RENEWALS: 691-4574
DATE DUE

WITHDRAWN
UTSA Libraries

TELECOM
Foreign Trade Engineering Company

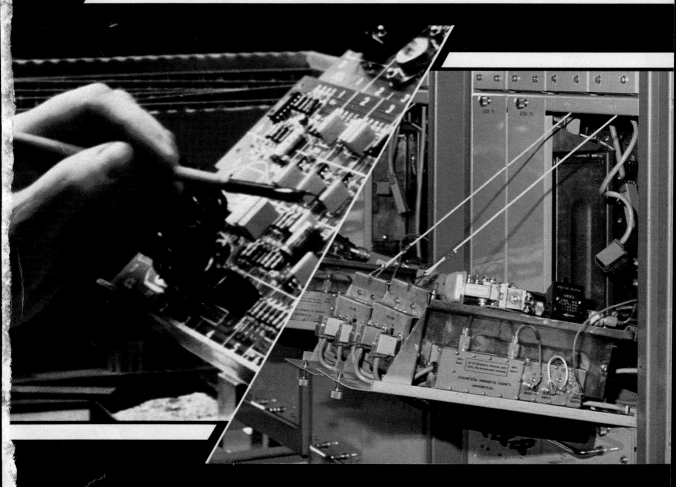

Major exporters and importers of electronic and communications equipment, elements, components and instruments for the communications industry.

Our engineering activities abroad include the supply of complete plant and equipment and our industrial enterprises cover study, design, delivery, operation and servicing.
We can also undertake the purchase and sale of licences, patents and appropriate equipment.

Telecom
17 George Washington Street,
Sofia, Bulgaria.
Telephone: 8-61-81
Telex: 22075 / 22076

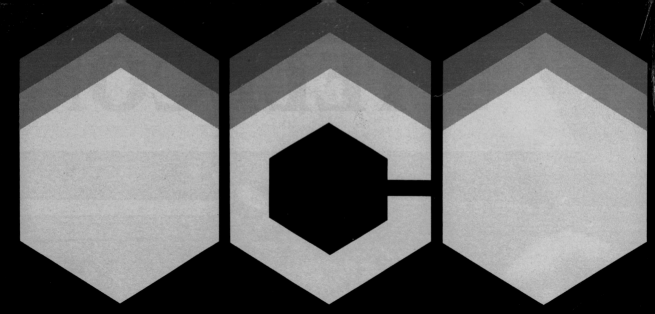

Export List:

- Soda Ash/Light
- Urea
- Ammonium Sulphate
- Ammonium Nitrate
- Triple Super Phosphate
- Sodium Nitrate
- Complex fertilisers
- Perocine / Zineb
- Thiosol
- Ammonium Bicarbonate
- Sodium Nitrate
- Minium
- Lead Glazing
- Zinc Sulphate
- Calcium Carbonate
- Borax - industrial
- Alpha Pennine
- Turpentine - industrial
- Urotropine
- Caustic Soda - liquid
- Ethyl Acetate
- Cyclohexanon
- Dimethyl Sulphate
- Standard Gases
- Neon
- Helium
- Catalyst - non-used
- Gloss Agents
- Pentaerytrite
- Tall Butyric Acid
- Phthallic Anhydrite
- Urea Glue - liquid
- Pyrolene
- Benzene
- Acetone
- Acetaldehyde
- Acryl Nitrile
- Monoethyleneglycol
- Liquid Chloride
- Chloric Lime - 33-35%
- Argon Gas
- Carbon Sulphide
- Calcium Carbide
- Hydrochloric Acid
- Sulphuric Acid
- Formic Acid
- Diethyleneglycol
- Ethylene Epoxide
- Orthoxylene
- Styrene Monomer
- Phenol
- Toluene
- Sodium Sulphide
- Propylene
- Dichlorethane
- Butanol
- Octanol
- Kerosene
- Paraffin
- Lubricating Oil
- Propane - Butane
- Normal Paraffin
- Sodium Tripolyphosphate
- Aniline Dyes
- Photopaper - black and white
- Cadmium Oxide
- Verin Antifoamer
- PVC Emulsion
- PVC Suspension
- Caprolactam
- Pheneplast
- Polyethylene - high pressure
- Polyethylene - low pressure
- Butadiene
- Microporous Separators
- Polyester Resins
- Methylmethacrylate
- Polyester Varnish for furniture
- Flooring
- Artificial Leather
- Synthetic Rubber
- Latex
- Rubber Hoses
- Automobile Tyres
- Deviplast Packing Paste
- Polyethylene Foil
- Polypropylene

For particulars, please contact:

CHIMIMPORT
2 St. Karadja Street,
Sofia, Bulgaria.
Telephone: 88-38-11/15
Telex: 22521 / 22522

MACHINOEXPORT ENGINEERING

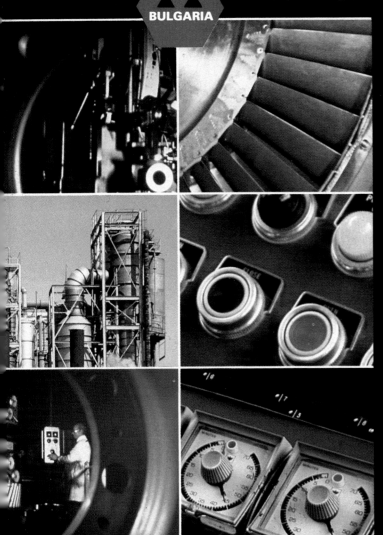

BULGARIA

A reliable service offering:

- Complete set of engineering services, study, design and construction of complete plant for the field of machine-building and woodworking.
- Technical and economic consultations
- Supply of technological equipment
- Designing of separate stages
- Construction and assembly
- Technological know-how
- Licences
- And any other related service

Machinoexport Engineering can implement Turnkey projects:

- Plants for the assembly or production of machine tools or woodworking machines
- Plants for the assembly or production of agricultural machines
- Enterprises for the treatment of material for the furniture industry
- Repair workshops for automobiles and agricultural equipment
- Repair plants for machine tools
- Plant repair workshops
- Light metal structures for a variety of industrial and commercial purposes

Machinoexport Engineering can supply:

- Machines for ELFA spark lamination
- Installations for ionic metal nitrification
- Centrispun pipes
- Installations for chemical nickelling
- Smelting and model equipment for ferrous and non-ferrous metals

Machinoexport Engineering - *your best bet!*

Machinoexport
5 Aksakov Street, Sofia 1000, Bulgaria.
Telephone: 88-53-21 Telex: 23425 / 23426

TECHNOEXPORTSTROY

State Corporation
TECHNOEXPORTSTROY is a Bulgarian state economic corporation with considerable experience in the international market and a considerable reputation for quality. Our activities include the design, assembly and construction, with authors' and investors' control, of residential buildings, hospitals and out-patients clinics, drug storehouses, theatres and public halls, sports facilities and stadiums, beach and holiday recreation sites, water supply and sewage plants, airfields, plants and factories, silos and all aspects of rail and road travel.

We also have the facilities to handle:
- Hydroenergy projects.
- The mechanical and electrical assembly of installations.
- Structures and equipment.
- Urbanization, regional and territorial planning.
- Land consolidation and topographic geodetic surveys.

Our staff of highly experienced engineers, mechanics, economists, townplanners and other specialists, work with some of the most sophisticated modern machinery available. Together we have completed over 450 projects in Asia, Africa and Europe, working to the highest standards and the tightest schedules.

Notable successes include:
- International airfields in Baghdad, Iraq, Tripoli and Sebkha in Libya.
- National theatre, hotel, stadium and other sports facilities at the Hotel Nigeria in Aden, South Yemen.
- Oil refinery, silo, out-patients clinics, hospitals and drug storehouses in Libya.
- Artificial Lakes in Syria and Morocco.
- Railway line, cement plant, bridges and a paper and pulp plant in Syria.
- Sports complexes in Libya and Tunisia.
- Residential buildings in Libya and the Federal Republic of Germany.
- Land reclamation in Morocco.

In 1975 TECHNOEXPORTSTROY was awarded the prize of Recognition and Prestige from the UN Institute in Geneva. And in 1977 we won the gold medal of the Milan Chamber of Commerce, Industry, Trades and Agriculture.

PO Box 786,
11 Antim I Street, Sofia,
Bulgaria.

Telephone: 87-85-11
Telex: 22128

BULGARIAN WINE IS *YOUR* WINE

Major exporter of wines and beverages.

VINIMPEX

19 Lavelaye Street,
Sofia, Bulgaria.
Telex: 22467

Import-Export
3 Positano Street,
Sofia, Bulgaria.
Telex: 22091
Cables: INDIMPORT

Twice winner of the
Gold Mercury
international prize.

industrialimport

Our wide range of exports include:

- Woollen Garments - suits, coats, trousers, overcoats, topcoats, dresses, skirts.
- Cotton Garments - shirts, blouses, pyjamas, dresses, raincoats, overalls, bed linen.
- Woollen, cotton and synthetic knitwear.
- Woollen, cotton and silk fabrics and blankets.
- Machine-made Persian-type carpets.
- Sports goods.
- Shoes and handbags.
- Sports and tourist appliances.
- Synthetic and artificial fibres.
- Rayon, cord and batting.
- Silicate products - blown and pressed glassware.
- Lighting fixtures.
- Sanitary porcelain.
- Faience tiles.
- Glass - ordinary, reinforced, section and constructional.

COUNTRIES OF THE WORLD
INFORMATION SERIES

INFORMATION BULGARIA

And Nation Shall Speak Peace unto Nation

COUNTRIES OF THE WORLD
INFORMATION SERIES

Also in this Series:

Information USSR
Information Hungary

INFORMATION BULGARIA

A short encyclopaedia of the People's Republic of Bulgaria

Edited by
The Bulgarian Academy of Science

Translated by
Sofia Press Agency

Pergamon Press

OXFORD · NEW YORK · TORONTO · SYDNEY · PARIS · FRANKFURT

U.K.	Pergamon Press Ltd., Headington Hill Hall, Oxford OX3 0BW, England
U.S.A.	Pergamon Press Inc., Maxwell House, Fairview Park, Elmsford, New York 10523, U.S.A.
CANADA	Pergamon Press Canada Ltd., Suite 104, 150 Consumers Road, Willowdale, Ontario M2J 1P9, Canada
AUSTRALIA	Pergamon Press (Aust.) Pty. Ltd., P.O. Box 544, Potts Point, N.S.W. 2011, Australia
FRANCE	Pergamon Press SARL, 24 rue des Ecoles, 75240 Paris, Cedex 05, France
FEDERAL REPUBLIC OF GERMANY	Pergamon Press GmbH, Hammerweg 6, D-6242 Kronberg-Taunus, Federal Republic of Germany

Copyright © 1985 Pergamon Press Ltd.

All Rights Reserved. No part of this publication may be reproduced, stored in a retrieval system or transmitted in any form or by any means: electronic, electrostatic, magnetic tape, mechanical, photocopying, recording or otherwise, without permission in writing from the publishers.

Photographs by Sofia Press, Bulgarian Photographs and the Bulgarian Encyclopaedia

First edition 1985

Library of Congress Cataloging in Publication Data
Information Bulgaria.
(Countries of the world information series)
Translated from Bulgarian.
Includes bibliographies.
1. Bulgaria—Dictionaries and encyclopedias.
I. Bŭlgarska entsiklopediiā (Firm). Editorial
Committee. II. Series.
DR53.I54 1985   949.7'7'00321   84-26544

British Library Cataloguing in Publication Data
Information Bulgaria: a short encyclopaedia of
the People's Republic of Bulgaria.—(Countries
of the world information series)
1. Bulgaria
I. Series
949.7'703   DR55
ISBN 0-08-031853-3

Typeset by Burgess & Son Ltd, Abingdon, U.K.
Printed in Great Britain by A. Wheaton & Co. Ltd., Exeter, U.K.

CONTENTS

Introduction	ix
List of Maps	xi
List of Map Facsimiles	xiii
A Note on Transliteration	xiv
The Dates in the Text	xv
Weights and Measures Conversion Table	xvi
List of Abbreviations	xvii
Glossary	xxi

PART I: GENERAL INFORMATION

General Information	3
Geological Structure	5
Topography	13
Mineral Resources	24
Climate	30
Water	34
Soils	45
Vegetation	49
Fauna	56
Protection of the Environment	62
National Parks	65
Population	72
The Bulgarian Language	77
District Centres	81
Selected Bibliography	135

PART II: HISTORY

The Bulgarian Lands in Antiquity	139
The Middle Ages in the Bulgarian Lands	145
The Bulgarian National Revival	177
Bulgaria in the Period of Capitalism	193
Bulgaria in the Age of Socialism	233
Chronological Outline	256
Rulers	265
Governments	266
Selected Bibliography	269

PART III: SOCIO-POLITICAL STRUCTURE OF THE PEOPLE'S REPUBLIC OF BULGARIA

Socio-political Structure of the People's Republic of Bulgaria	275
The Constitution	275
The National Assembly	280
The Supreme Court	284
The Bulgarian Communist Party	287
The Bulgarian Agrarian People's Union	299
Public Organizations	304
Selected Bibliography	318

PART IV: THE ARMED FORCES

The Armed Forces	323
The Ground Forces	328
The Anti-Aircraft Defence and the Air Force	329
The Navy	329
Selected Bibliography	337

PART V: THE ECONOMY

The Development of the Economy	341
Industry	358
Construction	399
Agriculture	408
Forestry	435
Game Farming	438
Fish-Farming Economy	439
Transport	440
Communications	452
Domestic Trade and the Co-operative Movement	459
Foreign Trade and Economic Co-operation	467
The Fiscal and Monetary System	481
Tourism and Resorts	492
Bulgarian Foreign Trade Organizations	517
Bulgarian Foreign Trade Representations	531
Selected Bibliography	539

PART VI: SOCIAL POLICIES

Living Conditions of the People	543
Public Health Services	550
Physical Education, Sports and Tourism	568
Rest and Recreation	594
Social Security	603
Selected Bibliography	610

CONTENTS

PART VII: EDUCATION

Education	613
General Development and Structure	613
Pre-school Upbringing	618
Secondary Education	620
Teacher Training	628
Higher Education	632
Facilities and Management of Education	635
Selected Bibliography	638

PART VIII: SCIENCE

The Development of Science Prior to 1944	641
Scientific Policy. The Organizational Structure and Management of the Scientific Units	647
Basic Trends and Achievements of the Natural and Technical Sciences	661
Basic Trends and Achievements of Social Sciences and Humanities	677
Selected Bibliography	684

PART IX: CULTURE

The Development of Culture and Cultural Policy in Socialist Bulgaria	687
Folk Art (Folklore)	691
Literature	698
Fine and Applied Arts through the Ages	740
Architecture	753
Music	765
Theatrical Arts	780
Cinema	795
The Circus	801
Amateur Art Activities	805
International Cultural Co-operation	810
Selected Bibliography	821

PART X: CULTURAL AND EDUCATIONAL ESTABLISHMENTS

The Library Clubs	825
Libraries	828
Archives	834
Museums	836
Selected Bibliography	847

PART XI: MASS MEDIA

Newspapers and Magazines	851
Broadcasting	860
The Bulgarian Telegraph Agency	864
Book Publishing	866
Printing and Publishing	870
Book Distribution	872
The Sofia Press Agency	873
Bulgarian Photography	875
The Copyright Agency	877
Hemus Foreign Trade Organization	878
Union of Bulgarian Journalists	879
Selected Bibliography	880

PART XII: FOREIGN POLICY OF THE PEOPLE'S REPUBLIC OF BULGARIA

Foreign Policy of the People's Republic of Bulgaria	883
Bulgaria and the Other Socialist Countries	885
Bulgaria and the Other Balkan Countries	891
Bulgaria and the Developing Countries	896
Bulgaria and the Capitalist Countries	900
Bulgaria and the Struggle for Peace, Security and Co-operation	905
Bulgaria's Participation in the United Nations and in Other International Organizations	909
Selected Bibliography	916

APPENDIX: BULGARIA IN FIGURES

Bulgaria in Figures	919

INTRODUCTION

Information Bulgaria is a comprehensive and up-to-date account of the People's Republic of Bulgaria, with a historical background and indications for future development. The book is the collective work of a large team of specialists from various disciplines and spheres of activity, and consists of articles assembled and edited by the Bulgarian Academy of Science, and translated into English by experts from Sofia Press. Their English-language text has been edited by Pergamon editorial staff.

The book covers all major aspects of the physical, political, social, economic and cultural life of the country, with major sections also on geography, natural resources, conservation, agriculture, industry, tourism, trade and commerce. It contains copious statistical information and many maps, and is lavishly illustrated with black and white and colour photographs. This is the Bulgarian view of Bulgaria and as such the book makes available, for the first time in English between the covers of a single volume, much information on Bulgaria which has been previously widely scattered or completely inaccessible.

The purpose of *Information Bulgaria* is to increase awareness, among opinion-makers, scholars, businessmen and the general English-speaking public throughout the world, of the enormous advances made by the People's Republic of Bulgaria in the forty years of its existence. During that time it has become an important, and in some fields a leading member of CMEA, and has also developed increasingly active relations with a number of countries outside the Socialist bloc, especially in Western Europe, the Middle East, North Africa and Nigeria. *Information Bulgaria* is also intended to inform all those potentially interested in developing commercial and cultural links with Bulgaria of the means of doing so.

As Publisher of *Information Bulgaria,* I hope that this volume will lead to a greater understanding and appreciation of the Bulgarian people and their beautiful country. I would like to express my thanks to Mr Todor Zhivkov, General Secretary, Central Committee of the Bulgarian Communist Party, and President of the State Council of Bulgaria, and to Mr Milko Balev, First Secretary, Central Committee of the Communist Party of Bulgaria, for the assistance they have given to Pergamon Press with the compilation of this book.

ROBERT MAXWELL
Publisher

LIST OF MAPS

Physical
Administrative
Agriculture
Industry
Geological
Climate
Soils
Vegetation
Bulgaria in the Seventh–Eleventh Centuries
Bulgaria in the Twelfth–Fourteenth Centuries
The Struggle of the Bulgarian People against Ottoman Domination (1396–1878)
Bulgaria after the Liberation (1878)
The September 1923 Anti-fascist Uprising
The Armed Anti-fascist Resistance (1941–44)
Bulgaria's Participation in the War against Nazi Germany (1944–45)

Maps reproduced by courtesy of the Bulgarian Encyclopaedia

LIST OF MAP FACSIMILES

Part of the Tabula Peutengeriana, showing the Roman provinces in the Balkans
An Anglo-Saxon map, one of the oldest maps to mention the Bulgarians and to depict their State
Fragment of a map of Eastern Europe from Hieronymus's essays featuring Bulgaria
Military chart of the late fourteenth century
Detail of a map of the world by Fra Mauro (1459)
Printed map by Gerardus Mercator (1634), Amsterdam
Copy of ethnographical map of the South Slavonic countries, compiled by MacKenzie and Irby, 1867
Reduced copy of maps of the Bulgarian autonomous regions determined by the Constantinople Conference (1876–77)
The Bulgarian Regions Determined by the Constantinople Conference (1876)

Map facsimiles reproduced by courtesy of the National Library, Vienna; the British Museum, London; the National Library, Paris; the Doges' Palace, Venice; and the Algemeen Rijksarchiv, The Hague.

A NOTE ON TRANSLITERATION

For several reasons, the transliteration of names used in this book is not entirely consistent, though we have in fact aimed at systematism within the constraints of the subject, proceeding from the criterion of easy intelligibility to the English-speaking reader.

The chequered history of the 'Bulgarian lands' is clearly reflected in the language, especially in the names of places which have, at different periods of their history, been at least Roman, Byzantine Greek, Turkish and finally Slav. Within the Slav family they may at various times have been Serbian, Macedonian or Bulgarian; and they may at other times have been subject to the influence of Romanian or Hungarian. Moreover, in transliterated form they may sometimes have become familiar to Western eyes first of all in spellings determined by the language of diplomacy of the times – French. In any one of these guises, spellings may have become so hallowed that it would be outlandish to reject them; but they cannot nevertheless be taken as models.

We have therefore been unable to be entirely logical either in adopting a system of letter by letter **transliteration** from cyrillic to latin script, or in using a system of phonetic **transcription** that would give the reader a close approximation of pronunciation. To attempt the latter would in any case have necessitated the use of a number of symbols and marks that would have been quite inappropriate in a book of this kind.

Thus for cyrillic ж we use -zh-, while allowing certain widely-accepted forms with -dj-; and for cyrillic х we use -h-, except in Russian names, when we use -kh-. We have used a double consonant -ss- for unvoiced cyrillic с, but prefer -ts- to -tz- and -u- to -ou-. The letter -u- thus represents both cyrillic ъ and cyrillic у, and though this is not an ideal solution, it has seemed to us that the gain in legibility has outweighed the loss of nuance.

Our hope is that the system we have adopted both avoids the sort of convoluted spellings that the untrained eye cannot absorb, and at the same time gives at least a little assistance to the reader who wishes to be able to use the names orally or, perhaps most importantly, to refer to them in other sources, including those given in cyrillic.

А	— A	Е	— E	К	— K	П	— P	Ф	— F	Щ — SHT
Б	— B	Ж	— ZH	Л	— L	Р	— R	Х	— H	Ъ — U
В	— V	З	— Z	М	— M	С	— S	Ц	— TS	Ь — —
Г	— G	И	— I	Н	— N	Т	— T	Ч	— CH	Ю — YU
Д	— D	Й	— I	О	— O	У	— U	Ш	— SH	Я — YA

THE DATES IN THE TEXT

The dates in this volume follow the calendar used in Bulgaria at the time of the events referred to. The Gregorian calendar (or 'new style') was introduced into Bulgaria on 1 April 1916, the date of 1 April ('old style') becoming 14 April. Before that the Julian (or 'old style') calendar was used. The other European countries adopted the new style in 1582. In the seventeenth century the difference between the Julian and Gregorian calendars was ten days, in the eighteenth – eleven days, in the nineteenth century – twelve days, and from 1901 to 1 April 1916 it had become thirteen days.

WEIGHTS AND MEASURES CONVERSION TABLE

Length
1 metre (m) = 1.09 yards (yd) = 3.28 feet (ft) = 39.37 inches (in)
1 kilometre (km) = 0.62 miles = 0.54 nautical miles

Area
1 sq. m = 0.000247 acres
1 hectare (ha) = 10,000 sq.m = 2.47 acres

Weight
1 kilogramme (kg) = 2.20 pounds (lb)
1 gramme (g) = 0.04 ounces (oz)
1 ton (t) = 1,000 kg = 2,200 pounds (lb)

Liquid Measures
1 litre = 1 cu. dm = 1.7598 pints = 0.22 gallons (UK) = 0.26 gallons (US)
1 cu. cm = 0.04 fluid ounces (fl oz; UK)

Temperatures
$n\ °C = (1.8n + 32)\ °F$
100 °C = 212 °F
50 °C = 122 °F
37.8 °C = 100.04 °F
10 °C = 50 °F
0 °C = 32 °F
−17.8 °C = 0 °F
−25 °C = −13 °F
−40 °C = −40 °F
−50 °C = −58 °F
−100 °C = −148 °F

LIST OF ABBREVIATIONS

AENOC	Association of European National Olympic Committees
AFIAP	Artiste de la Fédération Internationale des Artistes Photographis
AGT	*avtogruzovye tovary;* heavy-duty truck loads
AIC	agro-industrial complex
ATA	temporary admission (carnets)
ATE	automatic telephone exchanges
BAP	Bulgarian Agrarian Party
BAS	Bulgarian Academy of Science
BAU	Bulgarian Agrarian Union
BCAO	Bulgarian Circus Artistic Organization
BCCI	Bulgarian Chamber of Commerce and Industry
BCCS	Bulgarian Central Charity Society
BCP	Bulgarian Communist Party
BEP	Board for Environmental Protection
BHAU	Bulgarian Hunting and Angling Union
BNB	Bulgarian National Bank
BOC	Bulgarian Olympic Committee
BPA	Bulgarian People's Army
BRCC	Bulgarian Revolutionary Central Committee
BSCRC	Bulgarian Secret Central Revolutionary Committee
BSDP	Bulgarian Social Democratic Party
BSDU	Bulgarian Social Democratic Union
BSFS	Bulgarian Union for Physical Education and Sport
BTA	Bulgarian Telegraph Agency
BTS	Bulgarian Tourist Society
BTU	Bulgarian Tourist Union
BWP	Bulgarian Workers' Party
BWSDP	Bulgarian Workers' Social Democratic Party
BWSDP (l-ws)	Bulgarian Workers' Social Democratic Party (left-wing socialists)
BWU	Bulgarian Workers' Union
BZNS	*Bulgarski Zemedelski Naroden Suyuz*
CC	Central Committee
CCITT	International Consultative Committee for Telephone and Telegraph
CCU	Central Co-operative Union
CEP	Committee for Environmental Protection
Cibal	International Documentation Centre for Balkan Studies
cif	cost, insurance and freight

CISAC	Conference of Societies of Authors and Composers
Cinti	Central Institute for Scientific and Technical Information
CMEA	Council for Mutual Economic Assistance
CNC	computer-numerically controlled
CPSU	Communist Party of the Soviet Union
CRMC	Central Revolutionary Military Committee
CSKA	Central Sports Club of the Army
DAIU	District Agro-Industrial Unions
DAO	Defence Assistance Organization
DEFA	Deutsche Film AG, a film studio in the GDR
DNA	Deoxyribonucleic acid
DPYU	Dimitrov People's Youth Union
DRO	Dobrudja Revolutionary Organization
EBU	European Broadcasting Union
ECE	Economic Commission for Europe
Ecosoc	Economic and Social Council
ECSC	European Conference on Security and Co-operation
EEC	European Economic Community
EFIAP	Excellence de la Fédération Internationale des Artistes Photographis
Euronet	European Regional Network for Scientific and Technical Information
FAO	Food and Agriculture Organization
FIAP	International Federation of Photographic Art
fob	free on board
FRG	Federal Republic of Germany
GATT	General Agreement on Tariffs and Trade
GDP	gross domestic product
GDR	German Democratic Republic
GNP	gross national product
GP	general practitioner
GRT	gross registered tonnage
GWTU	General Workers' Trade Union
HQ	headquarters
IAC	industrial agrarian complexes
IAEA	International Atomic Energy Agency
IBRD	International Bank for Reconstruction and Development
ICAO	International Civil Aviation Organization
ICE	internal combustion engine
ICSU	International Council of Scientific Unions
IDA	International Development Association
IFC	International Finance Corporation
ILO	International Labour Organization
IMARO	Internal Macedonian-Adrianople Revolutionary Organization
IMCO	Intergovernmental Maritime Consultative Organization
IMF	International Monetary Fund
IMO	International Maritime Organization
IMRO	Internal Macedonian Revolutionary Organization
IMRO(u)	Internal Macedonian Revolutionary Organization (united)

LIST OF ABBREVIATIONS

	INMARSAT	International Organization for Maritime Telecommunications via Satellites
	INSEA	International Society for Education through Art
	IOC	International Olympic Committee
	ISSC	International Social Sciences Council
	ITU	International Telecommunication Union
	KI	single canoe
	MCL	Medical Commission on Labour
	MW	medium wave
	n.a.	not available
	NAIC	National Agro-Industrial Complex
	NAIU	National Agro-Industrial Union
	Nato	North Atlantic Treaty Organization
	NOVA	National Liberation Insurgent Army
	OAD	Organization for Assistance in Defence
	OAU	Organization of African Unity
	PCC	Political Consultative Committee
	PLIA	People's Liberation Insurgent Army
	PRB	People's Republic of Bulgaria
	PTTS	post, telegraph and telephone station
	PU	Popular Union
	PVC	polyvinyl chloride
	PYU	People's Youth Union
	RNA	Ribonucleic acid
	RPC	research and production complexes
	RSFSR	Russian Soviet Federative Socialist Republic
	SCBC	Secret Central Bulgarian Committee
	SEC	State Economic Corporations
	SMAC	Supreme Macedonian-Adrianople Committee
	SPC	scientific and production complex
	SSR	Soviet Socialist Republic
	SW	short wave
	Swapo	South West African People's Organization
	TIR	Transport International Routier
	TT	telephone and telegraph
	TUY	Tourist Union of Youth
	UCN	International Union for the Conservation of Nature and Natural Resources
	UNCTAD	United Nations Conference on Trade and Development
	UNDP	United Nations Development Programme
	Unesco	United Nations Educational, Scientific and Cultural Organization
	Unicef	United Nations International Children's Emergency Fund
	Unido	United Nations Industrial Development Organization
	UPU	Universal Postal Union
	USW	ultra-short wave
	VITIZ	Higher Institute of Theatrical Art
	WHO	World Health Organization
	WIPO	World Intellectual Property Organization

WMO	World Meteorological Organization
WOT	World Organization on Tourism
WP	Workers' Party
WYU	Workers' Youth Union
WWER	water-water energy reactor
ZMM	Machine-Tool Plants Groups

GLOSSARY

(Meanings and forms listed here are those used in the text)

aba	coarse wollen cloth
abadzhis	artisans and traders
bashi-bazouk	mercenary of Turkish irregulars, notorious for pillage and brutality
basilei	kings
beylerbey	governor-general
boyar	nobleman
chardak	broad-roofed balcony
chetnik	member of band or group opposed to Turkish occupation
chiflik	big farm, estate
chitalishte	communal cultural centre
chokoi	big landowner
chorbadjii	middlemen between State and population
dedets	senior
desyatuk	tithe tax collected in goods
dromos	ante-rooms
dzhelep	large-scale sheep breeder
firman	a royal decree
haiduk	member of band or group opposed to Turkish occupation, idealized by the masses as freedom fighters and protected by them
hamam	bath
horo	chain-dance
ipdjambazi	tightrope walkers
kadi	judge
kale	citadel
kashkaval	yellow cheese
kaza	administrative centre
klirik	serf to the Church and its estates
Knyaz	Prince
komitat	military administrative area
konak	town hall
kratema	chant
kuker	mummer
maazas	commercial buildings and storehouses
medrese	religious college attached to mosque
melikyane	lease of state-owned property as a source of income for life
naroden	people's, popular

nufuzes	male population over sixteen years old
opolchenie	voluntary corps
parik	serf to the local landlord
peperuda	butterfly
prust	vestibule
rakia	plum brandy
rayah	masses unenlightened by the Qur'ān
salhani	a type of factory
sanjak	administrative district
sirene	white brine cheese
soba	bedroom and parlour
spahi	rulers
spartakiad	major competition in several sports
survakar	first-footer
tegulas	ancient Greek tiles
teke	shrine
tell	habitation mound
thema	region
vilayet	region, province of the Ottoman region
voivode	military commander
yeniceri	janizaries
yunak	heroic
zasluzhil	honoured

PART I

GENERAL INFORMATION

Hadji Nikoli's Inn in Veliko Turnovo

The village of Zheravna in the Sliven district

Clock tower in Botevgrad

Pashov House in Melnik

Kuyumdjioglu's House in Plovdiv

National Revival buildings and clock tower in Tryavna

Ancient theatre in Plovdiv

Tsarevets fortress in Veliko Turnovo

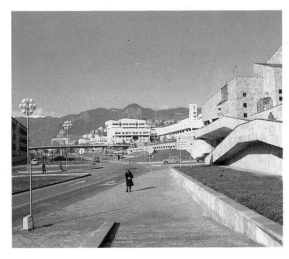

Smolyan town-centre

Church of the Holy Trinity in Svishtov

Bridge across the Yantra near Byala

Restored fortress gate of the first Bulgarian capital, Pliska

Palace of Sports and Culture in Varna

Sofia Central Railway Station

Hotel Vitosha – New Otani in Sofia

Monument to the Unknown Soldier, near the Church of St. Sophia in Sofia

Mausoleum to Russian and Romanian soldiers killed near Pleven in 1877

Lyudmila Zhivkova People's Palace of Culture in Sofia

GENERAL INFORMATION

The People's Republic of Bulgaria is a socialist state in South-eastern Europe. It is situated in the eastern part of the Balkan Peninsula between the meridian of 22°21′ East to the west (passing north-west of the Kitka peak in the Doganitsa mountain), the meridian 28°36′ East to the east (Cape Shabla on the Black Sea,) latitude 41°14′ North to the south (the Veikata peak in the Gyumurdzhinski Snezhnik mountain in the Rhodopes) and latitude 44°13′ North to the north (the mouth of the river Timok). The distance between the country's northernmost and southernmost points is about 330 km, and 520 km between its easternmost and westernmost points. Bulgaria has a total area of 110,911.5 sq. km (including inland and border river water), or 22 per cent of the area of the Balkan Peninsula. To the north Bulgaria borders on Romania. The 609 kilometre-long northern border passes along the middle of the River Danube from the mouth of the river Timok to the town of Silistra, then crosses the Dobrudja low country to reach the Black Sea at the Romanian village of Vama Veke. To the east Bulgaria borders on the Black Sea; the 378 kilometre-long border stretches between Vama Veke to the north and the mouth of the river Rezovska to the south. The country's southern border is 752 km long – 259 km shared with Turkey and 493 km shared with Greece. The border with Turkey starts from the mouth of the river Rezovska, following the ridges of the Strandja mountain and the Derventski hills to the river Maritsa east of the village of Kapitan Andreevo. This is where the border with Greece starts, then crosses the Rhodopes and Slavyanka mountains and reaches Mt. Tumba in the Belassitsa mountains, where the borders of Bulgaria, Greece and Yugoslavia converge. The western border (with Yugoslavia) is 306 km long; it follows the ridges of the Ograzhden, Maleshevska, Vlahina, Ossogovo, Milevska and Ruy mountains, crosses the western Balkan Range and, following the course of the river Timok, reaches the Danube. The total length of Bulgaria's state borders is 2,245 km, of which 1,181 km are land borders, 686 km river borders and 378 km sea borders. Bulgaria has a population of 8,929,332 (1982). The country is divided into 28 administrative-economic districts comprising 300 municipalities and settlement systems (as of 31 December 1983).

As of 31 December 1983 the settlement network includes a total of 5,385 inhabited localities, of which 228 are cities, 4,435 are villages, and 722 are hamlets and other types of settlements. The status of the territorial units has been determined by legislation. The capital Sofia, the largest city and centre of the Sofia settlement system, has a population of 1,082,315 (January 1983). The populations of the other district centres on the same date were as follows: Plovdiv – 367,195; Varna – 295,038; Russe – 178,920; Burgas – 178,239; Stara Zagora – 141,722; Pleven – 135,899; Sliven – 100,637; Shumen – 99,642; Tolbuhin – 98,857; Pernik – 94,859; Haskovo – 87,639; Yambol – 86,216; Gabrovo – 80,901; Pazardjik – 77,830; Vratsa – 73,014; Veliko Turnovo – 64,985; Blagoevgrad – 64,442; Vidin – 60,877;

The National Anthem

Silistra – 57,670; Kurdjali – 55,762; Kyustendil – 54,657; Mihailovgrad – 54,240; Razgrad – 51,761; Lovech – 49,754; Turgovishte – 48,022; Smolyan – 39,688.

The official language is Bulgarian – the oldest written Slav language. The monetary unit is the lev, equal to 100 stotinki. The national holiday is 9 September. The national anthem is called the 'Mila Rodino' (Dear Homeland). The President of the State Council is Todor Zhivkov. The highest state distinctions are the honorary titles Hero of the People's Republic of Bulgaria, Hero of Socialist Labour, and the Order of Georgi Dimitrov.

The state crest is round, with a lion rampant on a cogwheel in the middle set on an azure field. The field is bordered on either side by ears of wheat entwined with the national tricolour band; above the lion is a red five-pointed star and at the base, on a red band, stand the years of the foundation of the Bulgarian State (681) and the victory of the socialist revolution in Bulgaria (1944).

The state flag is a tricolour – white, green and red, arranged horizontally with the state emblem in the upper left-hand corner of the white field.

Bulgaria is a member of the world socialist community. It is a founding member of the Council for Mutual Economic Assistance (CMEA – 1949) and of the Warsaw Treaty (1955). It has been member of the United Nations since 1955 and maintains diplomatic relations with 117 states (on 30 June 1983) and commercial relations with 120 states (1982).

GEOLOGICAL STRUCTURE

The earth's crust in Bulgaria is formed of different magmatic, sedimentary and metamorphic rocks. It is thinnest in Northern Bulgaria and in the area of the towns of Varna and Burgas (about 30 km) but it becomes thicker in the south-west reaching up to 48 km in the Western Rhodopes and the Pirin mountains. The oldest are the magmatic and metamorphic rocks in Southern Bulgaria. Sedimentary rocks outcrop mainly in Central, Northern and Western Bulgaria. They are deposited in water basins which, during the different ages of the Paleozoic, Mesozoic and Cenozoic eras, covered parts of the region. Large blocks have been formed from the tectonic movements; faulted, folded and overthrust structures were formed and different magmatic processes occurred.

Bulgaria within the Reach of the Alpine-Himalayan Orogeny. The geological events during the past two geological eras, the Mesozoic and Cenozoic, bear the deepest imprint in the structure of Bulgaria's earth crust. The Balkan Peninsula space at that time was fully included in the most mobile belt of the planet – the Tethis geosynclinal belt spreading from the Atlantic to Indonesia. The structures of the Alpine-Himalayan orogeny originated as a result of the tectonic processes in the vast Tethis geosyncline to which the Stara Planina (Balkan) mountain range also belongs. The Balkan Peninsula has a marked symmetrical structure. The vast Thracian Central Massif (a micro-continent), formed of the oldest rocks outcropped in the Balkans, is in the middle part of the peninsula and simultaneously in the middle of the orogeny. To the south-west the massif is surrounded by the southern Dinaric branch of the Alpine-Himalyan orogeny and to the north-east by the northern, Carpathian orogenic branch. They form the Adriatic and Moesian Platform on which the Dinaride folds rest, namely the Balkanides and the South Carpathians.

The Thracian Central Massif is formed of archaic and proterozoic metamorphic rocks in which magma has intruded repeatedly. The massif is included in the Tethis geosynclinal belt as a micro-continent, i.e. as a previously consolidated segment of continental earth crust. The continuous Thracian massif elevation was connected with processes resulting in its denudation. As early as in the Early Paleozoic, two rift zones were established: the Kraishtides and the Vardar, where movement still continues today. The Dardan Massif was singled out between these rift zones. Quite narrow, the Dardan Massif and its two adjacent rift zones form a larger structural unit – the Kraishtide-Vardar lineament. It is not only an axis of tectonic symmetry in the Balkans but was also a paleogeographic barrier in the past three geological eras. The geological processes developed separately and in a different manner on both sides of the lineament. This is related to the magmatism, the sedimentation conditions and the water-basin types, the character and type of the geological formations, and the formation periods of the separate geosynclinal beds, etc. The formation of the Kraishtide and the Vardar rift zones results in the separation of the Thracian Central Massif into three sectors: Pra-Rhodope (including the Rhodope Massif, Sredna Gora and Strandja Mountains), the Pelagonic-Thessalian-Cyclade and the Dardan sectors. The Pra-Rhodope Massif was eroded many times during geological evolution. Of particularly great importance was a large depression in the western part of the Massif, the so-called Sredets Amphitheatre where oceanic rocks (Tribal-paleo-ocean) were formed in the oldest Paleozoic era. Later it experienced numerous faultings and magmatism. In the eastern part of the Rhodope Massif, on the other hand, was formed the so-called Sub-Rhodope zone – a zone of ruptures, ample magmatic activity and deep sea sedimentation in the Jurrassic and Early Cretaceous. Of particular importance was the tectonic-magmatic activity in the Late Cretaceous – in the northern strip of the Pra-Rhodope Massif. Many rift geosynclinal troughs were formed in that part of the Massif where thick sedimentary-volcanogenic complexes were accumulated and magma flowed profusely. This is the Sredna Gora structural zone folded by the end of the Late Cretaceous and closely annexed to the Balkanides – the northern branch of the Alpine-Himalayan orogeny. The subsequent erosion of the massif associated with ample magmatism occurred in the Late Eocene and in the Oligocene. The main field of fracturing and sedimentation was in the Eastern Rhodopes. New, vast areas of the Pra-Rhodope Massif sank in the Pliocene and the Quaternary, thus bringing about the origination of the Eastern Thracian and Aegian depressions. All this was against the background of elevation of the Massif as a whole.

The Rhodope Massif in its present form, as a part of the large Thracian Central Massif, is bounded by the Maritsa deep fault (parallel to the Maritsa river valley) to the north, and by the Strymon deep fault (along the valley of the Strymon river) to the west. On the east and south the massif continues beyond Bulgaria's borders. The Rhodope Massif was formed mainly of old metamorphic rocks: gneiss, gneiss-shales, amphibolites, migmatites, marbles, etc. The metamorphites are divided into nine strata, the lower three being conventionally related to the Archean, and the remainder to the Proterozoic. Several magmatic rock generations outcrop widely. Of particular interest are the so-called South-Bulgarian Granites. The total thickness of the Proterozoic rocks is estimated to be 8–10 km. The large Rhodope anticlinal and synclinal structures were formed of the Proterozoic rock complexes. The synclinal structures in the Late Eocene and the Oligocene were additionally ruptured by longitudinal faults. Overthrusts moving to the south were formed along the fault planes. The Sub-Rhodope zone was activated and the Eastern Rhodope Depression occurred. It was formed by thick sedimentary, sedimentary-volcanogeneous and magmatic complexes. Many of the faults in the Rhodope Massif in the Oligocene were canals along which large masses of acid magma were raised to the surface. The Rhodope rhyolites were formed after the setting of the magma. A prolific sulphide hydrothermal ore formation was connected with this magmatic activity. The field of tectonamagmatic and metallogenous mobilization included the Rhodopes, and to the north-west it was a strip up to 100 km wide crossing the whole Balkan Peninsula diagonally – to Slovenia (in Yugoslavia) and Carinthia (in Southern Austria). It is called the Transbalkan Strip.

The Moesian Platform can be regarded as the diametical opposite of the Thracian Massif. It is a stable lithospheric plate of Pre-cambrian consolidation. During the whole Phanerozoic it was built as a platform – a plate with a tendency to sink. The Moesian Platform occupies a large area in the lower stream of the Danube. Its northern boundary crosses immediately south of the South Carpathians and its southern rests on the Balkanides. Geologically the Platform is an enormous arch: its northern end sinks quite steeply to the north, and the southern one is less steep. The line of the highest elevated parts of the arch lies on the parallel – to the west it sinks to the Iron Gates, and to the east it rises. Lesser structures are the Dobrudja, the North-Bulgarian Uplift, the Balsh-Optash rises in the northern part of the Platform, and Strehaja in its western part. The North-Bulgarian Uplift is the most elevated structure in the platform in Bulgaria. Its end region has diameters ranging from 40 to 60 km. Its abutments are much larger, especially in the north-west. The Lom Depression is the most significant of the platform negative structures and like the above-mentioned uplifts is a perpetually changing structure. Marine sediments from 10 to 12 km thick were accumulated on its folded foundation. The Moesian Platform basement outcrops to the surface only in Northern Dobrudja. In Bulgaria's region it is at least 2–3 km from the surface. It is made of greatly folded Pre-cambrian metamorphic rocks. The shaly complexes of the Paleozoić overlie them transgressively, and further up there are mainly arenaceous and calcareous rocks forming the Platform superstructure. After the superstructure four carbonate platforms are situated one on top of the other: the Middle Upper Devonian (1,500–2,500 m thick), the Triassic (having a number of smaller structures important from the point of view of oil geology), Cretaceous (650–950 m thick) and the Late Cretaceous carbonate plate. The youngest sediments – the Neogene (being over 1,000 m thick) – were deposited mainly in the Lom Depression. To the east the Moesian Platform continues to the Black Sea. In fact, it is only the western part of the vast Ponto-Caspian Plate which also includes the Black Sea depression, the Georgian and Azurbaijan platforms and Southern Caspian. From the point of view of plate tectonics, the Ponto-Caspian platform is a primary lithospheric plate of sub-oceanic crust type. It has played an important role in the formation of the structures of the northern branch of the Alpine-Himalayan orogeny (South Carpathians, Balkanides, North Anatolian mountains, the Little Caucasus mountain system and the Elbrus mountain, as well as Caucasia and the Crimea).

Formations by erosion of Pliocene sands mixed with Lower Quaternary deposits near the town of Melnik

Folded Morphotectonic Structural Districts in Bulgaria. *The Balkanides* (the northern branch of the Alpine-Himalayan orogeny in the eastern part of the Balkans), the Sredna Gora districts and individual sectors of the Kraishtides, the Dardan Massif and the Southern Carpathians are included in the region of Bulgaria. The Moesian Platform and the Rhodope Massif are consolidated earth crust areas. As stable masses they control the development and the formation of the structural zone situated among them.

The Balkanides include three longitudinal zones – the

Balkan (Stara Planina) Mountains, the Fore-Balkan and the Transition Zone. The Sredna Gora mountain zone can also be included in them, taking a broader view.

The Stara Planina (Balkan) Structural Zone consists of three sectors: Western (represented by the Berkovitsa anticlinorium), Middle (the Shipka anticlinorium) and Eastern (the Luda Kamchia synclinorium). Both anticlinoria have quite identical structures. Because of the considerable structural elevation they have been deeply denuded and their Paleozoic cores are quite widely outcropped. The mantle is formed of the sedimented complexes of the Permian (partially), the Triassic, the Jurassic and the Lower Cretaceous. Folded sediments of Late Cretaceous and Paleogenic age transgressively positioned also outcrop in the Shipka anticlinorium. From north and south both anticlinoria are surrounded by deep faults – greatly fractured strips, several kilometers wide. The Cis-Balkan fault is to the south, and the Stara Planina (Balkan) front strip is to the north. Sredna Gora granites have been overthrusted to the north along the plane of the Cis-Balkan Deep Fault. This is the Stara Planina granite overthrust, an isolated part of which is the so-called tectonic klippe, about 30 km long, and which also forms the Botev Peak Massif (the highest peak in Stara Planina). Luda Kamchiya synclinorium, which includes the greater part of Eastern Stara Planina, has been complicated by many fold structures. The synclinorium joint sinks eastwards. The whole synclinorium also gradually extends in the same direction. The thick flysch sediments of the Late Cretaceous and Paleogenic age are quite characteristic of the Eastern Stara Planina mountain. In the Late Eocene, the Lyuliakova depression was filled by molassoide sediments. A unique structural zone about 200 km long (the so-called Kotel strip) lies north of the Luda Kamchiya synclinorium including the northern branches of Eastern Stara Planina. This has a diversified geological composition. East of the town of Kotel the strip structure is fairly simple – representing a synclinal structure pressed to the north. The western strip part, however, is filled by narrow folds and imbrications. All these, including the whole Kotel strip, are moved to the north.

The Fore-Balkan is enclosed between two deep faults which almost divide the whole country horizontally. The southern, which is called the Stara Planina front strip, is a deeply disintegrated earth crust strip on which the Stara Planina structural zone overrides the Fore-Balkan. The northern fault – the Fore-Balkan Deep Fault – is the boundary between the Balkanides and the Moesian Platform. Along its length the whole Fore-Balkan zone is divided into two longitudinal strips: the Actual Fore-Balkan to the South and the Transition Zone in the North. The western sector of the Actual Fore-Balkan is formed of the Belogradchik anticlinorium which is the most elevated part of the Fore-Balkan. The Paleozoic complexes of its core are widely outcropped in it. The Belogradchik anticlinorium overrides the Southern Carpathians to the north, and on the western part of the Moesian Platform. The Teteven anticlinorium fills the Middle Fore-Balkan and overrides to the north – to the greatest extent in its western part where its elevation is the highest. Its structure has been complicated by numerous less folded formations. The Eastern Fore-Balkan (formed of the Preslav anticlinorium) is relatively the lowest. The two flysch complexes – Tithonian-Lower Cretaceous and the Lower Middle Eocene are of particular interest. Their deposition shows the mobility of the geosynclinal basin during those ages. The flysch complexes are, however, separated by Upper Cretaceous carbonate platform sediments.

The Transition Zone is assumed to be the outermost Balkanide belt. To the south it borders with the Brestnitsa-Preslav flexure developed on a deep fault, and to the north with the Fore-Balkan Fault (the Balkanide front line). Most of it (east of the Iskur River Valley) was developed on the southern abutment of the All-Moesian Uplift. The structures filling the zone are linear keeping the direction of the Balkanide structure but being significantly smaller.

In the Sredna Gora Zone the Archean metamorphic complexes and the South Bulgarian granitoids, are typical of the Pra-Rhodope massif. The most commonly found ones are, however, the Upper Cretaceous sedimentary and magmatic rocks. The sedimentary rocks are formed in the longitudinal geosynclinal troughs, and the magmatic rocks are formed as a result of magma penetration along deep faults. The Sredna Gora Zone is bounded by the Cis-Balkan Deep Fault to the north, and the Maritsa Fault to the south. The Sredna Gora structure zone continues to the west in Yugoslavia, and to the North Pontiac mountain to the east. In Bulgaria, the Etropole and Tvurditsa diagonal line divide the Sredna Gora Zone into

The Stone Forest. Rock columns amidst Eocene sands near Varna

three sectors. The Western Sredna Gora sub-zone is formed on a deeply depressed point of the primary earth crust – the so-called Sredets Niche (Amphitheatre). Thick Paleozoic complexes, developed under geosynclinal conditions, were deposited in it. The Triassic sediments lying transgressively on a very diversified substratum in the West and Central Sredna Gora Zone have platform features, and geosynclinal features in the Eastern Sredna Gora sub-zone. In the region of the Strandja and Sakar mountains and the Svety Iliya hills they have been metamorphosed to a different degree. The Jurassic sediments in the Strandja Mountain have also been altered to a different degree. A large anticlinorium and a vast synclinorium are located in each of the three sectors of the Sredna Gora Zone. The Svoge anticlinorium and the Sofia complex trough are spread into the Western Sredna Gora, the Sredna Gora anticlinorium in the Central Sredna Gora, and the Upper Thracian tectonic trough south of it; the Strandja anticlinorium rises high to the south, and the vast Burgas synclinorium is filled by thick Upper Cretaceous sedimentary and igneous rocks to the north. The three large positive structures in the Sredna Gora Zone are highly elevated and overriden to the north. Their northern profusions have been transformed into flexures out of which large rock mass has been transported away forming overthrusts, this overthrusting before the Sredna Gora anticlinorium can be traced from the town of Zlatitsa to the town of Tvurditsa. The granites of the Sredna Gora anticlinorium core override the Upper Cretaceous and Eocene sediments of Stara Planina (Balkan) Mountain, the Cis-Balkan Deep Fault serving as an overthrust plane. Apart from the Upper Cretaceous sedimentary and igneous rocks, the plutonic massifs formed during the main structure formation – between the Late Cretaceous and the Paleogenic – are quite characteristic for the Sredna Gora Zone (Laramidian orogeny). These are the joint plutonics of the Vitosha and Plana mountains, near Plovdiv, the Monastirski Vuzvishenya hills, the Strandja Mountain, near Bourgas, etc. The repeated occurrence of faulting and the frequent change of lengthening and shrinking stresses is typical for the development of the Sredna Gora Zone. Geosynclinal troughs were formed during the fault formation in the Late Cretaceous when flysch sediments were also deposited. They were subjected to folding in the Laramidian orogeny (about 135 million years ago). A second general fragmentation – riftogenesis occurred in the beginning of the Late Eocene, and lower molass sediments were deposited in the newly formed troughs. They were folded during the general earth crust shrinking in the Pyrenean orogeny (about 70 million years ago). The next elongation occurred in the beginning of the Pliocene. It was connected with the formation of new basins (East Maritsa, Upper Thracian, Sofia valley, etc.) where fresh water upper molass sedimentation developed. These features in the Sredna Gora Zone tectonic development characterize it as a rift zone formed on the continental earth crust.

Southern Carpathians. The mountain ends of the Carpathian mountain system are located in North-western Bulgaria. In that part of the country east-facing folds are observed. They are formed of the Sinaja Cretaceous and Upper Cretaceous sediments, the so-called Krainen type, typical of the Southern Carpathians.

Kraishtides Structural Zone. It occupies a large part of South-West Bulgaria. The deep longitudinal faulting in north-south direction ($150°–170°$) is characteristic of it. The process of gradual faulting was continuous, traced since the Pre-cambrian era. Geosynclinal troughs were formed on numerous occasions as a result of the faulting. This rupturing contributed to the mosaic erosion of the Kraishtides Zone. The blocks formed in the tectonic process were of a specific type which brought about an exceptional complication of the geological structure. The Kraishtides Structural Zone bears the symptoms of a rift zone. A repeated alteration of elongation and shrinking stresses was also typical of it. Normal faults were formed under the effect of the elongation stresses along which dyke troughs originated. As a result of the shrinking the trough sediments were folded. Convergent folds formed during the transformation of the normal faults into reverse faults are also characteristic.

A small part of the *Dardan Massif* is also found in the country's south-west regions. It includes the Belassitsa Mountain, the Malashevska Mountain and the Ograzhden Mountain. This area is formed of Archaic metamorphites developed in multiple plication, the direction of the fold structures being parallel to the Kraishtides Structural Zone.

Rock formations of Triassic sandstone and conglomerates near Belogradchik

Fault Network. In Bulgaria the earth's crust is a mosaic of different sized blocks separated by faults. Some of the faulty structures are deep, i.e. they break the whole earth crust apart. They are conductors of magma which have erupted on several occasions during different geological ages. Usually the faults occur in zones, several km wide, where the earth crust is broken and diverse tectonic forms have originated (normal faults, reverse faults, overthrusts, folds, etc.) of different sizes but orientated in the direction of the deep fault. Some of the deep faults have also played a paleographic role, separating the water basins or provinces in these basins as a result of block shifting.

The structural zones in the Balkanides were formed by deep faults approximately in the direction of the parallel. They all, together with the faults bordering the Moesian Platform to the south (the Fore-Balkan Deep Fault) and the Rhodope Massif to the north (the Maritsa Deep Fault) form the Balkanide lineament. The Kraishtide Lineament intersects the whole Balkan Peninsula diagonally. Of great importance are the diagonal faults of north-eastern direction (30°–45°), especially the Etropole and Tvurditsa line dividing the lithosphere into three blocks: West Bulgarian, Central Bulgarian and Eastern Bulgarian. They had different behaviour in the geohistorical evolution. For example, sedimentary complexes of different type and varying thickness were formed at different times, including different magmatic rocks. The structural behaviour of the individual block is also different, as well as the time of the structure formation. The smaller blocks formed also had specific behaviour.

Magmatic Activity. The magmatic activity in Bulgaria which was quite diversified and intensive, especially in the older geological epochs, can be divided into three major stages: early, medium and late. *The Early Stage* is related to the formation and consolidation of the crystalline foundation widely developed in Southern Bulgaria (Rhodope Massif, Dardan Massif, Kraishtides, Sredna Gora Zone, the Sakar and Strandja Mountains). This stage occurred in the Pre-Cambrian and ended at the beginning of the Paleozoic. The magmatism in the early stage was of regional character and it marks a predomination of granite intrusions, whose formation was closely connected with the regional metamorphism. The oldest magmatic rocks have been completely altered – from regional metamorphism they have become gneisses and amphibolites. The large-sized serpentine bodies in the Rhodope Massif are also associated with old deep faultings of the earth crust. The crystalline foundation consolidation was concluded by the imbedding of acid magma, whose solidification formed the South Bulgarian granites – enormous bodies (batholiths) in the Western Rhodopes, the Rila, Pirin, Sakar and Ossogovo mountains and the Sredna Gora Zone. Despite the large size of the intrusive bodies they are monophase and are formed mainly of biotitic granites. The South Bulgarian granites are accompanied by numerous and thick aplite-pegmatitic veinous rocks often forming considerable pegmatitic fields. The country rocks where the granites have intruded have undergone intensive recrystallization. This re-crystallization was accompanied by the formation of a thick network of quartz-feldspar veins or wide magmatized zones. Magmatism ended by the implantation of single veins of granite-porphyries and diorite porphyrites.

A rather different situation is found for the rocks of the igneous-sedimentary complex, the so-called diabasophyllitoid (tribalic) formation, of probable Riphean-Cambrian age after which leukogranite and diorite bodies intruded – the so-called Strymon Diorite Formation. They were developed in the reach of old tectonic zones in the crystalline foundation – the Kraishtide Lineament Zone and the Dubna Strip.

To the earlier magmatism are related the diabases of the diabasophylloitoid complex outcropped in the cores of the Berkovitsa, Belogradchik and Shipka anticlinoriums. Diabase volcanism is of submarine character and is accompanied by intensive sedimentation of argillites, graywacks, arkoses, small-fragment conglomerates, impure quartzites and carbonate rocks. Volcanism rarely has acid character, and the more acid rocks rarely occur (quartz-porphyries and porphyrites). Diabase volcanism has related to it some iron, titanium and manganese ores.

The ore-formation significance of magmatism in the early stage is difficult to assess. It has been established that the serpentine bodies in the Eastern Rhodopes contain chrome-nickel, and the South Bulgarian granites contain iron and tungsten ores.

In the *Middle Stage* the magmatic activity was related to the development of the Paleozoic and Mesozoic superstructure in the reach of narrow geosynclinal zones (Stara Planina, Kraishtides, Sredna Gora). The volcanic and intrusive activities had a complex, multiphase character, and the metamorphic processes were limited only around the intrusions – in narrow contact zones. There are no thick aplitic-pegmatitic vein rocks.

In the early stage of the Hercynian structure formation the diabasephyllitoid formation was folded intensively and magma was intruded in it. After its setting the rocks of the so-called Stara Planina calcium-alkaline formation were established. It was found that the magma was introduced in four successive impulses. The Stara Planina granitoids are bearers of important iron and polymetallic ores (lead, copper, zinc, silver and gold). The magma of the so-called Stara Planina potassium-alkaline formation was embedded in several places in the Balkanides. These are small intrusive bodies and veins of extremely rare rocks rich in alkaline feldspars, alkaline pyroxenes and amphiboles having high potassium content. The magmatism was revived during the Stephanian and the Permian. It resulted in the pouring of lava of different chemical content into small water basins or on the earth surface in the area of the towns of Belogradchik, Berkovitsa, Vratsa, Botevgrad, up to the town of Sliven. Thus igneous rocks and their tuffs were formed. Of generous distribution are the quartz-porphyries in the Stara Planina (Balkan) mountains (in the area of the town of Sliven, Tvurditsa and Karlovo).

In the Mesozoic the magmatism moved to the south – in the Sredna Gora Structural Zone. Magmas spread amply in the water basins simultaneously with the deposition of the

Upper Cretaceous sediments forming andesites, dacites, trachyandesites (latites) and trachytes, and together with them – tuffs, tuffites, and tuffobreccias. In the late stages of volcanism residual magmas (lava) spread of predominantly trachybasalt composition and high potassium content. The Upper Cretaceous volcanism was the bearer of manganese, copper and pyrites ore formations. The intensive fold formation (Laramidian orogeny) from the Late Cretaceous to the Paleogene in the Sredna Gora Structural Zone was accompanied by a series of different composition and structure intrusions introduced in cracks. These are the so-called Laramidian plutonics. The early phases of the intrusions are represented by gabbro and gabbrodiorites, the middle ones by diorites, monzodiorites and diorites, and the late by granites, granitosyenite and syenites. Larger Laramidian plutonics are found in the mountains of Vitosha, Plana, the Monastirishte hills, near Plovdiv and elsewhere. The number of small-size plutonics in the Strandja Mountain and near Burgas is considerable. Important copper-molybdenum and iron-ore formations are related with the Laramidian bodies.

The *Late Stage* of the development of magmatism in Bulgaria began in the Paleogene and affected only the Rhodope Massif and the Kraishtide Structural Zone. It was associated with intensive elongation and ruptures of the earth crust where many depressions were formed in the Rhodope Massif. The magma spilling happened either in the water basins (e.g. in the East Rhodope Depression and the valley of the Mesta River where molass sediments were also formed) or on the surface as was the case in the middle parts of the Rhodopes. In the Eastern Rhodopes the Paleogene magmatic rocks are quite diversified: basalts, andesites, rhyolites accompanied by the respective pyroclastic rocks (tuffs, tuffobreccia and tuffites). The igneous rocks in the central and western parts of the Rhodopes, as well as in the Kraishtides are uniform rhyolites, rhyodacites, delenites which demonstrate the uniform acid lava nature. Volcanism in the Eastern Rhodopes is also accompanied by intrusive activities; small bodies of gabbro-monzonitic-granite composition have been embedded. The Rhodope polymetallic occurrences – mainly lead-zinc, often with copper, silver and gold ores are genetically connected with the volcanic and intrusive activities. The magmatic activity died down in the Neogene with the general lifting of the earth's crust. The residual basalt lava which poured in the deep tectonic zone intersecting North Bulgaria, Stara Planina and the Eastern Rhodopes along the line of the towns of Svishtov-Kurdjali was an act of volcanism. Several small volcanic foci continued to operate in the Neogene in the valley of the Strymon near the town of Petrich. Trachyandesitic lava and pyroclastites poured out from them. That was the last repercussion of active volcanism in the Tertiary.

Geological Structure

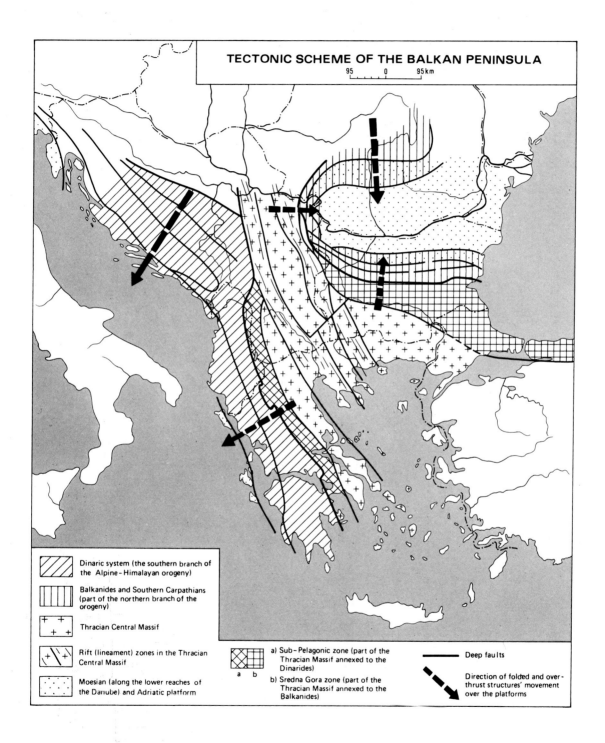

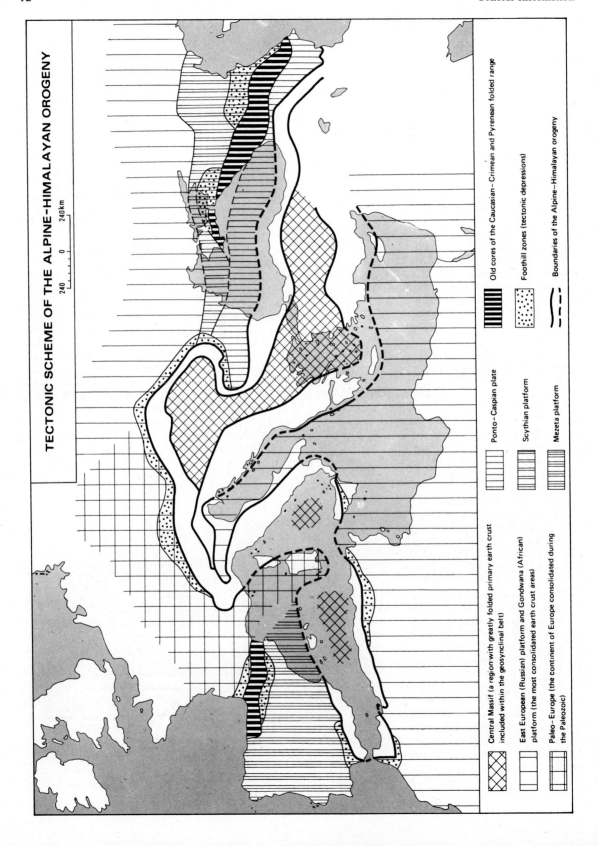

TOPOGRAPHY

The present-day topography of Bulgaria was formed as a result of different endogenic and exogenic processes. It was formed mainly in the Neogene and the Quaternary periods and corresponds to the so-called neotectonic stage in the development of the topography. It was characterized by the manifestations of differentiated epeirogenetic movements of the earth's crust of different rates and signs. A very important role in the formation of the individual morphostructural units was played by the large deep faults such as Fore-Balkan, the North Balkan, the Maritsa, the Struma (Strymon) and the Mesta faults. They include deep parts of the earth's crust, formed during different stages of Bulgaria's paleogeographic development and border the large morphographic units of the country. The characteristic feature of the topography is that the basic morphostructural units alternate in the north-south direction and spread mainly from west to east, in which direction their height also decreases. The topography is quite diverse: in relatively small regions of the country there are vast lowlands, plains, mountains, knob-and-basin topography, basins, deep gorges and river valleys. The relief is clearly divided in the vertical plane. Five altitudinal zones are markedly present. The lowland zone (0–200 m) occupies 34,881.7 sq. km, or 31.43 per cent of Bulgaria's area; that of the plains and hilly land (from 200–600 m) occupies 45,498.1 sq. km, or 41.0 per cent, the low mountains (600–1,000 m) – 16,911 sq. km, or 15.3 per cent, the medium-high mountains (from 1,000–1,600 m) – 10,899.9 sq. km or 9.77; and the high mountains (over 1,600 m) – 2,797.1 sq. km, or 2.5 per cent. Over two-thirds of Bulgaria is in the altitudinal zones under 600 m. In other words Bulgaria consists mainly of lowland, plains and hilly land. The country's average height above sea-level is 470 m.

The spread of the large morphostructural units from west to east corresponds to the continuous structural development of the lands in the eastern part of the Balkans. This is one of the reasons for dividing Bulgaria into four separate geomorphological regions.

The northernmost geomorphological region is the *Danubian Hilly Plain* developed on the Moesian Platform, an old peneplain land covered by practically horizontal Mesozoic and Tertiary sediments. In the Neogene period there were slight and slow elevations in the eastern parts and there was sinking around Varna and west of the Vit River. The southern boundary of the plain is along the contact line with the first mountains of the Balkan Range, also called the North Fore-Balkan Fault. It is about 500 km in length from the Timok River to the Black Sea. To the west it is up to 20 km wide and up to 120 km in the east in the Dobrudja. Its area is 31,522.6 sq. km, and its average height above sea-level is 178 m. The Danubian Hilly Plain is characterized by a hilly plateau-type relief which morphographically is divided into three sub-regions.

The Western Sub-region lies between the Timok and Vit rivers. In the Lom lowland vast flat interfluves prevail between the low courses of the Danube tributaries Archar, Lom, Ogosta, Skut, Iskur and Vit, from 100–150 m high, and less frequently from 200–250 m. The average height of the Sub-region is only 130 m. The flat relief of the watersheds descends towards the Danube and is slightly broken up. The tributaries of the Danube have assymetrical valleys with steep right-hand slopes and sloping left-hand slopes. In the western and south-western peripheral part of the Lom Plain the flat watershed dividing the plateaux rises to 200–250 m, and, at some places rivers have cut into deep canyon-type valleys for example Topolovets, Vidbol and others. The Zlatiata locality is mainly a plain.

The Middle Sub-region of the Danubian Hilly Plain is situated between the Vit and Yantra Rivers. In its profile the wide and meandering valleys of the Osum and Yantra rivers and their tributaries dominate, between which there are flat elevations usually 150–200 m high namely the Pleven heights reaching 317 m. The flat elevations are assymmetrical having steep western and sloping eastern slopes. Because of the development of wide river valleys, the mean height of this Sub-region of the Danubian Hilly Plain is small – 138 m. The steep western slopes of the elevations are quite indented. The river valleys also have slight indentation.

The Eastern Sub-Region is in general terms a flat plateau deeply intersected by the tributaries of the Danube. It includes the Popovo-Samuil heights; the Shumen plateau which has the highest point in the Danubian Hilly Plain (502 m); the Turnov Dyal peak; the Ludogorie; the Dobrudja; and others. They coincide with the summit of the North Bulgarian flat uplands. The flat watershed, which consists of strata eroded into landforms, occupies vast territories in the Ludogorie Plateau. It gradually rises from the west, north and east from 250 to 340–450 m.

The tributaries of the Danube which in most cases have irregular water drainage, flow in deeply intersected canyon-type valleys. The Dobrudja profile is a low, poorly eroded plateau which gradually decreases in height towards the Black Sea coast. The mean height of the Eastern Sub-Region is 204 m.

The narrow land strip along the Danube between the Timok River and Silistra is called the *Bulgarian Danubian Riverside*. It is separated from the Danubian Hilly Plain by a clearly marked slope.

The lowlands in the Bulgarian Danubian Riverside occupy about 900 sq. km. Most of this is part of the flood plain of the Danube. In the past the lowlands were flooded and swamped by the high waters of the Danube. The marshy land was drained and cultivated into arable land and meadows.

In its nothern part the Danubian Hilly Plain is covered by loess and loess-like formations ranging between 10–50 km in width. They are from 10–15 to 80–100 m thick. The loess cover gets thinner to the south, becoming loess clay. The loess cover in the Danubian Hilly Plain greatly reduces the contrasts in the ancient pre-loess relief. The loess depressions, loess funnels, and other loess features are characteristic variations in the landscape.

The linear relief of the *Stara Planina (Balkan Range) Region* lies south of the Danubian Hilly Plain, being an epigeosynclinal mountain range. To the north it is bounded by the Cis-Balkan Deep Fault. It is 25,986.2 sq. km and its average height is 523 m. The linear orientated relief in this region is determined by the structure and lithology. It is divided into two sub-regions – the Fore-Balkan and the Stara Planina (Balkan Range).

The Fore-Balkan which is the foothills of the Balkan Range has a hilly relief. Its stepped folded elevations are divided by structural and erosion depressions and have been deeply ruptured by the laterally crossing rivers. It is divided into three regions. The *West Fore-Balkan* lies up to the Vit River. The longitudinal faults were the reason for the intensive erosion which forms many elevations that slope down to the east – Babin Nos (1,108 m), Shiroka Planina, Milin Kamuk, and others. This Fore-Balkan region relief is quite diverse because of the wide structural depressions. Many of the tributaries of the Danube run along and across the Lom, Ogosta, Skut and other rivers. The erosion of the basic structures and their remodelling are the reason for the fairly small heights of this Fore-Balkan region – being 332 m.

The river Erma gorge in Jurassic limestone

The Middle Fore-Balkan, between the Vit and Stara Reka rivers has more marked relief. The stepped folded longitudinal elevations form a group. The Vassiliovska Planina mountain (with its highest peak Vassiliov, 1,490 m), and the Elena heights are in the south. To the north are the Dragoitsa, Strazhata and Kurshevo (583 m) elevations, the Devetak plateau, and others. The laterally running rivers – Vit, Ossum, Rositsa and Yantra intersect the mountain elevations by deep and narrow gorges, and their longitudinally running tributaries form wide anticlinal and synclinal valleys. The longitudinal elevations gradually increase in height to the south. This feature, as well as the great width of the Middle Fore-Balkan, greatly contributes to its average height being a considerable 420 m.

In *the Eastern Fore-Balkan* the hilly relief of the Gerlovo and Rishi valley prevails. In its western part there is the laterally formed Lisa Planina elevation (Golyman Sakar

Stara Planina from Mount Shipka

Topography

The river Krichimska

Peak, 1,054 m) and the Slannic hilly area; the Preslav-Dragoevo mountain and many low elevations intersected by the lower courses of the Luda Kamchia and Eleshnitsa rivers lie to the north and east. The prevailing low elevated hilly relief is the reason for the small average height of the Eastern Fore-Balkan (296 m).

A wide karst relief with many caves, chasms, grikes and a saxicolous cornice has occurred in the limestone rock complexes of the Fore-Balkan, and is rich in karst waters. They surface as numerous cold fresh-water springs.

Stara Planina (the Balkan Range) is the longest mountain range in Bulgaria. It was known to the ancient geographers by the name of Haemus. Its Turkish name, Balkan mountain, dates back to the beginning of the 19th century and the Balkan Peninsula was named after it. The Balkan Range lies in an arch-form crossing all the middle part of Bulgaria from the Timok river to the Black Sea (Emine Cape). Whilst the middle part of its ridge is the watershed between the Danube and the Maritsa rivers, its western part is intersected by the Iskur River in a gorge and is totally in the Danubian drainage area. It is divided into three regions.

The Western Balkan Range lies between the Belogradchik and the Zlatitsa passes. Its arch-like ridge consists of several separate sectors: the Sveti Nikola mountain (1,721 m), the Chiprovtsi-Berkovitsa mountain (Midjur peak 2,168 m), Koznitsa, and others. The northern mountain slope rises steeply above the longitudinal valleys of the upper courses of the Lom and Prevalska Ogosta rivers along a system of parallel normal faults. East of the Iskur Gorge, which is a very beautiful natural phenomenon, lies the Murgash sector of the mountain. North of it is the deep and fruitful Botevgrad valley, and east of it is the Etropole valley. The southern slope of the Western Balkan Range is less steep and is made of three deep karst surfaces: Koznitsa, Ponor and Zaburdo. This slope descends in steps along fault lines to the Sofia plain. The Western Balkan Range has a mean height of 849 m.

The Middle Balkan Range is situated between the Zlatitsa and the Vratnik passes. This is the narrowest but the most

The foothills of the Central Stara Planina near the Dryanovo Monastery

The river Iskur gorge at Lakatnik

massive part of the mountain. Its ridge is carved with deep saddles (Ribaritsa, Troyan, Shipka). Around the Pass of the Republic, the Middle Balkan Range sharply decreases in height to 680 m. The Zlatitsa-Teteven mountain (Vezhen peak, 2,198 m), the Troyan-Kalofer Mountain, (Botev peak, 2,376 m, the highest in the Balkan Range), the Shipka-Tryavna mountain (Ispolin peak, 1,524 m), and the Elena-Tvurditsa mountain (Chumerna peak, 1,536 m) stand out. The ruptured southern slopes of the Middle Balkan Range are very steep. The northern slopes are slanting and bounded by a number of orographically divided longitudinal tectonic-erosion depressions developed in the uppermost course of the Cherni Vit, Beli Vit, Cherni Osum, Vidima, Rositsa, Yantra, Elenska Reka and Stara Reka rivers. Only isolated parts of the northern slope are steeper. The steep slopes of the Kozya Stena and the Severen Djendem are quite impressive. Rounded (near Botev and Paskal peaks) and flat (near Revnets peak) ridges are outlined in the highest sectors of the Middle Balkan Range at a height of 1,800–1,900 m which are in sharp contrast with the steep faulted slopes under them. The rounded peaks of this highest part of the Balkan Range rise 200–400 m above them. Because of its high elevation in the Neogene-Quaternary its mean height is 961 m.

The Rhodope mountains

The Eastern Balkan Range lies between the Vratnik pass and the Black Sea. It is considerably wider than the other two mountain sectors. From the longitudinal valleys of the Luda Kamchiya and Hadjiiska rivers it is divided into two mountainous and hilly ranges. To the north lies the Kotel-Vurbitsa mountain (Razboina peak, 1,128 m). It has a steep northern normal fault slope ending in the steep slopes of Zlosten and Yurushka Stena. Its ridge is narrow and bare. East of the Luda Kamchia gorge, the Kotel-Vurbitsa mountain is split by the longitudinal valley of the Droinitsa River forming the Kamchia Mountain (ending at the Black Sea near the cape of Sveti Atanas in a high cliff), and the steeply sloping Emine mountain (ending at the Black Sea at the Cape of Emine). These two mountains together are called the Mator mountain. The ridge west of the Luda Kamchia gorge is high. East of it the region acquires a low mountainous and hilly profile. A second row of mountainous and hilly elevations rises to the south, in an west-east direction, under the common name of the Udvoi mountain, Rather higher are the Sliven mountain (Bulgarka peak, 1,181 m), Stidovo mountain (Ushite peak, 1,011 m) and Grebenets mountain (Gavanite peak, 1,034 m). East of the Songurlare valley are the low Karnobat mountain and the Aitos mountain. The southern slopes of this mountainous to hilly series are placer-type, steep and very well outlined. The Luda Kamchia depression lies between the north and south mountain series and is formed in a longitudinal synclinal structure. The Luda Kamchia and Hadjiiska rivers flow through it. In many places the flysch sediments are susceptible to erosion. The low-mountainous and hilly relief of the Eastern Balkan Range is related to its lower elevation in the Neogene and for this reason its average height in only 385 m.

The morphometric data for the Balkan Range system emphasizes the relief differences between the Balkan Range and the Fore Balkan. Forty-one per cent of the Balkans and 90 per cent of the Fore-Balkans are in the lowland and hilly belt (being under 600 m). Fifty-five per cent of the Balkan Range and only 10 per cent of the Fore-Balkans can be classified in the low mountainous and medium-high mountainous belt. Because of these differences in the vertical development of their relief, the average height of the Balkan Range is twice that of the Fore-Balkans. In the heights and elevations of the European part of the Alpine-Himalayan orogen these hypsometric differences in the parts of the Balken Range's orogenic region indicate the existence of moderate differentiations in their elevation during the Neogene-Quaternary.

South of the Balkan Range lies the third largest geomorphological unit in Bulgaria, called the *Transition Mountain-Depression Region.* It has greatly differentiated mountain-depression relief developed over medium and small block fractured epigeosynclinal and epiplatform morphostructures resulting from the transition from the Alpine geosynclinal structures to the Rila-Rhodope intermediary massif. To the north this region is bounded by the Fore-Balkan Fault, and to south by the North Rhodope Fault, which moved significantly in the Neogene-Quaternary. The differentiated neotectonic movements were the main factor in the formation and outlining of its complex relief which resembles gigantic mosaics. The region has an area of 30,707.4 sq. km, and an average height of 402 m. Linear relief shapes are predominent in the epigeosynclinal parts of the Transition Region (Kraishte, Sredna Gora, Strandja and the Sub-Balkan depression), and elongated and isometric block forms in the epiplatform parts (Ihtiman, Sredna Gora and Sakar Planina mountain). The

Topography

significance of the practically meridional faults becomes greater to the west in the Kraishte and the Ihtiman Sredna Gora mountains. The transistory region is divided into four sub-regions.

The West Mountain-Depression Sub-Region is situated west of valley of the Topolnitsa river. It is characterized by the longitudinal ridges running mainly from north-west to south-east. The many sharply outlined mountains: Konyavo (Vidin Peak, 1,487 m), Golo Burdo (Vetrushka peak), Strazha (Lyubash peak), Vitosha (Cherni Vruh peak, 2,290 m), Liulin (Dupevitsa peak, 1,256 m) Viskyar (Mechi Vruh peak, 1,077 m), Zavalska mountain (Kitka peak, 1,181 m) and the mountains in the Kraishte (Golemi Vruh peak, 1,481 m) have a mosaic structure and are divided by longitudinal depressions (Kyustendill, Radomir, Pernik, Trun and Ihtiman). In these parts of the Transition Region there are close-patterned joint blocks. The Sub-Region is drained by the upper courses of the Struma (Strymon), Iskur and Maritsa Rivers and their tributaries. The general tendency of elevation during the Neogene-Quaternary is the reason for the average height of the Sub-Region being 802 m.

The Sredna Gora Sub-Balkan Sub-Region has a much more clearly expressed linear distribution of the morphological relief elements. The Sushtinska Sredna Gora (Golyam Bogdan peak, 1,604 m) and Surnena Gora (Bratan peak, 1,236 m) mountains gradually diminish in height in the east and finally flatten out in the direction of the Tundja river. The southern slopes of these mountains (parts of the Sredna Gorge ridge) descend to the Upper Thracian Lowland in steps along fault lines. The northern part of the Sub-Region is occupied by the Sub-Balkan val-

Mount Malyovitsa in the Rila mountains

Mount Vitosha

The middle reaches of the river Maritsa

this corridor is simply cut off by the Cis-Balkan fault. The lesser elevation of the Sredna Gora orogen and its formation by the Balkan Range by means of the Sub-Balkan valleys is the reason for the relatively lower average height of the Sub-Region (495 m).

The Upper Thracian-Middle Tundja Sub-Region runs parallel in the south to the Burgas lowland, which is occupied by lowlands and hilly lands along the middle courses of the Tundja and Maritsa rivers. To the north it is bounded by the southern faulted slopes of the Sredna Gora and the eastern-most parts of the Balkan Range, and to the south by the northern faulted slopes of the Rhodope, Sakar and Strandja mountains. The Upper Thracian Lowland (with an average height of 166 m) is the largest plain in Bulgaria and the Balkans. It is the northern part of the historic province of Thrace. The Chirpan hills divide the plain into the Pazardjik-Plovdiv and the Stara Zagora plains which are alluvial formations. The lower part of the Sub-Region also includes the Yambol-Elhovo alluvial plain, and the Pliocene aggraded plains situated in the north. Above the plains rise the hilly elevations Manastirski Vuzvishenia, the Hissar-Bakadjik drop structure and the Sveti Ilia heights (416 m). The Sub-Region has an average height of 195 m.

To the south is the hilly *Sakar-Strandje Sub-Region*. It includes the Sakar mountain (Vishegrad peak 856 m), the northern branches of the boundary-forming Derventski hills (Gurenbair peak, 556 m) and the Bulgarian part

A panoramic view of the Strandja mountains

The Sredna Gora mountains

leys – the Sarantsi, Kamartsi, Zlatitsa-Pirdop, Karlovo, Kazanluk, Tvurditsa and others – forming a west to east chain. The Karlovo and Kazanluk valleys are known by the romantic name of the Rose Valley where three-quarters of the attar-yielding roses in Bulgaria are found. The Sub-Balkan valleys are separated by lateral mountain-drop structures such as the Gulubets, Koznitsa, Strazhata and Mezhdenish elevations. The northern boundary of

of the Strandja mountain (Gradishte peak, 710 m). The Sub-Region is separated from the southern parts of the Yambol-Elhovo plain and the Bourgas lowland by flexure and fracture slopes. The step-like southern slopes of the Sakar mountain reach the hilly relief by the Maritsa River.

'The Colt' ridge between Mounts Banski Suhodol and Kutelo in the northern Pirin mountains

The flat ridges are dominating elements in the relief of the Strandja mountain and the Dervent heights.

In general terms, the Sakar-Strandja Sub-Region is a flat arched elevation cut by the Strema gorge of the Tundja river.

The Sakar mountain dome stands out in the western part of the arch. The average height of the Sub-Region is 166 m.

The relative sinking of the Upper Thracian-Middle Tundja Sub-Region and the slight elevation of the Sakar-Strandja Sub-Region in the Neogene-Quaternary outline the southern and the eastern parts of the Transition Region as a faulted intermontane depression with a mean height of 185 m, while the Transition Region has a lower mean height (402 m) in comparison with its neighbouring regions.

South of the North Rhodope Fault is the fourth largest, geomorphological region in Bulgaria, the epiplatform *Rila-Rhodope Massif*. The structural disintegration of the Massif in the late Alpine period and the neotectonic stage caused the marked disintegration of its relief as a result of differential block movements. The mountainous relief of the Region is part of the vast Macedonian Thracian Massif of which the Ossogovo-Belassitsa mountain group, Rila and the major part of Pirin and the Rhodope mountains are in Bulgaria. The area of the Rila-Rhodope Massif is 22,771.6 sq. km, and its average height 896 m. Differentiated vertical movements in the Neogene-Quaternary caused considerable differences in the relief type of the different parts of the Massif. Nearly meridional faults divide the Massif into four sub-regions.

The mountains of the *Ossogovo-Belassitsa* (Ossogovo-Middle Strymon *Sub-Region* rise west of the Strymon (Struma) Fault, most of which is in Yugoslavia. The mountains of Ossogovo (Ruen peak, 2,251 m), Vlahina mountain (Kadiitsa peak, 1,924 m), the Maleshevska mountain (Ilyov peak, 1,803 m), Ograzhden (Markovi Kladentsi peak, 1,525 m) and Belassitsa (Radomir peak, 2,029 m) are prominent in this region. The Ossogovo and Ograzhden mountains have quite broadly developed flat drainage ridges. The eastern slopes of the mountains descend steeply along step-like faults to the graben valleys of the Struma (Strymon) and Strumeshitsa rivers. Of particular interest is the boundary mountain of Belassitsa, a horst-shaped block formed between two parallel normal faults before the early Pliocene. The average height of the Ossogovo-Belassitsa Sub-Region is 806 m.

The *Rila-Pirin Sub-Region,* which contains the highest mountain in Bulgaria, Rila (Mussala peak, 2,925 m, the highest in Bulgaria and the Balkans) and the Pirin mountain (Vihren peak, 2,915 m), rises betweeen the Strymon and Mesta fault valleys. Rila is a raised arched horst formation where two fault systems occur – a concentric and a radial one. The concentric system is formed of normal faults along which thermal mineral springs occur in places. The radial system of normal faults has

A view of the Pirin mountains

Mount Mussala and Lake Aleko in the Rila mountains

determined the orientation of the river valleys. Rila has an Alpine relief. There is good evidence of the Würm Glaciation – kettles, rock bars, cirques, moraines, glacial lakes (of 233 lakes, 189 are from the Ice Age, and poetically called 'the blue eyes of the mountain'). Many chalets and hotels, convenient roads, ski-lifts and drag-lifts have been built to serve the tourist industry. The Rila Monastery is sited on this mountain. Pirin rose as a high mountain during the Upper Tertiary and the Quaternary along peripheral faults where there are mineral water springs. The mountain has marked Alpine relief with numerous Würm Ice-age forms: kettles, cirques, moraines, glacial lakes (176 in Pirin). Erosional relief of an advanced form and raised in the Quaternary above the snow-line also partially reworked by glacial activity has been preserved along the top of the mountain. The high elevation of Rila and Pirin in the Neogene-Quaternary is the reason for the mean height of this Sub-Region being 1,285 m.

The *West Rhodope Sub-Region* is a vast mountainous land. Its main relief elements are the thick ridges of the Velika-Viden Sector (Srebren peak, 1,901 m) to the west, and the Perelik-Prespa sector (Golyam Perelik peak, 2,191 m, the highest in the Rhodopes) to the east. Quite high and wide ridges – Alabak, the Batak mountain (Golyama Siutka peak, 2,186 m), Chernatitsa (Golyam Persenk peak, 2,091 m) spread meridionally to the north of these ridges. To the east the Rhodope watershed ridge is divided into two by the Arda River valley, south of which is the boundary Arda's sector. Mountain ridges, 1,450–1,650 m high, occupy 25 per cent of the area of this Sub-Region. Above them rise the domes of the highest peaks, and below them lie deep river valleys. The Rhodope mountains are remarkable for the amazing beauty of their natural formations such as the Devil's Bridge on the Arda river, the Rock Bridges and the Trigrad gorge. National Holiday Resorts such as Pamporovo, Byala Cherkva and the spa centres of Velingrad and Narechen have been set up. Convenient roads have been built. The raising of the Sub-Region in the Neogene-Quaternary is the reason for its mean height of 1,098 m.

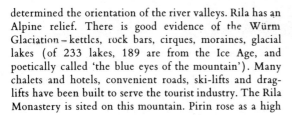

The view from the Pirin mountains

Topography

Mount Vihren, the highest in the Pirin mountains, and Mount Kutelo

The rocky shoreline of the Black Sea near Sozopol

The *East Rhodope Sub-Region* includes the hilly and lowland relief along the course of the River Arda and east of the Rivers Kayliika and Vurbitsa. The highest parts of the Sub-Region are the frontier ridges of Snezhnik and Muglenik, north of which the Strumni Rid and Iranski Rid strike meridionally. The Gorata hilly elevation strikes from west to east in a northerly direction going down steeply to the Maritsa valley. Deep gorges and vast valley-widenings have been carved by the rivers into the elevations. The lower elevation of the East Rhodope Sub-Region in the Neogene-Quaternary period is also the reason for its present smaller height – 329 m.

The Bulgarian parts of the Rila-Rhodope Region emphasize the predominantly high mountainous character of their greatly-raised relief. The hypersometrical development of the Sub-Regions illustrates their specific orographic conditions – more than half of the area of the West Rhodope Sub-Region is occupied by a medium-high mountain belt due to the mass occurrence of high-levelled surface terrains, and 30 per cent of the Rila and Pirin area is included in the high-mountainous belt because of the higher elevation of surfaces of the same age.

The *Bulgarian Black Sea Coast,* because of its specific climatic features and relief, although not being a separate, homogeneous, geomorphological region, takes the shape of a stripe of different widths along the Black Sea. It developed within the reach of the eastern parts of the Danubian Hilly Plain, the Balkan Range and the Transition Region. It strikes from Cape Kartal in the north to the River Rezovo in the south. To the west its boundary is conventionally determined by the climatic effect of the Black Sea. Two large bays (the Varna and Burgas) and numerous inlets are cut into the shore. The capes Kaliakra, Emine and Maslen Nos protrude furthest into the Black Sea. The shore is particularly picturesque near Cape Kaliakra. South of Cape Galata and along the Balkan Range coast the shore is steep, high and quite indented. Near the Bay of Burgas it becomes much lower. There are several small rocky isles in front of the Bay of Sozopol which are the only ones off the Bulgarian Black Sea Coast. In many places there are long and white sandy beaches. At Nessebur the beach widens and the winds have formed small dunes up to 20 m high. The miniature beautiful sand

A fiord-like inlet south of Sozopol

The Zmiiski Isle south of Cape Maslen

desert of Kumluk, the only one in Bulgaria, was formed there. It is formed of sand from sea accumulation. In the Middle Ages Kumluk was a place of solitude and contemplation for Hesychasts.

The Bulgarian Black Sea Coast is a first-class national and international holiday resort. A special programme has been approved for environmental protection, restoration and improvement on the Black Sea-Bulgarian Coast which will guarantee the protection of the specific landscape features at a time of fast development of its natural resources. The national and international holiday resorts of Albena, Zlatni Pyassatsi, Druzhba and Slunchev Bryag are to be found on the Black Sea Coast.

Modern Morphogenetic Processes. The exogeneous processes connected with local physico-geographical conditions have been of great importance for the creation of the modern landscape. Weathering, erosion and accumulation are determined by hydroclimatic and plant vegetation factors and are manifested under specific structural, lithological and neotectonic conditions. In the mountainous zones over 2,300 m there are long periods of temperature fluctuations around freezing point (0°C; frosty days) and under 0°C (freezing days), a total ranging from 200–270 days a year. A characteristic feature of this climatic boundary which is also of morphological significance is the fact that above it the number of freezing days ranges from 100 to 180 and of the frosty around 100. This creates conditions for the development of lasting cryogenic morphological processes caused by the low temperatures. In this belt weathering plays a considerable part. Moraines, rockslide terraces, etc., are formed. These forms, along the flat domelike peaks in the Rila and Pirin, have an insignificant total area. A subcryogenic belt developed below the cryogenic one, where the number of frosty days goes up to 120, and there is a sharp reduction of the freezing days (under 60). This also sets up good conditions for frost weathering during transitional seasons. Herbacious vegetation and coniferous shrubbing – dwarf pines – are characteristic of this belt, which descends to a height of 1,900–1,800 m. Grass strips and steps along the high mountainous pastureland are typical of its morphology. Further below is the forest belt, where the chemical and biogenic weathering are also of great importance together with the physical weathering.

Modern thickening of the slope sediments is observed concurrently with the weathering processes in the mountain foothills, which is typical both for the normal fault slopes (the southern foothills of the Balkan Range, etc.) and for the regions made of easily destroyed bogs and tuffs (the plateau foothills in the eastern part of the Danubian Hilly Plain and the tuff heights in the Eastern Rhodope Sub-Region). The finds of cultural remains of the Middle Ages at a depth of more than 10 m under the surface of these sediments (deluvial fans), the bogging of cultivated land and the piling of individual trees to the base of their crowns show that intensive modern deposits exist in the foothills of the mountains.

Because of the continuous cultivation of arable land and the soil structure destruction, there is soil erosion in places in Bulgaria. It has developed in different degrees depending on the slope and human activities. Zones having small-scale river and ravine networks (under 1.5 sq. km) include the flat land in the Danubian Hilly Plain, the Valley kettles in the Transition Zone, i.e. zones where the erosion gullying has been hampered by the lack of gradients. Zones having large densities of valley and plain networks (over 2.0 sq. km) have developed in the Rila-Rhodope Massif and in the Fore-Balkans. Deflation (wind erosion) appeared in many places establishing unstable microforms and limited, insignificant changes in the relief. This is caused by frequent, very strong winds with a velocity of 5–12 m per second, the droughts and the human disruption of the vegetation and top soil. Technogenic processes are the most recent and the most significant human relief formation factor in Bulgaria. A large number of favourable and unfavourable man-made forms have been formed in connection with mining of coal and other ores or minerals, road-building, irrigation and other types of civil engineering works such as mining tips, foundation pits in quarries, excavations for modern motorways and railway lines, etc. The lands affected by great relief distortions are subject to systematic reclamation. Tips are levelled off, the excavations are filled, and others are used as water reservoirs.

MINERAL RESOURCES

Coal. Following generally accepted international standards and according to the degrees of carbonification all types of coal are to be found in Bulgaria: lignite (soft brown), hard brown (matt and shiny), black (grades: long-flame, gas, fat, coking, lean, sintering) and anthracite. With just one exception they have a high ash and sulphur content which in some cases demands extra treatment. The lion's share of the principal deposits contain lignite coal – some 76.8 per cent (1980). About 40 coal areas, conventionally grouped in 15 coal basins, have been established. In most of the deposits the coal-bearing formation is thin (the maximum being about 3,200 m in the Dobrudja Basin). The coal sediments are viscous in form, and the coal-beds (with the exception of the Dobrudja Basin) are few in number, thin to medium-thick and limited in area (in the Pernik and Pirin Basins under 30 m thick, and under 40 m in the Sofia Basin). The coal-beds in the Svoge and Balkan Basins, which are to be found within the Balkanides – an active tectonic zone – are greatly folded and in some places fractured or ground. The coal-beds are deeply buried (the Dobrudja Basin), lie near the surface, or are outcropped (the Smolyan, West Maritsa, Elhovo, Chukurovo and other basins). Most of the Bulgarian deposits were formed under marsh-like conditions (the Pernik Province, the East Maritsa and Sofia Basins). A small part has been formed under littoral conditions (Burgas Basin) or in a tectonic trough and local bogland (Cis-Balkan Province). The largest number of basins in Bulgaria are to be found in zones of active tectonic manifestations (within the Balkanides, the Sredna Gora Zone and the Kraishtides) but there are also deposits which have features of the most compressed parts of the earth's crust (in the Moesian Platform – the Dobrudja and Lom Basins). Most of the coal is autochthonous. Allochthonous coal (formed of plants or transported from another place) is found only in the Dobrudja Basin, and the coal of the Burgas Basin is of a mixed autochthonous-allochthonous type. Five stages – five maximum degrees of carboniferous conditions were established, of which the late Paleozoic and Neogene are of the greatest significance. The Late Paleozoic is most clearly expressed in Late Carboniferous of about 250 million years ago, in the region of the West Balkanide Zone (Svoge Basin) and the most easterly part of the Moesian Platform (the Dobrudja Basin). There were also favourable conditions for the accumulation of coaly sediments in the Jurassic (mainly in the Early Liasic) in the West Balkan Province, and the Late Cretaceous (in the Cenomainian-Turonian) – in the West Sredna Gora Zone, the Balkan and Strandja Provinces. The Paleogene coal-forming stage (Late Eocene-Oligocene) is a typical feature of the Bulgarian lands, when peat which had become transformed into hard brown coal was deposited within the Rhodopes, the Sredna Gora Zone and the Kraishtides (the Pernik, Bobov Dol, Pirin, Burgas and other basins). The youngest, the Neogene stage (Miocene-Pliocene), is of the greatest economic importance for Bulgaria. At that time favourable conditions for the formation of lignite coal were established in the large fresh-water troughs – in the East Maritsa, West Maritsa, Sofia, Lom, Kyustendil Basins, and elsewhere.

Coal has a domestic use as fuel; it is also a major source of power: over 80 per cent of the electricity produced is from coal-fired power stations. In addition to heating and the generation of electricity, coal is also used in the processes of gasification and carbonization and for the production of high-temperature coking coal (black coal). Brown (lignite) coal is used as fuel either in its raw form or after dehydration and briquetting.

Oil and Gas. Relatively small gas, gas-condensate and oil deposits have been prospected in Bulgaria. Naturally consistent and hard bitumens (malt, asphalt, asphaltitite, etc.) have also been established. Oil gas found in different deposits is composed of the lower grades of the methane homological series. In the steam-phase condition, higher carbohydrocarbons are dissolved in the gas condensates (liquid under normal pressure and temperatures). The oil in most cases consists of hydrocarbons where gaseous and hard hydrocarbons ('paraffins') of different quantities, and fewer porphyrins, resins, and asphaltenes are dissolved.

Since 1947 large-scale surveys have been made in this country on the basis of complex geologo-geophysical methods. The oil and gas deposit near Tyulenovo village, Tolbuhin district, was struck in 1951, the Dolni Bubnik gas and oil, the Gigen oil, and the Chiren gas condensate deposits in 1962–1963, and the deposits in the central part of Northern Bulgaria including numerous very small deposits of no industrial importance were discovered in 1975. The oil and gas deposits in Bulgaria have complex geological structures and are small in area. Their water-retaining properties are poor and they lie deep below the surface. Industrial deposits (1983) are: the Burdarski Geran, Dolni Lukovit, Dubnik and Tyulenovo (gas and oil), the Chiren, Devetaki and Pisarovo (gas condensate). These deposits are mostly in Northern Bulgaria. The central part of the Moesian Platform (the area of the Pleven gravitational maximum) is of greatest interest. They are found at a depth of 140 m (at Tyulenovo) to 4,000 m (at Devetaki). The depth of the industrial deposits is from 100 m to 4,690 m. The geological age of the established deposits is from Mesozoic to Paleogene. The lithological composition of the reservoir rocks also varies widely – from carbonates to terrigeneous. The prospecting of new deposits of oil and gas is basically oriented towards the Mesozoic and Tertiary formations in Northern Bulgaria and the Black Sea shelf where there are promising oil and gas structures.

Ores. According to international standards the Bulgarian ore deposits are medium-type (few in number), small in size (several dozens) and non-industrial (several hundred types of ore manifestations). The ferrous and non-ferrous ores are of primordial importance, with those of the noble metals ranking second.

Of the ferrous metal ores in Bulgaria, the most important are the deposits of iron ore (Kremikovtsi). The deposits of manganese ores are of secondary importance and those of chromium and titanium are insignificant. The non-ferrous metal ores (copper and polymetallic ores) form a number of quite important deposits for industry in the Rhodope Zone (the Madan ore field), the Sredna Gora Zone (Medet, Elatsite, Asarel, Chelopech and others), and the Stara Planina (Balkan) mountain (Sedmochislenitsi, and others). The non-ferrous metal ores are mainly used for the production of noble (precious) and rare metals and dispersed elements (molybdenum, tungsten, bismuth, cadmium, rhenium, etc). The separate deposits of gold, tungsten and molybdenum ores are insignificant. The formation of deposits of fluorite, barytes, sulphur (pyrites) and some other industrial non-ore minerals are most closely connected with the ore formation processes.

Of the endogenic ore deposits, the magmatic deposits (containing insignificant concentrations of iron, titanium-iron and chromium ores), pegmatite deposits (a source of mainly non-ore minerals), the skarn deposits (a secondary source of iron and copper ores) and albitite-greisen ore outcrops (being only of mineralogical importance) are represented. The hydrothermal deposits are the basic source of ferrous, non-ferrous, noble and rare metals. The deposits of hydrothermal-sedimentary genesis (manganese pyrites and polymetallic-pyrites), the sedimentary deposits (iron) and the placer deposits (gold, iron) are of secondary importance.

Metallogenic Development. Bulgaria is included in the Mediterranean metallogenic belt – a planetary metallogenic unit of approximately equatorial orientation, about 18,000 km long and under 1,000 km wide – the contours of which coincide with the boundaries of the Tethys geosyncline. Its metallogenic development includes the interval from the Archaen to the Quaternary but the Alpine metallogenic age is of primodial importance (for Bulgaria too). Within the Caledonian-Hercynian and Alpine age the ore formation processes were cyclic, distinguishing between an early, middle and late stage of development.

Pre-Paleozoic Metallogenesis. The Archaean and Proterozoic rock complexes form a vast area in Southern Bulgaria (the so-called Pra-Rhodope Massif). Geosynclinal vulcanogenous-sedimentary rock complexes with potential metallogeny mainly of siderophillic elements (iron, nickel, cobalt, platinum, gold) predominate. It is possible that as a result of deep erosion, part of the deposits were washed away and another part liquidated during metamorphic processes. So far small ore formations (including non-ore minerals) whose future application is uncertain have been established. Iron and manganese bearing igneous sedimentary layers have been formed in the base of the Proterozoic rock complexes. By metamorphic processes they have become magnetite-haematite quartzes (jaspilite) and in manganese rocks (gondites) but are of no practical value. Other Pre-Cambrian ore formations are of magmatic genesis. These are slightly altered chromatite ore formations in serpentinized ultra-basic massifs and insignificant copper-nickel and titanium-iron mineralizations (in the area of Topolovgrad and Blagoevgrad). As a result of hydrothermal and metamorphic effect magnesite, megmetite, asbestos, vermiculite and other mineralizations were formed in the ultrabasites.

Celedonian-Hercynian Metallogenic Age. According to some authors, in the early stage of the age a rift occurred. Other authors believed an eugeosynclinal movement took place. An ocean-type earth crust started to take shape gradually from south to north. Thus, in the beginning (in the Upper Ryphean) in the Kraishtides the Kraishtide diabasophyllitoid complex was formed, and later in the Cambrian-Ordovician the Stara Planina (Balkan) mountain diabasophyllitoid complex in the zone of the present-day Western and Central Stara Planina mountain appeared. Iron-bearing (magnetite and haematite-magnetite) layers, very likely of the metamorphic origin, were established in the Kraishtides among chlorite schists and shales. Of primordial importance were the ore-formation processes related to basic volcanism in the Stara Planina Zone. Hydrothermal-sedimentary iron-ore (syderite) and manganese-bearing layers, as well as hydrothermal veins and metasomatic pyrite deposits with little copper have been formed. Of the regional low-degree metamorphism the magnanite ores have been transformed into gondite. The iron and copper-ore deposits underwent deformations and mineralogical alterations also took place but these ores are of no practical importance. Granitoid magmatism developed in the middle and late stages of the Caledonian-Hercynian age. The oceanic crust in the area of the Stara Planina Zone and the northern sector of Kraishte was granitized. On the basis of data from lead-isotope model age determination magmatism had two maxima: from 360 to 320 million years (Early and Middle Carboniferous) and from 280 to 240 million years (Permian). The first maximum is related genetically to the formation of gold quartz vein deposits in the Berkovitsa and Trun areas. Very likely some late magmatic iron-titanium (magnetite-ilmenite) mineralizations in gabbrovian rocks (near Chiprovtsi), hydrothermal vein quartz-molybdenum ore formations (near Bov-Sofia district, and near Trun) and marble deposits were also formed at that time. Hydrothermal metasomatic and vein medium- to low-temperature silver-lead and polymetallic deposits and ore outcrops were formed between the two maxima bearing no relation to concrete magmatic processes. Genetically more diversified ore-formation processes are related to the second maximum magmatism. Part of the syderitic layers which were subjected to granite intrusion thermal effects were transformed into high quality magnetite ore (near Martinovo village, Mihailovgrad district). Skarns are formed when they come into contact with marble. They are a favourable medium for metasomic processes – for substitution from hydrothermal high-temperature molybdenite, sulphide-arsenide and other mineralizations. Hydrothermal quartz-vein gold and polymetallic ores are also formed, as well as secondary hydrothermal (re-deposited) iron-ore mineralizations. In the Chiprovtsi ore zone, several ore-formation intrusions occurred in the course of the Paleozoic (a unique metallogenic phenomenon); later in the Alpine metallogenic age barytes veins were also formed.

In the Pra-Rhodope Massif (including the Rila-Rhodope area and parts of the Kraishtide and Sredna Gora Zone) granite magma was intruded, thus giving rise to South Bulgarian granites (some of the batholitic type). The continental type of earth-crust which had already formed was the magmatism medium. The formation of chalcophyllic and lithophyllic elements was related to magmatism. The subsequent erosion processes were quite intensive and for this reason insignificant skarn and pegmatite mineralizations mainly of non-metallic minerals resulted. In the Carboniferous and the Permian, basalt-andesite magma was intruded in parts of the Western Fore-Balkan. Magmatism is of an insular-arch type. Genetically, copper and lead-zinc ore outcrops were related to it. The ore formations in the Gramatikovo ore field are iron sulphide (pyrite, copper-pyrite and copper-zinc). They are associated with basalt-rhyolite geosynclinal volcanism and have hydrothermal sedimentary genesis. The ore formations are regionally rather poorly metamorphosed. Their age was determined differently – in the interval from the Paleozoic to the Jurassic.

Alpine Metallogenic Age. The ore-formation processes occurred on numerous occasions in the Mesozoic and the Cenozoic within Bulgaria. Many of them were significant by virtue of their scope and have determined the metallogenic distinctions of the different morphotectonic units. In the early Alpine stage (Triassic-Jurassic) deposits and ore outcrops of iron and a few manganese ores were formed, although in South-east Bulgaria they were poorly metamorphosed. The manganese ores were limited to the boundary between the Lower and Middle Triassic but their genesis has not been clearly identified. The iron ores were formed predominantly during the Jurassic but in the Haskovo district hydrothermal-metasomatic haematite-magnetite deposits were established among the Middle Triassic limestones which had subsequently been metamorphosed. It can be assumed that it was related to the Triassic magmatism outcrops. Of greater economic importance are the Jurassic epicontinental sedimentary iron ore layers in the Troyan ore district. The southern areas of the Moesian Platform were on many occasions subjected to tectonic transformations from the early to the middle Alpine stage (Triassic-Jurassic-Cretaceous), and the structures of the Balkanides were gradually formed. Outcrops of Triassic and Jurassic magmatism have also been established but the metallogenic phenomena were not related to them, but to the origins and multiple activization of faults (in the west-north-west direction) in the region of the Western Stara Planina (Balkan) mountain and the northern periphery of the Western part of Sredna Gora Zone. The ore formations are hydrothermal and low-temperature; they are composed of sulphide copper-lead-zinc, iron, carbonates, oxides and vein barytes. The Stara Planina front strip in the sector from the state boundary with Yugoslavia, south of Belogradchik to the Botevgrad area, is a relatively well-defined metallogenous zone. The southernmost zone can be traced along the line Slivnitsa–Kremikovtsi village (today a suburb of Sofia) and is characterized by iron deposits and ore outcrops. The main metal deposit occurred in places where the faults of approximate west-east orientation are intersected by a deep fault of north-north-westerly direction. In this way the deposits in the Iskur ore district – Kremikovtsi, Sedmochislenitsi, Plakalnitsa, Sokolets, Izdremets, and others were formed. The largest ore accumulations in the Iskur ore area are of the layer type (stratiform) and were mainly formed by the replacement of Middle Triassic limestones. The exact age of these ore formations has not been established. The ore formation is multi-stage and it is quite likely that it had interrupted, between the Middle Triassic and the Late Cretaceous. The formation of barytes veins was the last phase of the Sredna Gora Zone metallogenic development.

The disintegration of the Pra-Rhodope Massif was concluded in the Middle Alpine Stage (Cretaceous), whereas that of the southern part of the Moesian Platform continued. Granitoid magma was intruded amongst the territory of the Rila-Rhodope Massif. According to some geological chronometric determinations granitoid magmatism continued to the Early Paleogene. Hydrothermal, mainly veinous, high to middle-temperature tungsten, molybdenum and fluorite molybdenum ores were formed in connection with magmatic processes. The Sredna Gora tectonomagmatic and metallogenic zone was formed in the Late Cretaceous. The metallogenous processes were closely connected with the type and development of magmatism. Igneous-intrusive in form, magmatism has basic to medium, and in places, acid composition. The igneous rocks are calcium-alkaline or potassium sub-alkaline to alkaline. They were formed under rift, and according to some authors under island-arch conditions.

Sidero-chalcophyllic elements are predominant in the ore deposits. Copper is of primordial importance, with iron, gold, silver, sulphur, molybdenum, zinc, lead, selenium, bismuth, antimony, rhenium, barium, etc. being of secondary importance. The largest deposits are in the Panagyurishte ore zone – Medet, Elatsite, Asarel, Chelopech, etc. Hydrothermal-sedimentary iron-manganese ore formations and hydrothermal, predominantly metasomatic medium- to low-temperature massive copper pyrite (having an increased gold content) deposits are genetically connected with medium-acid calcium-alkaline igneous rocks. Their formation was accompanied with hydrothermal alterations near the ores forming zones of alunitic and diaspore quartzites. The magmatic titanium-iron (ilmenite-magnetite) and skarn iron (magnetite) ore formations were the result of basic magma intrusions. Copper and iron-copper (chalcopyritic and magnetite-chalcopyritic) ores in skarns and hydrothermal vein sulphide copper to copper-gold-poly-metal deposits are associated with the range of base, medium-acid and acid plutonic rocks mainly in the eastern part of the Sredna Gora Zone. The largest copper and molybdenum-copper and vein-intrusive (copper porphyry) deposits are genetically associated with plutonic rocks of medium-acid to acid calcium-alkaline to sub-alkaline potassium composition.

It is quite likely that at the end of the Late Cretaceous many vein baryte deposits and ore outcrops were formed in large areas – in parts of the Sredna Gora, Stara Planina and Kraishtid metallogenic zone which are not associated with magmatism outcrops. In some of them there were also fluoride and mercury mineralizations. The late Alpine stage includes the period from the Oligocene to the modern geological age. Weathering crusts of cobalt-nickel mineralizations were formed during the Oligocene on ultrabasic massif in the Rhodope region but they are of no industrial importance. At the same time the Varna Metallogenic Zone was formed in the Varna recess and the Kamchia depression. The deposits are volcanogenic-sedimentogenic manganese-silicate, carbonate and oxide. After Lutetian, the most significant metallogenic event was the formation of active tectonomagmatic and metallogenous zone spreading along the North-Western diagonal of the Balkans – from Slovenia to the Rhodope region (including its sectors in Greece) and Kraishte, intersecting several older tectonomagmatic and metallogenous regions and zones. In Bulgaria it includes a large area of the Rhodope Region and the Kraishtide Zone, and ore seams depend on its development to a large extent. From the Oligocene to the Miocene, and in individual places also during the Pliocene, ore formation was preceded by volcanic activities. Volcanism either had andesite-latite-rhyolite development (mainly in the Eastern Rhodopes) or is of rhyolite type (mainly in the central parts of the Rhodope Massif) and contributed to the formation of vulcanogenous-sedimentary rock complexes. The polymetal deposits were of primordial importance during the Late Alpine stage. They are zinc-lead having additional varying proportions of silver, copper, gold, cadmium, bismuth, antimony, etc. The grouping of the deposits into zones of ore fields in close proximity (Central Rhodope Ore Zone, the Giushevo Ore Zone, the Mesta Ore Zone) and ore fields in isolation (ore nodes) is characteristic – mainly among the volcanic dome structures (East Rhodope Ore Region). The polymetallic deposits are of the Miocene age, although this has not been proved for all of them. Tungsten-molybdenum mineralization outcrops in the south-west region of the Giueshevo Ore Zone. Fluorite deposits (in the Mihalkovo Ore Field, in the Palat occurrence, etc.) and antimony ore outcrops have also been formed. They are hydrothermal, vein, low-temperature and are associated with the last volcanic movement which took place during the Pliocene.

The last metallogenic phenomena (mainly during the Quaternary) were due to weathering processes. Ore concentrations predominantly of gold, copper, iron and to a certain degree silver, etc. originated in an exceptionally large number of places but according to present-day assessment most of them are insignificant. Placer gold deposits outcropped in the flood plains of the rivers and their tributaries in the Western regions of the Stara Planina Mountain (Ogosta seams), and in the Trun and Sofia districts, etc.

Metallogenic Units. Metallogenous zoning is based on the close relation between ore-formation processes and the development of tectonic structure. In Bulgaria the contours of the largest metallogenic units (metallogenic zones and regions) greatly coincide with the delineation of the morphotectonic zones and regions. They are distinguished on the basis of the rock's composition, the genesis and age of the deposits.

The Stara Planina (Balkan) Mountain Metallogenic Zone includes West and Central Stara Planina and parts of the Western Fore-Balkan. The Caledonian-Hercynian metallogenesis is almost fully developed and forms iron, gold, silver and lead ores. The Alpine metallogenesis is only partially outcropped and is characterized by the formation of copper, polymetallic, iron ores and barytes. The Sredna Gora Metallogenic Zone is geologically related to the Sredna Gora Zone. It is characterized by volcanogenous and plutonogenous, predominantly copper, and partially iron and manganese metallogenesis and dates from the Late Cretaceous. In the Rhodope Metallogenous Region (its boundaries coincide with the boundaries of the Rhodope region), late Alpine metallogenesis prevails, and mainly lead-zinc ores were formed. The Kraishtide Metallogenous Zone is a linear structure within the boundaries of the Kraishtides. The development of Hyrcynian (gold), Middle Alpine (baryte) and young Alpine (lead, zinc, fluorite) ore formations can be traced along a seam running from north to south. The Varna Metallogenous Zone is the westernmost boundary of the Eurasian manganese province. The formation of metallogenous units of a lower order (ore regions and zones, ore nodes and fields) was predetermined by deep faults. Quite often the faults are common to two or three metallogenous zones and regions forming metallogenically active lineaments and nodes.

Of the *metallogenically active lineaments and nodes* in Bulgaria the Sredets Ore Arch and the Thracian Ore Network are of the greatest importance. The Sredets Ore Arch is located on the north-eastern and eastern peripheries of a block of earth crust to a depth of 40 km and having maximum magnatic saturation. The ore zones running in a north-north-westerly direction are situated around the block where the largest ore deposits are found in the Stara Planina, Sredna Gora Zones and the Rhodope Region. The Bor-Majdanpek Ore Zone in Yugoslavia, the Iskur Ore Region, the Panagyurishte Ore Zone and the Central Rhodope Ore Zone range from north-west to south-east. In the western part of West Stara Planina the ore zones run in a west-north-westerly direction. The Thracian Ore Network includes isolated ore fields (nodes) in the south-eastern sector of the country including deposits of the eastern parts of the Rhodope Region and the Sredna Gora Zone. The nodes lying in a network of faults of north-western (120°–130°) and north-east (60°–70°) direction are metallogenically active. Part of the Transbalkan Zone of Post-Lutetian tectonomagmatic and metallogenic activities falls within Bulgaria. This zone stretches from Slovenia to the southern part of the Rhodope Region (including its branches in Greece) and Kraishte and determines the late Alpine metallogenesis in the Kraishtide Zone and the Rhodope Region.

Non-ore Minerals. Over 60 types of non-ore industrial minerals have been established in Bulgaria out of which about 40 are used in industry and civil engineering. Minerals which are of great importance to the country include: kaolin, clay, gypsum, barytes, limestones, marls, dolomites, marble, pegmatites, vein quartz, quartz sands, pearlite and zeoloites, rock and sea-salt, asbestos, magnesites and so on.

Kaolin. Kaolin is karstic in form (grikes, karst vortex) aming the Early Cretaceous limestones between the cities of Varna and Russe (Vetovo village, Russe district, Senovo village, Razgrad district, near Kaolinovo). Kaolin with a ratio of 1:4 is mixed with pure quartz sand and contains the minerals kaolinite, a little halloysite, hydromica and quartz. It is thought that the initial material of which kaolin was made was arkosic sand, transported and deposited in the karst formations and weathered later. The *Bentonite Clays* were formed during the Oligocene (near Dimitrovgrad and Kurdjali, and in the Stara Zagora district), the Miocene (in the Varna, Tolbuhin and Pleven districts) and the Pliocene (in the Kyustendil district, the Kraishte mountainous area). Industrial deposits of *refractory and fine ceramic clays* are found mainly in the Early Jurassic continental sedimentary rocks in the Kraishte and Zaburdo mountains, in the Balkan Mountain Range (near Etropole) and the Fore-Balkans, as well as in the Sarmatian alluvial-boggy sediments near Pleven. The ceramic clays near Pleven consist mainly of hydromica and contain less kaolinite and montmorillonite. The Early Jurassic clay layers number from 1 to 10, and they are generally from 1–5 m thick although occasionally layers 8 or 10 m thick are found. Clays formed in the Quaternary are nowadays widely used in civil engineering and household ceramics. A large *gypsum* deposit among the sedimentary rocks of Miocene (Baden) age is prospected in the Vidin district (near the villages of Koshava, Slanotrun and Pokraina). The gypsum layer-like corpus is 3–30 m thick and the average gypsum content in the material is 87 per cent. There are also gypsum deposits in the Stara Zagora and Yambol districts (of the Pliocene age) and elsewhere. *Barytes* is found in many deposits in Stara Planina, Sredna Gora, Kraishte and the Rhodopes. In the Kremikovtsi iron-ore deposit near Sofia barytes occurs as a by-product, forming niches and corpora from 3–50 m thick. *Limestones, marls and dolomites* occur practically in all geological formations. They are widely outcropped in Northern and North-east Bulgaria, in the Stara Planina Mountain Range and the Sredna Gora Zone. There are limestone quarries for facing stone and sculpture in the districts of Belogradchik, Vratsa, Russe, for the cement and lime producing industries in the Pernik, Vratsa, Lovech districts, near Devnya and elsewhere. Dolomites are quarried near Pernik, in Kyustendil district, in Belovo, and Yambol district, and marls (for the cement industry) in the Lovech district (Zlatna Panega village), Vratsa district, near Dimitrovgrad and in the Pleven district. Marble quarries are mined in the Blagoevgrad district (near the villages of Petrovo and Ilindentsi) and near the towns of Berkovitsa, Velingrad, Malko Turnovo, etc. The marble deposits (mainly of the Pre-Cambrian-Paleozoic age) are white or coloured, and often have diverse and beautiful patterns; they are broadly used as building and decorative material. *Pegmatites* – a source of feldspars and mica – exist in the Sredna Gora Zone (town of Strelcha) and the Plana mountain. Small *vein quartz* deposits occur in the areas of Yambol and Sliven (in the Sveti Iliya Hills). Industrial *pearlite* deposits were found amongst the rhyolites of Late Oligocene in the Eastern Rhodope (near the town of Djebel, Kurdjali and in Haskovo district). The clinoptiolite (zeolite) in the Kurdjali district is a basic rock-forming mineral of igneous tuffs several hundred metres thick and is a subject of industrial mining. *Quartz sands* are found among the Paleogene, Neogene, Pleistocene and Holocene formations – in the districts of Varna, Pleven, Shumen, Sofia and elsewhere. A 3,900 m high saline boss is found near the town of Provadia; the *rock-salt* (halite) was deposited in the Permian; the average halite content is 71.9 per cent. Fresh water is forced through boreholes from 600 to 1,800 m deep, and the saline solution produced (310 g per litre concentration) is pumped to the surface. A salt body over 700 m thick was found near the town of Omurtag at a depth of about 3,100 m, and its age is Upper Triassic. Sea-salt is manufactured near the towns of Burgas and Pomorie by sea-water evaporation. Small occurrences of antophyllitic and tremolitic asbestos have been found in the Kurdjali district (near the villages of Avren, Golyamo Kamenyane and Fetikler) and in the district of Blagoevrad. *Fluorite* is produced in the Smolyan district (Mihalkovo village) and in Blagoevgrad district (Palat village). The fluorite mineralizations are hydrothermal, vein-type, formed in the Tertiary.

Mineral Resources

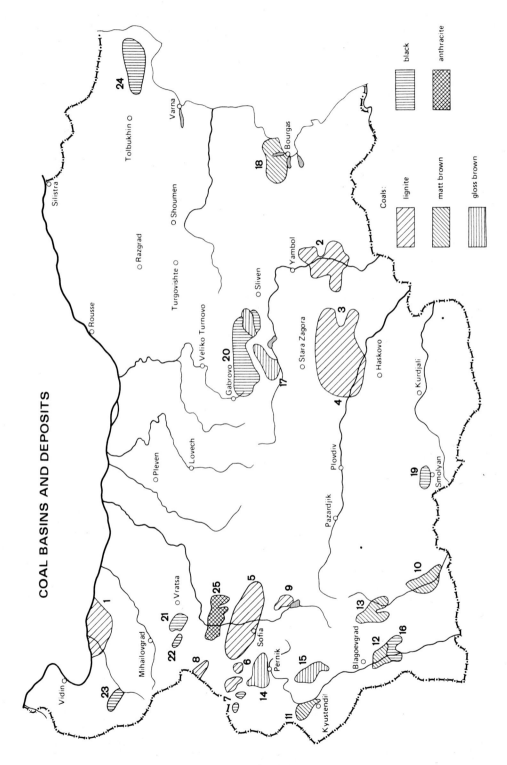

COAL BASINS AND DEPOSITS

1-Lom, 2-Elhovo, 3-East Maritsa, 4-West Maritsa, 5-Sofia, 6-Beli bryag, 7-Aldomirovtsi, 8-Stanyantsi, 9-Choukourovo, 10-Gotse Delchev, 11-Kyustendil, 12-Oranovo-Simitli, 13-Razlog, 14-Pernik, 15-Bobov dol, 16-Pirin, 17-Nikolaevo, 18-Bourgas, 19-Smolyan, 20-Balkan, 21-Gorno Ozirovo, 22-Gorna Luka, 23-Belogradchik, 24-Dobroudja, 25-Svoge.

CLIMATE

Bulgaria's climate is determined by the factors of atmospheric circulation patterns, the country's surface and geographical position. Situated in the southern part of the moderate climatic zone of Europe, Bulgaria comes under both the indirect influence of the Atlantic Ocean to the west and the direct influence of the continental parts of Central and Eastern Europe to the north-west and north-east. The Black Sea to the east and the Mediterranean (the Aegean included) to the south affect the country's climate locally. There is also occasional influence from the desert and semi-desert regions of Northern Africa and Asia Minor, as well as from the Arctic Ocean during the winter months. The climatic influence of the Atlantic Ocean is caused by the Icelandic cyclones. Occasionally the Azorian anti-cyclones carry humid and vertically unstable sea air. The Mediterranean extends its climatic influence through the Mediterranean cyclones and partially through the sub-tropic anti-cyclones. The cyclonal influence is stronger during the winter months (abundant snow falls) and the anti-cyclonal during the summer months (substantial rises in temperature, and dry weather). The Black Sea influence is felt all the year round, although this is limited to a narrow strip along the coast in Eastern Bulgaria. The sea air constitutes 45 per cent of all air currents flowing into the country. The continental climatic influence is slightly stronger due to Bulgaria's geographical position with regard to Central and Eastern Europe, from where the continental air comes in. Such air gathers over the Balkan Peninsula too, whose relief is also an important factor of Bulgaria's climate. The Dinaric Alps and the Carpathians prevent oceanic currents moving eastwards, thus favouring the formation of continental air masses over the Balkans. The continental influence is all the stronger owing to the fact that the north-eastern parts of the country are mainly lowlands, from where continental air, constituting 55 per cent of all air masses, flows in unobstructed.

The movement of air masses over Bulgaria is largely determined by the so-called active atmospheric centres in Europe, both cyclonal and anti-cyclonal. Most important among the cyclonal centres are the Icelandic low-pressure system (all the year round) and the Mediterranean low-pressure system (during the winter months), both of which account for the movement of humid air currents towards the Balkan Peninsula and the great changeability of the weather. The anti-cyclonal centres extend their influence mainly through the East European high-pressure system during the winter months and the Azorian high-pressure system throughout the year, as well as partially through the Siberian anti-cyclone. The anti-cyclones cause rare and scanty rainfall; in winter they bring cold weather with temperature inversions on the land surface and radiational mists, in summer sunny, often dry and hot weather. Local anti-cyclones are a slight factor influencing the weather during the winter months.

The Mediterranean cyclones affect the weather in Bulgaria directly when they pass over the country and indirectly when passing to the north-west or the south of it. Cyclones over Bulgaria usually take three main courses: a north-easterly (course Vb), an easterly (course Vc) and south-easterly (courses Vd and Ve). Whenever the Mediterranean depressions move north-eastwards and reach the central Danubian lowland a south-westerly air current is formed over Bulgaria. When the southerly winds reach the Bulgarian mountains a foehn occurs, which is particularly strong on the northern side of the Balkan Range and the adjoining hilly country of the Danubian Plain. Foehns often occur in the Sofia Valley, especially on the northern side of the Vitosha and Lyulin Mountains at a speed sometimes exceeding 30 metres per second. The combination of foehn, cyclonal activity and cold frontal depressions coming from the west gives rise to violent storms. Mediterranean cyclones passing directly over Bulgaria (course Vc) bring abundant snowfalls during the winter and rainfall during the rest of the year. In spring these cyclones are accompanied by thunder and torrential rains and occasional devastating hailstorms. The cyclones passing to the south of Bulgaria on courses Vd and Ve largely determine the winter weather. They condition the formation of a stream of cold air which penetrates the country from the north-east and gives rise to strong to stormy winds. Snowfalls are frequent, occasionally forming large drifts in the north-eastern parts of the country. During the late spring and summer, when the cyclones on courses Vd and Ve reach the Black Sea and move northwards to the Bulgarian Black Sea coast, conditions are created for heavy rainfalls. Such circumstances caused a record rainfall of 342 mm on 21 August 1951 in the region of the Varna Bay.

The climatic conditions within the country vary from place to place owing to local topography. Sixty per cent of Bulgaria's territory consists of hilly country, medium and high mountains. Mountains influence the overall climate, while local climates are formed in the valleys. The Balkan Range, for instance, blocks the passage of cold air masses into Southern Bulgaria and of warm air masses into Northern Bulgaria, thus affecting the temperatures, rain- and snowfalls and winds accordingly. Mean temperature to the south of the Balkan Range are 0.5° to 1°C higher than those in Northern Bulgaria. Conversely, rainfalls in the northern foothills of the Balkan Range are more abundant compared to those in the valleys in the southern foothills of the mountains: Vratsa – 851 mm as against 640 mm in Sofia, Gabrovo – 881 mm as against 656 mm in Kazanluk. The distribution of the strongest winds is also reversed: southerly winds prevail in the northern side of the Balkan Range and northerly winds prevail in the valleys at the southern foothills of the mountain. The strong winds in the northern foothills of the Balkan Range are foehn-like and are accompanied by a sharp rise in temperatures during the winter months. Occasionally the warm spells caused by the foehn may spread far into the hilly Danubian plain. In such cases the maximum temperatures in January may rise to well over 10°C (Knezha – 20°C, Pleven 19.4°C). Similar warm spells are experienced in the northern foothills of other mountains: in Sofia – 15–17°C, and in Petrich and Kyustendil – 17–19°C.

Depressions have a characteristic effect on the climate. During the winter months there occur temperature inversions and lasting radiational fogs, even with thin inversional clouds at a height of 200–300 m. Owing to the retention and the additional radiational cooling of the air in some depressions extreme minimum temperatures have been recorded: Trun –38.3°C, Sevlievo –35.4°C, Breznik –34.9°C, while on Mount Mussala peak the lowest recorded temperature is –31.2°C. The temperature inversion on the land's surface, the radiational fogs and the quiet spells are factors increasing air pollution with solid and gaseous substances in such places. This is the case with the valleys of Sofia and Pernik. Conversely, inclined terrains have cleaner air and more sunshine, as well as no temperature inversion. (The average number of foggy days in Sofia is 33 per year as against nine in Bankya).

Solar radiation is of major importance for the climate. The actual duration of sunshine throughout the year is between 2,000 and 2,400 hours in non-mountainous areas, with a minimum in December and a maximum in July. In terms of calories the daily duration of sunshine throughout the seasons shows a marked increase from winter to summer. Thus in January in Northern Bulgaria it is 110–120 cal per sq. cm per day and in July – 600–630 cal per sq. cm per day.

As well as varying a great deal in the horizontal and vertical direction temperatures in Bulgaria have a well-defined yearly course. The average annual air temperature for the country is 10.5°C. The average annual temperatures in the Danubian plains range from 10°C to 12°C. The corresponding figures for the upper Thracian plain, the Black Sea coast and the southern-most regions are 12°C and 14°C. In the course of the year the temperatures rise from January to July. In Vidin for instance, the average temperature in January is –1.6°C and 23°C in July the annual temperature range being 24.6°C, in Plovdiv respectively: 0.2°C and 23.6°C, in Petrich: 1.6°C and 25.1°C, in Michurin: 3.2°C and 23.1°C. Inland the spring temperatures are higher than those along the Black Sea coast and in the autumn the situation is reversed: in Varna the average temperature in April is 9.8°C and in Pleven 12.7°C, the average October temperature is 13.6°C in Varna and 12.8°C in Pleven. The highest temperature in Bulgaria was recorded in July 1916 in Sadovo (45.2°C) and the lowest in January 1947 in Trun (–38.3°C). The greatest temperature fluctuations have been recorded in January, in Knezha –35.5°C and +20.0°C, in Plovdiv –31.5°C and +19.4°C; in Petrich –27.8°C and +18.0°C. The lowest and highest average monthly temperatures on mountain peaks have been recorded in February and August respectively. Since 1940, the coldest winter was experienced in 1942 and the warmest in 1948. The mean January temperature in Knezha in 1942 was –12°C, compared to the average of –3.2°C. The mean minimum temperatures in January reflect adequately the intensity of the cold spell in a given place or region. In the Danubian plain and the higher valleys these temperatures vary between –4.5°C and –7.5°C, whereas in the valleys of Struma and Arda they are between –0.2°C and –2.0°C. The areas around Knezha, Trun and Ihtiman form what are known as islands of cold with mean minimum temperatures respectively: –8.2°C, –8.4°C and –7.8°C. The areas around Ivailovgrad, Michurin and Sozopol have the highest mean minimum temperature (–0.5°C). The sum total of the average daily temperature for the vegetation period, the average daily temperature being above 10°C, is comparatively high: in Petrich it is 4.385°C, in Stara Zagora 3.927° and in Vidin 3.680°. The duration of the plant growing period in the southern parts of the country is about seven and a half months and in the northern parts – about six and a half months. Bulgaria's climate is also characterized by warm spells during the winter months and cold spells during the summer months. Particularly strong warm spells were experienced in January of 1936 and 1948 and strong cold spells in mid-May 1943 and in May 1952.

Cloudiness in Bulgaria varies throughout the year with a maximum in December and a minimum in August. In the mountains the cloud mass drops relatively in the winter and increases in the summer. The cloud mass is of particular importance for the effectiveness of solar radiation. Rain falls from two basic types of clouds – cumuliform rainclouds and stratiform rainclouds. The amount and distribution of rainfall is largely governed by atmospheric circulation and topography. The mean annual rainfall is between 416 mm and 650 mm in non-mountainous areas and between 650 mm and 1,193 mm in the semi-mountainous and mountainous parts of the country. The northern Black Sea coast and the western parts of the Upper Thracian Valley get the smallest quantity of rain. In the semi-mountainous areas the highest amount of rain falls in some places at the northern foot of the Balkan Range (Vratsa, Gabrovo), border areas in the Strandja Mountains (Gramatikovo, and Malko Turnovo in Burgas district), and the border areas in the Eastern Rhodopes (Zlatograd). Rainfall is well distributed throughout the year. In northern and central Bulgaria February is the driest month, while June and, less frequently, May are the wettest. In the southern parts and on the southern Black Sea coast the wettest months are November or December and the driest is August. The seasonal distribution of rainfall is more defined: in the Danubian hilly plain and the high valleys rainfalls reach a maximum in summer, whereas in the southern parts and the southern Black Sea coast the most rain falls in winter. In Knezha, for instance, the sum total of summer rainfall is 190 mm as against 80 mm in winter, and in Petrich 103 mm as against 204 mm in winter. In the high mountain areas the maximum rainfall is in spring. The annual distribution of rainfall in percentages is as follows: rain up to 550 mm falls in some 20 per cent of the country's territory, between 550 and 650 mm falls over some 40 per cent, between 650 and 800 mm falls over about 23 per cent and above 800 mm over about 17 per cent (mountainous areas mainly). Research has shown that abundant daily rainfalls do occur in Bulgaria. Most important of the precipitation in solid form are snowfalls. The snow cover stays from 10 days (on the Black Sea coast) up to over 200 days in the high mountains. In the lowlands snowfalls can occur between 10 October and 15 May, although regular snowfalls occur from early December to late March. In the southern regions this period is much shorter. The other form of solid precipitation, hail, falls most frequently between May and August, being most harmful and frequent in June and early July. The mean annual precipitation on the territory of Bulgaria is 74,000 million cu. m.

In Bulgaria, winds vary a great deal in direction and speed. Although westerly winds prevail, in practice winds may blow from all directions. In the winter easterly winds are appreciably more frequent. The speed of the winds also varies widely: from 0 to 40 m per sec. The strongest winds are those on the south side of the Balkan Range, as well as the winds blowing along mountain ridges (foehns), the summer gales, and the northerly winds blowing down to the Black Sea coast in the winter. Their average speed in non-mountainous areas is not very high – from 1.5 to 4.0 m per sec. The Black Sea coast and some valleys in the southern foothills of the Balkan Range are an exception with a speed of the winds between 4 and 6 m per sec. Winds blowing along mountain ridges have the highest average speeds – 8 to 11 m per sec. Frequent lulls are also experienced (in 50 per cent of the cases observed). Strong destructive winds (tornadoes) occur rarely and are not typical of the Bulgarian climate.

Climatic conditions in Bulgaria provide favourable opportunities for the development of the national economy. The duration of sunshine is a vital factor, especially heat resources in the lowlands, characterized by large sum totals of the biologically active temperatures – from 3,600° to 4,400° – thus affording the heat needed for the growth of thermophilic crops. In some plains the humidity coefficient is rather low, especially during the summer months, occasionally falling below 1.0 and even below 0.6 (measured by N.M.Ivanov's coefficient) which involves land improvement measures, such as snow accumulation studies, rational utilization of rain-water, afforestation and anti-erosion activities. Due attention should be paid to the snow cover in the mountains and its effect on the water-level of the rivers and transportation. In many regions the weather is characterized by torrential rains, hailstorms, strong dry winds and frosts, which entails various contingent and routine methods of combating them.

In Bulgaria the following elements of the climate could be regarded as climatic resources: sunshine and solar heat (heliothermal resources); precipitation and snow cover (pluvial and nival resources); sunshine, moderate humidity, favourable temperatures, light winds for leisure and sport (recreational resources); sunshine, rainfall, temperature, etc. for agriculture (agroclimatical resources); winds (anemometric and wind power resources).

The Black Sea and the mountain climates for instance, favour leisure and sporting activities. The climate in Bulgaria is favourable for the cultivation of most grain crops, fruits, vegetables and industrial crops, characteristic of the climate in the southern part of the moderate climatic zone of Europe. The distribution of solar heat, sunshine and rainfall during the warm months provides optimum conditions for the cultivation of some subtropical crops, annuals in particular. The mountain climate affords favourable conditions throughout the year for medical treatment, leisure and sport.

Bulgaria's varied climatic resources are utilized at present in certain branches of the national economy (mainly in agriculture) and in sport, tourism, balneological treatment, leisure and recreation.

Owing to its geographical position and relief, Bulgaria has a variety of local climates which make it extremely difficult to divide the country into climatic zones. Its territory lies in the southern part of the moderate climatic zone and is indirectly influenced by the Atlantic Ocean. In most general terms its climate can be described as moderately continental, with some Mediterranean influences in the southernmost parts of the country.

Bulgaria has several types of climate: moderately continental, transitional continental, transitional Mediterranean and mountainous. It is inappropriate to speak of a Black Sea type of climate, since the influence of the Black Sea is limited to a very narrow dry strip along the coast in Eastern Bulgaria. The term 'transitional Mediterranean' is provisional and applies only to Bulgaria. The mountainous type of climate is characteristic of altitudes over 1,000 m, where the landscape changes owing to some weather constants: average daily temperatures below freezing in January, a snow cover remaining throughout the winter months, a lower annual temperature range up to 20°C, a relative drop in the average cloudiness in the winter months, and a lack of the radiational thermal inversions characteristic of other parts of the country. Bulgaria's territory is divided into five climatic zones each with a clearly-defined characteristic climate, of perennial weather and some of its basic elements: moderate continental, transitional continental, transitional Mediterranean, mountainous continental, transitional Mediterranean, mountainous and a zone of Black Sea climatic influence. The moderate continental zone comprises Northern Bulgaria and the high valleys in the western parts of Southern Bulgaria. It has a continental-type rainfall distribution, and the highest annual temperature amplitudes. The transitional continental zone comprises the northern half of Southern Bulgaria, including the Upper Thracian plain. Its climate is of a milder continental type, determined largely by the protection of the Balkan Range and the unobstructed influx of Mediterranean air masses from the south. The transitional Mediterranean zone comprises the southernmost border areas of Bulgaria. It is characterized by the mildest continental climate and a Mediterranean-type rainfall distribution. The mountainous zone includes altitudes over 1,000–1,200 m with relatively low temperatures, abundant rainfall and a lasting snow cover. Altitude and the correlation between the mountain ranges and air circulation are the main factors determining this type of climate. The zone of the Black Sea climate influence comprises a narrow strip of land (10–30 km). It is characterized by a modified continental climate. Rainfall to the north of the Balkan Range has a continental distribution, while to the south it is Mediterranean.

WATER

The physical and geographical conditions of Bulgaria have given rise to three genetically-related sources of water – ground waters, lakes, and rivers. The genetic links between, and the formation of, water resources depend on the combination of natural and anthropogenic conditions throughout the country.

The bulk of Bulgaria's water resources is formed in the middle heights of the Rhodopes and the Balkan Range. This area is only 23.7 per cent of the country's total, but one half (48.2 per cent) of its water resources are concentrated there. Bearing in mind these mountains' average elevation above sea-level, their run-off volume per unit of area is quite high, 10.67 and 12.26 litres per sq. km respectively. The ground waters in the Rhodopes and the Balkan Range make up a large percentage of the total water resources (33–40 per cent). This indicates a natural regulation of a sizeable volume of waters which is chiefly caused by dense afforestation. The sparse afforestation in the Eastern Rhodopes reduces ground water-feed to 23 per cent.

The alpine areas of the Rila and Pirin mountains are of great hydrologic and economic significance. They rank third in volume of water resources and have the highest water yield per unit of area – the average run-off volume per unit of area is 15.64 litres per sq. km. The ground waters in these areas form 42 per cent of the country's water resources. This high degree of natural regulation of run-off is due to the regulating effect of mountain talus, the dense forests and numerous natural alpine lakes located in the catchment areas. A large part of the declivity run-off is formed by snow reserves. The large number of lakes has a low water volume. Together with some small dams, they make up only 1.2 per cent of the country's water resources.

The country's medium and low mountain relief (the Fore-Balkan, the Balkan Range, the Sredna Gora, the Rhodopes, etc.) yields some 70 per cent of its water resources. The bulk of it is formed in the Rhodopes and the Balkan Range where water yield per unit of area is highest.

Ground waters and the artificially regulated run-off constitute some 28–45 per cent of the country's water resources. This percentage is highest in the Fore-Balkan because of substantial karst formation. The Fore-Balkan also contains the bulk of the country's lake water – 48.6 per cent.

The total water reserves of Bulgaria's plains and hilly areas (the hilly Danubian Plain, the Upper Thracian Valley and the Strandja-Sakar region) are estimated at 3.327 billion cu. m and constitute only 0.16 per cent of the country's water resources.

Lake Okoto and Mount Vihren in the Pirin mountains

Water

The Smolyan lakes in the Rhodope mountains

Though widely spread, mountain screes in Bulgaria are of smaller hydrological significance. The springs located at the foot of the mountains are usually seasonal. The bulk of ground waters is collected in the talus cover of the cirque slopes in Rila and Pirin. It is released by the thaw of thick snow cover and feeds the cirque lakes.

Karst Waters. Karst waters are a major part of Bulgaria's water resources. Some 10 per cent of the country's area is covered by karst terrains and the total annual volume yielded by some 89 of the bigger karst springs is 687 million cu. m. The average capacity of the springs located in the western part of the hilly Danubian Plain is 20–76 litres per sec. The Sarmatian limestones in the Dobrudja Plateau, and especially near the Black Sea coast, have a higher water yield. The biggest springs are located near the village of Obrochishte and the towns of Balchik and Kavarna. Another spring with a capacity of 400 litres per sec is found under the level of the Shabla lake. The largest karst springs in Bulgaria are found in Early Cretaceous and Late Cretaceous limestones. The Devnya springs (3,000 litres per sec) and the Glava Panega spring (2,500 litres per sec) are of this type. The areas of Triassic and Jurassic karst limestones are smaller and scattered in the western part of the Transition Zone and its subzone of Strandja-Sakar. Some bigger springs of this type are: the Dokuzaka spring (335 litres per sec) and the springs near the village of Polska Skakavitsa (109 litres per sec) and Bosnek (80 litres per sec). Paleogene limestone is restricted to the Chirpan Heights and the Eastern Rhodopes. The largest spring originating from such limestone is Khalkabunar in the Chirpan Heights (207 litres per sec). Proterozoic karst marble is widespread in the vast Rila/Rhodope mountain region and forms some artesian basins. The largest karst springs in the Rhodopes are: Khubcha (1,282 litres per sec), Tekira (1,424 litres per sec) and Kleptuza (420 litres per sec), etc.

Head-Waters. Head-waters are related to some artesian structures. The largest artesian synclinal and graben structures are located in the Upper Thracian Valley, the Sofia Valley, and in the depressions of Lom, Varna, and Lower Kamchia. The head-waters of these structures have several aquifers. Head-waters also drain along the deep fractures of some smaller inner- and intra-mountain artesian basins. The waters of all thermal mineral spas in the hydrothermal regions of Sredna Gora, Kraishte, the Western Balkan Range, and the Rila-Rhodope Massif are of this type. Fresh water, both ground and subterranean, make up the country's drinking and domestic water supply. Their total mineral content is up to 1,000 mg per cu. dm. There are three hydrogeological regions – Moesia, the Balkanide and the Rilo-Rhodope region. Every region consists of some subregions in which one or another type of water yield prevails.

The Moesian Hydrogeological Region covers the territory of Northern Bulgaria between the Danube River and the southern frontier of the Fore-Balkan. The region comprises two distinct artesian basins – the Northern Bulgarian and the Varna basins. The subterranean run-off is oriented towards the Danube river to the north and the Black Sea to the east. The aquifers are of considerable thickness and horizontal length. Ground waters are located predominantly among Middle Triassic, Late Jurassic, Cretaceous, and Sarmatian limestones and dolomites. The ground water in the alluvial deposits of the Danube river and its right-hand tributaries are of great significance, too. All aquifers excluding the youngest ones are fed by the infiltration of precipitation waters. They come to the surface in the Fore-Balkan, the Balkan Range, and the North Bulgarian Swelling. The subterranean waters are drained by the river network. One part of the waters of the Upper Jurrasic/Early Cretaceous and Sarmatian aquifers in the Varna artesian basin are drained into the Black Sea and Romania by some submarine springs. The total annual run-off (the natural reserve) of ground waters in the Moesian Region is estimated at about 2,050 million cu. m (65 cu. m per sec). Karst waters yield 47 cu. m per sec; the waters in the alluvial deposits – 15 cu. m per sec, and fracture waters – 3 cu. m per sec. The Late Jurassic/Early Cretaceous aquifer has the highest annual run-off of 915 million cu. m (29 cu. m per sec). The bulk of this aquifer's reserve is in the Varna basin. The waters of the Sarmatian and the Late Jurassic/Early Cretaceous aquifers, as well as of the Pliocene aquifer in the Lom valley and the alluvial deposits among the Iskur, Vit, Yantra, Kamchia and other rivers are extracted mainly by bore holes.

The Balkanide Hydrogeological Region. The Balkanide Hydrogeological Region occupies the central part of Bulgaria. It is bordered by the southern frontier of the

The seven Rila lakes

Fore-Balkan to the north, by the Rila-Rhodope Massif to the south, by the Black Sea to the east and by the Socialist Federal Republic of Yugoslavia to the west (the most eastern parts of the Southern Carpathians). Its subterranean waters are linked to the aquifers of a number of artesian basins and to the weathered crust of some solid, metamorphic and sedimentary rocks that form the mountains in the region. With the exception of the ground waters in the alluvial deposits of the Maritsa, Tundja, and Iskur river valleys and in the karst limestones and dolomites of the Triassic and the Later Cretaceous layers, the aquifers of the Balkanide Region are characterized by a relatively low water yield. They are fed through the infiltration of precipitation waters and drain into the river network and many springs. The annual subterranean run-off totals some 2,430 million cu. m (77 cu. m per sec). However, it is distributed extremely irregularly over the regions and aquifers. The reserve of unconfined ground waters in the alluvial deposits (38 cu. m per sec) and of karst waters in the Middle Triassic and the Cenozoic aquifers is the largest one. The subterranean run-off in the region of Western Thrace is the highest one. The alluvial deposits of the Maritsa river are its main aquifer. The Eastern Balkan Region has the poorest water yield. The volcanogenic-sedimentary complex is characterized by the poorest water yield while the karst dolomite limestones in the Western Balkan Range and the proluvial deposits along the Assenovitsa river to the north of the town of Assenovgrad have the highest water yield.

The waters of the alluvial deposits along the Maritsa river in the Upper Thracian Valley and of the Striama river in the Karlovo valley are intensively used for irrigation and water supply.

The Rhilo-Rhodope Hydrogeological Region. The Rhilo-Rhodope Hydrogeological Region includes the Ossegovo-Belassitsa mountain group and the Rila, Pirin and Rhodope mountains. Its ground waters are linked to the

The 'Kidney Lake' in the Rila mountains

weathered crust of the Precambrian metamorphites and the karst marbles of the Proterozoic, Quaternary and Pliocene deposits, which form some artesian basins, as well as to the alluvial deposits in the Arda, Mesta and Struma river valleys. The total annual underground run-off in the region is 1,290 million cu. m (41 cu. m per sec). The greatest share of this consists of fracture waters in the metamorphic rocks and the granitoids (21 cu. m per sec) and karst waters (15 cu. m per sec). The highest water yield is obtained from the Proterozoic and karst marbles; the lowest from the Priabonian tuffogenic rocks in the Eastern Rhodopes. The total mineral content of ground waters in Bulgaria ranges from 50–100 to 1,000 mg per cu. dm reaching higher values in some regions. The average value for the country is 582 – 14 mg per cu. dm. The waters with a mineral content from 500 to 1,000 mg per cu. dm are most widespread. As regards their anionic composition, Bulgarian waters are predominantly hydrocarbonate. Chloride waters are found in some small areas (around Burgas and Plovdiv) and sulphate waters – in the area of Panagyurishte, Nova Zagora, and Pleven. Sodium waters are most often found in rhyolites and dacites, magnesium waters – in gabro, serpentinites, and amphibolites. Most of the country's waters are calcium ones. In terms of hardness, they are classified into two types: medium hard (3–6 mg equivalent or 8.4–16.8° German) and hard (6–9 mg equivalent or 16.8–25.8° German). Medium hard waters are typical of the Fore-Balkan and the areas of Sredna Gora and Kraishte; hard waters – of the northern part of the Moesian region and the regions of Burgas, Strandja-Sakar and Eastern Maritsa. Some very soft waters (less than 3.0 mg per cu. dm) are found in the Rila/Rhodope and the West Balkan regions. Subterranean waters also contain some active biological components like iodine, fluorine, manganese, iron, etc. Their quantity depends largely on the general mineral composition of the water and their rock environment.

Every year about 2 million tons of dissolved substances are carried away by ground waters in Bulgaria. The annual average per square kilometre is 20 tons ranging between 1.5 and 100 tons in different areas and rocks. The smallest amount of salts is carried away from plains and the largest from mountainous regions. The use of chemicals in agriculture and the draining of industrial waste into the rivers has resulted in pollution of ground waters (mainly the ground and karst ones) and necessitate the taking of appropriate precautionary measures.

Lakes. Bulgaria has a total of 336 lakes most of which are small in size. Their total area is 95 sq. km. Bulgaria's natural lakes are littoral, high-mountain, karst, tectonic, landslide and riverside lakes. The littoral and glacial lakes play an important part in the water balance of the country.

The littoral lakes are the lakes of greatest economic importance for the country. Their total water volume is 219 million cu. m. It was the Quaternary epeirogenic sinking of the Black Sea coast which brought about the formation of these lakes. The lagoon lakes are formed in shallow sea bays enclosed by sand spits and are in direct or indirect connection with the Black Sea (Alleppu, Arkutino [an official reserve] Pomorie lake, etc). The liman lakes are actually river mouths under the sea (the Beloslav lake, Burgas lake [part of which is an official reserve], Varna lake, Durankulak swamp, Ezerets lake, Shabla lake, etc.). *The high-mountain lakes* were formed during the Glaciation in the Pirin and Rila mountains, at an altitude of 1,900–2,400 m. They are situated in cirques usually encircled by small moraines and outcrops. Some of them occur in terraces and very often the upper lake flows into the lower one. Bulgaria has about 260 high-mountain lakes – the Seven Rila lakes, Mussala lakes, Urdini lakes, etc. in Rila, and in Pirin the Popovi lakes, Bunderitsa lakes, Vassilash lakes and Vlahina lakes. There are only a few *karst lakes* which are found in stagnated ponors. They were formed as a result of the solution of limestone by atmospheric waters. Lakes of this type are the Rabisha lake (tectonic conditions also contributed to its formation), Devetashki lakes and the Suhotoezero lake in Dobrudja near the village of General Kolevo (Tolbuhin district) – this lake is shallow and runs dry in summer. The number of *tectonic lakes* is small. They lie in tectonic troughs – Kupen lake (below the Kupena peak in the Balkan Range, at an altitude of 1,945 m; the snow of the Quaternary period has modified the lake greatly), Panichishte (in the northern part of Rila, at an altitude of 1,375 m – it lies in a plain oval depression), Skalensko lake (near the Mokren Pass in the Eastern Balkan Range, now dammed).*The landslide lakes* occur in depressions formed after earth slips. Some of them are found on the Black Sea coast and in the Rhodopes. Kashliar and Balchishka Tuzla are situated in terraces between the Shabla Cape and Varna. The Smolyan lakes and Chair lakes are in the Rhodopes. *The riverside lakes* fill old erosion-worn forms, mainly former meanders. Their water level depends on the water level of the rivers. The lakes on the flood plains of the rivers Iskur, Yantra, Kamchia, Ropotamo, Tundja and Maritsa are of this type. Most of the lakes along the Danube, with a total of about 3,000 ha, have been drained. Only the Sreburna lake remains and is an official reserve. There are two basic types of *swamp* in Bulgaria: *low-land (eutrophic) swamps and highland (oligotrophic) swamps*. The eutrophic swamps occur mainly in low-lying places and rarely on the watershed levels – degraded lakes and old river-beds. They are fed by the rain, river and ground waters rich in mineral salts. The latter determine the nature of the vegetable life in the swamps. The swamps on the Black Sea coast, along the Danube, and on the banks of the inland rivers are of this type.

The oligotrophic (highland) swamps are situated in regions with a cool and damp climate such as is found in the mountain regions. They are fed mainly by rainfall water containing few mineral salts. Swamp vegetation is typical in these areas. Such swamps occur in the Rhodopes and Vitosha – there are many peat-bogs on the plateau, the waters of which flow down to the outskirts of the plateau forming torrents. Some are to be found in Rila too.

To increase the area of arable land, for some health and other reasons, today most of the swamps in Bulgaria are drained, the run-off share of their water balance being increased by hydrotechnical means (dykes, draining channels, pump stations). Some swamps are kept as breeding-pools and fisheries, because of the abundance of nutrients in them. The littoral swamps, after being treated, have been developed into salt-pits (the Pomorie and Burgas salt-pits are located in the place of one-time swamps). The mud of the Pomorie swamp and Balchishka Tuzla is used for medical treatment and in the chemical industry.

River waters form the bulk of the water resources in Bulgaria. The distribution of the rivers was much influenced by the geomorphological development of the country and its climatic conditions. The central position of the Balkan Range is not favourable for the formation of big catchment basins, and also affects the direction of river flows. As a result, the greater part of the Balkan ridge is the main watershed between rivers in Northern and Southern Bulgaria. From the Baba peak in the western Balkan Range the main watershed turns south, taking in the upper Iskur river valley. From the eastern Balkan Range it is diverted to the south and runs along the heights surrounding the Burgas plain on the west; then, following the Strandja mountains it reaches the border with Turkey. Thus the territory of Bulgaria is divided into two run-off regions: the Black Sea region with a run-off to the Black Sea (57 per cent of the country's territory) and the Aegean region with a run-off to the Aegean Sea (43 per cent). The rivers in the Black Sea region fall under two sub-regions – one comprising the rivers flowing into the Black Sea via the Danube and the other including the rivers flowing straight into the sea. The Rila mountains are the main river water source, the deepest and longest rivers (the Maritsa, Iskur and Mesta rivers) rising there. The average annual run-off volume in Bulgaria is 18.87 billion cu. m (the rainfall is estimated at 74 billion cu. m). The Black Sea run-off region has 7.89 billion cu. m (a mere 42 per cent) and the Aegean region – the remaining 10.98 billion cu. m (about 58 per cent).

The river network varies in density because of the variety in relief, the geological structure and the climatic conditions. It is most dense (over 2 km per sq. km) in some places on the northern slopes of the Balkan Range, in the western border mountains (Ograzhden, Vlahina and Ossogovo) and in the Eastern Rhodopes. The low-mountain and hilly regions of the country, occupying the larger part of it, have an average river network density from 0.5 to 2.0 km per sq. km. Least dense is the river network (0.5 km per sq. km) in the littoral plains and the level inter-valley areas in the Danubian Plain, the Dobrudja Plateau, the Upper Thracian Valley, the Burgas lowland and the Sub-Balkan valleys. Some parts of Dobrudja have a seasonal river run-off only.

All rivers, except the Timok, Danube, Velleka, Strumeshnitsa and Erma, have their source in Bulgaria, mainly in the high mountains. The catchment areas are comparatively small, the bigger ones varying between 200 and

The river Iskur

The Iskur valley between the Lozen mountains and Mount Vitosha

20,000 sq. km. Bulgaria has approximately 560 rivers of over 2.3 km in length.

Rivers in the Black Sea Run-Off Region which are Tributaries of the Danube. The Bulgarian tributaries of the Danube, except those which have their source on the Dobrudja plateau, have similar features. Most of them rise in the Balkan Range, with the exception of the Iskur which rises in Rila. The Dobrudja rivers display different features. They rise in the Ludogorie area and very often do not reach the Danube because of water loss.

Bulgaria's longest river is the Iskur (368 km long). It has its source in the central part of the Rila where the swift and clear waters of the rivers Cherni, Levi and Beli Iskur meet. Leaving the Samokov valley, the Iskur river runs along the Pancharevo Gorge. North of the Sofia valley it forms the long and picturesque Iskur Gorge crossing the Balkan Range. The average density of the tributaries in the large river basin is 1.1 km per sq. m. The biggest tributaries are the rivers Malki Iskur, Blat, Palakarpia, and Zlatna Panega. They have an average annual run-off of 54.5 cu. m per sec. High water in Rila comes during May–July and in the Danubian Plain – during March–June. Low water is during August–October. The Beli Iskur, Iskur and Pancharevo dams are located on this river.

The Yantra river, 285.5 km long, is the Danube's biggest tributary rising in the Balkan Range. It runs along a deep, narrow valley until it arrives in Gabrovo, forming beautiful gorges in the Fore-Balkan areas. In the Danubian Plain the river has a wide and sandy bed. Its course is calm and meandering. Its biggest tributaries are the rivers Rossitsa, Lefedjira, Belitsa and Studena. The river's average annual run-off is 44.4 cu. m per sec. The river level shows considerable seasonal fluctuations. In the Danubian Plain the river's bed has been corrected.

The Ossum river (314 km long) is the third longest tributary of the Danube. It rises in the Balkan Range, forms a deep gorge in the Fore-Balkan and runs across the Danubian Plain where it is wide and meandering. Its biggest tributaries are the rivers Beli Ossum, Komanska, Suha, and Dripla. The average annual run-off is 14.85 cu. m per sec.

The Ogosta and Vit rivers are other significant tributaries rising in the Balkan Range and they have many features similar to those of the Yantra river. Their sources are on the northern slope of the Balkan Range; then they meander across the Danubian Plain, their wide, shallow river-beds filled with alluvial deposits.

The Topolovitsa, Vidbol, Archar, Lom, Tsibritsa and Sju Skut rivers are tributaries of the Danube which rise in the Fore-Balkans and run across the western part of the Danubian Plain. The greater part of their reaches are in the plain. The Fore-Balkans being very close to the Danube the rivers were not able to develop bigger catchment basins. The lowest reaches of the Lom and Tsibritsa rivers are largely fed by spring waters. The Danube tributaries – the Russenski Lom, Krapinets and Suha rivers – which rise in Dobrudja, are very different from the others. Their catchment areas are oval-shaped. The river-beds are situated in regions made up of Cretaceous and Sarmatian limestone with patches of loess. Since the limestones are karsted, the loess and soil cover does not prevent seepage of the river waters. The

The Studena Reservoir at the foot of Mount Vitosha

The river Kamchia

Russenski Lom river is 196.9 km long, with an average annual run-off of 5.6 cu. m per sec, and part of its basin is beyond the loess-covered area. This river reaches the Danube without losing too much water, while the majority of the Dobrudja river valleys are dry and fill with water only during torrential rain.

Rivers Flowing straight into the Black Sea. The rivers flowing straight into the Black Sea collect their waters from the Ludogorie area, the Eastern Balkan Range, the Bakadjitsite Heights and Strandja Mountains. From north to south the major rivers are: the Batova, Provadiya, Kamchia, Dvoinitsa, Aheloi, Aitoska, Russokastro, Sredetska, Fakiiska Reka, Izvorska Reka, Ropotamo, Diavolska Reka, Karaagach, Veleska and the Rezovo. Their total catchment area makes up 15 per cent of the country's territory.

The Kamchia (244.5 km long) is the largest river running straight into the Black Sea. It comprises the rivers Goliama Kamchia (198.5 km long, and arguably the main one) and Luda Kamchia (200.9 km long, with an average annual run-off of 8.8 cu. m per sec). It has a wide valley in its lower reaches which, in geomorphological terms, is a liman; the banks are low and covered with ash and elm forests. It is known by the name of Longoz. On entering the Black Sea the river forms an under-water delta, and its major tributaries are the Eleshnitsa and Komoludere rivers. The average annual run-off is 25.3 cu. m per sec. In some places the Kamchia river bed has been corrected. The Kamchia resort complex is located at the river's mouth.

Rivers in the Aegean Run-off Region. The Aegean run-off region has a larger variety of physical and geographical conditions and regimes of river run-off than the others, including highland and medium-high mountains, valleys, and lowlands. The region is divided into two: its western part includes high mountains and lowlands, crossed by the Struma and Mesta river valleys; the eastern part comprises some medium-high mountains and the Upper Thracian Valley, the sub-Balkan valleys and the river system of the Maritsa and its big tributaries the Tundja and Arda.

The *Struma,* 290 km long bordering Greece, is the fifth biggest Bulgarian river. It rises in the Vitosha mountain, collects the waters of some large karst springs and runs through the Studena dam. It forms the deep picturesque gorges of Krakra, Zemen, Skrineno, Kresna and Rupel. Major tributaries are: the Svetlia, Trekliano, Dragovishtitsa, Cherman, Rila, Vlahina and Pirinska Bistritsa rivers. The average annual run off is 82.5 cu. m per sec.

The Mesta river, flowing out of the Cherna Mesta river rises in the Eastern Rila Mountains. Its length in Bulgarian territory is 126 km. It runs through the Razlog valley, the deep and beautiful gorge of Momina Klissura and the Gotse Delchev valley. Its biggest tributaries are the Bela Mesta, Belitsa, Iztok, Retizhe and Dospat rivers. The average annual run-off is 25 cu. m per sec.

The Maritsa river is the biggest Bulgarian river in the Aegean run-off region, its length within the territory of Bulgaria being 321.6 km. Its source is the Maritsa Lakes in the Rila Mountain. Its upper reaches pass through a deep narrow valley, covered by thick coniferous forests. Through the Momina Klissura gorge the river flows into the Upper Thracian Valley. Its most important tributaries are the Chepino Stara Reka River, the Harmanli, Vucha, Arda, Topolinitsa, Luda Yana, Striama and Tundja rivers and the Chepelare river. Its average annual run-off is

111.00 cu. m per sec making it the river with the highest water content in Bulgaria. Its water is used in several hydro-electric power waterfalls.

The Tundja is the biggest tributary of the Maritsa and the second longest river in Bulgaria – 349.5 km. It rises in the Balkan Mountain. Its steep reaches cut through a deep valley. Between the Derventski hills and the Sakar Mountain the river has formed the deep and picturesque Strema Gorge. Its main tributaries are the Tuzha river, the Popovo river, the Assenovska Reka river, the Mochuritsa and Kalnitsa rivers. Its average annual run-off is 37.7 cu. m per sec. The Georgi Dimitrov and Zhrebchevo dams have been constructed to regulate the waters of this river.

The Arda river, which is the other significant tributary of the Maritsa, rises in the Western Rhodopes and is the longest river in the Rhodope Mountains. It is 241.3 km long. Its upper part cuts through a deep narrow valley with small valley widenings. The middle reaches of the river have also formed numerous valley widenings which are linked by deep narrow gorges. The Dyavolski Most (Devil's Bridge) gorge is strikingly picturesque. Before leaving the territory of Bulgaria the river flows through the long meandering Kamildol gorge. Its biggest tributaries are the Vurbitsa, Krumovitsa, Cherna Reka, Malka Arda and Perperek rivers. Its average annual run-off is 76.6 cu. m per sec. The Kurdjali, Studen Kladenets and Ivailovgrad dams have been constructed along its middle reaches.

Formation of the River Run-off. Owing to the varied landscape of Bulgaria the quantitative ratio between the stable (constant, made up of underground waters) and unstable (inconstant, made up of rainfall or melting snow) components of river run-off varies widely. In the karst areas the percentage share of the stable component is very high. It is high in all the rivers of the Dobrudja region (58 per cent in the Kanagyrol river, 62 per cent in the Batova river), in a number of small rivers in the Fore-Balkan region (54 per cent in the Zlatna Panega river) and in the rivers in the Balkan Mountains (70 per cent in the Iskrets river), Pirin (75 per cent in the Byala Reka river and the Rhodopes (51 per cent in the Chepelare river). Also high is the percentage share of the stable component in the forested mountain areas of the Rhodopes, the Balkan Mountains, Rila, and Pirin Mountains but in these latter regions it does not exceed the percentage of the unstable component. The stable component contributes 45 per cent of the run-off that the Maritsa river receives in the locality of the town of Belovo, 48 per cent of the run-off that the Eninska Reka river collects from the southern slope of the Balkan Mountains and 44 per cent of the run-off of the Kanina Reka river in the Western Rhodopes.

In the low-lying and undulating woodless areas of Bulgaria the unstable component supplies much more water to the rivers than the stable one. It is 61 per cent of the run-off of the Lom river in the Danubian Plain and 72 per cent, 62 per cent and 63 per cent, respectively, of the run-off of the Skut, Russenski Lom and Dryanovo rivers in the Fore-Balkan region. The percentage share of this component is high mainly in the small basins limited to one type of natural landscape. As the area of the river basin increases, and other landscapes with run-off forming at different times are included, the fluctuations in run-off are naturally levelled. This is exemplified by the basin of the Maritsa river, the unstable run-off component of which is 55 per cent near the town of Belovo, before the river enters the Upper Thracian Valley, and then is lowered to 52 per cent at the town of Pazardjik and to 38 per cent at the town of Svilengrad.

The Batak reservoir in the Rhodope mountains

In Bulgaria the territorial distribution of the total river run-off expressed on the basis of the level of run-off water is as follows. In the mountain belt located higher than 1,000 m above sea level in Rila, Pirin, the Central Balkan and the Rhodope Mountains (in the latter it includes the upper reaches of the Arda river) the level of run-off water rapidly increases from 500 to 800 mm, reaching 1,000 mm in the uppermost parts of Rila and Pirin. For the medium-high mountainous areas in the rest of Bulgaria the level of run-off water ranges between 300 and 500 mm, and for the low mountainous and hilly areas it is between 150 and 300 mm, becoming less than 150 mm in the Upper Thracian Valley and the Danubian Plain. The lowest values for the level of run-off water have been recorded for the Dobrudja Plateau, a consequence of its karst relief.

Rila and Pirin have the highest water yield per unit area but the annual volume of the run-off from these mountains is not great because of their comparatively small area. The Danubian Plain and the Upper Thracian Valley, irrespective of their vast area, have the smallest supply of river water owing to their lowest water-yield.

River Regime. The great variety of types of landscape in Bulgaria accounts for the varied conditions of river run-off formation and for the type of fluctuation in annual run-off. One of the main characteristics of the river regime is the phasic nature of river fluctuations. Daily fluctuations in run-off looked at over the course of a year make periods of high and low water easily identifiable. These periods represent the major phases in the river regime with respect to the nature of the fluctuations. Climatic conditions are the main factor determining the existence of phases in the river regime. In the areas of sub-Mediterranean climate, the phase of high water occurs in winter and the low-water phase during the summer and autumn months. In the remaining parts of Bulgaria high water occurs in spring, while the phase of low water again occurs in summer and autumn but is of lesser duration.

Both the winter and spring high water periods are caused by rainfall and snow water. In the temperate continental climate of Bulgaria the phase of high water begins with the snow melting in the southern river basins as early as March and is subsequently sustained by water from the snow melting in the higher mountainous water catchment areas, to which later on rainfall water is added. During the phase of high water rivers receive more than 50 per cent of their annual run-off volume. In the high-mountainous water-catchment areas the water volume at high water is not great, because the period of the snow melt and hence of high water is short, lasting for only a month and a half, while in summer and autumn these areas receive considerable rainfall which increases the river water volume at low water. In the northern part of the Danubian Plain the water volume at high spring water is also low. This is accounted for by the low amount of water supplied by snow melt and spring rainfall.

In the Danubian Plain the maximum in river run-off occurs as early as the first half of March owing to the early snow melt in this area. The phase of high water being intermittent, the period of maximum run-off is not clearly defined. In the Fore-Balkan region the period of maximum run-off is more sharply defined and the duration of the spring high-water phase is shorter. A secondary maximum occurs in June which is related to maximum rainfall which falls in this month.

The maximum in March is a result of melting snow and spring rainfall. Because of the higher elevation of the Balkan Mountains the snow-melt is delayed and the run-off maximum is shifted to April and together with the secondary maximum in June, caused by the abundant rainfall typical of this month, a prolonged high-water phase results. The rivers in the Western Rhodopes and in the western subzone of the Transition Zone have run-off regimes similar to those of the Balkan mountain rivers. This similarity can be attributed to the similar medium-high mountainous physico-geographical landscape found in both areas. The April maximum is shifted to May and continues even into June in the basins with northern slopes. The high water period includes these two months and is sustained almost entirely by the intensive snow-melt. The secondary minimum run-off of the rivers in the Balkan Mountains and the Rhodopes during the winter becomes the main maximum owing to a prolonged delay in snow melt and a lack of surface run-off. In the low-lying parts of the Transition Zone the high-water phase is clearcut and occurs from January to March when the river run-off rises to the maximum. In South-east Bulgaria the high-water phase occurs in winter and the maximum run-off is in January.

The run-off regime of rivers sustained to a large extent or predominantly by karst waters falls within a separate category. Waters formed in deep karst are characterized by insignificant fluctuations and hence provide a naturally regular run-off to the rivers they supply. On the other hand, rivers sustained by shallow karst water run-off have great fluctuations in the annual distribution of their run-off.

Of the physico-chemical characteristics of river run-off (temperature, ice formations, muddiness and chemical composition) it is the chemical composition of river water that has recently attracted great attention in relation to the considerable anthropogenic changes detected. Hydrochemical studies of water in the Bulgarian rivers have shown that in the major part of it hydrocarbonate, calcium and sulphate ions predominate, i.e. that river waters are hydrocarbonate-calcium-sulphate in composition. In separate smaller areas hydrocarbonate-sodium (calcium or chloride) and hydrocarbonate-sulphate ions predominate in river waters. Mineralization is highest in the Danubian Plain, the eastern Upper Thracian Valley and the Burgas Valley and is related mainly to the cultivated areas. The surface run-off of rainfall water leaches the chernozems and some other soils and on reaching rivers increased the mineral content in their water. In the western Upper Thracian Valley, the valleys in the Transition Zone and the low and medium-high mountainous areas of Bulgaria the mineral content has been estimated at 200 to 300 mg per litre. In these areas the river-water mineralization is to a large extent accounted for by the vast areas of cultivated land, the presence of rocks, containing more readily soluble minerals. The lowest mineral content of river water has been found in the mountain areas, owing to the extensive forests there and the rapid run-off of rainfall down the steep slopes.

Hydrological Zoning of Bulgaria. Climatic conditions are the main factor in the formation and the regime of river run-off. Relief and other landscape features only further specify the impact of this basic factor. On the basis of the two climatic types existing in Bulgaria (sub-Mediterranean and temperate continental) two main hydrological zones, divided into regions, have been identified within the country.

The zone where the mediterranean climate influences river run-off includes the south-eastern part of Bulgaria, covering 27.7 per cent of its total area. This is the zone in which 20 per cent of the country's water resources are formed (3,810 billion cu. m). It comprises four different landscapes (regions).

1. The Eastern Balkan Mountain region. Its area is not larger than 2,575.6 sq. km. The winter run-off is either negligibly higher than or equal to the spring run-off. The low-mountainous landscape, the scarce afforestation and the low rainfall yield a low run-off volume per unit area (3.93 litres per sq. km) accounting for the low annual volume of the water resources – in this region 0.297 billion cu. m.

2. The Upper Thracian Valley region includes almost the whole of the Valley, the Burgas lowlands, Sakar Mountain, the Manustir, Sveti Iliya and Bakadjik Heights and some of the sub-Balkan valleys. This is the largest hydrological region in Bulgaria, covering an area of 18,433.2 sq. km. The water yield per unit area is as low as 2.94 litres per sec per sq. km. The small rivers which rise predominantly in the low mountains and undulating highlands receive very little water when flowing through the lowlands. This explains the comparatively low annual volume of the water resources – 1,436 billion cu. m – of this large region.

3. The Strandja region includes only the higher parts of the Strandja Mountain containing the Valeka river basin and the uppermost portions of the Fakiiska Reka river catchment basin. This is the smallest region with an area of 1,882.2 cu. km. Here, however, the rainfall is considerably higher, evaporation losses are relatively low and the run-off volume per unit area is high, amounting to 9.28 litres per sec per sq. km. This results in there being a significant volume of water resources (0.513 billion cu. m) yielded by this small area.

4. The Eastern Rhodope region includes the entire catchment basins of the Arda and Byala Reka rivers. Its area is 5,616.6 sq. km. The rainfall in this region is even higher and the run-off volume per unit area is among the highest for the country, reaching 13.30 litres per sec per sq. km. These conditions result in considerable water resources of 1,564 billion cu. m in this area.

The zone where the continental climate influences river run-off comprises 72.3 per cent of Bulgaria's area. Within this zone the landscape is considerably more varied and consequently the conditions for the formation and the regime of river run-of are more diversified. Two subzones have been defined.

A. The undulating plain subzone with a low increase in snow-water supply to the rivers. The comparatively low amount of rainfall received and the considerable evaporation result in low run-off volumes per unit area (2.84 litres per sec per sq. km on average). Because of these conditions the subzone yields a total volume of water resources of 4,885 billion cu. m, which is more than three times lower than the volume produced by the mountain subzone.

1. The Western Danubian Plain/the Western Fore-Balkan region covers an area of 9,259.1 sq. km. The low run-off volume per unit area (2.37 litres per sec per sq. km) and the morphography of the region permit the formation of small rivers, the Ogosta river being the only exception. The annual volume of water resources amounts to 0.745 billion cu. m.

2. The Central Danubian Plain/the Central Fore-Balkan region comprises an area of 12,206.2 sq. km. The run-off volume per unit area is comparatively high – 4.11 litres per sec per sq. km. The annual volume of water resources total 1,585 billion cu. m.

3. The Eastern Danubian Plain/the Eastern Fore-Balkan region has an area of 10,806.2 sq. km. It is characterized by an increase in the hydrological effect of karst processes as a result of which the run-off is naturally regulated (the stable component prevails over the unstable one) to a degree not found in any other region within the subzone. The run-off volume per unit area is 3.20 litres per sec per sq. km and the annual volume of water resources totals 1,090 billion cu. m.

4. The Dobrudja region covers an area of 12,156.6 sq. km. Hydrographically it consists of two parts – one with run-off and one without run-off. Karst processes are highly intensive causing large amounts of water to infiltrate into the ground and subsequently to flow underground towards the Danube or the Black Sea. Owing to this, the average run-off volume per unit area of 0.43 litres per sec per sq. km is the lowest one recorded in Bulgaria and the total water resources scarcely reach 0.198 billion cu. m.

5. The Sredna Gora Mountain – the Fore-Balkan region has an area of 7,837.1 sq. km, a run-off volume per unit area of 4.68 litres per sec per sq. km and a total volume of water resources of 1,267 billion cu. m.

B. Mountain subzone. Its mountainous nature creates very favourable conditions for the formation of river run-off. The average run-off volume per unit area is 11.19 litres per sec per sq. km. Although this subzone covers an area which is about 3.5 times smaller than the total area of Bulgaria, it has an annual volume of water resources of 10,803 billion cu. m which is more than one half of the country's water resources.

1. The Balkan Mountain region is the largest one in the subzone, stretching over an area of 9,265.8 sq. km. After the Rila/Pirin mountain region it has the second largest run off volume per unit area – 13.50 litres per sec per sq. km. This, together with its vast area, makes it the region with the greatest amount of water resources totalling 3,944 billion cu. m.

2. The Western Rhodopes region covers an area of 6,148.3 sq. km. The considerable amount of rainfall received, the vast denuded surfaces and the slow snow-melt yield a comparatively high run-off volume per unit area – 11.53 litres per sec per sq. km – and hence one of the country's most significant volumes of water resources – 2,355 billion cu. m.

3. The Kraishte region covers an area of 5,708.2 sq. km. The comparatively lower height of the mountains in this region and the extensive area of valley-like fields have resulted in a relatively low run-off volume per unit area – 5.09 litres per sec per sq. km. Consequently the annual volume of water resources is low, amounting to only 0.918 billion cu. m.

4. The Vlahina/Belassitsa region is the smallest one in this subzone and one of the smallest within the country,

covering an area of 2,902.4 sq. m. The mountainous areas receive high amounts of rainfall and despite considerable evaporation the water yield per unit area is high, reaching 8.61 litres per sec per sq. km. The annual volume of water resources is estimated at 0.798 billion cu. m.

5. The Rila/Pirin Mountain region. It also includes the Vitosha Mountain both because of its proximity and the similarity of hydrological conditions. The total area of this region is 6,189.9 sq. km. It has the most favourable physical and geographical conditions for the formation of run-off and hence the country's highest water yield per unit area has been measured here – 14.23 litres per sec per sq. km. The annual volume of its water resources is one of the highest, totalling 2,788 billion cu. m.

SOILS

There are numerous types of soil in Bulgaria, since they have developed under the influence of various soil-forming factors: climate, bed rocks, vegetation, relief, water resources and the age of the territory.

The 1974 composition of the soils within the country into geographical groups provided three soil zones, which include seven subzones (belts), 28 districts and 63 regions.

Geographical Soil Zones of Bulgaria (1974)

Zones and Subzones	Area (sq. km)	Percentage
Forest-steppe zone of Northern Bulgaria	45,740	41.2
Danubian subzone	23,170	20.9
Danube Plain and hilly Fore-Balkan subzone	15,930	14.3
Fore-Balkan subzone	6,640	6.0
Xerothermal zone of South Bulgaria	43,520	39.2
Central Bulgaria subzone	30,260	27.3
Southern Bulgaria subzone	13,260	11.9
Mountain zone	21,730	19.6
Forest belt	20,020	18.3
Woodless belt	1,710	1.3

The forest-steppe zone of Northern Bulgaria covers the Danube Plain and the Fore-Balkan and has an average height of 100–600 m. It has developed under the influence of the temperate continental climate under forest-steppe and forest vegetation cover. The predominating soil types are the chernozems and the grey forest soils. The zone is divided into three subzones.

The Danubian subzone includes the northern, major part of the Danube Plain, the height of which is mainly between 100 and 250 m. Loess is the main parent material here and the Fore-Balkan subzone comprises the southern part of the Danube Plain and the northern part of the Fore-Balkan, the height of which is 200 to 600 m, occasionally reaching 800 to 900 m. The main soil type is the grey forest soil derived from Cretaceous and Quaternary deposits. In the Fore-Balkan subzone, covering the southern part of the Fore-Balkan with an average height between 400 and 800 m but occasionally reaching 1,000 to 1,300 m, the parent rock material of soils consists mainly of Cretaceous sediments and partly of sediments from the Triassic, Jurassic and other periods. The light-grey (pseudopodzolic) forest soils are the main soil type.

The Xerothermal zone of Southern Bulgaria covers the flat and hilly regions in Southern Bulgaria, the height of which does not exceed 700 m. It has developed under the influence of the transitional continental and transitional Mediterranean climates under forest, shrub and meadow vegetation, including representatives of Mediterranean flora. The cinnamonic forest soils and the smolnitsas are the most frequently occurring soils in this zone.

Two subzones have been identified within it. The Central Bulgarian subzone lying south of the Balkan Mountains includes the following areas: the Kraishteto, the Fore-Balkan valleys, the Sredna Gora Mountain, the southern part of the Eastern Balkan Mountain and the Upper Thracian valley. Granites, granodiorites, schists, limestones and some other materials serve as parent rocks in the plain and hilly regions, while in the valleys the parent rock is formed by Pliocene lacustrine sediments and Quaternary deposits. The main soil types are the smolnitsas and the cinnamonic forest soils. Along the river valleys diluvial and alluvial meadow soils are found, but these are less common.

The Southern Bulgarian subzone lies to the south of the Central Bulgarian subzone but it is not a compact one. It comprises the part of the Struma river valley located to the south of the village of Krupnik (Blagoevgrad district), the part of Mesta river valley stretching to the south of the town of Yakoruda, the Eastern Rhodopes, the Sakar Mountain, the Dervent Upland and the Strandja Mountain. The parent rocks within this subzone show great variety. The soil cover consists of cinnamonic, forest soils, diluvial soils, diluvial meadow soils and alluvial meadow soils. The Strandja Mountain is the only place in Bulgaria where yellow soils occur.

The mountain zone covers the mountain areas located higher than 700–800 m above sea level. It has developed in a non-uniform mountain climate under forest and meadow vegetation cover with pronounced altitudinal zonality. Brown forest soils and mountain meadow soils predominate. This zone comprises two belts. The forest belt stretches across all the Bulgarian mountains, including the deciduous (beech), coniferous (fir and spruce) and mixed forests growing between 800 and 1,700–2,000 m above sea level.

Irrespective of the variety in the parent material, the latitude and the ecological characteristics, mineral soils and brown forest soils of large particle size predominate. The woodless (sub-alpine) belt covers the woodless areas in the Balkan Mountains and in the Vitosha, Rila, Pirin and Ossogovo ranges located above the upper forest boundary; that is, higher than 1,800–2,000 m. The parent material is made of the weathered products of granites (in Rila), granodiorites and crystaline schists (in the Balkan Mountains and Ossogovo) and limestones and marble (in Pirin). Skeletal and fragmental mountain meadow soils, rich in organic matter, predominate in this belt.

In Bulgaria the USSR genetic classification is used. The basic taxonomic unit is the *soil type* (as in the Federal Republic of Germany) which corresponds to the *soil series* (polypedon) of the morphometric classification of the USA and the *soil group* in the French system of genetic process classification. The main soil types occurring in Bulgaria are given in the table on page 48.

The chernozems cover almost all the western and central parts of the Danube Plain, the Dobrudja Plateau and a great part of the Ludogorie region. They have developed on loess under the influence of the temperate continental climate and underlie meadow-steppe or forest-steppe vegetation. Calcareous, typical leached and degraded chernozems have been distinguished within this type. The chernozems have a stable, granular and crumby structure, high water-holding capacity and good water permeability.

In terms of texture they are classified as sandy loams, loams and silt loams. The humus content ranges between 2.5 and 5 per cent.

The soil reaction is slightly alkaline (pH of 7.2–8.6) in calcareous and typical chernozems and neutral or slightly acid (pH of 6.5–7.0) in leached and degraded chernozems.

The soils contain sufficient amounts of nitrogen, but negligible amounts of phosphorus and potassium. The natural fertility of chernozems is high and the greater part of them (about 86 per cent) is under cultivation. Grains (wheat, maize, barley, beans, peas and soya), commercial (sunflower, sugar beet, broad-leaved tobacco) and orchard crops, as well as vines, are grown on the areas covered by this soil type.

The grey forest soils occur in the south-eastern part of the forest-steppe zone of Northern Bulgaria. They cover vast areas of the Fore-Balkan and Ludogorie regions and smaller territories in Dobrudja and the north-western Danube Plain. These soils form the transition between the chernozems stretching along the Danube and the brown forest soils of the Balkan Mountains. They have developed in a cool temperate continental climate under deciduous forest vegetation (oak, hornbeam, linden, maple, etc.). The grey forest soils have a fairly good structure and favourable water properties. In texture they range from loams to sandy loams.

The humus content is about 3–5 per cent in virgin soils and about 1.5–2.5 per cent in cultivated soils. The soil reaction is slightly-to-moderately acid. The contents of nitrogen, phosphorus and potassium are comparatively low, hence the fertility of these soils is not high. The use of appropriate agrotechnics and fertilizers permits the successful cultivation of many winter (wheat, barley) and spring (maize, beans, peas) crops on grey forest soils.

The smolnitzas (chernozem-smolnitzas) occur together with the cinnamonic forest soils in the valleys and low fields of Central and Southern Bulgaria, in the Thracian Valley, the lowlands of the Tundja Hilly Region, the Burgas Valley and some low fields of South-west Bulgaria. They have been formed on flat sides of low elevation, with insufficient surface water run-off and under hydrophilic meadow and meadow-swamp vegetation. The smolnitzas are heavy, colloid-rich soils with a comparatively good water-resistant structure, high water-holding capacity and low aeration. In texture they are loams. The humus content is over 5–6 per cent in non-cultivated soils and between 2.5 and 3.5 per cent in cultivated soils. The soil reaction ranges from slightly acid to slightly alkaline. The soils have sufficient supplies of nitrogen and potassium, but are low in phosphorus. They are very fertile and are used for growing valuable grain and commercial crops (wheat, barley, maize, beans, peas, sunflower, cotton, etc.).

The cinnamonic forest soils occur at a height of 700 to 800 m in the areas to the south of the Balkan Mountains, in the Tundja Hilly Region, around the Sredna Gora Mountains, in the Eastern Rhodopes, in the valleys of the Struma and Mesta rivers and in the Sofia Field. They have developed in an arid or semi-arid subtropical to transitional continental climate with a Mediterranean influence under xerophytic deciduous forests (mainly of oak and hornbeam). These soils are subdivided into typical and leached. They are characterized by a stable crumbly or crumbly-granular structure, low water permeability and poor aeration. They are heavy-textured soils, which are hard to cultivate. The humus content varies between 2 and 5–6 per cent, occasionally reaching 8–9 per cent. Their reaction is slightly alkaline, neutral or slightly acid. These soils contain negligible amounts of nitrogen, moderate amounts of phosphorus and considerable amounts of potassium. Their fertility is not high. When not irrigated, they are suitable for the growing of vines, tobacco, wheat, barley, roses, lavender and some other crops, but irrigated they are used for establishing orchards or fields of maize and cotton. The pseudopodzolic soils are common in the foothills and rolling regions of Northern and Southern Bulgaria, which rise to about 800 or 900 m above sea level. They have been formed in a transitional continental climate under deciduous forest vegetation (mainly oak, hornbeam, etc.). The pseudopodzolic soils have a loose structure and poor physical properties: on drying they form a thick crust, become compacted and develop cracks which make their cultivation difficult. They contain 1–2 per cent of humus in the arable layer and low quantities of nitrogen and phosphorus. Their reaction is acid (pH of 4–5).

The natural fertility of these soils is low, and because of this only 13 per cent of this area is cultivated.

They can be used successfully for the growing of more tolerant crops, such as rye, oats, clover, tobacco and grass fodder crops, and more rarely of plum-trees, blackcurrants and raspberries.

The yellow soils are found on the slopes of the Strandja Mountain in the catchment area of the river Veleka at a height of 300–400 m, in a wet sub-tropical climate. These slopes are put mainly under oriental beech, mixed with a winter oak, hornbeam and some other tree species. The yellow soils have a nutty-crumbly structure and optimum permeability. In terms of texture they are sandy loams and loams. The humus content is about 2–2.5 per cent, the soil reaction is slightly acid to acid (pH of 5.5–6). Their natural fertility is not high. These soils are mainly used for forestry. Some of the area is tilled and sown with arable and fodder crops.

The brown forest soils occur high in the mountains between 800 and 2,000 m above sea level. They have developed in a temperate climate with abundant rainfall under deciduous (mainly beech), mixed and coniferous forests. Within this soil-type dark brown forest soils and secondary grassed brown forest soils have been identified. They have a nutty structure, low water-holding capacity, high water permeability, and good aeration. The content of humus is 4–7 per cent in the uncultivated areas and decreases to 2–3 per cent in the cultivated ones.

The soil reaction is acid (pH of 4.5–6). The nitrogen status is unfavourable and the supplies of phosphorus and potassium are low. These soils have low fertility. They are used for growing potatoes, raspberries, strawberries, oats, rye, etc. and for pastures and meadows.

The dark-coloured forest soils (mountain forest dark-coloured soils) occur in the upper part of the central mountain belt and in the lower part of the sub-alpine belt (height between 1,600–1,700 and 2,200 m) in the Rila, Rhodope, Pirin, Balkan, Vitosha, and Sredna Gora mountain ranges. They mainly underlie high-mountain forest vegetation (spruce, white fir, dwarf pine, sometimes white pine, beech and some other trees). These soils are subdivided into peaty and chernozem-like. They have a fine granular structure and average-to-high water permeability. These soils are sandy loams to loams in texture. They have a high humus content (10–20 per cent) and an acid reaction (pH of 4–5).

The content of nitrogen is high, whilst phosphorus and potassium are present in low quantities. The unwooded plots on dark-coloured forest soils are planted with potatoes, raspberries and black currants.

The mountain meadow soils are found in the highest areas above the tree line (lying higher than 1,800–2,000 m and higher than 2,500 m) on the ridges of the Pirin, Rila and Balkan mountain ranges.

They are not common but can still be found in the Ossogovo, Belassitsa, and Vitosha mountains and in the Rhodopes. These soils are covered by high-mountain (alpine) grass vegetation. Soddy, chernozem-like and peaty soils have been identified within this soil type. They are loose soils with a fine granular structure, high water-holding capacity and poor permeability. In texture they are sandy loams. They have a high content of humus (between 10 to 20–25 per cent) and an acid reaction (pH of 4–4.5). The mountain meadow soils contain comparatively large quantities of nutrients but the greater part of them cannot be utilized and are used for high-mountain pastures.

The meadow soils are alluvial soil deposits of limited occurrence. They are found on river floodplains and over-floodplain terraces, where they overlie alluvial or diluvial deposits (alluvial meadow soils) or on piedmont diluvial drags (diluvial meadow soils). The alluvial meadow soils occur in the Danubian lowlands and in the valleys of the rivers Lom, Ogosta, Iskur, Vit, Yantra, Rossitsa, Maritsa, Tundja, and Struma. Their structure varies from gravelly to sandy-loam. These soils are loose, do not form a crust and have low water-holding capacity and high water permeability. They contain 1.5 per cent of humus and considerable amounts of nutrients; their reaction is slightly acid to alkaline. These soils are characterized by high fertility and are suitable for the growing of an appreciable variety of crops and especially of vegetables.

The diluvial meadow soils occur in the Fore-Balkan valleys and in the foothill regions of the Rhodopes, Rila and Pirin mountain ranges. Their structure is granular. The texture of these soils varies widely and hence the water-holding capacity and the water permeability of this soil type differs from soil to soil. The content of humus varies between 1 and 3.5 per cent and the soil reaction is alkaline, neutral or slightly acid. These soils have only slight nitrogen and phosphorus content. They are suitable for the growing of orchard fruits, fodder and commercial crops (mainly tobacco).

The rendzinas (humus calcareous soils) are a soil type, the formation of which is related exclusively to the abundant presence of calcareous rocks (limestones, marble, marbly limestones, etc.) and is independent of bioclimatic conditions. They occur in limited areas surrounded by chernozems and grey forest soils in Northern Bulgaria, by cinnamonic forest soils and smolnitzas in Southern Bulgaria and by brown and dark-coloured forest soils and yellow soils in the mountain areas. The humus carbonate content in rendzinas is very

The Main Soil Types in Bulgaria – Area in Hectares and Percentages

Soil Type	Area (m. ha)	Percentage
1. Chernozems	2.30	21.0
2. Grey forest soils	1.80	16.3
3. Smolnitzas	0.60	5.3
4. Cinnamonic forest soils	2.80	25.3
5. Pseudopodzolic soils	0.33	3.0
6. Yellow soils	0.03	0.2
7. Brown forest soils	1.64	14.9
8. Dark-coloured forest soils	0.10	0.9
9. Mountain meadow soils	0.17	1.5
10. Meadow soils	0.73	6.5
11. Rendzinas	0.27	2.4
12. Boggy soils	0.07	0.6
13. Saline soils	0.02	0.02

high, which accounts for their pronounced crumbly-granular structure and high water permeability. In texture they are loams to silt loams. The humus content ranges from about 5–7 per cent to 10–12 per cent in virgin soils and from 2.5–4 per cent in cultivated soils.

The soils' reaction is neutral to alkaline. The amount of nitrogen and phosphorus in them is significant, while the potassium content is high. The shallow rendzinas are not suitable for farming, while the thicker ones can be used successfully for growing vines and stone fruit-trees (apricot-trees, cherry-trees and almond-trees).

The boggy soils occur in some Danubian lowlands (Ostrovo, Cherno Pole, Belene), around the lakes along the Black Sea coast (Varna Lake, Pomorie Lake, Burgas Lakes) and in the Upper Thracian Valley (in the swamps near Sadovo and Straldja). They have developed near lakes, swamps and rivers as a result of flooding and lack of run-off. These soils are characterized by an accumulation of high amounts of undecomposed or partially decomposed organic matter (peat). After drainage and some other improvements these soils can be used for farming, mainly for the growth of commercial crops.

The saline soils contain high amounts of water-soluble mineral salts which adversely change the soils' physical and chemical properties and lower soil fertility. Solonchaks and solonetzes have been distinguished within this soil type. The solonchaks contain more than 1 per cent of water-soluble salts and cover small areas of the drained swamps near the Danube and the Black Sea. The solonetzes have a cation-exchange capacity, including more than 20 per cent of exchangeable sodium. They are found predominantly in the districts of Plovdiv, Sliven and Yambol, where they are used for pastures and ranges. These soils can be improved through drainage and chemical amelioration.

VEGETATION

The vegetation of Bulgaria is characterized by a diversity determined by the varied ecological conditions in the country and the combination of several climatic influences: moderate continental, transitional continental, transitional Mediterranean and mountainous; by the relief (formation of transitional and micro-climates); and inanimate factors: soil, subterranean rocks, hydrological and other related conditions. All this ensures the formation of varied ecotypes and of vegetation formations (phytocoenoses), respectively ecosystems (biogeocoenoses) realized in the process of a long evolution of the vegetation cover from the Tertiary period to this day. Important changes have recently occurred in the vegetation cover under the influence of human activity.

The floral elements constituting the country's vegetation cover are also very varied in their biology, ecological demands and origin. Over 3,500 higher plant species are known at present. Some of these plants have a **cosmopolitan** distribution, such as *Phragmites australis*, *Typha augustifolia* and *Typha latifolia*, *Pteridium aquilinum*, *Medicago polymorpha*, *Poa pratensis* and a number of other plants, predominantly aquatic and marshy species, as well as a large number of weeds and ruderal plants.

Some vegetation is **circumpolar** (around the tropical zone, their habitats are scattered), such as *Agrostis tenuis*, *Millium effusum*, *Poa nemoralis*, *Trifolium pratense*, *Rubus idaeus*, *Nymphaea alba*, and many others; the majority of them are aquatic and hygrophile, as well as ruderal plants.

The group with **Eurasian** distribution is the most commonly found. Among them there are tree species such as *Alnus glutinosa*, willow *(Salix fragilis)*, white poplar *(Populus alba)*, aspen *(Populus tremula)*, birch *(Betula pendula)*, white pine *(Pinus silvestris)*, and others, a number of shrubs, such as *Juniperus sabina*, *Tamarix ramosissima*, *Frangula alnus*, *Berberis vulgaris*, and so on but all perennial herbaceous plants, such as *Dichantium ischaemum*, *Agropygon pectiniforme*, *Astragalus onobrychis*, *Brisa media*, *Galium odoratum*, *Dictamnus albus*, *Medicago falcata*, a number of species belonging to the genera *Artemisia*, *Campanula*, *Carex*, *Centaurea*, *Euphorbia*, *Echinops*, *Festuca*, *Inula*, *Jurinea*, *Lathyris*, *Phlomis*, *Vicia*, *Trifolium*, *Veronica*, *Silene*, *Stipa*, *Scabiosa*, and many others.

Comparatively fewer are the species with **European** distribution: fir *(Abies alba)*, spruce *(Picea abies)*, *Fraxinus oxycarpa*, *Quercus robur*, lilac *(Syringa vulgaris)*, primrose *(Primula veris)*, *Mellitis melissifolium*, species belonging to the genera *Verbascum*, *Trifolium*, *Thymus*, *Centaurea*, *Euphorbia*, and many others.

Mediterranean species spread predominantly into the southern part of the country, such as: *Quercus coccifera*, *Phyllirea latifolias*, *Juniperus excelsa*, *Juniperus oxycedrus*, *Pistacia terebinthus*, *Cistus incanus*, *Cistus salvifolius*, *Asparagus acutifolius*, many herbaceous plants, and especially therophytes such as *Psilurus incurvus*, *Brachypodium distachions*, *Scleropoa rigida* from the cereal plants, and representatives of the genera *Trifolium*, *Medicago*, *Trigonella*, etc. from the leguminous plants.

Endemic species are much less widespread. They characterize the phytocoenoses formed by them and the vegetation cover in general. Some of these species occur as dominants of the plant formations and the vegetation cover of which they consist. The more important Balkan endemic species for the vegetation cover of Bulgaria are: *Sesleria comosa*, *Festuca valida*, *Festuca penzesii*, *Chamaecytisus absinthioides*, *Genista rumelica*, *Pinus peuce* and *Aesculus hippocastanum*; the Bulgarian endemics are: *Festuca riloensis*, *F. pirinica*, *F. balcanica*, *F. stojanovii*, *Primula deorum*, *Astragalus aitosensis*, and *Convolvulus suendermanii*.

As territory of Bulgaria was not completely frozen during the Glacial period (only the highest parts of the Rila and Pirin Mountains were covered with ice), vegetation was less affected than in Central and Northern Europe, where many species disappeared during the Glacial period. Migration from south to north started only after the last glaciation, from the Balkan Peninsula and Bulgaria as well. This suggests that vegetation in Bulgaria and in the Balkan Peninsula has an older, more primary, **relict** nature. Some species have even been preserved in Bulgaria from as far back as the Tertiary period, such as: *Haberlea rhodopensis*, and *Ramonda serbica*, both species belonging to the tropical family *Gesneraceae*. Other Tertiary relicts are: *Primula deorum*, *Pinus peuce*, *Aesculus hippocastanum*, *Pinus heldreichii*, *Fagus orientalis*, *Rhododendron ponticum*, *Laurocerasus officinalis*, *Daphne pontica*, and *Platanus orientalis*.

The zonal vegetation in the country is composed of deciduous forest xerotherms (adapted to a dry and warm climate), with abundant Mediterranean elements in the southernmost parts of the country, which decreases rapidly to the north, as there are few species in North Bulgaria. Mountain massifs which occupy about a quarter of the country's territory affect the distribution of the

The Iskur gorge near the Cherepish Monastery in the Western Stara Planina

Vegetation

plant species and the vegetation cover. In fact, the relict character of the Bulgarian vegetation has been preserved above all in the mountainous regions, where the conditions have existed for a long time for the growth of old floral elements that were and continue to be the source of distribution and enrichment of the flora and vegetation in Bulgaria and in the neighbouring countries.

The vertical distribution of vegetation is in vegetation belts typical of the country.

In the **Alpine belt** (above 2,300–2,500 m above sealevel) the vegetation consists mainly of herbaceous plant formations (phytocoenoses), to a lesser extent of phytocoensoses consisting of low shrubs and bushes, or of groups of lichen. Groups and formations of *rock vegetation* are widespread on the rocks and screes, which abound in the Alpine and partly in the Sub-alpine belts, particularly in the region of the glacial formations (circuses).

On silicate terrains on the rocks, the predominant species in most areas are: *Saxifraga pedemontana, S. bryoides, S. sancta, S. androsacea, S. exarata, S. retusa, S. carpatica*, as well as the species *Geum bulgaricum, R. incomparabilis, Allium victorialis, Agrostis rupestris*, and others. In some places there are lichen groups. On the screes there are groups of *Geum reptans, Poa cenisia, Ranunculus crenatus, Oxyria digyna*, and others. The predominant species on calcareous rocks are *Saxifraga oppositiofolia, S. sempervirens, S. spruneri, S. luteoviridis* and *S. ferdinandi-coburgi*. Other widespread species include *Sesleria klasterskyi*, edelweiss (*Leontopodium alpinum*), *Androsace villosa* and *Papaver degenii*. Some groups occur on silicate and calcareous rocks and screes: *Saxifraga adscendens, S. paniculata, Armeria alpina*, and related plants.

On good and poor soil, mainly on the mountain peaks and ridges of Rila and Pirin and to a lesser extent of the Balkan Mountain, Vitosha, Ossogovo and the Rhodopes, acidophilous plants occur: *Sesleria comosa, Carex curvula, Festuca riloensis, F. oiroides, Agrostris rupestris* and *Juncus trifides*, whereas near the snow-drifts there are formations of *Alopecurus gerardii* and *Omalotheca suoina*. Silicate terrains are characterized also by *Salix herbacea, S. retusa, S. hastata, Empetrum nigrum, Vaccinium uliginosum*, and others.

Formations of *Elyna bellaedii, Festuca pirinica, Sesleria korabensis*, etc. occur on calcareous terrains predominantly in the Pirin Mountain, on relatively poorly developed soils. Formations of *Carex kitaibelliana, Dryas octopetala, Salic reticulata*, etc. are found mainly on calcareous and less on silicate terrains.

In the **Subalpine belt** (from 2,000–2,200 to 2,300–2,500 m above sea level) the original vegetation, not affected by human activities, consists of phytocoenoses formed mainly of *Arctoalpine floral elements*, as in the Alpine belt. Shrub formations prevail, consisting mainly of dwarf pine *(Pinus mugo)*, which cover extensive areas in some places in the Rila and Pirin Mountains, whereas in the other mountains – the Balkan Range, Vitosha, the Rhodopes and Belassitsa – only isolated groups or single specimens have been preserved as relicts of the distribution of dwarf pine shrubs in the past. Shrub formations of *Alnus viridis* have grown up in humid areas among the dwarf pine vegetation. In some places in this belt there are

Forest vegetation in the foothills of Mount Vitosha

An ancient spruce (56 m high) in the Parangalitsa reserve in the Rila mountains

also original plants of blueberries *(Vaccinium uliginosum)*, though in most cases this plant's habitat grew as a result of human influence. Plants consisting of the relict species *Rhododendron myrtifolium*, which have also begun to grow in other areas under the influence of human activities, have limited distribution in the Eastern Rila Mountain and in the Central Balkan Range.

Due to the great variety in the relief, especially in the areas of the glacial formations (circuses), habitats with different humidity arose in the Rila and Pirin mountains ensuring the emergence of various formations, especially of hygrophile plants, and in the glacial lakes, aquatic plants as well. The majority of the hygrophile plants are original. Apart from the *Alnus viridis* formation occurring on the humid slopes, mainly along the streams, the most frequent formations in the peat-bogs consists of *Salix lapponum* in the Rila and Vitosha mountains and of *Salix waldsteiniana*; herbaceous formations of *Carex acuta, C. echinata, C. rostrata, Eriphorus sp. div., Trichophorum caespitosum, Molinia coerulea, Deschampsia caespitosa, Luzula alpinopilosa, Hydronardaea coenoses*, Sphagnum peat-bogs, as well as the endemic plant formations for the Rila mountain, consisting of *Primula deorum*.

Aquatic phytocoenoses occur in the glacial lakes of Rila and Pirin, with predominance of *Ranunculus aquatilis, Callitriche palustris, Sparganium angustifolium*, as well as the fern-like species *Isoetes setacea*.

In the past, most of the dwarf-pine shrubs were destroyed either by cutting them to produce charcoal, or by burning them to increase the Sub-alpine pastures. Secondary scrub and herbaceous formations appeared in the place of the dwarf-pine formations. Part of the secondary formations had limited distribution in the rocky regions or in superhumid areas near the Sub-alpine lakes and mountain streams, and after the anthropogenic destruction of the dwarf-pine formations which are more competitive under Sub-alpine conditions, shrubs and grasses occupied larger and sometimes even vast territories in this belt. Some of the phytocoenoses of these secondary formations are very resistant and play an important role in the water turnover and soil retention. The formation of *Juniperus sibrica* occupies the largest space of scrub vegetation, whereas the formations of *J. pygmaea, Chamaecytisus absinthioides, Vaccinium myrtillus, V. vitisidaea, V. uliginosum, Bruckenthalia spiculifolia* and *Acrostaphylos uva-ursi* are less well distributed. The largest areas are covered by the secondary herbaceous formations of *Nardus stricta*, followed by the Balkan endemic species *Festuca valida*, and of *F. paniculata, F. amethystina, F. nigrescens, Poa media*, and *Bellardiochloa violacea*. Calcareous terrains are covered by the phytocoenoses of *Festuca penzesii* and *F. pipinensis* which is similar to it, as well as by shrub formations of *Daphne oleoides*, and *D. kosaninii, D. cneorum*. In some places in the Sub-alpine belt, mainly among the dwarf pine vegetation, there appear groups of single trees of spruce *(Picea abies), Pinus peuce, Sorbus aucoparia, Salix caprea*, and others.

In the **coniferous forest belt** (from 1,300–1,600 to 2,000–2,200 m above sea level), formations of *Picea abies, Pinus peuce* (endemic for the Balkan Peninsula) and *Pinus silvestris* predominate, while on the calcareous terrains of Pirin and Salvyanka there is also a *Pinus heldreichii* formation. As a result of wind damage, cuitting or burning, part of the forests were destroyed and replaced by secondary forests of birch *(Betula pendula)*, aspen *(Populus tremula), Alnus incana* or shrubs of dwarf pine *(Pinus mugo), Juniperus sibirica, Chamaecytisus absinthioides, Vaccinium sp. div*, etc., which have usually penetrated from the Sub-alpine belt. Relict formations of the shrub-like *Potentilla fruticosa* are characteristic of the Rhodope Mountains.

Secondary herbaceous formations appeared in the place of the former coniferous forests, consisting of *Agrostis tenuis, Festuca rubra, Nardus stricta, Pteridium aquilinum Calamagrostis arundinacea, Epilobium angustifolium*, and others. Calcareous terrains are usually covered by vegetation formations of *Convolvulus suendermanii* and *Centranthus kellereri*, which occurs lower, in the beech belt and in some places there are also formations of *Festuca hirtovaginata*. In some places, mainly in the Rhodopes and partly in the Rila and Pirin, some of the lands in this belt have been converted to arable land, where mainly potatoes and to a lesser extent flax, oats, and other crops are grown.

In the **beech forest belt** (from 1,000–1,200 to 1,300–1,700 m above sea level) the original vegetation consists of deciduous forests with a predominance of *Fagus silvatica* formations. Particularly characteristic are beech forests with undergrowth of the evergreen *Laurocerasus officinalis*, which are widespread in the Balkan Mountains. Fir forests *(Abies alba)* occur occasionally. Mixed forests of beech with *Ostrya carpinifolia*, etc. are also found in some rocky valleys, predominantly on calcareous terrains. Where the destroyed beech forests have been replaced by secondary formations, these consist almost of the same species as the formations typical of the coniferous belt, but there are also herbaceous formations, predominantly of *Agrostis tenuis, Festuca rubra, F. balcanica, F. dalmatica, Nardus stricta, Pteridium aquilinum* and *Cynosurus cristatus*, whereas *Festucopsis sancta* and *Avenula compacta* formations are widespread on calcareous terrains as primary formations. Part of the territory of the beech belt has also been turned into arable land, where mainly potatoes are grown, less frequently oats and flax, and only rarely barley, rye, and other cereals.

In the **hornbeam and oak forest belt** (from 500–700 to 1,000–1,200 m above sea-level) the original vegetation is also *deciduous forest*, in many places relict in nature. *Quercus* forests are most widespread: the *Quercus dalechampii* formation and *Carpinus betulus* woods, with a characteristic, beautiful, brightly blossoming, ephemeral spring grass cover of *Scilla bifolia, Erythronium dens canis, Hepatica nobilis, Anemone ranunculoides, Isopyrum thalictroides*, species of genus *Corydalis*, etc. Mixed deciduous woods are also spread in some places on the rocky slopes, often with a prevalence of *Acer hyrcanum*. A large part of the formations in the vegetation cover of this belt are relict in character, specific for the Balkan Peninsula, for Southern Europe and for other regions. Formations of *Ostrya carpinifolia, Pinus nigra* and *Tilia tomentosa*, whose phytocoenoses are particularly characteristic of North-eastern Bulgaria (Russe,

Silistra and Razgrad districts, etc.), are spread over this belt predominantly in Southern Bulgaria, often on calcareous terrain. *Fagus moesiaca* formations are typical of the lower parts of the mountain slopes, predominantly in the Fore-Balkan region, reaching 1,000 m above sea-level.

There are other relict formations such as the *Castenea sativa* formations which occupy large areas in the northern foothills of the Belassitsa Mountain and of the Berkovitsa part of the Balkan Mountains, as well as in some parts of the Salvyanka and Pirin Mountains. The relict woods of *Fagus orientalis* found in Bulgaria (in Strandja and the Eastern Balkan Mountain) are a unique phenomenon in Europe. In Strandja they occasionally have an undergrowth of evergreen shrubs of *Rhododendron ponticum* and often of *Laurocerasus officinalis, Ilex aquifolium* and *Vaccinium arctostaphylos*. Other South-euxinic elements in these woods are *Trachystemon orientale*. Also of eastern origin are *Quercus polycarpa* formations, frequently found in the vegetation cover of Strandja and the Eastern Balkan Mountains. *Quercus hartwissiana* is also to be found in the Strandja woods, consisting predominantly of oaks. It has a primitive leaf structure, which is also a unique phenomenon in Europe. Of particular interest are the woods in the Preslav part of the Balkan Mountain, consisting of *Aesculus hippocastanum* – a relict and endemic species for the Balkan Peninsula. These forests are also a unique phenomenon for Europe, because they have been preserved since the Tertiary period.

The original vegetation in this belt has been strongly influenced by human activities, mostly due to the proximity of settlements. Many of the woods have been destroyed, to be replaced by secondary formations which have sprung up in their place, or the areas have been turned into arable land. More important among the secondary formations are the *Corylus avellana* scrubs which also penetrate into the beech belt, and occasionally into the coniferous belt. Scrubs consisting of *Juniperus communis* are less widespread and so are the grass formations – *Agrostis tenuis, Pteridium aquilinum, Festuca valesiana*, and to some extent *Festuca panciciana* and *Festuca stojanovii*, whereas calcareous terrains are covered by formations of *Sesleria latifolia, S. rigida*, etc. The formation consisting of the Balkan endemic species *Genista rumelica* and *G. lydia*, appearing predominantly in the place of *Quercus* forests destroyed in the past, is a secondary formation of predominantly characteristic of southern rocky slopes. In Strandja there are *Calluna vulgaris* formations, unique for Bulgaria, which are generally more widespread in the Holarctic region. *Erica arborea* which is very widespread in Southern and South-eastern Europe and Asia Minor occurs only in the Strandja region of Bulgaria.

In the **oak belt** (below 500–700 m above sea-level) the original vegetation also consists of deciduous forests. It is characterized above all by xerothermal oak formations of *Quercus cerris, Q. frainetto, Q. pubescens* and *Q. virgiliana*, which are widespread in hilly regions throughout the country (the Danubian, Dobrudja and Tundja hilly regions). Formations of *Fraxinus ornus* and *Acer monspessulanum* should also be included in this belt. Formations where walnuts predominate or are present occur naturally in some places at the foot of the moutain slopes, mainly in rocky and talus localities. In fact, xerothermal vegetation, above all xerothermal oak formations, is generally characteristic of the zonal vegetation in Bulgaria. In Bulgaria the deciduous summer vegetation is the most southern-type of green forests bordering on sclerophyllous evergreen forests of Mediterranean vegetation, some of which penetrate into the southernmost regions of the country.

Most of the original vegetation has been destroyed and it is being replaced mainly by cereal crops, vineyards, orchards, and some southern crops, such as tobacco and cotton. Part of the forests represent in fact shrubwood, most of them very thinned out and exhausted. Many of the forests have a substantially changed composition and structure. Others have been naturally substituted by less productive and less stable forest or scrub and grass formations. *Carpinus orientalis* gradually penetrated and settled in many places. On more strongly eroded terrains it has completely or almost completely replaced the original vegetation, creating its own *C. orientalis* formation.

In some places, particularly in the eastern part of the country (mainly in the Tundja hilly region), formations of *Paliurus spina-christi* have appeared, occupying rather large areas in some regions and determining the appearance of the landscape. Other shrubs, e.g. *Jasminum fruticans*, usually find refuge in the *Paliurus spina-christi* formations in Southern Bulgaria and along the Black Sea coast, making up independent formations in some places. Formations of grass vegetation have also established themselves among the *Paliurus spina-christi* bushes. In this way complexes of scrub and grass formations have been created. There are lilac (*Syringa vulgaris*) formations also in some places on rocky terrains, mainly on calcareous soils. Part of the lilac formations are original, part of them have spread to replace the original forests destroyed by man. *Cotinus coggygria* formations appearing on calcareous terrains as secondary formations are less frequent. The secondary grass formations consisting of *Chrysopogon gryllus, Dichantium ischaemum* and *Poa bulbosa*, are widespread whereas mainly in Southern Bulgaria there are also ephemeral formations consisting of *Brachypodium distachyon, Psilurus Incurvus, Vulpia ciliata* and *Vulpia bromoides*, a number of *Trifolium* species and *Medicago*.

Mediterranean vegetation has also penetrated into the southern part of Bulgaria, mainly along the Struma valley, the Black Sea coast and the Upper Thracian Lowland. In most cases it replaced the original xerothermal forest vegetation which had been destroyed. The evergreen scrubs of *Quercus coccifera* which have spiky thick leaves, *Phyllirea latifolia, Juniperus excelsa* and *Juniperus oxycedros* are examples of this. *Quercus coccifera* formations exist in several places along the Struma valley only; there is one single formation in the Gotse Delchev region, while *Phyllirea latifolia* formations are more frequent along the Struma valley, in the Eastern Rhodope Mountains, Strandja and the southern Black Sea coast, reaching the town of Aytos to the north. *Juniperus excelsa* formations occur mainly in the Kresna gorge (the Struma valley) and to some extent in the Rhodopes (the valleys of the rivers Vucha and Chaya). *Juniperus oxycedros* is more widespread along the valleys of the rivers Struma and Mesta, in the Rhodopes and in some regions on the southern slopes of the Balkan Mountains and along the Black Sea coast. *Pistacia terebinthus* formations are also of Mediterranean origin; they cover large areas in Sliven District (the southern slopes of Sredna Gora Mountain and the Balkan Range). There is a limited distribution of formations of the Mediterranean species *Cercis siliquastrum* (in the Patleina region, in the Preslav part of the Balkan Range and elsewhere), *Ozyris alba* and *Cistus salvifolius*. *Cistus incanus* formations are more widespread in South Bulgaria.

The Sub-Mediterranean vegetation consists of deciduous formations which are relatively thermophilic and drought-resistant. Usually there are Mediterranean elements in them. Sub-Mediterranean vegetation occurs mainly in the southern and eastern (coastal) parts of the country. It consists of plant formations of *Quercus pubescens*, some *Q. frainetto* and *Q. cerris*, *Fraxinus ornus*, *Acer monsspessulanum*, and their derivative forest formations of *Carpinus orientalis*, scrubs of *Paliurus spina-christi* and *Cotinus coggygria*. A large number of grass formations of *Dichantium ischaemum*, *Chrysopogon gryllus* and *Poa bulbosa* also occur, as well as ephemeral formations.

The heavily wooded banks of the lower reaches of the river Kamchia

Mainly on calcareous terrains in this belt and in some plain regions, in the place of xerothermal oak forests, in Northern and South-eastern Bulgaria there is also secondary **steppe vegetation.** Very large areas of steppe vegetation occur over the exposed calcareous hills in South-western Bulgaria – in the Golo Burdo, Chepan, Zemenska and other mountains. Steppe vegetation consists mainly of grass phytocoenoses, above all of *Bromus riparius*, *Festuca stojanovii*, *F. pseudovina*, some *Thymus* species, *Artemisia alba* and *A. austriaca*, some *Jurinea* species, less frequently formations of the low shrubs of *Amygdalus nana*. Steppe vegetation can also be said to include *Astragalus angustiolius* formations, occurring predominantly in South-western Bulgaria; *A. aitosensis* formations on the hills around the town of Aytos, and *A. thracisus* in the Upper Thracian lowlands.

In the **lowlands** the original vegetation has been almost completely destroyed, with only a little remaining in some places. It used to consist of *Quercus pedunculiflora*, *Ulmus minor* and *Fraxinus oxycarpa*, and to a lesser extent of *Acer campestra* and *Quercus robur*, while along the rivers it consisted of formations of *Alnus glutinosa*, *Salix* species, *Populus alba* and *Populus nigra*, *Salix purpurea* shrubs, *Tamarix ramosissima* and *T. tetrandra*.

Where the original vegetation has been destroyed in the lowlands, valleys and along the rivers, the areas have been turned predominantly into arable lands where mainly vegetable crops, and fruit trees are grown. Some of the lands have been turned into meadows with a predominance of formations of *Festuca pratensis*, *Poa silvicola*, *Alopecurus pratensis*, *Elymus repens*, *Agrostis stolonifera*, *Lolium perenne* and less frequently *Phleum pratense*, *Hordeum secalinum*, *Bromus commutatus* and *Poa pratensis*. Pastures of *Cynodon dactylon* have been formed in moderately moist and semi-moist areas, rapidly covering temporarily abandoned land.

Riverside forests occur in some places in the eastern part of the country – along the Black Sea coast (the rivers Batova, Kamchia, Ropotamo, etc.) and partly along the rivers Tundja and Struma, with a predominance of *Ulmus minor*, *Fraxinus oxycarpa*, some *Quercus pedunculiflora*, and an abundance of lianas – *Similax excelsa*, *Periploca graeca*, *Clematis vitalba*, *Vitis vinifera* and blackberry. Along the periphery of those forests the lianas form almost impenetrable 'curtains'. Riverside forests are an original phenomenon in South-eastern Europe and they are common in Bulgaria. They require a specific ecological regime, consisting mainly of a periodic flooding by river water, i.e. a change of the water regime and of the supply of nutrients in the soil. These forests are exotic and they are an unusual element in Bulgaria's vegetation.

Aquatic and riverside vegetation consists of algae and higher plants – hydrophytes. Approximately 3,500 algal species (out of the 4,000 species in Bulgaria) occur in the vegetation of freshwater reservoirs. There are more than 70 species of higher water plants. Three species of them are fern-like: *Isoetes setacea*, *Salvinia natans* and *Marsilea quadrifolia*, the rest being flowers: *Ceratophyllum demersus*, *Potamogeton natans*, *Nuphar luteum*, *Nymphaea alba*,

Wolffia, Lemna minor, Nymphoides flava, Valisneria, Microphillium, Trapa natans, Spirodella, Marsilea quadrifolia, Elodea canadensis, as well as representatives of terrestrial species: *Sparganium polyedrum, Ranunculus*, etc. Two species of Zostera marina are typical of salt water areas.

Riverside vegetation, referred to also as marsh and bog vegetation, is represented by hydrophytes. Most of them are higher plants, more than 350 species being floral. The lower plants are mostly mosses, peat mosses being more important among them. The riverside vegetation consists mainly of herbaceous species: *Schoenoplectus lacustris, S. triqueter, S. tabernemontani, Typha angustifolia, T. latifolia, T. laxmanii, Phragmites australis, Cyperacea, Carex* species, *Juncus maritimus, Leucojum aestivum*, and buttercup. Riverside trees include willows, poplars, some species of *Alnus* and *Ulmus* which, intertwined with creeping and climbing plants, form the riverside forests.

Halophytic grass vegetation occurs in some places in the country, mainly in South-eastern Bulgaria (the districts of Burgas, Sliven, Yambol and some parts of Stara Zagora). This vegetation is found in saline soils, with formations of *Salicornia europea, S. ramosissima, Sueada maritima, S. altissima, S. heterophylla, Camphorosma monspeliaca, C. annua, Aeluropus litoralis, Puccinellia convoluta, P. distans, Crypsis aculeata, Limonium latifolium, L. gmelinii, L. vulgare* and *L. meyeri* prevailing, while less saline soils are covered with phytocoenoses of *Crypsis alopecuroides, Juncus gerardii, Elymus elongatus, Hordeum geniculatum, Phacelurus digatatus* and *Atriplex hostata*.

Psammophytic vegetation occurs along the coastal sands of the Black Sea and less frequently elsewhere. It consists of various formations, with the following species predominating: *Cionura erecta*, the herbaceous species – *Ammophylla arenaria, Leymus arenaria, L. racemosus, Festuca vagnata, Galilea micronata, Pancratium maritimum, Peucedanum arenarium, Trachomitum venetum, Jurinea albicaulis, Artemisia maritima* and *A. campestre, Centaurea arenaria, Corispermum nitidum, Glautium sp. div.*, etc. Groups of *Elymus pycnanthus* and *E. farcatus, Eryngium maritimum* and *Cakile maritima* grow on the saline coastal sands. In the sandy areas in the interior of the country there are groups of *Chenopodium botrys* and some *Melilotus alba* and *M. officinalis*.

Groups of southern herbaceous plants are found in the southern part of the Struma valley and along the southern Black Sea coast (to the south of the Balkan range), predominantly on wet sands. Among these plants is the highly decorative *Erianthus ravennae* (reaching a height of 2–3 m) which in tropical countries forms a considerable part of vegetation formations.

Ruderal and weed vegetation, referred to also as *cultigenic or anthropophytic*, consists mainly of mobile floral species, some of which have been introduced into the county by man. Such species are, for instance, the recently imported *Galinsoga parviflora* and *G. quadriradiata, Azolla filiculoides* and *A. caroliniana*, which spread rapidly and occupy more and more territories. Some species have been imported as cultivated or park trees, such as *Ailanthus altissima* and *Zynzyphus jujuba*, both originating from East Asia. Both species found favourable conditions in Bulgaria and turned wild, forming in some places their own groups or even formations. *Ailanthus* occurs in many places throughout the country, while Zinzyphus is found only in the warmer regions, e.g. along the Black Sea coast not further north than Nessebur, on Djendem Tepe in Plovdiv and in the Struma valley near Katuntsi village. These are firmly established wild-growing foreign species.

Great attention is paid in modern society to the problem of the rational utilization of natural vegetational resources. This is related to the problem of preserving plant species and their formations, as well as the vegetation cover of the country as a whole. Important strides forward have already been made in this respect, a consistent system has been elaborated and rational measures have been taken for protection of the environment.

FAUNA

Compared to most European countries Bulgaria's wildlife is characterized by a great variety of species determined by the country's zoogeographic position. Bulgaria's terrestrial fauna belongs to the Palaearctic part of the Holarctic zoogeographic region and is in the border zone between the Euro-Siberian and Mediterranean zoogeographical sub-region. This explains the presence of thermophilic species side by side with cold-resistant ones. Another prerequisite for the variety of species is the variety of biotopes as a result of the large number of altitudinal zones in the topography. The widest range of species is found in the plains, the number decreasing with the height. The palaeoclimatic changes, chiefly in the Glacial periods of the Quaternary Period, were an important factor contributing to the varied fauna. The Bulgarian lands were not completely glaciated, and northern animals survived after the recession of the glaciers finding favourable living conditions in the high parts of the mountains (glacial relicts), whereas the warmer regions (even during the maximum glaciation, during the Würm) saved a small number of Tertiary relicts, survivors from the abundant thermophilic Tertiary fauna. Thermophilic animals from South-west Asia penetrated via 'land bridges' existing in the past between the Balkan Peninsula and Asia Minor. The diversity of factors which formed the country's fauna could explain the comparatively high percentage of endemics with certain groups of Invertebrata: nearly all representatives of the spelaean fauna, part of the Tertiary relicts, and the high mountain species.

We have grounds for believing that Bulgaria's fauna comprises many more than the recorded species. The number of the known, partially known and probable groups of animals is approximately as follows: Invertebrata – unicellular – 3,000; sponges – 30; Coelenterata – 40; worms – 1,000; Mollusca – 500 (of them – chitons – 2, mussels – 70, snails – 430); Crustacea – 600; Arachnida – 2,100; Myriapoda – 180; insects – about 30,000 and Vertebrata – 659 species, of them Cyclostomata – 2, fish – 177, amphibians – 16, reptiles – 35, birds – 337, mammals – 92.

The geographic location conditions the existence of varied zoogeographic elements despite the small territory.

Cosmopolitan species. The cosmopolitan species are found almost everywhere. In Bulgaria their number is small and the relatively highest percentage is concentrated among the unicellular. Of the Vertebrata there are single cosmopolitan species: *Falco peregrinus* (peregrine falcon), *Tyto alba* (barn owl), *Callinula chloropus*. In the past century human activity led to the cosmopolitan distribution of some synanthropic species: *Rattus rattus* (ship rat), *Rattus norvegicus* (common rat), *Mus musculus* (house mouse), *Musca domestica* (house fly), *Blatta orientalis*, (Oriental cockroach), *Blattella germanica* (German cockroach), and many parasitic species: unicellular, worms, Acaridae and insects in man and domestic animals.

Endemics. The endemics are Invertebrata. They are encountered in systematic groups in the main, whose representatives are either not very mobile or are species in restricted habitats with specific living conditions. Nearly all Bulgarian spelaean animals are endemics – 120 species (out of a total of 130 species), of different systematic groups. The percentage of endemics in subterranean waters is high: nine species of Batinelacea, eleven species of Isopoda. In open water endemics are as follows: Copepoda – fifteen species, Hydrocarina – eighteen species (ten fresh and eight salt water), Ostracoda – five species and water snails – seven species. Among terrestrial animals the greatest distribution of endemics is: Oniscopoidea – 28 species, Arachoidea – sixteen species, and of the insects – among the Trichoptera (caddisflies) – fifteen species (their larvae live in fresh waters), the Hemiptera (true bugs) – fifteen, the Diptera (two-winged flies) – fifteen.

Relict Animals. The relict animals in Bulgaria are of different origin and of different age. The spelaean animals in the main are Tertiary relicts. They are survivors of the once thermophilic fauna in Southern Europe during the Tertiary Period. Such are some genera Oniscoidea Langidae and Coleoptera (beetles) and all cave-dwellers. Typical relicts are the marine Oniscoidea of the genus Sphaeromides, and are of the fauna found in the subterranean sandy water-bearing layers. Many species of animals, mainly insects, survivors from the cold spells during the Quaternary Period, called glacial relicts inhabit the high parts of the Rila, Pirin and other Bulgarian mountains. Some of these species are distributed as far as North Europe, others are inhabitants of the high mountains in Central and Southern Europe, Bulgaria included. Among them are grasshoppers, beetles and many species of butterflies. The Rila Mountains are the richest in glacial relics. Steppe relics in Bulgarian fauna are: *Mesocricetus auratus* and *Sicista subtilis*. The slightest changes in the climate of the Strandja Mountains during the Quaternary Period saved South Pontic relics and this proves the connection with the fauna of Asia Minor.

Holarctic Species. The Holarctic species are few in number: *Cervus elaphus* (red deer), *Canis lupus* (wolf), *Anas Platyrhynchos* (mallard), *Corvus corax* (common raven), *Sterna hirundo* (common tern).

Palaearctic Species. The Palaearctic species in Bulgarian fauna are comparatively plentiful: *Vulpes vulpes* (fox), *Martes nivalis* (marten), *Apodemus sylvaticus* (yellow-necked mouse), *Accipiter nisus* (sparrowhawk) *Dendrocopos major* (great spotted woodpecker), *Alauda arvensis* (skylark), *Bufo bufo* (common toad).

Euro-Siberian Fauna. Euro-Siberian fauna is found in the Euro-Siberian sub-region of the Palaearctic zoogeographic region. It (together with representatives of the Central European fauna) is the main constituent of Bulgarian fauna comprising over 50 per cent of the species. They dominate the entire territory of the country except for some of the southernmost regions and their relative number is highest in the mountains. Euro-Siberian fauna consists of cold-resistant species and for most of them the southern limit of distribution passes through Bulgaria. The general distribution is ubiquitous and the more cold-resistant ones live in isolation in coniferous woods in the higher mountains: *Tetrao urogallus* (capercaillie), *Nucifraga caryocatactes* (nutcracker), *Parus ater* (coal tit), *Clethrionomys glaereolus* (bank vole); and in the high mountain meadows and rocky terrains: *Lacerta vivipara* (common lizard), *Vipera berus* (adder, or viper), *Tichodroma muraria* (wall creeper). The following are widespread in open terrain: *Microtus arvalis* (orkney vole), *Arvicola terrestris* (water vole), *Perdix perdix* (partridge), *Natrix natrix* (grass snake); and in woods – *Sciurus vulgaris* (red squirrel).

Some Euro-Siberian fauna which originated in the western regions of the Euro-Siberian zoogeographic sub-region is designated as *Central European fauna*. They are distributed in Europe except for the most northern, western and southern territories of the Continent. The Central European and Euro-Siberian fauna are found together in Bulgaria. Typical Central European species are: *Martes martes* (pine marten), *Glis glis* (the edible, fat or squirrel-tailed dormouse) *Muscardinus avellanarius* (doormouse), *Parus cristatus* (crested tit), *Bombina bombina* (fire-bellied toad), *Triturus cristatus* (crested newt).

Arctic-Alpine and Boreal-Alpine Fauna. The Arctic-Alpine and Boreal-Alpine fauna are the most cold-resistant and there are fewer of them than any other category. There are about 60 Arctic-Alpine species in Bulgaria, of which over 90 per cent are insects (chiefly butterflies and beetles). Their habitats are the Alpine meadows and, for some, the woods in the highest mountains: *Erebia lappona* over 2,400 m, *Titanio schrankiana* over 2,250 m, *Ceratophyllus borealis*, *Agabus solieri*, *Salda littoralis* (of the Hemiptera – true bugs), *Aeropedulus variegatus* (grasshopper) over 2,200 m. The lower limit of the separate species is 1,300. The richest Arctic-Alpine fauna is that of Rila (over 70 per cent), followed by that of Pirin. The number of species in the Balkan Range, the Rhodopes, Slavyanka and Mount Vitosha is approximately the same. *Turdus torquatus* (ring ouzel), *Bombus lapponicus* and *Hadena mailardi* are widespread. The Arctic-Alpine animals in the Bulgarian mountains originate from species which in pre-Glacial and Glacial times inhabited the tundra, or the mountains in the moderate altitudes. During the glaciation periods, they moved to the Central European tundra belt where they mixed; with the warming of climate, some of each species moved to the North, and some went into the mountains.

Steppe Fauna. There are few steppe fauna in Bulgaria. A small part of the south-west boundary of the extensive Eurasian steppe zone passes through Bulgaria. This peripheral position of the country explains the lack of characteristic steppe species. In Bulgaria only Dobrudja is a primary steppe and it is the only Bulgarian habitat of: *Otis tarda* (great bustard), *Otis tetrax* (little bustard) and *Sicista subtilis*. Other species like *Cricetus cricetus*, *Mesocricetus auratus*, *Putorious eversmani* (polecat) and some birds typical of Dobrudja penetrated the Danubian Plain but did not reach South Bulgaria. Some species of southern origin form in South-east Bulgaria a second region with steppe fauna: *Myomimus personatus*, *Cricetulus migratorius*. The steppe fauna is an insignificant part of Bulgaria's fauna. The composition and distribution are stongly influenced by human activity. Some larger species (the bustards) are threatened with extinction. On the other hand, agricultural activity led to the deforestation of vast territories and re-formed them as secondary steppes. The prevalence of cereal crops there attracted the *Citellus citellus* (hamster), *Vormela peregusna*, *Spalax leucondon* and a number of steppe insects from all over the country. A number of birds encountered near the water place in the steppes find suitable conditions in the coastal and some Danubian lakes and swamps.

Pontic Fauna. Pontic fauna is found in the easternmost parts of the country. The aquatic animals originated in the closed late Tertiary and Early Quaternary basins (Pontic, Old Pontic) of which the Black Sea has remained. These species now mainly inhabit places with lower salinity, where the conditions are close to the ancient ones (predominantly fresh water and semi-saline) basins. Today Pontic fauna is distributed in the Caspian and Azov-Black Sea basins, and some of the species are exclusively in the Sea of Azov or in the Black Sea, or in the semi-saline Black Sea sections (lakes, swamps, limans, river mouths). Some of the fresh water fauna in inland basins and in the Danube also have Pontic origin. There are some Invertebrata Pontic elements existing among some groups of worms, snails, mussels and crustaceans. Of the Pontic Vertebrata we should mention some species of fish, chiefly of the Acipenseridae family (sturgeons): *Acipenser guldenstadti*, Clupeidae (herring family): *Spratella sprattus phaleria*, *Caspialosa kessleri pontica*, and mainly Gobiidae (gobies) – relict species, *Gobius cephalarges*, etc. The Pontic fauna includes the terrestrial East Mediterranean species whose distribution is confined to the Black Sea coast. Of the Vertebrata we encounter *Phasianus colchicus* (ring-necked pheasant), (in its original region of distribution), *Phalacrocorax pygmeus* (pygmy cormorant), *Accipiter badius brevipes* (Levant sparrow-hawk), *Lacerta practicola pontica*, *Lacerta taurica*, *Natrix natrix* (grass snake), etc.: some inhabit only the eastern regions of the country, others are widespread in the plains.

Mediterranean Fauna. The Mediterranean fauna is the most thermophilic. This complex of animals which originated in the Mediterranean sub-region of the Palaearctic zoogeographic region ranks second in Bulgaria in terms of number. There are many of this species, designated as Holo-Mediterranean distributed in this sub-region. In Bulgaria such are *Sylvia cantillans* (sub-Alpine warbler), *Oenanthe hispanica* (black-eared wheatear) and other birds; *Clemis caspica*, *Malpolon monspessulanus* and other Vertebrata and many insects and spiders. The East Mediterranean (Ponto-Mediterranean) species are numerous but distributed only in the eastern parts of the Mediterranean region. Typical of these in Bulgaria are the birds *Hippolais olivetorum* (olive-tree warbler), *Lanius nubicus* (masked shrike), *Sitta neumayer* (rock nuthatch), most of the reptiles such as *Ophiops elegans*, *Flaphe situla*, etc. The Invertebrata are also well represented in this fauna. A small number of East Mediterranean species are encountered only around the Black Sea and constitute the terrestrial Pontic fauna. The typical Mediterranean fauna consists of thermophilic animals, and for this reason it is best represented in some of the most southerly areas (the valleys of the rivers Struma and Maritsa, the southern Black Sea coast). As we go north the number and variety of species gradually decrease, but certain relict habitats (the valley of the river Russenski Lom, the Balkan Mountains around Sliven) give shelter to some typical Mediterranean species.

Sub-Mediterranean Fauna. The sub-Mediterranean fauna is part of the Mediteranean fauna which comprises species distributed both in the Mediterranean zoogeographic sub-region and in the areas nearby. The sub-Mediterranean fauna consists of expansive species with a Mediterranean origin. Many of them reached some parts of Central Europe after the Glacial periods and certain species went as far as North Europe. In Bulgaria the sub-Mediterranean species inhabit places at an altitude up to 500 or 600 m, and in South Bulgaria and along the Black Sea coast the number of species is greater. There exist also sub-Mediterranean species which penetrated into the east from the centres of species formation and reached Central Asia. The fauna in question comprises examples of all systematic groups; mammals are well represented by some bats, and birds by a diversity of species of the order of Passeriformes (Passerines).

Migration of Animals. The composition, variety and number of animal organisms are subjected to seasonal changes connected with cyclic phenomena in their biological functions, or to spontaneous changes caused by other factors.

Migration of Insects. Some large species of swift flying butterflies able to cover great distances migrate such as *Daphnis nerii* which flies via Bulgaria on its way to the South; *Acheronita atropos* arrives in Europe from Africa flying across the Mediterranean Sea; *Protoparce convolvuvi* populates large areas; *Pyrameis cardui* flies from the east to the west coast of North Africa but in Bulgaria flies across shorter distances. The *Locusta migratoria* migrated *en masse* in the past.

Migration of Fish. Fresh-water fish perform the shortest migrations. They inhabit the middle and lower parts of rivers and migrate to the upper and middle reaches to reproduce there: *Vimba vimba*, *Leuciscus cephalus*, *Lucioperca lucioperca*, *Perca fluviatilis*, *Barbua barbus barbus*, etc. The number of these species is large. The migrant fishes live and reproduce in waters with different degrees of salinity. The only species inhabiting fresh waters and reproducing there is the *Anguilla vulgaris* (eel) which migrates farthest and spawns in the Sargasso Sea. The other migrant fish inhabit the Black Sea but reproduce in fresh waters. These migrations are characteristic of the Acipenseridae (sturgeon), except *Acipenser ruthenus* whose habitat is the Black Sea but which migrate to the Danube for reproduction. By autumn the adults return to the sea and by the end of the year the small fry move to saline waters. Migrant Black Sea fish spawning in the Danube are: *Caspialosa kessleri pontica*, *Caspialosa caspica nordmanni* (which reproduce in other rivers too). The migrant fish spawn mainly along the Bulgarian bank of the Danube. Some Black Sea fish with a different distribution in the different seasons perform more intricate migrations. *Engraulis encrasicholus* spends the winter along the Anatolian Black Sea coast (sometimes, but rarely, near the Crimean Peninsula and in the Sea of Marmora). It passes along the Bulgarian shores in April–June and in the autumn when it goes to places to spend the winter. The migrations of *Pelamis sarda* (belted bonito), *Pomatomus saltatrix*, *Trachiurus trachiurus* (scad) are similar in nature and time. *Scomber scombrus* (mackerel) has almost the same route but spawns in the Sea of Marmora where it spends the winters.

Migration of Birds. The country's geographic position determines characteristic features in the migration of birds. Most of the Bulgarian species are migrant birds divided into different categories and degrees of passage. Anatinae (ducks), Anserinae (geese), Porzana (crakes) *Archibuteo lagopus*, *Corvus frugilegus* (rook) which in summer live in the tundra and taiga of the Scandinavian Peninsula and Siberia spend the winter in Bulgaria;

Columba palumbus (wood pigeon), *Scopolax rusticola* (woodcock) and *Otis tarda* (great bustard) are partial migrants. A large number of bird species fly across Bulgaria to reproduce in the northern regions, mainly in East and Central Europe. Most of the Gaviidae (divers), *Gavia arctica* (Arctic loon), *Philomachus pugnax* (ruff), *Pocoides tridactylus* (three-toed woodpecker), *Calidris alpina* (dunlin), *Alauda arvensis* (skylark), *Lullula arborea* (woodlark), *Charidrius morinellus* (plover), *Numenius arquata* (curlew), etc. fly when migrating across Bulgaria. A large number of migrants brood in Bulgaria: Hirundinidae (swallows), Apodinae, *Luscinia megarhynchos* (nightingale), Ciconiidae (storks), *Platalea leucordia (spoonbill), Oriolus oriolus* (golden oriole), *Cuculus canorus* (cuckoo), *Caprimulgus europeus* (nightjar), *Merops apiaster* (bee-eater), Acrocephalus (warblers), etc. The storks which spend the winter in South Africa have the longest migratory route and it takes them a month and a half to two months to fly there. The nightingale, golden oriole, cuckoo, *Coracias garrulus* (roller), bee-eater, etc. spend the winter in Central Africa; pelicans, cranes and other song-birds, in West Asia and the Arabian Peninsula. The chief migratory route goes via Bulgaria and two migratory routes were established across Bulgarian territory as early as antiquity: Via Pontica – along the Black Sea coast, and Via Aristotelis across the valley of the river Struma. A large number of places which are stop-overs (the forests) of the migrants are protected, particularly along the Black Sea.

Migration of Mammals. They usually roam (nomadically). The bats migrate over large distances. Specimens marked in the USSR and caught in Bulgaria helped establish the passage of three species of Chiroptera: *Vespertilio pipistrellus* – 1,150 km; *Vespertilio nathusii* – 1,950 km; *Nyctalus noctula* (noctule, or great bat) – 2,347 km, one of the longest passages in the world.

Nature Regions of Bulgaria. There are seven nature regions in Bulgaria. The fauna in each comprises different zoogeographic elements.

The *north plain region* is the habitat of widespread Euro-Siberian and Central European plain species. Typical steppe elements are encountered in the eastern parts and west of Dobrudja their number decreases. The fauna over the greater part of the arable area is not varied. Popular species are: *Lepus europeus* (brown hare), *Microtus arvalis* (orkney vole), *Perdix perdix* (partridge), *Coturnix coturnix* (quail), *Lacerta viridis* (green lizard), etc. This region is also the habitat of: *Canis lupus* (wolf), *Apodemus agrarius* (field mouse), *Citellus citellus, Merops apiaster* (bee-eater), *Coracias garrulus* (roller), *Corvus frugilegus* (rook), etc. The Danubian swamps and islands are the habitats of: *Pelicanus crispus* (Dalmation pelican), *Platalea leucordia* (spoonbill), *Plegadis falcinellus* (glossy ibis), and many species of herons and ducks, *Phalacrocorax pygmaeus* (pygmy cormorant), *Sterna hirundo* (common tern), and other wading birds. Some of them build their nests in Bulgaria only in the Sreburna reserve. *Bombina bombina, Pelobates fuscus* and other amphibians are characteristic of the humid places.

The woods in the plains (chiefly in Ludogorie) are the habitat of *Sus scrofa* (wild boar), *Capreolus capreolus* (roe deer), *Streptopelia turtur* (turtle dove), *Oriolus oriolus* (golden oriole) abd *Lacerta praticola* (forest lizard). Characteristic of Dobrudja are the Otidae (bustards), *Cricetus Cricetus, Mesocricetus auratus, Sicista subtilis* and *Putorius eversmanni* (steppe polecat).

Euro-Siberian and Central European forest fauna predominate in the *Balkan Range region*. The animals in the Balkan Range belong mostly to species associated with beech woods. Native inhabitants of the oak belt are distributed in the eastern parts of the Balkan Range and its foothills. The highest parts in the West and Central Range are meadows and pastures where high mountain and Arctic-Alpine elements are found. Common species in the beech zone are: *Salamandra salamandra, Glis glis* (the edible, fat, or squirrel-tailed dormouse); of the insects – *Rosalia alpina* (beetle), the butterfly and *Aglai tau* (Tau Emperor Moth). In light and sparse forests the oak zone is a suitable habitat for: *Cervus elaphus* (red deer), *Capreolus capreolus* (roe deer), *Meles meles* (badger), *Certhia brachydactyla*, and *Coronella austriaca* (smooth snake). *Alectoris graeca* (rock partridge – chukar) and *Pyrrhocorax graculus* (Alpine chough) build their nests in the rocky places and karst areas. *Eremophila balcanica* (Balkan lark), *Anthus spionoletta* (water pipit) and some insects have their homes in Alpine meadows and marshlands.

The ordinary plain and mountain Euro-Siberian and Central European species dominate in the *south-west mountainous and hilly region*. Sub-Mediterranean species are also common and single species of Arctic-Alpine fauna live in the highest parts of rhe mountains Ossogovo and Belassitsa. There are no species exclusively characteristic of this region. It is the habitat of *Apodemus microps* (forest mouse), *Muscardinus avellanarius* (dormouse), *Lacerta agilis* (sand lizard), *Rana graeca* (frog), *Martes martes* (pine marten), *Apodemus flavicollis* (yellow-necked mouse), some Picicae (woodpeckers), Turdidae (thrushes), and *Elaphe longissima;* of the insects, some Isophya, Scolytidae (bark beetles), Cereambycidae (Longhorn beetles), *Lucanus cervus* (stag beetle), *Lymantria dispar* (gipsy moth) are widespread in the oak and mixed forests of the region.

The *Struma-Mesta region* is inhabited by the greatest number of thermophilic species. Mediterranean and sub-Mediterranean elements prevail throughout the Struma area of the region. Thermophilic Euro-Siberian and Central European species dominate north of Simitli and in the valley of the river Mesta. Common among the Vertebrata are *Lacerta erhardii riveti, Telescopus fallax, Elaphe situla, Lanius nubicus* (masked shrike), *Sylvia cantillans* (sub-alpine warbler), *Sylvia hortensis* (Orphean warbler), *Monticola solitarius* (blue rock thrush), and other birds; *Apodemus mystacinus;* among the Invertebrata – *Iris oratoria, Paranocaracis bulgaricus, Nemoptera sinuata* and the butterfly *Rethera komorovi*.

The fauna of the Rila-Rhodope region is Euro-Siberian. The inhabitants of Rila, Pirin and the Western Rhodopes are those associated with coniferous forests. Most of the Arctic-Alpine species to be found in Bulgaria live in the highest parts of the above three mountains and Vitosha and Slavyanka. Sub-mediterranean species exist in the river valleys of the Thracian region. They are widespread in the Eastern Rhodopes. *Hirundo daurica* (red-rumped swallow), *Procerus scabrosus*. Typical Mediterranean species are found in the lower parts (the courses of the rivers Arda and Byala Reka): *Ophiops elegans*. The diversity of zoogeographic species – from most thermophilic typical Mediterrnanean types to most cold-resistant Arctic-Alpine with the added presence of endemics creates the richness of the Rila-Rhodope fauna. The Rila-Rhodope region has the greatest number of endemic species in Bulgaria. They are common in the higher parts of the mountains; they are characteristic of the region and have not been found anywhere else. The number of coniferous forest inhabitants only found in this region is large: *Parus ater* (coal tit), *Parus cristatus* (crested tit), *Prunella modularis* (hedge sparrow – dunnock), *Aegolius funereus* (Tengmalm's owl – Boreal owl), *Glaucidium passerinum* (pygmy owl), *Sorex minutus* (pygmy shrew), *Pitymys dacius* and *Plectotus auritus* (long-eared bat). Typical inhabitants of the coniferous belt and often going beyond its upper limit are: the bear, *Clethrionomys glareolus* (bank vole), *Tetrao urogallus* (capercaillie), *Dyrocopus martius* (black woodpecker), *Picoides tridactylus* (three-toed woodpecker), *Loxia curvirostra* (crossbill), *Nucifraga caryocatactes* (nutcracker), *Vipera berus* (viper), *Lacerta vivipara* (common lizard) and *Rana temporaria* (common frog). *Rupicarpa rupicarpa* (wild goat), *Microtus nivalis*, *Tichodroma muraria* (wall creeper) and most of the species of Aquilidae (eagles) and Gypidae (vultures) inhabit the rocky areas and screes devoid of vegetation in the highest parts of the mountains.

The majority of animals in the *Thracian region* are plain species of northern and sub-Mediterranean origin. Species of southern origin have penetrated the southern parts of Eastern Thrace: *Cricetulus migratorius*, *Myomimus personatus* and *Microtus guentheri*. Characteristic of the region are *Mycromis minutus*, *Emberiza melanocephala* (black-headed bunting), *Phasianus colchicus* (pheasant, ring-necked pheasant), *Passer hispaniolensis* (Spanish sparrow), *Alectoris chucar*, *Falco vespertinus* (red-footed falcon), the serpents *Eryx jaculus* and *Coluber najadum*, of the Invertebrata – *Lathrodectus tridecimguttatus*, *Saga natolie* and *Empusa fasciata*.

Thermophilic Euro-Siberian and Central European species predominate in the greater part of the *Black Sea region* and sub-Mediteranean and typical Mediterranean species prevail south of Sozoplo. The mild climate favours the movement into the region of some sub-Mediterranean species north of the Balkan Range. Typical of the region (though small in number) are the Pontic elements: *Lacerta praticola*. Many bird species (divers in the main) live only or mainly around the Black Sea: *Pelecanus onocratulus* (white pelican), *Tadorna tadorna* (shelduck) and other ducks; *Phalacrocorax carbo* (great cormorant), *Recurvirostra avosetta* (avocet), *Himantopus himantopus* (black-winged stilt, black-necked stilt), *Charadrius alexandrinus* (Kentish plover, snowy plover), most of the Laridae species (gulls), *Haliaeetus albicilla* (white-tailed eagle) and *Panurus biarmicus* (bearded titmouse). *Canis aureus* (jackal)

inhabits the southernmost parts of the region. Of the reptiles most typical are *Gymnodactylus kotschyi*, *Ophisaurus apodus*, *Lacerta trilineata*. Certain insects like Cicindelinaa and *Acrotylus longipos*, and some flies are commonly found on the beach.

The distribution of animals in the *fresh water basins* is usually not connected with the terrestrial regions. Bulgaria's fresh-water fauna belongs entirely to the Mediterranean sub-region of the Holarctic zoogeographic region. Of the fresh-water Invertebrata the unicellular, some worms, Entomostraca and insects are common. Crustaceans and Rotatoria prevail in the plankton; some worms and larvae of insects exist on the bottom biocoenoses (benthos). Amongst the most common fish are Cypriniformes; those of the families Acipenseridae (sturgeons), Cobitidae, Percidae and single species of other families like *Salmo trutta* (sea trout), *Esox lucilus* (pike), *Silurus glanus* (sheat-fish), *Cottus gobio*, are fewer. Of the other Vertebrata we should mention the amphibians frog and newt, and the reptiles *Clemmys caspica* and *Emys orbicularis*.

The Black Sea fauna consists of three basic components. About 10 per cent are autochthonous organisms of the Pontic fauna. There are also 20 per cent of species of fresh water origin, tolerating waters of higher salinity, which are localized in the areas around river mouths, semi-saline lakes and other regions of lower salinity. The most common species of fresh water origin are in the groups Olygochaeta, Rotatoria, Cirripedia, Copepoda and Ostracoda. Of the fish in the Black Sea the Acipenseridae, Cypriniformes and *Salmo trutta labrax* are of fresh water origin. Most numerous (about 80 per cent) are the sea species that entered the Black Sea during its late connection with the Mediterranean Sea. They are post-Glacial immigrants which came to dominate over the autochthonous fauna with the increase of salinity. They are representative of all animal groups inhabiting the Black Sea: *Aurerial aurita*, *Pilema pulmo*; crustaceans – *Balanus*, *Idothea tricuspidata*, *Carcinus maenas*, *Diogenes varians*; molluscs – *Cardium edule*, *Nassa reticulata*; the fish *Acanthias acanthias*, *Mugil cephalus*, *Scomber scrombus* (mackerel), *Mullus barbatus ponticus*, *Trigla lucerna* (tub gurnard), *Crenilabrus tinca*, Blenniidae (Blennies), *Belone acus*, *Scorpaena porcus* and *Trachinus draco* (greater weever), of the mammals – three species of dolphins and *Monachus monachus*. The diversity is increased by some rare fish occasionally penetrating from the Mediterranean Sea, sea turtles, etc. The species brought into the sea by man and forming part of its permanent composition are an insignificant percentage. Most of the Black Sea fauna is in fact a poorer Mediterranean fauna (about 3.5 times). The basic causes for the poor Black Sea fauna are the poisonous hydrogen sulphide under the 150–200 m surface layer, the low salinity (15–18 per cent) and the lower temperature.

Wildlife formation in natural conditions is a slow process. The rapid changes in the quantitative and qualititative composition of the fauna in recent centuries have been due exclusively to human activity. The changes in the vegetation cover have the strongest influence on the fauna. The gradual decrease of forests which in the past covered the greater part of Bulgaria's territory has restricted the forest fauna. In the mountains and some hilly regions wildlife is comparatively well preserved. The substitution of deciduous forests with coniferous and high-stemmed trees with dwarfish trees led to changes in the composition of species. The reclamation of vast areas in the valleys caused the extinction of certain species, and increased the density of pest species. Combating pests with chemicals made the fauna poorer. The drainage of most swamps along the Danube and inland destroyed many good nesting sites for wading birds: pelican, herons, *Platalea leucordia* (spoonbill), *Plegadis falcinellus* (glossy ibis) and *Cygnus olor* (mute swan). The urbanization of the Black Sea coast and the pollution of the environment reduced the possibilities for nesting in most swamps and lakes along the coast. The fauna in basins close to large industrial regions, Beloslav Lake, Mandra Lake and some rivers, has been seriously affected. Urban areas have poor fauna too.

Intensive hunting led to the extinction of some species of Bulgarian fauna: *Bos primigenius*, *Castor fiber* (beaver), *Lynx lynx* (lynx); and *Lyrurus tetrix* (heat cock), *Monachus monachus*, *Gypaetus barbatus* (bearded vulture) are threatened with extinction. The number and distribution of the bear, wolf, jackal, wild goat, raven, *Grus grus* (crane), bustards, most vultures and eagles, some wading birds, reptiles and amphibians, and most of the sturgeons have been strongly reduced.

Man's direct or indirect participation enriches the fauna with new species by acclimatization. *Dama dama* (fallow deer) was restored to the fauna of Bulgaria. Other acclimatized animals are *Ondatra zybetica*, *Myocastor coypus* (nutria), *Ovis musimon* (mouflon) and other animals; *Salmo irideus*, *Salvelinus fontinalis*, *Ctenopharyngodon idela*, *Hipopthalmychtis molitrix*, *Gambusia affinis* and other fish. Some species entered accidentally and acclimatized themselves: *Nyctereutes procyonoides*, *Lepomis gibbosu*, *Amiuris nebulosus*, and some insects harmful to plants – *Phyloxera vastatrix*, *Quadraspidiotus perniciosus*, *Leptinotarsa decemlineata*, and *Hyphantria cunea*.

Great care is taken to protect and enrich the fauna of Bulgaria. The network of reserves in national parks is being examined. A large number of species, valuable, rare or threatened with extinction are protected by law. The hunting and angling seasons are strictly regulated. Systematic measures have been taken to increase game. Game breeding farms have been created, the number of harmful predatory animals is controlled and in winter the game is fed. The stocking of dams, rivers and lakes with fish and the maintenance of clean water helps to preserve stocks of fish in Bulgaria.

PROTECTION OF THE ENVIRONMENT

A system of natural-scientific, technically productive, economic, administrative and legislative measures has been set up with the aim of conserving and transforming the environment, as well as maintaining and raising its productivity, and using natural resources for the benefit of mankind. The preservation of the environment is a problem closely related to the protection of nature. Its historical development is determined by socio-economic, political, cultural, demographic and other factors leading to spoiling of the environment. Prior to Bulgaria's liberation in 1878, human influence on the environment was characterized mainly by the clearing and burning down of forests for fuel and building material, as well as the transformation of wooded areas into arable land and pastures. This caused erosion of terrains and the depletion of water sources. With the onset of industrialization there began a process of pollution, in particular of rivers. The development of the wood-processing industry entailed an intensive devastation of forests in certain regions, without any accompanying measures for their restoration.

The problem of environmental protection came to the fore in the years of socialist construction, owing to Bulgaria's development as an industrial-agrarian state, the intensification of agriculture and demographic changes. The need to make good the consequences of the senseless exploitation of natural resources in the past added to the importance of the problem. Within the socialist structure there exist prerequisites for purposeful control of the interaction between society and nature. Environmental protection is a problem of political importance in the party and state policy of building a developed socialist society. Under the Constitution of the People's Republic of Bulgaria the conservation and protection of the environment and natural resources is an obligation of state bodies, economic and public organizations and of every individual citizen. The Theses, approved by the Twelfth Congress of the Bulgarian Communist Party (1981) pay special attention to the problem, setting out the following tasks: 'To introduce into production low-waste or altogether waste-free technologies, as well as systems for recycling water supplies and scientific and technical achievements that could be instrumental in *protecting the environment and in the comprehensive solution of ecological problems*. Special attention should be devoted to land conservation and improving the fertility of the land through systematic and comprehensive measures for combating water and wind erosion, swamping and salination of the land and by stepping up reclamation and irrigation of spoiled areas. The optimum use should be made of above-ground and underground waters and care should be taken to protect them from pollution. The sea, as a national resource, should be utilized more comprehensively. Steps should be taken to improve the natural surroundings in urban and industrial areas, to bring down the level of noise, to reduce air pollution from motor transport and some industrial activities, to speed up the building of filtering installations for waste waters, to put urban waste to better use. The nation's awareness of ecology should be improved'. Environmental protection constitutes a section in the Unified Plan for the country's socio-economic development. The utilization of natural resources is regulated by the following bills adopted by the National Assembly: a Law on Hunting (1948), a Law on Forests (1958), a Decree for the Conservation of the Country's Landscape (1960), a Law on Fishing (1961), a Law on the Prevention of Air, Water and Soil Pollution (1963), a Law on Environmental Protection (1967) and Rules for its Application (1969), a Law on Waters (1969), and a Law on the Protection of Arable Land and Pastures (1973).

Programmes and directives are issued by the State Council of Bulgaria. There are also a number of acts and regulations on environmental protection issued by the Council of Ministers and some other authorities.

The supreme organs of state power are involved in the protection of the environment. Thus, for instance, the National Assembly has a Permanent Committee for environmental protection, the State Council of Bulgaria has a Council for the conservation and reproduction of the environment, and there is an autonomous supra-departmental body with the Council of Ministers – the Committee for Environmental Protection (CEP). The tasks of the CEP are implemented through its local organs – the sixteen district boards for environmental protection (BEP). Efforts are made to encourage and control the regenerative capacity of the environment. Game and fish farms are being set up with a view to revitalizing and enriching natural resources, and the natural water basins are stocked with local and acclimatized fish species. Afforestation is a form of combating erosion and replanting the cleared forest-land. Timber is being imported from the USSR as a measure for conserving the Bulgarian forests

The purification station at Slunchev Bryag

handed over to the court. A nation-wide movement for nature conservation has been launched in the country following a decision of the Secretariat of the Central Committee of the Bulgarian Communist Party of 1970. It is sponsored by CEP with the National Council of the Fatherland Front, which co-ordinates, organizes and directs the public movement aimed at preventing the negative effect of the scientific and technical revolution on the environment, for its protection, regeneration and enrichment, for promoting an educated attitude to natural resources. The committee supervises the implementation of plans for the exploitation and planting of woodland, the use of land, the protection of waters, air and soil from pollution, the management of protected areas and sites. It also has an extensive programme for the popularization of its work. The district and local committees for environmental protection do extensive organisational and educational work aimed at widening the scope of the social movement and involving an increasing number of people in environmental protection measures and in widening their scope; they expand the network of educational programmes popularizing the study of ecological problems.

and containing the clearing operations. The scientific and technological revolution has created problems connected with the appearance of a number of new pollutants – chemical substances, smoke, dust and radiation. These bring about ecological imbalance, and harm agriculture, affecting the crops, plants and animals. Furthermore they prove harmful to the health of people. With a view to gearing economic activities to the needs of environmental protection CEP issues instructions binding for all state and economic organizations. The ministries and departments are accountable for the conservation and improvement of the environment, as well as for the rational use and preservation of natural resources in the branches and spheres of activity of which they are in charge. The People's Councils manage, co-ordinate and control environmental protection. There is a set of expedient measures and sanctions aimed against those who are responsible for acts of pollution or have caused harm to the national economy. No plant, industrial enterprise or other project can be commissioned unless filtrating installations are built and the requisite environmental protection measures are implemented. Low-waste and waste-free technologies are introduced into all branches of the economy, as well as systems for recycling the water supply and water consumption norms. Control is exercised over all construction plans for industrial enterprises, airports, quarries and other sites and if they do not contain an adequate solution to the environmental protection problem, they are not approved and no finance is provided for them. The control extends to all projects polluting the air, water and soil; quarries, mines and construction sites spoiling the natural lie of the land; to the building of purification installations and the reclamation of areas, affected by the construction of industrial projects. In the event of violation a writ is issued against the offenders or the respective materials are

An oil cleaner in Varna harbour

A large number of research and design institutes are involved in the development and application of modern methods and technologies for the protection of the environment. Some 82 per cent of the research activities, both basic research and practical developments, are carried out in departments of the Bulgarian Academy of Science. The scientific and co-ordinating centre for the conservation and regeneration of the environment with the Bulgarian Academy of Sciences deals with ecological problems and the economic aspects of environmental protection, and co-ordinates, on a national scale, basic research in its programme 'Scientific foundations of the protection and regeneration of the environment'. The State Committee for Science and Technological Progress co-ordinates and monitors research and design activities connected with environmental protection, the development and introduction of waste-free, and low-water technologies, as well as technologies involving a closed cycle of water consumption and the building of filtering installations. A Research Centre for the protection of the environment and water resources has been set up with the Committee for Environmental Protection for the solution of current problems of the conservation and regeneration of the environment. Bulgaria engages in active international co-operation in the sphere of ecology, especially with the socialist countries within the framework of the CMEA (Council for Mutual Economic Assistance). The CMEA member-countries pool their efforts in joint research aimed at finding prompt and effective solutions to current problems and the timely application of these solutions. Bulgaria takes part in organizations, committees and other UN bodies directly or indirectly concerned with environmental matters, as well as in other world and international organizations and unions. There are bilateral conventions with Bulgaria's neighbouring countries for the protection of waters from pollution.

Nature Conservation. There were isolated attempts at nature conservation in Bulgaria as early as the late nineteenth century, but the results of these nature protection activities were insignificant. The indiscriminate clearing of forests for pastures by cutting or burning them down, the expansion of arable areas, and the draining of swamps has brought about unforeseen ecological changes which spoiled the countryside. Owing to ill-advised land improvement measures the only habitats of rare and typical representatives of the flora and fauna were destroyed or depleted, as were the richest habitats of some species, which brought about a sharp drop in their overall number.

After 9 September 1944 special care was devoted to the problems of nature conservation. Public ownership of the natural resources makes possible a judicious control of their conservation and use. New natural sites and natural resources were discovered. They had to be tapped without damaging the natural surroundings. Environmental matters assumed particular importance in the context of the scientific and technical revolution. Stringent measures were adopted aimed at preventing an ecological imbalance. Holidaymaking and tourist activities expanded entailing the need to preserve the natural surroundings, their educational and aesthetic value. The consistent implementation of the environmental protection policy involved a large number of measures and the setting up of an administrative structure for their control. Norms and regulations have been issued concerning the use of natural resources and the protection of the environment. Respective administrative bodies were set up for the management and control of these activities. In 1960 a Decree for the Protection of Nature in Bulgaria was issued which was the first document with a new modern approach to the problems of protecting nature. It was later replaced by a Law on the Protection of Nature (1967) and Regulations for its implementation (1969). The document specifies the protected natural sites: national parks, reserves, beauty spots, historic sites, plant and animal species. The national parks include large areas with rich and varied wildlife and places of outstanding natural beauty. The nature reserves are protected areas designated for the preservation of animate and inanimate natural objects in their natural surroundings. The protected areas are places with a characteristic landscape. Construction work is permitted in them according to a special plan taking into account the natural scenery. They are used as places of recreation and tourism, but the use of the land, the waters and other natural resources is under control. The natural sites include places of geological, palaeontological, botanical, scientific, cultural, historical, and natural-scientific value, such as: caves, rock formations, springs, waterfalls, lakes, centuries-old forests, and groups of or solitary centuries-old trees. The historical sites are areas, localities and regions where the natural surroundings in which historic events have taken place are preserved or in which there are historic buildings or monuments. The protected plants and animals are valuable and rare species threatened by extinction or preserved in sufficient numbers for the needs of science or the economy. There are in Bulgaria 3,397 protected sites, including: 10 national parks, 93 reserves, 1,977 natural objects (including 1,547 centuries-old trees), 59 protected localities, 777 historical sites, 63 plant species and 418 animal species.

Nature conservation being an important branch of environmental protection is a question of political importance for the State. In approaching matters of environmental protection the Bulgarian Government abides by international conventions on the conservation of the flora and fauna. The current approach, and the existing organization of the activities in the sphere of environmental protection, create prerequisites for the compensation of the negative effect of the scientific and technological revolution on the environment.

NATIONAL PARKS

National Parks are large protected territories with ecosystems which are still in their virgin state or little affected by human actitivity, and contain rare or threatened plant and animal species in their natural habitat, as well as geomorphological formations, or nature scenes of great beauty. The park may also include protected sites of other categories, such as wild life or game reserves, unique scenery, protected nature zones, historical sites, etc. They have a special status under the 1967 Ecology Act and the 1969 Code regulating its enforcement. When a territory is proclaimed a national park, the authorities take appropriate action to discontinue or curtail, within the shortest time possible, all forms of human activity affecting the environment, and introduce rules and regulations to ensure the efficient preservation of the existing ecological, geomorphological, and aesthetic values. Given the strict observance of certain rules, visits to the park are possible for recreation, for aesthetic enjoyment, for the contemplation of nature, for scientific exploration and research, etc. Funds are allocated to maintain park rangers and other personnel to supervise and manage the parks. National parks are surrounded by buffer zones – strips of territory outside the bounds of the park where visits by the public are kept within limits. There are hotels, camp-sites, car park and other tourist facilities. The purpose of the buffer zone is to prevent the public from interfering with the natural life of the area, and to absorb the noises of normal human activity which might have bad effects. The standards for a national park stipulated by Bulgarian law, are similar to those set out in the definition for a national park adopted by the Congress of the International Union for Ecological Protection (New Delhi, 1970). Internationally-agreed requirements stipulate, as one of the most important functions of national parks, the preservation of their natural diversity, and the absence of man's influence on the ecosystems and the genetic stock, whilst recreation, rest, and tourism are allowed within certain limits.

The first Bulgarian national park was Vitosha, so designated in 1934. It is situated close to the capital, on the mountain of the same name, spread over its most picturesque parts, and has on its territory a mountain station for natural exploration and research, and two preserves – Bistrishko branishte (the Bistritsa reserve) and Torfeno branishte (the Peat reserve). The Golden Sands was designated a national park in 1943. It is situated

Mount Vitosha

north of Varna, in a coastal area, and includes forest plantations (part of the Hachuka forest complex), and erosion gullies in the localities Aladja Monastery (where a twelfth century cave hermitage of the same name is situated), Golden Sands (an international seaside resort), and Bezhanata. The architectural and ethnographic park, the Etar museum, was designated in 1967; it is situated near Gabrovo, spreads over to include the tributaries of the Yantra River – Sivek and Strashka, and incorporates picturesque mountainous country. Ropotamo was designated a national park in 1962; it is situated 15 km south of Sozopol and includes the estuary of the River Ropotamo which flows into the Black Sea, and has in its territory swamp forests and vegetation conglomerates of Mediterranean, near-Asiatic, euxinic, endemic, and relict species. It also contains four nature reserves – Zmiiski Ostrov (Snake Island), Morski Pelin (Sea Mudwort), Arkutino and Vodnite Lilii (the Water Lilies). Russenski Lom was declared a national park in 1970. It is situated 20 km south-east of Russe, spreads over a part of the valley of

Water lilies in the Ropotamo National Park

the Russenski Lom River with canyons and rock features that impart a distinct character to the landscape, and the rock-face churches near the village of Ivanovo. Steneto was designated in 1963. It is located in the Troyan-Kalofer mountain (the middle section of the Balkan Mountain Range), and includes ancient forests, a cave canyon, rich and diversified fauna. On its territory is the Steneto reserve. Sinite Kamuni (the Blue Stones) was designated in 1980. It is situated north of Sliven, in the Sliven part of the Balkan Mountain Range, and includes the Blue Stones rocky massif with its picturesque rock formations and ancient beech woods. Shumensko Plateau was designated in 1980. It is situated above Shumen, includes an interesting landscape with precious animal and plant species and ecological concentrations, and has the Bukaka (the Beechwood) reserve on its territory. Pirin was proclaimed in 1962. It comprises areas of the Pirin Mountain from Vihren Park down to about 1,000 m above sea-level, and includes unique ecosystems, rare animal and plant species, a large number of lakes, caves and waterfalls, and has on its territory the reserves Bayuvi Dupki (the Bayuv Holes) and Djindjiritsa. Kobaklaka was proclaimed in 1951. It is situated near Tolbuhin, and includes a plain forest with exotic plant species.

Reserves. Reserves are a category of protected virgin territories of natural locales (featuring distinct scenic beauty, or plant and animal species (or their concentrations), valuable for science or threatened with extinction. The reserves are complex natural entities complete in themselves and form a part of the country's national wealth. The Ecology Act guarantees the inviolability of these territories by prohibiting any human action impairing their pristine wildness. No economic activity whatsoever is allowed in them – exploitation of natural resources, building, recreation, and the like. Human intervention in the preserves is allowed only in the cases of massive proliferation of a calamitous insect pest, and then only after a careful and thorough consideration by experts and the issue of an official order. Systematic scientific observations are conducted in the reserves on animate and inanimate nature in a natural environment, with special permission co-ordinated with the Bulgarian Academy of Science on the basis of a pre-submitted and agreed scientific programme. The principal purpose of these nature reserves is to preserve the wealth of wildlife, to explore scientifically the regularities in the development of nature, to control and trace the structure, functioning and dynamics of relatively undisturbed ecosystems in the absence of human interference; preserves are also used for monitoring purposes, as well as for education. On the basis of various criteria Bulgarian reserves have attained international recognition. Out of the 93 existing reserves ten are on the United Nations list of reserves and equivalent national parks. On the initiative of Unesco a specialized category of biosphere reserves was set up in 1977 in which may be included national parks or other categories of protected wild territories, as well as locales with a characteristic culture of land cultivation, which preserve typical ecosystems from the separate biotic regions of the world. The biosphere reserves serve as banks of inviolate ecosystems and as etalones for ecological monitoring. Having seventeen biospheric reserves, Bulgaria holds the world's second place (after the United States). As from 1983, with the Srebarna reserve and the Pirin national park, Bulgaria has been included in the convention for the preservation of the world's cultural and wildlife heritage. The biospheric reserves are Srebarna,

The Pirin mountains: Mount Todorin Vruh from Mount Vihren

National Parks

The Pirin mountains: the view from the Bayuvi Dupki and Djindjiritsa reserves

'The Concert Hall' of the Sueva Dupka cave in the tectonic limestones of the western foothills of the Stara Planina

Kamchia, Chuprene, Boatin, Tsarichina, Steneto, Djendema, Bistrishko Branishte, Parangalitsa, Maritsa Lakes, Bayuvi Dupki – Djindjiritsa, Alibotush, Mantaritsa, Kupena, Dupkata, Chervenata Stena, and Uzunbodjak. The most important Bulgarian reserves are Sylkossia (1931), Parangalitsa (1933), Bayuvi Dupki (1934). The largest are Uzunbodjak – 2,385.8 ha, Djendema – 1,775 ha, Parangalitsa – 1,580 ha, Maritsa Lakes – 1,509 ha, Bayuvi Dupki – 1,449.7 ha.

Protected Areas. Protected areas are a category of natural sites protected for the preservation of nature and wildlife in isolated corners of the land, where the beauty and individual character of the scenery are safeguarded. These areas contain plant varieties, rare for the country's flora or threatened with extinction, or sanctuaries of rare typical animal species. The statute referring to this protected ecological unit is relatively freer. No special permits to visit them are required. Certain forms of human activity which do not impair nature and do not inflict irreparable damage on the separate objects of protection – the concentrations of biological species or elements of the typical landscape, etc., are permitted. Natural beds of rare beautiful flowers are protected by seasonal wardens or voluntary guards (the so-called green patrols). Some protected areas can act as buffer zones around reserves or national parks. They are suitable places for the preservation of isolated fragments of representative ecosystems and of surviving pockets of threatened wildlife.

Unique Natural Sites. Unique natural sites are natural phenomena, locales with a geological, palaeontological, botanical or other particular aspect, which present a scientific, cultural, historical, or aesthetic interest. Sites from inanimate nature, proclaimed to be unique nature spots, relate to topographical freaks (sand dunes, rock mushrooms, beautiful rock formations, stone cataracts, moraines, earth or stone pyramids, rock bridges, canyons,

Vertical strata known as the 'Cart Rails' in the Iskur river gorge in the Western Stara Planina

The Djendema reserve in the Central Stara Planina

overhanging rocks) as well as caves, peculiar areas, hydrographic landmarks (springs, waterfalls, lakes, swamps and marshes, meandering rivers, estuaries), or fossil clusters. Unique natural sites of animate nature are ancient woods, groups or individual specimens of ancient trees, localities of rare or interesting bushes, grasses, plants, animals, successfully-introduced exotic types of trees or bushes, etc. Unique natural sites also include, besides the unique object under protection, the environment in which it exists. On such sites the law prohibits all kinds of human activity which will disturb the natural set-up or change the optimum ecological parameters of the environment of the object protected.

The most remarkable geomorphological and rock formations in Bulgaria are the abrasion mushrooms near the village of Ravda (Burgas district), carved out by the surf; the Belogradchik rocks (monumental massifs of rocks, eroded into freak forms, spread over a 30 km-long and 3 km-wide strip in the Western Balkan Mountain Range); God's Bridges (a rock bridge 25 m high and 14 m thick) near the village of Lilyache, Vratsa district, spanning the Lilyachka river; Vratsata (a high, narrow, rocky gorge near the town of Vratsa); Djuglata, or the Camel, (a rocky formation near the village of Tserovo, Sofia district); the Dobrovan stone mushrooms (near the village of Kozichino, Burgas district); the Dorman rocks (or Dorman holes, situated in Dubrash, the Western Rhodopes, near the village of Mesta, Blagoevgrad district); the Erma Zhdrelo (a picturesque gorge near the town of Trun); the Zemen pyramids (beautiful rock shapes several hundred metres tall); Kaleto (a gorge in the ridge near Reselets, Pleven district, some 300 m long, and nearby there rise beautiful rocks, 100 m tall); Kayaliite (rock formations in Dubrash, the Western Rhodopes, near Kayaliiski Peak – resembling fortresses, mushrooms, animals); Kostenurkata (the Tortoise) – a rock formation near the village of Dobromir, Burgas district; Kupenat (a 30 m tall rock formation near the River Mesta, 18 km from Gotse Delchev); the Kurdjaliiski stone mushrooms; the Kutina pyramids (in Sofia district); the Lakatnik rocks; the Lion's Head (a rock sphinx in Ropotamo National Park); the Melnik pyramids (eroded sedimentary rocks, 100 m tall, near the town of Melnik); Nastanska Mogila, a rock formation near the village of Nastan, Smolyan district; Orlitsata (a rock formation near the village of Medven, Sliven district); Pobiti Kamani (erect cylindrical columns

The Rock Bridges in the Western Rhodopes

The Melnik sandstone pyramids

of eroded rocks, 3 m in diameter and 5 m in height); Ritlite, vertical rock walls 60 to 80 m high, in the Iskur gorge; the Rock Bridges (Smolyan district); the Stobski pyramids (in the foothills of Rila, Kyustendil district); the Trigrad gorge (with its marble rockface 350 m sheer off the ground, in Smolyn district); Halkata, a unique rock formation near Sliven, etc.

Also under protection are 46 waterfalls, proclaimed to be unique natural sights. The best known of them are Boyana Fall, Big Leap Fall, Goritsa, Kademliya, Kosten Kamak, Pruskaloto, Raisko Pruskalo, Samokovishteto, Skakavets, Skakavitsa, Skakaloto, Skaklya, Suvacharsko, Suchurum and Haiduk Falls, etc. Among the interesting hydrographic landmarks, proclaimed to be remarkable nature sights, are the Springs (a karst source, out of which the River Medvenska flows – the left tributary of the Luda Kamchiya in the Eagle's Rock locality) and Black Springs (a cave with a big karst source in the lands of the village of Medven, Sliven district).

Eighty-four caves are under protection as remarkable nature sights, among them the Bacho Kiro, the Gradeshnishka Cave, Duklata, Ledenika, Ledenitsata, Lednitsata, Lepenitsa, Magura, Mishin Kamak, Morovitsa, Ponora, Snezhanka, Saeva Dupka, Temnata Dupka, Haidushka Peshtera and Tsarevets.

Other remarkable nature sights are the locality of the ancient monastery Sedemte Prestola (the Seven Thrones); the dolmen 'Kamennata Kashta' (Stone-House) in the lands of the village of Ostar Kamach, district of Haskovo; Opanski Bair (in the land of the village of Opanets, Pleven district), a major find of Tertiary fossils; the Fossil Find (in the district of Plovdiv; finds the fossils of giant proboscidian mammals of past geological times; the site has an area of 9,100 ha), etc.

Remarkable nature sights from animate nature are the ancient wood Kuriyata (in the lands of the village of Galabets, Burgas district); the ancient oak wood Snezhanskata Kuriya (in the land of Goren Chiflik, Varna District); Bozhur Polyana (a large natural bed of wild peonies in the land around the village of Parvomaitsi, Veliko Turnovo district); the Kleptuza forest (in the land of Velingrad; the region features extremely beautiful Rhodope landscapes); Manzul (an ancient beech wood in the Manzul locality near Panagyurishte); the ancient oak wood Genchov Orman (between the villages of Iskur and

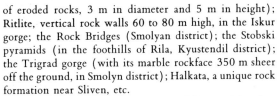

The Stone Mushrooms – part of the Kurdjali pyramids in the Eastern Rhodopes

The Trigrad gorge in the Southern Rhodopes

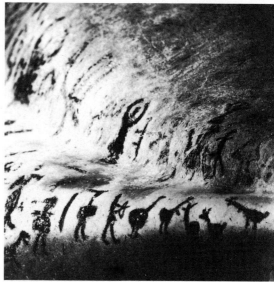

Paintings in the Magura Cave in the western foothills of the Stara Planina

Krushovene, Pleven district, where some oak trees are over 500 years old, and measure 6 m in girth); the Lugut-Drumkata (a large natural bed of ordinary wild peonies in the area of the village of Bozhuritsa, Pleven district); the ancient beech wood Kushbunar (in the locality Sinite Kamani near Sliven); Livadite (a large natural bed of Rhodope tulips in the area of the village of Sivino, Smolyan district); the ancient beech wood Sakaradja (in the area of the village of Chavdar, Sofia district); the ancient beech wood of Bratiya (in the land of the village

A waterfall near Vishograd in the Central Stara Planina

of Chavdar, Sofia district); the ancient oak wood Tulovskata Kuriya (in the area of the village of Tulovo, Stara Zagora district); the ancient oak wood Vetrenskata Kuriya (in the area of the village of Vetren, Stara Zagora district); the ancient oak wood Chirpanskata Kuriya (in the area of the town of Chirpan); the oldest tree in Bulgaria, the Baikusheva Mura (Baikushev's white fir) in the Bunderitsa locality, the Pirin mountain; the tree is 16 metres tall, measures 5.72 m in girth, and is 1,200 years old; the plane tree in Gotse Delchev (25 m tall, measures 7.6 m in girth, and is 500 years old); the oak tree in the Tulpan locality (near the village of Zabornovo, Burgas district) which is 24 m tall and 5.3 m in girth; the Peshtera poplar (26 m tall and 11 m in girth – the thickest poplar in Bulgaria), etc.

A total of 1,977 sites and objects have been proclaimed unique natural sites; 1,574 of them are ancient trees.

Protected Plants and Animals. This is a category of protected species, and includes rare species, plants and wild animals, threatened with extinction, economically useful, or being of special interest to science and education. Their care is the basic activity within the campaign for the preservation and reproduction of the natural flora and fauna within the general protection of the ecology. The Ecology Act absolutely prohibits the killing of protected species (the case is slightly different with those partially protected). It is only for scientific reasons and with special permission that the gathering or hunting of protected species is allowed.

There are 63 species of protected plants, 21 species of trees and bushes and 37 grass plants. Besides these, four species are partially protected, in certain areas, at certain times of the year. Prohibitions apply to the cutting,

National Parks

The Baikushev white fir, the oldest tree in Bulgaria

picking, uprooting, destroying, gathering, transferring, trading with, or exporting overseas in fresh or dry condition protected plant species. Under control are the wild species of economic importance, particularly medicinal plants, which would easily become extinct or would diminish if the established rules concerning their gathering were not observed.

Protected animals number 418 species, 105 cave-dwelling species, 12 invertebrates and 301 vertebrates, of which 8 are amphibian, 16 are reptiles, 234 are birds and 43 are mammals. In addition to these, two species are partially protected. The cave-dwelling fauna, represented by worms, crustacae, centipedes, arachnida and insects includes exceptionally rare species: a large number of species and even genera are indigenous only to Bulgaria. The invertebrates under protection are represented by insects, among which there predominate endemites, relicts, rare and useful species, severely imperilled by the widespread use of insecticides. Vertebrates are most severely threatened with extinction; those protected include a large number of useful or vanishing species. The two types of Bulgarian tortoises – the pike-thigh and the pike-tail – are partially protected in certain regions, seasons, and age-groups. Species with drastically reduced populations or threatened by extinction are protected by the Hunting and Fishing Act: e.g. grouse, bears, badgers, chamois; other game (fowls and mammals, as well as fish) are permitted for hunting and angling within defined limits.

Red Book of the People's Republic of Bulgaria. The publication of regional red books marks a rational initiative in the preservation of the genetic stock. The precise specification and updating of the national problems of ecological protection will yield better results if inventories are made of extinct or rare plants and animals. Thus, increasing importance is attached to the national red books. A decree of the Council of Ministers commissioned the Bulgarian Academy of Science, more particularly its Scientific Co-ordination Centre on the Preservation and Reproduction of the Environment, to compile and publish a *Red Book* of the People's Republic of Bulgaria. It includes a total of 763 objects of protection: 574 plant species and 189 animal species. Bulgaria is among the few countries in the world with a national *Red Book*.

The Bachkovo waterfall in the Rhodope mountains

POPULATION

Data regarding the population in the Bulgarian lands prior to the Liberation in 1878 are rather scanty and are mainly from the second half of the 19th century. They refer to the Bulgarian population within the then boundaries of the Ottoman Empire. In 1836, 1844 and 1866 the Turkish administration made attempts at estimating and carrying out a census of the population in its European domains. In 1874–75 a somewhat more successful census of the male population over sixteen years old (the *nufuzes*) was carried out in the Danubian *vilayet* (region), which includes the *sanjaks* (adminstrative districts) of Tulcha, Varna, Russchuk, Turnovo, Vidin, Nish, and Sofia. In connection with the great interest in the Turkish empire manifested by many European states at the time of their capitalist development (the search for new markets and capital investment areas and for political influence), many travellers, scientists, journalists, and military officers studied the natural resources, the population, the economy and the socio-economic life of Turkey. In using these and other data, taxation registers and materials, they strove to correct the imperfect attempts of the Turkish administration to estimate and take census of the population. In a very general way, those were the first statistical materials on the population of the lands inhabited by Bulgarians. Their estimate of the number of Bulgarians in 1850 and on the eve of the Liberation from Ottoman rule, in Moesia, Thrace and Macedonia was between four and six million. According to the Ukrainian historian Yuri Venelin, in the 1830s the number of all Bulgarians was 2,545,000. At about the same time the French geologist, Ami Boue, gives the figure of 4,500,000. On the basis of literary sources, Konstantin Jirecek computed the number of Bulgarians, irrespective of domicile and religion, to have been about 5,500,000.

The results of retrospective studies show that the probable population of what are approximately today's territorial boundaries of Bulgaria amounted to about 1,179,000 in the year 1500, that in less than two centuries the population doubled and rose to 2,463,000 and to 4,315,000 by the end of 1900.

After the Liberation in 1878, thirteen regular censuses of the population of Bulgaria were taken (the first census in the Principality of Bulgaria was carried out in 1880), in Eastern Rumelia, in 1884, for the purpose of establishing the exact number of the total population and the number of Bulgarians within the country's boundaries. For the

Population in the Fifteenth–Nineteenth Centuries within the Approximate Territorial Boundaries of Present-Day Bulgaria

Years	Number of population in approximately today's boundaries of Bulgaria	Probable degree of deviation – people plus and minus
1500	1,179,000	89,000
1550	1,386,000	108,000
1600	1,631,000	131,000
1650	1,990,000	143,000
1700	2,257,000	176,000
1750	2,463,000	183,000
1800	2,745,000	124,000
1850	2,984,000	134,000
1900	4,315,000	

period of 1880–1980, the population of Bulgaria shows an annual gradual smooth increase, which was interrupted substantially only during the wars (the Balkan War 1912–1913, the Inter-allied War 1913, the First World War 1914–1918 and to a considerably lesser degree during the Second World War, 1939–1945).

On 1 January 1982, the population of Bulgaria was 8,905,600. It constitutes 6.8 per cent of the population of the Balkan Peninsula, 1.3 per cent of that of Europe and 0.22 per cent of the population of the earth. The increase in population during the contemporary period (from 1944 to 1980 it increased by 1,963,300), can be characterized as more rapid than that of Austria, Belgium, Great Britain, GDR, Hungary, Czechoslovakia, but slower than that of the population of the USSR, the USA, Italy, Holland, Norway, Poland, Romania and some other countries. The total increase in the population is due mainly to its natural growth. Foreign immigration was reflected only in certain periods, mainly up to 1952.

A more exact idea of the dynamism in the increase of the population is given by the censuses started at the end of the nineteenth century.

The high coefficient of natural growth in the past was accounted for by the high birth-rate and high death-rate. The period of highest birth-rate, about 40 births per thousand of the population (1905 – 43.5 per thousand), continued up to the Balkan War in 1912. The high birth-

Population of Bulgaria 1880–2000

Year	Population (Number)
1880[1]	3,155,000
1890	3,762,000
1900	4,315,000
1910	4,980,000
1920	5,037,000
1930	5,997,000
1940	6,624,000
1950	7,273,100
1960	7,905,500
1970	8,514,900
1980	8,876,600
1990[2]	9,445,000
2000	9,821,100

[1] The data for the years up to 1940 are adduced from comparison with today's territory of Bulgaria. (From 1880 to 1940, the territorial boundaries of Bulgaria were changed 6 times.)
[2] The data for the years 1990 and 2000 are prognostic.

The Population of Bulgaria According to the Censuses 1880–1975

Years	Number of population (at the end of the year)	Average annual relative growth for the periods of 1881–1975 (in %)	
1	2	3	
1880[1]	2,007,919	1881–1887[1]	1.3
1884[2]	975,030	1885–1887[2]	0.6
1887	3,154–375	1888–1892	1.0
1892	3,310,713	1893–1900	1.6
1900	3,744,283	1901–1905	1.5
1905	4,035,575	1906–1910	1.5
1910	4,337,513	1911–1920	1.2
1920	4,846,971	1921–1926	2.1
1926	5,478,741	1927–1934	1.3
1934	6,077,939	1935–1946	1.3
1946	7,029,349	1947–1956	0.8
1956	7,613,709	1957–1965	0.9
1965	8,227,866	1966–1975	0.6
1975	8,272,771	1881–1975	1.2

[1] Principality of Bulgaria.
[2] Eastern Rumelia.

rate was accounted for by the very great proportion of the rural population (on the average, about 80 per cent). In the urban population, a regular birth-rate was observed.

During the wars (1912–18) the birth-rate was lower than the death-rate. The two post-war demographic compensations in 1914–15 and 1920–24 did not cover the tremendous losses in men (about 570,000). A stable birth-rate was established from about 1935 to about 1950 – on average between 22 and 24 per thousand, i.e. about half that of the preceding period.

A new upswing in the birth-rate is observed from 1935 to 1950, when it rose to 24 to 25 per thousand. After 1950, entirely new qualitative sides of the socio-economic factors began to act. The birth-rate gradually declined and reached its lowest level (13.9 per thousand) in 1982 (14.4 per thousand in the towns, 13.1 per thousand in the villages); from 1966 to 1977 the proportion of births increased.

Bulgaria fell among the countries with the lowest birth-rate. From among the European socialist countries only Hungary (1980) had a lower birth-rate. After 1950, the stream of intensive migration from the villages to the towns strongly disrupted the age pyramid of the rural population. This led to a drop in the birth-rate in the villages, as a result of which, from 1966, it became lower than that in the town. Irrespective of this, the higher fertility among the women in the villages was preserved. The territorial differences in the birth-rate had a stable character. The general tendencies for it to drop embraced all districts. For 1980–82 the district of Kurdjali had the highest birth-rate (about 22.2 per thousand), the district of Vidin, the lowest birth-rate (10.4 to 10.8 per thousand). In Bulgaria, a policy of encouraging larger families was implemented and after 1956 it assumed a systematic character.

On the average per year for the period 1976–80 there were 43,823 more live births than deaths. This is also the average annual natural increase for the population for this period (approximately three live births and two deaths every twelve minutes). The death rate coefficients for all age groups decreased. The greatest decrease was achieved in child mortality.

From about 150 to 160 per thousand at the end of the nineteenth century and the first decades of the twentieth century, child mortality fell to about 120 per thousand in the first years of the people's rule and to 22 to 24 per thousand for 1975–78. For 1982 child mortality was 18.2 per thousand.

The natural increase of the population for 1983 was 2.7 per thousand (birth-rate 13.9 per thousand, death-rate 11.2 per thousand). It was different for the different groups of the population. After 1960, the natural growth of the population in the towns has continuously increased; in the villages it has continuously decreased.

In different parts of the country the birth-rate, the death-rate and the natural increase show considerable deviations from the general development of the country; this is especially marked among the rural population where there are strong tendencies towards depopulation and falling behind in the natural increase.

The migration movements up to 1878 were characteristic of the whole period of Ottoman rule. They were

Population Increase

Natural increase of the population in the towns Indicators	1926	1940	1945	1950	1955	Years 1960	1965	1970	1975	1980	1982
Absolute increase (in inhabitants)	12,671	5,962	10,487	30,425	27,229	27,986	31,778	48,852	58,150	41,511	35,358
Relative increase (per thousand)	11.4	4.1	6.2	15.5	11.5	9.6	8.5	11.0	11.4	7.5	6.2
Natural increase of the population in the villages											
Indicators											
Absolute increase (in inhabitants)	96,650	49,556	52,882	78,012	55,789	48,431	27,043	12,798	−3,456	−11,271	−11,485
Relative increase (per thousand)	22.4	10.1	10.0	14.8	10.9	9.8	6.1	3.2	−1.0	−3.3	−3.6

connected with the political and socio-economic conditions in the Ottoman Empire. They were particularly marked during the time of the Austro-Turkish wars in the sixteenth–eighteenth centuries, and especially during the time of the feudal disorders and anarchy in the Ottoman Empire, and the Russo-Turkish wars in the eighteenth–nineteenth centuries. The Bulgarian population made unsuccessful attempts at liberation from Ottoman rule (Turnovo Uprising 1598, Chiprovo Uprising 1688, uprising in Western Bulgaria 1737, uprising in Vidin district 1774), which led to the mass emigration of the Bulgarian population to the Danubian Principalities of Wallachia, Banat, Moldavia, Bessarabia, to South Russia, Hungary, Serbia, etc. After the withdrawal of the Russian forces and crushing of the Kresna-Razlog Uprising (1878) some 60,000 refugees from Macedonia and the Adrianople region emigrated to Bulgaria. New waves of Bulgarian immigrants arrived after the Ilinden (St. Elijah's Day) and Preobrazhenie (Transfiguration) Uprising (1903) – by the end of 1903 alone some 30,000 refugees from Macedonia and the Adrianople region found shelter in Bulgaria. From 1901–10 a great number of Turks emigrated from the country.

The Inter-allied War of 1913, the First World War 1914–18 and consequent changes in the state boundaries led to considerable migrations. The number of refugees from Thrace, Macedonia, Dobrudja and the Western frontier regions increased considerably.

In accordance with the 'Voluntary Migration Convention', 1919, and more particularly under the 'Mollov-Kafanderis' Agreement, of 1927, a great part of the Bulgarian population from Aegean Macedonia and Western Thrace emigrated to Bulgaria. According to official data for the granting of refugee loans to migrants whose occupation was agriculture, there were 253,067 migrants.

From the beginning of the twentieth century and especially during the world economic crisis of 1929–33, many Bulgarians went to earn a living in Western Europe, North America (Canada, the USA), South America (Argentina and other countries) and Australia. A great number of them remained there permanently.

Almost everywhere outside the boundaries of Bulgaria, the Bulgarians preserve their native language and national traditions and in most of the places have their own schools and their own periodicals and Press. They maintain contacts by correspondence with relatives and close friends in Bulgaria and many of them visit the country.

The correlation between the sexes in Bulgaria during the period of 1880–1980 was always within the limits of the favourable demographic proportions, with the exception of the war and the immediate post-war periods. The demographic compensation (late marriage and childbirth) after the First World War was incomplete and failed by far to compensate for losses from the wars. After the Second World War it was considerable and led to an increase in the birth-rate and of the natural growth.

On average, for the last 25–30 years the correlation between live-born girls and boys was 100:105(106), between dead women and men it was 100:110 and between living women and men at the end of 1980 it was nearly 100:100. A certain exception was observed between the urban and rural populations – for every thousand men in the towns, there were 1,010 women; in the villages for every 1,000 men, 995 women (1976). Women in the villages are of a higher average age and have a higher average life expectancy.

In 1900–1095, the average life span was 42 years, for 1974–76, it was 71.31 years (men – 68.68 years, women – 73.91 years).

For the average life span Bulgaria holds one of the leading places in the world. At the end of 1975, the population above the age of 60 constituted 16.3 per cent of the total population as against 7.8 per cent for 1934.

The urban population in the period 1880–1980

continuously increased its absolute and relative share compared to the rural population. Because of the agrarian character of the economy during the capitalist period, the rural population was considerably more numerous than the urban. During this period, the urban population increased very slowly – from 18.8 per cent in 1887 it rose to 21.4 per cent in 1934, or in 47 years it marked an increase of nearly 2.6 points.

Change in the Correlation between the Urban and the Rural Population

Years	Urban population Absolute number	%	Rural population Absolute number	%
1887	593,547	18.8	2,560,828	81.2
1892	652,328	19.7	2,658,385	80.3
1900	742,435	19.8	3,001,848	80.2
1905	789,689	19.6	3,245,886	80.4
1910	829,522	19.1	3,507,991	80.9
1920	966,375	19.9	3,880,596	80.1
1926	1,130,131	20.6	4,348,610	79.4
1934	1,302,551	21.4	4,775,388	78.6
1946	1,735,188	24.7	5,294,161	75.3
1956	2,556,071	33.6	5,057,638	66.4
1964	3,703,600	45.3	4,473,900	54.7
1965	3,822,824	46.5	4,405,042	53.5
1970	4,509,800	53.0	4,005,100	47.0
1972	4,785,200	55.7	3,809,300	44.3
1973	4,917,200	56.9	3,730,200	43.1
1975	5,061,087	58.0	3,666,684	42.0
1977	5,283,100	59.9	3,539,500	40.1
1980	5,546,000	62.5	3,330,600	37.5
1982	5,735,800	64.2	3,193,500	35.8

A turning point in the correlation between the urban and the rural populations came under Socialism (socialist industrialization, urbanization, changes in the character of agriculture, collectivization, industrialization of agriculture, tendency towards equalization of the village with the town), which made it possible for many former villages to be given the status of towns. Estimates for the future indicate a still greater increase in the urban population at the end of the century.

With respect to the territory, the population lives in 5,379 inhabited places (on the average there are 48 inhabited places per 1,000 sq. km. The status of inhabited places in Bulgaria is determined and changed by a law decreed by the State Council. The inhabited places are urban and rural. On 31 December 1982 in Bulgaria there were 227 towns; 4,425 villages; 407 neighbourhoods; 306 hamlets; 7 industrial and mining settlements; 5 railway stations; and 2 monasteries. The villages were most numerous and the greatest number of people lived in the towns.

Bulgaria is a comparatively densely-populated country. From 1880 until today it has continuously increased its territorial density. Today it is more than twice as great as in 1900 – at the end of 1981 its territorial density was 80.3 people per sq. km. Compared to the average density of the Earth (33 people) and Europe (70 people) it is more densely populated.

Bulgaria is also more densely populated than are the neighbouring countries of Turkey (59 people) and Greece (74 people), but has less than the average density of Yugoslavia (88 inhabitants) and Romania (94 inhabitants). Its average population density in 1982 was 80.5 inhabitants per sq. km.

From a geographical point of view, the present population density shows great variations. The mountainous regions possess the lowest density – in the Balkan Range, Rila, Pirin, Western Rhodopes, Kraishte and Strandja Mountains there are almost uninhabited places and localities, in which the density of the population rarely surpasses 30 inhabitants per sq. km.

Density of the Population of Bulgaria

Year	Average number of inhabitants per sq. km	Average annual increase (in inhabitants per sq. km)
1881	29.7	—
1887	32.7	0.50
1892	34.4	0.34
1900	38.9	0.56
1905	41.9	0.60
1910	45.0	0.62
1920	47.0	0.20
1926	53.1	1.01
1934	58.9	0.72
1946	63.5	0.38
1956	68.8	0.53
1960	71.4	0.65
1970	76.8	0.54
1972	77.5	0.35
1975	78.7	0.40
1977	79.5	0.41
1980	80.0	
1982	80.5	

The density of the population typically decreases with the increase of the altitudes above sea level. The altitude belt of 500–700 m, with a comparatively small area where the city of Sofia is situated, is an exception with one million inhabitants.

Among Bulgarian citizens there exists solid spiritual and political unity, unharmed by second language and ethnographic differences created in the course of the socio-economic development.

In the People's Republic of Bulgaria the Constitution guarantees to all citizens freedom of conscience, and the majority of the people are free of religious prejudices. The believing Bulgarians are mostly Eastern Orthodox, a small number are Muslims, Catholics and Protestants. The Church is separate from the State.

Substantial changes in the social composition of the population have resulted from the nationalization of industry, the expropriation of the lands of the large landholders and of the country's industrialization. A leading place is held by the workers, who moved from 17 per cent at the end of the capitalist period to 60.6 per cent in 1975, followed by the co-operative farmers – 14.4 per cent (prior to 1944 the rural population constituted 70 per cent of the population of Bulgaria).

The relative share of the employees has considerably increased – from about 5 per cent in 1944 to 23.2 per cent in 1975.

The interior migration movements are closely connected with the socio-economic development of Bulgaria. After the Liberation in 1878 the relative hunger for land, a result of the great agrarian overpopulation under the conditions of an extensive agriculture, was particularly strong. It was this that determined the constant migrations from the mountainous regions to the fertile land – the Danubian hilly plain, the upper Thracian lowland, other plain regions and fields. Urbanization, although at slower rates, was connected with the migration stream from the villages to the towns. It was considerably more active after the First World War. Because of lack of land the peasants, not only from the mountainous regions, but also from the plains, migrated to to the towns, where they became workers, employees, and craftsmen. The character of the interior migration movements was changed as a result of the socialist transformations – they orientated themselves mainly from the villages to the towns, from the small to the large towns, from the smaller villages to the larger villages. The drastic changes in the socio-economic conditions in Bulgaria under Socialism have led not only to changes in the character of the internal migration processes, but also in their geographical orientation – from the villages to the towns, but mainly to the chief industrial centres and the new and rapidly developing industrial settlements (Dimitrovgrad, Madan, Rudozem), to the ore-mining regions which are being industrialized, to Sofia, the larger towns, to the places with a rapid development in means of productivity. The more even distribution of the means of productivity and of production capacities, the consolidation of farming, the legal restrictions on settlement in certain large towns and the development of transport towards 1970, all resulted in the migration streams subsiding. The decrease in the lasting migration streams made it possible to increase day-to-day travelling, and cultural and shopping trips; in 1975 the number of people travelling to work was 622,716 (14.0 per cent of travellers), two thirds of them being from the villages to the towns.

In Bulgaria the lower limit of the working age of the population is between 15 and 16 years, the upper, 60 years of age for men and 55 for women. The densely-populated regions with the greatest number and largest towns have the greatest labour resources and the greatest production possibilities of the population. The sparsely-populated rural regions with a high average age have the most limited production possibilities for their labour resources. The data for 1975 show a marked reduction in the lowering rate of economic activity of the population, typical of the years after 1946. A strong impact on the economic activity of the population has been produced by the involvement of women in production. The tendency towards equalizing the two sexes as well as the sharp drop in the lower and the higher groups are typical of the economic activity during the years of socialist development. In the urban population it is considerably greater than among the rural population.

The distribution of the economically-active population into spheres and branches of employment in the national economy show (1948–1975) very great changes. The relative share of people employed in the unproductive sphere is increasing. From a territorial aspect, there are comparatively large regions (especially among the rural population), with a relatively small share of economic activity of the population. It is greatest in the rapidly developing towns, the outer districts of the large towns, the industrial regions, the territories with a pronounced composition of people of advanced age but with favourable production conditions for temporary inclusion in work, and territories which have attracted temporary migration streams. The population in the agricultural regions has the least economic activity with stongly-marked migrations during the years of socialist construction and has limited possibilities for the reproduction of labour resources and for temporary enlistment for work of the numerous population above the normal working age. The activitiy coefficient in Kraishte, and in certain regions of the Central Balkan Range, is below 40 per cent and in certain places, even below 30 per cent. The western and central parts of the district of Kurdjali, the Ludogorie and the Eastern Balkan Range have comparatively low economic activity, but they have a comparatively high relative share of population of below working-age with real possibilities therefore of considerably increasing their labour resources.

Economic Activity of Population according to Censuses (1934–1975)

Year	Total		In towns		In villages	
	Number	%	Number	%	Number	%
1934	3,346,105	55.1	515,751	39.6	2,830,354	59.3
1946	4,034,730	57.4	702,418	40.5	3,332,312	62.9
1956	4,150,207	54.5	1,134,229	44.4	3,015,978	59.6
1965	4,267,798	51.9	1,859,131	48.6	2,408,659	54.7
1975	4,447,784	51.0	2,659,935	52.6	1,787,849	48.8

THE BULGARIAN LANGUAGE

The language of the Bulgarian people belongs to the southern branch of the Slavonic languages of the Indo-European language group, which also includes Serbo-Croatian and Slovene. The history of the Bulgarian language – the earliest written Slavonic language – helps establish a number of phenomena and regularities within the evolution of all Slavic languages. Its development comprises three main periods: Old Bulgarian (ninth–tenth centuries), Middle Bulgarian (twelfth– fourteenth centuries), and Modern Bulgarian (fifteenth century and later), which were preceded by a pre-literal period.

The beginnings of the Old Bulgarian period date back to the creation of an alphabet (the Glagolitsa) and literature in the Bulgarian vernacular (862–863) by the Thessaloniki monks, Cyril and Methodius. The invention of the Glagolitic alphabet (41 letters) constituted an original creative act, and though the author of the Cyrillic alphabet, Constantine Cyril the Philosopher, drew on his knowledge of Greek lettering and of some Oriental scripts, Glagolitic was a new graphic system and ingenious creation, exactly adapted to the phonetic peculiarities of the Old Bulgarian language. In the 9th–11th century the Glagolitsa was mainly used in the places closely related to the immediate activity of Cyril and Methodius and their disciples and followers – Moravia, Pannonia, and Bulgaria. Twelve surviving Glagolitic manuscripts are known: the Assemanius Gospel (late tenth or early eleventh century), one of the most beautiful Old Bulgarian Glagolitic monuments; the Zograph Gospel *Codex Zographensis* (second half of tenth century), one of the oldest monuments of old Bulgarian culture; the Marianus Tetraevangelia, *Codex Marianus* (late tenth and early eleventh centuries), whose rubrics and miniatures are instructive about the illumination of Bulgarian books of the time; the Clozianus manuscript (eleventh century); the Sinai Psalter and Sinai Euchologium (second half of eleventh century); the Sinai Service Book and the Rila Glagolitic Sheets (end of eleventh century); the Grigorovichev Sheet; the Ohrid Sheet; the Boyana Palimpsest; and the Zograph Palimpsest. In the ninth–eleventh centuries yet another script was created in Bulgaria – the Cyrillic alphabet. It includes the 24 letters of the Greek titular code lettering to which several new signs have been added (Б, Ж, Щ, Ц, Ч, Ш, Ъ, Ь, Ы, Ѣ, Ю, Ѧ, Ѫ, Ѩ, Ѭ) for the sounds peculiar to the Old Bulgarian tongue. The Cyrillic script is used by Bulgarians, Russians, and Serbians. The Romanians have also used it. Nine Cyrillic Old Bulgarian manuscripts survived: the Sava Gospels (beginning of the eleventh century) – an important moment for the study of the earliest development of the Old Bulgarian language and the illumination of manuscripts; the Codex Suprasiliensis (first half of eleventh century) – the most voluminous of Old Bulgarian manuscripts; the Hilendar Sheets (middle of eleventh century); the Enina Apostle (second half of eleventh century), which reflects fundamental phonetic and grammatical peculiarities of Old Bulgarian and is the oldest known copy of *The Acts of the Apostles;* the Undolski Sheets, a Macedonian Cyrillic sheet (end of eleventh century); the Triod Fragment (end of eleventh century); the Zograph Sheets; and the Novgorod (Kupriyan) Sheets. Testimony to the evolution of the language and the graphic appearance of the alphabet is borne by the surviving inscriptions. The Old Bulgarian Inscriptions alone number more than 40. The best known are the

Facsimile of the Gospel of Assemanius, end of tenth-beginning of eleventh century

Inscription of Chirgubil Mostich, tenth century

Inscription of Chirgubil (Proto-Bulgarian chieftain's title, cf. the Slav 'boyar') Mostich (middle of tenth century), the Dobrudja Inscription (943), Samuil's inscription (993) and the Bitola Inscription of Tsar Ivan Vladislav (1015–16). Information on Old Bulgarian has also been obtained from later copies of original and translated works of the Old Bulgarian period, from archaic onomastic data (geographical names, etc,) and the like.

Modelled on the Bulgarian of the Thessaloniki region, the East-Bulgarian dialect, the Old Bulgarian literary language, even at its very earliest, featured an enviable literary polish, strict observance of the linguistic canon, lexical wealth, and grammatic precision. Phonetically it displays some distinct differences from its contemporary Slavic tongues, but in vocabulary and grammatical structure it is very close to them. The phonetic system of Old Bulgarian consists of 11 vowels, а, е, о, н (*i* sound), оу (*u* sound), ъ, ь, ы, ѣ (broad *e*), ѧ ([ę], nasal *e*), ѫ ([ǫ], nasal *o*); two syllabic vowels – *r* and *l*; 25 consonants – б, в, г, д, ж, з (д͡з sound), з, з′ (soft *z*), к, л, л′, м, н (*n* sound), н′, п, р, р′, с, с′, т, х, ц, ч, ш and iota (*j* sound); and two consonantal combinations – шт and жд. All syllables were open, ending in a vowel, e.g. рыба, край, день.

Later, but still in the Old Bulgarian period, in consequence of the dropping of ъ and ь, there are also closed syllables (ending in a consonant). In certain positions consonants are palatal (soft). Morphologically, Old Bulgarian possesses a rich case and declension system: seven cases for nouns (nominative, genitive, dative, accusative, instrumental, locative, vocative) and three numbers (singular, plural and dual); adjectives have two forms – simple (добър) and complex (добърын); there are three genders (masculine, feminine and neuter).

Verbs have two aspects (perfective and imperfective), three persons (first, second, and third), three numbers, five conjugations, three moods (indicative, imperative and conditional), seven verbal tenses (present, aorist past imperfect, pluperfect, past indefinite, future in the past, future perfect), invariable verbal forms (infinitive and supine) and variable normal forms (participles). The syntactic structure is very well developed, and the lexical system offers great synonymic wealth. The Old Bulgarian literary language emerged and developed at a time when a Bulgarian ethos and a Bulgarian ethnic and linguistic awareness were being shaped within the bounds of the Bulgarian State, in the geographical regions of Moesia, Thrace, and Macedonia. The language was used for state and political needs, in church and in school, in scholarly and cultural life. During the 'Golden Age', in the Preslav and Ohrid academic schools, a rich original and translated literature was created by Old Bulgarian writers Kliment of Ohrid, Konstantin of Preslav, Naum of Ohrid, John the Exarch, Monk Hrabr, Presbyter Kozma, etc. As the pioneer written language of the Slavonic tongues, Old Bulgarian attained an international function; for centuries it played the role of literary language for a number of Slavonic, and in some cases non-Slavonic languages in Eastern Europe. From the tenth century onwards the language transcended the bounds of Bulgaria and spread to Russia, Serbia and the Wallachian Moldavian princedoms. In these countries Old Bulgarian mutated with time under the influence of the local vernaculars, and on the basis of it a number of variant forms (redactions) developed: a Russian, a Serbian variant, and later a Romanian redaction. On the basis of its international functions, some foreign scholars have wrongly termed Old Bulgarian 'Old Slavic' or 'Old Church Slavonic'. But it is more correct to term the language Old-Bulgarian, since traditionally a language is defined according to its ethnic affiliation, not its functions.

The Middle Bulgarian period is the second phase of the historical development of the Bulgarian spoken and literary language. Owing to the rigid traditionalism of the scholars of the day, the numerous literary monuments of the period offer only limited possibilities to trace with any precision the gradual evolution in spoken Bulgarian from a synthetic to an analytic system in the noun form. Some monuments, such as 'The Trojan Parable' (fourteenth century) reflect the changes in the living language: the two jer-vowels ъ, ь in a weak position (before syllables which do not have a jer in them, and also in final position) have already ceased to be voiced: the nasal vowels ѧ [ę], ѫ [ǫ] came to have similar pronunciation or gradually merged into a secondary jer and became interchangeable; the tendency to depalatalise the consonants appeared; the case system continued its slow disintegration. The use of prepositions became more widespread, and the use of the articles became established. This is a typical peculiarity of the Bulgarian language, which distinguishes it from the other Slavonic languages. The back vowel ы was equalized with the front vowel и, some participles and participle constructions died out, the 'да'-construction came in to replace the disintegrating

Facsimile of the Dobreisha Gospel, first half of the thirteenth century

infinitive and supine, the use became stabler of the complex form of the future formed with the modal verb хощѫ 'want', which undergoes an evolution resulting in the future tense particle ще, typical of Modern Bulgarian.

The period witnessed the creation of national hymnography and hagiography. Among the especially valuable written monuments are the Bologna Psalter (late twelfth and early thirteenth centuries); the Dobrian's Minaios (first half of thirteenth century) which contains one of the oldest sticherons exalting the Bulgarian saint, John of Rila; the Orbel Triod (second half of the thirteenth century); Dragan's Minaios (second half of the thirteenth century), unique among the extant minaea manuscripts by the number of Old Bulgarian hymnographic works, etc. Among the monuments of thirteenth century Bulgarian medieval art and scholarship are the illuminated Dobreisha Gospel, whose language echoes some idiosyncrasies of contemporary popular speech, the Vratsa Gospel, and the Pogodin Psalter.

In the thirteenth-fourteen centuries and the first half of the fifteenth century there were scholars and writers working in the capital Turnovo and the monasteries of Mount Athos, united by the ideological and aesthetic concepts of the Bulgarian representatives of Hesychasm. Under the guidance of Patriarch Euthymius of Turnovo they created a rich hagiographic and hymnographic literature. In accordance with the spirit and taste of the late Middle Ages in Europe, the literary Bulgarian language in the Turnovo Academic School artificially returned to the norms of Old Bulgarian, and continued to influence the literary languages of the other Slavonic nations, becoming the third classical literary language of medieval Europe, after Latin and Greek.

The works of Euthymius of Turnovo, Yoasaph of Bdin and Grigori Tsamblak are models of exquisite literary language and style. The literary Bulgarian of the period was far removed from the contemporary spoken language.

The Modern Bulgarian period is the third phase of the historical evolution of the spoken Bulgarian language. It has all the characteristic features of the Bulgarian of today. In the beginning of the seventeenth century an attempt was made in the Bulgarian lands to create a literary language on the basis of the spoken tongue. This is the language of the Bulgarian damascenes, collections of religious didactic tracts which first appeared at the end of the sixteenth century and continued to appear until the middle of the nineteenth century.

In the Modern Bulgarian period, which coincided with the Ottoman rule, the traditional language of the Old Bulgarian and Middle Bulgarian periods also used as a literary, written language, but without the nasal vowels.

Church Slavonic, i.e., the Russian variant of the Bulgarian literary language from the tenth to the fifteenth centuries, also became widespread. Thus in a variety of ways the language of Cyril and Methodius lived through the centuries and conveyed the literary and language traditions from the time of Cyril and Methodius to the modern Bulgarian literary language.

Modern literary Bulgarian appeared on the scene in the second half of the eighteenth century, the period of the Bulgarian National Revival. Three schools appeared in the first period of its development: (1) The Church Slavonic School. Its adherents (Hristaki Pavlovich, Konstantin Fotinov) adopted Church Slavonic as the basis of Modern Bulgarian, and since they did not know the system of this classical, dead language, they filled the gaps with elements of popular speech. This school had no special impact on the structure of modern literary Bulgarian. (2) The Slav-Bulgarian School. Its champions (Neophyte Rilski, Neophyte of Hilendar-Bozveli) thought that modern literary Bulgarian should be based on colloquial usage and Church Slavonic, and that the language should be purged of intrusive Greek and Turkish words by borrowing, where need be, from the richness of Church Slavonic and Russian. The school had a tangible influence on the development of modern literary Bulgarian. (3) Modern Bulgarian School. Its advocates maintained that the literary language must be based on the system of folk dialects. This idea was upheld, in both theory and practice, by eminent scholars such as Peter Beron, Vassil E. Aprilov, Ivan Bogorov and Naiden Gerov. Modern literary Bulgarian was formed mostly on the basis of the dialects of the Central Balkan Region and the Sredna Gora mountain, but also includes traits peculiar to the western and south-western parts of the country. Its major distinctions are: the eleven vowels of Old Bulgarian have decreased to six, the jer-vowels ъ and ь have vanished from final position, the broad vowel ѣ has assumed dual pronunciation (-я, -е), the old case forms have disappeared entirely, the role of the prepositions has increased, the comparative degrees are formed with the aid of the particles по- and най-, the definite article has come to be widely used and a new mood (inferential) has come into use to indicate events not witnessed. After the Crimean War (1856–1958) the basic styles of the literary language, those of popular science, art, journalism and adminstration, set into clearly defined moulds. The standards of the literary language and of orthography became clear, and the sources and ways and means of enriching the literary language increased. Owing to the close, centuries-old Russian-Bulgarian cultural and historical contacts, literary Bulgarian in its early period, as well as later, tended to echo the rich and mature literary Russian. Up to the beginning of the 19th century, literary Russian in both lexical content and phrasal structure was under the strong influence of Church Slavonic. This lexical penetration of Bulgarian by Russian had three effects: it enriched the cultural vocabulary of the Bulgarian language, ousted the intrusive Turkish words, and set a model of modern reformation of the literary language. The Bulgarian word stock is being substantially enriched and supplemented by the international cultural vocabulary, the foreign borrowings coming via Russian or being taken directly from the respective West European languages. This process has continued to the present day. The opposition of Ivan Bogorov, Naiden Gerov, Alexander Todorov-Balan and Stefan Mladenov to an undue rate of borrowing has safeguarded against a heavy contamination of modern literary Bulgarian with foreign lexical intrusions, but in general the purist stance has been moderate. In the second period of the development of modern literary Bulgarian (1878–1944) the processes of standardization began, seeking to establish unity and stability in both literary and colloquial usage: preserved as the orthographic and spoken standard was the pure value of unaccented vowels (добре, not дубре) and of the accented á in final position in the case of feminine nouns (главá, not главъ). In line with the strong central and eastern character of the language, it assumed softness of some verbal endings (нося - носят) and mobility of the vowel ъ when adjacent to р, л: (гръб - гърбав, мълча - мълквам) as well as some other typical characteristics. At the end of the nineteenth century the West Bulgarian pronunciation became the rule (ме, те instead of the East Bulgarian ма, та). The development of the literary language during this period was accelerated by the writers Ivan Vazov, P. R. Slaveikov, A. Konstantinov and K. Velichkov, as well as by the Bulgarian linguists Lyubomir Miletich, Benyu Tsonev, Stefan Mladenov and others. The language of *belles-lettres* and poetry attained high peaks of artistic elegance, expressiveness and colour.

The third period of the development of modern literary Bulgarian began after 9 September 1944 and is connected with the profound changes in private and public ways of living under the conditions of Socialism. Owing to the migration from village to towns and industrial centres, processes began in which the traditional speech habits of the different dialect speakers underwent changes, and their viability and stability was weakened. Some of the styles of modern literary Bulgarian (scientific, journalistic, administrative) have become lexically and structurally richer, and the vocabulary has expanded on the basis of public, political, scientific, technical and terminological neologisms or borrowings. Modern literary Bulgarian is supradialectical not only functionally, but phonetically and formally as well. Its distinctive feature is that it is national in usage and uniform in grammatical and stylistic features. The orthographic reform of 1945 widely introduced the phonetic principle. It brought written language closer to contemporary educated pronunciation. This facilitated assimilation of spelling rules and correct pronunciation by the learner. The letters with no phonetic value were dropped from the alphabet. Linked with the Modern Standard Bulgarian literary language are also two non-standard forms: the Banat written regiolect (in the Banat province) and the Skopje written regiolect. The problems of modern literary Bulgarian are the concern of the Institute for Bulgarian Language at the Bulgarian Academy of Science, of the Chairs of Cyrilomethodian Studies and Modern Bulgarian Language at the University of Sofia, of the Chairs of Modern Bulgarian Language at the Universities of Plovdiv and Veliko Turnovo and at the Higher Pedagogical Institutes in Shumen and Blagoevgrad.

DISTRICT CENTRES

Sofia. Sofia is the capital of the People's Republic of Bulgaria, the country's main political, economic, scientific, cultural and transport centre, and largest city; it is a resort, a spa and a tourist centre of national and international significance. Its population in 1982 was 1,082,315. Bulgaria's central political and administrative bodies are concentrated here, being the seat of the main state institutions and bodies of state power: the National Assembly, the State Council, the Council of Ministers, the Court and Prosecutor's office, the Sofia People's Council, the Central Committee of the Bulgarian Communist Party, and the central bodies of all political and public organizations. It is the centre of the Greater Sofia Settlement System (incorporating twelve urban municipalities in the actual city, as well as three towns and 34 villages) and the centre of the Sofia district.

It is situated in the western part of Central Bulgaria – in the southern part of the Sofia valley and at the foot of the Vitosha and Lyulin mountains at an altitude of 550 m above sea-level. Second to Madrid among European capitals in altitude, it has a moderate continental climate and mineral-water springs.

Sofia is one of the oldest cities in Europe. Traces of neolithic and chalcolithic settlements have been excavated in the city. A Thracian settlement existed since about the eighth century BC near the thermal springs in the central part of the present-day city. It was part of the kingdom of the Odrisae in the fifth century BC. A Thracian tribe called the Serdi inhabited the area in the first century BC. The settlement was taken over by Rome in 29 BC and formed a part of the Roman province of Thrace as the centre of a strategia (district) in 46 BC. The Romans called it Serdica, after the Serdi. During the reign of Emperor Ulpius Trayanus (98–117) it was granted an internal autonomous administrative status with a municipal council and a People's Assembly. The city, re-named Ulpia Serdica, grew quickly and was fortified by walls. Its strategic location on the crossroads of the main roads from Central Europe to Byzantium and from the Danube to the Aegean was of great importance for its development. In the third century AD the city became the provincial centre of Inner Dacia and minted its own coins. It reached its peak during the reign of Constantine the Great (306–337) and became a rich and prosperous city. It was the venue of the Serdica Ecumenical Council in 343, which was attended by representatives of the whole Christian world of the time. At the fall of the fourth century AD it was included within the boundaries of the Eastern Roman Empire. The city was intensively urbanized during the reign of Emperor Justinian (527–565) and Tiberius Constantinus (578–582). In 809 it was conquered by Khan Krum (803–814) and incorporated into the Bulgarian State under the Slav name of Sredets (or Triaditsa, in Byzantine chronicles). It was a strong fortress and an important strategic and administrative centre of the First Bulgarian State. The city was conquered by Byzantium in 1018. It was liberated by Tsar Assen I in 1194. From the twelfth–fourteenth centuries, it was a centre of highly developed craftmanship (gold-working, iron-working, weapons, pottery, etc.) and a busy trading centre. In the late fourteenth century, the city was first referred to as Sofia (after the Church of St. Sophia), and this name was eventually officially established. After resisting fiercely, Sofia fell under Ottoman rule in 1382. It became the capital of the beylic of Rumelia. After the Crimean War of

The coat of arms of the city of Sofia

1853–1856 Sofia entered a period of gradual decline, lost its key administrative status and was downgraded to a centre of a *sanjak* in the Danubian Vilayet. The Bulgarian population took an active part in the struggle for an independent Bulgarian church. A revolutionary committee was set up by Vassil Levski in 1870. The city was turned into a Turkish military camp during the Russo-Turkish war of Liberation of 1877–1878. It was liberated by the Russian troops on 4 January 1878. Sofia's key geographical location favoured its proclamation as capital city of the Principality of Bulgaria on 3 April 1879 (3 April was declared the Day of the Capital in 1979). Its population grew rapidly (primarily as a result of immigration), particularly when in 1888 the city was linked to the Vienna-Istanbul railway line. Sofia was the centre of the struggle of the Bulgarian working class against Capitalism. A society of the Bulgarian Social Democratic Party was founded in 1892, and in 1901 it became the seat of the Central Committee of the Bulgarian Workers' Social Democratic Party. It was also the centre of the Bulgarian people's anti-fascist struggle. During the Second World War it was the seat of the central leaderships of the Bulgarian Communist Party, the Workers' Youth Union, the Bulgarian All-People's Students' Union, the National Committee of the Fatherland Front and the General Staff of the National Liberation Insurgent Army. It was the centre of the First Insurgent Zone. The first Popular Democratic Government of the Fatherland Front was formed in Sofia on 9 September 1944.

By the turn of the century several industrial enterprises had appeared – a distillery, a brewery, ceramics, metal, furniture and leather factories and the first hydro-electric power station. The first power supply began in 1900, and the first tram-line started operating in 1901. Two hundred and three industrial enterprises were operating in 1926, and 432 in 1939. Textile, food and mechanical enterprises, metal-working and rubber industries, predominantly using imported raw materials, were in existence before 1939. By 1944, 37 per cent of the country's industrial capacity was concentrated in Sofia, and the first industrial estates were formed. Industry was disrupted during World War II, when the city suffered heavy damage in the bombing raids of 1944.

Swift industrialization, a substantial increase in population and size and high rates of housing construction characterized the city's development after 9 September 1944. Industrial enterprises of key importance to the national economy were constructed. Today Sofia is the country's main industrial centre. The capital leads in the ferrous metallurgy, mechanical engineering, chemical, rubber, textiles, food, printing, and publishing industries. Its major enterprises include the L. I. Brezhnev Metallurgical Combine; the D. Ganev Non-ferrous Metals Works; the D. Blagoev Insulated Cables Works; the Kliment Voroshilov Electric Appliances Plant; the Communications Appliances Plant; the Transformer Manufacturing Plant; the Vassil Kolarov Electric Machinery Plant; the Electrical Engineering Plant; the 6 September Electric Truck Plant; the Sredets Electric and Motor Plant; the G.

Housing construction in Sofia

Dimitrov Rolling Stock Yard; the Machine Tools Plant; the Automobile Repair Plant; the G. Kirkov Boiler Factory; the A. Ivanov Refrigerator Factory; the Elevator Manufacturing Plant; the Soft Drinks Plant; the Rodopa Meat Packing House; the Malchika Research and Production Combine for Sugar Products; the Sofiisko Pivo Brewery; the 8 March, the E. Thaelmann and Bulgaria Cotton Textile Mills; the N. Ivanov Woollen Textile Mills; the V. Ivanov Hosiery Factory; the Pharmaceutical Chemicals Combine; the Aroma Plant; the D. Toshkov Chemical Combine; the 9 September Footwear Factory; the Vitosha Tailoring Factory; the Medical Equipment Plant; the Cellulose and Cardboard Plant; the Sofia Tram Construction Yard; the Plant Construction Combine; the H. Smyrnenski Construction Materials Plant; the Combine for Reinforced Concrete Panels and Articles; the Narodna Republika Plastics Processing Combine; the Moskva Unit and Section Furniture Group; the G. Dimitrov Tyre Plant; the Bulgarplod, Mlechna Promishlenost, Bulgarsko Pivo, Vinprom and Izot State Economic Corporations; the Bulgartabak, Pirin and Resprom Commercial and Industrial Corporations; and the D. Blagoev and G. Dimitrov Printing and Publishing Houses.

Sofia is the centre of the Sredets Agro-Industrial Complex and of a Pedigree Cattle-Breeding Complex which meet the needs of the large consumption centre. They specialize in the production of meat, dairy products and vegetables.

The capital is the country's biggest railway junction. It is crossed by international railway lines and by the Western Europe-Istanbul international highway via Belgrade. There are four railway stations. The Central Station (one of the most up-to-date and beautiful buildings in Sofia) has trains going to Western Europe, the Soviet

Union, Turkey and Greece, as well as for the domestic railway service. There are daily bus services to Nish, Skopje, Kavadartsi, Ohrid, Strumitsa and Istanbul. There are four domestic bus terminals. The Balkan Airlines' domestic services have daily fights to Varna, Burgas, Vidin, Silistra, Turgovishte, Stara Zagora, Haskovo and Plovdiv. More than 30 regular and charter lines connect Sofia with major cities in Europe, Asia and Africa. The city transport network consists of 20 tram, 9 trolley-bus and over 200 bus routes. The demands of modern communications have necessitated the reconstruction of the road network and main outlet highways, the construction of communication facilities – flyovers and underpasses, and construction work is underway on Sofia's underground.

Sofia is the country's major cultural and educational centre boasting long-standing traditions in literature and education since the early years of Ottoman domination. The first monastery schools existed by the end of the 15th century. The Sofia School of Literature was opened in the 16th century. A mutual instruction school was opened in 1839, a boys' grade school in 1857, a girls' school in 1859 and a mixed secondary school in 1879. A Bulgarian Literary Society was founded in 1869 (which became the Bulgarian Academy of Sciences in 1911). The first Bulgarian higher educational establishment – the present-day Kliment Ohrid Sofia University – was inaugurated in 1888. In the 1981–82 academic year, Sofia had twelve higher educational establishments, including Sofia University; the Medical Academy; the K. Marx Higher Education Institute; the Higher Institute of Fine Arts; the Bulgarian State Conservatory; the K. Sarafov Higher Institute of Theatrical Art; the St. Kliment of Ohrid Theological Academy; the Higher Institute of Agricultural Sciences and several Engineering Institutes. There are also five semi-higher institutes, scores of general education schools, secondary vocational colleges, technical high and art schools and foreign language schools. Sofia has the biggest health-care establishments, national medical and scientific institutes, institutes of hygiene, dietology and occupational diseases, the National First Aid Station, the Pirogov Emergency Medical Aid Institute, the National Obstetrics Clinic, hospitals, polyclinics, specialized dispensaries, and child-care establishments. It is also the seat of the Bulgarin Union of Physical Education and Sports, the Bulgarian Hikers' Union and the Bulgarian Field Sports Society.

The square outside the Lyudmila Zhivkova People's Palace of Culture in Sofia

Old architectural Sofia

Sofia is a scientific and cultural centre: the Bulgarian Academy of Sciences with integrated centres for mathematics and mechanics, physics, chemistry, biology, pedology, philosophy and sociology, history, language and literature with their own scientific research institutes; the Cyril and Methodius National Library; the Ivan Vazov National Theatre; Sofia National Opera House; Operetta Theatre; Puppet Theatre; Drama Theatres; eighteen national and four city professional ensembles; dozens of amateur groups; 77 community centres; the G. Dimitov Central Pioneer Children's Palace; a Youth Centre; seven art galleries; a botanical garden; a Zoo and 34 museums; a natural history museum; a revolutionary movement museum; an ethnographical museum; an archeological museum; a museum of Bulgarian-Soviet friendship; a Bulgarian literature museum; a science, agricultural and ecclesiastical archeological museum; as well as the museum houses of political and cultural figures. It also houses Bulgarian Television, Bulgarian Radio, and the Bulgarian News Agency. Sofia hosts the Sofia Music Weeks; the

Festival of European Humour; the May Literary Festival; the International Competition for Young Opera Singers; and the International Competition of violin-making. Sofia is the seat of: the International Federation of Philosophical Societies; the International Federation of Sports Acrobats; the Banner of Peace International Children's Assembly; the World Centre for Technical Information and Urbanism; the World International Trade Union of Food, Tobacco, Hotel and Affiliated Workers; the European Weightlifting Federation; the European Federation of Sports Psychology and Physical Activities; and the International Committee for Immunology of Reproduction.

The present city centre was fully reconstructed after 9 September 1944. Major post-war construction includes: the Mausoleum of Georgi Dimitrov; the buildings of the Central Committee and the State Council; the Central Department Store; the Council of Ministers; the Monument to the Soviet Army; the Common Grave; the Boyana State Residence; the L. Zhivkova People's Palace of Culture; the Monument to the Unknown Soldier; the Monument to V. I. Lenin; the Banner of Peace Monument; the V. Levski, Universiada and Festivalna Sport Stadiums and Halls; the V. Zaimov, L. Tolstoi, Mladost and Lyulin residential areas; the Hristo Botev Students' Town; the Hotels Moskva, Novotel Evropa and Vitosha-New Otani; and the West, North, South and Liberty Parks. Mount Vitosha was declared a national park and features cable-car services.

A view of the Church of St. Sophia and the Alexander Nevsky Memorial Cathedral in Sofia

A view of Sofia. To the left is the Orbita Youth Complex, and in the centre the de luxe Hotel Vitosha-New Otani

In 1976 the areas of ancient Serdica and medieval Sredets were declared an architectural-historical reserve. About 700 buildings, several places of historical interest and urban areas contributing to Sofia's architectural and historical development were declared cultural monuments by 1983. The old architecture in the central parts of the city is being restored and incorporated into the modern city. Preserved architectural monuments: the remains of a Roman palace, thermae, basilicas, administrative buildings; the Byzantine Rotunda of St. George and Church of St. Sophia; the Ottoman Boyuk Mosque, Banya-Bashi Mosque and the Black Mosque. Restoration work is under way on the ruined churches and monasteries in and around the city – the Boyana Church, the Church of St. Petka Samardjiiska, the Church of St. Nedelya, the Dragalevtsi, Kremikovtsi, Kurilo and Guerman Monasteries. Post-Liberation monuments include the National Assembly (1884), monuments to the Russian liberators – the Alexander Nevski Memorial Cathedral (1904–12), the Russian Monument (1882), the Doctors' Monument (1883), the Monument to the Liberators (1907) and the Monument to Vassil Levski (1880–1895).

Sofia's coat of arms was designed in 1900 for Bulgaria's participants in the Paris Exhibition. It represents a shield divided into four fields: the upper left-hand field portrays the goddess Tyche, the custodian of the city, wearing a crown representing the city walls. The upper right-hand field depicts the Church of St. Sophia, after which the city was named. The lower left-hand field depicts Vitosha Mount and the lower right field a small temple to Apollo the Healer, symbolizing Sofia's thermal springs around which the first settlements appeared. The lion portrayed in the centre is a feature of the national coat of arms, a crown in the uppermost part in the form of a fortress wall symbolizes the city's strength and steadfastness over the ages. The motto, 'Ever Growing, Never Old', written on a ribbon encircling the lower half of the coat of arms; was added in 1911.

Plovdiv

Plovdiv. Plovdiv is a district town in Southern Bulgaria situated 156 km south-east of Sofia. The administrative, economic and cultural centre of Plovdiv district, and the second city in size and importance after Sofia, its population was 367,195 in 1982. It is situated by the river Maritsa on and around some of the Plovdiv Hills, in the Pazardjik-Plovdiv area of the Upper Thracian Plain at an average altitude of 160 m above sea level. The climate is transitional continental.

A general view of Plovdiv

Plovdiv is one of the oldest settlements in Bulgaria. Settlements from the neolithic, chalcolithic, the Bronze and the Early Iron Ages have been discovered. In 342 BC the settlement on Trimontium Hill was conquered by Philip II of Macedon (359–336 BC) who expanded it and turned it into an urban centre; it was named Philippopolis after him (a Thracian translation of Pulpudeva), and parts of the town walls still stand today. In 45 AD it was captured by the Romans. In the second and third centuries, it experienced an economic, political and cultural boom, and after the fourth century it played a leading role in the administration of the province of Thrace. It became the headquarters of the General Assembly of Thracians and a metropolis (the district centre of an administrative territory). From 88 AD coins were minted here. Under the Romans the population increased and the city expanded beyond the boundaries of the three hills. The town's waterworks were an impressive system of aqueducts, which brought water from the northern slopes of the Rhodopes. Remains have been discovered of residential areas, mosaics, reservoirs, the eastern and southern parts of the forum, an ancient theatre and a stadium, fortifications, Roman baths and others. In Roman times the city was called Trimontium.

In the mid-third century Philippopolis was destroyed several times by attacks. In 395 it became part of the Eastern Roman Empire. In the eighth century the Byzantine Emperor, Justinian I (527–565) restored and strengthened the town's fortifications. In the seventh century the Slavs settled permanently in the area of the city. In the reign of Khan Malamir (831–836) Plovdiv was included within the boundaries of Bulgaria and was given the Slavonic name of Pupuldin or Puldin. In the early eleventh century the town fell under Byzantine rule. After its liberation it was again included within the boundaries of the Bulgarian State (1186–1396). During the Second Crusade (1189–92) the city was destroyed. It was restored in c. 1198 by Ivanko, the Governor of Plovdiv district. In 1205 Tsar Kaloyan captured Plovdiv from the crusaders with the help of the Bulgarian Paulicians. Large-scale construction and restoration work was carried out under Tsar Ivan Assen II (1218–41). During the thirteenth and fourteenth centuries, medieval Plovdiv grew, expanded and strengthened. From the eleventh-fourteenth centuries it was a major administrative, political, religious and cultural centre.

A street in Old Plovdiv

It strongly resisted the Ottoman conquest, but was finally captured in 1364. From the fifteenth-seventeenth centuries it supplied the Turkish army and the population of the capital with agricultural products. Its main products were rice, sesame, and woven woollen fabrics. It developed as a manufacturing and commercial centre. During the National Revival Period it was an important economic, cultural and political centre. There was an upsurge in the development of handicrafts (over 50) and trade. The 18th century saw the development of homespun tailoring, inn-keeping, rug-making, soap-making, furriery, dyeing, cotton printing and so-forth; the town became a major trading centre. The Plovdiv merchants did business far beyond the boundaries of the Ottoman Empire – Egypt, India, Central and Western Europe – selling wheat, barley, rye, maize, oats, beans, flax, rice, tobacco, oil, fruits, silkworm eggs, and handicrafts. Many Plovdiv families had commercial houses and agencies in Constantinople, Odessa, Vienna, Manchester and Calcutta, carrying out the entire wholesale export and import. The hills were built up with rich houses, erected by Bulgarian builders, which were notable for their workmanship, design, external appearance, rich internal design and wonderful harmony of colours. The narrow cobbled streets followed the line of the terrain. Plovdiv's houses are a major achievement of Bulgarian National Revival architecture.

It is probable that Plovdiv had a monastery school as early as the beginning of the seventeenth century. In the late eighteenth and early nineteenth centuries there were two or three monastic schools. In 1849 and 1850 three Bulgarian mutual instruction schools were established. In 1850 a boy's grade school was set up – the Bishopric Sts. Cyril and Methodius School (here Cyril and Methodius Day was celebrated for the first time on 11 May 1851). In 1861 a secular school for girls was set up, which became a grade school in 1866.

Along with the establishment of a Bulgarian school system, Plovdiv was one of the few Bulgarian towns with a great number of foreign schools. The town's ethnic composition accounted for the presence of Greek, Turkish, Jewish, and Armenian schools, while the great interest of western countries in the town as a large economic and cultural centre accounted for the presence of foreign educational establishments.

The patriots of Plovdiv fought in the struggle for ecclesiastical and national independence and political freedom. In 1861 the Bulgarian parish (established in 1850) officially broke with the Greek Patriarchate and established a Bulgarian Church Board.

Many churches were built, today declared monuments of culture: the Church of St. Marina; St. Nedelya (with chapel); the old Church of St. Petka (eighteenth century); St. Dimitrius (1830) with the chapel of Sts. Ciril and Yulita (18th century); Sts. Konstantin and Elena; The Assumption (nineteenth century) with the chapel of St. Nicholas (1835) and a bell-tower (built in 1880 in the honour of the Russian liberators); St. Haralampii (1874); St. Kevork and the Catholic Church of St. Ludowig (1861). Book publishing made its appearance to meet the needs of education and culture. In 1869 Vassil Levski established a local revolutionary committee here. On the night of 3/4 (15/16) January 1878, Plovdiv was liberated.

After the Treaty of Berlin, in 1878, Plovdiv became the capital of Eastern Rumelia. It was here that on 6 September 1885 the Unification of Eastern Rumelia and the Principality of Bulgaria was announced.

By 1885 about twelve manufacturing enterprises were operating. An important role in the recognition of Plovdiv as a commercial centre was played by the First Bulgarian

A view of Old Plovdiv with Hissar Kapia and houses of the National Revival period

Agricultural and Industrial Exhibition and the Plovdiv Chamber of Commerce and Industry (established 1885). During the capitalist period Plovdiv developed as an industrial, transport and trade centre, second in importance after Sofia, with a dominant food industry.

In 1892 an organization of the Bulgarian Social Democratic Party was established. During the armed antifascist struggle in 1941–44 Plovdiv was the centre of the second Insurgent Operative Zone.

Today Plovdiv is the country's second largest industrial city after Sofia. The most important industries are food and beverages, mechanical engineering, metal working, electrical engineering and electronics, and these have the following factories and plants: an Amalgamated Memory Devices Factory; a typewriter factory (the only one in the country); the Rekord Motor Trucks Combine; a combine for woodworking machines and furniture assembly lines; an electrical appliances plant; an induction motors plant; the A. Ivanov Plant for Motor Trucks Spares; the Dimiter Blagoev Non-ferrous Metals Combine; the Kocho Tsvetarov Instrument-Building Plant; the G. Grozev Bolts, Nuts and Wire Factory; the 1 May Cannery; the Rodopi Tobacco Factory; the Vinprom Wine Distillery; the Vassil Kolarov Sugar Plant; the Kamenitsa Brewery; a Soft Drinks Factory; a meat-packing house; the Maritsa Cotton Textile Plant; the Bulgaria Men's and Ladies' Wear Factory; the A. Yurukov Knitwear Factory; the P. Tchengelov and K. Chistemenski Footwear Factories; the Pulpudeva Tannery; the Agria Insecticide Plant; the Alen

The Trakia housing estate in Plovdiv

Mak Plant; the Tchaika Handbag Factory; the Druzba Glassware Factory; the Trakia Amalgamated Pottery Combine; the Napreduk Furniture Factory; the Rodina Paper and Packaging Plant; and others. There is also an Institute on Non-ferrous Metallurgy, and a Canning Industry Institute. The city organizes the spring and the autumn International Plovdiv Fairs.

Plovdiv is the centre of an Agro-Industrial Complex and the Fruit and Vegetable Research and Production Complex. All crops are grown there and stock breeding is highly varied. It has a Company for Agricultural Chemicals; the Maritsa Vegetable Cultures Institute; an experimental variety-testing station; a Fruit-Growing Institute; an Institute for Fresh-Water Fish Breeding; and an Institute for Tobacco and Tobacco Products.

Plovdiv is the largest road and rail junction in Southern Bulgaria, which connect it with Sofia, Burgas, Svilengrad, Karlovo, Panagyurishte and Assenovgrad.

The city has five higher educational establishments; eleven technical high schools; secondary schools; general education schools; health-care centres; seven child-care centres; sports and tourist clubs; sixteen community centres; a National Library; a theatre; an opera; a State Philharmonic Orchestra; a puppet theatre; the Trakia Folk Song and Dance Ensemble; the Angel Bukoreshtliev Vocal Group; a Broadcasting centre; the Hristo Danov State Publishing House; an archaeological museum; ethnographical museum; a Museum of the National Revival and the National Liberation Struggles; a Museum of the Revolutionary Movements; a Natural History Museum; the Yordanka Nikolova Museum House; an art gallery; a General Trade Union Cultural Centre; a youth club; a pioneer club and nine trade union cultural centres. Plovdiv hosts the International TV Drama Festival; the International Festival of Chamber Music; the International Open-air Painting Session for Artists of the Socialist Countries; the International Photography Exhibition; the Alyosha Festival of Russian and Soviet Songs; the Festival of Bulgarian Short Films; the Days of the *Literaturnaya Gazeta* newspaper; Days of Bulgarian Literature on the theme 'The working class and literature'; the Days of Bulgarian Music; the Days of Short Adult Puppet Plays with international participation, and exhibitions by national artists. New architecture includes the Kamenitsa, Druzhba, Izgrev and Trakia residential areas, as well as various public buildings.

There are over 200 cultural monuments in the old quarter of Plovdiv. In 1936 it was declared an architectural and historical preservation area, and restoration and reconstruction work is underway. In 1979 Plovdiv was awarded the European Gold Medal for the Protection of Cultural Monuments.

Varna. Varna is a district city in North-eastern Bulgaria, 469 km north-east of Sofia, and is the administrative, economic and cultural centre of the Varna district with population of 295,038 in 1982. It is situated on the Bay of Varna on the Black Sea, east of lake Varna, at an altitude of 3–20 m above sea-level. It has a moderate continental climate strongly influenced by the Black Sea. In 1950 it was declared a national seaside resort.

Varna is one of the oldest settlements in Bulgaria. Artefacts dating back to the middle and late Palaeolithic, Mesolithic, Chalcolithic and Bronze Age have been discovered in the town and its vicinity. Roman thermae (second century), a Roman bath (third–fourth centuries), an early Christian basilica with fourth century mosaics, necropolises and remains of dwellings, churches and other buildings of the ninth–tenth and thirteenth–fourteenth centuries have been found in the town, as well as an early Christian basilica (fifth–sixth centuries), an early Bulgarian rampart (the eighth-century Asparuh wall), and an old Bulgarian monastery (tenth–eleventh centuries) in the area around the town.

Present-day Varna was founded in the sixth century BC as an ancient Greek colony by the name of Odessos. In the third and second centuries BC it enjoyed great economic prosperity and minted its own coins. In the seventh century it was given the Slavonic name of Varna. In the late eighth century the town became part of the First Bulgarian Kingdom; after the adoption of Christianity in 864 it became an important religious centre. In c. 971 Varna fell under Byzantine rule. At the end of the twelfth century the town was captured by Tsar Assen I (1187–96) and in March 1201 it was liberated by Tsar Kaloyan (1197–1207). In the thirteenth and fourteenth centuries it was a stronghold and major commercial port of the Bulgarian State, visited by Venetian and Genoese ships. In 1389 it fell under Ottoman rule and in 1399 it was ravaged by the Tartars. The town gradually became an administrative centre and the strongest fortress on the western Black Sea coast. At the end of the first half of the nineteenth century there was a noticeable upsurge in Varna's economic and cultural life. A revolutionary committee was set up by G. S. Rakovski's followers in 1861. The Bulgarian population in Varna took part in the struggles for the recognition of the Bulgarian Church. The Bulgarian church community, founded in 1860, united the Bulgarian population and directed the struggles. A mutual instruction school was opened in 1860. On 27 July 1878, the Varna fortress was ceded to the Russians and the town was liberated from Ottoman rule.

A socialist party organization to guide the struggles of the Varna workers was set up in 1893. In 1901, on the instructions of the Central Committee of the Bulgarian Workers' Social Democratic Party, Georgi Bakalov organized the Varna-Odessa clandestine channel to smuggle in Lenin's newspaper *Iskra*. Internationalist fighters from Varna maintained both the Varna-Odessa and the Varna-Sevastopol clandestine party channels. During the armed anti-fascist struggle of 1941–44 Varna was the centre of the Tenth Insurgent Operative Zone.

The town's economic structure is based on industry, transport and tourism. The main industries are mechanical engineering, textiles and food. The major enterprises are the G. Dimitrov Shipyards; a ship-repair yard; the V. Kolarov Mechanical Engineering Plant; an electrical appliances plant; a ship-fitting plant; the Purvi Mai Cotton Mill and Textile Plant; the Druzhba Factory; the Yanko Kostov Cannery; the Varnensko Pivo Brewery and the Chernomorski Ribolov Fisheries; and there are institutes of hydrodynamics, ship-building and fish resources.

Varna is the centre of an agro-industrial complex, whose main fields are viticulture, fruit-growing, poultry-farming, and dairy cattle-breeding.

Varna is a major transport centre. The Russe-Varna line, the first railway in Bulgaria, was opened in 1866. In 1882 a bill on the construction of a port was passed. Today the town maintains transport links with many towns and cities at home and abroad. Sea routes connect Varna with towns along the Bulgarian coast and with the USSR, Romania, Turkey and other sea states. The seat of the Bulgarski Morski Flot Shipping Company is in Varna. Varna has links with 22 airlines in the world and eight Bulgarian towns. It is on the Russe-Varna and Sofia-Varna railway lines. The construction of a deep-sea channel between lake Beloslav and lake Varna, which connected Varna with the industrial town of Devnya, greatly increased the freight turnover of Varna's station port. A ferry is in operation between Varna and Ilychovsk. There is a special institute for maritime transport.

Varna – a view of the pedestrian zone

The town has an Institute of Maritime Studies and Oceanography; three institutions of higher education; a semi-higher institute; ten technical high schools; many schools of general education; seventeen community centres; libraries; two theatres; an Opera; an open-air theatre; a District Art Gallery; a symphony orchestra; a broadcasting centre; a District History Museum; a Natural History Museum; a planetarium; an aquarium; health-care centres; sports, tourist, and field sports societies; and eleven sports centres.

Many national and international cultural, scientific and political events (congresses, symposia, conferences, meetings) are held in Varna. They include: the Varna Summer International Music Festival; the International Ballet Competition; the Varna Pop Song Competition; the May Choir Festival; the Bulgarian Film Festival; the Red Cross and Health Film Festival; the Bulgarian Puppet Theatre Festival and the Golden Sands International Rally.

The Varna skyline is dominated by the Church of the Holy Virgin, built between 1883 and 1886 and designed by the architect Maas of Odessa. The Euxinograd Palace was built in 1886, the Renaissance-style High School for Girls in 1901–1902, and the new port in 1906. In accordance with the 1946–1956 town plan new residential areas, the Palace of Culture and Sport, the Hotels Varna and Cherno More, the Party Building, the Observatory, the Students' Town, and many other buildings were erected, turning Varna into a modern and beautiful town. The Golden Sands and Druzhba international resorts north of the town also feature interesting architectural designs.

Tourism is of great importance to the economy of Varna. Wide sandy beaches stretch along the coast from Varna up to the Golden Sands international resort; many modern hotels and restaurants, rest homes, camping sites and holiday houses have been built there. Varna also boasts good sea-water pools.

A general view of Varna

Russe. Russe is a district town in North-eastern Bulgaria, 331 km north-east of Sofia, and is the administrative, economic and cultural centre of the Russe district. In 1982 its population was 178,920. Situated on the high terraced bank of the Danube, east of the island of Matel which joins with the mainland and of the estuary of the river Russenski Lom, it lies in the eastern part of the Danubian plain at an average altitude of 46 m above sea-level. Its climate is moderate continental.

Flint implements from the Upper Palaeolithic period have been found on the Danubian bank near Russe. There is a Chalcolithic habitation mound in the town. A Thracian settlement was founded on this mound at the end of the Bronze Age and during the Early Iron Age. Another fortified Thracian settlement was established along the river Russenski Lom. During the first century, during the reign of Emperor Vespasian (69–79), a Roman castle was built, which was known as Sexaginta Prista (Lat. 'Sixty ships'). It was an important fortress and home port of the Lower Danube fleet, and existed until the settlement of the Slavs who built their own fortress there – the Golyamo Yorgovo fortress. In 1388 it fell under Ottoman rule. The name 'Russi' is mentioned by merchants from Dubrovnik and in Romanian sources. Its present name was first mentioned in the early fifteenth century. During the sixteenth and seventeenth centuries, private monastery schools were opened in the town. A secular school was founded in 1820 which in 1838 became a school of mutual education (the Lancaster type of school). A boys' grade school was founded in 1858 and a girls' school in 1865. For the needs of the first grade school Angel Hadjioglu financed the publication of the *Bulgarian Primer* in Moscow in 1844, and in 1843 A. Hadji Rousset published in Strasbourg one of the first geographical maps of Bulgaria. During the National Revival period the citizens of Russe took an active part in the struggle for national education. In 1864 the Greek bishop Synesius was forced to leave the town. In 1871 a revolutionary committee was set up in the home of 'Baba' Tonka Obretenova, which served as a link between the Bulgarian Revolutionary Central Committee (BRCC) in Bucharest and the network of internal revolutionary committees. In 1874 the Russe committee became the central committee for the interior of the country. The first capitalist enterprises were founded before the Liberation of 1878. The town's proximity to the Danube and the trade relations established with a number of countries in Central Europe helped trade and industry to flourish. The Russe-Varna railway line, the first in Bulgaria, was built between 1864 and 1866. From 1864 until the Liberation in 1878 Russe was the seat of the Danube Vilayet (a province of the Ottoman empire) – the Tuna Vilayet. After the Liberation it was the largest town in the Principality of Bulgaria, being an important commercial, industrial and port centre. The foundations of a Bulgarian national fleet were laid; one of the first credit banks and joint-stock companies was established. The biggest machine factory in Bulgaria was founded in Russe in 1907. A number of printing houses were also set up here (it was in 1891 that Marx and Engels' Communist Manifesto was issued for the first time in Bulgaria). The first Bulgarian naval school – the Naval Technical Academy (later transferred to Varna) was founded in 1881 and in 1883 the *Obraztsov Chiflik* Agricultural School, the first of its kind in the Principality of Bulgaria, was set up. In 1885 the first Bulgarian technical society was also founded here. The *Svetlina* (Light) and *Komuna* (Commune) socialist circles were set up in Russe in 1892 and 1893 respectively. An organization of the Bulgarian Workers' Social-Democratic Party was formed in 1896. During the armed anti-fascist struggle of 1941–44 combat groups operated in the town. Having crossed the Danube the first Soviet military forces entered Russe on 8 September 1944.

Today Russe is a major industrial centre. The leading industries are mechanical engineering, metal-working, chemicals, food and textiles. There are a large number of industrial enterprises in the town: the Heavy Machine-Building Plant; the G. Dimitrov Combine for Agricultural Machines; the Ivan Dimitrov Shipyard; the Vassil Kolarov Rolling Stock Plant; the Zhiti Plant for Wire and Wire Articles; the Zita Machine Tool Plant; the Cherven Plant for Fodder and Mill Machines; the S. Karadzha Machine-Tool Plant; a plant for small tractors; the N. Kirov Plant for Electrical Installation Materials; the Plant for Electrical Insulation Materials; the Plant for Printed Circuits; a Communications Equipment Enterprise; the P. Karaminchev, G. Genov, Dunarit, and L. Tadzher Chemical Works; the Trud Plant for Refractory Materials; the Plant for Ferro-Concrete Pipes; the Plant for Pre-Stressed Pipes; the Plant for Perlite Articles; the Prefabricated Panel Combine; the 9 September Leather Plant; the Dunavska Koprina, Yuta and V. Piskova Textile Mills; the Arda

The Liberty Monument in Russe

Plant for Women's Ready-Made-Clothes; the Fazan Hosiery Plant; the A. Atanassov Furniture Factory; the Plant for Tubular Furniture; the Aerated Soft Drinks Plant; the Dunavia Cannery; the I. Dechev Bakery; a Poultry Combine; the D. Blagoev Combine; the Dunav Printing and Publishing House and others. Russe has also a number of research institutes: eight research, design and construction institutes, the Institute of Agricultural Machine-Building, and Institutes of Agriculture, Chemistry, River Transport, Engineering, Welding, Water Problems and Technology of Machine-Building.

Russe has its own agro-industrial complex which has to meet the demands of a large number of consumers and of the local food industry. It engages in viticulture, fruit-growing and calf- and pig-farming. The Dunav Game Farm breeds pheasants, hares and hunting dogs and there is also the *Obraztsov Chiflik* Institute of Seedology, Seed Production and the Growing of Leguminous Plants.

The town is a significant communications centre. It has two passenger, two freight railway stations and two marshalling yards, two bus stations, a river port and an airport. The Bridge of Friendship between Russe and Giurgiu, the largest bridge on the Danube, is built on two levels – one for road and one for rail transport. It connects northern Bulgaria with three main international high-ways – Budapest-Brasov-Bucharest-Russe; Warsaw-Lvov-Bucharest-Russe and Moscow-Kiev-Ungeni-Bucharest-Russe. Railway lines connect Russe with Sofia, Varna, Silistra, Gorna Oryahovitsa and Podkova. Since 1983 two tourist ships have been running along the Russe-Belgrade-Budapest-Bratislava-Vienna-Passau-Russe route. The port of Russe is one of the major Danubian freight and passenger ports and is also Bulgaria's largest river port. It is the seat of the Bulgarian River Shipping Company, and of the Dunay trans-Bulgarian-Soviet Transport Association; the Interlichter and other shipping companies of the USSR, Hungary and Czechoslovakia have their agencies there.

Russe has a Polytechnic, a college and thirteen technical high schools, as well as high schools; schools of general education; nursery schools; crèches; two sport clubs; a tourist and a field sports society; a sport centre; a yacht club; nine community centres; nine trade-union houses of culture; a pioneer and a youth centre; a Drama Theatre; a Puppet Theatre; a symphony orchestra; an Art Gallery; an exhibition hall; the Rodina and Dunavski Zyutsi Choirs; the N. Kirov Song and Dance Ensemble; libraries; a Radio and TV centre; a District History Museum; the Baba Tonka Museum House; the Z. Stoyanov Museum House; and a Museum of Transport and Communications. Various cultural events are held in Russe, such as the March Music Days International Festival; the Golden Rebec Folk Festival; the H. Vitanov Literature Festival; and days of military-patriotic literature held under the motto 'Working Class, Forward!'.

Several specimens of pre-Liberation architecture have been preserved, such as the Lom Public Baths, the Church of the Holy Trinity (1632), Armenian Church of the Virgin Mary, St. Paul's Catholic Church, the St. George's and All Saints Churches, and several old houses featuring wood-carving of the Tryavan school. Even before the Liberation in 1878 Russe followed a consistent town-planning policy which imposed a number of requirements quite modern for the times; sound construction, conformity with the overall plan, rich façade decorations, etc. Streets with kerb stones, pavements, Viennese street lamps, and European-style public houses and restaurants appeared for the first time in Bulgaria. After 1878 magnificent forms of neo-Baroque and neo-Renaissance architectural styles as well as elements of eclecticism appeared in the town; whole complexes were built exhibiting an extravagance of styles typical of Russe and unique in Bulgarian architecture. Although the town did not have endorsed town-planning until 1890, it exhibited a harmonious development, with a clear town-planning concept, a unity of styles, large-scale building, and overall integrity. The first lift in Bulgaria, operated by human power, was installed in a building erected in 1896. Several residential areas were established in different parts of the town. The greater number of the buildings there consisted of solid two-storey houses – the mansions of the rich Russe aristocracy; some of these stately homes stand in small parks facing the Danube. The façades of the buildings are richly decorated. One hundred and fifty buildings have been declared monuments of culture.

There are some notable examples of modern architecture in Russe; the Hotel Riga, the Pantheon of Bulgarian National Revival Heroes, and the Zdravets, Vuzrazhdane, Charodeyka, Iztok, Rodina and Druzhba housing estates. Many monuments have been erected to the memory of eminent figures of the National Revival politicians and men of culture.

The Hotel Riga in Russe

The National Revival Park, the Youth Park, the Prista Park, and the Lipnik Park provide excellent facilities for rest and recreation.

Russe is the birthplace of Elias Canetti – Nobel Prize winner for literature.

The Sava Ognyanov Drama Theatre in Russe

Burgas. Burgas is a district town in South-east Bulgaria, 385 km east of Sofia, and is the administrative, economic and cultural centre of the Burgas district. Its population in 1982 was 178,239. Situated by the Burgas Bay on the Black Sea coast, its average altitude is 17 m above sea-level. It stands in the plain of Burgas, and to the north, to the west and to the south-west are the Atanassovsko, Burgas and Mandrensko lakes. It has a transitional continental climate with a strong Black Sea influence. In 1953 Burgas was declared a national seaside resort, while the Burgas mineral springs (to the north-west of the city) were proclaimed a national spa.

Because of its crossroads situation and the convenient sea port, the town developed rapidly. Later on, fishermen from the Constantinople area settled here. The origin of the town's name can be traced to the Greek words pyrgos and burgos, meaning fortification or tower (because of the numerous fishing towers in the vicinity). In the second half of the seventeenth century Burgas developed extensive trade in timber, foodstuffs, sea salt, etc. In the early nineteenth century, because of the Russo-Turkish wars, the town acquired major strategic importance. It was freed from Ottoman rule at the end of January 1878. The town underwent rapid development, growing into an important

The centre of Burgas

On the site of the present-day city were Chalcolithic, Bronze Age, Thracian (sixth-third centuries BC) and Roman settlements. The first written reference to Burgas was made by the eighth-century Byzantine poet Manuel Philes and in medieval Turkish chronicles (fourteenth century). Later references to Burgas were made by the Turkish historian Hadji Kalfa (seventeenth century). He gives the town two different names: Borgos and Pyrgos.

sea port and commercial centre with large numbers of hired workers. The first socialist circle was organized in 1894.

The town's first industrial enterprises (machine-building and food processing) did not come into existence until after 1900. The growth and development of the town was also due to its rail link with Sofia (completed in 1890) and the opening of an enlarged and modernized port in

1903. On the eve of the Second World War Burgas had about 40 major factories and workshops. Trade was also highly developed. Its inhabitants took part in the armed anti-fascist struggle of 1941–44. Combat units were established in 1941, as were underground Fatherland Front committees.

Today Burgas (along with Varna) is Bulgaria's biggest seaport, equipped with new and specialized quays and with rail links to the interior of the country. Industry is represented by a number of plants, some of them of national importance: the Petro-Chemical Plant (along with its oil-processing and petro-chemical research institutes), the Ilia Boyadjiev Shipbuilding Yard; the Cherveno Zname Rolling-Stock Yard; the V. Kolarov Electric Cable Plant; the Okeanski Ribolov and Chernomorski Ribolov Fisheries; the Slavyanka Fish Factory and the Gocho Ivanov Timber Works.

The town is the centre of an agro-industrial complex which produces grain, fodder and industrial crops, vegetables, and perennials and also engages in poultry-farming and cattle-breeding.

In Burgas there are a Higher Institute for Chemical Engineering; two semi-higher institutes and seven technical high schools; general high schools; child-care establishments; a district hospital; sports, tourist and field sports societies; seven community centres; libraries; two theatres; an opera house; an open-air theatre; a district art gallery and a local history museum. Regular events include the International Sunny Beach Music Festival (the ten-day Symphony Music and International Folklore Festival held in Burgas). An International Symposium on sculpture and National Friends of Sea Art Exhibition are also held here.

The new urbanization plan divides the town into five zones: dockland, industrial, administrative-commercial, resort and housing zones. Extensive housing construction (the P. R. Slaveikov, Tolbuhin, Izgrev and other housing estates) and industrial building have determined the city's architectural appearance. The most outstanding architecture monument is the Church of Sts. Cyril and Methodius (1894–95) built of andesite and marble. There are also several monuments to outstanding political and cultural figures.

Tourism represents an important sector of the town's economy. Burgas is developing as a large seaside resort, with its wide sandy town beaches. Modern sea-water baths have also been built. Tourists pass through Burgas on their way to almost all southern Black Sea resorts.

The Tolbuhin housing estate

Burgas from the sea

Stara Zagora. Stara Zagora is a district town in Southern Bulgaria, lying 231 km to the south-west of Sofia, and is the administrative, economic and cultural centre of the Stara Zagora District. Its population in 1982 was 141,722. It is situated in the south-eastern foothills of the Surnena Gora Mountains, at an altitude of 200 m above sea level and has a transitional-continent climate. To the north-west are the Stara Zagora Mineral Baths, a national spa resort since 1967.

The Opera House

The Dimiter Blagoev residential quarter

In the environs of the town of Stara Zagora lies the Bereket Mound, inhabited during the Neolithic, Chalcolithic and Bronze Age; later a Thracian fortress and a sanctuary were built there. Near this mound remnants of pre-historic dwellings have been found. Chalcolithic ore mines, Thracian tombstones, and a large necropolis have been discovered. On the side of the present-day city the Thracians built the town of Berbe in the fifth century BC. In the early second century AC the Romans built the town of Augusta Trajana (the name was preserved until the third century), whose population was predominantly Thracian. Roman roads connected the ancient city with Europe, Asia, the Black Sea and the Danube. Remnants of Roman baths have been discovered near the present mineral baths. In the mid fifth century the town was sacked by the Huns. In written evidence of that time it is referred to as Verea. At the end of the sixth century it was razed to the ground by Avars and Slavs. In the later half of the eighth century the city was rebuilt by the Byzantine Empress Irina, who gave it the name of Irinopolis, but in the following centuries Byzantine and Latin chronicles refer to it as Verea or Vero, while Bulgarian documents from the thirteenth century refer to it as Borui. In 1370 the town fell under Ottoman rule. During the National Revival it developed rapidly as a large centre of agriculture, crafts (nearly fifty different crafts were developed), trade and learning. 1841 saw the opening of a boys' grade school and 1863 of a girls' grade school, at which Vassil Levski, Raina Knyaginya, Dimiter Blagoev and others studied. The population took part in the struggle for an independent church, and national liberation. In 1869 Levski formed a revolutionary committee, which became a district committee in 1872. The town was the centre of the Stara Zagora uprising in 1875. During the Russo-Turkish War of Liberation of 1877–78 it was liberated twice, on 10 July 1877 and on 1 January 1978. On 19 July a Russian Army under General Gurko and a Turkish Army under Suleiman Pasha fought a battle for the town; the Bulgarian Volunteer Corps also fought in it. After the Russian retreat the town was burnt down and destroyed. After the Liberation in 1878 it was a district town of Eastern Rumelia, and after the Unification in 1885, it became district and county centre. 1897 saw the formation of the Kaval Music Society, which in 1925 became the first amateur opera group, and in 1946 it became the first opera theatre outside the capital. 1893 saw the formation of a local Bulgarian Workers' Social Democratic Party organization. During the armed struggle against Fascism of 1941–44 it was the centre of the Fifth Revolutionary Zone.

Stara Zagora is a large industrial centre with advanced food, textile, chemical, and mechanical engineering industries. It has the following industrial plants: a Fertilizers Combine; a Combine for Memory Devices; the Svetlina Lighting-Fixtures Works; the Mebel Furniture Factory; the Natalia Plant; the Zagorka Breweries, the Meat Packing House; the Enev Cannery; a grain-fodder works; a dairy; the Sluntse Tobacco Factory; a poultry-plant; a creamery; an industrial building enterprise and the Beroe Mechanical Engineering Plant. The town also has a Research Institute for Food Processing Machines.

Stara Zagora is the centre of an agro-industrial complex covering a large area which specializes in grain, viticulture, vegetables and stock-breeding. It also has a Research and Production Centre for Poultry Breeding, and an Institute for Cattle- and Sheep-breeding. It is a junction of the Sofia-Plovdiv-Burgas and Russe-Podkova railway lines. It is a major trading centre.

The town has two institutes of higher education; one college; technical high schools; specialized and general high schools; kindergartens and nurseries; health care establishments; sports, tourist, and field sports clubs; a community centre; libraries; a theatre; an opera house; a symphony orchestra; art gallery; an open-air theatre; a stadium; an observatory; a District History Museum; the Geo Milev House-Museum; and a radio broadcasting station. An autumn literature festival is organized in Stara Zagora.

After the Liberation in 1878 the town was quickly restored and reconstructed according to the town-planning scheme of the Czech architect Lubor Bayer. In 1895 the setting out of Ayazmo Park, today named the Lenin Park, was begun. There are remnants of an ancient mosaic, an ancient administrative building, Roman Baths, and the 15th-century Eskai Mosque. There are monuments in memory of those who died in the Russo-Turkish War of 1878, and in the anti-fascist struggle, and a monument to the Soviet Army. The modern housing estates are the Kazanski, Gradinski, and Trite Chuchura housing estate. The eastern part of the town has a memorial to 'The Defenders of Stara Zagora'.

The Lenin Park

Pleven. Pleven is a district centre in the central part of Northern Bulgaria. Situated 174 km north-east of Sofia, it is the administrative, economic and cultural centre of the Pleven district, with a population of 135,899 in 1982, and at an altitude of 100–160 m above sea-level, situated on the river Tuchenitsa (the right-hand tributary of the river Vit), west of the Pleven Heights and in the central part of the Danubian Plain, it has a temperate continental climate.

In Pleven and in the Kailuka Wooded Park, remains of the Neolithic Age, Chalcolithic settlements, and of Thracian settlements of the Moesians and Triballi tribes have been found. In the Kailuka area, the Romans founded the Storgosia posting-house in the first–second centuries. Later on a fort was erected to protect the posting-house. In the Strazhata area to the north of the castle, an early Christian necropolis has been unearthed, one of the biggest in Bulgaria, dating from the fourth and fifth centuries. The fortress was destroyed, probably during the Avar and Slav invasions. In the Kailuka was the site of a Bulgarian settlement during the Middle Ages, (ninth–eleventh centuries), which existed until the end of the fourteenth century as Kamenets castle. The town traded with Byzantium. In the Middle Ages it was a crafts and commercial centre; it traded with Dubrovnik and Brasov, and had a cattle market. In the late fourteenth century it was conquered and destroyed by the Turks. In the sixteenth century it was an export market, mainly via the port of Nicopolis. To meet the requirements of the local population and the large Ottoman army, furriery, shoe-making, braziery, the trade in wool, hides, tobacco, cereals, wine and brandy developed. The cattle market's name spread throughout the Ottoman Empire. Pleven turned into a major administrative and garrison town.

The people of Pleven were among the first to begin the struggle for church independence: in 1825 they lodged a protest against the new 'bishop's tax' imposed by the Patriarchate. After the Crimean War of 1853–56, they fought against the unruly and self-willed Bishop Dorotheus and removed him, and after him his deputy Paissi. In the sixteenth century the town had five monastery schools. In 1840, the first Bulgarian secular girls' school was opened, and in 1890 the first Bulgarian viticultural and vinicultural high school. In 1869, Vassil Levski set up one of the first revolutionary committees in the country. During the Russo-Turkish War of Liberation (1877–78), epic battles were fought near Pleven between the Russian and Romanian troops and the Turkish army of Osman Pasha.

After the Liberation in 1878, the town was an administrative, transport and industrial centre. In 1892, the best equipped and most extensive wine cellars in the Balkan peninsula were built (the oldest collection of wines in the country has been kept there since 1893). The first industrial enterprises were established after the Liberation and also with the processing of agricultural produce. An organization of the Bulgarian Workers' Social-Democratic Party was established in 1894. In December 1899, the constituent congress of the Bulgarian Agrarian Party (BAP) was held here and its newspaper *Agrarian Defence* was published. Here the BAP organized the country's first Party schools for training cadres. Pleven was a centre of the anti-fascist uprising of June 1923. During the years of the armed anti-fascist struggle of 1941–44, the town was a centre of the Eleventh Operative Insurgent Zone.

Today Pleven is one of the country's most developed industrial towns. Along with the traditional food-processing, building materials, textiles and knitwear industries, new key industries have been set up: mechanical engineering, oil-processing, chemicals, etc. Main industrial enterprises include Petro-Chemical Works; the Nikola Vaptsarov Mechanical Engineering Plant; the Chemical Engineering Plant; the Tractor-Repair Plant; the Karlo Lukanov Electric and Motor Trucks Works; the Plant for Aluminium Casts; the Deseti Dekemvri Mechanical Engineering Plant; the Ilinden Plant for counter-pressure casting equipment by the A. Balevski method; the Steel Mill; the Vladimir Zaimov Cement Factory; the Purvi Mai Ceramics Works; the Prefabricated Housing Elements Factory; the General Ivan Vinarov Glass Factory; the Vuzhod Fire-Proof Clay Materials Plant; the Georgi Kirkov Cannery; the Tobacco Industry Enterprise; the Wine Factory; the Kailuka Brewery; Poultry Factory; a meat-packing plant; the Vassil Kolarov Rubber Footwear plant; the Assen Halachev Textile Combine; the Sanya Knitwear Factory; the Moesia Clothing Plant and the Republika Furniture Factory. It also has a Research Design Institute on Materials Technology, the Forging Press, Power-Generating Plastics Processing Machinery Institute

A new housing estate in Pleven

The District Youth Centre

and the Casting and Casting Technologies Research Institute.

It is the centre of two agro-industrial complexes, whose traditional field is viticulture. It has excellent research facilities for the development of agriculture – the Viticulture and Viniculture Institute, and the Fodder Institute, both being of national significance.

Pleven is an important transport centre with heavy traffic and a large road transport organization, and an important junction on the Sofia-Gorna Oryahovitsa railroad, with a branch-line from Yassen station to Cherkovitsa. It is also a major road junction, with roads to Sofia, Russe, Svishtov, Livech, Troyan, Teteven and Nikopol.

Pleven has a Higher Medical Institute; two semi-higher institutes; six technical high schools; high schools; healthcare establishments (in 1865, Bulgaria's first civil hospital was built here); child-care establishments; sports, tourist and field sports societies; six community centres; libraries; a youth centre; a trade-union cultural centre; a pioneer children's centre; a drama theatre; a puppet show theatre; an opera; a symphony orchestra; an art gallery; the House of Books; the Sparta Sports Centre; the Museum of the Liberation of Pleven in 1877; the District History Museum; a group of houses of the National Revival Period; a Field Sports Museum; and houses of famous people. It holds the Katya Popova Laureate Day's International Musical Festival; the March Festival of Literature; the June Festival of Public-Spirited Poetry; the National Festival of Soviet Folk Dances; and the Druzhba International Amateur Film festival. The new Storgozia and Druzhba housing estates are under construction, and there is an original waterfall in the city centre.

Architecture of the pre-Liberation period includes the old bridge over the river Vit near Pleven (built in 1865-66 by Kolyo Ficheto of Tryavna), and the Church of St. Nicholas (built 1834, probably on the site of an older church, featuring icons by Dimiter Zograf and N. I. Obrazopisov). The Churches of Sveta Parashkeva (1862), and the Holy Trinity (1893) have also been declared monuments of culture.

The Monument of the Unknown Soldier

Pleven is one of the towns with the greatest number of monuments built by the Bulgarian people in honour of the Russian liberators: the Mausoleum of Russian and Romanian soldiers; the Skobelev Museum-Park; the Monument of Liberty to commemorate Russian soldiers who died in the second attack against the city; eight fraternal common graves in the neighbourhood, and the Panorama of the 1877 Liberation of Pleven.

Sliven. Sliven is a district town in Eastern Bulgaria, 279 km east of Sofia. The administrative, economic and cultural centre of Sliven district, its population was 100,637 in 1982. It is situated in the southern foothills of a mountain in the eastern part of the Balkan Range at an altitude of 260 m, and has a transitional continental climate. The Sliven mineral springs which were declared a national spa in 1967 are located south-west of the city.

The town of Sliven emerged in Late Antiquity as an emporium. Its development was facilitated both by an ancient road passing the town between Kabile (7 km from Yambol) and Nicopolis and Istrum (today the village of Nikyup, Veliko Turnovo district) and by the town's location. Remains of late Roman fortifications, fortress walls, the foundations of a three naved basilica and a Roman necropolis have been found. Along with the Zagore region, Sliven was included within the boundaries of the Bulgarian State in the reign of Tervel in 705. Sliven was referred to as Istilifunus by the Arab traveller Idrisi (twelfth century). During the period of the Second Bulgarian State (1186–1396) Sliven was a cultural centre. Some 24 monasteries (today in ruins) were built in its vicinities. The town was known as 'Little Athos'. In the mountains around Sliven there are ruins of a medieval castle. During the Ottoman invasion the medieval town and castle were destroyed and the monasteries were burnt down. In the seventeenth century Sliven developed as a town of handicrafts and the centre of a *kaza* (administrative district) and in the late eighteenth century as the centre of a *sanjak*. Many varied crafts developed during the National Revival period. The town specialized in the production of rifles and guns, and of agricultural, craftsmen's and domestic ironmongery (a hundred workshops turned out a total of 12,000 rifle barrels per year). Over 2,700 looms and 70 mills produced 176,000 rolls of cloth. There were 66 water-mills, 984 workshops and 35 inns. In June the town held one of the biggest fairs in Bulgaria. In 1834 Dobri Zhelyazkov (Fabrikadjiata) set up the first textile enterprise in European Turkey, which in 1883 set up a textile school. Sliven's inhabitants participated in the struggle for national and ecclesiastical independence; in 1859 they expelled the Greek Bishop; and in 1871 the Bishopric of Sliven was set up. Sliven was also the centre of a vigorous *haiduk* movement against the Ottoman oppressors. It was known as the town of one hundred *voivode*. The revolutionary committee, set up by Vassil Levski in 1871, became a district committee. During the April 1876 Uprising Sliven was the centre of the Second Revolutionary District. During the Russo-Turkish War of Liberation (1877–1878) 800 workshops and 100 houses were reduced to ashes. On 4 January 1878 Sliven was liberated by Russian troops. After the Liberation, Sliven became one of Bulgaria's industrial textile centres. Some fifteen textile mills were operational by 1892. In 1896 one of the first workers' strikes broke out there. In 1892 the first Bulgarian textile trade union was set up. A social democratic organization was set up in 1891. During the armed anti-fascist struggle of 1941–44 the Sliven Party district was included within the Sixth Insurgent Operative Zone, with its centre in the town of Yambol.

Sliven is a large industrial centre. Its leading industrial sectors comprise food and beverages, mechanical engineering, textiles and glass, their major factories and plants being the Pobeda Works; the Dinamo Plant; the Lenin Electro-Vacuum Plant; the Commercial Furnishings Factory; the Amalgamated Grain and Fodder Enterprise; a soft drinks factory; a meat-packing plant; a dairy; the S. Ivanov Plant; the Spectur Factory; the L. Kalaidjiev Plant for Spare Parts; the D. Slavov Ceramics Factory; a construction and installations plant; the V. Kolarov Economic Combine; the Dikotex Woollen and Cotton Textile Mills; the D. Zhelyazkov Hosiery Plant; the Balkanbas Mining and Energy Complex; and the Institute of Light Sources and Quartzes.

The town is the centre of an agro-industrial complex specializing in viticulture, fruit growing and stock-breeding. There is an Institute for Animal Diseases. Sliven is on the Sofia-Burgas railway line.

There is a semi-higher institute in Sliven, as well as high school; secondary vocational and comprehensive schools; child- and health-care establishments; sports, tourist, and field sports societies; ten community centres; a theatre; a puppet theatre; a trade union cultural centre; a House of Literature; libraries; a district history museum; house-museums of famous people and an art gallery. The city is the venue of the Sliven Fires Cultural Festivals and the traditional May Cultural Festivals.

Of the National Revival Period there are the mutual instruction school in the courtyard of the Church of St. Sophia, the clock tower (1808), and the Churches of St. Demetrios (1831) and St. Nicholas (1934). Of particular interest is the wood-carved iconostasis in the Church of St. Demetrios ascribed to the Debur school of woodcarving. The typical Sliven house is open and symmetrical; 36 houses have been declared monuments of culture. In 1959, when Sliven became a district centre, large-scale housing and public building was begun. Housing construction is concentrated mainly in the D. Grouev residential area, and the Iztok and V. Zaimov housing estates. There are monuments to figures of the National Revival Period. A cable-car links the town with the Karandila Mountain Resort in the Sinite Kamuri locality, and there are four town parks for rest and recreation.

A general view of Sliven, with the Sinite Kamuni in the background

Shumen. Shumen is a district in North-eastern Bulgaria, lying 381 km east-north-east of Sofia and is the administrative, economic and cultural centre of the Shumen district. Its population in 1982 was 99,642. Situated in the eastern foothills of the Shumen plateau at an average altitude of 180 m above sea-level, its climate is moderate continental.

The Shumen plateau was inhabited both in the Copper Age (fifth–fourth millenium BC) and in the Bronze Age (third-second millenium BC). A Roman town with imposing public buildings and houses, surrounded by high, soundly fortified walls, emerged on the ruins of the fortified Thracian settlement during the second century BC. In the early Byzantine age it was still an important manufacturing, administrative and cultural centre as well as a major stronghold in the province of Lower Moesia. In the sixth and seventh centuries the fortress was captured by the Slavs. Soon after the Slav-Bulgarian State was founded, the town became a component part of the Pliska-Madara-Shumen-Khan Omurtag's camp system which was a safe barrier against the aggressive aspirations of the Byzantines. Later on, during the time of the Second Bulgarian State, Shumen fortress retained its important defensive role. In the eleventh and twelfth centuries it was known by the name of Misionis.

In 1388 Shumen was seized by the Turks. As a result of the irreparable damage the fortress was abandoned and during the first half of sixteenth century the town grew in a picturesque valley surrounded by the high Shumen plateau. In the fight against Russia, the Turks made the town a stronghold which was part of the Russe-Varna-Silistra-Shumen-fortress system. The town developed as a centre of many crafts (copperworking, leather tanning, silk production, weaving and the manufacture of firearms) and as a cultural centre. During the National Revival Period the public-minded intelligentsia of Shumen turned the town into a torch-bearer of Bulgarian national culture. The Hungarian and Polish revolutionaries, led by Lájos Kossuth, who immigrated after the defeat of 1848 revolution, had a great influence on the development of the national culture and consciousness of the numerous Bulgarian craftsmen and merchants. The two thousand foreigners introduced new ideas and ways of life in Shumen. The balls, concerts, musical evenings, theatre performances, etc., which they organized, brought the Bulgarian young people and the intelligentsia out of the haze of the oriental lifestyle.

In 1851 the Hungarian Mihai Safran formed the first Bulgarian orchestra, and in 1856 one of the first community centres in the country was set up. This, along with the establishment of a community centre in Lom, the first play performed there, 'Mihal Mishkoed', a comedy staged by Sava Dobroplodni in 1856, marked the beginning of the Bulgarian National Theatre. Dobri Voinikov, a writer of the Bulgarian Revival, was born in Shumen. In 1859 he started polyphonic choir singing in Bulgaria.

After the Liberation the town remained a typical crafts centre. There were a few small industrial enterprises, among them the country's first brewery (established in 1882).

From 1886 to 1887 Dimiter and Vela Blagoeva worked as teachers in Shumen. They organized one of the first socialist groups in the country. Vassil Kolarov (1877–1950), an outstanding figure of the Bulgarian and international workers' and communist movement, and a close associate of Georgi Dimitrov, was born here. Shumen was the centre of the Eleventh Insurgent Operative Zone during the armed anti-fascist struggle of 1941–44.

Today Shumen is a big industrial centre. The leading industries are mechanical engineering, food and beverages, tailoring and woodworking. The main plants and factories include the Madara Lorry Works; a repair works for wheeled tractors, and a spare parts factory for agricultural machines; the S. S. Stamenov Aluminium-Processing Plant; the E. Markovski Ship Fittings Plant; the Avgoust Popov Amalgamated Enterprises for Furniture Units and Sections; the P. Volov Chemical Works; a dairy factory; a meat packaging-house; the Amalgamated Lime Factories; foresty enterprise; a canning factory and the Shumensko Pivo Brewery. Shumen is the centre of an agro-industrial complex and a livestock-breeding centre, specializing in pig-fattening, viticulture and fodder crops. There is a Pig-breeding Research Institute; a Buffalo-Breeding Research Station; a regional sheep-breeding station; and a Sugar Beet Institute. It is the junction of the Sofia-Varna main line and the Karnobat branch line as well as the centre of good road transport to Russe, Varna, Silistra, Razgrad, etc.

Shumen has a Higher Teachers' Training Institute; a semi-higher institute; six technical high schools; secondary vocational and general education schools; kindergartens and nurseries; health care establishments; sports, tourist and field sports societies; a drama theatre; an operetta theatre; a puppet theatre; an art gallery; ten community centres; libraries; a philharmonic orchestra; a District History Museum; the L. Kossuth, P. Volov, D. Voinikov and V. Kolarov museum houses, and a branch of the Archeological Institute with the Bulgarian Academy of Science.

Remains of the Shumen fortress are preserved west of the town; and the Tombul Mosque of 1744, the old bazaar and the Clock Tower of 1740 are preserved as monuments of architecture. A group of buildings of the National Revival Period is to be found near the Church of the Ascension. Other sights are the Soviet Army Monument, the common grave of those who perished in the anti-fascist struggle, and the 1,300 years Bulgaria Memorial Complex. The Shumen Plateau has been declared a national park.

A general view of Shumen

Monument to Khan Asparuh

Tolbuhin. Tolbuhin is a district town in North-eastern Bulgaria. Situated 512 km north-east of Sofia, it is the administrative, economic and cultural centre of the Tolbuhin district, and its population in 1982 was 98,857. Situated on the Dobrudja plateau, the eastern part of the Danubian Plain, at an average altitude of 200 m above sea-level, it has a moderate continental climate.

Remnants of a Roman settlement, dating from the third or fourth century, have been found on the territory of the town. In the fifteenth century a settlement named Hadjioglu-Pazardjik grew there, to become a busy centre of trade and crafts in the seventeenth, eighteenth and nineteenth centuries. It was captured by Russian troops during the Russo-Turkish Wars. After 1878 it developed as a grain dealing and crafts centre. The Bucharest Peace Treaty in 1913 left it under Romanian domination. It was the centre of the Dobrudja national liberation struggle and headquarters of the Central Committee of the Dobrudja revolutionary organization. The Craiova Treaty of 1940 returned it to Bulgaria.

The town's main industries today are food and beverages, mechanical engineering, footwear, textiles and furniture; its major industrial enterprises include the Kalorima Works; a plant for trailers and containers; a grain-fodder combine; a vegetable oil factory; a fodder plant; the Dobrudjanska Mebel Furniture Factory; the Dobrich Footwear Factory; the Dobrudjanska Textil Factory; the Rodina Factory; the Orlov Works; a dairy and a poultry combine.

It is the centre of a rich agricultural region with two Agro-Industrial complexes and Pedigree Livestock-Breeding Centre. They specialize in grain, industrial and fodder crops and in pig-breeding and poultry-farming. The town is a transport centre, with two railway stations and on the inter-section of the roads to Varna, Silistra, Balchik, General Toshevo and others.

In Tolbuhin there are: a semi-higher institute; five technical high schools; vocational and general education schools; child- and health-care centres; sports, tourist and field sports' societies; a community centre; libararies; a drama theatre; folk song and dance group; a brass band; the Dobrudjanski Zvutsi Choir; a District History Museum; ethnographical Museum; the Y. Yovkov House-Museum; an art gallery; an open-air theatre and a stadium. Every year the Yovkov Turzhestva spring literary festivals are held there.

An area of houses dating from the National Revival has been preserved, while the town in the centre has been redeveloped, now featuring modern administrative, commercial and cultural buildings. There are monuments to the Soviet Army, to those who died in the Bulgarian Patriotic War of 1944–45 and a Khan Asparuh Memorial.

Pernik. Pernik is a district centre in Middle Western Bulgaria, 31 km south-west of Sofia. The administrative, economic, and cultural centre of the Pernik district, its population was 94,859 in 1982. It is located in the Pernik valley of the river Struma at an average of 700 m above sea-level. The Lyulin Mountains rise to the north-east, and the Golo Bardo Mountains to the south-west. It has deposits of brown coal and a moderate continental climate.

was incorporated into the Bulgarian State. Pernik castle played an important political, administative, military, economic and cultural role.

Pernik and its vicinities were a major farming and livestock area from the fifteenth–nineteenth centuries. 1876 and 1877 saw the first primitive extraction of glossy brown coal for the needs of the Turkish Army. In 1891 the Pernik coalfields were declared state property. The

Monument to Krakra Pernishki

Four Early Stone Age settlements, each built on the site of the previous one, have been discovered in the Pernik area. During the Copper Age there was another settlement about 200 m south-west of them. It also existed at different times during the Bronze and Iron Age. One Late Stone Age settlement that continued into the Copper and Bronze Ages has been uncovered on Krakra Hill. The Early Iron Age saw the foundation of a Thracian settlement on that same hill. In the early fourth century BC it became a stong fortress maintaining relations with some regions of Thrace, Macedonia, Illyria and the Aegean coast. The fortress was later destroyed by the Celts probably in the early third century BC. The ruins of several Roman, an Early Byzantine and a Slav settlements are also located in that area. In the early ninth century the latter

Pernik State Mining Enterprise was established, as were the first Bulgarian state-owned pit and a workshop for mining implements. 1895 saw the construction of a first coal separation installation in Bulgaria. The St. Ann's Pit, the largest in Bulgaria, was opened in 1927. The country's first mining rescue agency was founded in 1928.

The Pernik coalfields turned Pernik into the main source of energy for the country's developing industries, and so conditions for the rise and development of a workers' revolutionary movement were created. Under the direction of G. Dimitrov the first Marxist circle was set up in 1906, the workers of Pernik staged their first political strike, and the Pernik Miners' Union was founded. The first Bulgarian Workers' Social Democratic Party (left-wing) organization was set up in 1908. The first class organization of Bulgarian miners, the Colliers' Union which united all colliery workers in Bulgaria was founded in 1909 on the initiative of Pernik miners. A number of combat groups operated in the town during the armed anti-fascist struggle of 1941–44.

At present, Pernik is a large centre of coal-mining and heavy industry, with iron and steel, mechanical engineering, the coal and electrical engineering predominating. Its major industrial enterprises include the Georgi Dimitrov Mining and Generating Combine; the Lenin Iron and Steel Works; a rectifier plant; the Struma Plant for Heavy-Duty Machine Tools; the Krakra Pernishki Welding Machines Works; the Ferromagnets Plant; the Kristal Glassware Factory; the Pektin Works; a tailoring factory; the Stoyan Mihailov Metallic Structure Works; the E. Georgieva Knitwear Factory; and the Institute of Heavy Machine Engineering.

Pernik is a key railway junction with a large goods turnover due to its developed industries and close links with Sofia. The town has a big network of undergound and overground narrow-gauge and normal lines. The international Sofia-Kulata-Thessaloniki-Athens highway passes through the Daskalovo quarter, as do the Sofia-Kyustendil-Gyueshevo and the Trun roads.

The town has three technical high schools; comprehensive schools; child- and health-care establishments; two tourist and one field-sports societies; two sports centres; nine community centres; libraries; six trade-union cultural centres; a theatre; a district history museum; a Museum of Mining; an art gallery and a pioneer children's centre. It is the venue of the May Festival of Culture, the November Festival of Music, and the National Festival of Mummers and *Survakar*. A number of modern housing estates have been built. The Monument of Georgi Dimitrov and the Pernik Miners' Revolutionary Struggle consisting of an impressive statue of G. Dimitrov and a memorial hall built half-underground stands in the town's central square.

A general view of Pernik

Haskovo. Haskovo is a district town in Southern Bulgaria, 234 km south-east of Sofia. It is the administrative, economic and cultural centre of the Haskovo district, with a population of 87,639 in 1982. The town is situated on the river Haskovo in the hilly region of the district, at an average of 150 m above sea level, and has a transitional continental climate. To the west of the town are the Haskovo mineral springs, declared a national spa in 1952.

The monument to the victims of the 1912–13 war

A view of Haskovo

Remains of middle Neolithic and Thracian settlements have been unearthed in the area of the town. In the Middle Ages it was an administrative, economic and military centre. In the ninth century it was a fortified town surrounded by 2.30-m thick walls and had a garrison. The inner city consisted of houses, utility buildings, workshops and water reservoirs. Outside the walls lay the outer city. In the late fourteenth century the town developed as an important centre at the crossroads of major trade routes. During Ottoman rule Haskovo was a town of flourishing crafts, situated west of the nearby Uzundjovski Fair. 1836 saw the opening of the first monastery school. In 1869 Vassil Levski founded the first revolutionary committee. After the Liberation the town developed as a tobacco-producing centre. The first Bulgarian Workers' Social Democratic Party organization was set up in 1904. During the anti-fascist struggle of 1941–44 Haskovo was the centre of the Seventh Insurgent Operative Zone.

The town's industrial development is represented by three main industries – mechanical engineering, food and beverages, and textiles, concentrated in the following enterprises: a chemical engineering combine; an instrument building plant; the Mlada Gvardia Tobacco Machine Works; a machine tool factory; the Avtomatika Plant; an electric motor plant; a plant for low voltage equipment; the N. Terziev Factory; the Z. Dimitrov Textile Works; the M. Gogov Factory; the Mir Plant; a meat-packing plant; a dairy combine; a brewery; a wine factory; the Trakia Tobacco Factory; the I. Gebeshev Oil Refinery; a food and beverages combine and a poultry-dressing combine.

Haskovo is the centre of two agro-industrial complexes specializing in cereals, tobacco, cotton, viniculture, vegetables, sheep and cattle-farming. There is a quail farm near the city. It is a major highway intersection of the roads for Plovdiv, Harmanli, Kurdjali, and Dimitrovgrad, and a major station on the Russe-Dimitrovgrad-Kurdjali-Podkova railway line.

Haskovo has technical high schools; special and general high schools; kindergartans and crèches; as well as a number of health-care establishments; sports, tourist, and field sports societies; thirteen community centres; libraries; a drama theatre; a district history museum; an art gallery; and a Museum of Urban Life in the Post-Liberation Period. Haskovo is the venue of the Southern Spring Days of Literature. Its places of historical interest are the churches of the Virgin Mary (1857) and of the Archangel Michael (1861), featuring rare wood-carved iconostases; the Eski Djumaya Mosque (1635); the Kirkov School built in 1882; as well as a number of Revival Period houses. There are also ruins of the Medieval Fortress. Other sights are the monuments to those who died in the anti-fascist struggle and to the revolutionary Captain Petko the *Voivode,* who fought under Garibaldi. There is a picturesque deciduous forest to the north-west of the city, now called Victory Park. Another park, the Klokotnitsa Park, commemorates Tsar Ivan Assen II's victory over Theodore Comnin in 1230 at the Battle of Klokotnitsa.

Yambol. Yambol is a district town in South-eastern Bulgaria, 304 km south-east of Sofia. The administrative, economic and cultural centre of a district, its population in 1982 was 86,216. Situated on the Tundja river, at an average altitude of 100 m above sea level, it has a transitional continental climate.

There are remains of a Neolithic settlement, as well as a Roman and Medieval fortress. Seven km north-west of the town stand the remains of Kabile, an ancient Thracian town which appeared in the late second millennium BC. Yambol is an ancient settlement. From the eleventh–fourteenth centuries it was a stongly fortified town called Diampolis, Dimpolis, or Hyampolis, playing an important role in the confrontations between Bulgaria and Byzantium. An inscription from 1357 mentions it by the name of Dulbilino. After a long siege in 1373 Yambol was captured by the Turks. Under Ottoman rule the town developed mainly crafts and trade. A monastery school was opened in 1832 and a revolutionary committee was established in 1873. After the liberation from Ottoman rule it was an agricultural and craftstown. The construction of the Plovdiv-Burgas and Yambol-Elhovo railway lines greatly helped its development. During the armed anti-fascist struggle of 1941–44, Yambol was the centre of the Sixth Insurgent Operative Zone.

The town's main industries are primarily textiles, mechanical engineering, food-stuffs, construction and tailoring. Among the town's main industrial enterprises are the G. Kalchev Automobile Repair Plant; the Tundja Cotton Textile Plant; a plant for polyester silk; the D. Dimov Chemical Works; the Ivan Tenev Plant for Hydraulic Elements; the S. Zlatev Plant; the Sila Agricultural Machinery Plant; a ceramics factory; a pre-fabricated housing construction plant; a soft drinks factory; a packaging-house; a winery; and the Yantisa Centre for New Commodities and Fashion-Wear. Yambol is the centre of an agro-industrial complex with fertile lands specializing in viticulture; fruit-growing; market-gardening; wheat; maize; sunflowers, and sugar beet. There is an irrigation experimental centre and the Tundja Pheasant and Deer Game Farm. Yambol is an important centre of communications, and has road links with Sliven, Burgas, Elhovo, Nova Zagora and other towns.

Yambol houses a number of semi-higher institutes; technical high schools; specialized and general education schools; child-care and health establishments; mineral baths; sports, tourist and field sports unions; a drama theatre; an operetta; a puppet show theatre; a chamber orchestra; a Thracian folk song and dance ensemble; six community centres; public libraries; a sports hall and a planetarium. Regularly held here are the Sunrise Over Tundja Days of Literature and the Golden Diana Review of Chamber Choirs.

The centre of the town has a sixteenth-century covered **bazaar** which has been restored and incorporated into the modern town's architecture, the fifteenth-century Eski Mosque, and St. George's Church (1737) with a valuable

A general view of Yambol

The restored sixteenth-century bazaar

wood-carved iconostasis. Just outside Yambol are the Borovets and Vassil Kolarov Parks. Yambol is a starting-point for outings to the Bakadzjitsite, the remains of a Roman fortress, and the Monastery of St. Spass with the Alexander Nevski memorial church, built as a token of gratitude to the liberating Russian soldiers.

Gabrovo. Gabrovo is a district centre in Northern Bulgaria, lying 220 km north-east of Sofia, and is the administrative, economic and cultural centre of the Gabrovo district. Its population in 1982 was 80,901. Situated on the northern slopes of the Shipka mountain in the central Balkan Range on the river Yantra and its tributaries Sinkevitsa, Panicharka and Zhulteshka river at an altitude of 390 m it has a moderate-continental climate influenced by the mountains.

Remains of the Late Stone Age and the Copper Age have been discovered in the area of Gobrovo, and there are also Thracian tumuli, the remains of ancient roads and Roman and medieval fortresses and churches. The town first emerged, probably in the early fourteenth century, on the major trade routes from the Carpathians via Svishtov to Constantinople and grew to become a major commercial and crafts centre particularly during the late eighteenth and nineteenth centuries. The inhabitants of Gabrovo are famed for their strong national spirit, having taken part in the 1856 Uprising led by Dyado Nikola, in the Turnovo Uprising of 1862 and having fought in the detachments of Hadji Dimiter and Stefan Karadja and under Hristo Botev. In 1868 Vassil Levski set up a revolutionary committee here which prepared an uprising with great vigour. During the April 1876 Uprising Tsanko Dyustabanov's detachment was formed there. During the Russo-Turkish War of Liberation it was Gabrovo that supplied the Russian Army and the Bulgarian volunteers with ammunition and food throughout the battles to hold the Shipka Pass in August 1877.

The National Revival Period was marked by the development of handicrafts. In the late eighteenth and early nineteenth centuries iron-forging (farriery, axe-making, cutlery, balance-making, gun-making, watch-making and shoe-making), leather-working, pottery, silkworm-breeding (silk stuffs and clothes) and some other crafts developed rapidly. In the early nineteenth century wool-braiding flourished, later on lace came to be turned out in large quantities; in 1860 a workshop for lace braids opened in the town, which gradually became a factory. In 1870 there were some 800 fulling mills here, and 36 workshops for woollen braids. Retailers and trade companies made their appearance. The products by Gabrovo craftsmen were well received on markets outside the Ottoman Empire. Gabrovo dealers traded with Russia, Romania, Bosnia, Austria-Hungary, Germany, Italy, France, Asia Minor, and had branches in Bucharest, Brasov, Braila, Galati, Moscow, Nezhin and Odessa. After the Liberation from Ottoman domination the capital they had amassed, as well as their great craftsmanship, facilitated the rapid transfer to factory production. Trade and industrial enterprises and joint-stock companies appeared. In 1882 the first spinning mill opened in Gabrovo, and was followed two years later by the Alexander Worsted Spinning Mill (in 1884–85 both factories began turning out machine-spun woollen materials); in 1887 a leather factory opened up; and in 1888 a coal deposit started to be worked in the foothills of Mount Bedek in the higher parts of Gabrovo.

On the eve of the Balkan War of 1912–13 the town boasted some 27 workshops and factories.

Gabrovo was the cradle of new Bulgarian education. A monastery school was opened in 1625, and in 1835 the first new Bulgarian secular school was opened – the Gabrovo mutual education school, which in 1872 became a secondary school. In 1886 *Rossitsa,* the first Bulgarian socialist newspaper, was issued in Gabrovo. A Workers' Party organization was founded there in 1928. The population played an active part in the armed anti-fascist resistance of 1941–44. Along with the town of Gorna Oryahovitsa Gabrovo was the centre of the Eighth Insurgent Operative Zone.

Today Gabrovo is one of the country's industrial centres with textiles, machine-building and metal-working, leather and fur, shoe-making and food and beverages industries. The Gavril Genov Woollen Textiles Combine is situated there, together with the Dobri Kartalov Woollen Knitwear Works; the Vassil Kolarov Cotton Fabrics Combine; the Podem Hoisting Equipment Plant; the Bolshevik Instrument-making Works; the Yantra Plant for Textile Machinery and Equipment; an Industrial Electronics Plant; a Printing Appliances Plant; the Petko Denev Household Appliances Plant; the Dimitrov Blagoev Leather-working Plant; the Surp i Chuk Shoe Factory; Kapitan Dyado Nikola Plastics Plant; the Nezavissimost Furniture Factory; the K. Stoev Insulation Materials Plant and the Institute of Instrument-making which develops intensive fruit- and vegetable-growing and livestock breeding.

Gabrovo is a town with heavy traffic, being the junction of roads from Sevlievo, Veliko Turnovo and Tryavna. It is connected with the road-network of Southern Bulgaria by the Shipka Pass, and has regular coach transport to the regional centres and all villages in the district, to many district centres and to the capital. The Vurbanovo branch line links it with the country's rail network.

In the town there are a Higher Institute of Mechanical and Electrical Engineering; three technical high schools; a number of secondary and comprehensive schools; integrated child-care establishments; sports, hiking and field sports societies; a stadium; health-care establishments; thirteen community centres; a trade-union house of culture; a drama theatre; an open-air theatre; many libraries; a youth centre; a pioneer-centre; a children's art school; exhibition halls and a district History Museum. The town also features the National Museum of Public Education, the Museum of Architecture and Construction, and the House of Humour and Satire, where the National Humour and Satire Festival is held. Gabrovo is the birthplace of a great number of heroes of the national-liberation struggles who fell in the April 1876 Uprising.

The town's oldest part is the May Day Square with the clock-tower dating from 1833. The National Revival architecture is represented by the Aprilov Water-Fountain, built in 1762, the Church of the Holy Virgin built in 1865 and famed for its abundant wood-carving and icons by the Tryavna painters Tsanyo Zahariev and Tsanyo

Gabrovo

The clock tower in Gabrovo

Simeonov, and the V. Aprilov Secondary School, built by the Tryavna master-builder Gencho Kunev. Numerous bridges span the Yantra river. On the Gramadata Rock in the middle of the river stands a monument to Racho the Blacksmith, the legendary founder of the town. Many other monuments commemorate outstanding political figures, and there are scores of sculptures, memorials and common graves erected in memory of the Russian soldiers who fell in the Russo-Turkish War of Liberation of 1877–1878. A house-museum has been set up in memory of the child-partisan Mitko Palauzov. An original ethnographical museum-park has been built in the Etur town district, displaying the architecture, lifestyle and crafts typical of the National Revival Period.

Gabrovo is the starting point for tourist routes to the Shipka areas of the Balkan Range, for the village of Bozhentsi (an architectural preserve) and for the Shipka-Buzludje National Museum-Park, where there are monuments to the Russo-Turkish War of Liberation (1877–1878), to the foundation of the Bulgarian Workers Social Democratic Party (1891) and to partisan struggle over the 1941–44 period.

Modern building development in Gabrovo

Pazardjik

Pazardjik. Pazardjik is a district town in Southern Bulgaria, 120 km south-east of Sofia. It is the administrative, economic and cultural centre of the Pazardjik district and its population in 1982 was 77,830. The town is situated on the river Maritsa in the Pazardjik-Plovdiv region of the Upper Thracian plain at an altitude of 205 m above sea-level and has a transitional continental climate.

The town was founded in 1485, although remains of Thracian and Medieval settlements have been unearthed there. In the seventeenth and eighteenth centuries it flourished as a port and warehouse on the river Maritsa where grain, rice, wine, timber from the Rhodopes and iron from Samokov were stored. The river was then used as a water route, along which goods were ferried on rafts to Adrianople and Constantinople.

Its situation at a crossroads made the town a major trading centre and the venue of big trade fair – the Marashki Fair. In the early nineteenth century the main occupations in the town were market gardening, crop-growing and stock-breeding. The crafts of leather-tanners, coppersmiths, tailors, blacksmiths, farriers, broadcloth weavers, rug-makers, and dyers, and the trades of building and wheel-making also flourished here. The town was a producer of saltpetre. In 1855 Pazardjik took part in the Paris Exhibition with goods manufactured by its craftsmen. The foundations of education were laid in the seventeenth century with the opening of a monastery school, and in 1847 a mutual instruction and a grade school were opened. In the second half of the nineteenth century Pazardjik developed as a major cultural centre. As a result of the intense ecclesiastical struggle in 1859 the town won independence from the Greek Patriarchy. Its citizens played an active part in the revolutionary struggles for the country's liberation from Ottoman domination. In 1869 Vassil Levski founded a revolutionary committee here. The town was liberated by the detachment under General Gurko on 2 January 1878. In 1894 a Bulgarian Workers' Social Democratic Party organization was founded in Pazardjik. During the armed anti-fascist struggle of 1941–44 Pazardjik was the centre of the Third Insurgent Operative Zone.

Today Pazardjik is a major industrial centre with well developed food, mechanical engineering, metal working, chemical, paper and cellulose and electronics industries.

The house of Stanislav Dospevski

The Metropolitan Church of the Holy Virgin

Its industry includes the Konstantin Russinov Rubber Combine; the Metodi Shatarov Storage Battery Plant; the D. Banenkin Magnetic Discs Plant; the Metalik Machine-Tool Works; the Trakia Combine for the manufacture of cardboard, pasteboard and packaging; the Maritsa Cannery; the Vela Peeva Hemp Textiles Factory; the P. Abadjiev Furniture Factory; the P. Korchagin Children's Clothing Enterprise; the Stoyan Popov Construction Equipment Plant and the Woodworking Institute.

Pazardjik is the centre of two agro-industrial complexes, specializing mainly in vegetables (with the largest hothouses in the country), fruit, tobacco, grapes, cereals and fodder. It also has an experimental station for agricultural chemicals.

The town lies on the international Belgrade-Sofia-Istanbul highway and railway line. It is connected via Panagiurishte with the Sofia-Burgas motorway and via Srednogorie and Etropole with the Sofia-Pleven-Russe and the Sofia-Varna motorway. Major roads link Pazardjik with Peshtera, Batak and Dospat to the south and with towns of South-western Bulgaria via Velingrad.

Pazardjik has a branch of the Plovdiv Medical Institute; a semi-higher institute; three technical high schools; secondary schools; general education schools; health establishments; kindergartens and crèches; sports, tourism and field sports societies; four community centres; libraries; a youth centre; a trade-union centre; a pioneer children's centre; a drama theatre; an amateur operetta; a district history museum; the museum-house of Stanislav Dospevski; and an art gallery. It is the venue of the Winter Music Week's Festival, the April Days of Literature, and the national review of the amateur operettas.

Of architectural interest are the town's National Revival Period buildings – the eighteenth century Metropolitan Church of the Virgin Mary with its remarkable iconostasis, carved by the master wood-carvers of Debur, as well as the chapels of St. Petka (1856), the Holy Archangel (1860), Sts. Constantine and Helena (1868–70), and numerous houses, built in the style of the Plovdiv period houses. After 1959 the town centre was redeveloped, the new buildings being adapted to the architectural style of the National Revival Period. New residential quarters appeared, and there are numerous monuments built to commemorate outstanding political and cultural figures of the National Revival Period, and revolutionaries who died in the 1923 Anti-fascist Uprising and during the armed anti-fascist struggle of 1941–44.

Vratsa. Vratsa is a district centre in North-western Bulgaria. Situated 116 km north-east of Sofia, it is the administrative, economic and cultural centre of the Vratsa district with a population of 73,014 in 1982. Located in the north-eastern foothills of the Balkan Range by the picturesque Vratsata Gorge on the river Leva Reka, at an altitude of 370 m above sea-level, it has a temperate-continental climate.

settlement at the beginning of the Second Bulgarian State. It developed rapidly as an administrative and educational centre. The Vratsa Gospel was written here during the thirteenth century. Under Ottoman rule the city became a centre of handicrafts and learning. In the late eighteenth century Bishop Sophronius of Vratsa worked actively in the area to encourage the process of awakening of the Bulgarian national spirit. In 1824 the people of Vratsa

A street in Vratsa

Remains of the late Paleolithic and Neolithic Ages have been found in the caves around the city, while traces of Thracian settlements and burial mounds have been discovered in the vicinity. Excavations on the Mogilan Mound in Vratsa in 1965–66 unearthed three stone tombs of Thracian rulers of the fourth century BC with a rich collection of ritual objects (over 500 pieces of gold, silver, bronze, etc.). Two medieval Bulgarian necropolises (twelfth–fourteenth centuries) and remains of the medieval town wall have also been uncovered. Other finds include the treasure of jewellery and silver coins dating back to the time of Tsar Ivan Shishman (fourteenth century) and Sultan Bayazid I (fourteenth–early fifteenth centuries).

The town is thought to have originated as a fortified

began the struggle for church and national independence. During the preparations for the April Uprising of 1876, Vratsa was the centre of the Third Revolutionary District. Eight men of Vratsa fought in the detachment of Hristo Botev. During the Russo-Turkish War of Liberation of 1877–78 a cavalry squadron was formed in Vratsa, thus marking the beginning of the Bulgarian cavalry. On 28 October 1877, Russian troops liberated the town. In 1894 a local group of the Bulgarian Workers' Social Democratic Party was set up. During the preparation of the September 1923 Anti-fascist Uprising the Bulgarian Communist Party District Committee as well as the military command were located in Vratsa. The people of Vratsa took part in the armed anti-fascist struggle of 1941–44. Vratsa was the centre of the Twelfth Insurgent Operative Zone.

The Meschii Tower

There are two defence towers of the National Revival Period, which are commonly dated to the late seventeenth century. By their composition and architectural design they represent fortified dwellings – the Kurtpasha, or Serapion Tower in the town centre and the Meshchii Tower north-west of the town centre which is bigger, twice as tall, with a clearly defensive function, converted in the late nineteenth century into a clock-tower and fire watch-tower. Also well-preserved are the Church of the Ascension (constructed probably before 1840) and St. Nicholas Church (completed in 1862) which have remained largely unchanged, with icons painted by Stanislav Dospevski and wood-carvings, as well as one-and two-storeyed houses of the National Revival Period typical of the region, and the Stone Fountain built entirely of travertine during the eighteenth century.

The Hristo Botev Monument and Drama Theatre

Its leading industries at present are chemicals, mechanical engineering and metal-working, food, textiles and wood-working, with the following industrial enterprises: a chemical works; the Veslets Cast-Iron Works; a machine-tools repair plant; a medical equipment factory; a dairy; a meat-packing plant; the Vratitsa Cotton Textile Factory; the Yordan Lyutibrodski Silk Factory; the Okolchitsa Hemp and Linen Factory and the Dub Furniture Factory.

The town is the centre of an agro-industrial complex in a rich agricultural region specializing in viticulture and fodder. Stock-breeding is well-developed. It has a Sericulture Research Institute, a Swine Disease Institute and an ornamental plants nursery. Vratsa is on the intersection of the Vidin-Lom-Mezdra and Vratsa-Oryahovo (the nearest Danubian port) highways.

Vratsa has one semi-higher institute, four technical high schools; general high schools; child- and health-care establishments; sports, tourist and field sports societies; three community centres; libraries; a district history museum; a drama theatre; an art gallery and a philharmonic orchestra.

In 1976 the city acquired a new big architectural complex known as the Palace of Culture, incorporating a theatre and an opera, the district history museum, an art gallery and a Bulgarian-Soviet friendship centre; there is a monument to Hristo Botev and an ethnographical group of buildings of the National Revival Period. North-east of Veslets Hill a high-school students' village is under construction. Vratsa is a starting-point for walks in the Western Balkan Range.

Veliko Turnovo. Veliko Turnovo is a district town in the central part of Northern Bulgaria, 241 km north-east of Sofia. The administrative, economic and cultural centre of the Veliko Turnovo district, its population was 64,985 in 1982. Situated in the Fore-Balkan northern foothills of the Balkan Range, it is built up the steep sides of the Turnovo Gorge on the river Yantra on the Tsarevets, Trapezitsa, and Sveta Gora hills. The fine natural scenery is enhanced by the deep meanderings of the river Yantra. Its climate is moderate continental.

Veliko Turnovo is an ancient town. Remains of the Late Palaeolithic, Neolithic, Chalcolithic, Bronze and Iron Ages have been unearthed there. The Tsarevets and Momina Krepost hills have remains of fortress walls, gates and towers dating back to the early Byzantine period and the period of the Slav migration. During the period of Byzantine rule (eleventh–twelfth centuries) it was an important town. The brothers Peter and Assen, local feudal barons, rose in rebellion against Byzantine rule in November 1185. After the foundation of the Second Bulgarian State Turnovo became its capital. The town first grew on the slopes of Tsarevets and Trapezitsa but gradually the strip of land running along the Yantra river between the two hills was also populated and called Nov Grad (New Town). Turnovo developed as a major centre of crafts (forging, iron-working, weaving, tanning, pottery and gold- and silver-working), trade and culture. Access to the stongly fortified royal city on the hill of Tsarevets was through three gateways and a drawbridge. At the foot of Tsarevets was the quarter for foreign merchants and craftsmen. The town became famous for its beautiful architecture. Many palaces and churches, richly decorated with mosaics and mural paintings typical of the Turnovo school of arts, were built then. In Bulgarian and foreign (Byzantine and western) sources, Turnovo is mentioned as 'The Great City of Turnovo' ('Magna civitas Trinov') 'regal capital city', 'second to Constantinople', and so on. Turnovo was the centre of medieval Bulgarian coin-making. The Bulgarian, Byzantine, West European and Turkish medieval coins found during excavations, testify to flourishing internal and foreign trade. Literary schools, which were the centres of organized educational and literary activities, were founded in and around the town during the thirteenth and fourteenth centuries. In the literary schools of Kilifarevo, headed by Theodosius of Turnovo, and in the school of Patriarch Euthymius of Turnovo in the Monastery of the Holy Trinity, many monks and clerics were taught and passed on their learning.

In July 1393, after a three-month siege, the capital of the Second Bulgarian State was seized by the Ottoman Turks and reduced to ashes. Much of the population was massacred, and those who survived were driven away.

In the seventeenth and eighteenth centuries many Bulgarians from the surrounding villages and from villages in the Balkan Range, settled in Turnovo, where they set up their own communities. Various crafts developed (homespun manufacturing and tailoring, silk trade and silkworm breeding, dyeing, rug-making, gold- and silver-working, etc.) and the town became an important trading centre. In the first half of the 19th century well-to-do Bulgarian artisans and merchants gradually assumed the leading role in the town's economic and social life. Turnovo became once again a stronghold of the Bulgarian national spirit. Its people took part in the preparation of a number of uprisings, such as the 1598, 1686 and 1700 Turnovo uprisings, the Velcho Conspiracy of 1835, and 1856 uprising organized by the famous rebel 'Dyado' Nikola, and the 1862 Turnovo uprising. During the National Revival Period Turnovo was an educational centre. In 1869 a revolutionary committee was set up here by Vassil Levski, and Turnovo was the centre of the First Revolutionary District of the 1876 April Uprising.

On 25 June 1877, the Russian forces led by General I. V. Gurko liberated the town from Ottoman rule. There was an upsurge of political activity in the town. On 10 February 1879, a Constituent National Assembly was called in Turnovo which drafted and adopted the Turnovo Constitution. From 1890 to 1894, while D. Blagoev was living and working in Turnovo, the town became the centre of the socialist movement. The first socialist meeting in Bulgaria was called in the vicinity of Turnovo on the initiative of D. Blagoev and N. Gabrovski in April 1891. After the Buzludzha Congress held in 1891 Turnovo was, for a certain period, the seat of the General Council of the Party. The first party publication, the *Rabotnik* (Worker) newspaper was published in Turnovo from 1892–94. The population of Veliko Turnovo took an active part in the anti-fascist struggle of 1941–44.

The first industrial enterprise in the town was a small factory for silk spinning opened in 1861. Later several small enterprises of food, textile and leather industries were founded. Today the main industry is the food industry, followed by machine-building and the textile industry. The biggest industrial enterprises include: the T. Saraliev Cereals; Fodder and Food-Oil Combine; a packing plant; the Balkan Brewery; the E. Popov Machine Tools Plant; the Television and Radio Sets Factory; the Plant for Memory Devices; the Electric Hoists Factory; the V. Levski Factory for Unwoven Textiles; the V. Mavrikov Textile Mills; the D. Georgiev Factory for Ready Made Children's and Young People's Clothes; the D. Blagoev Timber Combine and the Pobeda Furniture and Upholstery Plant, etc. The town is the centre of an agro-industrial complex specializing in dairy-cattle breeding, vegetable growing, fodder plants, viticulture and fruit-growing.

Near the town is the intersection of the Sofia-Varna and Russe-Stara Zagora-Podkova motorways. A road network connects Turnovo with the towns of Svishtov, Pavlikeni, Elena, Gorna Oryahovitsa, Strazhitsa, etc. The trans-Balkan Russe-Stara Zagora-Podkova railway line passes through the town. Gorna Oryahovitsa Station connects Veliko Turnovo with the Sofia-Varna railway line.

Veliko Turnovo has two institutes of higher education; technical high schools; many schools of general education; child-care establishments; five community centres; libraries; a drama theatre; an open-air theatre; a district history

The House with the Monkey

museum; an art gallery; a broadcasting studio; sports, tourist, and field sports societies and stadium. Every year the National Review of History in the Arts is held in the town, as is the Republican Festival of Tourist Films, a Children's Song Review, and a Festival of Amateur Arts.

There are many old churches in Veliko Turnovo and its vicinity. On the banks of the Yantra stands the church of St. Dimiter built in 1185 where the uprising of Peter and Assen against Byzantine rule was proclaimed; the church of the Holy Martyrs built in 1230 by Tsar Ivan Assen II to commemorate his victory over Despot Comnensus of Epirus is near the village of Klokotnitsa; the small Church of St. George which stands at the foot of Trapezitsa, was also built during the reign of Tsar Ivan Assen II, and was repaired and painted in 1612; the church of Sts. Peter and Paul of the late thirteenth and early fifteenth centuries, famous for its stone ornaments and frescos dating from between the thirteenth and seventeenth centuries. Wonderful specimens of Bulgarian architecture have been preserved from the National Revival Period. There are whole quarters which still retain the atmosphere of the last century. The buildings, with their unique and bold designs, stand like ornaments one above the other on the slopes, overhanging the cliffs and abysses of the river Yantra, creating a fabulous townscape. The narrow and crooked cobbled streets afford the occasional picturesque glimpses of the Tsarevets, Trapezitsa and Sveta Gora Hills. The town's architectural beauty was greatly enhanced by Kolyo Ficheto, a master-builder of the National Revival Period who in 1836 completed the Church of St. Nikola, begun by the master-builder I. Davdata. Kolyo Ficheto built the house of Nikola Koyov, also known as the 'house with the monkey' (1849); the inn of Hadji Nikola (1858); the Church of Sts. Cyril and Methodius (1861); the House of Sarafka (1861); Sts. Constantine and Helena's Church (1872); the Turkish Konak (town-hall; 1875); the house of Etem Bey; and others. New housing complexes rise in the western part of the town. The old town area, Tsarevets and Trapezitsa, were declared a historical and architectural preservation area in 1965. The Monasteries of the Holy Trinity and Transfiguration are in the vicinity of Veliko Turnovo. Tourism plays an important role in the town's economy; its beauty and its numerous places of historical and architectural interest attract many Bulgarian and foreign tourists to the old Bulgarian capital.

The Old Town and river Yantra

The fortress on Tsarevets Hill

Blagoevgrad. Blagoevgrad is a district town in south-west Bulgaria, 101 km south of Sofia. Named after Dimiter Blagoev, founder of the Bulgarian Workers' Marxist party, it is the administrative, economic and cultural centre of a district with a population of 64,442 in 1982. It is situated in the eastern part of the fertile Blagoevgrad valley in the foothills of the south-western slopes of the Rila Mountains at an average altitude of 410 m above sea level. The river Bistritsa, a tributary of the Struma, flows through the town. It has a transitional-continental climate with Mediterranean climatic influence. In the north-eastern part of the city there are thermal mineral springs, and in 1963 Blagoevgrad was pronounced a spa town.

The Trade Union Centre

In ancient times a Thracian settlement called Skaptopara appeared near the mineral springs. In the first years of Ottoman rule another settlement appeared here as a crossroads fortification and market place. A monastery school was opened in the sixteenth century. In the second half of the eighteenth and the early nineteenth century the town developed as a major crafts and trade centre, growing tobacco and rice. The treaty of Berlin of 1878 left the town under Ottoman rule. The *Edinstvo* (Unity) Committee was set up here and played an important part in the preparation and execution of the Kresna-Razlog Uprising of 1878–79. The population also took part in the Ilinden-Preobrazhenie Uprising of 1903. The town was freed from Ottoman rule on 5 October 1912 during the Balkan War. The first socialist party organization in the district was established in 1913. During the September 1923 Anti-fascist Uprising the town was a revolutionary centre. During the armed anti-fascist struggle of 1941–44 the town was the centre of the Fourth Insurgent Operative Zone and the headquarters of the District Committee of the Bulgarian Workers' Party and the Workers' Youth Union.

Blagoevgrad is one of the towns with the highest industrial development in the country today. Its leading industries are food and mechanical engineering. Its major enterprises include the Pirin Tobacco Factory; the Pirinsko Pivo Brewery; a construction elements plant; a plant for measuring-instruments; a plant for mechanical computer units; a loudspeaker factory; the Gotse Delchev Textile Mill and the Dimo Hadjidimov Wood-working Factory. The town is the centre of an agro-industrial complex engaging mainly in tobacco, vegetable, and fruit growing, and stock-breeding. It also has a branch of the Institute of Agricultural Management and Organization, as well as establishments for hybrid pig-breeding and agricultural improvement and anti-erosion measures.

The city's main transport communications with the rest of the country are along the Sofia-Blagoevgrad-Kulata-Thessaloniki (Greece) international road and railway. Another road connects Blagoevgrad with the town of Delchevo in Yugoslavia.

Blagoevgrad also has a training college for primary school teachers; four technical high schools; general education high-schools; child-care establishments; a district hospital; a spa treatment centre; one sports, one field sports and two tourist societies; two community centres; a youth centre; a drama theatre; a district history museum; a broadcasting studio; and the Pirin State Folk Song and Dance Ensemble. Every year the Alen Mak Political Song Festival is held in the town.

Architecture of the Bulgarian National Revival includes the Church of the Holy Virgin, built in 1844, with its original architecture and precious wood-carvings, as well as several houses in the Varosha district and the G. Izmirliev Museum-House. There are a number of monuments to those who died for Bulgaria's liberation from Ottoman rule, in the Balkan War of 1912–13, in the anti-fascist resistance of 1923–44 and the Bulgarian Patriotic War of 1944–45.

The town has two parks: the Bachinovo and the Loven Dom parks. The city is a starting-point for outings in the south-west Rila mountains.

Vidin. Vidin is a district centre in North-western Bulgaria, situated 199 km north-west of Sofia, and is the administrative, economic and cultural centre of the Vidin District. Its population was 60,877 in 1982. Located on the right bank of the Danube River, in the Vidin Plain, its altitude is about 30 m above sea-level. The low bank of the river is protected by a dyke and concrete hydraulic engineering installations. The climate is moderate continental.

The remains of a large prehistoric settlement of the Copper Age and of Bronze Age settlements and a necropolis have been found in the vicinity of the town. Vidin is one of the oldest towns on the Danube and was founded more than 2,000 years ago as the Roman fortress of Bononia, which was built on the site of a Thracian settlement. Under the name of Bdin in the First Bulgarian State (681–1018), it became the centre of an administrative region and independent episcopate. In 1003, after an eight-month siege, it was conquered by the Byzantine Army. At the time of the Second Bulgarian State (1186–1396) it developed as a powerful stronghold on the State's north-western border, and withstood the attacks of Kumanians, Tartars and Magyars. In the late thirteenth century Bdin was the centre of an independent feudal principality under the Despot Shishman. When the ruler of Vidin, Mihail Shishman, was proclaimed Tsar in 1323 the city was re-integrated into the Bulgarian State. From 1371 Vidin was the capital of the Kingdom of Vidin. It flourished as a major commercial city and port and as an educational and literary centre. The Vidin literary school was established during the reign of Ivan-Stratsimir (late fourteenth century). A *tetraevangelie*, copied in Vidin during the fourteenth century, is today in the British Museum in London as well as in the *Bdin Collection* (1360). In 1396 Vidin was captured by the Turks. In 1404 the first uprising broke out against Turkish rule. During the Ottoman occupation Vidin was a *sanjak* (Turkish administrative district) centre, well-fortified by the Turks. Between 1794 and 1805 it was the seat of the independent feudal ruler of the area, Osman Pazvantoglu. Sophronius of Vratsa worked in Vidin from 1800 to 1803 compiling the First and Second Vidin Collections. During the late eighteenth and early nineteenth centuries the town developed as a major economic centre. It was famed for its master coppersmiths, potters and goldsmiths. In 1864 the Vidin community organized a struggle for Bulgarian enlightenment and an independent Bulgarian church. In 1868 Antim I was appointed Metropolitan Bishop of Vidin. During the Russo-Turkish Liberation War the Turks surrendered Vidin Castle to the Russian forces commanded by General Borevski on 7 April 1878.

After the Liberation in 1878 Vidin developed as a city of handicrafts and small industries. Dimiter Blagoev taught in Vidin from 1887–90. He founded the first teachers' association in the country and the Teachers' Association Newspaper (1889–90) and set up a workers' socialist study circle. In July 1898 a socialist party organization was created. During the September 1923 anti-fascist uprising hundreds of Communists and Agrarians were imprisoned and many of them massacred. The inhabitants of Vidin took an active part in the armed antifascist resistance of 1941–44.

The first factories in Vidin started appearing only during the 1920s. When Vidin became a district centre (1959) and the Chemical Combine was constructed in 1968, the ratio between agriculture and industry changed.

The Chemical Combine incorporates the Vidin Polyamide Fibre plant and the Vida Tyre Plant. The complex includes a repair shop, a thermo-electric power station, a research institute for industrial chemistry and a research and development centre. Other factories include the Vida Shirt Factory; the Dunav Tobacco Factory; a cannery; a dairy factory; a meat packing plant; a wine factory; a soft drinks factory; a dairy research and production complex; the Georgi Dimitrov Pump Factory; a metal-cutting tool plant; the Kosta Yordanov Household Pottery Factory; a ceramics factory; the Bdin Factory for Furniture Units and Packaging.

Vidin is the centre of an agro-industrial complex in a fertile agricultural area. The main products are cereals and fodder, cattle, and pigs. The region specializes in vegetable, grapes and fruit growing. It also has a viniculture and viticulture research complex; a dairy institute; an experimental station for cattle-breeding; a pig-breeding department and a district veterinary control centre.

The river Danube connects Vidin with the other Danubian countries; there is a ferry linking Vidin with Kalafat, in Romania. The port has become very active in recent years in the import and export of goods. The town and district of Vidin benefited greatly when it was linked with the national railway network in 1923 – the Mezdra-Boychinovtsi-Vidin line, a branch of the Sofia-Pleven-Varna main line; the Vidin-Koshava local line started to

The cruciform barracks (1798)

The medieval fortress Baba Vida

operate in 1969. Vidin is connected by road to Bregovo, Kula, Sofia and Lom.

The city has two technical high schools; high schools; kindergartens and crèches; health-care establishments; sports, tourism and field sports societies; a sports centre; a stadium; a marina; a drama theatre; a district history museum; a cultural centre; an art gallery; a library; and a pioneer children's club. The Vidin Summer Theatre Days are held in the Baba Vida Fortress, now a museum. Preserved old architectural monuments include fortifications and the Stambol, Telegraf and Enichar Gates. The old part of the city contains the sixteenth-century Church of St. Pantaleimon and the seventeenth-century Church of St. Petka, both single-naved churches built half underground. According to the inscription in the courtyard, the Church of St. Petka was painted in 1633, of which some frescoes have been preserved. Other architectural monuments of interest are the Turkish *konak* (town-hall;

second half of the eighteenth century), the mosque and Pazvantoglu's library, and the 'Cruciform Barracks' (1798; restored). The old houses were built in the Western Bulgarian style with some influence from Europe and the mid-19th century architecture of Plovdiv. After the Liberation in 1878 it was mainly the town centre that was affected, St. Demitrius's Cathedral (built in 1885–89), the *Tsvyat* Community Centre, and the Mausoleum of Antim I.

Many monuments speak of the town's heroic past. The beautiful riverside gardens around the Baba Vida Fortress contain a memorial to those who died in the anti-fascist resistance. One hundred and sixty-four officers and soldiers of the Third Ukrainian Front, who fell in the Second World War on Yugoslavian territory, were buried in the Soviet cemetery. The town's well-preserved historical monuments attract many Bulgarian and foreign tourists.

Silistra. Silistra is a district town in North-eastern Bulgaria, situated 443 km north-east of Sofia and is the administrative, economic and cultural centre of the Silistra district, which is a part of the Dobrudja plain. Its population in 1982 was 57,670. Situated on the right bank of the Danube, it is right on the Romanian border at an average altitude of 30 m above sea-level, and has a moderate-continental climate.

In the vicinity of Silistra there are remains of a Thracian settlement on whose site the ancient town of Durostorum grew up. During the third and sixth centuries it was a fortified town of great military and economic importance, and an outpost on the road along the Danube. When the Bulgarian State was founded it was incorporated within its borders. From the eighth–tenth centuries it had the Slav name of Druster; it became a large, well-fortified Bulgarian medieval administrative and economic centre. During the early ninth century it was a centre of an administrative province – a *komitat*. After the conversion to Christianity in 864 and the recognition of the independence of the Bulgarian Church it became the seat of the Bulgarian Patriarch. During Byzantine rule it was an administrative centre of the *thema* Paristrion. In 1413 it fell to the Ottomans. In the fourteenth century it was called Silistra; later on it was built up as a fortress, and was repeatedly besieged by Russian troops during the Russo-Turkish wars. In 1667 a monastery school was founded, in 1860 a grade school, and in 1869 a girls' school. In 1869 a revolutionary committee was set up. The citizens of Silistra took an active part in the 1877–1878 Russo-Turkish war of liberation. In 1894 a Bulgarian Workers' Social Democratic Party (BWSDP) organization was set up. After the Inter-Allied War of 1913, Silistra came under Romanian rule. In 1919 the Bulgarian Communist Party (BCP) organization was set up. From 1925 to 1940 the struggle of the Silistra Communists against the national economic and cultural oppression of the Romanian *chokoi* (big landowners) was directed by the Dobrudja Revolutionary Organization. Silistra was restored to Bulgaria (together with the Southern Dobrudja) by force of the 1940 Treaty of Craiova. The anti-fascist struggle of 1941–1944 was led by the Communist Party organization.

Silistra's industry is dominated by mechanical engineering and metal-working, food and wood-working industries: it has a computer technology plant; an office machines plant, the Combine for Multi-purpose and Cutting Machines; the Stomana Steel Works; the machine-tool plant; the Captain Mamarchev Industrial Rubber Goods Plant; the Chrome Manufacturing Plant; the Lipa Furniture Factory; the Dobrudja Cotton Textiles Plant; the Kamushitt Combine; a construction and assembly plant; the Nektar Cannery; and the Wood-Working Combine where timber from Komi in the USSR is processed.

It is a centre of an agro-industrial complex with well-

A general view of Silistra

developed viticulture, fruit-growing, vegetable-growing, stock-breeding and fishing. It is a port on the Danube and the last stop on the Samuil-Silistra railway.

In Silistra there are: a semi-higher institute; three technical high schools; general schools; kindergartens; crèches; health centres; two lecture halls; a drama theatre; an art gallery; a history museum (in the Medjidi Tabia fortress); an ethnographical museum and monuments to anti-fascists. Still preserved are remnants of a late-classical tomb, an ancient and medieval fortress and settlement, and a necropolis. A modern town centre and a riverside park were laid out. The Golden Ear of Wheat Literary Festival is held here.

Kurdjali. Kurdjali is district town in Southern Bulgaria, 262 km south-east of Sofia. The administrative, economic and cultural centre of the Kurdjali district, its population in 1982 was 55,762 inhabitants. Situated in a small valley in the middle course of the river Arda at an altitude of 240 m it is near the Kurdjali and Studen Kladenets reservoirs. The climate is transitional Mediterranean, slightly influenced by the mountains.

sor – many were killed, while others fled into the mountains or adopted Islam. A census of 1800 registered a population of 230. Under the Treaty of Berlin of 1878 Kurdjali became part of Eastern Rumelia, but after 1885 it was again under Turkish rule. Gradually the town developed as an administrative and economic garrison centre. In 1913 the Bucharest Peace Treaty returned Kurdjali to Bulgaria. After the Balkan War of 1912–13

The town centre

A Neolithic settlement and two Thracian burial mounds have been unearthed not far from the town. Some 5 km east of Kurdjali, on the rocky hill of Hissar Altu, stands the Muniak (Moniak) Fortress. During the time of the First (681–1018) and Second (1186–1396) Bulgarian States a medieval Bulgarian settlement, which developed as an economic and administrative centre of the Ahridos region, existed here.

At the end of the fourteenth century and the beginning of the sixteenth century the eastern parts of the Rhodopes were assimilated by the Ottoman Empire. The population waged a heroic struggle against the oppres-

and especially after the First World War many Bulgarian refugees from Thrace, the nearby villages and other parts of the country settled there. In 1919 the Communists established a party organization in the town. During the armed anti-fascist struggle of 1941–44 the BCP's Regional Committee formed combat groups for sabotage action here.

Before the town's liberation from Ottoman rule, Kurdjali was a craftmen's village with a great number of homespun manufacturers, tailors, shoe-makers, potters and coppersmiths. Soon it became a centre for trade in the Rhodope tobaccos. The tobacco industry developed mainly with financial backing from foreign tobacco companies. Between 1930 and 1934 prospecting began in the mining concessions of the Kurdjali region. Between 1939 and 1941 German capital financed the digging of mine galleries and the construction of an ore-crushing installation, a small washing plant for lead and zinc concentrate, a small blast-furnace, and a diesel-electric power station. There was a cable-car to the major ore deposits.

Today ore-mining and ore-dressing occupy first place among the town's industry, followed by tobacco processing, mechanical engineering and textiles. The main enterprises are the Georgi Dimitrov Lead and Zinc Works; the Rodopi Research Production Centre for the mining and dressing of non-metalliferrous mineral raw materials; the Komsomolets Mechanical Engineering Works; the Pnevmatika Mechanical Engineering Works; the Bistrets Reseaerch Production Centre for Water Purification Equipment; the Kapitan Petko Voivoda Automobile Repair Works; a tobacco sweating factory; the Orphei Knitwear Factory; a meat packaging-house and the Breza Furniture Factory.

Kurdjali is the centre of an agro-industrial complex specializing mainly in tobacco, stock-breeding and fruit.

The town is on an important railway junction. It is on the railway line of Russe-Stara Zagora-Dimitrovgrad-Podkova. Ores, metals, trass and tobacco leave Kurdjali by rail and agricultural products arrive.

Kurdjali has a teachers' training college; four technical high schools; general education schools; child- and healthcare establishments; sports, tourist, and field sports societies; a sports centre; reservoir marinas; tourist facilities; rest houses; four community centres; libraries; a drama theatre; a cultural centre; a Book House; the District Art Gallery; the District History Museum; a symphony orchestra; and a seismic station.

Many cultural festivals take place in Kurdjali – the Screen of Friendship Film Festival, May Days of the Theatre, district amateur arts festivals, the Bulgarian Book Week, and the Lights over the Rhodopes' Days of Literature and Art. In the centre are a monument to the victims of the 1912–18 wars and one to Georgi Dimitrov.

West of the small river of the town, at the bottom of Borovets Hill, is the old part of the town with its narrow winding alley ways and low houses.

After the First World War Kurdjali expanded eastwards because of the influx of Bulgarian refugees from Thrace and the interior of the country. The new town plan of 1973 is changing Kurdjali into a beautiful modern town; since 1975 the Vuzrojdentsi Housing Estate has been under construction on the right bank of the river Arda.

Kyustendil. Kyustendil is a district centre in Southwestern Bulgaria, 90 km south-west of Sofia. It is the administrative, economic and cultural centre of the Kyustendil District with a population of 54,657 in 1982. The town is situated in the southern parts of the Kyustendil Valley, near the river Banshtitsa (the right-hand tributary of the river Struma), 500 m above sea level. Kyustendil has a transitional continental climate. In 1950 it was declared a national spa; its mineral waters have a temperature of 73°C, a slight odour of hydrogen sulphide and an acrid taste, owing to the combination of hydrocarbonate-sulphate-sodium, sulphide, fluoric and silicon salts.

A general view of Kyustendil

The earliest known inhabitants of the Kyustendil area were the Thracians of the Danteleti tribe, who inhabited the upper reaches of the Struma. Reference to this ancient settlement can be found in the writings of the ancient Greek writer Theopompus (fourth centry BC), the Roman Pliny the Elder (first century AD) as well as the Roman historian Titus Livius (first century BC–first century AD). Its ancient name is Pautalia (from the Thracian *pote* meaning spring). Until 270 Pautalia constituted part of the Roman province of Thrace, and later the province of Inner Dacia. During the fourth century the fortress on the Hissarluka Hill was built, reinforced during the reign of the Byzantine Emperor Justinian I (527–565), to protect the town from the frequent attacks of Slavs, Avars and other barbarian tribes. After 553 the name Pautalia is no longer mentioned. In a charter issued by the Byzantine Emperor, Basil II, in 1019 Kyustendil is mentioned as Velbuzhd – assumed to be the name of a Slav tribal chief. Under Byzantine rule the town became a major religious and adminstrative centre.

The town was incorporated into the Bulgarian State by Tsar Kaloyan (1197–1207). In 1282 the Serbian King Milutin conquered Velbuzhd. The Bulgarian Tsar, Mihail III Shishman, fell in battle against the Serbian King, Stefan Dechanski, on 28 July 1330. About 1355 Velbuzhd and the surrounding area were incorporated into the semi-independent feudal Velbuzhd principality under the despot Deyan. In 1372 Velbuzhd fell to the Turks, although it preserved a comparative independence until 1395. The citizens made their last attempt to throw off the Ottoman yoke in *c.* 1427–1428. In the years of the Ottoman domination Velbuzhd was the centre of a *sanjak* (administrative district), incorporating the muncipalities of Radomir, Dupnitsa, Petrich, Melnik, Doiran, Tihves, Veles, Struga, Strumica, Radovis, Shtip, Kratovo, and Vranja.

During the National Revival Period the town developed rapidly. The crafts and trade flourished. The citizens of Kyustendil took an active part in the struggles for an independent church. After the foundation of the Bulgarian exarchy in 1870 the town became a metropolitan centre.

The first monastery school was opened in 1821 in the courtyard of the Church of the Virgin Mary. In 1849, on the foundations of the old school a new one was built which stands to this day. In 1872 a revolutionary committee was set up in the town. Kyustendil was liberated on 29 January 1878. After the liberation Kyustendil retained the character of a centre of crafts and trade. Craftsmen's guilds were formed and in 1912 the first spinning mill for worsted yarn was opened, financed by French capital. Several tobacco warehouses, distilleries, mills, an oil refinery, a cannery, carding workshops, and a wood-working workshop were set up, financed by local and foreign capital.

A socialist society was founded in the town in 1893. During the Soldiers' Uprising of 1918 a detachment seized the Kyustendil Army Headquarters (24 September). In 1920, the First and Second Kyustendil Communes were established as a result of municipal election victory. A Workers' Party organization was set up in March 1927. The citizens of Kyustendil played an active part in the armed anti-fascist struggle of 1941–44.

Mechanical engineering, metal processing, textiles, food and beverages, and wood processing are the town's main industries. Main enterprises include the Emil Shekerdjiiski Condenser Plant; a kitchen furnishings and electrical equipment complex for public catering establishments; a transformer factory; the S. Ludev Metal Working Plant; the Velbuzhd Worsted Yarn Factory; the Marek Woollen Knitwear Factory; the D. Blagoev Cannery; a tobacco sweating factory; a grain fodder plant; the D. Kalyiaski Factory for furniture, pacakaging and shaped wood; and the Ilyo Voivoda Factory.

Kyustendil is the centre of an agro-industrial complex, specializing in fruit growing (with an Institute for Mixed Plantation Fruit-Growing Nurseries, etc.), tobacco, and market-gardening (flower and vegetable-growing in hot houses).

The town lies on an intersection of the following

highways: Sofia-Kyustendil-Skopje, Kyustendil-Bosilegrad-Vranja (Yugoslavia), Kyustendil-Blagoevgrad, and Kyustendil-Stanke Dimitrov-Samokov-Plovdiv. Buses connect Kyustendil with all towns and villages in the district.

The town centre is a resort area, which includes the springs: the Chifte Banya (1912), Dervish Banya (1566), and Alay Bania (built on the site of an old Turkish building); a balneological complex (opened in 1966) with a polyclinic, the Miners' Rest House with a balneological clinic, and balneosanatoriums. There are open-air mineral water pools. The composition of the water here is recommended for the treatment of disorders of the bones, joints and peripheral nervous system, as well as gynaecological and light heart diseases. The Hissarluka hill outside the town has been made into a park covering 130 ha, with forests of black and white pine, spruce, fir, oak and lime (the first trees were planted in 1891). In the town there is a sports centre with a covered sport hall and sports, tourism and field sports societies. Kyustendil has several health-care establishments; child-care establishments; four technical high schools; high schools; general education schools; a Pioneer Children's Centre; five community centres; a drama theatre; an art gallery and a local history museum.

Its sights include: St. George's Church in the Kolusha quarter; the Pirkova Tower of the National Revival Period (sixteenth–seventeenth centuries); the Fetih Sultan Mehmed (1531) and Ahmed Bey (1575) Mosques; a wall left from the Devehani Caravanserai (1606); the Churches of the Virgin Mary (1816) and of St. Demetrius (1866); as well as houses of the National Revival Period, with impressive architectural layout and interior decoration. The remains of the ancient and medieval town of Pautalia Velbuzhd in the centre of Kyustendil have been declared an architectural and archaeological reserve.

A general view of Kyustendil

Milhailovgrad. Milhailovgrad is a district town in North-western Bulgaria, 114 km north of Sofia. Named after Hristo Mihailov, a party functionary and leader of the September 1923 Anti-fascist Uprising, it is the administrative, economic, and cultural centre of the Mihailovgrad district. Its population was 54,240 in 1982. Located in the broad valley of the river Ogosta in the western foothills of the Balkan Range at an average altitude of 160 m, it has a moderate continental climate.

Monument of the Eternal Flame

Remains of a Thracian settlement have been found on the Kaleto Hill. The Roman town of Montana was also located at the bottom of the hill. Near the town's big spring there was a shrine and a temple to Diana and Apollo. In the third century a fort was erected on the hill to station troops which was used until Late Antiquity. The ruins of Roman villas have been found on the outskirts of the town. Around three of them the remains of various buildings have been found, and early Christian basilica of the fourth century with an apse added to it and a hypocaust. After settling in Montana, the Slavs gave it the name of Kutlovitsa. Another village bearing the same name was later built nearby, and the old one came to be called Golyama Kutlovitsa. When Bulgaria fell under Ottoman rule, the inhabitants of Golyama Kutlovitsa were either driven away or massacred, and the village was destroyed. It was gradually re-populated. The villagers took part in the 1688 Chiprovtsi Uprising. On 3 November 1877 the Russian troops liberated the village from the Ottomans.

The post-Liberation period saw the further development of crafts and the rise of woodworking, ceramics, foodstuffs and rubber enterprises.

The first group of the Bulgarian Social Democratic Workers' Party in the town, now called Ferdinand, was set up in 1902. After World War I, the party organization was headed by H. Mihailov, Z. Popov, and A. Grekov. The town was the north-western centre of the September 1923 Anti-fascist Uprising and the headquarters of the Central Revolutionary Military Committee (CRMC). On 23 September a workers' and peasants' government was set up there. Many soldiers were taken prisoner and the insurgents seized one canon, several machine guns, rifles, and ammunition. The revolutionary army established its own hospital and a District Military Revolutionary Committee to direct the uprising in the Vratsa district. G. Dimitrov and V. Kolarov, members of the CRMC, arrived on 24 September to lead the uprising. Rebels from the town and the nearby villages were organized in a single revolutionary military formation, the Ferdinand Battalion, commanded first by G. I. Russinov and later by Ivan Mihailov. The battalion was ordered by the CRMC to take part in the Battle of Boichinovtsi. On 26 September it was disbanded. On 28 September fascist units entered the town. A Workers' Party organization was set up in 1928. During the armed anti-fascist struggle of 1941–44, the town was the centre of the Twelfth Insurgent Operative Zone.

Today the town is a big industrial centre. The main industries are mechanical engineering, metal-working, food and beverages, building materials, textiles, tailoring, and wood-working. Mechanical engineering is represented by: the Mir Building Machines Works; the Analitik Works for Analytical Devices and Apparatus; the Pretsiz Works for Measuring Instruments and Devices; the Druzhba Agricultural Machinery Plant; the Electroakoustika Works; the 23 Septemvri Fitting Works and the Avtomatika Works. The town also has a dairy; a meat packaging plant; a poultry processing and packing plant and a farm for eggs; a floor-tile factory; a works for bricks and roof-tiles; an enterprise for the extraction and processing of stone materials; a prefabricated construction elements works; the H. Mihailov Ladies' Wear Factory; the Montana Factory for Worsted Yarn; the I. Dimitrov Footwear Factory; the G. Damyanov Works and the Yavor Factory.

Mihailovgrad is the centre of two agro-industrial complexes, which grow grain crops and perennial plants (apples and vines). There is an agricultural construction enterprise, a pig farm, a District Livestock Selection Centre, a pedigree supply enterprise under the District Agro-Industrial Union, the State Agricultural Chemicals Research Station and the enterprises for reclamation and erosion control.

The town has road links to Sofia, Lom, Vidin, Berkovitsa, Chiprovtsi, Vratsa, Oryahovo and, through Boichinovtsi Station, to the Mezdra-Vidin line.

Mihailovgrad has three technical high schools; two high schools and general education schools; child- and healthcare establishments; a sports society; a field sports society; a stadium; a youth centre; three community centres; libraries; a drama theatre; an amateur operetta theatre; an art gallery; a Trade-Union Cultural Centre; and a

Museum of the September 1923 Uprising, part of it incorporating the house of H. Mihailov. The town holds the April Festival of Culture and awards a literary prize. There is a sculpture at the southern entrance of the town to commemorate the September 1923 Anti-facist Uprising. Another monument in the town centre contains the eternal flame to the victims of the Uprising. The train which transported the rebels from Boychinovtsi to Berkovitsa on 24 and 25 September is preserved. The house, where the main rebel group met on the night of 22 September in the Parta area 2 km north-west of the town, is also preserved. An obelisk on Kaleto Hill marks the spot where the signal for the uprising was given. The spots where the September rebels fell in Bugaza, Ivanovo Selo, Pchelina, Slaninshteto and the Zhivovtsi Halt areas have been declared historic spots.

The Ogosta Hotel in Mihailovgrad

Razgrad. Razgrad is a district town in North-east Bulgaria, 375 km north-west of Sofia. It is the administrative, economic and cultural centre of the Razgrad district, and had a population of 51,761 in 1982. The town stands on the banks of the River Beli Lom, which crosses the Ludogorie Plateau in the eastern part of the Danubian Plain, at an altitude of 150–200 m above sea-level and has a moderate-continental climate.

The Clock Tower (1764)

Remains of a paleolithic settlement, one of the largest in South-eastern Europe known so far, have been found in the region of Razgrad. Habitation mounds of the Chalcolithic and the Bronze Age, as well as ruins of a Thracian settlement which later developed into the ancient town of Abritus, have been discovered outside the town. Archaeological excavations brought to light Bulgaria's largest coin treasure of late antiquity, consisting of 835 Roman solidi weighing a total of 4 kg. On the ruins of Abritus there appeared a medieval Bulgarian settlement which existed from the seventh to the tenth centuries. During the time of the Second Bulgarian Kingdom it was known under the name of Hrusgrad or Hruzgrad. In 1388 the town fell under Ottoman rule.

In 1573 Razgrad became the administrative centre (*kaza*) of a large district. It developed into a centre of trade and handicrafts; a large section of its population were cartwrights, potters, cutlers, furriers, gardeners and merchants. The sixteenth and seventeenth centuries saw a trade colony of Dubrovnik (Ragusa) in the town and the town was mentioned as an exporter of hides and furs. Two janizary corps and other Turkish troops were stationed in the town. A monastery school was set up in 1667. In 1872 Angel Kunchev founded a revolutionary committee there. On 28 January 1878 the thirteenth army corps, under the command of Prince Dondukov-Korsakov, liberated the town. A socialist group was formed in 1890, which in 1892 became an organization of the Bulgarian Social Democratic Party. The people of Razgrad fought in the armed anti-fascist struggle of 1941–44.

Razgrad's leading industries are food, porcelain, pharmaceuticals and mechanical engineering. Main enterprises include an antibiotics research and production plant; the Himik Factory; a machine-tool plant; the Druzhba Mechanical Engineering Plant; the D. Stefanov Glass-ware and Porcelain Plant; a meat-packing firm; a grain and fodder enterprise; a poultry farm; the Progress Woodworking enterprise; the Mir Children's Toys Factory and the Orel Knitwear Factory.

Razgrad is the centre of an agro-industrial complex in a fertile agricultural region. Grain and fodder crops predominate. Stock-breeding is well-developed, with a cattle- and sheep-breeding station, a district veterinary medical centre, and a hybrid pig-breeding enterprise.

The town has two technical high schools; general education schools; kindergartens and crèches; health care establishments; an Institute of Contagious and Parasitic Diseases (successor to the Ospenia Institute founded in 1881 with the town's hospital, where the first Bulgarian laboratory for the preparation of anti-variola vaccine to be provided to the Principality of Bulgaria and Eastern Rumelia was set up); as well as sports, tourist and field sports societies; five community centres; a theatre; an art gallery; a district history museum and the Abritus Archaeological Museum. The Bulgarian Women's Poetical Horizons Days of Poetry dedicated to Stanka Nikolitsa Spasso-Elena (one of the first Bulgarian women of letters) are regularly held in Razgrad, as is the Dimiter Nenov Music Festival.

During the National Revival Period Razgrad was a centre of handicrafts and trade, with a flourishing building trade. The lead domes and tall minarets typical of Turkish religious and civil architecture dominated the appearance of the town. The houses are of the Dobrudja type, and mostly asymmetrical. In 1860 construction started on the church of St. Nicholas. Today this large, three-naved building has been declared a monument of culture. In 1764 a clock-tower was erected in the town's high street. Other sites declared monuments of culture are the Ahmed Bey Mosque (1442), the Ibrahim Pasha Mosque (1614–16), the building of the high school (constructed in 1885) and the Demir Baba *teke* (shrine) near town. Today the town's centre consists of blocks of flats and modern public buildings. The Vassil Levski and Orel Housing Estates are under construction. On the outskirts of the town is the Pchelina wooded park.

Lovech. Lovech is a district centre in Northern Bulgaria, 167 km north-east of Sofia. The administrative, economic and cultural centre of a district, its population in 1982 was 49,754. It is situated on both banks of the river Ossum, in the spur between the northern foothills of the Balkan Range and the Danubian plain, at an average altitude of 190 m above sea-level, with a moderate-continental climate.

Palaeolithic, Neolithic, Chalcolithic and Bronze Age remains have been found in the caves near the town (the Tabashka, and Levski Caves). Remains of a fortress wall, the foundations of churches, residential and commercial buildings are preserved on Hissarya Hill. Lovech stands on the foundations of the Thracian settlement of Melta. At a later period the Romans established a roadside station here under the name of Presidium-Melta on the road from the Danube to the Aegean via Philippopolis. The town has had its present name since the end of Byzantine rule (1018–1186), its old Bulgarian name being Lovuts. Lovech was an important stronghold protecting the approaches to Turnovo during the 1185–87 uprising of Peter and Assen against Byzantine rule. After a three-month siege, Byzantium was compelled to sign the Peace Treaty of Lovech (1187), which formally recognized the Second Bulgarian State. Lovech was a centre of brisk trade, maintaining commercial relations with Dubrovnik and other commercial towns during the fourteenth century. Lovech came under Ottoman domination in 1393. The nearby Monastery of the Holy Virgin was a major literary centre in the 16th century. Lovech was twice seized by Russian troops during the Russo-Turkish war of 1806–12. Lovech established itself as an economic and educational centre in the eighteenth and nineteenth centuries. Various crafts developed – ironmongery, watch-making and tanning. The town was called *Altun Lovech* (Golden Lovech). The population had a keen national awareness. The struggle against the Greek exarchate for an independent Bulgarian Church began in 1839. In 1869 V. Levski set up one of the country's first revolutionary committees here, and Lovech became a leading centre of the Internal Revolutionary Organization in 1870. During the Russo-Turkish Liberation War of 1877–78 a small Cossack detachment seized the town with a sweeping attack on 5 July 1877. The detachment retired on 14 July under pressure from a Turkish brigade headed by Rifat Pasha. On 22 August 1877 a Russian detachment broke the Turkish fortifications and entered the town.

Tanning was still the town's main craft after the Liberation: there were 21 small and medium enterprises in the town, producing sole-leather, upper-leather, suitcases, bags and gloves. A number of factories were set up – a tobacco factory; a basket-weaving factory; a factory for paints and varnishes; a factory for milk powder and canned food; and large mills. The town's connection with the Sofia-Varna railway line through the Levski railway station and the establishment of a connection with the Danubian port of Svishtov (1921) were of great importance for its advancement. A socialist group was founded in the town in 1893, and a party organization in 1902. The Hristo Kurpachev Partisan Detachment operated in the region of the town during the years of the armed anti-fascist struggle (1941–44).

Mechanical engineering, tanning, furniture-making and knitwear are the town's leading industries. Its industrial enterprises include the Balkan Mechanical Engineering Works, which specialize in the production of parts for transport machine-building; the Elprom Works for Manual Electrical Instruments and Electric Motors; a Tubular Furniture Factory; a tempered cast iron works; the Melta Cannery; the Serdica Milk Powder and Children' Food Factory; a meat-packing plant; the Velour Factory for Soft Leather Processing and Leather Clothing Production; the T. Katzarov Works for Furniture Units and Upholstered Furniture; a Pre-Fabricated Panel Works with a Housing Construction Department; and the S. Edrev Child and Baby Knitwear Factory.

Lovech is the centre of an agro-industrial complex specializing in grain production, livestock breeding and perennial plants; it houses a Station for Biological and Integrated Plant Protection in Bulgaria's mountain regions, which is an enterprise for soil improvement and erosion combat.

The town is on the road from Svishtov via the Troyan pass to Plovdiv. Lovech has regular bus services to the towns and villages in the district as well as to Sofia, Pleven, Vratsa, Gabrovo and Stara Zagora. It is connected to the Sofia-Varna railway line through the town of Levski.

View of Lovech, with the covered bridge built by master-builder Kolyo Ficheto

Lovech has four technical high schools; a high school; schools of general education; health and child-care establishments; sports, tourist and field sport societies; a sports centre; three community centres; libraries; a district history museum; a theatre; and a Vassil Levski Museum. Every year the town holds the Lilac Music Festival.

A street in the Varosha district of Lovech built during the National Revival Period

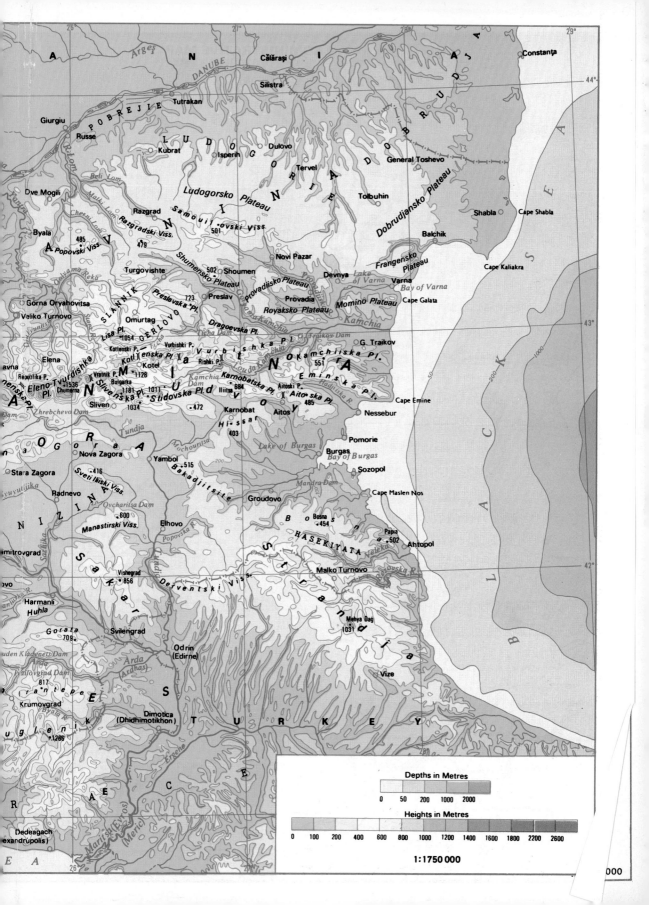

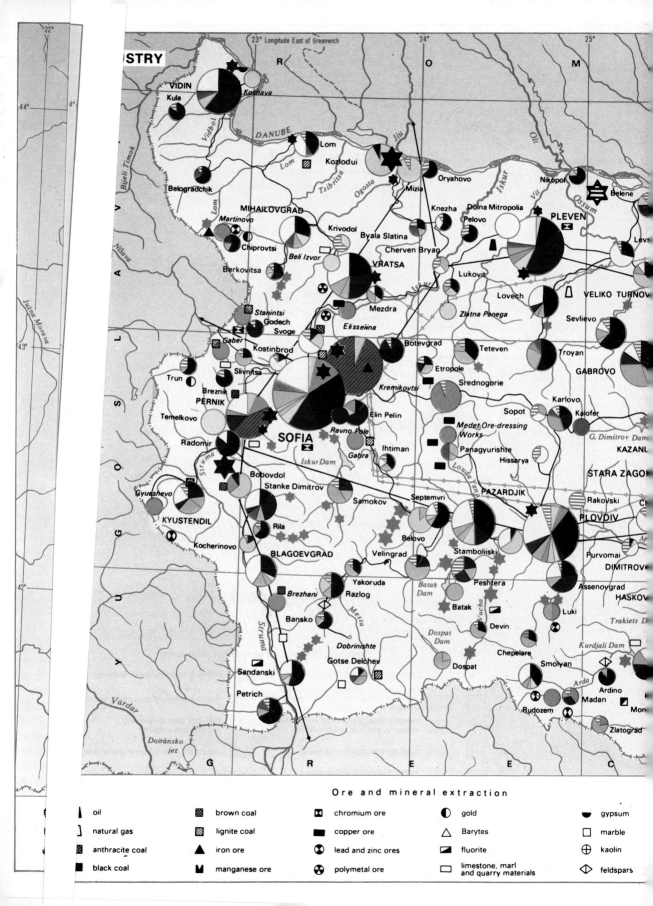

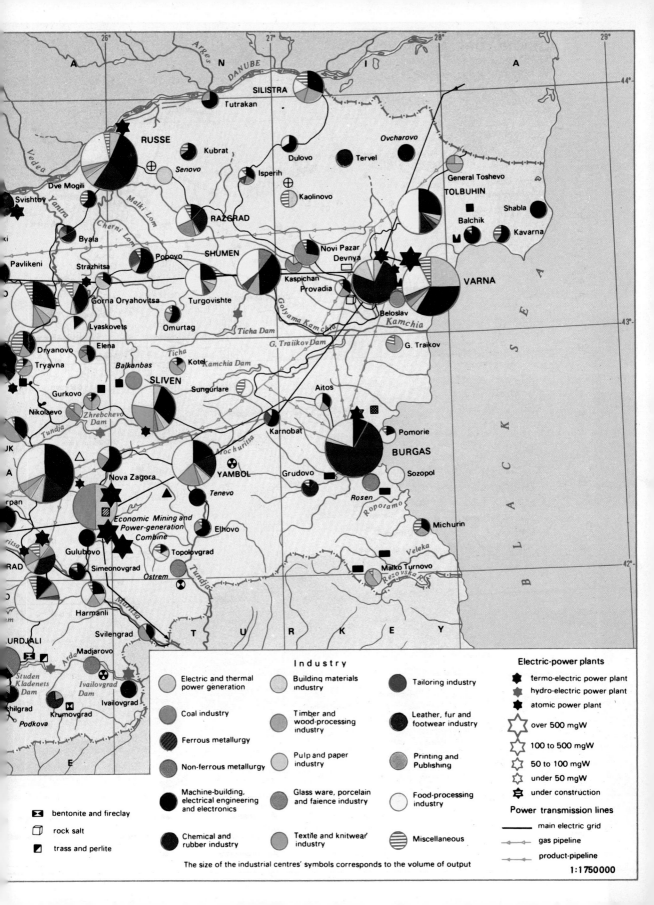

GEOLOGICAL MAP

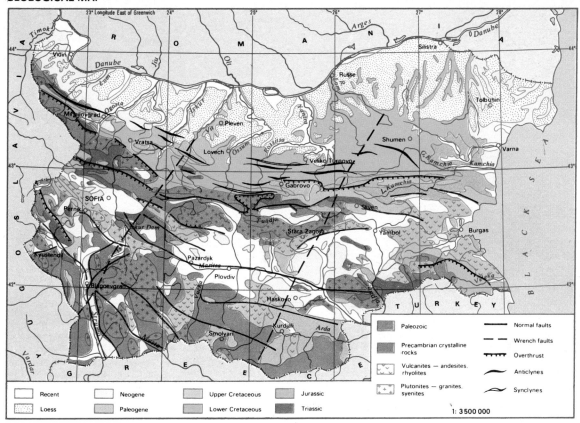

CLIMATE MAP

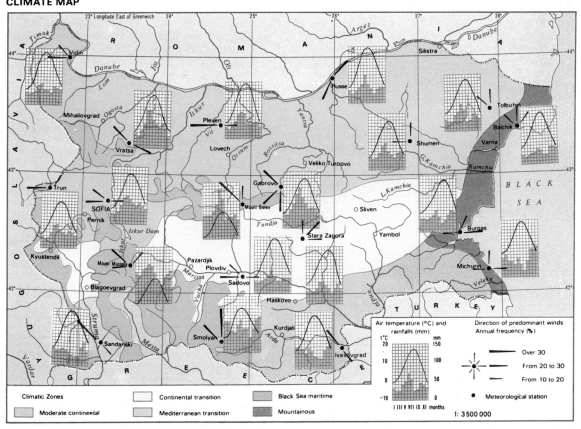

SOILS

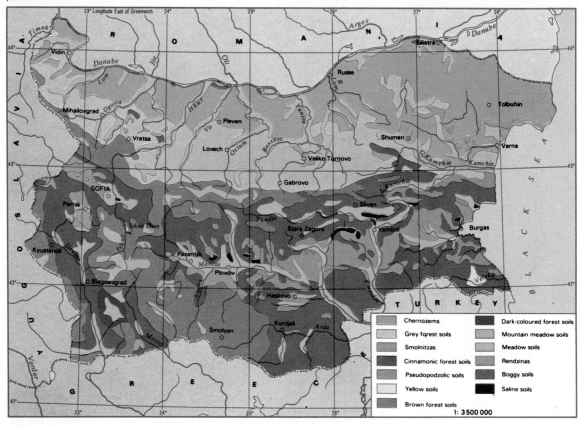

VEGETATION

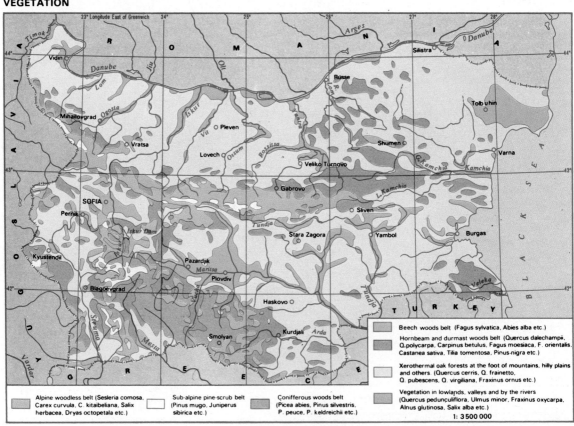

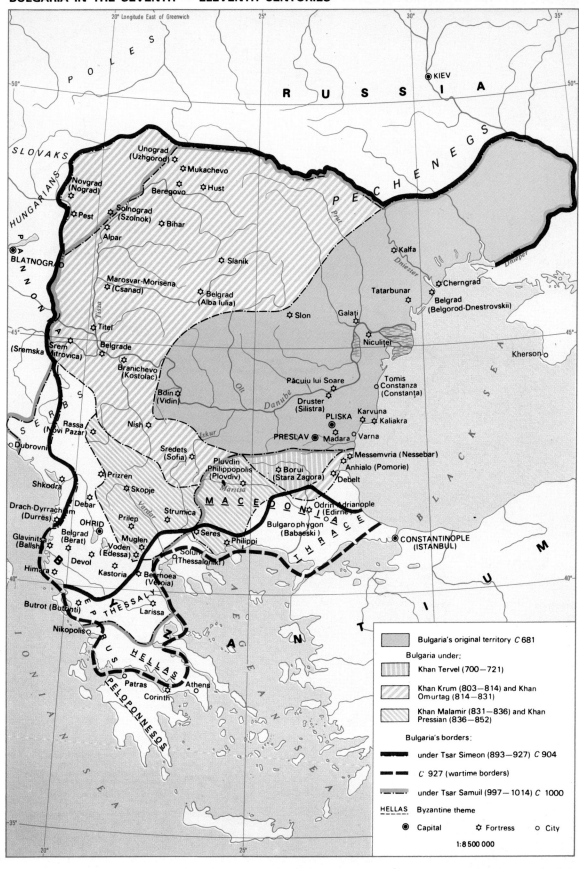

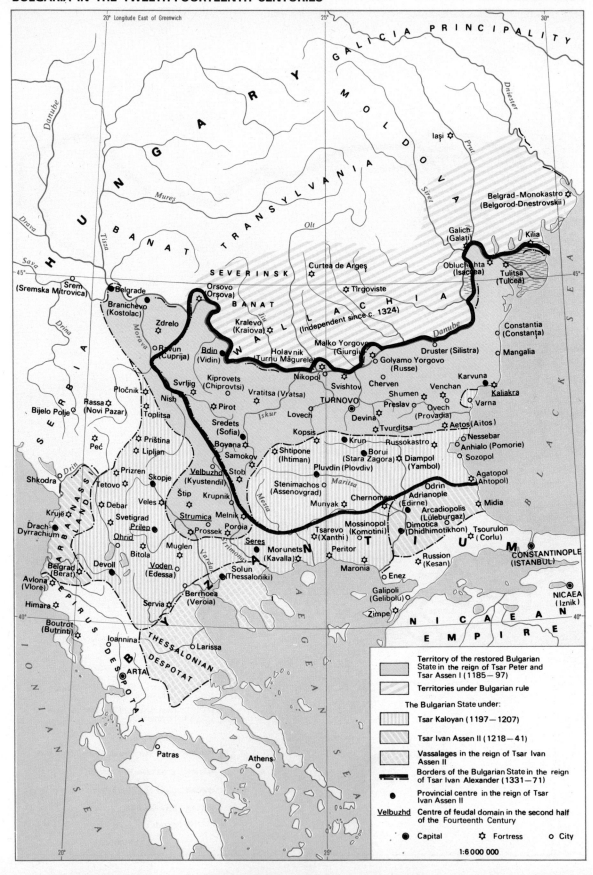

THE STRUGGLE OF THE BULGARIAN PEOPLE AGAINST OTTOMAN DOMINATION (1396–1878)

Legend:
- Uprising led by Konstantin and Fruzhin (1404)
- Tŭrnovo Uprisings (1598 and 1686)
- Chiprovtsi Uprising (1688)
- Uprising led by Karposh Voivode (1689)
- Haiduk detachment
- Braila riots (1841–1843)
- Proclamation of Bulgarian autocephalous church
- Secret Central Bulgarian Committee
- Route of F. Totyo's detachment (1867)
- Route of P. Hitov's detachment (1867)
- Route of H. Dimitŭr and S. Karadja's detachment (1868)
- Local revolutionary committee set up by V. Levski
- Centre of the internal revolutionary organization
- Centre of the Bulgarian revolutionary émigrées
- Bulgarian Revolutionary Central Committee
- Revolutionary movements in 1875
- Giurgiu Revolutionary Committee
- April Uprising (1876)
- Uprising's centre
- Region of mass participation
- Route of G. Benkovski's flying squad
- Route of H. Botev's detachment (1876)

1 : 3 500 000

BULGARIA AFTER THE LIBERATION (1878)

- – – – Bulgaria's borders according to the Treaty of San Stefano March 3 1878
- ——— Borders of the Principality of Bulgaria and Eastern Rumelia according to the Treaty of Berlin (1878)
- 6.IX.1885 Proclamation of the Union (September 6. 1885)
- ········ Demarcation line between the Ottoman Empire and the Balkan countries by virtue of the London Peace Treaties of 1913
- ——— Bulgaria's borders according to the Peace Treaties of Bucharest and Constantinople (1913)
- –·–·– State borders according to the Peace Treaty of Neuilly (1919)
- ░░░ Territories taken from Bulgaria after the First World War (1919)
- 1940 Southern Dobrudja, returned to Bulgaria under the Treaty of Kraiova (1940)
- –··–··– The state borders after the Second World War under the Paris Treaties (1947)
- 1 USSR

1 : 3 500 000

THE SEPTEMBER 1923 ANTI-FASCIST UPRISING

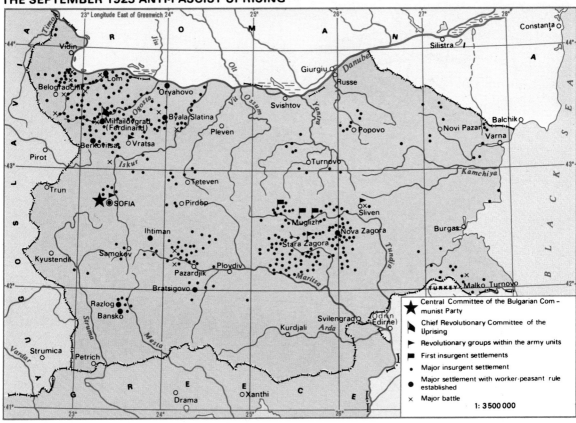

THE ARMED ANTI-FASCIST RESISTANCE (1941—44)

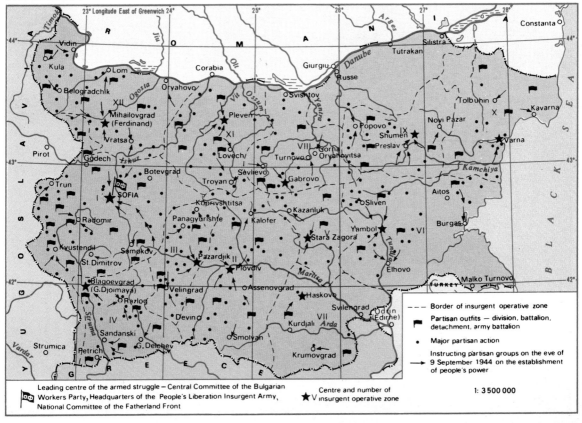

BULGARIA'S PARTICIPATION IN THE WAR AGAINST NAZI GERMANY (1944—45)

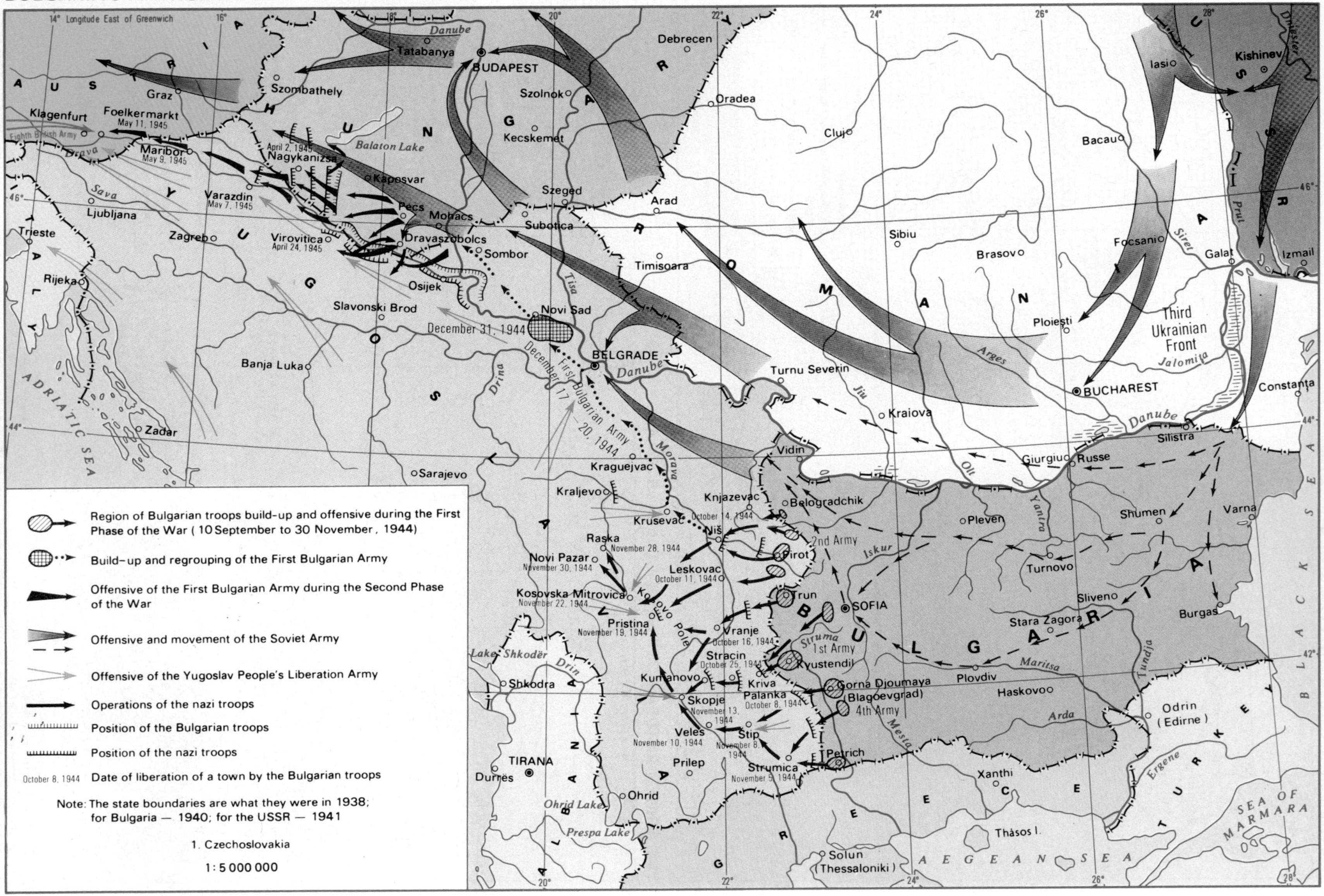

Region of Bulgarian troops build-up and offensive during the First Phase of the War (10 September to 30 November, 1944)
Build-up and regrouping of the First Bulgarian Army
Offensive of the First Bulgarian Army during the Second Phase of the War
Offensive and movement of the Soviet Army
Offensive of the Yugoslav People's Liberation Army
Operations of the nazi troops
Position of the Bulgarian troops
Position of the nazi troops
October 8, 1944 Date of liberation of a town by the Bulgarian troops
Note: The state boundaries are what they were in 1938; for Bulgaria — 1940; for the USSR — 1941
1. Czechoslovakia
1 : 5 000 000
14° Longitude East of Greenwich
BUDAPEST
BELGRADE
BUCHAREST
SOFIA
TIRANA
Third Ukrainian Front
First Bulgarian Army December 17 — 20, 1944
December 31, 1944
2nd Army
1st Army
4th Army
Fighth British Army
ADRIATIC SEA
AEGEAN SEA
SEA OF MARMARA
BLACK SEA
Foelkermarkt May 11, 1945
Maribor May 9, 1945
Varazdin May 7, 1945
Virovitica April 24, 1945
Nagykanizsa April 2, 1945
Knjazevac October 14, 1944
Leskovac October 11, 1944
Raska November 28, 1944
Novi Pazar November 30, 1944
Kosovska Mitrovica November 22, 1944
Pristina November 19, 1944
Vranje October 16, 1944
Stracin October 25, 1944
Palanka October 8, 1944
Skopje November 13, 1944
Veles November 10, 1944
Stip November 8, 1944
Strumica November 5, 1944
Gorna Djoumaya (Blagoevgrad)
Odrin (Edirne)
Solun (Thessaloniki)

Stratesh Hill on the outskirts of the town, is a park including a Zoo, an open-air theatre, and a tourist hostel.

The town's National Revival Period quarters (the Varosha and Drustene) are situated on either sides of Hissarya Hill on the right bank of the river Ossum. The narrow, steep cobbled streets are preserved. The houses are mainly of two storeys, with stone cellars and half-timbered oriels, stone-walled courtyards and big wooden porches. The houses, where Vassil Levski and other revolutionaries worked, are preserved. The Varosha quarter has been declared an architectural-historical reserve. By 1821 there were seven small churches in Lovech. The Churches of the Assumption (with magnificent iconistases, of the Tryavna school, by the master wood-carver Peter) and the Church of St. Nedelya (with mural paintings by the brothers Nencho and Nahum Iliev of Debur) were constructed in 1843. Nikola Fichev built the only covered bridge in the country, entirely of wood and with stone foundations in 1872–74.

The bridge was burnt down in 1925, was rebuilt of reinforced concrete in 1931 and reconstructed in 1981, on the model of the original. The Marx-Engels, Lenin and 9 September housing estates have been built. An experimental residential area is being built in the Cherven Briag region. An avenue of Bulgarian-Russian Comradeship-in-Arms is being built in the Stratesh Park. The White and the Black Monuments have been erected there in memory of the Russian soldiers who fell for the town's liberation during the Russo-Turkish War of Liberation. Lovech is the birthplace of the first Bulgarian cosmonaut Georgi Ivanov.

A general view of Lovech

Turgovishte. Turgovishte is a district town in Northeastern Bulgaria, 340 km north-east of Sofia, the administrative, economic and cultural centre of Turgovishte district with a population of 48,022 in 1982. It is situated on the banks of the Vrana river in the northern foothills of the Preslav mountains at an alititude of 150 m above sealevel. The climate is moderate continental. Turgovishte was declared a spa in 1963.

The Museum of Ethnography, in the House of Hadjiangelov, built during the National Revival period

There are ruins of a chalcolithic habitation mound within the area of Turgovishte precincts (known as the Polyanitsa culture – the name derives from the locality where the habitation mound was discovered) in addition to ruins of a Byzantine fortress dating back to the fifth–sixth centuries, and a Bulgarian settlement of the twelfth–fourteenth centuries. Turgovishte was founded during the sixteenth century. In the late eighteenth and early nineteenth centuries Turgovishte was a major economic centre with developed crafts and trade. The Eskidjumaya Fair was traditionally held there. A school was opened in 1846. After the Liberation in 1878 Turgovishte developed as a county agricultural, craftsmen's market and administrative centre. During the armed anti-fascist struggle of 1941–44 Turgovishte was in the Ninth Insurgent Operative Zone; the M. Petrov Partisan Detachment operated in the region.

Turgovishte's main industries are food and beverages, mechanical engineering and wood-processing, the major plants being: the Zora Refrigeration Machine-Building Plant; the N. Kamov Electrical Navigation Equipment Plant; a mineral water and soft drinks bottling factory; a wine-factory; the Energia Storage Plant; the Granit Ceramics Factory; the S. Delchev Wood-Processing Works; the Staria Dub Plant; the Fibreboard Plant; the Boks Tannery; the Septemvri Factory and the Rhodopa Works. Turgovishte is the centre of an agro-industrial complex, which specializes in viniculture, viticulture, market-gardening and fruit growing; it also has a cattle-breeding and sheep-breeding experimental station. Turgovishte is on the Sofia-Varna railway line.

There are technical high-schools in Turgovishte in addition to vocational high schools of general education; child- and health-care establishments; sports, tourist, and field sports societies; three community centres; libraries; a theatre; a puppet theatre; a district history and ethnography museum; a stadium; and the Mizia Folk Song and Dance Ensemble. Turgovishte is the venue of the May Days of Culture.

The city centre has been architecturally developed and four urban zones are being laid out. Notable old buildings include a church built in 1851 by masters of the Tryavna School and houses of the National Revival Period in the Varosha Quarter which have been declared monuments of culture. The Loven Park has various recreation amenities and facilities.

Smolyan. Smolyan is a district centre in Southern Bulgaria situated 258 km south-east of Sofia and the administrative, economic and cultural centre of the Smolyan District. Its population in 1982 was 39,688. It is situated on the river Cherna, a tributary of the river Arda in the Western Rhodopes, at an altitude of 900 m above sea-level. It has a mountain climate with Mediterranean climatic influence. In 1963 the town and its surroundings were declared a mountain health resort.

The new town centre

The locality was populated in ancient times by the Thracian Koilaleti tribe. In 11 BC, the area was conquered by the Romans. In the seventh century the area was settled by the Slavonic tribe of the Bulgarian group – the Smolyans. It was included within the Bulgarian State of Khan Pressian (836–852). At the beginning of the eleventh century it was re-conquered by the Byzantines. During the reign of Tsar Kaloyan (1197–1207) it was a possession of Despot Slav, in 1246 John III Ducas Vatatses conquered it (together with a great number of Bulgarian settlements in the Rhodope area). In 1254–55 the population of the Rhodopes rebelled and, with the help of Tsar Mihail II Assen's (1246–56) troops, it again joined the Bulgarian State; in 1343 it was part of the Ruler Momchil's possessions. Smolyan and its region fell, after stubborn resistance, to the Ottomans after 1372. In the mid-seventeenth and early eighteenth centuries the mass conversion of the population to Islam was carried out. In 1515, Smolyan and the upper reaches of the river Arda were presented by Sultan Selim I to the royal doctor Aha Chelebi, the whole area being given the same name. The population had a stong national consciousness. In the seventeenth and eighteenth centuries, *haiduk* detachments operated throughout the region. In the late eighteenth and throughout the nineteenth century, trade and crafts developed – homespun tailoring, brasswork, etc. During the nineteenth century Bulgarian schools and community centres were set up. The struggle for church independence began in 1860; by 1870 the church service was already performed in Bulgarian. In 1876, Smolyan joined the Bulgarian Exarchate. During the Russo-Turkish Liberation War of 1877–78, it was liberated by the Combined Caucasian Cossack Brigade led by P. A. Cherevin. Under the Berlin Treaty of 1878 the town was returned to the Sultan's power. It was liberated on 12 October 1912 by the 21st Sredna Gora Infantry Regiment commanded by Col. V. Serafimov. In 1919, Bulgarian Communist Party groups were founded. The people of Smolyan participated in the anti-fascist struggle of 1923–44 and a partisan detachment was formed.

Leading branches of the district centre's industry are mechanical engineering, mining, food-processing and wood-working. It has an insulated conductors plant; the Komuna Contact Elements Plant; an instrument plant; a vehicle repair plant; a food-processing plant; the Rhodope Fabrics Factory; the Sluntse enterprise; a meat packaging-house and others. It is the centre of an agro-industrial complex producing potatoes, fodder, tobacco, perennial plants and with developed stock-breeding. It is also the district centre for engineering and development.

The town is at the junction of two important highways: the Plovdiv-Smolyan-Rudozem, and Krichim-Devin-Smolyan-Kurdzhali roads. Transport is by bus.

Smolyan has one semi-higher institute; three technical high schools; general educational schools; kindergartens and crèches; medical establishments; sports, tourism and field sports societies; three community centres; libraries; a drama theatre; a District History Museum; an art gallery; an observatory with a planetarium; a District Youth Centre and a District Pioneer Children's Centre.

In the town's three quarters there are a great number of Rhodope houses from the National Revival Period (eighteenth century–1878) standing either in groups or separately. St. George's Church (1858; with murals from 1888, featuring carved wood altar gates and a spire built in 1904), the Church of the Holy Spirit (1891), and Brahom Bey's Konak (town-hall) are all to be seen in Smolyan. The greatest number of architectural monuments are in the Raikovo quarter. The Cheshit Mahala in the Dolno Raikovo quarter is a typical Rhodope group of houses. In the Gorno Raikovo residential area are the Home of Pangalov – one of the pinnacles of Rhodope architecture – the Shakir Konak, the Ali Bey Konak and others. There are other valuable architectural monuments in the Ustovo quarter. The triple-naved basilicas of St. Nicholas Thaumaturge (1836, rebuilt in 1883), built by the master Stoyu Dimov, featuring an icon of Sts. Cyril and Methodius by Stanislav Dospevski in 1865, and an icon by Dimiter Zograph (1838); the Dormition of the Virgin built in 1865 by the master V. Dimov, with murals (1866); and the triple-naved basilicas of St. Nedelya

(1836, by the masters H. and V. Kissyov, painted 1882) and St. Theodore Stratilates (built 1836, with murals 1871–72) are in Raikovo. The groups of buildings are matched by the picturesque Rhodope-stone bridges, which are either arched or vaulted.

New public and residential buildings have basically changed and modernized the town's appearance. The general development plan of 1970 is aimed at merging the town's three districts. Several groups of buildings are emerging – public buildings, cultural buildings, commercial buildings and hotels.

Smolyan is developing as a tourist centre. The Gorge of the river Cherna, the Smolyan Lakes, the panoramic road to the Pamporovo Resort Complex make it the central feature of the whole series of places of interest in these parts.

A general view of Smolyan

SELECTED BIBLIOGRAPHY

Atlas Narodna republika Bulgaria (Atlas of the People's Republic of Bulgaria). Sofia, 1973.
BESHEVLIEV, V. *Die protobulgarischen Inschriften.* Berlin, 1963.
BESHKOV, Vl. et P. BERON *Catalogue et bibliographie des Amphibiens et Reptiles en Bulgarie.* Sofia, 1964.
BONCHEV, E. *Problemi na bulgarskata geotektonika* (Problems of Bulgarian Geotectonics). Sofia, 1971.
Demografiya na Bulgaria (Demography of Bulgaria). Sofia, 1974.
DIMITROV, D. *Klimatichnite resursi na Bulgaria* (The Climatic Resources of Bulgaria). Sofia, 1974.
DRENSKI, P. *Ribite v Bulgaria* (Fish in Bulgaria). Sofia, 1951.
Entsiklopediya 'Bulgaria' (Encyclopaedia 'Bulgaria'). Vol. 1, Sofia, 1978.
Fauna na Bulgaria. Bezgrubnachni (The Fauna of Bulgaria. Invertebrates). Sofia, 1963.
Fizicheska geografiya na Bulgaria (Physical Geography of Bulgaria). Ed. Z. Gulubov. Sofia, 1977.
Flora na Narodna republika Bulgaria (The Flora of the People's Republic of Bulgaria). Vol. 1, Sofia, 1963.
Geografiya na Bulgaria v 3 toma. Tom 1. Fizicheska geografiya (Geography of Bulgaria in 3 Vols, Vol. 1. Physical Geography). Sofia, 1982.
Geologiya i neftogazonosnost severnoi Bolgarii (The Geology and the Oil and Natural Gas Deposits in Northern Bulgaria). Moscow, 1976.
Geologiya i neftogazonosnost na severoiztochna Bulgaria (The Geology and the Oil and Natural Gas Deposits in North-eastern Bulgaria). Sofia, 1981.
Gramatika na suvremenniya bulgarski knizhoven ezik v 3 toma (A Grammar of Modern Standard Bulgarian in 3 Vols) Sofia, 1982–83.
IVANOVA, V. *Opazvane n okolnata sreda* (Preservation of the Environment). Sofia, 1981.
Izsledvaniya iz istoriyata na bulgarskiya knizhoven ezik ot minaliya vek (Studies in the History of Standard Bulgarian in the 19th Century). Sofia, 1979.
KEREMIDCHIEV, G. *Borba za knizhoven ezik in pravopis* (The Struggle for a Standard Language and Spelling). 2nd edn. Sofia, 1943.
Klimatichen atlas na Narodna Republika Bulgaria (A Climatic Atlas of the People's Republic of Bulgaria). Sofia, 1956.
Klimatichen Spravochnik za NR Bulgaria (A Climatic Reference Book of Bulgaria). Vols 1–4. Sofia, 1978–82.
KOINOV, V., T. GYUROV and B. KOLCHEVA. *Pochvoznanie* (Soil Science). Sofia, 1980.
Magmatizum i Metallogeniya Karpato-Balkanskoi oblasti (Magmatism and the Metal Deposits of the Carpatho-Balkan Area). Sofia, 1983.
MARKOV, G. *Bozainitsite v Bulgaria* (Mammals in Bulgaria). Sofia, 1959.
Mineralite v Bulgaria (Minerals in Bulgaria). Sofia, 1964.
MINKOV, M. *Naselenie i osnovni sotsialni strukturi* (Population and Basic Social Structures). Sofia, 1976.
MIRCHEV, K. *Istoricheska gramatika na bulgarskiya ezik* (Historical Grammar of the Bulgarian Language). 3rd edn. Sofia, 1978.
MLADENOV, St. *Geschichte der bulgarischen Sprache.* Berlin, 1929.
MLADENOV, S. *Suvremenniyat bulgarski knizhoven ezik i narodnite govori* (Modern Standard Bulgarian and the Bulgarian Dialects). Sofia, 1943.
Nashi rezervati i prirodni zabelshitelnosti (Reservation and Natural Sights in Bulgaria). Vols 1–3. Sofia, 1968–74.
Neftogazonosnost na Predbalkana [Mezhdu rekite Ogosta i Yantra] (Oil and Natural Gas Deposits in the Foothills of the Balkan Range [Between the Rivers Ogosta and Yantra]). Sofia, 1980.
Opazvane i vuzproizvodstvo na prirodnata sreda (Protection and Reproduction of the Environment). Sofia, 1981.
Opazvane i vuzproizvodstvo na prirodnata sreda. Sbornik ot dokumenti na Durzhavniya suvet na NRB (Protection and Reproduction of the Environment. Collection of Documents Issued by the State Council of the People's Republic of Bulgaria). Sofia, 1982.
PATEV, P. *Ptitsite v Bulgaria* (Birds in Bulgaria). Sofia, 1950.
PENCHEV, P. G. *Obshta hidrologiya* (General Hydrology). 4th edn. Sofia, 1975.
PENKOV, M. *Pochvite v Bulgaria – opazvane i podobryavane* (Soils in Bulgaria – Their Preservation and Improvement). Sofia, 1983.
PESHEV, T. and N. BOEV *Fauna na Bulgaria. Grubnachni* (The Fauna of Bulgaria. Vertebrates). Sofia, 1962.
POPOV, V. *Chudni kutove iz nashata rodina* (Wonderful Spots around Our Homeland). Sofia, 1970.
POPOV, K. *Po nyakoi osnovni vuprosi n bulgarskiya knizhoven ezik* (On Some Basic Problems of Standard Bulgarian). Sofia, 1973.
Pochveno-geografsko rayonirane na Bulgaria (Soil and Geographical Zonation of Bulgaria). Sofia, 1974.
Ribite v Cherno more. Bulgarsko kraibrezhie (Black Sea Fishes of the Bulgarian Coast). Varna, 1963.
RUSINOV, R. *Istoriya na novobulgarskiya knizhoven ezik* (History of Modern Standard Bulgarian). V. Turnovo, 1976.
SELISHCHEV, A. M. *Staroslavyanskii yazik* (The Old Slavonic Language). Parts 1–2. Moscow, 1951–52.
SPIRIDONOV, Z. *Oazisi na divata priroda* (Oases of Wild Nature). Sofia, 1977.
STAINOV, P. *Zashtita na prirodata. Pravni izsledvaniya* (The Protection of Nature. Legal Studies). Sofia, 1970.
STEFANOV, B. and B. KITANOV. *Kultigenni rasteniya i kultigenna rastitelnost v Bulgaria* (Cultigen Plants in Bulgaria). Sofia, 1962.
STOYANOV, N., B. STEFANOV and B. KITANOV *Flora na Bulgaria* (The Flora of Bulgaria). 4th rev. and enlarged edn. Parts 1–2. Sofia, 1966–67.
TOSHKOV, M. and N. M. VIHODTSEVSKI *Zashtiteni prirodni obekti* (Protected Natural Sights). Sofia, 1971.
TSONEV, B. *Istoriya na bulgarskii ezik* (History of the Bulgarian Language) Vols 1–3. Sofia, 1934–37.
YOVCHEV, Y. *Osnovi geologii i polezni iskopaemie teritorii NR Bulgaria* (Foundations of Geology and Ores and Minerals in the Territory of the People's Republic of Bulgaria). Sofia, 1965.
YOVCHEV, Y. *Polezni iskopaemi* (Ores and Minerals). Vol. II. 1. Sofia, 1956.
Zapazvane i povishavane na pochvenoto plodorodie (The Preservation and Improvement of Soil Fertility). Sofia, 1974.

PART II

HISTORY

THE BULGARIAN LANDS IN ANTIQUITY

The Palaeolithic and Neolithic Ages. The Bulgarian lands of today were amongst the first cradles of human life on the European continent. The earliest traces of the presence of man go back to the Lower Palaeolithic or Old Stone Age: and comprise mainly flint implements (flake and blade tools) discovered in the region of the Rhodope mountains and in Northern Bulgaria. However, traces of old cultures in the territory of present-day Bulgaria dating back to the Middle Palaeolithic Age (from about 100,000–40,000 BC) are greater in both number and variety. These were discovered mainly in caves (the Devetashka Cave in the Lovech district, Samuilitsa-II near the village of Kunino in the Vratsa district, the Bacho Kiro Cave near the Dryanovo Monastery, and many others) and in old settlements in Northern Bulgaria (near the village of Beloslav in the Varna District and the village of Musselievo in the Pleven district, in the Kremenete region, not far from the town of Batak, and elsewhere). A great number of caves from the Upper Palaeolithic Age (from about 40,000–10,000 BC) have been found and explored, and many settlements uncovered on the northern slopes of the Stara Planina (the Balkan Range), in the valleys of the rivers Iskur, Vit, Ossum, Yantra, and elsewhere. Relics from this period provide evidence of the spiritual advancement of man. This was the time when a variety of primitive beliefs and rituals appeared – animism, fetichism, totemism, magic and others. The population increased considerably.

The culture of the Upper Old Stone Age in the Bulgarian Lands differs in some respects from its development in Central and Western Europe, mainly in the shape of tools and the techniques employed in their making. It has more features in common with the culture of the Mediterranean community and its kindred cultures in South-east Europe. The Middle Stone Age (Mesolithic) in the Bulgarian lands goes back to 10,000–7,000 BC. The development of productive forces was given a new impetus; implements were now made of stone, wood and bone. Mesolithic finds have been mainly uncovered in the surface layers of the quicksands in the Pobiti Kamuni (Stone Forest) locality, near the villages of Beloslav, Strashimirovo, Slunchevo, Banovo and the town of Devnya (Povelyanovo quarter) in the Varna district, and some other settlements in North-east Bulgaria.

Cult column from the Polyanitsa locality, near Turgovishte. Neolithic Age. National Archaeology Museum, Sofia.

The New Stone Age (Neolithic), 6,000–5,000 BC and Copper-Stone Age (Chalcolithic), 4,000–3,000 BC, witnessed a marked rise in the culture of the farming and stock-breeding communities on the territory of modern Bulgaria. The most significant remnants of the settlement patterns of that period are the *tells* (habitation mounds), which are formations of different layers of culture resulting from successive settelements on one and the same spot. There are several hundred such mounds, most of them in South and North-east Bulgaria. During the Chalcolithic Age material and spiritual culture reached a high level. Settlements were built following certain urbanization plans and some of these were fortified. The means of production developed still further; the making of stone and bone tools was improved and copper was introduced (excavations in the Stara Zagora district bear evidence of copper processing). The rich finds in the necropolis in the town of Varna (gold ornaments and other objects) and the abundance of objects evidencing the development of religious beliefs combine to present a picture of one of the most advanced civilizations in Europe at that time, contemporaneous with the culture of Ancient Egypt and Mesopotamia. The ethnic origin of the population inhabiting the Bulgarian lands, however, has not been established with certainty.

The Thracians. The transitional period to the Bronze Age (3,000–2,000 BC) witnessed fundamental economic, ethnic and cultural changes on the territories of modern Bulgaria. New ethnic groups appeared. Bronze began to be produced and was used as the basic material in the making of tools and weapons. Trade and cultural contacts were formed with tribes and peoples from Asia Minor, the middle reaches of the Danube, the northern Black Sea coast and mainland Greece. In that epoch the territories of present-day Bulgaria played an important role as the link between the cultures of Asia Minor and Central Europe. The bearers of the Bronze Age culture in these territories were the Thracians, earliest written records of whom are found in Homer's *Iliad*, which gives a description of the participation of Thracian chieftains and their legions in the Trojan War (twelfth century BC). The leaders of their tribes are described in Homer's epic as possessing the same material culture as the Hellenic *basilei* (kings) and using the same title. The ethnogenesis of the Thracians is a complex and prolonged process, the result of the merging of migrant elements particularly at the end of the Copper-Stone Age (Chalcolithic) with autochthonous Indo-European strata in ancient Thrace. The eleventh century BC is considered the beginning of the Iron Age in ancient Thrace, the time when the formation of the Thracian ethnic community was completed.

After the great ethnic migrations and changes at the end of the second and the beginning of the first millennium BC, which affected the Balkan Peninsula, Asia Minor and the Aegean world, the Thracians settled in the central, eastern and south-eastern regions of the peninsula. There were about 90 Thracian tribes. The more numerous amongst them were: the Odrysae (originally inhabiting the lands

Gold ring from a burial mound unearthed at Ezerevo, near Plovdiv, with a Thracian inscription. National History Museum, Sofia

along the lower reaches of the River Hebros, nowadays called the Maritsa, and the valley of the Ergin, today's Ergene River); the Bessi (mainly inhabiting the western, mountainous parts of ancient Thrace, and later on, together with some of the sub-branches, moving over to the region of Bessica, which occupied a large part of the present-day districts of Pazardjik, Plovdiv and Stara Zagora, with the settlement of Bessapara as major centre); the Serdi (inhabiting the lands of today's Sofia district, with Serdica – present-day Sofia – as the major tribal centre) and the Maidae (inhabiting the lands along the middle reaches of the Strimon, nowadays the Struma river, and its valley between the Kresna and Rupel gorges (Spartacus, the organizer and leader of the greatest rebellion of slaves in the ancient world, came from the ruling tribal strata of the Maidae); the Getae (inhabiting the lands between the lower reaches of the Danube and the Stara Planina, and the Asamus river, the present-day Ossum river, to the west) were among the first to form permanent settlements, becoming farmers (mostly stock-breeders and horse-breeders in particular); the Dentheletai (inhabiting the lands along the upper reaches of the Strimon river, from the Ossogovo and Ruen mountains up to the Skombros mountain, today Vitosha, and to what is known as Znepole); the Astea (inhabiting the regions between the settlements of Apollonia and Salmydessos; the tribal centre and, later on, the major town of the Thracians in the first century, was Blizye – the present-day town of Vize) and many others. According to the

Bronze chandelier with an image of a dancing satyr, from Mezek, near Haskovo. Fifth-fourth centuries BC. National Archaeology Museum, Sofia

ancient Greek historian Herodotus (sixth century BC) the Thracians were the second largest people on earth, after the Hindus.

In the seventh–sixth centuries BC the first Greek town-colonies founded by European and Asia Minor Greeks appeared on the Black Sea coast of Thrace. The most important colonies were: Apollonia (present-day Sozopol), Messambria (Nessebur), Odessos (Varna), Dionysopolis (Balchik), Bizone (Kavarna), Tomis (the present-day Romanian town of Constanta) and others. The Greek colonies were usually founded on the sites of old Thracian settlements. At first they were temporary trade centres for the exchange of goods with the Thracian population, but later they turned into city-states, which became busy centres of flourishing trade and crafts. They grew into major cultural centres and were the first places in Thrace where the slave system.took root.

Towards the end of the sixth century BC socio-economic development in Thrace reached such a level that conditions were created for the appearance of a state system. Small, separate Thracian states appeared as early as the second half and the end of the sixth century. In the seventies of the fifth century BC the most powerful Thracian tribe, the Odrysae, founded the vast, long-lasting Odrysae Kingdom. Later, some of the regions to the south and south-east were also incorporated, and to the north, in all probability, also the land along the upper and middle reaches of the Tonzos river (Tundja), and the territory to the north of Hemus (Stara Planina), as a result of which the Odrysae Kingdom stretched as far as the reaches of the Histria river (Danube). The Kingdom gradually gained new territories also in the west and north-west: to the west it bordered on the Strimon river; to the north-west, on the upper and part of the middle reaches of the Oskios river (Iskur); to the south-west, on the lands of the town of Abdera, not far from the estuary of the Nestos river (Mesta); and to the south-east, on the environs of the town of Byzantium (Istanbul). The Odrysae Kingdom incorporated the most populous and socially, economically, culturally and politically most advanced Thracian tribes, between which, however, there were differences in levels of development. Certain features in the tribal relations still prevailing among the Odrysae were an obstacle to the establishment of a centralized government in the Odrysae Kingdom. The intrigues and revolts organized by individual representatives of the tribal aristocracy played a negative role and undermined the unity and stablility of the state.

The first known ruler of the Odrysae Kingdom was Teres (from between 480 and 460 to 440 BC), who distinguished himself by his statesmanlike wisdom and foresight. He tried to impose his rule over all Thracian tribes on the territory between the lower reaches of the Danube and the Aegean Sea, and between the Black Sea and the Strimon river, and to expand the lands of the Odrysae Kingdom northwards. In order to strengthen his friendly relations with the rulers of the Scythian Kingdom, Teres married a Scythian princess. Teres's son, Sitalkes (440–424 BC), maintained good relations with the

Athens of Pericles and made efforts to conclude a military alliance, directed against ancient Macedonia. He married a high-born Greek lady from Abdera, whose relatives had a great deal of influence in the ruling circles in Athens. All this led to the conclusion of a military treaty between the Odrysae Kingdom and Athens, under which Sitalkes was to attack Macedonia with a large army and thus assist Athens in the implementation of her plans concerning the Chalcidice Peninsula and Macedonia. Judging by the scale of the preparations Sitalkes made for this campaign, it may be concluded that his goal was to expand the Odrysae Kingdom to the west and gain the territories between the Strimon and the Axios river (Vardar), where the population was predominantly Thracian.

Clay seal from a habitation mound at Karanovo, near Sliven. Late Chalcolithic Age. National Archaeology Museum, Sofia

The armed conflict between Macedonia and Athens broke out on the eve of the Peleponnesian War (431–404 BC). Under the terms of the treaty the Odrysae ruler was obliged to attack Macedonia, penetrate deep into its territory and meet the Athenian fleet on the shores of the Gulf of Strimon. In all probability the Athenians were to make a landing there and wait for Sitalkes, but they violated the treaty and did not send him the promised support, for, according to Thucydides, they were frightened by the destructive power of the Thracian army. Sitalkes started negotiations with the Macedonian King and stated some of the reasons for his campaign. The military operations and political moves of the Odrysae ruler did not bring him the desired results, as the campaign was cut short.

Sitalkes's successor, Seuthes I, did not possess the foresight and military talents of his predecessors; he therefore fell under the sway of Macedonia, became allied through marriage with the Macedonian dynasty and simply carried out its policies, which affected the interests of the Odrysae Kingdom in the west. The Kingdom lost its political influence among the Scythians and the Hellenic cities of Athens and Sparta. Towards the end of the Peloponnesian War the Odrysae Kingdom split into two parts, the Upper and Lower Kingdoms. In the fourth century BC Kotys I (c. 383–360 BC) succeeded in restoring the unity of the Kingdom. During that period the Thracian state dealt a serious blow to the Athenian rulers and gained influence along the Aegean seacoast. During the rule of Kotys's son, Kersebleptis (359–341 BC), the Odrysae Kingdom broke up into three. The Macedonian King, Philip II, taking advantage of the weaknesses and division of the Odrysae Kingdom, organized several forays into Thrace and conquered part of its territories. After the death of Philip II the Thracians rose in revolt and defeated the Macedonian garrisons. At the beginning of his reign Philip's son, Alexander of Macedon, launched a punitive operation against the North Thracian tribes in what is present-day Northern Bulgaria. In 331–330 BC the Thracians staged a rebellion against Macedonian rule, which ended, however, in failure. During this period the Odrysae ruler, Seuthes III, waged several wars against the Macedonians, managing to preserve the independence of his possessions, which approximately covered the territory of the present-day Stara Zagora district and part of the districts of Sliven, Haskovo and Yambol, and founded the town of Seuthopolis (near the present-day Kazanluk) as the capital of his kingdom.

The development of farming, stock-breeding, mining (in particular of gold), of various crafts and of trade with the Hellenic world, created favourable conditions for an economic upsurge in the fifth–fourth centuries BC, which was a period of great political and military advancement for the Odrysae Kingdom. Remnants have been uncovered from that time of well-built, fortified cities, such as Philippopolis, and the rectangular and domed tombs of high-born Thracians have been revealed. Thracian toreutics reached a very high level, as can be seen from the well-preserved silver and gold treasures, consisting of horse tackle, appliqué work, silver and gold vessels, richly ornamented with patterns of the plant and animal world and motifs associated with the Thracian religion. An example of the most significant monuments of Thracian art is provided by the mural-paintings of the Kazanluk Tomb, dating back to the fourth and the beginning of the third centuries BC. This is on the Unesco list of the world's natural and cultural heritage, which is an indication of the high level of the pictorial arts in the Odrysae Kingdom.

Thracian art is the result of independent development and has its specific characteristic features, as can be seen in the Thracian way of life, Thracian mythology, the genres which the Thracian rulers required and the taste of the Thracian craftsmen. Thracian culture was a bridge between East and West. The spiritual life of the Thracians included ancient beliefs and cults which can be classified into the following three groups: worship of the Sun (the solar cult), belief in the immortality and transmigration of the soul (chthonic cults) and the cult of regenerating nature (the orgiastic cults). The Thracian language belongs to the Indo-European family; from it have been preserved about 250 words, 500 geographic names of rivers, mountains and settlements and 1,000 personal names.

Macedonian penetration into Thrace (which continued until the year 281 BC) retarded economic development in the area. At the same time, the Thracian lands were economically and politically integrated into the Hellenistic world, which set in motion certain syncretic processes between the Thracian and Greek cultures. The third and second centuries were a period of decline; a great part of the Thracian population emigrated, and fertile areas of Thrace fell under the domination of Macedonia, Pergamum, Egypt and Syria. In 279 BC the Celtic tribe Gaelatae settled in the lands to the south of Stara Planina and founded the Celtic Kingdom, which was wiped out about 212–211 BC as a result of a mass uprising of the Thracian population. The Odrysae Kingdom was restored, though it was unable to attain its earlier political and military power.

About the middle of the second century BC Rome conquered Macedonia and the Hellenic cities. The Thracian Kingdom and some tribal unions inhabiting the lands of what is present-day Northern Bulgaria did not fall under Roman domination and for two centuries waged a struggle for the preservation of their independence. In 45 BC all the Thracian tribes to the south of the Danube were conquered by the Romans, and Thrace was turned into a Roman province. About 15 AD present-day Northern Bulgaria and the Dobrudja region were incorporated into the Province of Moesia, which in 86 AD was divided into Upper and Lower Moesia.

The establishment of Roman rule in the lands to the north and south of Stara Planina (Lower Moesia and Thrace) created favourable conditions for the introduction and consolidation of the slave system. Many cities were built, becoming centres of the ancient material and spiritual culture: Serdica (Sofia), Philippopolis (Plovdiv), Augusta Trajana (Stara Zagora), Pautalia (Kyustendil), Ratiaria (the present-day village of Archar, in the Vidin district), Oescus (near the village of Gigen, in the Pleven district), Nove (near Svishtov), Sexaginta Prista (Russe), Durostorum (Silistra), Montana (Mihailovgrad), Abritus (near the town of Razgrad), Marcianopolis (near Devnya), Nicopolis ad Istrum (near the village of Nikyup in the Veliko Turnovo district), Nicopolis ad Nestrum (near the present-day town of Gotse Delchev) and the old Hellenic colonies Odessos, Messambria and Apollonia. The economic upsurge of these lands continued well into the second and the first half of the third century, and a vast network of roads was built.

The Barbarian invasion – part of the incursions of the Barbarian tribes into the territory of the Roman Empire known as the Great Migration of Peoples – reached Thrace

Bronze helmet of the type referred to as Thracian, found at Kovachevitsa, near Blagoevgrad. Fourth century BC. National History Museum, Sofia

and Lower Moesia in the middle of the third century. As a result of the economic and political changes in the Roman Empire the situation in Thrace at the end of the third century was considerably changed. The provinces of Thrace and Lower Moesia were merged to form the diocese of Thrace, with the following six provinces: Thrace, Lower Moesia, Scythia, Rhodope, Hemimont and Europe, which embraced the East Balkan lands to the east of the rivers Mesta and Vit. Thrace became the hinterland of the new capital, Constantinople. In the fourth century Thrace became one of the richest and most powerful regions of the empire; the development of crafts gained new impetus; many of the towns, which were fortified as part of the defence system of the capital, were consolidated and flourished; many monumental public and religious buildings were erected.

In the seventies of the fourth century Thrace was seriously affected by the incursions of the Gothic tribes; the Roman legions were defeated near Adrianopolis (Edirne) on 9 August 378. These events had a great effect upon the social, economic and political development of Thrace; the old slave-owning system was dealt a serious blow, the new form of exploitation – the colonus system –

Clay tumbler with two handles, of the type referred to as Thracian, from Mihalich, near Haskovo. Early Bronze Age. National Archaeology Museum, Sofia

was consolidated and towards the end of the fourth century the Thracian coloni were already bound to the land. After the division of the Roman Empire into the Eastern and Western Empires (395), Thrace remained within the boundaries of Byzantium, the Eastern Roman Empire. In the fifth century the area was the target of devastating incursions by the Huns and Ostrogoths, which affected some of the regions.

With the passage of time the ethnic composition of the population of the Balkans started to change. The north-western parts of the Balkan Peninsula were dominated by the Illyrians and the southern parts by the Greeks, while the whole central and eastern part was populated by Thracians. As a result of the continued processes of Romanization and Hellenization in the preceding centuries, part of the Thracian population, particularly in the town centres, lost its ethnic identity, while the Thracian peasants in their vast majority still spoke their own language, professed their own religion and observed their own customs. Alongside the older population, in the sixth century the peninsula was also inhabited by other ethnic groups, which had penetrated the region during the time of the incursions or had been settled there as a result of Byzantium's policies, as separate federates, whose number however was comparatively small. The situation changed considerably when the Slavs and the Proto-Bulgarians came to the Balkan Peninsula and linked their destinies and history for ever. Their appearance in the lands of ancient Thrace marked the beginning of a new epoch in its history.

THE MIDDLE AGES IN THE BULGARIAN LANDS

The Slavs in the Balkan Peninsula. The Slavs belong to the Indo-European linguistic and ethnic community. In the first milennium BC they inhabited the central part of Eastern Europe, between the rivers Oder, Visla and Dnieper. The second half of the first milennium BC marked the beginning of the migration of the Slav tribes from their lands of origin, their territorial expansion towards the central part of Europe and of their division into distinct groups. Roman writers of the first and second centuries mentioned the Slavs by the collective name of Venedi. The attempts of the Roman Empire to conquer the Slav tribes living at the periphery of the Slav territories failed. However, the contacts with the Roman Empire and the Germanic tribes contributed to the development of the Slavs politically, economically and culturally. The period between the second and fourth centuries saw the final separation of the Slavs into tribal groups, distinguished by their collective names: the Antae for the Eastern Slavs, the Venedi for the Western Slavs, and Slavenae for the Southern Slavs. Towards the middle of the fifth century the Slavenae moved southwards to the Carpathian mountains and settled in the ancient provinces of Pannonia and Dacia, to the north of the Danube. At the end of the fifth and beginning of the sixth centuries the Slavenae and the Antae, together with the Proto-Bulgarians, started their numerous inroads into the central parts of the Balkan Peninsula. Their raids of the first half of the sixth century were characterized by the employment of organized military units operating in most cases on a large front. In its effort to stop them, Byzantium accompanied intensive military operations and large-scale building of fortifications by recourse to tactical stratagems, using the differences between the individual Slav tribes, attracting representatives of the Slav aristocracy into the administration of the empire, using other peoples (such as the Gepidae and the Avars) as their allies. During the second half of the sixth century the Slav incursions became difficult to resist; they penetrated deep into the area south of the Danube, reaching the Adriatic, Thessaly and the environs of Constantinople. During the first quarter of the seventh century, taking advantage of the fact that the Byzantine Empire was engaged in wars with Persia, the Slavs intensified the process of mass colonization. Within a very short time the Balkans radically changed their ethnic composition: they were Slavonicized. The Slavs not only settled in Moesia, Thrace, Illyricum and the Rhodope region, but also reached Thessaly, Greece itself and the Peloponnesos. Some of them settled in the Aegean and Adriatic islands. Along the two banks of the Danube, between the Carpathians and Stara Planina, were the seven Slav tribes and the Severians; the valley of the Timok river was inhabited by the Timochanae; the valley of the Morava river – by the Moravians; the valley of the Struma river – by the Strymonians; the lands to the southwest and north-west of Thessaloniki – by the Sagudatae and Dragovitae respectively; the lands near Mount Athos – by the Rinhinae; the lands round the Ohrid and Prespa Lakes – by the Bersitae (Bursyaks); Thessaly – by the Velegezitae; Epirus – by the Vayunitae; Peloponnesos – by the Ezertzae and Milingae, and Helles – by the Vihitae. The settlement of the Slavs in the Balkan provinces of Byzantium is one of the most significant processes in the history of South-east Europe in the early Middle Ages. A radical change took place in the composition and character of the population of the Balkans, and the Slavs turned into a major political factor in the history of the Peninsula. The settlement of the Slavs established the free village community as the main socio-economic unit, which served as the basis for the transition from Antiquity to the Middle Ages, and to feudalism. As a form of social organization it proved its vitality and stability. The settlement of the Slavs also marked the beginning of a cultural and political symbiosis of the Slavonic and Byzantine world, which brought about the formation of a new type of medieval European civilization.

The Proto-Bulgarians. The Proto-Bulgarians are of Turkic origin. Their language is related to the group of Western Turkic languages as an independent branch. Their ancient homeland is the Altai. The first mention of the ethnic name Bulgarians occurs in the Anonymous Chronograph (334 AD). The Hun invasion of the Bulgarian lands in the second half of the fourth century forced some of the Proto-Bulgarians to move on to Armenia, and others to settle in Pannonia and the Carpathian valleys. At the end of the fifth century the Proto-Bulgarians were Byzantium's allies in fighting the Ostrogoths. Later they made several incursions into the Balkan Peninsula. Throughout the first half of the sixth

Coin minted by the Thracian ruler Seuthes III. National History Museum, Sofia

century the raids of Slavs and Proto-Bulgarians recurred in periodic succession. By the end of the sixth century three strong Proto-Bulgarian tribal unions had formed – the Kutriguri, who inhabited the steppes west of the Don; the Utriguri – in the steppes between the Don and the Kuban, and the Onoguri (Onogunduri) – in the area beyond the Kuban and in the Caspian region. In 567-568 the Western Turks conquered the Proto-Bulgarians and incorporated their lands into their Khaganate. In 632, after a prolonged struggle, the Proto-Bulgarians, led by Khan Kubrat of the Unogunduri, regained their freedom and built the powerful military and tribal alliance 'Great Bulgaria', whose borders were, to the east – the Kuban, to the west – the Dnieper, to the north – the Donets, and to the south – the Sea of Azov and the Black Sea. Its capital was Phanagoria on the Taman Peninsula. After the death of Khan Kubrat (in the mid-seventh century), Great Bulgaria disintegrated under pressure from the Hazaras, who came from the east. Some of the tribes remained in their original lands and accepted Hazari domination. Others drifted to the confluence of the Kam and the Volga and formed the independent state of Bulgaria. Two other groups moved towards Pannonia and the Ravenna area.

The Founding and Consolidation of the Bulgarian State. A part of the Proto-Bulgarians, led by Khan Asparuh, settled in the lands north of the Black Sea – from the Dnieper to the Danubian delta (the Ongul area). Thence, for a period of several years, he proceeded to raid the Byzantine lands in today's Dobrudja. It was probably at this time that he entered into an alliance with the surrounding Slav tribes. In 680 the Proto-Bulgarians of Khan Asparuh defeated the Byzantine troops which invaded the Ongul, gave chase as far as the Danube, and then in turn invaded the Empire's Balkan domains. In a mere few months Byzantium lost its territory in the north-eastern part of the Balkan Peninsula. Khan Asparuh then negotiated an alliance with the Slav tribes in Moesia and Dacia – the seven Slav tribes and the Severians – with which he founded the Bulgarian State. The alliance of Slavs and Proto-Bulgarians, which brought them to statehood, was the ultimate result of their previous inter-relationships in the situation actually prevailing at the time. In their newly-founded state the Slavs and Proto-Bulgarians dwelt in territorial and ethnic independence, keeping their local self-government. From the very outset the Bulgarian State emerged as a well organized political and military power. In 681 a joint force of Proto-Bulgarians and Slavs defeated the Byzantines in Thrace for the second time. The Byzantine Emperor, Constantine IV Pogonatus, was compelled to conclude a peace treaty with the Proto-Bulgarian Khan in the summer of 681. The Empire was then treaty-bound to pay annual tribute to Bulgaria. This amounted to a virtual legal recognition of the new Bulgarian State. In 685 the Slav tribe of the Timochani, dwelling in the Timok valley, was freed from the oppression of the Avar Khaganate and incorporated into Bulgaria. The founder of the Bulgarian State, Khan Asparuh, died in a battle with the Hazaras in 700, north of the Danubian estuaries.

In the last quarter of the seventh century a body of Proto-Bulgarians, led by Kuber out of Avar domination in Pannonia, settled among the Slavs in Macedonia, mostly in the Keramissichian (Bitola) plain. In 685, acting in alliance with the Slav tribe of the Dragoviti, Kuber's Bulgarians made an attempt to seize Thessaloniki. The Proto-Bulgarians of Asparuh and Kuber were closely allied.

The founding of the Bulgarian State constitutes an important event in the millennia-old history of Europe. Its emergence on the political scene largely altered the course of events in South-east Europe. It made for the political

The Middle Ages in the Bulgarian Lands

Khan Kubrat and his Sons. Painting by Dimiter Gyudjenov. National Art Gallery, Sofia

Relief of a horseman cut into a stone slab. Pliska. National History Museum, Sofia

and ethnic unification of the Slav tribes and the Proto-Bulgarians in the Balkan Peninsula. The founding of the state marked the start of the formation of the Bulgarian nation, which became the outpost of Slavdom in its struggle against the expansion of Byzantines, Avars and Franks.

Khan Asparuh, who was descended from the House of Dulo, a Proto-Bulgarian clan, was succeeded after his death by Khan Tervel (700–721), also a descendant of the Dulos and presumed to have been a son of Asparuh. In 705 a Bulgarian army assisted the deposed Byzantine Emperor, Justinian II, to regain the throne. In recompense Khan Tervel received the title of Caesar (second only to that of the Byzantine Emperor) and, in addition, the region of Zagora, which was a key strategic location. In 708 the Emperor marched on Bulgaria with the aim of re-capturing Zagora, but in the battle of Anchialo (Pomerie) the Byzantine army suffered a complete rout. In 711 and 716 Khan Tervel invaded Thrace and reached the walls of Constantinople. During the second campaign, Emperor Theodosius III was forced to conclude a peace treaty with the Bulgarians; the treaty defined the state border between the two countries, committed Byzantium to pay an annual tribute to Bulgaria, arranged for the mutual repatriation of political refugees, and regulated bilateral trade relations. In 717–718 the Bulgarian troops extended assistance to the Byzantine capital, besieged by Arabs, crushing them and rescuing Constantinople. The victory of the Bulgarians over the huge Arab army played an important role in building the prestige of the young Bulgarian State.

After the long reign of the Dulo dynasty, Khan Kormisosh (721–738) ascended the throne of Bulgaria and for most of his reign continued the policy of the House of Dulo, maintaining relations with Byzantium.

About the middle of the eighth century a political crisis erupted in Bulgaria in consequence of the internecine struggle among the Proto-Bulgarian aristocracy over the succession. In the course of about fifteen years the throne was occupied for brief spells by representatives of the houses of Volkil and Ughain (Vineh – 753/4–760; Telets – 760–763, Sabin – 763–766, Umor – August–September 766, Toktu – 766–767, and Pagan – 767–768). Byzantium took advantage of these rivalries to embark on a succession of military operations lasting, with brief intermissions, for twenty years. Constantine V Copronymus launched nine campaigns against the Bulgarian State, with the aim of destroying it, but had his plans thwarted by staunch Bulgarian resistance.

After prolonged defensive battles against Byzantium the Bulgarian State again assumed the offensive under Khan Telerig (768–777). In 774 he sent a force of 12,000 strong to the area inhabited by the Slav tribe of the Berzites (west of Sofia) to recruit their support in the wars against Byzantium and to incorporate them into Bulgaria. This was an extension of the policy of integrating the Slav tribes and lands into the bounds of the Bulgarian State. The domestic political crisis was gradually overcome and on the ascension to the throne of Kardam (777–803) the internal fighting stopped. Bulgarian foreign policy turned its attention toward the southern and south-western Slav tribes. In 785 the Byzantines invaded Thrace and began to rebuild towns and fortresses. In 789 the Bulgarian troops penetrated the valley of the Struma, inhabited by the Slav tribe of Strimonians, and defeated the Byzantine troops. In 791, another war broke

Inscription of Tsar Ivan Vladislav, 1015–16. Bitola

out between Byzantium and Bulgaria; Khan Kardam led his troops to Adrianople and, near the citadel of Provat (north-east of the city), sent the Byzantines fleeing in disarray. In July 792, near the border fortress of Markella, the troops of Khan Kardam dealt a crushing blow to the Byzantines; the Bulgarians captured the whole train of the Byzantine army, incuding the royal tent and treasure hoard. Emperor Constantine VI had to sign a peace treaty under which Byzantium was bound to pay annual tribute to Bulgaria. In 796 Byzantium attempted to violate the peace, but the resolute countermoves of Khan Kardam forced it to abide by its treaty obligations of 792.

The foreign policy thus established was even more consistently pursued by Khan Krum (803–814). He made deft use of the defeat of the Avar Khaganate by the Franks and in 805 annexed part of its territory, Transylvania and the Carpathian region. Also incorporated into the Bulgarian State were a number of Slav and Proto-Bulgarian tribes which had then been under Avar rule. Part of the Avar now came under the power of Bulgaria, which also included in its territory the Belgrade area, Srem (Sirmium, today's Sremska Mitrovitsa), Transylvania and the Hungarian plain between the Tisza and the Danube. A common border was established with the Franks and their penetration into the region of the Middle Danube was halted. The expansion of Bulgaria evoked a sharp reaction from Byzantium. In 807 the Byzantines violated the peace and in a retaliatory move Khan Krum sent Bulgarian troops down the Struma valley and defeated the Byzantine army, seizing its train with 1,100 pounds of gold. In 809 he captured the important fortress of Serdica, as a result of which Bulgaria gained control over the strategic road running through the heart of the Balkan Peninsula (Constantinope-Plovdiv-Sofia-Nish-Belgrade): this gave access to Macedonia and the Aegean – regions inhabited by Slavs of the Bulgarian group. To divert attention from the south-western regions and to strike at the heart of Bulgaria, Emperor Nicephorus I Genik launched a campaign in 811 against Bulgaria, overran it and burnt its capital, Pliska. But on their march back the Byzantines ran into the ambush laid for them by Khan Krum and were completely destroyed in the Veregava pass (Vurbitsa pass), and Emperor Nicephorus was killed (26 July 811). Despite the victory gained, Khan Krum asked for peace and for a renewal of the treaty of 716. A categorical refusal by Byzantium led to a reopening of hostilities. In the following year Khan Krum took the fortress of Develt (near today's village of Debelt in the Burgas district), besieged and captured Messembria (Nessebur), a fortified city on the Black Sea, and extended his possessions over the Black Sea coast south of Stara Planina down to Sozopol. The war was renewed in 813, and in the battle waged in that year at Versinikia near Adrianople the Byzantines suffered another defeat and the Bulgarian

The Constructional Genius of Khan Omurtag. Painting by Boris Angelushev

troops reached the walls of Constantinople. On the order of Khan Krum they dug a huge moat in front of the Golden Gate, up to the sea, and tightened their grip round the city. The Byzantine Government opened negotiations for peace, during which the Byzantines treacherously attacked the Bulgarian ruler, but succeeded only in killing his retinue. In retaliation for the Emperor's bad faith Khan Krum raised the siege, ordered the destruction of all military installations throughout South-eastern Thrace, took Adrianople and returned his army to Bulgaria loaded with spoils. In the winter of 813–814 Khan Krum sent a 30,000-strong army to Eastern Thrace to harass the Byzantines and prevent them from building their defences. He also started preparations for a new march against Constantinople; 5,000 carts and 10,000 bullocks were to transport the sophisticated machinery for the siege of the Byzantine capital. At the height of these preparations Khan Krum died of a heart attack (13 April 814).

The military feats of the Bulgarians during the reign of Khan Krum brought about the incorporation into the Bulgarian State of new areas south of the Balkan Range, inhabited by a large Slav population. The victories also raised the prestige of Bulgaria among the Slav tribes in the Balkan Peninsula and in the whole of Europe. To deal with Bulgaria, the Byzantine Emperor had to solicit the aid of the Frankish Emperor Charlemagne. Khan Krum integrated the Slav aristocracy in the ruling hierarchy, and by introducing universal written laws equalized the Proto-Bulgarian and the Slavic element. This accelerated the process of Slavonicization of the Bulgarian ethno-political community and the formation of a united Bulgarian nation.

Khan Krum's son Omurtag (814–831) tried to continue the offensive against the Empire, but his troops were defeated. He opened peace negotiations and at the beginning of 815 concluded a thirty-year peace treaty with Byzantium, which legalized the territorial expansion of Bulgaria into Thrace, and a policy of good neighbourly relations with Byzantium was maintained throughout his reign. In 823 Bulgarian troops aided the Byzantine Government in suppressing the peasant rising led by Thomas the Slav. Having established peace with Byzantium, Khan Omurtag turned his attention to his north-eastern and north-western borders. Around the year 820 a Bulgarian army repulsed an invasion (presumably Hungarian or Khazar) at the river Dnieper. A war broke out (827–829) between Bulgaria and the Frankish Empire, caused by the secession of the Slave tribes of Timoks, Branichevs, and Abodrits from the Bulgarian State. Through successful military operations and diplomatic negotiations, Khan Omurtag defeated the Franks and restored the unity of his country. The territorial expansion and development of feudal relations called for a better state organization, so Khan Omurtag abolished the union

The Middle Ages in the Bulgarian Lands

Prince Boris I (with sword). Fresco from Santa Maria in the Monte Celio Church, Rome

structure and established a centralized state government. The country was divided into a military administrative areas, called *komitats*. Khan Omurtag engaged in large-scale construction; the capital city of Pliska, destroyed by the Byzantines, was restored; a new fortress and palace were erected, as well as a ceremonial hall on the river Ticha (Kamchiya) near the village of Khan Krum in the Shumen district, and a large palace on the Danube. Many inscriptions and memorials have been preserved, testifying to this large-scale construction and to the activities of eminent military commanders. Khan Omurtag persecuted Christianity and its preachers, because he believed that the propagation of Christianity among the people would undermine the foundations of the State and create conditions for the infiltration of Byzantine influence.

Khan Omurtag had three sons – Enravotha (Voin), Zvinitsa and Malamir. After his death, contrary to the rules of succession, his youngest son Malamir ascended the throne (831–836). During his reign Khan Malamir maintained good relations with the Franks, signing a peace treaty which settled the relations between Bulgaria and the Frankish Empire for nearly twenty years. In 832 Byzantium broke the thirty-year peace and invaded the **territory of Bulgaria**. The Bulgarian troops commanded by Khan Malamir and his close aide Kav-Khan Isbul drove them back and mounted a counter-offensive in Eastern Thrace, taking the fortress of Provat (near Adrianople),

and Philippopolis. Malamir continued his father's cultural and construction policies, building a large water-main in the capital. Like his father he persecuted the Christians, and even condemned to death his own elder brother, Enravotha, for having adopted Christianity.

After his death, Khan Malamir was succeeded to the throne by Pressian (836–852), son of his brother Zvinitsa. In 837 the Slav tribes in the areas of the Rhodope mountains, Aegean Thrace and South Macedonia rose against Byzantine rule. The Bulgarian troops, headed by Kav-Khan Isbul, gave support to the Smolyans and defeated the Byzantines in a battle near the Philippi fortress. Thus the Slav tribes in the south-western parts of the Balkan Peninsula were incorporated into the Bulgarian State. The territory they inhabited, which had until then been called 'Slavinii' (Slav lands) by Byzantine authors, was now given the Proto-Bulgarian name of Kutmichevitsa (newly-annexed land); in later sources it is referred to as 'the third part of the Bulgarian Kingdom'. Within a very short time the Bulgarian's succeeded in consolidating their positions in the newly conquered lands, making them an integral part of their state. In 839–842 Khan Pressian was forced to enter a war with the Serbs, instigated by Byzantium.

The Power of the Bulgarian State. The incorporation of the new Slav tribes into the emerging Bulgarian ethos

Christianization of the Bulgarians. Miniature from the Bulgarian translation of the Chronicle of Manasses. Fourteenth century. Vatican Library

determined its Slavonic character. The defence of the Slav ethos and its uniting round the Proto-Bulgarians was the first major mission of the early-medieval Bulgarian State, which became a strong centripetal force for the Slav tribes in the Balkan Peninsula in their struggle against Byzantine expansion. At the same time the Bulgarian State was politically active in Central and Eastern Europe through its relations with the Avar Khaganate, the Frankish Empire, the Pannonian Principality and the Khazar Khaganate. Bulgaria became a serious obstacle for the nomadic tribes and peoples (Khazars, Hungarians, Turks, Pechenegs, etc.), invading from the north and north-east. It brought political balance and calm in South-east Europe.

Khan Pressian's successor to the throne was his son Boris I (852–889). The Bulgarian-Byzantine conflict, which continued into the beginning of Boris's reign, was settled with the conclusion of the treaty of 856, by force of which Byzantium acknowledged the conquered Bulgarian lands. On the north-western border the Bulgarians were forced to fight on two fronts against the German Kingdom and the Croatian Kingdom.

Towards the middle of the ninth century all the prerequisites existed for Bulgaria to join the Christian world. Being a heathen state, Bulgaria stood aside from the development of European spiritual and material culture based on Christian ideology. The adoption of a general and obligatory religion would guarantee the religious and ideological unity needed in the Bulgarian State by the ruler and the feudal aristocracy in order to consolidate the ideological basis of their power. It was also necessary for the amalgamation of Slavs and Proto-Bulgarians. It was first envisaged that Christianity would be adopted from the Western European Church, and in 862 Boris I and King Ludwig of Germany concluded an agreement which, among other things, settled the conversion. In order to break the union and prevent further *rapprochement,* Byzantium, in coalition with Greater Moravia, Croatia and Serbia, declared war on Bulgaria in 863. The Bulgarian troops were defeated, and by the terms of the peace treaty Bulgaria was obliged to adopt Christianity from Constantinople. The first to be baptized, late in 863, were the Bulgarian envoys to the Byzantine capital. In the spring of 864 a special mission arrived in Pliska. Boris was baptized with the Christian name of Mihail, and along with him other Bulgarian dignitaries were also baptized. Then began the baptizing of the people, which took about half a year. The adoption of Christianity abolished the heathen religious structure; the temples were turned into churches or destroyed, and Christian ones built on their sites.

Gold cross found at Pliska. Ninth–tenth centuries. National History Museum, Sofia

archate in Constantinople resulted in the ultimate association of its history and fate with the Christian East and with the Byzantine-Slav political, religious and cultural community. The adoption of Christianity and the creation of an independent Bulgarian Church accelerated the process of ethnic amalgamation.

Boris I sent many young Bulgarians, including his son Simeon, to be educated in Constantinople. In 886 he received—in Pliska—Kliment, Naum and Angelarius, pupils of Cyril and Methodius, who had been driven out of Great Moravia. Together they elaborated a broad plan for educative and literary work. The Ohrid and Preslav Literary Schools were set up and many distinguished Bulgarians were sent to monasteries to engage in literary work; among them were his brother Doks and his son Todor Doksov. In 889 Boris I relinquished the throne to his son Vladimir (889–893), took holy orders and retired to a monastery to follow more closely the work of the men of letters. As ruler, Vladimir tried to change the domestic and foreign policy of Bulgaria; he struck up an alliance with the German King Arnulf against Great Moravia,

Tsar Simeon (893–927). Painting by Dimiter Gyudjenov. National Art Gallery, Sofia

A large part of the *boyars* (noblemen) and the people opposed the conversion, and in 865 a rebellion broke out in the ten *komitats* (regions) of the country. It was suppressed by Boris I and fifty-two boyars and their families suffered the death penalty. The Bulgarian ambition to establish an independent church ran counter to Byzantine policies and the negotiations with Patriarch Photius in Constantinople, opened by Boris I in 864, failed. Bulgaria resumed its political alliance with the Germans and established contacts with the Roman Church. The negotiations with the Pope were more fruitful; Boris I received an answer from Pope Nicholas I to his one hundred and fifteen questions and the papal consent, in principle, to the establishment of an independent church. Bulgaria's temporary acceptance into the bosom of the Roman Church made Byzantium more compliant to Boris's demands concerning the Bulgarian Church problem. In 870 the Eighth Ecumenical Council in Constantinople decided that the Bulgarian Church diocese should pass under the jurisdiction of the Patriarchate of Constantinople, but should be a separate autocephalic archbishopric. The Roman clergy left Bulgaria and were replaced by the Byzantine ones, headed by Joseph, Archbishop for Bulgaria, consecrated by the Patriarch of Constantinople. The alliance of Bulgaria with the Partri-

tried to return heathen ways and persecuted the Christians. His actions threatened the existence of the Old Bulgarian script and literature in Bulgaria. Boris I left the monastery to participate in the dethronement of Vladimir and in the state convention held in Preslav in 893, which made the far-reaching decisions to put Simeon on the throne, move the capital from Pliska to Preslav, establish Old Bulgarian as the official language and replace the Byzantine clergy by Bulgarian. Then Boris I retired to the monastery for life. Soon after his death he was canonized by the Bulgarian Orthodox Church.

The new Bulgarian ruler, Simeon (893–927), received the best of education at the Magnaur Academy in Constantinople. He returned home about 886, took holy orders and was preparing to be the head of the Bulgarian Church. Instead of becoming a senior clergyman and man of learning, however, he became a statesman and military commander. Immediately after his ascension to the throne, relations between Bulgaria and Byzantium worsened. The Byzantine Emperor, Leo VI the Philosopher, moved the market for Bulgarian goods from Constantinople to Thessaloniki (894). This exposed Bulgarian tradesmen to the arbitrary actions of the local authorities. Simeon's intercession with the Emperor to change this order met with a blunt refusal. In response the Bulgarian ruler broke the peace with Byzantium, invaded Thrace and defeated the Byzantines at Adrianople. Byzantium allied with the Hungarians and incited them to attack Bulgaria. In 895, carried across the Danube by the Byzantine fleet, the Hungarians invaded North-east Bulgaria and besieged the Bulgarian army, with Simeon at the head, in the fortress of Mundraga. To gain time the Bulgarian ruler conducted lengthy negotiations with the Byzantine diplomat Leo Chyrosfactus. In 896, with the help of the Pechenegs, he defeated the Hungarians and drove them into the lands north-west of the Carpathians. Then the Bulgarian army resumed its war with Byzantium, dealt it a decisive blow at Bulgarophygon and forced the Empire to sign a peace treaty (896). In 904 another treaty legalized the territorial expansion of Bulgaria into South Macedonia and South Albania, the border with Byzantium passing twenty kilometres north of Thessaloniki. The peace gave the Bulgarian people opportunities to concentrate on intensive construction and cultural development. Through the actions of the Byzantines, the end of 912 saw a new change for the worse in Byzantine-Bulgarian relations. A Bulgarian campaign in Eastern Thrace met with no resistance, and with Simeon at the head, the army reached the walls of Constantinople and besieged it (913). Forced to negotiate, the Byzantines organized a brilliant reception for Simeon, declared him Tsar *(basileus)* and arranged for his daughter's marriage to the Byzantine heir to the throne. The following year Byzantium broke the peace treaty, and the hostilities were resumed. The Bulgarian troops entered the territory of the Empire, besieged and took Adrianople. The military actions continued for several years, and in 917 Byzantium formed an anti-Bulgarian coalition with the Pechenegs, Hungarians and Serbs. Tsar Simeon let the Byzantines cross the Thracian border and concentrated his troops in the plain between Anchialo and Messembria (Nessebur). On 20 August 917, along the banks of the river Acheloi, the Bulgarian army led by Tsar Simeon inflicted a heavy defeat on the Byzantines, which caused the end of the coalition. Pursuing the Byzantines, the Bulgarians reached the vicinity of Constantinople and once more defeated them at Katasirti. The attack on the Byzantine capital was postponed, for Tsar Simeon was forced to neutralize Serbia which, being an ally of Byzantium, was hostile to Bulgaria. After this victory (918) he dispatched an army which defeated the Byzantines in Thrace and penetrated deep into Hellas. In 920 the Bulgarian army reached the Dardanelles and in 921 defeated the Byzantines at Pigi ('Life-giving spring', present-day Balikli) in the vicinity of Constantinople. Tsar Simeon was once again unable to complete the campaign because of unrest among the Serbs. In 922 his troops re-entered Byzantium and recaptured Adrianople. At the same time he agreed with the Arab Khalif, Ubaydallah al-Mahdi, to combine forces for actions against the Empire, but the alliance was thwarted by Byzantine diplomacy. In 923 Tsar Simeon and his troops were again at the walls of Constantinople. The Byzantines started negotiations and at a meeting, on 9 September at Kosmidion (in the upper part of the Golden Horn), Tsar Simeon and Emperor Romanus Lecapenus concluded a truce. In the following year the Tsar was forced to mount a further action against Serbia, which resulted in its incorporation within the boundaries of the Bulgarian State. In the latter half of 926, following diplomatic talks, Pope John X sent two emissaries to Bulgaria to crown Simeon and bestow Patriarchal rank on the head of the Bulgarian Church. In a further war with the Croatians the Bulgarians were defeated. Tsar Simeon died on 27 May 927, after a heart attack.

During the reign of Tsar Simeon the Bulgarian State reached the summit of its political grandeur and might. Its territories bordered on three seas – the Black Sea, the Aegean and the Adriatic – and it embraced the larger part of the Balkan Peninsula. Bulgaria was the stronger side in the conflicts with the Byzantine Empire. With his political ideas, cultural activities and military triumphs Tsar Simeon set an example for the medieval Bulgarian tsars to follow.

The military and political achievements of the Bulgarian State during the reign of Tsar Simeon were accompanied by the flourishing of Bulgarian culture, in what is now referred to as the Golden Age. The development of Bulgarian letters, architecture and the arts during this period had a significant influence on other Slav countries. The greatest cultural achievement during Simeon's reign was the development and enrichment of Slavonic literature. The initiator of literary activity was the Tsar himself. A circle of eminent men of letters was formed in Preslav, the most distinguished of them being Monk Hrabr, John the Exarch and Constantine of Preslav. Art and architecture flourished and many fortresses, churches and monasteries were built. The capital city of Preslav became the centre of an impressive building programme.

In the latter half of the ninth century and the first quarter of the tenth, events of extreme importance took place in Bulgaria, leading to significant changes in its historical development. The conversion of the Bulgarian

The Middle Ages in the Bulgarian Lands

Tsar Simeon Defeating the Byzantines at Aheloi. Illumination from the Madrid Manuscript – a chronicle by John Scylitses, twelfth–thirteenth centuries. Madrid National Library

people to Christianity and the setting up of an autonomous Bulgarian church, as well as the obligatory use of the script and literature turned Bulgaria into the foremost Slav literary centre. The major result of the brilliant upsurge of the Bulgarian State was the lasting formation and consolidation of the Bulgarian ethos on vast territories of the Balkan Peninsula (Moesia, Dobrudja, Thrace, the Rhodope region, Macedonia and the lands beyond the Danube).

Bulgaria from the Second Quarter of the Tenth Century to the Beginning of the Eleventh Century. The death of Tsar Simeon was followed by a change in the situation in the Balkans. All states and tribes (Serbs, Pechenegs, Hungarians and others), which had previously suffered from its military power were now openly hostile. Tsar Simeon's successor to the thone, his second son Peter (927–970), did not share his eminent father's ambitions. During the early years of his reign the chief adviser and the effective head of state was his uncle Georgi Sursuvul, a champion of a peaceful policy towards Byzantium. The two of them held secret negotiations with the Byzantine Emperor Romanus Lecapenus and in 927 concluded a thirty-year peace treaty, under which the latter acknowledged the Bulgarian conquests up to 912, the title of Tsar for Peter I and of Patriarch for the head of the Bulgarian Church; it also provided that the Byzantine Government should resume its payment of annual tax, stipulated in the treaty of 896. To consolidate the treaty, Tsar Peter I married the Byzantine Princess Maria who, in honour of the treaty, took the name of Irina. A group of boyars discontented with the pro-Byzantine policy of the Tsar organized two plots against him – one in 928, a coup led by his brother Ivan, and another one in 930, led by the other brother Mihail. At the same time Serbia broke away from Bulgaria.

The initial reaction to the policy of Tsar Peter was an expression of internal class struggles among the Bulgarian boyars. The first half of the tenth century saw an intensification of the process of social and class differentiation. It sharpened the differences and opposition between the privileged secular and ecclesiastical aristocracy on the one hand, and the dependent population, burdened with many taxes and duties, on the other. The inability of the supreme state power to cope with foreign invasions and protect the population suggested that the State was demoralized and on the verge of ruin. The crisis also pervaded the Bulgarian Church, and the corruption which set in early among the higher clergy and monks undermined the church's popularity. The abyss between the preachings and deeds of the clery, between the Christian ideal and reality, and the dissemination of different heretic teachings among the black clergy and the people intensified the social contradictions in Bulgarian society.

A reaction to social injustice and corruption in the official church in the tenth century was the spread of hermitism, which raised the ideal of escape from reality and praise of the spiritual perfection of human nature. The most eminent representative of hermitism under the reign of Tsar Peter I was Ivan Rilski (John of Rila) who was posthumously canonized by the Bulgarian Orthodox Church.

A serious threat to the bases of the existing social order and the rule of the feudal aristocracy and church was Bogomilism, the teaching of the priest Bogomil, a contemporary of Tsar Peter. With its simplicity and sharp, logical criticism of social vices it gained popularity among the broader strata of Bulgarian society and soon became a mass

Obverse — Reverse
Seal of Tsar Peter I. Tenth century.

movement against the Tsar, the boyars and the higher clergy.

The views of the Bogomils were related to those of the Paulicians and Messalians, which existed in the Byzantine Empire and penetrated Bulgarian territories mainly after the conversion. Bogomilism was based on the dualistic concept of life. The Bogomils considered that the universe had two distinct principles – Good (God) and Evil (Satan). They held that heaven and the human soul were the creation of Good, and the external world and the human body – the creation of Evil. There existed a sharp struggle between the two principles, and Good would win over Evil. As a result of their dualistic views the Bogomils had a negative attitude towards temporal power, the Tsar and the boyars, declaring them 'serfs of Satan' and considering that all those who worked for them were anathema to God. Thus they completely undermined the hold of the existing social order on the country. They preached that the clergy were useless as a mediator between God and man, a thoroughly parasitic stratum, and 'serfs of Satan'. The symbols, rites and rituals of the church were completely rejected, and their services were held in ordinary houses or in the open (since God was everywhere), and consisted of reading and interpreting the New Testament.

The Bogomils declared themselves against war and murder, preached asceticism and reconciliation, rejected marriage, forbade the killing of any living creatures except snakes, etc. They were organized in communities, called brotherhoods, each of which had a leader, called *dedets* (senior). There were three categories of Bogomils – 'perfect', 'believers', and 'listeners'. The 'perfect' ones were initiated into all details of the teaching, obliged to observe all its canons and to perform leading and organizational functions.

The wide popularity of the teaching not only caused anxiety in the Church, but also disturbed Tsar Peter I, who asked the advice of the Patriarch in Constantinople on how to cope with the heretics and how to punish them. A merciless struggle against the Bogomils began: they were declared heretics, subjected to persecution and torture, and imprisoned as criminals dangerous to the State.

In 934 the Hungarians started regular onslaughts on the Bulgarian lands, as a result of which Bulgaria lost a large part of its territories beyond the Danube. It was not until 965 that Tsar Peter signed a peace treaty with the Hungarians, under which they were not to invade the Bulgarian lands, and the Bulgarians were to allow them a free passage for their attacks on Byzantium. This worsened relations between Bulgaria and Byzantium. In 967 the Byzantines broke the peace, refused to pay their annual duty and launched a campaign against Bulgaria, which was interrupted by a mutiny in the Byzantine army.

In 968 Byzantine diplomacy incited Prince Svyatoslav Igorevich of Kiev to war with Bulgaria. The thirty thousand-strong Bulgarian army which went to meet him could not offer serious resistance and the Russian troops took some eighty fortresses. Because of a threat from the Pechenegs against Kiev, Svyatoslav withdrew from the Bulgarian lands, but the following year he started a new campaign against Bulgaria. Tsar Peter I sent his sons Boris and Roman to Constantinople with a mission to concert the actions of Bulgaria and Byzantine against Prince Svyatoslav. Meanwhile Tsar Peter fell ill, took monastic vows and on 30 January 970 died. He was later canonized by the Bulgarian Orthodox Church. The Government fell temporarily into the hands of the nobility, who sought the help of Byzantium and the Pechenegs.

Tsar Peter's two sons returned to Bulgaria and in

Relief of a lion, found at Stara Zagora. Tenth–eleventh centuries. National History Museum, Sofia

970–971 Boris II occupied the throne. He concluded a treaty with the Russians against Byzantium, and in 970 Bulgarian and Russian troops commanded by Prince Svyatoslav invaded Northern and Eastern Thrace, reaching as far as Adrianople. In 971 the Byzantine Emperor, John I Tsimisces, launched a campaign against Bulgaria, and in April of that year the Byzantines conquered the capital of Veliki Preslav, despite the heroic resistance put up by Bulgarians and Russians. Svyatoslav and his troops were besieged in Drustur (present-day Silistra). After a three-month siege the Russians concluded a truce with the Byzantines and withdrew (Svyatoslav was later killed by the Pechenegs at the estuary of Dnieper). After the withdrawal of the Russians the Byzantines took the rest of the Bulgarian fortresses and the North-eastern parts of the country fell *de facto* under Byzantine rule. Boris II and Roman were captured.

The Bulgarian lands west of the river Iskur preserved their political independence and the traditions of the Bulgarian State and administrative system. After the conquest of Veliki Preslav and Drustur, the Bulgarian Patriarch Damyan came to live in those parts, establishing his temporary residence in Sredets. The supreme head of the Church therefore temporarily represented royal power as well, symbolizing the independence and unity of Bulgaria. The government of the free Bulgarian lands was in the hands of the brothers David, Moses, Aaron and Samuil. After the death of John I Tsimisces (976) the brothers launched military actions against the Byzantine Empire to regain the conquered Bulgarian lands. They were active on a large front and in different directions. In 976 Byzantine rule in the north-eastern part of Bulgaria was overthrown, but in that year David was murdered by the Wallachians, Moses was killed at the siege of Ser, and Aaron was accused of collaboration with the Byzantines and killed on Samuil's order.

In 977 Tsar Boris II and Roman fled Constantinople and made for Bulgaria. As a result of confusion at the border, Boris was murdered; Roman went to Vidin, where he was greeted as Tsar (977–997). Later he established his seat in Skopje. During his rule the power was *de facto* in the hands of his first military commander, Samuil.

In 977–983, under the leadership of Samuil, the Bulgarian army seized control of Thessaly and conquered Larissa, and on 17 August 985, they defeated the Byzantine army which had penetrated Bulgaria through the Klissura Pass. In 991 Samuil again fought against Byzantium and in this campaign Tsar Roman was taken captive. In 996 Samuil launched a campaign for Thermopylae, but on his way back he was surprised and put to rout at the river Sperchios. After the death of Tsar Roman (997), and for lack of any direct successors of Simeon's family, Samuil, who in addition to being Roman's first military commander also had close ties with the previous dynasty, was crowned Tsar. It was probably at that time that Samuil moved the capital, first to Prespa and later to Ohrid. In 998, seeking to terminate Byzantinum's collusion with the Serbs in the Bulgarian rear, Tsar Samuil attacked and conquered the regions of Zeta, Dukla and Trebine. At that time brisk diplomatic efforts brought about recognition by the Papal seat of the title and the authority of Tsar Samuil, the signing of a treaty with Hungary and the subsequent marriage of his son Gavril-Radomir to the daughter of the Hungarian King Geza. At the beginning of the eleventh century Bulgarian-Byzantine relations worsened; in 1001 Byzantium recaptured North-eastern Bulgaria and in 1002 Thessaly; in 1003–1004 the Byzantines besieged and conquered Vidin. In order to distract them Tsar Samuil launched a campaign into Eastern Thrace and seized Adrianople. The Byzantine Emperor Basil II succeeded in conquering Skopje, but not far from Pernik he was repulsed and suffered heavy losses. Tsar Samuil undertook two campaigns against Thessaloniki. The military actions lasted until the year 1013 without any decisive battle being fought. In order to arrest the Byzantine offensive Tsar Samuil erected a broad wall blocking the gorge near the present-day village of Kyluch in the Blagoevgrad district. Emperor Basil II's attempts to overcome the wall proved futile, but the Byzantine troops succeeded in taking the Bulgarians by surprise in the rear, on the slopes of Belassitsa mountain. On 29 July 1014, the Bulgarian army concentrated there was besieged and on the orders of Emperor Basil II the 15,000 Bulgarian soldiers taken prisoner were blinded. For this unheard-of cruelty Basil II acquired the name of 'Bulgaroctonus' (the Bulgar-Slayer).

The blinded prisoners were sent back to the Bulgarian Tsar in Ohrid, and at the sight of this tragic procession Tsar Samuil suffered a heart attack. He died two days later on 6 October 1014.

On 15 October 1014 the son of Samuil, Gavril-Radomir (1014–1015) was crowned Tsar. Soon afterwards Basil II undertook a new offensive against the Bulgarians. The Byzantines attacked Bitola, reduced the palaces of Tsar Gavril-Radomir to ashes, and then proceeded to conquer Prilep, Shtip, Ostrovo and Voden. Byzantine diplomacy succeeded in splitting the Bulgarian aristocracy, and as a result of its intrigues Tsar Gavril-

The Soldiers of Tsar Samuil Besieging Salonika and Killing its Governor, Duke Gregory Tharonites. Illumination from the Madrid Manuscript

Radomir was slain by his cousin Ivan-Vladislav while game-hunting. Upon ascending the throne, Tsar Ivan-Vladislav (1015-18) moved the capital from Ohrid to Bitola, where he re-erected the fortress. In the specially placed inscription there he called himself 'autocrat Bulgarian', 'a Bulgarian by birth', noting further that the rebuilt fortress would serve as a 'refuge, a salvation and a shield' for the Bulgarians. In 1015 the *voivode* (military commander) Ivats defeated the Byzantines in the battle of Bitola and this forced Emperor Basil II to withdraw his troops from Bulgaria. The campaigns of the Byzantine detachments which had set out to seize Serdica and Strumitsa had also failed. In 1018 Tsar Ivan-Vladislav besieged the Adriatic stronghold of Drac and was killed outside its walls. After his death the collapse of the Bulgarian State, with all its perilous consequences, was inevitable.

The Bulgarian aristocracy split into two opposed groups. The first, which stood for complete and immediate surrender to the Byzantine Emperor, was headed by the Queen, the Patriarch and the Governor of the Inner Province. The group which stood for struggle to the bitter end, was headed by Pressian, the successor to the throne and the *voivode* Ivats and Nikulitsa. In March 1018 Emperor Basil II launched his last campaign for the conquest of Bulgaria. Most of the boyars surrendered their castles voluntarily to obtain titles and posts as Byzantine nobles in exchange. The resistance against the Byzantine conquerors was continued by the heir to the throne who, together with his brothers Alussian and Aaron, retreated into the mountain Tomor; by the *voivode* Ivats, who fortified his position in the stronghold of Promista, again in the Tomor mountain; the *voivode* Nikulitsa and by the two *voivode* Elemag and Gavra, who made an abortive attempt to make the people take up arms against the Byzantines. The last fighter in independence – the *voivode* of the Srem fortress, Sermon, was killed by the Byzantines through a trick. To enable him to overpower Bulgaria completely, Emperor Basil II Bulgaroctonus collected together the surviving Bulgarian troops and sent them, together with their commanders, to Asia Minor, to fight in the conquest of Armenia.

The Bulgarian People under Byzantine Rule. In the seized Bulgarian lands the Byzantines imposed their own military and administrative system. The conquered territories lying between the Danube and the Stara Planina were included in the *thema* Paristrion (the sub-Danubian region), with its original seat at Preslav, later moved to Drustur; the lands from the Prespa Lake and the river Drim to the river Struma, together with the Sofia plain, formed the thema Bulgaria, with Skopje as its capital; the towns of the outlying Bulgarian north-west (Belgrade, Branichevo, Sirmium – present-day Sremska Mitrovitsa, etc.), conquered in 1019, formed an independent unit which was called 'thema of the Danubian towns', with Sirmium as its capital; the lands lying to the north of the Danube delta formed the thema of 'Western interfluve'; and the territories to the south of the Stara Planina were incorporated into the existing Byzantine themas of Thrace, Strymon, Thessaloniki, etc. The Bulgarian Black Sea coast and the Rhodope regions emerged as separate themas under different names.

The introduction of the new military and administrative organization in the conquered Bulgarian lands was accompanied by important changes of both demographic and political character. Part of the boyar aristocracy, mostly the relatives and the descendants of the last Bulgarian Tsar Ivan-Vladislav, were sent into exile. Some of the exiled Bulgarian nobles remained in Constantinople, but most of them were banished to the outlying Byzantine provinces of Armenia and Georgia.

The Byzantine taxation system was introduced in an even more severe form than in the other themas of the Empire. The Bulgarian peasants suffered gravely from the arbitrariness of the Byzantine officials during tax collection and when brought to court.

Large-scale secular and clerical land-owning became

The Insurgent Bulgarians Proclaim Peter Delyan Tsar. Illumination from the Madrid Manuscript

more widespread, as did the exemptions. The number of serfs (*pariks* and *kliriks*) rose. Along with the corvée and the rent in kind, a rent in cash was introduced from 1040. The ecclesiastical autonomy of the Ohrid Archbishopric, granted by Emperor Basil II, was gradually restricted, and the representatives of the higher Byzantine clergy mounted a campaign of Hellenization. The feudal and the clerical bondage was aggravated by the foreign invasions: in the mid-eleventh century the lands between the River Danube and Stara Planina were exposed to the ruinous incursions of the Pechenegs, and later of the Uzi; the Hungarians raided the lands from the north-west; the Normans, coming from Italy, made several incursions into Macedonia, and the First (1096) and Second (1147) Crusades caused a great deal of destruction and hardship.

The very first decades of Byzantine bondage saw the Bulgarian people rising in a tenacious struggle for the restoration of state independence. The worsening of the economic plight of the Bulgarian population provoked widespread popular discontent. In 1040 a rebellion broke out led by Peter Delyan, the son of Tsar Gavril-Radomir. With an army recruited in the Trans-Danubian lands, he crossed the Danube at Belgrade, where he was proclaimed Tsar, and called upon the Bulgarians to rise in arms. Soon the rebels reached Nish, invaded the valley of the river Vardar and seized Skopje. At the same time a rebellion led by Tihomir broke out in the Drac region. The two insurrectionist armies met at Skopje, where Tihomir was murdered in a conspiracy and Peter Delyan remained sole leader of the rebel forces. One of the armies moved to the south-west and soon the whole of the Ohrid and Prespa regions was liberated. The other army, led by Antim, cut southwards down to the Peloponessus, where the poorest stratum of the Greek population joined the rebels. To the north-east the rebellion spread up to Sofia. The main army of insurgents, under the command of Peter Delyan, advanced south towards Thessaloniki, inflicting crushing defeats on the Byzantines on their way. The rebellion spread over a large territory and assumed a liberating and anti-feudal character. At the peak of the rebellion Peter Delyan's cousin Allusian, son of Tsar Ivan-Vladislav, escaped from Byzantium and joined the insurgents. Having faith in his cousin, Peter Delyan put him at the head of a 40,000-strong army and sent him against

Thessaloniki, but Allusian acted carelessly and his troops were routed. Upon his return to the rebel camp he committed a new betrayal: he blinded Peter Delyan during a feast and hurried back to Byzantium. The rebellion, however, endured. The Byzantine Emperor, Michael IV the Paphlagonian, sent Norman mercenaries against the Bulgarian rebels, who suffered a number of heavy defeats. The desperate resistance of the Bulgarians lasted for a year and it was not until 1041 that the boyar Botko, the defender of the Boyana fortress near Sofia, was forced to surrender. Peter Delyan was taken prisoner and sent to Constantinople. In the territory it covered and in the number of rebels who fought in it, this was the biggest Bulgarian uprising against Byzantine domination.

A quarter of a century later, in April of 1066, a protest movement against the excessive taxes and duties broke out in Thessaly. The movement was organized by the Bulgarian population in alliance with the Wallachians.

The centre of the rebellion was the town of Larissa. To ensure the success of the Liberation movement they had started, the organizers sought to win over to their side the boyar Nikulitsa Delphinus, Governor of Larissa and grandson of Samuil's *voivode* Nikulitsa. At first Nikulitsa Delphinus refused to take part in the rebellion, since his sons were in Constantinople, and he even informed the Emperor about the plans. Very soon, however, he was compelled to take the lead. The rebellion broke out in June in Larissa and soon spread to Trikkala, Pharsalus and other places. A detachment was sent to seize and destroy the fortress of Kythros on the Aegean coast (some 80 km south of Thessaloniki). The main insurrectionist army headed north and conquered Servia (on the border between Thessaly and Macedonia). The Byzantine Emperor's promise to spare the lives of the rebels if they stopped fighting undermined their morale and soon Nikulitsa Delphinus gave himself up.

The defeat of the Byzantines by the Seljukian Turks at Mantsikert in Armenia in 1071 and the incursion by the Hungarians into the north-western parts of the Balkan Peninsula stirred up the Bulgarians anew. The initiative of the rebellion and its organization was the work of the Bulgarian boyars of Skopje, who were headed by Georgi Voiteh, a descendant of an old aristocratic family. The uprising was to break out in the autumn of 1072 in Priznen, and Skopje was to be its centre. In order to win the favour of Prince Michael of the Zetas, the rebels proclaimed his youngest son Konstantin Bodin a Bulgarian Tsar under the name of Peter. Together with Konstantin Bodin a band of 300, led by the commander Petrila, joined the rebels, and in the first battle the Byzantine army was routed and its commander Damian Dalasin taken prisoner. Skopje was conquered and Georgi Voiteh appointed its Governor. This success won the rebellion more supporters and helped to spread it over almost the whole of Macedonia. Konstantin Bodin, at the head of a detachment, moved north-east and reached Nish. Another detachment under the command of Petrila roused the population of the Polosk region (Tetovsk) to revolt, captured Ohrid and Devol and reached Kostur, where he was taken by surprise and defeated. After this blow Petrila abandoned his army to find refuge in Serbia. From Kostur the Byzantine army advanced northward, reached Skopje and conquered it. For a short time Georgi Voiteh went over to the side of the Byzantines, but was soon to regret his decision, inform Konstantin Bodin of his shifted loyalty and return to the Bulgarian ranks. The decisive battle took place in the southern part of Kossovo Pole. A betrayal decided the outcome of the fighting in favour of the Byzantines, since the Serbian-employed Byzantine commander deserted Konstantin Bodin and thus settled the outcome of the battle in advance. The Byzantine army succeeded in gaining control of the whole territory of the insurgents (1073) and the leaders of the rebellion were captured. On the way to Constantinople Georgi Voiteh died of wounds he had received, while Konstantin Bodin was thrown into a dungeon.

In 1074 a rebellion broke out in the sub-Danubian regions, and an attempt by the Byzantine Emperor Michael VII Ducas to quell the unrest by appointing a Bulgarian governor of the area failed. The boyar Nestor, who was appointed to the post, sided with his compatriots and assumed leadership of the popular unrest. There are no records as to how long the unrest in the sub-Danubian regions lasted, but for many years the populace there refused to acknowledge Byzantine power and supported the Bulgarians to the south of the Stara Planina in their struggle against Byzantium.

Four years after this uprising the Bulgarians made two new unsuccessful attempts, one from Plovdiv by Leka (a member of the Paulician religious sect) who tried to organize the population in the region of Sredets, and one, simultaneously, by the Bulgarian population in the town of Messembria (the present-day Nessebur) where Dobromir also attempted to organize a revolt.

Between 1084 and 1086 the Paulician rebels in Thrace maintained their resistance. An immediate cause of their uprising was the massacre of the Paulician notables by Byzantine Emperor Alexius I Comnenus, who summoned them under a false pretext to Mossinopol (the region of the present town of Gumurdjina) and ordered them to be murdered . The Paulician rebels were led by Travul, a former courtier in Constantinople, who fled the royal city followed by a number of adherents. The rebels took the Belyatovo fortress (north of the modern town of Plovdiv) and made it their centre of operations. They managed also to conclude a treaty with the Pecheneg tribes. The uprising gradually covered the entire Thrace region. In 1086 a large force under the Great Domesticus Gregorius Baccuriani was sent against the rebels, but the Byzantine army was defeated in the battle of Belyatovo, where both Baccuriani and the Commander Nikolai Vrana were killed. The Emperor then sent new forces and Travul considered it wiser to withdraw, together with the Pecheneg warriors, to the regions north of the Balkan Range.

Another form of resistance against Byzantine oppression was the spread of social and religious teachings and movements among the Bulgarian population, especially of

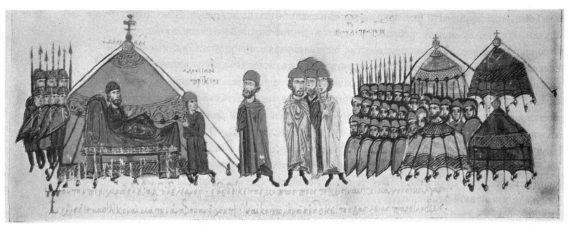

The Camp of Peter Delyan in Ostrovo and the Arrival of Allusian. Illumination from the Madrid Manuscript

Assen and Peter. Painting by Peter Gyudjenov

the Bogomil teaching and the Paulician religious ideas. The Bogomil teaching had its followers in Thrace and in the Rhodopes, and the town of Plovdiv and the Draguvitia settlement became Bogomil centres.

Their leaders and followers participated actively in the rebellions and revolts against Byzantine rule. Born on Bulgarian soil, the Bogomil teaching later spread to other countries, was disseminated in Asia Minor in the eleventh century, and early in the twelfth century found very wide support among the population of the Byzantine capital itself. Emperor Alexius I Comnenus subjected the Bogomils to systematic persecution. In the second half of the eleventh and the beginning of the twelfth century, seclusion had become the preferred form of resistance and many of the hermits (of which best known are Prochor Pchinski, Gavril Lesnovski and Yoakin Ossogovski, later canonized by the Bulgarian Orthodox Church) set up unpretentious monasteries which later developed into centres of learning and culture for the Bulgarians.

The oppressive Byzantine rule, which had lasted for 169 years, had grave consequences for the Bulgarian people, who were deprived of the opportunity to direct their own historical destiny and thwarted in their political and cultural development. The Bulgarian population was subjected to extermination by both the Byzantine authorities and the numerous invaders of the Bulgarian lands. The evolution of Bulgarian nationhood also suffered significant setbacks because of the Hellenization of part of the population, the falling away of Bulgarian territories from the main body of the Byzantine Empire, and other factors. The Bulgarians, however, managed to manifest their perseverance and their resistance to foreign influences by preserving their ethnic identity and assimilating huge masses of such ethnically different tribes as the Pechenegs, Kumanians and others who had migrated to the Bulgarian territories. Byzantine rule proved particularly harmful in that it disrupted the Bulgarian tradition of statehood and obliterated the institutions created in the Early Middle Ages.

Resurgence and Power of the Bulgarian State. In the second half of the twelfth century a profound internal crisis was imminent in the Byzantine Empire, which had repeatedly suffered crucial military defeats, In 1183 the Hungarians invaded the Bulgarian territories and took Sredets. Of particular danger for the Empire was the invasion of the Normans, whose fleet headed for Constantinople after the seizure of Drac and Thessaloniki. Centrifugal and separatist trends became prevalent in Byzantium, and the plight of the Bulgarians was rendered almost unbearable by the constantly increasing financial and political burden and the inability of the Byzantine State to stave off foreign invasions. When, in the spring of 1185, the Bulgarian population from the Balkan Range and the Danube regions was forced to pay additional tax on the occasion of the royal marriage of Emperor Isaac II Angelus, a spontaneous revolt erupted. The two brothers Assen and Peter, boyars of the town of Turnovo, headed the revolt and turned it into an organized uprising against foreign domination. The uprising was officially declared at the end of 1185 in Turnovo, on the day the St. Demetrius Church was consecrated in the town. The rebels pronounced Peter Tsar of the Bulgarians (1185–1187), and within a short time Byzantine rule over the northern regions of Bulgaria was liquidated. The rebels proceeded to take the passes in the Balkan Range and finally entered the Thracian lowland.

For several months they waged successful operations against the Byzantine troops and the Emperor was forced to replace three of his commanders. The fourth, Alexis Vrana, revolted against the Emperor and laid siege to Constantinople. In June and July 1186 Emperor Isaac II Angelus himself led his troops into Northern Bulgaria. To avoid unnecessary casualties, Peter and Assen withdrew with their rebel army to the north of the Danube, while the commanders of the strongholds who were left behind

Gold seal-ring of Tsar Kaloyan, found at Veliko Turnovo. Twelfth–thirteenth centuries. National History Museum, Sofia

surrendered without resistance. The Emperor took this as proof of a great victory and returned to Constantinople. But in the autumn of 1186 Peter and Assen crossed the Danube at the head of the combined forces of rebels and Kumanians, with whom they had made an alliance, wiped out the Byzantine strongholds throughout the country and once again entered the Thracian lowland. Attempts by the Byzantine units to act as they had the previous summer were foiled, and the Bulgarian and Kumanian units kept the Emperor's army off the Balkan Range. In the winter of 1186–87 the Emperor sent his army to Sredets and in the early spring of 1187 advanced towards Turnovo from the west. The Byzantine troops were checked in Lovech and laid a three-months' siege to the town. The period proved sufficient for the Bulgarian troops to surround them. The Emperor was forced to sign a peace treaty which was a *de facto* recognition of the restored Bulgarian State. The youngest brother of Assen and Peter, Kaloyan, was sent as hostage to Constantinople, to guarantee the observance of the treaty.

In 1189–90 the Third Crusade, headed by the German Emperor Friedrich Barbarossa, passed through the Balkan Peninsula. The Bulgarians promptly seized the opportunity to strengthen their position and the prestige of their newly restored state. Tsar Peter abdicated voluntarily in favour of his younger brother Assen (1187–96) who had doubtlessly been a better statesman and military commander. Peter himself became Governor of the North-eastern Bulgaria and the Dobrudja region, with Preslav as its centre.

Once relieved of his anxieties about the Crusaders, the Byzantine Emperor organized a campaign to conquer the Bulgarian capital of Turnovo and laid an unsuccessful siege to the town. While letting the Byzantine troops withdraw, the Bulgarians laid an ambush for them in the Trevna pass in the central part of the Balkan Range and dealt a crushing blow on the invaders.

In the following years hostilities against Byzantium concentrated on the territories south of the Balkan Range and covered a wide front, from the river Struma to the Strandja mountain. Sredets and the adjacent regions were liberated, and the Byzantine troops suffered another crushing defeat in the battle at Arcadiopolis, the present-day town of Luleburgas, in Eastern Thrace (1194). The Bulgarians continued their offensive and reached the Ser region (1195) taking over enemy strongholds on their way. Byzantine power in Eastern Thrace and the Aegean area was seriously threatened. Considerable success was also scored in the Bulgarian military offensive against the Hungarians, as a result of which the Belgrade and Branichevo territories were recovered.

The intrigues of Byzantine diplomacy incited part of the Bulgarian nobility to oppose the firm, efficient rule of the Bulgarian Tsar, Assen I. A conspiracy was organized against him and in 1196 he was murdered by his cousin Ivanko. Having succeeded to power, Ivanko proved unable to keep it and, when a siege was laid to Turnovo by Assen's brother Peter, fled to Constantinople. There he was promoted to the post of Governor of the Plovdiv region, having already been admitted to the Byzantine court. Peter was again crowned Tsar of the Bulgarians (1196), only to be murdered less than a year later (1197) by disgruntled boyars. He was succeeded by the youngest brother, Kaloyan (1197–1207), who had managed to escape from Constantinople and get back to Bulgaria in 1196.

The first task of the new ruler was to put an end to the discontent and cessationist aspirations. Acting firmly, he succeeded in overcoming the boyars' opposition within a short time. Thanks to his tact as a diplomat he also secured the support of the Governor of the Plovdiv region, Ivanko; of Dobromir Chris, who had previously renounced the power of Turnovo and was at the time ruler of Vardar Macedonia; and of the Governor of the Rhodope area, Ioanitius Spiridonaki. Thus Northern Thrace and part of the Rhodopes and Macedonia were quickly annexed to Bulgaria, and in 1201 Tsar Kaloyan conquered Varna, the only Byzantine stronghold north of the Balkan Range. In 1202 he signed a peace treaty with the Byzantine Emperor Alexius III Angelus, by virtue of which all territorial acquisitions of Tsar Kaloyan were recognized by Byzantium. In 1203 the Bulgarian army defeated the Hungarian troops who had invaded the north-eastern Bulgarian lands. Negotiations with the Roman Church were motivated by the will to reaffirm international recognition of the position and status of the Bulgarian State. They had started as early as 1199, and grew most intensive in 1202–4, when ambassadors and letters were repeatedly exchanged between Tsar Kaloyan and Pope Innocent III.

In the autumn of 1204 the negotiations were crowned with success and Tsar Kaloyan was given the title of 'King' and the right to mint coins. Archbishop Vassili of Bulgaria was given the title of Primate of the Bulgarian Church, which the Pope considered equal to that of Patriarch. Early in November 1204, the Papal Nuncio, Cardinal Leo, solemnly crowned the Bulgarian Tsar. Yet in practice, although it recognized the supremacy of the Pope, the Bulgarian Church preserved its independence. The Union had a formal, rather political and diplomatic character, and did not essentially affect Bulgaria's affiliation to the Eastern Orthodox Church.

In 1204 a substantial change took place in the political situation in the Balkan Peninsula. During the Fourth Crusade, Constantinople was captured by the western knights. The Byzantine Empire collapsed and for some time disappeared from the historical stage. The Latin Empire was established in its stead. Its first Emperor Baldwin I and his entourage made no secret of their aggressive plans toward Bulgaria. The attempts of Pope Innocent III to avert the imminent conflict between Bulgaria and the Latin Empire produced no results. In the summer of 1204 the Latins advanced across Thrace and subjugated a number of towns. The Byzantine aristocracy from Thrace sought the help and protection of Tsar Kaloyan. The Latin knights set out on a campaign, led by their Emperor, and a decisive battle took place on 14 April 1205 near Adrianople, where the Bulgarian forces, led by Tsar Kaloyan, routed the knights and captured Emperor Baldwin I. The whole region of Eastern Thrace, with the exception of Constantinople and some adjacent towns, fell into Bulgarian hands. Tsar Kaloyan headed for the lands lying between the Maritsa and Mesta rivers, and Serrai was captured. The Bulgarian successes frightened the Byzantine aristocracy, who renounced their alliance with Tsar Kaloyan. At this turn of events the knights were able to rally from the heavy blow they had suffered. In July 1207 the new Latin Emperor Henri and the Thessalonian King Bonifacius of Montferrat agreed to unite their forces and attack Bulgaria. On the way back to Thessaloniki, in a battle with the Bulgarians in the Rhodopes, Bonifacius of Montferrat was killed. Tsar Kaloyan besieged Thessaloniki, but under the city walls, just before the decisive storming of the city, he fell prey to a conspiracy (October 1207), and the Bulgarians lifted the siege.

The reign of Tsar Kaloyan brought to a natural conclusion the Bulgarian people's struggle for liberation and aspirations toward unification. It formed a most significant chapter in Bulgarian history. Thanks to Kaloyan's talent as a statesman, general and diplomat, within a short time after its restoration the Bulgarian State became a major military and political force in the Balkan Peninsula.

At the bottom of the conspiracy against Tsar Kaloyan were his wife and his sister's son, Boril. Tsar Boril (1207–18) usurped royal power and started persecuting Kaloyan's relatives and adherents, a number of whom were destroyed. Others seceded from the State (Sebastocrator Strez, brother of Boril, ruled in Vardar Macedonia, having as its centre the stronghold of Prossek; Despot Alexis Slav established independent rule over a region in the Rhodope Mountains, with a centre at the Tsepina fortress). Yet others fled from the country. The two sons of Tsar Assen I, Ivan Assen and Alexander, who had legal rights to the throne, fled for their lives to the Russian principality of Galicia. The unity of the Bulgarian State, achieved at the cost of precious sacrifices and much bloodshed, began to crumble.

Tsar Boril continued the anti-Latin policy of his

The Victory at Klokotnitsa, 1230. Painting by Mito Ganovski and Boicho Grigorov. National Military History Museum, Sofia

predecessor. In the spring of 1208 he invaded Thrace with a Bulgarian and Kumanian army. In a battle fought near Plovdiv the Bulgarians suffered a defeat which brought about a consolidation of the Latin coalition to which Despot Mihail of Epirus, Despot Alexis Slav and Sebastokrator Strez were attracted. The allies waged a war on a broad front and achieved certain successes, but as a result of diplomatic efforts of the Pope their actions were paralysed. In 1213–14 Tsar Boril concluded peace treaties with the Latin Empire and the Hungarian kingdom, consolidating them with dynastic marriages.

In the meantime, inside the country opposition to and dissatisfaction with the rule of the usurper Boril grew rife. They found expression not only in official opposition involving a section of the boyars, but also in the wide acceptance of Bogomil teaching among the mass of the people. This compelled the Tsar and the higher clergy to hold a council against the Bogomils in the capital, Turnovo, in February 1211. A few years later an insurrection against Boril's rule broke out in the Vidin region; this was crushed with the assistance of Hungary. The Bulgarian State was in a position of acute crisis in its domestic policy.

In 1217 Ivan Assen returned to Bulgaria at the head of Russian and Kumanian battalions to win back in battle his captured throne. He was joined by representatives of the feudal aristocracy and backed by the mass of the people.

Isolated, Tsar Boril fortified himself in Turnovo. After a nearly seven-month siege the capital opened its gates. Boril was removed from the throne, and in the spring of 1218 Ivan Assen II was proclaimed Tsar of Bulgaria.

The new Bulgarian ruler was distinguished for his exceptional qualities as statesman and diplomat, which he maintained on numerous occasions during his reign. In 1219 Tsar Ivan Assen II concluded a peace treaty with the Hungarian King Andrew II, as a result of which the Belgrade and Branichevo regions were returned to Bulgaria, and the Bulgarian Tsar married the Hungarian Princess Anna-Marie. He concluded a peace treaty with the Despot of Epirus and regained a part of the lost territories. At first Ivan Assen II concluded peace also with the Latins, had his daughter Elena betrothed to the minor Emperor Baldwin II, became the latter's guardian and established his political influence over the Latin Empire. Through skilful diplomatic activity Ivan Assen II restored the political might of the Bulgarian State and restored to it a great part of the territories it had lost under Boril. In March 1230 the Despot of Epirus, Todor Comnenus, violated the peace treaty and unexpectedly entered Bulgaria, but on 9 March 1230, near the fortress of Klokotnitsa (Haskovo district) Tsar Ivan Assen II won a brilliant victory over the Epirus troops. Todor Comnenus himself was made captive, together with his whole army, and taken to Turnovo. The common soldiers were allowed

Obverse　　　　　　　　　　　　　　　　　　Reverse

Gold coin from the reign of Tsar Ivan Assen II. National Archaeology Museum, Sofia.

to go back peacefully to their homes. This highly humane gesture by the Bulgarian Tsar became known all over the Balkan Peninsula and made a stong impression on his contemporaries. The Bulgarian forces continued their victorious march, and within a short time Bulgarian rule was restored over the Rhodope region, in the Aegean region and in Macedonia. The Arbanassian lands (Albania) were also included within the boundaries of the Bulgarian State. As previously, at the time of Tsars Simeon and Samuil, Bulgaria once again reached the Adriatic Sea. The Epirus-Thessaloniki empire of Todor Comnenus collapsed. The Serbian Kingdom became vassal to Bulgaria.

As a result of the victory of Klokotnitsa and the ensuing developments the Bulgarian State gained a hegemony in the Balkan Peninsula. The Hungarian advance into Bulgarian territory in 1231–32 was repelled by the forces led by the brother of Tsar Assen II, Sebastocrator Alexander. The alliances with Hungary and the Latin Empire were disrupted and Union with the Roman Church was in fact liquidated.

In the resulting situation, with the open hostility of the Catholic world, Tsar Ivan Assen II was compelled to change the State's foreign policy towards a return to the bosom of the Eastern Orthodox Church and an alliance with the Nicean Empire. In 1235 a military-political alliance was concluded between Bulgaria and Nicea, directed against the Latin Empire. At the same time, with the agreement of all European Patriarchs, the Bulgarian Patriarchate was restored. The Bulgarian side ceded to the Nicean Empire the region between the Maritsa and Gallipoli on the Balkan Peninsula. The armies of the two States besieged Constantinople by land and by sea, and to save the Latin Empire from annihilation the Roman Pope called Hungary and other Catholic states to a Crusade against Tsar Ivan Assen II and the Nicean Emperor, John III Dukas Vatatses, proclaiming them 'heretics and schismatics'.

In 1236 Ivan Assen II again changed the foreign policy of the Bulgarian State. Realizing that assisting the Nicean Empire could disrupt the established *status quo* in the Balkans, he sought *rapprochement* with Hungary and the Latin Empire, and renewed relations with the Pope. In the last years of his reign he established flexible relations, depending upon precise circumstances, with Latins, Niceans and Hungarians. In 1241 Bulgaria beat off an attack on its frontiers by the Tartars in their impetuous onslaught against Central Europe.

The great territorial expansion of the Bulgarian State after 1230 had a favourable effect on its economic and cultural development. Its authority was considerably enhanced, and its ties with other countries and peoples broadened. The Bulgarian Patriarchy acquired great importance in the clerical and literary life of the Slav people. A great deal was done for the urban development of the capital Turnovo, situated in a beautiful natural setting on the banks of the Yantra river. The capital was made up of several fortresses: Tsarevets, Trapezitsa, Sveta Gora, Novgrad and Devingrad, the main one being on Tsarevets Hill. The name of Tsar Ivan Assen II is associated with the construction of many churches, fortresses, the Palace on Tsarevets Hill and other works. It was in his reign that the minting of coins – gold, silver and copper – began in Bulgaria. The churches and monasteries were generously endowed with lands and granted privileges. Crafts and trade flourished, and the towns gained

Obverse   Obverse

Gold seal of Tsar Konstantin Assen. Thirteenth century. National Archaeology Museum, Sofia.

prosperity. Literature, the arts and crafts blossomed. Men of letters, artists, and monks flocked to Turnovo from all parts of the Balkan Peninsula. The transference of the relics of saints, popular among the Balkan peoples (Ilarion of Muglen, John of Rila, Petka Epivatska, Mihail Voin, Philoteya and others) to the Turnovo churches enhanced the glory of the Bulgarian capital. The long reign of Tsar Ivan Assen II is justifiably considered one of the most successful and fruitful periods in the history of Medieval Bulgaria.

The Decline and Renaissance of the Bulgarian State. After the death of Tsar Ivan Assen II, his son, a minor, from his marriage with the Hungarian Princess Anna Maria ascended the throne as Kaliman I (1241–46). In 1242 Bulgaria was invaded by the Tartars. Devastated and plundered by the hordes of Khan Batu, it was forced to pay taxes to the Golden Horde.

A palace coup led to a change on the throne when Tsar Kaliman I was poisoned by Ivan Assen II's second wife Irina. Her son, who was also under age, succeeded to the throne as Mihail II Assen (1246–56). During the regency the leading role was played by his mother and Sebastocrator Peter, who virtually reigned as deputy in the western half of the kingdom. The Nicean Empire took advantage of the difficulties of the Bulgarian State and the strife in the palace to occupy Serrai, Thessaloniki, the Rhodope region and Adrianople. At the same time Southern and Western Macedonia, Epirus and Albania were severed from the Bulgarian State. Under these circumstances the regency was forced to make peace with the Nicean Emperor and promise to assist him in winning back Constantinople (1247). The Hungarians occupied the Belgrade and Branichevo districts.

In 1253 the Bulgarian foreign policy again became active. A successful war was fought against Serbia, in close alliance with Dubrovnik, and in 1254 large-scale military operations were launched against the Nicean Empire for the liberation of the territories it had occupied. However, in the autumn of 1256 King Mihail II Assen was assassinated in a plot masterminded by his cousin Kaliman. When he took power, Kaliman married the widow of Mihail II Assen (daughter of the Hungarian vassal Prince Rostislav Mihailovich) and was proclaimed Tsar (1256–1257). A civil war broke out in Bulgaria, during which Tsar Kaliman II was killed. There were three claimants to the throne, engaged in internecine strife: Prince Rostislav Mihailovich (father-in-law of the murdered Tsar), Despot Mitso (son-in-law of Tsar Ivan Assen II) and the Boyar Konstantin Tih (Governor of Skopje district). The Turnovo boyars elected Konstantin Tih (1257–77). One of the other claimants, Mitso, fortified his positions in the Southern Bulgarian regions, with Messembria as their main centre. Tsar Konstantin Tih conducted a policy of *rapprochement* with the Nicean Empire; a peace treaty was concluded in 1257 and the Tsar married Irina, daughter of the Nicean Emperor Todor II Lascarid and granddaughter of Ivan Assen II, for which reason he also assumed the name of Assen. In 1259–61 he fought a long war, with varying fortunes, against the Hungarians, and in 1260–63 he crushed Boyar Mitso. In 1263, Byzantium intervened in the internal strife in Bulgaria and captured the Black Sea towns of Messembria and Anhiallo, but in 1265, in alliance with the Tartars, Tsar Konstantin Assen attacked Byzantium. He succeeded in freeing Boyar Yakov Svetoslav from his dependence on the Hungarians, gave him the title of Despot and helped him strengthen his positions in the district of Vidin. In 1269 the Tsar married the Byzantine princess Maria – niece of Emperor Michael VIII Palaeologus. Under the terms of the marriage settlement Byzantium undertook to return the fortresses of Messembria and Anhiallo to Bulgaria as dowry. After three years of procrastination, Byzantium refused to carry out its pledges.

Konstantin Assen invaded Thrace but the Empire sent

the Tartars against him, and from 1272 onwards they invaded and plundered the Bulgarian lands every year. The Tsar was compelled to seek the help of the King of Naples, Charles I of Anjou, and a political alliance was concluded between them.

The frequent wars and internal strife made the tax burden on the Bulgarian people unbearable. Their plight worsened as a result of the foreign invasions by Tartars, Hungarians and Byzantines. The task of defending state and people from enemy attacks, arbitrary actions and social injustice proved more than royal power or the feudal boyars could shoulder. The devastating Tartar incursions became the scourge of the Bulgarian population. Subjected to merciless exloitation by the Tsar and boyars, condemned to ruin and subjugation by foreign invaders, and left utterly defenceless, the Bulgarian peasantry sought salvation and a way out of its plight by its own resources. An anti-feudal uprising of Bulgarian peasants, led by Ivailo, himself a peasant, broke out in North-east Bulgaria in the spring of 1277. At first it was directed at the boyars in an effort to secure the supplies and weapons the rebels needed, and also to take vengeance on them for their misdeeds in the past, but later the insurgents stood up against the invading Tartar bands, routed them and drove them beyond the Danube. These victories considerably enhanced the influence of the rebel army, which was joined by large numbers of peasants; entire regions renounced the central government and placed themselves under the command of the peasant leader. In the autumn of 1277 the insurgent army advanced on the capital, Turnovo, and won a brilliant victory in which the Tsar was killed. The rebel army seized the boyars' estates and townships one after the other, reached Turnovo and besieged it. Byzantine diplomacy again took advantage of events in Bulgaria and the forces of Emperor Michael VIII Palaeologus invaded, bringing along with them the claimant to the Bulgarian throne, the grandson of Tsar Ivan Assen II – Ivan Assen III. In this complicated political context Queen Maria and the Turnovo boyars decided to open the gates of Turnovo and recognize Ivailo as Tsar. The united forces of insurgents and boyars marched against the Byzantine army coming from the south, and battles with the Emperor's superior forces continued unabated in the summer of 1278 in the region south of the Balkan Range. Ivailo displayed great courage, the advance of the Byzantine army was checked and its way to the mountains barred. Byzantine diplomacy won the Tartars over to its side, and led by Kassim Beg they invaded Northern Bulgaria. Ivailo marched with a part of his army to the north but was besieged in Druster. At the same time the Byzantines landed fresh troops in Varna, advanced on Turnovo and, spreading rumours about Ivailo's death, succeeded with the assistance of some disgruntled boyars in placing Ivan Assen III on the throne. Ivailo routed the Tartars and marched back to the Bulgarian capital. Byzantium sent two armies in succession to help the threatened Ivan Assen III, but in the summer of 1279 Ivailo routed these armies in the Balkan mountain passes. Seeing that he was unable to keep the throne for himself, Ivan Assen III robbed the royal treasury and fled to Constantinople. The boyars elected one of their own number to be the new Tsar – Georgi Terter I (1279–92). The war between the insurgents and the boyars lasted nearly a year, until exhausted in countless battles and poorly armed, the people's army was defeated. In 1280 the uprising was put down and Ivailo sought refuge with the Tartars, where, during a feast, he was beheaded on the orders of Khan Nogay, as a result of Byzantine intrigues. The memory of Ivailo lived on in the people's minds and on two occasions later false Ivailos came to the fore to lead the people's masses against the Turks invading their lands from Asia Minor.

By its scope and significance as a powerful social movement of anti-feudal character, the uprising headed by Ivailo could be compared to that of Dolcino in Italy (1303–07), that of Wat Tyler in England (1381), the Jacquerie in France (1358) and the peasant uprising in Catalonia (fifteenth century). It has precedence over them in time, and was the only anti-feudal peasant uprising in Europe whose leader took the throne, though for a short time only.

The high degree of feudalization of the Bulgarian social-political structure, the acute social crisis at the top and at grass-roots level, and the weakness of royal power in the second half of the thirteenth century brought about the political fragmentation of the country and the emergence of a number of feudal divisions: the brothers Durman and Kudelin held the Belgrade and Branichevo districts, with Zhdrelo fortress as its centre; the Vidin district was governed by Despot Shishman; the Sub-Balkan district, with Kopeie (near Karlovo) as its centre, by the brothers Smilets, Voyssil and Radoslav; the Krun district by Despot Eltimir.

In spite of these unfavourable circumstances Tsar Georgi Terter I conducted an active foreign policy from the very start of his reign. In 1281 he established close diplomatic relations with the King of Naples, Charles I of Anjou, and joined the anti-Byzantine coalition formed by him, but Byzantine diplomacy foiled the Tsar's plans and he had to steer the foreign policy of the Bulgarian State along new lines. Peace treaties with Byzantium and the Serbian Kingdom were concluded in 1284.

In the following year the Khan of the Golden Horde, Nogay, invaded the Bulgarian lands, razing to the ground everything in his path. Tsar Georgi Terter I was compelled to acknowledge himself his vassal, send his son Todor-Svetoslav as hostage with the Tartars and give his daughter in marriage to Nogay's son, Chaka. The previous taxation dependences now became direct subordination to the Golden Horde. Tartar domination over Bulgaria inflicted a damaging blow to the authority of the Tsar and deepened the crisis in the country. The squabbles between the different boyar clans were clearly instigated and exploited by the treacherous policy of Byzantium and Nogay.

In 1292 Tsar Georgi Terter I lost the confidence of his suzerain, Khan Nogay, and was compelled to give up his

Patriarch Euthymius Bidding Farewell to the People of Turnovo. Painting by Boris Denev

throne and seek political asylum in Byzantium. Nogay placed his obedient servant Boyar Smilets on the vacated throne (1297–98). His brief, featureless reign is one of the saddest chapters in the history of medieval Bulgaria. The Tartar hordes roamed unhampered, pillaging and plundering the country. Tsar Smilets looked passively on the annexation of the districts held by Durman and Kudelin to Serbia. Nor did he react to the Serbian advance on the Vidin despotat, which led to the expulsion of Despot Shishman. It was the intervention of the Tartars that restored Bulgarian rule over that territory. After the death of Smilets, power passed temporarily into the hands of his wife, who was of Byzantine descent. Intrigues and internal strife flared up again in the Bulgarian State.

In the late thirteenth century a sharp power struggle broke out within the Golden Horde, in which Khan Toktu emerged the victor. Nogay was killed. His son Chaka, accompanied by Todor Svetoslav, sought refuge in Bulgaria and in 1300 seized power.

Shortly afterwards, Todor Svetoslav engineered a plot against Chaka and killed him.

Seasoned by long years of exile, Todor Svetoslav (1300–21) shouldered a grave political legacy. Feudal separatism was rife. From the start he took firm measures to bring into line his political adversaries among the boyars and high clergy. He imposed the death penalty on the Patriarch of Turnovo, Yoachim III, suspected of treason.

Attempts by Byzantium to use the opposition emerging in Bulgarian society for its own ends and dethrone Tsar Todor Svetoslav yielded no results. He made an alliance with his uncle Eltimir, the Despot of Krun. An attempt by Smilets's brother, Radoslav, awarded the rank of Sebastocrator by the Byzantine Emperor, to occupy the Sub-Balkan region also failed, and he was taken prisoner along with thirteen Byzantine notables. Tsar Todor Svetoslav opened negotiations with the Byzantine Emperor and exchanged the captives for his father, Georgi Terter I, who was held prisoner in Byzantium.

In less than a year Todor Svetoslav stabilized his position on the throne, overcame internal boyar opposition, repelled the intervention of Byzantine and *émigré* boyar forces and achieved partial centralization of the fragmented Bulgarian State, torn by internal strife. These changes cleared the way to overcoming the crisis in Bulgaria.

After decades of defensive policy *vis-à-vis* Byzantium, the Bulgarian State passed onto the offensive and restored its sovereignty over a number of towns and fortresses in Northern Thrace, Zagora and the Black Sea region south of the Balkan Range. In 1304, in a decisive battle in the valley of Skafida (today the Fakiiska river), the Byzantines sustained a crushing defeat. Military operations were carried right into Eastern Thrace. In early 1307, the two belligerent sides concluded a peace treaty, recognizing the restoration of Bulgarian sovereignty over the liberated territories.

Bulgarian foreign policy – considerably reactivated and freed from Tartar tutelage – went beyond the framework of relations with the Byzantine Empire. Commercial and political contacts with Venice were strengthened, in spite of the sharp deterioration of relations with the Republic of Genoa. During 1317–21 relations with Serbia also improved and the alliance between the two countries assumed an anti-Byzantine character.

After the death of Tsar Todor Svetoslav, his son Georgi Terter II (1321–22) succeeded him. He took advantage of the power struggle in Byzantium between Andronicus II and Andronicus III, and immediately after his ascension to the throne launched a campaign in the south to liberate the Plovdiv district and the Rhodopes from Byzantine rule. He succeeded in taking Plovdiv and many fortresses in the Rhodopes, and headed for Adrianople, but in late 1322, at the height of the war, he unexpectedly died. He left no heirs (his death brought the Terter dynasty to an end), and this fact precipitated a temporary crisis in the court, during which the Byzantine Emperor Andronicus III invaded Northern Thrace and occupied Plovdiv.

The Bulgarian State under the Shishmans. The interregnum and the altercations among the boyars over the succession to the throne lasted from December 1322 to May 1323, when finally they elected as Tsar the Despot of the Vidin Principality, Mihail Shishman (1323–30). This marked the beginning of the Shishman dynasty in Bulgaria.

Tsar Mihail III Shishman ascended the throne at a time when Bulgaria and Byzantium were at war and brought this war to an end, restoring Bulgarian sovereignty over Northern Thrace and the Black Sea coast. The Byzantine Government proposed peace, one of the conditions being the marriage of the Tsar to the sister of the Byzantine Emperor Andronicus III – Todor, the widow of Todor Svetoslav. The royal marriage took place in the summer of 1324 and thereby Mihail Shishman became the legitimate heir to the Terter dynasty and was officially recognized by Byzantium. As the dynastic struggles between Andronicus II and Andronicus III continued, Tsar Mihail Shishman sought to expand Bulgaria's territory. Initially he was in an alliance, made in 1327, with Andronicus III, but in early 1328 started talks with Andronicus II and promised to support him. After suffering defeat by the forces of Andronicus III (March 1328), Andronicus II sought help from the Bulgarian Tsar, who concentrated large forces on the Bulgarian-Byzantine border and sent a 3,000-strong detachment of mounted troops to Constantinople with the intention of taking the Emperor's Palace and occupying the Byzantine capital. By tactical manoeuvres Andronicus III managed to secure the withdrawal of the Bulgarian force from Constantinople; he then emerged the victor in the Byzantine internecine war and made peace with Mihail III Shishman, sealed with the Treaty of Peace and Alliance, which was directed against Serbia (1329). In the spring of 1330 the Tsar began military and diplomatic preparations for a war against Serbia. His brother Belaur, the Despot of Vidin, and his cousin Ivan-Alexander, the Despot of Lovech, also engaged in intensive war preparations. As his own preparation, the Serbian King Stefan

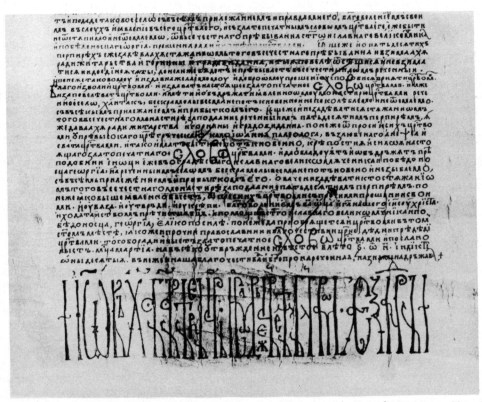

Facsimile of the final part of the endowment deed given by Tsar Ivan Alexander to the Zograph Monastery on Mount Athos in 1342

Dechanski hired a Catalan army. The battle between the Bulgarian and Serbian forces took place near Velbuzhd (today Kyustendil) on 28 July 1330. The Serbian army launched a surprise attack on the Bulgarian camp, which was not ready for action on account of the one-day truce declared at the request of the Serbians. A heavy defeat was inflicted upon the Bulgarian army and Tsar Mihail III Shishman was himself wounded in the battle and died a few days later. The Serbs entered Bulgaria but encountered Bulgarian reserve forces under the command of Belaur. After a round of negotiations the Serbian King abandoned his plans for advance into Bulgaria. Under pressure from Serbia, the Bulgarian throne was ascended by Ivan-Stefan (1330–31), son of Tsar Mihail III Shishman from his first marriage to Anna (Anna-Neda), daughter of the Serbian King Stefan Milutin.

In February 1331 two Turnovo boyars – the Protovestiarius Raksin and the Great Logoteth Philip – headed a group of boyars who carried out a coup. The Boyar council elected Mihail III Shishman's nephew – the despot of Lovech, Ivan Alexander (1331–71) – as Tsar.

In the spring of 1332 Tsar Ivan Alexander made an alliance with the Serbian King Stefan Dushan and gave his sister Elena in marriage to him. In that same year he defeated the Byzantines in the battle at Rusocastro (1332) and restored to Bulgaria the towns in the Eastern part of the Sub-Balkan valley it had lost in 1330.

Despite the successes scored in the state foreign policy, the process of feudal disintegration of the Bulgarian territories and formation of independent regions continued: Despot Deyan ruled in Velbuzhd district; Protosevast Hrelyo in the region enclosed between Strumitsa, Stip and Stob, with the Rila Monstery as its centre; the descendants of the Terters – the brothers Balik, Todor and Dobrotitsa – were the feudal lords of Dobrudja, with the fortress of Karvuna (today the town of Kavarna) as its main centre.

After the death of Emperor Andronicus III a long internecine war broke out in Byzantium between the claimant to the throne, John Cantacuzene, and the regency of the Emperor John V Palaeologus, a minor. Bulgaria, Serbia and the other small, independent Balkan states, as well as the Seljuk Turks, were gradually involved.

In 1341 the Bulgarian forces began incursions into Eastern and Western Thrace. To win the Bulgarian Tsar over to their side, the Regency ceded to Bulgaria many towns and fortresses in Northern Thrace and the Rhodopes. This created great difficulties for the claimant to the throne, John Cantacuzene. His situation was worsened as a result of the formation of a new independent

The Middle Ages in the Bulgarian Lands

Facsimile of the final part of the endowment deed given by Tsar Ivan Alexander to the St. Nicholas Monastery in Mraka, Radomir Region, in 1347

Bulgarian district in the Rhodopes and Western Thrace, under the sovereignty of Momchil. The later exploited the differences between John Cantacuzene and the government in Constantinople to establish an independent realm under his own rule, with Tsarevo (Xanti) as its centre, and Peritor (near the bay of Bistonis on the Aegean) its main fortress. With the assistance of the Seljuk Turks, John Cantacuzene defeated Momchil in battle at Peritor (7 July 1345). At that time certain changes in the Bulgarian royal court affected to some extent the political situation in the country. In 1345 Tsar Ivan-Alexander sent his first wife Todora off to a monastery and married the Jewess Sarah, who on her conversion to Christianity also adopted the name of Todora. To deprive his son from his first marriage, Ivan Stratsimir, of the right to succeed him to the throne, he made him the ruler of Vidin district which was also declared a kingdom. Thus the process of disintegration of the Bulgarian State was accelerated.

The middle of the fourteenth century saw the rise of a new great power in the Balkan Peninsula and in Europe – the Ottoman State. During the first half of the fourteenth century the Osman and the Seljuk Turks were often used by the Byzantines as allies. Their pillaging raids in Thrace, the Rhodopes and Zagora had already become a common occurrence. The threat of Turkish conquest at that time appeared most real to Byzantium and in 1351 Emperor John Cantacuzene sought the support and assistance of the Bulgarian and Serbian kings, but his request fell on deaf ears. One of the most significant results of the lack of understanding between the Balkan states and their disunity was the permanent settlement of the Osman Turks under the leadership of the son of Sultan Orhan-Suleyman in the Balkan Peninsula. In 1352 they conquered the Fortress of Zimpe, and in 1354 the strategically important fortress of Gallipoli.

The growing Osman threat, the insecurity of the population in the peripheral areas and intensive feudal exploitation led to the wide dissemination of the Bogomil teaching and of other heretical, social-religious teachings and movements within the confines of Bulgaria. The secular and Church authorities took serious measures to eradicate them; Church conventions on this problem were held in Turnovo in 1355 and 1360.

In spite of the weakening of the State, cultural life in Bulgaria was still on the upsurge. The Turnovo art school marked further progress in its development. Some of its best works were the murals in the Turnovo churches, in Boyana, Zemen and Ivanovo, a village near Russe. The art of illumination flourished, the most brilliant examples of this school being the miniatures in the London Gospel (the Gospel of Ivan-Alexander), the Vatican copy of the Chronicle of Manasses and the Tomich Psalter. Literature reached unprecedented heights in its development. Along with church literature appeared works on historical and secular subjects. A literary school – the Turnovo school of letters – carried out a great volume of work and gained wide popularity.

In the thirteenth and fourteenth centuries Bulgaria was among the major Slav and European states renowned for their culture.

In 1364 the Ottoman armies under Sultan Murad crossed the Southern Bulgarian border and took over the towns of Stara Zagora and Plovdiv. Alarmed by the Ottoman incursion, the Bulgarian Tsar began fortification works in all fortresses north of the Balkan Range and chiefly in the capital Turnovo. In the second half of the fourteenth century the Hungarian Kingdom launched offensive actions against the Balkan Peninsula. In the spring of 1365 the Hungarian army invaded the Vidin Kingdom, seized its capital and took possession of its entire territory, capturing Tsar Ivan Stratsimir, together with his family. In 1369 the armies of Tsar Ivan Alexander of Turnovo, assisted by those of Dobrotitsa, the Despot of the region of Dobrudja, and the Wallachian chieftain Vladislav I Vlaiku, liberated the Vidin Kingdom from the Hungarians and restored Ivan Stratsimir to his throne.

In 1366 the knights of the Savoy Duke, Amadeus VI, headed by the Duke himself, attacked the Bulgarian strongholds on the Black Sea coast and seized the towns of Sozopol, Anchialo, Messembria, Emona and Kosyak. In the spring of 1367, following prolonged negotiations, the knights withdrew, surrendering the strongholds south of the Balkan Range to the Byzantine authorities.

In 1369 the Turks captured Adrianople and moved their capital to that city. Wedged in the territories of the Balkan countries, the Ottoman state bordered on the Turnovo Kingdom in the north, the Sea of Marmara in the south, the Constantinople region in the east, and the Principality of Uglesh, whose capital was Serrai, in the west.

Faced with the strong threat of Ottoman conquest, the politically and territorially disintegrated Bulgaria had also to cope with the fruit of the inconsistent and often shortsighted policy of Tsar Ivan Alexander, who died on 17 February 1371, after 40 years of rule.

The campaign against the Turks organized by the Despot of Serrai, Uglesha and by King Vulkashin of Prilep, and the decisive battle that followed on 26 September 1371 at Chirmen (the village of Ormenon in present-day Greece), brought victory for the Turks. In 1372, on the pretext that the new Tsar in Turnovo, Ivan Shishman (1371–95), had violated the treaty, the Ottoman Turks invaded the Turnovo Kingdom. The situation became critical for Turnovo. Tsar Ivan Shishman was compelled to sign a peace agreement which made him and his Kingdom vassals to the Ottoman Sultan, and to give his sister, Kera Tamara, in marriage to Sultan Murad (1375). To avoid attacks on themselves, other Bulgarian rulers – Tsar Ivan Stratsimir, the Dobrudja Despot Dobrotitsa, the Velbuzhd Despot Konstantin Dragesh and the Prilep ruler Krali (King) Marko – also recognized the sovereignty of the Ottoman Turks. The Byzantine Empire also became vassal.

In 1378 the Ottoman Turks violated the peace agreement with the Turnovo Kingdom and continued their advance to the north-east, capturing the strongholds of Stipone (the present-day town of Ihtiman) and Samokov. After years of brave resistance, Sofia fell to the Turks in 1382, and then Nish and Pirot in 1385. The Turks were now wedged firmly between Bulgaria and Serbia. The Serbian King Lazar made an attempt to form a Christian Defensive Alliance against the Ottoman Turks, including King Tvrdko of Bosnia, Tsar Ivan Shishman of Turnovo and the new Dobrudja ruler, Ivanko, but the other Bulgarian rulers, Tsar Ivan Stratsimir, Despot Konstantin Dragesh and Marko kept their status of loyal vassals of the Ottoman Turks, and the lack of unity hampered the anti-Ottoman resistance.

In 1387 the united armies of the Bulgarians, Serbs and Bosnians defeated the Ottomans at Plochnik, but the Turnovo Kingdom and the Dobrudja Despotat became the victims of Ottoman revenge, when in 1388 Commander Ali Pasha advanced through the passes in the Eastern Balkan Range and captured by storm many Bulgarian strongholds. Tsar Ivan Shishman was again forced to acknowledge his vassalage to the Ottomans and to surrender the Druster fortress.

Having thus dealt with the rebellious Bulgarian vassals, Sultan Murad launched a campaign against Serbia and Bosnia. King Lazar met the Turks at Kossovo Pole on 15 June 1389. His army included Bulgarian, Bosnian and Wallachian units. Despite the initial victories of the Christian Alliance forces (Sultan Murad himself was killed in battle), the Ottomans finally defeated them. Their advance could no longer be restrained.

In 1393 the Ottoman Turks renewed their large-scale offensive against the Balkan states. Chelebi, son of Sultan Bayazid I, headed the army against Bulgaria, and after a surprise attack laid siege to Turnovo. The defence of the Turnovo fortress was led by Patriarch Euthymius. On 17 July 1393, after three months of courageous, stubborn resistance, Turnovo fell.

The aristocracy was massacred, and the clergy and the population were driven into exile or sold as slaves. Tsar Ivan Shishman fortified his positions at Nikopol. Supported by the Hungarians, the Wallachian Prince, Mircho the Elder, fought a battle against the Ottomans on 17 May 1395 in the Rovine plain, which was won by the Turks. Among those killed in this battle were the Prilep ruler Marko and Despot Konstantin Dragash, who had fought on the side of the Turks. On the pretext of breach of promise by Tsar Ivan Shishman and Despot Ivanko to lend him support Sultan Bayazid attacked their armies and conquered them. Tsar Ivan Shishman was captured and imprisoned, to die in captivity only a year later. The Velbuzhd Despotat and the Prilep Kingdom were also conquered. The victories of the Ottoman Turks had a tremendous impact on the states in Central and Western

Haiduks Lying in Ambush. Painting by Dobri Dobrev. National Military History Museum, Sofia

Europe. The appeal by King Sigismund of Hungary to organize a Crusade against the Ottomans was supported by the Pope, and in response knights from all over Western Europe joined his troops. The allied European forces concentrated in Hungary and advanced into the Balkans. Tsar Ivan Stratisimir destroyed the Ottoman garrison stationed in his capital and also joined the Christian armies. However, although numerous, the armies lacked homogeneity and unity of action and command. This led to their crucial defeat by the Turks on 25 September 1396 in the battle at Nikopol. Soon afterwards the Vidin Kingdom was also conquered and Tsar Ivan Stratsimir was led into captivity. This brought to an end the independent development of the medieval Bulgarian State.

Bulgaria was the first victim of the Ottoman expansion in the mainland of the Balkan Peninsula. The resistance it put up enabled the other Balkan and European states to organize their own defence against the Ottoman conquerors at this crucial period in history.

The Bulgarians in the Fifteenth to Seventeenth Centuries. The Ottoman conquest destroyed the Bulgarian State institutions and the independent Bulgarian Church, and disrupted the normal development of medieval Bulgarian nationhood. The Bulgarian lands were included in the Rumelia *beylerbey* and divided into several *sanjaks* (districts). The Bulgarian population was totally deprived of the right to participate in the local and central bodies of government. With the destruction of the Bulgarian feudal class, the population was socially levelled out on the principle of general absence of rights and poverty. As an entity it was included in the category of *rayah,* subjected to all forms of exploitation in the feudal system of production. The Bulgarians, like all non-Muslim subjects of the Empire, were deprived of civil rights. Many were killed, thousands of families were compelled to emigrate, the strongest and healthiest boys were taken forcibly away from their families and educated as fanatic Muslims to become *yeniceri* (Janizaries) – soldiers of the Sultan. Those Bulgarians who remained in their country

St. George of Yanina in Haiduk Attire. Icon from Bansko. National Art Gallery, Sofia

Church of St. Naum, Ohrid

were subjected to religious and national discrimination. There followed numerous attempts at resistance to Ottoman methods of government, but the laws of subordination to the local feudal ruler were imposed in the Bulgarian lands. The new type of relations led to a high degree of feudal exploitation of the Bulgarian population, which was forced to pay numerous taxes to the central authorities and to the local *spahi* (rulers).

The type of economic relations and political regime imposed by Ottoman rule retarded Bulgarian economic development. The social and political status of the Bulgarians was regulated by Ottoman feudal law, based on the *shari'a* a (Muslim religious legislation) and the secular legislation. The policy of control and pressure was carried out by the local authorities, of which the *Kadi* (judges) was the most influential. From the end of the sixteenth century the advanced feudal development clashed with the principle of centralized government of the Ottoman Empire. The wish to establish unconditional and complete ownership of land by the feudal class resulted in specific, parasitic forms of land ownership such as the *chiflik* (farm) ownership and the *melikyane* (lease of state-owned property as a source of income for life). This promoted separatist trends among the local feudal lords and resulted in increased feudal exploitation of the Bulgarian population, in which corruption and lawlessness became a daily occurrence in the Bulgarian lands. The process reached its culmination at the end of the eighteenth and the beginning of the nineteenth centuries, when a prolonged period of anarchy began in the Balkan Ottoman dominions, and especially in the Bulgarian lands.

Despite the burden of foreign rule the Bulgarians succeeded in preserving their ethnic identity and distinguished themselves from their conquerors in their moral and ethical values, their religion and their way of life. They preserved their national consciousness and sense of unity with all the people in the traditionally Bulgarian territories of Dobrudja, Moesia, Thrace and Macedonia. Their common history, language, way of life, material culture and Eastern Orthodox religion were factors of tremendous importance for the preservation of Bulgarian nationality.

In the period of Ottoman rule the Bulgarians resisted foreign oppression by an intricate system of forms and methods of opposition in which anti-feudal and liberation trends were intertwined. The most stubborn, permanent and mass form of anti-Ottoman resistance, the Haiduk movement, emerged as early as the fifteenth century as an essentially anti-feudal, popular movement. It was a peasant movement which spread to all the Bulgarian lands and continued through the centuries of Ottoman rule. The Haiduks were organized in small units, which took action against individual representatives of the foreign authorities, small Ottoman military units, stronghold garrisons etc. Their actions destabilized foreign authority in the Bulgarian lands and kept up the spirit of the Bulgarians.

Hundreds were haiduk chieftains, such as Chavdar, Lalush, Angel, Momchil, Strahil, and others, including also women–Elenka, Sirma, Rumyana–whose deeds were glorified in numerous folk songs and legends and live on in the memory of the Bulgarian people.

The earliest attempts to cast off Ottoman rule were made in the first decades of the fifteenth century. In 1404 the two heirs to the Bulgarian throne, Konstantin (son of the Tsar Ivan Stratsimir) and Fruzhin (son of Tsar Ivan Shishman) took part in the anti-Ottoman coalition of King Sigismund of Hungary. They also organized the first Bulgarian uprising against the Ottomans over the entire area of North-western Bulgaria.

In 1443 and 1444, under the leadership of King Wladislaw III Jagiello of Poland and the Hungarian Commander Jan Hunyadi, two crusades were organized against the Turks. The crusaders were joined by Bulgarian, Serbian, Bosnian and Albanian volunteers. The successes in the first crusade of the Christian armies raised the hopes of the Bulgarian and other Balkan peoples for an early liberation of the Balkan Peninsula from the Turks. In the fierce battle of 1444, however, the Ottoman army of Murad II defeated the crusaders near the town of Varna.

During the sixteenth and the seventeenth centuries a number of uprisings against the Ottoman rulers broke out in the Bulgarian lands, including an uprising in Turnovo in 1598, organized by the merchant Todor Balina, Bishop Dionysius Rali of Turnovo and Pavel Djorbjich of Dubrovnik. One of the Turnovo citizens, believed to be a direct descendant of the Shishman dynasty, was proclaimed Tsar of the Bulgarians. The uprising spread all over the Turnovo region and was supported with military actions by the Wallachian *voivode,* Michael Bravul, in Northern Bulgaria. The Ottoman authorities sent out a regular army against the rebels and defeated them. Thousands of them were killed. About 60,000 Bulgarians emigrated to the territories north of the Danube.

In 1686 the Bulgarian population, encouraged by the war of the Holy Alliance of Austria, Poland, Russia and Venice against the Ottoman Empire, again rose in arms. The revolutionary preparations were lead by Prince Rostislav Stratsimirovich, who considered himself descendant of the Bulgarian Tsar Ivan Stratsimir. Rostislav managed to secure the support of the Russian Patriarch Joachim. Heavy fighting ensued and the rebels were defeated near Turnovo and the smaller town of Gabrovo. Rostislav Stratsimirovich himself was wounded and found shelter in Rila monastery.

Only two years later, in 1688, another armed rebellion broke out with the town of Chiprovets as its centre. It was headed by Georgi Peyachevich and Bogden Marinov, and was much better organized than the previous ones. The rebels were divided into units. They adopted offensive military tactics and succeeded in capturing the local administrative centre of Kutlovitsa (today Mihailovgrad). But the rebellion did not spread beyond this area and was drowned in blood by the Turks. Many thousands of Bulgarians emigrated to the north and settled in Banat.

The very next year, 1689, the population of the regions between Kyustendil and Skopje in Western Bulgaria rose up in arms. This was actually a mass peasant uprising, headed by *Voivode* Karposh, who was proclaimed king. The rebel forces acted jointly with the Austrian troops and managed to remain in power for a year, but were crushed by an 18,000-strong Ottoman army, including Tartar units.

All revolutionary movements in the sixteenth and seventeenth centuries were connected with the wars against the Ottoman Empire by European states. In the eighteenth century the Russo-Turkish wars gave a strong impetus to the anti-Ottoman resistance movement. The number of haiduk units was constantly increasing, and so was the number of people joining them. Regional peasant riots became frequent.

The efforts of the Bulgarian people in the fifteenth to eighteenth centuries to throw off the Ottoman oppression and restore the independence of the Bulgarian State are of particular historic significance. The Bulgarians accumulated combat experience and preserved their ethnic identity and national awareness, despite their losses.

During the nineteenth century the uprisings increased in number, while the organizations and strategy of the anti-Ottoman resistance was continually improved.

Throughout the period of Ottoman rule the Bulgarians had preserved the basic values of Bulgarian medieval culture, despite the annihilation of a great many cultural centres and of the Bulgarian intellectual élite. Bulgarian culture acquired a profoundly democratic character. The cultural and literary traditions, which had suffered a severe setback, were gradually revived in the folklore, in which genres of high aesthetic value emerged and developed. The monasteries became cultural and educational centres, which took care to preserve, copy and disseminate the medieval manuscripts, and promoted the creation of new, original Bulgarian literature. Works which were new in spirit and content appeared during the seventeenth century. The learned men among the Bulgarian Roman Catholics also contributed to the creation of new Bulgarian literature, permeated with the Renaissance ideas and a truly patriotic spirit. The medieval traditions in painting, architecture and wood-carving were gradually developing in line with modern trends.

At the beginning of the eighteenth century, the socio-economic processes in Bulgarian society brought about a marked activation of economic life. Significant changes likewise occurred in the social structure and in political and cultural life. The period of the Bulgarian National Revival had started.

THE BULGARIAN NATIONAL REVIVAL

Social and Economic Development. In the beginning of the eighteeth century the first signs of a new historical process appeared – the Bulgarian National Revival was beginning. This is the period when capitalist relations emerged and developed on Bulgarian soil, when a new class – the Bulgarian bourgeoisie – appeared and consolidated its positions. It is the period of the formation and the strengthening of the Bulgarian nation, of national self-consciousness. It was then that the national ideology of the Bulgarian people was created. This was the time when the idea of the Bulgarian national, liberating, anti-feudal, bourgeois-democratic revolution matured, and when the revolution was prepared and carried out. Its main purpose was the abolition of the foreign political, economic and spiritual yoke and the re-establishment of an independent Bulgarian State. During the Revival a sweeping movement for national liberation grew apace, in three main forms: a movement for a new, secular Bulgarian education and culture; the struggle for national independence of the Bulgarian Church; and a movement for political liberation. The development of the process of revival falls into three periods: from the beginning of the eighteenth century to the mid-twenties of the nineteenth; from the second quarter of the nineteenth century to the Crimean War (1853–56); and from the Crimean War to the Liberation from Ottoman domination (1878).

The defeats and the territorial losses which the Ottoman Empire suffered in the Austro-Turkish and the Russo-Turkish Wars at the end of the seventeenth century, throughout the eighteenth century and in the first third of the nineteenth century caused deep changes in its military-feudal structure. Landownership was no more granted by virtue of office but gradually became hereditary. In 1826 the janizary corps was liquidated and a regular army was created. The 1832–34 land reform abolished the feudal-*spahi* system in the Ottoman State, and the State became the main feudal lord and exploiter, together with the local hereditary feudal lords who exploited the labour of the population in the vassal lands. In order to meet the enormous expenses of the numerous wars, to support the new state army and to cover the arrears of the treasury, the Ottoman power raised the tax burden systematically and excessively. The Bulgarian population was suffering under the growing economic oppression of the State and the local feudal lords. Being the main productive regions in the Empire, the Bulgarian provinces supplied Istanbul and other parts of the Empire with agricultural produce, meat and meat products, woollen materials and furs, and provided food and clothes for the Turkish regular army. In the Bulgarian provinces sheep-breeding played an important role, satisfying the needs of the domestic market in the Ottoman Empire in meat and meat products, furs and woollen materials. The demand for these goods stimulated the development of trade and handicrafts, the rise of a significant Bulgarian commercial stratum, the gradual turning of the old Bulgarian crafts into small bourgeois enterprises, and the emergence of manufacture. Some specifically Bulgarian trades flourished: manufacture of *aba* (coarse woollen cloth), braiding for decorating clothes, leather-working, etc. The products of the Bulgarian artisans were in great demand not only on the Bulgarian market and Istanbul, but also in far-away parts of the Empire, as well as on the Eastern markets. Exports to Europe consisted mainly of farm produce: cotton (from Macedonia), cattle (from the north-west and the south of Bulgaria), grain (from the Danubian Plain), silk, furs and wool. In the eighteenth and the early nineteenth centuries when the position of the Bulgarian commercial bourgeoisie was not yet strong enough, the main centres of commercial activity were the local fairs. Traditional annual fairs took place in Uzundjovo (now in the district of Haskovo), Serrai, Eski Djumaya (now Turgovishte) etc. Later, the biggest merchants established their own firms in Russia, the Danubian principalities, Austria, Britain and other places. The trade firms of the Robev brothers, Evlogi and Hristo Georgiev, the Tapchileshtov brothers and the Geshov family conducted brisk and large-scale export-import trade on a contemporary European level. Usury, which had been widespread, gradually became less significant. Joint-stock companies were established, which engaged also in credit activity. The richest merchants abroad became the first Bulgarian bankers. Handicrafts grew into small industries which, during the Revival period, were still manufacturing industries. In 1834 the first Bulgarian factory for woollen textiles was opened in Sliven. Later other textiles factories were opened, as well as factories for silk, alochol, furs, etc. Many of them were short-lived because their owners could not hold out against competition and because of the unstable situation in Turkey. Social and property differentiation was visible in the

Paissii of Hilendar. Painting by K. Denchev

Bulgarian villages and towns. The Bulgarian bourgeoisie strengthened its position and by the middle of the nineteenth century was already stratified into several groups. The uppermost stratum – the *haute bourgeoisie* – was composed of rich merchants who made large-scale transactions on the Bulgarian, Ottoman and European markets; rich landowners, suppliers of cattle and of animal products; tax collectors and owners of manufacturing industries and factories. Economically this was the most powerful part of the Bulgarian bourgeois class. Another group was the *chorbadjii* – people who functioned as middlemen between the State and the Bulgarian population. A part of the *haute bourgeoisie* depended upon the Ottoman economic structure and the Turkish authorities for its material prosperity. This determined the conservative position of this part of the Bulgarian Revival bourgeoisie in the liberation movement and its shunning revolutionary methods in the struggle for increased rights. The Bulgarian *haute bourgeoisie*, with the exception of the small Turkophile group, wanted also the liberation of the country from Ottoman rule, which impeded and restricted its own development and undermined its security, but its representatives preferred this aim to be achieved by peaceful means, through the diplomatic steps of the Great Powers and above all, through the military intervention of Russia. The merchants, artisans and farmers of moderate means formed the stratum of the middle and petty bourgeoisie – a rather dynamic class. Some of them who grew richer gradually joined the upper stratum; others were overpowered by the competition of the European industrial goods, which started penetrating deeper and deeper into the Ottoman Empire and from the constantly increasing taxation burden. Many went bankrupt. The petty bourgeoisie and the middle class played an active role in the liberation movement. They felt most painfully the implications of foreign oppression, European competition and the instability of every production activity in the Ottoman Empire. These two strata, together with the small-holders, property owners and the unpropertied classes in town and village, formed the mainstream of the revolutionary movement. Parallel with the development of capitalist trade and the crafts, with the appearance of manufactured goods and factories, as well as with the emergence of capitalist elements in

The Bulgarian National Revival

agriculture – the *chifliks* (big market-oriented grain-producing farms) and large-scale landownership in the Bulgarian village – came the first use of hired labour. The small producers were ruined, in the villages many peasants were left without land, and this process – together with that of proletarianization of the petty bourgeoisie in the towns – developed more intensively than the process of consolidation of the bourgeoisie as a class and of the new capitalistic type of industrial production. The Ottoman rule, with all its social characteristics – the regime of national discrimination, the lack of a legal order and the lack of any protection whatsoever of the property and life of the Bulgarians – delayed the development of capitalist relations on Bulgarian soil.

The Emergence of an Ideology of National Liberation. The Bulgarian nation of the Revival period was formed through the preservation of the national spirit during the centuries of foreign yoke in the geographic regions inhabited by Bulgarians – Moesia, Dobrudja, Thrace and Macedonia. A real manifesto of the newly-born Bulgarian nation was *The Slav-Bulgarian History* (1762) written by the ideologist of the Bulgarian Revival – the monk Paissii of Hilendar (1722–73). Reminding his countrymen of the glorious past of medieval Bulgaria, Paissii called on them to engage in a fight for self-determination, to become aware of their nationality and to form a separate ethnic community in the Ottoman Empire with its own language, past and goals, differing from those of the Greeks who professed the same faith, and from the Serbs, their Slav brothers by blood. The monk of Hilendar outlined the historical tasks of the Bulgarian people – to achieve spiritual liberation from the Constantinople Greek patriarchate and to frustrate its efforts to denationalize the Church and the School, as well as to achieve political liberation from the Ottoman domination.

At the same time *The Slav-Bulgarian History* gave the first impulse to the national revolution. Enthusiastic Bulgarians took turns to copy the book, which was circulated among the people as a holy relic. Some fifty copies of it are known to have existed. A passionate disciple of Paissii of Hilendar and the first to copy *The Slav-Bulgarian History* was Sophronius of Vratsa (1739–1813), a scholar-teacher and public figure. In 1806 in Rimnik (Wallachia) he published the first book printed in the New Bulgarian language, *Nedelnik*, a collection of Sunday sermons, which was circulated widely. In the eighteenth and the beginning of the nineteenth centuries the enhanced Bulgarian national consciousness found expression in various political and diplomatic actions, in different revolutionary movements, etc. In Bucharest at the beginning of the nineteenth century, Sophronius of Vratsa together with Bulgarian dignitaries from Vratsa, Teteven and other Bulgarian towns, created the first Bulgarian political circle or centre. He helped the first Bulgarian embassy in Russia to fulfil its mission. At the same time he formulated more clearly the aspirations of the Bulgarians towards political

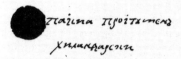

Facsimile of *The Slav-Bulgarian History*, with the signature of Paissii of Hilendar

freedom through the help of Russia. Bulgarians took part in the Serbian rebellion for national liberation (1804–13) and in the Greek national-liberation uprising (1821–29). Bulgarian volunteers fought together with the Russian army in the Russo-Turkish wars in the second half of the eighteenth and the first third of the nineteenth centuries. In 1811 during the Russo-Turkish war of 1806–12, the first Bulgarian independent military unit was formed, called *Bolgarskoe zemskoe voisko* (Army of the Bulgarian Land). Many Bulgarians took part in the armed struggle against the Ottoman Empire in the Russo-Turkish war of 1828–29. During the siege of Silistra the detachment of Georgi Mamarchev (1786–1846) displayed great courage. In the south-eastern part of Bulgaria there were many popular armed actions, and at the same time preparations were being made for a liberation uprising.

The cohesion of the nation was badly affected by the mass emigration of Bulgarians to the Danubian principalities and Russia caused by the intolerable conditions of servitude in the period of feudal anarchy and the aftermath of the Russo-Turkish wars. Considerable numbers of Bulgarians settled in Russia, Wallachia, Moldavia and Serbia, where they founded new settlements or compact colonies in different towns. This badly affected the demographic development of the Bulgarian people and weakened its unity and powers of resistance. At the same time the emigration supported and organized numerous revolutionary actions and initiated the establishment of significant cultural and printing centres, where newspapers were published, and where eminent publicists, scholars and men of letters worked. A Bulgarian colony of many thousands was formed in Constantinople in the thirties and forties of the nineteenth century. It engaged in various public and cultural activities and gradually became the main centre of the Bulgarian National Revival and of legal Bulgarian public movements and organizations, and the most important cultural centre of the Bulgarians.

Movement for National Enlightenment and Culture. The first significant manifestation of the process of Revival was the movement for Bulgarian national, secular education and culture – for the cultural and national independence of the Bulgarian people. In comparison with the remaining Balkan provinces of the Ottoman Empire, the position of Bulgaria was the most unfavourable. It had no direct ties with the advanced European countries; it was nearest to the capital of the Empire – Constantinople; it was the most densely populated by Turkish colonists; and was under the ecclesiastical rule of the Constantinople ecumenical patriarchy. All of this impeded the cultural development of the Bulgarian people, which suffered unheard of national and religious discrimination, oppression and depredation by its Ottoman masters, and had no secular schools, periodical Press or literature of its own. The only educational centres during the Ottoman domination were the schools which functioned at monasteries and churches. It was there that a small number of Bulgarian youths received the most elementary education. During the Revival Period, with the emergence and rise of the Bulgarian merchant and industrial bourgeoisie and with the maturing of the Bulgarian nation, the need for a secular education grew stronger. At the same time the necessity was felt to associate Bulgarian national culture with the achievements of the modern world. Thanks to the more favourable geographical location of Greece, Revival processes there started earlier. Secular schools appeared not only on Greek soil, but in many Bulgarian towns where there were compact groups of Greek residents such as Melnik, Plovdiv, Sozopol, Pomorie and Adrianople. These played a positive role in disseminating secular knowledge among the Bulgarians and in developing education and literacy in Bulgaria, but at the same time they became an instrument of the Greek Patriarchate in Constantinople in its endeavour to deprive the Bulgarian people of the feeling of nationhood in the spirit of the Great Hellenic nationalistic aspirations. In his *Slav-Bulgarian History* Paissii of Hilendar underlined the necessity of resisting this phenomenon, which endangered the mainstay of national consciousness. For him education was a powerful means of enhancing Bulgarian national consciousness and of carrying out the programme for Bulgarian national liberation. At the beginning of the nineteenth century Bulgarian men of letters were already aware that it was impossible to familiarize the Bulgarian people with the achievements of the contemporary European civilization without raising their general cultural level. The first to express the idea of creating a modern Bulgarian school was the prominent Bulgarian scientist and encyclopaedist Dr Peter Beron (1800–71). He outlined the first programme for the application of modern methods in Bulgarian education. In 1824 he published *The Fish Primer,* in which he underlined how important it was to teach the vernacular in the Bulgarian secular school, to apply the practice of mutual instruction and the phonetic method. He insisted that girls also should go to school and that teaching should relate to the capabilities of the children, etc. The first Bulgarian secular school was opened in Gabrovo in 1835 by Vassil Aprilov (1789–1847), a learned Bulgarian of the Revival period, with the help of some Bulgarian merchants. Its management was entrusted to the monk, Neophyte of Rila (1793–1881), one of the most prominent men of letters of this time. He composed teaching tables in accordance with the new system, and textbooks of Bulgarian grammar and Bulgarian language. Following the example of the Gabrovo school, by the time of the Crimean War (1853–56) there were already over 200 mutual instruction schools in different parts of Bulgaria. The first secular school for girls was opened in 1840 in Pleven, and by 1856 there were about 35 schools for girls. The young Bulgarian intelligentsia, who had studied abroad, mainly in Russia, gave an impulse to the development of the New Bulgarian educational process. In the mid-forties the first graduates returned home from Russia and opened grade schools. In these, the pupils were grouped in different grades according to their age and knowledge. The teaching proceeded in the form of separate lessons under the supervision of the teacher and with the active participation of the pupils. The first grade schools for boys were opened in 1845–46 in Elena, Kalofer, Pazardjik and Koprivshtitsa, while the first grade school for girls was opened in 1856 in Shumen. The national character of the

The seal of the Gabrovo school, 1832

New Bulgarian educational process was manifested in the way it was financed and governed. Schools were entirely supported by the Bulgarian population, through the personal donations of representatives of the different social strata in Bulgarian society, and were governed by boards of trustees. Thus from its very beginning education in Bulgaria was truly democratic. Bulgarian youth enjoyed the right and the opportunity to study free of charge, irrespective of social background. Many Bulgarians were

The Bulgarian National Revival

Bulgarian students in Moscow, 1861

sent to study abroad at the expense of Bulgarian municipalities, monasteries and rich fellow-countrymen, or received scholarships from the Russian Government or from Slav charities. After the Crimean War (1853–56) social progress was reflected in the rapid development of education. The school network was enlarged, and pedagogical theory and practice were updated. On the eve of the Liberation from Ottoman rule (1878) there were more than 2,000 schools in Bulgaria, since they were opened even in the smaller settlements and less developed regions. Bulgarian emigrants ran more than forty Bulgarian schools in Romania; the first Bulgarian secondary school was opened in Bolgrad (Bessarabia) in 1858. Secondary general education schools were opened in Gabrovo (1872) and Plovdiv; specialized schools – in Shtip (1869), Veles (1857), Svishtov (1873) and in other places. The importance of the role of education in the social life of the country was reflected in the proclamation of 11 (24) May as a national holiday – the day for glorifying the work of the Slav 'Enlighteners', Cyril and Methodius.

Through the school system and the education Bulgarians received at home and abroad, contemporary European knowledge and ideas penetrated Bulgarian society and facilitated the familiarization of the Bulgarian people with the achievements of world civilization, spreading among the masses and raising them out of medieval illiteracy and ignorance. The development of schooling and of national education by Bulgarian financial means (often impeded by the Turkish authorities and the systematic resistance of the Constantinople Patriarchate) was recorded by the foreign travellers and diplomats who visited Bulgaria, as an important achievement of the enslaved Bulgarian people.

As a result of the successful development of education, Bulgarian national consciousness was further enhanced, and this provoked the Ottoman authorities to make fresh attempts to impede the political advance of the Bulgarian people. In the sixties the Governor of the Danubian *vilayet* (a province of the Ottoman Empire), Midhat Pasha, drew up a project for merging the Bulgarian and the Turkish schools, which failed totally because of the resistance of the Bulgarians and their now firmly established national consciousness. Greek propaganda and the efforts of the Greek Patriarchate in Constantinople to appoint Greek teachers and foster a Hellenistic spirit in the Bulgarian schools were particularly strong in Macedonia and many parts of Thrace. Most active in opposing Greek influence were the brothers Dimiter (1810–62) and Konstantin (1830–62) Miladinov, who perished in prison, Grigor Purlichev (1830–93), Kuzman Shapkarev (1834–1909), Yordan Djinot (1818–82), and Georgi Dinkov (1839–76). The enhanced national consciousness and individual character of the Bulgarian people, achieved through the Church and the school, were obstacles to the expansionist aspirations of the bourgeois Serbian State, which in the middle of the nineteenth century made no secret of its territorial pretensions towards Macedonia. Therefore in the sixties, and particularly in the seventies of the nineteenth century, Serbian teachers were appointed and textbooks supplied by the Serbian authorities, who tried in this way to oust the Bulgarian national school and influence the consciousness of the Bulgarians inhabiting the frontier regions and Macedonia. In addition the Serbian ruling circles tried unsuccessfully to destroy the national character of the Bulgarian schools and those of the Macedonian Bulgarians through their Serbian

The first Bulgarian teachers' rally, Plovdiv, 1870

schools. The Bulgarian schools in the western and southwestern Bulgarian regions preserved their national spirit and developed simultaneously, in parallel with, and in collaboration with, the schools in the remaining parts of Bulgaria.

In this period the foundation of other cultural and educative establishments were laid. In 1856 the first communal cultural centres *(chitalishtes)* appeared in Svishtov, Lom, Shumen, etc. They grew in number very quickly, not only in the towns but in the villages, too, becoming the centres of lively cultural and educational activities: theatrical performances were staged there, lectures on popular scientific and historical patriotic themes were held, celebrations were organized, museum objects and folklore materials were displayed there. They also helped the schools in their activity. The works and studies of the scientists and scholars of the Revival, such as Dr Peter Beron, Spiridon Palauzov (1848–72), Marin Drinov (1838–1906), Dr Ivan Seliminski (1799–1867) and others, laid the bases of the separate branches of Bulgarian science. In 1869 in Braila (Romania) the first Bulgarian literary society was established, and this was the archetype of the Bulgarian Academy of Science. The rapid development of education determined the formation of a New Bulgarian written and spoken language, and stimulated the development of literature and art. In the first half of the nineteenth century the foundations of New Bulgarian printing were laid.

In the decades following the Crimean War Bulgarian national literature blossomed. Petko Rachev Slaveikov, Lyuben Karavelov, Hristo Botev, Vassil Drumev, Dobri Voinikov and Ivan Vazov created mature works of fiction, poetry and drama. The Bulgarian periodical Press underwent a remarkable development; periodicals were published (mainly in Constantinople and in Romania), stimulating discussion of basic problems concerning the developments of the Bulgarian nation, its struggle for religious, national and cultural independence, which had their ideological bases worked out in them. The *émigré* revolutionary Press, and particularly the newspapers, edited and published by Georgi Rakovski, Lyuben Karavelov and Hristo Botev – *Dunavski Lebed, Svoboda (Nezavissimost)* and *Zname* – played an important role in the dissemination of ideas of liberation, revolution and democracy and in the psychological preparation of the people for the national revolutionary struggle.

The Bulgarian national spirit found its expression also in fine arts (Zahari Zograph, Stanislav Dospevski, Nikolai

The Bulgarian National Revival

Pavlovich), and in the original works of the national master-builders (Alexi Relits, master Milenko, Nikola Fichev, etc.), who built remarkable public buildings, bridges and houses in the Revival style.

The Struggle for an Independent Bulgarian Church. The movement for the restoration of the independent Bulgarian Church and the removal of the supremacy of the Patriarchate of Constantinople was a specific manifestation of the Bulgarian national revolution, and of the growing consciousness of the Bulgarian nation as a separate entity within the Ottoman Empire and among the Balkan peoples. When Bulgaria fell under Ottoman domination the Bulgarian Church was subordinated to the Constantinople Patriarchate, which was recognized by the Ottoman Government as the sole representative of all Christian peoples within the Ottoman Empire, the Bulgarians included. Taking advantage of their privileged position, the Greek clergy gradually ousted all the Bulgarian bishops and priests, gaining complete superiority in Bulgaria. The Bulgarians were replaced by Greeks, and Greek became the only language used for church services. All clerical ranks could be bought, and corruption spread far and wide, while the Bulgarian people had to shoulder the financial burden. The Greek clerical authorities also made repeated attempts to assimilate the Bulgarians and destroy their feeling of national identity. The Constantinople Patriarchate, obsessed with chauvinism, preached the great idea of the Greek bourgeoisie of a pan-Hellenic rebirth within the boundaries of the former Byzantine Emperor. The Bulgarian people were actually under dual domination – the Ottoman political oppression and the oppression of the Greek clergy.

The first open conflicts of an ecclesiastic and national nature flared up during the twenties and thirties of the nineteenth century in towns such as Vratsa, Skopje and Samokov. The Bulgarian population not only objected to the predatory, malpractising policies of the Greek clergy, but insisted that they be replaced by Bulgarian bishops. The Bulgarians of the largest diocese, that of Turnovo, with the great Bulgarian patriot and religious leader Neophyte Bozveli (c. 1785–1848) at its head, waged a firm struggle against the Greek rulers throughout the thirties and forties.

From the mid-forties onwards the struggle for an independent Bulgarian Church gained stature as a rather better organized movement with a clear-cut programme. Prominent among its national leaders were Neophyte Bozveli and Ilarion Makariopolski (1812–75). They sent repeated messages to the Sultan containing not only demands for the reinstatement of the Bulgarian bishops but a programme for cultural and national self-determination, calling for the establishment of Bulgarian secular schools, the development of Bulgarian books, literature

Facsimile of the *Firman* proclaiming the Bulgarian Exarchate, 28 February 1870

The first Bulgarian Synod. *Left to right:* Ilarion Makariopolski, Panaret of Plovdiv, Exarch Antim I, Ilarion of Kyustendil

and the like, as well as for civil administrative autonomy to equal the status of the remaining nationalities within the Empire.

The Patriarchate of Constantinople scornfully rejected the Bulgarian demands and launched a campaign against the movement, banishing its leaders to Mount Athos, but its adherents refused to abandon their cause. In 1849 they succeeded in obtaining the Sultan's permission to build their own church in Constantinople and to install a Bulgarian bishop there. In essence this signified that for the first time the Bulgarians gained official recognition as a separate national entity within the Empire. The Constantinople Church gathered together an impressive congregation, led by outstanding representatives of the Bulgarian bourgeoisie.

In its turn the Bulgarian community in Constantinople gradually emerged as leader of the movement for an independent Bulgarian Church, which was steadily growing both in scope and in numbers. The community launched numerous initiatives, expressing the will of the Bulgarian people to be rid of Greek clerical oppression, and acting as official representatives of the Bulgarian people before the Ottoman authorities. The Bulgarian emigrants in South Russia, Wallachia, Moldavia and elsewhere also participated actively in the movement. Following the Crimean War, the fight for an independent Bulgarian Church reached the proportions of a powerful national movement, involving the whole Bulgarian nation. On 18 February 1856 the Ottoman Government issued a new decree on the reformation of the Empire – the *Hattihumayun* – promising equal civil rights to all subjects, and though these were only promises, the Act provided a pretext for the Bulgarians to claim religious and civil autonomy. Within the year a petition on behalf of 6,400,000 Bulgarians was delivered to the Sultan demanding that they be allowed to appoint their own church and administrative head. The Ottoman Government again rejected the demands but the struggle went on. Representatives of different Bulgarian towns settled in Constantinople and the campaign of petitions for an independent Bulgarian Chruch was resumed.

The deaf ear the Supreme Porte and Patriarchate of Constantinople turned to their legitimate claims finally convinced the Bulgarians that they had no alternative to open defiance of Greek religious domination. Prompted by the Bulgarian community in Constantinople, Ilarion Makariopolski solemnly declared the independence of the Bulgarian Church from the Patriarchate during the Easter service in the Church of St. Stefan on 3 April 1860. This act aroused unprecedented enthusiasm amid the Bulgarian population. Many villages forced their clerics to relinquish their pledge to the Patriarchate and addressed Ilarion

Makariopolski as head of the Bulgarian Church. By the end of the 1860s the Patriarchate had totally lost its hold on Bulgaria. Everywhere the Greek clergy was purged and the Church was almost exclusively in the hands of Bulgarian ecclesiastics. However, the Turkish Government wanted to prolong the Greek-Bulgarian conflict indefinitely and was delaying official sanction of the situation. The untiring revolutionary activities of the Bulgarians during the 1860s finally brought the crisis to a head. The Supreme Porte was forced to issue a *firman* (a royal decree) on 28 February 1870 decreeing the founding of an independent Bulgarian Exarchate that encompassed many of the Bulgarian regions. It included fifteen sees, and the opportunity for that number to grow was left open should a plebiscite in other Bulgarian territories show a two-thirds majority in favour. The establishment of the Bulgarian Exarchate as an all-national institution more or less drawing together ethnically the Bulgarian nation and the appointment in 1872 of the first Bulgarian Exarch Antim I (1816–88) marked a significant victory for the Bulgarian people.

Born as a result of the awakening national consciousness, the religious struggle carried this consciousness to further heights, thus bringing about national consolidation of the Bulgarian people. As a peaceful form of struggle, targeted not on the Sultan's rule, but rather on foreign Greek influence and the de-nationalizating intentions of the Patriarchate, it enjoyed the support of the large majority of the Bulgarian people. It spread to all corners of the Bulgarian lands and through the numerous mass campaigns, organized by its leaders, brought the Bulgarians from all regions in touch, and enabled them to realize that they were bound by their national interests in an integral whole. In the course of the religious struggle public figures emerged who were to play an important part of the nation's development up to the liberation from Ottoman domination and in political, public and cultural life afterwards. The crucial effect of the movement for an independent Bulgarian Church was that it managed to make the Bulgarian national problem a matter of international concern. The Great Powers invested a large amount of diplomatic effort in the independent Bulgarian Church issue and in that way the existence of a hitherto unknown nation different from the Greek and the other Balkan nations, cherishing its own natural aspirations and goals, was suddenly brought to the notice of the Bulgarian governments and public.

Thus the religious issue had a political side which occupied an important place in the Eastern policy of the Great Powers. For over ten years it was a central issue in which the antagonistic interests of the powers clashed. Each one of them was actively involved in the Bulgarian-Greek conflict, Russia, France and Great Britain in particular.

The National Revolutionary Movement. The struggle of the Bulgarian people for political freedom, for the restoration of the destroyed Bulgarian State and the liquidation of the despotic feudal Ottoman system is

Vassil Levski

the third major component of the Bulgarian Revival. During the second quarter of the nineteenth century a wave of popular liberation campaigns swept the Bulgarian lands. An anti-Ottoman plot, known as *Velchova zavera,* was organized in 1835 in the Turnovo region; liberation and anti-feudal uprisings broke out in the Berkovitsa and Pirot regions in 1834–37, in 1841 in the Nish region and in 1850 – in the Vidin, Lom and Belogradchik regions.

Bulgarian emigration in the Danubian principalities, assisted by Serbian and Greek emigrants, was responsible for the organization of three liberation uprisings occurring in 1841, 1842 and 1843 in Braila. The example of the other free Balkan peoples who had won their political and state independence with Russian help, and the contacts the Bulgarians had with the governments of these newly-liberated states, put the Bulgarian revolutionary movement into a totally different position. As a consequence of its successful wars against the Ottoman Empire Russia grew to be the most influential power in South-eastern Europe. Bulgaria's sovereign neighbours attracted the Bulgarain political emigrants, who gathered together in the Danubian principalities and in Serbia, and this created favourable conditions for their patriotic activities to marshal the people's forces from outside. Bulgarian emigration centres were also set up in Southern Russia. In every instance of uprisings, attempted or performed during the second quarter of the nineteenth century, the Bulgarians were hoping for

Russian intervention in favour of their liberation, and for support and assistance from Serbia and the Danubian principalities.

From the forties onwards the Bulgarian question began to draw the attention of European diplomatic circles. Public opinion in Russia showed a marked increase in sympathy towards the Bulgarian people, but the explicit hostility of the Russian Emperor Nikolai I towards even the slightest suggestion of revolution determined his Government's resolve not to get involved in the Bulgarian liberation struggles. The 1841 uprising in the Nish region, which broke out at the time of the Egyptian crisis, combined with the growing contradictions of the Great Powers over the Eastern Question, stirred up considerable interest about Bulgaria among the important European countries. They sent several missions to Bulgaria to study the situation there. The Bulgarian Question became a public and diplomatic concern in Europe. The Bulgarians themselves did their best to plead their case before the European governments. During the forties Alexander Exarch (1810–21) launched several campaigns in defence of his people, sending memoranda to the Governments of France, Britain, Austria, Prussia and Russia, and engaging in negotiations on behalf of the Bulgarian people with the French, British and finally the Russian Governments. The Bulgarian national liberation movement occupied a special place in the policy conducted by Serbia. They turned it to their own ends, namely to liberate the Serbian territories that had remained under Ottoman rule and to secure for themselves a hegemonic role among the Southern Slavs – the *Nachertanie* programme. They also developed

Georgi Rakovski

Panayot Hitov

an appetite for the annexation of certain Bulgarian regions to Serbia. The numerous Polish emigrants in the Balkans and in Constantinople also played an active part in the Bulgarian liberation movement. They backed the Bulgarian demand for the independence of the Bulgarian Church from the Constantinople Patriarchate and later took an active part in the Uniate movement and in France's attempts to convert Bulgaria to Catholicism, as well as in the attempts by Britain to penetrate it through Protestantism.

The Crimean War of 1853–56, Russia's defeat and the Ottoman Empire's growing economic and political dependence on the West were of paramount importance for the Bulgarian liberation movement. Russia's enemies – Britain, France and Austria – took up important positions in the Ottoman Empire, stepped up their expansionist policies in regard to its markets and enslaved it financially. Unable to compete with the imported goods, certain Bulgarian crafts declined. In the meantime, however, the production of industrial crops in Bulgaria and the development of Bulgaria's trade within the Empire and abroad were given an impetus by the Empire's livelier business relations with the West. All this led to the growth and consolidation of the Bulgarian bourgeoisic and the development of Capitalism in the Bulgarian lands. The rift in the class differentiation of Bulgarian society became wider. Turkey's efforts to overcome its financial difficulties by shifting the burden onto the enslaved peoples through heavier taxation and plunder were met with an ever-growing indignation on the part of the Bulgarians. Developing Capitalism in Bulgaria resulted in a sharp

mission it had to play, was now ready to devote its efforts to the political liberation of the country.

A period of ideologically-based nation-wide revolutionary struggle was naturally ushered in. The sporadic and isolated armed outbursts and other anti-Ottoman actions of the kind that took place in the first half of the nineteenth century were a thing of the past. The failure of the Ottoman Government to live up to its promises for reforms, proclaimed in the *Hattisherif* (1839) and the *Hattihumayun* (1856) provided ample enough proof that it was ill-equipped to carry out any reform in the existing

Filip Totyu

Hadji Dimiter

contradiction with the feudal syatem and with the lack of legal order and security. The Bulgarian nation became firm in its conviction that it could not but regard the overthrow of foreign oppression, the liquidation of the antiquated economic and social structure and the foundation of an independent Bulgarian State as a necessary prerequisite for the economic, political and cultural progress of the Bulgarian peoples and as a major historical priority of the day. Thus the Bulgarian nation entered the decisive stage of its liberation movement.

The nation which during the fight for an independent Bulgarian Church and new secular education and culture had clearly revealed its individual features and grasp of the

regime of oppression and exploitation, as well as testimony to its internal weaknesses and irreversible decay. The Bulgarians realized that if they were to win their freedom their one recourse was a purely Bulgarian uprising. During this period political thought matured and the political organizations of the major social strata of the population began to emerge. Political liberation of Bulgaria was set as the key goal of the struggle which spread throughout the Bulgarian lands and involved the emigration in Romania, Russia, Serbia and the Bulgarian colony in Constantinople. Various social and political forces were active in the liberation movement, and their views are recorded in the periodicals and in a number of programme documents of the Revival Period. The Russophiles among the *haute bourgeoisie* in Romania, Russia, Constantinople and at home were represented by the Benevolent Society (founded in 1854 as a Central Bulgarian Trusteeship) and by the Bulgarian Trusteeship in Odessa. These placed their hopes for Bulgarian liberation chiefly in Russia's intervention, either diplomatic or military, once the international situation was right. They objected on principle to all revolutionary methods. Heterogeneous in social composition and inconsistent in its political outlook was the Secret Central Bulgarian Committee (SCBC), 1866–68. In 1867 it put forward a programme for

Stefan Karadja

political dualism with the Ottoman Empire, formulated in a memorandum to the Sultan and the governments of the great Powers in a vain attempt to reorientate the Bulgarian liberation movement toward the West-European governments. Despite its failings, the SCBC played an important part in bringing the Bulgarian claims home to the European governments and public.

Revolutionary ferment in anticipation of a general national uprising was the order of the day. Inspired by the revolutionary ideas prevalent throughout Europe, the leaders of the Bulgarian liberation movement never failed to spot what was pertinent in them to the interest of their national cause. They established personal and organizational contact with the revolutionary and national movements in the Balkans, in Russia and elsewhere in Europe. Russian revolutionary democratic thought, represented by the works of A. I. Hertzen, A. G. Chernyshevsky, N. A. Dobrolyubov and others, exerted a tremendous influence on the Bulgarian revolutionary movement. The emigrants in Russia and Romania kept in touch with the Russian revolutionaries. The Italian national liberation movement, the work of G. Mazzini and Giuseppe Garibaldi, also gained tremendous popularity among the Bulgarians, so much so that many went out to join it as volunteers. The Bulgarian revolutionaries also made a careful study of the Eastern policy of the Great Powers and of their contradictions and elaborated their tactics with regard to international relations. Realistically assessing the peculiarities of Russian autocracy, they came to the conclusion that Russia's political stake in the Balkans could well promote the cause of Bulgarian liberation.

The new phase in the national revolutionary movement is closely related to the views and work of Georgi Stoikov Rakovski (1821–67) – the founder of the Bulgarian revolutionary ideology and of the organized national liberation struggle of the Bulgarian people. During the late fifties and the sixties he elaborated the tactics, already tried out by other national movements, of rallying and training armed detachments on free territories outside Bulgaria, to be sent to the homeland. There, they were to turn the Balkan Range into a centre of the revolution, which was then to spread throughout the country. The armed uprising was to be headed by a provisional revolutionary administration, whose statutes Rakovski drew up in 1862 and called the Provisional Bulgarian Administration, with himself at the head. In that year, with the permission and assistance of the Serbian Government, Rakovski organized in Serbia a military group – the First Bulgarian Legion – made up of Bulgarian revolutionaries, which was to cross over into Bulgaria. The Legion joined in the battles fought between Serbs and Turks. The diplomatic settlement of the Serbo-Turkish conflict led to the disbanding of the Legion. Serbia helped the liberation movements of the neighbouring Southern Slav peoples so as to liberate and unite the Serbian regions which were still under Turkish rule. The revolutionary crisis sweeping the Balkans in 1868–69 created favourable conditions for the transfer of detachments to Bulgaria. In the beginning of 1867 Rakovski founded, in Bucharest, a centralized leadership of the Bulgarian armed detachment movement – the Supreme Bulgarian Secret Civil Administration – and prepared a document setting out for the detachments a programme which reflected the knowledge and experience accumulated by the great theoretician, politician and strategist of the Bulgarian national revolution. Rakovski believed in independent revolutionary actions rather than in concerted actions by the Balkan states, In the spring of 1867 Lyuben Karavelov (1834–79), another great leader of the national revolution, set up, in Belgrade, a Bulgarian committee for the organization of detachments of Bulgarian emigrants and refugees in Serbia.

The planned massive liberation campaign scheduled for

1867 did no more than rally the Second Bulgarian Legion in Belgrade (prematurely disbanded in July 1868) as a military training school for commanders in the impending uprising, and transfer to Bulgaria the detachments led by Panayot Hitov (1830–1918) and Filip Totyu (1830–1907). Organized and armed by the Supreme Command and the Benevolent Society the detachments entered Bulgaria in the spring of 1867 with the task of checking the political orientation of the people and its preparedness for an uprising. After heavy battles with the Turkish forces the survivors returned to Serbia. There the Bulgarian Committee rallied a big detachment at Zaichar (July 1867), but the Serbian authorities prevented it from crossing into Bulgaria. The detachment movement reached its climax in the summer of 1868, when the detachment headed by Hadji Dimiter (1840–68) and Stefan Karadja (1840–68) crossed over into Bulgaria and performed feats of heroism. Its task was to organize an uprising in Bulgaria and set up a provisional government in the region of the Balkan Range. Fierce battles were fought in the Turnovo region, where Stefan Karadja was seriously wounded and captured by the Turks, and also on Mount Buzludja (in the Balkan Range), where almost all of them were killed, along with the *Voivode* Hadji Dimiter.

The feats performed by the detachments in 1867–68 displayed the revolutionary spirit of the Bulgarian people, its eagerness to fight for its freedom, its readiness for self-sacrifice and its ardent patriotism. Like the Second Bulgarian Legion, they too aroused deep sympathy in Europe and focused the attention of European diplomatic circles on the Bulgarian Question and the fate of the Bulgarian people. But the short existence of the detachments, and the drastic measures taken by the Turkish authorities against the rebels and peaceful population, proved that it was impossible to organize an uprising in this way.

In 1869 the Bulgarian revolutionary movement embraced the idea of setting up an all-Bulgarian organization with two centres – one in emigration and one inside the country – which was to create a network of secret committees in Bulgaria, to prepare the people for the uprising. The next stage in the development of the national revolutionary movement is chiefly associated with the work of Lyuben Karavelov and especially Vassil Levski (1837–73), the most distinguished champions of the new ideological approach. After 1869 L. Karavelov formed a revolutionary committee in Bucharest, with himself as leader and ideologist. Meanwhile, Vassil Levski was active inside the country, setting up a network of revolutionary committees, in hundreds of villages and towns, which became the backbone of the internal revolutionary organization. With his revolutionary and democratic ideas most clearly revealed in his draft statute of the Bulgarian Revolutionary Central Committee (BRCC) and his revolutionary, organizational and propaganda work, unparallelled in magnitude and fruitfulness, Levski made his mark as a brilliant strategist, organizer, ideologist and leader of the Bulgarian national-liberation revolution. In 1870 a general assembly was convened in Bucharest and the BRCC was formed, adopting the programme and statutes drawn up by Levski and Karavelov, and revolutionary activities were in full swing by 1871–72. But the disclosures that followed the ill-advised hold-up of the Turkish bullion waggon in the

Facsimile of the Statutes of the Bulgarian Revolutionary Central Committee, published in Geneva, 1870

Arabakonak Pass, wilfully carried out in September 1872 by a group of revolutionaries without Levski's knowledge, dealt a heavy blow to the organization. Levski's execution in February 1873 and the destruction of the revolutionary network disrupted the activities of the BRCC. In late 1874 Hristo Botev (1848–76), a staunch revolutionary democrat, upholding the ideas of Utopian Socialism and Materialism, took up the leadership of the emigrants' revolutionary democratic wing.

The Eastern crisis of 1875–78, unleashed by the uprisings in Bosnia and Herzegovina, facilitated the rallying of the revolutionary forces. The committee with Hristo Botev at the head decided to rouse the people to rebellion in September 1875. Short-term, hasty preparations and inadequate co-ordination of action restricted the scope of the uprising. Fighting was limited mainly to the Stara Zagora region, and insurgent detachments were organized in Shumen and Russe. The uprising in Stara

Zagora and the other attempts at rebellion in the autumn of 1875 were quickly crushed by the Turkish authorities but did nevertheless contribute to the aggravation of the Eastern crisis.

Discouraged by the failure, the Bucharest BRCC ceased to operate. But before the year was out a new leading body was formed by a group of younger revolutionaries in the Romanian town of Giurgiu (today's Gyurgevo). The Giurgiu revolutionary committee took up the cause of the BRCC and prepared a plan for a new uprising in the spring of 1876. To facilitate preparations, the country was divided into several revolutionary districts, to which the Giurgiu committee sent emissaries to organize the revolutionary actions. Assisted by emigrants and scores of local revolutionaries, they began to restore the old revolutionary network of the BRCC, added new committees and started preparations for the uprising. In April–May 1876 the April Uprising broke out before it had been adequately prepared. Armed poorly, if at all, and heavily outnumbered by the bashi-bazouk hordes and regular Turkish army, the rebels in the Panagyurishte revolutionary district displayed unparalleled heroism in defending Panagyurishte, Batak, Bratsigovo, Perushtitsa and other parts. In the Turnovo revolutionary district, the detachment led by Priest Hariton engaged in a nine-day battle with Turkish troops at the Dryanovo monastery. The insurgents from the Gabrovo-Sevlievo region succeeded in resisting Turkish attacks near the villages of Kravenik, Novo Selo, Batoshevo and others for eight successive days. A unique example of undaunted courage was given by Hristo Botev who, at the head of a two hundred-strong detachment set up in Romania, was fighting not far from the Danubian town of Kozlodui (Vratsa district) on 17 May. Relentlessly pursued by the Turkish troops, the detachment was defeated in the Vratsa mountains and Hristo Botev was killed. The Pianets region in West Bulgaria also rose up in arms. The April uprising was suppressed with unheard-of brutality. Over 30,000 men, women and children were killed, hundreds of towns and villages were burnt, and thousands of people were gaoled or exiled. But despite the military defeat the uprising had an enormous political significance. It placed the Bulgarian Question at the centre of attention of European diplomatic circles and spurred on Russia's activity on the Eastern and Bulgarian issues. The surviving Bulgarian patriots had no intention of abandoning the national struggle under the new circumstances. A new revolutionary organization centre was set up in Bucharest, entitled the Bulgarian Central Charity Society (BCCS). Financially assisted by the Slav charity committees in Russia, it sent Bulgarian volunteers to Serbia to take part in the Serbo-Turkish War of 1876, thus continuing the Bulgarian people's struggle against the oppressor. The BCCS also launched preparations for a new uprising in Bulgaria, but the atmosphere of intolerable anguish and the lack of funds in the country foiled this plan.

After the uprising, the Bulgarian liberation movement embarked upon a new stage of development. The Bulgarian emigrants in Romania and Russia, and the Bulgarian colony in Constantinople were actively engaged in acquainting the European public with the hardships and current plight of the Bulgarian people, and with their unshakeable resolve to fight for their freedom. The campaign was taken up by the whole Bulgarian population as an act of self-defence intended to win over the European governments to the Bulgarian national cause. This was the first initiative to involve all strata of the Bulgarian bourgeoisie. In the existing favourable international situation even the Russophile *haute bourgeoisie* in Romania and elsewhere responded to the common call for political liberation. They sent a delegation to several European countries to lay the Bulgarian demands for political independence before the respective govern-

The standard of the insurrectionists from Panagyurishte: *Svoboda ili Smurt* (Freedom or Death)

ments. A public movement in support of the Bulgarian people began to gain momentum in Europe and the Balkans, and above all in Russia.

The Russo-Turkish War of Liberation of 1877–78 and the Restoration of the Bulgarian State. The solution of the Bulgarian Question was conditioned by Russia's reorientation of its policy from a peaceful settlement of the Eastern crisis to a war with the Ottoman Empire. Before proceeding to military action, Russia made one more diplomatic move. The Great Powers, and Britain in the first place, were hoping to delay a solution of the pressing problems in the Balkans and to prevent Russia's military intervention. A conference of representatives of the Great Powers was convened in December 1876 in Constantinople. Bulgarians from all over the country, the BCCS and the Constantinople community sent numerous statements and petitions to the conference, openly demanding the restoration of the independent Bulgarian State within its ethnic territories of Moesia, Dobrudja, Thrace and Macedonia and submitting projects for the

The Bulgarian National Revival

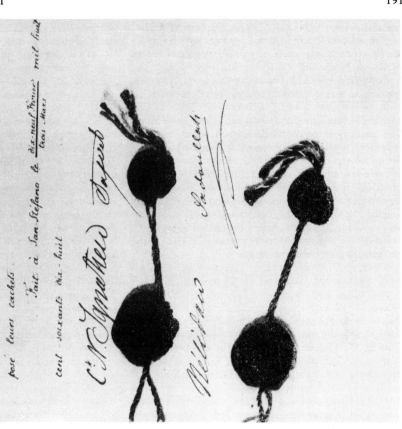

Facsimile of the first and last pages of the San Stefano Peace Treaty, 1878

political structure of the future Bulgarian State. The most radical and democratic were the demands put forward by the Bulgarian people's rally, organized in Bucharest by the BCCS in November 1876. In this way, though not formally represented, the Bulgarian people was nevertheless able to make its presence felt at the Constantinople conference by having its demands and ideas about its own state laid before the Great Powers and the Supreme Porte. The decisions of the conference envisaged the creation of two autonomous Bulgarian regions approximately within the boundaries of the Bulgarian Exarchate. But the actual outcome was nil on account of the Ottoman Government's stubbornness and the diplomatic ploys of Britain, which encouraged the Ottoman Government to reject the London Protocol granting autonomy to the Christian peoples in the Balkans, signed by the Great Powers on 19 March 1877. Thus, armed conflict between Russia and the Ottoman Empire became inevitable.

Russia declared war on the Ottoman Empire on 12 April 1877, and the Bulgarian people took an active part in the Russo-Turkish War of Liberation (1877–78). Thousands of patriots joined the Bulgarian volunteer forces and fought shoulder to shoulder with the Russian army, displaying excellent combat ability and peerless heroism. On 15 June 1877 the Russian army crossed the Danube and liberated Svishtov. After hard-won victories near Pleven and Stara Zagora, and the heroic defence of the Shipka Pass, the Russian troops went on the offensive. The Bulgarian population helped the liberators also through active reconnaissance work during the crossing of the mountain passes. They set up armed detachments which protected the population, cared for wounded and sick soldiers, provided food and fodder for the army, etc. The Russian soldiers and Bulgarian volunteers performed miracles of heroism during the crossing of the unscalable ridge of the Balkan Range in severe winter conditions. The general offensive of the Russian army in the south started on 1 January 1878. Edirne was captured and the road to Constantinople was clear. The Ottoman Government was forced to seek peace, and following lengthy negotiations the peace treaty was signed in San Stefano on 19 February (3 March) 1878. Thus after long years of struggle, the heroic April Uprising and the unswerving will of the Bulgarian people to fight for liberation, this war finally won back Bulgaria's hard-earned freedom.

The San Stefano Peace Treaty of 1878 defined the state frontiers according to ethnic principles and included approximately the territories as outlined by the Constantinople Conference. The formation of a large Bulgarian State and Russia's increasing influence in the Straits, resulting from its victories on the battlefield, were unacceptable to its rivals. Thus on 1 June 1878 the Berlin Congress was convened and succeeded in dwarfing the results obtained with such an enormous military effort and loss of life. By virtue of the Berlin Treaty, Bulgaria was dismembered. North of the Balkan Range was the autonomous Principality of Bulgaria, including the Sofia district; Eastern Rumelia, in the south, remained under the Sultan's direct political and military control, gaining nothing but administrative autonomy; Macedonia and the Edirne region of Thrace remained under the rule of the Ottoman Empire; Serbia got Nish, Pirot and Vranja, and Romania – Northern Dobrudja. The unfair solution of the national problem resulted in hostility and misunderstanding among the Balkan peoples and led to serious conflicts and national catastrophes. Nevertheless, the restoration of the Bulgarian State was of exceptional importance for the Bulgarian nation.

The decisions of the Berlin Congress were met with angry indignation by the Bulgarian people. Protest letters from all over the country were sent to the Governments of the Great Powers and the Supreme Porte, openly expressing the people's will to fight for a united Northern and Southern Bulgarian and for the liberation of Macedonia and the Edirne region of Thrace. On the initiative of former revolutionaries a chain of *Union* committees was founded, whose purpose was to organize armed resistance against the implementation of the Berlin Treaty and the struggle for the liberation of Macedonia and the unification of the Bulgarian lands. Under their guidance, military volunteer detachments were formed in Southern Bulgaria to ensure that Eastern Rumelia stayed free and Bulgarian. The struggle of the Bulgarians in Macedonia culminated in the Kresna-Razlog uprising of 1878–79, which was fought for the liberation of that area and its reunion with free Bulgaria. The insurgent operations started on 5 October 1878 in the Kresna region and spread to the Bansko-Razlog valley. In less than a month the insurgents liberated over 40 villages and towns on both banks of the Struma river and in the Razlog region. The uprising took place at a time when the international situation was extremely complicated and unfavourable. It was put down with great brutality by the Ottoman troops, who received no diplomatic support whatsoever. The democratic-minded public in the world saw in it a continuation of the Bulgarian people's struggle against Ottoman oppression.

Although it did not liberate the Bulgarian lands completely, the Russo-Turkish War of Liberation stands out as an epoch-making event in Bulgarian history. It played the role of a bourgeois-democratic revolution, eliminating Ottoman feudalism and all it stood for, together with foreign political oppression in the Bulgarian lands. The Principality of Bulgaria and Eastern Rumelia set out along the path of free capitalist development.

BULGARIA IN THE PERIOD OF CAPITALISM

The Establishment of the Bulgarian National State. The restoration of the Bulgarian State after a period of five centuries presented the Bulgarian people with the possibility of independent national development. A new state system was built, with the help of Russia, following the pattern of modern bourgeois states. Already during the Russo-Turkish War of Liberation a special office for civil administration of the liberated Bulgarian lands was set up, headed by *Knyaz* (Prince) Vladimir A. Cherkasky, functioning in co-operation with the High Command of the Russian army. Its task was to lay the foundations of the Bulgarian political and legal system. With the assistance of local patriotic forces, new state institutions, bourgeois in nature, replaced the overthrown Turkish administration in the liberated regions. City ruling councils were established as elective and collegiate bodies of local self-government. Judicial and police authorities became consolidated; and measures were taken for legalizing the land reform carried out spontaneously in the liberated regions. The aid of prominent Bulgarian men of culture, including Prof. Marin Drinov, Naiden Gerov and Todor Burmov, was enlisted in this activity. Under the San Stefano Peace Treaty the establishment of the new order in Bulgaria and the control of its implementation were to be entrusted to a Russian Imperial Commissioner for a period of two years. Under Article 7 of the Berlin Treaty the term of the Provisional Russian Administration in Bulgaria was cut to nine months. Prince Alexander Mikhailovich Dondukov-Korsakov was appointed Russian Imperial Commissioner in the Principality of Bulgaria, while General A. D. Stolypin was appointed Governor-General of Eastern Rumelia. The main body of the supreme central power in the Principality was the Administrative Council, which had seven departments and provided the prototype of the intended government.

Headed by Prince Dondukov, the Council took positive steps toward the consolidation of the Bulgarian State and issued a number of normative acts. The foundations were laid for the major institutions of the Bulgarian state apparatus, which provided the basis of the bourgeois-liberal system in Bulgaria. Almost everywhere the local power was in the hands of liberal and democratic functionaries. The Bulgarian armed forces were set up and compulsory military service was introduced for men from twenty to thirty years of age. Military training was conducted by Russian officers in the spirit of Russian military traditions. A school for the training of officers was opened in Sofia in November 1878. Legislation was radically transformed, under the leadership of the Russian jurist Sergei Lukyanov.

Bulgaria's administrative system was completed and regular elections for administrative councils were held. Marin Drinov carried out an educational reform, partly on the basis of certain aspects of the local school system existing prior to the Liberation and partly according to some of the main provisions of the Russian educational system.

The Administrative Council put in great efforts to settle agrarian relations stemming from the acute refugee problem and played a great role in transferring big landowners' farms to the peasants.

The new state institutions and system which replaced the Turkish feudal order were the same for Northern and Southern Bulgaria, which was a prerequisite for their future unification.

A major task of the Russian Provisional Administration was to prepare a draft of a Constitution of the Bulgarian State. This was carried out by the Russian jurist Sergei Lukyanov. The first Bulgarian parliament – the Constituent National Assembly – was convened in Turnovo on 10 February 1879. It worked out and adopted the first constitution of Bulgaria (16 April 1879), called the Turnovo Constitution. The constitutional discussions were accompanied by clashes over the Principality's state system between two political groupings holding conflicting views – the Liberal Party, which was an exponent of the interests of the petty and middle bourgeoisie, and the Conservative Party, which represented the *haute bourgeoisie* of the merchants and money-lenders. The approved constitution was bourgeois-democratic in character; universal suffrage for men of age was introduced, as well as wide local self-government and a one-chamber parliament; broad civil rights and freedoms were granted. The Principality was declared a hereditary and constitutional monarchy ruled by a *Knyaz* (Prince). Sofia was designated capital of Bulgaria. The Turnovo Constitution was a historic acquisition for the Bulgarian people. It consolidated the political and economic changes in Bulgaria and facilitated the Bulgarian people's democratic and progressive development. Although limited in nature, for it did not exhaust the possibilities for democratic transformation, the constitution was the focus of bitter class and

Prince A. M. Dondukov-Korsakov and the first Administrative Council, 1878

political struggles throughout the period of bourgeois rule. The reactionary forces strove to abolish it, or to change it in a conservative spirit, while the popular masses defended it as a guarantee of their freedom and advance.

On 7 April 1879, the first Great National Assembly, convened in Turnovo immediately after the Constituent Assembly was prorogued, having elected *Knyaz* of Bulgaria the Prussian Prince Alexander Battenberg (1857–93), nephew of the Russian Empress.

Parallel with the establishment of the Principality of Bulgaria, the foundations were laid of the administrative system of the autonomous province of Eastern Rumelia.

According to the Berlin Treaty, the drafting of the province's Constitution was entrusted to an International Commission of representatives of the states which had taken part in the Berlin Congress. Thanks to the efforts of the Russian delegates, the Constitution (adopted on 14 April 1879) guaranteed the political autonomy of Eastern Rumelia and its bourgeois-parliamentary rule. Alexander Bogoridi (1822–1910), a Bulgarian who held high administrative posts in the Turkish state apparatus, was appointed Governor-General of Eastern Rumelia. A Directorate, comprising five members, became the executive body of the province, while its legislative body was the Regional Assembly.

Eastern Rumelia's dependence was shown in the Sultan's right to adjourn the Regional Assembly at the request of the Governor-General, to approve the members of the Directorate, to station troops in the province and to guard its borders, and in the province's obligation to pay an annual tribute to the Ottoman State. In practice, this dependency proved to be to some extent only nominal.

Having fulfilled its basic obligations as regards the organization of the Principality of Bulgaria and Eastern Rumelia, the Provisional Russian Administration turned over the entire state apparatus to the Bulgarians by June 1879. Supreme power in the Principality was solemnly handed over to Prince Alexander I (25 June 1879), and in Eastern Rumelia to Governor-General Alexander Bogoridi (19 May 1879). In July 1879 the Russian troops left Bulgaria.

After the Liberation profound socio-economic changes, particularly in agrarian relations, occurred in Bulgaria, and a spontaneous land reform was carried out. Already during the war and afterwards the Bulgarian peasants had seized the lands of Turkish landowners and beys who had left the country. Later, the emigrating Turks began selling their lands and about a quarter of the arable land was bought up by Bulgarians, especially peasants.

Turkish large-scale farming was done away with. Petty farming and farm production became dominant in the Bulgarian villages. Although slowly, it was gradually brought into line with up-to-date requirements. The transformation in farming improved to a certain extent the plight of the peasant masses, but at the same time intensified the process of their stratification. Part of the petty and middle farmers began losing their lands and were proletarianized. Another part seized more lands; gradually a considerable stratum of well-to-do peasants became established and the peasant bourgeoisie took shape. Yet the majority of peasants in Bulgaria were petty and middle commodity producers.

The dominant social class in Bulgaria was the petty bourgeoisie, including the petty urban artisans and tradesmen and the peasants who owned small plots of land. They comprised the overwhelming majority of the Bulgarian people. The *haute bourgeoisie* – rich tradesmen, money-lenders, contractors, etc. – was small in number and economically weak. Considerable means were concentrated in the hands of a small number of people, but the young Bulgarian bourgeoisie found it more profitable to invest the accumulated capital in trade and usury rather than in the development of the national industry. Therefore capitalist urban production developed at a slow pace.

In the first five years following the Liberation, only 23 industrial enterprises were set up; they were semi-artisan in nature. The poorly developed industry could not absorb the proletarianized peasants and craftsmen. Bulgaria remained a poorly developed agrarian country with prevailing light industry and petty farm production and a comparatively low level of productive forces. Capitalist development was also hampered by the unequal trade contracts concluded between the western states and Turkey and imposed on Bulgaria under the Berlin Treaty.

The political life of the Principality in the early post-Liberation years was determined by the struggle between the Liberal Party, whose ideal was the existing petty-bourgeois relations and the Turnovo Constitution, and the Conservative Party, which stood for a strong monarchic power and restriction of popular rights. On his accession to this Bulgarian throne Prince Alexander I at once sought to impose reactionary amendments to the Constitution. The first Bulgarian Government included representatives of the Conservative Party, headed by Todor Burmov (5 July–24 November 1879). From the very beginning the Prince and the Government began violating the Constitution and schemed to elect a National Assembly that would ensure lawful amendment of the country's major laws. During the election campaign the Liberal Party, representing the petty and middle bourgeoisie, raised the slogan of safeguarding the Turnovo Constitution and developing the democratic principles enshrined in it. Two groupings had taken shape by 1879 – a moderate one headed by Dragan Tsankov (1828–1911) and an intransigent one, headed by Petko Karavelov (1843–1903) and Petko Rachev Slaveikov. Their differences amounted to setting the limits of democracy in the Constitution; the moderates regarded it as being extreme and were willing to amend it in a more conservative spirit; the intransigents believed it to be in line with political conditions in the country, containing the prerequisites for progressive development.

In the National Assembly elections held in the autumn of 1879 the conservatives suffered total defeat. Alexander I dismissed the Assembly, and a new Government was appointed, again comprising members of the Conservative Party, headed by Bishop Kliment (Vassil Drumev,

Petko Karavelov

24 November 1879–24 March 1880). In the new parliamentary elections the Liberal Party won an even more convincing victory and the Prince had no choice but to entrust it with the country's Government. The Liberal Governments of Dragan Tsankov (March– November 1880) and of Petko Karavelov (1880–81) pursued a policy that corresponded with the national interest and took measures which promoted state development. The Second National Assembly passed a number of bills which were to guarantee the implementation of the democratic principles enshrined in the Turnovo Constitution. One of the most important documents issued under Liberal Party rule was the Law on the People's Voluntary Corps, in which all men under the age of forty were to serve, while its command was entrusted to elective bodies and appointed persons. In its foreign policy the Liberal Government upheld the country's independence, strove for national unification, opposed Austro-Hungarian and German aspirations and steered a course of close co-operation with Russia.

Prince Alexander I and the conservatives did not become reconciled to Liberal Party rule and waited for the right internal and external conditions to launch an offensive.

Taking advantage of Russia's difficulties after the assassination of the Russian Emperor Alexander II on 1

March 1881, Prince Alexander carried out a *coup d'état* on 27 April 1881 in alliance with the conservatives. He described his intention to abdicate, should the Great National Assembly fail to accept his proposals for a way out of the deadlock, as he called it. Petko Karavelov's Government was overthrown and replaced by a Provisional Government under the Russian General, Kazimir Ernrot, Minister of War (27 April–1 July 1881). The Second Great National Assembly, elected by terror and malpractices, was composed almost entirely of the Prince's adherents. It voted the extraordinary powers demanded by the Prince empowering him to rule the country without a constitution for a seven-year term. Upon expiry of this term the Great National Assembly was to be convened to revise the Constitution. The regime thus established was in fact a dictatorship of the Prince and the *haute bourgeoisie*. The European powers, Russia included, approved of the 27 April coup. A State Council was constituted, whose members were appointed by the Prince. A new electoral law was worked out on the basis of the two-stage system, and suffrage was greatly restricted. The constitutional freedoms of meetings and the Press were likewise considerably restricted through special bills. Liberals were persecuted and many of them were interned or went into exile. The leaders of the intransigent liberals (P. Karavelov, P. R. Slaveikov and others) went to Eastern Rumelia. Being imposed through outside interference and violence, the regime did not enjoy the support of the people.

As early as the autumn of 1881 there were differences of opinion between the conservatives and the Russian representatives in the Government. On 1 January 1882 the Prince formed a new Government made up of conservatives only. A strong popular movement against the regime broke out, and the Prince was once again compelled to seek Russia's support in order to strengthen his position. A new Government was formed, headed by the Russian General Leonid Sobolev (23 June 1882–7 September 1883) and the Russian General Alexander Kaulbars was appointed Minister of War. However, soon afterwards the Prince fell into disagreement with the Russian representatives and thus deprived his regime of outside support. Assisted by Dragan Tsankov, who agreed to amend the Constitution in a conservative spirit, Alexander I overthrew General Sobolev's Government. He decreed his decision to restore the Constitution and to submit it for amendment by the National Assembly. A Cabinet of moderate liberals and conservatives under Dragan Tsankov was formed (7 September 1883–29 June 1884) and on 5 December 1883 the Third Ordinary National Assembly passed a bill on constitutional amendments. Alexander I gave up his powers. The intransigent liberals headed by P. Karavelov and P. R. Slaveikov declared themselves for the complete restoration of the Constitution, and won a decisive victory in the elections for the Fourth National Assembly. It was from their midst that P. Karavelov formed a Government (29 June 1884–9 August 1886) which restored the Turnovo Constitution in full. The contradictions between moderate and intransigent liberals became more acute, and late in 1883 the Liberal Party split into two separate and hostile parties – the Karavelov and the Tsankov parties. The collapse of the extraordinary powers regime also gravely affected the Conservative Party, which left the political stage.

The Unification of the Principality of Bulgaria and Eastern Rumelia. The Constitution of Eastern Rumelia had established a comparatively complicated administrative order in that artificially founded, small state. The aim of the western states was to neutralize Bulgarian ethnic supremacy through a greater involvement of the Turkish and Greek minorities in the Government. In May 1879 Alexander Bogoridi formed the first Government of Eastern Rumelia (Directorate) and appointed Gavril Krustevich (1820–98) its Chief Secretary (Prime Minister) and Director for Internal Affairs. The first elections for a Regional Assembly held on 17 October 1879 brought victory to the Bulgarian population, with forty Bulgarian deputies as against sixteen of non-Bulgarian origin. The Bulgarians won an even greater victory in the elections for the Standing Committee (the Assembly's standing body), filling the total of ten seats. Ivan Evstratiev Geshov (1849–1924) was the first president of the Regional Assembly and of the Standing Committee. Notwithstanding the election results, Eastern Rumelia developed unevenly. The Ottoman Government deliberately impeded its state and economic advance. The inter-party alliances and clashes in the region also acquired a markedly Bulgarian character. Two parties took shape – the Popular (Conservative) Party, which represented the interests of the *haute bourgeoisie* and had pro-Russian leanings, and the Liberal (also called the Unification, or State) Party – of the middle and petty bourgeoisie, with anti-Russian leanings. The key posts in the province's management were filled by representatives of the *haute bourgeoisie,* most influential among them being the Geshov family of merchants and landowners. Bulgarian nationality became established as dominant in the province and this fact was manifest in the province's independent status, despite the Sultan's right to occupy with his troops certain points of strategic importance. Following the settlement of the national question in Eastern Rumelia, the struggle for state and political unification with the Principality of Bulgaria came to the fore. The realization of this aim, however, was linked with complex political problems on a Balkan and all-European scale, as well as with violating one of the most unjust provisions of the Berlin Treaty.

In late 1884 and early 1885 there was an upsurge in the popular movement for the liberation of Macedonia and the Adrianople Region of Thrace, and for the unification of Northern and Southern Bulgaria. Early in 1885 the Bulgarian Secret Central Revolutionary Committee (BSCRC) was set up in Plovdiv, and undertook the organization of the unification. Zahari Stoyanov (1850–1889), a pre-Liberation revolutionary figure, was the founder, ideologue, and actual leader of the committee. Its statutes and programme were based on the

Members of the Bulgarian Secret Central Revolutionary Committee and activists, 1885

programme of Vassil Levski's revolutionary organization. The Committees declared as its supreme goal 'the final liberation of the Bulgarian people, both through a moral revolution and arms'. It was pointed out that the Bulgarian people had no aggressive intentions against any people whatsoever, and that it aspired after a Balkan peoples' federation, based on equality and national sovereignty. In June 1885 the committee had its composition renewed, including functionaries who stood for the immediate unification of Eastern Rumelia with the Principality alone. The liberation of Macedonia was to be the task of a later stage in the unification of the Bulgarian people. The unification movement was also joined by representatives of Prince Alexander I, whose aim was to stabilize his prestige after the abortive extraordinary powers regime and to enhance his influence abroad through a successful political campaign. The tactical plan of action was specified in August and a great portion of the officers in the Eastern Rumelian army were enlisted for the cause. The military troops of the Principality were set in full fighting trim.

The unification was effected quite expeditiously between 3 and 6 September 1885, following exactly the plan drawn up in advance. On 6 September Eastern Rumelia was proclaimed part of the Bulgarian State. Gavril Krustevich, the region's Governor-General from 1884–85 offered no resistance. The BSCRC members set up a Provisional Government headed by Dr Georgi Stranski (1847–1904). Major Danail Nikolaev (1852–1942) was appointed Commander-in-Chief of the armed forces. The Bulgarian troops entered former Eastern Rumelia, jointly with the government representatives of the Principality.

The Unification stirred up European politics. Events in Bulgaria were in the limelight of international relations, since they violated the Berlin Treaty and disturbed the unity of the Ottoman Empire. The Russian Emperor declared himself against the Unification, mainly because of his dislike for Prince Alexander I. The Russian officers were withdrawn.

The Ottoman Government sent a protest note to the Great Powers, but it did not dare to start a war, since Russia and the other Great Powers warned it to refrain from hostile actions against Bulgaria. In view of Bulgaria's

Arch erected in Plovdiv to welcome the victors of the Serbian-Bulgarian War, 1885

expansion, Serbia and Greece put in territorial claims to the Empire as well as to Bulgaria. A conference of Ambassadors of the Great Powers to settle the Eastern Rumelia Question was held in Constantinople late in October 1885. Under the pretext that the balance of power in the Balkans had been disturbed, the Serbian King Milan, encouraged and financially backed by Austria-Hungary, ordered the Serbian troops to attack Bulgaria on 2 November 1885. Considerable Serbian forces advanced towards Sofia and Vidin. In the first battles the Bulgarian army, though small and lacking in experienced commanders, put up a brave resistance and after prolonged battles at Dragoman withdrew to a defensive line at the approaches to Sofia and Slivnitsa. The Vidin fortress was beseiged by Serbian troops. In the meantime thousands of Bulgarian volunteers and the Bulgarian units guarding the Bulgarian-Turkish border were quickly transferred to the front line. After hard battles lasting three-days, the Bulgarian army, led by young captains and majors, won a decisive victory. The Serbian élite corps was defeated, and the main Bulgarian force advanced towards Serbia. The siege of the Vidin fortress was raised. The Serbian troops at the northern part of the Bulgarian-Serbian border also retreated. The road to Belgrade was open. It was only through Austro-Hungarian intervention that Serbia was saved from utter military defeat. The Bucharest Peace Treaty (19 February 1886) restored the pre-war *status quo*. Bulgaria's victory in the war sealed the Unification. After prolonged diplomatic bickering the Bulgarian-Turkish agreement (the Tophane Act) was signed on 5 April 1886 by which the Great Powers recognized the Unification. The Bulgarian Prince was entrusted with the government of Eastern Rumelia. In actual fact the two regions were united into a single state, having a common army, common National Assembly, etc.

The unification became a fact on the political map of the Balkans and Europe. The Ottoman domination over Southern Bulgaria, though a formal one, was finally done away with, and a considerable portion of the Bulgarian nation was re-united. The Bulgarian State now had greater opportunities for economic, political and spiritual progress. The blood shed in the battlefield and clever diplomacy helped carry through the first stage of the Bulgarian people's unification programme after the Liberation.

Bulgaria at the Turn of the Century. After the Unification the struggle between tsarist Russia and its West-European adversaries for domination of Bulgaria intensified. Russian diplomacy sought to remove Alexander I, while the western states supported him. Encouraged by its military victories and diplomatic successes, part of the Bulgarian bourgeoisie steered a course of complete break with Russia and closer economic ties with Austria-Hungary and the other big western states. The Bulgarian bourgeoisie was divided into Russophiles and Russophobes. Dragan Tsankov was the leader of the extreme Russophile group; the other part of the Bulgarian bourgeoisie, closely linked with West-European capitals and striving to detach Bulgaria from Russian influence, was headed by Stefan Stambolov (1854–95). The two camps waged bitter struggles for power and for determining the country's foreign policy. There was growing opposition to the Prince among the officers, who plotted to depose him. The Prince was regarded by Russophile propaganda as the sole cause of the difficulties experienced by the Principality. The conspirators were headed by high-ranking officers – the chief of the Military School, Major Peter Gruev (1855–1942), Captain Anastas Benderev (1859–1946), Captain Radko Dimitriev (1859–1918) and others. On the eve of 9 August 1886 the Struma regiment and the cadets from the Military School seized the Palace and the government departments in Sofia. The Prince was forced to abdicate, and immediately afterwards he was driven out of the country. The plotters, however, were isolated from the people and the political parties, and displayed political naïvety and inconsistency in their actions. Their attempt to form a Provisional Government of prominent political figures (P. Karavelov, S. Stambolov, D. Tsankov) failed. On 9 August a Russophile Government under Bishop Kliment was established. Lt.-Col. Sava Mutkurov (commander of the Plovdiv garrison) and Stefan Stambolov (Chairman of the National Assembly), who headed the Russophobe

Radko Dimitriev

Anastas Benderev

opposition, declared themselves against the coup. A quick counter-coup was carried out on 11 August. The Provincial Government was proclaimed traitorous and was outlawed, and S. Stambolov cabled Prince Alexander inviting him to return to the throne. On 17 August the Prince came back to Bulgaria, but since the Russian Emperor disapproved of his staying in the country, on 25 August he abdicated, having appointed a regency including S. Stambolov, P. Karavelov (who was soon replaced by Georgi Zhivkov) and Sava Mutkurov. Stefan Stambolov was in complete control and power was usurped by the Russophobe bourgeoisie. The political crisis in the country was exacerbated, and the officers who had taken part in the conspiracy emigrated. The groupings and factions within the country finally took shape according to their attitude towards tsarist Russia. Through its special envoy, General Kaulbars, the Russian Government made an attempt to change Bulgaria's foreign policy, but failed. In November 1886 Russia broke off diplomatic relations with Bulgaria, and the crisis in Bulgaria grew more acute. The opposition outside the country stepped up its activity, and in 1886 the Russophile officers in exile set up a revolutionary committee in Bucharest. In February 1887 they staged rebellions in Russe and Silistra, which were brutally suppressed, and government terror mounted.

After lengthy consultations conducted by Bulgarian emissaries with the West-European countries, the Prussian Prince Ferdinand Saxe-Coburg-Gotha (1861–1948), an Austrian army officer, was nominated candidate for the Bulgarian throne. On 25 June 1887 the Third Great National Assembly elected Ferdinand Prince of Bulgaria. According to the Berlin Treaty, the elections had to be endorsed by the Turkish Government. France, Germany and Russia declared the election unlawful, and the Ottoman Empire followed suit. After he was sworn in, Ferdinand I entrusted Stefan Stambolov with the formation of a new government.

The seven-year term of Stambolov's rule (1887–94) was characterized by accelerated capitalist development, benefiting from the expanded domestic market after the Unification. The process of initial accumulation of capital intensified. Favourable conditions were created for the development of manufacture, mainly the textile and the food industries. During the Stambolov rule a total of 80 new factories were opened, while many of the old ones were reconstructed and expanded. Credit societies and over 200 joint-stock companies were founded. The main foreign trade partners were Austria-Hungary, Britain, France and Germany. Capitalist development prompted the consruction of new railway lines, roads, ports, telegraph lines, etc. The Government resorted to foreign loans to finance the expanding railways. The People's Liberal Party (Stambolovists) which represented the big commercial and industrial bourgeoisie introduced protectionism. Trade agreements on the increase of import duties were signed with Britain and other West-European states in 1889–90. The first Bulgarian economic exhibition was staged in Plovdiv in 1892. Large-scale construction and urbanization of Sofia and other cities was launched. The foundations of national institutions of culture and education were laid. The first Higher School (today Sofia University) was opened in Sofia in 1888, and a unified system of education was introduced.

The Government made it its goal to promote the cultural and educational consolidation of the Bulgarian population in the European part of the Ottoman Empire, as a preliminary step towards the future unification campaign.

In 1890 Stambolov managed to get the consent of the Ottoman Government for the nomination of Bulgarian bishops to the dioceses of Skopje and Ohrid.

As the big commercial and industrial bourgeoisie was not numerous, the Government of Stambolov lacked a broad social base. In order to remain in power it resorted to harsh police terror. The opponents of the regime, including the young socialist movement, were subjected to brutal persecution and reprisals. The Russophobe policy of the Government also antagonized broad social circles from the sphere of the upper middle class, clergy and intelligentsia.

The main enemies of Ferdinand and Stambolov's Government among the bourgeoisie were the factions of the Liberal Party – the adherents of Tsankov and Karavelov (what was called the underground opposition) but they could not organize the people's resistance and overthrow the regime. The discontent found expression in conspiracies and terrorist actions, which stepped up government reprisals. The police regime and the tense political situation hindered further capitalist development of foreign trade, and therefore a part of the commercial and industrial bourgeoisie opposed the dictatorship of Stambolov. By its extremely intransigent attitude toward Russia, the regime obstructed international recognition of the Bulgarian Prince. Ferdinand made skilful use of the broad social and political movement against the Government in order to get rid of Stambolov and concentrate power in his own hands.

The Government formed by Konstantin Stoilov (19 May 1894–18 January 1899) consolidated the domination of the *haute bourgeoisie*. At the end of 1894, after the foundation of the Popular Party, an independent Populist Government came to power. Representing the interests of bankers, businessmen, industrialists and big landowners, the populist regime carried on the policy of encouraging and furthering industry and capitalist development. A law was passed on the encouragement of local industry, new trade contracts were concluded, and customs duties on imported industrial products were increased. Railroad construction developed at a hectic pace, supported by a special law (1895). The number of joint-stock companies

Dimiter Blagoev

grew. The Law on Land Tax (1895) did away with the tithe on agricultural products left over from the time of Ottoman domination, and introduced taxation on land estates, thus giving an impetus to capitalist progress in agriculture. The measures taken by the Populist Government required vast funds, collected by way of direct and indirect taxes, which were a heavy burden on the masses. In the sphere of foreign policy the Government pursued normalization of relations with Russia (restored in 1896) and stabilization of the international standing of Bulgaria. But the enormous spending in construction and other spheres encumbered the country financially and discredited the Government completely. The continuous aggravation of the financial and economic crisis, the failures of the regime, and the growing discontent of the masses were used by the bourgeois opposition parties and Ferdinand I to topple the Government (January 1899).

At the end of the nineteenth century, after the resignation of K. Stoilov, several governments came to power in quick succession. Ferdinand I took advantage of the weakness of bourgeois policy, and gradually consolidated his position and established a personal regime of his own. The Radoslavov Administration (1899–1901) gave *carte blanche* to its partisans to appropriate public funds and was thus instrumental in bringing about the establishment of Ferdinand's personal regime. Looking for a way out of the financial crisis and of covering the rocketing costs of maintaining the regime, the Government adopted in 1900 the Law on the Tax of Land Products for 1900–1901, by virtue of which land tax was replaced by tithe in kind. The massive agrarian movement and peasant rebellions against the introduction of the tithe (1900) precipitated the fall of the discredited Government.

The rapid capitalist development brought about changes in the class structure of Bulgarian society and the correlation of political forces. The new commerical and industrial bourgeoisie increased in number and consolidated its economic and political positions. The working class extended its ranks. The petty and rural bourgeoisie accounted for the vast majority of the people. Despite the sharp polarization of class forces, it was the most numerous social class. The changes in the class structure caused changes in the political forces as well. The former Conservative Party ceased to exist. The Liberal Party split up into factions (1883–84) which gradually took shape as independent parties representing the interests of different strata of the bourgeoisie – the Liberal Party led by Vassil Radoslavov (1854–1929), the Progressive Liberal Party led by Dragan Tsankov (1828–1921) and Stoyan Danev (1857–1949), the Democratic Party of P. Karavelov and Alexander Malinov (1867–1938) and the People's Liberal Party of S. Stambolov and Dimiter Petkov (1858–1907). In 1894 the Popular Party was founded under the leadership of K. Stoilov and I. E. Geshov.

The end of the nineteenth century also saw the formation of the working class. The extremely hard conditions of life and work, cruel exploitation by the bourgeoisie seeking quick profits, and the lack of labour legislation incited the workers to fight for an improvement of their living standard against the arbitrariness of the masters and for the formation of trade unions. The Bulgarian Typographic Society – the first workers' professional organization in Bulgaria – was set up in 1883. The first attempts of workers to organize strikes date back to the beginning of the 1880s.

On 20 July 1891, on the initiative of D. Blagoev (1856–1924), the founder of Socialism in Bulgaria and the greatest proletarian revolutionary and theoretician of Marxism in the Balkans at the turn of the century, the Bulgarian Social Democratic Party (BSDP) was founded and stood at the head of the struggle of the working class for social and political liberation. The BSDP was the first Marxist Party in the Balkans. At the opening stage of capitalist development it guided the nascent workers' movement toward scientific Socialism, Marxism, and directed its economic battles to the ultimate objective – Socialism, which was to be achieved through uncompromising class struggle. From the very beginning of its existence, the BSDP fought the opportunist ideological trends whose supporters believed that there were none of the prerequisites for the dissemination of Marxism and for the formation of a proletarian party in the underdeveloped capitalist society of Bulgaria at that time. In 1892 the opponents of the foundation of such a party set up a social democratic organization with an opportunist orientation – the Bulgarian Social Democratic Union (BSDU).

In 1894 the BSDP and the BSDU merged under the name of Bulgarian Workers' Social Democratic Party (BWSDP) but the controversies within the party continued. Two trends developed – the revolutionary Marxist trend, led by D. Blagoev and the reformist-opportunist one, led by Yanko Sakuzov (1860–1941). They were engaged in a persistent ideological and political struggle. Under the influence of the BWSDP propaganda the workers' movement intensified. Strikes were staged relatively often: there were sixty in 1891–1900. The first workers' trade unions appeared; 74 such societies were formed in the 1894–1900 period.

The agrarian movement in the country also gained momentum. At the end of 1899 a congress held in Pleven laid the foundations of the Bulgarian Agrarian Union. Conceived as a professional and educative farmer's organization, the Union gradually grew into a political party which adopted the estate principle in the formation of its policy. The foundation of the Bulgarian Agrarian Union (BAU) and its establishment as a political organization was an important event in the political life of the country, making it possible to organize the peasants' struggle against the reactionary regimes, to involve them in political life and to create the prerequisites for the unity of the working people in the struggle for democracy. As of 1904 Alexander Stamboliiski (1879–1923) became the recognized ideologist and leader of the BAU.

In the first decade of the twentieth century the development of Capitalism in Bulgaria made considerable progress. The amassed capital was invested in industry, signs of concentration appeared, and the first cartel was formed. Several large banks were set up with foreign and local capital. Agriculture acquired more and more the characteristics of a commodity economy, the developing domestic market and bigger agricultural exports being conducive to that end. Poor and middle-bracket peasants increasingly came to ruin; the number of these deprived of land or in possession of small plots rose all the time (130,000 in 1910). A sizeable rural bourgeoisie came into being. Nevertheless, the rural economy was small-scale, fragmented and unproductive. Bulgaria remained an economically backward agrarian country with a moderately developed, predominantly light industry; this made it dependent on foreign capital, invested mostly in big banks and trade.

The economic upsurge affected the development of the working class and caused changes in its structure. The number of industrial workers increased, and the working class played a relatively significant role in the economic and political life of the country.

The Marxist trend in the BWSDP led by Dimiter Blagoev fought to identify the activity of the party with the aspirations of the working class, to make it a guiding force of the Bulgarian proletariat. The revolutionary Marxists, called 'left-wing' socialists were forced to carry out sharp ideological and theoretical polemics with the opportunist trend (the 'common-cause' socialists) who stood for class co-operation with the bourgeoisie. As a natural result of the deep theoretical, tactical and organi-

The Buzudja Congress, 1891 Painting by K. Buyukliiski and P. Petrov

zational contradictions between the two trends came the purge of the opportunists, the 'common cause' socialists from the party, sanctioned by the Tenth Congress of the BWSDP (1903). The BWSDP ultimately emerged as a revolutionary Marxist party of the working class under the name of Bulgarian Workers' Social Democratic Party (left-wing socialists) – BWSDP (l-ws). A new period in the party history was initiated – the period of left-wing Socialism. The Bulgarian Marxists took their place in the left wing of the European socialist movement, whose objective was to preserve the revolutionary character of Marxism. The Party stayed true to its historical mission – to organize the working class, to defend uncompromisingly its interests, to lead the revolutionary struggle against Capitalism, for Socialism. The 'common-cause' socialists, also called 'right-wing' socialists, continued their existence organized in a separate party – the opportunist Bulgarian Workers' Social Democratic Party, concerned mainly with the interests of the petty bourgeoisie. The BWSDP (l-ws) started to play a greater role as organizer and leader of the working class. Its influence in the sociopolitical life of the country increased. In the elections for deputies to the National Assembly in 1913 it won 54,297 votes and eighteen seats in the National Assembly. In 1910 the BWSDP (l-ws) won the municipal elections in Samokov and in 1912 – in Sliven, and the first communes were established.

The fast growth of the working class, the difficult conditions under which it was placed, and the activities of the revolutionary Marxists determined a new upsurge in the workers' movement. A struggle for the elaboration of workers' legislation was organized. The General Workers' Trade Union (GWTU) was set up in 1904, under the leadership of the BWSDP (l-ws). The total number of strikes carried out under the leadership of GWTU during the period 1904–11 amounted to 550 (210 in 1910 alone). Significant manifestations of the workers' struggle were the miners' strike (1906), the general strike of railway workers (1906–7), the strike of the Sliven textile workers (1908), the strike of the tobacco workers in Varna, Shumen, Russe and Plovdiv (1910) which was a complete success, and the general printers' strike (1910–11), in Sofia. The mass-political movement among the workers for the elaboration of social legislation intensified by the end of 1909 and in 1910.

The party split, the fruitless inter-factional struggle and the insufficient influence of the bourgeoisie among the masses gave an opportunity for Ferdinand I to establish himself as a decisive factor in leadership and to consolidate his personal regime. Petko Karavelov's Government

Gotse Delchev

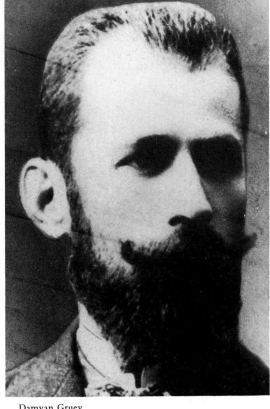

Damyan Gruev

(February–December 1901), which inherited a critically serious financial and economic situation in the country, tried to find a way out of signing a foreign financial loan, but the government proposal was defeated by a vote of deputies in the National Assembly. On 21 December 1901 a Cabinet of the Progressive Liberal Party, headed by Dr Stoyan Danev (1857–1949), was formed. Meanwhile a certain liberalization was introduced in the methods of home rule. Positive changes took place in all spheres of life. Bulgaria became one of the most important actors on the Balkan political scene. Laws bringing about the stabilization of the state-administrative structure were adopted, local industry was encouraged, and commercial contracts with European states were signed. The modernization of the army continued.

In May 1903 a Government of the People's Liberal Party (the second Stambolov regime), which ruled the country for five years, was formed.

The party expressed the interests of the Bulgarian industrial bourgeoisie, linked to the West-European world. Through its legislative initiatives the regime contributed to the progress and modernization of the economy, which was in the interest of the young industrial bourgeoisie. Two foreign loans were contracted for modernization and equipment of the army. At the same time, its home policy was anti-democratic, consolidating further Ferdinand's personal regime. The right to strike was restricted, and some professional organizations were banned.

The fast growth of capitalist industry was based on cruel exploitation of the workers and on lower prices for the women's and children's labour used. Indirect taxes were increased twice during the period 1903–7. The disatisfaction and discontent of the masses intensified, and strikes and protest demonstrations broke out in the country. On 3 January 1907, at the inauguration of the National Theatre in Sofia, students and railway workers on strike booed Ferdinand. The party in power resorted to extreme measures; the University was closed, the teachers dismissed and the students drafted into the army. Reactionary laws restricting the right of workers and employees to strike were passed and reactionary amendments were made to the Law on the Press. The remaining parties, taking advantage of the Government's difficult position, formed an Opposition – the so-called patriotic bloc (Democratic, Popular, Progressive-Liberal, Radical-Democratic and Broad-Socialist) and embarked on a loud campaign against Ferdinand's personal regime, demanding popular rule. The Prime Minister, Dimiter Petkov, was shot in the street in February 1907.

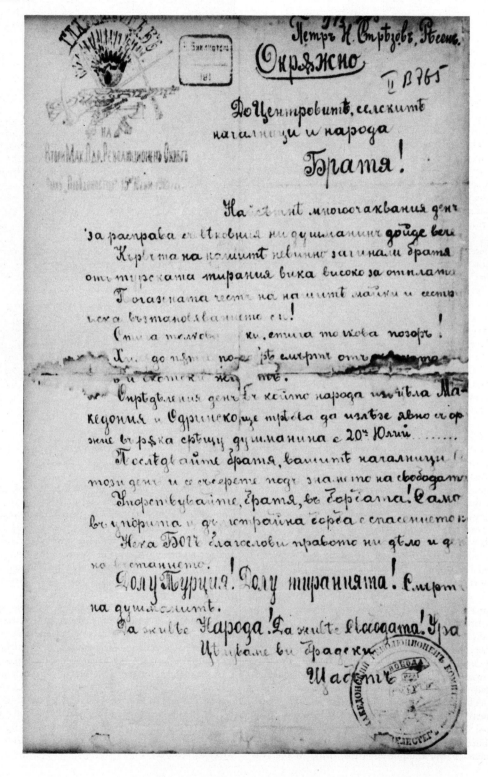

Facsimile of the appeal of the General Staff of the Bitola Revolutionary District, July 1903. 'Long live the people! Long live freedom!...'

Ilinden-Preobrazhenie Uprising. Despite the differences in their foreign policy orientation, the various political groupings of the bourgeoisie shared the same view on the issue of uniting the Bulgarian lands and the liberation of the Bulgarians in Macedonia and the Adrianople region, which were still under Ottoman rule by virtue of the Berlin Treaty of 1878.

Despite the fact that Article 23 of the Berlin Treaty envisaged the introduction of reforms in the Ottoman Empire and the granting of internal administrative autonomy to the regions inhabited by non-Turkish populations, the Ottoman regime did nothing in this respect. The severe national, socio-economic and religious contradictions which gave rise to the Bulgarian national liberation movement during the period of National Revival, were not eliminated. Therefore the Macedonian and Adrianople Bulgarians pooled their forces in a joint liberation action, and the free Bulgarian State encouraged their revolutionary struggle. The activities of the Bulgarian Exarchate, which included the Ohrid, Skopje and Veles dioceses, in promoting the church-school cause and the enhancement of educational and cultural work, helped in raising the national consciousness of the Bulgarian population in the Ottoman European regions, and in protecting the Bulgarians from the anti-national ambitions of the neighbouring countries. Many detachments continued to operate in the two provinces after the protest actions against the decisions of the Berlin Congress and after the failure of the Kresna-Razlog Uprising (1878–79). In October 1893 in Thessaloniki, the foundations of a revolutionary organization were laid – the Internal Macedonian-Adrianople Revolutionary Organization (IMARO). Its committee network extended quickly to many areas in Macedonia and the Adrianople region. The ideologists and leaders, Damyan Gruev (1871–1906) and Gotse Delchev (1872–1903) did an enormous amount of organizational work. The Statute of the IMARO stipulated that the goal of the organization was to unite all dissatisfied elements in Macedonia and the Adrianople region, no matter what nationality, in order to gain complete political autonomy for the two provinces through revolution. In terms of its national composition the IMARO was a Bulgarian organization. It relied mostly on the Bulgarian population in Macedonia and the Adrianople region, and its cause was a natural continuation of the Bulgarian national revolution from the National Revival Period. The Bulgarian farmers and craftsmen, led by the intelligentsia, constituted the major driving force of the national liberation movement. The organization's institute of detachments was set up under the leadership of Gotse Delchev. Agitational organizing detachments were set up in all regions, to carry out revolutionary propaganda among the population and serve as a nucleus at the outbreak of the uprising.

The legal organization of the liberation movement in the Principality was headed by the Supreme Macedonian-Adrianople Committee (SMAC), founded in Sofia in 1895. The committee assisted IMARO in setting up the detachment's institute and in supplying it with armaments and funds. Gradually sharp differences of opinion on the issue of the tactics and joint leadership of the revolutionary movement arose between the two organizations. In September 1902 SMAC announced an uprising in the region of Gorna Diumava (Blagoevgrad) in order to provoke the interference of the Great Powers. Despite its rapid suppression, the uprising nevertheless had a strong impact in Europe. The situation in Macedonia and the Adrianople region became tense. The terror inflicted by the Turkish Government acquired unprecedented dimensions, and the dissatisfaction of the population slowly grew. The congress of IMARO, convened in Thessaloniki in January 1903, adopted a decision on an uprising. The leader of the revolutionary organization – Gotse Delchev – was killed in a battle at this crucial moment. The concrete plan of the uprising was elaborated at a congress of the Bitola revolutionary district, headed by D. Gruev, and a directorate was elected, consisting of D. Gruev, Boris Sarafov and Anastas Lozanchev. The uprising broke out on St. Elijah's Day, 20 July 1903 in the district of Bitola. The insurgents seized power in Bitola, Ohrid, Kostur, Lerin and other districts, and revolutionary power was established in the town of Krushevo under the leadership of the socialist Nikola Karev (1877–1905), which in practice implemented the revolutionary-democratic and internationalist principles of IMARO.

The uprising in the Adrianople region broke out on Transfiguration Day, 6 August. Revolutionary-democratic power was proclaimed in the liberated areas of the Strandja region. Insurgent actions started by the end of August in Ser revolutionary district, the detachments of IMARO operating together in this area with the detachments of SMAC. Guerrilla battles were waged in the revolutionary districts of Thessaloniki, Skopje and Struma.

The Ottoman Government took fast and cruel measures for the suppression of the uprising. An eighty-thousand-strong army was sent against the insurgents of Bitola revolutionary district, and a forty-thousand-strong army was concentrated in the Strandja region. On 19 September the directorate adopted a decision for the uprising to be interrupted. Armed clashes continued for more than three months throughout the territory of Macedonia and the Adrianople region. According to IMARO data, during that period 239 battles took place, and the 26,400 insurgents who participated in them fought against a 350-thousand-strong regular army and the bashi-bazouk. 205 villages were set on fire and more than 70,000 persons remained without shelter. Loss of life came to over 5,000, and 15,000 were imprisoned. About 30,000 people left their native regions, finding refuge in free Bulgaria.

The Ilinden-Preobrazhenie uprising was the most heroic manifestation of the struggle for national liberation of the Bulgarians in Macedonia and the Adrianople region. The uprising was widely reported abroad and activated international policy. Reforms were envisaged at a meeting between the Russian Tsar and the Austro-Hungarian Emperor, held in Mursteg (1 October 1903), but they were of an extremely restricted nature. The liberation struggle in the two areas continued, and armed

Representatives of the Bulgarian and Ottoman sides at the signing of the armistice at Chataldja, November 1912

detachments operated continuously, fighting the Turkish troops.

The Government of Racho Petrov (1903–6) signed an agreement with Serbia for unity of action in improving the situation of the Christian population in the European province of the Ottoman Empire. However, it did not come into force because of the contradictory views of the two States on the Macedonian issue. The autonomy of Macedonia, regarded as a step toward its unification with Bulgaria, remained a fundamental principle of Bulgarian policy, while Serbia strove to expand its territory to the detriment of the Bulgarian lands in the Ottoman Empire.

Bulgaria on the Eve of, and during the Wars (1912–18). By the beginning of 1908 the internal political life of the country was marked by a certain stability. The new Government of the Democratic Party, headed by Alexander Malinov (16 January 1908–16 March 1911) repealed the Stambolovists' extraordinary measures with respect to the university, the teachers and the students in order to calm down the country. The Democratic Party, expressing the interests of the industrial bourgeoisie, consistently carried through undertakings for the consolidation and preservation of the political and economic domination of the bourgeoisie. Bulgaria witnessed an economic upsurge; in terms of absolute growth rates the Bulgarian economy made the greatest headway compared with the economy of the other Balkan countries.

A. Malinov's Cabinet continued the traditional protectionist policy; the Law on the Encouragement of Local Industry was extended; the road and railway network was further constructed; commerce developed actively; banking was promoted; foreign investments (Belgian, French, Austro-Hungarian and German) penetrated the country through the banks and the joint stock companies.

The foreign policy of the Democratic Party was focused on *rapprochement* with Russia. The Russian strategy for the establishment of a union of Balkan States (between Bulgaria and Serbia mainly) as an instrument against Turkey and as a barrier against Austro-German expansion in the Balkans corresponded to the basic line of Bulgarian foreign policy. The liberation of Macedonia and the Adrianople region of Thrace, being a historically substantiated policy for the attainment of the Bulgarian national ideals, was supported by the public at large. Under these conditions, A. Malinov's Government embarked on a course of preparation for a war with the Ottoman Empire. The first step in its implementation was the declaration of Bulgarian independence, casting off vassal relations with the Ottoman Empire which, though only formal, restricted Bulgaria's rights as an equal partner in its foreign political activities. The outbreak of the 1908 Young Turkish Revolution created favourable conditions.

The Bulgarian State was officially proclaimed independent on 22 September 1908 in Turnovo. By a special manifesto to the Bulgarian nation, Bulgaria became a fully

Bulgarian soldiers on Tumba peak on the Belassitsa mountain, World War I

independent and sovereign state, enjoying equal rights in international relations.

On the following day Austria-Hungary annexed Bosnia and Herzegovina. The Proclamation of Bulgarian independence and the annexation were among factors threatening the outbreak of an international conflict, the Balkan Crisis (1908–9), which greatly exacerbated Bulgarian-Turkish relations. The conflict was averted through the active mediation of Russia and the Great Powers came to recognize the new political status of Bulgaria. The Bulgarian State was then faced with the immediate task of tackling the national problem through national unification. The sorry plight of the Bulgarians in Macedonia and the Adrianople region, still under Ottoman domination, and the need for accelerated economic development gave grounds for war with the Ottoman Empire. In March 1911, Ivan Evstratiev Geshov, leader of the Popular Party, formed a coalition Government with the Progressive Liberal Party led by Dr Stoyen Danev. The first home-policy undertaking of the new Cabinet was to convene the Fifth Great National Assembly (June–July 1911), which voted to amend the Turnovo Constitution. King Ferdinand was authorized to conclude treaties with foreign countries without consulting the National Assembly. The Geshov Administration undertook military and diplomatic preparations for a war with Turkey, signing treaties with Serbia, Greece and Montenegro (1912), and forming an anti-Ottoman alliance of Balkan States. However, the Bulgarian Government made the mistake of signing not a general agreement, but separate treaties with each of the Governments of the allied States. This made it easier for the other members of the alliance to negotiate separately, later, at the expense of Bulgaria. The treaty with Serbia stipulated that Macedonia should be divided into two zones. The undisputed zone, covering the territories south-east of the Kriva Palanka-Ohrid line, was given to Bulgaria irrespective of whose troops would first set foot on it; the fate of the disputed zone, covering the territories north-west of that same line, was to be decided later either by bilateral negotiations or through the arbitration of the Russian Emperor.

The territorial provisions in the treaty with Greece were also not clearly stated. The treaties bound Bulgaria to fight the Ottoman Empire with the largest military contingent and in the main theatre of war. By the end of September 1912, the Balkan alliance had completed its preparations for the war, and on 5 October 1912 the allies declared war on the Ottoman Empire.

The bulk of Bulgarian forces were concentrated on the decisive front, in Thrace. At the very beginning of military operations the Bulgarian troops broke the resistance of the Ottoman army and captured Buleburgaz and Bunanissar. The Ottoman army retreated to the Chataldja position. Some other Bulgarian contingents advanced

Bulgarian artillerymen firing at British and French aircraft in World War I

towards the Rhodopes and Southern Macedonia and reached the Aegean Sea. The allied armies captured Kossovo, Northern Albania, Vardar, Macedonia, Epirus and Southern Macedonia. Facing a military defeat, the Ottoman Government asked for an armistice, but the Bulgarian side missed the opportunity of ending the war successfully and concluding an advantageous peace treaty. Following the abortive attack against the Chataldja positions, Ferdinand I accepted the proposal for an armistice, but the situation had changed to Bulgaria's disadvantage. Operations were renewed after the 1913 Young Turkish coup in Istanbul. The offensive of the Ottoman troops at Chataldja, Bulair, and Sarköy was repulsed. On 13 March, the Bulgarian troops achieved the considerable feat of seizing the fortress of Adrianople. Skodra and Yanina were captured almost at the same time.

The peace treaty between the Ottoman Empire and the States of the Balkan Alliance was signed in London on 17 May 1913. The Empire ceded to the allies all territories to the west of the Midye-Enos line, as well as the Aegean islands. This meant the end of Ottoman national oppression and the feudal remnants in the liberated regions were eliminated. In distributing the recovered territories, however, the allied States broke into heated disputes and a military conflict became imminent. Serbia and Greece signed a pact against Bulgaria, and were also joined by Montenegro. The Serbian and Greek authorities in the occupied parts of Macedonia resorted to terrorist measures against the Bulgarian population there, and the position of Bulgaria was further complicated by the stand adopted by Romania, which under the pretext of upset balance of power after the war, made claim to Southern Dobrudja.

In spite of all warnings issued by Russia, the Government formed on 1 July 1913 and headed by the extreme Russophile, Stoyan Danev, did not take any measures to prevent military conflict. The Bulgarian army was transferred to the west to fight its former allies, and on 16 June 1913 Ferdinand ordered the Bulgarian troops to start operations against the Serbian and Greek forces that had occupied Macedonia. Thus the Inter-Allies War (the Second Balkan War) broke out. On that very day the Serbian and Greek forces went on the offensive, joined by Montenegro, Romania, and the Ottoman Empire. Pressed on all sides, Bulgaria was forced to surrender. The 1913 Bucharest peace treaty gave to Serbia and Greece the larger part of Macedonia and to Romania – Southern Dobrudja. Bulgaria retained a small part of Macedonia and Aegean Thrace. The Istanbul treaty, signed by Bulgaria and the Ottoman Empire in 1913, stipulated that Bulgaria should also be deprived of Eastern Thrace and Adrianople. Instead of being nationally united, Bulgaria emerged from this first national catastrophe with a plundered and ruined economy: she was compelled to pay reparations to the amount of 700 million golden levs; a total of fifty-five thousand were killed in military operations and another 105 thousand were wounded; the damage to the economy amounted to two billion levs. The refugee question then came to the fore once again in all its gravity, as thousands

Mutineers of the 36th infantry regiment. Southern front, July 1918

of refugees fled from the Macedonian regions left under Serbian and Greek administration and settled in Bulgaria. The BWSDP (l-ws) led a popular campaign to punish those responsible for the defeat, and the parliamentary group of the party proposed in the National Assembly that they should be brought to trial. The proposal, however, was rejected.

The struggle of the Bulgarians under Serbian and Greek occupation administration continued. In the summer of 1913 uprisings broke out in the Tihves, Ohrid-Debar and other regions. IMARO also maintained its revolutionary activities. The coalition Government of Dr Vassil Radoslavov, formed on 4 July 1913, considered national unification possible only through a new war against Serbia and Greece and relied on the support of the central European States. This policy enjoyed the full backing of the monarch.

In order to consolidate its position, the Government dissolved the National Assembly. The parliamentary elections in November 1913 were won by the Opposition, but instead of resigning, the Cabinet of V. Radoslavov dissolved the newly-elected National Assembly and fixed a date for new elections in the beginning of 1914. By means of terror and falsifications it managed to win a marginal parliamentary victory, and the Opposition's lack of unity enabled it to stay in power. During the one-month crisis following the assassination of the Austrian Archduke in Sarajevo, Bulgarian society was divided into several camps on the issue of foreign policy. The Government Coalition proclaimed neutrality, sympathizing at the same time with the Central Powers. The group of pro-*Entente* parties like the Popular, the Progressive Liberal, and the Democratic Parties, did not believe that war was the only way of satisfying national aspirations, but should participation in the war become inevitable, it would declare itself in favour of joining the *Entente*. The petty-bourgeois parties (the BAU, the Radical Party, the 'right-wing' socialists) rejected the war as a means of attaining national ideals. The BWSDP (l-ws) declared itself against the two imperialist blocks and for Bulgaria's non-interference in the war. It saw the establishment of a democratic federation of Balkan States as a solution to the national question in the Balkans.

The Government of V. Radoslavov proclaimed neutrality but went on pursuing the policy of gradual *rapprochement* with the Central Powers. On the eve of the war, it achieved a loan from the German banking concern Disconto Gesellschaft amounting to 500 million levs, and thus actually bound Bulgaria to the Central Powers. It did its best to assist Germany and Austria-Hungary in transporting military supplies and in sending military experts to the Ottoman Empire. The *Entente* countries, too, exercised pressure to win Bulgaria over to their side. They made offers of territorial gains which the Government found insufficient. Through the French capitalist, de Closier, they succeeded in attracting prominent opposition figures, but the scandal which broke out over de Closier's trade company was used by the Government to square its accounts with its political opponents. To resist the calls of the Government and opposition parties to join in the war, the left-wing socialists pursued an active anti-war campaign. Dimiter Blagoev, leader of the BWSDP (l-w s) was authorized to reject the proposal by the Russian social democrat G. V. Plekhanov to ally Bulgaria to the *Entente*. The appeal to take the side of the Central Powers made by the German social chauvinist Parvus to the Bulgarian Government also met a definite refusal. Thus, the Bulgarian left-wing socialists drew a clear-cut demarcation line between their stand and the social-chauvinistic positions adopted by the leaders of the Second International during the war.

In August 1915 Bulgaria and Germany signed a treaty and a secret agreement supplemented with a military convention between Germany, Austria-Hungary, and Bulgaria. Bulgaria was placed under an obligation to start military operations against Serbia and was guaranteed the return of both the disputed and undisputed zones in Macedonia and the Morava river valley. Austria-Hungary and Germany agreed that, if Greece and Romania should engage in hostilities, Bulgaria would regain the lands it ceded to them under the Bucharest Treaty of 1913. The Government and monarch hastily decided that the Central Powers' victories in the summer of 1915 meant a decisive turning-point of the war, thus proving shortsighted and incompetent to orientate themselves in the complicated and dynamic politico-military situation. The adventurous foreign policy of King Ferdinand and his personal bias in favour of the Central Powers took the upper hand. The leaders of the pro-*Entente* parties condemned the policy of the Government at a private meeting with Ferdinand and pin-pointed the greater chances of a victory for the *Entente*. The leader of the BAU, Alexander Stamboliiski, later sentenced to life imprisonment, behaved most audaciously at this meeting.

On 1 (14) October 1915 Bulgaria declared war on Serbia and the other *Entente* States. According to the convention signed with Germany, the latter was to take command of the Bulgarian armed forces. Three armies numbering more than 800 thousand, were mobilized. The First and the Second army moved against Serbia; the Third remained as a reserve in the vicinity of Pleven. Pressed by the Austro-Hungarian army to the north and subjected to energetic pressure by the Bulgarian troops to the east, the Serbian army could not but retreat. The Bulgarian troops liberated Shtip, Veles, Kumanovo, Skopje and Ohrid, and cut off the Serbian divisions' retreat towards Thessaloniki. The defeated Serbian army retreated across the Albanian mountains to the Adriatic Sea. On the southern (Thessa

Bulgarian and Russian soldiers fraternizing. Northern front, 1918

loniki) front the second army was engaged in battles with British and French troops which had both numerical and techno-military superiority. The Bulgarian detachments, however, managed to utterly defeat the enemy at Krivolag and were able to develop their offensive. For political reasons, the German High Command stopped the Bulgarian offensive, which had serious consequences for the Bulgarian position at the southern front.

Operations started getting bogged down in positions along the front line from Orfan Bay on the Aegean to Albania. In the summer of 1916 the High Command made an abortive attempt to break through the *Entente* positions in southern Macedonia. Trench warfare continued until the autumn of 1918.

In the autumn of 1916 Bulgarian troops were drawn into the war against Romania. After bloody battles they succeeded in liberating Dobrudja. Later, the front line settled along the reaches of the Seret river and in the Danubian delta, where more trench warfare began. The prolonged military operations dealt a heavy blow to the Bulgarian economy: some industries deteriorated; devaluation and unemployment assumed monstrous proportions; agriculture also declined sharply owing to the lack of able-bodied men and the requisitioning of cattle.

The plunder of Bulgaria by its German ally and the unpopularity of the war against Russia, coupled with high prices and famine, provoked wide popular discontent. The first signs of unrest appeared in the rear; starving women, children and old people were part of it. Unrest also flared up in the army, which the left-wing socialists directed against the war and for peace and social justice.

The victory of the Great October Socialist Revolution in Russia in 1917 lent a powerful impetus to the world revolutionary process. The left-wing socialists enthusiastically welcomed the victory of the October Revolution and sided with its cause. The party's activities at the front and in the rear centred on popularizing the slogans of the revolution and appealed to the Bulgarian working people to follow the examples set by their Russian brothers.

Influenced by the October Revolution, the general crisis underway in Bulgaria and all other belligerent powers gradually assumed a revolutionary character.

Having lost the confidence of the people, no bourgeois party proved capable of taking Bulgaria out of the crisis. The party of the left-wing socialists stood at the head of the movement for peace and radical reform in state administration and economic management, and against the high cost of living and famine. Left-wing socialist propaganda resulted in the setting up of soldiers' committees which engineered mutinies in a number of army units at the front. Fraternizing of Russian and Bulgarian soldiers started at the front along the Seret river; by early March 1918 military courts sentenced approximately 40,000 soldiers and 800 officers, 450 of them to death.

The rising revolutionary wave forced the Government and the palace to look actively for a way out of the crisis. After the victory of the October Revolution, the Soviet Government proposed that all belligerent powers should start negotiations for a just and democratic peace without any annexations and reparations. The *Entente* States declined the offer of the Soviet Government, which was compelled to negotiate only with the Central Powers. The Government of V. Radoslavov took part in the 1918 Brest-Litovsk talks and fully supported the German and Austro-Hungarian imperialist claims. The treaty with Romania signed in Bucharest on 7 May restored Southern Dobrudja (with some corrections of the frontier) to Bulgaria. Northern Dobrudja was proclaimed a condominium of the allies.

In 1918 the situation in Bulgaria deteriorated sharply. Its industry, transport, agriculture and crafts were completely disrupted, and the army was starving, shoeless and exhausted. On 21 June 1918 a Coalition Government consisting of democrats and radicals and headed by Alexander Malinov was reformed. It did not, however, live up to expectations that it would take Bulgaria out of the war. In September 1918 the Anglo-French army went on the offensive on the Southern front, and their main actions were directed at the units of the Bulgarian army. Greatly superior in terms of men and ammunition, the *Entente* troops broke through the Bulgarian defences at Dobro Pole (15–18 September). The retreating troops rebelled. On 23 and 24 September they reached Berovo, Pehchevo, and Tsarevo Selo and formed insurgents' detachments. On 24 September the insurgents seized the chief operational

Alexander Stamboliiski

Striking transport workers, 1919–20

headquarters of the army located in Kyustendil and on 27 September, in Radomir, declared Bulgaria a republic. Released from prison, the BAU leaders Alexander Stamboliiski and Raiko Daskalov (1886–1923) took control of the Soldiers' Mutiny. The insurgents headed for Sofia to overthrow the monarch and the bourgeois government, but were defeated by loyalist forces assisted by German troops. The Soldiers' Mutiny was the first attempt to overthrow the monarchy and establish a democratic republican Government in Bulgaria.

On 29 September 1918 Bulgaria concluded the Thessaloniki armistice with the *Entente* States: its army was demobilized, the country was occupied by *Entente* troops and some 100,000 Bulgarian soldiers were left as hostages in Macedonia. King Ferdinand I was compelled to abdicate (3 October) and left the country. His son Boris III (1918–43) ascended the throne.

The Post-War Revolutionary Crisis and the BAU Government. Striving to suppress the revolutionary movement and to preserve its shaken political supremacy, the bourgeoisie tried to expand the social base of its power. On 17 October 1918 A. Malinov was asked to form a new Government. Besides radicals and democrats, it also included representatives of the BAU, the Social Democratic Party, the Progressive Liberals and the Popular Party. Later, it was headed by the populist leader Todor Todorov. The left-wing parties won a landslide victory in the parliamentary elections of 17 August 1919. The BAU received 180,000 votes, and the Bulgarian Communist Party (BCP) – 120,000. A new Coalition Government of the BAU, Popular and Progressive Liberal Parties, headed by Alexander Stamboliiski, was formed on 6 October 1919. Threatened by the revolutionary upsurge of the working people, the bourgeoisie was temporarily forced to yield power to the BAU, hoping to set the peasants against the workers in an irreconcilable duel, to render their alliance impossible and, ultimately, to preserve the capitalist system in the country.

Having emerged victorious from World War I, the *Entente* States forced the defeated countries into accepting peace treaties, which had been worked out beforehand, dictating the terms. The Treaty with Bulgaria was signed in the Paris suburb of Neuilly on 27 November 1919. The predatory imperialist conditions of the Neuilly peace treaty drove the country to its second national catastrophe. Bulgaria suffered considerable territorial losses: forced to yield Western Thrace to Greece, it was deprived of an outlet on the Aegean. Serbia took the country's westernmost districts of Tsaribrod, Bosilegrad and the border areas along the Timok and Strumitsa rivers. Southern Dobrudja was given to Romania. Heavy reparations to the amount of 2,250 billion golden francs were imposed on Bulgaria. Thousands of tons of coal and timber and scores of thousands of cattle were handed over to the victorious states. Bulgaria was deprived of the right to raise a regular conscript army, being allowed only to organize a small volunteer army, border guards and gendarmerie units, numbering up to 33,000. A regime of severe persecution of the Bulgarian population was introduced in the neighbouring, conquered lands. Thousands of Bulgarian families had to leave their native regions and look for shelter and subsistence either in Bulgaria or in foreign countries. After World War I, the number of Bulgarian refugees from Macedonia, Thrace, Dobrudja, and other outlying areas to the west totalled 500,000. The refugee question was among the most heavy aftermaths of the second national catastrophe.

After the war Bulgaria slid into a grave economic and political crisis. Economic life was totally disrupted: both industrial and agricultural production plummeted; trade exchange was in a state of havoc. Exhausted by the tremendous war expenditures, the finances of the State were also affected by the rapid devaluation of the lev. Prices soared. The state of economic ruin proved most disastrous for the working people, who also bore the burden of paying the heavy reparations. Compared with the pre-war period, the wages of factory and office workers increased only four-or five-fold, while the cost of living rose fifteen- to twenty-fold. Popular hatred of the bourgeoisie and the monarchy grew, since they bore the responsibility for the national catastrophe.

The revolutionary movement gained momentum in late 1918 and early 1919. The BWSDP (l-ws) proved the only political party capable of consistently and faithfully defending the interests of the working people, organizing and leading their economic and political struggle, lending it a definite class and anti-capitalist character. It organized mass popular actions against rising prices and famine, and in support of political amnesty for the imprisoned participants in the anti-war movement and the return of the Bulgarian divisions held as hostages by the *Entente*. Protest rallies and demonstrations were held in towns and villages. The influence of the BAU in public life grew. Expressing the democratic aspirations of the working peasants, the BAU leadership made an effort to curb the monarch's influence in the country's political life, showing at the same time the hesitation typical of petty bourgeois democracy.

The revolutionary crisis in Bulgaria created conditions for a rapid growth of the power and influence of the BWSDP (l-ws). Led by Dimiter Blagoev it participated in the founding of the Communist International in March 1919, and in contrast to many other parties, joined the International as a united party without inner struggles or factions. The Twenty-second Party Congress (May 1919) changed the party's name to Communist and adopted a Programme Declaration based on the main principles of Leninism.

In the summer of 1919 the revolutionary movement in the country intensified. Mass strikes were organized by the tobacco workers in Stanimaka (present-day Assenovgrad), Gorna Djumaya (Blagoevgrad), and Dupnitsa (Stanke Dimitrov), as well as by the sugar factory workers in Gorna

Raiko Daskalov

Oryahovitsa, Sofia and Russe, the textile workers in Gabrovo, and others. On 1 July Pernik miners went on a general strike. A total of 146 strikes led by the BCP (l-ws) and the GWTU were declared in 1919.

The most notable manifestation of the mass movement in the summer of 1919 was the protest action organized by the Party on 27 July 1919 against high prices, hunger and unemployment. On 7 December 1919 municipal elections were held and the BCP won in half the district towns and in many county centres and other small towns as well as in dozens of villages. This was the beginning of the Party's broad municipal activity, realized through the organization of town and village communes. The post-war revolutionary crisis reached its culmination during the transport workers' strike (1919–20) which spread all over the country. In support of the transport workers a one-week general political strike was declared on 29 December 1919. The transport workers' strike ended in failure because of disunities among the democratic forces. At the same time the bourgeoisie succeeded in using the BAU and the Coalition Government of Alexander Stamboliiski against the strikers.

After the end of the transport strike the National Assembly was dissolved and a date for new parliamentary elections was set. In the elections the BAU won a majority of the votes (followed by the BCP) and on 21 May it constituted an Independent Government headed by

Alexander Stamboliiski. By its composition and ideology the BAU was a petty bourgeois party. Alexander Stamboliiski evolved the 'theory of estates', according to which society was divided into estates, and not into classes. The main estate was the 'peasantry' which was subject to the plundering of the 'townspeople'. Stamboliiski was aware of the fact that the peasantry was not a socially homogeneous mass, but he believed that it would seize political power through the peasants' estate organization and would perpetuate small-scale and middle-scale property through reforms.

After the war the BAU became the most numerous party, and its influence predominated in the countryside. Its supporters were small and mainly middle peasants and comprised the basis of the left democratic wing in the party headed by Alexander Stamboliiski and R. Daskalov. The well-to-do peasants, though a minority in the party, had a considerable influence in the local party leading bodies. They formed the social basis of the right wing, which incited the BAU to unite with the bourgeoisie and oppose the Communist Party.

The heterogeneous social composition and petty bourgeois ideology of the BAU determined the vacillating, inconsistent and contradictory policies of the independent Agrarian Government.

The BAU was an anti-monarchic and republican party. The international situation as well as the situation in Bulgaria in post-war Europe made it impossible to question the position of the monarchy, and this is why the Agrarian Government attempted at least to do away with the conditions favouring the existence of the personal regime. The Tsar was stripped of one of his most important prerogatives – the office of Commander-in-Chief of the army. This function was assumed by the Government. At the end of his rule Stamboliiski was preparing a new constitution under which the tsarist power was to remain purely representative.

The Agrarian Government sent out to completely discredit the bourgeois parties in order to frustrate their return to power. Legal proceedings were started against the Ministers of V. Radoslavov's Government, which had been in office during World War I. In 1922 a referendum was held to indict the Ministers of the Governments of I. E. Geshov, Dr S. Danev and A. Malinov, which had held office during the Balkan War and at the end of World War I. Thus the bourgeois parties were deprived of leadership and their leaders were arrested and brought to court.

The Agrarian Government had a dual attitude towards the Communist Party, regarding it as a natural ally in the struggle against bourgeois reaction and at the same time a dangerous rival and adversary in gaining influence over the masses. The bourgeois reaction did all it could to intensify the disagreements between the two parties of the working people.

Alexander Stamboliiski at the Genoa Conference, 1922

The Agrarian Government carried out a number of reforms which affected the interests of the bourgeoisie. Foreign trade in cereals and cereal products was taken away from the capitalists and put into the charge of a specially established state body – the Consortium. New systems of additional and graduated taxation of the capital and property of the bourgeoisie were introduced. A law was passed for the expropriation of big urban residential property for the needs of the State and society. Labour service was introduced for all citizens who had to work for a certain period on big construction sites such as railways, roads, etc. The Law of Land Ownership limited the ownership of arable land to thirty hectares. The expropriated and state land, put together, formed a state fund from which land was given away to the landless peasants and small holders. The reforms carried out by the Agrarian Government affected considerably, though only partially, the interests of big business.

The foreign policy of the Agrarian Government was peaceful, aiming to take the country out of international isolation. Bulgaria was admitted to the League of Nations. The endeavours of Stamboliiski's Government were aimed at a *rapprochement* with France and understanding with the neighbouring countries, the Serbian-Croatian-Slovenenian Kingdom in the first place, with which it signed the Nish agreement in 1923. The controversial issues between the two countries were settled mainly in favour of Yugoslavia. This caused the discontent of the leadership of the Internal Macedonian Revolutionary Organization (IMRO) whose right wing took the upper hand and entered into an alliance with the bourgeois reaction for a struggle against the agrarian regime.

An agreement was signed with the countries which had won the war for paying off Bulgaria's reparation debt. The Agrarian Government maintained friendly relations with Poland and Czechoslovakia, and made steps toward a *rapprochement* with Soviet Russia. Under the pressure of the Western States the Government allowed the White Guard of Baron Wrangel, expelled by the Red Army, to settle in Bulgaria, but at the same time it permitted the Russian Red Cross mission to organize the return of the White *émigrés* to Russia.

During the period of agrarian rule the BCP was consolidated both ideologically and organizationally. In the municipal elections in 1920 it won 65 village and 32 town municipal councils (communes), with the help of which it carried out many undertakings in defence of the immediate interests of the working people. The GWTU and the other workers' trade organizations intensified their activities. In November 1920 the left wing of the BWSDP joined the BCP; the GWTU united with the militant ranks of the reformist trade unions.

The bourgeoisie, having preserved its dominating position in the economy, regarded the continuation of BAU rule as a threat to its class domination and gradually launched a political offensive, in order to consolidate its forces. In 1922 all the bourgeois parties joined to form a reactionary political grouping named the Constitutional Bloc. However, the bourgeois parties did not have a broad social base. They did not believe that they would restore their power by parliamentary means and began preparations for a coup with the assistance of the army. The influence of Italian Fascism penetrated the bourgeois parties, and the reactionary army officers also took their side. The underground Military League was formed in 1919. Its legal front was the fascist organization *Naroden Sgovor* (Popular Alliance) formed in 1922. In the beginning of 1922 the bourgeoisie was about to launch a decisive offensive. Representatives of the Constitutional Bloc negotiated with the headquarters of Wrangel's army, which had all its equipment and military organization well preserved, on common actions against the agrarian regime. The conspiracy of the reaction was revealed by the BCP and the Government was forced to expel Wrangel's generals and partially disband the White Army. In the autumn of 1922 the parties of the bloc tried to organize 'rallies' against the BAU in different settlements, but this was frustrated by the unity of action between communists and agrarians.

In the parliamentary elections on 22 April 1923 the BAU won a considerable majority in the National Assembly. The Government made the wrong assessment that the bourgeois parties had been defeated and turned its blows against the BCP. The aggravated relations with the BCP facilitated the bourgeoisie in its conspiracy to seize power. The immediate organizers of the *coup d'état* were the Military League and the Popular Alliance, but representatives of all bourgeois parties were involved in the conspiracy. On the night of 8–9 June 1923 the conspirators, supported by the monarch, carried out a *coup d'état* in the capital. The Agrarian Ministers were arrested. The Tsar appointed a Fascist Government headed by Prof. Alexander Tsankov (1879–1959) comprising representatives of the Military League, the Popular Alliance, the Constitutional Bloc, the National Liberal Party and the Social Democratic Party.

Scores of thousands of agrarians and communists rose spontaneously against the *coup d'état:* in the district of Pazardjik under the leadership of Alexander Stamboliiski; in the districts of Pleven, Turnovo, Shumen, Varna, Karlovo and in many other regions of the country. In Pleven district the uprising took on a really mass and organized nature. More than 100,000 people rose in revolt across the country. In many places communists and agrarians acted in unity. However, although it disapproved of the *coup d'état,* the Central Committee (CC) of the BCP, because of its doctrinaire attitude toward the BAU and its Government and because of lack of Marxist-Leninist maturity, took the mistaken stand of neutrality. The June 1923 anti-fascist uprising was brutally crushed, and on 14 June, Alexander Stamboliiski was captured and murdered.

The *coup d'état* was a crushing blow to the democratic movement in the country. It replaced the traditional bourgeois democracy by a reactionary dictatorship, a more efficient form of defence of the bourgeois domination under the new conditions. The Government of A. Tsankov (1923–25) revoked the gains of the working people. It had no sound social basis among the masses and

Proclamation of worker-peasant rule in Knezha, 24 September 1923

Communists and Agrarians arrested in the suppression of the September 1923 uprising

increased its pressure on the bourgeois parties to unite. In August 1923 the Popular Alliance, the United People's Progressive Party and part of the democrats and radicals formed a reactionary political grouping called *Democratichen Sgovor* (Democratic Alliance) as the social and political force of the 9 June regime.

The September 1923 Anti-Fascist Uprising. The onslaught of Fascism caused a wave of indignation among the working people. The new revolutionary upsurge was headed by the BCP whose leadership, under the pressure of the masses and after the critical assessment of the Comintern, gradually rectified the erroneous policy of 9 June. Vassil Kolarov (1877–1950), who was Secretary-General of the Comintern at that time, returned to Bulgaria. With his assistance a meeting of the CC of the BCP was convened from 5–7 August 1923 which set as the Party's main task 'an all-round preparation for a mass armed uprising' in the name of worker-peasant government. For the first time in its history the BCP took a concrete decision for the immediate preparation of an armed uprising and the establishment of a worker-peasant rule. To implement its new policy the party leadership appealed to the leadership of the BAU, the Social Democratic Party, the IMRO and some public and trade union organizations for unity of action in the common struggle. The proposal was accepted only by the militant left-wing sections of the BAU, which was steadily re-establishing its organization. The articles on the United Front by Georgi Dimitrov (1882–1949), published at the end of August and the beginning of September 1923, were an important contribution to the political preparation of the uprising. They argued the need for uniting all progressive and democratic forces against Fascism. The military and technical preparation for the uprising was being carried out along with the political. A Military-Technical Committee, set up by the CC of the BCP, worked out a plan of action and on 20 September the extended CC fixed the date for the general armed uprising, which was to break out on the night of 22–23 September. A General Military-Revolutionary Committee was elected, composed of Vassil Kolarov, Georgi Dimitrov and Gavril Genov (1892–1934).

The insurgents' actions began prematurely, on the night of 13–14 September in Muglizh (Stara Zagora district). On the night of 19–20 September the uprising was proclaimed in the whole of the district. The insurgents launched an attack on Stara Zagora and Chirpan but, in spite of their heroism, they failed to capture them. They also attempted an attack on Kazanluk.

The uprising was most successful in Nova Zagora, where its leader was Petko Enev (1889–1925). The insurgent army gained control in the town (except the barracks) and established worker-peasant rule both in the town and in all the villages of the country. Worker-peasant rule was established in most of the villages of Stara Zagora, Chirpan and Kazanluk counties. On the night of 22–23 September the uprising spread to the rest of the country. It was in North-west Bulgaria that the uprising involved the largest number of people and was best organized, with the town of Ferdinand (Mihailovgrad) as a centre, where the General Military-Revolutionary Committee established its quarters. The insurgents seized the towns of Ferdinand, Byala Slatina, Berkovitsa and Oryahovo, and won a victory at the railway station of Boichinovtsi. The workers and peasants in the district of Vidin (in particular in Lom and the county) and in Belogradchik county rose in arms, together with the insurgents in Vratsa. Heavily outnumbered by the enemy, the insurgents in North-west Bulgaria had to retreat and cross the Yugoslav frontier. Fierce battles were fought at the railway station of Saranbei (today the town of Septemvri) and in Bratsigovo. 59 villages rose in the

Gavril Genov

district of Pazardjik and in 40 of them worker-peasant rule was established. In taking the town of Razlog the insurgents were supported by the soldiers of the local army unit. In Sofia, due to betrayal, the police succeeded in crushing the military-revolutionary committee, but in the region of Ihtiman and Dolna Banya the insurgents were very active. The uprising spread also to parts of Samokov, Sofia and Pirdop counties, and to a number of counties in the other regions. In Burgas district, mainly in Karabunar (called Grudovo today) and the county, the fighting was particularly fierce. Burgas was attacked by a large insurgent detachment. In the districts of Shumen, Veliko Turnovo and Pleven it was only the population of a few villages that rose in arms.

For the first time in the history of Bulgaria, during the

Insurrectionists from Lessichevo, near Pazardjik, condemned to death

University students demonstrating against the Alexander Tsankov regime

Communist students in Vratsa prison, 1927

September 1923 anti-fascist uprising the people, up in arms and led by the BCP, set about establishing worker-peasant revolutionary rule. Its form and functions were clearly defined in North-west Bulgaria where a considerable part of the territory was liberated – from the Danube to the Balkan Range and from the river Iskur to the Bulgarian-Yugoslav frontier. In all places, as soon as the fascist power was overthrown, local revolutionary committees were set up as organs of worker-peasant power. All revolutionary committees, from those in the villages to the General Military-Revolutionary Committee, comprised communists and agrarians. The revolutionary authorities introduced strict law and order, protected the life and property of every citizen, and extended humane and generous treatment to the enemy prisoners.

The September anti-fascist uprising of 1923 – the first organized anti-fascist uprising in modern history, ended in defeat. The Fascist Government crushed it cruelly, taking savage reprisals on the defeated insurgents. Thousands of communists, agrarians and non-affiliated anti-fascists were killed without charge or trial. The main cause for the defeat of the uprising was the insufficient bolshevization of the BCP. The most opportune moment for crushing the reaction – 9 June, when the masses rose spontaneously in large numbers against the still uncertain rule of the conspirators – had been missed. Nevertheless, the great number of victims cemented the anti-fascist unity of the Bulgarian people and created strong anti-fascist traditions. The uprising was a turning-point in the development of the BCP. The Party gained experience in the organization and leadership of the masses in the armed struggle. During the uprising the tactics of the united front were applied in practice. As Georgi Dimitrov stated, 'We regard the anti-fascist popular uprising of September 1923, which was organized and headed by the Bulgarian Communist Party, as the turning point in the Party's development from left-wing Socialism to Bolshevism.' (G. Dimitrov, *Works,* Vol. 14, Sofia, 1955, p. 240)

The heroic struggle of the Bulgarian workers and peasants under the leadership of the BCP had wide repercussions abroad, especially in the neighbouring Balkan countries and in Germany. The example of the party of the Bulgarian proletariat and the lessons drawn from its struggle enriched the experience of the world communist movement.

Between Fascism and Bourgeois Democracy. After the uprising the Government of the Democratic Alliance consolidated its positions. It was in full command of the state machine, the police and the army. Yet the issue of power was still pending, and there was fierce strife between the various groups within the Democratic Alliance. Its foreign policy sought closer contacts with Britain and Italy but it failed to overcome Bulgaria's international isolation. Relations with Yugoslavia remained tense even though the Bulgarian rulers observed the burdensome and unjust clauses of the Treaty of Neuilly and adhered to the Nish Agreement of 1923.

To make its position 'lawful' the Government held parliamentary elections in November 1923. The BCP and the left wing of the BAU participated in the elections with a joint ballot paper. In spite of the terror and the restrictive measures against the worker-peasant bloc, which was given the opportunity to contest the elections with candidates of its own in only 12 out of 72 constituencies, the communists and agrarians scored a major success; in a number of counties they won more votes than the Government Coalition – Varna, Provadia, Preslav, Belene, Gabrovo, Gorna Oryahovitsa, Svishtov, Sevlievo, Turnovo and Sliven. In the struggle to preserve its power, in January 1924 the Alliance Government promulgated a State Defence Act under which the BCP, the Bulgarian Young Communist League, the GWTU and some other organizations were outlawed.

Despite the conditions of underground existence and savage terror, the BCP began to re-establish its organizations. A new CC, headed by Stanke Dimitrov (1889–1944), was formed. On the initiative of Georgi Dimitrov and Vassil Kolarov, who had escaped abroad, a BCP Bureau-in-Exile was formed in Vienna (1923–24) and helped the CC in the country to re-establish the Party on an underground basis. In early 1924 a Labour Party was founded as a legal organization of the BCP. The Party established close relations with the left wing of the BAU headed by Petko D. Petkov (1891–1924). The first underground conference of the BCP – the Vitosha Party Conference – was held on 17 and 18 May 1924. It consolidated the party forces and kept up the policy of an armed uprising. The Party broadened its activities for the building up of a united anti-fascist front, and an Action Committee of communists and left-wing agrarians was set up to direct the joint activities. The BCP's work among the mass organizations of the working people and among the refugees, especially from Macedonia, was further improved. The Party established close links with the left wing of the IMRO and made an attempt to win over the whole organization to the side of the united front. The leadership of the IMRO (Todor Alexandrov, Alexander Protogerov and Peter Chaulev), disappointed with the foreign policy of Tsankov's Government and pressured by the masses, signed a Manifesto of the IMRO in May 1924 in Vienna calling for the unification of all revolutionary forces of the national-liberation movement of the Bulgarians in Macedonia and for co-operation with the progressive parties and organizations of the Balkan peoples. The Government exerted pressure on the leadership of the IMRO, forcing them to renounce the Manifesto, and took advantage of the murder of T. Alexandrov on 31 August 1924 to stage a ruthless massacre of the leaders of the IMRO left wing in September of that year.

The second half of 1924 and the beginning of 1925 were marked by a process of partial stabilization of Capitalism paralleled by an ebb in the revolutionary movement. The party leadership in the country failed to appreciate the need to renounce the policy of armed uprising. An ultra-leftist deviation emerged within the BCP (l-ws), its representatives substituting the action of partisan units (1924–25) for the legal forms of mass struggle. Vassil Kolarov and Georgi Dimitrov who were at the head of the Party's leadership in exile, instructed that the policy of armed uprising should be renounced. However, the leaders of the BCP's military organization (1920–25) gave in to fascist provocations and retaliated for government terror with terrorist actions. On 16 April 1925 a bombing incident in St. Nedelya Cathedral served as the pretext for crushing the anti-fascist movement. Martial law was proclaimed throughout the country. There followed a wave of mass murders of communists, agrarians and non-affiliated anti-fascists. Kosta Yankov and Ivan Minkov (leaders of the Party's military organization), Dimiter Gruncharov (the leader of the agrarian left wing), and Ivan Manev (political secretary of the BCP), all lost their lives. The anti-fascist poets Geo Milev, Hristo Yassenov and Sergei Rumyantsev, the publicist Josef Herbst and many others were murdered without charge or trial. Thousands of anti-fascists were court-martialled and many were executed. A great number of communists and agrarians had no choice but to emigrate. The events of April 1925 brought about an abrupt change in the country's political situation. The anti-fascist parties suffered a grievous defeat.

The terrorist regime of Tsankov's Government not only met the resistance of the masses inside Bulgaria, it also aroused the wrathful indignation of the European progressive public at large. A large-scale campaign against the atrocities of the Bulgarian reaction was mounted abroad. Committees for struggle against the white terror in Bulgaria were set up in Paris, Berlin, Vienna and in many other places.

Romain Rolland, Maxim Gorky and many other writers spoke out in defence of the Bulgarian people. Henri Barbusse visited Bulgaria and wrote his book *Les Bourreaux;* Marcel Willard wrote *What I Saw in Bulgaria.* The attempt by the Tsankov Government to use the bombing as a justification for mass murder failed. Tsankov's Cabinet became an embarrassment to both the bourgeoisie and the monarch.

To prevent the complete defeat of the ruling Democratic Alliance, the reactionary bourgeoisie resorted to a political manoeuvre. On 4 January 1926 a second Cabinet of the Alliance was formed at the head of which was Andrei Lyapchev (1926–31). It continued A. Tsankov's policy though it used different, more flexible means. A conditional and partial amnesty was declared in January 1926. At the same time new trials were being staged against the participants in the revolutionary movement.

The time when A. Lyapchev's Government held office coincided with a period of temporary and partial stabilization of Capitalism, when the bourgeoisie consolidated its economic and political positions. A characteristic feature of Bulgaria's development in that period was the intensified process of merging of industrial and banking capital. The nation's economy was in the hands of nine major financial enterprises, five of them foreign. Capitalism in

Striking miners from the Tvurditsa colliery, 1931

Bulgaria was established by two foreign loans contracted by A. Lyapchev's Government: the refugee loan and the stabilization loan, which helped consolidate the economy but on the other hand made the country much more economically and politically dependent on foreign capital. An agreement was concluded with Greece (1927) by virtue of which Aegean Thrace lost a large part of its Bulgarian population as a new wave of refugees arrived in Bulgaria.

After the second blow it sustained at the hands of fascist reaction in April 1925 the BCP abandoned the policy of armed uprising and concentrated on strengthening its relations with the masses. In the summer of 1925 the legal Independent Workers' Trade Unions were set up under the Party's guidance to take the place of the GWTU, which had been outlawed. The workers' Press was restored. The *Rabotnichesko Delo* newspaper started publication in 1928. The creation in 1927 of the Workers' Party (WP), the legal form of the underground BCP, and in 1928 of the Workers' Youth Union (WYU), the legal form of the underground Komsomol, was the greatest success of the revolutionary workers' movement.

In 1929 the capitalist world fell into the grips of an economic crisis of unprecedented scope and duration, which affected industry as it did agriculture, credit and trade. It did not spare Bulgaria and its effects were specifically reinforced by the burdens imposed by the Neuilly peace treaty, the exploitation of working people by foreign monopoly capital, etc. The weakness of Bulgarian Capitalism was also a contributory factor. Bulgaria remained a petty bourgeois country with an underdeveloped Capitalism, its agriculture fragmented, with scanty resources and primitive equipment. Therefore the crisis affected primarily agriculture, with the small farmers the hardest hit. Industrial production began to drop fast, and a number of enterprises cut back or discontinued production. The country's trade balance was affected. The export and import of goods decreased. The crisis led to changes in the structure and system of Bulgarian Capitalism. Concentration, on a scale unprecedented in the history of Bulgarian Capitalism, was carried out in the field of industry and credit, and banking and industrial capital merged. A. Lyapchev's Government sought a way out of the crisis by shifting its burdens onto the shoulders of the working people. It resorted to wage cuts and raised the price of industrial goods. Thousands of workers were dismissed, and the plight of the working class and the peasants therefore created conditions for an intensification of the revolutionary movement. The strike movement was on the upsurge; a total of 6,000 tobacco workers from Haskovo, Plovdiv, Stanimaka and Kurdjali went on strike in 1929. The influence of the WP grew.

The deepening of the economic crisis and intensification of the revolutionaryy movement compelled the bourgeoisie to re-group its political forces. With the Democratic Party of A. Malinov taking the lead, a new bourgeois political grouping of the opposition parties, the Popular Bloc, was set up consisting of the BAU headed by Dimiter Gichev (1893–1964), the National Liberal Party of Georgi Petrov and the Radical Party. In the parliamentary elections of June 1931 the Popular Bloc won a majority of the votes – 600,000 and 150 seats. The WP, in alliance with the left wing of the BAU, formed a worker-peasant bloc which won 31 seats. The dictatorship of the Democratic Alliance was overthrown after eight years of rule, and a crushing blow was delivered to the most reactionary circles of the Bulgarian bourgeoisie. On 29 June 1931 a Government was formed headed by A. Malinov, replaced on 31 October by Nikola Mushanov (1872–1951). The Popular Bloc's aims as it came to power were the abolition of the dictatorship, the reinstatement of the constitutional regime and the improvement of the economic situation of the people. Because of its heterogeneous composition, two tendencies manifested themselves within it: a reactionary and a democratic one. The struggle between them went on throughout its period in office.

The Government effected a certain internal democratization. The role of the National Assembly increased and civil liberties expanded. More favourable conditions were created for legal activity by the communist movement. Political amnesty was granted. Some measures were taken to improve the lot of the working masses, but the reactionary legislation inherited from the Democratic Alliance Government was not abolished and the policy against the workers' movement continued. The State Defence Act was not repealed, and monarcho-fascist officers continued to rule the army. The BCP was subjected to cruel persecution: the secretary of the CC of the BCP, Nikola Kofardjiev (1904–31); the secretary of the CC of the WP, Petko Napetov (1880–1933); and other Party functionaries were killed.

Despite the overwhelming electoral victory of the WP in the capital in the municipal elections (1932) the authorities did not allow the lawfully elected communist councillors to take office in the Sofia Municipality. The workers' elected representatives were suspended.

The inability of the Popular Bloc to guarantee calm and normal development exacerbated the crisis. The struggle between the parties of big business and of the middle and petty bourgeoisie grew sharper. The contradictions within the Popular Bloc multiplied, and a re-alignment of the class-determined political forces in the country began. The reaction was preparing another *coup d'état*. A. Tsankov's adherents in the Democratic Alliance organized a party of a fascist type, called the National Social Movement, which was an exponent of the interests of the extreme bourgeois reaction. Other organizations of a fascist type also appeared. The ideological (later 'political') circle Zveno, constituted in 1927, occupied a special place in the right wing of the bourgeoisie. It was joined by a group of men, dissatisfied with the Government of the Alliance and headed by Dimo Kazassov (1886–1980), who were former functionaries of the Democratic Alliance, socialists, radicals, democrats and officers of the reserve. Anti-communism was the dominating element in Zveno's ideology; it rejected the bourgeois democratic form of government and praised strong 'supra-party' authoritarian bourgeois power of the fascist type. In January 1934 the majority of the Zveno members joined the Popular Social Movement of A. Tsankov while the remaining part, headed by Kimon Georgiev (1882–1969), continued the independent existence of the organization. Zveno was a small grouping, weak in its social activity. The Military League, which was again preparing a violent change of Government, established contacts with it.

Characteristic of the bourgeois camp was political disunity, the existence of a large number of political factions, unions and groups, and lack of authoritative, influential political parties.

Under the new political conditions the influence of the WP grew quickly and it became the legal party of the Bulgarian proletariat. The Independent Workers' Trade Unions, the Workers' Youth Union and other revolutionary mass organizations scored organizational success, and the parliamentary group of the Party worked actively to defend the direct interest of the working people. 1931 saw a fresh upturn in the strike movement, which spread to different sectors of industry but was strongest in the textile industry (Yambol, Sliven, Varna). The revolutionary upsurge gripped the masses of refugee Bulgarians from Macedonia, Thrace, Dobrudja and the Western frontier lands. The Internal Macedonian Revolutionary Organization (united) – IMRO(u) and the Dobrudja Revolutionary Organization (DRO) were all active, and the BAU left wing, headed by Lazar Stanev (1897–1938), was also active in helping to build a united front.

In 1933 the national socialists led by Hitler came to power in Germany and this gave a strong impetus to reaction and Fascism in all countries. To rout the German Communist Party and the anti-fascist movement, and to deliver a blow to the international communist movement, the nazis in Germany engineered one of the most notorious provocations of the century – the Reichstag Fire. The Reichstag was set ablaze on 27 February 1933, and the nazi propaganda trumpeted that the fire was the work of 'international Communism' and that it was the signal for an uprising and take-over of power by the communists. Thousands of German communists were arrested and interned. The Bulgarian communists Georgi Dimitrov, Blagoi Popov and Vassil Tanev, who were staying in Germany, were arrested and accused of having organized the fire. At the trial in Leipzig in 1933 Georgi Dimitrov, the accused became an accuser of German Fascism. He unmasked the real incendiaries, the nazis, and delivered the first moral and political blow at Fascism. A wide international movement was launched in defence of the accused communists, and a protest against the methods of Fascism started in Bulgaria in which not only communists but also broader democratic circles took part. Under the pressure of progressive public opinion the world over, the fascist court was compelled to acquit G. Dimitrov and the other accused communists. The Popular Bloc Government refused to let G. Dimitrov return to Bulgaria and, having been granted Soviet citizenship, he went to the USSR.

Demonstration outside the German Legation in New York in defence of Georgi Dimitrov, 1933

After the Reichstag Fire trial G. Dimitrov became the indisputable leader of the BCP; under his leadership the Party started to overcome its sectarianism and dogmatism.

From Military-Fascist to Monarcho-Fascist Dictatorship. On the night of 18–19 May 1934 the Military League and Zveno, with the help of the army, carried out a *coup d'état* and the Popular Bloc Government was overthrown. The coup of 19 May was one of the most sinister events, entailing the gravest consequences in the history of modern Bulgaria. It began a process that reinforced the monarcho-fascist dictatorship, concentrated power into the hands of the monarch and his clique of Generals, high-ranking officials and political careerists and divested the people of their elementary rights and liberties. The new Cabinet was headed by the Zveno leader Kimon Georgiev and composed chiefly of representatives of the Military League and Zveno. The bourgeois non-fascist parties and the other bourgeois parties condemned the coup but resigned themselves to it. The BCP characterized it as a fascist coup but through errors of secretarian dogma was unable to build a united front and oppose it. The manifesto issued by the perpetrators of the 19 May coup proclaimed the establishment of a 'national supra-party power'. They reorganized the state administration on a fascist basis, disbanded Parliament and dissolved all political parties and the IMRO. All the democratic rights and freedoms of the citizens were abolished. In order to create a broader social support, the Government set up state trade union organizations: the Bulgarian Workers' Union (BWU) and the Agrarian Association of Farmers. The entire socio-political life, culture and education were heavily centralized. The strongest blow was delivered to the revolutionary workers' movement.

In the economic sphere the regime made an attempt to introduce some state monopolies, thus partially infringing upon the interests of monopoly capital. The Government's foreign policy was orientated towards France. Diplomatic relations with the USSR were established on 23 June 1934.

Lacking a wide social base and political support, the regime was eroded by deep contradictions among the ruling clique. The monarch was afraid of losing the crown, while the *haute bourgeoisie* was displeased with some of the Government's economic reforms. The monarch managed to win over some of the Military League leaders with royalist inclinations and on 22 January 1935 engineered the resignation of the Government. The new Government was headed by General Pencho Zlatev (1881–1948). This marked the beginning of a series of transitional Cabinets which paved the way for the establishment of an open monarcho-fascist dictatorship. The Government of Georgi Kyoseivanov (1884–1960), former chief of the Royal Chancellery, was formed on 23 November 1935, and a Cabinet reshuffle in October 1936 ended the process of the transformation of the military-fascist dictatorship into a monarcho-fascist one. Monarchism and Fascism merged into a single system and King Boris started playing the part of a Bulgarian Führer. He took over all aspects of management of the country: appointed 'civilian' governments, commanded the army and police, and directed foreign policy. The open monarcho-fascist dictatorship relied on the most reactionary circles of the *haute bourgeoisie*, on the nationalistic and militaristic organizations, and on the armed forces, which were purged of anti-royalist commanders. Unlike nazi Germany and fascist Italy, the monarcho-fascist dictatorship in Bulgaria failed to create a mass, monolithic party of a fascist type. Fascist and nationalistic groups continued to exist, and new ones were set up, but they did not have a wide social foundation and were hostile towards each other. The establishment of fascist supremacy in the political life of the country met with the hostility of the masses, but their forces were disunited. The historic decisions of the Seventh Congress of the Comintern, in 1935, for building a united workers' front with broad popular support in opposing Fascism and war, became the BCP's programme of action. Under the guidance of G. Dimitrov, who was elected Secretary-General of the Comintern, the Party resolutely overcame leftist sectarianism. In 1936 the Sixth Extended Plenum of the CC of the BCP (l-ws) elaborated the decisions of the Seventh Congress of the Comintern in conformity with the Bulgarian conditions. The task of building a popular anti-fascist front was given top priority. The Party worked actively in state trade unions, community centres, co-operative, educational, sports, and women's groups, and temperance and other mass organizations. In many of these the Party gained wide influence and introduced a definite political content. The WP and the WYU, which had been disbanded after 19 May 1934, were re-established on an underground basis. The BCP organized and led a mass strike movement. Within three months in 1936 there were 150 strikes with over 35,000 participants (the tobacco workers in Plovdiv and in all Bulgaria, the textile workers in Gabrovo, Sliven, etc). The Party became an initiator and organizer of the unification of all democratic non-fascist forces in Bulgaria in a broad, popular, anti-fascist front against the onslaught of Fascism and the danger of a new war.

The popular movement and the mounting danger of fascist aggression in Europe promoted the activity of the bourgeois parties' leaders. In May 1936 an opposition coalition was formed called *Petorka* (the Five), consisting of the BAU right-wing 'Vrabcha 1', the Radical Party, the Social Democratic Party, the National Liberal Party and the Democratic Alliance. The Five declared themselves in favour of restoring the constitutional parliamentary regime. The BCP supported the democratic and anti-fascist actions of the Five and at the same time strengthened its co-operation with all anti-fascists, and led the mass, popular struggle. The General Youth Front, comprising all anti-fascist youth organizations, was a major achievement of the popular front movement in Bulgaria. The 'Nationwide Petition to the King', drawn up by the Five, with which the WP and the other democratic parties associated themselves, became a platform for uniting the democratic forces in Bulgaria. Popular constitutional committees were set up in connection with the municipal elections in 1937 and all upholders of democracy and the Constitution participated in them. In the elections in March 1937 all democratic parties joined together, voting under the banner of restoring the Turnovo Constitution. The Democratic Association formed at the beginning of 1938 (the WP, the Five, the BAU left-wing 'Alexander Stamboliiski' and the Democratic Party) which voted in the elections for the Twenty-fourth National Assembly with a common ballot and won 60 seats out of 160, was a fresh success in building the popular front. On the eve of World War II the Bulgarian democratic community unanimously declared against the fascist aggressive policy as a major threat to world peace.

The BCP was most consistent in its anti-fascist line. Under the onslaught of Fascism it became a decisive national force. In 1938 the CC of the BCP decided that the two parties of the working class – the BCP and WP – should merge into a one-class, revolutionary party called the Bulgarian Workers' Party (BWP). The reorganization was completed in 1940. The Party was strengthened organizationally and politically and expanded its relations with the masses. The successful pursuance of Dimitrov's policy showed that it had mastered the Leninist strategy and tactics of revolutionary struggle and finally become a party of the new, Leninist type.

With the establishment of an open monarcho-fascist dictatorship the penetration of German Imperialism into Bulgaria was intensified. Exports to Germany, which in 1936 constituted 48 per cent of the total, grew to 67.8 per cent in 1939, and imports increased from 61 to 65.5 per cent respectively. The policy of protectionism with regard to the national industry ended and Bulgaria gradually became an agrarian and raw material back yard of Germany. The country's international isolation and the internal political instability of the monarcho-fascist regime compelled the ruling circles on the eve and at the beginning of World War II to abide by a policy of relative non-commitment. At the same time they sought protectors among the Great Powers to help them solve Bulgaria's territorial problems. Particular preference was given to nazi Germany, which cited the unjust nature of the 1919 Treaty of Versailles as a justification for its aggressive policies. Bulgaria's economic link with German Imperialism and the similarity of the political regimes in the two countries were of essential importance for the pro-German orientation of Bulgaria's foreign policy.

The rightist legal opposition, represented by the Bulgarian bourgeoisie connected with British and French monopoly capital, tried to orientate Bulgaria's foreign policy toward Great Britain and France, but like the government circles it had anti-Soviet leanings. The left

Students, surrounded by police, demonstrating for the restoration of the Turnovo Constitution, 24 May 1936

democratic forces headed by the BWP continued their consistent struggle against Fascism, for peace and neutrality, in defence of national independence and for signing a pact of mutual assistance with the USSR. The BWP upheld this position during the election campaign for the Twenty-fifth Ordinary National Assembly (December 1939–January 1940). The Party participated in the elections in a united front with the left democratic forces.

Bulgaria during World War II. At the outbreak of World War II the Bulgarian Government declared that it would follow a policy of neutrality. In January 1940 the Government of G. Kyoseivanov concluded a trade agreement with the Soviet Union which was highly advantageous to Bulgaria. After the elections for the Twenty-fifth National Assembly G. Kyoseivanov was removed from the Government. On 15 February 1940 a new fascist Government, headed by Prof. Bogdan Filov, was formed; King Boris III became the uncontrolled and unchallenged ruler of the country's home and foreign policy. Measures were taken to build up the fascist state machine still further. A Law on Civil Mobilization was adopted. The police apparatus was strengthened. At the end of 1940 the National Assembly adopted a Law on the Defence of the Nation which introduced nazi anti-semitic and racist laws in Bulgaria. Under the Law on Organizing the Bulgarian Youth a state youth organization was set up after a nazi pattern. The new Government declared that it was going to continue the policy of neutrality. On 7 September 1940 an agreement was signed with Romania on the strength of which South Dobrudja was returned to Bulgaria.

After the formation of the Tripartite Pact, nazi pressure for involving Bulgaria in the war on the side of the fascist aggressive coalition was stepped up, especially after the stationing of a 500,000-strong German army in Romania. In a personal message to King Boris III, the King of England warned him that if nazi forces were allowed into Bulgaria it would be transformed into a theatre of war. At the same time the Italian dictator, Benito Mussolini, invited Bulgaria to join in the forthcoming aggression against Greece. The proposal was rejected by the King. Bulgaria's neutrality made it possible for the Greek command to withdraw a great part of its forces from the Bulgarian-Greek frontier and to concentrate them against the aggressor. The military failures of the Italian forces in Greece provoked still greater pressure upon Bulgaria on the part of nazi Germany.

In November 1940 the Soviet Union, a neutral country at the time, proposed to Bulgaria the signing of a pact of mutual assistance. Although this was exceptionally advantageous to Bulgaria, the King and the Government rejected it. This gave rise to a broad, mass movement, guided by the BWP, in support of the Soviet proposal. The BWP took a stand against Bulgaria's participation in the war, either on the side of Germany or on the side of Britain and France. The Party raised as a main slogan in the mass struggle: 'Peace, neutrality and alliance with the USSR!'

The leading circles of the bourgeois and petty-bourgeois parties, constituting the bourgeois non-fascist Opposition, also took a stand against the pro-German orientation.

In the beginning of 1941 the Government of B. Filov gave its consent, in principle, for Bulgaria's joining the

A German train destroyed by sabotage

Tripartite Pact, which was officially formulated on 1 March 1941. The involvement of Bulgaria in World War II on the side of the fascist aggressive coalition was an objective result of the policy of monarcho-fascism, but at the same time this involvement took place under the conditions of indisputable coercion – the ruling circles found themselves faced with the choice of either allowing the free entry of the German forces grouped on the Romanian bank of the Danube, or of having them invade the country by force. On 1 March 1941, the vanguard units and staff of the Twelfth Germany Army passed through the northern Bulgarian frontier.

Although the Bulgarian territory was used by the nazi army in its aggression against Yugoslavia and Greece, the Bulgarian army did not take part in the military operations against the two neighbouring States. After the military defeat of Greece and Yugoslavia, the nazi command entrusted the Bulgarian Government with the task of introducing a Bulgarian Administration in Macedonia, the Morava river valley and Aegean Thrace. The fictitious 'national unification' became a main slogan of the ruling fascist bourgeoisie, with which it justified its alliance with Germany. After the treacherous attack of nazi Germany on the USSR (22 June 1941), the Bulgarian ruling circles steered clear of direct participation in the military operations in the East. The diplomatic relations with the USSR were preserved. Bulgaria's military participation in the world conflict consisted in the dispatching of an occupation corps to Serbia and in the declaration of a 'symbolic war' on Britain and the USA.

Immediately after the nazi invasion of the USSR the Politburo of the CC of the BWP adopted a course of armed struggle (24 June 1941) against the monarcho-fascist dictatorship and the nazi presence in Bulgaria. The main forms of struggle were mapped out and the organs of leadership were set up. The course adopted by the Party was a natural continuation of the long anti-fascist struggle, which in the new situation, reached its highest form – an armed struggle – with a popular uprising as a long-term prospect. The Party applied new methods of struggle. Already in the summer of 1941 it proceeded to set up partisan units and a network of combat groups for sabotage and subversive actions. The mass resistance against the political and economic activities of the Government was stepped up, and sabotage actions were carried out in important sectors of production serving the nazi army. The work of the Party among the army was expanded with a view to attracting it to the side of the people's struggle. To organize and lead the armed struggle, a central military commission was set up in the CC of the BWP, headed by Hristo Mihailov (1893–1944), as well as military commissions with the district party committees. A home front was established, which had an increasingly palpable effect on the military and political situation in the

Todor Zhivkov greeting partisans from the Fourth Sofia Insurgent Brigade, 9 September 1944

country. Characteristic of the Bulgarian resistance from its very beginning was its marked social and class character. Not only was it guided by the Communist Party, but its composition was essentially communist. Communists, members of the WYU and sympathizers from the broad circles of the working class and the peasants took part in the resistance. The bourgeois Opposition rejected all forms of armed struggle.

On 25 November 1941, the Government of B. Filov attached Bulgaria to the anti-Comintern pact and on 12 December 1941 it broke off diplomatic relations with the USA and declared war on Britain and the USA. Although it was a 'symbolic war' Sofia and other towns were subjected to devastating air raids by the Anglo-American air forces (14 November 1943–17 April 1944), in which 11,718 buildings, 1,700 industrial enterprises, 40 schools and 10 hospitals were destroyed. Thousands of civilians perished under the ruins; the country's economy was disrupted – whole industrial branches had to cut back or completely discontinue their production; the supply of food to the population became difficult – basic goods disappeared, the prices of consumer goods rose, the black market and inflation ran high. The bourgeoisie took advantage of the war to pile up tremendous profits. Class contradictions were exacerbated, and the political and economic diktat on the part of Germany grew stronger. At German request, in September 1942 the Soviet consulate in Varna was closed down. On 22 February 1943, an agreement was signed with Germany for the deportation of 20,000 Jews to Poland, but thanks to the active resistance of the progressive public, organized by the BWP, the Bulgarian Jews were saved from annihilation. The monarcho-fascist Government stepped up its terror against the anti-fascist movement. Many people's fighters were killed; thousands of patriots – communists and non-Party members – were thrown into prisons and concentration camps, or interned. A heavy blow at the Party was the betrayal in its CC (1942). Several members and associates of the CC and of the Central Military Commission were arrested, sentenced to death and executed, including Anton Ivanov (1884–1942), Secretary of the CC, Anton Popov (1915–42), Atanas Romanov (1911–42), Nicola Vaptsarov (1909–42), Peter Bogdanov (1908–42), Georgi Minchev (1907–42) and Tsvyatko Radoinov (1895–1942). The anti-fascist General Vladimir Zaimov (1888–1942) was also sentenced to death and shot.

In 1942, at the initiative of the Party's Bureau-in-Exile under G. Dimitrov, the BWP proceeded to build a mass, anti-fascist organization – the Fatherland Front. In the Programme of the Fatherland Front, announced over the Hristo Botev radio station (17 July 1942), first of all were included democratic, national-liberation tasks through which the Party came close even to the most vacillating

Welcoming partisans at Belovo Station

anti-fascist circles. The anti-fascist struggle was unfolded under the banner of the Fatherland Front. Local, county and regional committees were formed and in August 1943 a National Committee of the Fatherland Front was set up, which was composed of representatives of the BWP, the BAU 'Pladne' and Zveno, whose leaders after 1936 adopted a leftist position and took an anti-fascist stance. Later, they were joined by leftist social-democrats and independent intellectuals. The defeats inflicted on the nazi army in 1943 at Stalingrad and the Kursk salient gave rise to confusion and consternation in Bulgarian fascist circles. At the same time, the people's resistance was strengthened and entered a new stage. By a decision of the CC of the BWP in March–April 1943 the partisan units were brought together in a unified National Liberation Insurgent Army (NOVA) headed by a General Staff. The country was divided into twelve insurgent operative zones. The partisan movement assumed a mass character, a great number of soldiers passed over to the side of the partisans and several soldier battalions were formed.

The development of the military operations unfavourable to Germany, and the capitulation of Italy (1943) intensified the contradictions among the ruling élite in Bulgaria. On 28 August 1943 King Boris III died of a heart attack and his son, who was under age, was proclaimed King Simeon II. This aggravated the political crisis in the country still further. In violation of the Constitution, the National Assembly elected (9 September 1943) a council of regents consisting of Prince Kiril of Preslav, Bogdan Filov and Gen. Nikola Mihov. On 14 September 1943 a new Government was appointed, headed by Dobri Bozhilov. The ruling circles made use of the army to crush the resistance movement. A special, repressive, armed force – the gendarmerie – was formed. The widely developed resistance did not allow the fascist ruling élite to send Bulgarian units to the Eastern Front against the USSR.

A decisive change in the disposition of power in the world set in during the first half of 1944. The Soviet army marched onward irresistibly and reached the frontiers of Romania, while British and American units landed in France. The impending defeat of nazi Germany deepened the contradiction among the Bulgarian ruling élite. In May 1944 a prolonged ministerial crisis arose. On 1 June 1944, a Government was formed under Ivan Bagryanov, and in order to prevent the collapse of the dictatorship, it attempted to sign a truce with Britain and the USA, at the same time taking steps to crush the internal revolutionary front. In July–August a 100,000-strong force of army and police was pitted against the partisans. The NOVA repelled the fascist offensive and was preparing for a counter-offensive. Towards the end of August 1944 one national liberation division, nine partisan brigades, more than forty partisan detachments, numerous bands and hundreds of combat groups with a total of more than 30,000 fighters were operating in the country. About 200,000 helpers and supporters also took an active part in the struggle. Compared with the other satellite States, the anti-fascist armed resistance in Bulgaria was unfolded on a very wide scale. With the advance of the Soviet army towards the Bulgarian frontiers the anti-fascist struggle flared up with fresh force. A revolutionary situation was coming to a head. On 26 August 1944 the CC of the BWP issued a circular letter to the leaders of the Party organizations in which the preparation for an armed uprising for overthrowing the monarcho-fascist dictatorship and setting up people's democratic rule was set as a most urgent and immediate task.

Taking into account the total military and political impasse facing nazi Germany, its own inability to cope with the internal partisan front and the danger which the impetuous advance of the Soviet army posed for the existing regime, Bagryanov's Government decided to switch over to 'strict neutrality' in the anti-Soviet war. The aim of this decision was to prevent the entry of the Soviet army into the country and to bring to a successful conclusion the negotiations which had started in Ankara and Cairo with Great Britain and the USA for Bulgaria to come out of the war with their help. The Soviet Government refused to recognize the proclaimed neutrality, considering it as insufficient proof of change in the policy of monarcho-fascist Bulgaria. The fascist dictatorship was no longer able to hold on to its position. In order to preserve the capitalist system in the country the bourgeois ruling circles resorted to their final resource in saving the monarchy and the bourgeois opposition system. On 2 September 1944 a Government was formed of what had until then been the legal Opposition, headed by the rightist agrarian Konstantin Muraviev. Taking part in this Government were the BAU 'Vrabcha 1', the Democratic and the Popular Party. In its declaration of 4 September the new Cabinet announced that it was going to restore the bourgeois democratic liberties and that its foreign policy would follow a line of neutrality towards the German-Soviet war and of achieving an armistice with Britain and the USA. Evaluating the formation of Muraviev's Government as an attempt on the part of the Bulgarian bourgeoisie to deceive the masses and avert a people's uprising, the CC of the BWP decided that the struggle should be continued until its overthrow. The main blow of the anti-fascist resistance was now directed not only against the monarcho-fascist bourgeoisie, but also against that non-fascist bourgeoisie which had passed over to counter-revolutionary positions. The Government of K. Muraviev disbanded certain fascist bodies and organizations and proclaimed a political amnesty, but in fact it did not do away with the system of dictatorship. In its foreign policy it also took inconsistent and indecisive steps: neutrality in the Soviet-German war was confirmed, but its unconditional observance was not secured; at the same time negotiations with the Anglo-American representatives in Cairo were stepped up; and the question of breaking off relations with Germay was settled. In this situation, on 5 September 1944, the Soviet Government declared that it was breaking off all relations with Bulgaria and would hence-forward be in a state of war with it. This was a powerful blow against the government policy, whose main objective was not to allow the entry of the

Welcoming the Soviet Army in Sofia, September 1944

A political General Strike, Plovdiv, 1944

Red Army into Bulgaria. The Government's ambition to win Anglo-American support was also frustrated; the negotiations which had opened in the summer of 1944 in Cairo were suspended. It was only now that the Government began to act more decisively. It broke off diplomatic relations with Germany and with the pro-nazi regimes and decided to declare war on Germany, with effect from the evening of 8 September.

The declaration of war by the Soviet Union served as a turning-point in the development of the Bulgarian revolution. The time had come to make a bid for power. The armed struggle grew into a mass uprising. On 5 September 1944 the Politburo of the CC of the BWP, in conjunction with representatives of the NOVA General Staff, adopted a plan for the uprising. It was decided that the main blow would be dealt in Sofia on the night of 8–9 September, with supporting blows in Pleven, Plovdiv and other large towns. Instructions about the preparation and carrying out of the uprising were also given by the Party's Bureau in Exile.

On 6 September the uprising had already spread to different parts of the country. Partisan detachments proceeded to capture towns and villages, where they set up Fatherland Front rule. On 7 and 8 September in Silistra, Pleven, Varna and elsewhere the working people attacked the prisons and liberated the political prisoners. In Sofia and other towns strikes, demonstrations and meetings were organized as well as attacks on strategic and other major objectives. When on 8 September units of the Third Ukrainian Front stepped onto Bulgarian soil the uprising was at its height. The Soviet forces were welcomed by the Bulgarian people as liberators and no resistance was offered by the military units. The revolutionary upheaval in the country and the entry of the Soviet army allowed the Fatherland Front to pass to a direct seizure of political power. A large part of the military units of the Sofia garrison passed to the side of the uprising, as did the Minister of War. According to the plan of the CC of the BWP and the NOVA General Staff, the main blow of the uprising started after midnight, on 9 September in Sofia. The insurgent strike forces, grouped into two streams, directed their attack on the Ministry of War. The army units captured the Ministry, and the revolutionary forces consisting of partisans, members of combat groups, and other anti-fascists and Fatherland Front members, led by Todor Zhivkov, were concentrated near the Ministry of War in full fighting trim. On 9 September Sofia was completely controlled by the insurgent forces. State power was wrested from the hands of the bourgeoisie and monarcho-fascism and passed into the hands of the working people. The first people's Democratic Government of the Fatherland Front, headed by Kimon Georgiev, was formed. The triumph of the Fatherland Front in Sofia was a signal for revolutionary actions everywhere. On

9 and 10 September the monarcho-fascist Government was overthrown in the whole country, and power was taken over by the Fatherland Front committees.

The 9 September Uprising succeeded as a logical result of the development of the Bulgarian people's anti-fascist struggle and as the highest form of its manifestation. With the victory of 9 September a whole era of the history of the Bulgarian people came to an end – the era of class exploitation – and the foundation was laid for the Bulgarian socialist revolution.

BULGARIA IN THE AGE OF SOCIALISM

The Development of the Political System of Society. Radical changes took place on 9 September 1944 in the composition of the bodies of state power in Bulgaria. The first Popular Democratic Government was composed of representatives of the four Fatherland Front parties – the BWP (communists), the Bulgarian Agrarian Party (BAP), the Zveno Popular Union (PU) and the BWSDP. The new Government was composed of four communists, four agrarians, four Zveno members, two social-democrats and two independent intellectuals. Kimon Georgiev, the leader of the Zveno PU was appointed Prime Minister. A new Council of Regency was formed of two non-party functionaries (Venelin Ganev and Tsvyatko Boboshevski) and one communist (Todor Pavlov).

On 17 September 1944, at a rally in Sofia, Prime Minister Kimon Georgiev announced the Programme of the Fatherland Front Government, known as the Second Fatherland Front Programme. It was a broad, far-reaching programme of action involving all the democratic forces in the country and ensuring the further strengthening of their unity.

The organization of the socialist State in Bulgaria proceeded in three main stages: dismantling the bourgeois-fascist apparatus for the coercion of the working people, restructuring another part of the state apparatus, and establishing new bodies for popular rule. This process began on 9 September 1944 and continued to the middle of 1948 when the organization of the State was brought into compliance with the Constitution of the People's Republic of Bulgaria, adopted late in 1947.

The abolition of the fascist state machine involved first of all the police, gendarmerie, intelligence and counter intelligence forces, prisons and concentration camps, courts martial, regional reforming squads and some other reactionary state bodies and institutions. The citizens' political rights and freedoms were restored and publication of newspapers propagating fascist ideology was suspended. Reactionary laws and police restrictions on political life were rescinded.

The Twenty-fifth Ordinary National Assembly and all fascist organizations were dissolved, and organs of state and municipal administration were purged of adherents to Fascism and counter-revolution. At the same time, new state bodies were established: the people's militia, and revolutionary intelligence and counter-intelligence. The functions of the district, county and municipal bodies of government were radically altered. The judiciary and the prosecutor's office were completely restructured. A People's Court was instituted to try those guilty of Bulgaria's involvement in the world war against the allied nations and of the crimes perpetrated during it. Brought to justice were the Regents and Ministers of the Governments in power from the 1 January 1941 to 9 September 1944, deputies to the Twenty-fifth National Assembly and other civilians and military.

The organization of the state system of people's democracy was inextricably bound up with the activities of the National Fatherland Front Committee and of local Fatherland Front Committees. The National Committee of the Fatherland Front exercised political control over the activities of the supreme state bodies, made recommendations about important national issues, guided the development of the revolutionary process and followed the realization of the major undertakings and changes in society.

Similar functions with regard to the local bodies of power were performed by the respective Fatherland Front Committees, through which the political unity of the people was strengthened and the alliance of the working class, the working peasants and the people's intelligentsia was consolidated.

In view of the urgent task of Bulgaria's immediate participation in the war against nazi Germany, there was a revolutionary restructuring of the army, from which the

Kimon Georgiev announces the programme of the Fatherland Front, 17 September 1944

Georgi Dimitrov, Marshall Fyodor Ivanovich Tolbukhin and Sergei Semyonovich Biryuzov at Sofia Station, 22 February 1946

most reactionary pro-fascist elements were removed. Outstanding officers, anti-fascists and servicemen loyal to the Fatherland Front Government, were promoted to command posts. The units of the People's Liberation Insurgent Army, of which the People's Guard was formed, joined the army forces, and a total of 450,000 men were mobilized within a very short time. In fact Bulgaria had entered World War II on the side of the anti-nazi coalition on 9 September 1944, and on 17 September 1944 the Bulgarian People's Army (BPA) was placed under the operative command of the Third Ukrainian Front. By its active participation in the final stage of World War II, the Bulgarian people made a contribution of its own to the defeat of Nazism.

The Government took important steps towards the economic stabilization of the country and the establishment of public ownership in the overall economic activity. The property of fascist war criminals, and of factory-owners and tradesmen who had been in the service of the nazis, was confiscated. The public (state and co-operative) sector was developed as the economic base of popular rule. Also included in this sector were railway transport, trade, credit and other economic sectors owned by the bourgeois State prior to 9 September 1944. The State became the owner of the coal mines and other mining enterprises, electric-power stations, arms factories and large trade enterprises. The state-owned enterprises laid the beginnings of the socialist sector in the economy. The public sector included the great majority of the co-operatives. A great many co-operative farms were formed. As a result of the confiscations, mixed state and private enterprises were set up. The small-scale commodity-production sector, including hundreds of thousands of small farms and craftsmen's workshops, was quite large, but the bulk of industry remained under private capitalist ownership.

The BWP (communists) maintained the thesis that the broader the social and political sections involved in it, the smoother and wider the transition to Socialism would be. Hence it conscientiously and genuinely fought to strengthen the Fatherland Front and engaged in careful, patient effort among the working people. At the same time, it was clear that the Party would hold leading positions in society if it increased its membership and rallied closely together its ranks, extended its influence over the Fatherland Front Committees, the mass organization and the bodies of state power, established itself as a principal and prestigious partner among its allies, and strengthened its links with the masses. And, indeed, the Party proved equal to these high requirements.

Among the mass organizations for the consolidation of the people's democratic power, the Workers' Youth

Union (WYU), with a membership of 420,000 (early in 1945), and the GWTU, with a membership of about 300,000 were of utmost importance. Hence the BWP (communists) which had a membership of over 250,000, devoted great efforts to their organizational and ideological advancement, to strengthening the youth and trade-union movements in the country, and to the active involvement of the youth and trade unions in building the new life.

The deepening of the revolutionary process in the country brought about certain centrifugal tendencies in the other Fatherland Front Parties – the BAP, the Zveno PU and the BWSDP. Several wings emerged in these parties – left, right and centre – and acute intra-party strife broke out on the issue of attitude to the BWP (communists) as a leading political factor in the country, as well as to the prospects of Bulgaria's socialist development.

The formation of political opposition forces which early in September 1945 were legalized in independent political parties, was closely connected with the preparations for holding the first parliamentary elections after 9 September 1944. The opposition decided to boycott them. At the climax of the pre-election campaign, on 4 November 1945 the leader of the BWP (communists), Georgi Dimitrov, returned to Bulgaria after twenty-two years of exile. His pre-election speeches delivered on 6 November *(To a Brilliant All-out Victory in the Elections)* and on 15 November *(To a Victory of the People over Reaction and Ill-wishers of New Bulgaria)* had a great mobilizing effect and contributed in the largest measure to the overwhelming success of the Fatherland Front in the elections for the Twenty-sixth Ordinary National Assembly.

On 18 November 1945, 84.80 per cent of the electorate turned out at the polls (though voting was not compulsory), of whom 88 per cent voted for the Fatherland Front.

Those elected to the Twenty-sixth Ordinary National Assembly included 94 communists, 94 agrarians, 44 Zveno members, 31 social-democrats, 11 radicals, and 2 independents, a total of 276 deputies; among them were 18 workers, 51 farmers, 15 women and 12 young people under 30 years of age. Vassil Kolarov, member of Politburo of the CC of the BWP (communists) was elected Chairman of the National Assembly, while Georgi Traikov (BAP) and Peter Popzlatev (Zveno PU) were elected Deputy Chairmen. During its short term in office the first Parliament of the people's democratic Bulgaria carried out an enormous amount of work in the area of legislation and endorsed about 330 statutes.

The second Fatherland Front Government, headed by Kimon Georgiev, was formed on 31 March 1946 and was composed of five communists, four agrarians, four Zveno members, two social-democrats, one radical and one independent. Traicho Kostov (BWP communists) and Alexander Obbov (BAP) were appointed to the newly instituted posts of Deputy Premier.

Early in 1946 the Fatherland Front political parties had the following membership: the BWP (communists) – 413,225; the BAP – 152,788; the Zveno PU – 31,111; the BWSDP – 29,039 and the Radical Party – 5,595. The opposition parties, i.e., the BAP (N. Petkov) had a membership of 53,531, the BWSDP (united) – 3,020, and the Democratic Party – 1,607. These figures show that the number of communists was about twice that of the members of the other Fatherland Front parties whose claims for 'equal power' were based neither on the real correlation of forces nor on their actual influence among the masses. The BWP (communists) continued to make concessions to its allies with regard to their representation in the Fatherland Front committees and in government bodies though it undoubtedly held leading positions in the country's political life.

In the referendum held on 8 September 1946 the Bulgarian people said a categorical 'No' to the monarchy and resolute 'Yes' to the people's republic. Out of the total of 4,509,354 voters, 4,132,105 (or 91.63 per cent) took part in the referendum. The number of invalid ballot papers was 123,690 (or 2.99 per cent). Of the valid ballot papers 3,833,183 were in favour of a people's republic (95.63 per cent) and 175,234 (4.37 per cent) of monarchy. In this way the people expressed its genuine will on the issue of cardinal importance for the country – the issue of the form and character of state Government. A week later, at a ceremonial session the National Assembly adopted the following decision on the basis of the vote: 'Today, 15 September 1946, the Twenty-sixth Ordinary National Assembly, acting on the will of the Bulgarian people expressed by rare unanimity in the referendum held on 8 September 1946, and in accordance with Article 5 of the Law on Plebiscite for the abolition of monarchy, etc., proclaims Bulgaria a people's republic'.

The functions of Head of State were taken over by the provisional (until the adoption of the new Constitution) supreme state organ, named the Presidency of the Republic, instituted by the National Assembly. It was composed of the Chairman and Deputy Chairmen of the National Assembly. Hence Vassil Kolarov in his capacity of Chairman of the Twenty-sixth National Assembly became the provisional President of the Republic.

In the elections for a Great National Assembly in 1946, about 4,250,000 people (94.18 per cent) of all voters took part. The victory of the Fatherland Front over the opposition forces was indicative of the people's confidence in the policy of the Fatherland Front Government – about 3,000,000 votes (70.10 per cent) and 366 seats in the Great National Assembly of all 465 seats. The votes given for the BWP (communists) amounted to 53.16 per cent of all votes, for the BAP – 13.23 per cent, for the BWSDP – 1.87 per cent, for Zveno PU – 1.65 per cent and for the Radical Party – 0.19 per cent.

For the opposition parties about 1,200,000 votes (28.35 per cent) were given, which secured them 99 seats in the Great National Assembly: 90 for the BAP (N. Petkov), 8 for the BWSDP (united) and one for the Independent Intellectuals. This result reduced to naught the assertion spread at that time that the opposition had behind it the majority of the Bulgarian people and that the Government of the Fatherland Front was not a representative Government.

Georgi Dimitrov signing the roll-call vote for the Constitution of the People's Republic of Bulgaria, 4 December 1947

The results obtained from the elections sanctioned and reaffirmed in a convincing manner the leading role of the BWP (communists) and opened the way to thoroughgoing revolutionary changes. The fact that the Communist Party had a majority in the highest organ of state power assured the adoption of a Constitution based on socialist principles. Up to the signing of the Peace Treaty with the victorious powers, the policy of the Party and the Government had to conform with the as yet unsettled international situation of the country, but the road for the socialist reconstruction of society had definitely been cleared.

As a party having won a majority in the National Assembly the BWP (communists) had all the legal grounds to form an independent Government of communists, but it did not avail itself of this right. In the interest of consolidating the alliance among the workers, peasants and people's intelligentsia, the BWP (communists) decided that the other Fatherland Front parties should again be represented in the new Government. On 22 November 1946 the Chairman of the CC of the BWP (communists) and founder of the Fatherland Front, Georgi Dimitrov was elected as Prime Minister and formed the third Fatherland Front Government. It included ten representatives of the BWP (communists), five of the BAP, two of the Zveno PU, two of the BWSDP and one independent.

The third Fatherland Front Government found itself confronted with severe problems – the signing of a Peace Treaty, the adoption of a new Constitution, the elaboration of a two-year plan for the country's development, etc. But as the decisive posts in commanding positions in the State were occupied by communists and Fatherland Front members, the BWP (communists) was able still more firmly and confidently to follow the road opened up by the 9 September people's uprising.

At the end of 1946 and the beginning of 1947, the opposition groups stepped up their activity. Some of the leaders of the old bourgeois parties also reappeared on the political stage. Supported by certain imperialist circles they strove to organize a common front of all reactionary and anti-communist forces in the country. This compelled the BWP (communists) and the other revolutionary forces, after the signing of the Peace Treaty between Bulgaria and the victorious powers in the war (6 February 1947), to take action to defeat the reactionary opposition in the country.

The Thirteenth Plenum of the CC of the BWP (communists) held on 14 October 1947 decided that the process of socialist transformation of society should be accelerated and the leading role of the Party in the system of people's democracy should be strengthened. The practical programme for passing to a new stage in the development of the socialist revolution included the transformation of the Fatherland Front into a unitary socio-political organization; a re-working of the draft-Constitution in a socialist spirit; nationalization of the large capitalist estates and the expansion of the state and public sector in the national economy; consolidation of the worker-peasant alliance; deepening of the friendship and co-operation with the Soviet Union and with the people's democratic countries, etc. Soon afterwards the National Committee of the Fatherland Front and the leaderships of the political parties and mass organizations expressed their solidarity with these decisions and adopted them.

On 4 December 1947 the Great National Assembly unanimously adopted the new republican Constitution, popularly called the Dimitrov Constitution, which legally consolidated the historic gains of the Bulgarian people after the victory of the 9 September people's uprising, reaffirmed the people's democratic form of government and secured the necessary conditions for the building of Socialism in Bulgaria. The Constitution of the People's Republic of Bulgaria of 1947 was socialist in character, which guaranteed society's successful transition from Capitalism to Socialism and reaffirmed socialist social relations and the further development of the new social system.

At the end of 1947 and during the first half of 1948 a restructuring of the state organization was carried out in accordance with the new Constitution. A Presidium of the National Assembly was elected on 9 December 1947, and on 11 December the fourth Fatherland Front Government was formed, with Georgi Dimitrov at the head. The new Council of Ministers included fourteen representatives of the BWP (communists), five of the BAP, two of the BWSDP, and two of the Zveno PU. Early in 1948 a State Planning Commission, a Commission for State Control, a Committee for Science, Art and Culture, a Public Prosecutor's Office of the People's Republic of Bulgaria and several other new state organs and institutions were set up. A number of Ministries were restructured, planning departments and services were set up and large economic associations were formed. A Supreme Court and a Chief Public Prosecutor of the Republic were elected and changes were introduced in the system of justice. With the creation of the People's Councils as local organs of power and of local self-government the first stage in the development of the Bulgarian socialist State was completed.

At the end of 1947 the youth unions were united into a single Youth Organization. On 2 and 3 February 1948, the Second Congress of the Fatherland Front was held. It took decisions on the reorganization of the Fatherland Front into a unitary socio-political organization whereby the elements of coalition were liquidated, the alliance of the working people from town and country was consolidated, and the leading role of the working class, headed by the Communist Party, was enhanced.

The restructuring of the political system of society also involved the political parties. The BAP at its Twenty-seventh Congress (28–29 December 1947) adopted the construction of Socialism as its practical task and recognized the leading role of the Communist Party in the country, and in October–November 1948, it rejected the 'town-versus-country' principle of organization and declared itself to be a class organization of poor and middle working peasants. In the summer of 1948 the BWSDP was merged with the BWP (communists), and early in 1949 the Zveno PU and the Radical Party disbanded themselves and terminated their activity. Thus in socialist Bulgaria two friendly parties – the BWP (communists) and the BAP continued in existence and worked together.

The Fifth Congress of the BWP (communists) was held from 18–25 December 1948. It was attended by 1,045 delegates representing 496,000 party members. The Congress approved the political activity of the Party after 9 September 1944, and charted its general line for the years to come – a line of building a socialist society along a path of socialist industrialization of the country, reorganization of agriculture on socialist lines, and revolutionary changes in the sphere of culture. A decision was taken that the Party should be named Bulgarian Communist Party (BCP) and its central bodies were elected. Georgi Dimitrov was elected as General Secretary of the CC of the BCP.

The Fifth Party Congress worked out the programme for the period of transition from Capitalism to Socialism and introduced clarity on a number of important theoretical questions, specifically concerning the periodization of the history of the BCP, the character of the 9 September Uprising and certain objective standards, as well as specific features of socialist construction in Bulgaria. Georgi Dimitrov developed, with great insight and maturity of creative thought, his conception of people's democracy as a specific form of proletarian dictatorship – as the rule of the overwhelming majority of the people, the leading role in which is played by the working class and its communist vanguard. He pointed out that the political system of people's democracy was intended to secure the development of the country along the road to Socialism, that the People's Democratic State was a State of the period of transition from Capitalism to Socialism, that people's democracy and Soviet power differed only in form and some specific features, but not in their class essence and substance.

In connection with the further development of the political system of people's democracy, the Fifth Congress of the BCP set the tasks of steady consolidation of the leading position of the working class in all fields of life; of strengthening the worker-peasant alliance; of continued improvement and perfection of the state apparatus; and of enhancing the role of the BCP as the leading and motivating force in society.

In the autumn of 1949, changes were effected in the administrative and territorial division of the country, whereby the existing municipalities and counties made up fourteen districts (after 1951 – twelve, and the district of the city of Sofia). On 5 May 1949 elections were held for County and Municipal People's Councils, and on 18 December the same year – elections for a First National Assembly and District People's Councils.

Rally to mark the end of the Seventh Party Congress, Sofia 1958

After Georgi Dimitrov's death (2 July 1949) a Government was formed headed by Vassil Kolarov (20 July 1949). After his death on 23 January 1950, the posts of General Secretary of the CC of the BCP and Prime Minister of the People's Republic of Bulgaria were assumed by Vulko Chervenkov, whose name is connected with what is called the personality cult, which led to certain deviations from the general line followed by the Fifth Party Congress and to violation of the Leninist principles of party leadership and norms of party life. The personality cult and its negative consequences were an outgrowth alien to the Party and an untypical and temporary phenomenon. That was evident from the brave and consistent struggle which the BCP began in order to eradicate this alien phenomenon in its ranks.

The April Plenary Session of the CC of the BCP (2–6 April 1956) condemned and rejected the personality cult and the methods of work connected with it, and elaborated a number of undertakings for the strictest observance of the principles of collective leadership, for developing inner-party democracy, for strengthening socialist legality, and for a reorganization of ideological work. In the area of organizational questions, the Plenum decided that the First Secretary of the CC of the BCP, Todor Zhivkov, should become head of the Politburo and proposed to the National Assembly that Vulko Chervenkov be relieved of the post of Chairman of the Council of Ministers and appointed as its Deputy Chairman, while the member of the Politburo of the CC of the BCP, Anton Yugov, be appointed Prime Minister. Mapping out the main lines and directions of the future socio-economic, political and cultural development, the April Plenum provided new conditions and possibilities for securing the all-round progress and ascending socialist development of the People's Republic of Bulgaria.

The elaboration and practical realization of the political policy of promoting the gains of Socialism in Bulgaria and making the country a modern industrial and agricultural State with a high development of culture and science, which was confirmed at the April 1956 Plenum of the CC of the BCP, as well as the elaboration of the Party Programme endorsed by the Tenth Congress of the BCP in 1971 for building a developed socialist society and the adoption of a new Constitution, also in 1971, are inseparably linked with the name of Todor Zhivkov.

As long-standing First Secretary and (since 1981) General Secretary of the BCP CC, Prime Minister and (since 1971) President of the State Council of the People's Republic of Bulgaria, as a worthy heir and successor to the cause and the traditions bequeathed to the Party by Dimiter Blagoev and Georgi Dimitrov, Todor Zhivkov has emerged as an outstanding theoretician, politician, organizer, ardent patriot and internationalist, an outstanding functionary of the international communist and workers' movement, and as a man of great personal charm. In his reports to the Seventh (1958), Eighth (1962), Ninth (1966), Tenth (1971), Eleventh (1976) and Twelfth (1981) Party Congresses, and to CC plenums, and in the decisions of other Party forums, many vital questions of Marxist-Leninist theory and of Bulgaria's and the BCP's practical activities were developed further. Todor Zhivkov's speeches, reports, articles and interviews published at home and in over forty foreign countries in many different languages, in particular his theoretical formulations and practical approaches since the Twelfth Congress (1981), touch on many fundamental questions concerning the country's past and present, the BCP's strategy and tactics in building Socialism, the tasks facing the Party and the major topics of our times.

Todor Zhivkov has played a vital role in confirming the policy of peaceful coexistence between states with different social systems and promoting peace and understanding among the peoples of the Balkans, Europe and the world. He is a tireless fighter for the further all-round development and promotion of friendship between the Bulgarian and Soviet peoples, of closest economic, political and cultural co-operation and mutual assistance among all countries of the world socialist community and of furthering the unity and cohesion of the international communist and workers' movement.

The April 1956 Plenum and the policy adopted at it are neither the result of the chance combination of factors and events, nor the result of subjective whims. They were, at a particular historical juncture, a response to a great necessity, a conclusion reached from the general activities of the Party and development of the country, and the only possible alternative, dictated by historic necessity, for the further and accelerated movement of society along the road of Socialism, an alternative that was in line with objective laws, with the interests of the people and the programme tasks of the BCP.

At the April Plenum the Party set out neither to change the general policy of building Socialism, pursued until then, because it was the right policy, nor to belittle the achievements made in the course of socialist construction. The underlying historical significance of the Party's struggle was to fully restore and strictly adhere to the Leninist principles of leadership and norms of party life, and to bring the constructive deeds of the people into full correspondence with the objective laws of Socialism and comprehensively resolve outstanding problems on the basis of the creative application and development of Marxist-Leninist theory. The decisions of the April Plenum were not of a temporary or transient nature. The political course elaborated on the basis of these decisions was one of consistently enriching and developing the Party's policy of building a new society, taking strict account of changes in the situation at home and abroad, through the development of revolutionary theory and the experience of socialist construction.

In the post-April 1956 period cardinal questions were solved concerning the democratization of socio-political life and the broad involvement of the masses in management. A number of functions hitherto carried out by state bodies were now entrusted to public organizations. Essentially new forms of management were evolved in the sphere of production, combining the features of state and public management. The volunteer squads of working people as well as comrades' courts were set up, and measures were taken to strengthen public control by the masses. The role of the public organizations themselves was promoted. Subsequently the principle of public-state and state-public management was elaborated and applied in many spheres of life as a new socio-historical mechanism for strengthening the creative role of the millions and ensuring their direct participation in management.

All these processes are being implemented under conditions of growing improvement in the socialist state as the prime means of building the new society. Through the mechanism of socialist democratization, the socialist system provides the working people with real rights in management at all levels. In the process of building mature Socialism, the Bulgarian socialist state is gradually acquiring the features of a socialist State of all the people, which means a growing social basis, increased participation in state management by the masses, and increasing democracy.

An important factor in making the millions the immediate collective and individual masters of their own history is the application of the new system of management based on the economic approach. This approach gives even wider scope for the activity of the working people and its further development. It also provides a new formulation of public ownership which enables and likewise requires the collective and the individual worker to show greater concern for the protection and development of the wealth society has entrusted to him and to take an active and responsible part in its management.

The April policy line is, therefore, a creative, innovative and vital policy because it is a response to and realization of a social requirement and historic necessity. This is a necessity involving all main areas of the life of society. The root question is to provide conditions for the socialist social system to reveal and develop all its essential features, and its vast potential, in order to unleash the creative energy of the masses and to raise the country's material and cultural life to new heights.

A new administrative reform was carried out in Bulgaria at the beginning of 1959, aimed at bringing the management of the national economy closer to production, and at raising the responsibility, activity and number of initiatives of the local Party, state and economic organs, and of all working people engaged in the fulfilment of the economic and cultural goals. The 13 districts and 117 counties then in existence were abolished and 30 new administrative and economic districts were set up instead (the number was later reduced to 28). The basic administrative unit within the district became the municipality including the territory of the unified co-operative farms. This new administrative division of the country signified an all-round reorganization of the state leadership and economic management, ensuring the expansion of democratic centralism of leadership, the elimination of

Workers' parade on the 25th anniversary of the 9 September Uprising, 1969

certain redundant links in it and the increase of its effectiveness.

The Fourth National Assembly (19 November 1962) elected the First Secretary of the CC of the BCP, Todor Zhivkov, Chairman of the Council of Ministers of the People's Republic of Bulgaria.

Voting for the new Constitution. Sofia, 1971

The historic Tenth Congress of the BCP (1971), which adopted the Programme of the BCP for the construction of the advanced socialist society in Bulgaria, was a major event in the development of the Party and the country. The Programme formulated the principal theoretical, ideological and strategic party views and principles of the further progress of socialist society and of its conversion into an advanced social system. It elucidated the basic principles of the advanced socialist society considered to be the highest and final stage of the development of Socialism as the first phase of the formation of Communism.

The profound and progressive changes which had occurred in all walks of life gave rise to the necessity of working out a new Constitution. On 15 March 1968 the National Assembly elected a Commission to prepare a draft, and after a nationwide discussion, the new Constitution of the People's Republic was adopted by a referendum on 16 May 1971, proclaimed by the Fifth National Assembly on 18 May 1971 and published in *Durzhaven Vestnik* (the State Gazette). The new Constitution reflects and legislatively consolidates the successes of the Bulgarian people achieved during the construction of the new social system.

The Programme of the BCP on the construction of the advanced socialist society serves as the theoretical basis of the new Constitution of the People's Republic of Bulgaria. Its entire content is imbued with the basic objective of this Programme: 'Everything in the name of man, everything for the good of man'. The Constitution has preserved all stipulations of the 1947 Constitution characteristic of Socialism concerning the social, political and economic structure of the country and the rights and freedoms of the citizens. In addition, it takes into account the further development of these relations under the new conditions. The Constitution reflects not only the state structure but also the unified system of the social management of society.

Under the new Constitution a State Council of the People's Republic of Bulgaria has been set up, the highest permanently functioning organ of state power linking the adoption of decisions with their fulfilment. On 7 July 1971 the First Session of the Sixth National Assembly elected the First Secretary of the CC of the BCP, Todor Zhivkov, as President of the State Council of the People's Republic of Bulgaria; it also elected a member of the Politburo of the CC of the BCP, Stanko Todorov, as Chairman of the Council of Ministers.

The Seventh National Assembly passed a law (1 December 1978) on the further improvement of the country's territorial and settlement structure. This law amended and supplemented the law on the People's Councils. The settlement systems, which are now organizational forms of local self-government, have proved their efficiency as basic political, social, economic and administrative units.

Public organizations are an integral part of the political organization of the socialist society. They reflect and uphold the specific interests of the various sections of the population, engage the working people in socialist construction and in the management of society, work for raising the level of their socialist consciousness, linking the Party with the masses still more closely, rallying them round the policy of the Party and mobilizing them for its implementation. Their role in the social management system is continuously growing.

Relations between the State and the public organizations have been greatly modified. The public organizations have taken on the fulfilment of many state functions and tasks. These are: the National Council of the Fatherland Front, the Central Council of Trade Unions, the CC of the Dimitrov Young Communist League and the Managing Board of the Central Co-operative Union. They have been granted the right to legislative initiative, their control over many state bodies and institutions being increased considerably. Co-operation between these public organizations and the State has evolved new forms. Their participation in the solution of major social, political, economic, cultural and educational tasks has been growing even more active, as has their participation in the formation of state power organs, in taking decisions pertaining to the affairs of the State and in implementing major tasks set by the State.

Public organizations have strengthened their independent role, improving the forms and methods of their work and extending the democratic principles of their structure and their all-round activities.

Chain dances performed by amateur ensembles in the 9 September parade, Sofia, 1976

Their material base has been considerably improved, as has their sphere of competence on a wide range of questions. Their role in implementing the party and state policy has also been widened.

The Twelfth Congress of the BCP (31 March–4 April 1981) analysed the main results of the activities of the Party in implementing the tasks underlying the Seventh Five-Year Plan (1976–80) and adopted theses on the social, economic and cultural development of the People's Republic of Bulgaria during the Eighth Five-Year Plan (1981–85) and up to 1990. The Congress took the decision to launch an all-round intensification of the national economy and of the other spheres of public life, to go ahead with the further development of socialist democracy and to further raise the material and cultural standards of the population. Todor Zhivkov was elected General Secretary of the CC of the BCP on 4 April 1981.

Elections for the Eighth National Assembly were held on 7 June 1981. Out of the 400 deputies elected, 271 are members of the BCP, 99 are members of the BAP and thirty are non-party people. The percentage of deputies belonging to no party, of women, of workers, of engineering specialists and of workers in the field of culture has also increased. Thirty-nine per cent of the composition of the National Assembly has been changed.

On 16 June 1981 the First Session of the Eighth National Assembly re-elected the General Secretary of the CC of the BCP, Todor Zhivkov, President of the State Council of the People's Republic of Bulgaria, while Grisha Filipov, member of the Politburo of the CC of the BCP, was elected Chairman of the Council of Ministers of the People's Republic of Bulgaria.

The structure and activities of the state apparatus have been constantly improving at the present stage of the development of Bulgarian society. Provisions have been made to eliminate redundancy in the activities of state organs, to introduce new and up-to-date methods of management and consistently to apply the Leninist principles of organization and work of the socialist State.

The socialist State expresses and upholds the interest of the broad masses, being a basic form of the rule of the people. The leading role of the working class is carried out through the socialist State. The State helps consolidate the alliance of workers, farmers and other kinds of working people and the participation of the masses in the management of society. Democracy is a characteristic feature, a distinguishing trait and the most important principle in the organization and functioning of the Bulgarian socialist State. The involvement of all working people in the running of the State has led to the consolidation of the socialist State and the successful solution of the tasks facing socialist construction. For this

Twelfth Party Congress, Sofia, 1981

reason the policy of the BCP in the field of social management and state construction has been primarily directed towards the expansion of socialist democracy in the activities of the State. An eloquent example of this is the introduction of the public-state and the state-public principle in the management of culture, science and education and in control activities. The principle is also being applied in the sphere of material production, through the creation of the National Agro-Industrial Union, economic associations, etc. A number of other activities, exercised by state organs, have been re-assigned to public organizations or to state-public organs.

The role of the socialist State continues to grow. This fact has determined its place within the political system of society and within the practice of socialist construction. It has assumed great significance as the principal organizer and regulator of the economic processes in society. In addition, the State guides the development of education, science and culture, it conducts a peaceful foreign policy, organizes the country's defence and facilitates the all-round integration of the countries belonging to the world socialist community.

The constant expansion of the role of the State has been determined by the rapid rate of socialist construction, by the growth of its scale and scope, by the complex tasks confronting society in its further progress and by the increasing role of the entire system of social management.

The BCP plays the leading role within the political system of the socialist society in the People's Republic of Bulgaria. It ensures the political guidance of all spheres of public life, as it does unity within the political system of Socialism, determining the main trends in the development of Bulgarian society along the road of Socialism. The BCP enjoys such a position because it is the party of the major class – the working class – because it expresses the historic objectives and interests of this class, because it possesses the richest amount of experience in leading the masses and because science and policy constitute an organic unity in its activities. In its relations with the state and public organizations the Party makes a strict distinction between its own functions and theirs. The Party does not undertake the task of solving all problems stemming from the management of society, neither does it interfere grossly in the activities of state organs or public organizations. It ensures the political leadership of society, as it raises the scientific level of its management. At the present stage of the country's development – the stage of building the advanced socialist society – the leading and guiding force of the BCP has increased still further, turning into a basic principle of the organization and activities of the political system of society in the People's Republic of Bulgaria.

Building the Material and Technical Basis of Socialism. During the first three years after 9 September 1944, the attention of the Communist Party, the National Committee of the Fatherland Front and the People's Democratic State was focused, above all, on the settlement of the country's international position, the political consolidation of the Fatherland Front Government and the restoration of the national economy, which had been ravaged during the Second World War. During that period measures were taken to limit capitalist private ownership and partially to expropriate the bourgeoisie. The problems connected with the elimination of capitalist ownership were solved in an all-embracing fashion later on.

The process of restricting capitalist ownership was carried out through state-supervised regulation of and control over private capitalist enterprises and privately-held commerce, through public control over capitalist production and through high taxation. These measures were aimed at meeting the population's essential needs and barring the way of profiteers.

State-sponsored regulation included not only the output but also the supply of privately-held capitalist enterprises with raw materials and other auxiliary material. The prices of goods manufactured by privately-held capitalist enterprises and the wages of workers and employees were also placed under the control of the State. Price formation and the introduction of work quotas limited capitalist profit-making. In addition, it was the State that determined to which organizations and enterprises the goods manufactured should be sold, thus preventing speculation by property owners and merchants and the additional enrichment of the country's industrial and commercial bourgeoisie.

Control over the activities of the capitalist enterprises was also effected by the public, primarily by the trade union committees in the different enterprises, whose basic task was to help increase production, save materials, improve the quality of output and strengthen labour discipline. To achieve this, numerous recommendations and suggestions which factory-owners were obliged to take into consideration were put forward at the production meetings at enterprises. The trade unions also took part in setting the prices of goods.

The expropriation of the Bulgarian bourgeoisie was carried out during the course of several years. Socialist nationalization was one of the prime tasks connected with the abolition of private capitalist ownership but it was not the only means of establishing public ownership over the basic means of production. In addition to nationalization, confiscations, the introduction of state-owned monopolies, the purchase of capitalist-owned enterprises and other kinds of bourgeois-owned property, the transformation of joint state- and capitalist-owned enterprises into socialist ones were carried out with a view to abolishing capitalist ownership.

The nationalization of privately-owned industries and mines was effected at the end of 1947. It brought about a decisive dominance of the socialist sector in the country's industry. The radical reorganization of the economy provided the necessary conditions and prerequisites for effecting large-scale and concentrated industry and for setting up powerful industrial complexes in the country.

The first steps toward industrializing the country were taken during the reconstruction period, and especially through the successful implementation of the Two-Year

The nitrogen fertilizer plant at Dimitrovgrad

The Georgi Dimitrov dam

Economic Plan (1947–48). During the First Five-Year Plan (1949–53) socialist industrialization became the mainstream in the struggle of the BCP for building Socialism and the most important condition for turning Bulgaria from an agrarian-cum-industrial country into an industrial-cum-agrarian one. The implementation of the Second Five-Year Plan was launched at the beginning of 1953. During that period, however, due to the violation of the basic requirements of socialist industrialization, its rates of development were slowed down and a number of major branches of heavy industry such as machine-building, energy generation, the chemical industry, and ferrous and non-ferrous metallurgy lagged behind unjustifiably in their establishment. Industrial rates of development were half those during the First Five-Year Plan.

In the period between April 1956 Plenary Session of the CC of the BCP and the Seventh Congress of the Party (1958), an enormous amount of work was done in Bulgaria to overcome the slower rates of socialist industrialization. The efforts of the Party and State were directed, above all, towards the restoration of the development of heavy industry as the first priority, the correct allocation of the productive forces, the increase in labour productivity, the accelerated introduction of the achievements of scientific and technological progress, and more active participation in the international socialist division of labour, etc. Economic relations between Bulgaria and the Soviet Union were a decisive factor in the development of the country's socialist industrialization.

The BCP also found the most suitable form of reshaping Bulgarian agriculture which corresponded to the specific Bulgarian conditions, thus facilitating the mass involvement of the working peasants in the construction of Socialism. Co-operative farms were set up as a stabilized form of agriculture, which was the only possible form of mechanizing and modernizing Bulgarian agriculture with the aid of machine and tractor stations, of improving the living standards of the Bulgarian population and of directing the rural economy towards Socialism.

Ploughing up old field-boundaries for the first co-operative farms

By means of the co-operative farms, the construction of which began immediately after 9 September 1944, labour and the means of production were socialized. The ownership of the pooled land was maintained and a rent was received in accordance with the size of the land. This rent was an economic category characteristic of the transitional period from Capitalism to Socialism in Bulgaria. It was bound up with the process of the gradual transformation of private land-ownership into public, and did not essentially change the socialist nature of co-operative farms.

The year 1950, when about 1,000 new co-operative farms were set up with nearly 350,000 peasant households, and co-operative land increased from 13.6 per cent at the beginning of the year to 15.1 per cent at the end of the same year, proved to be a turning-point in the co-operation of agriculture. The total number of co-operative farms reached the figure of 2,501, which was almost half of all peasant households in the country.

The 1956 April Plenary Session of the CC of the BCP ushered in a new era in the socialist transformation of Bulgarian agriculture. Only three months after the Plenum, and in order to implement its decisions, a joint session of the CC of the BCP and the Council of Ministers of the People's Republic of Bulgaria (6–7 July 1956) was held, which discussed and confirmed by a special decree a number of measures to further strengthen co-operative farms, expand and render rural economy produce less costly, and stimulate the material incentive of co-operative farmers and farms. Stressing the fact that co-operative

The Maritsa-East thermo-power plant

farms united 77 per cent of the country's peasant households and 75 per cent of the arable land, and following the recommendation of Todor Zhivkov, First Secretary of the CC of the BCP, the decree set the task of completing co-operation in agriculture in two to three years. The principle of the voluntary nature was to be strictly observed and the attention of the Party and the state organs was drawn to the requirement that new co-operative farms were to be set up only after a most careful consideration of the natural and economic conditions of each village and after a most serious political and organizational preparation.

The correct policy of the State in this sphere quickly brought about positive results in agriculture. The Fifth National Conference of Co-operative Farms (held in December 1957) noted that 3,158 co-operative farms had been founded and that 86.5 per cent of the arable land had been co-operated. The co-operative farms proved to be a form of co-operation which most successfully and fully combined the personal and public interests of the peasants. This encouraged not only the poor peasants and those with smallholdings but also the peasants of medium-size land holdings to join the co-operative farms. About 92 per cent of all the land subject to co-operation had been included in the co-operative farms by the middle of 1958.

The victory of Socialism in the Bulgarian countryside signified the total establishment of socialist production relations and public ownership over the means of production in all spheres of the national economy (in its two forms: the state one and the co-operative one). The necessary conditions were created for introducing new and up-to-date technology, for developing widespread improvement in construction, for increasing agricultural output and for raising labour productivity.

After the April 1956 Plenum of the CC of the BCP, socialist industrialization became the main trend in the country's economic policy – a policy of accelerated development of the productive forces and improvement of the people's living standards. Over this period the policy line of the BCP and its CC, headed by Todor Zhivkov, a line of consistent, accelerated industrialization, rapid development of heavy industry, and especially of the machine-building, power-generating, chemical and metallurgical industries, gained great impetus and yielded impressive results.

The policy of the country's accelerated development was given wide support by the population. Socialist activity was launched in all sectors of the national economy. Various initiatives and ideas were advanced by the work-force, and movements for high-quality products and exemplary socialist labour developed widely. In the early 1960s Bulgaria achieved tremendous success in its economic development, with the support of the international socialist division of labour, and economic co-operation with the Soviet Union and the other socialist countries.

The main tasks set for industry during the Third Five-Year Plan period (1958–62) were fulfilled two years ahead of schedule. During that period adequate preconditions were created for the further accelerated construction of the material and technical basis of Socialism in Bulgaria, which provided a graphic illustration of the cogency of its economic policy. At the end of 1958 an amalgamation of the co-operative farms was carried out: the existing 3,457 co-operative farms were merged into 972. This created favourable conditions for the mechanization of agriculture on a larger scale and for specialization of agricultural production.

The country's accelerated socio-economic development continued during the Fourth Five-Year Plan period (1961–65). Industry was definitely being established as a dominant sector of the economy in the People's Republic of Bulgaria. Moreover, it was heavy industry (machine-building, power-generation, metallurgy and the chemical industry) that were most strongly developed. Over the same period agriculture likewise showed rapid rates of development. Through specialization and concentration of agricultural production, and by means of its mechaniza tion and electrification, by the end of the fourth fifth-year plan period agricultural output increased by 23 per cent over the 1956–60 period. Through the average yields of a number of crops and the productivity of some farm animals Bulgaria reached the front rank of all countries.

The industry of socialist Bulgaria also continued developing as the most dynamic sector of the national economy during the Fifth Five-Year Plan (1966–70). Over this period alone it generated 48.8 per cent of the national income and developed at rates much higher than those in some advanced countries. The establishment of the two agro-industrial complexes (AIC) as large-scale, economically-powerful socialist organizations with a high degree of production concentration and specialization in the various sectors, marked a new stage in the socialist development of agriculture from 1970 onward.

Over the 1971–75 period efforts were channelled to further development of the productive forces, streamlining the structure of the national economy, its overall intensification and growing efficiency, and boosting scientific and technical progress. By the end of 1975 the volume of industrial and agricultural output rose by 55 and 17 per cent, respectively, as compared to 1970. The ratio of industrial to agricultural output changed from 77.1:22.9 to 81.2:18.8.

During the Seventh Five-Year Plan (1976–80) Bulgaria's economy developed even more rapidly. Industry provided the necessary conditions for the further intensification and introduction of the advances of techno-scientific progress in all sectors of the economy. The material and technical basis of agriculture expanded. In agricultural production Bulgaria is among those with the highest levels of concentration and specialization in the world today, and ranks first among the Council for Mutual Economic Assistance (CMEA) member-states in the export of farm produce. Transport, communications and other sectors of the national economy have likewise been developed further.

In the 1971–80 period alone the Bulgarian people actually built a new, second Bulgaria. This is illustrated by the fact that fixed capital in the country grew from 33,000 million in 1970 to about 77,000 million levs in 1980, and capital investments reached 52,000 million levs, or 1.6 times more than in the five preceding Five-Year Plans together. Compared to 1939, total industrial output in 1980 increased 74 times and agricultural output 2.55 times. The ratio between industrial and agricultural output in 1980 was already 84:16. Modern socialist Bulgaria produces in less than three days an amount of industrial products equal to that for the whole of 1939.

Today the People's Republic of Bulgaria, which ranks 101st in territory and 61st in population in the world, holds first place in the per capita production of electric trucks, soda ash, cigarettes and tobacco, and is among the leading producers of electronic items. It ranks first in the Balkans in the per capita production of electricity, ferrous ores, brown and lignite coal, synthetic and man-made textile fibres, wheat, sunflower seeds, sugar and other products.

The Lenin Metallurgical Works in Pernik

Transformations in the Sphere of Culture. The profound changes in Bulgaria's cultural life in the age of Socialism took place simultaneously with the socialist industrialization of the country and with the socialist reconstruction of agriculture. The main aim of the revolutionary transformations in the sphere of culture was the reorganization of overall cultural activity on Marxist-Leninist foundations, the formation of socialist consciousness and a highly developed system of intellectual life, communist education of the working people and popularization of the cultural wealth of society. The victory of the socialist revolution in Bulgaria in the sphere of culture meant inheriting the most valuable cultural achievements of the past, raising the cultural standards of the people, providing wide access to education and science, and general ascending development of culture and art.

In the initial years of the socialist revolution the efforts of the BWP (communists) and of the people's democratic rule were devoted to the eradication of the fascist ideology from the country's social, political and cultural life. The fascist organizations were dissolved, and the cultural institutions and organizations were purged of people holding fascist views. Fascist laws in the sphere of culture were revoked, publications with fascist and reactionary orientation were suspended and the organs of fascist propaganda were eliminated. Essential changes took place in the structure of the cultural institutions and organizations: a Ministry of Propaganda was set up (after 1945, Ministry of Information and Arts, and after 1948, Committee for Science, Art and Culture), the Ministry of Public Education was reorganized, and a number of cultural unions were restored and new ones were formed. New textbooks, containing no reactionary ideas, were published and the libraries and bookshops were purged of pro-fascist literature.

The Party and the people's democratic regime dealt successfully with the problems of the ideological-theoretical reorientation of the old intelligentsia, for which a number of favourable prerequisites were at hand – the strong democratic traditions established in the cultural life of society, the roots of a large part of the intellectuals in bourgeois Bulgaria in the people, the emergence and development of a robust proletarian culture in the age of Capitalism, the close links of the Communist Party with the creators of cultural values, the influence of Soviet culture, etc. This is why almost the entire country's intelligentsia rallied around the cause of the Fatherland Front Government and was most active in the process of socialist transformation in the sphere of culture. At the same time, measures were taken to create a new socialist intelligentsia by attracting young people from the working class and working peasantry to the higher, semi-higher and secondary educational establishments.

The cultural policy of the Fatherland Front eliminated illiteracy, broadened the network of educational and cultural institutions and led to a radical breakthrough in

A farm-machinery station in Strazhitsa

the development of science and the arts. Marxism-Leninism was established as the leading ideology of Bulgarian society, exercising a powerful influence over the country's general development and the intellectual development of the individual. On the basis of this new ideology a thorough reappraisal was made of the country's cultural heritage, and the guidelines for the development of Bulgarian culture – national in form and socialist in content – were outlined.

In the first half of the 1950s the development of Bulgarian culture was to a greater extent hampered by the personality cult. The April 1956 Plenum of the BCP CC was the beginning of a new stage in the development of Bulgarian socialist culture. Public education and Bulgarian scholarship received wide scope for development. A real breakthrough was made in literature and the arts. Amateur art activities became a mass movement; libraries, museums and community centres became irreplaceable centres of enlightenment and culture; the ideological and artistic level of the mass media was raised.

The April 1956 Party policy united the various generations of artists on the positions of Marxism-Leninism and around the method of socialist realism. It inspired deep faith in the young artistic intelligentsia. The Leninist principles of the Party and popular character of art was confirmed. The Party's appeal, 'Closer to the People, Closer to Life!' launched by Todor Zhivkov in 1958, resulted in greater artistic reflection of modern themes, an increased social and educational role of art and the creation of major works of art which re-create the wealth and variety of socialist life.

In May 1967 the First Congress of Bulgarian Culture was held; it elected a Committee for Art and Culture and confirmed the public-state principle in cultural activities. District, town-district and municipal councils for art and culture were elected, involving outstanding cultural figures and representatives of public organizations. Socialist democracy was broadened in the sphere of culture. Joint activities by cultural unions and state and public bodies and organizations increased. Greater opportunities were created for integration between the arts, science, education and the social sphere.

The Third Congress of Bulgarian Culture, held in May 1977, discussed and decided on fundamental questions of the make-up, tasks and activities of the Committee for Culture and improving the activities of the National Artistic Creativity, Cultural Activities and Mass Media Complex. The expansion of the public principle in the management of culture brought about the creation of a Council of Presidents of Cultural Unions and other co-ordinating councils of the Committee for Culture.

The foremost tasks in the cultural policy of the Party

and State of recent years have been those of nationwide aesthetic education, the many-sided and harmonious development of the individual and the fight against outdated prejudices in people's minds. To this end an Integrated National Programme for Popular Aesthetic Education has been drawn up and is being put into practice. Its aim is to awaken and constantly develop the creative aptitudes of the individual. In addition, a special comprehensive programme was drawn up and effected for marking the thirteen-hundredth anniversary of the foundation of the Bulgarian State.

Socialist Bulgaria's culture has achieved new heights in the sphere of science and education as well as in the sphere of the arts and the mass media. This can be seen clearly in the gradual transformation of science into a force for production; the reconstruction of the system of education in the spirit of the BCP's strategic course at the stage of building mature Socialism; the greater ideological and aesthetic maturity of literature and art; the orientation of Bulgarian artists towards the modern theme; the expansion of the public-state principle in cultural management; the enhanced role of the Press, publishing and broadcasting; intensified integration between science, education, art and the mass media; and closer ties between culture and material production.

Modern Bulgaria has a talented socialist intelligentsia which is closely bound to the policy of the BCP. Its contribution to the development of socialist culture and its unflagging efforts to raise the people's cultural level, to establish the socialist way of life and to form the new man, are outstanding.

Decisive steps have been made to expand the facilities available to Bulgarian culture. One example is sufficient to illustrate this: the Lyudmila Zhivkova People's Palace of Culture in Sofia, which was opened on the eve of the Twelfth Party Congress in 1981.

The flowering of Bulgarian culture at this stage of building mature Socialism has become a major factor in the development of society and an objective and vital necessity for all-round social advancement and the versatile development of man. Culture is becoming increasingly involved with the questions of socialist construction, helping to build a material and technical basis appropriate for developed Socialism, improving socialist social relations and raising the socialist consciousness of the working people.

The April 1956 party policy in the sphere of culture is distinguished by adherence to principle; consistency; creative quests and a stimulating atmosphere; respect and trust for men of culture; and a scientific approach to the problems of cultural development and the ideological front in the country. This policy has played an exceptionally beneficial role in developing the talent and creative forces of all generations of educationalists, scientists, artists and cultural figures, in making possible the great achievements won in all spheres of culture and the contribution made by socialist Bulgaria to the world's cultural treasure-store. The achievements of socialist Bulgaria are clear proof of Georgi Dimitrov's idea: 'In the sphere of culture there are no small or large nations. There are no adequate or inadequate people. Each people, however small it may be, is capable of making its own contribution to the treasury of world culture.'

Raising the International Prestige of Bulgaria. The foreign political situation of Bulgaria after the 9 September 1944 Uprising was extremely unstable and unsettled. As a result of the short-sighted policy pursued by the monarcho-fascist regime, Bulgaria found itself in almost complete political isolation. It had been in a state of war with the anti-Hitlerite coalition, and since 8 September 1944, also with Germany. Its diplomatic relations were severed with ten other countries, among which were France, Italy and Poland. Bulgaria started on its socialist development under extremely unfavourable international conditions; therefore one of the fundamental tasks of the people's democratic rule, established on 9 September 1944, was to defend the country's national interests and to settle its international situation.

The Programme of the Fatherland Front Government, declared on 17 September 1944, expressed the readiness and resolution of the new power to maintain 'cordial friendship with the Soviet Union', to establish 'friendly relations with the Western democracies' and to work for the 'final and fraternal agreement among the Balkan States on all contentious issues'.

In response to the request of the Bulgarian Government a Peace Treaty was signed in Moscow on 28 October 1944, between the USSR, Great Britain and the USA, on the one hand, and Bulgaria, on the other. The Fatherland Front Government expressed its full approval of the treaty and made sincere efforts to ensure its strictest observance. Indicative of this is that the Allied Control Commission situated in Bulgaria to monitor the fulfilment of the Treaty (1944–47) had no protests or complaints in this respect.

By immediately joining the anti-fascist coalition, Fatherland Front Bulgaria distanced itself from the policies of the monarcho-fascist Governments which ruled the country prior to 1944, and manifested its readiness to contribute to the establishment of new international relations in the Balkans, in Europe and in the world, based on just and democratic principles.

During the two periods of the Patriotic War of Bulgaria (1944–45) the Bulgarian People's Army suffered more than 32,000 dead and wounded, while the economic losses of Bulgaria amounted to more than 160,000 million levs, tantamount to about two thirds of its national income in 1945. The main slogan of the Fatherland Front at that time was: 'Everything for the front, everything for victory!'. The Fatherland Front Government was fully aware of the fact that Bulgaria's active participation in the final victory over fascist Germany would be of decisive significance for its standing at the forthcoming peace conference, and for continuing and intensifying the process of revolutionary transformations in society. Although the claims of the Fatherland Front Government for the recognition of the Bulgarian army as a co-belligerent

Vassil Kolarov reading the Bulgarian Government's statement on the draft peace treaty. Paris, 1946

army, and of Bulgaria as a co-belligerent state, were not recognized, the Government justifiably expected that Bulgaria's participation in the war would undoubtedly bring to an end its international isolation and help improve the country's international position.

During the war the Fatherland Front Government took a number of foreign policy steps which resulted in a gradual thaw in the country's international relations. The Bulgarian Government received the political missions of the Soviet Union and of Yugoslavia, and sent its political representatives to these countries. Diplomatic relations were restored with France (October 1944), Italy (January 1945) and Yugoslavia (May 1945). Bulgaria for its part recognized the Polish Provisional Government of National Unity (July 1945) and expressed its wish to establish diplomatic relations with it. New Ministers Plenipotentiary were appointed to Romania, Turkey, Switzerland and Sweden.

After the war against Germany ended, Bulgaria intensified its foreign policy activities. In August 1945, diplomatic relations were restored with the Soviet Union and Poland; with Czechoslovakia in October 1945; and with Albania in November 1945. The US Government, for its part, gave its consent to the appointment of General Vladimir Stoichev as Bulgaria's political representative to the US.

In 1946 and early in 1947 the Fatherland Front Government steadily pursued the policy of consolidating the country's international position and of establishing good neighbourly relations in the Balkans. It severed diplomatic relations between Bulgaria and Franco's Spain (April 1946), expressed readiness to restore diplomatic relations with Hungary (December 1946), recognized the Austrian Republic (May 1946) and the independence of Syria and Lebanon (January 1947), etc. During this period, as well as during the war, Bulgaria's interests in the countries with which it had no diplomatic relations were defended by Switzerland.

The Peace Treaty with Bulgaria was signed in Paris on 10 February 1947. By virtue of the Paris Peace Treaty, Bulgaria preserved its frontiers of 1 January 1941, recognizing the terms of the Craiova Treaty of 1940 concerning the return of Southern Dobrudja to Bulgaria, but Bulgarian claims for an outlet to the Aegean Sea were not gratified. The Bulgarian armed forces were limited to 65,500 men and the northern part of the Greek-Bulgarian frontier was demilitarized. Allied (Soviet) troops had to be withdrawn from the country within three months. Reparations to the amount of 70 million US dollars were imposed, to be paid by Bulgaria to Yugoslavia and Greece in kind over a period of eight years. No claims on

Georgi Dimitrov addresses a rally on the return of the Bulgarian delegation from the USSR. Sofia, 1948

Germany, for compensation of damage or losses sustained during the War, were granted. Bulgaria was obliged to guarantee to all allied nations most favoured nation status in trade, industry and shipping for a period of eighteen months. Control over the enactment of the Peace Treaty for the period was assigned to the Heads of Diplomatic Missions of the USSR, Great Britain and the USA in Sofia.

The Bulgarian Government's request that Bulgaria be granted the status of a co-belligerent country was not granted even in the final version of the Peace Treaty. It was pointed out in the Preamble to the Peace Treaty that Bulgaria had actively participated in the war against Germany, and it was maintained – in spite of historical facts – that this applied also to the period after the signing of the Armistice with the Governments of the USSR, Great Britain and the USA, i.e. after 28 October 1944.

The Paris Peace Treaty, though containing a number of harsh clauses, consolidated the international position of Bulgaria. The national independence and state sovereignty of the country were preserved, while its international relations acquired new dimensions. This provided even more favourable conditions and possibilities for the expansion of the revolutionary process in the country. The Government was now able to carry out in practice its Fatherland Front Programme with fewer difficulties, and the Communist Party could further consolidate its leading role in society.

The signing of the Peace Treaty in 1947 marked the beginning of a new stage in the foreign policy of Bulgaria. On 26 July 1947, Prime Minister Georgi Dimitrov sent a request to UN Secretary General Trygve Lie for Bulgaria's admission to membership of the United Nations Organization, expressing the Bulgarian Government's full agreement with the fundamental principles embedded in the UN Charter and the readiness of the Bulgarian Government to fulfil all obligations ensuing from the future membership of the People's Republic of Bulgaria in this organization. On 15 September 1947, the Allied Control Commission discontinued its activities, and by mid-December all Soviet troops were withdrawn from Bulgaria. The Governments of Great Britain and the USA recognized the Fatherland Front Government and restored diplomatic relations with Bulgaria. Diplomatic relations were also restored with Belgium, Austria, Egypt, Hungary and Holland.

In spite of the difficulties existing in relations with Greece and Turkey, Bulgaria steadfastly strove towards consolidation of peace and good neighbourly relations in the Balkans, proceeding from the principled standpoint that the interests of all Balkan nations required that the Balkan Peninsula be in every way transformed from 'a powder keg and a hotbed of war' into a 'sound mainstay and powerful factor of peace and democracy not only in this part, but also in the rest of the world'. An important factor in the creation of the new political climate in the Balkans was the victory of the people's democratic revolutions in four Balkan countries: Bulgaria, Romania, Yugoslavia and Albania.

At the end of 1947 and early in 1948 Bulgaria signed treaties of alliance with the rest of the people's democratic countries in the Balkans: Yugoslavia, Albania and Romania. In this way a sound front was created against the reactionary attempts of some Western forces, including the ruling circles in Greece and Turkey, to hamper the development of the revolutionary process in the people's democracies, to interfere in their internal affairs and threaten them with aggression.

A Treaty of Friendship, Co-operation and Mutual Assistance between the People's Republic of Bulgaria and the Union of Soviet Socialist Republics was signed in Moscow on 18 March 1948. It finally endorsed the fraternal relations between the two countries and opened up new vistas for Bulgaria's all-round development along the path of Socialism. In the ensuing months Bulgaria signed similar treaties with the rest of the people's democracies: the Czechoslovak Republic (23 April 1948), the Polish Republic (29 May 1948) and the Hungarian Republic (16 July 1948).

The signing of the treaties for friendship and mutual assistance between the USSR and the people's democracies, as well as between the people's democracies themselves, established new principles in international relations – the principles of socialist internationalism – and created the necessary conditions and prerequisites for the consolidation of the international prestige of the countries belonging to the world socialist system, for the formation of the international socialist community. The expulsion of the Yugoslav Communist Party from the Cominformbureau was followed by stagnation in Bulgarian-Yugoslav relations, notwithstanding the readiness of the People's Republic of Bulgaria to adhere to the Treaty of Friendship, Co-operation and Mutual Assistance between the two countries and to maintain friendly relations with Yugoslavia.

In September 1948, Bulgaria renewed its application for admission to membership of the United Nations Organization. It was pointed out in a letter sent by the First Deputy Prime Minister and Minister of Foreign Affairs, Vassil Kolarov, to the UN Secretary-General that the Constitution of the People's Republic of Bulgaria of 1947 endorsed the democratic institutions and democratic order in the country, that the Bulgarian Government pursued a policy of peace and co-operation with all nations, and 'fulfils in good faith the numerous and heavy obligations of military, economic and other natures, imposed on Bulgaria by the Peace Treaty'. Once again Bulgaria's request was not granted.

Toward the end of 1948, Bulgaria established relations with yet other countries of the world. On 12 November it restored diplomatic relations with Argentina, and on 30 November a declaration was published concerning the recognition of the People's Democratic Republic of Korea and of the State of Israel, later affirming the readiness of the Bulgarian Government to establish diplomatic relations between Bulgaria and these two countries.

In the 1945–48 period Bulgaria's foreign trade increased about 3.5 times. The number of countries with which Bulgaria maintained economic relations reached 35. Trade agreements were signed with 22 of them. The Soviet Union ranks first in the country's foreign trade, followed by Czechoslovakia and Poland. The foreign trade relations of Bulgaria were of great significance both for the development of the country's national economy and for its active participation in international economic co-operation, as well as in the international socialist division of labour which was beginning to take shape at that time.

In 1949 Bulgaria and some other socialist countries founded the CMEA. This enabled it to develop its economy on the basis of planned socialist development in close co-operation with the other socialist countries. Together with them, Bulgaria vigorously opposed the formation of the North Atlantic Treaty Organization (NATO) in 1949 and the Paris Agreement of 1954, emphasizing its opposition to the division of the world into military blocs and alliances. After the coming into force of the Paris Agreements, which led to the Federal Republic of Germany's incorporation into NATO, in 1955 Bulgaria stabilized its international security by signing the Warsaw Pact. This is the first mutilateral political union of the socialist countries, which involved Bulgaria even more actively in important international initiatives concerned with peace and security. In the following years, together with the other member-states of the Warsaw Treaty, Bulgaria proposed on several occasions that steps be taken towards simultaneously halting the actions of the North Atlantic and Warsaw Treaties or at least towards the simultaneous dissolution of their military organizations. In its attempts to help improve the international situation and to bring about effective disarmament, between 1955 and 1960 Bulgaria unilaterally reduced its armed forces on several occasions.

Bulgaria's admission to membership of the UN in 1955 was one recognition of the democratic and peace-loving nature of its foreign policy.

The April 1956 Plenum of the BCP CC marked a breakthrough in the country's foreign policy by establishing the collective method of party and state leadership in matters pertaining to the country's international relations. With its innovatory approach to basic foreign-policy questions, the plenum outlined a consistent and clear-cut new course. It gave a strong impetus to the country's all-round development and closening ties with the USSR and the other socialist countries and contributed to the development of socialist construction. This created even better prerequisites for Bulgarian diplomacy's active contribution to international relations. At the BCP's following congresses, particularly the Tenth, Eleventh and Twelfth Congresses (1971, 1976 and 1981), the 1956 policy line was clearly expressed, confirmed and devel-

A Bulgarian and a Soviet soldier in the battle on the Drava

oped. Much credit for the development of the country's foreign policy goes to the General Secretary of the CC of the BCP of the People's Republic of Bulgaria and State Council President Todor Zhivkov, with his direct participation in the formation, leadership and practical implementation of the country's foreign policy, his theoretical works and his appraisals of general and particular international problems.

The first treaties of friendship, co-operation and mutual assistance linking Bulgaria with the other socialist countries expired in the 1960s. In order to reflect the changes that had occurred, new treaties were signed which reflected more fully and accurately the higher stage of mutual relations. Over this period Bulgaria signed new bilateral treaties of friendship, co-operation and mutual assistance with Poland (6 April 1967), the USSR (12 May 1967), the GDR (7 September 1967), Romania (19 November 1967), Czechoslovakia (26 April 1968) and Hungary (10 July 1969). The treaties guaranteed the countries' independence and security and the unconditional respect of each other's sovereignty, equality and non-interference in each other's internal affairs. During these years, the CMEA increasingly became an organization that actively aided all-round economic integration, including production.

Bulgaria's foreign policy has achieved major successes in normalizing the country's relations with the Balkan countries, in strengthening mutual trust and developing co-operation among them. As early as the second half of the 1950s there was already a trend towards the normalization and improvement of relations with Yugoslavia, which had deteriorated after 1948. Progress was also made in relations with Turkey. In 1954 diplomatic relations with Greece, which had been broken off during World War II, were re-established. This was the beginning of continual attempts to resolve in the interest of both parties the many problems of bilateral relations which had remained unsolved since the war. Bulgaria proposed drafts for non-aggression treaties to Greece and Turkey and again underlined that Bulgaria had no territorial pretensions to any Balkan or other state. During the 1960s, through the active efforts of Bulgaria and the other states, a process of gradual *détente* in international relations started to develop and strengthen in the Balkan region.

Bulgaria plays an active role in the work of the Political Consultative Committee of the Warsaw Treaty member-states and in the other bodies of the organization. Bulgaria, along with the other member-countries, is the co-sponsor of many proposals and intiatives for strengthening international peace and security, halting the arms race and introducing effective disarmament.

Together with that of the other socialist countries,

Welcome for the Bulgarian army in Nish, Yugoslavia

Bulgarian diplomacy has played an energetic role in the process of turning the international relations of the 'cold war' into peaceful coexistence among states of different social systems, a process which began in the late 1960s and developed during the 1970s. Both at state and at sociopolitical level Bulgaria joined in the efforts to provide the necessary conditions for convening an all-European Conference. The People's Republic of Bulgaria participated in all stages of the Conference on Security and Co-operation in Europe, and many of its proposals have been reflected in the Helsinki Final Act of 1975.

Of particular importance for the improvement of Bulgaria's relations with the developed capitalist states are the high-level meetings with the leaders of these countries.

In the significant complication of the international situation and the serious threat to international peace and security that has arisen as a result of growing politico-military confrontation and the beginning of a new round in the arms race, Bulgaria is decisively opposed to the deployment of new US medium-range missiles in Europe and the view of the European continent as the possible field of a 'limited thermo-nuclear conflict'.

Bulgaria pays particular attention to the situation in the Balkans, which continues to be relatively calm. Particular progress has been made in relations among the Balkan countries. This has allowed the Bulgarian Government to present a broad and concrete programme for the further development of relations of peace and co-operation with these countries, for improving the climate of mutual trust and for making the Balkans a zone of lasting *détente*.

Together with the other socialist member-states of the Warsaw Treaty, Bulgaria continues to do all that is necessary to promote peace, understanding, trust and co-operation with all nations, to help curb the arms race, to bring about measures for effective disarmament and to establish *détente* between states with different social systems for the peaceful development of the world.

In the era of Socialism, Bulgaria has made historic gains in all spheres of life. It has built a modern industry and achieved large-scale concentration and specialization in agriculture. The social structure has been radically changed, the Bulgarian people's moral and political unity is strengthening and its market and cultural standards are rising.

Of great significance for Bulgaria's advancement along the road to Socialism is the all-round assistance of the USSR and its co-operation with and mutual assistance to the remaining countries of the socialist community. The national economy is developed on the basis of closest integration, specialization and industrial co-operation within the system of the CMEA. Bulgaria plays an active part in carrying out the Comprehensive Programme for Socialist Economic Integration among the CMEA member-countries adopted in 1971.

CHRONOLOGICAL OUTLINE

(Including events not described in the text)

470s BC The Odrysaean Kingdom was founded in Thrace with Teres as its first ruler.

45 BC Thrace was turned into a Roman province.

Early first century AD Present-day Northern Bulgaria and Dobrudja were incorporated into the Roman province of Moesia.

86 AD Moesia was divided into Upper and Lower, bordered by the Kiabros river (present-day Tsibritsa).

395 AD The Roman Empire was divided into Eastern and Western. The Thracian diocese remained in the Eastern part (Byzantium).

Late fifth and first half of sixth centuries First raids of Slavonic tribes into the present-day Bulgarian lands.

466–467 and 474 Fresh Hun invasions into Thrace.

480 An alliance was established between Byzantium and the Proto-Bulgarians.

493–499 Mass Proto-Bulgarian invasions of Thrace.

First half of sixth century Joint actions of Bulgarians and Slavs against Byzantium.

Sixth century, last decades Early Slavonic settlement in the Balkan peninsula.

Early seventh century Slavonic tribes settled in the Balkan peninsula.

630s The Proto-Bulgarian tribal union of Great Bulgaria was founded.

Mid-seventh century Proto-Bulgarian groups headed by Khan Asparuh settled in the Ongul area between the lower courses of the Dnieper and Danube rivers.

680 Proto-Bulgarian troops defeated the Byzantine army in the Ongal and settled in the Balkan peninsula. They formed an alliance with the seven Slavonic tribes and the Severians.

681 Peace treaty with Byzantium. The Bulgarian State was recognized.

686 A Proto-Bulgarian tribe headed by Kuber settled in the Ceramisian plain (today the Bitola plain).

705 A Proto-Bulgarian and Slav army undertook a campaign to Constantinople and aided Emperor Justinian II in his attempt to restore the throne. The Zagora area was annexed to Bulgaria.

716 The first commercial treaty was signed with Byzantium.

717–718 Bulgarian troops defeated the Arabs besieging Constantinople and this saved the capital of Byzantium.

792, July Bulgarian troops defeated the Byzantine army at the Markella fortress.

805 The Bulgarian army destroyed the Avar Khaganate.

809 The Bulgarians seized Sredets (present-day Sofia).

811 The Byzantine invaders were routed in the battle at the Vurbitsa Pass.

812 Bulgarians seized Messambria (present-day Nessebur) and conquered the Black Sea coast.

813 The Bulgarian army headed by Khan Krum defeated the Byzantine troops in the battle at Versinikia, took Adrianople and reached the walls of Constantinople.

815 The thirty-year peace treaty was signed between Bulgaria and Byzantium.

827 and 829 Hostilities between Bulgaria and the Frankish Empire.

838 The Rhodope mountains, the territories between the Struma and Mesta rivers, and the greater part of Macedonia were annexed to Bulgaria.

842–843 The Bulgarian State incorporated more territories of Macedonia inhabited by Slavs from the Bulgarian group (the Ohrid, Bitola and Devol areas).

864 Christianity was adopted as a state religion.

866, November Pope Nicolas I sent *Knyaz* Boris I answers to questions he had posed him. Roman mission arrived in Bulgaria.

868, 20 February Pope Adrian II met Cyril and Methodius in Rome and consecrated the Slavonic letters devised by them.

870 An independent Bulgarian Church was established.

886 The disciples of Cyril and Methodius went to Bulgaria. The beginnings were laid for organized dissemination of the Slavonic alphabet in the Bulgarian lands.

893 The Preslav Convention was held and a decision was taken to put Knyaz Simeon on the throne, to move the capital from Pliska to Preslav, to establish the Old-Bulgarian language as official literary language in Bulgaria, and to replace the Byzantine clergy by Bulgarian church servants.

896 Bulgarian army headed by Tsar Simeon heavily defeated the Byzantine troops at Bulgarophygon (Eastern Thrace).

904 A Bulgarian-Byzantine treaty was concluded, providing for Bulgaria's territorial expansion. The border with Byzantium was shifted as far as twenty kilometres north of Thessaloniki.

Chronological Outline

913 Bulgarian army headed by Tsar Simeon besieged Constantinople.

917, 20 August Byzantines routed in the battle of Acheloi.

917, early in September Byzantine army freshly defeated at Katasirti near Constantinople.

923 Tsar Simeon again besieged Constantinople.

924 Tsar Simeon undertook a campaign against Serbia and annexed it to Bulgaria.

927 A thirty-year peace treaty was signed between Bulgaria and Byzantium.

Tenth century, first half The socio-religious doctrine of Bogomilism spread in Bulgaria.

971 North-eastern Bulgaria fell under Byzantine rule. The State was preserved in the south-western territories under the brothers David, Moses, Aaron and Samuil.

976 North-eastern Bulgaria shook off Byzantine rule and state unity was restored.

986 Byzantines routed in the battle of Trayanovi vrata (Trajan Gates).

996 Campaign of Tsar Samuil to Thermopylae. The Bulgarian army was defeated by the Byzantines in the battle near the Sperchios river.

1014 The Bulgarian army was defeated in the Belassitsa battle. Fifteen thousand Bulgarian soldiers were blinded under the decree of Byzantine Emperor Basil II.

1018 Bulgaria fell under Byzantine rule.

1040 Uprising headed by Peter Delyan.

1066 Uprising in Thessaly headed by Nikulitsa Delphinus.

1072 Uprising headed by Georgi Voiteh.

1074 Uprising headed by Nestor in the Danubian region.

1078 Rebellions against Byzantine domination in Nessebur and Sofia headed by Dobromir and Leka.

1084–86 Paulician rising in Thrace.

1096 Crusaders of the First Crusade passed through Bulgaria.

1111 Bogomil preacher Basilius was burned on the stake in Constantinople.

1147 Crusaders of the Second Crusade passed through Bulgaria.

1185–87 Liberation uprising organized by brothers Peter and Assen.

1190 Byzantine campaign to Turnovo defeated in the Trevna pass.

1202–04 Fourth Crusade. The Balkan territories of Byzantium were conquered by the Crusaders. The Latin Empire was established (1204).

1204, autumn Tsar Kaloyan concluded a union with Rome.

1205, 14 April The Bulgarians headed by Tsar Kaloyan defeated the Latins at Adrianople.

1211 An Ecclesiastical Convention was held in Turnovo condemning Bogomils.

1230, 9 March Epirus army was defeated by the troops of Ivan Assen II in the battle of Klokotnitsa.

1241 Bulgaria repelled the border inroads of the Tartars during their raids of Central Europe.

1253, June A commercial and military treaty was signed between Bulgaria and Dubrovnik.

1277–80 A peasant anti-feudal uprising headed by Ivailo broke out. The Tartars driven away to the north of the Danube.

1277 Ivailo won a victory over the royal army. Tsar Konstantin Assen was killed.

1304 The Byzantine army was defeated in the battle near the Skafida river. Byzantine rule was shaken off in the territories between Stara Planina (the Balkan Range) and Mount Strandja.

1332, 18 July Byzantine army defeated in the battle at Rusocastro.

1345, 7 July Momchil defeated by a strong Byzantine-Turkish army in the battle at Peritor.

1347 A commercial treaty was signed between Bulgaria and Venice.

1365–69 The Vidin Principality was seized by Hungarians and proclaimed Vidin Bandom.

1366 The Black Sea towns were raided by the fleet of Amadeus VI of Savoy.

1371, 26 September The Christian army headed by Vulkashin and Uglesha were routed by the Ottoman Turks in the Chirmen battle.

1371–78 Thrace and the Aegean were conquered by the Ottoman Turks.

1382 Sofia fell under Ottoman rule.

1387, 27 May A commercial treaty was signed between the Dobrudja despot Ivanko and the Republic of Genoa.

1388 The Ottoman Turks seized Shumen, Ovech (present-day Provadia), Madara etc.

1393, 17 July The Turnovo Kingdom fell under Ottoman rule.

1395 The Dobrudja Principality fell under Ottoman rule.

1396, 25 September The Western Crusaders under the command of Hungarian King Sigismund, subsequently joined by Bulgarian Tsar Ivan Stratsimir, were defeated by the Ottoman Turks in the battle at Nikopol. The Vidin Kingdom fell under Ottoman rule.

1404 An uprising broke out, headed by Konstantin and Fruzhin.

1412 Risings in Vidin, Provadia, Madara, Veles and the Ovchepole area (near the town of Shtip).

After 1417 An anonymous author compiled the *Bulgarian Chronicle*.

1443, 3 November Christian army headed by Jan Hunyadi in which Bulgarians also fought clashed with Ottoman troops at Nish.

1444, 10 November The Christian coalition (Hungarians, Poles, Czechs, Wallachians and Bulgarians) clashed with Ottoman army at Varna. The Christian troops suffered defeat.

Sixteenth–seventeenth centuries Mass forcible conversions of Bulgarians to Islam.

1598 Outbreak of the Turnovo uprising organized by Todor Balina, Dionisius Rali, Pavel Djordjich and the Sorkochevich brothers.

1686 Outbreak of the second Turnovo uprising headed by Rostislav Stratsimirovich and Saveli Dubrovski.

1688 Outbreak of the Chiprovets uprising headed by

Georgi Peyachevich, Bogdan Marinov, Ivan Stanislavov and Luka Andrenin.

1689 Outbreak of the Karposh uprising (led by *Voivode* Karposh) spreading to the areas between Kyustendil and Skopje. An abortive rebellion in the Razgrad-Provadia area.

1761 Franciscan monk Blasius Kleiner wrote a history of Bulgaria in Latin.

1762 Paissii of Hilendar completed *The Slav-Bulgarian History*.

1824 Peter Beron published the *Riben bukvar* (the Fish Primer).

1829 An abortive uprising in the Sliven area organized by Georgi Mamarchov.

1829-30 Mass emigration of Bulgarian population from Eastern Bulgaria to the Ukraine, Bessarabia and Wallachia.

1832-34 Land reform was carried out in Turkey. The *spahi* system was abolished.

1834 The first Bulgarian wool factory was established in the town of Sliven.

1834, December–January 1835 Outbreak of an anti-feudal liberation uprising by the Bulgarians in the Pirot and Nish areas.

1835, 2 January The first mutual instruction school was opened in Gabrovo.

1835, 4-5 May An anti-Ottoman conspiracy was organized in the Turnovo area (known as Velchova zavera).

1836, 7 May Rising in the Berkovitsa area.

1836, August Rising in the Pirot and Belogradchik area.

1837, 14 November Rising in the Berkovitsa area.

1840, November The first Bulgarian Girls' School was opened in Pleven.

1841, 6 April Peasants anti-feudal uprising in the Nish area.

1841, 13 July First rebellion in Braila.

1842, 10 February Second Braila rebellion.

1843, September A third abortive rebellion of Bulgarian immigrants in Braila.

1844, April The first Bulgarian journal *Lyuboslovie* (Philology) was launched.

1846 The first grade school was opened in Koprivshtitsa.

1846, 20 April The first number was published of the first Bulgarian newspaper *Bulgarski Orel* (Bulgarian Eagle).

1849, 9 October The Bulgarian Memorial Church of St. Stefan was opened in Constantinople. A Bulgarian governing body was set up with the Memorial Church.

1850, 29 May A mass liberation and anti-feudal uprising broke out in North-western Bulgaria.

1851, 11 May First Celebration of the Day of Slav Enlighteners Cyril and Methodius at the Eparchial School in Plovdiv.

1853-56 The Crimean war broke out. Bulgarian voluntary units took part in combined operations with the Russian army.

1856, 12 May Liberation uprising in North-western Bulgaria.

1856, 29 July Liberation uprising in the Turnovo and Gabrovo area.

1856 The first reading clubs were established in the towns of Svishtov, Lom and Shumen.

1859, 3 May The Bolgrad Secondary School 'Sts. Cyril and Methodius' was opened (the first Bulgarian secondary school).

1860, 3 April Easter Action. The Bulgarian Church seceded from the Greek Patriarchate.

1862 The first Bulgarian legion was organized in Belgrade.

1862, 15 June Georgi Stoikov Rakovski set up the Provisional Bulgarian Headquarters in Belgrade, the first organizational centre of the Bulgarian national revolutionary movement.

1866, 20-27 March The Bulgarian Secret Central Committee was founded in Bucharest.

1866 G. S. Rakovski set up the Supreme Bulgarian Secret Civil Headquarters in Bucharest.

1867, 28 April A *haiduk* band led by Panayot Hitov crossed the Danube at Tutrakan.

1867, spring Lyuben Karavelov founded the Bulgarian Committee in Belgrade.

1867, 17 May A *haiduk* band led by Filip Totyo crossed the Danube at Svishtov.

1867, 20 July Another *haiduk* band was formed in Zaichar to pass into Bulgaria.

1867, autumn–spring of 1868 A Second Bulgarian legion was formed in Belgrade.

1868, June A Bulgarian society was established in Bucharest.

1868, 6 July A detachment led by Hadji Dimiter and Stefan Karadja crossed the Danube and set off for the Balkan Range.

1868, 11 December–24 February 1869 Vassil Levski set up the first revolutionary committees in Bulgaria.

1869 The first Bulgarian pedagogical school was opened in Shtip.

1869, 26-30 September The first Bulgarian Literary Society was founded in Braila (present-day Bulgarian Academy of Science).

1869, autumn The BRCC was founded in Bucharest.

1870, 27 February A Sultan *firman* was issued on instituting a Bulgarian Exarchate.

1872, 16 February The first Bulgarian Exarch Antim I was elected.

1872, 27 December Vassil Levski was captured by the Ottoman authorities.

1873, 6 February Vassil Levski was hanged in the vicinity of Sofia.

1875, 12 August A General Assembly of the BRCC was convened and a decision was taken on making preparations for an uprising in Bulgaria.

1875, 16 September Liberation uprising in the Stara Zagora, Shumen and Russe areas.

1875, November–December Meeting of the Giurgiu revolutionary committees at which the decision was taken to declare an uprising on 1 May 1876.

1876, 20 April The April Uprising was declared in Koprivshtitsa.

1876, 17 May The detachment led by Hristo Botev

disembarked on the Bulgarian coast at Kozlodui, Vratsa district.

1876, 18 June–20 October The Serbo-Turkish war began. Mass participation of Bulgarian volunteers.

1876, 10 July The BCCS was set up in Bucharest.

1876, August Bulgarian volunteer battalions were formed to take part in the Serbo-Turkish war.

1876, 17 November A Bulgarian volunteer unit was formed under the Commander-in-Chief of the Russian army.

1876, 11 December–8 January, 1877 A conference of the Ambassadors of the Great Powers was held in Constantinople on resolving the Eastern crisis.

1877, 12 April Russia declared war on Turkey. The Russo-Turkish war began.

1877, 17 April The Bulgarian volunteer force was formed consisting of six batallions.

1877, 15 June The Russian troops crossed the Danube at Svishtov.

1877, 19 July The Battle at Stara Zagora.

1877, 9–11 August Heroic defence of the Shipka pass by Russian soldiers and Bulgarian volunteers.

1877, 28 November Pleven was taken by the Russian army.

1877, 23 December Sofia was liberated.

1877, 27–28 December Heavy battles for the fortified Ottoman camp in the Shipka-Sheinovo area. The enemy army under the command of Veisel Pasha was defeated.

1878, 19 January Armistice was concluded between Russia and Turkey in Edirne (Adrianople).

1878, 19 February The San Stefano peace treaty was signed, ending the war.

1878, 1 June–1 July The Berlin Congress was held.

1878, 1 July The Berlin treaty was signed. Bulgaria was split.

1878, 26 November The Military School was opened in Sofia for training military commanders.

1878, 5 October–July 1879 The Bulgarian population in the Kresna-Razlog area rose in protest against the unjust decisions of the Berlin Congress.

1879, 10 February The Constituent Assembly was opened in Turnovo to draft a Bulgarian constitution.

1879, 13 March Alexander Bogoridi was appointed Governor-General of Eastern Rumelia.

1879, 14 April The Constitutional Statute (Constitution) of Eastern Rumelia was signed.

1879, 16 April The Constituent Assembly adopted the first Bulgarian Constitution (Turnovo Constitution).

1879, 17 April–16 June First Great National Assembly. Alexander Battenburg was elected Bulgarian *Kynaz* (Prince).

1879, 5 July The first Bulgarian Government formed with Todor Burmov as Prime Minister.

1879, 19 July Prince Alexander I issued a decree on setting up Bulgarian diplomatic missions in Constantinople, Bucharest and Belgrade.

1881, 27 April Coup d'état instigated by Prince Alexander I.

1881, 1 July The Second Great National Assembly voted the proposals of the Prince on suspending the Constitution and the prerogatives he demanded for a period of seven years.

1883, 2 February The Bulgarian Typographic Society was founded in Sofia – the first workers' trade organization in Bulgaria.

1883, 22 December The Law on Promoting Popular Industry was published.

1883, December The extraordinary powers regime ended.

1884, April Gavril Krustevich was appointed Governor-General of Eastern Rumelia.

1885, 10 February The BSCRC was founded in Plovdiv for carrying out the unification of Eastern Rumelia and the Bulgarian Principality.

1885, June The first issue of the *Suvremenni Pokazatel* Journal (Contemporary Indicator), the first socialist journal in Bulgaria, was published.

1885, 6 September The unification of Eastern Rumelia with the Bulgarian Principality was proclaimed.

1885, 24 October–24 March 1886 The Conference of Representatives of the Great Powers in Constantinople on Eastern Rumelia was held.

1885, 2 November Serbia attacked Bulgaria. The Serbo-Bulgarian War broke out.

1885, 5–7 November The Slivinitsa Battle. Victory of the Bulgarian army.

1885, 15 November The Bulgarian army took the town of Pirot.

1885, 7 December An Armistice was signed between Bulgaria and Serbia.

1886, 19 February The Bucharest Peace Treaty, putting an end to the Serbo-Bulgarian War, was signed.

1886, 24 March Representatives of the Great Powers in Constantinople signed the Bulgarian-Turkish Agreement recognizing the Unification of North and South Bulgaria.

1886, 9 August Prince Alexander I was dethroned.

1886, 11–18 August A counter-coup for the restoration of Prince Alexander I was carried out.

1886, 16 August A body of regents (Stefan Stambolov, Petko Rachev Slaveikov and Ivan Stranski) was appointed.

1886, 26 August Alexander I abdicated.

1886, 27 August–25 June 1887 Government of the regents Stefan Stambolov, Petko Karavelov (to 15 October 1866), Georgi Zhivkov (from 15 October 1866) and Sava Mutkurov.

1886, 6 November Russia broke off diplomatic relations with Bulgaria.

1887, 16–19 February Mutinies of Russophile army-officers in the towns of Silistra and Russe broke out.

1887, 25 June Prince Ferdinand of Saxe-Coburg-Gotha ascended the Bulgarian throne.

1887, 20 August–19 May 1894 The regime of Stefan Stambolov was established.

1888, 1 October A Higher Course in Pedagogy was started at the First Boys' Secondary School in Sofia which was transformed on 1 January 1889, into a School of Higher Learning (today's 'Kliment of Ohrid' University of Sofia).

1891, 20 July The Constituent Congress of the BSDP was held on the Buzludja Peak (the Balkan Range).

1891, September Dimiter Blagoev's book, *What Is Socialism and Can It Thrive in Bulgaria?*, the first original Bulgarian Marxist work, was published.

1892, 15 August The First Bulgarian Agrarian-Industrial Exhibition and Agrarian-Industrial Fair in the town of Plovdiv were held.

1892, 1 November The first issue of the *Rabotnik* (Worker) newspaper, the political organ of the BSDP, was published.

1893, 23 October The foundations of IMARO were laid.

1894, 3 June The Popular Pary was constituted.

1894, 4 October The first issue of the *Sotsialist* newspaper (Socialist), organ of the BWSDP was published.

1894, 5–8 November The Central Trade Union of Printers was founded.

1895, 28 January The Local Industries Encouragement Act was enforced.

1895, 19–28 March The SMAC was constituted.

1896, 2 February Diplomatic relations between Russia and Bulgaria were restored.

1896, 12 February Prince Ferdinand's ascension to the Bulgarian throne was recognized by Russia.

1896, 22 September The Democratic Party was founded.

1896, 9 December The first commercial treaty between Bulgaria and Austria-Hungary was signed.

1897, January The first issue of the *Novo Vreme* (New Times) journal, a theoretical organ of the BWSDP, was published.

1897, 16 February A commercial treaty between Bulgaria and Serbia was signed.

1897, 28 February A commercial treaty between Bulgaria and Italy was signed.

1897, 23 May A commercial treaty between Bulgaria and France was signed.

1897, 2 July A commercial treaty between Bulgaria and Russia was signed.

1897, 12 July A commercial treaty between Bulgaria and Great Britain was signed.

1897, 5 September The first issue of *Rabotnicheski Vestnik* (Workers' Newspaper), organ of the BWSDP was published.

1899, 11–13 November The Progressive Liberal Party of Dragan Tsankov was founded.

1899, 28–31 December The Constituent Congress of the BAU was held in Pleven.

1900, April–May Peasants riots broke out against the *desyatuk* (a tax collected in goods, amounting to one-tenth of the produce) in the villages of Trustenik, Russe district, Durankulak (present-day Blatnitsa) and Shabla, Tolbuhin district.

1902, 23 September A rebellion in Gorna Djumaya district (present-day Blagoevgrad district) broke out, organized by the SMAC.

1903, 2–4 January A Congress of the IMARO was held in Thessaloniki. A resolution for the organization of an uprising in Macedonia and the Adrianople region was adopted.

1903, 5 May–16 January 1908 The Second Regime of Stambolov.

1903, 6–12 July The Tenth Congress of the BWSDP was held in Russe; the left-wing socialists formed a revolutionary Marxist party, called the BWSDP (l-ws).

1903, 20 July The Uprising of St. Elijah's Day and Transfiguration Day broke out in the Bitola revolutionary district.

1903, 5–6 August The beginning of the uprising in the Adrianople revolutionary district.

1903, 28 August An *aide mémoire* of the IMARO was sent to the diplomatic envoys of the Great Powers in Sofia.

1903, 14 September The beginning of the uprising in the Ser revolutionary district.

1903, 3 October An *Entente* between Russia and Austria-Hungary on a draft for reform in Macedonia was signed in Mursteg, Austria.

1904, 26 March A Bulgarian-Turkish agreement about the normalization of relations between Bulgaria and Turkey was signed.

1904, 21–22 July The Constituent Congress of the GWTU was held in the town of Plovdiv.

1908, 6 September Alexander Malonov's Government expropriated the railway lines of the 'Eastern Railways' Company on Bulgarian territory.

1908, 22 September Proclamation of Bulgaria's independence.

1909, 6 April A Bulgarian-Turkish Protocol was signed with which Turkey recognized Bulgaria's independence.

1910, July–Februrary 1912 The Samokov Commune, the first municipal commune in Bulgaria, was constituted.

1912, 29 February A Bulgarian-Serbian Treaty for Friendship and Alliance was signed, laying the foundations of the Balkan Alliance.

1912, 16 May A Secret Treaty for Defensive Alliance was signed between Bulgaria and Greece.

1912, 17–18 August The Constituent Conference of the Union of Workers Social Domocratic Youth was held in the town of Russe.

1912, 26 September Montenegro opened hostilities against Turkey.

1912, 3 October Turkey broke off diplomatic relations with Bulgaria.

1912, 4 October Turkey declared war on the Allies.

1912, 5 October Bulgaria and Greece declared war on Turkey.

1912, 5 October–1 April 1913 Balkan War I (between Turkey and the states of the Balkan Alliance).

1912, 7 October Serbia declared war on Turkey.

1912, 5–12 October The Bulgarian army defeated the major units of the Turkish Eastern Army (the Lozengrad operation).

1912, 14–20 October The Bulgarian army dealt a heavy blow to the Turkish army (the Luleburgas-Bunarhissar operation).

1912, 20 November A truce was signed between Turkey and the Allies.

1913, 10 January The coup of the Young Turks was carried out in Constantinople; hostilities were resumed.

Part of the Tabula Peutengeriana, showing the Roman provinces in the Balkans. The map was first compiled during the second century, and last revised in the fourth century. This specimen (the only one surviving) is a parchment copy dating from the thirteenth century, now in the National Library, Vienna.

An Anglo-Saxon map, one of the oldest maps to mention the Bulgarians and to depict their State. Dating from the early eleventh century, it is now in the Manuscript Department of the British Museum Library, London.

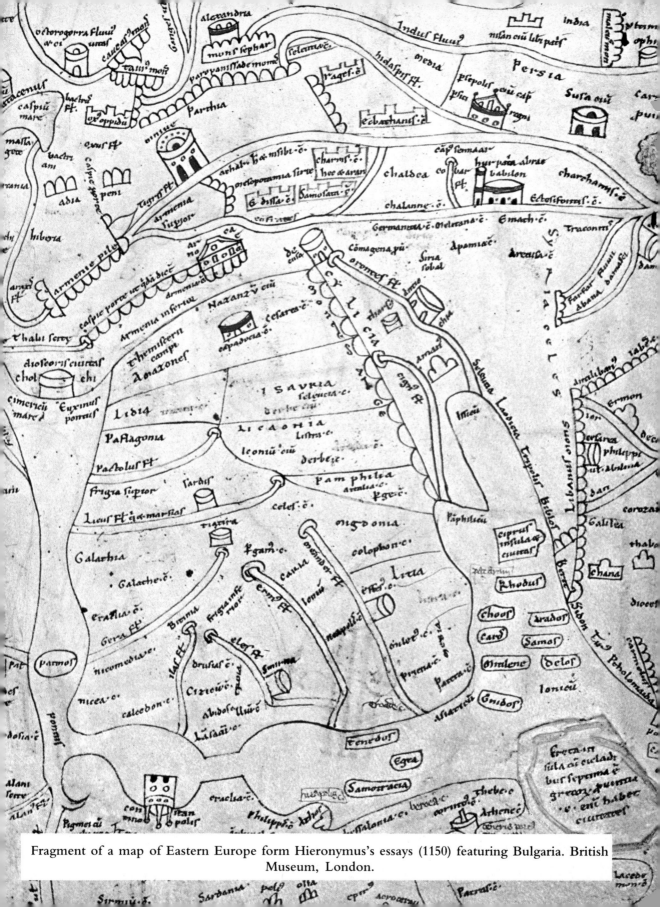

Fragment of a map of Eastern Europe form Hieronymus's essays (1150) featuring Bulgaria. British Museum, London.

Detail of a map of the world by Fra Mauro (14..), 'Bulgaria'.

9), with Bulgaria designated by the words 'Zagora' and
oges' Palace, Venice.

Copy of ethnographical map of the South Slavonic countries

th Slavonic countries, compiled by MacKenzie Irby, 1867.

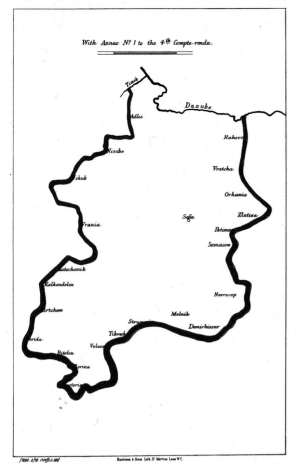

Reduced copy of maps of the Bulgarian autonomous regions determined by the Constantinople Conference (1876–77). Reproduction from Correspondence Respecting the Affairs of Turkey, N13 (1878).

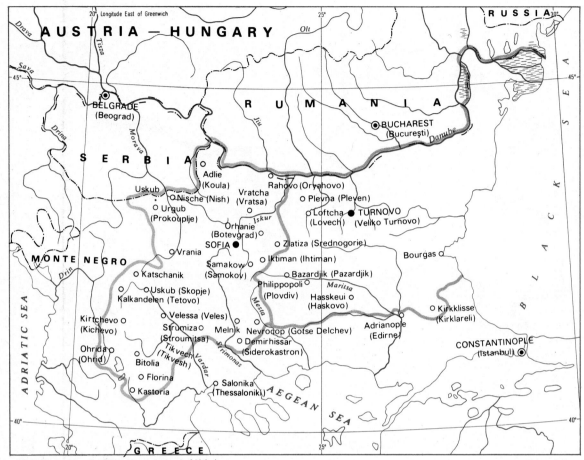

*Complier's error-should be Urgub instead of Uskub

The Bulgarian regions determined by the Constantinople Conference (1876–77) represented on an ordinary map.

1913, 12–13 March The Bulgarian army seized the fortress of Adrianople (the Edirne operation).

1913, 17 May The Peace Treaty between Bulgaria, Serbia, Greece and Montenegro on the one hand, and Turkey on the other, was signed in London.

1913, 19 May A secret Serbian-Greek military treaty directed against Bulgaria was concluded.

1913, 16 June–18 July Balkan War II (between Bulgaria and its allies joined by Turkey and Romania).

1913, 28 June Romanian troops crossed the Bulgarian frontier in the Dobrudja Region.

1913, 30 June Turkish army units crossed the Midye-Enos frontier line.

1913, 18 July An Armistice for the cessation of hostilities was signed.

1913, 28 July The Peace Treaty of Bucharest was signed between Bulgaria on the one hand, and Greece, Serbia, Romania and Montenegro on the other.

1913, 16 September The Treaty between Bulgaria and Turkey was signed in Istanbul.

1914, 28 July Austria-Hungary declared war on Serbia: the beginning of World War I.

1915, 24 August A Bulgarian-Turkish Agreement was signed on the rectification of the Bulgarian-Turkish frontier.

1915, 24 August The Bulgarian-German Treaty for Friendship and Alliance, the Military Convention between Germany, Austria-Hungary and Bulgaria, and the secret Bulgarian-German Agreement were signed.

1915, 1 October Bulgaria declared war on Serbia and joined in World War I on the side of the Central Powers.

1915, 3 October Great Britain and France declared war on Bulgaria.

1916, 21 March An act was enforced for the introduction of the Gregorian Calendar in Bulgaria.

1916, 1 April The new calendar was introduced in Bulgaria: 1 April became 14 April.

1916, 1 September Bulgaria declared war on Romania; the Third Bulgarian army started its offensive in the Dobrudja.

1917, 2 July Greece declared war on Bulgaria.

1918, 7 May A Peace Treaty with Romania was signed in Bucharest; Southern Dobrudja was given back to Bulgaria.

1918, 15–18 September The joint British-French army made a break-through in the Bulgarian front at Dobro Pole.

1918, 24 September–2 October The Soldiers' Mutiny broke out; Bulgaria was proclaimed a republic in the town of Radomir on 27 September.

1918, 29 September An Armistice was signed between Bulgaria and the *Entente* in Thessaloniki.

1918, 3 October King Ferdinand I abdicated in favour of his son, Boris III.

1919, 2–6 March The BWSDP (l-ws) took part as co-founder in the Constituent Congress of the Communist International in Moscow.

1919, 25–27 May The Twenty-Fourth Congress of the BWSDP (l-ws) was proclaimed the First Congress of the BCP (l-ws); it adopted a Programme Declaration whose underlying principles were the principles of Leninism.

1919, 27 November The Peace Treaty of Neuilly was signed; Bulgaria lost Aegean Thrace, Southern Dobrudja, the Western fringes, and the Struma region.

1919, 7 December Elections for municipal councils, in which the BCP polled the majority of votes in 22 towns and 65 villages, and municipal communes were organized.

1919, 26 December–19 February 1920 The general transport workers' strike was carried out.

1919, 29 December–3 January 1920 A general political strike was organized by the BCP in support of the transport workers.

1920, 28 March Elections for the Nineteenth Ordinary National Assembly; the BAU polled the majority of votes (followed by the BCP).

1920, 21 May–9 June 1923 Independent Government of the BAU.

1922, 19 November A Referendum for the trial of war-criminals guilty of the national catastrophes of 1913 and 1918 was carried out.

1922, 6 December Vassil Kolarov was elected Secretary-General of the Communist International.

1923, 23 March An agreement was concluded in Nish between Bulgaria and the Serbo-Croatian-Slovenian Kingdom for the settlement of issues related to the safety of the Bulgarian-Serbian frontier.

1923, 9 June A military-fascist *coup d'état* was effected by the Military League and the *Naroden Sgovor* (Popular Concord) Party.

1923, 9–14 June An uprising of the people's masses against the organizers of the coup broke out.

1923, in the night of 22–23 September The beginning of the September anti-fascist uprising, the first organized anti-fascist uprising in the world, which was cruelly suppressed by the fascist Government.

1924, 27 January The Labour Party, as a legal organization of the BCP, was founded.

1924, 2 April The BCP, the Labour Party, the Bulgarian Young Communist League, the GWTU and other organizations were banned under the State Defence Act.

1924, 7 May Dimiter Blagoev, the leader of the BCP, died in Sofia; his funeral (which took place on 11 May) turned into a powerful anti-government demonstration.

1924, 17–18 May The Vitosha Underground Conference of the BCP approved of the revolutionary course of the Party taken in September the previous year, and took a decision to continue the policy of organizing an armed uprising.

1925, 16 April A bombing incident in St. Nedelya's Church in Sofia was perpetrated by representatives of the extreme-left faction of the military organization of the Party. It was followed by mass arrests and massacres of communists, agrarians, and other non-affiliated anti-fascists, without charge or trial.

1925, July The beginning was laid for the foundation of the Independent Workers' Trade Unions.

1925, 6 October A committee for the protection of the victims of white terror in Bulgaria was set up in Paris.
1925, autumn The IMRO(u) was founded.
1925, September The DRO was founded in Vienna.
1926, 21 December Andrei Lyapchev's Government concluded foreign loans for relieving the Bulgarian refugees from Macedonia, Thrace, Dobrudja and the Western fringes.
1927, 20 February The Constituent Conference of the WP, a legal organization of the BCP, was held.
1927, 5 March The first issue of the *Rabotnichesko Delo* (Workers' Cause) newspaper, organ of the WP, was published in Sofia.
1927, 9 December An agreement was signed in Geneva between Bulgaria and Greece, under which Aegean Thrace lost a large part of its Bulgarian population.
1928, 24 May The Constituent Conference of the WYU, a legal organization of the Bulgarian Young Communists League, was held in Sofia.
1930, 30 March The Bulgarian All-National Students' Union was founded on the initiative of the Communist Party.
1931, 21 June Elections for the Twenty-second Ordinary National Assembly were carried out, won by the Popular Bloc.
1931, 29 June–19 May 1934 Government of the Popular Bloc.
1931 Mass strike action; 198 strikes were declared.
1932, 25 September The WP won the municipal elections in Sofia, cancelled by the Popular Bloc.
1932 262 strikes were declared.
1933, 21 September–23 December The Reichstag Fire Trial in Leipzig against Georgi Dimitrov, Blagoi Popov and Vassil Tanev.
1933 277 strikes were declared.
1934, 5 January An International Committee was set up in Paris for the release of Georgi Dimitrov and all the anti-fascists imprisoned in Germany.
1934, 19 May A *coup d'état* was perpetrated by the Military League and the Zveno political circle.
1934, 14 June The activities of the political parties in the country were banned.
1934, 23 June Diplomatic relations between Bulgaria and the USSR were established.
1935, 20 June The first underground issue of the *Rabotnichesko Delo* newspaper came out in Sofia.
1935, 20 August Georgi Dimitrov was elected Secretary-General of the Communist International.
1936, the beginning of February The Sixth Extended Plenum of the CC of the BCP, which worked out a platform for the creation of a popular anti-fascist front and approved of the Dimitrov course of action taken by the Party.
1936, May The political group of the non-fascist bourgeois Opposition, known as *Petorka* (the Five) was formed.
1936, July An agreement was reached between the WP and the Social Democratic Party for the establishment of a united front in their common struggle for the restoration of the Turnovo Constitution, etc. The General Youth Front and the Central Youth Constitutional Committee were founded.
1936, December The 'Five', some other democratic parties and the WP submitted an all-national petition to the King, which became a platform for the unification of the democratic forces in Bulgaria.
1937, January An agreement was reached between the democratic parties for joint participation in the municipal elections in March 1937. Constitutional committees were set up throughout the country.
1938, the beginning of A democratic coalition of the eight non-fascist parties – the WP, the 'Five' the BAU, 'Alexander Stamboliiski' wing and the Democratic Party – was formed for joint participation in the elections for the Twenty-fourth Ordinary National Assembly.
1938, March Elections for the Twenty-fourth Ordinary National Assembly; the Democratic Coalition won 60 seats.
1938, March–1940 The BCP and the WP merged into a one-class, revolutionary party under the name of the Bulgarian Workers' Party (BWP).
1939, 24 December–28 January 1940 Elections for the Twenty-fifth Ordinary National Assembly; the Democratic Coalition polled 28 per cent of the votes.
1940, 5 January A Bulgarian-Soviet Trade Agreement was signed in Moscow.
1940, 7 September An agreement between Bulgaria and Romania was signed in Craiova; Bulgaria regained Southern Dobrudja.
1940, 27 November–February 1941 A nationwide movement was organized by the BWP, in support of the Soviet Proposal (20 September 1939) for signing a Treaty for Friendship and Mutual Assistance (the Sobolev Campaign).
1941, 1 March Prime-Minister Bogdan Filov signed a Protocol at Vienna for Bulgaria's accedence to the Tripartite Pact.
1941, 6 March The CC of the BWP issued a Declaration against Bulgaria's accedence to the Tripartite Pact.
1941, 22 June Nazi Germany treacherously attacked the Soviet Union. The CC of the BWP launched an appeal for irreconcilable struggle against the fascist invaders.
1941, 24 June Decision of the Politburo of the CC of the BWP to start an armed struggle against the monarcho-fascist dictatorship and the nazi presence in Bulgaria.
1941, end of June The first partisan group in Bulgaria was set up in the region of Razlog.
1941, 23 July First broadcast of the Hristo Botev radio station, organized by the Bureau-in-Exile of the CC of the BWP.
1941, July Formation of a Central Military Commission with the CC of the BWP.
1941, 12 December The Bulgarian Government declared war on Great Britain and the USA.
1942, 17 July The Hristo Botev radio station announced the Programme of the Fatherland Front elaborated by the Bureau-in-Exile of the CC of the BWP.

Chronological Outline

1942, 31 December The first number of the Fatherland Front newspaper was published.

1942 A total of 136 Underground committees of the Fatherland Front were set up.

1943, March Decision of the Politburo of the CC of the BWP, organizing the partisan groups into NOVA, and dividing the country into twelve insurgent operative zones.

1943, 10 August The National Committee of the Fatherland Front was formed.

1943, 9 September–9 September 1944 A Regency Government was formed by Bogdan Filov, Prince Kiril and Gen. Nikola Mihov.

1943, 14 November–17 April 1944 Air raids by the British-American Air Forces on Sofia and other towns.

1944, 25 April The Chavdar partisan group was reorganized into the first partisan brigade in Bulgaria.

1944, 26 August A meeting of the CC of the BWP adopted Circular No. 4 on making immediate preparations for popular armed uprising.

1944, end of August The National Committee of the Fatherland Front issued a Manifesto to the Bulgarian people to take immediate action to form a Fatherland Front Government.

1944, 5 September USSR declared war on the monarcho-fascist Government of Bulgaria; Georgi Dimitrov issued a Directive with last instructions on the preparation for the uprising; a joint meeting of the Politburo of the CC of the BWP and representatives of the General Staff of the NOVA adopted the plan of the uprising.

1944, 6–8 September Fatherland Front rule was established in a large number of towns and villages. Prisons were taken and political prisoners were set free.

1944, 8 September Soviet army divisions crossed the Bulgarian-Romanian border; a joint meeting of the Politburo of the CC of the BWP and members of the Fatherland Front National Committee and the General Staff of the NOVA elected the members of the first Fatherland Front Government.

1944, 9 September A popular anti-fascist uprising broke out and the monarcho-fascist dictatorship was overthrown; the first Fatherland Front Government was formed.

1944, 9 September–15 September 1946 A new regency was formed by Todor Pavlov, Venelin Ganev and Tsvyatko Boboshevski.

1944, 10 September Decree of the Council of Ministers on forming a People's Guard and People's Militia.

1944, 14 September The offensive of the nazi troops was held back in the Kula area.

1944, 17 September The Fatherland Front Government declared its Programme.

1944, 18 September The first issue of the now legalized *Rabotnichesko Delo* (Workers's Cause) newspaper published.

1944, 30 September The Council of Ministers passed a Decree on putting on a people's trial of the culprits for Bulgaria's involvement in the Second World War and the crimes connected with it.

1944, September–November First stage of the Patriotic War of Bulgaria. The BPA under the operative command of the Third Ukranian Front of the Soviet army operated in the present-day Federal Republic of Macedonia and Serbia.

1944, 12 October The Council of Ministers passed a Decree abolishing all the reactionary laws of the fascist Government.

1944, 28 October An Armistice was signed in Moscow between the Government of Bulgaria and the Governments of the USSR, Great Britain and the USA.

1944, December–May 1945 The second stage of the Patriotic War of Bulgaria: The First Bulgarian army operated within the Third Ukranian Front on the territory of Yugoslavia, Hungary and Austria.

1945, 9–12 March The first Congress of the Fatherland Front was convened.

1945, 14 March A Bulgarian-Soviet Trade Agreement was signed in Moscow.

1945, 16–20 March The Constituent Congress of the GWTU was held.

1945, 14 August Diplomatic relations between Bulgaria and the Soviet Union were re-established.

1945, 4 November Georgi Dimitrov returned to Bulgaria after a twenty-two-year exile.

1945, 18 November Parliamentary elections for the Twenty-sixth Ordinary National Assembly were held, with 88 per cent of the votes for the Fatherland Front candidates.

1946, 12 March The Land Ownership Act was passed (Agrarian reform).

1946, 8 September A referendum for the abolition of the monarchy was held, with 95.63 per cent of the votes for the Republic.

1946, 15 September Bulgaria was declared a People's Republic.

1946, 27 October Elections for a Great National Assembly were held, the BWP (communists) wining 53.16 per cent of the votes.

1947, 1–5 February The first National Conference of Co-operative farms was convened.

1947, 10 February The Peace Treaty with Bulgaria was signed in Paris.

1947, 10–16 March A monetary reform was carried out.

1947, 1 April The Great National Assembly passed an Act introducing a two-year plan in the national economy.

1947, 27 November A Treaty of Friendship, Co-operation and Mutual Assistance with Yugoslavia was signed.

1947, 4 December The Great National Assembly adopted the Constitution of the People's Republic of Bulgaria.

1947, 16 December A Treaty of Friendship, Co-operation and Mutual Assistance with Albania was signed.

1947, 21–22 December The Constituent Congress of the Union of the People's Youth was convened.

1947, 24 December The Great National Assembly adopted a Decree on the nationalization of private industrial and mining enterprises.

1947, 26 December The Great National Assembly adopted a Decree on the nationalization of banks and reorganization of banking.

1948, 16 January A Treaty of Friendship, Co-operation and Mutual Assistance with Romania was signed.

1948, 2–3 February The Second Congress of the Fatherland Front was held, which adopted the Statute and Programme for the construction of Socialism; the Fatherland Front was transformed into a unified social and political organization: its coalition structure was abolished.

1948, 18 March A Treaty of Friendship, Co-operation and Mutual Assistance with the USSR was signed.

1948, 23 April A Treaty of Friendship, Co-operation and Mutual Assistance with Czechoslovakia was signed.

1948, 29 May A Treaty of Friendship, Co-operation and Mutual Assistance with Poland was signed.

1948, 16 July A Treaty of Friendship, Co-operation and Mutual Assistance with Hungary was signed.

1948, 11 August The BWSDP joined the BWP (communists).

1948, 18–25 December The Fifth Congress of the BWP was held: it adopted the general party line for the building of Socialism; the BWP (communists) changed its name to BCP.

1948, 29 December The Great National Assembly adopted the First Five-Year Economic Plan for 1945–53.

1949, 5–8 January Bulgaria participated in establishing the CMEA.

1949, 15 May The first election for People's Councils, People's Judges and jurymen were held.

1949, 23 May Decree of the Presidium of the Great National Assembly was passed on instituting Dimitrov Prizes for outstanding achievements in the field of science, art and literature, as well as for inventions and concepts.

1949, 2 July Georgi Dimitrov died.

1949, 18 December Elections for the First National Assembly were held.

1950, 23 January Vassil Kolarov died.

1950, 8–10 June The Third National Conference of the BCP was held.

1950, 25 December A Law on the Defence of Peace was adopted.

1952, 10 May A monetary reform was carried out.

1953, 20 December Elections for the Second National Assembly were held.

1954, 25 February–3 March The Sixth Congress of the BCP was convened: it adopted guidelines for the Second Five-Year Plan of 1953–57.

1955, 14 May The Warsaw Treaty of Friendship, Co-operation and Mutual Assistance was signed.

1955, 14 December Bulgaria became a member of the UN.

1956, 2–6 April The April Plenary Session of the CC of the BCP opened a new stage in the development of the Party and the country.

1957, 22 December Elections for the Third National Assembly were held.

1958, 2–7 June The Seventh Congress of the BCP was held. It adopted guidelines for the Third Five-Year Plan of 1958–62.

1959, 17 January The CC of the BCP took a decision on establishing thirty new administrative-economic districts.

1962, 25 February Elections for the Fourth National Assembly were held.

1962, 5–14 November The Eighth Congress of the Bulgarian BCP was held. It adopted guidelines for the development of the national economy during the 1961–80 period.

1966, 27 February Elections were held for the Fifth National Assembly, People's Councils, People's Judges and jurymen.

1966, 14–19 November The Ninth Congress of the BCP was held. It adopted guidelines for the Fifth Five-Year Plan of 1966–70.

1968, 1 July Bulgaria signed the Treaty on Non-Proliferation of Nuclear Weapons.

1970, 21 October The Council of Ministers adopted Provisional Rules and Regulations for the AIC.

1971, 20–25 April The Tenth Congress of the BCP convened; it adopted a Programme for the Construction of a Developed Socialist Society in the People's Republic of Bulgaria, guidelines for the Sixth Five-Year Plan of 1971–75, and amendments and supplements to the Party Statute.

1971, 16 May A Referendum on the new Constitution of the People's Republic of Bulgaria was held.

1971, 18 May The new Constitution of the People's Republic of Bulgaria was adopted officially at a Ceremonial Meeting of the Fifth National Assembly.

1971, 27 June Elections for the Sixth National Assembly were held.

1971, 7 July A State Council of the People's Republic of Bulgaria was set up; Todor Zhivkov was elected as its President.

1973, 13 March The Council of Ministers and the Central Council of the Bulgarian Trade Unions passed a joint decree on the introduction of a five-day working week and reduced working hours.

1974, 20–22 March A National Party Conference on increasing productivity was convened.

1975, 30 July–1 August The People's Republic of Bulgaria signed the Final Act of the All-European Conference on Peace and Security in the World in Helsinki.

1976, 22 March–2 April The Eleventh Congress of the BCP convened; it adopted the Guidelines for Social and Economic Development of the People's Republic of Bulgaria during the Seventh Five-Year Plan of 1976–80.

1976, 30 May Elections for the Seventh National Assembly.

1978, 21–22 April National Party Conference on the improvement of the socialist organization of labour and the planning principle in economic management.

1978, 1 December The Seventh National Assembly passed an Amendment to the People's Councils Law; the settlement systems were established as the basic political, socio-economic and administrative units.

1981, 31 March–4 April The Twelfth Congress of the BCP convened. Theses for the socio-economic development of the People's Republic of Bulgaria during the Eighth Five-Year Plan of 1981–85 and up to 1990 adopted.

1984, 22–23 March A National Party Conference was held and a long-term programme was adopted on improving quality in all spheres of social life.

RULERS

Khan Asparuh	(680–700)
Khan Tervel	(700–721)
Khan Kormisosh	(721–738)
Khan Sevar	(738–753/4)
Khan Vineh	(753/4–760)
Khan Telets	(760–763)
Khan Sabin	(763–766)
Khan Umor	(August–September 766)
Khan Toktu	(766–767)
Khan Pagan	767–768)
Khan Telerig	(768–777)
Khan Kardam	(777–803)
Khan Krum	(803–814)
Khan Omurtag	(814–831)
Khan Malamir	(831–836)
Khan Pressian	(836–852)
Knyaz Boris I	(852–889)
Knyaz Vladimir	(889–893)
Tsar Simeon	(893–927)
Tsar Peter I	(927–970)
Tsar Boris II	(970–971)
Tsar Roman	(977–997)
Tsar Samuil	(997–1014)
Tsar Gavril-Radomir	(1014–15)
Tsar Ivan-Vladislav	(1015–18)

* * *

Tsar Peter II	(1185–87 and 1196–97)
Tsar Assen I	(1187–96)
Tsar Kaloyan	(1197–1207)
Tsar Boril	(1207–18)
Tsar Ivan Assen II	(1218–41)
Tsar Kaliman I	(1241–46)
Tsar Mihail II Assen	(1246–56)
Tsar Kaliman II	(1256–57)
Tsar Konstantin Assen	(1257–77)
Tsar Ivailo	(1278–79)
Tsar Ivan Assen III	(1279)
Tsar Georgi Terter I	(1279–92)
Tsar Smilets	(1292–98)
Tsar Chaka	(1300)
Tsar Todor-Svetoslav	(1300–21)
Tsar Georgi Terter II	(1321–22)
Tsar Mihail III Shishman	(1323–30)
Tsar Ivan-Stefan	(1330–31)
Tsar Ivan-Alexander	(1331–71)
Tsar Ivan Shishman	(1371–95)
Tsar Ivan-Stratsimir	(1371–96)

* * *

Prince Alexander I	(17.04.1879–26.08.1886)
Ferdinand I of Saxe-Coburg-Gotha, Prince	(25.06.1887–22.09.1908)
and King	(22.09.1908–3.10.1918)
Tsar Boris III	(3.10.1918–28.08.1943)
Tsar Simeon II	(28.08.1943–15.09.1946

GOVERNMENTS

From the Liberation of 1878 to 9 September 1944

1. Todor Burmov (5.7.1879–24.11.1879)
2. Metropolitan Kliment (Vassil Drumev) (24.11.1879–24.3.1880)
3. Dragan Tsankov (24.3.–28.11.1880)
4. Petko Karavelov (28.11.1880–27.4.1881)
5. Johann Kazimir Ernrot (24.4.–1.7.1881)
6. No Prime-Minister (1.7.1881–23.6.1882)
7. Leonid Sobolev (23.6.1882–7.9.1883)
8. Dragan Tsankov (7.9.1883–29.6.1884)
9. Petko Karavelov (29.6.1884–9.8.1886)
10. Provisional Government headed by Metropolitan Kliment (9–12.8.1886)
11. Provisional Government headed by Petko Karavelov (12–16.8.1886)
12. Vassil Radoslavov (15.8.1886–28.6.1887)
13. Konstantin Stoilov (28.6–20.8.1887)
14. Stefan Stambolov (20.8.1887–19.5.1894)
15. Konstantin Stoilov (19.5.–9.12.1894)
16. Konstantin Stoilov (9.12.1894–18.1.1899)
17. Dimiter Grekov (18.1–1.10.1899)
18. Todor Ivanchov (1.10.1899–27.11.1900)
19. Todor Ivanchov (27.11.1900–9.1.1901)
20. Racho Petrov (9.1.–19.2.1901)
21. Petko Karavelov (19.2–21.12.1901)
22. Stoyan Danev (21.12.1901–5.5.1903)
23. Racho Petrov (5.5.1903–22.10.1906)
24. Dimiter Petkov (22.10.1906–26.2.1907)
25. Dimiter Stanchov (27.2.–3.3.1907)
26. Peter Gudev (3.3.1907–16.1.1908)
27. Alexander Malinov (16.1.1908–5.9.1910)
28. Alexander Malinov (5.9.1910–16.3.1911)
29. Ivan Evstratiev Geshov (16.3.1911–1.6.1913)
30. Stoyan Danev (1.6.–4.7.1913)
31. Vassil Radoslavov (4.7.–20.12.1913)
32. Vassil Radoslavov (23.12.1913–21.6.1918)
33. Alexander Malinov (21.6.–17.10.1918)
34. Alexander Malinov (17.10.–28.11.1918)
35. Todor Todorov (28.11.1918–7.5.1919)
36. Todor Todorov (7.5.–6.10.1919)
37. Alexander Stamboliiski (6.10.1919–21.5.1920)
38. Alexander Stamboliiski (21.5.1920–9.6.1923)
39. Alexander Tsankov (9.6–22.9.1923)
40. Alexander Tsankov (22.9.1923–4.1.1926)
41. Andrei Lyapchev (4.1.1926–12.9.1928)
42. Andrei Lyapchev (12.9.1928–15.5.1930)
43. Andrei Lyapchev (15.5.1930–29.6.1931)
44. Alexander Malinov (29.6–12.10.1931)
45. Nikola Mushanov (12.10.1931–7.9.1932)
46. Nikola Mushanov (7.9.–31.12.1932)
47. Nikola Mushanov (31.12.1932–19.5.1934)
48. Kimon Georgiev (19.5.1934–22.1.1935)
49. Pencho Zlatev (22.1.–21.4.1935)
50. Andrei Toshev (21.4.–23.11.1935)
51. Georgi Kyoseivanov (23.11.1935–4.7.1936)
52. Georgi Kyoseivanov (4.7.1936–14.1..1938)
53. Gerogi Kyoseivanov (14.11.1938–23.10.1939)
54. Georgi Kyoseivanov (23.10.1939–15.2.1940)
55. Bogdan Filov (15.2.1940–11.4.1942)
56. Bogdan Filov (11.4.1942–14.9.1943)
57. Dobri Bozhilov (14.9.1943–31.5.1944)
58. Ivan Bagryanov (1.6.–1.9.1944)
59. Konstantin Muraviev (2–8.9.1944)

Since 9 September 1944

1. (60) Kimon Georgiev (9.9.1944–31.3.1946)
2. (61) Kimon Georgiev (31.3.1946–23.11.1946)
3. (62) Georgi Dimitrov (23.11.1946–10.12.1947)
4. (63) Georgi Dimitrov (11.12.1947–20.7.1949)
5. (64) Vassil Kolarov (20.7.1949–18.1.1950)
6. (65) Vassil Kolarov (19.1.–1.2.1950)
7. (66) Vulko Chervenkov (1.2.1950–15.1.1954)
8. (67) Vulko Chervenkov (16.1.1954–17.4.1956)
9. (68) Anton Yugov (17.4.1956–13.1.1958)
10. (69) Anton Yugov (14.1.1958–15.3.1962)
11. (70) Anton Yugov (16.3.–19.11.1962)
12. (71) Todor Zhivkov (19.11.1962–11.3.1966)
13. (72) Todor Zhivkov (12.3.1966–7.7.1971)
14. (73) Stanko Todorov (8.7.1971–15.6.1976)
15. (74) Stanko Todorov (16.6.1976–16.6.1981)
16. (75) Grisha Filipov (since 16.6.1981)

Governments

Vassil Kolarov

Georgi Dimitrov

Todor Zhivkov

SELECTED BIBLIOGRAPHY

ANGELOV D. *Les Balkans au Moyen Age: la Bulgarie des Bogomiles aux turcs.* London, 1978.
ANGELOV, D. *Bogomilstvoto v Bulgaria* (Bogomilism in Bulgaria). 3rd enl. and rev. edn. Sofia, 1969. 1st edn., 1947.
ANGELOV, D. *Obrazuvane na bulgarskata narodnost* (Formation of the Bulgarian Nation). Sofia, 1971.
ANGELOV, D. *Obshtestvo i obshtestvena misul v srednovekovna Bulgaria, IX–XIV vek* (Society and Social Thought in Medieval Bulgaria, Ninth–Fourteenth Centuries). Sofia, 1970.
Aprilskoto vustanie 1876 (The April Uprising of 1876). Vols 1–3 (Documents). Ed. A. Burmov. Sofia, 1954–56.
Arhiv na Vuzrazhdaneto (Archives of the National Revival Period). Ed. D. Strashimorov. *Dokumenti po politicheskoto vuzrazhdane* (Documents on the Political Revival). Vol 2. *Dokumenti po Suedinenieto* (Documents on the Unification of Eastern Rumelia and the Bulgarian Principality). Sofia, 1908.
ARNAUDOV, M. *Tvortsi na bulgarskoto vuzrazhdane* (Architects of the Bulgarian National Revival). Vol. 1. *Purvi vuzrozhdentsi* (The First Enlighteners). Sofia, 1969.
Assimilatorska politika na turskite zavoevateli. Sbornick ot dokumenti za pomohamedanchvaniya i poturchvaniya (The Assimilatory Policies of the Ottoman Conquerors. Collection of Documents on Forcible Conversions to Islam and Turkicization. Fifteenth–Nineteenth Centuries). Comp. and ed. by P. Petrov. 2nd edn, Sofia, 1964. 1st edn, 1962.
BEROV, L. *Dvizhenieto na tsenite na Balkanite prez XVI–XIX vek i evropeiskata revolyutsiya na tsenite* (Price Fluctuations in the Balkans in the Sixteenth–Nineteenth Centuries and the European Price Revolution). Sofia, 1976.
BEROV, L. *Ikonomicheskoto razvitie na Bulgaria prez vekovete* (Bulgaria's Economic Development through the Ages). Sofia, 1974.
BESHEVELIEV, V. *Bulgarisch-byzantinische Aufsätze.* London, 1978.
BESHEVLIEV, V. *Prabulgarski epigrafski pametnitsi* (Proto-Bulgarian Epigraphy). Sofia, 1981.
BESHEVLIEV, V. *Purvobulgarski nadpisi* (Early Bulgarian Inscriptions). Compl. rev. and enl. edn., Sofia, 1979. German translation: *Die Protobulgarischen Inschriften.* Berlin, 1963.
BLAGOEV, D. *Prinos kum istoriyata na sotsializma v Bulgaria* (Contribution to the History of Socialism in Bulgaria). Sofia, 1956. Numerous edns.
BLAGOEV, D. *Biografia* (Biography). Sofia, 1979.
BOZHILOV, I. *Tsar Simeon Veliki (893–927). Zlatniyat vek na Srednovekovna Bulgaria* (Tsar Simeon the Great [893–827]. The Golden Age of Medieval Bulgaria). Sofia, 1983.
BOZHINOV, V. *Bulgarskata prosveta v Makedonia i Odrinska Trakia, 1878–1913* (Bulgarian Letters in Macedonia and Adrianople, Thrace, 1878–1913). Sofia, 1982.
BOZHINOV, V. *Politicheskata kriza v Bulgaria prez 1943–44* (The Political Crisis in Bulgaria in 1943–44). Sofia, 1957.
BOZHINOV, V. *Zashtitata na natsionalnata nezavisimost na Bulgaria. 1944–47* (The Defence of Bulgaria's National Independence, 1944–47). Sofia, 1962.
Bulgaria prez vekovete (Bulgaria through the Centuries). Sofia, 1982.
Bulgaro-rumunski vruzki i otnoshenia prez vekovete. Izsledvania XII–XIX vek (Bulgarian-Romanian Ties and Relations through the Centuries. Studies, Twelfth–Nineteenth Centuries). Sofia, 1956.
Bulgaro-Suvetski otnosheniya i vruzki. Dokumenti i materiali (Bulgaro-Soviet Relations and Ties. Documents and materials). Vols. 1–2. Sofia, 1976–81.
Bulgarskata komunisticheska partiya. Dokumenti na tsentralnite rukovodni organi (The Bulgarian Communist Party. Documents of the Central Governing Bodies). Ed. I. Yotov. Vols. 1–5. Sofia, 1972–81.
BURMOV, A. *Izbrani proizvedeniya* (Selected Works). Vols. 1–3. Eds. Zh. Natan, et al. Sofia, 1968–76.
BUZHASHKI, E. *Dimiter Blagoev i pobedata na marksizma v bulgarskoto sotsialistichesko dvizhenie*, 1885–1903 (Dimiter Blagoev and the Victory of Marxism in the Bulgaian Socialist Movement, 1885–1903). Sofia, 1960.
CHICHIKOVA, M. *Sevtopolis* (Seuthopolis). Sofia, 1970.
CVETKOVA, B. *Les institutions ottomanes en Europe.* Wiesbaden, 1978.
DAMYANOV, S. *Frantsia i bulgarskata natsionalna revolyutsia* (France and the Bulgarian National Revolution). Sofia, 1968.
DAMYANOV, S. *Frenskata politika na Balkanite, 1828–53* (French Policies in the Balkans during the 1928–53 Period). Sofia, 1977.
DANOV, H. *Drevna Trakiya* (Ancient Thrace). Sofia, 1969. Published in German: *Altthrakien*, Berlin, 1976.
DECHEV, D. *Die thrakischen Sprachreste.* Wien, 1957.
DIMITROV I. *Anglia i Bulgaria (1938–4). Navecherieto i nachaloto na Vtorata svetovna voina* (Great Britain and Bulgaria 1938–41. The Eve and Beginning of World War II). Sofia, 1983.
DIMITROV, I. *Bulgaro-italianski politicheski otnoshenia 1922–43* (Bulgarian-Italian Political Relations between 1922 and 1943). Sofia, 1976.
DIMITROV, I. *Bulgarskata demokratichna obshtestvenost, fashizmut i voinata 1934–39* (The Bulgarian Democratic Public, Fascism and War, 1934–39). Sofia, 1976.
DIMITROV, I. *Burzhuaznata opositsia v Bulgaria, 1939–44* The Bourgeois Opposition in Bulgaria, 1939–44). Sofia, 1969.
DOINOV, D. *Kresnensko-Razlozhkoto vustanie 1878–79* (The Kresna-Razlog Uprising of 1878–79). Sofia, 1979.
Dokumenti za bulgarskata istoriya (Documents on Bulgarian History). Vols. 1–6. Sofia, 1931–51.
Dokumenti za Bulgarskoto vuzrazhdane ot arhiva na Stefan I. Verkovich (Documents on the Bulgarian Revival Period from the Archives of Stefan Verković). 1860–93. Comps. D. Veleva and T. Vulov. Sofia, 1969.
Dokumenti po obyavyavane na nezavisimostia na Bulgaria – 1908. Iz tainiya kabinet na Knyaz Ferdinand (Documents on the Declaration of Bulgarian Independence – 1908. From the Secret Files of Knyaz Ferdinand). Comps. T. Todorov and E. Statelova. Sofia, 1968.
DRINOV, M. *Izbrani suchineniya* (Selected Works). In 2 volumes. Ed. and pref. by I. Duchev. Sofia, 1971.
DUYCHEV, I. *Bulgarsko srednovekovie* (Medieval Bulgaria). Sofia, 1972.

DUYCHEV, I. *Iz starata bulgarska knizhnina* (Excerpts from Old Bulgarian Letters). Books 1–2. Sofia, 1940–44.

DUYCHEV, I. *Medioevo bizantino-slavo. Vol. 1–3. Saggi di storia politica e culturale.* Roma, 1965–71.

DUYCHEV, I. *Slavia Orthodoxa. Collected Studies in the History of the Slavic Middle Age.* With a Preface by I. Sevchenko. London, 1970.

Edinodeistvieto na bulgarskiya narod s drugite balkanski narodi v antifashistkata borba, 1940–45 (Common Actions of the Bulgarian and other Balkan Peoples in the Anti-Fascist Struggle, 1940–45). *Dokumenti i materiali* (Documents and Materials). Eds. N. Todorov, et al. Sofia, 1974.

Evreiski izvori za obshtestveno-ikonomicheskoto razvitie na balkanskite zemi (Jewish Sources on Socio-Economic Development in the Balkans). Comp. and transl. with notes by A. Hananel and E. Eshkenazi. Vols. 1–2. Sofia, 1958–60.

FOL, A. *Demografska i sotsialna struktura na Drevna Trakia prez I hilyadoletie pr.n.e.* (Demographic and Social Structure of Ancient Thrace in the First Millenium BC). Sofia, 1970.

FOL, A. *Politcheska istoriya na trakite. Kraya na hilyadoletie do kraya na peti vek predi nashata era* (A Political History of the Thracians from the End of the Second Millenium to the End of the Fifth Century BC). Sofia, 1972.

GANDEV, H. *Bulgarskata narodnost prez XV vek* (The Bulgarian Nationality in the Fifteenth Century). Sofia, 1972.

GANDEV, H. *Problemi na Bulgarskoto vuzrazhdane* (Problems of the Bulgarian Revival Period). Sofia, 1976.

GENCHEV, N. *Frantsia i bulgarskoto duhovno vuzrazhdane* (France and the Bulgarian Spiritual Revival). Sofia, 1979.

GENOV, G. *Mezhdunarodni aktove i dogovori zasyagashti Bulgaria* (International Acts and Treaties concerning Bulgaria). Sofia, 1940.

GENOV, Ts. *Osvoboditelnata voina 1877–78* (The Liberation War of 1877–78). Sofia, 1978.

GEORGIEV, V. *Burzhoaznite i drebnoburzhoaznite partii v Bulgaria 1934–39* (The Bourgeois and Petty Bourgeois Parties in Bulgaria in 1934–39). Sofia, 1971.

GOTSEV, D. *Natsionalno-osvoboditelnata borba v Makedonia, 1912–15* (The National Liberation Struggle in Macedonia, 1912–15). Sofia, 1981.

Grutski izvori za bulgarskata istoriya (Greek Sources on Bulgarian History). Vols 1–11. Sofia, 1954–83.

GYUZELEV, V. *Bulgarskata durzhavnost v aktove i dokumenti* (The Bulgarian State System in Acts and Documents). Sofia, 1981.

GYUZELEV, V. *Knyaz Boris Purvi. Bulgaria prez vtorata polovina na XIX vek* (Knyaz Boris I. Bulgaria during the Second Half of the Nineteenth Century). Sofia, 1969.

GYUZELEV, V. *Srednovekovna Bulgaria v svetlinata na novi izvori* (Medieval Bulgaria in the Light of New Sources). Sofia, 1981.

HRISTOV, H. *Agrarnite otnosheniya v Makedonia prez XIX i nachaloto na XX vek* (Agrarian Relations in Macedonia in the Ninteenth and early Twentieth Centuries). Sofia, 1964.

HRISTOV, H. *Agrarniyat vupros v bulgarskata natsionalna revolyutsia* (The Agrarian Problem in the Bulgarian National Revolution). Sofia, 1976.

HRISTOV, H. *Osvobozhdenieto na Bulgaria i politikata na zapadnite durzhavi 1876–78* (The Liberation of Bulgaria and the Policy of the Western Countries in 1876–78). Sofia, 1968.

HRISTOV, H. *Paisii Hilendarski. Negovoto vreme, zhiznen put i delo* (Paisii of Hilendar. His Time, Life and Work). Sofia, 1972.

Ilindensko-Preobrazhensko vustanie ot 1903 godina (The Uprising on St. Elijah's Day and on Transfiguration Day in 1903). Eds. H. Hristov, et al. Sofia, 1983.

IRECHEK, K. *Istoriya na bulgarite* (History of the Bulgarians). Ed. V. N. Zlatarski. Sofia, 1929.

IRECHEK, K. *Knyazhestvo Bulgaria. Negovata povurhnina, priroda, naselenie, duhovna kultura, upravlenie i noveisha istoriya. Chast 1. Bulgarskata durzhava* (The Bulgarian Principality. Its Area, Nature, Population, Spiritual Culture, Government and Recent History. Part 1. The Bulgarian State). Trans. by E. Karavelova. Plovdiv, 1899.

Istoriya, izkustvo i kultura na srednovekovna Bulgaria. Sbornik statii (History, Art and Culture of Medieval Bulgaria. Collection of Papers). Comp. by V. Gyuzelev. Sofia, 1981.

Istoriya na antifashistkata borba v Bulgaria, 1939–44 (History of the Anti-Fascist Struggle in Bulgaria, 1939–44). In 2 Vols. Vols. 1–2. Eds. K. Vasilev, et al. Sofia, 1976.

Istoriya na Bulgaria (History of Bulgaria). In 14 Vols. Vols. 1–4. Ed. D. Kossev, et al. Sofia, 1979–83.

Istoriya na Bulgaria (History of Bulgaria). 2nd rev. edn. in 3 Vols. Comp. and ed. by D. Kossev, et al. Vols. 1–3. Sofia, 1961–64.

Istoriya na Bulgarskata kumunisticheska partia (History of the Bulgarian Communist Party). Ed. in chief R. Avramov. 4th edn. Sofia, 1981. 1st edn. 1970.

Istoriya na otechestvenata voina n Bulgaria 1944–45 (History of the Patriotic War of Bulgaria 1944–45). In 4 Vols. Vols. 1–3. Sofia, 1981–83.

Istoriya na profsuyuznoto dvizhenie v Bulgaria (History of the Trade Union Movement in Bulgaria). Sofia, 1980.

Istoriya na Srubsko-bulgarskata voina 1885 (History of the Serbo-Bulgarian War of 1885). Sofia, 1925.

ISUSOV, M. *Politicheskite partii v Bulgaria 1944–48* (Political Parties in Bulgaria in 1944–48). Sofia, 1978.

IVANOV, Y. *Bogomilski knigi i legendi* (Bogomil Books and Legends). Sofia, 1970. Reproduced from the 1925 edn. French translation: *Livres et legendes bogomiles (Au sources du catharism).* Paris, 1976.

IVANOV, Y. *Bulgarite v Makedonia. Izdirvaniya i dokumenti za tyahnoto poteklo, esik i narodnost s etnografska karta i statistika* (The Bulgarians in Macedonia. Studies and Documents on their Genealogy, Language and Ethnic Identity with Ethnographic Maps and Statistical Data). 2nd enl. edn. Sofia, 1917. 1st edn. 1915.

IVANOV, y. *Bulgarski starini iz Makedonia* (Bulgarian Historical Monuments in Macedonia). Sofia, 1970. Reproduced from the 1931 edn.

IVANOV, Y. *Severna Makedonia. Istoricheski izdirvania* (Northern Macedonia. Historical Studies). Sofia, 1906.

Izvori za istoriyata na Trakia i trakite (Sources on the History of Thrace and the Thracians). Vol. 1. Comp. by V. Velkov, Z. Gocheva, V. Tupkova-Zaimova. Sofia, 1981.

Izvori za starata istoriya i geografiya na Trakia i Makedonia (Sources on the Ancient History and Geography of Thrace and Macedonia). 2nd enl. edn. Sel. and transl. by G. Katsarov, et al. Sofia, 1949. 1st edn. 1915.

JURUKOVA, J. *Coins of the Ancient Thracians.* Oxford, 1976.

KACAROV, G. *Beiträge zur Kulturgeschichte der Thraker.* Sarajevo, 1916.

KESYAKOV, B. *Prinos kum diplomaticheskata istoriya na Bulgaria* (On the History of Bulgarian Diplomacy). Vol. 1. 1878–1925. Treaties, Conventions, Agreements, Protocols and other Documents and Diplomatic Acts. Pref. by S. Balamezov. Sofia, 1925.

KIRIL PATRIARCH BULGARSKI *Bulgarskata ekzarchia v Odrinsko i Makedonia sled Osvoboditelnata voina 1877–78* (The Bulgarian Exarchate in the Adrianople Region and Macedonia after the Liberation War of 1877–78). Vol. 1. 1878–85. Books 1–2. Sofia, 1969–70.

KIRIL PATRIARCH BULGARSKI. *Graf N. P. Ignatiev i bulgarskiyat tsurkoven vupros. Izsledvane i dokumenti, 1864–72.* (Count N. P. Ignatiev and the Bulgarian Church Question. Studies and Documents, 1864–72). Sofia, 1958.

KIRIL PATRIARCH BULGARSKI. *Katolicheskata propaganda sred bulgarite prez vtorata polovina na XIX vek* (Catholic Propaganda among the Bulgarians in the Second Half of the Nineteenth Century). Vol. 1. 1859–65. Sofia, 1962.

KIRIL PATRIARCH BULGARSKI. *Prinos kum bulgarskiya tsurkoven vupros. Dokumenti ot avstriiskoto konsulstvo v Solun* (On the Bulgarian Church Question. Documents from the Austrian Consulate in Thessaloniki). Sofia, 1961.

KIRIL PARTRIACH BULGARSKI. *Prinos kum uniatstvoto v Makedonia sled Osvoboditelnata voina, 1879–95. Dokladi na frenskite konsuli v. Solun* (On Uniatism in Macedonia following the Liberation War, 1879–95). Reports of French Consuls in Thessaloniki). Sofia, 1968.

Selected Bibliography

KIRIL PATRIARCH BULGARSKI. *Suprotivata sreshtu Berlinskiya dogover. Kresnenskoto vustanie* (The Resistance against the Berlin Treaty. The Kresna Uprising). Sofia, 1955.

KLEINER, B. *Istoriya na Bulgaria sustavena v 1761 g.* (History of Bulgaria. Comp. in 1761). Pref. by I. Duychev. Sofia, 1977.

KOLEDAROV, P. *Politcheska geografia na srednovekovnata bulgarska durzhava. Ch. 1 Ot 681 do 1018 g. Chast 1* (Political Geography of the Medieval Bulgarian State in 681–1018. Part 1). Sofia, 1979.

KOLEVA, T. *BKP i mezhdunarodnoto kumunistichesko dvizhenie, 1914–24* (The Bulgarian Communist Party and the International Communist Movement, 1919–24). Sofia, 1973.

KOSSEV, D. *Kum istoriyata na revolyutsionnoto dvizhenie v Bulgaria prez 1867–71* (Towards the History of the Revolutionary Movement in Bulgaria in 1867–71). Sofia, 1958.

KOSSEV, D. *Rusiya, Frantsia i bulgarskoto osvoboditelno dvizhenie 1860–69* (Russia, France and the Bulgarian Liberation Movement. 1860–69). Sofia, 1978.

KOSSEV, D. *Septemvriiskoto vustanie 1923* (The September Uprising of 1923). 2nd edn. Sofia, 1973. 1st edn. 1954.

KOSSEV, K. *Bismarck, Iztochnivat vupros i bulgarskoto osvobozhdenie, 1856–78* (Bismarck, the Eastern Question and Bulgaria's Liberation 1856–78). Sofia, 1978.

KRAOHUNOV, K. *Diplomaticheska istoriya na Bulgaria, 1886–1915. T. 1. Velikite durzhavi i Bulgaria prez 1886–87* (A Diplomatic History of Bulgaria 1886–1915. Vol. 1. The Great Powers and Bulgaria in 1886–87). Sofia, 1928.

Latinski izvori za bulgarskata istoriya (Latin Sources on Bulgarian History). Vols. 1–4. Sofia, 1958–81.

LISHEV, S. *Bulgarskiyat srednovekoven grad* (The Bulgarian Medieval Town). Sofia, 1970.

Makedonia. Sbornik ot dokumenti i materiali (Macedonia. Collection of Documents and Materials). Eds. D. Kossev, et al. Sofia, 1978. English translation: *Macedonia, Documents and Materials*. Sofia, 1978. Russian translation: *Makedonia. Sbornik dokumentov i materialov*. Sofia, 1980. French translation: *La Macedonie. Recueil de documents et materiaux*. Sofia, 1980.

MARKOVA, Z. *Bulgarskoto tsurkovno-natsionalno dvizhenie do Krimskata voina* (The Bulgarian National Movement for an Independent Church prior to the Crimean War). Sofia, 1976.

MATEEV, B. *Dvizhenieto za kooperativno zemedelie v Bulgaria pri usloviyata na kapitalizma* (The Movement for Co-operative Farming in Bulgaria under Capitalist Conditions). Sofia, 1967.

Mezhdunarodni otnosheniya i vunshna politika na Bulgaria sled Vtorata svetovna voina. Sbornik ot studii i statii (International Relations and Bulgaria's Foreign Policy after World War II. Collection of Studies and Articles). Eds. M. Isusov et al. Sofia, 1982.

MIHAILOV, G. *Trakite* (The Thracians). Sofia, 1972.

MIHOV, N. *Contribution a l'histoire du commerce bulgare. Documents officieles et rapports consulaires.* Vols. 1–6. Sofia, 1941–70.

MIHOV, N. *Naselenieto na Turtsia i Bulgaria prez XVIII i XIX vek. Bibliografsko-statisticheski izsledvaniya* (The Population of Turkey and Bulgaria in the Eighteenth and Nineteenth Centuries. Bibliographic and Statistical Studies). Vols. 1–5. Sofia, 1915–68.

MITEV, Y. *Suedinenieto 1885* (The Unification of Eastern Rumelia with the Bulgarian Principality in 1885). Sofia, 1980.

MUSHMOV, N. *Monetite it pechatite na bulgarsite tsare* (The Coins and Seals of the Bulgarian Tsars). Sofia, 1924.

MUTAFCHIEV, P. *Istoria na bulgarskia narod* (A History of the Bulgarian People). Vols. 1–2. Sofia, 1943.

MUTCHIEVA, V. *Agrarnite otnosheniya v Osmanskata imperiya prez XV–XVI vek* (Agrarian Relations in the Ottoman Empire in the Fifteenth–Sixteenth Centuries). Sofia, 1962.

MUTCHIEVA, V. et S. DIMITROV. *Sur l'état du système des timars de XVIIe–XVIIIe siècle.* Sofia, 1968.

NAUMOV, G. *Rabotnicheskata partiya v Bulgaria, 1927–39* (The Workers' Party in Bulgaria, 1927–39). Sofia, 1980.

NIKOV, P. *Vuzrazhdane na bulgarskiya narod. Tsurkovno-natsionalni borbi i postizheniya* (The Revival of the Bulgarian People. Achievements in the Struggle for National and Church Independence). Sofia, 1971. 1st edn. 1930.

OGNYANOV, L. *Voynishkoto vustanie ot 1918* (The Soldiers' Mutiny of 1918). Sofia, 1978.

Osmanski izvori za istoriyata na Dobrudja i Severoiztochna Bulgaria (Ottoman Sources on the History of Dobrudja and North-Eastern Bulgaria). Comp., transl. and ed. S. Dimitrov. Sofia, 1981.

Osvoboditelna borba na bulgarite v Makedonia i Odrinsko, 1902–4 (The Liberation Struggle of the Bulgarians in Macedonia and the Edirne region, 1902–4). *Diplomaticheski dokumenti* (Diplomatic documents). Eds. N. Todorov, et al. Sofia, 1978.

Osvobozhdenieto na Bulgariya. Materiali ot yubileina mezhdunarodna nauchna sesiya v Sofia (The Liberation of Bulgaria. Proceedings of the International Jubilee Scientific Session held in Sofia). Eds. H. Hristov, et al. Sofia, 1982.

Osvobozhdenieto na Bulgaria ot tursko igo 1878–1958. Sbornik statii (The Bulgarian Liberation from Ottoman Domination, 1878–1958. Collected Papers). Eds. D. Kossev, et al. Sofia, 1958.

Otechestvenata voina na Bulgaria 1944–45. Dokumenti i materiali (The Patriotic War of Bulgaria 1944–45. Documents, Materials). Vols. 1–4. Eds. P. Panayotov, et al. Sofia, 1978–82.

PAISII HILENDARSKI. *Istoriya slavyanobolgarskaya sobrana i narezhdana Paisiem yeromonahom v leto 1762. Stukmil za pechat po purvoobraza i uvod Y. Ivanov* (Slav-Bulgarian History compiled and written by Hieromonk Paisii in the year 1762. Prepared for publication after the original and prefaced by Y. Ivanov). Sofia, 1914. First copy by Sophronius in 1765. Preface, Modern Bulgarian Version and Commentaries by B. Raikov. Sofia, 1972.

PALAUZOV, S. *Izbrani trudove* (Selected Works). In 2 vols. Eds. V. Gyuzelev and H. Kolarov. Vol. 1: *Izsledovaniya po istoriya na Bulgaria i evropeiskiya prez Srednovekovieto* (Studies on the Medieval History of Bulgaria and South-East Europe). Sofia, 1974.

PALESHUTSKI, K. *Makedonskiyat vupros v burzhoazna Yugoslavia, 1918–41* (The Macedonian Problem in Bourgeois Yugoslavia, 1918–41). Sofia, 1983.

PANDEV, K. *Natsionalnoosvoboditelnoto dvizhenie v Makedonia i Ordrinsko, 1878–1903* (The National Liberation Movement in Macedonia and the Adrianople Region, 1878–1903). Sofia, 1979.

PANTEV, A. *Anglia sreshtu Russia na Balkanite, 1879–94* (England against Russia in the Balkans 1879–94). Sofia, 1972.

PASKALEVA, V. *Bulgarkata prez Vuzrazhdaneto* (The Bulgarian Woman during the Revival Period). Sofia, 1964.

PENEV, B. *Istoriya na novata bulgarska literatura* (History of Modern Bulgarian Literature). Vols. 1–4. Sofia, 1976–78.

PETROV, P. *Obrazuvane na bulgarskata durzhava* (The Foundation of the Bulgarian State). Sofia, 1981.

PETROV, P. *Subdonosni migove za bulgarskata narodnost. Kraya na XIV vek 1912 godina* (Crucial Events for the Bulgarian Nation. End of the Fourteenth Century–1912). Sofia, 1975.

PETROV, S. and H. KODOV. *Starobulgarski muzikalni pametnitsi* (Old Bulgarian Monuments of Music). Sofia, 1973.

PETROVA, D. *Bulgarskiyat zemedelski naroden suyuz i Narodniyat front, 1934–39* (The Bulgarian Agrarian Union and the Popular Front during 1934–39). Sofia, 1967.

PETROVA, D. *BZNS prez perioda na ikonomicheskata kriza, 1929–34* (The Bulgarian Agrarian Union during the Economic Crisis of 1929–34). Sofia, 1979.

PETROVA, D. *BZNS v kraya na burzhoaznoto gospodstvo v Bulgaria 1939–44 g.* (The Bulgarian Agrarian Union at the end of the Bourgeois Rule in Bulgaria 1939–44). Sofia, 1970.

PETROVA, S. *Devetoseptemvriiskata sotsialisticheska revolyutsia 1944 g.* (The Socialist Revolution of 9 September 1944). Sofia, 1981.

Polozhenieto na bulgarskiya narod pod tursko robstvo. Dokumenti i materiali (The Plight of the Bulgarian People under Ottoman Domination. Documents and Materials). Comp. and ed. N. Todorov. Ed.-in-chief H. Hristov. Sofia, 1953.

PRIMOV, B. *Bugrite. Kniga za Bogomil i negovite posledovateli* (The Bougres. A Book about Priest Bogomil and his Disciples).

Sofia, 1970. Published in French: *Les Bougres. Histoire du pope Bogomile et de ses adeptes.* Paris, 1975.

Problemi na Bulgarskoto Vuzrazhdane (Problems of the Bulgarian Revival). Eds. K. Sharova, et al.

Problems of the Transition from Capitalism to Socialism in Bulgaria. Eds. M. Isusov, et al. Sofia, 1975.

RADEV, S. *Stroitelite na suvremenna Bulgaria* (The Builders of Modern Bulgaria). In 2 Vols. 1–2. Pref. and ed. by P. Zarev. 2nd edn. Sofia, 1973. 1st edn, 1940.

RADUNCHEVA, A. *Doistoricheskoe izkustvo v Bolgarii* (Prehistoric Art in Bulgaria). (In Russian). Sofia, 1973. Published in English: *Prehistoric Art in Bulgaria from the Fifth to the Second Millennium BC.* Oxford, 1976.

Russia i osvobozhdenieto na Bulgariya (Russia and the Liberation of Bulgaria). Ed. V. Gyuzelev. Sofia, 1981.

SILYANOV, H. *Osvoboditelnite borbi na Makedonia* (The Liberation Struggles in Macedonia). In 2 Vols. 1–2. Preface L. Panayotov. Sofia, 1983.

SIRKOV, D. *Vunshnata politika na Bulgaria 1938–41* (Bulgarian Foreign Policy in 1938–41). Sofia, 1979.

SNEGAROV, I. *Istoriya na Ohridskata arhieskopiya-patriarshiya* (A History of the Archbishopry and Patriarchate of Ohrid). Vols. 1–2. Sofia, 1924–32.

SPIROV, N. *Preobrazhenskoto vustanie* (The Uprising of Transfiguration Day). 2nd enl. edn. Sofia, 1983. 1st edn. 1965.

STATELOVA, E. *Diplomatsiyata na Knyazhestvo Bulgaria, 1879–86* (The Diplomacy of the Bulgarian Principality in 1879–86). Sofia, 1979.

STATELOVA, E. *Iztochna Rumelia (1879–85). Ikonomika, politika, kultura* (East Rumelia in 1879–85. Economy. Policies, Culture). Sofia, 1983.

TODOROV, N. *Balkanskiyat grad XV–XIX vek* (The Balkan City in the Fifteenth–Nineteenth Centuries). Sofia, 1972. Russian translation: *Balkanskii gorod XV–XIX vekov.* Moscow, 1976.

TODOROV, N. *La ville balkanique sous les Ottomans, XVe–XIXe siècle.* London, 1977.

TODOROV, N. and V. TRAIKOV. *Bulgari uchastnitsi v borbite za osvobozhdenieto na Gurtsiya, 1821–28. Dokumenti* (Bulgarian Participation in the Struggles for the Liberation of Greece, 1821–28. Documents). Sofia, 1972.

TODOROVA, T. *Diplomaticheska istoriya na vunshnite zemi na Bulgaria, 1878–1912* (A Diplomatic History of the Foreign Loans of Bulgaria, 1878–1912). Sofia, 1971.

TODOROVA, T. *Obyavyavane nezavisimosta na Bulgaria prez 1908 g. i politika na imperialisticheskite sili* (Bulgaria's Declaration of Independence in 1908 and the Policies of the Imperialistic Powers). Sofia, 1960.

TONEV, V. *Dobrudja prez Vuzrazhdaneto* (Dobrudja during the Revival Period). Sofia, 1973.

Turski iztochnitsi za bulgarskata istoriya (Turkish Sources on Bulgarian History). Vols. 1–6. Sofia, 1964–77.

TRAIKOV, V. *Ideologisheski techenia i programi v natsionalnoosvoboditelnite dvizhenia na Balkanite do 1878 godina* (Ideological Currents and Programmes in the National-Liberation Movements in the Balkans up to 1878). Sofia, 1978.

TSVETKOVA, B. *Haidukstvoto v bulgarskite zemi prez XV–XVIII vek. Studiya i dokumenti* (The Haiduk Movement in the Bulgarian Lands in the Fifteenth–Eighteenth Centuries. Introductory Study and Documents). Vol. 1. Sofia, 1971.

TSVETKOVA, B. *Pametna bitka na narodite* (Memorable Battle of the Peoples). 2nd compl. rev. and enl. edn. Varna, 1979. 1st edn, 1969. Published in French: *La bataille memorable des peuples.* Sofia, 1971.

TUPKOVA-ZAIMOVA, V. *Byzance et les Balkans à partir du VIe siècle. Les mouvements ethniques et les Etats.* London, 1979.

TUPKOVA-ZAIMOVA, V. *Dolni Dunav – granichna zona na Vizantiiskiya zanad. Kum istoriyata na severnite i severoiztochnite bulgarski zemi, kraya na X–XII vek* (The Lower Danube – the Border Area of Western Byzantium. Towards the History of the Northern and North-Eastern Bulgarian Lands, the End of the Tenth–Twelfth Centuries). Sofia, 1976.

Ustanovyanvane i ukrepvane na narodnodemokratichnata vlast. Septemvri 1944–May 1945. Sbornik dokumenti (Establishment and consolidation of the People's Democratic Rule. September 1944–May 1945. Collection of Documents). Comps. B. Mateev, et al. Sofia, 1969.

VAKLINOV, S. *Formirane na starobulgarskata kultura, VI–XI vek* (Formation of Old Bulgarian Culture, Sixth–Eleventh Centuries). Sofia, 1977.

VANCHEV, Y. *Novobulgarskata prosveta v Makedonia prez Vuzrazhdaneto do 1878* (New Bulgarian Letters in Macedonia from the National Revival Period to 1878). Sofia, 1982.

Varna 1444. Sbornik ot izsledvaniya i dokumenti v chest na 525-ta god. ot bitkata krai Varna (Collection of Studies and Documents Commemorating the 525th Anniversary of the Battle near the Town of Varna). Sofia, 1969.

VELIKI, K. and V. TRAIKOV. *Bulgarskata emigratsiya vuv Vlahia sled Ruska-Turskata voina 1828–29. Sbornik ot dokumenti* (Bulgarian Immigrants in Wallachia after the Russo-Turkish War of 1828–29. Collection of Documents). Sofia, 1980.

VELKOV, V. *Gradut v Trakia i Dakia prez kusnata antichnost, IV–VI vek.* (Towns in Thrace and Dacia in the Late Antiquity, Fourth-Sixth centuries). Sofia, 1959.

VELKOV, V. *Roman Cities in Bulgaria.* Amsterdam, 1980.

VENEDIKOV, I. *Istoriya na Srubsko-Bulgarskata voina, 1885* (History of the Serbo-Bulgarian War, 1885). Sofia, 1910.

VULKOV, G. *Bulgarskoto opulchenie. Formirane, boino izpolzuvane i istoricheska sudba* (The Bulgarian Volunteer Corps. Formation, Military Operations and Historical Fate). Sofia, 1983.

Vunshnata politika na Bulgaria. Dokumenti i materialii (Bulgaria's Foreign Policy. Documents and Materials). Vol. 1, 1879–86. Eds. N. Todorov, et al. Sofia, 1978.

Vuoruzhenata borba na bulgarskiya narod protiv fashisma 1941–44. Dokumenti (The Armed Struggle of the Bulgarian People against Fascism, 1941–44. Documents). Comp. M. Ereliiska, et al. Sofia, 1962.

Vuzpomenatelen sbornik po sluchai stogodishninata ot osvoboditelnata Rusko-Turska voina 1877–78. Dokumenti sbornik s podbrani materiali ot arhivite i bulgarskiya vuzrozhdenski pechat, 1876–78. Istoriya izvori. May 1876–November 1878 (Collection Commemorating the Centenary of the Russo-Turkish Liberation War 1877–78. Documentary Selections from Archives and the Bulgarian Press during the National Revival Period 1876–78. Historical Sources, May 1876–November 1878). Comp. by V. Tileva, et al. Sofia, 1979.

ZHECHEV, N. *Braila i bulgarskoto kulturno-natsionalno vuzrazhdane* (Braila and the Bulgarian National Cultural Revival). Sofia, 1970.

ZHIVKOV, T. *Biografichen ocherk* (Biography). Sofia, 1981.

ZHIVKOVA, L. *Kazanlushkata grobnitza* (The Kazanluk Tomb). Sofia, 1974. Published in German: *Das Grabmal von Kazanluk.* Recklinghausen, 1973; Russian: *Kazanlikskaya grobnitsa.* Moscow, 1976; English: *The Kazanluk Tomb,* Tokyo, 1982.

ZHIVKOVA, L. *Chetveroevangelieto na tzar Ivan Aleksandur* (The Tetraevangelia of Tsar Ivan Alexander). Sofia, 1980. Published in German: *Das Tetraevangeliar des Zaren Ivan Alexander.* Recklinghausen, 1977.

ZLATARSKI, V. *Istoriya na bulgarskata durzhava prez srednite vekove* (The History of the Bulgarian State in the Middle Ages). Ed. P. Petrov. 3rd edn. Vols. 1–3. Sofia, 1970–72.

ZLATEV, Z., B. MATEEV and V. MIGEV. *Bulgaria v epohata na sotsialisma* (Bulgaria in the Socialist Era). Sofia, 1981.

PART III

SOCIO-POLITICAL STRUCTURE OF THE PEOPLE'S REPUBLIC OF BULGARIA

SOCIO-POLITICAL STRUCTURE OF THE PEOPLE'S REPUBLIC OF BULGARIA

The political system of the People's Republic of Bulgaria comprises interconnected state and public organizations and collective bodies. It is characterized not only by a high degree of organization, but also by unity of purpose and action.

Its main components are the Bulgarian Communist Party (BCP), the Bulgarian Agrarian Party (BAP), the socialist State and its apparatus, the public organizations (including the cultural unions), and the work-forces at enterprises. The BCP is the leading force in society and the State.

All the revolutionary transformations in the country after the Second World War, which brought about the triumph of Socialism in Bulgaria, have been carried out under the guidance of the Communist Party, and this has affirmed the Party as guiding force. The Communist Party guides the working of the state bodies and organizations but does not itself carry out their functions. On the basis of scientific knowledge of the objective laws of social development, the Party works out the general policy and the strategy for success, determined by the interests of the people and by its ultimate goal – the building of a classless communist society, which will be a voluntary association of men free from social oppression and exploitation, with fully developed personalities and fully-realized potential, in a context of highly developed production and of social activities organized according to the requirements of science.

An important component of the political system of socialist society is the State and its apparatus. It is a State of the working people from town and country, led by the working class.

By the referendum of 8 September 1946, the monarchy was abolished and a republican form of state government established.

The main trend in the evolution of the socialist State in the People's Republic of Bulgaria is toward the constant development and improvement of democracy, the active participation of the working people in the governing of the State, economic and cultural advancement, the improvement of the work of the state apparatus and increased popular control over its activity. The State constantly expands its social basis and is increasingly becoming a State of all the people. Its goal is the building of the communist society.

The Constitution. The Constitution is the fundamental document which affirms the social and state system and on the basis of which government of the country is conducted. The constitutional development of socialist Bulgaria is connected with two Constitutions. The first, adopted in 1947, affirmed the socialist principles in the country's socio-political and economic life and created the legal prerequisites for socialist development. This Constitution was socialist in character, reflecting the specifics of the transitional period from Capitalism to Socialism and containing certain elements of the old social and legal system, such as private ownership.

The 1947 Constitution played an important role in the consolidation of the socialist social system.

The substantial changes in socio-political life and in the country's economy during the 1950s and 1960s made it necessary to draw up a new constitution and in March 1968 the National Assembly elected a constitutional commission, consisting of 78 national representatives and presided over by Todor Zhivkov, which worked out an appropriate draft. The draft was published for nationwide discussion, in which more than three million citizens took part. The constitutional commission took into consideration the 14,000 proposals put forward, introduced a number of improvements in the draft and presented it to the National Assembly. In May 1971, the Assembly decided that a Referendum should be held. By a vote of all the people, cast on 16 May 1971, the new Constitution was adopted and entered into force on 18 May 1971.

In the People's Republic of Bulgaria all power stems from the people and belongs to the people. The people are the source and subject of state power. State power is implemented in two ways: through freely elected representative bodies, and directly. The representative bodies are the National Assembly and the People's Councils. They are elected directly by the voters. No-one can be officially included in the composition of the National Assembly or in the People's Councils by appointment or through inherited privilege. The composition of the Assembly is decided on the basis of universal, equal and direct suffrage by secret ballot. The representative bodies are composed of the political representatives of the people and express its will and sovereignty.

The representative bodies constitute the foundation of state power and of the entire state apparatus, and the remaining state bodies are directly or indirectly derived from them. The National Assembly elects the State Council, the Government and the Supreme Court of the People's Republic of Bulgaria; it also elects the Chief Public Prosecutor of the Republic.

All other state bodies are correspondingly dependent upon the representative bodies and carry out their activity under their guidance and control.

State power is implemented through the forms of direct democracy. A special Act on referendums determines in what cases and form the nationwide and local referendums shall be held, and how the consultative forms of direct democracy shall be used.

The public organizations are an important form of social organization. They are voluntary mass associations with defined membership, and through them specific interests of the citizens – which may have a personal or social character – are realized. The Constitution regulates the right of citizens to set up organizations for political, professional, cultural, artistic, scientific, religious, sports and other purposes not concerned with the economy. For joint economic activity, the citizens may also unite into co-operatives. The public organizations enlist and unite different strata of the population to express and defend their specific interests. They are an important manifestation of the creative initiative and self-government of the people.

The Constitution provides for the possibility for public organizations to fulfil state functions transferred to them with their consent. To the trade unions, for instance, has been delegated the right of control over safety at work.

The Constitution also allows the combining of activities performed by state bodies with activities performed by public organizations. Thus the possibility is envisaged for the National Assembly to form public state bodies. Such bodies are, for instance, the Committee for Culture, the Committee for State and Public Control and the Committee for Scientific and Technical Progress. The possibility is thus created for a wide cross-section of people who are not civil servants to take an active part and have a decisive say in the management of the respective fields of state life, on the principle of performing a public duty.

In the political system of the socialist society an important place is also occupied by the work-forces. The forms and order of their participation in the management of the economy and social life are determined in special laws. A work-force is an association of people who perform common tasks and interact in the process of work. They are interconnected by the common goals of their labour, by its results and remuneration, and by their mutual relations in the work process. The work-force is the principal social component of any economic or other organization.

The socio-economic structure of the People's Republic of Bulgaria covers the organization of economic life, the system of economic relations, the character and forms of ownership, the ultimate goal of social production, the forms of organizing economic activity, and the principles of distributing the social product and managing the national economy.

The basis of the socio-economic structure is socialist ownership of the means of production. It excludes the exploitation of man by man; since all citizens have the same interest in the means of production, the possibility of exploitation of some people by others is removed.

The goal of social production is to meet the growing material and cultural needs of the people as fully as possible, and to secure the people's well-being and the country's progress.

The socialist economic organizations, built on the principle of socialist ownership of the means of production, are the basic forms of organizing economic activity.

Labour is a basic socio-economic factor, and the distribution of the social product is carried out according to the quantity and quality of the labour invested by every member of society.

The socio-economic structure is developing steadily and is directed by the socialist State in a planned and organized manner.

The national economy is developing as part of the world socialist economic system and, above all, of the Council for Mutual Economic Assistance (CMEA), carrying out, within the latter's framework, a constant deepening of economic, scientific and technical co-operation and economic integration with the other member-states.

According to the Constitution the following kinds of ownership exist: state (of all the people), co-operative, ownership by the public organizations, and personal. Demarcation is according to who holds the right of ownership, and the purpose for which the objects of ownership are used. Socialist property is that which belongs to the State or to organizations formed voluntarily by the citizens. This property is also called public, because its holders are big public collectives and it is used to satisfy common needs of all the people or of the collectives who form a given socialist organization. Personal property is that of the individual citizen, which serves to meet the needs only of the owner and the members of his or her family. There are also mixed kinds of public ownership – state and co-operative, state and public, and co-operative and public.

There is also property belonging to international economic organizations. Organizations of this type have been set up by the member-countries of CMEA for joint economic activity.

The Constitution does not forbid the existence of ownership by foreign citizens and firms in Bulgaria; special laws lay down the order, extent and the forms of such property.

State ownership is the basis for the development of all other forms of public ownership. It is the highest form of socialist ownership and represents a unified fund.

The State can own all kinds of property, unrestricted in form and quantity. A number of most important items are declared by the Constitution to be exclusively state property: these include the plants and factories, banks, the underground riches, the natural sources of energy, nuclear power, pastures, waters, roads, the railway, water and air transport, the means of communication, radio and television.

The State assumes rights of ownership in different ways. Initially, state ownership appeared as a result of nationalization. Now new state property is created through planned construction and industrial activity. The State places its property under the management of the work-forces of economic and other organizations. The right of management of the state economic and other organizations empowers them in their own name to possess, use and dispose of the property entrusted to them by the State, and also of the means acquired in the process of their economic activity. In exercising these powers the state organizations take into consideration what the State has formally established. As managers, state organizations enter into contractual relations with each other, or with co-operatives, public organizations and citizens. Part of the profit realized as a result of this management remains at the disposal of the organization. This is reflected in the wages of the work-force, which has an interest in the better management of state property. The greater the income the state organization derives as a result of managing such property, the greater the proportion of it that is set aside for remuneration of labour.

Several state organizations may unite and create joint facilities, buildings, holiday houses, etc.

Some state-owned property is placed at the disposal of co-operatives, public organizations or citizens, to be managed by them. Such property may comprise, for example, farm lands, pastures, forests, waters, quarries, machinery, etc. The State can place such property – gratis or in return for payment – at the disposal of co-operative or public organizations. The State also grants citizens land for personal use or for housing construction.

The co-operative is the second most significant form of socialist ownership. Co-operatives are voluntarily established public organizations for joint economic activity. The co-operative as a juridical person has the right of co-operative ownership. There are as many individual holders of the right of co-operative ownership as there are co-operatives. In co-operative ownership there is no unified fund. Every co-operative has its own right to possess, use and dispose of the objects.

Property owned by agro-industrial complexes, and by producer, consumer and housing construction co-operatives, is co-operative property, as also is that of the district co-operative unions, of the Central Co-operative Union and of their enterprises.

Inter-co-operative organizations also possess co-operative property. Every co-operative may own property which it needs for its activity as permitted by law. The right over co-operative property encompasses rights over fixed and floating capital, registered trademarks, licences, claims, etc.

The members of the co-operatives themselves exercise the right of co-operative ownership by the respective co-operative bodies.

Public organizations acquire their right to ownership mainly through the purchase of objects with the money received as payments from the members or as a result of economic activity. The socialist State helps the creation of ownership of public organizations by placing at their disposal pieces of its own property, gratis or in return for payment, or by granting them the right to build on state-owned land.

As owners, public organizations themselves exercise their rights through their own system. Outside bodies cannot interfere with the activity of these organizations or with the exercise of the right to ownership. In exercising their right, public organizations must take into consideration their own interests, those of the members, and the public interest.

Land is a basic natural resource and one of the most important means of production. In the People's Republic of Bulgaria the land belongs to the State, the co-operatives, the public organizations and the citizens. Most farm lands and all forests and pastures are state-owned. The state places these lands at the disposal of agro-industrial complexes, co-operatives and forestry enterprises and also of citizens, free of charge. Agricultural organizations and citizens are obliged to make the fullest possible use of the lands thus placed at their disposal and to take due care of them. In cases provided by law, buildings can be erected on them.

Citizens may own real and movable pieces of personal property for meeting their own family members' needs, as well as petty tools for production and ancillary activities, and the produce of these activities. Personal ownership cannot be used to the detriment of the public interest or for the purpose of obtaining unearned income. The citizens' right to personal property is expressed in their ability to possess, use or dispose of immovable and movable property intended to meet their consumer needs. Citizens of the People's Republic of Bulgaria are subject to the right to personal property: foreign citizens who have a permanent domicile in the country may also possess property on an equal footing with Bulgarian citizens. The objects of the right to personal property may be both immovable and movable items used for meeting the citizens' needs: among these are motor cars, dwellings, holiday villas, household accessories, articles for personal use, etc.

Families comprising both spouses and their children who have not attained majority (unless they have married before coming of age), as well as citizens who have not created a family of their own, can own as their personal property the following immovables: a dwelling, a villa, a garage, a studio.

The rights of the person and citizen are one of the most important elements in the relations between society, the State and the individual. The individual enjoys his rights in order to ensure his material well-being and cultural development, as well as to take part in the political, economic and cultural life of society.

One of the main principles upon which the legal status of the citizen is based is the equality of rights. All citizens of the People's Republic of Bulgaria are equal before the law. No privileges or restrictions of rights are allowed on the basis of nationality, origin, religion, sex, race, education, social or material status. This equality manifests itself in the equal rights and equal obligations of the citizens.

Citizens enjoy a wide range of rights. The Constitution, however, deals only with the basic rights, which determine the fundamentals of the juridical position of the person and citizen in society. They ensure his participation in economic, political and cultural life, guarantee his freedom and inviolability, and contribute towards the further improvement of social relations in defence of the interests of the individual and society.

The Constitution contains a special section (Chapter III) devoted to the basic rights and obligations of the citizen. Other parts of the text (Chapters, I, II and IV) also deal with the fundamental rights of the citizen. The stipulations of the Constitution are of tremendous importance. They constitute an important part of the text of the Fundamental Law and the entire legal system of the country but do not exhaust the content of the legal status of a person in society, since the citizens also enjoy other kinds of rights of a normative character within the legislation of Bulgaria.

The basic groups of constitutional rights of the citizens are the political, economic, social and cultural rights and democratic freedoms.

The political rights ensure participation in the political life of the country, in the exercise of state power, and in the government of the State and society.

The electoral right enables the citizen to choose the representative and other elective bodies of the State, as well as to be elected to them. Suffrage is universal and enables all citizens who have reached a certain age to vote in elections. All Bulgarian citizens of eighteen years and over can vote and be elected, regardless of sex, nationality, race, religion, education, occupation, official or social position and property status, with the exception of those placed under full judicial disability. Persons who cannot consciously guide their actions are placed under full judicial disability by court procedure.

Suffrage in the People's Republic of Bulgaria is equal. Every Bulgarian citizen can participate in elections on an equal footing with other voters. The vote of one elector is equal to that of any other. Every elector has the right to only one vote, and votes only once in a given election.

Constituencies are defined on the basis of an equal number of inhabitants.

Suffrage is direct. Citizens vote directly for the candidates whom they wish to elect as national representatives or people's councillors, or members of other elective bodies. Voting is secret, and candidates who have received more than one half of the votes in the constituencies are considered elected.

The Constitution stipulates that the national representatives and councillors shall be responsible to and report to their electorate. Elected persons can in principle be removed before expiry of the term for which they were elected. If a national representative or people's councillor fails to justify the confidence of his electors, he can be removed. A request for removal can be made by the Party and public organizations or by at least one-fifth of the electorate in the respective constituency. The decision to remove an elected representative is taken directly by the electors in the manner in which the voting itself was affected – by secret ballot. A new election is held for the place of the national representative or people's councillor who has been removed.

The right to remove an elected representative is an expression of genuinely popular rule, an active means of public control over the activity of the people's representatives.

The Constitution, however, forbids the formation of organizations oriented against the socialist system and the rights of the citizens, or for the propagation of a fascist or other anti-democratic ideology.

It stipulates that citizens have the right to submit requests, complaints and proposals. This right is implemented in the manner established by law.

The socio-economic rights of the citizen secure participation in the distribution and utilization of material benefits in accordance with the principles of Socialism. These are the right to a job, the right to leisure, the right to social security, a pension and assistance, as well as the right stemming from certain social obligations of the State and society towards the citizen. Every citizen has the right freely to choose his profession. The State secures the right to a job by developing the socialist socio-economic system. Labour is remunerated according to its quantity and quality. The right to leisure is implemented through diminishing the working hours without reducing labour remuneration and without infringement of other rights.

The State and the public organizations secure the leisure of the working people through a well developed network of holiday houses, clubs, houses of culture and other places for rest and recreation.

Of great significance among the social rights of the citizen is the right to security, pension and assistance. This is connected with such social problems as incapacity for work, sickness, accident, maternity, invalidity, old age and death, as well with the upbringing of children. In the cases established by law, citizens receive pensions, indemnities and other social aid. Social security covers all workers, office employees and agricultural workers. To secure it, insurance premiums have been established which are calculated as a percentage of salary. The premiums are paid by the enterprises, institutions, co-operative farms and private persons, in cases when the latter are the employers. It is forbidden to deduct the insurance premiums from the remuneration of workers and employees.

In Bulgaria old-age pensions are received by all men who have reached 60 years of age after 25 years of service, and by all women who have reached 55 years of age after 20 years service. Some categories of workers and office employees, who have worked in more unfavourable conditions may retire on pension earlier. Workers and office employees who have been incapacitated for work, permanently or for a prolonged period because of sickness or industrial accident, receive disability pension.

Sick workers and office employees receive an indemnity in cash during their sickness. A lump sum is paid to workers and office employees when a child is born as well as in other cases established by law. Persons covered by the social insurance scheme take part in the management of this insurance.

Mothers enjoy particular protection and care by the State, and by the economic and public organizations. They are guaranteed maternity leave, both before and after the baby's birth, with salary or wages preserved. Free obstetric and medical aid, maternity homes, pain relief during labour, free medicines during pregnancy, etc. are secured. Considerable relief and assistance is provided in the raising of children.

Minors, the disabled and elderly people who do not have relatives or have remained without help from their relatives, are under the protection of the State and society. For this purpose a special category of social pension has been provided.

Of great significance among the social tasks set for the State is the protection of marriage and the family.

The State devotes comprehensive care to the health of the people by organizing therapeutic, prophylactic and other health establishments. Every Bulgarian is entitled to free medical aid.

Preventive medical care is of particular importance; state and public organizations disseminate health education and culture, and encourage the development of physical education and tourism. Cultural rights hold a major place in the system of basic rights. In the forefront is the right to free education in all kinds and all levels of educational establishment.

The State encourages education and all-round improvement in working conditions at the educational establishments, grants scholarships and encourages particularly gifted students. Conditions are being created for the introduction of compulsory secondary education.

The Constitution devotes particular care also to the advancement of science, art and culture.

The Constitution regulates the freedoms of the citizens, both personal and socio-political. Personal freedoms include the freedom and inviolability of the person, freedom of personal and family life, the inviolability of the home, secrecy of correspondence, and freedom of conscience and religion. No one can be detained for more than 24 hours without an order from the court or Public Prosecutor. Offences and corresponding punishments are established only by law. A law establishing the penalty for a given act or increasing the penal responsibility cannot have retroactive force. Punishment is personal, corresponding to the crime, and can be imposed only by established courts of law. Every citizen has the right of defence against illegal interference in his personal and family life and against encroachments upon his honour and good name. Without the consent of the dweller, no one can enter a home except in the cases and under the conditions provided for by law. The Constitution also establishes the secrecy of correspondence, telephone conversations and telecommunications.

Every citizen has freedom of conscience and of religion. Taking up a public or state post, and the exercise of political and other rights are not dependent upon the profession of any religion. In the People's Republic of Bulgaria the Church is separate from the State. It is forbidden to abuse the Church and religion for political ends and to form political organizations on a religious basis.

The Constitution establishes a number of socio-political freedoms – freedom of speech, of the Press, of assembly, of meetings and demonstrations – which make it possible for citizens to participate actively in political and public life. It not only proclaims the political freedoms, but also stipulates that they are secured by guaranteeing the citizens the necessary material and other conditions for this purpose.

It also provides the right of asylum to foreign citizens persecuted for defending the interests of the working people, participation in a national-liberation struggle, progressive political, scientific and artistic activities, or fighting against racial discrimination or in defence of peace.

The right of asylum is not a subjective right of foreign citizens, but gives the State the right, according to assessment by competent bodies, to give shelter to every non-Bulgarian citizen persecuted abroad for his progressive beliefs and activities.

Side by side with the basic rights, the Constitution also establishes basic obligations of the citizens, thereby creating a balance in their legal status. This unity removes the possibility of abuse of rights.

The basic obligations established by the Constitution are:

(1) The obligation to work. Labour is a basic right of every citizen, a matter of honour and duty towards society. For this reason every able-bodied citizen is obliged to do socially useful work in accordance with his capabilities and qualifications. Citizens are obliged to protect and multiply socialist property as an inviolable foundation of the social system in the country.

(2) The defence of the country is a supreme duty and a matter of honour for every citizen. High treason and betrayal of the homeland are the gravest crimes against the nation.

(3) Every citizen is obliged to co-operate in the preservation and consolidation of peace. Abetment to and propaganda of war are qualified as grave crimes against peace and mankind and are forbidden and

punishable. The Penal Code of Bulgaria provides special punishment for crimes of this kind.

The rights and liberties provided in the Constitution are subject to material, political and legal guarantees. In practice, the different kinds of guarantee act in conjunction and their joint operation assures the reality of rights and liberties. The material guarantees stem from socialist ownership and the socialist mode of production and are objective in character. The political guarantees for the rights of the person and citizen in Bulgaria are factors directly connected with the socialist political system and stem from it. The essence of socialist democracy presupposes self-government by the working people, development of their activity and initiative, immediate and decisive participation of the masses in the government of the State, in the management of the national economy and in the advancement of culture. The legal guarantees are implemented through the function of the socialist State in defending the legal order, through the laws and the activities of the courts, the Public Prosecutor's Office, and the organs of State and public control. The legal status of the citizens in society is not determined restrictively. With the development of the socialist social relations, with the further perfection and deepening of socialist democracy and socialist humanism, the rights and liberties of the citizens are constantly developed and expanded.

The state bodies of the People's Republic of Bulgaria are organized as a unified system, a unified mechanism for the implementation of state power. At the base of the system lie constitutional principles which stem from the nature of the socialist system. The people are the only holders of state power. They implement this power directly (through referendums and other forms of direct democracy), or through representative bodies elected and authorized by them. This also determines the unique place of the representative bodies (the National Assembly and the People's Councils) in the management of society. They are formed by the electorate through universal, equal and direct suffrage by secret ballot and are accountable to and contolled by the voters. Their activity is guided by the interests of the working people. The electors can, through their national representatives and councillors, give assignments to the representative bodies and remove those whom they have elected.

The National Assembly elects the Council of Ministers, directs and controls its activity. The same dependence also exists between the People's Councils and their executive committees. As champions of the people's will, the representative bodies form, guide and control the activity of all other state bodies. The representative bodies implement state power in its entirety. The National Assembly decides the most important issues in the government of the country and regulates the activity of the whole state apparatus. The remaining state bodies implement only certain state activities. The powers of the judicial bodies operate only in the administration of justice. The ministries are each authorized to direct only a certain branch of the national economy or to implement a certain state function (planning, control, etc.).

Democratic centralism is the guiding organizational and functional principle for the state apparatus. It contains the requirement for the widest possible participation of the working people in the formulation and implementation of state policy, for taking into account their interests and developing their initiative. On the other hand, it presupposes the obligatory character of the unified state policy for all bodies and officials. From this stem also the requirements for subordination and accountability by the state bodies from lower levels upwards, and the mandatory nature of the decisions of the superior organs for the inferior ones. Democratic centralism also presupposes collective discussion of the problems, on the one hand, and one-man management and personal responsibility in the execution of decisions, on the other.

A basic principle of the socialist system is socialist democracy. It is manifested in the relations between the people and the state bodies, i.e. in the formation of representative bodies; in public control over the activitiy of the entire state apparatus; in the different forms of participation of the working people in the preparation and adoption of government decisions and in the organization of their implementation, and in the creation of conditions for the exercise of the socio-political rights of the citizens – to unite into political, professional, cultural, co-operative and other organizations. Socialist democracy permeates the entire process of government. Particularly rich are the forms of the working people's participation in the preparation and adoption of government decisions. It is also manifested in the formation of the state bodies and in the methods of implementing state power.

For the implementation of these varied state functions, basic sub-systems of state bodies are also set up. The representative bodies fashion state policy and implement the general guidance of the socio-economic processes, the development of the country as a whole and of the individual settlement systems. The executive and administrative bodies (the Council of Ministers, ministries, committees, executive committees of the People's Councils, etc.) carry out the executive and administrative activity of the State in the different spheres of social practice. The judicial bodies are called upon to implement the administration of justice by the State, to apply the sanctions determined by the law to transgressors of the socialist legal order and to settle legal disputes between state bodies and citizens, between citizens and between economic organizations. The Public Prosecutor's authorities supervise the exact and uniform application of laws by all officials and citizens.

The National Assembly. The National Assembly is the supreme representative body and overall seat of state power, which expresses the will of the people and their sovereignty. It is elected by the whole electorate, and the Constitution determines its composition of 400 national representatives. It consists of representatives of all social strata elected by universal, equal and direct suffrage by secret ballot, and is answerable to its electorate. In all its activity concerning the government of society, it expresses

The National Assembly building, Sofia

the will of the people and is guided solely by their interests.

The leadership of the National Assembly is composed of a Chairman and Vice-Chairmen, elected from the ranks of the national representatives. The Chairman presides over the sessions of the Assembly, co-ordinates and assists in the work of the permanent commissions and national representatives, and represents the Assembly in relations with the parliaments of other states and with citizens and public organizations.

The National Assembly sets up subsidiary organs – permanent and *ad hoc* commissions; the permanent commissions facilitate the preparation of laws and control their observance by ministries and other state bodies. As the overall seat of state power, it guides the other state bodies, but does not carry out operational tasks. The authority of the National Assembly manifests itself in determining the status of the state bodies, appointing the managers of the central state bodies, regulating social relations and controlling the activities of the state apparatus. It exercises supreme guidance over home and foreign policy, and determines the tasks and organization of the State Council, the Council of Ministers, the People's Councils, the courts and the Public Prosecutor's Office. Included in its authority is the right to set up, close down, merge and rename ministries and other departments with the rank of ministries, and to decide other important problems of the structure of central and local state bodies.

The National Assembly is the supreme organizer of the planned management of society. Through the adoption of unified plans for the country's socio-economic development and the state budget, as well as through reports on their fulfilment, it directs and controls the country's economic development. It is the only legislative body in the People's Republic of Bulgaria. It adopts and amends the Constitution and the laws, whereby it regulates the most important social relations and determines the conduct of the state bodies, the public organizations and the citizens.

Within its authority are the declaration of war and the conclusion of peace, changes in national borders, etc.

As the overall seat of state power, it exercises supreme control over the activity of all state bodies and over the observance of the Constitution and laws, both by state bodies and by the public organizations and the citizens.

These acts have supreme legal authority. This is manifested in their binding force for all state bodies, public organizations and citizens. Through this mechanism the National Assembly carries out its policy on the main questions of the management of the socio-economic processes.

The laws adopted by the National Assembly regulate

Georgi Dimitrov during the sessions of the Twenty-Sixth National Assembly, December 1945, Sofia

lasting social relations of fundamental significance. In elaborating laws, the Assembly is assisted by the whole state apparatus and by the public organizations. The right to legislative initiative has been delegated to the State Council, the Council of Ministers, the Permanent Commissions of the National Assembly, the national representatives, the Supreme Court and the Chief Public Prosecutor. The Fatherland Front, the trade unions, the Dimitrov Young Communist League and the Central Co-operative Union are vested with the same right in questions related to their own activity.

The organizational activity of the National Assembly in the government of the country is manifested in the construction of the system of state bodies as a whole and of the different sub-systems, and the definition of their authority. This is done first of all through the Constitution and then through the special organizing acts, such as the People's Councils Act, the Structure of the Courts Act, and the Public Prosecutor's Office Act.

The National Assembly forms the State Council, the Council of Ministers and the Supreme Court, and elects the Public Prosecutor of the Republic. As the overall seat of state power, each newly elected Assembly forms the central state bodies through which it will carry out the policy of the State. Every year the Council of Ministers reports on its overall activity to the supreme body. The National Assembly monitors the activity of the Supreme Court and of the Public Prosecutor's Office. It has become common practice to hear reports on the work of individual ministries during its sessions. It can revoke acts of the State Council and the Government when they are at variance with the law or are incorrect.

The Party House in Sofia. On the left is the Council of Ministers; on the right is the State Council

The State Council links the adoption of key decisions on the management of society and the organizing of their implementation. It ranks second in importance in the system of state bodies, its task being to assist the National Assembly and to implement directly complex state management in the country's government.

The State Council is elected by the National Assembly from among the national representatives, and is answerable to it. Its mandate depends directly on that of the National Assembly. Every newly elected National Assembly elects a new State Council, and changes in its composition may be effected at any time.

Within the framework of the Constitution, the State Council organizes and controls the execution of the basic tasks stemming from the laws and decisions of the National Assembly, exercises overall guidance and control over the activities of the Council of Ministers and other state bodies, and takes decisions and carries out executive and administrative activity on basic questions of state administration. On the other hand, it can itself draft and organize the drafting of bills by other leading bodies.

The spheres of authoritiy of the State Council are divided into three groups:

(1) It conducts a number of major organizational and political activities connected with the functioning of the representative system as a whole and of the National Assembly in particular. It decides the date of elections for the National Assembly and People's Councils, and of referendums; calls the National Assembly to sessions; decides on which of the draft laws submitted to the Assembly a nationwide discussion should be held, etc.

In certain cases the Constitution invests the State Council with the right to decide matters within the competence of the National Assembly when the latter is not in session. In urgent cases, the State Council makes changes by decree or supplements individual stipulations in the laws. On the proposal of the Chairman of the Council of Ministers, it dismisses and appoints members of the Government. The State Council declares a state of war in the case of armed attack against the country or when an international mutual defence obligation must be urgently fulfilled. It appoints or dismisses the commander-in-chief of the armed forces. Its decisions on all these questions are subject to approval by the National Assembly during its first following session.

(2) The State Council is also vested with a number of permanent powers, concerning the government of the country. It issues decrees and other legal acts on fundamental questions, resulting from acts and decisions adopted by the National Assembly, gives interpretations obligatory for all laws and formal decrees, and exercises general guidance of the defence and security of the country; appoints and dismisses the members of the State Defence Committee, appoints and releases from duty the senior commanding staff of the armed forces and awards high military ranks; represents Bulgaria in international relations; appoints, recalls and dismisses from duty the diplomatic and consular representatives of the People's Republic of Bulgaria in other countries, at the proposal of the Council of Ministers; ratifies and renounces international agreements, establishes diplomatic and consular ranks, sets up and closes departments which do not have the status of ministries and appoints and dismisses their heads; exercises control over the activity of the Council of Ministers and of the heads of ministries and other departments; exercises control over the strict observance of the laws and other acts of the National Assembly and of the acts issued by it; revokes incorrect and unlawful acts of the Council of Ministers, heads of ministries and other departments, People's Councils and their executive and administrative bodies, and exercises the right of pardon; remits collection of debts from insolvent debtors to the state institutes; awards orders, medals and titles of honour; sets up and closes down administrative territorial units, changes the limits of municipalities, districts of towns and of the administrative districts, and determines their administrative centres; grants, restores and revokes Bulgarian citizenship, and grants asylum. In implementing some of these powers the State Council is assisted by auxiliary bodies such as the commissions on awards, pardons, honorary titles, the granting of asylum, and the remission of bad state debts.

(3) The third kind of authority of the State Council can be exercised only in case of war and provided that it is impossible for the National Assembly to be called. In these cases the State Council performs most of the functions of the National Assembly (including its legislative powers), issues decrees with which laws can be amended or repealed, or unregulated matters can be legally regulated. It adopts the consolidated plans for socio-economic development and the Budget, as well as the reports on their application, and elects and releases from duty the Council of Ministers, the Supreme Court and the Chief Public Prosecutor. The State Council submits these decrees for approval by the National Assembly at its following session.

The Council of Ministers is the supreme executive and administrative body of state power. It is elected by the National Assembly, which can at any time replace its membership entirely or in part. The Council of Ministers conducts its business under the guidance and control of the National Assembly or, when the latter is not in session, of the State Council. It is answerable to the National Assembly and is obliged to report to it every year. The constitutional status of the Council of Ministers determines its place in the hierarchy of the state bodies, after the National Assembly and the State Council.

The Council of Ministers is made up of the Prime Minister, the Deputy Prime Ministers and members – Ministers and heads of central departments with the rank of ministries. Members of the Government are not required to be national representatives. It heads and manages the whole system of executive and administrative bodies in the country, and its tasks are to carry out the main guiding decisions in the country's government. It has the right of legislative initiative. It submits most of the draft laws or drafts for the amendment of existing ones.

Another kind of authority of the Council of Ministers

relates to the management of the whole system of executive and administrative bodies. It administers, co-ordinates and directly controls the activity of the ministries and other departments, and of the executive committees of the People's Councils. The forms of administration and control may vary considerably through normative and directive acts, instructional and control conferences, exchange of information and experience, consideration of reports on and discussion of the work of individual ministries, associations, executive committees of District and Municipal People's Councils, etc.

The Council of Ministers organizes the implementation of foreign and home policy and of the social and cultural activity of the State. It is also entrusted with the organization of the country's economic activity. It approves normative acts regulating the economic mechanism and its application in the different sectors of the national economy; approves programmes for the development of individual economic sectors; allocates the plan for the country's socio-economic development among the ministries, departments and People's Councils; introduces changes in the approved current plans, and approves technical and economic reports on the construction of large facilities. It also decides all major questions concerning Bulgaria's economic, scientific and technical co-operation with other countries.

An important element of the legal status of the Council of Ministers is its authority in the field of co-ordination and control. It co-ordinates the activity of all executive and administrative bodies as a whole and also, when necessary, the activity of individual ministries, departments or executive committees of the People's Councils.

As controlling body, the Government has the right to repeal unlawful and incorrect acts of Ministers, heads of other central departments and the executive committees of People's Councils.

The Council of Ministers adopts decrees, orders and decisions, which regulate social relations at the organizational and executive level. This determines their legal force, which follows that of the law and of the decrees of the State Council.

A characteristic feature of the structure of the Council of Ministers is the presence of a Bureau, which functions as an inseparable part. It is made up of members of the Council of Ministers, to assist the supreme executive and administrative body in its operative and administrative activity. Its tasks therefore have a functional orientation. This also determines the composition of the Bureau, which includes the Deputy Chairmen of the Council of Ministers and the heads of the functional ministries and departments.

The Bureau of the Council of Ministers carries out some of the functions of the Council of Ministers, especially those connected with the operational management of the economic and social processes. These are mainly organizational, co-ordinating and control functions. It gives advance consideration to drafts of important government documents. It also approves a programme for the development of such individual sectors as do not affect the main proportions of the national economy, takes measures for the proper functioning of the economic mechanism, decides current questions of an economic and financial character, matters related to economic, scientific and technical co-operation, etc.

The ministries and other departments (committees) each organize the execution of state policy in a certain sector of the national economy or the implementation of a certain state function. They are constituted by the National Assembly. Ministers and chairmen of committees with the rank of ministries are elected by the National Assembly and are members of the Govenment.

The activity of a branch ministry is limited to that branch. The acts that branches issue generally have a binding force only with regard to their subordinate organizations and institutions.

The situation with the functional ministries is different. They are concerned with the whole state executive apparatus in order to create unity of purpose in its activity. Thus, for instance, as a result of planning the tasks and resources of the country are distributed between the branch bodies to attain certain goals in the development of society as a whole, or in the separate branches. The purpose of the central functional state bodies is to guide economic, social and cultural development by regulating, controlling and co-ordinating the activity of state bodies.

An essential element in the legal status of the ministries is the principle of one-man management. At the head of the ministry or committee stands a Minister or his equal, who is responsible to the higher state bodies for overall activity of the body managed by him. He organizes the work of the leading officials in the ministry (Deputy Ministers, Directors, etc.), and guides and controls the activity of the whole system. He carries out his guiding functions both directly and through legal acts issued by him – rules and regulations, orders, instructions, etc.

In the administration of a ministry, the one-man principle is combined with the collective principle. In every ministry a college is formed, its composition being determined by the Council of Ministers. It is headed by the Minister, Deputy-Ministers and General Secretary; certain outstanding specialists are usually appointed as members also. The college discusses and takes decisions on the fundamental issues of the ministry. The collective principle in the management of the ministry helps the taking of competent decisions on complex and important questions. The decisions are signed by the Minister and, in case of disagreement, the Minister acts at his own discretion and refers controversial issues to the Council of Ministers.

The Council of Ministers may take branches of the administration under direct management by forming commissions, councils, general directorates and administations which do not have the rank of ministries. There are also branch or functional bodies, which manage a certain state activity on a state-wide scale.

The Supreme Court. In the People's Republic of Bulgaria justice is administered by the courts. The judicial

The District Court in Russe

system includes the regional and district courts, tribunals and the Supreme Court. Every court has its own seat – a corresponding population centre, and judicial region – the territory in which it conducts its activity. The courts apply the laws strictly and fairly towards all citizens and juridical persons. All citizens of Bulgaria are equal before the law.

Judges and jurymen are elected. In fulfilling their functions they are independent and obey only the law. Only Bulgarian citizens with legal education can be elected judges.

The Supreme Court is the highest judicial body; it exercises supreme judicial supervision over the activity of all courts and ensures exact and equitable application of the laws. It also supervises the activity of the special jurisdictions, unless otherwise stipulated by the law. The judges and the jurymen of the Supreme Court are elected by the National Assembly for a term of five years.

The Supreme Court consists of a president, deputy presidents, presidents of departments, judges and jurymen. There are three divisions: criminal, civil and military, which are themselves divided into sections and court sessions. The judges form the plenum of the Supreme Court.

As court of first instance, the Supreme Court hears cases of great significance. As an appellate court, it hears complaints and protests against judicial acts by other judicial bodies. It also plays a supervisory role – examining cases made for the repeal of acts concerning all courts in the country and the special jurisdiction that have entered into force.

The general assembly of the division of the Supreme Court issues interpretative decisions which serve as guidelines in the work of the courts and the special jurisdictions. The plenum of the Supreme Court issues interpretative decrees which are obligatorily observed by the courts and special jurisdictions, as well as by administrative bodies whose acts are subject to judicial supervision.

The Supreme Court has the right to submit bills to the National Assembly.

It reports to and is responsible for its activity to the National Assembly and, between sessions, to the State Council.

The District Court consists of a president, deputy presidents, judges and jurymen. The judges are elected by the National Assembly for a period of five years and the jurymen are elected on the basis of universal, equal and direct suffrage by secret ballot, also for five years. Between sessions of the National Assembly, district court judges are elected by the State Council. Collectively, the judges of the District Court form its general meeting. The District Court examines and decides – as court of first instance – civil and criminal suits in the cases determined by law, and – as appellate court – the sentences and judgements of the regional courts, against which appeals have been made.

The Regional Court is the basic, first instance court in the Bulgarian judicial system. It acts within the framework of a certain territorial region, determined by the State Council, and consists of a president, judges and jurymen. The president, deputy president and judges are elected by the National Assembly and – between sessions – by the State Council, for a term of five years. The jurymen of the Regional Court are elected directly by the electorate, also for five years. The Regional Court is competent to hear all civil and criminal cases, at first instance, except those which are entrusted by law to another court. Within the limits set by the law it also examines administrative cases based on appeals and protests against acts of special jurisidictions and other bodies.

The Public Prosecutor's Court of the People's Republic of Bulgaria is a system of state bodies whose basic task is to defend the established social and state system and the rights and lawful interests of the citizens and of state economic and public organizations. It performs its functions along the following basic lines: it supervises exact and equitable observance of the law by the ministries and other departments, the local state bodies, the economic and public organizations, officials and citizens; it combats crime and violations of the law perpetrated in the country, instigates proceedings or takes measures to do so against the guilty person; guides and supervises the application of the law by the bodies for preliminary investigation – the investigating judges and officers, in accordance with the rights with which it is vested; takes part in hearing cases by the courts and other bodies administering justice and supervises the serving of the punishments and other corrective measures imposed on persons found guilty; takes measures for the repeal of unlawful acts and restoration of violated rights; carries out or organizes an investigation of crimes and other legal violations and elaborates measures for overcoming them and for the removal of the causes and conditions that give rise to them; and takes an active part in the perfection of the country's legislation.

The Public Prosecutor's Office is a unified and centralized system. It consists of a Chief Prosecutor's Office, and District and Regional Prosecutor's Offices; the Military Prosecutor's office is also included in the system.

The Chief Public Prosecutor is the highest official, who exercises supervision directly and through his subordinate prosecutors for the exact and uniform observance of the law by ministries, other departments, local state bodies, economic and public organizations, officials and citizens. The Chief Public Prosecutor is elected by the National Assembly for a period of five years. He can also be removed before the expiry of his term. He is responsible to the National Assembly and accounts for the activity of the Public Prosecutor's Office before it and – between sessions – before the State Council. He appoints and dismisses the other public prosecutors and has the right of legislative initiative. The main tasks, functions and competence of the Chief Public Prosecutor are regulated by the Constitution of 1971 and by the Public Prosecutor's Office Act.

Public Prosecutors in Bulgaria are independent and act only on the basis of the law. They seek the co-operation of the state bodies, public organizations and work forces and rely on the active assistance of the citizens. The orders of the Public Prosecutor are mandatory for officials and citizens.

The Municipal People's Council in Kazanluk

Bulgarian citizens with higher legal education can be appointed Prosecutors.

The People's Councils are local representative bodies of state power and popular self-government. Their establishment was stipulated by the 1947 Constitution and they constitute an important part of the system of state bodies. According to the 1971 Constitution the country is divided into districts and municipalities. The organs of state power in these administrative territorial units are the District and Municipal People's Councils. They are elected for a period of two and a half years on the basis of universal, equal and direct suffrage by secret ballot. In addition, in every population centre in the municipalities, with the exception of their administrative centres, the population elects its own local self-government body, the mayoralty, which consists of a mayor and three to seven councillors.

From 50 to 150 councillors are elected to the Municipal People's Councils and from 70 to 150 to the District People's Councils; the People's Council of the capital, Sofia, has 250. To direct the operational activity every People's Council elects an executive committee of its own, consisting of five to seventeen members. It also elects, from among its own members, the standing commissions, which are its auxiliary bodies. As local bodies the People's Councils have a territorially limited competence within the respective administrative territorial unit. They pursue state policy in their own area, develop activities for the fulfilment of general state tasks and decide questions of local significance. Within the limits of their competence they guide the development of the economy, public health, social and communal services, and cultural and educational activities. They guide, co-ordinate and control the activity of the economic organizations and institutions within their area. They safeguard the observance of public order, legality, and the protection of the rights and interests of the citizens, to preserve socialist property and strengthen the country's defensive capability. The People's Councils elaborate, cost and adopt plans for social and economic development in accordance with the state Plan and Budget, and organize their fulfilment. They adopt decisions, orders, rules, regulations and instructions. The hierarchically superior People's Councils guide and control the activities of the inferior ones, and revoke unlawful or incorrect acts and actions by these or their executive commitees. Disputes between Ministers or heads of other departments and the executive committee of a District People's Council are settled by the Council of Ministers. The People's Councils rely on the initiative and wide participation of the population and the political, professional and other public organizations. At least once a year, in a manner determined by the law, they make a report of their activity to the electorate.

The executive committee is part of the system of the People's Council. It connects the representative body which adopts decisions with the people implementing the decisions; the representative with the professional executive bodies of state power. It is formed by the representative body from among the councillors. Only a councillor can be a member of the executive committee.

Only some of the members of the executive committee are salaried by the Council – the president, the deputy presidents and the secretaries. The remainder – specialists, political and public workers – participate as a public duty.

The executive committee is a local executive and administrative body with general authority. The latter determines its functional subordination to the hierarchically higher executive committee and to the Council of Ministers. The executive committee organizes the fulfilment of the tasks originating from the decisions of the People's Council and from the acts of the National Assembly, the State Council and the Council of Ministers.

In managing specific branches the executive committee observes the normative acts of the respective ministries. It takes decisions and gives instructions on the basis of the law and other normative acts.

For the immediate, operational guidance of economic and social activities the People's Councils form specialized branch and functional executive and administrative bodies – administrations, departments, directorates. The executive committee conducts general guidance of these bodies, co-ordinating and controlling their activity.

The State Awards of the People's Republic of Bulgaria are honorary titles, orders and medals and other awards for contribution or distinction conferred by the State Council. They were instituted on the basis of the 1947 and 1971 Constitutions and are legally regulated by decree of the State Council.

The conferring of State Awards takes place in a ceremony in accordance with the protocol approved by the State Council. The mass information media popularize the names and examples of the laureates.

Public Organizations. Public Organizations in Bulgaria are voluntary self-governing alliances, founded and governed according to the principles laid down in the 1971 Constitution and according to a charter endorsed by each organization. They represent an essential element in the political structure of socialist society. The cultural unions, state and other associations of citizens are public organizations.

Mass organizations like the Fatherland Front, the Bulgarian Trade Unions, the Dimitrov Young Communist League and the co-operative unions are of great importance in a socialist society. The Constitution stipulates the right of citizens to form organizations with political, professional, cultural, artistic, scientific, religious, sports and other non-economic purposes. They can form co-operative unions for carrying out mutual economic activities. Public organizations draw various strata of the population into the process of socialist construction, represent and protect their individual interests and work to raise the level of their socialist consciousness. They unite the personal interests of the citizens with their participation in the life of society in the pursuit of socially significant ends; increasingly they assist state bodies in the fulfilment of their tasks. In accordance with the Constitution, public organizations, like state bodies, strive to make citizens' rights a reality.

The state and the public organizations propagate health education and culture, encourage sports and tourism, and take special care of the health of children and young people. Working mothers in Bulgaria enjoy the special protection of the state, economic and public organizations. At the current stage of building the developed socialist society the importance of public organizations in Bulgaria has grown considerably, and the State relies increasingly on their assistance.

With the broadening of democracy in the working of state bodies, the role of public organizations in the social management of the country expands. Social management is the principal line of developing socialist democracy in Bulgaria. It is determined mainly by the growing socio-political unity of society, by the process of development of the dictatorship of the proletariat into a State of all the people. The forms, means and methods of interaction between the public organizations, the state bodies and the work forces are being perfected. The functions and rights of public organizations are being developed and enriched. They have steadily come to aid the State to a growing extent in forming its organs, through nominating national representatives, participating in collective state bodies, developing them on a public basis, and creating public-state bodies. They contribute towards the implementation of control over the activity of state organs, the taking up of some of their functions by public organizations, and the exercising of the right to legislative initiative.

As a guiding force in society and the State, the BCP takes political leadership of all public organizations and movements, in accordance with the development of the socialist State and the construction of the developed socialist society.

The Bulgarian Communist Party. The Bulgarian Communist Party is a vanguard detachment of the working class, the leading political force in the entire socio-political life of the People's Republic of Bulgaria. The most conscious section of the working class, farm labourers and the people's intelligentsia have been organized in the Party on a voluntary basis. (Statute of the BCP, Sofia, 1981, p. 5). It is guided by the teaching of Marxism-Leninism and its ultimate goal is the construction of communist society.

The BCP is built on the principle of democratic centralism. Its supreme body is the Congress and – between Congresses – the Central Committee. The Central Committee elects the Politburo and the Secretariat from its own composition to guide political and organizational work between meetings of the Plenum. District, city, regional and municipal Party organizations have been set up, corresponding with the administrative division of the country. The primary party organization or cell is the basic organizational unit. On 1 January 1984, the Party numbered 892,000 members. The *Rabotnichesko Delo* (Workers' Cause) newspaper and the *Novo Vreme* (New Times) theoretical journal are the press organs of the Central Committee; its other publications include the journal *Politicheska Prosveta* (Political Education), *Partien Zhivot* (Party Life), and *Politicheska Agitatsia* (Political Agitation) and the newspaper *Ikonomicheski Zhivot* (Economic Life). Todor Zhivkov is General Secretary of the Central Committee (he was First Secretary from 1954–81). The members and candidate-members of the Politburo of the Central Committee of the BCP are: Todor Hristov Zhivkov, Georgi (Grisha) Stanchev Filipov, Dobri Marinov Djurov, Yordan Nikolov Yotov, Milko Kolev Balev, Ognyan Nakov Doinov, Pencho Penev Kubadinski, Peter Toshev Mladenov, Stanko Todorov Georgiev, Todor Iliev Bozhinov, Chudomir Assenov Alexandrov, Andrei Karlov Lukanov, Georgi Ivanov

Atanassov, Georgi Yordanov Momchev, Grigor Georgiev Stoichkov, Peter Georgiev Dyulgerov, Stanish Bonev Panayotov, Dimiter Ivanov Stoyanov, and Stoyan Dimitrov Karadjov.

Scientific services to the Central Committee of the BCP are rendered by the Academy of Social Sciences and Management, the Institute of Bulgarian Communist Party History and the Institute of Public Administration.

The Academy of Social Sciences and Management at the Central Committee of the Bulgarian Communist Party is a centre for training specialists in the field of social sciences and public administration, and for research. The Academy was founded in 1969 on the basis of the Higher Party School (in existence until then) at the Central Committee of the BCP, the Institute of the Organization of Management at the Council of Ministers and the Centre for Further Training of Management Personnel at the Ministry of Labour and Public Welfare. The Academy maintains close contacts with related institutes and centres in the USSR and the other socialist countries, and engages in joint research projects, symposia, seminars and other measures to exchange experience. It publishes *Scientific Studies* (since 1970) in four sections: 'Philosophy', 'Economics', 'History' and 'Public Administration'.

The Institute of Bulgarian Communist Party History at the Central Committee of the Bulgarian Communist Party was founded in Sofia in 1953. It co-ordinates nationwide research projects on problems of the history of the BCP under Capitalism and Socialism, and the history and contemporary problems of the Balkan and international communist and workers' movements. It publishes collective and individual scientific papers, monographs, collections of documents, etc. The Institute maintains ties with related scientific institutes in the USSR and other countries, organizes joint research projects, and takes part in bilateral and multilateral symposia and conferences. Its scientific publication is *News of the Institute of Bulgarian Communist Party History* (since 1957).

The Institute of Public Administration at the Central Committee of the Bulgarian Communist Party was founded in 1969 within the framework of the Academy of Social Sciences and Management. In 1975 it separated from the Academy of Social Sciences and Management to become an independent Institute. It engages in research, both theoretical and practical, and in teaching aspects of public administration. The Institute maintains ties with related institutes and centres in the USSR and in other socialist countries, and engages in joint research projects.

The emergence of the Communist Party in Bulgaria was a historically inevitable result of the development of Capitalism, the birth of the working class and the working-class movement, and the dissemination of Marxism at the close of the nineteenth century. The BCP dates back to 1891. On the initiative of Dimiter Blagoev, the founder of Socialism in Bulgaria, the Constituent Buzludja Congress was held on the peak of Buzludja in the Balkan Range on 20 July (2 August) 1891. The first congress was attended by fifteen delegates. Socialist societies from the towns of Turnovo, Gabrovo, Sliven, Stara Zagora,

Commemorative plaque where the Constituent Congress of the BSDP was held on Mount Buzludja in 1891

Kazanluk and other places united into the Bulgarian Social Democratic Party (BSDP). The Congress adopted the Programme and Statute of the Party based on the principles of scientific Socialism, elected a leading body – the General Council, and in order to propagandize the ideas of scientific Socialism and the programme demands of the Party, it decided to start the publication of a periodical and a series of pamphlets, the 'Bulgarian Social Democratic Library'. Dimiter Blagoev's book *What is Socialism and Can it Thrive on Bulgarian Soil?* (1891), the first original Marxist work in Bulgaria, was the first issue of the series. It was decided to use the monthly *Den* (Day; 1891–96) published by Yanko Sakuzov, for the purpose of socialist propaganda until the publication of a weekly Party organ. The BSDP was created in the struggle against the populist-utopian, nihilist and other ideological trends, the representatives of which held the view that no conditions existed in Bulgaria for the dissemination of Marxism or for the existence of a socialist movement under the underdeveloped capitalist relations of the time. Opposing the opportunists who stood against the creation of the Party, most of the socialist societies, headed by Dimiter Blagoev, called the Second Congress of the BSDP in Plovdiv in August, 1892. The congress reaffirmed the decisions of the Buzludja Congress to openly announce the formation of the Party and to start the publication of the Party organ *Rabotnik* (Worker; 1892–94). The decision that the Party should engage in propaganda among industrial workers and the other working people was of paramount significance. In the political struggle the Party was to act independently in defence of the short-term and long-term interests of the working class. The members of the BSDP were called 'partyists'. In 1892 the opportunist-led societies formed an organization of their own, which was called the Bulgarian Social Democratic Union. Its members were called 'unionists'. Sharp polemics developed between the partyists and the unionists. On the pages of the *Rabotnik* newspaper the partyists drew attention to the process of capitalist development, which had started in Bulgaria, and to the need for a Marxist party of the proletariat. Proceeding from the position of economism, the unionists rejected the necessity for such a party. On the pages of their organ the *Drugar* (Comrade) newspaper (1893–94), they upheld the view that the socialist movement in Bulgaria had to be initiated through the formation of workers' societies designed to fight for economic gains. In the course of that struggle, they asserted, the workers themselves would arrive at the stage of political action. Using more solid argumentation, the partyists refuted the views of the unionists, giving a thorough explanation of the leading role of the BSDP in the workers' movement, and, without giving up the economic struggle, they stressed the primary importance of the political struggle. In 1894 the partyists and unionists were united into one single party called the Bulgarian Workers' Social Democratic Party (BWSDP). Two trends took shape within the Party: the revolutionary-Marxist one, headed by Dimiter Blagoev, and the reformist-opportunist one, led by Yanko Sakuzov. Upholding the Marxist position, Dimiter Blagoev and his fellow-workers Gavril Georgiev and Georgi Kirkov exposed the attempts of the opportunists to replace Marxism with petty bourgeois views, to make the Party engage in large-scale activities among the peasant and urban petty bourgeoisie, and to force upon the Party the conciliatory tactics of adaptation to the interests of the petty bourgeoisie minimum programme propaganda, election compromises and press propaganda outside Party control. The decisions of the Fourth Congress (1897), in particular the decision to start the publication of the *Rabotnicheski Vestnik* (Workers' Newspaper; 1897–1939), designed to raise the level of class consciousness of the workers and reflect and guide their struggle, constituted a considerable achievement in the drive for closing party ranks among the working class and for improving the organizational state of the BWSDP. The first issue of *Novo Vreme* magazine (January 1897), edited by Dimiter Blagoev, was published before the Congress. The *Rabotnicheski Vestnik* newspaper and *Novo Vreme* magazine played a major role in the ideological struggle of revolutionary Marxism against opportunism. At the insistence of the Marxists, the Fifth Congress of the BWSDP (1898) adopted a decision to centralize the party Press. That decision was based on the view that the Party had to be proletarian in terms of its composition and nature. Differences between the two trends flared up still further during the emergence of the peasant movement and during the mass peasant riots (1900) against the introduction of agricultural taxation in kind, which had been in existence during the Ottoman bondage. At the Seventh Congress of the BWSDP (1900) the Marxists resolutely condemned the repressions to which the Government had subjected the peasants. But they rejected the attempts of the opportunists to re-direct the Party's activities to the countryside, and to open its doors wide to craftsmen and peasants possessing small landholdings. Resolutely fighting for the preservation of the proletarian nature of the Party, the revolutionary Marxists, however, paid no heed to the progressive aspect of the peasant movement or to the revolutionary potential of the peasantry. After the Seventh Congress opportunism assumed a new character and became a faction within the BWSDP. Its ideologist Yanko Sakuzov started the publication of the magazine called *Obshto Delo* (Common Cause; 1900–05), in which he expounded the thesis of uniting all 'productive social strata' in order to implement the 'democratic principles' of reforms which allegedly led to Socialism. In the so called 'productive social strata', Sakuzov placed the peasants first, then the craftsmen and the commercial and industrial stratum (i.e. the capitalists). The workers came last. Through the common cause programme, the Bulgarian opportunitists, called *obshtodeltsi* after their publication and programme, rejected the Marxist view that class differences under Capitalism were irreconcilably antagonistic, and declared themselves for co-operation with the bourgeoisie. The Marxists, or left-wing socialists as they were called, resolutely opposed Sakuzov's opportunist theory. In his celebrated articles *Pro domo sua* (1900), 'Marxism or Bernsteinism' (1901), 'Opportunism or Socialism' (1902) and in other publications, Dimiter Blagoev exposed the views of the Bulgarian opportunists as an attempt to transplant Bernstein's revisionism onto Bulgarian soil and to replace Marxism with reformism. In his argument, Blagoev used Lenin's work *Chto Delat'* (*What is to be Done?;* 1902) and Lenin's newspaper *Iskra* (Spark; 1900–03), which for a time was imported to Russia through Bulgaria. At the Eighth Congress (1901) and the Ninth Congress (1902), in the party Press and in their day-to-day work, the Marxists fought relentlessly against opportunism and in support of upholding scientific Socialism. This was a struggle for the very nature, role and tasks of the Party and for the future of the revolutionary movement in Bulgaria.

The Party got rid of the opportunist elements in 1903, which saved it from the danger of degenerating into a petty bourgeois reformist organization. This came about as a natural result of the profound theoretical, tactical and organizational differences between the Marxist and the opportunist trends. The action was approved by the Tenth Congress of the BWSDP held in the town of Russe in 1903. In this way the Party proved itself to be a revolutionary Marxist party of the working class under the name of the BWSDP, known at the time and in scientific circles as the party of 'narrow' socialists. The Congress introduced democratic centralism as the basic organizational principle. The *Rabotnicheski Vestnik* newspaper became the Party's central organ and the *Novo Vreme* magazine its theoretical organ.

Free of the opportunists, the Party directed its main efforts to organizing and educating the working class in the spirit of scientific Socialism. The General Workers' Trade Union (GWTU) was formed under its guidance in 1904. The Party organized and led a large number of strikes to improve the living standard of the working people and to gain political rights for the working class. It also engaged in extensive work among young people; the Union of Workers' Social Democratic Youth was founded under its leadership in 1912. The role of the Party in the country's public and political life grew in importance simultaneously with the growth and the increasing ideological maturity of the working class. In 1908, in the National Assembly elections, the Party obtained 2,500 votes, while in 1913 it obtained 54,297 votes and 18 seats in the National Assembly. Representatives of the Party were elected to many municipal and district councils. In 1910, in the town of Samokov, and in 1912 in Sliven, the BWSDP received most of the votes and communes were set up. The Party displayed remarkable qualities as a revolutionary and Marxist Party during that period. It waged a relentless struggle against the opportunists (right-wing socialists) and against their attempts to subordinate the interests of the working class to those of the bourgeoisie. The Party purged its ranks of the centrist, anti-Party groupings of the anarcho-liberals (1905) and the progressists (1908). Led by Dimiter Blagoev, the Party unmasked all opportunist trends, such as Yanko Sakuzov's followers, the supporters of Bernsteinism, centrism, etc. within the international socialist movement. The Party's characteristic features were: profound devotion to the ideas of Marxism and proletarian internationalism, unshakable confidence in the strength of the working class, class intransigence toward the bourgeoisie, and self-imposed, iron discipline within the Party ranks. All this made it an original, revolutionary Marxist Party in the international workers' movement, similar to the Bolshevik Party in Russia. The Party waged a courageous struggle against the Balkan and the Inter-Allied Wars (1912–13), and against the First World War (1914–18). It firmly defended the position of proletarian internationalism and raised the slogan of peace and a Federative Balkan Republic. Its parliamentary group voted against the war credits. The BWSDP condemned the stand of the opportunists of the Second International, who advocated the defence of the bourgeois homeland, as an act of betrayal of the interests of the proletariat. In his articles *Magister dixit* (1914), *Plekhanov and Parvus* (1915), *Kautsky and Plekhanov* (1915) and *The International and the War* (1915) Dimiter Blagoev criticized the views of the leaders of the Second International and insisted on the creation of a new International on the basis of revolutionary Marxism. This was a valuable contribution to the struggle against opportunism and social chauvinism on an international scale. The Party took an active part in the Zimmerwald Movement, welcomed the victory of the Great October Socialist Revolution of 1917 with great enthusiasm and launched large-scale activities at the front-line and within the country to popularize its slogans.

Despite its revolutionary virtues, during the period from 1903 to the October Revolution in 1917 the BWSDP had not as yet become a party of the new Leninist type. It differed from the Bolsheviks on a number of basic problems in the theory, strategy and tactics of the revolutionary struggle. Although its activities infused the masses of soldiers at the front-line with revolutionary ideas, for the above-mentioned reason the Party was unable to head the Soldiers' Mutiny in 1918, which was defeated.

The BWSDP embraced Leninist principles under the influence of the October Revolution. Unlike many other parties, it entered the Communist International, being one of its co-founders. At its Twelfth Congress, held in 1919 and declared as the First Congress of the BCP, it was given its new name and adopted a Programme Declaration, stressing that the Great October Socialist Revolution had ushered in the era of the downfall of Capitalism. The seizure of political power by the proletariat was defined as an immediate, practical task, and the armed uprising was characterized as the means of achieving this end. Demands for a dictatorship of the proletariat in the form of Soviet power, and for the creation of a people's militia and of a Red Army, were also advanced. The demand was also put forward for turning the means of production and exchange into public property, for an all-round defence of workers, peasants and small-property owners. In spite of some imperfections, the Programme Declaration was a major step forward in the development of the Party along the road to Leninism. In the conditions of the post-war revolutionary crisis of 1919–23, the BCP established itself more strongly than ever as the vanguard of the working people in their struggle against the onslaught of capital, and became a major political force. In 1922 it numbered 38,036 members. The GWTU, numbering about 30,000 members, the Bulgarian Communist Youth Union with a membership of about 14,000, and other mass organizations, such as the Central Women's Commission – numbering 4,500 members, the *Osvobozhdenie* (Liberation) Co-operative, with a membership of 70,000, and the Union of Former Prisoners-of-War and Hostages numbering 16,521 members, acknowledged the leadership of the Party. The circulation of the *Rabotnicheski Vestnik* was about 30,000. The Party organized and led a sweeping strike movement against exploitation, starvation and misery. It organized a 55-day strike of transport workers (December 1919–February 1920) and the one-week General Political Strike of 1919. The Party defended the cause of Soviet power in Russia with great determination. It organized a large-scale operation to help the victims of the Volga Famine (1921–22) and directed the struggle to drive the White Russian counter-revolutionaries of the Wrangel faction out of Bulgaria. In its role of revolutionary vanguard, the BCP defended the working people and rapidly grew in prestige and influence throughout the country. In 1920 it scored considerable successes in the elections – its parliamentary group numbered 42, second after the Bulgarian Agrarian Party (BAP). In the same year, the Party won elections in 32 town and 65 village municipal councils, which turned into communes. In 1922 it had 3,623 municipal and 115 district council members, and 1,496 trustees of schools. The Third Congress of the BCP, held in 1921, adopted a resolution on the agrarian question based on Lenin's teaching of the

May Day celebration, Sofia, 1905

alliance of workers and peasants as a necessary condition for the victory of the revolution. The Fourth Congress, in 1922, adapted the tactics of the Comintern for a united workers' front to suit Bulgarian conditions. The GWTU offered to joint forces with the reformist and 'neutral' trade unions against capitalist exploitation. In 1923 the Party raised the slogan of a government of workers and peasants, and in its struggle against the onslaught of Fascism, the Party (left-wing socialists) took practical steps for united action with the BAP (for example, by participation in a common bulletin in the 1922 referendum concerning the trial of those responsible for the two national catastrophes). However, the Party was still unable to regard the democratic organizations as its allies in the struggle against the reactionary bourgeoisie.

A fascist dictatorship was established in Bulgaria after the military and fascist coup of 9 June 1923. In opposition to the coup, the working people, under the leadership of the local Communist and Agrarian Party officials, took part in an armed rising in a number of places in the country – the Pleven, Karlovo, Shumen, Pazardjik districts, and elsewhere. At that point, however, the Central Committee took the false position of 'neutrality'. In the conditions of grave political crisis following the coup, the sound Marxist nucleus in the Party leadership, headed by Vassil Kolarov and Georgi Dimitrov, assisted by the Executive Committee of the Communist International, brought about a change in the strategy and tactics of the BCP in an effort to unite anti-fascist forces for an armed uprising to establish a worker-peasant government. The Session of the Central Committee of 5–7 August 1923 instructed the Party to start immediate political and military preparations for an armed uprising. The BCP called on the Bulgarian Agrarian Union (BAU) and the other non-fascist parties and organizations to unite into a common front. The Party explained that Fascism was not only anti-communist but also anti-popular. The world's first anti-fascist uprising, organized and led by the BCP, and headed by Vassil Kolarov and Georgi Dimitrov, broke out in Bulgaria in September 1923. Despite the heroism of the insurgents, the anti-fascist September Uprising failed for a number of objective and subjective reasons. However, it became a turning-point in the process of the Leninization of the Party. For the first time in Bulgarian history, under the guidance of the BCP, the insurgent people attempted to establish worker-peasant power. Indeed the Party demonstrated its readiness and ability to use all revolutionary means, including armed uprising, to achieve political power. The tactics of a united labour front, based on an alliance of workers and peasants, were put into effect during the uprising.

After the defeat of the uprising, the BCP was officially

Members of the BWSDP at its Sixteenth Congress, Varna, 1909

outlawed under the fascist Law on the Defence of the State, 1924. In extremely clandestine conditions the Party went on with its ideological, strategic and tactical reorientation towards Leninism, uniting reliable party members on the basis of the September line of the BCP. The Vitosha Clandestine Conference of the BCP, 1924, rallied the sound party forces, condemned the right-wing liquidating elements that had appeared within its ranks, and took the decision to go ahead with preparations for an armed uprising.

A process of temporary and partial stabilization of Capitalism and an ebb in the revolutionary movement was noticeable at the end of 1924 and the beginning of 1925. The party leadership in the country did not pay sufficient heed to the fact that the situation made it imperative to change the line of preparing for an armed uprising. An ultra-leftist wing sprang up within the BCP, the representatives of which substituted legal forms of mass activity with actions by partisan detachments and punitive acts, thereby breaking its links with the masses. Using the bombing in the St. Nedelya Cathedral (April 1925) as a pretext, the fascist authorities launched a massacre of the leaders and followers of the communist anti-fascist movement. The Moscow Conference of the BCP, 1925, instructed the Party and the worker's movement to work closely with the masses and to replenish their strength for a future revolutionary upsurge. The Vienna Extended Plenum of the Central Committee of the BCP (left-wing socialists), 1926, elaborated on this line of tactics in greater detail. The Second Regular Clandestine Conference of the BCP, held in Berlin in December 1927–January 1928, denounced the right-wing opportunists and leftist sectarians who had combined in an unprincipled bloc in their struggle against the leadership of the Party. The BCP campaigned to restore the party organization underground, and to build up revolutionary trade unions and unite the working class within them. It was successful in using various legal means to re-establish its links with the working people and to organize their battles against the onslaught of capital and the fascist reign of terror. The Workers' Party (WP), which was formed in February 1927 and was used by the BCP for legal work, with the *Rabotnichesko Delo* newspaper as its press organ, played a major role in re-establishing the Party's links with the working people in town and countryside. The WP was a class revolutionary organization of the proletariat which gradually expanded and embraced about 30,000 members. It exercised wide influence on the working people, becoming a genuine organizer and leader of the class struggle. Through the WP, the Independent Workers' Trade Unions and the Workers' Youth Union, the BCP headed the new revolutionary upsurge of the masses in

Dimiter Blagoev and other Bulgarian delegates at the Second Balkan Social Democratic Conference, Bucharest, 1915

Bulgaria and led the strike movement involving thousands of workers from 1929 onward. Despite the atrocious police terror in the legislative elections of 1931, the WP obtained 170,000 votes and 31 parliamentary seats, while in the municipal elections of 1932 it received 230,000 votes, and in the number of votes cast in Sofia surpassed all the bourgeois parties put together. Nevertheless, the bourgeois authorities did not hand over the Sofia municipality to the communists.

Due to the heavy blows inflicted on the Party and its members after the September Uprising of 1923 and the April events of 1925, and because of errors in the struggle against the right-wing and left-wing deviations, the leftist sectarians – who pursued a mistaken line – dominated the leadership of the Party at the Second Extended Plenum of the Central Committee of the BCP held in Berlin in 1929. Their dogmatic assessments, methods of leadership and extreme left-wing slogans impeded the revolutionary upsurge of the working class and working peasants during the Depression of 1929–34 and after, hampering the construction of a united front and leading to the Party's detachment from the masses. The left-wing sectarians became a serious obstacle in the way of party rearmament with the principles of Leninism.

The sound Marxist-Leninist nucleus within the Party, headed by Georgi Dimitrov and Vassil Kolarov, waged a relentless fight against the leftist sectarians. During the Reichstag Fire Trial of 1933, through his courageous conflict in the fascist court, Georgi Dimitrov became a standard-bearer in the struggle against Fascism. His victory in Leipzig was also a serious blow to the Bulgarian leftist sectarians. Under the leadership of Georgi Dimitrov, the sound forces of the Party began to overcome leftist sectarian deviation and dogmatism.

The coup of 19 May 1934 strengthened the fascist dictatorship, and the WP, Workers' Youth Union and Independent Workers' Trade Unions were subsequently banned. The BCP was unable to mobilize the working class and working people in mass resistance to the onslaught of fascist dictatorship, and this fully revealed the impotence of the leftist sectarian leadership and the erroneousness of its policy. The decisions of the Fifth Extended Plenum of the Central Committee of the BCP, held in 1935, marked a considerable step forward in vanquishing leftist sectarian deviation, achieving a breakthrough in the life of the Party. The Plenum renounced the line followed by the leftist sectarian leadership before and after the coup of 19 May and set the construction of a broad united front against rising Fascism as the principal political task of the Party. The Plenum made personnel changes in the Central Committee, electing officials of the WP and the Independent Workers' Trade Unions, who had broad experience in the mass work of the Party.

Georgi Dimitrov delivering the concluding speech at the Fifth Party Congress, 1948

The historic decisions of the Seventh Congress of the Comintern, 1935, on the construction of a united workers' broad Popular Anti-Fascist Front against Fascism and war, became the programme of action of the BCP, too. Under the leadership of Georgi Dimitrov, who was elected General Secretary of the Comintern, the Party resolutely overcame the leftist sectarian deviation. The Sixth Extended Plenum of the Central Committee of the BCP, 1936, was a turning-point in that respect. It made concrete the decisions of the Seventh Congress of the Comintern, in accordance with Bulgarian conditions, setting the formation of the Popular Anti-Fascist Front as the Party's prime task. The Plenum paid special attention to the work of the Party among the working people, and to the expansion of its influence on the masses. Convinced followers of the new line were elected to the leading bodies of the Party, which guaranteed its successful implementation. Carrying out the policy mapped out by the Sixth Extended Plenum under difficult underground conditions, the BCP and the WP scored great successes in their ideological and organizational consolidation and in their struggle to re-establish links with the masses. The Party carried out extensive work among the working class, the peasants, youth, women and intellectuals. It also began to work actively in the state-sponsored syndicates, in the co-operative, educational, sports, temperance societies, women's and other mass organizations. In many of these organizations the Party gained wide influence, enabling it to use them to educate the working people and defend their interests. In 1936–40 the BCP organized and led a mass strike movement, achieving a certain degree of unity of action with the leaders of some other non-fascist organizations and parties – such as the *Pladne* BAP, the *Vrabcha* BAP and the BSDP – in the parliamentary and municipal elections. The same was also true of other initiatives. The Party set up constitutional and popular-front committees in defence of the democratic rights and economic interests of the working people. The underground conditions in which the communist movement was forced to function rendered useless the existence of two working-class parties: the Bulgarian Communist BCP and the WP. Following a decision of the Central Committee of the BCP (1938), the two parties merged into one class revolutionary party under the name of the Bulgarian Workers' Party (BWP). Its reconstruction was fully completed in 1940. The Party was consolidated politically and organizationally, and it rapidly expanded its ties with the masses as it prepared for the great events to come. The implementation of the new line showed that the Party had mastered Leninist strategy and tactics, having become a party of a genuinely new, Leninist type. On the eve of the Second World War and during the first year after it had started, the BCP vigorously opposed Bulgaria's involvement in it. It carried out a large-scale operation, known as the Sobolev Action, in support of acceptance of the proposal by the Soviet Government to conclude a pact of friendship and mutual assistance between the Soviet Union and Bulgaria, to save the country from another national catastrophe.

On 24 June 1941, two days after nazi Germany's treacherous invasion of the Soviet Union, the Politburo of the Central Committee of the BWP adopted the line of organizing an armed anti-fascist liberation struggle against the nazi invaders and their Bulgarian collaborators. The decision was of historic significance for the destiny of the Bulgarian nation. A Central Military Commission was set up at the Central Committee of the BWP, as were military commissions, attached to the district Party committees. The First partisan detachments and combat and sabotage groups were also formed. The new line was being carried out under the unfavourable influence of a number of internal political factors. No revolutionary situation existed in Bulgaria after 22 June 1941. Bulgaria was not overrun and occupied by the nazi troops, but was an ally of nazi Germany. When the ruling fascist top clique allied with the Axis Pact, the bourgeois authorities had a well-organized state apparatus, a strong police force and an army, which they used in their struggle against the anti-fascist movement. The Government skilfully played on the national sentiments of the Bulgarian people: the overwhelming majority of the bourgeoisie rallied round the fascist Government's nationalist policy, and a considerable part of the petty bourgeosie, as well as certain sections of the intelligentsia, were also influenced by it. The masses of the people, however, regarded the struggle of the Soviet Union with open sympathy, and hated the fascists. A process began which led to the emergence of a revolutionary situation. In determining its line, the BCP was not guided only by the internal situation or temporary political factors, but subordinated its tactics to the main strategic goal – the defeat of Bulgarian and German Fascism – by combining national and international tasks and adopting firm internationalist positions. The Party pointed out that the future of the Bulgarian nation and that of all other nations enslaved by Fascism would be determined by the outcome of the Great Patriotic War of the Soviet Union against nazi Germany, and that the efforts of the Bulgarian nation would facilitate the speedier defeat of nazi Germany and its liberation. Under the leadership of the BCP, the formation of the Fatherland Front was put under way in 1942, uniting the working class, the working peasants, the people's intelligentsia and all the patriotic and democratic forces of the Bulgarian people in a struggle against Fascism. The Fatherland Front continued and further developed the Party's tactics on building up a united and broad popular anti-fascist front. Its programme, drawn up by the Bureau-in-Exile of the BWP under the guidance of Georgi Dimitrov included, above all, democratic all-national liberation tasks which could bring the Party closer to even the most hesitant anti-fascist circles. The partisan movement in Bulgaria grew in scope as a result of the victories of the Soviet Army on the Eastern Front, and of the growing discontent of the masses with the policy of the Government. In March and April 1943 the Central Committee of the BWP divided the country into twelve insurgent operational zones, thus laying the foundations of the People's Liberation Insurgent Army. Its partisan units and combat groups carried out thousands of acts of sabotage, inspiring the people to fight the fascists, and keeping the fascist authorities in a state of constant tension. The BWP engaged in large-scale activities aimed at demoralizing the fascist army. The nationwide resistance did not allow the fascist Government to send a single soldier to the Eastern Front to fight against the Soviet Union. The National Committee of the Fatherland Front was created in August 1943. It included representatives of the BWP, the Alexander Stamboliiski BAP *(Pladne)* and the *Zveno* political grouping. The social democratic and independent intellectuals joined the National Fatherland Front Committee later. The setting up of the National Committee gave an impetus to the growth of the anti-fascist struggle on a nationwide scale. On 26 August 1944 the Central Committee of the BWP issued the historic circular letter No 4 to the leadership of the Party organizations, in which it set the preparation of an armed uprising to overthrow the monarcho-fascist dictatorship and the establishment of a popular democratic regime as the most urgent and immediate task. The last Bulgarian bourgeois Government, which did not break off the alliance with nazi Germany, was formed on 2 September 1944, and on 5 September the Soviet Government was forced to declare war on Bulgaria. That same day the Politburo of the Central Committee of the BWP and the Main Headquarters of the People's Liberation Insurgent Army worked out the plan of the uprising. Strikes and demonstrations, accompanied by armed clashes with the police, broke out throughout the country on 6 and 7 September and the armed uprising was in full swing when Soviet army units entered Bulgarian territory on 8 September. The entry of the Soviet Army paralysed the counter-revolutionary forces and saved the country from civil war. Rallied round the Fatherland Front, under the leadership of the BWP and with the decisive assistance of the Soviet Army, the Bulgarian people overthrew the reactionary dictatorship and established the popular democratic regime on 9 September 1944. The triumphant uprising paved the way for the construction of Socialism and marked the beginning of the socialist revolution in Bulgaria.

Since its very inception, the popular democratic regime established on 9 September 1944 has been a regime of the working class in alliance with the other working people in the towns and countryside under the leadership of the Party. It is essentially a dictatorship of the proletariat in the form of a people's democracy. After 9 September 1944, the BWP (communists) became the ruling and guiding force in the socio-political life of the country. It worked vigorously for the consolidation of the popular regime, directed the destruction of the old bourgeois-fascist state apparatus and the construction of a new state apparatus, and organized and guided the participation of the Bulgarian people in the war on the side of the freedom-loving nations headed by the Soviet Union to finally rout nazi Germany. The Eighth Extended Plenum of the Central Committee of the BWP (communists), 27 February–1 March 1945, worked out the Party's line of further consolidation of the popular democratic regime and adopted measures aimed at the Party's ideological and organizational consolidation. The Party already had its own organizations in all towns, enterprises and in almost all villages of the country. In January 1945 it numbered 254,140 members (on 9 September 1944 its membership was about 25,000). The Plenum adopted new Party Statutes appropriate to the new situation. The Eighth Plenum played the role of a congress in terms of the importance of its decisions. Under the leadership of the BWP (communists), revolutionary measures were consistently carried out to overcome the resistance of the reactionary bourgeoisie and create conditions for large-scale construction of Socialism in Bulgaria. Property acquired illegally during the war was confiscated, an agrarian reform was carried out, foreign and domestic trade passed under state control at an ever increasing pace, a nationwide referendum was held on 8 September 1946 on the abolition of monarchy, and on 15 September Bulgaria was proclaimed a People's Republic. Vassil Kolarov was elected President of the Provisional Presidency. The BWP (communists) received more than half the votes in the elections to the Great National Assembly in 1946. The leading role of the BWP (communists) was confirmed by Parliament. A Government was formed, headed by Georgi Dimitrov, and on 4 December 1947 the National Assembly adopted the first Constitution of the Republic, which legislatively confirmed all the revolutionary gains of the working class and working people, achieved as a result of the historic victory on 9 September 1944. By its nature, it was a Constitution of Socialism under construction. The nationalization of capitalist-owned industry, mines and banks took place on 23 December 1947. The private capitalist sector in foreign trade and wholesale commerce was abolished and privately-owned retail trade was quickly curtailed, too. The process of abolishing private ownership and eliminating the capitalist class in the towns was completed, the dictatorship of the proletariat was fully established, and all the preconditions were assured for planned socialist construction.

The Fifth Congress of the Party, held in 1948, at which it was renamed the Bulgarian Communist Party, was of historic significance to the Party and the country. The political report of the Central Committee, delivered by Georgi Dimitrov, offered a Marxist-Leninist evaluation of the history of the Party, with a detailed, scientific analysis of the main periods in its development. At the same time the Congress revealed the nature of the Uprising of 9 September as a profoundly revolutionary act which laid the foundations of the socialist revolution in Bulgaria. It also defined the concept of people's democracy as a form of dictatorship of the proletariat and mapped out the general policy of the Party – the construction of Socialism in Bulgaria. The Congress adopted the First Five-Year Economic Plan (1949–53), which constituted a large-scale programme of industrialization, socialist reconstruction, co-operation and mechanization of rural economy. A new Party Statute was adopted, reflecting the new stage in the struggle for Socialism, promoting the consolidation of the unity of party ranks, raising the quality of work and closer ties with the masses. The Fifth Congress elected Georgi Dimitrov General Secretary of the Central Committee of the BCP. After the Congress, the Party mobilized the people for the country's socialist industrialization, for the co-operativization of rural economy and for the implementation of the plan for the country's cultural rebirth. The First Five-Year Plan was fulfilled in four years, with the selfless assistance of the Soviet Union. The Sixth Congress of the BCP, held in 1954, adopted the Second Five-Year Economic Plan (1953–57) for the further construction of Socialism. The April Plenum of the Central Committee, 1956, resolutely condemned the personality cult which had temporarily appeared in the BCP at the time when the Party was headed by V. Chervenkov (1950–56) and marked a new stage in the development of the Party. The decision of the April Plenum played a major role in the full restoration of Leninist norms of Party life and improvement of the methods of Party and State leadership, and in the promotion of the activities and initiatives of the people in socialist construction. The Seventh Congress (1958) entered the history of the Party and the country as the Congress of victorious Socialism in Bulgaria. It stressed that as a result of the Party's correct Marxist-Leninist policy the exploiting classes had been eliminated, socialist production relations had been established in industry and agriculture, and the moral and political unity of the working class, co-operative farmers and people's intelligentsia had been forged; in other words, that the transition from Capitalism to Socialism had been completed. The Seventh Congress adopted Directives on the Third Five-Year Economic Plan (1958–62), setting the further construction of socialist society, with the priority development of heavy industry, as the central task. The January Plenum, 1959, which adopted a decision for an accelerated development of the national economy, for the reorganization of the state and economic leadership, and for a new administrative division of the country, also played an important role in the development of the Party. The Third Five-Year Economic Plan was fulfilled in three years. The Central Committee adopted the Fourth Five-Year Economic Plan (1961–65). The Eighth Congress, 1962, reaffirmed the correctness of the political line conducted by the Party between the Seventh and Eighth Congresses, adopting directives on the development of the People's Republic of Bulgaria for the 1961–80 period. Projections for more effective state and economic leadership, raising the material and cultural standards of the people, were mapped out. The Ninth Congress, 1966, confirmed the correctnesss of the April line of accelerated economic development and scientific and technological progress, and of increasing the strength and fostering the creative initiative of the working class, co-operative farmers and socialist intelligentsia in the drive for the further construction of socialist society in Bulgaria. The directives on the Fifth Five-Year Plan (1966–70) were also approved. The key task was the intensification and modernization of the national economy, raising labour productivity and effectiveness. The Congress gave a high assessment of the consistent and highly principled activities of the Communist Party of the Soviet Union and of the other fraternal parties toward restoring unity in the international communist and working class movement on the basis of Marxism-Leninism and proletarian internationalism.

A number of Plenums of the Central Committee held after the Ninth Party Congress were of great importance in promoting party theoretical thinking and elucidating the prospects of the People's Republic of Bulgaria in its struggle for all-round construction of a socialist society. The July Plenum, 1968, discussed problems of perfecting public administration, which were rendered increasingly urgent by the dynamic development of socialist society and by the scientific and technological revolution. Measures were mapped out affecting the overall, scientific system of public administration. The September Plenum, 1969, considered the problems of speeding up the processes of the integration of production, of the further application of the achievements of scientific and technological progress, of the automation of production and of perfecting the system of management of the national economy. The April Plenum 1970, discussed problems of the further concentration of agriculture on an industrial basis. It defined the agrarian and industrial complexes as a new and improved form of concentration in agriculture, and drew up a plan for the further mechanization and chemicization of agricultural production in order to achieve high labour productivity and cheaper agricultural production.

The Tenth Congress, 1971, which adopted the Party Programme on the construction of the advanced socialist society in Bulgaria, marked a new stage in the development of the Party and the country. The Programme offers a profound, theoretical definition of the advanced socialist society as the highest and final stage in the development of Socialism – a stage in the transition from Capitalism to Communism. It is a contribution to Marxist-Leninist theory about the construction of socialist society under specifically Bulgarian conditions. The Tenth Congress also reaffirmed the Directives on the development of the People's Republic of Bulgaria during the Sixth Five-Year Plan (1971–75), whose main social and economic task was to ensure comprehensive satisfaction of the growing material and cultural needs of the people, and to raise the level of their socialist consciousness on the basis of the total utilization of the results of the scientific and technological revolution, of the rise in labour productivity and of the economic progress of the country. The basic principles of the new Constitution were also discussed and certain amendments and supplements to the Statutes of the BCP were adopted.

The activities of the Party after the Tenth Congress were focused on the construction of the advanced socialist society as 'the principal and immediate historic task of the BCP' (Programme of the BCP, Sofia, 1971, p. 41). A number of Plenums of the Central Committee were held after 1971 in the spirit of the Tenth Congress and of the Party Programme adopted by it. The October Plenum, 1971, dealt with the reconstruction of the sciences and higher education, outlining certain steps designed to perfect the organization and management of scientific activity. The December Plenum, 1972, prepared a complex theoretical programme aimed at a gradual increase in the material and cultural standards of the nation. The November Plenum, 1973, discussed problems of agriculture, mapping out major steps for raising standards to the level of up-to-date requirements. The February Plenum, 1974, adopted decisions designed to raise the Party's ideological work to the level of the tasks set by the Tenth Congress and its Programme. It mapped out the general tendencies of future ideological work to educate the working people in the spirit of Socialism and involve them in the construction of the advanced socialist society. The National Party Conference of 1974 debated problems related to higher labour productivity, pointing out the main factors and reserves for raising productivity, as well as ways and means for their most effective use.

The Eleventh Congress, 1976, summed up the experience of the Party after the Tenth Congress, enriching and making concrete the formulations in the Party Programme on the future construction of the advanced socialist society. The Congress adopted the *Main Guidelines for the Country's Social and Economic Development* during the Seventh Five-Year Plan (1976–80), whose main task was the overall intensification of the national economy, and improving the efficiency and quality in all spheres of social activity. The July Plenum, 1976, discussed measures for the consistent fulfilment of the decisions of the Eleventh Congress. It mapped out practical ways for the consistent application of Leninist principles in social policy, of observing a strict regime of economizing on material and time, and of achieving a high rate of effectiveness in the utilization of the country's labour, material, financial and currency resources. The National Party Conference of 1978 discussed questions connected with perfecting the socialist organization of labour and planned management as a decisive factor in the intensification of the economy, in achieving a high rate of effectiveness and high quality, and in the further construction of the advanced socialist society.

The decisions of the Twelfth Congress of the BCP, 1981, are a considerable step forward towards the advanced socialist society. The Congress evaluated the main results of the Party's activities aimed at implementing the Party Programme, and the Seventh Five-Year Plan. It mapped out the basic trends of the country's economic policy – a decisive line of intensifying the national economy and of consistently applying the economic approach in all spheres of public life. In addition, it discussed a number of problems of social and cultural policy, and outlined measures for perfecting the political system. The Congress confirmed the theses on the work of the Party, on the social, economic and cultural development of the People's Republic of Bulgaria during the Seventh Five-Year Plan (1976–80) and on the tasks to be implemented during the Eighth Five-Year Plan (1981–85) and up to 1990. The Congress also adopted a resolution on the 25th anniversary of the April Plenum of 1956, and adopted amendments to the Party Statutes, including the restoration of the post of General Secretary of the Central Committee of the BCP. The January Plenum of the Central Committee of the BCP, 1984, discussed and adopted the memorandum of Todor Zhivkov, General Secretary of the Central Committee of the BCP, to the Politburo of the Central Committee of the BCP *On Some Urgent Questions Concerning the Application of the Economic Approach and the Improvement of Management* which analysed vital problems of the further social and economic development of the country, of increasing the role of the Party, of furthering socialist democratic principles and the creative initiative of the working people. The memorandum continued and enriched the theoretical formulations and practical approaches advanced by Todor Zhivkov after the Twelfth Congress. The National Party Conference of 1984 discussed the question of improving the quality of life in all spheres of society, and adopted a *Long-Range Party Programme for the Improvement of Quality.*

The successes scored by the BCP in the country's socialist construction are due, above all, to its consistent Marxist-Leninist policy, the creative utilization of the experience of the Communist Party of the Soviet Union, the concrete application of the Marxist-Leninist teaching about the objective processes of socialist revolution, the fraternal and selfless assistance rendered by the Communist Party of the Soviet Union and by the Soviet Union to the BCP and Bulgaria, and co-operation with the other socialist countries.

The BCP is part of the international communist and workers' movement and has made a definite contribution to its theory and practice. It took an active part in the Conferences of representatives of Communist and Workers' Parties in Moscow (1957, 1960 and 1969), the Conference of European Communist and Workers' Parties in Berlin (1976), the Conference of Communist and Workers' Parties in Prague (1977), the Meeting of European Communist and Workers' Parties for peace and disarmament in Paris (1980) and other meetings. The BCP has given its whole-hearted support to the documents adopted by these conferences. It has expressed its solidarity with the measures taken to strengthen the unity of the international communist movement and of all anti-imperialist forces for the preservation of world peace. The BCP contributes actively to the consolidation of the world socialist community, maintaining ties of friendship with the fraternal Communist and Workers' Parties from the socialist and capitalist countries.

Congresses, Conferences and Major Plenums of the Central Committee of the Bulgarian Communist Party.

I. Congresses of the Bulgarian Social Democratic Party
First Congress of the BSDP, 20 July (2 August) 1891, Buzludja Peak.
Second Congress of the BSDP, 20 August (1 September) 1892, Plovdiv.
Third Extraordinary Congress of the BSDP, 3–6 July (15–18 July) 1893, Turnovo.

II. Congresses and Conferences of the Bulgarian Workers' Social Democratic Party
First Congress of the BWSDP, 3–7 July (15–19 July) 1894, Sofia.
Second Congress of the BWSDP, 27–31 July (8–12 August) 1895, Sofia.
Third Congress of the BWSDP, 21–25 July (2–6 August) 1896, Sofia.
Fourth Congress of the BWSDP, 13–18 July (25–30 July) 1897, Kazanluk.
Fifth Congress of the BWSDP, 13–17 July (25–29 July) 1898, Yambol.
Sixth Congress of the BWSDP, 11–18 July (23–30 July) 1899, Gabrovo.
Seventh Congress of the BWSDP, 9–14 July (21–26 July) 1900, Sliven.
Eight Congress of the BWSDP, 22–28 July (3–9 August) 1901, Pleven.
Ninth Congress of the BWSDP, 28 July–3 August (10–16 August) 1902, Turnovo.
Tenth Congress of the BWSDP, 6–12 July (19–25 July) 1903, Russe.
Eleventh Congress of the BWSDP, 18–20 July (30 July–1 August) 1904, Plovdiv.
Twelfth Congress of the BWSDP, 31 July–4 August (12–16 August) 1905, Sofia.
Thirteenth Congress of the BWSDP, 31 July–4 August (12–16 August) 1906, Sliven.
Fourteenth Congress of the BWSDP, 8–11 July (21–24 July) 1907, Pleven.
Fifteenth Congress of the BWSDP, 20–23 July (1–4 August) 1908, Gabrovo.
Sixteenth Congress of the BWSDP, 19–22 July (31 July–3 August) 1909, Varna.
Seventeenth Congress of the BWSDP, 11–14 July (23–26 July) 1910, Sofia.
Eighteenth Congress of the BWSDP, 3–7 July (15–19 July) 1911, Plovdiv.
Nineteenth Congress of the BWSDP, 15–17 August (27–29 August) 1912, Russe.
Twentieth Congress of the BWSDP, 29 June–1 July (11–13 July) 1914, Sofia.
Twenty-first Congress of the BWSDP, 10–12 August (22–24 August) 1915, Sofia.
First Conference of the BWSDP, 16–17 September 1917, Sofia.
Second Conference of the BWSDP, 22 September 1918, Sofia.

III. Congresses, Conferences and Major Plenums of the Central Committee of the Bulgarian Communist Party
First Congress of the BCP, Twenty-Second Congress of the BWSDP, 25–27 July 1919, Sofia.
Second Congress of the BCP, 31 May–2 June 1920, Sofia.
Third Congress of the BCP, 8–10 May 1921, Sofia.
Fourth Congress of the BCP, 4–7 June 1922, Sofia.
First Clandestine Conference of the BCP, 17–18 May 1924, Vitosha Mountain.
Moscow Conference – First Extended Conference of the Central Committee of the BCP, end of July–beginning of September 1925, Moscow.
First Extended Plenum of the Central Committee of the BCP, August 1926, Vienna.
Second Extended Plenum of the Central Committee of the BCP, August–September, 1929, Berlin.
Fifth Extended Plenum of the Central Committee of the BCP, January 1935, Sofia.
Sixth Extended Plenum of the Central Committee of the BCP, 5–6 February 1936, Sofia.

Seventh Plenum of the Central Committee of the BWP, January 1941, Sofia.

Eighth Extended Plenum of the Central Committee of the BWP (communists), 27 February–1 March 1945, Sofia.

Sixteenth Plenum of the Central Committee of the BWP (communists), 12–13 July 1948, Sofia.

Fifth Congress of the BCP, 18–25 December 1948, Sofia.

Sixth Congress of the BCP, 28 February–3 March 1954, Sofia.

April Plenum of the Central Committee of the BCP, 2–6 April 1956, Sofia.

Seventh Congress of the BCP, 2–7 June 1958, Sofia.

January Extended Plenum of the Central Committee of the BCP, January 1959, Sofia.

November Plenum of the Central Committee of the BCP, 28–29 November 1961, Sofia.

Eighth Congress of the BCP, 5–14 November 1962, Sofia.

Ninth Congress of the BCP, 14–19 November 1966, Sofia.

July Plenum of the Central Committee of the BCP, 24–26 July 1968, Sofia.

Tenth Congress of the BCP, 20–26 April 1971, Sofia.

October Plenum of the Central Committee of the BCP, 4–6 October 1971, Sofia.

December Plenum of the Central Committee of the BCP, 11–13 December 1972, Sofia.

November Plenum of the Central Committee of the BCP, 28–29 November 1973, Sofia.

February Plenum of the Central Committee of the BCP, 7–8 February 1974, Sofia.

National Party Conference, 20–23 March 1974, Sofia.

Eleventh Congress of the BCP, 29 March–2 April 1976, Sofia.

July Plenum of the Central Committee of the BCP, 1–2 July 1976, Sofia.

National Party Conference, 21–22 April 1978, Sofia.

Twelfth Congress of the BCP, 31 March–4 April 1981, Sofia.

November Plenum of the Central Committee of the BCP, 29–30 November 1982, Sofia.

January Plenum of the Central Committee of the BCP, 3 January 1984, Sofia.

National Party Conference, 22–23 March 1984, Sofia.

The Bulgarian Agrarian People's Union (BZNS) or Bulgarian Agrarian Party (BAP) is the political party of co-operative farmers. In alliance with and under the guidance of the BCP, it participates in the government of the country and in socialist construction. It is one of Europe's oldest democratic peasant parties. Its supreme organ is the Congress, which elects the Ruling Council which, in turn, elects the Standing Board. A Supreme Union Council may be convened between congresses, if deemed necessary. The district, municipal and local agrarian organizations are formed according to the country's administrative division. At its 34th Congress (1981) the party had a membership of 120,000. Its press organ is the *Zemedelsko Zname* newspaper. The Secretary of the BAP (since 1974) is Peter Tanchev.

The members of the Standing Board of the BAP are: Alexi Ivanov Vassilev, Angel Dimitrov Angelov, Dimiter Ivanov Karamukov, Nikolai Georgiev Ivanov, Yanko Markov Spassov, Boyan Georgiev Parvanov, Dimiter Haralampiev Dimitrov, Ivan Nikolov Kaludov, Kamen Ivanov Kamenov, Lyuben Stoyanov Stefanov, Milena Asenova Stamboliiska, Pando Vulev Vanchev, Radoi Petrov Popivanov, Svetla Raikova Daskalova, Stoyan Tonchev Stoyanov and Angel Ivanov Yordanov.

The BAP came into existence as a result of the discontent of the working peasants with the policy of the Bulgarian bourgeoisie and with capitalist exploitation. It was constituted at a congress held in Pleven (28–31 December 1899) as a professional and educational organization of farmers. Its founders included Tsanko Tserkovski, Yanko Zabunov and D. Dragiev. After the peasant riots of 1900 against the introduction of taxation on agricultural crops in kind – the tithe – and the first participation of the BAP in the parliamentary elections (January 1901), its Third Congress (1901) declared the Union to be a political organization. Alexander Stamboliiski became the ideologist and leader of the BAP from 1904 onward. The Seventh Congress (1905) adopted the first BAP Programme developed by A. Stamboliiski in the spirit of agrarianism and the town-versus-village principle, turning it into a fully-fledged political organization. Its membership consisted of representatives of various strata of Bulgarian peasantry, predominantly poor and middle peasants. Its varied social composition and town-versus-village ideology accounted for its contradictory development. From the early period of its existence it developed as a democratic organization, one of the world's most leftist agrarian organizations, but during various periods in its history it was pushed by the right-wing forces within it to adopt reactionary positions. It developed further in a state of continued strife between its left-wing, democratic forces and the right-wing forces, which upheld the interests of the rural bourgeoisie. The BAP's struggle against the bourgeoisie and its parties, and against the autocratic regime of Ferdinand I, was conducted from petty bourgeois positions, but it did help peasants tear themselves away from the political influence of the bourgeoisie and participate in progressive social struggles. On the eve of the Balkan War (1912–13) the BAP took a stand against militarism and against the amendment of the Turnovo Constitution in favour of the monarch. However, its parliamentary group voted for the war credits for the Balkan War. (The BAP declared itself against Bulgaria's entry into the First World War (1914–18) and for neutrality, but after the country's entry into the war, the agrarian deputies voted for the war credits demanded by the Government.) BAP activists took part in the anti-war movement in the army and the whole country, despite the resistance of the right-wing forces. Although it did not embrace the socialist ideas of the Great October Socialist Revolution, it supported the slogan for peace without

A rally during the Seventeenth Congress of the Bulgarian Agrarian Party, Sofia, 1922

annexations and reparations, regarding with approval some of the Soviet Government's reforms. Many agrarians took part in the Soldiers' Mutiny (1918), and Alexander Stamboliiski and Raiko Daskalov assumed the leadership of the mutineers.

During the war years (1912–18) the BAP extended its influence, becoming the country's largest party (77,278 members in 1919). At its Fifteenth Congress (1919) the left-wing, headed by Alexander Stamboliiski, was victorious. After the Congress the extreme right-wing members headed by D. Dragiev were expelled from the party and formed the Stara Zagora Agrarian Union (Dragiev's followers). Covert right-wing elements (M. Turlakov, K. Tomov and others), however, stayed on. The petty-bourgeois ideology of BAP, and its political platform, fully crystallized after the First World War. The non-scientific, reactionary town-versus-village theory was adopted, according to which it was not class contradictions but the differences between town and village populations that mattered most, and the leading role in society belonged to the larger, peasant population. The BAP proclaimed its slogan for 'independent agrarian government' which was to defend 'the interests of all peasants'. It declared for 'a third road' i.e., not only against the bourgeoisie and its parties, but also against the working class and the Marxist Party. Without rejecting the capitalist system, the BAP stood for restricting the dominance of the *haute bourgeoisie*, for eliminating the vices of Capitalism, for the protection of small and middle-size private property and the promotion of the co-operatives.

After the war, the BAP was enlisted in the bourgeois coalition Governments of A. Malinov (October–November 1918) and T. Todorov (November 1918–October 1919). In October 1919 it formed a coalition Government, headed by Alexander Stamboliiski, with the Popular Party and the Progressive Liberal Party, which was hostile to the revolutionary movement and crushed the transport workers' strike (December 1919–February 1920). In the Parliamentary elections in March 1920, the BAP won a majority and formed an independent Agrarian Government, headed by Alexander Stamboliiski. It made a serious attempt at steering the country's foreign and domestic policy along a democratic road and imposing a petty-bourgeois solution to the post-war revolutionary crisis. Without abolishing the domination of the *haute bourgeoisie* in the national economy, the Agrarian Government tried to restrict it through reforms, to defend the small and medium landed property, and to improve the plight of the working peasant masses (an agrarian reform, a law on state-co-operative monopoly of trade and export of grain by consortium, progressive income tax on labour service, and others). The foreign policy of the Agrarian

Government was aimed at fostering peaceful co-operation with all countries, including Soviet Russia.

Affected by the reforms carried out by the BAP, and fearing the growing influence of the BCP, the bourgeoisie engineered a coup to regain power. Reactionary and fascist organizations were set up (Military League, *Naroden Sgovor* [Popular Alliance] and others) which carried out political and terrorist acts. The attitude of the BAP to the BCP was hesitant and contradictory. At the initiative of the BCP, unity of action was to some extent achieved in the struggle against the onslaught of reaction. The BAP Government facilitated the action organized by the BCP in aid of the victims of the Volga Famine (1921–22). The BCP and the BAP acted in parallel to foil the action of the Constitutional block in Turnovo, voting with one ballot-paper in the referendum on bringing to trial those responsible for the national catastrophe (1922). Under the influence of the right-wing forces, at the end of 1922 and in early 1923, the Agrarian Government subjected the BCP to persecution, while at the same time its opposition to big business slackened.

The disunity between communists and agrarians facilitated the military fascist coup on 9 June 1923, which brought down the Agrarian Government. The BAP leader, Alexander Stamboliiski, was savagely murdered, as also were other leading figures and many progressive BAP activists. In opposition to the coup, under the guidance of local communists and agrarian militants, the working people rose in arms in many places across the country. In August 1923, having drawn a lesson from the defeat, the left-wing forces of the BAP, headed by P. D. Petkov and Ts. Avramov, accepted the agreement on a united front for armed struggle against Fascism and the establishment of worker-peasant power which had been proposed by the BCP. During the September 1923 Anti-fascist Uprising the foundations were laid of the militant alliance of the BCP and the sound forces within the BAP. After the defeat of the Uprising, the BAP was in a state of organizational disarray and had no consistent political line, but its influence had not diminished. The agrarians participated in conjunction with the communists in the parliamentary elections in November 1923. The BAP wing of the united front, headed by P. D. Petkov, D. N. Gruncharov, N. Petrini and others, stood for an active struggle against Fascism, and in August 1924 signed an agreement with the Central Committee of the BCP on starting preparations for a new armed uprising in the name of a democratic worker-peasant government. It resolutely condemned the attempts made by K. Tomov, M. Turlakov and others to do away with the militant progressive traditions of the BAP. Many outstanding BAP officials were murdered during the fascist reign of terror in April 1925. Three currents emerged in the BAP – right, left and centre. The right wing wavered between opposition to Fascism and co-operation with the fascist Government. The centrists opposed the fascist Government, yet were not in favour of united action with the BCP, but of an independent BAP policy under the slogan 'Neither Left, Nor Right'. The left-wing fought for a leftist orientation of the BAP and for alliance with the BCP. The situation within the BAP grew even more complicated during the partial stabilization of Capitalism (1924–29) and during the Great Depression (1929–34). At the end of 1926 it split up. The extreme right wing headed by K. Tomov was disowned and emerged as the 'Orange' BAP (K. Tomov) which gradually turned fascist and in January 1934 merged with A. Tsankov's fascist 'National Social Movement'.

Thirty-Fourth Congress of the Bulgarian Agrarian Party, Sofia, 1981

The *Vrabcha-1* BAP was set up (1926–44) with the centrists playing the main role. Organizationally and politically split into two wings, it opposed the ruling fascist Democratic Alliance and struggled against it. The *Vrabcha-1* BAP did not accept the united front and the left wing, headed by L. Stanev, alone co-operated with the BCP. The leaders of the *Vrabcha-1* BAP, D. Gichev, G. Yordanov and K. Muraviev, associated the overthrow of the fascist regime, restoration of bourgeois democracy and solution of the economic crisis with a policy of conciliation with the non-fascist bourgeois parties. In 1931 the *Vrabcha-1* BAP entered the Popular Block coalition and its Government (1931–34). Its leadership clashed with a considerable part of its rank-and-file membership because the Government did not meet the needs of the working peasants, which had grown acute in the Depression. A new complex regrouping of forces took place within the BAP, and the left wing merged organizationally with the BCP. The Alexander Stamboliiski BAP (*Pladne*) was formed in March 1932, in opposition to the *Vrabcha-1* BAP. In

August 1933 it merged with other opposition groupings, forming the United BAP ('Alexander Stamboliiski' and *Vrabcha-1*). The extreme right wing within *Vrabcha-1* BAP formed a new Union: *Vrabcha-1* BAP-Georgi Markov *(Serdika)*. Before and after they had joined the Popular Block Government, *Vrabcha-1*, 'Alexander Stamboliiski' and the United BAP had a hostile attitude to the BCP. For this reason and because of the incorrect attitude taken by the left-wing sectarian leadership of the BCP during the Depression, the united front was weakened and co-operation between communists and agrarians was effected solely on a local basis.

After the coup of 19 May 1934 all parties, including the BAP, were banned. The United BAP disintegrated into the 'Alexander Stamboliiski' BAP (*Pladne*), *Orach*, *Selski Glas* and other groupings. The two main wings *Vrabcha-1* and 'Alexander Stamboliiski' BAP opposed monarcho-fascism. The small groupings like *Orach* and *Selski Glas* evolved to the right and their officials collaborated with the fascists. On the eve of the Second World War the United *Vrabcha-1* and 'Alexander Stamboliiski' BAP became more democratic. It stood for the restoration of the Turnovo Constitution, for the abolition of the State Defence Act, in support of the strike movement, etc. The *Vrabcha-1* faction, however, preferred co-operation with the right-wing forces of the non-fascist opposition. 1936 saw the formation of the so called 'Five', including the BAP, National Liberals, right-wing Radicals, right-wing Socialists and Lyapchev's block, which stood for the restoration of the Turnovo Constitution and bourgeois democracy. The 'Alexander Stamboliiski' wing stood for co-operation with the BCP. The two BAP wings took part in the popular-front movement led by the BCP (1936–39). During the Second World War the BAP remained divided: the 'Alexander Stamboliiski' and *Vrabcha-1* BAP. Both wings stood against Monarcho-Fascism and its anti-popular home and foreign policy allied with nazi Germany. The 'Alexander Stamboliiski' BAP worked together with the BWP (communists) to find a democratic way out of the revolutionary crisis, adopted the Programme of the Fatherland Front and participated in its formation. Agrarians formed the second largest force in the Fatherland Front after the communists. Their participation in the Fatherland Front and the adoption of its Programme and methods of struggle by the 'Alexander Stamboliiski' BAP constituted an important point in the ideological and political evolution of the Agrarian Party. Its left-wing forces took part in the armed anti-fascist struggle from 1941–44 and in the popular uprising on 9 September 1944, but despite the efforts made by the BCP to draw the *Vrabcha-1* faction into the Fatherland Front and the anti-fascist struggle, it remained aloof. Together with the parties of the bourgeois opposition, it formed a counter-revolutionary Government headed by K. Muraviev, (2–8 September 1944) which was the last Bulgarian bourgeois Government.

After the victory of the 9 September Socialist Revolution, the BAP was restored as a united organization, and became a co-ruling party and main ally of the BCP in governing the country. At its National Conference (May 1945) the BAP declared itself in support of the Fatherland Front and in alliance with the BCP. The right-wing anti-communist forces in it, led by G. M. Dimitrov (Gemeto), came out in opposition to co-operation with the BCP and the Fatherland Front policy (late 1944–early 1945). They were defeated by the joint efforts of the Fatherland Front and the sound forces within the BAP. In the summer of 1945, the remainder of the reactionary elements in BAP, headed by N. Petkov, left the Fatherland Front, forming the opposition BAP (Nikola Petkov). D. Gichev's opposition group joined in at the end of 1945. The BAP (Nikola Petkov) formed the nucleus of the counter revolutionary bourgeois opposition in the country. A right-wing group, headed by A. Obbov, which upheld the town-versus-village premise of the BAP and opposed co-operation with the BCP, remained in existence in the Fatherland Front Agrarian Party. Despite the Opposition's attempts to impose a reactionary policy on the BAP, the latter worked for the consolidation of the people's democratic system. In the summer of 1947 the opposition BAP was disbanded, while A. Obbov's group was removed from the BAP. The 27th Congress of the BAP (December 1947) adopted a course towards a renovation of its ideology and policy. In 1947 Georgi Traikov was elected Secretary (until 1974). The final all-round ideological and political reorganization took place after the conference of the Supreme Council, 31 October–1 November 1948, when it broke with the town-versus-village principle and decided to build itself on the class principle as an organization of the poor and middle peasants, adopting the construction of Socialism as its Programme task, and recognizing the leading role of the BCP in the country's socio-political life. The BAP became a new type of agrarian organization which has been successfully participating in socialist construction in alliance with and under the leadership of the BCP. Since the April 1956 Plenum of the Central Committee of the BCP, the role of the BAP in socio-political life has increased; its joint activity with the BCP has grown in scope and substance. The BAP's contribution to the victory of the co-operative system in the Bulgarian countryside (1948–58) was most significant. Its 32nd Congress (October 1971) approved a programme in which it adopted the construction of the advanced socialist society in Bulgaria as its immediate historic task.

The BAP participates actively in the solution of the major economic, political, social and cultural problems of the country. Its 34th Congress (May 1981), outlined the tasks to be carried out by the BAP in fulfilment of the basic goals in the country's development during the eighth Five-Year Plan and up to 1990, and defined its place in the effort to intensify rural economy. The session of its Supreme Council (17 March 1983) discussed the central organizational, political and ideological tasks of augmenting the local agrarian organizations' contribution to the social and economic progress of the country at the present stage.

The BAP carries out broad and fruitful international

Georgi Traikov

activity centred round the struggle for peace, security and co-operation. It maintains ties with about 120 peasant, kindred and other democratic parties and movements, and on the basis of bilateral and multilateral co-operation makes a contribution of its own to the development of the political, economic, techno-scientific and cultural contacts of the People's Republic of Bulgaria with the other countries, to the greater unity of the forces of the international democratic agrarian movement, and to the strengthening the anti-imperialist front and reaffirmation of the policy of peaceful co-existence. The BAP has sponsored and participated in many international forums of democratic peasant and kindred parties and movements – the First Meeting of Representatives of Agrarian Organizations in Sofia (1962); the Peasant Parties' Meeting in Sofia (1967); the Meeting of Europe's Three Oldest Agrarian Parties (1968); the International Meeting for Peace, Co-operation and a Better Life for the Working Peasants of 37 peasant parties and organizations in Sofia (1971); the Peasant Parties' Meeting in Moscow (1973); the International Conference during the 23rd BAP Congress, with the participation of 81 foreign delegations (1976); the First Conference of the European Agrarian and Democratic Public in Helsinki, attended by 35 peasant parties, centre, left-wing liberal, radical and other democratic parties from 23 countries (1976); the Meeting of Peasant and Kindred Democratic Parties and Organizations in Havana, attended by 39 delegations (1977); the Sessions of the Preparatory Committee for holding a meeting of peasant parties in Panama (1977), Costa Rica (1978) and Venezuela in 1979; the International Conference for Solidarity with the Palestinian Peasants and the Arab People of Palestine, in Damascus, with 44 delegations taking part (1978), and in Baghdad, with 181 delegations (1980); the International Meeting-Dialogue on *Détente* on the occasion of the 80th anniversary of the founding of the BAP, attended by 92 delegations (1980); the International Meeting-Dialogue 'for *Détente,* for Peace and Social Progress' held after the 34th BAP Congress and attended by 112 foreign delegates (1981); the Second European Peasant Meeting in Helsinki, with the participation of representatives of 41 parties and organizations from 25 countries of Europe, the United States and Canada (1983), and others.

PUBLIC ORGANIZATIONS

The Fatherland Front – 'the most popular social and political organization and a nation-wide movement, the embodiment of the nation's unity, of the unity of the working class, the peasants and the people's intelligentsia ... is the expression of the joint efforts of communists, members of the Agrarian Party and non-party working people. It is the greatest mass public and political stronghold of the BCP and the people's power, an important factor in the involvement of the working people in socialist construction and social management, as well as a genuine school for communist, patriotic and international education of the people'.

Under the leadership of the BCP, the Fatherland Front combines the efforts of the public organizations and movements included in it for the fulfilment of common tasks; it is the initiator and organizer of events of national and local significance, and of broad political campaigns. According to the Statute of the Fatherland Front (Sofia, 1977, p. 6–7) membership can be individual and collective; collective members are the Bulgarian Trade Unions, the Dimitrov Young Communist League, the Bulgarian Red Cross and other public organizations. The guiding principle in the organizational structure is democratic centralism. The Congress, which is convened once every five years, is the supreme body; it elects by secret ballot a National Council of the Fatherland Front (since 1974 Pencho Kubadinski has been President) and a Central Control and Auditing Commission. The local organizations are the basic units of the Fatherland Front; the leadership of a local organization is elected annually at the general meeting. By 31 December 1982 it was 4,279,609 strong.

The Fatherland Front was founded on the initiative of the BWP as a broad, popular, anti-fascist movement of the patriotic and democratic forces of the Bulgarian people. It is the continuation and further development of the Popular Anti-fascist Front and is the highest achievement of the Party's united-front tactics. The name Fatherland Front was first mentioned in a broadcast of the Hristo Botev radio station on 7 May 1942. On 17 July 1942, the radio station proclaimed the Programme of the Fatherland Front which had been discussed and adopted by the Party Bureau-in-Exile, with the active participation and under the guidance of Georgi Dimitrov (1882–1949). 'The supreme duty of the Bulgarian people, of the army and patriotic intelligentsia', it was emphasized in the Programme, '... is to rally into a mighty Fatherland Front to save Bulgaria'.

The Programme consisted of twelve items, covering the most important questions of policy, economy and ideology and representing a broad, democratic, anti-fascist and anti-imperialist platform. It raised problems of a national liberation, home-policy and economic nature, and assigned respective tasks. The first part of the Programme reflected changes in the foreign policy orientation of Bulgaria, the demand that the country should not be dragged into the nazi war which was so disastrous to the Bulgarian people, that the alliance with nazi Germany and her allies should be broken off, that Bulgaria should join the struggle of the freedom-loving peoples headed by the Soviet Union and aiming at the utter defeat of the German imperialists, and that the export of foodstuffs and raw materials to Germany should cease. The greater part of the Programme discussed problems of home policy, economy and ideology, concerning mainly the national interests of the country. The problems revealed the necessity for unity between the revolutionary and patriotic forces in joint action for the salvation of the fatherland. What the Programme envisaged for the democratization of social life was to free the political and military men persecuted and imprisoned for anti-fascist activity, to repeal all anti-national and fascist laws, to tear the army away from the pernicious influence of the monarcho-fascist dictatorship, ensuring civil rights to military men, to ban the fascist-organizations, to eradicate the fascist ideology and racial hatred, and to put an end to the practice of violating the national honour of the Bulgarian people. Economic development was viewed as closely dependent upon the possibility of preserving the national wealth and the nation's labour from any foreign encroachment.

In this section, the Programme took up the cause of defending the interests of the working urban and village populations by guaranteeing the right to live and work in normal conditions.

The programme was based on democratic principles and represented a broad platform for uniting all social strata. The main and most imminent tasks to be solved were the overthrowing of the anti-national Government and the establishment of a genuinely national Government of the Fatherland Front, which would guarantee the independent political and economic development of Bulgaria. The

Programme did not include problems of social policy. It was '... the practical nation-wide democratic platform of our Party during the war for the liberation of our country from Fascism and the German occupation.' (Dimitrov, G. *Collected Works,* Volume 14, Sofia, 1955, p. 266)

The practical development of the Fatherland Front was carried out by the BWP. The Central Committee of the Party entrusted Tsola Dragoicheva with responsibility for the committees of the Fatherland Front in the country. The concrete work was done by the Central Commission headed by Kiril Dramaliev (1892–1961); at the first meeting, held underground in Vitosha in August 1942, problems concerning the building up of the Fatherland Front were discussed. Pancho Angelov (1873–1956), Ivan Pashov (1881–1955), Mincho Neychev (1887–1956) and many others worked on the commission, entrusted with special tasks. In August 1942 the Central Committee of the BWP published a declaration in the newspaper *Rabotnichesko Delo,* which was outlawed at the time, approving the Progamme and appealing to the leaders of the BAU, the Social Democratic party, the Democratic and Radical parties, the *Zveno* political circle, the Military Union and all economic, cultural, military and mass organizations to combine their efforts to build the Fatherland Front. The Central Committee of the Workers' Youth Union also appealed to the youth masses and their political organizations to rally their forces in the joint struggle against Fascism and in the implementation of the Programme tasks. The first underground committees of the Fatherland Front were set up; in a few months (by the end of 1942) their number exceeded 100. The local committees carried out mass, political propaganda with newspapers of their own, leaflets and other materials. In the autumn of 1942, the underground Fatherland Front Press came into existence, and the first issue of the newspaper *Otechestven Front* (Fatherland Front) came out in December.

The popularization of the Programme and the foundation of committees enhanced the mass character of the most efficient form of armed struggle in 1941–44, namely the partisan movement. Of particular importance to further development was the Directive of the Central Committee of the BWP of August 1943, which outlined the character, basic motives, organization and forms of struggle of the Fatherland Front and emphasized the necessity of creating a directing centre for the anti-fascist movement. Despite the wavering of some bourgeois leaders and sabotage by others, with the assistance of the Central Committee of the BWP such a centre was set up. It included the National Committee of the Fatherland Front, which comprised the *Pladne* left wings of the BAU 'Alexander Stamboliiski', the *Zveno* political circle, the BWSDP and other mass organizations which had accepted the Programme. On 10 August 1943 the National Committee was set up with, as members: K. Dramaliev (from the BWP), Nikola D. Petkov (1893–1947; from the BAU '*A. Stamboliiski*') and Kimon Georgiev (1882–1969; from *Zveno*). In September the same year, Grigor Cheshmedzhiev from the BWSDP and Dimo Kazassov (1886–1980; from the independent intellectuals group) also sat on the committee.

The initial, constructive stage in the development of the Fatherland Front ended with the setting up of the National Committee. During the second stage there was a rapid expansion of the committees, which became accessible to the masses and consolidated their links with the partisan detachments. Active support was lent to the armed struggle – food and clothes were provided, medical assistance was rendered to underground revolutionaries, sick and wounded partisans were given shelter, and the partisan ranks were reinforced with new men. The Fatherland Front committees in Sofia and a number of surrounding districts in Plovdiv and its environs, in Vratsa and district, in Varna, Smolyan and many other towns and villages carried out activities on a large scale. The influence of the Fatherland Front amidst the soldiers and officers of many garrisons increased greatly.

The local, district, and town committees of the Fatherland Front played an active role in exposing the falsity of fascist propaganda and giving faithful information about political events at home and abroad. These committees issued leaflets, bulletins and fly sheets: the Plovdiv District Committee published the newspaper *Glas na Otechestvenya Front* (Voice of the Fatherland Front), the newspaper *The Patriot* was published in the northern district of Sofia, the local committee in the village of Poruchik Geshanovo (today's village of Geshanovo), Tolbuhin district, published the newspaper *Otechestvo* (Fatherland). At the end of November 1943, the National Committee published an appeal to lend open moral support to the partisan movement. The first issue of an information bulletin was prepared. The National Committee showed greater readiness to engage in enterprises during the spring and summer of 1944 – two supplements to the bulletin were issued as well as a manifesto and appeals to the Bulgarian people and army for bold actions directed at seizing power. During the summer of 1944 the Fatherland Front played an important part in rallying and mobilizing all the anti-fascist forces. A network of new district and local committees covered the whole country (7,000 in almost 5,000 towns and villages). During the first days of September 1944, the activities of the National Committee were almost legal; it declared itself openly to be against the Government's anti-national policy, it gave the regents a warning and demanded that they should surrender power to the Fatherland Front which, under the leadership of the BWP, was the only force capable of leading Bulgaria out of the crisis. The anti-imperialist, anti-fascist and democratic slogans underlying the Programme of the Fatherland Front constituted the platform of the victorious popular uprising of 9 September 1944. The first people's democratic Government of the Fatherland Front was formed, with Kimon Georgiev as Prime Minister and a Government including representatives of the BWP, the BAU *Pladne*, *Zveno*, the Social Democratic Party and two non-party members. On 9 September, Fatherland Front Bulgaria entered the war against fascist Germany. With the

Meeting in support of the Fatherland Front Government, Sofronievo (near Vratsa)

assistance of the National Committee a government programme was drawn up and announced on 17 September. The programme specified and further elaborated the tasks set in the first Programme. The Government, led by the BWP, undertook a number of tasks to consolidate the people's power, to liquidate the vestiges of Fascism, to bring the Patriotic War to a victorious end, to abolish the monarchy, to implement the programme for the democratic reorganization of the country, and to create the necessary conditions for socialist development.

On 20 September 1944, the National Committee issued a circular letter which contained instructions about setting up Fatherland Front committees all over the country and detailed directions concerning the structure, rights and obligations of the committees, the interrelations between the party and state and administrative bodies, the mass organizations, etc. Organizational, Mass Activity, Military and Public Relations departments were set up within the National Committee; youth and women's commissions were set up within the district committees. The number of the committees grew in a very short time. In October 1944 the Declaration of the four Fatherland Front parties was issued, outlining concrete measures for consolidating militant unity in the ranks of the Fatherland Front. The principle of parity of representation in the committees was elucidated on the basis of equal obligations and equal engagements in overcoming the hardships caused by the war, the damage and shameful heritage of the past. The declaration specified the fact that the committees were not bodies having governing authority, they were a mass political support of the Government and popular bodies which served as a link between the people and the Government without, however, replacing the state and administrative authorities which were being set up. The documents of the BWP (communists) and the Fatherland Front emphasized the significance of the committees as bodies of public and mass control. The Fatherland Front parties, which were joined by the Radical Party of Stoyan Kosturkov (1886–1949), continued to play an active role in the political life of the country, but the BWP (communists) was recognized as the leading force in the political system of the people's democracy. A deep ideological and political differentiation, which later ended in the splitting up of the parties, arose in the BAU and the BWSDP. A radical democratic group and a right-wing group were set up in *Zveno*. During the summer of 1945 some of the leaders of the BAU, of the Social Democratic Party and of right-wing and centre groups, tried to undermine the inner unity of the Fatherland Front, and to discredit the leading role of the BWP (communists) both in the Front and in the fulfilment of the Programme. They left the Fatherland Front, raised demagogic slogans and sabotaged the undertakings and initiatives of the Government.

The First Congress of the Fatherland Front (9–12 March 1945) clarified the functions, internal political and organizational structure of the committees. The Fatherland Front had now to focus on an even wider mobilization of the broad masses for the victorious end of the Patriotic War, and for the consolidation of the people's power and the national economy, which had been disrupted and plundered by the German nazis. The problem of holding parliamentary elections at which all united-front parties should put forward a common ticket and a joint election platform, was given due consideration. The Congress exposed the aspiration for power of the anti-popular opposition. The parliamentary election campaign after the Congress ended in complete victory for the Fatherland Front (1946). The union between the workers and peasants grew closer and the democratic and patriotic forces rallied more firmly round the Communist Party, whose leading role was universally recognized.

The consolidation of the people's democratic power made it possible to rid the country completely of the economic basis of Capitalism. The agrarian reform of 1946 was carried out, capitalist enterprises and banks were nationalized (December 1947), and a new republican Constitution was drafted and adopted (4 December 1947). Conditions were created for proceeding to the all-out construction of a socialist society. The Second Congress of the Fatherland Front (2–3 February 1948) adopted a new programme and a new organizational statute. The coalition elements in its structure were removed and the Fatherland Front became a united, popular, social and political organization, with the tasks of socialist construction reflected and clearly formulated in the renewed programme. On the basis of the principle of democratic centralism, the Congress elected the new Fatherland Front bodies – the National Council and the Central Control and Auditing Commission. After the Second Congress the Fatherland Front furthered the consolidation of the unity of the Bulgarian people, engaging in enormous socio-political, educational and instructive activities. In 1949 *Zveno* and the Radical Party ceased independent activities and merged into the Fatherland Front; in 1950 the General Agrarian Trade Union and the Bulgarian National Women's Union joined the ranks too, as their programmes completely coincided with that of the Front. The Social Democratic Party merged into the BCP in 1948.

During the 1950s, the process of the involvement of the broad masses in the Fatherland Front and the education and re-education of its numerous members was slowed down by misconceptions which had penetrated the party leadership, namely that the historical role of the Fatherland Front had already become irrelevant and that after the elections of the People's Councils as local bodies of the people's democracy, it had to merge into them. In the documents of the Third Congress of the Fatherland Front (28–29 May 1952) these misbeliefs and erroneous trends were confirmed, and the Fatherland Front was declared to be the first and direct assistant of the People's Councils. Cases of merging between the leadership of the Fatherland Front and the Councils became more frequent, council plans were declared to be those of the Front organizations, the Front activities acquired a cultural and educational character. As a result, the political importance of the Fatherland Front as the main transmitter and stronghold of the party was greatly decreased. The April Plenum of the Central Committee of the BCP in 1956 condemned these misconceptions of the role, place and importance of the Fatherland Front in the building of Socialism. In a resolution of the Central Committee of the BCP of 17 January 1957, G. Dimitrov's assessment of the functions of the Fatherland Front was restored – the most popular, self-governing socio-political organization, the genuine embodiment of the union of the working class, peasants and the people's intelligentsia, the main aim of which is to educate the people in a patriotic and socialist spirit and to be the mass support of the people's power.

The resolutions of the Fourth Congress (11–12 February 1957) marked a new stage in the development of the Fatherland Front, during which there was a decisive improvement in the activity of the organization: the social and control functions were recognized as highly significant; mass ideological, political, educational and economic activities were carried out on a large scale; measures were outlined for improving the style and methods of work of the governing bodies. The Congress adopted a resolution concerning the consolidation and development of the Fatherland Front as the most popular self-governing public and political organization, and as the most popular public support of the people's power. The April 1956 party policy adopted at the Seventh Congress of the BCP (1958), ensured the necessary conditions for the rapid reorganization of the Fatherland Front, in conformity with the changes in the administrative and territorial division of the country, as well as with the process of amalgamation of the co-operative farms. Changes were made in the structure of the National Council, and in the structure and tasks of the district and newly-founded local committees. The interaction between the Fatherland Front and the People's Councils was enhanced, so that the voluntary labour of the population to help agriculture could be put to a more useful purpose; and other objectives were envisaged, such as afforestation and environmental protection, sanitation of built-up areas, gathering of medicinal herbs, and of scrap and waste materials. The Law of the Comrades' Courts (1961) enhanced the role of the Fatherland Front in the implementation of efficient public control, and the committees were entrusted with the task of organizing and guiding the Comrades' Courts which functioned in town areas, villages, industrial enterprises and other places. The Fifth Congress of the Fatherland Front (14–16 March 1963) supported the resolutions of the Eighth Congress of the BCP (1962) and approved them as the guiding principle in the work of the Fatherland Front. The organization strove to develop socialist democracy, public forms of activity, and the involvement of working people in the management of the country; it assumed a number of functions of the state bodies – Comrades' Courts, volunteer detachments, public control, and many others, the final objective being the mobilization of labour of the broad masses of the

Georgi Dimitrov and Todor Zhivkov at a meeting of the Fatherland Front, 1948

population and the building of the material and technical basis of Socialism. The Sixth Congress of the Fatherland Front (15–17 May 1967) mapped out measures for the implementation of the resolutions adopted at the Ninth Congress of the BCP (1966). The main task was to complete the reorganization of the Front in order to improve the forms and methods of work among the different social strata. Of primary importance for the processes of promoting the activity of the Front were the resolutions of the July 1968 Plenary Session of the Central Committee of the BCP for improving the system of social management and delineating the functions of the socialist State and the mass organizations. With the application of the new methods and means, the Fatherland Front acquired features characteristic of a nation-wide movement, and the delimitation between its members and the rest of the population was abolished. A number of organizations (though preserving their organizational independence and names) merged into the Fatherland Front; the Committee of the Bulgarian Women's Movement, the National Peace Committee, the Committee of Bulgarians Abroad, the National Temperance Committee, etc.

The Programme of the BCP, adopted at its Tenth Congress, 1971, highlighted the perspectives for the development of the Fatherland Front as 'the most popular public support of the Party and the people's power... In its activity it should apply forms of work which are intrinsic to a nation-wide movement'. The Constitution (1971) states: 'The Fatherland Front is the embodiment of the union of the working class, the peasants and the people's intelligentsia. It is the public support of the people's power, a mass school for patriotic and communist education of the population and for the involvement of the working people in the management of the country' (Article 11). The Fatherland Front National Council was granted the right to legislative initiative. The Seventh Congress of the Fatherland Front (20–22 April 1972) further developed the formulation for the transformation of the Fatherland Front from an organization into a movement and from a movement into an organization. The Congress favourably assessed the results achieved in the fulfilment of the tasks of socialist construction. The Eighth Congress of the Fatherland Front (14–16 July 1977) made a detailed analysis of the many-sided and beneficial activities of the Fatherland Front in having turned the Party imperative of the formation of the socialist personality into reality, with the consolidation of the ideological and political unity of the working class, agricultural workers and scientific and creative intelligentsia around the resolutions of the Eleventh Party Congress (1976) and the Seventh Five-Year Economic Plan for increasing confirmation of the public principle in the work of the Fatherland Front and the mass organizations and cultural unions which had merged within it, for giving opportunities to the people's amateur performing and art activities, and for raising the international prestige of the country. The Ninth Congress of the Fatherland Front (21–23 June 1982) gave an account of the results of the rich and versatile activity after the Eighth Congress and described in detail the trends and forthcoming tasks in the social, economic, socio-political and cultural spheres. The resolution outlined the role and functions of the organization as a mass support of the Party and Government for the further development of socialist democracy, for a better and more complete use of the rights to legislative initiative granted to it, for control over the management of the representative and economic bodies, for improving the system of self-sufficiency, sanitation and urbanization of rural settlement systems and built-up areas. Especially important are the tasks relating to ideological and educational work in the residential areas through more intensive individual activity among the populace, in order to solve the problems of the family and moral and aesthetic education, to lead the struggle against anti-social manifestations, to raise the cultural and living standards of the population and the level of public services, and to further the patriotic and international education of the coming generation. The Congress adopted amendments to the Statute of the Fatherland Front, in conformity with the changes which had come about in the organizations, and a declaration which supported the aspirations of the governments and peoples of the socialist countries to ease international tension, further disarmament and consolidate peace and security.

The Dimitrov Young Communist League (DYCL or Komsomol) is a mass voluntary socio-political organization of Bulgarian youth, expressing its general interests and aspirations. It is the first assistant of the BCP in its work for the communist education and all-round development of the younger generation, for the latter's full and highly efficient social realization in the construction of the developed socialist society and the defence of its achievements. The DYCL is a reserve of the Party and an active participant in the overall life of the country, an important pillar of society in its further development.

The League's principal task is to work for the formation of a Marxist-Leninist world outlook among youth, to educate young men and women in the spirit of the people's revolutionary and labour traditions, of socialist patriotism and proletarian internationalism, of affection and loyalty to the Soviet Union, the Communist Party of the Soviet Union and the Lenin Komsomol, of fraternal solidarity with the peoples of the socialist community and full support of the democratic forces in the world.

The Komsomol provides opportunities for the manifestation of the all-round interests and abilities of the younger generation and is an important factor for their self-fulfilment in the fields of labour, education, political life, art and culture, physical education, sport and tourism, and recreation. The Youth League has major achievements as an initiator and political organizer of the movement for the techno-scientific creativity of youth.

In conformity with the Constitution of the People's Republic of Bulgaria the DYCL takes part in the development of socialist democracy, involves youth in the management of socialist society, and through its Central

Todor Zhivkov with a group of children during the First Congress on Public Education, Sofia, 1980

Committee exercises legislative initiative. The Komsomol has the right to submit for discussion to the Party, state and other bodies all questions pertaining to the improvement of the work of the economic organizations, departments, educational establishments, cultural and research institutes; it exercises public control of the work of the state and economic bodies and mass organizations among youth. According to the Statutes of the DYCL, every citizen of the People's Republic of Bulgaria from the age of fourteen to thirty can be a member of the Komsomol. In 1982 it had a membership of 1.5 million. The press organs of the Central Committee of the DYCL are the *Narodna Mladezh* (People's Youth) newspaper (since 1944), and the *Mladezh* (Youth) magazine (since 1948). The First Secretary of the Central Committee of the DYCL is Stanka Shopova (since 1981).

The League works under the guidance of the BCP, with democratic centralism as a basic organizational principle. Its supreme ruling body is the Congress, which the Central Committee convenes at least once every five years; in the interim between congresses, the League's activity is directed by the Central Committee. District, town, city-district and municipal organizations have been formed in accordance with the administrative division of the country. The main organizational units are the societies, set up on the territorial-production principle in all educational establishments, enterprises, offices, construction sites, forestry estates, co-operative farms, military units, villages and neighbourhoods where there are at least three members.

The DYCL originates from the Workers Social-Democratic Youth Union established in 1912, renamed Bulgarian Communist Youth Union in 1919. It is also the successor to and continuator of the revolutionary traditions of the Workers' Youth Union and the Bulgarian National Students' Union. After the victory of people's democratic power on 9 September 1944, the Workers' Youth Union launched a drive for the creation of a united people's youth league. The first step in its implementation was the establishment in 1945 of a Unified All-Student Youth Union and a General People's Union of University Students. The congresses of the two largest youth unions – the Workers' Youth Union and the Agrarian Youth Union – were held concurrently from 1 to 3 November 1947. In his letter of greetings to both these congresses Georgi Dimitrov recommended the establishment of complete unity of the younger generation by setting up a unified youth organization. The young welcomed this Party recommendation, and 4,970 unified youth societies with a total membership of 456,000 were formed by the end of November 1947.

The Constituent Congress of the People's Youth Union (PYU) was held on 21 and 22 December 1947. It adopted the Statutes and elected the Union's leadership. After the congress all political and mass youth organizations officially terminated their existence. In his speech to the Congress Georgi Dimitrov appealed to the youth, which was already united, to rally around the Fatherland Front and join as actively as possible in the effort to develop the national industry, to achieve the industrialization and electrification of the whole country, and to build the foundations of the socialist society. A vivid manifestation of the League's activity during that period was the mass movement of Bulgarian youth for participation by voluntary labour in the building of Socialism – the youth-brigade movement; some 600,000 young people took part in national, district and local youth brigades in the period 1946-50.

After the Fifth Congress of the BCP (1948), which charted the general line for building the economic and cultural foundations of Socialism, and adopted the first Five-Year National Economic Plan (1949-53), the PYU focused its activity on the implementation of the Party's programme. The fourth extended plenum of the Union's Central Committee (February 1949) defined the young people's socialist education as the Union's pivotal task. The decisions taken at the National Conference of Work-Teams (May 1949) promoted the younger people's participation in socialist emulation. In the very first year of the Five-Year Plan period many young workers introduced innovations in production, by which they helped boost labour productivity and cut down production costs. In the villages, youth was actively involved in the foundation and consolidation of the co-operative farms, and many youth teams obtained unprecedented yields in agriculture. After Georgi Dimitrov's demise (2 July 1949), the Plenum of the Central Committee of the PYU held in this connection on 9 July decided to name the organization the Dimitrov People's Youth Union (DPYU). The Second Congress (21-24 February 1951) outlined the new tasks of the Union: to assist the BCP even more energetically in the fulfilment of the Five-Year Plan within four years, to take an increasingly active part in the industrialization of the country, to contribute to the social reorganization of agriculture and to the training of young specialists in this field, to improve the work conducted among pioneer children and secondary school students, etc. Young people made a conspicuous contribution in the nationwide drive to fulfil the First Five-Year Plan ahead of schedule, with thousands of shock-workers, front-rankers in production, innovators and Heroes of Socialist Labour emerging from their ranks.

A fresh upsurge in the Union's activity began after the April 1956 Plenum of the Central Committee. The Union's members advanced a number of initiatives in production, designed to boost labour efficiency, upgrade the young workers' skills and improve working conditions. The youth-brigade movement was given fresh impetus. With all its activity, the DPYU contributed to the victory of Socialism and to the country's economic and cultural upsurge.

After the Seventh Congress of the BCP (1958), which is now known in national and party history as the congress of victorious Socialism in Bulgaria, the Union faced increasingly important tasks of improving young people's communist education and promoting their wider participation in the process of labour. At the proposal of the Party Central Committee, the Union's Fourth Congress

(27 November–1 December 1958) renamed it the Dimitrov Young Communist League. The delegates to the Congress therefore declared it the Ninth Congress of the Dimitrov Young Communist League (DYCL or Komsomol), thus stressing its succession to the Bulgarian Communist Youth Union, the Worker's Youth Union and the DPYU. Todor Zhivkov's speech at the Congress, which outlined the principal tasks of young people in the drive for Communism, became a strategic platform of the Komsomol's work for the education of the young at the new stage of accelerated socialist construction.

The period following the congress was characterized by the Komsomol's still more intensive work for the fulfilment of the Third Five-Year Plan (1958–62) ahead of schedule. In response to the Party's appeal, young people actively joined in the construction of the biggest industrial projects. The DYCL assumed patronage over the key industrial projects of the Third Five-Year Plan period – the Kremikovtsi Metallurgical Combine, the Nitrogen Fertilizer Plant near Stara Zagora, and the Maritsa-East Thermo-Power Plant – as well as over stock-breeding. An important step toward the overall accomplishment of the DYCL's tasks was the involvement of youth in the movement for commercial labour. The League advanced the appeal 'Study and Work, Work and Study' and initiated a drive for the mastery of science and technology. In 1961–62 alone, some 100,000 young people raised their educational level while at their jobs. Over 400,000 young workers in industry, construction and transport improved their engineering training, while more than 40,000 young agricultural workers mastered the operation of farm machinery. Each year more than 500,000 young people study the theory of Marxism-Leninism in the political education study-groups. The Tenth DYCL Congress (25–28 April 1963) considered the League's activity in the light of the Eighth Party Congress decisions and approved important decisions concerning the communist education of youth. It specified power engineering, metallurgy, and the chemical and mechanical engineering industries as the key sectors of Komsomol activity.

The theses about work with youth and the Komsomol elaborated by Todor Zhivkov, General Secretary of the Central Committee of the BCP, and the decisions of the December 1967 Plenum were of paramount significance in increasing the Komsomol's role and prestige in the overall economic, political and cultural life of the country. It became the chief organizer of youth sports, tourist, amateur art activities and those related to patriotic-military education and techno-scientific progress – a genuine exponent of the general interests and aspirations of young people. The Eleventh DYCL Congress (10–14 January 1968) analysed the results of the League's activity in the period following its Tenth Congress and drafted the guidelines for its future activity, the fundamental task being the broad involvement of youth in socialist construction. Particular attention was attached to the ideological and political elevation of youth, to the patriotic and internationalist education of the young generation. To this end the Congress adopted decisions on the improvement of its organizational work and perfection of the League's structure. The Tenth Congress of the BCP (1971), which approved a new Party Programme for the construction of an advanced socialist society, was a landmark in the development of the Komsomol. The Twelfth DYCL Congress (25–28 May 1972) set young people the prime task of active participation in the construction of the material and technical basis of the developed Socialism. The Kozlodui Atomic Power Plant, the Bobov Dol Thermo-Power Plant, the Antonivanovtsi and Krichim hydro-power projects and the Devnya industrial complex were built under the patronage of the League in the period between the Twelfth and Thirteenth Congresses. The Eleventh Congress of the BCP (1976) and the July 1976 Plenum of the Central Committee gave a highly favourable assessment of the Komsomol's activity and outlined its major tasks in the fulfilment of the Seventh Five-Year Plan. The Thirteenth DYCL Congress of 9–11 May 1977 adopted decisions guaranteeing the fulfilment of the Party assignment for an improvement of the quality and efficacy of the overall work for the communist education of youth, for rallying it around the party policy, and for its active involvement in the construction of the developed socialist society.

The general strategy and policy of the Party in its activity among youth and the Komsomol throughout the construction of the developed socialist society was elaborated in Todor Zhivkov's Letter (18 July 1978) to the Central Committee of the DYCL.

In response to the letter a nationwide youth movement was launched for the accomplishment of the tasks and guidelines charted in the letter. The movement was centred around the Komsomol, which was the representative of the Party before Youth, and of Youth before the Party, State and society. The function of the Komsomol as chief co-ordinator of those activities of the state, economic and public bodies and organizations which provide conditions for the development and education of youth was extended further. Inspired by its motto 'Study and Work, Buoyancy and Daring', the Komsomol began work on a wide scale and with new qualitative criteria for the comprehensive and effective self-realization of youth, for its general education and professional training on a contemporary level, for keeping abreast of the topmost achievements of scientific and technical progress, for the ideological, moral and aesthetic education of each young man and woman.

The Fourteenth DYCL Congress (25–27 May 1982) set the prime task of increasing the League's organizing and political role in all spheres of youth activity, and in the life of society in general, so that young people could fully accomplish their highly responsible tasks in the overall work and struggle for the fulfilment of the decisions adopted by the Twelfth Congress of the BCP. The Congress stressed the necessity of carrying through the renovative changes, the beginning of which was marked by Todor Zhivkov's Letter to the Central Committee of the DCYL, so as to bring the level of the

Komsomol organizations' work up to the requirements and criteria set out in the letter.

Since the Fourteenth Congress, the deeds, creative efforts and enthusiasm of the Bulgarian Komsomol members, of all young men and women, have been linked with the realization of the principal youth initiatives such as 'Scientific and Technical Progress and Leading Experience – a Domain of Youth Daring' and 'Worthy Offspring of the Class', and with the 'Memory' international relay race dedicated to the 40th anniversary of the socialist revolution in Bulgaria and to the 40th anniversary of the victory over German Nazism and Japanese Militarism. The Komsomol has assumed patronage over key industrial and agricultural sectors, and has mobilized the young people for devoted labour at such youth projects of national significance as the Third Metallurgical Base at Debelt, the Kozlodui Atomic Power Plant, the Maritsa-East Thermo-Power Plant, the Belmeken-Sestrimo Hydro Power Scheme, the Pernik Industrial Complex, the reconstruction and modernization of the Leonid Brezhnev Iron and Steel Works, the Chemical Works in Dimitrovgrad and the Copper Dressing Plant in Srednogorie. A whole region of the country which is to develop at particularly rapid rates by decision of the Party and the Government has been proclaimed a 'Republic of Youth' by the Komsomol.

The DYCL is a member of the World Federation of Democratic Youth (November 1945) and the International Union of Students (August 1946) since their foundation and has been energetically participating in their activities. It maintains relations with over 250 national, regional and international youth and students' organizations.

Orders: People's Freedom (1941–44), first class (1947); People's Order of Labour, gold (1949); three times Order of Georgi Dimitrov (1958, 1971, 1982).

The DYCL is authorized by the BCP to exercise direct leadership over the *Septemvriiche Dimitrov Pioneer Children's Organization,* which is a mass, autonomous organization of children in the People's Republic of Bulgaria.

With its specific forms and methods it works actively for the attainment of the ultimate goal – the upbringing of a many-sided and subsequently an all-round personality, an active builder of Socialism and Communism. The Septemvriiche Dimitrov Pioneer Children's Organization is a reserve of the Dimitrov Komsomol. In its work it is guided by the example and experience of the All-Union Lenin Pioneer Children's Organization. The underlying principles of the Septemvriiche are: communist ideas and socio-political orientation; unity and continuity of the three generations; autonomy and self-government on the basis of democratic centralism combined with competent pedagogical guidance; taking account of the age and individual peculiarities of children; voluntary membership and active participation in the organization's work; systematic, consistent and continuous activity.

Members of the Septemvriiche Organization are children in the nine–fourteen age range. The ceremony of admission takes place on 23 September, the day of Bulgarian pioneer children.

The main documents defining the requirement for membership of the organization are the solemn pledge and the laws of the Dimitrov pioneer. The organization educates children in a spirit of fidelity to the party cause, ardent patriotism and socialist internationalism, developing their interests and a communist attitude to labour, working for their harmonious development and preparing them for active participation in the country's social and political life.

The ideas of the 'Banner of Peace' movement expressed in its motto 'Unity, Creativity, Beauty' have developed widely and enriched the activity of the organization which provides broad opportunities for the expression and development of the creative faculties and endowments of the young.

The pioneer children's organization is a continuator of the children's communist movement in Bulgaria and was founded shortly after the socialist revolution. In the autumn of 1944, on the initiative of Georgi Dimitrov, the Politburo of the Central Committee of the BCP took a decision on setting up a children's organization with the Workers' Youth Union. Many prominent educationists, public figures and scientists were involved in the accomplishment of this task. Various organizations of children already existed, attached to other youth organizations. In 1945 these organizations joined together in the nationwide children's organization *Septemvriiche*. In 1949 it was renamed Dimitrov Pioneer Children's Organization, after Georgi Dimitrov. Its structure is built on the educational-territorial principle, consisting of district, city, city-district and municipal organizations. The pioneer children's company is the primary unit of organization. Its governing body is the company assembly, which can be either of a working or of a ceremonial nature. The company's executive body is its Council. As of 1982, the Organization had a membership of about one million. A Central Council of the pioneer children's organization at the Central Committee of the DYCL provides effective management by tackling its problems, preparing documents in connection with its activity and looking after the training and qualification of pioneer leaders. The Komsomol works for the communist education of children in close co-operation with the school and the family, jointly with party, trade-union, cultural, sports and other public and state organizations.

The Dimitrov Pioneer Children's Organization is a contingent of the international children's democratic movement and maintains relations with over 60 children's democratic organizations the world over. It is a founding member of the International Committee of Childrens' and Adolescents' Organizations and is an active participant in the international children's campaigns, competitions, contests, Olympiades and seminars sponsored by the Committee.

Orders: People's Republic of Bulgaria, first class (1959); Georgi Dimitrov (1969).

Pioneer children from Burgas in the 9 September parade, 1969

Public Organizations

Congresses of the Dimitrov Young Communist League

First (Constituent) PYU Congress – 21–22 December 1947, Sofia.
Second DPYU Congress – 21–24 February 1951, Sofia.
Third DPYU Congress – 13–15 May 1954, Sofia.
Fourth DPYU Congress – 27 November–1 December 1958, Sofia (Ninth DYCL Congress).
Tenth DYCL Congress – 25–28 April 1963, Sofia.
Eleventh DYCL Congress – 10–14 January 1968, Sofia.
Twelfth DYCL Congress – 25–28 May 1972, Sofia.
Thirteenth DYCL Congress – 9–11 May 1977, Sofia.
Fourteenth DYCL Congress – 25–27 May 1982, Sofia.

The Bulgarian Trade Unions. The Bulgarian Trade Unions are 'a mass, self-governing socio-political organization of the working class, the peasants, the employees and all working people, who are organized on a voluntary basis, irrespective of their political affiliation, nationality, race, sex and religious convictions' (from the *Statute of the Bulgarian Trade Unions,* Sofia, 1982, pp. 4–5). The basic units are the local trade union organizations. The entire structure of the Unions is based on the branch-production principle, in combination with the territorial principle. They work under the political guidance of the BCP. In the process of social homogenization of society, they enlarge their functions from mass organizations of the working class to mass organizations of all working people.

The Bulgarian Trade Unions conform to the principle of democratic centralism. Congress is the supreme guiding body; the Central Council functions between congresses, and between the Plenums of the Central Council the work is executed by the Bureau of the Central Council and the Secretariat of the Central Council. The President of the Central Council of the Bulgarian Trade Unions is at present Mr Peter Dyulgerov (elected in 1981). At the Eighth Congress of the Trade Unions in 1982 the membership was announced as 4,000,000. The Bulgarian Trade Unions have their own newspaper, *Trud* (Labour).

The professional associations of Bulgarian workers, the Syndicates, were formed immediately after Bulgaria's Liberation from Ottoman rule in 1878. Following attempts to set up professional organizations among the printers, the Bulgarian Typographic Society was formed in 1883, in Sofia. Other professions followed suit and formed their own associations. These, however, were of a purely subsidiary nature and concerned themselves solely with the workers' financial interests. In the 1880s, in many of the district centres, Teachers' Associations were formed with both educational and strictly professional tasks. The activities of the Teachers' Association in the town of Vidin, set up in 1889 and headed by socialists under the direct guidance of Dimiter Blagoev, were particularly important. Socialist propaganda and the creation of the BSDP in 1891 helped the emerging working class to adopt a class-conscious approach to its problems. At the beginning of the 1890s united workers' associations and professional syndicates appeared. In 1894, the Central Printing Workers' Syndicate was formed, and in 1895, the newly-created Bulgarian Teachers' Union began the dissemination of populist, reformist and revolutionary ideas. In the first decade of the twentieth century, two trends, revolutionary and reformist, were already discernible in the trade union movement. On the initiative of the BWSDP (left-wing) the revolutionary professional and united syndicates formed, in 1904, the United Workers' Syndicate, whose basic principles were militant class struggle and close ideological and organizational relations with the revolutionary-orientated Social Democratic representatives. In the same year, the United Free Workers' Syndicate was formed. Under right-wing influence, the latter took a reformist stand and adhered to the principle of ideological and organizational neutrality. The intensification of the class struggle and the developments in the United Workers' Syndicate led to significant, positive changes in the organizational structure of the trade union movement, chiefly owing to the work of Georgi Kirkov, Georgi Dimitrov and other eminent workers' representatives. From 1907–10 centralized trade unions had been set up for artisans and industrial workers. At the same time the United Free Workers' Syndicate merged with the syndicates of the *Proletarii* anarcho-liberal socio-democratic union to form the United Workers' Syndical Union. In 1909 the reformist trade unions of the state and municipal employees also united in a Union of Unions. Both the revolutionary and the reformist wings maintained international contacts. On the eve of, and during the First World War of 1914–18, strong left-wing trends were obvious in the United Workers' Syndical Union. In the post-war revolutionary crisis and under the impact of the October Revolution of 1917, the left-wingers gradually came closer to the revolutionary trade unions. In September 1920 the majority of the members of the United Workers' Syndical Union merged into the United Workers' Syndicate. In 1919–22 the professional

The building of the Central Council of Bulgarian Trade Unions, Sofia

unions organized a great number of strikes, of which the large-scale Transport Workers' Strike (1919–20) was the longest. Georgi Dimitrov, who was Secretary of the United Workers' Syndicate, contributed greatly to the adoption of Leninism by the Bulgarian Trade Union movement in the period after the Great October Socialist Revolution. The United Workers' Syndicate became a mass organization; at the beginning of 1920 it united eighteen trade union organizations with 335 local sections and a membership of 30,061 and was the largest syndicate in Bulgaria at that time. The United Workers' Syndicate also participated in the constitution of the Red International of the Trade Unions, the Profintern.

The onset of fascist dictatorship in Bulgaria in 1923 put an end to all legal forms of revolutionary workers' organizations. The BCP combined the legal forms of working class struggle with underground action and in 1925 formed the Independent Workers' Trade Unions.

organizations to defend the interests of the working class. After the invasion of the Soviet Union by nazi Germany on 22 June 1941, the revolutionary trade union activists organized acts of sabotage on production destined for the nazis, and shared the burden of the struggles of the working class against the anti-national policy of the fascist Governments. Many trade union activists joined the partisan movement. Prominent members of the revolutionary trade union movement took part in the organization of mass political strikes and in the People's Uprising on 9 September 1944.

Immediately after the 1944 socialist revolution in Bulgaria, 32 mass branch trade union organizations were set up on a voluntary basis. They merged into the United Workers' Trade Unions. The new trade union organizations were characterized by unity of action and together actively participated in strengthening the government of people's democracy and in efforts to restore the country's economy. They were particularly efficient in establishing

The Gavril Genov Chemical Works, Russe

Workers relaxing at the District Council of Trade Unions, Pernik

The new professional organization strove to achieve unity of action and to defend the immediate interests of the working people. They participated in the great strikes of the textile and tobacco workers, and in other events. The 'Revolutionary Profposition' was created to send activists out among the reformist and state organizations of teachers, railway workers, post office employees, etc. When, in mid-1930, the fascist Government set up its own state-professional associations and made them unite in a Bulgarian Workers' Union, the Independent Workers' Trade Unions ceased to exist. The cadres and members of the revolutionary trade union organizations began to work, under the guidance of the BCP, among the workers and employees organized in the Bulgarian Workers' Union. They made use of the state professional

forms of workers' control and incentives for patriotic work emulation in industry and transport. The United Workers' Trade Unions also established close relations with the trade union organizations in the Soviet Union and the people's democracies, and participated in the formation of the World Federation of Trade Unions. At the Third Trade Union Congress in 1951 the Central Council of the Trade Unions was constituted as the leading body of the trade union organizations. At the Seventh Congress of the Trade Unions in 1972, the Central Council of the Trade Unions changed its name to the Central Council of the Bulgarian Trade Unions.

After the Fifth Congress in 1948, the Trade Unions became one of the basic factors in the struggle of the working class for building a socialist society. They became the initiators of different activities for the achievement of the goals of the Five-Year Economic Plans. They adopted and applied the experience accumulated by the Soviet working class in the organization of socialist emulation, in the socially-orientated basis of production and in the creation of the material and technological basis of Socialism.

The Trade Unions are, at the same time, representatives and defenders of the many-sided interests of the working people in the different state, economic, and state-and-public bodies. Their socio-economic, cultural and educational functions have gradually been extended, and at present they play the role of a school for management, a school for administration, and a school for Communism. A year after its historic April Plenum in 1956, the Central Committee of the BCP held yet another Extraordinary Plenum, dedicated solely to the state of the Trade Unions and the future guidelines for their development.

The Trade Unions increased their efforts to assist the realization of the Communist Party's programme after the Tenth (1971), Eleventh (1976) and Twelfth (1981) Congresses of the BCP.

The formulations of Todor Zhivkov, General Secretary of the Central Committee of the BCP and President of the State Council, concerning the Unions' share in the construction of the mature socialist society were of particular significance for their further development. In his speech to the Ninth Party Congress 1982, as well as in a number of statements and lectures, Todor Zhivkov further developed his formulation of the role of the Unions as a public guarantor of the realization of the country's economic and social policies: as the active driving force behind the target-plans in the field of economics; as assistants to the workers' collectives in the execution of their duty as sole owners of socialist property and collective managers of their respective production units; and as executors of the new tasks and functions in labour and labour relationships, as reflected in the *Party Theses on a new Labour Code*. The Bulgarian Trade Unions confirmed their important place in the political system in the People's Republic of Bulgaria, their role as champion of the Party's cause, enjoying the mass support of the working people; they are a decisive factor for the development of socialist democracy. This has made it possible for the Trade Unions to agree to take responsibility for some state functions. They have accordingly assumed certain legislative functions, have deputies in the National Assembly and participate in the work of high-level, responsible state and economic bodies. The Trade Unions adopt agreements and sign collective labour contracts defending the interests of the working people. They control safety regulations and organize the rest, recreation and short-term leaves and holidays of the workers and employees; and they control, in the public interest, the retail prices of consumer goods. They also guide the work of the conciliation commissions whose members are elected to act in the first instance in trying to solve labour disputes. The Trade Unions participate at all levels in the working out of the Five-Year and Annual Plans for the socio-economic development of the country;

A holiday house for workers from the Sofstroi Building Organization, Michurin

they stimulate the production activities of the working people by organizing socialist emulation for high efficiency and quality in all spheres of work; they play a large part in the development of the self-initiative, innovators' movement, and in the mass dissemination and adoption of positive experience accumulated in the production sphere. They are also active in defining labour norms and remuneration. Basing their actions on the workers' efforts to achieve the goals of the socio-economic development plans, the Trade Unions are active in organizing the defence of the workers' basic rights, such as trade and public services, food, housing, working and protective clothing, admission of children to kindergartens and crèches, allocation and use of public funds and in particular, of the specialized funds for social, cultural and everyday needs set up in individual enterprises. They participate in decision-making on the right to work and on the workers' and employees' pension rights upon retirement. One important function is to work actively for the education of the working people for the comprehensive development of the individual, and the Trade Unions take an active part in efforts to raise qualifications and promote the education of the workers. They possess large material resources and technological facilities, as well as cadres for the implementation of their tasks and functions.

SELECTED BIBLIOGRAPHY

Socio-Political Structure of the People's Republic of Bulgaria

ANANIEVA, N. *Control Function of the National Assembly.* Sofia, 1974.
ANDREEV, M. *History of the Bulgarian Bourgeois State and Law. From the Liberation from Ottoman Rule, 1878, to the Great October Socialist Revolution.* Rev. edn. Sofia, 1975.
Constitution of the People's Republic of Bulgaria. Sofia, 1971.
Georgi Dimitrov and the Bulgarian People's Democratic State. Ed. Y. Radev. 2nd rev. edn. Sofia, 1982.
MILANOV, Zh. *State Power and State Government.* Sofia, 1976.
NACHEVA, S. *State Procedural Norms.* Sofia, 1983.
PAVLOV, S. *Penal Procedure of the People's Republic of Bulgaria.* 3rd edn. Sofia, 1979.
RADEV, Y. *State Representative Bodies in Bulgaria (1944–47).* Sofia, 1965.
SLAVOV, P. *The Bulgarian Socialist State. Essence, Laws, Stages and Prospects.* Sofia, 1975.
SPASSOV, B. *Basic Rights and Obligations of the Citizens in the People's Republic of Bulgaria. Problems of Socialist Law and Legality.* Sofia, 1978.
SPASSOV, B. *La Bulgarie.* Paris, 1973.
SPASSOV, B. *Koinoniki kai kratika organosi tis laiki demokratia Bulgaria. Matafrasi Kostas Saparas.* Bulgarian Books, Athens, 1977.
SPASSOV, B. *Konstitutsia i Narodnoe Predstavitelstvo v Narodnoi Respublike Bolgarii.* Progress, Moscow, 1977.
SPASSOV, B. *The Legislative Activity of the National Assembly.* Sofia, 1971.
SPASSOV, B. *The National Assembly. Organization and Legislative Activity.* Sofia, 1980.
SPASSOV, B. *Ordinamento costitutionale della Republika popularle Bulgaria.* Rome, 1982.
SPASSOV, B. *Questions of the New Constitution.* Sofia, 1973.
SPASSOV, B. and G. ZHELEV. *State Law of the People's Republic of Bulgaria.* 2nd edn. Parts 1–2. Sofia, 1979.
STALEV, Zh. *Bulgarian Civil Procedure Law.* 3rd rev. edn. Sofia, 1979.
State, Law, Government. Sofia, 1977.
State and Legal System in the Construction of a Developed Socialist Society in the People's Republic of Bulgaria. Sofia, 1975.
STOICHEV, S. *Acts of the National Assembly.* Sofia, 1974.
STOICHEV, S. *The Development of Direct Democracy.* Sofia, 1980.
STOICHEV, S. *The Electoral System of the People's Republic of Bulgaria.* Sofia, 1977.
TADJER, V. *The Right of State Socialist Ownership in the People's Republic of Bulgaria.* Sofia, 1975.
Theoretical and Practical Problems of Social Development. Sofia, 1978.
Theses of the Twelfth Congress of the Bulgarian Communist Party. Sofia, 1981.
Theory of the Socialist State. Sofia, 1978.
VULKANOV, V. *Bulgarian Citizenship.* Sofia, 1978.
VULKANOV, V. *Legal Status of the National Representaive.* Sofia, 1976.
ZHELEV, G. *The Essence of the Representative System of the People's Republic of Bulgaria.* Sofia, 1964.
ZHELEV, G. *The National Assembly according to the New Constitution of the People's Republic of Bulgaria.* Sofia, 1971.
ZHIVKOV, T. *On the New Constitution of the People's Republic of Bulgaria. Report to the Sixteenth Session of the Fifth National Assembly, 7 May 1971. – Selected Works.* Vol. 19. Sofia, 1976.
ZHIVKOV, T. *Report of the Central Committee of the Bulgarian Communist Party to the Twelfth Congress and the Forthcoming Party Tasks.* Sofia, 1981.
ZHIVKOV, T. *Twelfth Congress of the Bulgarian Communist Party and the Further Construction of Mature Socialism. Problems, Tasks, Approaches.* Sofia, 1982.

The Bulgarian Communist Party

The Bulgarian Communist Party, Resolutions and Decisions. Vols. 1–5. Sofia, 1965.
Documents of the Central Leading Bodies. Vols. 1–5. Sofia, 1972–81.
GEORGIEV, G. *Selected Works.* Sofia, 1971.
History of the Bulgarian Communist Party. Sofia, 1951–55.
KIRKOV, G. *Selected Works.* Vols. 1–2. Sofia, 1950–51.
KOLAROV, V. *Selected Works.* Vols. 1–3. Sofia, 1954–55.
Programme of the Bulgarian Communist Party. Sofia, 1971.
Statute of the Bulgarian Communist Party. Sofia, 1981.
ZHIVKOV, T. *Problems and Approaches in the Construction of Advanced Socialism in the People's Republic of Bulgaria.* Sofia, 1984.
ZHIVKOV, T. *Selected Works.* Vols 1–29. Sofia, 1975–83.
ZHIVKOV, T. *The Twelfth Congress of the Bulgarian Communist Party and the Further Construction of Advanced Socialism.* Sofia, 1982.

The Bulgarian Agrarian Party

Alexander Stamboliiski. His Life, Work and Legacy (Messages of Greetings, Reports and Communications Read at the International Scientific Conference on the Occasion of Alexander Stamboliiski's Birth Centenary. Sofia, 6 and 7 June 1979.) Sofia, 1980.
DIMITROV, G. *Collected Works.* Vols. 7–14. Sofia, 1953–55.
GENOVSKY, M. *The Agrarian Movement and Leninism.* Sofia, 1948.
GENOVSKY, M. *The Road of Agrarian Unification. A Historical Survey.* Sofia, 1947.
KOSSEV, D. 'The Peasant Movement in Bulgaria at the End of the Nineteenth Century. Foundation of the BZNS and the Attitude of the Bulgarian Workers' Social Democratic Party to the Agrarian Question' *Istoricheski Pregled* (Historical Review), No. 5, 1949.
KOZHUHAVOV, K. *The Bulgarian Agrarian People's Union in the Past and the Present Tasks.* Sofia, 1956.
KOZHUHAVOV, K. *Pages from the Most Recent History of the BAP.* Sofia, 1969.
PEKAREV, Y. *History of the Agrarian Movement in Bulgaria. The Bulgarian Agrarian People's Union.* Vol. 1. Varna, 1945. Vol. 2, Dobrich, 1947.
PETROVA, D. *The Bulgarian Agrarian People's Union and the Popular Front, 1934–39.* Sofia, 1967.

PETROVA, D. *The BZNS at the End of Bourgeois Domination in Bulgaria, 1939–44*. Sofia, 1970.
STAMBOLIISKI, A. *Selected Works*. Sofia, 1979.
TANCHEV, P. *The BAP Always Faithful to the Party, the People and Socialism*. Selected Works in two volumes. Sofia, 1981.
TANCHEV, P. *The BAP and the Construction of the Advanced Socialist Society*. Selected Works. Sofia, 1983.
TISHEV, D. *The Joint Work of the BCP and the BAP in Socialist Construction*. Sofia, 1969.
TISHEV, D. *The Unity of Action between Communists and Agrarians in the Struggle against Fascism*. Sofia, 1967.
TRAIKOV, G. *Articles, Speeches, Reports, 1922–68*. Sofia, 1968.
TRAIKOV, G. *Seventy Years of Bulgarian Party*. Sofia, 1970.
ZHIVKOV, T. *The Party is Faithful to its Allies. Speeches, Reports and Articles*. In two volumes. Sofia, 1976.

The Fatherland Front

BONEV, V. *Za edinniya, Narodniya i Otechestveniya front* (On the United, Popular and Fatherland Front). 2nd rev. and enl. edn. Sofia, 1974.
DIMITROV, G. *Suchineniya* (Works). Vols. 11–14. Sofia, 1951–55.
DRAGOICHEVA, T. *Povelya na dulga* (The Call of Duty). Vols. 2–3. 2nd edn. Sofia, 1980.
Istoriya na antifashistkata borba v Bulgaria 1939–44 (History of the Anti-Fascist Struggle in Bulgaria 1939–44). Vols. 1–2. Sofia, 1976.
ISUSOV, M. *Politicheskite partii v Bulgaria 1944–48* (The Political Parties in Bulgaria 1944–48). Sofia, 1978.
KUBADINSKI, P. *Otechestveniyat front – front na tsyaloto otechestvo* (The Fatherland Front as a Front of the Whole Country). Sofia, 1982.
PETROVA, S. *Devetoseptemvriiskata sotsialisticheska revolyutsiya 1944* (The Socialist Revolution on 9 September 1944). Sofia, 1981.
SHARLANOV, D. *Otechestveniyat front i sotsialisticheskata revolyutsiya* (The Fatherland Front and the Socialist Revolution). Sofia, 1975.
SHARLANOV, D. *Suzdavene i deinost na Otechestveniyat front. Yuli 1942–Septemvri 1944* (The Creation and Activities of the Fatherland Front. July 1942–September 1944). Sofia, 1966.
ZHIVKOV, T. *Isbrani suchineniya* (Selected Works). Vols. 1–3. Sofia, 1975.
ZHIVKOV, T. *Otechestveniyat front – nashe istorichesko zavoevanie* (The Fatherland Front – Our Historical Achievement). Sofia, 1982.

The Dimitrov Young Communist League

DIMITROV, D. *Koga i kak beshe suzdadena Dimitrovskata pionerska organizatsiya 'Septemvriiche'* (When and How the Dimitrov Pioneer Children's Organization 'Septemvriiche' was Created). Sofia, 1969.
DKMS v rezolyutsii i resheniya na kongresite, konferentsiite i plenumite na TK. (The Dimitrov Young Communist League in the Resolutions and Decisions of the Congresses, Conferences and Plenums of the Central Committee). Vols. 1–4. Sofia, 1973–78.
Istoriya na mladezhkoto revolyutsionno dvizhenie v Bulgaria (History of the Revolutionary Youth Movement in Bulgaria). 2nd enl. edn. Sofia, 1972.
Vinagi gotov. Dokumenti i resheniya za deinosta na DPO 'Septemvriiche' (Always Ready. Documents and Decisions about the Activities of the Dimitrov Children's Pioneer Organization 'Septemvriiche'). Eds. D. Shopova and R. Marinova. Sofia, 1975.
ZHIVKOV, T. *Nyakoi osnovni problemi na rabotata s mladezhta i Komsomola. Izbrani suchineniya* (On Some Basic Problems of Work with the Young and the Komsomol). Selected Works. Vol. 14. Sofia, 1976.
ZHIVKOV, T. *Uchenie i trud, zhizneradost i druznovenie. Pismo do TK na DKMS* (Study and Work, Vitality and Daring. Letter to the Central Committee of the Dimitrov Young Communist League). Sofia, 1978.

The Bulgarian Trade Unions

History of the Bulgarian Trade Union Movement. Sofia, 1980.

PART IV

THE ARMED FORCES

THE ARMED FORCES

The armed forces of the People's Republic of Bulgaria defend the revolutionary achievements of the working people, the independence, sovereignty and territorial integrity of the nation. Together with the armed forces of the other socialist countries they guarantee the security of the socialist community and the preservation of peace.

They differ essentially from the armed forces of the capitalist states. They reflect the nature of the socialist State and social order and embody the common interests of the working people and of the socialist State, the moral and political unity of the people, socialist patriotism and internationalism. They are the successors and continuers of the progressive military and revolutionary traditions and virtues of the Bulgarian army and Bulgarian people throughout the ages.

From the very first days of its existence, the Bulgarian State, formed in 681, had set up a military organization of its own. The respective armed services followed the feudal pattern: cavalry, infantry and supplies. The armed forces flourished during the reign of the Bulgarian tsars who were also renowned for their military competence and exploits – Krum, Simeon, Samuil, Kaloyan, Assen I and Ivan Assen II. In the struggle of the Bulgarians against the centuries of Ottoman rule (1396–1878), typically Bulgarian forms of military organizations, such as the *haiduk* and *chetnik* units, were born. Bulgarian armed units participated during this period in the uprisings of the neighbouring peoples against the Ottoman oppressors, as well as in the Russo-Turkish wars in the eighteenth and nineteenth centuries. In 1862 the Bulgarian revolutionary Georgi Rakovski organized the First Bulgarian Legion in Belgrade. In 1867 Russian support was secured to set up a military college in Belgrade to train young Bulgarians for the future armed struggle against Ottoman rule. This became known as the Second Bulgarian Legion. On the eve of the Russo-Turkish war for the liberation of the Bulgarian people (1877–78), the Russian Command organized Bulgarian volunteers into a *Bulgarsko opolchenie* (Bulgarian voluntary corps). The volunteers formed three brigades and six battalions, numbering 12,000 men at the end of the war. This volunteer force was put under the direct command of General Nikolai G. Stoletov and participated in military operations and in the battles for the Bulgarian town of Stara Zagora and the historical places of Shipka and Sheinovo.

After the Liberation of Bulgaria from Ottoman rule in 1878, the volunteer units formed the backbone of the armed forces of the new Bulgarian State. As early as 1878 a Bulgarian local army was set up with Russian assistance and existed till 1879. On 26 November 1878 a military school was opened in Sofia. In 1880 the Conscription Bill made military service compulsory. The young Bulgarian army, trained by Russian officers and by *opolchenie* Commanders, participated actively in the unification of Eastern Rumelia (Southern Bulgaria) with the Principality of Bulgaria in 1885. In the Serbo-Bulgarian War of 1885 and the Balkan War of 1912–13, the Bulgarian army demonstrated mass heroism.

The modern Bulgarian People's Army is a successor to the underground armed units organized by the Bulgarian Communist Party (BCP) during the resistance against the bourgeoisie and Fascism (1941–44).

The predecessors of the modern Bulgarian armed forces were the revolutionary units formed during the September 1923 armed anti-fascist uprising. The revolutionary army of 1923 consisted chiefly of infantry, although cavalry, artillery and supplies were also available. The army was under the command of the Supreme Military and Revolutionary Committee whose members were competent military experts with adequate experience, loyal to the revolutionary cause. The revolutionary army showed a very high morale and excellent fighting capabilities and was exemplary in its revolutionary discipline, endurance, comradeship and humanity. The revolutionary army of 1923 is the forerunner of the modern Bulgarian People's Army; therefore, in the tradition of historic continuity, 23 September, the actual date of the 1923 anti-fascist uprising, is nowadays celebrated as the Day of the Bulgarian People's Army.

During the Second World War (1939–45) the BCP organized a large-scale armed resistance by the anti-fascist forces against the nazi invaders and the monarcho-fascist dictatorship. Partisan units, for active armed resistance, and underground sabotage groups were formed. A Central Military Commission at the BCP Central Committee, and military commissions at party district committees were set up to organize and guide the armed resistance against Fascism.

The increasing number and intensified activities of the partisan units and sabotage groups in the spring of 1943 heralded a new stage in the anti-fascist resistance. This

Forcing a crossing

necessitated certain structural changes in the partisan units and sabotage groups. To this end, a comprehensive plan for military operations was worked out by the BCP Central Committee.

Bulgaria was divided into twelve insurgent operating zones, and the partisan units were reorganized to form, along with the other active anti-fascist groups and units, the People's Liberation Insurgent Army (PLIA). In 1943 the Central Military Commission was transformed into PLIA Headquarters (HQ). Each zone was put under the military command of the local HQ while the immediate political guidance was taken over by the district committees of the BCP and representatives of its Central Committee.

At the beginning of September 1944 the PLIA consisted of 54 partisan units (one division, 10 brigades, 36 detachments, 7 soldiers' partisan battalions and detachments), hundreds of sabotage groups, underground organizations and groups in the armed forces and over 200,000 supporters who assisted the anti-fascists by supplying them with food and shelter.

The PLIA was a really popular, revolutionary army defending the interests of the working people. Its character was determined by the very essence and image of the anti-fascist struggle, which was a social and class struggle by nature. Not only was it led by the BCP, in its overwhelming majority it consisted of communists. The BCP, which actually bore the brunt of the anti-fascist struggle, was the only party within the Fatherland Front organization, uniting all anti-fascist forces in the country at that time, to have its own organized military force (PLIA). The operations of PLIA played a significant role in disorganizing the German rear, stepping up the national liberation war and achieving the victory of 9 September 1944. The PLIA then became the revolutionary nucleus of the Bulgarian People's Army (BPA).

In the process of the formation of the armed forces of new Bulgaria, of particular importance was the work by the BCP among the military personnel of the bourgeois State, aiming to win the support of large parts of the army to the people's cause and transform the government military force into a real people's army.

As a result of this activity more than 3,000 servicemen joined the partisan detachments. There were even cases of entire units going over to them. In addition, underground anti-fascist organizations were operating in almost all army units.

The Party's consistent work in the army also resulted in a number of army detachments participating in the preparation and execution of the socialist revolution of 9 September 1944. The overwhelming majority of the soldiers enthusiastically welcomed the victory of the revolution and declared their support of the revolutionary cause. The soldiers' committees formed before the victory were granted legal status and new ones were set up. During the revolution soldiers' committees operated in 210 out of a total of 250 units of the Bulgarian army. They occupied barracks, arrested officers known to have committed crimes against the people, elected as commanding officers true representatives of the people, put themselves at the disposal of the local committees of the BCP and the Fatherland Front and carried out their instructions.

The Armed Forces

A tank column

In order to secure the support of the military units and progressive officers for the preparations for the armed uprising, the BCP had used not only its own influence and resources but also those of the Fatherland Front and chiefly of the *Zveno* political circle.

On 9 September 1944 the army had actually been won over from the fascist command and pledged its support to the cause of the Fatherland Front. Thus in the very first days of the revolution the links of the army with the then dominant Monarcho-Fascism were destroyed and its social function as a stronghold of bourgeois domination was eliminated; a new army was created, of an entirely new nature, and with new goals and tasks, a people's army, new in spirit and class characteristics. Revolutionary measures accelerated its further reconstruction: first and foremost, the problem of cadres was taken care of; the fascist High Command was replaced, Military posts were assigned, and the role of the leading element in the army was given to several hundred former partisans, political emigrants who had returned home from the USSR, and reserve officers who had suffered repression in the past for their progressive ideas, and who were now restored to their officer's posts and in some cases promoted to the rank of General. Intensive officer training courses were organized. A new force, the 40,000-strong People's Guard, was formed, joined by former partisans and other antifascists. The People's Guard helped change further the army's class character and affirm it as truly representative of the people. Immediately after the uprising the post of Assistant Commanding Officer, with specific political functions, was created. To this post were assigned people who had passed through the severe school of the antifascist struggle and had accumulated great political and revolutionary experience. A significant part in the revolutionary reconstruction of the army was played by the military departments set up by the central and local leading bodies of the Party and the Fatherland Front.

On the very day of 9 September Bulgaria joined in the war effort of the anti-fascist coalition. The Patriotic War that followed was for Bulgaria a natural continuation of the armed anti-fascist resistance against the nazi invaders and the monarcho-fascist regime in the country. In the immediate war preparations the Fatherland Front Government declared mobilization of 454,000 reservists, 286,000 of whom were sent directly to the front line. The just cause of this war gave encouragement to the Bulgarian people to employ all their moral strength and material resources in the name of ultimate victory over Fascism.

On 18 September the troops of the BPA who were assigned to take part in the military operations, were placed under the operational command of the Commander-in-Chief of the Third Ukrainian Front, Marshal Tolbukhin of the Red Army.

From September to November 1944 it fought on the fronts in Vardar Macedonia, Southern Serbia and Kosovo-Metohija. The task of the BPA was to join in the offensive of the Red Army and the Yugoslav People's Liberation Army, defeat the forces of the 'Serbia' German military formation, and cut the enemy communications, thus preventing withdrawal of the German troops from Greece. Bulgarian military units participating in this task included fourteen combined-arms units, an aviation division, the

On a reconnaissance mission

Danube military fleet, etc. In extremely bad weather, fighting a heavily armed and experienced enemy, the BPA carried out four successful offensives known in history as the Nish, the Stracin-Kumanovo, the Bregalnica-Strumica and the Kossovo offensives. As a result of its intensive operations, the Bulgarian Army, in close co-ordination with the Soviet Army and units of the Yugoslav National Liberation Army, cut the communication lines and checked the retreat of a considerable part of the nazi armies from Greece, and liberated Southern Serbia, Vardar Macedonia and Kossovo with their respective administrative centres of Nish, Skopje and Prishtina.

In the course of the war, the Supreme Command of the BPA and the Ministry of War in Sofia undertook changes in the organization, composition, equipment and management of the armed forces.

After the completion of the military operations in Macedonia, Southern Serbia and Kossovo-Metohija, and the shift of the centre of operations to the North of Yugoslavia and Hungary, Bulgaria's participation in the anti-nazi war continued. At the end of November the First Bulgarian Army was formed, with six divisions composed of two corps. It was again placed under the operational command of the Commander-in-Chief of the Third Ukrainian Front of the Red Army. In December 1944 the First Army units were concentrated in Northern Yugoslavia, west of Belgrade. From 22–29 December the Srem offensive was mounted, and in early January 1945 the Bulgarian troops marched into Southern Hungary. There, incorporated into the Balaton defensive operation of the Third Ukrainian Front against the German counter-offensive in Hungary in March 1945, the First Bulgarian Army succeeded, from 6–19 March, in carrying out the defensive operation on the river Drava. In the Mur offensive of 29 March–5 April, the Bulgarian troops were co-ordinated with the Soviet 57th Army and defeated the main forces of the German 68th Army Corps, took over the oil-rich region of Nagykanizsa and advanced on Cakovech. From 7–15 May the Bulgarian troops chased the retreating German units and on 11 May, east of Klagenfurt in Austria, met with units of the British Eighth Army advancing from the west.

In the Patriotic War the BPA inflicted heavy losses on the enemy, killing or capturing 96,000 men; the Germans also sustained huge losses in ammunition, equipment and vehicles. By its successful operations, the BPA had a considerable share in the ultimate defeat of nazi Germany by the Allied Powers. In battles with the nazi forces Bulgaria suffered about 40,000 casualties. In launching its Patriotic War Bulgaria both sided with the democratic nations fighting against nazi Germany and successfully defended its national interests. By participating in the liberation of Yugoslavia and Hungary the BPA also fulfilled its internationalist duty. The outcome of the Patriotic War strengthened the position of the popular Government in Bulgaria and helped create favourable conditions for the future course of the socialist revolution.

After the victorious end of the Patriotic War, the BPA went through two successive stages of development, from the end of the war to the mid-1950s, and from the mid-1950s to the present day.

The first period saw speedy reconstruction of the organizational structures and cadres, and of the equipment and training and educational system in the army. To carry out the strenuous programme of reshaping the army, in 1946, the BCP under the leadership of Georgi Dimitrov undertook a number of important measures. The Bill on Army Management and Control adopted by the National Assembly created the necessary conditions for the final defeat of the counter-revolutionary military conspiracies and helped purge the army of the remaining anti-national and fascist elements. The leading role of the Party and its Central Committee became the guiding principle of military policy. Intensive courses were organized to train representatives of the working people as commanding officers. The Fifth Congress of the BCP (1948) adopted guidelines for laying the foundations of the socialist society and defined the main tasks of the country's defence and training of the armed forces. Important party and government measures were consequently taken to strengthen the position of the BPA as mainstay and reliable defender of the socialist fatherland. The army units and formations were reorganized and supplied with advanced Soviet-made weapons and equipment. The instruction and training of troops was reorganized in accordance with new statutes and regulations based on Soviet military science and the experience accumulated in the Second World War. Within the army organization an orderly and comprehensive system of political structures was set up. After the Fifth BCP Congress, 1948, organizations of the BCP and the People's Youth League (today's Dimitrov Young Communist League; DYCL) were set up within the BPA.

The establishment, improvement and training of the armed forces during this period was done on the basis of the conventional means of warfare. By the mid-1950s it had become a modern army of a new, socialist type.

The development of the BPA in the second period has been influenced by important socio-political, economic and military-technological factors. Rapid economic development has enlarged the possibilities for production of weaponry, various types of military equipment, ammunition and military facilities. The improvement in communications has led to better operational manoeuvre capability. The social, class and political changes in society have strengthened the army personnel's moral and political unity, which has become an important factor and resource for guaranteeing the high combat capability and state of readiness of the armed forces.

Developments in the BPA during this period embrace its technical facilities and organizational structure, as well as the forms and methods of warfare, the system of training and instruction, and all other aspects of its make-up. Through the increased economic possibilities of the country and the fraternal help of the Soviet Union, as well as co-operation with the other socialist countries, the BPA receives additional supplies of modern military equipment and weaponry. Considerable improvement has also been registered with regard to the technical equipment of all armed services; motorization and mechanization have begun to play an increasingly important role, and the fire and striking power of the motorized-rifle formations has improved immensely. A scientific and technological revolution has been carried out in the BPA.

At the organizational level, the necessary changes have been introduced in the army structure to make it comply more fully with the requirements of modern warfare and operational, tactical and strategic concepts of the Allied Forces of the Warsaw Treaty member-states, and to guarantee a destructive blow at any possible aggressor. Profound consideration has been given to any possible conditions leading to an outbreak of hostilities, and to the warding off of enemy aggression, the management of the forces, etc. The whole system of operational, combat and political training of the armed forces has been reconstructed to conform with the requirements of modern warfare.

In the training of commanding and staff officers note is taken of the impact of the revolutionary changes in military science on the forms and methods of modern warfare and on the operations of the armed forces. Considerable attention is paid to the wider and comprehensive use by the military cadres of the latest developments in military science and in the instruction and training of the armed forces. The units and formations practise modern methods of warfare in most difficult conditions, training to demonstrate their ability against a powerful and experienced enemy.

As a result of the tremendous amount of work done under the leadership of the BCP Central Committee, the army is now organized, armed and trained on a modern basis.

Of decisive importance for the development of the BPA during the second period have been the new conditions

Pioneers

which took shape in the mid-1950s when the Warsaw Treaty Organization was established. The present member-states are Bulgaria, the GDR, Poland, Romania, the USSR, Hungary and Czechoslovakia.

The Warsaw Treaty Organization is a defence alliance whose chief task is to guarantee security to peace-loving states and to maintain peace in Europe. The treaty is open to all countries which are ready to contribute towards the cause of world peace and security. The member-states are guided by the UN Charter and by the principles of respect for the independence and sovereignty of all nations and of non-interference in their domestic affairs. As an active member of the Organization according to Articles 1 and 2 of the Treaty, Bulgaria has undertaken to refrain in its international relations from the use of force, to settle its international disputes by peaceful means, to participate in all international events aiming at the strengthening of peace and security and to work towards a general reduction of armaments and the prohibition of nuclear, hydrogen and other weapons of mass destruction. The BPA is an integral part of the armed forces of the Warsaw Treaty member-states. The army contingents included in the Organization's armed forces remain part of the BPA, are stationed in Bulgaria and are subordinate to the Bulgarian Ministry of National Defence. Their overall training and combat efficiency are provided by the BPA according to plans co-ordinated with the Joint Command.

The BPA participates in the joint military exercises of the Warsaw Treaty member-states and has played host to the Hemus exercises (1964), as well as to the Rhodope military exercises in 1967 and 1982 respectively. It thus contributes to increasing the defence capability of the Warsaw Treaty armies.

Apart from defence matters, co-operation among the allied countries also covers the promotion and strengthening of their political, economic and cultural relations. Parallel to this it has been envisaged that the Warsaw Treaty will lose its force as soon as a European security

Mine-throwers

treaty is signed. The Warsaw Treaty member-states have repeatedly confirmed their readiness to dismantle the Organization simultaneously with the dissolution of the North Atlantic Treaty Organization (NATO), and to prevent increased membership of the two politico-military blocs.

The People's Republic of Bulgaria participates in the activities of the Warsaw Treaty Organization and works for the attainment of its goals; its foreign policy supports the cause of peace, security and co-operation among the nations of the world. Such is the spirit of the constructive proposal by the General Secretary of the BCP and President of the State Council of the People's Republic of Bulgaria, Todor Zhivkov, for creating a nuclear-free zone in the Balkans and thus turning this powder keg of the past into a region of peace and good-neighbourly relations.

Supreme defence planning in Bulgaria and management of its armed forces are executed by the Central Committee of the BCP and by the supreme bodies of state power, i.e. the National Assembly, the State Council and the Council of Ministers. The National Assembly decides the questions of declaring war and concluding peace, declares partial and general mobilization, appoints and dismisses the Commander-in-Chief of the armed forces. Between its sessions these functions are taken over by the State Council which convenes the National Assembly on the earliest convenient occasion to make a statement on the decisions taken. In addition the State Council undertakes the general management of the country's defence, appoints and dismisses senior officers and gives promotions to high military ranks. The general development of the BPA is guided by the Central Committee of the BCP and the Council of Ministers, while its direct management is entrusted to the Ministry of National Defence.

The armed forces of the People's Republic of Bulgaria are divided into Ground Forces, Anti-Aircraft Defence, Air Force and Navy. Logistics also belongs here. The sevices are divided into arms branches, regular armed forces and special task forces. The armed forces of Bulgaria also include the Border Guard, the Labour Corps, and the forces of the Ministry of Transport.

The Ground Forces. The Ground Forces are most numerous and varied in their composition. They possess great fire and striking power, high manoeuvrability and independence of action. They also incorporate arms branches, special task forces and logistics services. The Ground Forces are equipped with advanced firearms,

Setting up a report centre

artillery, tanks and anti-tank facilities, all-purpose armoured cars, communication and other equipment. The motorized-rifle units are the most numerous; they are equipped with weapons against ground and air targets, including automatic rifles, artillery, tanks, armoured personnel carriers and aircraft. The tank units are the main striking force: equipped with tanks, self-propelled artillery, etc., they possess high manoeuvrability and resistance to the destructive power of nuclear weapons. The artillery units form the main firing power and are equipped with rocket and conventional artillery for a number of different purposes, mortars and other weaponry. The task of the anti-aircraft defence units of the Ground Forces is to cover the latter's formations from air strikes.

The Anti-Aircraft Defence and the Air Force. The Anti-Aircraft Defence and the Air Force comprise aviation, the anti-aircraft and radiotechnical formations and the respective logistics services. The anti-aircraft units are the mainstay of the country's air defence; the Air Force has supersonic aircraft; the radiotechnical units possess highly effective military equipment for the location, target destination and control of the means designed to destroy air targets.

The Navy. The Navy consists of surface units, aviation and coastal artillery. It also includes auxiliary and special-purpose vessels and logistic services.

There are special branches in each service, whose task is to provide the necessary conditions for carrying out combat operations (reconnaissance, engineers, gas troops, signalmen, etc.).

The Commanders of the Bulgarian People's Army are trained at higher military schools where they study as full-time cadets or by correspondence, admission depending on the applicant's performances at entrance examinations. The higher military schools may be opened and closed down by decision of the Council of Ministers; each has the rights of a juridical person, as well as its own name, banner, seal and stamp.

The first and only military school before 9 September 1944 was founded in Sofia on 26 November 1878 on the basis of the military training command formed in that year in Plovdiv. The Russian Emperor's commissary in Bulgaria, Prince A. M. Dondukov-Korsakov, contributed to its opening. A military academy was set up in 1912.

After the victory of the socialist revolution, military training was reorganized and a network of higher military schools was set up to meet the needs of the newly built

Anti-aircraft gunners

Guarding the water frontiers

armed forces. These comprised the Military Academy, higher military schools for training officers and secondary military schools for sergeants.

The Georgi Stoikov Rakovski Military Academy in Sofia is a tertiary educational establishment for the training of highly qualified command, political, engineering and scientific cadres for the army. The Academy uses the positive traditions of Bulgarian military skills, and the rich experience of Soviet military science and Soviet military academies. It is staffed by military cadres of outstanding political, scientific and pedagogical experience, well versed in the art of command. Admitted to the Military Academy are army officers who have graduated from higher military schools, served in the country's armed forces for a definite period of time and shown the necessary efficiency and command qualities.

The Vassil Levski People's Higher Military School in Veliko Turnovo is the successor of the first Bulgarian military school. Trained in it are officers of the command, combat engineering and technical staff for the ground forces who serve in motorized-rifle, border guard, signals, chemical, motorized, engineer and tank units.

The Georgi Dimitrov People's Artillery School in Shumen was set up in 1948. It trains commanding officers and technical staff officers for the following specialisms: terrestrial artillery, anti-aircraft artillery, sound-flash survey, geodesy, cartography, artillery armament, radio engineering, computer technology and automated control systems.

The Georgi Benkovski People's Higher Air Force School in Dolna Mitropoliya, in the Pleven district, was set up in 1945. Its predecessor was the aviation school establishment in Sofia in 1914. It trains officers for the air force and for the needs of civil aviation: pilots, navigators, staff officers and engineers handling the exploitation and repair of aircraft and aircraft equipment.

The *Nikola Yonkov Vaptsarov People's Higher Naval School* in Varna was orginally set up in 1881 as an engine operator training school for the Russe fleet. It graduates engineers and command officers in ship navigation and radar and communication engineering, ship designers, and training cadres for the navy and the civil fleet.

In flight

The *General Blagoi Ivanov People's Higher Military Construction School* in Sofia was set up in 1966 to train officers in the following specialisms: construction of industrial, housing and public buildings and construction technology.

The higher military schools have modern training equipment and facilities, with specialized libraries and their own publications. Training lasts four or five years, depending on the specialism chosen. The military establishments have been awarded the Order of the People's Republic of Bulgaria, First Class, while the *Georgi Sava Rakovski Military Academy* has also been awarded the Order of Georgi Dimitrov, the highest Bulgarian order.

Bulgarian army officers also undergo training at Soviet military schools.

Reservists are trained at the *Hristo Botev People's School for Officers of the Reserve* in Pleven, which was formerly known as the reservist training course of the military school in Sofia, set up in 1889. In 1901 a school for training infantry officers of the reserve was established, and in 1904 a school for artillery officers of the reserve was set up. The two schools merged in 1908 to form a school for officers of the reserve, which was based on Knyazhevo, near Sofia. The People's School for Officers of the Reserve is open to young people with higher and secondary education who do their military service and spend a specific period in army units serving as non-commissioned officers. This school has been awarded the Order of 9 September 1944, First Class with swords, for its activity in training officers of the reserve.

There are also secondary military schools for the training of non-commissioned officers. These are the *Georgi Izmirliev School* in Gorna Oryahovitsa, the *Anton Ivanov School* in Varna and the *Maestro Georgi Atanassov School* in Sofia. There are likewise secondary training schools for non-commissioned officers at the *Georgi Dimitrov People's Higher Artillery School* and at the *Georgi Benkovski People's Higher Air Force School*. Admitted to these schools are young people who have completed their eighth grade. Training lasts from two to four years. The graduates of these schools receive special secondary education and a proficiency degree in the respective specialism.

Over 70 per cent of Bulgarian army officers have higher education, received at either military or civil educational establishments, 55 per cent of them having graduated in engineering specialisms.

The training and education of officers and staff is the main task of the command and the political bodies.

Training is done according to study programmes worked out by the respective bodies of the BPA General Staff and approved by the Minister of National Defence. The academic year is divided into two periods, winter and summer. The aim is to provide officers and staff with a sound knowledge of the basic types of training (tactical training, firing instruction, flying training, naval training, drill), to cultivate firm habits and skills in handling advanced and sophisticated military equipment and weaponry, and high moral and fighting qualities of courage, self-denial and unshakable psychological stability. Training takes place in classrooms, on firing ranges and armour training areas, in the air and at sea, and in conditions very close to those of actual combat. Socialist emulation is part of the training.

On completing their academic year, men on military service who have excelled themselves in combat and political training are awarded distinctions by the Minister of National Defence, and star-performer military units are given interim banners and diplomas.

The methods of training take account of the sophisticated character of military art in the context of the scientific and technical revolution, the possibility of the potential enemy using mass-destruction weapons, and the need for troops to be adequately trained.

The educational courses aim at shaping the world-outlook of servicemen and commanders, and at their patriotic, internationalist, military, aesthetic, moral and physical instruction. The basic forms of education are classes in Marxist-Leninist theory and political science, mass political, cultural and sports activities, and individual work with the staff. Bulgaria's heroic past, the socialist present and the bright future of the Bulgarian people are the main sources of education.

For *the political education* of army personnel a system of political bodies and organizations has been set up to plan, control and carry out party and political tasks.

The party and political work is aimed at: implementing party and state policy within the BPA; rallying the officers and men around the Central Committee; mobilizing their efforts to increase the combat capability of the troops and fulfil the production programmes; raising the quality of military activity by encouraging the initiative, creativity and efficiency of the personnel and by improving its organization and executive abilities; constantly strengthening the ideology, morale and discipline of the army units; training the servicemen in a spirit of courage, endurance, resoluteness, heroisim and responsibility; stressing the best methods in educational and military activity in a spirit of emulation.

The guidance of party and political work is done by the BCP Central Committee through the Chief Political Department of the People's Army, which has the status of a Central Committee department. In order to ensure the application of the principle of collectiveness in the solution of key problems, the Central Committee set up a Bureau of the Political Departments of the People's Army, whose decisions are taken by a majority of votes and communicated as directives and instructions of the Head of the Chief Political Department.

Directives and instructions on basic aspects of party and political work are issued after being co-ordinated with the Central Committee and signed by the Minister of National Defence and the Head of the Chief Political Department; instructions on current matters of party and political work are issued by the Head of the Chief Political Department.

The political bodies in the armed forces work for the implementation of the Programme and the Statute of the BCP, the decrees and resolutions of the Party and the Government, the orders of the Minister of National Defence and the instructions of the Head of the Chief Political Department. They are called upon to strengthen the combat readiness of the armed forces. The instructions of the political bodies on matters pertaining to party and political work are mandatory for all lower commanders and political bodies and for party, trade-union and Young Communist League organizations.

The party organizations in the BPA are guided by the Programme and the Statute of the Party, by the decisions of the party congresses and Central Committee and by the instructions of the Chief Political Department. They are called upon to carry out the policy of the Party and are under the direct leadership of political bodies and party committees.

The Komsomol organizations in the BPA are an integral part of the DYCL. They assist the party organizations and support the commanders in carrying out party policy aimed at raising the combat readiness of the troops.

A culture-and-education system with important tasks also operates in the BPA.

The culture-and-education establishments help the Commanders, political bodies, BCP and Komsomol organizations in the cultural and aesthetical education of the troops.

The Central Home of the People's Army is its oldest culture-and-education institution, which has asserted itself as a centre for political, cultural and aesthetical education of the servicemen. It is from this Home that most of the smaller culture-and-education establishments have proliferated. The main activities of the Central Home are: propaganda and mass political work; ideological and aesthetical education; organizing the recreation and leisure of servicemen and their families; the patriotic and internationalist education of the servicemen; organizing amateur art activities in the army. The Central Home houses the central army library, which has a stock of more than 120,000 volumes. A group of part-time lecturers read more than 2,000 lectures and reports a year in the spheres of military art, science, literature and the arts. It is the ideological and methodological administrator of the Homes of the People's Army and garrison clubs. It assists in the administration of the amateur art groups in the units: garrison women's committee, young officers club, military stamp collectors' group, hiking society, literary circle, etc. The Central Home of the People's Army has been awarded the Order of the People's Republic of Bulgaria, first class.

The National Military History Museum collects, studies and stores relics from Bulgaria's heroic past and works for the education of the servicemen and citizens of Bulgaria in a spirit of national pride, patriotism and internationalism. The museum's beginning can be traced back to July 1916 when World War I was still continuing. It houses more than 150,000 exhibits, arranged in twenty-one collections, and is in charge of collecting, displaying and storing valuable articles. Its most important task is the cultural and educational work among servicemen, schoolchildren, university students and citizens. More than 20,000 people visit the musem annually. It is also engaged in research and has been awarded the Order of Cyril and Methodius, first class. All the museums and Halls of Soldiers' Glory in the units of the BPA are methodologically controlled by the National Military History Museum.

The Theatre of the People's Army is a successor of the front-line theatres which gave performances during Bulgaria's Patriotic War (1944–45). It was founded in 1950, and up to 1981 had staged 169 productions and given 10,014 performances attended by 7,220,000 spectators,

1,100 performances for servicemen free of charge, and 1,845 entertainment programmes in army units.

The Army Theatre is successfully upholding the achievements of the Bulgarian theatre on the international stage, giving guest performances in the GDR, the USSR and elsewhere. It is a regular participant in the Schiller Days International Theatrical Festival in Mannheim, Federal Republic of Germany. A club with a theatre company was set up six years ago. It has staged more than fifteen plays since then, and has been awarded the Order of the 9 September 1944, first class. There are three People's Artists, fifteen Merited Artists and three Dimitrov Prize winners in the company.

The People's Army Film Studio was set up in 1947 by a decree of the Council of Ministers. A successor of the Film Department whose crews took part in Bulgaria's Patriotic War (1944–45), by 1982 the studio had made more than 300 educational and more than 500 documentary films and had released more than one hundred editions of the *Army Screen* newsreel. At the end of the same year the Studio's stock amounted to 3,845 feature, 5,612 documentary and 667 educational films. Its distribution network ensures the screening of two films weekly in the units of the BPA. It is a regular participant in the international film festivals of the armies of the Warsaw Treaty member-states and a winner of the Order of the 9 September 1944, first class, with swords.

The Military Artists' Studio. In 1949 a workshop was set up by the National Military History Museum, and in January of 1955 this was re-established by order of the Minister of National Defence as a Military Artists' Studio. By 1981 the members of the Studio had created more than 300 paintings, 100 sculptures and 150 graphic works. Many of these works are among the best achievements of Bulgarian painting. The Military Artists' Studio attracts civilian artists to work on military and patriotic subjects; it organizes exhibitions, discussions and other events in the units.

The Representative Ensemble of the Bulgarian People's Army was founded in September 1944 for artistic, creative and performing activity among servicemen. It consists of a choir, an orchestra and a dance troupe. By 1982 it had given 3,000 concerts before military and civilian audiences. The ensemble is known far beyond the boundaries of Bulgaria and has given successful guest performances in the USSR, Yugoslavia, Poland, the GDR, the People's Republic of China, the People's Democratic Republic of Korea, Hungary, Algeria, etc.

The Army Variety Company, set up in 1962, gives more than 100 concerts a year in Sofia and other garrisons, with guest performances in the USSR, Czechoslovakia, Cuba, Poland, etc.

A group of military writers and composers was set up by the Ministry of National Defence in 1956 to replace the former Studio of Military Writers. The main tasks of its members is to write fiction and poetry for the army and to popularize the military-patriotic and the internationalist themes among the servicemen.

In addition to the central culture-and-education establishments, many more army homes, garrison clubs, unit clubs, Dimitrov rooms, museums, libraries, cinema halls and ensembles have been set up in the garrisons of the BPA.

Army homes and garrison clubs were set up in 1948 for the servicemen and their families. Army and public activists are employed as members of artistic councils, garrison women's committees, garrison commissions for cultural activity, etc. The clubs have at their disposal halls for performances, rehearsals and lectures, libraries with reading rooms, halls for exhibitions and games, children's rooms, parks and open-air theatres.

Unit clubs and Dimitrov rooms were set up by Central Committee decree in 1951. The unit clubs are complex cultural and educational establishments which incorporate halls for performances, games, rehearsals, lectures and music halls, photo laboratories, libraries, cinema halls, museums and Rooms of Soldiers' Glory. These establishments work directly among the troops. The Dimitrov rooms were set up in 1949 to carry out the cultural and educational activity of the Komsomol societies. They are organized in every company, battery and independent platoon.

Founded after the 9 September 1944, the culture-and-education establishments of the BPA are the inheritors of everything valuable created by the people in the sphere of art and culture, which they develop and enrich under the new conditions, putting it at the service of the country and the army. In the forty years of the popular regime these establishments have turned into centres of cultural, educational, and artistic work, into champions of the cultural policy of the Party in the army, contributing to the development and to the enrichment of the Bulgarian cultural treasury.

The military mass media are propaganda bodies of the Party in Bulgaria's armed forces. They assist Commanders and political bodies in the education and training of the troops.

Most publications are of the Ministry of National Defence; others – of the Chief Political Department and the General Staff of the People's Army; a third category comprises publications of the different services.

The Narodna Armiya (People's Army) daily is the newspaper of the Ministry of National Defence. It is the successor and the continuer of the *Narodna Armiya* newspaper published by the BCP in the 1920–23 period.

The newspaper assists the Ministry of National Defence in the communist education of the army and the people, in raising the combat capability and preparedness of the units, in strengthening their discipline and in increasing their vigilance, and in the building and development of the People's Army.

The newspaper carries topical military, political, domestic and international news and commentaries.

The Armeiski Komunist (Army Communist) magazine is a monthy publication of the People's Army Chief Political Department. It elucidates party decisions, highlights key aspects of Marxist-Leninist theory and combats ideologies alien to it, aids commanders, staffs, political leaders and Party and Komsomol organizations in their work on the military training of personnel and enhancing the combat capability of the army units and studies, summarizes and disseminates the positive experience amassed in the field of party, political and ideological work in the army and educates its personnel in the spirit of socialist humanism, patriotism and internationalism.

The *Armeiska Mladezh* (Army Youth) magazine is the Chief Political Department's monthly for army youth, which explains the BCP policy in the construction of Socialism and the country's defence, educates young soldiers in socialist patriotism and internationalism, helps inculcate in them lofty moral and combat virtues, treats problems of Komsomol work in the army, and highlights aesthetic and ethical aspects of army life routine.

The *Bulgarski Voin* (Bulgarian Soldier) magazine is a monthly publication of the People's Army Chief Political Department. Its main purpose is to contribute to the patriotic education of soldiers and people, its subject-matter reflecting the moral image and combat spirit of the modern army. Its issues, most diverse in genre and content, tell about the nation's history, socialist present, and future prospects. The magazine allots considerable space to fiction, and treats at length problems of literature and the arts. It also covers international developments and sports events, scientific news and various other topics of interest.

The *Armeiski Pregled* (Army Review) magazine is a specialized monthly release of the Ministry of National Defence intended for Commanding Officers. It discusses problems of the theory and practice of general combat and army training, the latest in military science and propagates the experience of renowned Commanders, as well as the results of socialist emulation.

The *Voenna Tehnika* (Military Technology) magazine is a specialized illustrated monthly of the Ministry of National Defence for engineering and technical personnel. It covers problems on handling and maintenance of armaments and equipment, provides information on the latest achievements in military technology and warfare, propagates the front-rank experience of the engineering personnel, the most recent achievements and innovations, and the results of the techno-scientific creativity of young people in the ranks of the BPA.

The *Serzhant* (Sergeant) magazine is a specialized illustrated monthly of the Ministry of National Defence intended for non-commissioned officers, publishing materials concerning their education and role in training soldiers, in maintaining discipline and statutory order in army units.

The *Krile* (Wings) magazine is a socio-political, techno-scientific and literary magazine for aviation and astronautics, a joint publication of the Ministry of National Defence, the Central Committee of the DYCL, the Organization for Assistance in Defence, the National Space Research and Exploration Committee and Balkan Bulgarian Airlines. Its main task is to popularize the policy and decisions of the BCP in the field of military and civil aviation, space research, air training, and aviation sports.

The *Grazhdanska Otbrana* (Civil Defence) magazine is a specialized illustrated monthly of Bulgaria's Civil Defence Staff covering problems of civil defence in popularly written, scientific and publicistic materials.

The *Patriot* Magazine is a monthly of the Central Council of the Organization for Assistance in Defence (OAD), covering the large-scale civil defence work carried out among the population, especially among the younger generation. It supports the district and municipal councils of OAD, the auxiliary defence clubs and committees, relating popular events and mastery in military-technical sports.

The *Voenno-istoricheski Sbornik* (Military History Review) magazine is a bi-monthly scientific publication of the BPA General Staff. Its purpose is to promote military historical science in Bulgaria and the patriotic and internationalist education of army personnel. It publishes studies on military history, on the revolutionary past and the national-liberation struggles of the Bulgarian people, and exposes the falsifications of the military and revolutionary history of Bulgaria, the Soviet Union and the rest of the socialist countries from the point of view of Marxist-Leninist methodology.

Broadcasts destined for the army by *the respective departments* of Bulgarian Television and Bulgarian Radio are other channels providing the BPA with information on topical issues, and matters of army training and education, popularizing the best achievements, marking the anniversaries of outstanding military and revolutionary events, organizing meetings between soldiers and veterans from the revolutionary movement and the wars, soldiers' parents and relatives, covering youth work-teams, amateur art and sports events taking place in the army and much else.

The *Army Publishing House* specializes in releasing fictional works of a patriotic spirit and manuals for the purposes of army training and instruction. Intended for the general public, its editions are wide-ranging in terms of themes and genres, and enjoy recognition both in the country and abroad. Over 60 of its titles have been published in large editions in nine languages.

Bulgaria's armed forces comprise also some special-purpose branches with specific peace- and war-time tasks; they are related to the BPA in one way or another.

Bulgaria's Civil Defence was established in 1962 to take over the Local Anti-Aircraft Defence set up in 1951. It is part of the country's integrated defence system, its chief tasks being to protect the population and the national economy from the horrible effect of modern weapons of mass destruction by carrying out general and obligatory training, early warning, timely evacuation, etc.; to organize the functioning of the national economy, if necessary; to organize rescue operations and restoration works in areas of wholesale destruction. Its task is also to combat natural calamities such as earthquakes, floods, hurricanes, fires, etc. To serve its needs, special units have been formed of citizens, 'civilly' mobilized in cases of emergency. The main activities of Civil Defence are laid down in the long-term, five-year and annual plans for the country's socio-economic development. The Civil Defence is under the general management of the Council of Ministers and is organized by the state, economic and public bodies on a territorial-production principle, covering the entire civilian population.

The Labour Corps is a military building organization participating actively in socialist construction and in strengthening the country's defence capability. Its chief purpose is to contribute to the implementation of the unified state socio-economic development plan. It has built thousands of kilometres of roads and railway lines, scores of large industrial power-generating, hydro-technical and various other facilities. One of its important tasks is to train skilled contruction workers. Labour Corps troops are given military and technical training, skills and habits; systematic ideological work is carried out among their ranks for their communist upbringing.

The Labour Corps is subordinate to the Ministry of Construction and Territorial Planning.

The Border Guards are a specialist trained force for guarding the state boundaries, set up in 1946 on Georgi Dimitrov's advice in connection with the complicated situation along the frontiers following the 9 September 1944 revolution in Bulgaria. As part of Bulgaria's armed forces, they have been organized under the direct leadership of the BCP.

Alongside protecting the national frontiers, the Border Guards combat smuggling, and jointly with the respective bodies of the Ministry of Public Health exert sanitary and veterinary border control. The Border Guards units have at their disposal modern equipment for frontier protection, ensuring early warning and timely action against violators. They act in co-operation with the local population in exemplary relations of high socialist morality and patriotism. Especially helpful have proved to be the voluntary groups for assisting the Border Guards. Various local industrial, cultural, educational and other organizations often take patronage over Border Guards units, and culture marches are organized to the frontiers – all this being conducive to improving the morale and political tempering of the Border Guards. The DCYL also carries out large-scale work among the Border Guards, patronage over which it assumed as early as 1947.

The Ministry of Transport Corps have the task of doubling and electrifying railroad tracks and building motor-ways, assisting the maintenance and efficient functioning of the railroad network, and training skilled executive and managerial personnel for rail transport. The Ministry of Transport Corps has constructed many railroads, a great number with double tracks; it takes part in the general overhauling and maintenance of railways, in the electrification of railroads, and in the construction of motorways.

The Organization for Assistance in Defence, of the Council of Ministers, was established in 1977 as a successor of the Voluntary Organization for Assistance in Defence which existed from 1951 to 1967, and the Organization for Military-Technical Training and Military Patriotic Education of Youth, of the DYCL (1967–77). It is set up along public-state lines.

The Organization for Assistance in Defence works to inculcate a patriotic spirit in working people and in youth, and helps young people prepare for military service. Through an extensive network of motor, amateur radio, avio-naval and marksmanship clubs it provides training in various technical disciplines, organizing competitions in model aircraft construction, gliding, ship-modelling, motorcycling, parachute jumping, sailing, motorboat racing, amateur radio operation, marksmanship and other sports. The Organization represents Bulgaria in the respective international sports federations, and takes responsibility for participation in world, European and Balkan championships, as well as in 'Friendship and Fraternity' competitions in these sports.

Popular participation in broad-scale defence is voluntary. The Organization for Assistance in Defence is set up on the territorial-production principle. Central and local clubs and federations for applied military and technical sports also function within the Organization.

As part of the socialist State the Armed Forces of the People's Republic of Bulgaria have been established and developed on the basis of the socialist social and state order. The material and technical base for their development is the country's advanced economy; the ideological and theoretical base is Marxism-Leninism and its theory of warfare and the army; the legal base is the Constitution of the People's Republic of Bulgaria. The principles by which the BCP and the Government are guided in the policy with respect to military matters are a correct reflection of existing social realities, organically related to the laws of war and armed struggle. The main guiding principle is the leading role of the Party through whose all-round organizational and ideological work the growing defensive might of the socialist homeland is ensured.

An important principle in developing the armed forces is the political education of the troops as part of their overall training. This principle is implemented through the systematic and specific party and political work carried out in the army. Moral and political unity, and high socialist consciousness as elements of the army are provided by a system of education and training based on Marxist-Leninist pedagogics, psychology and ethics.

Friendship, fraternal unity and mutual assistance with the Soviet Army and the armies of the other socialist community countries are a firm principle in developing the BPA.

Education in a spirit of socialist patriotism and internationalism is another principle of prime importance in an army of a socialist type.

Another major principle is the armed forces' indestructible unity with the Bulgarian people, conditioned by the fact that the army is an integral part of socialist society, established and developing upon its socio-political, ideological and moral foundations. The Bulgarian people and its armed forces have common socio-political objectives and goals. The people sends into the army educated young people, who return steeled young men, devout patriots and internationalists, having acquired various skills they will need in socialist construction.

An important principle also is the constant improvement of the organization of the different services and branches and their balanced and co-ordinated development.

No less important is the strict centralization which ensures maximum order and disciple, flexibility, prompt reaction and close co-ordination.

The need to co-ordinate the actions of great numbers of people, and for stringent discipline and perfect organization make undivided authority an indispensable principle in the development of the army.

The armed forces of the People's Republic of Bulgaria, as part of the armed forces of the Warsaw Treaty member-states, have as their main duty the protection of the sovereignty and territorial integrity of the State, and the security of the socialist community. They are a reliable shield of the socialist gains of the people and their peaceful labour.

Red alert

A naval coastal patrol

A military parade in Sofia

SELECTED BIBLIOGRAPHY

A Brief Military History of Bulgaria (681–1945). Sofia, 1977.
The Bulgarian Communist Party – Organizer and Leader of the People's Army. Collection. Sofia, 1963.
The Bulgarian Communist Party and the People's Army. Sofia, 1976.
The Bulgarian Communist Party in Resolutions and Decisions of the Congresses, Conferences and Plenums of the Politburo of the Central Committee. Vol. 4. Sofia, 1955.
The Bulgarian Communist Party's Work in the Army (1941–44). Documents and Materials. Sofia, 1959.
Bulgaria's Patriotic War (1944–45). Documents and Materials. Vols. 1–4. Sofia, 1978–82.
The Bulgarian People's Armed Struggle against Fascism (1941–44). Documents. Sofia, 1983.
Constitution of the People's Republic of Bulgaria. Sofia, 1983.
DIMITROV, G. *Political Report of the Central Committee of the Bulgarian Workers' Party (Communists) to the Fifth Party Congress*. Sofia, 1974.
DIMITROV, G. *Works*. Vol. 11. Sofia, 1954.
DJUROV, D. *On the Party's Military Policy*. Sofia, 1971.
Establishment and Consolidation of the People's Democracy. September 1944–May 1945. Collection of Documents. Sofia, 1969.
Establishment and Development of the Bulgarian People's Army 1944–64. Sofia, 1965.

Georgi Dimitrov on the Bulgarian People's Army. Sofia, 1959.
Georgi Dimitrov on War, the Army and the Protection of the Homeland. Sofia, 1972.
History of Bulgaria. Vols. 1–3. Sofia, 1961–64.
History of the Bulgarian Communist Party. Sofia, 1981.
History of Bulgaria's Patriotic War (1944–45). Vols. 1–3. Sofia, 1981–83.
History of Military Art. Vols. 1–3. Sofia, 1959–67.
KUKOV, K. *The April 1956 Plenum of the Central Committee of the Bulgarian Communist Party and the Development of the Army*. Sofia, 1976.
LENIN, V. I. *On War, Army and Military Science*. Moscow, 1965.
MARKOV, M. *State and Military Symbols, Rituals, Festivals and Celebrations*. Sofia, 1977.
MARX, K. and F. ENGELS. *Works*. Vol. 21, Moscow, 1961.
PETROV, T. *Bulgarian Orders and Medals*. Sofia, 1983.
SEMERDJIEV, A. *An Army of Bulgaria's Socialist Revolution*. Sofia, 1963.
SEMERDJIEV, A. *A Guard of the Homeland and Socialism*. Sofia, 1972.
The Warsaw Treaty – a Mainstay of Peace and Socialism. Sofia, 1981.
ZHIVKOV, T. *Thirty Years Since the Socialist Revolution Triumphed in Bulgaria*. Sofia, 1974.

PART V

THE ECONOMY

THE DEVELOPMENT OF THE ECONOMY

The socio-economic development of Bulgaria during the First Bulgarian State (681–1018) was characterized by a gradual process of feudalization. A great number of the free peasants were turned into serfs as a result of continuous rises in taxation and the obligatory purveyance for the army, the oppression and arbitrary rule of the feudal military aristocracy, and bad harvests and natural calamities. The development of the economy was slowed down by frequent wars. The landless peasants rented lands and farm equipment from the local landed aristocracy, in return for which they had to give from a tenth to a half of their crops and perform many different duties. The economic power and social privileges of the military and tribal aristocracy increased as a result of the exploitation of the peasants, the spoils gained from the numerous wars and the seizure of lands. The feudal system of large-scale landowning was probably already prevalent in the second half of the tenth century. The basic means of livelihood were farming, crop-raising being of primary significance (the major crop was wheat, followed by barley, rye, millet and oats). Vegetable gardening, vine-growing and stock-breeding were also well developed. A number of craftsmen developed their trade – iron and copper smiths, potters, builders, goldsmiths, etc. The Bulgarian State established its first trade contacts (with Byzantium they were regulated by a treaty of 716). Foodstuffs and handicraft products were exchanged. In the ninth and tenth centuries trade agreements were concluded with the ancient Russian State of Kiev.

Towards the tenth century commodity production began to develop. It was however restricted in scale, because of the small urban population and the barter system of farm produce exchange. The differentiation of crafts, penetration of simple commodity production into the feudal economy, and expansion of the home market contributed to the formation of the Bulgarian feudal town which gradually turned into a commodity production centre. Money penetrated into Bulgaria through the country's foreign trade relations and the wars.

With the introduction of Christianity in 864, the difficulties hitherto existing in the country's economic relations with the European states were overcome. Through its geographic location Bulgaria began to play an intermediary role in trade between Europe and Constantinople, and in the period of Byzantine domination (1018–1186) the transition to more mature feudal relations was accelerated. The size of the feudal rent increased, and a taxation ordinance of 1040 added a money rent. Large-scale farming was developing rapidly, and commodity production was penetrating into the villages.

During the Second Bulgarian State (1186–1396) the relative share of large-scale feudal farming increased, as well as the number of serfs. Nevertheless, in the thirteenth and fourteenth centuries the number of free peasants in Bulgaria was still very high as compared with the countries of Western and Central Europe. Under the reign of Assen II (1218–41), alongside the advancement in the political sphere, a new impetus was given to the country's economic life. The first Bulgarian coins were minted, the art of minting in Medieval Bulgaria being characterized by the distinction and originality of the iconographic material. A number of towns (Turnovo, Sofia, Plovdiv, Varna, Sozopol, Silistra, Vidin and others) became centres of crafts and trade. Big colonies of foreign merchants from Dubrovnik, Genoa and Venice settled in the busiest of these towns. The role of vine-growing in agriculture was gaining increasing importance in Thrace, along the Black Sea coast, in Macedonia and in other parts of the country. From the industrial crops it was flax, hemp, cotton, poppy and anise that were most widespread. Bee-keeping and silkworm-breeding were also developing. Stock-breeding was largest in scale on the big feudal farms of the tsar, the boyars and the monasteries. The period between the twelfth and the fourteenth centuries witnessed a considerable advancement in crafts. There were some fifty different trades, which testified to a higher degree of labour division as compared to the ninth and tenth centuries. Most of the craftsmen were free citizens, but there were also some who were bound to their lords. Towards the end of the thirteenth and the beginning of the fourteenth centuries the artisan's commodity production began to be partially replaced by the made-to-order system.

In spite of urban growth the social stratification of the town population had not yet been fully accomplished. Archaeological finds of coins and some documents from the period clearly show an advancement of domestic trade. Weekly open town-markets and annual fairs were organized. A limited stratum of Bulgarian merchants came into being. Foreign trade was the major sphere of investing trade capital and was regulated by special treaties. Bulgaria traded mainly with Byzantium, Venice, Genoa, Dubrovnik, Serbia, Hungary, Moravia and Wallachia. The development of money-commodity relations gave rise to the emergence of money-lenders' capital as well. Craftsmen and bondsmen increasingly fell into the power of rich money-lenders. The dominant form of feudal rent was in kind. The corvée rent was stipulated in terms of hectares to be tilled, while towards the fourteenth century it was calculated in terms of the number of days a serf had to work free of pay on his master's lands or for the country's ruler. The money rent, whose significance grew with the development of money-commodity relations, increased the degree of bondsmen exploitation.

The development of feudalism in Bulgaria was slowed down by the invasion of the Osman Turks. Ottoman domination led to stagnation in the country's economy, due to the large-scale destruction, the great number of war casualties and the lower stage of economic and social development of the invaders as compared to that of the subdued population. Most of the land was distributed by the sultans among the Turkish *spahis* in exchange for the right to collect part of the taxes due to the State (mainly in kind), which was a form of feudal rent in return for their military administrative obligations. Feudal exploitation and national oppression highly aggravated the plight of the Bulgarian population. Apart from the taxes the peasants paid, which numbered about 70, the latter were exploited through the big deliveries for the State, purchased at prices three times lower than the market prices.

Towards the middle of the fifteenth century economic life began to recover. The economic role of the towns grew and domestic trade turnover increased. Trade and crafts were regulated by the State. Many products, such as processed skins, homespun cloth, rugs and others, were in high demand on the foreign markets. The end of the sixteenth century saw the beginning of the decline of the Ottoman Empire, accompanied by intensified exploitation of the Bulgarian population by the feudal lords. The empire was in the grips of a financial crisis, prices went up, taxation increased sharply. Foreign trade relations were on the decrease. There was only limited expansion in the crafts industry and in the trade exchange between towns and villages. The arbitrary plunder of the peasants grew in scope. The so-called *chifliks* (big farms) were set up, extending the rights of the big landowners. These were dominated by the feudal mode of production, but the germs of capitalist exploitation were gradually emerging. The Central Government increasingly often levied additional taxes. Corruption and misappropriation on the part of the administration, and plundering by robber gangs and janizary garrisons led to further deterioration in the plight of the population and checked the development of the productive forces. In the seventeenth and the eighteenth centuries the volume of agricultural production increased at a very slow rate, from lack of proper conditions to sell the surplus produce at reasonable prices. The peasants could usually till only one fourth to one third of the arable land.

The consumer needs of the Ottoman feudal State raised the significance of the Bulgarian lands. The second half of the eighteenth century saw the differentiation of specialized production regions. The southern Bulgarian lands, where large flocks and herds were raised, became the centre of both the *dzheleps'* trade (the *dzheleps* were large-scale sheep-breeders and traders of small farm animals and were in charge of the supply of the big towns and cities in the empire and of the army) and of textile production. The *abadzhis* (artisans and traders in frieze coats and jackets – upper garments of men's folk costume, made of thick woollen homespun cloth) and the merchants of the town of Plovdiv carried their products to the Middle East markets; in this way Bulgarian goods reached as far as India. The *abadzhis* from Koprivshtitsa and Sliven took their products to the large markets of the Ottoman Empire. Raw and processed hides and skins were exported from the northern and north-eastern Bulgarian lands. Wheat, foodstuffs and meat were exported from the Dobrudja district and the eastern regions of the country through the port of Varna. Trade with countries beyond the Danube, in particular Romania, and with Russia also expanded. Many villages, especially those in the foothills of the Balkan and Sredna Gora mountains, became centres of crafts. Merchants from all over the Ottoman Empire, as well as from other countries, regularly attended the fairs in Uzundzhovo (the village of Uzundzovo, Haskovo district) and Eskidjumaya (near the present-day town of Turgovishte). The trade exchange between towns and villages rapidly increased, and a stratum of well-to-do Bulgarian craftsmen and tradesmen was evolved.

The second quarter of the nineteenth century saw the beginning of a new stage in the development of the economy in the Bulgarian lands. The agrarian reform of 1832–34 abolished the *spahi* institution. Most of the *spahis* had succeeded in transforming part of their land possessions into *chifliks* through the use of forced labour, sharecroppers and hired labourers. An act was passed in 1839 which guaranteed Bulgarians the right to private ownership of farm land. Gradually parts of the lands owned by the Turks, became the property of Bulgarians. The compulsory purveyances for the State were discontinued in the 1837–41 period. The introduction of free trade in farm produce led to an increase in the areas under crops and of commodity agricultural production, which speeded up the process of class and property differentiation in the village. A stratum of well-to-do peasants and *chiflik* owners emerged amongst the Bulgarians.

The limited imports of European manufactured goods to the Ottoman Empire and the growing needs of the standing Turkish army created conditions for a considerable rise in the production of the crafts industry in the Bulgarian lands. A process of differentiation therefore started among the craftsmen, too. On the basis of the existing trade and money-lenders' capital and handicrafts manufacture, the process of primary accumulation of capital and the formation of a commercial and industrial bourgeoisie began at the end of the eighteenth and the beginning of the nineteenth centuries. Many of the richer craftsmen searched for markets for their increased production outside the Ottoman Empire (in Russia, Romania, Austria and elsewhere). A number of large-scale merchants appeared in Gabrovo, Turnovo, Constantinople and elsewhere. Some of the rich craftsmen became middlemen, who mainly organized scattered, smaller workshops making frieze, homespun cloth and woollen braiding. The first modern industrial factories were opened in Sliven (1834), the village of Dermendere in the Plovdiv district (1847), Stara Zagora (1858–60), Karlovo (1873–74), Gabrovo (1882) and elsewhere.

The development of modern industrial enterprises before 1878 was rather limited, due to the lack of protectionist policies and of security for the subjugated population, to arbitrary taxation and to an insufficiently large home market. Communications and transport were partly improved through the construction of roads and bridges; the first railway lines (1866–74) and the first telegraph lines (1855) were also inaugurated. The 1860s marked the establishment of the first shareholders companies, which however were very few in number (only eleven up to 1878), short-lived, and engaged in relatively limited activity. The first loan companies appeared in such towns as Varna, Constantinople and Pleven. An impoverished semi-proletarian urban and rural population evolved as a separate stratum, providing the workforce necessary for capitalist manufacturing and trade, and cheap hired labour for farming and construction.

After the Crimean war of 1853–56 the economic situation of ordinary Bulgarian craftsmen deteriorated rapidly because of the increased import of cheaper and high-quality manufactured goods from Western Europe. The craft guilds were on the decline. The beginnings of capitalist development in the Bulgarian lands, the upsurge of national consciousness and the decay of the Ottoman Empire created the prerequisites for a national revolution.

The five-century long Ottoman yoke (1396–1878) hindered the development of productive forces in the Bulgarian lands. Bulgaria started on the road of capitalist development much later than most European countries. The Russian-Turkish Liberation War (1877–78) in fact played the role of a bourgeois democratic revolution and gave a strong impetus to the development of the productive forces and the establishment of capitalist production relations in the newly-founded State. Capitalism in Bulgaria developed, however, at a time when the capitalist system had already entered into its final stage, namely Imperialism. With the very first independent political steps the country took, it fell into economic dependence and ended as a raw-material supplier for the advanced European capitalist states, which impeded the development of its own productive forces.

Up to the late 1890s the political and economic life of the country was dominated by the petty bourgeoisie; the prerequisites for capitalist development were thus created, and the impoverishment of thousands of peasants resulted in the formation of the country's working class. The provisions of the Berlin Treaty (1878) placed Bulgaria in the position of a semi-sovereign State and therefore made it an easy prey to European finance capital. Taking advantage of the treaty's capitulatory conditions, foreign industrialists and merchants imported their goods free of tax and customs duties, thus making them more competitive than the products of the weakly developed Bulgarian industry. In their effort to oppose the capitulation privileges of their foreign counterparts, the Bulgarian commercial and industrial circles launched the idea of pursuing a protectionist economic policy. The agrarian-industrial convention, held in 1892 during the first big international fair in Bulgaria after the Liberation, drew up a programme to ensure the development of the Bulgarian economy along capitalist lines. A bill was enacted in 1895 for the encouragement of local industries, which guaranteed considerable privileges for the bigger enterprises, (duty-free import of machinery and raw materials, exemption from direct taxation, free construction sites for the building of factories, a 35 per cent reduction in tariff rates of the Bulgarian State Railway Lines, etc.). The foundations of a national industry (mainly light industry) were laid and the number of industrial enterprises rapidly grew – from 20 in 1878 to 345 in 1910. The postal communication network created immediately after the Liberation was now expanded and a network of big shareholder banks and insurance companies, predominantly with foreign capital, was rapidly built. In 1885 the Bulgarian National Bank (BNB) issued its first banknotes, and in 1897 the first foreign-trade agreements were concluded. Rural economy also made considerable advances: the import of farming machinery was increased and by 1910 the number of regular agricultural workers showed a three-fold increase as compared with 1900. The number of large capitalist farms also grew. The process of mass improvement of farmers subsided towards the end of the nineteenth century as a result of lower interest rates, decreased taxation (after the peasant riots against the reimposition in 1900 of the *desyatuk* – a tithe paid in kind, equivalent to one tenth of the peasants' crop) and a rise in the market prices of agricultural produce.

The foundations of the co-operative movement were laid. In 1890 the first Bulgarian co-operative was set up in the village of Mirkovo, Sofia district. The co-operative farming movement gained momentum. The volume of farm produce increased, thus preparing the conditions for the growth of exports – mainly grain, cattle and flour. As a rule imports exceeded exports, and substantial external, community and government loans granted under unfavourable conditions were needed to keep the country's balance of payments steady. Up to the wars of 1912–18 the bourgeoisie had not yet felt the need of government interference in the economy and had pursued a liberal economic policy.

The two national catastrophes, however, disrupted the country's economy and entailed perilous consequences – territorial changes with a disastrous effect on agriculture and foreign trade, enormous financial burdens under the Neuilly Peace Treaty of 1919, impoverishment of the people, and economic stagnation. The Government (1920–23) of the eminent leader of the Bulgarian

Agrarian Party (BAP), Alexander Stamboliiski (1879–1923), brought to power by the people, introduced some progressive reforms, such as the agrarian and tax reforms and established state monopoly on grain and other exports, but failed to destabilize the economic domination of the bourgeoisie. In 1919–23 the transition towards the monopolistic forms of Capitalism was speeded up. Cartels and syndicates appeared, mainly in industry. Foreign monopoly capital flooded industry. Financial groupings, substantial by Bulgarian standards of the time, formed around the foreign capital banks – the Bulgarian General Bank (1905), with French and Hungarian capital; the Balkan Bank (1906), mainly with Austrian and French capital; the Italo-Bulgarian Commercial Bank (1919), and the Credit Bank (1905), mostly with German capital. The majority of them remained in existence throughout the 1930s.

The economic crisis in Bulgaria at the end of World War I (1914–18) proceeded in a peculiar way. The reduced areas under crops, worn out tools and unfavourable weather conditions caused a decline in farm produce. As a result of the rapid devaluation of the national currency unit – the lev – the bourgeoisie sought ways to circulate capital by injecting it into industry. New industrial branches appeared and as early as 1921 industrial output exceeded the pre-war level by 41 per cent. In the period of temporary and partial stabilization of Capitalism (1924–29), the crisis vacillations in the economy were gradually abated by an economic recovery. In 1926 the pre-war volume of farm produce was reached, with the prevailing trend already pointing to intensive instead of extensive crops and to the promotion of poultry-farming. Industrial production increased three-fold. Capital was further concentrated and centralized. The volume of foreign trade considerably exceeded the pre-war level. The exports of tobacco, vegetables, fruits and the products of poultry-farming marked the most rapid growth. The lev was relatively stabilized and money circulation reduced under the suspended gold convertibility of the banknotes and the state monopoly on foreign currency operations.

The 1929–33 world economy crisis proceeded in Bulgaria with a number of peculiarities. The volume of industrial output was more or less unchanged because the decline in some industries was compensated for by larger output in others. Farm produce increased as a result of peasants' frantic efforts to make up for the decrease in purchasing prices. Foreign trade went down sharply because of the unfavourable prices for Bulgaria. As of 1931–32 trade was restricted to bilateral deals only, and these mostly with Germany and on a clearing principle or as compensation deals, which limited Bulgaria's appearance on the world market. Home trade and credits were in a state of crisis. The bourgeois State was forced to abandon its liberal economic policy and as of 1930 started interfering actively in economic life – by control over foreign trade, credit and insurance societies, direct intervention in the trade in agricultural staple goods, and regulation of worker-employer relations. In industry the concentration and centralization of capital increased. In 1934–41 there was a certain revival in the economy: in 1941 the country had 3,872 industrial enterprises of which 3,467 were small private capitalist ones (with a staff of 28 workers on the average), 130 state-owned enterprises (turning out 8–9 per cent of the overall industrial production) and 275 co-operative enterprises (turning out 6 per cent of the output).

During World War II (1939–45) Bulgaria was completely converted into a farming appendage of nazi Germany. The national economy was systematically robbed, mainly through trade, military supplies, the BNB's financing of the German mines in Bulgaria, support of the German troops, etc. In 1944 75.1 per cent of the imports and 89 per cent of the exports went to Germany, including Bohemia and Moravia. The economy of the country was on the verge of disaster – a steady decline in both industrial and agricultural output, rising inflation, soaring prices, and falling standards of living.

Throughout the period of Capitalism Bulgaria remained one of the most backward agrarian countries in Europe with poorly developed productive forces, anaemic industry – chiefly small-scale craft industry, parcelled up and low-productivity farming. In 1944 the arable land was owned by more than one million farmers; it was parcelled up into 12 million plots of less than 0.4 hectares each on average. The land was cultivated with primitive farming equipment (only three thousand tractors). 83 per cent of the country's active population was engaged in agriculture. Arable farming was the more developed branch; animal husbandry was low-productive. Though the dominant branch of the economy, agriculture generated only 50 per cent of the national income. Industry consisted mostly of small-size enterprises employing only 112,000 people in 1939 as against 146,000 actively employed craftsmen. The leading branches were the food and tobacco industries (50 per cent of the total industrial output) and textile manufacturing, while the branches of heavy industry were in a rudimentary state. In 1939 the ratio between agricultural and industrial output was 75.2 : 24.8; industry generated a mere 15 per cent of the national income. Agricultural output constituted 95 per cent, industry – only 5 per cent of the country's exports. The 1937 per capita value of industrial output (in US dollars) was 28 in Bulgaria, Belgium – 316, Great Britain – 399, Germany – 343, United States – 470, France – 210, Sweden – 272. The poor economic development of the country was accompanied by persistent unemployment (particularly great was the latent unemployment in agriculture). The low level of economic development was organically connected with the inadequate division of labour between the separate regions of the country. With the exception of the Sofia-Pernik area the economics of the regions had a predominantly agrarian character. Particularly backward were the southern regions (Pirin, the Rhodopes, and the Strandja) and some of the northern ones (mainly South Dobrudja).

By its nature the 9 September 1944 popular uprising was a socialist revolution which prepared the ground for radical reforms in the economy of the country. With the

The Development of the Economy

establishment of the people's democratic power there started a process of abolition of the economic foundations of Capitalism and a gradual asserting of socialist production relations. Conditions were created for the rapid development of productive forces and thorough changes in the structure of the economy. In the very first months production was placed entirely under the control of the workers and the State. Workers' control was exercised through the trade unions and the committees of the Fatherland Front. State control over production was exercised through a specialized economic planning body of the Council of Ministers – the Supreme Economic Council (1945–48), the ministries and the militia department dealing with financial abuses.

In almost all sectors and branches of the economy in which larger private capital was involved the principles of State Capitalism were applied. The quantity, kind and quality of industrial output were regulated through state plans and state-supply orders, the fulfilment of which was obligatory for the private enterprises also. The State regulated the circulation of goods and raw materials, controlled trade and prices, and took sanctions against private owners for any forms of profiteering. In 1946 labour quotas, obligatory for all enterprises, were introduced and the principle of paying workers according to the quantity and quality of labour was asserted. All this reduced the role played by proprietors and subordinated production to the interests of workers. Mixed state-and-private enterprises were set up (mostly in foreign trade) as a way of using private capital in the interests of the people's democratic State. The State held considerable shares in credit, construction, etc. Private initiative was also restricted by means of taxes. The tariff rates of taxes on the war-time profits was increased and a progressive-income tax was introduced in 1947.

In 1946 an agrarian reform was put into effect by passing a Land Ownership Law which restricted the rights of the big landowners, thus opening the way for a rapid socialist reform of agriculture. The co-operative farms, which appeared even before 9 September 1944 as a form of collective farming, were consolidated. Their number jumped from 110 at the end of 1944 to 1,100 in 1948. The co-operative farms were set up mostly by the poor and the needy, incorporating additional hectares from the state land stock. In this way small, private farming grew into large-scale, socialist agriculture. The specific economic conditions in the Bulgarian villages made it necessary to preserve private ownership of the land and the rent relations connected with it in the co-operative farms. The capitalist state-owned farms (50 in 1944) and the lands expropriated during the 1947–48 agrarian reform were the foundation on which 86 state-owned agricultural farms were created.

The State undertook measures to do away with the economic aftermath of World War II. German property was placed under the provisions of the Law on the Control of the Property of Enemy States' Nationals (1944); the property of those convicted by the People's Court was confiscated as well as the property obtained illegally or through speculation after 1935; the shares of foreign shareholders and owners were purchased at terms favourable for the State. The first foreign trade agreements concluded with the Soviet Union (1945) and the other socialist states played a major role in the restoration of the economy. In June 1946 the private insurance companies were nationalized and a *state insurance monopoly* was introduced. A monetary reform was carried out in 1947 in an effort to stabilize the lev.

A wheel excavator for opencast mining

The property of the upper, middle and a large portion of the petty bourgeoisie was finally nationalized with the Nationalization of Private Industrial and Mining Enterprises Law of December 1947 and with its amendment passed the following year. About 6,000 industrial enterprises were nationalized. The socialist reforms in crediting were facilitated by the existing state banks (the BNB, the Bulgarian Agriculture Bank and the Co-operative Bank). The private banks were nationalized in 1947 – banks were thus united and *state bank monopoly* established, laying the foundations of a socialist banking system. By the end of 1947 foreign trade was carried out almost entirely by specialized state-owned enterprises. The former state-capitalist property in transport (mainly railway transport) went into the hands of the socialist State. Road transport was still privately owned but systematic state control was exercised. In 1948 private vessels were nationalized. Large farming equipment was purchased in 1948 and handed over to the State Machine-and-Tractor Stations. Wholesale trade and a substantial part of retail trade passed into the state and the co-operative sector. The Central Co-operative Union, founded in 1947, reorganized the country's traditional co-operative trade and took charge of

The Bobov Dol thermo-electric power station

almost all trade in the villages and a major part of trade in the cities.

The consolidation of the public sector stimulated the development of the planned management of the economy. The working class and the Bulgarian Communist Party (BCP) acted in unity with the working peasants and the BAP. They strengthened their positions as the leading force in public and economic life. The economy developed under the leadership of the BCP and the socialist State. The main trends in the development of the national economy and the specific objectives to be achieved at each stage of socialist construction were outlined in resolutions adopted at the congresses and Plenary Sessions of the Central Committee of the (BCP). In 1947–48 a Two-Year Economic Plan for economic recovery and development was implemented. Its main objectives were to raise the living standard of the working people and to reach the pre-war level in all economic sectors. The Soviet Union contributed very much to the recovery of the Bulgarian national economy. It supplied large quantities of ferrous and non-ferrous metals, petrol products, machines, cotton, rubber, and granted considerable long-term credits, etc. The foundations were laid of industries for the production of capital goods. This was a qualitatively new moment in the country's economy. The first steps were taken in national machine-building (internal combustion engines, electric engines, machine tools, lathes, etc.). Electrical engineering released its first products. The construction of twelve thermal and sixteen hydro-power stations started as early as 1947. The Georgi Dimitrov, Studena, and Vassil Kolarov dams were built. The first unit of the Combined Chemical Works in Dimitrovgrad was built, followed by other heavy industry enterprises. Many young people joined the youth-brigade movement to take part in important national projects such as the building of highways and railways, industrial enterprises, etc.

At the beginning of 1948 thousands of small enterprises were merged into bigger economic units – state industrial enterprises and about twenty state corporations. The introduction of cost-accounting as a means of socialist management enhanced the role of accountancy and control and created favourable conditions for the full application of the principle of distribution of incomes according to the work done. By the end of 1948 the pre-war level of the economy was attained, the national income equalled that of 1939, the gross national product (GNP) was 25 per cent greater and industrial output had doubled. The supply of foodstuffs and consumer goods to the population was significantly improved. In spite of the extremely bad weather conditions in 1948, agricultural produce increased by 3 per cent over 1939. The proportion between industrial and agricultural output changed to 40:60. The share of industry in the GNP was 31.9 per cent, the number employed in industry doubled, hidden unemployment, characteristic of bourgeois Bulgaria, was reduced. The fulfilment of the Two-Year Plan created the necessary material and social prerequisites for building the economic foundations of Socialism, transforming Bulgaria from an underdeveloped agrarian State into an advanced industrial-agrarian one.

The Fifth BCP Congress (1948) was of historical importance for the development of the economy. In the main report, delivered by the prominent leader of the BCP and of the international communist and workers' movement, Georgi Dimitrov (1882–1949), the general policy of the Party for building the foundations of Socialism in Bulgaria was outlined on the basis of an analysis of the fundamental revolutionary changes. The First Five-Year Plan (1949–53) had as its main objective the improvement of the economy through electrification and industrialization, and through the collectivization and mechanization of agriculture. During that period industrialization was the leading line and the main instrument for the rapid construction of the material and technical basis of Socialism. Consequently, the development of heavy industry, and especially of electricity production, mining, metallurgy, machine-building and the chemical industry, was given priority. The plan was fulfilled in four years. Over 700 industrial enterprises were built, enlarged or overhauled and put into operation, including some big machine-building, chemical and textile plants and power-stations. The towns of Dimitrovgrad, Rudozem, Madan and Kurdjali became new centres of heavy industry, and the Vassil Kolarov, Rossitsa, Topolnitsa and Georgi Dimitrov dams were built.

Agricultural co-operation was accelerated, especially in 1950, when considerable numbers of middle peasants joined the co-operative farms, and co-operation spread to whole regions. The average arable land of one co-operative farm increased from 346 hectares in 1949 to 844 hectares in 1952. By the end of 1952 the co-operative farms covered 60.5 per cent of the arable land. The output of the co-operative farms and the state farms continued to grow; higher yields were obtained in crop-raising and stock-breeding. During the Five-Year Plan the national income increased by 8.4 per cent annually, industrial production by 20.7 per cent, the retail trade by 10.5 per cent, and the real annual incomes of the population by 8.4 per cent.

The objective of the Second Five-Year Plan (1953–57) was to proceed with industrialization and to further the development of socialist agriculture. Priority was given to the accelerated development of electrification, coal-mining, ferrous and non-ferrous metallurgy, as well as to the growth of consumer goods production. Capital investments amounted to 3,200 million levs. The first Bulgarian Lenin metallurgical plant, the Karl Marx Soda Plant in Devnya, and the superphosphates plant in Dimitrovgrad were put into operation; the first stage of the construction of the Lead and Zinc Plant in Kurdjali was completed, and some big hydro-power plants were built. In 1954 the bridge over the Danube connecting the Bulgarian city of Russe with the Romanian city of Giurgiu was completed. Export underwent positive structural changes and the share of capital goods in it was already prevailing. But during that period serious deviations from the policy of rapid industrialization adopted at the Fifth BCP Congress were committed. On the basis of an incorrect evaluation of the level of development already attained, the Sixth BCP Congress (1954) set the drastic improvement of the material and cultural standard of the working people as the main objective of the Five-Year Plan. The implementation of that unrealistic and badly timed objective adversely changed the consumption-accumulation ratio in the distribution of the national income. The relative share of the accumulation fund decreased from 23.8 in 1952 to 20.3 in 1954 and 14.3 in 1956. This limited possibilities of investing in the development of the national economy and in the expansion of the material and technical facilities and therefore slowed down the rate of socialist construction.

The decisions of the April Plenary Session (1956) were a turning point in the economy. At that session the unfavourable consequences of the retarded industrialization during the Second Five-Year Plan were admitted, subjectivism in solving the economic problem was condemned, decisions were taken which set the system of planned economic management on a scientific basis, investment activities were enhanced, and production underwent a profound reorganization, which created the necessary material prerequisites for a general economic upswing. The initiator of the new economic policy, known as the April 1956 line, was Todor Zhivkov (b. 1911), now General Secretary of the Central Committee of the BCP and State Council President of the People's Republic of Bulgaria. The subsequent period of economic and social prosperity in Bulgaria is closely associated with his name and his personal contribution.

In the early years of the Second Five-Year Plan, despite the good results achieved in the development of its material and technical basis, agriculture lagged behind the general economic development. At the January 1955 Plenary Session important measures were taken to improve planning and management in agriculture. The

The Belmeken hydro-electric power station

positive results achieved in enlarging the material and technical facilities of the co-operative farms, and the higher incomes of the co-operative farms, further stimulated socialist collectivization. The purchase prices for agricultural products were raised, which had the effect of a material incentive for the co-operative farmers; the prices of some industrial goods used in the co-operative farms were reduced, and co-operative farms were granted some financial concessions. By the end of 1957 the number of co-operative farms increased to 3,202, which tended 73.1 per cent of the arable land. During the Five-Year Plan period the average annual rate of growth of the national income reached 7.8 per cent, of industrial production – 13.2 per cent, of agriculture – 4.75 per cent, of retail trade – 14.9 per cent, and of the real incomes of the population – 7.1 per cent. By the end of the Second Five-Year Plan the socialist mode of production was already predominant, the share of the socialist sector in industrial output reached 98 per cent, in agriculture – 86 per cent, in retail trade – 99 per cent. These profound changes in the economy and in the class structure of society proved that Socialism was already victorious in the country, and that Bulgaria had become a socialist industrial-agrarian country with a system of large-scale collectivized and mechanized farming.

At the Seventh BCP Congress (1958) the completion of the socialist reorganization of the economy was confirmed and guidelines for the Third Five-Year Plan (1958–62) were adopted. The plan envisaged accelerated industrialization and primarily the development of heavy industry on a modern technological basis, improvement of the territorial structure of the economy and the further progress of agriculture through intensification. The main

The Dimiter Blagoev non-ferrous metal works in Plovdiv

the Medet Ore-dressing Factory, and the Semi-conductors Plant in Botevgrad were put into operation; the industrial complex Maritsa-Iztok was inaugurated, etc. The process of organizational and economic consolidation of the co-operative farms continued. Financial funds were created similar to those in the state enterprises. About 90 per cent of the main agricultural operations (ploughing, harrowing, sowing, harvesting) were mechanized. Technical reconstruction of railroad transport began, electric and diesel engines were introduced. During the Five-Year Plan period the average growth of the national income reached 6.7 per cent, of industrial produce – 11.7 per cent, of agriculture – 3.0 per cent, of retail trade – 7.3 per cent, of

The Elatsite ore-dressing works at Mirkovo; a mechanical shop

objectives of the Plan were fulfilled in three years. The line of gradually increasing the accumulation fund was still in force. The average rate of accumulation grew from 19.8 per cent to 26.1 per cent during the second five-year period. This created the possibility of implementing an impressive investment programme. Some 3,300 million levs were allotted for capital investment, 72.3 per cent of them intended for enlarging and improving the material basis of industry. Essential changes took place in the structure of industrial production – the share of electricity production, fuel production, ferrous and non-ferrous metallurgy, machine-building and chemical industry increased. The Georgi Dimitrov Copper-dressing Plant near Pirdop (now Srednogorie) was put into operation, and so were big new plants for cement-making, vessels for river navigation, batteries, etc. It was then that the construction of the Kremikovtsi Metallurgical Complex (now called Leonid Brezhnev) was started. In 1958 the collectivization of agriculture was completed and in 1959 the amalgamation of the co-operative farms was begun; their number was reduced from 3,290 (1958) to 972 (1959) with an average arable land of 4,190 hectares each. The main tasks facing agriculture were the increase of production concentration and specialization, the introduction of modern equipment and of industrial production methods, the development of co-operative property and the improvement of income distribution.

During the Fourth Five-Year Plan (1961–65) the economy continued to develop rapidly. Capital investments amounted to 8,400 million levs.

The Dimiter Blagoev Non-ferrous Metals Combined Works, the Plant for Asynchronic Electric Engines in Plovdiv, the Combined Petrochemical Works in Burgas,

the real incomes of the population – 4.6 per cent. The Eighth BCP Congress (1962) ratified the guidelines for the development of the national economy during the period 1961–80, which constitute a general programme for the economic and cultural development of the country. At the same time the Congress mapped out the ways for improving the management of the national economy. At the May 1963 Plenary Session of the Central Committee, guidelines for the implementation of these objectives were drawn up. In 1964 a process of perfecting the management of the national economy was instituted, on a basically new principle. Big economic units were organized – called the State Economic Corporations.

The Fifth Five-Year Plan (1966–70) made provisions for further accelerated economic development through intensification and modernization of all sectors, and increased labour productivity and economic efficiency in order to satisy more fully the material and cultural requirements of the people. The various sectors of material production developed on the basis of electrification of

The Development of the Economy

production processes, complex mechanization, renovation of technologies and extended automation. The decisions of the July 1968 and September 1969 Plenary Sessions played a major role in securing the attainment of the plan's targets. The average growth rate of the national income reached 8.8 per cent, of industrial output – 11 per cent, of agricultural output – 3.5 per cent, of retail trade – 8.6 per cent, of the real income of the population – 6.0 per cent. In 1970 the national income surpassed its 1965 level by 52 per cent; higher labour productivity accounted for 96 per cent of this growth. Capital investments amounted to 15,300 million levs. A considerable part of the production capacities in the Kremikovtsi iron and steel works was put on stream, as was the USSR-Bulgaria gas pipeline. The Medet copper smeltery and refinery was completed. New big enterprises were opened – for petrochemicals, chemical fibres, synthetic resins, plastics etc. and some of the existing ones were extended. Preparations were made for intensive concentration of agriculture and for setting it on an industrial basis.

The Kozlodui nuclear power station; central control room

The main trends in the development of the economy during the Sixth Five-Year Plan period (1971–75) were the broad implementation of the achievements of scientific and technical progress, further accelerated industrialization, consistent introduction of new industrial technologies in all economic sectors; raising productivity on the basis of the scientific and technical revolution, complex and increasing satisfaction of the growing material and spiritual requirements of the people and raising its socialist consciousness. At the Tenth BCP Congress in 1971 a new Programme was adopted, setting the objective of building a developed socialist society in the People's Republic of Bulgaria as the highest stage in the construction of Socialism. The decisions of the 1974 National Conference of the BCP envisaged the further increase of productivity, which was considered to be a cardinal problem in the building of a mature socialist society. In the period of the Sixth Five-Year Plan the economy developed at a high and stable rate. National incomes increased by 46 per cent, higher labour productivity accounting for 98.8 per cent of this growth. Capital investments amounted to 21,700 million levs. The first Bulgarian atomic power station, at Kozlodui, was put into operation; new capacities were introduced in metallurgy, machine-building, and chemical and light industry. Following the decisions of the April 1970 Plenary Session the development of agriculture entered a new stage. New large-scale economic organizations appeared – the agro-industrial complex (AIC), the scientific and production complex (SPC) and other agricultural organizations. By the end of 1982 there were 296 AICs and 105 other agricultural organizations in the country, included in the network of the National Agro-Industrial Union (NAIU; established in 1979). Various measures were taken for implementing the December 1972 Programme for raising the people's living standard. Minimum wages were increased, as well as pensions, scholarships, grants and social benefits, and maternity leave was extended. In addition, a five-day working week was introduced.

The Eleventh BCP Congress in 1976 adopted the guidelines for social and economic development up to 1990 and outlined the main tasks of the Seventh Five-Year Plan (1976–80) – the increasing satisfaction of the people's requirements through a dynamic and proportional development of the national economy, the rapid increase of labour productivity efficiency and quality through modernization, reconstruction, concentration and accelerated implementation of scientific and technical achievements and by raising the working people's consciousness and consolidating the socialist way of life. This objective was implemented in the conditions of deepening socialist economic integration. Important decisions for the development of the economy were taken at the 1978 National BCP Conference. The growth of the national income during the five-year period was the result of higher labour productivity. About 31,000 million levs were allotted for capital investments. The second stage of the USSR-Bulgaria gas pipeline and its first extension were completed. Industrial projects constructed during the period included a polyester silk plant in Yambol, an aluminium processesing combine in Shumen, an iron foundry in Ihtiman and a ferryboat line connecting the city of Varna with Ilichovsk in the USSR. The Varna thermo-electric power station and the Kozlodui atomic power station were extended, and many plants were modernized and reconstructed. The considerable extension of the material and technical basis of agriculture brought to it some of the characteristics of industrial production. Economic management was improved.

The Twelfth BCP Congress in 1981 defined the main tasks of the country's social and economic development under the Eighth Five-Year Plan (1981–85) and up to

1990. It was planned to continue the line of comprehensive satisfaction of the ever growing material, intellectual and social needs of the people, of the all-round development of the personality on the basis of the overall intensification of the national economy, of the accelerated implementation of the achievements of scientific and technical progess and front-rank experience, the gradual implementation of the new economic approach and its mechanism, and the further improvement of socialist social relations. Over the first two years (1981–82) of the Five-Year Plan 15,900 million levs were spent on capital investments, and a fixed capital amounting to 14,500 million levs was put into operation. The National BCP Conference of March 1984 adopted a long-term programme for the improvement of quality, the latter being regarded as the key economic problem.

In the period of socialist construction the economy developed at high and stable rates. In average annual growth rates of domestic product and national income Bulgaria occupies one of the leading places in the world. The structure, organization and management of the national economy have improved. Production concentration has intensified. Thanks to the high economic growth rates the gross domestic product (GDP) in 1982 was 21 times greater than in 1939, the national income – 13 times, industrial output – 81 times, agricultural output – 2.8 times. In the 1949–82 period the average annual growth rate of the domestic product amounted to 8.65 per cent, of the national income – to 7.75 per cent, of labour productivity – to 7.95 per cent. A large investment programme has been carried out with the aim of building the material and technical basis of different industrial branches, their technical reconstruction and modernization. Radical changes have been effected in the structure of the national economy. From a backward agrarian country under Capitalism, Bulgaria has turned into an industrial-agrarian socialist State with a developed industry and modern, highly concentrated agriculture. In 1982 industry accounted for 67 per cent of the domestic product and for 54 per cent of the national income; 36.3 per cent of the labour force was employed in industry.

The economic system of Bulgaria is socialist and is based on the public ownership of the means of production, ruling out the exploitation of man by man, and is developing in a planned way. The forms of property are state (people's) property, co-operative property, the property of the public organizations, and personal property. The state and co-operative forms of property are already very closely related and are still undergoing a rapid process of convergence and mutual enrichment, tending to merge into a single form of people's socialist property. The socialist State, acting on the authority of the people, is the real owner of both forms of property. The State manages the national economy and the other areas of social life on the basis of unified socio-economic development plans with the aim of increasing satisfaction of the constantly growing material and intellectual requirements of all members of society for the all-round development of the individual. Labour is the basic social and economic factor. The socialist principle of distribution of goods is applied 'from each according to his abilities, to each according to his work'. The social funds for meeting the needs of the people are constantly increasing.

A major economic reform was launched in 1964. After five years of experimentation in different sectors of the national economy the system was completed and put into practice (1969). It is known as the 'economic mechanism for management of the national economy' and includes different rules for planning, economic regulation, incentives, the organizational structure and the distribution of functions between the separate bodies responsible for economic management. Periodically the

The steel pipes plant in Septemvri

economic mechanism is changed and improved in accordance with the changes which have occurred in the general conditions and the factors of economic development, in accordance with the changes in the social objectives and plan targets. The main object of those changes is strengthening the scientific basis and improving the quality of the centralized planning decisions, the creation of a better economic climate favouring innovation, creativity and enterprise, and the application of economic constraints in the economic organizations, producing greater efficiency, higher quality and a better competitive position for Bulgarian goods and services in domestic and international markets. The economic mechanism is improved also by promoting democracy in decision-taking at all levels of management and at all stages of planning, and organizing effective implementation of the economic decisions adopted.

Bulgaria has been a member of the Council for Mutual Economic Assistance (CMEA) since 1949 and

The Development of the Economy

The Stamen Stamenov alluminium processing works in Shumen

The Madara lorry plant in Shumen

plays an active part in socialist economic integration. On the basis of the international socialist division of labour, the People's Republic of Bulgaria is involved in specialization and industrial co-operation arrangements for different types of production, and participates in international energy and transportation systems and in joint construction projects and international production organizations and enterprises. It also participates in the pursuance of concerted licence, trade and credit policies. The result of socialist economic integration is the achievement of an efficient sectoral and inter-sectoral structure of the national economy. The all-round economic, scientific and technical co-operation and integration with the USSR plays a key role in the building of the material and technical basis of Socialism in Bulgaria. With the scientific, technical and economic assistance of the USSR more than 300 enterprises (1977) and projects in different sectors of the national economy have been built in Bulgaria, the problems of the fuel energy and raw material basis have been resolved and the conditions for achieving higher rates of economic growth created. The economic relations of the People's Republic of Bulgaria with other states and its participation in the international division of labour are steadily broadening.

Table of Important Dates
1945
13 March. Decree by the Council of Ministers on co-operative farms.
14 March. Bulgarian-Soviet trade agreement signed in Moscow.
1946
 8 March. National Assembly endorsed a law on the confiscation of property acquired after 1 January 1935 through speculation or illegal means.
12 March. National Assembly endorsed a law on labour land ownership (agrarian reform).

1947–48 – Two-Year National Economic Plan
1947
10–16 March. Monetary reform carried out.
 9 June. Inauguration of the first domestic airline (Sofia-Burgas) by Balkan Bulgarian Civil Airlines.
 1 September. Opening of the first republican trade fair in Plovdiv – the eleventh to be held there.
12 September. First Bulgarian international airline (Sofia-Budapest) opened.
16 October. First furnace of the Vulkan Cement Works in Dimitrovgrad commissioned.
26 October. First power-line Kurilo-Plovdiv put into operation.
 5 November. The Republic Pass through the Central Balkan Range completed.
23 December. Great National Assembly endorsed a law on the nationalization of private industry and mining enterprises.
27 December. Great National Assembly endorsed a law on the nationalization of banks and the reorganization of banking.

1948
 5 January. First train travelled on the Pernik-Voluyak line.
March. Commissioning of the Kalin water-pumping and storage station on the Kamenitsa river.
16 May. Chemical plant for dry wood distillation opened near Kocherinovo.

The Chavdar bus works in Botevgrad

The Analitik plant in Mihailovgrad; assembly line for electronic stablilizing rectifiers

1949–52 – First Five-Year Plan

1949

5–8 January. Foundation of the CMEA. Bulgaria is one of the founder-members.
22 March. Trial operation began of the Sofia Thermo-Electric Power Station.
18 September. Completion of the Stanke Dimitrov-Bobov Dol railway line.
1 December. Lyaskovets-Zlataritsa railway line opened.

1950

3 February. By decree of the Council of Ministers and the BCP Central Committee the co-operative farms were separated from the general co-operatives and became independent production enterprises.
28 May. Construction of irrigation system in the Belene Lowlands completed.
4 November. Vehicle repair factory opened at Gara Iskur-Sofia.
5 November. Vehicle repair factory opened in Plovdiv.
24 December. Dimitrovgrad Asbestos-Cement Plant opened – the first Bulgarian plant for asbestos-cement pipes.

1951

4 February. The Vidima and Petrovo Hydro-Electric Power Stations commissioned.
20 March. Rationing for industrial goods abolished.
17 April. Commissioning of the Brushlyan Irrigation System in the Russe District.
5 November. Opening of the Dimitrovgrad Chemical Works.
18 November. Construction completed on the V. Kolarov Dam.
21 December. Opening of the Ernst Thaelmann Cotton Mills in Sofia.

1952

2 February. Opening of the pectin factory near the village of Tsurkva (today Daskalovo), Pernik district.

The Georgi Kostov DC electric motor and equipment works in Sofia

31 October. Lovech-Troyan railway line completed. Construction began on the Druzhba tourist resort near Varna – the first international seaside resort in Bulgaria.
18–25 December. Fifth BCP Congress, general line of the Party for building Socialism established.

The Development of the Economy

The memory devices plant in Veliko Turnovo

The computer works in Sofia; tuning ES-1035 all-purpose computers

27 April. First Bulgarian concentrated fodder plant in Pleven put into operation.
10 May. Monetary reform.
30 November. The S. Kiradjiev Cellulose Factory, the first in Bulgaria for sulphate cellulose, put into operation in Novi Krichim (today Stamboliiski).
21 December. Completion of the railway line south of the Balkan Range, including the Koznitsa tunnel, the longest in Bulgaria.

1953–57 – Second Five-Year Plan
1953
8 August. First steel emerged from the Lenin Iron and Steel Works, Pernik.
3 November. Lenin Electrical Porcelain Factory opened in Nikolaevo.
5 November. Lenin Iron and Steel Works, the first metallurgical plant in Bulgaria, put into operation.
11 November. Ore-dressing factory in the town of Rudozem started operation.
5 December. Official opening of the Studena Dam.

1954
1 May. First experimental Bulgarian television broadcast.
20 June. Bridge of Friendship opened over the Danube between Russe and Giurgiu.
28 August. K. Marx Soda Plant put into operation in Devnya.
5 September. Penicillin factory opened in Razgrad.
6 November. Construction work on the Alexander Stamboliiski Dam completed.

1955
1 September. First modern dry docks opened at the G. Dimitrov Shipyard in Varna.
5 September. Completion of the first stage of the Kurdjali Lead and Zinc Works.
6 September. G. Dimitrov Hydro-Power System put into operation.

1956
18 January. First telephone radio-relay line between Sofia and Plovdiv put into operation.
16 August. First metal mine opened in Kremikovtsi.
8 September. Iskur Hydro-Electric Power System put on-stream.
23 December. Lead workshop opened at the Kurdjali Lead and Zinc Works.
  Construction work began on the Zlatni Pyassatsi seaside resort near Varna.

1957
16 April. Opening of the Central Department Store in Sofia – the largest in Bulgaria.
20 April. Petrohan Power Scheme put on-stream.
4 May. First radiophonic radio-relay line opened between Sofia and Pleven.
9 August. Telex links established with the USSR and Hungary, thus including Bulgaria in the international telex service.
12 October. Superphosphate Fertilizer Factory opened in Dimitrovgrad.

1958–60 – Third Five-Year Plan
1958
2 January. Batak Hydro-Electric Power Station put on-stream – the first underground power station in Bulgaria.
15 June. Studen Kladenets Hydro-Power System, the first stage of the Ardino Power Scheme, put on-stream.
4 December. Devnya Cement Plant went into operation.
5 December. Maritsa Textiles Plant started operation in Plovdiv.
6 December. Georgi Damyanov Copper Works put into operation near Pirdop (today a town district of Srednogorie).
  Construction began of the Slunchev Bryag holiday resort at the seaside near Nessebur.

The electronic transforming elements plant in Sofia; X-ray crystal control

1959
8 August. Completion of the Ivan Dimitrov Shipyard in Russe.
6 September. Opening of the Batak Hydro-Power System.
7 November. Sofia Television Broadcasting Station started broadcasting regular TV programmes.
26 December. Official opening of Bulgarian TV.
29 December. Asbestos works in Kurdjali put into operation.

1960
1 January. Bulgaria's first international photo-telegraphic links opened with the USSR and Hungary.
14 March. Construction work began on the Kremikovtsi (today the L. I. Brezhnev) Iron and Steel Works with the construction of a metal structures workshop.
9 September. First turbo-generator of the Maritsa-Iztok 1 Power Station put into operation.
25 December. Metodi Shatorov storage battery factory opened in Pazardjik.
28 December. Opening of the Wilhelm Pieck cement plant outside Beli Izvor, Vratsa district.

1961–65 – Fourth Five-Year Plan
1961
7 May. Two VHF radio transmitters put into operation.
1 September. International Gentex service opened with Hungary, thus incorporating Bulgaria into the European Gentex network.
9 November. Construction completed of the first Bulgarian nuclear reactor, the IRT-100.
17 December. The Dyanko Stefanov glass factory opened in Razgrad.
24 December. The D. Blagoev Non-ferrous Metals Combine opened near Plovdiv.

1962
1 January. Monetary reform carried out.
22 March. Fourth group of turbines put on-stream at the Maritsa-Iztok 1 Thermo-Electric Power Station.
8 April. Druzhba Glass Factory opened in Plovdiv.
11 June. Construction began of the Maritsa-Istok 2 Thermo-Electric Power Station.
14 November. Machine Repair Factory, the first enterprise of the Kremikovtsi (today L. I. Brezhnev) Iron and Steel Works, put into operation.

1963
2 January. First Sofia-Bucharest TV relay line put into operation, thus linking Bulgaria with the international Intervision network.
21 April. Pulp and paper works opened in Bukyovtsi (today Moesia), Vratsa district.
28 April. First electrified Plovdiv-Sofia railway line opened.
29 April. Induction motor factory opened in Plovdiv.
25 May. The Topolnitsa Hydro-Power and Irrigation System went into operation.
8 June. Vratitsa Cotton Textile Works opened in Vratsa.
23 June. Nitrogen fertilizer plant opened near Stara Zagora.
29 June. Official opening of the first stage of construction of the Maritsa-Iztok Industrial Estate.
20 July. Second electrified railway line opened between Gorna Oryahovitsa and Russe.
27 August. First Bulgarian coke produced at the Kremikovtsi Iron and Steel Works.
5 October. Electric Truck Plant started operation in Sofia.
5 November. Opening ceremony of the Kremikovtsi Iron and Steel Works.
30 December. Petro-chemical works near Burgas put into regular operation.

1964
2 September. The Chlorine and PVC Factory, Sugar Factory and Cement Plant extension were opened near Devnya.
23 November. Russe Thermo-Electric Power Station put into operation.
11 December. The Ivailovgrad Hydro-Power System went into operation.
29 December. Official opening of the Medet Ore-Dressing Combine.

1965
21 April. The first steel emerged from the Kremikovtsi Iron and Steel Works.
29 April. Semi-conductor factory opened in Botevgrad.
5 September. The Gara Iskur Semi-Cellulose and Cardboard Plant started operation in Sofia.
25 December. A modern hothouse combine completed outside Pazardjik.
25 December. Factory for Protein Bioconcentrate Mixtures opened near Kostinbrod Station.

1966–70 – Fifth Five-Year Plan
1966
22 January. Prefabricated Units Works opened in Stara Zagora.
24 March. Chiren-Vratsa gas pipeline completed.

The Development of the Economy

10 July. TV and VHF relay station opened on Mount Botev.
30 October. Cement Factory opened near Zlatna Panega, Lovech district.
4 November. The first Moskvich 408 car assembled at the Balkan Plant in Lovech.

1967
11 April. The Craiova-Oryahovo-Boichinovtsi power-transmission line opened.
1 November. The Vladimir Zaimov Cement Works started operation near Pleven.
7 November. Slunchev Bryag TV Transmitter began operation.

1968
20 January. The Petrohan TV Transmitter began operation.
29 April. The Elektronika Plant opened in Sofia.
28 June. Vratsa Chemical Works completed.
2 November. Dimiter Dimov Synthetic Fibres Plant opened in Yambol.
7 December. Official opening of the electrified Sofia-Karlovo railway line.

1969
22 August. Opening of the Varna Thermo-Electric Power Station.
24 August. Official opening of the Black Sea resort of Albena.
26 August. TV transmitter in Kyustendil went into operation.
2 September. The Blagoi Popov Plant for special shaped steel went into operation in Pernik.
5 September. The Sofia-Mezdra double-track railway line came into regular use.
6 November. The Avram Stoyanov storage battery factory started operation in Mihailovgrad.
20 November. The Lift Factory opened in the Sofia suburb of Gara Iskur.

The Elektra plant in Sofia

The Vassil Kolarov electric machines plant in Sofia

1970
27–28 April. BCP Central Committee Plenum on the concentration and development of agriculture on an industrial basis.
5 June. A new polystyrene production installation was put into operation at the Burgas Petro-Chemical Works.
11 August. The first Bulgarian synthetic rubber produced at the Burgas Petro-Chemical Works.
3 November. Chemical plants near Vidin went into operation.
5 November. The Synthetic Rubber Factory, the Bulana Polycrylic Fibre Plant and an extension of the Petro-Chemical Works in Burgas went into operation.
6 November. The Strumni Rid television transmitter went into operation.

1971–75 Sixth Five-Year Plan
1971
25 March. The Pleven Petro-Chemical Works went into operation.
31 March. The Anton Ivanov Mechanical Engineering Works, the Typewriter Factory and the Hydraulic and Pneumatic Tool Factory went into operation in Plovdiv.
5 April. Completion of the Dimiter Ganev Non-Ferrous Metals Combine in Sofia.
11 April. The pharmaceuticals factory in Stanke Dimitrov went into operation.
23 May. Instrument-building factory opened in Pravets.
6 October. The factories for thermal measuring equipment and for printed circuits went into operation in Russe.

1972
17 March. The Tolbuhin TV relay station went into operation.
29 April. An automatic control centre went into operation in Sofia for the Sofia-Karlovo railway line.
30 July. Launching of the first 38,000-ton coal and ore carrier built in Bulgaria.

19 August. The Druzhba 400-kV power cable between the USSR and Bulgaria opened.
8 September. Automatic telephone line between Sofia and Moscow went into operation.
9 September. Bulgarian TV transmitted its first colour broadcast.
5 November. Mill 1700 for cold rolling went into operation, thus completing the technological cycle of the Kremikovtsi Iron and Steel Works.
11–13 December. BCP Central Committee Plenum adopted the December Programme for raising the people's living standards.

1973
13 March. A decree issued by the Council of Ministers and the Central Council of the Bulgarian Trade Unions on the introduction of shorter working hours and a five-day working week.
15 June. The Pravnik TV transmitter went into operation.
1 August. The first Bulgarian compound mineral fertilizers produced at the Mineral Fertilizer Plant in Devnya.
6 September. The Sviloza Chemical Works went into operation in Svishtov.
23 September. Opening of the Lom Electric and Motor Trucks Plant.
28 September. Videotelephone link between Sofia and Varna.
28 September. First group of turbines opened at the Belmeken Hydro-Electric Power Station.
30 November. Computer Memory Plant opened in Stara Zagora.
13 December. First power generated by the Bobov Dol Thermo-Electric Power Station.

1974
27 January. Bulgaria's largest production line for ammonia went into operation and completed the production cycle of the Devnya Artificial Fertilizers Plant.
20–22 March. BCP National Conference on the question of labour productivity.
24 July. First nuclear power from the Kozlodui Nuclear Power Station switched into the country's power grid.
1 September. Official opening of the USSR-Bulgaria gas pipeline in Vratsa.
4 September. Opening of the Kozlodui Nuclear Power Station, the first in Bulgaria.
5 September. Some of the largest projects of the Devnya Industrial Estate put into operation – the Mineral Fertilizers Plant, the Calcinated Soda Factory and the first stage of the Devnya-West Harbour.
6 September. Official opening of Sofia Central Station.
1 November. The Belmeken-Sestrimo Hydro-Electric Power System went into operation.

1975
27 March. Official opening of the Bobov Dol Thermo-Electric Power Station – the first major thermal power station entirely designed by Bulgarian experts.
12 June. The 200,000th silage harvester of the KIR-1.5 model emerged from the G. Dimitrov Agricultural Machinery Plant in Russe.
9 September. First regular transmission of Bulgarian TV's Channel 2.

5 November. The Antonivanovtsi hydro-power unit of the Vucha power station put into regular operation.
31 December. The TV transmitter near Shumen, the most powerful in the Balkans, started operation.

1976–80 – Seventh Five-Year Plan
1976
24 March. The Khan Asparuh, a 100,000-ton tanker and flagship of the Bulgarian fleet and shipbuilding, was launched.
27 March. The second reactor of the Kozlodui Nuclear Power Station went into operation.
10 July. The Maritsa Cannery was opened in Pazardjik – the largest in Bulgaria.
8 September. The Asparuh Bridge and navigational channel opened in Varna.
6 December. Construction begun on the Heavy Investment Mechanical Engineering Plant near Radomir.
10 December. Ground is broken for the joint Bulgarian-Romanian Heavy Mechanical Engineering Plant at Russe and Giurgiu.

1977
1 November. The Haskovo Brewery was opened.
2 December. The first 210 megawatt unit of the Maritsa-Iztok Thermo-Electric Power Station went into operation.
22 December. The modern Kailuka Brewery, fully mechanized and automated, opened in Pleven.

1978
February. The Sinite Kamuni TV Transmitter went into operation near Sliven.

The Podem electric hoist plant in Gabrovo

The Development of the Economy

- 21–22 April. National BCP Conference on questions of improving the socialist organization of labour and planned economic development.
- 14 November. Official opening of the Varna-Ilichovsk ferry service, one of the biggest railway ferries in the world.

1979
Commutation Elements Plant started operation in Smolyan.
Polyester silk processing factories went into operation in Dimitrovgrad and Yambol.
Construction begun on an Integrated Radio and TV Centre on Mount Vitosha.

1980
Contact Elements Plant went into operation in Devin.
4 September. A new factory for chlorovinylchloride and polyvinylchloride opened in Devnya.

1981–85 – Eighth Five-Year Plan
1981
- 2 March. Bulgaria's first quantometric laboratory opened at the G. Dimitrov Lead and Zinc Works in Kurdjali.
- March. The first automatic quasi-electronic telephone exchange tested.
- 22 March. Yana-Vitinya section of the Hemus Motorway completed.
- 13 May. Plant opened for special diesel-motor iron casting at the Veslets Pig-Iron Foundry in Vratsa.
- 5 September. The first Bulgarian polypropylene powder for mechanical engineering and the pharmaceutical industry produced at the Burgas Petro-Chemical Works.

1982
Opening of the Mirovo-Malo Konare section of the Trakia Motorway.

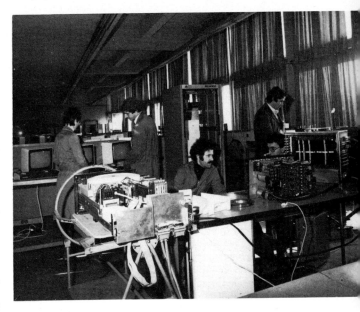

The Elektronika plant in Sofia

1983
- 18 February. The Elatsite Ore-Dressing Complex opened.
- 31 March. Purification station for sewage opened near Ihtiman – the country's first plant for purifying all waters of a complete settlement system.

1984
March. BCP National Conference on the problems of quality.

INDUSTRY

The beginnings of Bulgarian industry can be traced back to domestic crafts in ancient times – manual processing of farm produce, clay and other materials. With the increase in labour productivity crafts gradually separated from farming and were recognized as a separate trade within the village community, the tribe or larger unit. In the seventh and sixth centuries BC there were already a number of independent small artisans in the semi-urban colonies along the Black Sea, founded by immigrant merchants, and towards the fifth and fourth centuries BC crafts appeared in the first semi-urban settlements in the Odryssaean Kingdom in Thrace. The crafts industry developed, and in the second half of the eighteenth and the first half of the nineteenth centuries its scale rapidly grew, due to the expansion of the home market and the comparatively minor import of European manufactured goods. The beginning of the nineteenth century saw a rapid growth in the production of woollen braiding, concentrated mainly in the towns of Gabrovo, Kalofer, Sliven, Karlovo, Sopot and Kazanluk. In the 1860s and 1870s the number of woollen braiding mills reached nine to ten thousand, with an average annual output of 2,600 tons of woollen braid. Some other textile crafts were also well developed, for example, manufacturing of homespun, carpet-making, clothweaving, the making of goat hair rugs, fulling, textile-dyeing and carding. There were also a number of other trades, such as tailoring, leather and fur-dressing, iron-working and other metal-processing (axe- and knife-making and the farrier's and coppersmith's trades). A better-off stratum was gradually formed of craftsmen who looked for foreign markets in order to sell their increased production, both within the boundaries of the Ottoman Empire and beyond them, e.g. in Syria and India. Small, scattered, domestic private workshops appeared for the production of woollen textiles, furs and leathers, knives, ready-made clothes, etc. In the third quarter of the nineteenth century a number of centralized manufacturing enterprises were set up within Bulgarian territory for the production of carpets, blankets, silk-spinning, furnaces and implements for the smith's trade. Large-scale manufacturing did not spread over the whole of social production, but it brought about the ruin of many small craftsmen. In 1831 Dobri Zhelyazkov the Manufacturer (1800–65) set up his own textile workshop (the first textile mill in the European part of the Ottoman Empire) in the town of Sliven; in 1936, after the construction of new factory buildings, it became the 'State Woollen Textile Mill' – the first in Bulgaria. Prior to the Liberation in 1878 there had been other attempts to open factories, most of which, however, were short-lived and unsuccessful. The major factors which held back development of the industries were the lack of large amounts of capital and of security within the empire, the import of competitive manufactured goods from abroad, bad transport facilities and high taxes. Up to 1878 there were only 25 industrial enterprises, with a total work-force of 600–760 workers. During the second quarter of the

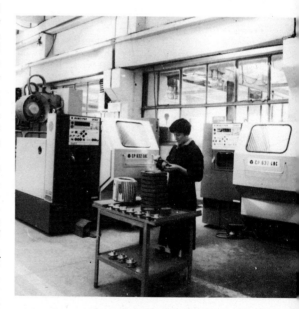

The Elprom factory in Troyan

nineteenth century the competitiveness of cheaper manufactured goods imported from Western and Central Europe increased, but this did not lead to the ruin of masses of craftsmen. Many new trades appeared which did not have to compete with the manufacturing world, for instance, the maintenance of new equipment in industries and transport, services for the town population (maintenance of the water-supply and sewerage systems, servicing

The Typewriter factory in Plovdiv

The Shesti Septemvri electric trucks plant in Sofia; electric trucks of the Druzhba type

of electrical appliances, the barbers' and hairdressers' trade, etc.).

The development of Bulgarian industry started after the Liberation in 1878, at a time when in other countries Capitalism was entering its highest stage of development – Imperialism. The strong influence of advanced capitalist countries held back the development of national industry. During the first two decades the development of capitalist manufacturing was rather slow because of the limited home market, the low import tariffs (5–14 per cent before 1895) and lack of industrial capital, and the use of the greater part of existing capital for money-lending and trade. In 1894 the country had 501 factories and workshops with a total of 5,732 workers, that is, an average of eleven workers per manufacturing unit.

After 1895 large-scale capitalist industry developed at an accelerated rate because of the increase in import tariffs (up to 65 per cent), expansion of the home market as a result of the construction of railway lines, seaports, roads, administrative and residential buildings in the towns, lower-interest credits offered by the newly set-up modern banking system, and the special privileges the larger industrial enterprises enjoyed under the Home Industry Encouragement Acts of 1895, 1905 and 1909. The Act of 3 March 1896 made it compulsory for all civil servants, members of Parliament and schoolchildren to wear home-manufactured clothes and shoes. In 1911 Bulgaria had 345 larger enterprises (employing more than ten workers) with a total work-force of 15,886 workers; home manufactured goods met 45 per cent of total requirements (as compared to 13 per cent in 1896). In the 1904–11 period the average annual growth in industrial production was 13 per cent (the estimate is based on information collected about enterprises subject to statistical control) as compared to 13 per cent in Japan, 12.5 in Serbia, 8.1 in Romania, 4.9 in the USA and France, 4.2 in Russia, 3.4 in Germany and 2.1 in Great Britain. The higher concentration and centralization of capital led to the appearance of monopolistic companies, particularly under the influence of the increased import of foreign capital.

After the 1912–18 wars Bulgarian industry managed, in a comparatively short time, to reach and even surpass its pre-war level of production. In 1921 there were 1,544 registered enterprises with a total of 55,000 workers. The comparatively rapid restoration of the economy was due to the inflation of the lev (the Bulgarian monetary unit), which compelled the Bulgarian bourgeoisie to invest its capital in industry; to the appearance of new branches of industry; and to increases in import tariffs and control over trade in foreign currency. Between 1921 and 1929 (1926 and 1929 in particular), the output of Bulgaria's capitalist industry almost doubled. After the period of stagnation during the world Depression 1929–33, Bulgarian industry continued to increase its output in the second half of the 1930s, but the rates of growth were lower than those in the 1920s. The average rate of growth in industrial production in bourgeois Bulgaria for the 1921–41 period was 6.5 per cent, as compared to 5.2 per cent in Japan, 4.9 in Germany, 4.5 in Romania, 4 per cent in Yugoslavia, 3.8 in France, 3.6 in Poland, 3.5 in Italy, 3.4 in Hungary, 1.1 in the USA, and 0.1 in Great Britain. Towards 1941 there were about 3,872 registered industrial enterprises, out of which 3,467 were privately owned, 130 state-owned (accounting for about 8–9 per cent of industrial output) and 275 co-operative enterprises (accounting for 6 per cent of industrial output). An enterprise had, on average, 26 workers and machine equipment of 110 hp. The number of artisans' workshops was still very high. The equipment and technologies used were still primitive, the major productive force being manual labour. Private investment predominated. The share of foreign capital invested between the two world wars was between 16 and 18 per cent. In 1939 industrial output accounted for a mere 15 per cent of the national income. As compared to 1931, industrial output in 1939 increased almost 1.8 times (in terms of comparable prices). In spite of this growth rate, however, Bulgarian industry was unevenly developed (the predominant branch was light industry accounting for 77 per cent of total industrial output), inefficiently structured (the basic divisions were the food and textile industries) and unequally distributed throughout the country. Machine-building accounted for only 2.4 per cent of the country's total industrial output, and the chemical industry, for 1.9 per cent (mainly volatile oils, dyes and varnishes, soap, matches and calcium carbide). The level of development was 1,000–1,600 per cent lower than that of the advanced industrial countries.

The Georgi Dimitrov shipyards in Varna

After the victory of the 9 September 1944 uprising a people's democratic Government was established, which provided favourable conditions for the development of industry. During the reconstruction period of 1945–47 industrial enterprises were still in the possession of private owners. Therefore a system of state and workers' control was set up to monitor and direct production. During that period the trade unions did much to improve the organization of labour. The State took measures to ensure normal conditions for production and to prevent attempts at sabotage engineered by private owners. In 1946 an attempt was made to introduce planning into the financial economy, including industry. A two-year economic plan was adopted for the 1947–48 period which gave a boost to industrial growth. On 23 December 1947 private industrial and mining enterprises were nationalized: 1,997 larger industrial organizations and 4,027 smaller ones (flour-mills, printing houses, soft drinks factories, carding workshops, etc.). Opportunities were created for the construction of an advanced industrial system. Early in 1948 socialist industry amounted to 91.7 per cent, of which 81.0 per cent was accounted for by state-controlled industry, 10.6 per cent by co-operative industry and 8.3 per cent by crafts and small-scale private industry. Economic aid from the Soviet Union and the labour upsurge towards the end of 1948 resulted in a two-fold increase in the total volume of industrial output as compared with 1939. Priority tended to be given to the production of capital goods. At the Fifth Congress of the BCP (1948) special importance was placed on industrialization as a crucial factor in socialist development. The Resolution of the Congress on the First Five-Year Plan for Economic Development (1949–53) stated: 'The major economic and political objectives of the Five-Year Plan are to lay the foundations of Socialism in Bulgaria by the industrialization and electrification of the whole country, and the introduction of a co-operative system and mechanization of agriculture'. (*Rezolyutsii na V kongres na BKP* [Resolution of the Fifth Congress of the BCP]. Sofia, 1949, p. 30). Bulgaria's economic backwardness and the inherited industrial system determined the specific problems facing socialist industrialization: within a historically short period of time to build up heavy industry as a basis for providing other branches of the national economy with adequate equipment and ensuring Bulgaria's participation in the international socialist division of labour via mutually profitable specialization and co-operation in different branches of industry.

As a member of the CMEA (since 1949) Bulgaria took an active part in the economic integration of the CMEA countries. The major task of the first two five-year economic development plans was to lay the foundations of Socialism through industrialization and electrification (during the First Five-Year Plan for 1949–53), as well as through promoting electrification, coal mining, non-ferrous metallurgy, and by constantly expanding the production of consumer goods (during the Second Five-Year Plan for 1953–57). In 1948–57, the average annual growth of industrial output was 16.1 per cent; for the production of capital goods – 18.8 per cent and for consumer goods – 14.4 per cent (see Table 1). Towards the end of the First Five-Year Plan (1952) the socialist sector accounted for 98.2 per cent of total output, of which state industry accounted for 85.9 per cent and co-operative industry for 12.3 per cent. The private sector produced 1.8 per cent. Towards the end of the Second Five-Year Plan the socialist sector almost kept its share, whereas the share of the co-operative sector rose to 16.8

per cent. Priority was given to ferrous metallurgy, the chemical and rubber industries, machine-building and metal-working, the building materials industry, the fuel industry, textiles and glass, and the porcelain and faience industries. Industry won itself an important place in the overall development of the national economy. Its relative share in the general structure of the domestic product rose to 55 per cent (as against 32 per cent in 1948) and to 41 per cent in the structure of the national income (as against 23 per cent in 1948). Bulgaria was transformed into an industrial-agrarian country with a modernized industry. Industrial enterprises were reorganized, developed and enlarged. From 4,000 in 1944, the number of industrial enterprises was reduced to 2,305 towards the end of the First Five-Year Plan. Of these 1,168 were state-owned and 1,137 were co-operative. By the end of the Second Five-Year Plan their number had dropped to 1,879, of which 980 were state-owned and 899 co-operative. Whereas in 1945 an average of 28 persons were employed in an industrial enterprise, of whom only 22 were workers, in 1957 the average number of industrial workers and employees in a single state-controlled enterprise rose to 430, and to 96 in a co-operative enterprise.

The heavy mechanical-engineering combine in Russe

The machine-tool plant in Sofia

In the period 1952–57 the total industrial output increased 216 per cent; in state enterprises it increased 209 per cent and in co-operative enterprises – 288 per cent. Alongside this, labour productivity also rose: by 48.2 per cent in 1952 as against 1948 and by 91.0 per cent in 1957 or by an annual average of 7.4 per cent. In 1949–57, some 1,847 million levs were invested in building up the resources and technological base of industry. Of these 622 million were utilized during the First Five-Year Plan and 1,225 million during the Second Five-Year Plan. A total of 1,537 million levs of fixed capital were put into operation.

The April Plenum of the Central Committee of the BCP (1956) raised the issue of surmounting the delay in the industrialization rate during the Second Five-Year Plan. A policy of accelerated industrialization was adopted with special priority for heavy industry as the basis of modern technology. In 1958–82 industry became the staple branch of the economy. The National Party conferences (1974, 1978, 1984) provided the basic guidelines for raising labour productivity, improving organization and the quality of production. Major issues were resolved such as: an increased use of domestic raw materials and development of the power industry at a highly accelerated rate; rapid development and specialization of machine-building, the chemical industry and metallurgy; a sensible territorial allocation of productive forces; fuller utilization of material, manpower and financial resources; approximating growth rates of production of capital goods to those of consumer goods; expanding production of consumer goods as a guarantee for raising the people's living standards; providing better quality and greater competitiveness of production; raising the profitability of production by overall intensification; and maintaining high and stable rates of economic growth. A concentration and specialization of industrial enterprises was also effected. In 1964 state economic corporations were launched as a new organizational form of branch management of industry. The main object in the management of plant concentration was the system of

Table 1: Indices and Structure of Total Industrial Output in 1957 by Industrial Sectors

Sector	Indices for 1957 Base year 1939 (=100)	Indices for 1957 Base year 1948 (=100)	Annual average growth in 1948–57 in %	Structure in % 1939	Structure in % 1948	Structure in % 1957
Total	782	386	16.15	100.0	100.0	100.0
Electric power and thermal energy	1,000	498	19.60	1.8	1.8	2.4
Fuel industry	572	311	13.45	4.6	4.2	3.4
Ferrous metallurgy (incl. ore extraction)	3,100	5,700	54.20	0.2	0.1	0.9
Machine building and metal-working	4,000	650	23.10	2.4	7.3	12.3
Chemical and rubber industries	1,800	808	26.10	1.9	2.0	4.1
Building materials industry	1,300	647	23.10	1.8	1.7	2.9
Timber and wood-processing industry	488	215	8.85	10.3	11.6	6.5
Pulp and paper industries	629	302	13.10	1.5	1.5	1.2
Glass and porcelain-faience industries	2,000	628	22.70	0.3	0.5	0.8
Textiles	775	344	14.70	19.8	14.3	12.8
Clothing industry		663	23.40		4.0	6.9
Leather, fur and shoe industry	1,000	350	15.00	2.1	2.8	2.6
Printing and publishing industry	348	211	8.65	1.7	1.4	0.7
Food industry	466	284	12.35	51.2	41.5	30.9

diverse yet interlinked production units (economic combines, economic corporations, echelons of specialized enterprises, economic complexes). The economic combine proved to be the most appropriate form of direct organization and management of production. Product-technological organization of production processes was also introduced, based on higher and more differentiated concentration and specialization. The complete systems of production were supplanted by a new type of production system in which the production cycle was completed outside the territory of the individual enterprise. The larger scale production led to increasing the size of production units, the rate of production specialization and the size of the series. New organizational forms of industrial management were introduced, such as corporations and associations uniting the interests of economic organizations in homogeneous and mixed branches. Through the CMEA, Bulgaria specialized in the production of over 600 items – electric trucks, electric hoists, tractors and some kinds of farming machinery (silage combines, tractor harvesters, etc.), machinery for the food industries, equipment for radio-electronics, instrument design, shipbuilding, the automobile industry, machine-tool manufacture, etc.

In 1982 the share of industry in the GDP increased to 67 per cent, and in the national income to 54 per cent. The volume of total industrial output increased ten times in the period 1957–82, the production of capital goods – fifteen times, and that of consumer goods – seven times (see Table 2). Heavy industry predominated, which created favourable conditions for the development of other branches of the national economy, as well as for the participation of Bulgaria in the international division of labour. The strategic branches of the national economy were further developed and strengthened, namely power generation, metallurgy, machine-building, metal-working (with its main sub-branches connected with progress in technology, such as mechanical-handling equipment manufacture, ship-building, metal-working, machine-building, hydraulics, instrument-making, production of equipment for automation, communication equipment, electronics, and computer equipment), chemical industry, building materials industry, etc. To meet the ever-increasing needs of the working people and improve their standard of living there was a constant expansion of production in the textile, food, clothing, leather, fur and shoe industries, and in other branches of light industry. The production of export goods also increased 310 per cent in the period 1970–82, of which capital goods increased 440 per cent and consumer goods 180 per cent. The high growth of industrial production as a whole and the difference in the growth rates by sectors brought about significant changes in the structure of industry. Manufacturing industry increased its share: in 1982 it contributed 92.2 per cent of the total output. The role of branches processing raw materials for industry also increased. In 1982 manufacturing industry used 61.1 per cent of industrial raw materials, compared to 54.3 per cent in 1970; the use of agricultural raw materials was 38.9 per cent and 45.7 per cent respectively. Extractive industry gave 3.6 per cent, and the electric power and thermal energy output was 4.2 per cent of the total output. The high growth rates of industrial production caused the transfer of labour from agriculture. In 1982 the relative share of those employed in industry was 36.3 per cent of all employed in the national economy, compared to 12.9 per cent in 1956. The average annual number of workers and employees in industry increased from 488,068 in 1956 to 1,401,997 in 1982, i.e. 290 per cent. Machine-building, electro-technical and electronic industries employ the largest number of people (17.4 per cent in 1982). Party and government policy is aimed at improving the technical, technological and organizational level of industrial production. In 1982 the real volume of fixed capital per unit of labour increased 600 per cent over 1957, with an average annual growth rate of 7.4 per cent; in 1961–82 the energy per production worker increased 530 per cent (an average of 7.9 per cent annually); the total installed capacity per worker increased from 2.7 kW to 14.4 kW (at an average of 7.9 per cent annually); the total of electrical power used as motive power and in technological processes per worker increased from 4,221 kWh to 19,388 kWh (an average of 7.2 per cent annually). In 1958–82 the productivity of labour increased on average 6.05 per cent annually, which is 430 per cent over 1957.

The instrument engineering works in Pravets

The total number of state and co-operative enterprises was 2,157 in 1982, 1,974 of them being state-owned and 183 – co-operative. In 1982 there were 1,386 thousand workers and employees occupied in them. The fixed capital of the state-owned and co-operative enterprises amounted to 31,000 million levs. Compared to 1957 they marked an increase of about 1,420 per cent with an average annual growth of 11.2 per cent. In the period between the Third and Seventh Five-Year Plans (1958–80) 33,500 million levs of capital investments were spent on their construction. During the same period new fixed capital was put into use of about 27,700 million levs. A large part of the fixed capital of industry was concentrated in electric power and thermal energy production, in the fuel industry, ferrous metallurgy, machine-building, chemical industry, building materials industry, and glass, porcelain and faience industry. The growth rates of fixed capital in those branches of industry producing means of production, are considerably higher than in those producing consumer goods. Compared to 1957, in 1982 fixed capital in heavy industry increased over 1,500 per cent (by an average of 11.2 per cent annually), in light industry – about 1,100 per cent (by an average of 10.1 per cent annually). Some 41.4 per cent of the total volume of investment in industry in 1982 was used for reconstruction and modernization. The resources and technology already existing in industry could thus ensure a further development of this branch, which is a major factor in the construction of a developed socialist society in Bulgaria.

The machine-tool plant in Sofia

Table 2: Indices and Structure of Total Industrial Output in 1982 by Industry Sector

Sector	Indices for 1982 Base year 1939 (=100)	Indices for 1982 Base year 1957 (=100)	Average annual growth rate 1958–82 in %	Structure in % 1982
Total	8,100	1,000	9.70	100.0
Electric power and thermal energy	14,800	1,500	11.45	4.2
Fuel industry	7,200	1,200	10.60	1.4[1]
Ferrous metallurgy (incl. ore-extraction)	95,000	3,100	14.70	3.4
Machine-building and metal-working	150,500	3,700	15.55	22.5
Chemical and rubber industries	64,000	3,600	15.40	8.2
Building materials industry	20,200	1,600	11.70	4.4
Timber and wood-processing industry	1,800	377	5.40	3.1
Pulp and paper industries	7,800	1,200	10.50	1.4
Glass, porcelain-faience industries	42,100	2,100	12.95	0.9
Textile / Clothing industry	4,500	578	7.20	5.8
Leather, fur and shoe industry	5,100	527	6.90	1.1
Printing and publishing industry	2,200	631	7.60	0.5
Food industry	2,500	563	7.20	27.1

[1] For coal-mining only.

The Elprom plant in Varna; washing machines

Under the Eighth Five-Year Plan (1981–85) industry continues to play the leading role in the development of national economy. It is developing at an accelerated pace; the rates of development of heavy and light industries are approximately the same; productivity is increasing; a balanced development of the different branches of the national economy is ensured; production profitability is growing; and production facilities are combining to concentrate and specialize with consequent improvements. Priority is being given to those branches determining the structure of the national economy; light industry production is on the increase.

1. The Coal Industry: turns out 1.4 per cent of the total industrial output (1982). Prior to the socialist revolution (9 September 1944) it was confined to a poorly developed coal industry and supply of firewood. The first steps in coal-mining as a private enterprise were made as early as 1852 in the central part of the Balkan Range (Tryavna mountain). Industrial coal-mining began in 1891 when the Pernik Coal Basin was nationalized and the largest coal mining enterprise at that time, the Pernik State Mines, was founded. The working of the Bobov Dol and

Table 3: Share of Industrial Sectors – Years and Sectors

Sector	1975	1977	1978	1979	1980	1981	1982
1	2	3	4	5	6	7	8
Electric power and thermal energy	2.2	2.3	2.3	2.3	3.7	3.8	4.2
Coal-mining	1.0	0.8	0.8	0.8	1.4	1.3	1.4
Ferrous metallurgy (incl. ore-extraction)	3.4	3.7	3.6	3.5	3.9	4.0	3.4
Machine-building, electro-technical and electronic industries	24.8	27.7	28.7	28.7	23.2	23.7	22.5
Chemical and rubber industries	7.6	7.8	8.0	8.3	8.3	8.2	8.2
Building materials industry	4.0	4.0	4.1	4.2	4.9	4.8	4.4
Timber and wood-processing industry	3.5	3.3	3.2	3.1	3.3	3.3	3.1
Pulp and paper industries	1.4	1.3	1.3	1.3	1.5	1.6	1.4
Glass, porcelain-faience industry	0.9	0.9	0.9	0.9	1.0	1.0	0.9
Textile and knitwear industries	7.8	7.5	7.5	7.5	5.1	5.1	5.8
Clothing industry	3.6	3.3	3.2	3.1	1.9	1.9	2.1
Printing and publishing industry	0.4	0.5	0.5	0.4	0.5	0.5	0.5
Leather, fur and shoe industries	1.7	1.6	1.4	1.4	1.2	1.2	1.1
Food industry	25.1	23.0	22.4	22.5	22.9	23.7	27.1

Kyustendil coal deposits began in 1891–93. Coal was mined most primitively, the pits were partially mechanized (mainly the Pernik Mines). In 1945–55 the geological deposits were further prospected and coal-mining was adapted to the needs of developing industries. Since 1960 modern and highly efficient mechanization, scientifically-based organization of labour and computer technology have been introduced. The total coal output increased to 33,550,000 tons (1982). Bulgarian coal has a high ash and sulphur content and low calorific value. Hard brown, soft brown (lignite), black and anthracite coal are produced. Lignite coal predominates (75.8 per cent of the total output in 1982). Open-cast coal-mining is used in the Maritsa-East Coal Basin (the largest lignite coal deposit in Bulgaria), the Chukurovo, Stanyantsi, Sofia and Gotse Delchev coalfields; underground mining methods are employed in the Maritsa-West Coal Basin. The brown coal deposits are small. The Pernik Coal Basin, where several underground and open-cast pits operate, is nearly exhausted. Most promising is the Bobov Dol Coal Basin (underground mining). Insignificant amounts of black coal are mined in the Balkan Range Coal Basin, and anthracite coal is obtained in the Svoge Coal Basin. Coal mining is organized by large mining and power enterprises, the largest being the Maritsa-East Combine in Radnevo (for lignite coal mining). Coal mining has a considerable share in the country's energy balance. Coal consumption is increasing in absolute terms, but in relative units it is decreasing. The greater part of coal is used for electrical power generation, and the remainder in industry, transport and for household needs. Prospective coal basins and occurrences are the Dobrudja (rich in black coking-coal), the Elhovo-Yambol, and the Balsha occurrence in the Sofia Coal Basin (rich in lignite coal). The geological conditions in these three Basins are extremely complex, and considerable technical and techological difficulties have to be overcome.

This industrial sector is managed by the Ministry of Energy and Raw Material Resources – the Energetika Corporation.

2. Power and Heat Energy Production accounts for 4.2 per cent of the total output of industry (1982). The beginnings of this sector were laid in 1900 with the commissioning of the first hydro-electric power station on the Iskur River at Pancharevo, near Sofia, with a capacity of 1,720 kW. It was designed to supply Sofia with electricity. Up to 9 September 1944, 174 small, local power stations had been built; 11 of them were state-owned, 34 belonged to municipal councils, 15 to the water

The 23 Septemvri plant in Mihailovgrad; moulding and casting shop

resource syndicates, 37 to the popular banks and co-operative societies, and 77 to various companies and private firms. In 1944 there were 117 electric power stations with a total capacity of 130.5 MW and an annual output of 310.8 million kWh. Of the hydro-electric power stations, the most powerful was the *Vucha* (7 MW); of the thermo-electric power stations – the *Kurilo* (15 MW); and of the Diesel generators – the *Madan* (1.2 MW). A total of 791 communities (13 per cent of the towns and villages and 39.7 per cent of Bulgaria's population) were supplied with electricity. Electric-power consumption was 42 kWh per capita (1938–39); in the same period the average electric-power consumption level in the world was 230 kWh per capita (500–800 kWh in the most advanced European countries). After the nationalization of industry in 1947 new power stations were built, while others were closed as uneconomic. In 1982, 234 electric-power stations with a total capacity of 9,499,000 kW were operational, of which 87 were hydro-electric (1,895,000 kW), 140 thermo-electric (5,844,000 kW) and one was nuclear (Kozlodui, 1,700,000 kW). The 1982 power output was 40,135 million kWh (152 times as much as in 1939), including 3,049 million kWh from the hydro-electric, 26,660 million kWh from the thermo-electric and 18,746 million kWh from the nuclear power stations.

Exports amounted to 2,711,467 thousand kWh (from 1974 to Turkey, from 1975 to Yugoslavia, and from 1982 to Greece). Since 1953 Bulgaria has ranked first in per capita electric power consumption in the Balkans, since 1960 it has reached the average world standard, and since 1970 it has ranked among the leading European countries (4,531 kWh of electric power per capita were generated in 1982). The principal power consumers are industry, followed by the communal services and households, agriculture, transport, and other sectors of the national economy. On average, these sectors are 98 per cent electrified and in some of them this figure has reached 100 per cent (ferrous metallurgy, the paper and pulp industry, printing, tailoring, cement, etc.). Central heating was introduced in 1950 following the commissioning of thermo-electric power stations. At present thirteen towns are centrally heated: Sofia, Pernik, Vratsa, Pleven, Veliko Turnovo, Gabrovo, Russe, Shumen, Razgrad, Plovdiv, Assenovgrad, Sliven and Yambol.

The Hidravlika hydraulic equipment works in Kazanluk; automated line

In 1982 the number of centrally heated flats per 1,000 flats in the towns with central heating was 249; the capacity of the heating turbines was 758,100 kW. A total of 13,178,000 million kilocalories of energy was produced, as against 476,000 million in 1950. In Sofia alone the length of the central-heating pipelines was 590 km, and 6,301,676 g.cal. of heat were consumed. This sector is managed by the Energetika Corporation, of the Ministry of Energy and Raw Material Resources.

3. Metallurgy: an industrial branch for the extraction and dressing of ores and for the manufacture of ferrous, non-ferrous, rare and precious metals and their alloys, this accounts for 3.4 per cent of the total industrial output (1982). It is managed by the Ferrous Metallurgy and Non-Ferrous Metallurgy corporations of the Ministry of Energy and Raw Material Resources.

Ferrous Metallurgy includes the mining and dressing of iron ores and the manufacture of cast iron, steel and rolled ferrous metals (rolled steel) and their products (pipes, wires, ropes, galvanized sheet-iron, tin-plated sheet-iron, etc.).

Iron has been mined in the Bulgarian lands since Roman times. In the Middle Ages, after Bulgaria fell under Ottoman domination in 1396, iron-ore mining was rather limited and primitive, developing mainly in the Rila, Pirin and Strandja mountains. Its greatest progress was in the eighteenth and nineteenth centuries. In Samokov special hammers driven by a water-wheel were used.

The foundations of modern ferrous metallurgy were laid after 9 September 1944, when the Lenin Metallurgical Combine was commissioned in Pernik (1953) and the Kremikovtsi Combine (today's Leonid Ilych Brezhnev Iron and Steel Works) in Sofia (1963). 1,617,000 tons of cast iron and ferrous alloys, and 2,584,000 tons of steel were produced in 1982. This sector uses Bulgarian and imported iron ore from the USSR. The annual iron output of 1,552,000 tons (1982) cannot satisfy Bulgaria's needs. The ore is mined in open-casts in the Kremikovtsi area and underground in the Martinovo, Krumovo and Dryanovo-1 mines.

Non-Ferrous Metallurgy includes the mining and dressings of non-ferrous ores and the production of non-ferrous metals. In Roman times, first century AD, the mining of lead, silver and copper ore was quite highly developed; gold mining developed mainly in the sixteenth century. Before 9 September 1944, mining was poorly organized and on a fairly limited scale, using mainly foreign capital. Of greatest importance was the Plakalnitsa copper mine near Zgorigrad, in the Vratsa district. A few small, private shareholding societies were founded after World War I (1914–18). The public sector was poorly developed. The Pirin shareholding society, founded in 1924, was the leading producer of lead and zinc ore; in 1941 it began mining the lead and zinc deposits in the Madan ore-field in the Rhodope mountains, financed by the Germans. Attempts were also made to mine copper ore near Burgas. After the nationalization of private industrial and mining enterprises in 1947, large-scale ore-dressing facilities were built: the Gorubso Mining Combine in Madan for mining of lead-zinc ores, operating the Madan, Rudozem, Erma Reka, Luki and Kurdjali ore-fields; the Georgi Dimitrov Mining Combine in Eliseyna, in the Vratsa district, and the Chiprovtsi State Mines (Iskur ore-field). Copper ores are mined underground in the Burgaski Medni Mining and Ore-Dressing Combine, which is based in Burgas (the Burgas and Strandja mountain ore-field) and in open pits; in the Medet Mining and Ore-Dressing Combine in Panagyurishte, and by open-cast and underground methods in the Panagyurski Mining and Ore-Dressing Combine in Elshitsa village, in the Pazardjik district. Copper production has expanded following the commissioning of the Elatsite Mining Enterprise in Etropole. Flotation and one gravitational ore dressing factories have been built for the processing of non-ferrous metal ores into concentrates. Since 1969 a plant for bacterial copper leaching has been in operation. Lead and zinc are produced at the Georgi Dimitrov Lead and Zinc Plant in Kurdjali, which has been operating since 1955, and at the Dimiter Blagoev Non-Ferrous Metal Combine in Plovdiv, which was commissioned in 1961; copper is produced at the Georgi Damyanov Combine, in Srednogorie, which was put into operation in 1958. The treatment of non-ferrous metals and alloys is concentrated in the Dimiter Ganev Non-Ferrous Metal Processing Factory in Sofia (turning-out section, sheet, strip and wire material); aluminium is produced at the Stamen Stamenov Aluminium Processing Combine in Shumen.

The communications equipment plant in Blagoevgrad

4. Mechanical and Electrical Engineering and Electronics Industries

are a key industrial sector, accounting for 22.5 per cent of the total industrial output (1982).

Historical and archaeological studies indicate that metal processing in Bulgaria existed as early as the fourth millenium BC, when copper, bronze and iron artifacts (decorations and tools) were made. Mechanical engineering proper commenced only in the eighteenth century on the basis of such crafts as fulling and watch-making, and in the first half of the nineteenth century, when the first textile machines were made in Sliven and Plovdiv based on prototypes brought from Russia. After Bulgaria's liberation in 1878 the crafts began declining. The first mechanical engineering works in Bulgaria was the Russe Repair Shop (called Fleet Arsenal), founded in 1879. The first registered metal working enterprise was the Bratya Sahakyan Iron Foundry in Sliven, founded in 1881. Facilities were set up for the repair of railway cars, locomotives and mining equipment, and for the manufacture of metal products such as wire and wire products, horse-shoes, household utensils, beds, cooking ranges, etc. By 1912 there were 37 small metalworking enterprises with a few workers each. The 1912–25 period was characterized by stagnation in the development of the mechanical engineering and metal-working industries. After 1926 private metal-working enterprises belonging to shareholding societies were founded: Zhiti in Russe, Veriga in Burgas, Chilichna Ruka in Sofia, Bratya Panovi in Plovdiv, etc. There were also several state-owned enterprises for the repair of transport vehicles and mining equipment, and several munitions factories. In 1939 the number of the metalworking and mechanical engineering enterprises with more than ten employees reached 94, and the total number of people employed in this sector was 4,400, with metal-processing and mechanical engineering accounting for 2.4 per cent of the total industrial output. During World War II (1939–45) the mechanical engineering and metal-processing industry was in a state of stagnation, and there was a decline in the number of enterprises. Bourgeois Bulgaria could not establish a well-developed mechanical engineering and metal-working industry, and the country's needs for machines were met entirely through imports.

After the victory of the 9 September 1944 Uprising the small mechanical engineering and metal-working enterprises became more active, and after the nationalization of private industrial enterprises and mines in 1947, the engineering and metal-working enterprises were integrated and became specialized, and the organization of production was improved. More production facilities were built and the range of products was extended. Mechanical engineering played an important role in the restoration and development of the national economy. In 1948 the number of people employed in state and co-operative-owned mechanical engineering and metal-working enterprises reached 25,754 (10.2 per cent of industry's total work-force); output increased 620 per cent over 1939 levels. The metal-working industry was the largest. In the 1939–48 period the production of cast-iron products increased 211 per cent, of stoves 168 per cent, of metal fasteners 183 per cent, and of metal utensils 371 per cent. The inventory of metal-working products became more varied, and there was an increase in mechanical output. Lack of experience and largely inadequate production facilities and equipment determined its character. The articles and machines were rather simple and of inferior technical standards. The first sophisticated individual machines (locomotive, motor car, excavator, etc.) were also made on the basis of foreign prototypes, as the necessary technical documentation and qualifications were lacking.

During the first two five-year periods (1949–57) the semi-artisan mode of production was reconstructed along industrial lines. Bulgaria received economic, technical and scientific assistance from the USSR and other socialist countries. The plants for more sophisticated engineering products (machine-tools, ships, signal and power equipment, etc.) worked on the basis of Soviet technical documentation. Soviet specialists rendered technical and scientific assistance, and Bulgarian workers and engineers improved their skills in the USSR. The organization of production and the theoretical training of technical personnel were improved, and planning and dispatch

The telephone and telegraph equipment plant in Sofia

control were introduced. In 1952 the total industrial output increased 1,800 per cent over 1939 levels, and nearly 300 per cent over 1948 levels. The range of engineering products increased greatly and they became more sophisticated. About 70 kinds of machines and metal-working products were turned out in 1952. Ships, milling machines, motor-driven drills, ball-bearing mills, tractor-driven combine harvesters and other types of agricultural machines were added to the production list, as also were freight machines and passenger railroad cars, trolley-buses, X-ray equipment, electric meters, automatic telephone exchanges, etc. By 1958 the volume of production had increased 5,000 per cent over 1939 levels. New machines such as new types of tractor-driven combine harvesters, water-turbines, weaving-looms and hydraulic presses went into production. The first export of machinery took place in 1956: 22 lathes, 70 diesel engines, 32 water-pumps, 2 threshing machines, 50 freight cars and ships worth a total of 7.3 million levs. Despite this success, the mechanical engineering and metal-working industries were lagging behind the requirements of the national economy. Their share in the country's total industrial output in 1956 still remained low – 11.5 per cent. The development of typical mechanical engineering sub-branches lagged behind. The amount of capital investment was comparatively small.

The decisions of the April 1956 Central Committee Plenum marked a new stage in the development of mechanical engineering, to which considerable material and financial resources were allotted. New sub-branches appeared (instrument engineering, electronics, mechanical-handling engineering, etc.); the inventory of products expanded. In 1964 the rate of development of the mechanical engineering and metal-working industry exceeded that of industry as a whole by 0.4 points. The mechanical engineering and metal-working industry became the leading sector in the national economy. Production specialization and territorial decentralization took place and new engineering enterprises appeared. The following five-year periods were characterized by an accelerated, extensive development of mechanical engineering. In 1975 the number of state-owned enterprises increased over 500 per cent over 1960 levels, reaching 580, and fixed capital increased 580 per cent. As a result of international specialization within the CMEA the production of new machines and equipment was begun. Of special importance were electric trucks, electric hoists and some types of electronic equipment. Co-operation and bilateral specialization with other socialist countries and the USSR in particular provided the basis for the production of commercial vehicles, engine-driven industrial trucks, ships, tractors and agricultural machines, stage regulators, communication equipment, electronic devices, computer hardware, etc. The manufacture of machine-tools, hydraulic and electronic components and shipbuilding also made considerable progress. Substantial changes occurred between 1960 and 1975 in the structure of mechanical engineering production, when the share of mechanical handling engineering, the production of hydraulic components and machine-tools, computer engineering, electronics, etc. increased, and the export potential of mechanical enginering rose. In 1975 the value of engineering exports reached 1,847.4 million currency levs. The mechanical engineering and metal-working industry became a leading sector and a basis for technological progress and investment, and its role in the formation of the national income continued growing. A stage of intensive development of the mechanical engineering and metal-working industry started after the Eleventh BCP Congress (1976) and the National Party Conference (1978). In 1982 the total industrial output of the mechanical engineering and metal-working industry went up 400 per cent over 1970 levels. A 1978 UN survey placed Bulgaria eighteenth in terms of volume of engineering exports (in US dollars) among 42 leading engineering countries. During the Eighth Five-Year Plan period (1981–85) the mechanical engineering and metal-working industry continued to develop at priority rates. The mechanical engineering industry up-dates its own facilities and supplies the basis of other national economy sectors with highly productive automated machinery, automatic production lines, systems of machines and automated factory shops.

This industry is managed by the Ministry of Mechanical Engineering – the Balkancar Scientific-Production and Commercial Corporation, Metal Casting, Machine-Tool Plants, Food and Tobacco Industry Machines, Household Equipment, Shipbuilding, Hydraulics, Elprom, Resprom, Isot, Instrument Engineering and Automation, and other state economic corporations, the Technology of Metals Scientific and Production Corporation, the combines for heavy engineering, electric-hoist engineering, chemical engineering, etc., the Scientific and Production Combine for Semi-Conductor Devices in Botevgrad, the Mechatronika Scientific and Production Combine, etc.

The Kliment Voroshilov communications equipment plant in Sofia; colour TV sets

The Automotive Industry includes the assembly of motor-cars and lorries and the manufacture of buses, internal combustion engines, trailers, containers, two-wheeled vehicles, generators, assemblies and parts. Basic enterprises: Madara Lorries Combine in Shumen (including: the Madara Lorries Factory in Shumen, the Automobile Superstructures Factory in Preslev, the Pliska Wagon Trailer Factory in Novi Pazar, the Automobile Axle Factory in Shumen and the Development and Research Centre in Shumen) – responsible for the assembly of Soviet GAZ-53A commercial vehicles (since 1967); Czechoslovak Skoda heavy lorries (since 1970) under the trade name of LIAZ-Madara (since 1983); the Trailer Factory in Elena and Semi-Trailer Factory in Tolbuhin; the Chavdar Bus Combine in Botevgrad, manufacturing public transport buses based on a Bulgarian design; the Balkan Amalgamated Works in Lovech – assembling Soviet Moskvich motor-cars; and the Vassil Kolarov Engineering Works in Varna, manufacturing Perkins diesel engines under a licence agreement with the British company.

Electrical Engineering. This turns out electrical machines, generators, equipment and materials connected with the production, transportation, transformation and use of electric power. The basic enterprises are: the Vassil Kolarov and Nenko Iliev Cable and Wire Factories in Burgas and Sevlievo respectively, and the plants for insulated cable in Smolyan and bare wire in Sofia; the electric motor works for asynchronous electric motors in Plovdiv, Elprom in Troyan and Vassil Kolarov in Sofia; for electric machines, the G. Kostov in Sofia for DC machines,

The Beroe Robotics Research and Production combine in Stara Zagora

a micro-electric motor plant in Teteven, Elprom in Lovech, and Elektra in Sofia; for battery factories, the Metodi Shaterov in Pazardjik, Energiya in Turgovishte, Start in Tolbuhin; for high-voltage equipment factories, the High Voltage Equipment plant in Tolbuhin and a transformer plant in Sofia; for Low Voltage Equipment, the Electrical Equipment Plant in Plovdiv, and Equipment Factories in Assenovgrad, Perushtitsa, Brezovo, etc. Other factories include the Iskra Low Voltage and Power Electronics plant in Sofia, Elektrik for Low Voltage Complete Equipment in Radomir, a plant for electrical insulation materials in Russe, the Naiden Kirov Plant for electrical insulation articles in Russe, Elprom for porcelain electrical insulators in Kubrat, Dinamo for motor-car starters and generators in Sliven, Avangard for custom-made equipment in Sevlievo, and the Lenin electrical porcelain factory in Nikolaevo.

Power Industry Engineering. This manufactures power production machinery, generators and equipment (steam and water boilers, steam, gas and water turbines, heat-exchangers, condensers, electrical turbine generators, etc.) for thermo-electric power stations, condensation electric power stations, hydro-electric power stations, nuclear power stations and piston engines (except those of the automobile and tractor industries). The basic enterprises are: the Georgi Kirkov Boiler Works in Sofia (the largest specialized power industry machinery enterprise in Bulgaria) for heating, water heating, industrial, power and green-house boilers; the Kalorima Factory in Tolbuhin, for heating boilers, torches, water heaters, generators, etc,; the Spartak Factory in Burgas, for combustion equipment (torches and burners for heavy- and light-fuel oils); the N. Y. Vaptsarov Factory in Pleven, for Francis and Pelton-type water turbines; and the V. Kolarov Factory in Varna, for Diesel engines (since 1973 producing engines on a licence agreement with Perkins).

Railway Engineering. This manufactures railway carriages, machinery and equipment for rail transport, and repairs rolling stock. The basic enterprises are: the G. Dimitrov Railway Works in Sofia (for freight cars, truck chassis, rail points, fasteners, etc.); the Cherveno Zname Carriage Factory in Burgas, for different types of open and roofed freight cars; the A. Zhdanov Wagons Works in Dryanovo, for passenger carriages; the V. Kolarov Locomotive and Wagon Works in Russe; the Railway Transport Automation and Telemechanics Factory in Sofia; the Karlovo Wagon Repair Works in Karlovo; etc.

Shipbuilding and Ship Repairing Industry: includes the building and repair work of ships and other vessels. The basic enterprises are: the Georgi Dimitrov Shipbuilding Yard in Varna (the largest in Bulgaria), the Ivan Dimitrov Shipbuilding and Ship Repair Yard in Russe, the Iliya Boyadjiev Shipbuilding Works in Burgas, the Varna Ship-repairing Works (the largest in Bulgaria); and the Dragni Nedev Ship Repair Yard in Burgas. The shipbuilding industry specializes in the manufacture of 25,000 and 38,000 ton bulk carriers, container carriers (for 400 containers), 2,000 and 5,000 ton river and sea-going

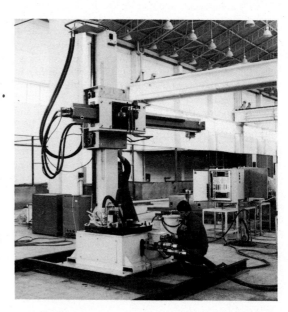

The Beroe Robotics Research and Production combine in Stara Zagora: an RB-232 industrial robot

tankers, reinforced concrete ship repair works, all-purpose dry cargo vessels, food carriers, sea-going tugs and other modern types of vessel. The first 100,000 ton tanker was built in 1977. Bulgarian is exporting vessels to the USSR, Poland, Cuba, Norway, Great Britain, Finland, the USA, Greece, Holland and other countries.

Mechanical Handling Equipment. Manufacture comprises

mechanical-handling machines for interrupted-operation electric-hoists (rope and chain), cranes (bridge, portal, bracket, tower-cranes), coilers and reelers, winches, elevators, stacking-lifts, lifts (passenger and cargo), electric trucks, engine-driven trucks, electric carriages, hand-operated trucks, etc.; and mechanical-handlers for continuous operation – conveyors with a driving element (belt, chain, elevators, pipe-line transport, etc.), and conveyors without a driving element (gravitation, roller, screw, inertia, etc.).

After 1948 the production of mechanical-handling vehicles for universal application was launched. Since 1956 Bulgaria has become specialized in the manufacture of mechanical-handling electrically-driven vehicles within the CMEA. Industrial trucks are central to the structure of production. Electric trucks are manufactured at the Shesti Septemvri, Dunav and Electric Truck Combines in Sofia and Lom, the Sredets Electric and Engine-Driven Trucks Plant in Sofia, and the Karlo Lukanov Plant, Pleven; engine-driven trucks are manufactured at the Record Plant, Plovdiv and the Sredets Plant, Sofia; electric rope hoists – at the Hoist-Building Works in Gabrovo; warehousing equipment (electric rope hoists, stacking-lifts, bracket cranes, etc.) – at the Yantra Warehouse Works in Gorna Oryahovitsa; cranes are manufactured at the Hristo Smirnenski Construction Equipment Works in Sofia, the Stoyan Popov Works in Pazardjik, etc.; grapples and mine-lifts – at the Struma Mechanical-Handling Works in Pernik; complex transportation systems (for coal and other bulk materials), rubber belt conveyors, etc. – at the Chervena Zvezda Mechanical Engineering Plant in Debelets; conveyor belts, mast elevators and electric winches – at the Mir Mechanical-Handling Plant in Mihailovgrad; fork-lift trucks – at the Petko Kunin Machine-Building Plant in Pavlikeni; elevators and blade-conveyors – at the Cherven Machine-Building Plant in Russe, chain-and-plate and screw conveyors – at the Purvi Mai Machine-Building Plant in Polski Trumbesh, etc.; lifts – at the works in Sofia and Stanke Dimitrov. In the manufacture of industrial trucks Bulgaria ranks third in the world, after the United States and Japan (it turns out about 20 per cent of world output); it ranks first among the socialist countries in the manufacture of electric hoists.

In 1982 more than 93 per cent of the output of electric trucks was exported to 32 countries; about 80 per cent of the output of engine-driven trucks was exported to 39 countries and more than 95 per cent of the output of stacking-lifts was also exported – mostly to the socialist countries.

Joint companies and firms have been set up with different capitalist countries; Sibicar (Italy), Technocar and Sofbim (France), Irioncar (FRG), Nordcartruck (Sweden), Nichibu Baleast (Japan), Rastem (the Netherlands), BKC Impex (Great Britain), etc. Technical co-operation has also been established and maintained with leading western firms through licences and exchange of technical expertise.

The Manufacture of Ventilation and Purification Equipment: includes the manufacture of fans and blowers of all-purpose and specialized designation, de-dusters, dust arrestors, silencers, purification stations (for water purification by biodiscs, and scrubbing of technological and ventilation gases by electric filters), sleeve filters, electric filters, surface and floating aerators, sand-blast equipment, mud-cleaning equipment, screw water conveyors, intake units with silencers, etc. The production of fans and ventilation equipment is concentrated in the Spartak (Burgas), Briz (Grudovo) and Klokotnitsa (Dimitrovgrad) engineering works, and the production of purification equipment in the Bistrets engineering works in Kurdjali.

The Georgi Dimitrov agricultural machine works in Russe

The Manufacture of Electronic Components: includes the manufacture of capacitors, resistors, ferromagnets, connectors, switches and other electronic elements in the plants in Sofia, Kyustendil, Smolyan, Pernik, Botevgrad, Aitos and Devin.

The Production of Computing and Office Equipment. Computing equipment has been manufactured since 1961 – external memory units, data tele-processing equipment and central processors (basic fields of Bulgaria's specialization in the Inter-Governmental Commission of the CMEA, including the USSR, GDR, Czechoslovakia, Poland, Hungary, and Bulgaria), microcomputer devices (invoice producing and accountancy machines, text processing machines, etc.), problem oriented complexes, electronic calculators, electronic cash registers, typewriters, etc.; registration equipment (since 1962) – registration (pending, standing, diagonal, filing); equipment for computer centres, libraries, desks and places of work; punching equipment, etc.; and registration equipment where data is mechanically retrieved on a semi-automatic principle. The basic manufacturers of computer equipment are the

Electronika Plant in Sofia, Orgtehnika in Silistra, and plants for computer equipment in Sofia, for memory devices in Stara Zagora, Plovdiv and Veliko Turnovo and for typewriters in Plovdiv. Office equipment is produced in the Samokov Registration Equipment Factory. Over 80 per cent of the products of the three basic lines of computer equipment is designed for export to the socialist countries for the assembly of computers in unified computer and unified minicomputer systems, established in 1974. Part of the registration equipment products are also exported to the socialist countries.

The Production of Instruments and Tools. Metal-working tools are made in the Bolshevik Tools Factory in Gabrovo, the Machine Tools Factory in Vidin and the Tools Factory in Dryanovo; wood-working tools – in the Tools Factory in Smolyan; abrasive tools – in the Dimiter Filipov Abrasive Tools Factory in Berkovitsa; measuring instruments – in the Measuring Tools and Instrument Factory in Blagoevgrad; equipment for operation under pressure – in the Pretsiz Factory in Mihailovgrad and the Factory for Custom-Made Equipment in Aprilovo village, Turgovishte district; pneumatic tools – in the Pnevmatika Engineering Works in Kurdjali; electric tools – in the Elprom Factory in Lovech, and elsewhere. Finished products are exported to the USSR, GDR, Poland, Cuba, France, Italy, Sweden, FRG and other countries.

The tractor plant in Karlovo

The Production of Timber and Wood-Working and Processing Machines. This sector manufactures saw-mills, feeders, standard wood-processing machines (band-saw), jointing machines, thicknessing machines, circular saws, joiners' lathes, milling machines, etc.), furniture-making machines (end-piece gluing machines, multilateral automatic drills, two-band polishing machines, two-side automatic edge gluing machines, etc.), automatic furniture-making machines, etc. The main woodprocessing machine factories are the Deveti Mai in Plovdiv, and plants in Yakoruda, Dospat and Svilengrad.

The Manufacture of Machines and Equipment for the Metallurgical Industry. This sector turns out machinery and equipment for the metallurgical industry (blast furnaces, steel foundries and coking installations for ferrous-alloy production, individual machines and equipment for rolled steel production), for the non-ferrous industry (lead-zinc works, production of black electrolyte copper, nickel and other non-ferrous metals), and foundry equipment. The basic enterprises are: the Struma Mechanical Engineering Works in Pernik, Chervena Zvezda in Debelets, Ilinden in Pleven, Stomana in Silistra.

The Manufacture of Machinery and Equipment for the Mining and Extractive Industries. This sector manufactures machines and equipment for underground works: drilling equipment (borelogs, drilling carriages, etc.), mine locomotives, equipment for ore and coal loading and transportation (conveyors, shaft underground machines, endless-rope machines, rope-changing machines, mining machine reduction gear, rubber conveyor belts, etc.), filling machines, platform hoists, grab buckets, mining drills, pneumatic hammers, etc.; machines and equipment for open-cast mining – complete conveyor systems for coal and earth moving works, winches, spreaders, etc.; crushing and grading equipment: crushers (jaw, conical, rotary, eccentric), crushing and grading installations, conveyors, sieves, etc.; ore-dressing equipment: crushers (ball, fan, hammer, roller), feeders, classifiers, vacuum filters, compressors, flotation machines, batching units, etc. The basic enterprises are: the Struma Engineering Works in Pernik, Pnevmatika in Kurdjali, Spartak in Burgas, Briz in Grudovo, Chervena Zvezda in Debelets and Komsomolets in Kurdjali.

The Manufacture of Medical Equipment, Apparatus and Instruments. The range of production of this industry includes X-ray equipment; physiotherapeutical and electronic equipment; dentistry equipment; sterilization, disinfection and washing equipment; laboratory and chemist shop furnishings; medical instruments and syringes; miscellaneous tools and devices. The basic enterprises are: the Electrical Medical Equipment Factory and Medical Apparatus Factory in Sofia, the Druzhba Medical Furnishings Factory in Varna and the plants in Topolovgrad. Part of the equipment (dentistry, phonocardioselector, Tonus-2 apparatus for diadynamic current treatment, cardiosignalizer, rheopletismograph, chemist shop furnishings, etc.) is exported.

The Manufacture of Metal Products. This sector produces assembled steel sections, steel casing and tubular scaffolding, greenhouse equipment, pallets and box-pallets, industrial and plumbing fittings, household plumbing equipment, sanitaryware products, metal binders, fasteners, wire and wire products, chains, railway building materials, heating radiators, steel packing, agricultural implements, household metal utensils, industrial vessels, cutlery, stoves, steel beds and couches, bed springs, etc. The basic factories are: the Gocho Grozev in Plovdiv, Zhiti in Russe, G. Dimitrov in Burgas, Ihtimanska Komuna in Ihtiman, Vuzhod in Popovo and Stoyan Buchvarov in Sevliero. Part of the output is exported to the USSR, GDR, Czechoslovakia, Hungary, Yugoslavia and other countries.

The Manufacture of Metal-cutting and Forging Machines and Equipment: includes lathes, milling machines; grinding machines; jig-boring machines; gear-cutters, shaping machines, slotter and broaching machines; metal cutters; special and specialized machines; automatic and semi-automatic process- and assembly-lines; automatic modules, machining centres; forging machines and equipment. The largest share of this production belongs to lathes, which are made in the Machine-Tool Plant in Sofia, the Large-Size Machine-Tool Factory in Pernik; the Machine-Tool Factory in Sliven; and the Mashstroi Machine-Tool Factory in Troyan. Drilling and boring machines are made by the Metalik Machine-Tool Factory in Pazardjik; transfer machines by the Stefan Karadja Machine-Tool Factory in Russe and the Machine-Tool Factory in Haskovo; cutters in the Machine-Tool Factory in Silistra; grinding machines in the Machine-Tool Factory in Assenovgrad; assemblies and parts for machine tools in the factories in Nova Zagora, Velingrad, Stara Zagora and elsewhere; forging machines in the N. Y. Vaptsarov Engineering Works in Pleven and the Forging-Machine Factory in Belene. Lathes are exported to the GDR, Poland, Romania, Hungary, the FRG, Italy, Turkey, Syria, Egypt and other countries.

The Manufacture of Castings and Forged Parts: includes castings of ferrous metals, cast iron (grey, inoculated, hydraulically solid, malleable), steel (carbon, alloyed and manganese alloyed); non-ferrous metal castings (bronze, aluminium, brass); press and stamp forgings of ferrous metals. Cast-iron has the highest share, manufactured by the Foundry in Ihtiman, the Veslets Foundry in Vratsa, the Factory for Malleable Cast-Iron in Lovech, the Progress Cast-Iron Foundry in Stara Zagora and the Foundry in Pleven. Steel castings are made in the Steel Foundries in Pleven, Pernik, Rakovski, Elektrometal in Sofia, the Dynamo Precision Steel Casting Factory in Sevlievo; non-ferrous castings – in the Aluminium Casting Factory in Pleven, the Stoyan Buchvarov Household Equipment Factory in Sevlievo, the Sakar Factory in Svilengrad; press moulds are made in the Surp i Chuk plant in Stara Zagora, the Madara Truck-Construction Works in Shumen, the Podem Electric Hoists Factory in Gabrovo and other engineering works.

Spare Parts Production. This range of manufacturing includes parts, assemblies, units and other elements required for the repair and maintenance of the machinery and equipment in industry, transport, communications and other sectors of the national economy, and of household appliances. The largest share in the range of production belongs to the spare parts for mechanical handling machines, motor vehicles and agricultural machines. In 1982 over 60 per cent of the machine-building and metal-working plants produced spare parts. About 30 per cent are exported.

The Production of Means of Automation: includes instruments and devices for the automation of production and technological processes; measuring and regulating instruments for electric and magnetic quantities; radio-measurement instruments; starting and regulating equipment for household and industrial application; machines for automatic technological process control; scales and balances; scientific research equipment for environmental parameter measurement and control. The basic enterprises are: the Belassitsa Instrument-engineering Works in Petrich, the Analitik Factory in Mihailovgrad, the Ferdinand Kozovski Factory for Actuators in Knezha, the Elia Factory in Nilopol, the Measuring Instruments and Meters Factory in Blagoevgrad, the Elektra Works in Sofia, the Automatic Weighter Works in Liyaskovets, the Instrument Engineering Works in Pravets, and the Ziti Works in Russe, the Kocho Tsvetarov Instrument-engineering Works in Plovdiv and a further plant in Koprivshtitsa. Part of the output is exported to the USSR, Czechoslovakia, Hungary, the GDR, Great Britain and other countries.

The Manufacture of Hydraulic and Pneumatic Equipment: includes axial piston pumps, gear pumps, gear motors, hydraulic power cylinders, brake cylinders, hydraulic and pneumatic distributors, valve throttles, hydraulic regulators, units, filters, hydro-servocontrol units, hydraulic systems for machine-tools and forging presses, hydraulic components for robots, hydraulic systems for automanipulators, hydraulic parts for steering units, hydraulic and pneumatic systems for mechanical handling equipment, pneumatic systems for robotics, control hydraulics, pneumatic machines and tools, rammers, etc. The basic enterprises are: the Hidravlika Engineering Works in Kazanluk, the Pnevmatika Works in Kurdjali, the Garant in Byala Slatina, the Budashtnost in Chirpan, the Boncho Shanov in Kazanluk, the Ivan Tenev Hydraulic Component Works in Yambol, factories in Elhovo and the village

The petro-chemical works in Pleven

The petro-chemical works in Burgas

of Tenevo near Yambol, the Control Hydraulics Works in Shipka, and the Mizia Accessories Works at Byala Slatina. Their products are exported to about 40 countries.

Refrigeration and Air Conditioning Equipment Production: includes domestic refrigerators, refrigerating chambers and installations, compressors and compressor-condenser generators, prefabricated cold stores and cold store equipment, air conditioners, burners and air heaters, etc. Refrigeration equipment is manufactured in the Anton Ivanov Works in Sofia, the Zora Engineering Works in Turgovishte and the Domestic Refrigerator Works in Slivnitsa.

Radio and Communications Engineering: produces radio electronic equipment, TV and wireless sets, low-frequency amplifiers, electro-acoustic converters, applied TV cameras, display monitors, TV switch stations, electronic equipment for farm-machine control and monitoring, transformers, stabilizers, collective aerial systems, industrial and technological electronic equipment and systems, communications equipment – automatic telephone exchanges, multiplex telephone systems, radio relaying systems, telephone sets, conference public address systems, switching parts and assemblies and radio telephones. The basic enterprises are: the Telephone Works in Belogradchik, the TV and Radio Set Works in Veliko Turnovo, the Electro-acoustics Factory in Mihailovgrad, the Loudspeakers Factory in Blagoevgrad, the Telephone and Telegraph Engineering Works in Sofia, the Telephone Equipment Factory in Bansko, the LV Relay Factory in the village of Banya in the Blagoevgrad district, the Plastic Parts Factory in Belitsa, and the Components Factory in Blagoevgrad. Over 50 per cent of the finished products are exported, mainly to the USSR.

Robot Engineering. This sector of industry produces industrial manipulators (gantry, standing and manually-balanced types) with different lifting capacities for the automation of auxiliary operations as separate units of machine-tools and of process and assembly-lines, including automated ones, of various designations; industrial robots, designed in Bulgaria for the automation of auxiliary operations in metal-cutting, forging, foundry and other machines, for painting, lacquer and thermo-insulation coating, and for other labour-intensive operations; industrial robots for the automation of basic and auxiliary processes of feeding metal-cutting, forging and foundry machines, and of glass and plastics manufacture, the application of lacquers, palletization and other processes, and on licence agreement with the Fanuc Company (Japan) for the automation of mechanical-handling operations of machine-tools, palletization, conveying and other processes; industrial robots for integrated machines; industrial manipulators and robots for automated process and assembly-lines, automatic technological modules, machine systems, etc. The basic producer is the Research and Production Combine of Robotics in Stara Zagora.

Heavy Investment Engineering. This sector of industry manufactures a complete range of equipment for the metallurgical, mining and thermo-power generating, civil engineering, cement, chemical and other industries. The basic enterprises are: the heavy mechanical engineering works in Russe and Radomir, the Struma in Pernik, Chervena Zvezda in Debelets, Mir in Mihailovgrad, N. Y. Vapsarov in Pleven, the Georgi Kirkov Boiler Plant in Sofia, and the plants for the production of machines for the chemical industry at Haskovo, Pleven and Provadia.

Agricultural Machine-Building: includes tractors and self-propelled chassis, soil penetration and cultivation machines, seed-drills, planting and fertilizer spreading machines, plant protection machines, water spraying machines and sprinkler installations, harvesters and grain-cleaning machines, devices for the mechanization of livestock-breeding and poultry farms, agricultural tools and implements. The basic enterprises are: the Georgi Dimitrov Mechanical-Engineering Works and the Tractor Assembly Plant in Russe, the Mechanical-Engineering Works in Kubrat, the Zavet near Razgrad, the Cherven Works in Russe, Purvi Mai in Polski Trumbesh, Slia in Yambol, Perla in Nova Zagora, Karnobatska Komuna in Karnobat, and the works in Dulovo. A part of their output is exported to the USSR and the other socialist countries.

The Manufacture of Machines for Civil and Road Construction: includes the mechanization of processes and the production of building materials, excavators, graders, bulldozers, scrapers, ditch diggers, road-building motor-driven rollers, complete technical equipment for brick and tile factories, cement factory installations, lime mixers, lime slaking machines, tesselated flooring machines, sand washers, parquet flooring polishers, vibrating screens, mortar mixers, asphalt mixers, building elevation cradles, painting machines, cradle hoists, tie-bars, tubular scaffoldings, equipment for house-building combines, etc. The basic enterprises are: the Hristo Smirnenski Crane Construction Factory in Sofia, the Petko Kunin Engineering

Works in Pavlikeni, Chervena Zvezda in Debelets, Mir in Mihailovgrad, Emil Popov in Veliko Turnovo, Struma in Pernik, the construction equipment works in the village of Glavanitsa near Silistra, in Aitos and Pazardjik.

Textile Machine-Building. This sector manufactures spinning machines (double-twisting machines, doubling machines, ring spinning frames, spindleless spinners, hollow-spindle Prenomit spinning machines); weaving equipment (shuttle and shuttleless weaving-looms); weaving-loom assemblies (including assemblies for automatic textile machines made in Klimovsk and Cheboksari in the USSR), and other textile machines and equipment. A basic enterprise is the Pobeda Mechanical-Engineering Works in Sliven, a part of whose products are exported to the USSR, Czechoslovakia, the GDR, Romania, Albania, Vietnam, Mongolia, Hungary and Nigeria.

The Manufacture of Machinery and Equipment for the Chemical Industry: includes complete process lines and installations and individual machines and equipment for the chemical industry. The basic enterprises are: the mechanical-engineering works in Haskovo, Pleven and Provadia, and the Plastics-Processing Machine-Building works in Levski. Part of the output is exported to the USSR, the GDR, Romania, Cuba, Iraq and Morocco.

The Manufacture of Machinery and Equipment for the Food Industry: includes machinery and equipment for the cannning industry; bottling and packaging installations, dairy equipment; machinery for the wine and spirits industry and filter press equipment; machines for the bread-making and the tobacco-processing industries, including complete installations for the food and canning industry. The basic enterprises are: the Cherveno Zname, Tacho Daskalov and Zheleznik in Stara Zagora, Belitsa in Kilifarevo, Khan Kubrat in Kubrat, Angel Peev in Chirpan, Vurbitsa in Mochilgrad, and the Avtomatika and Mlada Gvardia in Haskovo. Food industry equipment is exported to the USSR, the GDR, Mongolia, Poland, Romania, Hungary, Czechoslovakia, Cuba, Vietnam, the People's Democratic Republic of Korea, Yugoslavia, Greece, Iraq, Iran, Cyprus, Nicaragua, Pakistan, Sudan; tobacco industry equipment goes to the GDR, Poland, the USSR, Romania, Czechoslovakia and Iran.

5. The Chemical and Rubber Industry. Before 9 September 1944 the chemical industry was poorly developed. There were small private enterprises for the production of essential oils, perfumes, soap, spirits, varnish, dyes, calcium carbide, salts and pesticides; installations for oil processing, supplied by Romania; and for the production of lubricants (greases) and emulsion oils, of bakelite on the basis of formaldehyde resin and other chemical products. The foundations of the modern chemical industry were laid after the nationalization of private industrial and mining enterprises in 1947. The following sectors of the chemical industry developed: the production of inorganic chemical products, oil processing and oil chemistry, production of intermediate products of organic synthesis, chemical fibres, plastics, synthetic resins, and rubber, production of varnish and oil dyes, pesticides, synthetic dyes, chemical pharmaceuticals, perfume and cosmetic preparations, photo and film materials, reagents, detergents and cleansing reagents, and auxiliary materials.

The country produces all major inorganic chemical products: acids – hydrochloric, sulphuric, nitric, phosphoric; alkali, sodium, potassium, ammonium; salts – sulphates, nitrates, carbonates, and phosphates; sodas – soda ash and bicarbonate, chlorine, ammonia; mineral fertilizers – simple (nitrogen and phosphorus) and composite (containing nitrogen and phosphorus). The main producers are the industrial chemical plants in Dimitrovgrad, Devnya, Stara Zagora and Vratsa. The first large inorganic chemistry plants were built and put into operation in the period between 1950 and 1955 – the Chemical Combine in Dimitrovgrad (today the Industrial Chemical Combine) and the Karl Marx Chemical Works (today the Karl Marx Soda Plant) in Devnya. In 1963 the Nitrogen Fertilizer Plant (now the Industrial Chemical Combine) in Stara Zagora was commissioned; in 1966 – the Chemical Combine (now the Industrial Chemical Combine) in Vratsa. In 1969–80 another three chemical works of base chemistry were built up and commissioned: the Soda Ash Chemical Works (the biggest in Bulgaria and Europe, in operation since 1973), the Mineral Fertilizers Plant (in operation since 1974, today renamed the Plant for Composite Fertlizers and Plant for Phosphorus Fertilizers) and the Chlorine and Polyvinylchloride-2 Plant, in operation since 1980. In 1982 the total production was 458,976 tons of nitrogen fertilizers calculated as 100 per cent nitrogen; 238,932 tons of phosphorus fertilizers as 100 per cent P_2O_5; 1,458,700 tons of soda ash (as 98 per cent product); and 915,917 tons of sulphuric acid and other chemical products. The production of a considerable part of inorganic salts, pesticides and insecticides, chemi-

The Sviloza combined chemical works in Svishtov

cally pure substances and reagents is concentrated in the Angel Vulev Chemical Plant in Yambol, the Agrya Chemical Works in Plovdiv, the Krasnaya Zvezda Factory in the village of Vladaya in the district of Sofia, and the Himik Factory in Razgrad. Inorganic chemical products are exported to the USSR, Hungary, Czechoslovakia, Romania, Cuba, Vietnam, Yugoslavia, the FRG, Italy, Spain, Turkey, Greece and many other countries.

Several oil refining chemical plants have been built: the Leon Tadjer Oil Refinery in Russe was commissioned in 1953; the Oil Refinery in Burgas in 1965 and the Petro-Chemical Plant in Pleven for processing the product of the oilfield near the village of Dolni Dubnik, as well as oil supplied from the USSR. Different kinds of gasoline, diesel fuel, diesel fuel oil, mazut (black oil), mineral oils, liquid paraffins (from C_{10} to C_{17}), solid paraffins, kerosene, bitumen and city gas are produced. As regards oil processing, Bulgaria belongs to the group of countries with a developed petrochemical industry.

The Dimitrovgrad chemical works; cracking towers

The Petrochemical Industry. In 1964–67 the first petrochemical plant was commissioned, in Burgas. This plant, now called the Neftohim Petrochemical Combine, is the largest in Bulgaria, in which is concentrated the production of plastic materials and synthetic resins – high and low-pressure polyethylene, polystyrene and polypropylene. Its section in Shumen, the Panayot Volov Chemical Works, produces bakelite powder and phenolformaldehyde resin. The Industrial Chemical Combine in Devnya produces emulsion and suspension on polyvinyl chloride (PVC); the Dimitrovgrad Chemical Combine produces urea-formaldehyde resins; the Dimiter Toshkov Industrial Chemical Combine, Sofia, produces epoxies, phenol-formaldehyde and cresol-formaldehyde resins; the Gavril Genov Industrial Chemical Combine, Russe, produces unsaturated polyester, alkyd, alky-phenol and urea-formaldehyde resins. The plastic materials and resins produced are wholly processed in the enterprises of the *plastics processing industry,* the Peter Karaminchev Industrial Chemical Combine, Russe; the Kapitan Dyado Nikola Industrial Combine for plastic materials processing, Gabrovo; the Assenova Krepost Industrial Combine for Polymer Processing, Assenovgrad; the Botevgrad Industrial Chemical Combine and the Narodna Republika Industrial Combine for plastic materials processing, Sofia. Plastic materials are produced for household needs, for construction and industry. Among the plastic materials exported are plastic separators, for batteries and accumulators, artificial leathers and furs, pipes and other items. The production of synthetic chemical fibres started in 1968 with the commissioning of the Polyester Fibres Plant in Yambol, at present a section of the Dimiter Dimov Industrial Chemical Combine in Yambol. The production includes polyester fibres with the trade mark Yambolen and consists of staple fibres, cotton, woollen and linen, and converter cable. In 1978 a new, Polyester Silk Plant for textured and smooth silk, was put into operation, to cater for the needs of the textile industry and has an annual output of 6,500 tons. Polyamide fibres with the trade mark of Vidlon (textile consisting of silk, polyamide cord and technical silk) are produced at the Vidin plant, now called the Chemical Industrial Combine, built in 1966–70. They meet the needs of the home market, while part of the cord produced is for export to the USSR. Polyacrylnitrile fibres with the trade mark Bulana are produced in the Bulana Chemical Works, built in 1967–69 as part of the Petro-Chemical Combine in Burgas, and which is now a separate line at the Neftohim Industrial Chemical Combine in Burgas. Part of the polyacrylnitrile fibres are exported to the USSR, Poland, Yugoslavia, and other countries. Synthetic fibres are also produced at the Sviloza Industrial Chemical Combine in Svishtov, the Staple Fibre Plant, built in 1973, and the Synthetic Silk Plant, built in 1976. Staple fibres are produced (40,000 tons annually, including 14,000 tons of polyfibre and 26,000 tons of strengthened fibres) and also viscous silk (5,000 tons annually). The synthetic fibres meet mainly the needs of the home market. The total output of chemical fibres for 1982 was 105,215 tons.

The Pharmaceutical Industry and Industrial Micro-biology are among the most dynamic sectors of the chemical industry. The pharmaceutical industry began to develop in 1932 when a Production Laboratory for galena forms was set up, the Bulgarian Pharmaceutical Co-operative Society, called Galenus, in Sofia. Tablets with imported substances were produced. After 9 September 1944, the chemical pharmaceutical industry developed at an accelerated rate. Preparations were produced for the treatment of cardio-vascular, gastric and intestinal conditions; infectious, nervous, cancerous and other diseases; as well as a wide spectrum of antibiotics for human and veterinary medicine; chemical therapeutics, vitamins, spasmolytics, antipyretics, psychotropic, neuropic and biostimulating drugs. With the foundation of industrial microbiology in 1954 the production of antibiotics (tetracycline, oleandomycin, tetraolean and gentamycin), developed rapidly. There was also a rapid increase in the production of semisynthetic penicillins – ampicylin, tubocine, amopen, uromycin and cephalosporin antibiotics. In terms of production of antibiotics, of anti-TB preparations and phytochemicals, Bulgaria ranks among the first countries in Europe. Among the cardio-vascular medicinal preparations the best known are isodynite, nitrolong, chlophazoline, chlophadone, and stenopryl. In wide general use are the phytochemical preparations – nivaline, glauvent, alcyde V, vincapane, cratemon and gyrosital, which are prepared from the Bulgarian natural attar of roses, rosonol and maralite. Other pharmaceutical items are antiseptic plasters, cytoplast, sanplast, bandages, antimicrobal polyamide catgut, and antimicrobal polyamide sheets, and the effective skin substitute Pharmexplant. Over 300 medicinal preparations are produced to satisfy the needs of livestock-breeding – intra-mammary, intra-uterine, water soluble, deposit and aerosol preparations, and biostimulators; the antibiotics biovite, zinc-bacytracine and tylosine. The main enterprises of the chemical pharmaceutical industry of industrial and industrial microbiology are the Antibiotics Research and Production Combine in Razgrad and the Chemical Pharmaceutical Combines in Sofia, Stanke Dimitrov and Troyan; the Dressing Plant in Sandanski, the Plant for Microbial Preparations in Peshtera, and the fodder yeasts plants in Razgrad and Dolna Mitropolya. The Microbial Preparations Plant in Peshtera produces preparations for plant protection, dipel and the indispensable amino-acid L – lysine (bioconcentrate for livestock-breeding); the Razlog and Dolna Mitropolya plants produce fodder yeasts, an important protein component in the production of fodder mixes for stock-breeding, with a high content of protein – about 50 per cent. The production of alkaline protease (for detergents) and of pivosine (for brewing) was started in the microbial preparations shop in Botevgrad. Technologies for the production of enzymes for the food industry – alpha-amylase for bread-making, acid protease for wine-production and brewing and pectinase for fruit-juice production have also been developed.

The Essential-Oil Industry is one of the oldest industries in Bulgaria. Its beginnings were laid in the seventeenth century when the oil-bearing (oleaginous) rose was carried over to the Balkan Peninsula. Whereas in Ancient Persia and India only rose-water was obtained from the leaves of the oleaginous rose, the Bulgarian population developed an original technology for the production of rose oil by multiple redistillation on the basis of distillation equipment called alambics (copper cauldrons with a capacity of 120 litres), which consisted of several pieces of distillation equipment, called gyulpans. The production of gyulpans was interrupted in 1935. At the end of World War II there were 88 rose-distilleries in the districts of Karlovo, Kazanluk and Plovdiv. The annual output of rose-oil reached 4,000 kg (1903–5); later it varied between 1,000 and 2,000 kg, depending on the

The Sviloza combined chemical works in Svishtov

crop and the demand of the international market. Bulgaria has won recognition as a major producer of the highest quality rose-oil in the world and until 1950 had been its greatest exporter. After 1948 the old installations were replaced by new ones, and since 1975 the installations have been made of stainless steel with a capacity of 10,000 litres. Cultures such as lavender, seed of *fructus foeniculi* and pine cuttings are cultivated to be processed for several months or even for whole year. Peppermint and dill oil is produced exclusively for export, as well as other natural aromatic compounds – concretae (solids) – rose, lavender, resinoids based on resinous plants, absolutae based on concretae and resinoids, and extracts; from these several by-products are obtained, namely bases, compositions and essences. Anathol is produced, based on the *fructus foeniculi* seed, menthol – on peppermint, and hops extract – on hop cones. Synthetic aromatic products are obtained in small quantities. The Essential Oil Industry follows the guidelines supplied by the Bulgarska Rosa Research and Production Combine in Kazanluk.

The Perfume and Cosmetics Industry. The perfume and cosmetics enterprises, founded in Sofia before 9 September 1944, were the property of foreign firms Liszt, L. T. Piver, Mousson Lavendelle, Nivea, and Messeter. In 1892 Papazov founded (in Plovdiv) the first Bulgarian perfumery factory, P. Papazov, which took part in the 1892 Plovdiv Fair and was awarded a gold medal for phial-attar of roses. After 1921 the Nikola Chilov Works in Kostinbrod started the production of toilet soap and cosmetics. The Aroma Joint-stock Company was set up in 1931 at the head of the largest perfume and cosmetics enterprises in Bulgaria. After the nationalization of industry in 1947 all the perfume and cosmetics enterprises were amalgamated. Production was concentrated in the

The combined chemical works in Stara Zagora

Aroma Combine in Sofia and the Alen Mak Combine in Plovdiv. A factory for de luxe cardboard packing was opened in the village of Lessichevo, in the Pazardjik district, and a Decorative Cosmetics Plant in Rudozem. The following items are produced: eaux de toilette, perfumes, hair preparations (oil emulsions, nourishing creams, hair dyes, dye-stuffs, hair-sprays, permanent wave conditioners and shampoos), ladies', men's and children's cosmetics, body shampoos, lotions, depilatories, face packs, creams, toilet milks, gels, lipsticks, eye make-up, nail polish, shaving creams, tooth paste, and others. These cosmetics are exported to over twenty countries, mainly to the USSR. Other items of the chemical industry are the plant protection preparations produced in the Agrya Chemical Works in Plovdiv, photo and film materials at the Phohar factory in Sofia, synthetic dye stuffs at the chemical works in Kostenets, and detergents and cleansing reagents at the Industrial Plant for Household Chemical Products in the village of Ravno Pole, in the Sofia district.

The Rubber Industry includes the production of rubber shoes and technical wares. Before 9 September 1944 only rubber shoes and insignificant quantities of automobile and bicycle tyres were produced in the factory of the Bakish Joint-Stock Company in Sofia. After 1947, production was expanded through the reconstruction of the existing facilities and the construction of new plants. The Georgi Dimitrov Automobile Tyre Plant in Sofia specialized in the production of automobile and electric fork-lift truck tyres and inner tubes. New factories for technical rubber wares were built – the Technical Rubber Factory in Kula (1962) for the production of semerings, O-rings, sleeves, porous and solid sections, etc. as required by industry; the Kapitan Mamarchev Technical Rubber Wares Factory in Silistra (1965–66), mainly for automobile transport and household needs; the Pneumatic Tyres Plant in Vidin (1966–70), at present part of the Industrial Chemical Combined Works, for the production of light radial and heavy-duty diagonal automobile, tractor and electric fork-lift tyres, belts, and impregnated conveyor belt tissues; and the Latex factory in Byala (1968–70) and the factory in Madan (1972–75) for the production of technical rubber wares. Since 1973, after reconstruction, the Konstantin Russinov Rubber Wares Plant in Pazardjik has started producing rubber conveyor belts, rubber hoses, bicycle and moped inner tubes and tyres, soft collapsible containers for bulk materials, platten rollers for typewriters, etc. Since 1984 it has started recycling pneumatic tyres for heavy-duty automobiles, agricultural and building vehicles. In 1982 the output was 1,576,900 automobile tyres and 1,087,000 bicycle tyres and inner tubes. Compared to 1970, automobile tyre production has increased 290 per cent, and that of bicycle tyres – 210 per cent. After a number of reconstructions the production of rubber shoes has been concentrated in three industrial enterprises: the Vassil Kolarov Plant in Pleven; the Yako Dorossiev Plant in the village of Hadji Dimovo, in the Blagoevgrad district; and the Chavdar Plant in Sofia. All rubber footwear (boots, sandals), textile-rubber wares and materials for the shoe industry (synthetic solid leathers, ready-made pressed rubber and polyurethane soles, microporous rubber bands and all kinds of glues, heels, and frames), are among the items produced. In 1982, 14,281,000 pairs of rubber footwear were produced; the total output of the chemical and rubber industry has increased 300 per cent over 1970 levels. Institutional charge is taken by Ministry of the Chemical Industry; leading units include the industrial chemical combines in Burgas, Pleven, Devnya, Stara Zagora, Vidin and Vratsa.

The compound mineral fertilizer plant in Devnya

6. The Building Materials Industry. The production of building materials has been going on in Bulgaria ever since pre-historic times. Stone was the first to be extensively used, as evidenced by the Thracian tombs unearthed near the village of Duvanlii in the Plovdiv district, the village of Mezek in the Haskovo district, and Strelcha. As early as the fourth century BC the Thracians already knew all about and made use of burnt bricks. After the Roman conquest of Thrace (first century BC) hydraulic bonding agents, lime and Roman cement, obtained by baking a mixture of limestone and clay, were introduced in the construction of administrative and residential buildings. Ceramic manufacture made tremendous progress in the ninth and tenth centuries. Pottery centres appeared, providing decorative ceramics for use in architecture and the decoration of buildings. Bulgaria's fall under Ottoman domination (1396) led to a decline in the building materials industry. Urban expansion after the liberation in 1878 called for the local manufacture of a number of materials and articles that had hitherto been imported. In 1894 the first ceramic factory, Izida, was built in the village of Novoseltsi in the Sofia district (today Izida Works at the Elin Pelin junction) to produce stoneware pipes. In 1885–95 factories for solid bricks and tiles were built in the vicinity of several large Bulgarian towns (such as the Rabotnik Factory near Sofia, the Trud near Russe, the Trakia near Plovdiv, etc.). Construction work on the railway line running between Sofia, Mezdra and Roman (operating since 1897) brought about the excavation of the lime deposits near the villages of Vurbeshnitsa, Gorna Kremena and Tsarevets, in the Vratsa district, as well as certain granite and syenite quarries in the Vitosha mountain. The industrial production of Portland cement started in 1909, when the first cement factory, called 'Lev' then and Vladimir Zaimov State Cement Works today, was inaugurated in Pleven. In 1912 the Granitoid Cement Factory near the Batanovtsi railway station (now the Vassil Kolarov State Cement Works in Temelkovo) was opened. But up to World War II the output of building materials was not sufficient to meet the country's needs. In 1939 the production of bricks amounted to 65 million, of tiles and crest-tiles – 68 million, of cement – 225,000 tons.

After 9 September 1944 there was a reconstruction of the nationalized building materials plants, and new ones were added. In 1947 the Vulkan State Cement Works in Dimitrovgrad started operating, after the initial stage of its construction was completed. State cement works started functioning in following order: the Devnya in Devnya, 1958; Wilhelm Pieck in the village of Beli Izvor, Vratsa district, 1960; Zlatna Panega in the village of the same name in Lovech district, 1966; Vladimir Zaimov in Pleven, 1967. In 1960 the *cement* output was 700 per cent over 1939 levels, and in 1982 2,500 per cent over 1939 levels. Portland cement varieties and makes alone number thirteen. Six up-to-date cement plants produce a total of five million tons annually. The Vulcan and Wilhelm Pieck cement plants manufacture asbestos cement products, such as corrugated and flat-surfaced slates, asbestos cement

The Leonid Brezhnev iron and steel works in Sofia; a metallo-plastics shop

pipes and facing elements. *Ceramics* have an ever growing part to play in construction. Ceramic works with a high level of mechanized production were built in the 1965–80 period in the villages of Butovo and Ovcha Mogila in the Veliko Turnovo district, Dragovoshtitsa in the Kyustendil district, General Toshevo, Plovdiv, Kaspichan and Harmanli. The production of bricks in 1982 reached the 1,404 million mark, and tiles and crest tiles 49 million. There has also been a considerable increase in variety.

Ceramic flooring, as well as inner and outer wall facing tiles, were also produced in considerably larger numbers in the 1975–82 period at the ceramic flooring plant in Mihailovgrad, the Mizia State Ceramic Plants in Gorna Oryahovitsa, the Granit in Turgovishte, the Iskustvo in Troyan, and in Mezdra. By 1982 production figures were: flooring tiles – 1,092,000 sq.m, and facing tiles – 277,000 sq.m.

Growing industrialization in construction has stepped up the production of new, *highly-effective building materials*. The production of light-weight fillings for concrete and other binding matter was introduced with the construction in 1959 of the first pre-cast large-panel building in the A. Tolstoi housing estate in Sofia. Production in 1982 was: for mineral marl-wool – 11,200 tons, for ceramicite – 138,000 cu.m and for porous schist – 100,100 cu.m.

The production of perlite was first begun in Kurdjali, 1962; of ceramicite at the Al. Voikov State Ceramic Works in Novi Iskur, 1964; of porous sinter in Gulubovo, 1973; of porous schist at the Al. Voikov State Ceramic Works, 1976; of expanded vermiculites and vermiculite-based fire-resistant components at the Zavodproekt Joint Research and Design Institute of Building Materials, 1980. Mineral and glass wool and the articles made of them for the purpose of insulation, fire-proofing and packing are manufactured in Polski Trumbezh and at the Kosta Stoev Works in Gabrovo. Roofing waterproof bituminous materials have been manufactured at the Sofia plant for waterproof materials since 1962. Bitumen, hemp-fibre skin, voalite – re-inforced voalite included – are also being put into operation. The Gavril Genov Chemical Plant in Russe produces a large number of polymeric waterproof foils.

The production of *stone-facing materials* has grown considerably since 1966. It includes rock-shaped blocks and slates of limestone, marble, granite, gabbro, etc. Up-to-date, highly productive machinery and equipment for stone extraction and processing, such as wire-saw and chain stone-cutting machines, multi-hammer quarry bars and self-propelled drilling carriages, are used extensively. Stone quarries will be developed under the names of Murata near the village of Ilinden in the Blagoevgrad district; Chelyustnitsa, near the village of Chelyustnitsa, in the Mihailovgrad district; and Pravoslav, near the village of Pravoslav in the Stara Zagora district. The local white limestone quarried in Russe is turned into facing panels. The total amount of blocks extracted in 1982 reached 75,000 cu.m of which 41,080 cu.m were marble blocks. Most are processed at home, while the remainder are exported to the GDR, the USSR, Czechoslovakia, Hungary, Italy and Greece. The largest stone processing works operate in Mezdra, Berkovitsa, Sofia, Bratisigovo, Velingrad, Malko Turnovo, the villages of Chernomorets in the Burgas district, Strumyani in the Blagoevgrad district and Mramor and Granitovo in the Yambol district. The stone-chip waste obtained in the process is bound together with polyester resins or cement to make agglomerate tiles, for which a production line is operating in Mezdra. There were 896,700 facing panels manufactured in 1982, marble accounting for 669,200 of them. Owing to their highly decorative and functional quality, Bulgarian facing tiles are exported to the USSR, Hungary, Czechoslovakia, Poland, the GDR, the USA, Sweden and the Middle East.

The major share of *fire-resistant materials* is produced at the following plants: the Trud, turning out fire-resistant chamottes in Russe; the Mladen Stoyanov, also for fire-resistant chamottes at the Elin Pelin junction; the Simeon Ivanov, specializing in Dinas firebricks, in Sliven; the L. I. Brezhnev Iron and Steel Works, manufacturing calcined dolomite-based fire-resistant materials in Sofia; and the plant for unmolded fire-resistant materials and heat and acid resistant compounds of high silica, aluminium and magnesium content, in Radomir. The production processes here are highly mechanized and partially automated. The 1982 output of 170,914 tons of fire-resistant material was 130 per cent over the 1970 level. Some of these products are exported to Albania, Cuba, Mongolia and Czechoslovakia.

About 30 *non-metalliferous minerals* are mined and processed in Bulgaria. Most important among them are kaolin, bentonite, gypsum, quartz sands, perlite, dolomite, volcanic tuff, clays, trass, feldspars, fluorite, chalk and diatomite. There are some 50 open-cast and subterranean

The Narodna Republika plastics works in Sofia

The Alen Mak works in Plovdiv; a toothpaste production line

pits linked to mining flotation and processing plants. The largest among them are Vatia, in the village of Negushevo in the Sofia district and Dimiter Blagoev, in the village of Sernovo in the Razgrad district. Bulgarian exports of kaolin, quartz sands, bentonite, talc flour, pegmatite, fluorite, perlite, etc. go to the USSR, Poland, Hungary, the GDR, Romania, Yugoslavia, Italy and Greece.

The majority of the building materials and articles are manufactured by economic organizations at the Ministry

Table 4: Production of Building Materials and Articles

Name	Units of measure	Years of output 1975	1982
Cement	thousands of tons	4,358	5,614
Asbestos cement pipes (standard diameter 200/16)	thousands of metres	3,378	2,717
Asbestos cement panels	thousands of square metres	5,613	4,434
Bricks	millions in number	1,495	1,404
Tiles and crest tiles	millions in number	78	49
Lime	thousands of tons	1,590	1,776
Quarry yields	thousands of cubic metres	23,193	29,929
Fire-resistant materials	thousands of tons	150.9	170.9
Concrete pipes	thousands of metres	1,256	1,650
Basalt slabs	thousands in number	43,916	29,913
Stoneware pipes (standard diameter 150 mm)	thousands of metres	911	623

of Construction and Settlement Planning, such as the state economic organizations for cement and lime extraction, inert materials, extraction and processing of stone tiles, reinforced concrete structures and articles, etc. Enterprises of various authorities produce building materials and articles for finishing work and installations, among them sanitary ware, wall tiles, plate glass, fitted carpeting, polymeric floor-covering materials (Ruvitex, Dunapal), bitumen and bituminous compounds, glues, dyes, varnishes, resins, corrosion-resistant foils, polymeric rolled water-proof materials, wallpaper, joint-packing pastes, plastic building materials, polystyrene, plasticizers, fire-resistant materials, gypsum and plaster of Paris, woodwork, casing, radiators and the like.

7. The Timber and Wood-working Industry

The timber industry comprises wood-cutting, the initial processing of timber, its transportation to temporary storage facilities, loading centres, wood-working enterprises and the production of forestry items. Up to 1948 the cutting and initial processing of timber were done manually, the haulage and part of the transportation – by draught animals. In 1983 the degree of mechanization of the main processes of production in the wood-working industry was: felling – 98.5 per cent, haulage – 63.1 per cent, loading – 84.7 per cent, transportation – 100 per cent. The total network of forest paths used was about 17,000 km, of which 8,850 km had a gravel surface; some 6,578 km of public motorways (1983) were also utilized.

In late 1983 the average density of motorways running across the forests of the People's Republic of Bulgaria was 6.00 m per hectare of forestland. The average distance involved in the haulage of timber is 1.14 km, and 33.3 km in the process of transportation. In 1983 some 13,077 workers were employed in timbering and 3.03 million cubic metres of construction timber was produced, comprising coniferous and deciduous trunks for sawing, plywood, sleepers, mine props, cellulose materials, and timber for wood-panelling and technological processing, in addition to 890,740 cubic metres of firewood, 1,394 tons of charcoal and 1,550 tons of pine resins. In 1983 about 15 per cent of the timber was stored in batches in the wood cutting areas, about 80 per cent in temporary storehouses and 5 per cent in the central warehouses. Under an agreement between the USSR and Bulgaria the two countries have been co-operating in the production of timber since 3 December 1967. Activities on the Bulgarian side are conducted by the Durvodobiv i Stroitelstvo State Economic Corporation, established in 1975 in the Komi Autonomous Soviet Socialist Republic, with its seat in Usogorsk. Four industrial-forestry concerns have been built – in Usogorsk, Blagoevo, Mezhdurechensk and Gornomezensk. Houses, cultural establishments and on-site shops have been built. The production capacities for the felling, cutting out and loading of timber are estimated at 3.2 million cubic metres a year. Storehouses and workshops have been built and equipped with very efficient automated machines, and workshops for the processing of timber of little value and waste materials have also been established. Many roads – of which 380 km have an asphalt surface – have been constructed and are used to capacity throughout the year. There is also a motorway (108 km long) which is used to transport timber from Gornomezensk. During the period 1970–83 the volume of wood-cutting increased 650 per cent.

The Antibiotics Research and Production complex in Razgrad

The wood-processing industry incorporates the following lines of production: board-cutting, crating, plywood, veneer parquet-flooring, cask-making, wood panelling and consumer goods and articles. Before the Liberation in 1878 there were about 300 cutting-machines and wood-turning lathes. The first saw-frame factory was built in 1892 at the Belovo railway station – now the town of Belovo. Wood-working factories were also built in Burgas, Varna, Svishtov, Russe, Troyan and elsewhere. The first state owned saw-frame factory was built in 1927 and the first plywood factory in 1931. On 9 September 1944 there were many wood-working enterprises, but they were small in size and the employees were mainly artisans. After 1948 the existing factories were enlarged and modernized. New board-cutting workshops were opened and the plywood enterprises were reconstructed for production by the dry-hot method. A new plywood plant, the G. Damyanov, was built in the village of Govezhda in the Mihailovgrad district, and a big parquet-flooring workshop was opened at the Longoza State Wood-Working Plant in Cherven Bryag. Cask-making also took on an industrial form in the period 1961–70, when the old plants were mechanized. Twelve new plants for wood panelling were built in Slivnitsi, Troyan, Veliko Turnovo, Berkovitsa and elewhere.

The furniture industry was actually already flourishing prior to the country's liberation in 1878, but the first furniture enterprises came into being after that date. In 1939 furniture making held third place after the food processing and textile industries in terms of number of workers employed. Most furniture enterprises were small and had limited production capacities. The bigger ones – i.e. those with more than 40 workers employed – included the factories of the joint-stock companies Sofa, Mebel and Kotva in Sofia, Bor in Plovdiv, Karpati in Russe and Napreduk in Burgas.

After the nationalization of the industry in 1947, the small handicraft workshops were amalgamated, and their total number was reduced to fifteen. New workshops were built and new machines and technologies were introduced. Subsequently, sixteen new plants were built (in Troyan, Bansko, Cherven Bryag, Velingrad, Plovdiv, Stara Zagora, Shumen, Sofia and elsewhere) and six of the old ones were reconstructed (in Tryavna, Russe, Silistra, Veliko Turnovo, Yambol and Septemvri). Modern machines and aggregates and mechanized, automated production lines were introduced. Seventy per cent of the production facilities of the furniture industry were created during the period of the Fifth and Sixth Five-Year Plans (1966–70 and 1971–75 respectively). By 1982 furniture production had increased 300 per cent over the 1970 level, and sales per capita by 280 per cent. Nowadays both modern and traditional furniture are produced. Different furniture enterprises specialize in the production of different types of furniture. Chairs, for example, are produced at the Hemus Factory in Troyan and the Pirin in Bansko; the Nikola Vaptsarov Plant in Cherven Bryag and Vassil Sotirov in Velingrad specialize in kitchen furniture; single unit and modular furniture is chiefly produced at the Napreduk Plant in Plovdiv and the Moscow Amalgamated Enterprise in Sofia, the August Popov in Shumen, Mebel in Stara Zagora and Nezavisimost in Gabrovo. New, more effective materials are continually being introduced. A large percentage of the output is exported, both to the socialist countries (mainly to the Soviet Union) and to non-socialist countries such as the FRG, France, Great Britain, Belgium, the Netherlands, Switzerland, Finland, Kuwait, Jordan, Iraq, and Libya. This branch of industry is controlled by the timbering department at the Ministry of Forests and Wood-Processing and the Stara Planina and Mebel State Economic Corporations.

8. Pulp and Paper Industry. The pulp and paper industry had a 1.4 per cent share in the total volume of the country's industrial production in 1982. The first factory for the production of cardboard was built in the village of Kilifarevo, in the Turnovo district, in 1865 by Angel Popov but existed for only three or four years. A paper factory was built in Sofia in 1891 and this laid the foundations of the industrial manufacture of cardboard, and later, of paper. During the 1900–37 period another four paper factories were built – in Belovo, Kostenets, the Iskur Junction and Kocherinovo. The annual output was about 25,000 tons of paper and cardboard. Bleached and

The pharmaceuticals plant in Sofia

unbleached cellulose, wood fibre – both local and imported – and waste paper were used as raw materials.

After 9 September 1944 the oldest enterprises were reconstructed, modernized and enlarged. New modern plants were built in Novi Krichim – now the town of Stamboliiski – Moesia, the Iskur Junction, Plovdiv, Lukovit, Strazhitsa and Razlog in 1952, 1962, 1965, 1967, 1970, 1972 and 1974 respectively. Production is concentrated in the cellulose and paper combines in Moesia, Stamboliiski and Razlog; the combine for semi-cellulose and cardboard in Sofia; the Dimiter Blagoev Combine for Light-weight Paper, in Belovo; the Trakia Combine for Cardboard, Corrugated Cardboard and Packing Materials in Pazardjik; the Vassil Kolarov Combine for Paper and Paper Products in Kostenets; the plants for corrugated cardboard and packing in Lukovit and Strazhitsa; the plant for multi-layer cardboard in Nikopol, the state paper factories Iskur and Petko Napetov in Sofia, Chaika in Varna, Nikola Vaptsarov in Kocherinovo and Hadji Dimiter in Svoge, the Rodina Plant for Paper and Packaging in Plovdiv and the Suba Eneva Combine for Packing in Pavlikeni, amongst others. The biggest Forestry Complex in Bulgaria, which is situated in an area of 160 ha, is being built near the town of Silistra with the assistance of the Soviet Union and the other CMEA countries. It comprises a plant for wood fibre panels (built in the period 1973–76) and other plants and auxiliary units which are still under construction. They produce craft-cellulose in Stamboliiski and Moesia; straw cellulose in Moesia; semi-cellulose in Stamboliiski and Sofia; wood fibre in Kocherinovo, Kostenets and Moesia; wood-fibre panels in Silistra; paper for corrugated cardboard in Sofia, Stamboliiski, Razlog, Pazardjik and Plovdiv; corrugated cardboard in Sofia, Plovdiv, Lukovit, and Strazhitsa; multi-layer cardboard in Nikopol; other kinds of cardboard in Kocherinovo and Pazardjik; paper for writing and printing in Moesia and Sofia; thin wrapping paper in Belovo and Kostenets; sanitary and hygienic paper in Belovo and Kostenets; corrugated cardboard packing in Sofia, Plovdiv, Lukovit and Strazhitsa; cardboard packing in Pavlikeni, Svoge and Varna and paper bags in Stamboliiski. The factories are equipped with modern and highly efficient machines imported from the USSR, the GDR, Czechoslovakia, Austria, Finland, Italy and Great Britain. In the period 1970–82 the total industrial production of the pulp and paper industry increased 230 per cent, and the production of paper and cardboard articles 550 per cent. For the same period the stock of fixed capital increased from 154.2 million levs to 538.2 million levs. In 1982 there were 15,745 workers employed in this branch.

A rose-oil distillery

Coniferous and deciduous wood fibre (both local, and imported from the Soviet Union) is used for the production of cellulose, as well as straw and technological chips; local deciduous wood fibre from beech, poplar and birch trees is used for the production of semi-cellulose, and local coniferous wood is used for the production of wood pulp. Cellulose – both local and imported, chiefly from the Soviet Union – is used in the production of paper. Craft cellulose is exported to Yugoslavia, Romania and other countries; straw cellulose is exported to the GDR, Switzerland, Japan, Greece, and the FRG. Thin wrapping paper is exported to the USSR, Iran, Iraq, Pakistan and other countries. This sector of industry is controlled by the Ministry of Forests and Wood Industry and the Cellulose and Paper State Economic Corporation.

9. The Glass and Porcelain-Faience Industries

The glass industry produces glass containers for the food

Table 5: Output of the Pulp and Paper Industry by Years

Products	Unit of measure	1970	1975	1982
Wood fibre	thousands of tons	14.3	11.8	18.9
Cellulose	thousands of tons	77.3	179.0	212.3
Paper	thousands of tons	199.7	283.0	353.6
Cardboard / Soft Cardboard	thousands of tons	15.2	44.1	64.4
Corrugated cardboard	thousands of tons	94.3	161.5	184.9
Paper bags	millions of items	110.0	126.9	130.9

The Vidin chemical works; the tyre plant

processing and medical industry, construction glassware (window-glass, armoured glass, ornamental and profile glass), hollow glass tiles, glass packages for windows, water glass, household table sets of ordinary, potash and lead crystal and technical glass (quartz, electric bulbs, flat and curved hardened glass for automobile windscreens, mirrors, glass fibres, glass wool, and glass for electronics and other special purposes).

The oldest remnants of a glass workshop (c. sixth century) have been found in Odessos (Varna). During archaeological excavations glass workshops were also found in Preslav, the localities of Patleyna and Selishta (tenth century). A glass factory was built in the town of Sopot in the eighteenth century to make phials for attar of roses. There is also evidence that a small glass workshop existed in Kazanluk, with an annual production capacity of 5,000–6,000 phials for attar of roses. Bottles for balms, medicines, cosmetics and rose distillate were also made.

After the Liberation in 1878 glass factories were built in Samokov (1881) and near the village of Gebedje in Varna district, today Beloslav (1892). In 1931–32 the first glass factories for the production of bottles, glasses, cups, demijohns and other glassware were founded in Sofia, with Bulgarian, German and other foreign capital. In 1931–35 the Crystal Factory in Pernik started manufacturing flat window glass, with joint Bulgarian-Belgian capital. The Belfa Electric Bulb Factory in Sliven was built in 1933–34 with the assistance of German specialists. Plants built and commissioned after 9 September 1944 are the Dyanko Stefanov in Razgrad (1960), for flat glass; the Druzhba Plant in Plovdiv (1961) and General Ivan Vinarov Plant in Pleven (1966) for glass containers; and the Ninth of September Plant in Elena (1978) for hollow glass tiles and profile glass for the needs of construction. Old plants were reconstructed and overhauled and became specialized in the production of: glass containers, bottles and household table sets (the Stind Plant in Sofia); hollow glassware and flat glass, including hardened automobile windscreens (the Crystal Plant in Pernik); and hollow glassware and lead crystal (the Vassil Kolarov Plant in Sliven). Since 1960 the Kosta Stoev Plant for insulation materials in Gabrovo has been making equipment for the production of glass, basalt and marl-wood for thermal and sound insulation, and glass veil for fibre glass roofing material. A modern Soviet installation for insulation marl-wool was commissioned in 1973 in Polski Trumbesh.

Glassware is also produced by some local industry enterprises, the most outstanding of which is the Kamushit enterprise in Silistra (glass demijohns and lighting fixtures). The plants are highly mechanized.

Over 24,860,000 cu.m of flat glass and 31.3 million pieces of household glassware were produced in 1982.

The porcelain-faience industry produces ordinary and bone-china, white, coloured and decorated faience plates for interior tiling and sanitary-ware (wash-basins, toilet pans, bidets etc.). The production of clay vessels and utensils can be traced back to antiquity. The ancient peoples made ceramic articles of white and red clay, including bricks, pipes, cups, plates, vases, pots, etc. The first factory for fine ceramics, Izida, was founded at the Novosseltsi Junction (today Elin Pelin Junction) as long ago as 1894, gradually becoming expert in the production of ceramic piping, wall tiles, sanitary ware, terracotta tiles and porcelain. Some other porcelain-faience production enterprises were built before 9 September 1944: in the town of Novi Pazar in 1919, where the production of terracotta and wall tiles and household porcelain was introduced,

The Dimiter Blagoev tissue-paper works in Belovo

and in Kaspichan, for wall tiles and household porcelain; the faience ceramics factory was opened in Sofia in 1925, for wall tiles and fireproof glazed bricks for stoves, etc.; and in Vidin, the first specialized factory for household porcelain and faience articles, the Bolonia. The Ludogorski Porcelain Factory was built in the town of Razgrad in 1936.

After the nationalization of industry in 1947 the old enterprises were enlarged and renovated, and modern technology was introduced. The Izida plant specializes in the production of wall tiles (white, coloured and decorated) and modern porcelain articles (more than 300 kinds). The most modern factory in Bulgaria for the production of high quality household porcelain in a wide variety of forms, models and decorations was put into operation in the Kitka Combine at Novi Pazar in 1974. The Georgi Dimitrov plant in Kaspichan specializes in wall tiles and sanitary-ware. In 1947 the Bolonia factory was renamed Kosta Yordanov, and its line of production was confined to high-quality de luxe porcelain. As a result of co-operation with the Lomonosov plant in Leningrad, a special shop for bone-china was opened at the K. Yordanov factory in 1980, the only one of its kind in Bulgaria. The Razvitie factory, the only enterprise of its kind in Bulgaria for the production of china and porcelain for the needs of hotels, restaurants, holiday-houses and bars was opened at the village of Kaleitsa, in the Lovech district, in 1954. During the 1949–53 period the Lenin plant was built in the town of Nikolaevo, with Soviet assistance. It is the only plant manufacturing articles of high and low voltage electro-installation porcelain, and meeting the demands of the fast developing energy network and communications technology. The Khan Asparuh plant, the most modern one and with the highest degree of automation for the production of wall tiles, was opened in the town of Isperih in 1979. 42,206,000 articles of household porcelain were produced in 1982; 2,940,000 high voltage insulators and 2,133,000 low voltage insulators were made in 1982. During the 1954–83 period the glassware and porcelain-faience industry increased by about 4,500 per cent and 42,100 over the 1939 level.

The industry is under the control of the Ministry of Production and Trade in Consumer Goods – the Kvarts Economic Corporation.

10. The Textile and Knitwear Industries. The textile industry in Bulgaria is the oldest sector, dating back to the second quarter of the nineteenth century. It was preceded by manual domestic and manufactured textile production which developed in the late eighteenth and early nineteenth centuries. The main centres of homespun were Sliven and Kotel, and of woollen braids – Karlovo. The first textile mill for homespun stuff was founded in Sliven in 1836 by the Manufacturer Dobri Zhelyazkov. Another factory was opened in the village of Dermendere (today Purvenets, in the Plovdiv district) in 1847, to make woollen fabrics for the Turkish army. A factory for silk cloth and fabrics was opened in Stara Zagora (1860), for the spinning and processing of silk in Shumen (1850) and in the town of Turnovo, with a branch in Gabrovo (1861). From 1870–76 a number of other textile factories were founded in Sliven, Sofia, Karlovo, Panagyurishte, Russe and Silistra. After the Liberation in 1878 the wealthy manufacturers and dealers in wool, yarn, homespun and other domestic products started organizing factories, and by 1892 there were 26 textile factories out of a total of 90 industrial enterprises. Sliven, Gabrovo, Kazanluk, Tryavna, Karlovo, Samokov and other towns became the main centres of the textile industry. In the late nineteenth century about 73 per cent of the textile factories were owned by industrial companies or associations. During this period and in the early twentieth century the beginnings were laid of industrial production of cotton and woollen knitwear, concentrated in Gabrovo, Sliven, Russe and Sofia. There was a big boom in the textile industry after 1926, and by 1929 the number of textile enterprises in the country was 198. The textile industry was in possession of 20 per cent of the fixed capital, 30 per cent of the output and 42 per cent of the work-force of the industries being encouraged.

After the nationalization of industry (1947) the textile enterprises were reorganized and reconstructed; their number decreased 250 per cent (1947–49). During the First (1949–53) and Second (1953–57) Five-Year Plans the process of agglomeration of the textile enterprises continued; the imbalances existing within various sub-branches were eliminated; the production of the most widely used textile articles increased. During the 1956–65 period the industrial plants were reconstructed and renovated. The capital invested in the textile industry in 1964–65 was as much as that used in the First and Second Five-Year-Plan periods taken together. Compared to 1939, the volume of textile production in 1982 was 2,200 per cent higher, the textile industry now produces 700

The Stind glassware factory in Sofia

per cent more wool, about 1,100 per cent more cotton, and about 3,300 per cent more silk fabric. For the same period the production of cotton knitwear increased nearly 1,000 per cent, of woollen – more than 3,000 per cent, of stockings – 800 per cent; the production of non-woven textiles increased more than 22,500 per cent (1963–81). Reduced-spinning systems were introduced in worsted yarn spinnings; pneumo-mechanical spinning machines in cotton spinning; and unshuttled weaving machines in weaving. In finishing and dyeing, high-finishing textile machines for knitted cloth and articles were brought in: in knitted goods manufacturing – highly productive lace machines with electronic sampling devices, automatic flat knitting machines, warp knitting Raschel machines, etc.; and in the production of unwoven textile materials new technologies, (sew-knitting, dry-fleece forming, guide-needle tamping, tufting, etc.). The structure of the raw materials used was changing. The share of chemical fibres increased (Bulgarian-made polyester, polyacrilnitril, polyamid and viscose fibres and silks). In 1982 there were 41.1 sq.m of cotton fabrics, 4.2 sq.m of woollen fabrics, 4.0 sq.m of silk fabrics, 15.3 items of knitted fabric articles, and 4 sq.m of linen textile goods produced per capita. The greater part of the products of the textile industry is used by the tailoring industry for the production of ready-made clothes. In 1982, 20.3 million m of cotton textiles were exported to the USSR, Yugoslavia, the GDR, Switzerland, Austria and other countries, and 1,077,100 sq.m of carpet to the USSR, Lebanon, the FRG, Switzerland and other countries. The textile industry is under the management of the Ministry of Production and Trade in Consumer Goods – the Pamukotex, Vitex, Svilena, Ruen and Rila economic combines.

The cotton-textile industry produces yarns, sewing materials, fabrics and other cotton goods (39 per cent of the volume of the overall industrial production of the textile and knitwear industry, 1982).

The first Bulgarian factory for cotton yarns and fabrics was founded in Varna in 1899 by the Prince Boris Industrial Joint-Stock Company (founded in 1897), with a part of the capital supplied by British companies. Another four cotton-textile factories had been built in Sliven, Gabrovo, Yambol and Turnovo by 1920. After the First World War the Fortuna Joint-Stock Company was established in Sofia (1927) for the production of fine cotton fabrics, the Gloria Joint-Stock Company (1924) for sateen and other cotton fabrics, and the Bulgaria Joint-Stock Company for the production of cotton yarns (1928). The Bulgaria Joint-Stock Company for the production of sewing materials, etc. was founded in 1930 in Kazanluk. Entirely Bulgarian capital was invested in the establishment of a number of bigger cotton-textile mills, such as Textile in Varna, and Prince Kiril and H. Metev in Gabrovo. In 1929 there were 37 cotton factories with 4,692 workers in Bulgaria, operating with a capital of 13 million gold levs and a total output worth 23 million gold levs. 1937 saw a new increase and concentration of production in the cotton-textile industry, with 225 workers and 257 hp driving power on average per industrial enterprise. The production of cotton fabrics increased from 22 million m in 1934 to 34 million m in 1939. The import of cotton-type manufactured and semi-manufactured products decreased, while the import of cotton increased from 2,142 tons a year in 1928–30 to 100,600 tons in 1937–39. The production of raw cotton rose from 900 tons in 1929 to 10,000 tons in 1939. By 1939 the cotton-textile sub-branch already had a total of 87 spinning and weaving factories with 10,396 workers and engine driving power of 16,191 hp. The value of its production was estimated at 1,197 million levs, which amounted to 36 per cent of the total textile production. After the Second World War (1939–45) cotton textile production dropped by one third, and the number of the workers employed was reduced from 10,396 in 1939 to 3,651 in 1944.

To bring textile production back to its feet again 2,280 tons of cotton were imported from the USSR in 1945, and 16,000 tons in 1946, including 9,000 tons of material supplied by the client. More than 40 cotton spinning plants with 227,300 spindles and 215 cotton textile plants equipped with about 6,000 looms were nationalized in 1947. The existing enterprises were enlarged, extended and modernized. The Ernst Thaelman Cotton Spinning Mill in Sofia was commissioned in 1951, to be followed by the Balkan Cotton Spinning Factory in Gabrovo, in 1953, the Maritsa Cotton-textile Combined Works in Plovdiv in 1958, the Vratitsa Cotton Textile Combined Works in Vratsa in 1967 and others. Compared to 1939 the quantities of cotton yarns produced in 1982 were 740 per cent greater, and of cotton fabrics – more than 1,070 per cent. The quantities produced per capita in 1982 were 9.6 kg of cotton yarn and 41.1 m. of cotton fabrics, as against 1.8 kg and 5.4 m. respectively for 1939.

Industry

Table 6: Production of the Cotton-Textile Industry by Years

Products	Unit of measure	1939	1948	1960	1970	1982
Cotton yarns	tons	11,545	15,900	49,044	73,697	85,803
Cotton fabrics	thousand metres	34,121	58,263	218,389	318,806	366,569

The sub-branch has 38 enterprises, 21 of which are factories, ten plants and seven combined works. The bigger ones are the cotton textile combined works – Maritsa in Plovdiv, Vassil Kolarov in Gabrovo, Vratitsa in Vratsa, Purvi May in Varna, Tundja in Yambol, the cotton plants Zheko Dimitrov in Haskovo, Petko Enev in Nova Zagora, Osmi Mart in Sofia, and the cotton spinning plants in Bulgaria and Ernst Thaelman in Sofia.

The wool-textile industry is the oldest sector of the textile industry, manufacturing 45 per cent of the total industrial production of textiles and knitwear.

It started as a domestic production and eventually developed into a trade. The nineteenth century saw a boom in the manufacture of woollen fabrics, mainly homespun, frieze and, later, broadcloths and braids for the needs of the vast Ottoman Empire and for export. The major centres of wool textile manufacture include Sliven, Gabrovo, Tryavna, Kotel, and Panagyurishte. The first wool-textile manufacturing workshops were set up in the 1830s in Plovdiv and later in Turnovo and Shumen (1852), in Gabrovo (1860) and in Sliven (1870). In the period 1820–25 the primitive spinning and weaving devices were replaced by machines operated by water-power.

After the Liberation of 1878 wool-textile factories were opened in Gabrovo – in 1882 by I. Kalpazanov and in 1883–84 by the Alexander Joint-Stock Company, at present part of the Georgi Genov textile works; in Samokov – in 1886, recently merged into the Samokovska Komuna Textile Works; in Tryavna – in 1880, currently the Angel Kunchev Combined Works; in Sliven – in 1890–91 by the Andonov and Mihailov Joint Company, currently the Georgi Dimitrov Wool-Textile Combine, and elsewhere. In 1909 there were 32 factories with 2,343 workers and machinery of 3,335 hp. Over 90 per cent of their output went for export to Turkey. In 1937 the output of the wool textile industry doubled in comparison with 1930.

Following the nationalization of the industries in 1947 and as a result of reconstruction and new construction, the technical standards of production rose; the processes of concentration, specialization and co-operation were stepped up, bringing down the number of enterprises from 70 in 1947 to 33 in 1976. The average number of looms in a single plant went up from eight to ten in 1947 to 177 in 1976. In the wool textile industry in 1982 there were 1,919 looms (22 per cent automatic and 31 per cent unshuttled) and 168,078 worsted and 35,702 woollen yarn spindles (33,193 ring spindles and 2,509 spinning mules). The output of woollen yarn in 1982 was 38 million kg as against 840,000 kg in 1939, and of wool

The Vulcho Ivanov plant in Sofia

The Izida works in Elin Pelin

fabrics 37.3 million metres as against 5.3 million m in 1939. The relative share of worsted fabrics rose from 21 per cent in 1948 to 55 per cent in 1976. The 1982 per capita production was 4.2 m of wool fabrics, compared to 0.8 m in 1939 and 4.2 kg of wool yarns (0.1 kg in 1939). The major wool textiles combined works include: Georgi Dimitrov in Sliven, Georgi Genev in Gabrovo and Nacho Ivanov in Sofia, for the production of worsted yarn and woollen yarns and fabrics; Dimiter Blagoev in Kazanluk for worsted and semi-worsted yarns and fabrics; and Angel Kunchev in Tryavna and Samokovska Komuna in Samokov for yarn and worsted fabrics.

The silk-textile industry manufactures silk yarns and fabrics. There is data showing that the production of natural silk in Bulgaria goes back to the Middle Ages. The first silk weaving factory was built in 1858–60 in Stara Zagora by the French company Bonal Brothers. In 1861 Doino Viccenti from Bergamo, Italy, built a factory for silk spinning in Turnovo and the next year he opened a branch of the same factory in Gabrovo. After the Liberation of 1878 a silk spinning factory was built in Stanimaka (today Assenovgrad, 1895) which survived for five or six years. Other small workshops for silk spinning, weaving and dyeing of natural silk gradually appeared. Larger workshops were set up in the 1920–35 period for the manufacture of raw, natural silk. In 1921 the *Koprina* (Silk) Silk-worm Co-operative was founded in Svilengrad for the production of raw silk, or silk-worm seed. In 1927 the Filtis French-Bulgarian joint-stock company was established, with branches in Kazanluk (now the Rosa factory) for twisting – with some 3,600 spindles – and weaving – with about 42 looms – mainly crêpe fabrics; in Turnovo – for silk spinning with 64 guide pins (the modern silk spinning factory, Vassil Kolarov, has been built on its site) and in Sofia – the Feya factory (currently a workshop within the Bulgarska Koprina Combined Works) for high-quality jacquard fabrics. The Avedissyan filature with a moulinage workshop was opened in Russe for the unwinding and throwing of natural silk processed at the Mazlamyan weaving workshop (currently a section of the Dunavska Koprina Silk Textile Combine), as were the Razpopov brothers' filature and weaving workshop and D. Kolarov's factory Dunav. The silk textile enterprises in Sofia included: the joint-stock company Balkan Silk Factory (founded in 1917), the dyeing factory of the Brothers Landau Joint Company, set up in 1921 (currently a workshop of the Bulgarska Koprina Combined Works), the Mushanov brothers' silk spinning factory, and a couple of weaving factories (currently also workshops of the Bulgarska Koprina). The total output of the silk textile enterprises in 1939 consisted of 690 tons of dry cocoons, 92 tons of raw silk and 1.13 million m. of silk fabrics.

After 9 September 1944 the silk textile enterprises were nationalized and consolidated. New, more productive machines were gradually introduced into production, such as shuttled and unshuttled (hydraulic and combing) weaving machines, pre-weaving machines and machines for dyeing and finishing operations. Some of these were made in Bulgaria, others were imported – mainly from the socialist countries. The production of man-made silk and its blending into fabrics began after the 1950s. The 1960 output included 511 tons of dry cocoons, 187 tons of raw silk and 10.6 million m. of silk fabrics, in which the relative share of the natural silk yarn in the manufacture of silk fabrics was under 10 per cent. The silk textile industry is concentrated in the Svilena Industrial Combined Works in Karlovo, with branches in Karlovo – the Todossii Markov Combine (the largest in Bulgaria), in Sofia – Bulgarska Koprina, in Russe – Dunavska Koprina, in Kazanluk – the Rosa factory, in Svilengrad – Koprina, in Vratsa – Yordan Lyutibrodski, in Harmanli – Margarit Gogov, and the weaving workshop in Ivailovgrad. The branches cover the following main production activities: spinning of yarns from waste material in the cocoon production and silk spinning (flocksilk), throwing (moulinage) of natural and man-made silk for weaving and other purposes (sewing materials used in tailoring and embroidery, in surgery and for insulation purposes), weaving production and finishing-dyeing production. The combined works, processing polyester silk produced by the Dimiter Dimov Chemical Combined Works in Yambol, was commissioned in 1979 in Dimitrovgrad. The 1982 silk textiles output included 607 tons of dry cocoons, 172 tons of raw silk and 36 million m. of silk fabrics.

The hemp and linen industry manufactures hemp and linen goods for domestic and technical purposes. The industry began to develop after the Liberation of 1878, when the first rope-making workshops were opened, working on imported raw materials. In 1898 the Konop (Hemp) Joint-Stock Company was set up in Plovdiv for the manufacture of ropes, and in 1903 a rope-making factory was opened in Sofia by Ts. Payakov and I. Vuzharov, with

The Dimitrovgrad polyester silk processing works

fifteen workers and a fixed capital of 50,000 levs. After the First World War, the first industrial enterprises for the primary processing of linen and hemp were established in the village of Teke, now Obrochishte, in the Tolbuhin district; Vardim, in the Veliko Turnovo district; Byala Slatina, Pazardjik and elsewhere. In the secondary processing factories in Gabrovo, Varna, Russe, Pazardjik, Kazanluk and elsewhere the hemp and linen goods manufacture used both local and imported raw materials. The technological standards of these enterprises were rather low. The 1939 output included 967,000 m. of hemp-linen fabrics, 1,100 tons of yarn and 265,000 sacks. At the time of nationalization, in 1947, there were 36 hemp and linen enterprises for primary and secondary processing, which after realignment and technical reconstruction merged into seven in 1979. Currently the larger enterprises of the hemp and linen industry include: Vela Peeva in Pazardjik, for hemp spinning and the weaving of packaging and technical goods; Georgi Dimitrov in Provadia, for linen spinning and weaving; Rilski Len in Samokov, for primary processing of linen for spinning, weaving and finishing; and Yuta in Russe, for non-woven textiles. Other enterprises include the Georgi Kirkov factory in Kazanluk (for the production of bands, hoses and technical fabrics), Slavyanka in Tutrakan (for hemp fibres, yarns, strings and plates of hemp shive), and Okolchitsa in the village of Moravitsa in the Vratsa district (for hemp spinning and rope making), etc. In 1982, 9,877 tons of hemp and linen yarns and 1,794,000 linear m. of hemp and linen cloth were manufactured.

The production of non-woven textiles is a sub-branch of the textile industry, manufacturing non-woven fabrics by non-conventional methods which eliminate some of the labour-intensive processes involved in the manufacture of yarn (mixing, grinding and spinning), fabrics (pre-weaving and weaving) and knitwear (pre-knitting and knitting).

The foundations of the non-woven textile industry in Bulgaria were laid in 1962 with the importation of a machine Malimo-1600 (make No. 14010) from the GDR, installed at the Burya factory in Gabrovo. The three technologies, used for the production of non-woven fabrics, are as follows: sew-knitting on MALI machines imported from the GDR and ARA machines made in Czechoslovakia; dry fleece forming with additional chemical melding on machines imported mainly from West European countries (the FRG, Austria, Switzerland, Belgium); dry-fleece forming with additional guide-needle tamping on machines imported from socialist countries (the USSR, Czechoslovakia, Poland) as well as from some West-European countries (the FRG, Belgium, France, Austria), and tufting – on machines made in Great Britain. Some 70 per cent of the output is produced by sew-knitting technologies; the remaining 30 per cent are distributed among other technologies, tufting having a priority (up to 12 per cent). Over the 1963–83 period the output of non-woven textiles increased 275 times – from 200,000 sq.m in 1963 to 55 million sq.m in 1983. The principal manufacturers of non-woven textiles

The Liliana Dimitrova fine knitwear plant in Sofia; a Lipanit-type circular knitting machine

include: the Subi Dimitrov Comined Works in Sliven, using Schusspol-Malimo technology made in the GDR, Arahne, Aralup and Polara imported from Czechoslovakia, and Tufting made in Great Britain (mainly for upholstery); the Vassil Levski factory in Veliko Turnovo, for the production of non-woven textiles using the technology of dry-fleece forming with additional guide-needle tamping and chemical melding (upholstery and technical purposes); the Bolshevik factory in Troyan,

The shop for hand-made fleecy rugs in Kopilovtsi, in the Mihailovgrad district

using the sew-knitting technology and the Malivat machines (decorative and technical purposes); the Rodina factory in Tolbuhin, using the tufting technology (interior decoration); the Yuta Combined Works in Russe, using the technology of dry-fleece forming with additional guide-needle tamping and chemical melding (sewing materials and accessories for the clothing, footwear and sportswear industries, as well as multi-layer materials, materials for application in technology and backings to be coated with polyvinylchloride); the Samokovska Komuna Combined Works in Samokov, using sew-knitting technology and Malimo, Malipol and Voltex machines (application in the clothing industry and in interior decoration); the Orel factory in Razgrad, using the sew-knitting technology and Malimo machines (application in technology and packaging); the Balkan factory in Dragoman, using sew-knitting technology and Malivat and Voltex machines (linings and fabrics used in interior decoration).

The knitwear industry is a sector of the textile industry where, in the process of knitting, textile fibres are transformed into flat and hose textile goods (fabrics,

The Dimiter Blagoev printing works in Sofia

details, clothes and hosiery). Its output is 21.1 per cent of the total of the textile and knitwear industry (1982).

The knitwear industry in Bulgaria is the successor of handmade, home knitting, and later – of craft knitting. The founder of the knitwear industry was Hristak Momerin, who in 1890 set up the first knitwear enterprise in Bulgaria – a workshop in the town of Gabrovo for knitting men's, ladies' and children's outer garments and underwear. The small enterprise had an operating floorspace of 120 sq.m., twelve knitting looms, supplied by Germany, and 26 workers. After 1921 the enterprise was enlarged by his son Andrei Momerin and turned into a modern knitting mill, equipped with motor-driven knitting machines. By 1932 the mill had become the largest knitting mill in Bulgaria for fine knitted goods. In 1925 the Gabrovo and Triko companies merged into a trade joint-stock company, called Gabrovo, for the manufacturing and sale of knitted goods. The mill had four shops – knitting, weaving, finishing and sewing – and was serviced by 65 people. It produced different kinds of ladies', men's and children's outer garments and underwear. In 1921–22 the number of knitting mills in Gabrovo reached ten. It was during this period that the textile industry developed in Sofia. One of the numerous mills was the Nitra Textile Mill, owned by Nikola Rachev, specializing in ladies' and men's fine underwear of warp-knit, circular-weft knit and interlock fabrics. In 1924 D. N. Boyadjiev, in partnershp with the German producer Kalfmann, moved their hosiery factory from Germany to Sofia and called it the First Bulgarian Hosiery Factory Kabo. It produced ladies' stockings and men's and children's socks made of thicker cotton-yarn and artificial silk. The successor to the factory was the Kabo Stock Company, founded in 1930, which built a modern factory and installed new hosiery equipment – warp-knitting machines for fine stockings and socks with seams. In 1931 a cartel of hosiery factories was set up, which existed until 1936 and controlled 75 per cent of hosiery manufacture in Bulgaria. After 1936 new knitting mills were set up and new machines were imported. The production of knitwear and hosiery increased several times. The number of workers employed in some mills reached 150–200. During the nationalization period there were about 85 knitting mills and 45 hosiery factories. The old enterprises merged, expanded, were modernized and reconstructed. New factories and plants were built in the districts of Smolyan, Turgovishte, Blagoevgrad and Kurdjali. The production processes were mechanized and automated. In 1965–82 the output of knitwear increased by 250 cent and the output of hosiery by 160 per cent (see Table 7).

Bulgaria ranks first among the member-countries of the CMEA as regards the manufacture of knitwear per capita of the population (in 1982 – 15.3 items). The biggest investments in the knitwear industry were made between the years 1971 and 1980. Structure-determining enterprises are: the Dimitrovgrad Combined Works for the processing of polyester silk for tubular tricot and warp-knit fabrics of full polyester silk, and ready-made outer garments; the Proletarii Plant in Sofia, for ladies', men's and children's underwear, made of warp-knit and tubular tricot fabrics, curtains, lace, highly elastic cloth and underwear; the Burya Plant in Gabrovo, for ladies' knitwear made of warp-knit fabrics and printed textile fabrics for the needs of the sector; the Avram Stoyanov Plant in Tryavna, for ladies', men's and children's underwear made of tubular tricot fabric, for plush and ladies' tailored suits; the Orpheus Plant in Kurdjali, for tubular tricot plain jersey and cotton flannel fabrics and for the manufacture of ready-made sportswear, and ladies' and men's underwear; the Dobri Kartalov Plant in Gabrovo, for artificial fur, tubular tricot fabrics and outer garments, the Sanya Plant in Pleven, for knitted goods on

Table 7: Output of Knitwear by Years

Articles	Unit of measure	1965	1970	1975	1980	1982
Knitwear	thousand number	52,881	78,515	107,231	130,928	136,610
Hosiery (tights included; woollen included)	thousand pairs	38,488	39,253	53,327	63,182	68,173

hose flat knitting machines and their finishing, and the manufacture of ready-made clothes, as well as for the production, finishing and usage of fabrics knitted on circular-weft knitting machines; the Andrei Yurokov Plant in Plovdiv, for ladies' and children's knitted outer garments made on circular-weft knitting and hose flat-knitting machines; the Marek Plant in Kyustendil, for tubular tricot fabrics and the manufacture of men's ready-made knitwear; the Liliana Dimitrova Plant in Sofia, for luxury outer knitwear; the Rossitsa Plant in Sevlievo, for outer knitted garments, made of polyester silk, on automatic flat-knitting machines; the Stoyan Edrev Plant in Lovech, for babies' and children's outer knitwear on automatic flat-knitting machines; the Vulcho Ivanov Plant in Sofia, for stockings and tights manufactured on single cylinder hosiery machines; the Fazan Plant in Russe, for ladies' stockings, men's and childen's socks, and tights manufactured on double-cylinder hosiery machines; and the Dobri Zhelyazkov Plant in Sliven, for stockings, men's

The Sofia cigarette factory; king-size cigarette packaging installation

The 9 September footwear plant in Sofia

and children's socks and stockings made of natural and mixed materials on double-cylinder hosiery machines. Knitwear is exported to the USSR and other countries of Europe, Asia, Africa and Latin America.

11. The Tailoring Industry. Its output is 2.1 per cent of the total industrial output (1982). Before 9 September 1944 it was predominantly a craft industry, on a low technical level, for single or group orders. After the nationalization of industry in 1947, crafts' co-operatives and state industrial enterprises were set up for the needs of the tailoring industry. The inventory of articles produced was rather limited and the equipment used was poor. In 1962 the Rila State Economic Corporation was set up, incorporating all the dress-making and tailoring establishments of the state tailoring industry. New plants were built, such as the Vitosha Plant in Sofia, the Druzhba in Varna, the Mizya in Pleven, the Vida in Vidin, etc.; the fixed capital of the sector went up from 5.9 million levs in 1960 to 139.6 million levs in 1982, which is approximately a 3,040 per cent increase. For the same period the number of workers employed in the sector went up from 19,474 to 47,677 – a 240 per cent increase – and output increased 830 per cent.

The period of accelerated development of the tailoring industry started in 1970. Almost all dressmaking and tailoring establishments were enlarged, modernized and reconstructed. They were equipped with highly automated modern machines and numerical control devices, supplied by the FRG, the USA, Italy and other countries. New technologies for making clothes by pasting were introduced. The industrial enterprises were enlarged and specialized.

The total number of plants and factories in the sector is 27, incorporated in the Rila Industrial Corporation, Sofia, in the Ministry of Production and Trade in Consumer Goods. The leading enterprises of the tailoring industry are the Vitosha Plant in Sofia, specializing in the manufacture of men's suits and shirts, ladies' topcoats and tailored suits, men's and ladies' raincoats, children's and teenagers' clothes (4,000 workers are employed and the enterprise is the biggest on the Balkan Peninsula, exporting 29.3 per cent of its tailoring output); the Druzhba Plant in Varna, specializing in men's suits, jackets and trousers and ladies' topcoats and tailored suits (with 2,500 workers, and exports at 36.4 per cent of output); the Vida Plant in Vidin, specializing in men's shirts (with over 2,500 workers, and exports at 54.7 per cent of output); the Bulgaria Plant in Plovdiv – specializing in men's suits, men's and ladies' raincoats, and ladies' topcoats (with over 2,500 workers, and exports at 53 per cent of output); the Mizya Plant in Pleven – specializing in men's suits, jackets and trousers (with over 2,500 workers, and exports at 56.2 per cent of its tailoring output).

12. Printing. The beginnings of Bulgarian typography were laid in the sixteenth century, when printing shops were opened by Slav communities, Bulgarians among them, in Wallachia and Vienna. Before the Liberation in 1878 educated Bulgarian patriots also set up printing shops abroad. In 1828 N. Karastoyanov opened the first home-based one in Samokov. After 1878 the Bulgarian printers abroad moved back to their native lands. In 1881 the Bulgarian Government authorized the foundation of a state printing house, which was to grow over the next few years into the largest typographic establishment, both at home and in the Balkans; in 1944 it employed more than 1,500 workers. Quite a few private printing shops also appeared in Sofia, some of them used by the Communist Party Press. During the 1920–39 period a number of publishing and printing houses were amalgamated into bigger enterprises, while others, including Hristo G. Danov, Balkan, Polygraphia and Hudozhnik, broadened the scope of their activities. The first linotype composing, offset and intaglio printing machines and zincography shops were introduced. Joint-stock typographic companies, such as the Bulgarski Pechat (1918), Stopansko Razvitie and Utro-Dnevnik' (1922) and Mir (1924), came into being.

9 September 1944 found Bulgarian typography in a deplorable state. The people's democratic Government invested great efforts in its restoration and in providing the material foundation for the printed releases of the public and political organizations. On 23 December 1947 the larger printing houses were nationalized. Part of them were handed over to public and political organizations, while others started functioning as state-owned enterprises. In 1948 the State Polygraphic Corporation was founded to manage the nationalized printing houses. In 1950 a General Department of the publishing houses, the typographic industry and the distribution of printed matter was set up at the Council of Ministers in Bulgaria. In 1954 this was reconstructed into the Poligrafizdat Board at the Ministry of Culture. In 1954 the Dimiter Blagoev Combined Typography Works came into operation in Sofia; it now prints the bulk of the publishers' production. Further reconstruction and concentration was carried out in the typographic industry over the 1955–68 period. New printing houses were built, and up-to-date equipment and technology were provided. The Bulgarska Kniga State Corporation, based on the Poligrafizdat Board, was established at the Committee for Art and Culture in

The Malchika Research and Production complex in Sofia; an installation for the production of cracker biscuits

1968 and the Committee for the Press was formed in 1971. It undertook the agglomeration of printing capacities, establishing three combined typography plants in Sofia – the Georgi Dimitrov, Dimiter Blagoev and Decho Stefanov. Colour, offset and intaglio printing have been given prominence since 1970. New plants have been built, technology and equipment updated, electronic devices and photocomposition introduced, operative typography promoted as a means of reproducing scientific and technical information, automation and text-processing equipment advanced. The volume of typographic output has grown.

In a move to further improve the typographic industry a State Creative, Production and Economic Association 'Bulgarska kniga i pechat' (Bulgarian book-making and press publications) was founded at the Committee for Culture in 1982, its economic division including the Bulgarska Poligrafia State Corporation in Sofia with its 36 printing plants (seven of them in Sofia).

Table 8: Typographic Output (in thousand issues)

Publications	1970	1982
Newspapers (circulation)	816,720	917,688
Magazine (circulation)	45,125	61,648
Books (total print)	41,029	59,662

13. The Leather, Fur and Footwear Industries. The leather, fur, footwear and leather fancy goods industries contributed 1.1 per cent of the total industrial output in 1982.

Game-hunting for valuable furs, e.g. sable and others, was popular with the proto-Bulgarians who processed and exchanged them for other goods with neighbouring tribes. After the foundation of the Bulgarian State (1681) leather manufacture attained a high level of development. During the ninth and tenth centuries Russian leather, made of oxhide in Bulgaria, achieved widespread popularity abroad. The twelfth to fourteenth centuries period saw the formation in the towns (especially in Turnovo) of artisan communities of leather-workers and shoe-makers. After the Ottoman conquest of Bulgaria (1396) the majority of the tanneries were transferred to Gabrovo, which was to remain the chief leather-dressing centre up to the Liberation in 1878 and afterwards. The first tannery was founded in Gabrovo by N. Gutev in 1862, followed by others in Lovech, Shumen, Silistra, Vidin and Svishtov. A considerable quantity of shoes, some 200,000 pairs a year (1855–78), was also made in Gabrovo. Fur-dressing developed in Gabrovo, Troyan, Lovech, Shumen, Silistra, Vidin, Svishtov, Sofia, Plovdiv and elsewhere. In 1896 the first leather and fur-dressing factories were opened, using imported equipment and technology, and for the first time in Bulgaria machine-made shoes were produced, in Sofia. The first shoe factory was founded in Gabrovo in 1904. Up till 1928 the leather industry was developing at a more or less stable pace, when it was brought to a halt by the appearance of rubber shoes. There were 51 tanneries and 7 furriers in 1939. In 1947, 103 tanneries, 25 furriers and 18 shoe-factories were nationalized. Concentration, specialization, expansion, reconstruction and modernization were brought to the leather, fur and footwear industries. In 1982 there were seven tanneries, three furriers, fourteen shoe plants, five leather fancy goods, saddle and harness-making factories and two subsidiary workshops manufacturing shoe-lasts.

The leather industry is centred at the Dimiter Blagoev Works in Gabrovo, the largest producer of sole-leather in Bulgaria, also manufacturing leather from hides, and the only producer of pig-skin Russian leather and patent calf leather; the Deveti Septemvri Works in Russe, specializing in the dressing of raw hides; the Ahmet Tatarov factory in Sevlievo; Boks in Turgovishte; the leather factory in the village of Mitrovtsi, in the Mihailovgrad district; Todor Dokov in Etropole, manufacturing pigskin; and the Velur factory in Lovech – dressing small stock- and pig-skins and manufacturing leatherwear and gloves. The leather industry dresses annually some 22,000–25,000 tons of raw hide, 45 per cent of it imported. In 1982, the production of upper leather reached 539 million square decimetres, and of soft leather – 213 million square decimetres.

The major enterprises of the *fur industry* are: Bulgarska Kozhuharska Industria in Sofia, Pulpudeva in Plovdiv and Svoboda in Tolbuhin, where raw lamb and sheep skins are dressed. To provide furs for the industry, mink farms have been founded in Perushtitsa and the village of Bohot, in the Pleven district. Fur coats and articles made in 1982 amounted to 309,000. Most of these go to the USSR, Czechoslovakia, Poland, Romania, the FRG, Britain, Italy,

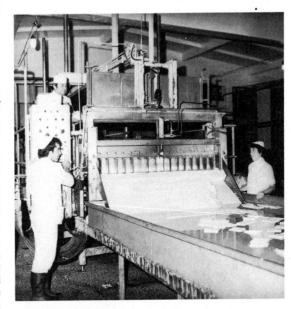

The dairy in Pazardjik; the white brine cheese shop

the Scandinavian countries and Japan. Fur exports in 1982 amounted to 20.3 million currency levs.

The leather fancy goods industry has three major factories – the Leda in Sofia, Chaika in Plovdiv and Prolet in Dimitrovgrad – using modern, highly productive equipment. A continuous production-line process is employed. Goods manufactured include: suitcases, attaché cases, bags, wallets, purses, gloves, etc. Nearly 70 per cent of production goes in exports to the USSR, Poland, Czechoslovakia, the GDR, the FRG, Holland, Sweden, Norway, Canada and Belgium.

The process of concentration and specialization in *the footwear industry* was, on the whole, completed by the mid-1970s. The Deveti Septemvri Shoe Works in Sofia specializes in the production of men's and boys' shoes; the Surp I Chuk works in Gabrovo – in men's shoes; the Peter Chengelov Works in Plovdiv – in ladies' shoes; the Vassil Muletarov Works in Peshtera – in men's and ladies' shoes; the Ilyo Voivoda Works in Kyustendil – in ladies' shoes; the Ivan Dimitrov Works in Mihailovgrad – in slippers; the Dobrich Works in Tolbuhin – in children's shoes; etc. The footwear industry makes use of highly effective equipment, brought from the FRG, Britain, France, Italy and others. In 1982, 19,171,000 pairs of shoes (slippers excluded) were manufactured, and 1,246,000 pairs – exported to the USSR, the FRG and elsewhere.

The industry is run by the Ministry of Production and Trade in Consumer Goods via the Pirin Commercial and Industrial Corporation.

14. The Food Industry. The output of the food industry accounts for 27.1 per cent of Bulgaria's total industrial output (1982 data).

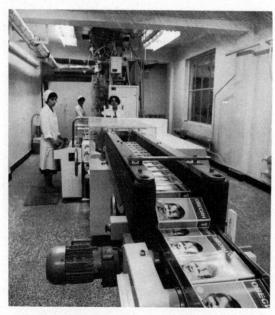

Production of baby foods and diet foods in Svishtov

The Bulgarian food industry dates from the middle of the nineteenth century. In 1853 an alcohol distillery was set up in the town of Radomir; in 1865 Sofia acquired a plum-brandy distillery. In 1882 a brewery was built in the town of Shumen and in 1880 a soft drinks factory was built in Plovdiv. During 1897 and 1898 a sugar refinery was built in Sofia with Belgian capital, and by 1899 the town of Varna already had one cannery. In 1909, the first wine-producers' co-operative was set up; it possessed one wine-cellar in the village of Suhindol. During the 1860s and 1870s small tobacco factories were opened in the towns of Plovdiv, Turnovo, Sliven, Sevlievo and elsewhere. Primitive dairies already existed in the districts of Haskovo, Harmanli, Kazanluk and elsewhere. Meat production and processing was carried out in the *salhani*, a capitalist type of factory that already existed before the Liberation of Bulgaria from Ottoman rule, for cattle slaughtering and meat production. The flour milling industry was well developed. Primitive water- and windmills existed all over the country.

Milk-processing into *sirene* (white brine cheese), *kashkaval* (yellow cheese), butter, and yoghurt – traditionally very popular in Bulgaria – has been known to the Bulgarians since the fifteenth century. Milk-processing enterprises were small-scale and organized on a non-industrial basis. As late as 1939 milk was still processed in 1,159 small dairies, of which 662 were private and 497 belonged to co-operative societies; the annual output was 73.9 million litres of milk (7.1 per cent cow's milk, 4.5 per cent buffalo's milk, 1.4 per cent goat's milk, and 87 per cent sheep's milk). Annual per capita consumption was far from satisfactory, amounting to only 27.3 litres of milk, 7 kg of *sirene*, 0.560 kg of *kashkaval* and 1.4 kg of butter. At the time of nationalization in 1947 there were 1,000 flour mills, nearly 1,700 bolters, and about 7,500 primitive water-mills in the country. Between 1925 and 1936, some 401 slaughter-houses were built, 306 of them in the villages. The only up-to-date slaughter-house was built in Sofia between 1930 and 1933. At the time of the nationalization meat was supplied by 102 slaughter-houses, of which only seventeen were equipped with appropriate technology and freezers, and meat-processing was carried out in 150 primitively equipped, private factories. The production of wine and other alcoholic beverages was in the hands of about 450,000 private producers, 20,000 private wholesalers and retailers and over 60 co-operatives (1947 data). Before the First World War, 95 per cent of vegetable oil requirements were satisfied by imports. The first oil-extraction and refining facilities appeared in the Olivia factory in the village of Yabliano, near Pernik, in 1923, and the Olio factory in Svishtov in 1924. At the end of 1944 there were 551 oil factories, of which only 76 had refineries. The sugar industry developed remarkably quickly; in 1913 sugar factories in the towns of Russe (Belgian-owned) and Gorna Oryahovitsa (Czech-owned) were put into operation, followed in 1914 by those in Plovdiv and the village of Kayalii, in the Burgas district, owned by French capital. In 1921 the five sugar factories merged to form the Sugar Factories Cartel, which was one of the largest and most stable monopolies in the country before the 1944 revolution. By the end of 1921 the sugar industry was the second largest industry in the country, after flour-milling. The tobacco industry also developed quickly. The first tobacco-shredding machines were imported and installed in Plovdiv in 1900, followed by the first cigarette-packing machines in 1905. Between 4,000 and 6,000 workers were employed in the tobacco industry. Tobacco companies were established in the towns of Plovdiv, Haskovo, Assenovgrad and Dupnitsa – now Stanke Dimitrov. In 1905 the Tobacco Factories Cartel was formed and in 1910 was renamed the United Tobacco Factories Cartel. In 1921 the Orientabako Joint-Stock Company was set up, with Italian capital; it dealt with tobacco growing, purchase, processing and selling. By the Second World War 52 new tobacco factories had been built (58 had already been built by 1922) and more than 320 companies were involved in the export of cured tobacco. In 1926–27 the Soft Drinks Cartel was set up, uniting 235 individual soft drink companies. The fish processing industry was poor: fishing was seasonal and done only in the coastal waters of the Black Sea, in the littoral lakes and in some inland lakes. Part of the catch was processed at the fish-markets in warehouses adapted for the purpose, as well in some small fish canning factories in the town of Burgas. The annual catch was about 4,000–5,000 tons (0.500 kg per capita), and the annual production of canned fish – about 500–800 tons.

Despite its primitive and small-scale nature, the food industry was the best developed industrial branch in the country before the socialist revolution. After 1944, the food industry continued its speedy development. The nationalization of industry in 1947 established state control over tobacco and alcohol manufacture and trade.

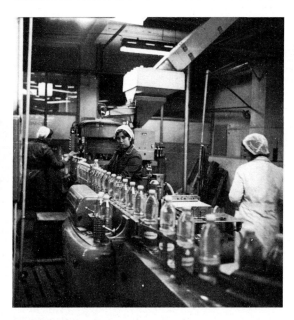

The Stoyan Syulemezov soft drinks plant in Sofia; bottling mineral water

Wine production was enlarged, restructured and modernized. Specialized production units (wine-cellars, *rakia* (plum brandy) distilleries, etc. for primary wine products) were commissioned, as well as installations for secondary wine-production (processing, stabilizing and bottling of wines, sparkling wines, grape brandy, vermouth, liqueurs, vinegar, tartaric acid, etc.). Large industrial enterprises were set up in the city of Sofia and in the towns of Pleven, Russe, Lyaskovets, Turgovishte, Preslav, Pavlikeni, Pomorie, Burgas, Perushtitsa, etc. Modern machines and equipment were introduced. Wine production from grapes, as opposed to other fruit, increased to 73.1 per cent in 1976. The supply of raw materials for the wine industry was reorganized along industrial lines. New varieties of grapes began to be cultivated, such as Cabernet, Riesling, Aligote, Tamianka,

The Hotel Nevrokop in Gotse Delchev

Pinot Chardonet, Rkatsiteli and others. The quality and range of wines was improved: popular wines on the local and international markets include: the red and white dry wines – Misket, Dimyat, Riesling, Rkatsiteli, Manastirska Isba, Cabernet, Mavrud, Gumza, Trakia, and Kadarka; the red and white sweet wines like Varna, Bisser, Turnovo; natural sparkling wines like Iskra, Lazur, and Biliana; the vermouths Cio Cio-San; white and red Vinprom; the aperitifs Roselli and Ropotama; and the cognacs Pliska, Preslav and Pomorie. 463,723,000 litres of grape wine, 76,943,000 litres of liqueur wines, and 33,177,000 litres of rakia were produced in 1982. Over 70 per cent of the output was exported to over 50 countries (1982 data). Large-scale *cigarette production* began with the concentration and specialization of the cigarette factories (from ten in 1948, to six in 1962). Handling was mechanized and new equipment, such as new types of tonga-baling machines, dust-removal and heating installations, mechanized flow-production lines for tobacco curing and handling, was introduced in the tobacco handling stores. New machines for rolling cigarettes and attaching filters, packing and cellophaning were introduced in the cigarette factories. After 1950 indoor curing facilities were provided to replace the seasonal curing process. The tobacco industry was concentrated in the tobacco-growing regions of the country. Curing halls were built in Plovdiv, Yambol, Haskovo, Blagoevgrad, Stanke Dimitrov, Kurdjali, Assenovgrad and elsewhere; new cigarette factories – in Blagoevgrad, Assenovgrad and Haskovo. In 1958 the production of filter cigarettes began, which now accounts for more than 90 per cent of the total production. Over 90 per cent of the processed tobacco varieties are cultivated within Bulgaria. The 1982 production figures were 124,835 tons of cured tobacco and 88,132 tons of tobacco products. The volume of exports of oriental tobacco was 63,266 tons for the USSR, Czechoslovakia, the GDR, Hungary, France, the FRG, Austria and other countries. In 1981–82 Bulgaria ranked second after the USA in cigarette exports. Tobacco was exported to 45 companies in about 30 countries in the world, and cigarettes to ten countries (1982).

Favourable weather and soil conditions and the availability of raw materials have been beneficial to the development of the *canning industry*. A number of big canneries have been built: Purvi Mai in Plovdiv, Georgi Kirkov in Pleven, Petko Enev in Stara Zagora, Maritsa in Pazardjik, Dunavia in Russe, Vitamina in Stamboliiski, and Brigada in Assenovgrad. To utilize raw materials more efficiently, smaller canneries were built within the agro-industrial complexes and the Central Co-operative Union, and new technologies and equipment were introduced. The canning industry started producing canned vegetables, meat and vegetables, tomato paste, processed and semi-processed vegetable and meat-and-vegetable products, baby foods, canned fruit, frozen vegetables, pulps, and semi-processed products (about 340 different products in all).

In both absolute and relative terms, the output of frozen fruit and vegetables, fruit and vegetable juices, canned meat-and-vegetable and vegetable foods, stewed fruits and peeled tomatoes increased. In 1982, some 315,039 tons of canned vegetables and 255,384 tons of canned fruit (apart from pulp) were produced. 256,981 tons of canned vegetables were exported to the USSR, the GDR, the FRG, Czechoslovakia, Poland, Great Britain (5,308 tons), Canada, Austria and elsewhere. In 1979 Bulgaria ranked third after Hungary and the USA in the production of canned foods per capita – 69.5 kg (1982).

Flour mills also underwent a period of merging, reconstruction and modernization. Grain and flour handling equipment was introduced, along with magnetic screening in the milling and packing machines. Additional cleaning machines, calibrated aspirators, hullers, graders, and pneumatic equipment were also introduced. Grain and fodder processing plants were set up in Burgas, Varna, Veliko Turnovo, Vratsa, Tolbuhin, and elsewhere, a flourmill in Sofia, amalgamated grain and fodder combines in the village of Dunavtsi and in Lom, Plovdiv and other towns. Rice-production was concentrated in the Pazardjik, Plovdiv and Yambol areas, and production of macaroni in the towns of Stara Zagora, Tolbuhin and Cherven Bryag. In 1982, some 1,334,000 tons of flour, 33,131 tons of rice and 18,197 tons of pastries were produced.

The milk-processing industry underwent a period of concentration and reorganization. In 1959 it was absorbed by the state sector, having previously been in the hands of consumers' co-operatives and People's Councils. In 1960 there were 1,470 dairies working on a strictly seasonal basis. Between 1960 and 1970, 38 large mechanized milk-processing plants (industrial dairies) were formed, supplemented by 60 smaller dairies. As a result of this process of concentration, by 1982 the milk industry consisted of 27 district enterprises comprising about 400 dairies. The industrial dairies are equipped with pasteurizing and cooling machines, equipment for souring cream, washing and filling machines, tank transport, etc. Automated production lines and new technologies were introduced. In 1982 some 23,036 tons of butter, 98,256 tons of *sirene*, and 25,297 tons of *kashkaval*, were produced; 22,050 tons of *sirene* and 3,465 tons of *kashkaval* were exported.

By 1982 *meat production and processing* had been concentrated in 28 meat processing plants with 57 subsidiaries. Medium-size (about 50 tons of output per shift) and large-size (over 50 tons of output per shift) plants predominate. In 1982, the large meat processing plants provided 68 per cent of the total output of the meat industry. Modern, efficient methods and up-to-date technologies are being used – mechanized meat-processing

The Elektronika plant in Sofia

lines, complex production lines for salami, frankfurters and skinless sausages, modern pork and beef sausage production technology, modern smoked salami production technology using the R-4 strain, and technologies for the use of animal and vegetable additives and by-products in salami-production. In 1982 some 490,951 tons of meat, 99,073 tons of meat products and 38,151 tons of canned meat were produced, and 53,702 tons of meat, including 31,670 tons of poultry, were exported.

The state-owned *fish and fish-processing industry* was established in 1947. Fishing was placed on an industrial footing, and by 1956, nineteen wooden-hulled drifters of between 80 and 240 hp had been built for the Black Sea Fishing Fleet. Fishing was mechanized, on the basis of Soviet experience in industrial fishing. In 1957 Soviet scientific bodies assisted in the organizing of the first scientific expedition to evaluate the fish reserves in the Black Sea. After 1975 the wooden-hulled drifters for the Black Sea Fishing Fleet were replaced by small trawlers, drifters, shoal detection boats and factory ships, all imported from the USSR. Seasonal fishing gave away to year-round fishing and this is now done with highly efficient equipment, including surface and bottom trawls. 1964 is considered to be the year in which deep-sea fishing began, when the Soviet Government delivered twenty deep-sea fishing trawlers and four cold-store factory ships. In 1982 the Ocean Fleet consisted of 25 trawlers and nine cold-store factory ships. Excellent onshore storage and processing facilities were built in Burgas and brought into operation in 1973. They housed mechanized port facilities, a refrigeration unit with a capacity of 10,000 tons of frozen fish, additional facilities for the production of ice and non-sterile fish products, a fish cannery (the Slavyanka combine), a factory producing deep-sea fishing equipment, a shipyard and a spare-parts shop, up-to-date storage facilities, a power plant and an administrative building for the Ribno Stopanstvo State Economic Association. In 1982 the Dragni Nedev Shipyards were built in the fishing port for the refits and running repairs on fishing trawlers, cold-store factory ships and support ships with a maximum displacement of 4,500 tons. High-capacity fish-machines for packing, washing, cleaning, etc., were installed in the fish-processing sector. The production process is organized on a continuous basis, with waste-free utilization of raw materials. Products include fresh and frozen fish, non-sterile fish products, semi-processed fish and delicatessen. In 1982 the catch was 139,645 tons, while canned fish production reached 12,375 tons. Part of this production is exported to Greece, Italy, Iraq, Syria, Egypt, Nigeria and other countries.

After 1947 the *sugar producing and processing industry* was radically modernized and expanded. New plants were built in the town of Lom in 1960, for the production of sugar and confectionery, and in Devnya in 1964, for the production of sugar. The existing factories were reorganized in branches: the Sugar Works in Gorna Oryahovitsa (the biggest in Bulgaria) and the Vassil Kolarov Sugar Works in Plovdiv, for sugar, sweets, sugar alcohol and other products; the K. Zlatarev Sugar Works in Dolna Mitropolya, for sugar and confectionery; the Svoboda Sugar Works in Kameno, for sugar; the Dimiter Blagoev Sugar Works in Russe, for sugar and enzymes, including yeast; two Research and Production Plants – Malchika in Sofia and Republika in Svoge – for chocolates and sweets; and the Geo Milev Works in Sofia for starch and grape-sugar. Factories are equipped with modern, efficient technologies – continuously operating diffusers, automatic centrifuges, up-to-date evaporation equipment, automated filtering and purification equipment, vacuum machines, etc. In 1982 some 399,000 tons of sugar and 114,135 tons of confectionery were produced, amounting to 44.7 kg of sugar and 12.8 kg of confectionery per capita.

After 1947 the *breweries* in Plovdiv, Shumen, Turnovo and Sofia were reconstructed and modernized and their capacity enlarged. New breweries were built in Stara Zagora, Mezdra, Burgas, Sofia, Varna, Blagoevgrad, Haskovo and Pleven. Production was mainly concentrated in the Sofiisko Pivo brewery in Sofia and in breweries like Shumensko Pivo in Shumen, Kamenitsa in Plovdiv, Balkan in Veliko Turnovo, Zagorka in Stara Zagora, Dunav in Lom, Ledenika in Mezdra, Cherno More in Burgas, Pirinsko Pivo in Blagoevgrad, Varnensko Pivo in Varna, Haskovsko Pivo in Haskovo and Kailuka in Pleven. Beer-bottling plants were built in Troyan, Russe, Turgovishte, Silistra, Tolbuhin, Yambol, Sliven, Kurdjali, Smolyan, and Petrich. The brewery sector has developed very quickly and between 1970 and 1982 the volume of production increased by 250 per cent. Production is based entirely on Bulgarian raw materials. Beer output in 1982 came to 547,669,000 litres. Bulgarian beer has been awarded a great number of gold and silver medals at

Karlovo Station

international fairs and competitions. At the 'World Selection' international competition for beer and soft beverages, organized by the Institut de Qualité in Brussels, Bulgarian brands have received both gold medals: Zagorka – De Luxe, Astica – De Luxe, Bolyarka –

A consumer services centre in the Lyulin housing estate in Sofia

Genuine, in 1981, 1982, 1983; Pleven – De Luxe and Shumensko Pivo – Special Brand, in 1982, 1983; Kamenitsa 100 – Special Brand in 1983; and silver medals: Zagorka – Special Brand in 1981 and 1982, and Veliko Turnovo – Special Brand in 1982.

In 1982 65 per cent of the *total output of soft drinks and mineral water* was produced by the Soft Drinks and Mineral Waters State Economic Association in Sofia, while the remainder was produced by companies run by district co-operative unions and the Ministry of Production and Trade in Consumer Goods. In 1958 and 1959 mechanized production lines were installed close to the natural mineral springs in Gorna Banya and Hissar, and later in Mihalkovo, for the production of aerated mineral water. After 1965, modern soft drinks enterprises were commissioned in Sofia, Plovdiv, Varna, Russe and Burgas. The degree of mechanization and automation is very high. The soft drinks are produced from fruit, vegetables, grain, and extracts of citrous fruits and medicinal herbs, with or without sugar, water, and carbon dioxide. Depending on the fruit content of the products, they are divided into two main categories: natural (fruit and vegetable) juices, and soft drinks based on citric and herbal extracts. The aerated soft drinks comprise herbal tonics, e.g. Coca-Cola, Pepsi, Tonic Water, Altai, Sun-Cola, and Breza; fruit drinks, e.g. Mandarin, Golden Orange, Pineapple, Apple and Raspberry; lemonade-type soft drinks, e.g. Etur, Cider etc.; aerated waters including Gorna Banya, Hissar, Michalkovo, Nevestino, Turgovishte and others; and low-calorie beverages with a reduced sugar content including Reneta, Buket and Limon. The annual consumption of aerated soft drinks and mineral water is 56 litres per capita (1982 data). The per capita consumption of natural mineral water is 1.4 litres, and of aerated mineral waters, 4 litres (1982 data). In 1982 some 118,371 tons of fruit juices were exported to the USSR, Poland, Cuba, the GDR, Austria, the FRG and elsewhere.

The *vegetable oil industry* is concentrated in fifteen industrial enterprises. These enterprises are either built in regions supplying the raw materials – the Mihailovgrad, Vratsa, Pleven, Shumen, Veliko Turnovo with two factories, Turgovishte, Varna, Tolbuhin, Silistra, Burgas and Stara Zagora areas – or in the regions where vegetable oil consumption is highest – in Sofia itself and the Sofia and Haskovo districts. The factories operate in consultation with the Vegetable Oils and Protein Research and Production Combine in Kostinbrod. They function throughout the year and process the entire harvest of sun-flower seeds, cotton seeds, and soya beans. Edible vegetable oils are produced in unrefined, refined, hydrated, bottled and bulk form. In the factories in Kostinbrod, Stara Zagora and Burgas, hydrated vegetable oils are produced for the food and soap industry. The factory in Tolbuhin produces margarine, and the Batunya Oil Factory in Boichinovtsi produces mayonnaise. Industrial vegetable oils from sun-flower and cotton seed are also produced, for the chemical, pharmaceutical, tanning, soap, coolants, printing, and other industrial sectors. By-products include sun-flower, soya, etc. husk residue, lecithin, alyphatic acids, mixed fodders, protein concentrates, active biological substances and others. The works in Burgas and Kostinbrod produce soap, and the factories in the village of Krushevo, near Veliko Turnovo and in Kostinbrod, produce bone-glue. In 1982 the production of vegetable oil was 167,777 tons, of which 161,757 tons were used for nutritional purposes and 6,020 tons by industrial concerns.

In 1982 the volume of production in the food industry was 170 per cent above the 1970 level, thus satisfying the needs of the country. The present level of development and current trends in the food industry are contributing to the increase in the population's standard of living.

The food industry is directed by the National Agro-Industrial Union, which includes the State Economic Corporations Vinprom and Bulgarplod, the Sugar Industry, the Fish Industry, Grain and Fodder, and the Vegetable Oils and Protein Research and Production Combine, and by the Central Co-operative Union, which includes the Soft Beverages and Mineral Waters Industrial Corporation.

CONSTRUCTION

The construction industry is erecting new buildings or extending, reconstructing and repairing the already existing buildings and facilities of production or non-production designation, including the fitting of machines and equipment. As a branch of the economy it started developing after the Liberation from Ottoman rule in 1878, when a large number of factories were built and extended, primarily with private capital. The State built a wide transport-communications network with funds raised by special taxation and foreign loans. The expansion of the cities gave a strong impetus to housing construction and town planning and urbanization. The greater part of the construction projects were carried out by private, capitalist building contractors. Construction in capitalist Bulgaria was characterized by a low degree of mechanization, hard working and living conditions for construction workers, lack of planning and squandering of funds and resources. State capital investments in this sector of the economy were insignificant – about 5.8 per cent of the national income a year on average for the 1926–44 period.

The rapid progress of the economy after 9 September 1944 considerably increased the investment capacities of the country. During the 1949–80 period, 84,665 million levs were invested in the national economy, of which 35,322 were in industry, 13,373 in agriculture, 8,102 in transport, 14,260 in housing and communal construction, and 13,608 in other sectors of the economy (see Table 1).

The increased investment led to a significant development of the entire economic sector of construction. In 1980 it accounted for 9.02 per cent of the GNP and 9.41

The dairy farm in Blagoevgrad

per cent of the national income; the building enterprises contributed 8.84 per cent of the total profit of the national economy, the relative share of wages was 12.28 per cent, and of those employed in this sector – 8.84 per cent of the nation's work-force.

A solid material and technical basis for the construction industry has been created. In 1976, the fixed capital in the

Table 1: Capital Investments by Plan Periods and Sectors (in million levs)

Periods	Total	Industry	Agriculture	Transport	Housing of communal construction	Other sectors and activities
1949–52	1,750	622	226	200	357	345
1953–57	3,243	1,225	572	247	713	486
1958–60	3,329	1,134	897	197	695	406
1961–65	8,390	3,556	1,772	531	1,421	1,110
1966–70	15,284	7,012	2,460	1,263	2,240	2,309
1971–75	21,736	8,882	3,326	2,266	3,404	3,858
1976–80	30,933	12,891	4,120	3,398	5,430	5,094

sector was 1,010 per cent higher than the 1960 level, with construction machines having the biggest share – 52.2 per cent. Industrial methods of construction have been introduced. A broad technical base has been built for the production of pre-cast, reinforced ferro-concrete units and sections, with plants, house-building combines, industrial construction enterprises, etc., distributed evenly across the country, as a result of which the main construction and assembly operations have been mechanized up to 90 per cent and over. Efficient, assembled-steel sections are increasingly widely applied in industrial construction, and extended pre-fab units in housing construction. The relative share of steel-concrete flats and houses in 1960 was 5.7 per cent, in 1965 – 36.3 per cent, in 1970 – 40.0 per cent, and in 1979 – 72.1 per cent. The degree of industrialization of housing construction has reached 45 per cent and of industrial construction – 80.0 per cent. A number of basic construction processes are performed at the highly-mechanized auxiliary plants of the construction organizations (concrete-solution units, reinforcement, shuttering, hardware workshops, etc.). Increasingly wide use is made of standard designs for identical buildings and facilities, or for parts of such (type sections), and of standard type construction elements and parts (type nomenclatures of reinforced concrete pre-fab units for housing and industrial construction), industrially mass-produced. No less widely used are the new and more effective construction materials – ceramicite, perlite porous agglomerate, glass wool, expanded polystyrene, polyvinylchloride and other synthetic materials. More advanced methods and technologies of construction are being developed and introduced – large panel sections, large-size shuttering, lifted slabs, climbing shuttering, etc. The industrialization of construction has increased economic efficiency in this sector. The main kinds of construction work have been mechanized: in 1978 the degree of mechanization of the basic construction and erection works had already surpassed 95 per cent. In 1960 the number of excavators was 226, and the amount of earth and quarry construction materials excavated was 17.9 million cu.m; in 1978 their number was 3,004, and the earth and quarry construction materials excavated – 149.6 million cu.m. In 1965, 462 tower and auto-cranes were used in erection work; by 1978 their number had increased to 3,000. As a result of mechanization the productivity of labour in construction more than doubled during the 1960–78 period, while the time terms of construction works were shortened. The mechanized, streamlined process of assembly and erection of the buildings and facilities from pre-fab sections has become typical of production organization in this sector. The large machines used in the basic construction and erection works are maintained and repaired by specialized economic organizations and enterprises working on a self-supporting basis and renting machines to the building organizations. The smaller machines and mechanical devices are exploited and maintained by units of the building organizations or directly by the work teams. The mechanization of construction has led to the rational and economical use of raw and prime materials and energy, with a reduction of the costs and improvement of the quality of construction and erection work and an upgrading of the skills of workers. Construction work is no longer seasonal in character. The annual average growth rate of the productivity of labour per worker in construction has increased. After 1970, as a result of the industrialization and mechanization of construction and erection work, the relative share of those employed in construction decreased, and amounted to 8 per cent of the total number of people employed in the national economy in 1976. Construction and erection works are performed mainly by state construction organizations – 92.5 per cent in 1976. Industrial construction is carried out by the organizations 'Industrial Construction', 'Assembly and Erection Work', 'Hydrostroi', 'Construction Mechanization', the 'Ministroi' Economic Corporation, the economic combines 'Metal Constructions', 'Steel-Concrete Construction and Parts', etc. The smaller construction projects are carried out by the works under the management of the respective District People's Council. Reconstruction, modernization and extension of buildings is done by the respective industrial enterprises, using primarily local resources and materials.

Housing, agricultural, commercial and communal construction, as well as that of the local infrastructure (culture, education, health, sports, tourism, etc.) is carried out by the general construction organizations – 'Sofstroi' in the capital city, and the district building organizations in the various districts. Specialized construction in the area of transport is carried out by the 'Transtroi' State Building Organization, and road construction by the Road Building State Department of Construction.

Industrial Construction. Immediately after 9 September 1944 industrial construction was carried out by private contractors and newly established construction co-operatives, lacking machines and equipment. During the period of the restoration of the economy (1947–48) a large number of industrial projects were built with the participation of national youth brigades and with whatever means were available to the respective ministries and central departments.

During the 1950s the first big, heavy industrial projects were built and commissioned with Soviet assistance in designing, supervision of erection work, delivery of machines and equipment, starting production and the training of personnel. These included the Combined Chemical Works in Dimitrovgrad, the Lead and Zinc Works in Kurdjali, the Karl Marx Soda Works in Devnya, the Stefan Kiradjiev Cellulose Plant in the town of Novi Krichim (today Stamboliiski), the Lenin Metallurgical Works in Pernik, the Nitrogen Fertilizer Plant in Stara Zagora, and thermo- and hydro-electric power stations and mines. The construction was carried out by Bulgarian organizations with the participation of Bulgarian designers. Some major heavy industrial projects were built in the 1960s – the Kremikovtsi Metallurgical Works (the

The clover-leaf junction on the Peyo Yavorov Avenue in Sofia

The Diana sports centre in Sofia

L. I. Brezhnev economic metallurgical works of today), machine-building plants, the petro-chemical works near Burgas, the first thermo-electric power stations in the industrial zones, and the power base of the state mining enterprise 'Maritsa-East' at the town of Radnevo. In the general plans there is a growing tendency to concentrate the plants in order to make the most rational use of the territory. Industrial buildings are erected on the basis of unification and typification of construction elements. The assembly type of construction has been gaining ground, and there are special plants producing units for this kind of construction. The buildings are designed in line with the latest requirements and trends of creating architectural ensembles, producing a strong impact with their impressive structure.

The 1960s and 1970s witnessed the beginning of scientific and technological achievements in industrial construction. Production capacities were increased in the engineering and chemical industries, in power development, metallurgy and electronics; atomic energy became part of the economy. The construction of large-scale industrial projects was launched, such as the chemical works in Vidin, Vratsa, Svishtov, Devnya, Yambol and Pleven; the mechanical engineering works in Varna and Sofia, the electronic chips works in Botevgrad and Sofia, the memory units works in Stara Zagora, the printed circuits works in Russe and many others; huge thermo-power stations and the Maritsa-East power complex, the Varna and Bobov-dol thermo-power stations, the Kozlodui atomic-power station and other enterprises of the heavy, light and food industries. Heavy investment, machine-building projects were initiated with the combined engineering works in Russe and Radomir. Industrial construction in the 1970s was characterized by the employment of new, effective materials, the wider use of metal structures and of larger prefabricated units. Industrial architecture is distinguished by an up-to-date approach to configuration and design. Notable among its achievements are the Anton Ivanovtsi electric-power station, the Kozlodui atomic power station, the Maritsa-East 3 thermo-power station, the chemical works in Devnya, the transistor electronics works in Botevgrad, the Elektronika works in Sofia, and the combined polyester silk works in Dimitrovgrad.

The successes in *hydroelectric power development* are particularly impressive. A relatively short period (1956–75) saw the completion of the Iskur hydro-power project (1956–57), whose five power stations have a total capacity of 226 MW and an annual output of 284 million kWh; the Batak hydro-power scheme (1958–59), the capacity of its three stations totalling 226 MW and its annual output – 796 million kWh; the Arda system (1958–64) with a total capacity of 280 MW and annual output of 600 million kWh; the Sandanska Bistritsa system (1969–72) with a total capacity of 55.5 MW and an annual output of 189 million kWh; the Belmeken-Sestrimo scheme (1973) with a total capacity of 755 MW and an annual output of 11,800 million kWh; and the Vucha scheme (1969–75) with a total capacity of 400 MW and an annual output of 752 million kWh.

Agricultural construction is also developing at accelerated rates. Outbuilding facilities are constantly built for the needs of plant-growing and stock-breeding, for storing and processing agricultural produce, artificial fertilizers and manures, and for sheltering and maintaining farming equipment and machinery. In the initial period of building the co-operative and state farms, and the machine and tractor stations (1944–50), these were restricted to cow-sheds, pigsties and poultry houses. Towards the middle of

Hotel Echo in Lagos, Nigeria, built by the Bulgarian Technoexportstroi State Corporation

the 1950s, the growth in number of the co-operative farms led to the intensification of construction work. Outhouse facilities increased both in size and variety. They now included ox-sheds, seed-depositories, fruit storage houses, repair shops, veterinary hospitals, etc. Nearly 50 per cent of these buildings follow standard architectural designs. Over the 1958–60 period, 27 per cent of Bulgaria's capital investments were allotted to agricultural construction. Large-scale pig, poultry, dairy, calf-fattening, ewe, and lamb-fattening farms with an industrially based production process came into being. Besides individual buildings, farm facilities are sometimes arranged in large compounds. Pre-cast reinforced concrete units are coming into much wider use in these constructions. Higher aesthetic criteria are set. The gap between technological and compositional methods in agricultural and industrial construction is being closed. Agro-industrial buildings are appearing. Large industrial dairy farms, housing from 500–2,000 cows, and pig farms, breeding 30,000–50,000 pigs a year are being built at agro-industrial complexes, research and production centres and the like. The achievements of Bulgarian agricultural architecture and construction are earning international acclaim. Various agricultural projects, designed by Bulgarian architects and sometimes constructed by Bulgarian builders, have been undertaken in more than 30 countries of Europe, Asia, Africa, and South America.

Newly-built facilities, as well as reconstruction and modernization, have also helped boost trade, public utilities and the tourist industry in Bulgaria.

An adequate commercial network is being built. Public utilities are being expanded and industrialized. There is a vast system of establishments for public services, industrial supply bases, workshops, parlours, reception centres engaging in more than 400 varieties of communal services, laundries and dry cleaners', household appliances, maintenance and repair shops, car service stations, housing construction and repair services, barber and hairdresser's shops, cosmetic parlours, etc. Fashion houses and centres offer elegant, made-to-measure clothes. There were 251 laundries, 2,293 barber and hairdresser's shops and 1,064 public baths in 1982. Large and properly equipped department stores and multi-service centres have been built, among them the Central Department Store and the Services Centre in Sofia, and the corresponding establishments in the district and larger towns. New administrative and residential buildings are erected complete with various shops, while the larger industrial enterprises and plants boast their own shopping and service centres.

The Hristo Kurpachev primary school in Sofia

Resort Construction. The variety of nature and climatic conditions is conducive to a large-scale resort construction programme, consisting of balneological, mud-curing and climatic (mountain and seaside) projects.

Bulgarian resorts are steadily growing in number and capacity. The seaside resorts of Druzhba (construction undertaken in 1949), Golden Sands (construction started in 1956), Sunny Beach (the largest in Bulgaria; construction started in 1958) and Albena (construction initiated in 1968) enjoy world-wide popularity.

Each of these resorts has an architectural configuration of its own, which is in harmony with its environment. The dynamic, original architectural composition of the Albena complex is particularly attractive, and its architects were awarded the Dimitrov Prize in 1971. The growing popularity of winter sports has accelerated the construction of mountain resorts, such as Pamporovo, Borovets and others. The original architecture of the Perelik Hotel Complex makes it a prominent feature in the Pamporovo resort. (See the chapter on Tourism and Resorts, pp. 492–514.)

Transport and road construction has also made tremendous progress. The network of existing railway lines is being reconstructed and modernized, and the traffic capacity of line routes and sections has been augmented. New tracks, such as those joining Pernik to Voluyak, and Lovech to Troyan have been laid. The first electric railway sections went into operation in 1963, and in 1984 the electric ring railway was completed. Railway tracks added up to a total length of 6,416 km in 1982.

The Bulgarian ports have also been expanded, reconstructed and brought up-to-date. They have also been agglomerated into complexes, such as the Burgas complex, including the six Black Sea ports of Burgas, Nessebur, Sozopol, Pomorie, Michurin and Ahtopol; the Varna complex, including the four harbours of Varna, Varna-West, Balchik and the Varna thermo-power stations; the Lom complex, with three ports on the Danube in the towns of Lom, Vidin and Oryahovo; the Russe complex, with five ports on the Danube in the towns of Russe, Svishtov, Silistra, Somovit and Tutrakan; and others. Modern airports, serving domestic and international flights, have been built in Sofia, Varna, Burgas, Silistra, Plovdiv, Turgovishte, Vidin and other towns. The Sofia, Varna and Burgas airports are international airway centres.

Modern asphalt and concrete covered roads and motorways have been built with petrol stations, campsites, motels, restaurants and rest-places abounding along them, a touch of decoration added by a profusion of extra-unconventionally shaped road signs; monumental shafts of stone at the entrance to villages, bearing their names; sculptures; fountains; rotundas; etc. Road construction is marked by a high level of mechanization. Sophisticated road and roadside facilities have been built. Bridge construction is particularly extensive. The Bridge of Friendship over the Danube (operating since 1954) is of tremendous importance. A number of pre-stressed arched and bowstring construction bridges spanning up to 70 m., as well as multi-crowned girder constructions with up to 40 m. spans each, have been built. A most notable achievement is the Asparuh bridge in Varna (1976).

Housing construction is among the top priorities of the Party's and Government's social policies. It is state-owned, to let or for sale to individual citizens and co-operatives on a single or group basis. Most of it is assigned to building organizations (about 80 per cent) and the rest is achieved by using local resources and materials, which goes for individual construction as well. Co-operative and individual construction are stimulated by the State, which provides materials, machinery and help from the building organizations. Pre-cast large panel construction is finding an increasingly wide application – from 19.2 per cent in 1967, it grew to account for 45 per cent of all newly-built homes in 1977. The production and assembly of pre-fabs have been industrialized so that in 1980 their capacity amounted to 55,000–60,000 new flats a year. The method of broad-surface shuttering is also gaining preference. Architectural and urbanization solutions in individual buildings and housing estates are steadily being improved. The disposition of the rooms is most functional and convenient. There is a tendency to partially or fully furnish new flats. Care is taken with the development of newly built areas. The residential quarters and complexes are self-sufficient (see 'Living Conditions', pp. 548–49).

Cultural Facilities Construction. Architecture has done a great deal for developing cultural facilities that will help preserve objects of cultural value and make them accessible to the people.

The Lyudmila Zhivkova People's Palace of Culture under construction in Sofia

In the first decade following 9 September 1944, construction amounted to restoring the destroyed, and completing the already started buildings, such as the Cyril and Methodius National Library, the Bulgarian Concert Hall and the National Theatre in Sofia, a library and a museum in Veliko Turnovo, etc. Some of the existing buildings were adapted to serve new purposes, as for instance the National Opera House in Sofia (the Alexander Stamboliiski Memorial House), the Opera House in Russe, etc. The construction of a limited number of new projects was initiated – the Cinema Centre in Boyana near Sofia, the Dimiter Blagoev Printing and Publishing House in Sofia, etc. After 1956 cultural construction started developing at a faster rate. Thus the pre-conditions were laid for the further flourishing of socialist culture. Eleven opera houses and theatres were built, as well as three district libraries, six museums and art galleries, many reading clubs, new production and editorial buildings of the radio, television and printing houses, etc. The National Theatre in Sofia was reconstructed (1977), the April Uprising memorial complex was erected, as well as the Home of Soviet Science and Culture in Sofia, the National Home of the Party on Mt Buzludja, the Lyudmila Zhivkova People's Palace of Culture in Sofia, and the 1,300 Years of Bulgaria Memorial Complex in Shumen.

Educational Construction. In the years of popular rule great successes have also been achieved in the construction of buildings for educational purposes. Many buildings were constructed to house unified secondary polytechnical schools. Educational construction was carried out using industrial methods, which made it possible to unify and typify to the maximum all the construction elements, thus ensuring flexible planning. The assembly method of construction is being widely used, such as the frame-and-panel beamless system and the lift-slabs method.

Kadin Bridge in the village of Nevestino in the Kyustendil district

Many schools, kindergartens and higher educational establishments were constructed. A Higher Food and Tobacco Institute was built in 1962 in Plovdiv. It was followed by the construction of the Veliko Turnovo Cyril and Methodius University (1967–75), the Karl Marx Higher Institute of Economics in Sofia (1973), and the Social Sciences and Public Administration Academy (1976).

A programme for the further construction and modernization of the material and technical basis of the unified secondary polytechnical schools and of the new type of technical vocational school has been drawn up and currently implemented. Educational and professional complexes are being set up for professional training and qualification. In 1982, 245 million levs of capital investment were spent on educational construction. Designing and building are based on normatives (number of students per unit of population). Schools and kindergartens form the centres of districts and microdistricts. The material and technical basis is changing not only quantitatively but qualitatively, becoming a substantial component of the environment for the harmonious development of the individual.

Health Care Construction. Health care is another sphere of brisk construction. Many hospitals, polyclinics, sanatoriums, prophylactoria and other medical establishments were built and fitted out with the latest equipment within a short period. The Pirogov Institute for Emergency Medical Aid in Sofia holds a place of

Construction

Table 2: Completed Projects of the Social Infrastructure

Projects	Number/Capacity	1970	1975	1980	1981
Completed homes	no.	45,656	57,151	74,308	71,419
General education	no.	73	29	22	34
	cap.	31,868	19,950	13,660	28,815
Crèches	no.	11	34	7	5
	cap.	835	4,920	1,470	960
Kindergartens	no.	37	124	97	88
	cap.	3,511	14,060	9,663	8,866
Holiday-homes	no.	56	16	6	9
	cap.	2,636	1,148	452	193
Incl. light constructions	no.	41	11	4	7
	cap.	876	348	64	93
Medical, polyclinic, sanatorial and resort establishments	no.	10	12	6	2
	beds	1,833	2,488	1,536	290
Hotels	no.	35	8	3	3
	beds	8,269	4,111	760	1,329
Water-supply network	km	2,312	1,355	723	625
Sewerage	km	329	234	192	206

prominence among them. The pharmacy network is growing (see Table 2 and the chapter on the Public Health Services, pp. 550–67).

The construction of office buildings constitutes one of the important branches of construction, marked by a new social orientation. The individual buildings are now being integrated in the entire architectural scheme of the respective town or village on the basis of detailed town-planning projects. This is a guarantee for creating fine architectural ensembles. Among the most notable office buildings in Sofia are those of the Machinoexport Foreign Trade Association, the Bulgarplod State Economic Corporation, Electroimpex, the Energoproekt research and design institute for power-plant construction, the Ministry of Transport, the Ministry of Light Industry, the special building with halls for the official signing of international agreements at the Boyana Residence of the State Council, etc. Other important office buildings are those of the City People's Councils in Varna, Pleven, Mihailovgrad, Vidin, and elsewhere. During the 1960s and 1970s the office buildings became part of the complex centres for public services, which are an organic component of the multi-purpose architectural ensembles.

The construction of office buildings is mainly financed by the State. Some special office buildings of plants, cultural unions, associations and institutions are also financed by the respective ministries, associations and unions.

The Construction of Sports Facilities. The concern for the physical development of the individual finds an expression in the improvement of facilities and equipment

The Hemus motorway; positioning a bridge with the help of helicopters

for physical education, sports and mountaineering. The construction of sports and hiking facilities contributes to this end. A great many stadiums, sports halls and grounds, sports equipment and mountain hostels have been built. Prioroty is given to indoor sports facilities which can be used all the year round. There are also sports training centres which are often combined with stadiums, such as the complexes in Stara Zagora, Russe, Pernik, etc. The Vassil Levski Stadium in Sofia, built in 1953, is the first modern stadium in Bulgaria. Quite remarkable, from an architectural point of view, are the sports auditoriums -- Festival Hall in Sofia (1968), the Palace of Sports and Culture in Varna (1968), the October Sports Hall (1977) and the September Winter Stadium (1973) in Sofia, and the indoor tennis courts in Sofia (1968) and Plovdiv (1982). (For further information see the Physical Education, Sports and Tourism chapter, pp. 568–93).

The construction projects abroad are of particular significance for Bulgaria. The country became involved in this type of construction project in 1956 with the building of a port in Latakia, Syria. In charge of such activities are the Technoexportstroi State Economic Corporation and the General Directorate 'Bulgarian Construction in the USSR'. The Technoexport specialist engineering organization exports complete plants and technological lines; the Agrocomplect engineering economic organization and the Bulgargeomin State Economic Corporation are involved in research, design, construction, delivery, operation and technical maintenance of projects related to agriculture and the food industry, complex geological prospecting work, engineering geological and hydrogeological works and the construction of mines. The Bulgarian construction organizations dealing in projects abroad have their own representative bureaux in about 25 countries. They employ a large number of engineers, architects, geologists, town-planners, technicians and builders and use modern machines and transport facilities. These enterprises are also making good use of the specialists and resources of the specialist construction and assembly organizations at the research and design institutes in Bulgaria. Bulgarian architects and building specialists regularly take part in international trade competitions. Their designs, which are up to modern architectural standards, are realized in practice within the set terms for quality and expediency. Bulgaria designs and builds in more than 40 countries in Europe, Africa, Asia and Latin America.

Research and design abroad covers the design of town-planning projects, the design of the infrastructure of housing, public, office, sports, hydro-technical, transport, agricultural, health, industrial and other kinds of construction, designer and investment control and technical assistance in compliance with the economic, social and cultural development of the respective countries. The Bulgarconsult Economic Organization was set up for the purpose, and all the design institutes and organizations and their numerous highly-qualified specialists are members of Bulgarconsult. Bulgarian architects and engineers have been winning important international competitions and tenders – the designs for the construction of buildings along the Mohammed V and Bourguiba Boulevards in the town of Tunis (1960), design for the central part of Berlin (1962), the town-planning design of the central part of Karlsruhe, FRG (1963), the design of the sports complex in Tunis (1964), the design for the San Francisco central square in the USA, the design for the construction of the central part of the capital of Mali, Bamako. Some of the designs which have already been realized include: in Syria (1957–70) – ten town-planning designs for the coastal towns of Latakia, Erihe, Djeble, Asharie, etc.; the Balme, Khabur and Asharie irrigation systems; the Balik and Kabur land-improvement systems for an area of 250,000 ha; office buildings and houses; hotels, irrigation facilities, bridges and roads in Iraq (1969–70) and in Guinea (1961–64); an Olympic-size stadium for 150,000 in Ghana (1965–68); the Labadi Biku resort and an administrative centre in the town of Accra; in Tunisia (1960–78) – hotels, a TV Centre in Bizerte, a national cultural centre, a national sports complex and a town-planning design for the town of Tunis; in Libya (1965–78) – a resort in the town of Sabrata and other projects in Zliten, Barche, Tobruk, Benghazi and Tripoli; district hospitals in the towns of Sirte, Bengashir, Tarhuna, Sabrata; poly-clinics; the Uaddi Aibda and Uaddi Ghan dams; water-supply systems and other projects. The Bulgarian designers won world reknown and recognition with the design for the National Theatre in the capital of Nigeria, Lagos (1973), which has a main hall for 5,000, a conference hall for 1,600, two cinema halls for 800 each and exhibition halls. Other notable achievements of Bulgarian architects and engineers are the designs for the irrigation system near the towns of Marrakesh, El-Jadida and Frih Ben Salam in Morocco; the Arib irrigation system; the Gakhal dam; the irrigation systems in the valley of Mitidja; schools, office and sports buildings in Algeria; a hotel in the capital of the People's Democratic Republic of Yemen; a pig-raising farm and a poultry farm in the Ouando district in the Congo; oil pipe-lines and reservoir installations in Iraq, different projects in Afghanistan, the People's Republic of Mongolia, Yugoslavia, the FRG, Lebanon, Cuba, Peru, etc. Awarded special prizes by international juries were the projects for the sewerage system in Damascus, Syria (1971); the designs for polyclinics in Libya (1974); the town-planning design for the town of Tunis (1965), and the design for the Municipal Hall in Al-'Ayn in the United Arab Emirates (1977).

The construction and assembly activities abroad involve the realization of almost all kinds of construction work. Among the major projects built by Bulgarian specialists are the National Theatre in Lagos, Nigeria (1973–75); the multi-purpose hall in Skopje; projects in Latakia, Syria; the Uaddi al-Hira (1975–77) and al-Djufra (1978) in Libya, sports stadiums in Tunisia (1965–67), in Benghazi and Tripoli (1965–68), in Iraq, and in the towns of Sebcha and Tripoli (1965–68) in Libya; in the town of Ilare, Nigeria; projects in Baghdad (1962–65); two in Syria, the Diala (1969–72) and Husseinia

(1972); irrigation systems in Iraq; the Balme, Khabur and Asharne irrigation systems in Syria; the Rastan (1959–61) and Meharde (1968–70) dams in Syria, the Nekor and Gru (1967–69) dams in Morocco; sewerage systems in Baghdad (1968–71) and Tripoli (1968–70); health establishments – polyclinics (1974–76), hospitals in the towns of Sirte, Ben Gashir, Al-Zahara, Zavia, Garian, Zliten and Nisurata in Libya; roads, and bridges – over the Euphrates (1967–69), the Sueira bridge over the Tigris (1969–72), the Rastan bridge over the Orontes (1972–74) in Syria, the Basra bridge over the Gyarmat Ali (1966–68) in Iraq; industrial and power projects – a thermo-electric power station for 740,000 kWh (1972–74) in Skikda, Algeria; a 220 kWh powerline in Libya; an oil project in Zavia (1971–73) and a nuclear research centre near Tripoli in Libya; an ore-dressing combine for the mining and processing of lead and zinc ores in Algeria; the assembly of two automated cement plants in Amel (1973–75) and assembly operations at the paper plant in Derizor in Syria. More than 80 large projects were completed and prepared for operation (turn-key) over the 1956–78 period, without claims during the past ten years.

Bulgarian construction and assembly workers, together with Soviet construction and assembly organizations, are working on industrial projects in the USSR. As a result of joint work over 140 large projects were built and put into operation in 1969–78 – the third stage of the Arkhangelsk and cellulose combine; the Lebedinsk and Mikhailov ore-dressing combines in the towns of Gubkin and Zheleznogorsk; the cement plant and the plant for silicate products in the town of Stari Oskov; the Groznensk gas-processing plant (all in the RSFSR); the plant for gas installations in Uzhgorod, in the Ukraine; the iodine plant in Nebit-Dag in the Turkmen Republic; gas-compressor and oil pump stations, cultural and services establishments, houses, etc. Bulgaria takes part in the construction of the Soyuz gas pipe line from Orenburg to the western frontier of the USSR; the Ust-Ilimsk cellulose plant; the Kiembaev Asbestos combine; different industrial projects; cultural and services establishments and housing estates at the Kursk magnetic anomaly, and in the construction and assembly of projects of the oil and gas industry.

Local administrative bodies, resident engineers and international consultant firms think highly of Bulgarian specialists and construction workers. This is due to the high quality of their designs and construction work and the short terms in which they materialize. The International Institute for Development and Prestige in Geneva awarded its 1975 prize to the Technoexportstroi State Economic Corporation for its contribution to 'the creation

Residential buildings and a kindergarten in the B-5 Zone of Sofia

of better life on Earth'. The Corporation has also been awarded the 1977 gold medal of the Chamber of Commerce, Industry, Crafts and Agriculture in Milan. The Bulgargeomin State Economic Enterprise has been awarded the Phoenicia international trade award for 1978. The Soviet Government values highly the work of the Bulgarian builders.

AGRICULTURE

Agriculture in capitalist Bulgaria was extensive and with low productivity. In 1934 there were 884,900 farms made up of 11,862 thousand plots of land with an average size of 0.37 ha. Of these, farms holding up to 10 ha. of land accounted for 92 per cent and those with over 50 ha. – 0.1 per cent. Each household had an average of thirteen to fourteen plots of land, which was cultivated mainly with draught animals and wooden ploughs, though over 22 per cent of farms did not own any draught animals. There were only 3,000 low-power tractors in the country, used mainly for threshing machines rather than for land cultivation. The cow was the main draught power.

Chemical products for agricultural purposes were not yet popular; fertilizers, herbicides and pesticides were practically unkown; irrigation was limited. Only 0.6 per cent of the arable land was irrigated. During the 1934–39 period the average wheat yield per ha. was 1,246 kg, barley – 1,300 kg, maize – 1,171 kg, sunflower – 830 kg, sugar-beet – 15,720 kg, apples – 2,670 kg, grapes – 4,460 kg, tobacco – 929 kg, etc. The average yield per cow in 1939 was 450 litres of milk, per sheep 1,511 kg of wool, while the annual average number of eggs per hen was 73.

The state of agriculture was further aggravated by the underdeveloped home market, the low level of industrialization, the lack of major consumer centres and the poor development of subscribed bank credit. Poor farmers could not sell their products themselves and thus fell victims to speculative trade capital and the private money-lenders. The relative agrarian overpopulation was developed to a high degree, mainly in the form of hidden unemployment. Sufferers from this trend were about 1 million poor peasants and farmers with medium-sized land holdings, who were exploited by the land owners and were victims of the bank, trade, and usury capital, and of the State. Tenant farming was widespread. In ten years (1920–30) the peasants' debts increased 1,200 per cent. The agrarian problem in Bulgaria grew out of proportion, and though the bourgeois Government sought to solve it through land consolidation and agrarian reforms it was in fact the large capitalist farms which profited from these measures. The process of ruination and impoverishment of most of the small and petty farmers continued. The peasants were seeking a way out of this disastrous situation by organizing farm associations and different co-operative alliances. In some regions they organized the so-called 'water-syndicates', and built land-improvement and irrigation facilities. A great number of credit and consumer co-operatives were set up, which proved to be a successful way of curbing exploitation to a certain degree, and of fighting the trade and usury capital. The co-operative movement began to develop: as many as 66 co-operative farms were organized, 29 of which continued to function until the 9 September 1944 revolution, regardless of the opposition of the bourgeois governments. The orientation of farmers towards the amalgamation of agriculture through producer co-operatives was a serious prerequisite for a faster and comparatively easier solution of the agrarian problem on the basis of collectivization after the victory of the socialist revolution.

Lavender plantation near Kazanluk

After 9 September 1944 the co-operative farms became the most suitable form of collectivization and socialist transformation of agriculture. The Bulgarian farmers made use of Soviet experience in this respect, adapting it to the specific features of agrarian relations in Bulgaria and observing the main principles – voluntary, gradual collectivization, democratic management, combining private and public interests and rendering active state assistance to the co-operative farms, such as political, economic and financial support. Socialist production relations in agriculture were consolidated by preserving private ownership of

Table 1: Collectivization of Agriculture

Year	Number of co-operative farms	Farmers in the co-operative farms (in thousands)	Percentage share of co-operative farmland from the total
up to 1944	29	1.7	—
1944	110	7	0.6
1946	480	41	3.7
1948	1,100	124	7.2
1950	2,501	502	51.1
1958	3,290	1,244	93.2
1960	932[1]	1,256	98.4

[1] The number of farms decreased because of amalgamation.

the land, which was carried out through rent. A specific approach, suited to Bulgarian conditions, was used in establishing the co-operative farms: the socialization of land not through a decree (land expropriation), but through gradually eliminating the rent against the land incorporated into the co-operative farm, which was done at the wish of farmers.

The most characteristic features of the co-operative farms as specific agrarian units in Bulgaria were the following: the means of production (except the land) were public property, of which there were two types – co-operative and state. The land remained the private property of the individual co-operative farmers, who were remunerated according to the quantity and quality of the work done. The distribution of incomes was based on two principles: according to the work done – not less than 70 per cent, and according to the private land-ownership rights – not more than 30 per cent. In the second half of the 1950s, at the initiative of the co-operative farmers, the share of the rent gradually decreased. Around 1960, following a decision of the general co-operative assemblies, the principle of distribution according to private land-ownership rights was suspended in almost all co-operative farms. Each household could set up its own farm on a plot of land, which according to the statute could be between 0.1–0.5 ha., and could raise an inseminated cow, a certain number of sheep and pigs and an unlimited number of poultry and beehives.

This type of co-operation and the principles upon which it was performed appealed to the Bulgarian farmers and they gradually began to join the co-operative farms. The first to join were the poor and petty peasants, and from 1950 onwards petty farmers began to join the co-operative farms in great numbers. In the very first years after their establishment yields went up by 20–30 per cent compared to that of private farms. The process of collectivization covered sixteen years.

The preparation and organization of the co-operative farms was carried out through a system of government acts. In 1945 the National Assembly passed the Law on Co-operative Farms, and in 1950 the Second National Conference of Co-operative Farms adopted a model Statute which served as the main law for their activities. On its basis, each co-operative farm had to prepare and adopt its own statute, reflecting the specific conditions of each community. A considerable part of the land expropriated from the big landowners under the Agrarian Reform Law of March 1946, was used for the setting up of state farms. These served as model socialist agricultural enterprises and supplied the co-operative farms with high grade seeds, seedlings and breeding animals. In 1956 the state farms were in charge of 3.7 per cent of the country's arable land. The State also established a great number of machine-and-tractor stations, which assisted production and provided technical maintenance to the co-operative farms. They played a decisive role in the creation and the development of an adequate material and technical base for large-scale farming.

The training of management cadres, university and agricultural college specialists and qualified farm workers greatly contributed to the development of farming. A system of research institutes and experimental stations was created; they worked on specific problems and in particu-

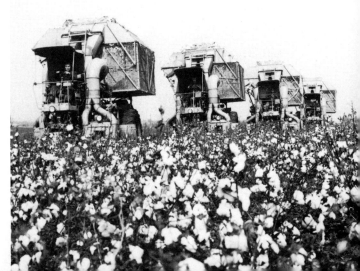

Cotton-picking with combine harvesters in Radnevo, in the Haskovo district

lar sectors, and were located in different parts of the country, according to the ecological features of the respective regions. On 9 September 1944 there were in Bulgaria 1,230 agronomists, some university graduates, 865 veterinary surgeons and 2,600 agro-technicians. Over the past few decades, the number of qualified specialists has greatly increased: in 1981 the total number of farm specialists was 56,500 of which 20,800 were university graduates: 2,500 engineers, 8,900 agronomists, 2,400 veterinary surgeons, and 3,400 livestock specialists. There were also 35,700 people with secondary special education, of whom 10,400 were technicians, 7,100 agronomy technicians, 4,800 livestock technicians and 4,500 assistant veterinary surgeons.

Sunflowers

The accelerated development of productive forces called for a change in both production and management structure of farming. The preparatory and organizational work for the merger of the co-operative farms started in the late 1950s. In 1959 there were 3,290 co-operative farms with an average of 1,153 ha. of arable land. They were multi-branch farms. The size of the different sectors and the bulk of homogeneous crops were small and did not allow the effective implementation of the achievements of science and technology and of the natural, economic and labour resources. Modernization of farming required a higher degree of concentration and specialization of agricultural production. The need to adapt the organization and production structure to the new requirements was recognized by the Party, the State and the members of the co-operative farms. The merger of the co-operative state farms was performed over a period of two years. Towards the end of 1960 there were 932 co-operative farms with an average of 4,266 ha. of land. In the 1960s the advantages of large-scale farming began to be felt most tangibly. An accelerated process of intensification of agricultural production was marked by large-scale application of fertilizers, irrigation, the creation and dissemination of new plant varieties and animal breeds, mechanization and automation, and other branches of scientific and technological progress. Considerable changes took place in the structure of the rural economy. The volume of agricultural production increased and its quality improved.

The merger process was conducted in an organized manner, taking into account the specific conditions and especially the type and level of specialization of agricultural production, terrain factors, population density according to regions and micro-regions, etc. The size of the separate merged farms varied widely (from 1,600 ha. to over 8,800 ha.); the majority of the farms had 4,300 ha. of land on average, according to their specialization and mechanization. The smaller farms were in regions specializing in the production of vegetables, tobacco and other crops which are more labour-intensive. The biggest farms were in regions specializing in grain production, where mechanization was high and labour intensity low.

The further development of the co-operative farms and of the national economy, accompanied by the rapid progress of science and technology, raised new problems and contradictions between the organizational and production structure of agriculture, and ways for their solution emerged. The development of agricultural production necessitated a further improvement in the effective use of all factor inputs and arable land, and the raising of quality as a prerequisite for meeting the population's

Combine picking broad-leaved tobacco at the Takia Agro-Industrial Complex, Plovdiv

Irrigation system at the Georgi Dimitrov Research and Production complex, Plovdiv

consumer needs of farm products, and for securing competitiveness in the world market.

The problem could be solved comprehensively through the industrialization of farming. This was made possible by the existing level of development of agricultural production at the beginning of the 1970s and the high degree of development and application of the achievements of science and technology. The country's economic development, and the new tasks facing farming, required a new type of co-operation between agriculture, the branches of industry processing agricultural products or supplying it with the means of production, and the respective research organizations. This meant organized guidance of the objectively developing process of agro-industrial integration.

Changes were effected in the organization of production and the management of agriculture, and in the sectors and activities directly involved in the servicing and development of farming. A new structural unit of agricultural organization was established – the agro-industrial complex (AIC). These are territorial production organizations with economic and social functions. According to the regulations, they have the rights, obligations and responsibilities of an economic organization in agriculture, and perform their activities according to the principles of self-accounting and self-support. Their internal organization and production structure is being continuously developed and perfected in compliance with the changes in the concentration and specialization of production. In the initial stage the main structural organization and production units were the co-operative and state farms and, in the second half of the 1970s, the specialized production and the production servicing enterprises, the branch farms and production sections.

The optimal combination between centralization and decentralization in the management and activities of the agricultural production units helps refine the internal organizational and production structure of the AICs. The AIC centralizes the management of the investment process, the consistent pursuit of a scientific policy, and

Irrigating a field at the Georgi Dimitrov complex, Plovdiv

the implementation of scientific and technological achievements in production through and beyond the investment process guarantees material and technical supply, the advantageous marketing of farm products, the regular maintenance of equipment, and the organization of the other servicing activities.

A new structural, organizational and production unit – the integrated team (comprising all villagers) which is both juridically and economically autonomous, started to develop in the early 1980s and has already become a dominant pattern. It is responsible for the management of public property, for the most efficient use of resources – land, manpower, raw and prime materials, agricultural technology and equipment, land improvement installations and all the stock and funds placed at its disposal. It uncovers and utilizes new opportunities for increasing production and raising its quality, and organizes accounting and supervision of the quality of labour and production. It takes decisions on such matters as providing incentive for workers who perform well and imposing sanctions on those who do not fulfil their duties conscientiously. It carries out social welfare and cultural programmes through the funds it has been authorized to raise in accordance with production results.

Toward the end of 1982 there were in Bulgaria 296 AICs with 12,431 ha. of arable land on average, and an average stock of fixed assets of 19 million levs, 134 tractors, the equivalent of 337 tractors reduced to 15 hp, and 48 combine harvesters, which produced 23 million levs worth of gross output.

Other territorial production complexes are formed parallel with the AICs – industrial-agrarian complexes (IAC) and research and production complexes (RPC). The IAC are set up in regions or in separate economic units specializing in the production of one or several products and capable of providing the basic part of the raw materials used by the food processing industry in the respective region. In these cases the enterprises of the food-processing industry are incorporated in the IAC.

The RPC, too, are organized in regions with specialized production, where there are regional research institutes or experimental stations for this kind of production. The better part of them operate in the field of vine-growing and wine-making. The various forms of structural, organizational and production units have inherited from the co-operative farms their democratic principles of management – election of the bodies of management and reporting to the work-force on the results of their activities, etc.

The production and management structure of agriculture is being developed in organic unity with the improvement of the structure of the economic systems related to agriculture. In the early 1970s the Ministry of Agriculture was incorporated in the Ministry of Food Industry, and thus the Ministry of Agriculture and the Food Industry came into being. In 1976 it was incorporated in the National Agro-Industrial Complex (NAIC) as a state economic system for the management of the production of finished foodstuffs. This economic system covers agriculture, the food-processing industry, farm machinery, production services for agriculture, the agricultural sciences and the establishments training cadres with higher, secondary and other qualifications. The Constituent Congress held in 1979 adopted a new form of management of agriculture and set up the National Agro-Industrial Union (NAIU) which incorporated the actitivities of the NAIC and those of public organizations involved in agro-industrial production. A new principle of organization and management, used for the first time in the sphere of material production, was applied in the formation of the NAIU. The economic and public organizations are collective members of the NAIU. Organized on the same principles are the territorial economic organizations in the form of District Agro-Industrial Unions (DAIU). These unite and co-ordinate on an economic basis the efforts and resources of the AICs,

Grain silos in Silistra

the enterprises of the food processing industry and the other organizations in the system of the NAIU.

All union bodies of management are elected (including the President, who holds the rank of Minister). The supreme body of management of the NAIU is the National Congress, which elects a Central Council of the Union and convenes once every five years. The Central Council elects an executive committee and a chairman responsible for the daily management of the Union. The executive committee issues all the normative documents for the management of the economic organizations and the members of the Union in conformity with the laws, the government decrees, the decisions of the Congress and the Union's Statute. There is also a Control Council, elected to supervise the economic and financial activity of the union, its managerial bodies, economic organizations and members.

There are several kinds of economic organization dealing in agriculture within the system of the NAIU. Most important are the so-called State Economic Corporations (SEC), whose functions differ widely. Some of them buy, process and market the products of the AICs, while others produce agricultural goods. Such are the Rhodopa State Economic Corporation, which buys, processes and markets meat and meat products; Bulgarplod, which purchases, processes and markets fruits and vegetables; Bulgarian Tobaccos – tobacco; Vinprom – grapes; and the Dairy Industry – milk, etc. Still others organize and provide services: Water Economy organizes the construction and exploitation of land-improvement projects and the use of water resources for irrigation; Repair of Farm Machinery, etc.

Research and production corporations dealing in the spheres of poultry, pig, cattle and sheep breeding, selected seeds and planting material, veterinary services, etc. were set up in the second half of the 1970s. A typical feature is the complete integration of production and science, which makes possible the accelerated introduction of the achievements of science and technology.

Since 1977, land has been made available to industrial enterprises, institutions, schools, etc. for setting up auxiliary farms on their territories. These are usually small, abandoned plots. The farm produce from these farms is used by their respective canteens; costs are thus reduced and the quality of the food served to workers and employees is improved.

Apart from the public farms, there are also personal farms. To many households these are a source of additional income and farm produce. The land given for personal use varies in size from 0.1 to 0.5 ha. per household, plus the courtyard. The personal farms are not allowed to use hired labour for land cultivation or for stock breeding. The state and managerial bodies of the agricultural organizations assist the personal farms in their development, supplying them with materials, seeds, fertilizers and young animals.

In 1982 the personal farms produced 977,775 tons of maize, 126,647 tons of wheat, 128,803 tons of barley, 1,959 tons of peanuts, 2,850 tons of sugar-beet, 146,205 tons of fodder-beet, 238,800 tons of tomatoes, 180,316 tons of melons and water melons, 416,530 tons of grapes, 279,273 tons of meat, 639,033 tons of milk, 1,392,907,000 eggs, 7,530 tons of honey, etc.

The development and improvement of the organizational-production, and organizational-managerial structure has resulted in a number of positive economic and social achievements: the socialization of production has been broadened, the standards of public ownership have been raised, the co-operative and state forms of ownership have been brought closer to each other and have become more inter-dependent, the process of creating a single unified form of ownership has continued, the social acquisitions of agricultural and industrial workers have been made equal by the introduction of the same principles for both categories of workers in the field of retirement pensions, sick leave, maternity benefits, etc. and by narrowing the gap between the annual labour remuneration of industrial workers and peasants.

The improvement of the organizational-production, and organizational-managerial structure in the system of agriculture makes it possible to make full use of comprehensive and economic methods in the management of agriculture. It finds an expression in the incentives provided for the introduction of scientific and technological progress as a comprehensive and most important factor for the intensive development of agricultural production, for harmonizing the interests of the individual and the work teams as a whole on a higher level of production development, for the advancement of social relations and for providing the necessary conditions for the more expedient and efficient management of public ownership which is the economic foundation of the socialist system of agriculture.

An important role in the further development of agriculture is assigned to the improvement of the material-technological base, which should correspond to the

Combine harvesters

Maize crop at the Kozlodui Agro-Industrial Complex

requirements of large-scale production and should take into account the systematically growing possibilities for the industrialization of agriculture parallel with the process of increasing agricultural production, raising the quality of production and increasing its economic efficiency.

The development of the material and technical base of agriculture is characterized by certain quantitative and qualitative indices which show that it is making progress in a comprehensive manner as a broad and sound basis for the long-term development of agriculture.

The main farming operations – ploughing, sowing, digging, harvesting, etc. – have been almost completely automated. The structure of power resources is being changed in two directions: the share of mechanical engines is increasing and the number of electrical engines is growing faster, both in relative and absolute terms. The increased use of electric power is most closely related to the introduction of industrial technologies in stockbreeding, and the broadening and improvement of storage facilities, which helps prevent the interruption of technological processes in some branches of plant-growing,

Agriculture

Table 2: Material Supply and Technical Equipment of Agriculture

Years Indices	1956	1970	1982
1. Fixed assets at initial cost (in million levs)	1,471	4,353	8,799[2]
2. Power resources (in thousand hp)	2,007	7,053	12,046
3. Share of mechanically-driven engines from the total (in per cent)	65.3	96.3	98.3
4. Tractors converted to 15 hp (number)	24,283	93,742	152,155
5. Combine harvesters of all kinds (number)	4,118	16,810	19,097
6. Irrigated areas (thousand ha.)	381.4	1,001.1	1,185.8
7. Irrigated area as a share of arable land (in per cent)	7.0	21.0	25.5
8. Chemical fertilizers supplied (in thousand tons pure content)	49	639	1,038
9. Plant protection chemicals (in tons)	8,548[1]	12,032	40,400
10. Electricity supplied (in million kWh)	91	675	1,139

[1] 1958.
[2] Not counting sheep, pigs and goats.

Vegetable seedlings at the Novi Krichim Agro-Industrial Complex

Vegetable hothouses in Pazardjik

stock-breeding, etc. Mineral fertilizers, plant-protection and veterinary preparations are in good supply. The land-improvement projects (dams, pumping stations, irrigation systems, etc.) guarantee high and stable yields from agricultural crops.

Agriculture is one of the key sectors of the national economy. Its development is characterized by two simultaneous processes. Its relative share in both the gross national product (from 59 per cent in 1939 to 13 per cent in 1982) and the net material product (from 65 per cent in 1939 to 19 per cent in 1982) decreases; the quantity of production increases and its structure improves.

The first process is due to the country's industrialization and the dynamic development of the other sectors of the national economy, especially industry, construction and transport. The second process is due to the intensification of agricultural production. The cultivated land has been reduced by about 200,000 ha., or 4.2 per cent, and the number of workers employed in agriculture has been cut 400 per cent. The increase in agricultural production and the changes introduced in its structure proceed according to a plan determined by the needs of society.

Table 3: Indices of Gross Agricultural Output (1939=100)

Years Indices	1948	1960	1970	1982
Gross agricultural output	103	152	211	284
Crop production	99	154	206	254
Animal production	109	149	222	344

Table 4: Structure of the Gross Agricultural Output (in per cent)

Years Indices	1939	1948	1960	1970	1982
1. Crop production	66.2	64.2	67.3	64.7	50.2
2. Animal production	33.8	35.8	32.7	35.3	49.8

Table 5: Production of Basic Agricultural Products per Capita

Products	1934–39[1]	1961–65	1971–75	1979–82
Wheat (kg)	287	273	362	466
Maize (kg)	142	198	290	318
Sunflower (kg)	23	42	51	50
Oriental tobacco (kg)	5.1	12.5	14.6	13.4
Sugar beet (kg)	21	178	198	174
Tomatoes (kg)	6	91	87	96
Apples (kg)	4	39	38	43
Grapes (kg)	73	125	122	122
Milk (litres)	101[2]	148	194	250
Eggs (number)	112[2]	167	204	272
Meat (slaughter weight; kg)	31[2]	48	67	89

[1] Southern Dobrudja included.
[2] 1939.

During the stage of building an advanced socialist society, the most urgent need is to supply the population with foodstuffs in compliance with a scientifically-based nutritive regimen. The possibilities in this respect, and the degree to which this task is accomplished, are determined by the per capita amount of agricultural products, produced and consumed.

Consumption of the main ration diet products, and especially of foodstuffs of animal origin, have in recent years been getting steadily closer to the nutritive regimen. Before the socialist revolution, agriculture was the sole source of foreign exchange revenue (99.6 per cent), and continued to dominate up to the beginning of the 1970s. It was only in the second half of the 1970s and the early 1980s that its relative volume diminished to within the range of one third to one fifth of the nation's total exports, on account of the greatly increased exports of industrial products. Nonetheless, the exports of agricultural products increased 50 times in absolute terms.

In a short time Bulgaria's agriculture attained a level of development which allows the simultaneous fulfilment of the three main tasks: to meet the population's needs for foodstuffs, to supply industry with raw materials, to establish itself as a major source of export products and hence, of foreign exchange revenue for the nation. It is of great importance for the steady development of the country's economy. In addition, the rural areas were the main source of manpower for the other sectors of the national economy.

In 1939, some 79.6 per cent of the active population in the country was employed in agriculture. In 1948 – 81.9 per cent, in 1956 – 70.1 per cent, in 1960 – 54.7 per cent, in 1970 – 35.2 per cent, in 1975 – 27.7 per cent, in 1980 – 23.8 per cent, and in 1982 – 22.3 per cent. The absolute and relative reduction of the work-force is chiefly due to the introduction of mechanization. Taking 1939 as a basis, one index of labour productivity in agriculture was 111.1 in 1956, 218 in 1960, 312.3 in 1965, 448.2 in 1970, 762 in 1980 and 886 in 1982.

The successful development of agricultural production in Bulgaria over the past forty years is closely associated with the scientific and technological revolution, progress in science and technology and the introduction of their achievements into agricultural production in various

Cucumber greenhouse at the Georgi Dimitrov Agro-Industrial Complex, Plovdiv

forms – new and more productive plant varieties and animal breeds, expansion and improvement of irrigation, improvement of the technical and technological level of production, development of the processes of internationalization of agricultural production and the application of national and foreign progressive experience, etc. The main and most substantial condition for the intensive development of agriculture was the relatively fast conversion from small-scale, parcelled-up, one-man farming of the past to the large-scale, collective, socialist farming and adequate exploitation of its advantages.

Crop Farming. The cultivation of crops is a major branch of agriculture, covering cereals, vegetables, fruit, viticulture, flowers and forage. It provides about half of the total agricultural output of Bulgaria. Fresh and processed plant products represent a considerable proportion of the food of the Bulgarian population, while also providing the fodder requirements for stock-breeding and a large part of the country's exports. Crops also provide a source of valuable raw materials for the food, textile, pharmaceutical, cosmetics and other industries.

Soil and climatic conditions in Bulgaria favour the

Picking peppers in Petrich

Tilling an orchard at Gavrilovo in the Sliven district

Mechanized bean-picking at the Georgi Dimitrov Research and Production Complex, Plovdiv

Mechanized pruning of fruit-trees at Purvenets in the Plovdiv district

growing of an impressive range of crops, including cereals, legumes, many species of fruit and vegetables, vines, tobacco and oil-bearing roses.

Ever since the foundation of the Bulgarian State in 681, cultivation has been a major source of livelihood for the Bulgarian people. During the early centuries of its existence mainly cereals were grown. During the Ottoman rule the Turks introduced new plants into Bulgarian agriculture: rice, cotton, maize, beans, tobacco, potatoes, sesame, etc. After the liberation of the country in 1878, sugar-beet, vetch, alfalfa, sainfoin and some other crops broadened the scope of Bulgarian agriculture. In the years following World War I, sunflowers and soya began to be grown and the areas under vegetables, orchard crops and vines were expanded. The socialist reconstruction of agriculture has resulted in the intensive development of plant growing. Nowadays production is increasing at rates that place Bulgaria among the world's most advanced countries. The intensification of agriculture, the introduc-

Picking pears in the Kyustendil district

tion of vast fields and new fast-growing and high-yield varieties, the use of irrigation and fertilizers, successful pest control, and the wide implementation of industrial technologies and mechanization are continuing to raise yields and total production. This is illustrated by the increase in staple crop yields. In 1939 the average yield of wheat was 1,310 kg per ha. and went up to 2,990 kg per ha. in 1970, and 4,634 kg per ha. in 1982. The respective figures for maize are 1,359, 3,727 and 5,484 kg per ha.; barley – 1,492, 2,891 and 4,077 kg per ha.; sunflower – 960, 1,458 and 2,010 kg per ha.; tomatoes – 20,493, 27,616 and 29,413 kg per ha.; and apples – 7,974, 7,237 and 12,612 kg per ha.

In 1982, the land in Bulgaria put to agricultural use amounted to 6,181,500 ha. of which 4,654,700 ha. were cultivated. Of this cultivated land, 3,817,800 ha. were under arable cultivation; 294,800 ha. under natural meadow; 214,600 ha. under complex seeded pastures and other land use categories; 327,500 ha. under fruit-trees and berry cultivation; and 1,526,800 ha. under pasture and common land. The largest share of total agricultural produce was contributed by cereals (31.9 per cent), followed by perennial crops (7.8 per cent) and vegetable crops (7.6 per cent).

Crop production develops in close collaboration with 45 research institutes and 138 experimental stations

Cherry orchard in blossom at the Belassitsa Agro-Industrial Complex, Petrich

carrying out specialist research relating to different natural and economic conditions. 728 varieties and hybrids have been created (613 since 9 September 1944) of 67 species of field, orchard, vine and other crops. Many of the Bulgarian varieties are regarded as supreme specimens of world plant-breeding and enjoy international recognition. Examples include the Sadovo 1, Ludogorka, and Russalka winter wheat varieties, the Knezha maize hybrids and various varieties of tomato, tobacco, cotton, etc. The testing and allocation of new varieties is carried out and an organization has been set up for variety testing, seed production and seed quality control.

As a result of long-term soil research, a plan has been elaborated for the application of suitable types and qualities of mineral fertilizers to individual zones and micro-zones. Soil maps have been drawn, which are used for the zoning of varieties and other forms of agro-technical measures..

Since ancient times crop farming has supplied the country with food and fodder. Out of the total area under cultivation, 82.3 per cent was given over to cereals in 1939 and 61.1 per cent in 1982, 9.2 per cent was given over to industrial crops in 1939 and 12.3 per cent in 1982 and 6.2 per cent was given over to forage crops in 1939 and 22.2 per cent in 1982. Despite the reduction of the area under cereals, the cereal problem has been solved by increasing the yields and the volume of production of these crops. In 1982 cereals accounted for 16.9 per cent of total agricultural output, industrial crops 11.6 per cent and fodder crops 3.4 per cent.

The cereals are subdivided into bread cereals (mainly wheat) and fodder cereals (mainly maize and barley). In 1982 some 2,286,700 ha. were sown with cereals and yielded 10,178,000 tons of cereal. Within this area 1,052,500 ha. were under wheat, 620,900 ha. under maize and 351,900 ha. under barley. Other cereals grown were rice, beans, soya and peas, the production of which was sufficient to meet the demands of the country. A large part of cereal production is used as raw material in the food and fodder industries. The varieties grown are high-yielding and predominantly of Bulgarian selection.

Industrial crops are used as raw material for the processing industries. The crops of greatest agricultural importance include sunflower, oriental tobacco, sugar-beet, and volatile oil-bearing plants. In 1982 they covered an area of 459,100 ha. and yielded production to the value of 982,400,000 levs. In Bulgaria the vegetable oil used for cooking is extracted from sunflower seeds. In 1982 some 250,000 ha. were sown with this crop and all the country's requirements for sunflower were fully met. Bulgarian tobacco is of rich quality and is much sought on the world market. First-rate Bulgarian varieties of tobacco have been produced. In 1982 oriental tobacco was grown on 88,100 ha. and yielded 126,000 tons; 63,266 tons of

Agriculture

Mechanized cherry-picking at the Georgi Dimitrov Agro-Industrial Complex, Plovdiv

oriental tobacco and 71,742 tons of cigarettes were exported. Sugar-beet is the main raw material for sugar production in Bulgaria. In 1982 some 60,500 ha. were sown with Bulgarian polyploidal varieties of sugar-beet and yielded 1,583,000 tons of tubers. Soil and climatic conditions in Bulgaria are particularly favourable for the growing of volatile oil-bearing plants. The Kazanluk oil-bearing rose is an exclusively Bulgarian plant and a result of these propitious conditions. It is used for the extraction of high-quality rose oil which is an important raw material of the perfume industry. The lavender and peppermint oils produced in Bulgaria have a high reputation on the world market. Fibre crops, such as cotton, hemp and flax are also grown.

Grain crops were cultivated in 1982 on 829,300 ha. and yielded produce to the value of 287,500,000 levs.

Vegetable production. The first references to vegetable growing on the Bulgarian lands dates from the first century BC. Frescos from the first and second Bulgarian kingdoms depicting onions, garlic and radishes have been preserved in several monasteries and churches. In the first half of the eighteenth century, Bulgarian horticulturists began leaving the country and seeking work abroad. The emigration of gardeners to Eastern and Central Europe involved more and more people, reaching as many as 15,000 persons. Bulgarian horticulturists created many varieties of cabbage, pepper, onions, aubergine and other vegetables, which are described in the catalogue of old Russian and Hungarian seed-selling firms and were noted for their high yield, nutritional qualities and flavour. These varieties had Bulgarian names and were recommended as being the best. During the first decades after the liberation of Bulgaria from Ottoman rule (1878), vegetables were grown mainly to meet the demands of individual households. After World War I the area under vegetable cultivation was expanded and vegetable production adjusted itself to market demands, gradually acquiring a pronounced commodity nature. Some of the products began to be exported and the foundation of the canning industry were laid. Before 1944 the production of vegetables was the work of 261,000 individual vegetable gardeners. A real boom got under way after 1944, especially after the establishment of large-scale socialist enterprises, such as co-operative farms, state farms and AICs. Large market gardens have been set up. The rapid development of vegetable production has been stimulated by favourable climatic conditions – which allow the cultivation of early, mid-early and late tomatoes, peppers, cucumbers and other vegetable crops; by a long Bulgarian tradition of vegetable growing; by the supply of valuable nutrients; and by the intensive use of land. Total production of vegetables and the gross income per unit area is 700–800 per cent higher than with cereals. In 1982 the vegetable crops (water-melons, potatoes and melons included) covered 4.4 per cent of the area under cultivation and accounted for 13.4 per cent of the total farming output of the country. The rapid development of vegetable production is characterized by an increase in the areas under cultivation, the restructuring of vegetable crops, the establishment of large greenhouse complexes, an increase in average yields and the introduction of new industrial technologies.

The total area under vegetables, water-melons, melons and potatoes by 1950 exceeded the pre-war level by 44 per cent and by 1974 had almost doubled. The relative share of tomatoes, green peppers, garden peas, garden beans and gherkins increased. During the Sixth Five-Year Plan (1971–75) tomatoes took up about 17 per cent of the total area under vegetable cultivation, peppers about 10 per cent and garden peas about 12 per cent. Greenhouse vegetable production was rapidly growing. By 1977 the area under glass-and-steel greenhouses amounted to some 700 ha. Large modern greenhouse complexes have been constructed, each with an area of 25 to 50 ha. under glass. The area under polyethylene non-heated greenhouses is also increasing rapidly. The average yields of the major vegetable crops (with the exception of onions) have gone up considerably. In terms of average yield of tomatoes and peppers Bulgaria ranks among the leading countries in the world. The total production of tomatoes over the period between 1971 and 1975 was double that of the period between 1953 and 1957. The production of green peas, French beans and gherkins multiplied many times. Over a

Mechanized hop-picking at the Anton Ivanov Agro-Industrial Complex, Velingrad

Vineyards at the Lenin Agro-Industrial Complex at Ivanovo in the Russe district

fifteen-year period (1960–75) exports of fresh vegetables increased by 300–400 per cent, tomato paste increased about 250 per cent, canned vegetables went up about 1,500 per cent. Since 1970 emphasis has been put on the export of canned vegetables.

Vegetable production in Bulgaria has embarked on a path of intensification and higher efficiency. Since 1966, vegetable gardens have undergone large-scale concentration and specialization, as a result of which some 250 to

Agriculture

300 large vegetable farms have emerged, occupying over 75 per cent of the area under vegetable crops and allowing individual stands to cover areas of more than 50 ha. Vegetable processing enterprises employing up-to-date industrial technologies are being established on the AICs. A process of integration is being effected between vegetable production and the canning industry. High yields have been obtained with early tomatoes of the Triumph variety (over 70 tons per ha.), green peppers of the Klurtovska Kapiya variety (60 to 70 tons per ha.) and some other vegetables. High yields have also been achieved with vegetables grown in greenhouse complexes: 120 to 130 tons per ha. of early tomatoes, 300 to 320 tons per ha. of cucumbers and 75 to 85 tons per ha. of peppers.

The yield and other characteristics of vegetable crops are related to the selection of more valuable varieties. Selection is orientated in three directions: selection of greenhouse tomatoes and cucumbers; selection of high-quality early field varieties for fresh consumption and export, and selection of high-quality simultaneously ripening varieties, with good biological characteristics permitting mechanized harvesting and mainly intended for canning. Over 70 new varieties have been produced, many of them hybrids of increased size, strength, etc.

Pruning vines

Harvesting grapes with a combine harvester

Picking grapes at Kovachevitsa in the Mihailovgrad district

Fruit-growing became a separate, minor branch of farming as early as the period of the Ottoman rule. Before the Liberation in 1878, some 4,800 ha., mainly of plum orchards and scattered walnut, apple, cherry and other trees were given over to fruit-growing. After the Liberation, nurseries and orchards were set up, managed by the agricultural colleges. Valuable European species and varieties were imported and spread throughout the country. Between 1880 and 1920 the area under orchards was doubled, reaching 10,400 ha. After World War I (1914–18) fruit-growing boomed and the export of fresh and semi-processed (pulped) fruits to Germany and other countries began.

Up to 1949 plums and apples were the dominant orchard crops, covering 95.6 per cent of the areas given over to orchards. Between 1950 and 1965 the area increased to 178,600 ha. The area under apple and plum trees continued to be the largest, although extensive stands of peach, walnut, apricot, cherry, morello-cherry and other trees were also established. The area given over to peach trees was increased very greatly indeed. The expansion of orchards and the increase in average yields is responsible for the increase of fruit production. In the first twenty years after 9 September 1944 the increase in fruit production was achieved by constantly extending the total area and the percentage of cultivable land given over to orchards. The following years saw the beginning of the restructuring and intensification of fruit-growing. The main objective was no longer the extension of the fruit-growing area at the former rates but the attainment of certain economic indices.

Scientifically-based zonation of orchard species has been established. Primary and secondary zones for the cultivation of each individual species have been identified. This is the basis on which the problems of the specialization and concentration of fruit production are being solved. The growing of the main orchard crops (apples, plums, peaches) has been concentrated in a limited number of zones. Substantial changes have been introduced in the system of planting and cultivation of trees in the new industrial orchards. The number of trees per area has increased and the varietal composition of individual species has been improved. The right balance is being sought between the different varieties, so that a constant flow of fruit for consumption and canning can be ensured. Modern industrial technologies are being put into operation in productive stands of 400–500 ha. and even of 1,000 ha. Complex mechanization is also being introduced and about 70 per cent of production processes have been mechanized. Plant protection, agro-technical techniques, and irrigation are widely used. In 1982 some 2,190,722 tons of fruit were produced, including 425,780 tons of apples, 105,873 tons of plums, 93,859 tons of peaches and 19,758 tons of strawberries. The total area under orchard crops was 125,756 ha. including 31,310 ha. under apples, 21,105 ha. under plums, 9,865 ha. under peaches and 3,361 ha. under strawberries. Agricultural science has greatly contributed to the successful development of fruit growing. Numerous scientific units

Mechanized spraying of vineyards near Loznitsa

Agriculture

(research institutes, complex experimental stations, experimental fields) have been established for the solution of theoretical problems arising from practical experience.

Selection and zonation have been carried out for many fruit varieties, including the Sheinovo, Sliven, Dryanovo, Probuda, Turgovishte, Plovdiv and Konkurent walnut varieties; the Markovo 11 almond variety; the Chervena Kurtovka, Elinpelinska, Rumyana, Chaya, Earliest Yellow and Oktomvriiska II peach varieties; Bilyana strawberry variety, and Bulgarski Rubin and Iskra raspberry varieties.

Viticulture has been practised by man since very ancient times. Its appearance on Bulgarian territory can be traced as far back as 3000 BC. Vine growing gained impetus during the First Bulgarian Kingdom (681–1018). The varieties grown during that period included Pamid, Mavrud, Shevka and the Broad Melnik Vine. During Ottoman rule some dessert varieties such as Afus-ali (Bolgar), Chaush, Kadun Parmak and Al Zeinel were introduced from Asia Minor. Before the appearance of phylloxera (1884–96) vines were grown only on their own roots. After the phylloxera crisis, and particularly after World War I, a rapid re-establishment of vineyards began by grafting vines onto phylloxera-resisting rootstock. The area under vines increased from 43,413 ha. in 1919 to 143,103 ha. in 1944. In 1927 the production of dessert grapes began. By the late 1930s Bulgaria had emerged as one of Europe's major dessert grapes exporters (50,000 tons per year). Under Capitalism, however, viticulture remained small-scale, scattered and technically backward.

Since 9 September 1944 vine-growing in Bulgaria has made enormous progress. Large-scale vineyards on well-chosen ground were set up. The first merging of vineyards was carried out. Most of them were located on the territories of co-operative and state farms and they were arranged into 4,000 vast stands of 25 to 35 ha. on average. The first scientific zonation of vine-growing in Bulgaria was carried out in 1962 on the basis of ecological criteria. New high-quality grape varieties – dessert (Cardinal, The Queen of Vines) and wine (Cabernet, Sauvignon, Merdeau, Rkatsiteli, Riesling, Uni Blanc, etc.) – were introduced. The production of rootstock increased (some 200 million pieces), as did that of engrafted rooted vines (some 100 million pieces).

In the early 1970s a quantitatively new stage began in Bulgarian viticulture. A transition was effected from the low growing of vines to the trellis-training of vines, stem height reaching 1.2–1.5 m. In 1982 the area under vines in Bulgaria was 186,980 ha. Wine grapes were grown on 148,350 ha. and dessert grapes on 18,637 ha. The area under dessert varieties had greatly diminished compared to 1968, since these varieties require more manual labour. The vineyards in the public sector have been grouped into 437 large tracts. The mechanization of the production process means that one worker can grow 8–9 ha. of vines. The trellis-training of vines requires varieties that are more resistant to cold, which has resulted in the widespread distribution of the Cabernet, Sauvignon, Rkatsiteli and Misket Otonel varieties. A re-evaluation of the zonation was made, suitable zones were selected, and the area for the cultivation of the local varieties Pamid, Mavrud, Gumza, Dimyat and Red Misket has been increased. Three vine-growing zones have been identified: an eastern zone, producing grapes for white wines and brandy; northern and southern zones mainly producing grapes for red wine; and dessert grapes.

In 1982 vineyards covered 3 per cent of the cultivated area in the country and 50 per cent of the area given over to perennial crops. They turn out some 5 per cent of total crop production and 57 per cent of total perennial crop production. In terms of per capita production of grapes Bulgaria ranks fourth in Europe. Viticulture is the basis of the wine industry which is responsible for 3 per cent of the total industrial output and 11 per cent of the food industry output. It is closely linked to the international market. Bulgaria ranks fourteenth in the world in terms of area under vines, but is fifth in the production of dessert grapes and second in their export. Bulgaria ranks thirteenth in the world for wine production and sixth for wine export. Over 70 per cent of the wine industry's output, and over 50 per cent of dessert grapes are exported. Bulgaria is first in Europe in production and export of vine seedlings. Bulgarian plant breeders have produced varieties which have earned a high reputation and are successfully grown inside the country. Among them are Danube Misket, Thracia Misket, Supur Ran Bolgar, Bulgaria, Mechta, 10 May, the seedless hybrid V-6, Sveznest, Mir and Brestovitsa (dessert grapes), and Buket, Bulgarian Riesling, Thracia Bisser, Mavrud, Branch 1, Pomorie Bisser, Plovdiv Malaga and Varna Misket (wine grapes).

Plant protection includes measures for pest, disease and weed control of agricultural plants. Before the socialist

Aerial spraying of orchards at Purvenets

Cultivation of the parasite *Trichograma,* used to combat pests

Spraying tomatoes with a Perla-3 spray

revolution the level of plant protection in Bulgaria was very low: it was limited in scope and yielded unsatisfactory results. Losses inflicted by pests amounted to 20–25 per cent of the total plant production. Since 1944 the most dangerous pests have been successfully controlled by using integrated agro-technical physico-mechanical, chemical and biological techniques. Predicting and signalling are an important feature of plant protection. Reliable predictions are made of the appearance and development of dangerous plant diseases and pests, of the areas that they may affect, of the severity of the problem and the expected damage. Modern prediction and signalling seek to determine optimal rates and timing (the period of the highest vulnerability to pests) for pesticide application. Plant quarantine is also used for pest control. It prevents the penetration and spread of unknown and dangerous pests from other countries and ensures containing the problem within the affected areas and its elimination.

In recent years an economic survey has been carried out of the major pests on agricultural crops. The integrated approach, a scientifically-based system of complex pest control, has been adopted. This involves a new range of chemical usage (low rates of pesticide application, low-volume and super low-volume spraying, the use of aerosols, seed disinfection, disinfection of empty storehouses, etc.), that are more effective and harmless. Weed control involves a special study of the effect of a number of herbicides on the useful soil microflora and on the biochemical and physiological changes in the plants being treated. There is wide use of biological methods of pest control, exploiting parasites (trichogram, prospaltella), predatory (perilus, phytoceilus) and entomopathogenic bacteria (entobacterin, disparin), fungi (boverin), antibiotics and antagonistic species. Methods that do not exercise a negative effect on the environment are being developed, for example, the method of ray sterilization and homosterilization, the genetic method, and the application of juvenile hormones and repellents. Thorough research is being carried out in the field of plant immunity. Plant varieties that are immune to diseases have been created and virus-free mother plantations have been established. Testing is carried out, too, for the selection of virus-free mother trees to enrich the range of varieties available for planting.

Stock-breeding is the second most important sector in Bulgaria's agriculture, after plant growing. It comprises sub-branches such as the breeding of cattle, bulls, sheep, goats, pigs, poultry, horses, hares, silkworms, bees, fish and game. The natural and economic conditions in Bulgaria are propitious for rearing various animal strains and breeds, and stock-breeding is a major occupation of the population in various parts of the country, in particular the semi-mountainous and mountainous regions. The development of multifarious crafts is based on stock-breeding, which provides the animal products for the population's needs in foodstuffs and clothing, and raw materials for the needs of the crafts, industry and trade. Various products such as wool, silk, honey, wax, and livestock were in demand and exported to foreign markets in Istanbul, Venice, Dubrovnik, etc., but during the

Agriculture

Ottoman rule there were no suitable conditions for the development of highly productive and advanced stock-breeding. After the Liberation in 1878, the State attended to the improvement of stock-breeding, and horse-breeding in particular. The first state stock-breeding farms were set up, in addition to research units which contributed to the import and breeding of high production pedigrees, to the improvement of the local breeds and breeding of new strains and to improved feeding and rearing. World War II stock-breeding met mainly consumer needs, such as those of the small farms for draught animal produce. During the war it suffered heavy damage: the number of arm animals decreased (pigs and goats excluded) and their productivity declined.

After 1944 intensive work began for the reinvigoration of stock-breeding. The implementation of the Two-Year National Economic Plan (1947–48) repaired the damage suffered by agriculture. The restructuring of stock-breeding on socialist foundations was guided and encouraged by numerous party and government decrees, and the decisions adopted at the April 1956 Plenum of the Central Committee of the BCP were particularly favourable. After the collectivization and amalgamation of co-operative farms had been completed, over the 1958–59 period, the

Rose harvesting at Rozino in the Plovdiv district

The Liliana Dimitrova dairy farm at Tsalapitsa, a branch of the Georgi Dimitrov complex, Plovdiv

public sector comprised 77.3 per cent of cattle, 77 per cent of pigs, 57.3 per cent of sheep, 29.2 per cent of fowl and the bulk of horses. The concentration in stock-breeding was coupled with radical structural changes, including the breeding of new strains. The number of horses, bulls and goats increased considerably, both relatively and absolutely. The number of oxen was reduced several times, thereby exerting an adverse effect on the relative share of cattle-breeding, despite the increased number of cows. The number of pigs and fowl and their relative share increased sharply. On 1 January 1983 the number of farm animals in thousands was: cattle – 1,782,600, sheep – 10,761,400, pigs – 3,809,800, and fowl – 42,853,100. Within the structure of the total agricultural output, stock-breeding accounted for 49.8 per cent in 1982. Cattle strains of various productivities were bred, with industrial cross-breeding of milk cows. Fine-fleeced and semi-fine-fleeced pedigree sheep were bred, thereby meeting the country's needs of merino. The premature weaning of calves and lambs was introduced in stock-breeding, whereas pig-breeding concentrated mainly on commercial production. The number of pigs bred for meat increased rapidly, the main stress being on hybrid varieties. The public sector was engaged in breeding only highly productive poultry breeds and hybrid layers and broilers; intensive broiler production was launched, in addition to industrial egg production. New breeds of pedigree horses were bred and improved for sport. Pedigree stock-breeding was stepped up by rearing valuable pedigree animals of all strains, and by artificial

Dairy farm at Razliv in the Sofia district

insemination. It covered 97 per cent of the total number of cows. A network of pedigree farms was set up, in addition to breeding centres. Fodder production was vastly increased, mainly by raising yields and expanding areas sown with grain and fodder crops and alfalfa, raising new crop varieties, large-scale adoption of ensilage and developing the fodder industry, thus meeting about 60 per cent of collective stock-breeding farms' needs of mixed fodder. A comprehensive fodder plan was established for the whole country in 1976, and the setting up of agro-industrial and industrial-agrarian complexes stepped up the concentration and specialization of stock-breeding production. Industrial methods and technologies were adopted and mechanized and automated stock-breeding complexes were set up including cattle-breeding, calf-fattening, pig-breeding, and sheep-breeding complexes and integrated poultry farms.

About 68 per cent of poultry meat, 62 per cent of veal, 36.2 per cent of eggs and 20 per cent of pork were produced on collective farms under industrial conditions. The average milk yield per cow amounted to 3,221 litres in 1982 (as against 450 litres in 1939); the average wool yield per sheep was 3,906 kg in 1982 (1,511 kg in 1939), mainly merino. The average laying capacity per hen was 204 eggs in 1982 (73 in 1939). A special decree of the Council of Ministers envisaged the commencement of measures aimed at the development of stock-breeding in the individual farms of co-operative members, workers and employees. As compared with the 1939 output, the production of milk rose 280 per cent in 1977, of cow's

Cattle grazing at Tsalapitsa in the Plovdiv district

milk alone over 600 per cent, of meat – 349 per cent, of eggs – 272 per cent, and of wool – 276 per cent. The public sector accounted for the bulk of stock-breeding production. Animal products were produced on the personal farms of co-operative workers. Milk accounted for 76.5 per cent of the total production in 1977 as against 5.0 per cent in 1950, meat for 61.2 per cent as against 3.2 per cent, eggs 49.3 per cent as against 0.7 per cent, and wool 71.2 per cent as against 9.6 per cent. In 1982, 91 kg per capita meat in slaughter weight were produced in addition to 264 litres of milk, 3,912 kg of wood, and 279 eggs. Marketing of stock-breeding production increased. In 1982 trade organizations purchased 74.6 per cent of the total output of milk, 74.9 per cent of meat, 57.5 per cent of eggs and 82.3 per cent of wool. Animal products intended initially to cater for the needs of the farms themselves now became commodity items.

Cattle breeding developed on the Bulgarian lands during the earliest historical periods. Cattle were bred by the Thracians, Slavs and proto-Bulgarians, mostly to act as beasts of burden and to provide food and clothing, as well as a subject of barter. In the Middle Ages and later, with the development of Capitalism, its importance grew. Before 9 September 1944, the private farmers rarely owned more than one to three cows, which yielded little and were mainly of local breeds (Bulgarian Grey Cow).

They were raised to provide a means of draught power, and to yield milk and meat. After the reorganization of agriculture on a socialist basis, the collective farms raised large, high-production herds of stud breeds. The reorganization influenced the structure of the cattle herd, and its breed composition. The number of cows increased by 28.7 per cent (1977) as against 1939, oxen numbers were reduced by 240 per cent, while the young cattle doubled; good conditions were created for the production of more milk and meat. At the very beginning of the co-operative farming system cross-breeding was organized on a mass scale, and this made it possible to raise new breeds of cattle within a comparatively short time – Bulgarian Cattle, Bulgarian Red Cattle and Bulgarian Siemental Cattle.

Dairy cattle-breeding has developed around the big towns and industrial centres where milk is supplied for direct consumption; dual-purpose cattle are reared in the remaining, more extensive areas of the country, while in some defined regions of wider grazing pasture beef cattle are preferred. Cattle-breeding accounts for 18.5 per cent of total agricultural produce. Its relative share of the total stock-breeding produce is 43.6 per cent in the collective farms, and 24 per cent in the private farms. As a result of the improvement of the breeds, especially through raising the quality of the feeds both the milk-yield and total

Agriculture

Sheep at the Agro-Industrial Complex at Senovo in the Razgrad district

cattle-breeding production increased in 1982. The yield of the collective farms was 3,221 kg milk per cow. Bulgaria is one of the first in Europe (for the collective farms it is 480 per cent) in the rate of the average milk-yield increase. Compared to 1939 the production of milk has been raised by 510 per cent and that of meat 400 per cent. Beef accounts for 19.6 per cent of the meat production; the number of head of cattle being 1,782,649 (1983). There are 28.8 head of cattle per 100 ha. of cultivated land, of which 11.4 are cows. Concentration and specialization of cattle-breeding brought about the introduction of industrial technologies, as well as the mechanization and automation of the production processes. Cattle-breeding complexes have been set up for 200–500 and up to 1,000–2,000 cows, as well as cattle-fattening complexes for 3,000–10,000 and up to 20,000 calves. Labour productivity in the complexes for industrial production of milk and meat has increased enormously. Breeding and improvement work is realized by the district breeding centres, in close co-operation and co-ordination with the teams of research workers. Their aim is to improve the existing breeds and create new ones – an original Bulgarian form for breeding work.

Sheep-breeding is the main livelihood for a large part of the Bulgarian population. The foundation of the Bulgarian State in 681 gave an impetus to the development of trade in wool which increased in the thirteenth and fourteenth centuries and became very brisk in the fourteenth–eighteenth centuries. A document of the archives of some merchants of Dubrovnik (1582) shows that the Bulgarians reared sheep of coarse fleece and fine fleece wool. The main breeds raised during the Ottoman oppression were Tsakel and Tsigai, but there were also some other breeds, developed mainly for wool. Later on, the flowering of the crafts (homespun tailoring, etc.) and trade in products (meat, cheese, etc.) tended to provide an increase in the number of sheep. In some regions local breeds (Karnobatska, Svishtovska, Blackface Plevenska, South Bulgarian sheep) were raised. To improve the local breeds some foreign breeds such as Merinofleisch, Fresian, Rambouillet, and Karakul were imported. After 9 September 1944 favourable conditions were created for the numbers of sheep to be increased, and for breeds to be improved and labour productivity raised. Big sheep-breeding farms were set up. In some indices Bulgaria ranks among the first in Europe and the world – in numbers of sheep (1,800–2,000), milk (over 47 tons) and wool yield (more than 5 tons) per 100 ha. of cultivated land. The total number of sheep in 1983 was 10,761,395. The breed composition of the flocks changed and their productivity was raised as a result of the new fine fleece and semi-fine fleece breeds raised through cross-breeding,

Pig-farm at Kozloduitsi in the Tolbuhin district

as well as through improved feeding and the introduction of artificial insemination. Two new fine fleece breeds were raised (Thracian sheep and North-East Bulgarian) for the plains, semi-fine fleece breeds were bred for the semi-mountainous regions, and the semi-coarse fleece breeds, Tsigai, for the mountainous regions. Bulgaria meets its needs for wool. About 50 per cent of the wool it produces is merino wool. In some areas lamb, weaned lamb meat and sheep's milk are the basic products. Dairy sheep-breeding is developed in some micro-regions of the plain and semi-mountainous areas. Sheep-breeding accounts for 14.0 per cent (1982) of the total farm produce. The average live-weight of the sheep has been considerably increased. Mutton accounts for about 13–14 per cent (over 70 per cent of young animals) of the total meat production. The wool, milk yield and fertility of the sheep has been raised. While the number of sheep has been reduced, the total wool production is almost 300 per cent greater than in 1939. Early weaning and fattening of the lambs, as well as intensification of breeding (the percentage of twinning being increased) has been achieved. Sheep-breeding production is improving as a result of the increase in number and raising of the average yield, full utilization of the existing and creation of new biological possibilities, and narrow specialization in the breeds – dual-purpose and for milk. Sheep-breeding units of different capacities, depending on the region and the purpose, have been set up, and industrial methods and technologies, with mechanization and automation of production processes, have been introduced. Breeding and improvement work is carried out by the district selection centres, in co-operation with the research workers' teams for the improvement of existing breeds and raising of new ones.

Pig-Breeding on the Bulgarian lands dates from the settlement of the Thracians. At first the White Long-Eared Pig was bred, then the East Balkan Pig. After the Liberation of 1878 the local breeds were improved by crossing them with boars of different breeds – Yorkshire, Berkshire, Mangalitsa and some others. Pedigree herds of the Yorkshire breed were raised on the farms of the Obraztsov Chiflik Agricultural School in Sadovo, in order to produce pedigree animals. Then the improved breed was crossed with the German White Noble Pig, and thus the Bulgarian White Pig came into being. Pork and lard pigs are raised, as well as the East Balkan, Dermanska and Mangalitsa breeds.

After 9 September 1944 the number of pigs was increased to 3,809,800 (1.1.1983) and the production of pork reached 488,000 tons. About 70 per cent of the pigs were reared on the collective farms. Priority is given to the pork pigs. Animals of foreign breeds are imported. Pig-breeding is developing at a comparatively high rate; numbers and the production of pork are over 500 per cent greater than in 1939. Pork accounts for more than 40 per cent of the meat produced in Bulgaria. The structure of meat production and the consumption of pork per capita has changed considerably: in 1982 pork accounted for 9.4 per cent in the structure of the over-all farm production.

Pig-fattening shed at Kozloduitsi in the Tolbuhin district

Pork production is concentrated in the pig-breeding complexes (there were some 35 pig-breeding complexes of different capacities in 1982). The technologies and organization of pig breeding are being constantly changed and improved. During the period 1960–79 the general fertility of the sows in the collective farms and the average growth per day rose, whereas fodder consumption per 1 kg of growth was reduced by 26.45 per cent. Approximately 50 per cent of the pigs fattened in the collective farms came from the Camborough hybrid. Large White, Landrace, Hampshire, Duroc and Petrain breeds were improved with a view to producing new genotype hybrids. 37 pure-bred (pedigree) and 22 reproduction herds were raised to provide 15–20 per cent of the hybrids produced.

The high efficiency of the pork production in the industrial farms is a result of the wide application of biotechnical methods – artificial insemination, early establishment of pregnancy, synchronization and stimulation of the oestrus and farrowing, etc. Early weaning of the pigs and their cage-battery raising are successfully practised.

Poultry-farming. Before the Liberation in 1878 all species of farm poultry, mainly of local breeds, were raised in Bulgaria. Poultry-farming was extensive and covered the needs of the family. The farms of the agricultural schools were the first centres of breeding-and-improvement work. Bulgarian eggs found a good market abroad, which encouraged the development of poultry-farming. Egg and egg-and-meat breeds were imported and crossed successfully with the local breeds. From 1935 onward the improvement of poultry-farming became more systematic. Incubators were created, and a control on the productive qualities and feeding was introduced.

Bee-keeping

Poultry factory at Tolbuhin

Substantial changes in poultry-farming were made after 9 September 1944. Collective poultry farms were set up, and big poultry houses with controlled micro-climate established. New technologies of breeding were introduced and are being constantly improved. The number of poultry is rapidly growing, especially of hens (about 400 per cent). Sanitation and feeding have been improved. The relative growth is highest with turkeys and lowest with geese. On 1 January 1983 Bulgaria had 42,853,108 fowls.

The average laying capacity of hens has almost doubled, and the overall production of eggs and poultry, and their consumption, are constantly growing. Their share in the public sector is quite considerable. In 1982 the relative share of the products of poultry-farming was 6.1 per cent of the total farm production. Compared to 1971, poultry-farming production rose by 41.5 per cent.

The production of eggs and poultry on an industrial basis was organized in 1965. Specialized state, inter-co-operative and state-co-operative enterprises were established for the production of eggs, and fattening of poultry (broilers). As a result of the broiler production which was developed, and the fast replacement of the layers, meat production grew rapidly. Great changes were made in the breed composition of the poultry to suit production purposes. The poultry breeds raised are: hens – Leghorn, New Hampshire, Sussex, Plymouthrock and Cornish, as well as various cross-breeds and hybrids; geese – local (Benkovska), Emden, Toulouse and some others; turkeys – Bronze, White Imperial, White Beltsville, Black Novozagorska; and ducks – Peking, Ruan, and Khaki Campbell. Many high-yield fowl lines and hybrid forms of egg breed and meat breed have been raised.

The fodder industry provides wholesome compounds for all species and classes of poultry, depending on their production purpose. Incubation methods and the existing incubator systems have been further improved, and some new ones have been introduced. The technologies of breeding (deep laying, cage and battery-cage breeding) are changing and being perfected. Poultry-farming is one of the sub-branches of stock-breeding in which the production processes – fodder delivery, watering, cleaning of the premises, collecting and sorting of eggs, slaughtering and poultry dressing, are mechanized and automated to the greatest extent. In a very short time Bulgaria has reached the standard of poultry-breeding achieved in developed countries in terms of scale and nature of production.

Egg and poultry production in Bulgaria is provided both by the national poultry farms for broilers and layers, and by the personally run farms. Now the tendency is to increase in absolute terms the quality and relative share of the production coming from the industrial enterprises and to further improve and specialize it.

Horse-breeding. During the First and Second Kingdoms, Bulgaria had a very good cavalry. Under Ottoman rule, horse-breeding fell into decline. The Liberation of 1878 found a great variety of local breeds raised in the country, and the State took immediate measures for the fast develoment of horse-breeding. A Supreme Council was formed to organize the improvement of existing breeds, the import of foreign breeds and the raising of new ones. The breeding and improving facilities are supervised by the State. The state stables and stud farms established at Kabyuk, Klementina and Bozhurishte help to develop and improve horse-breeding. The stud stations set up at the time of covering are provided with state-owned studhorses. The East Bulgarian, Plevenski and Dunavski horses, which are half-blood breeds, are the basis for the further breeding of horses. As a result of the rapid introduction of mechanization and motorized transport in agriculture the number of horses decreased abruptly (it was 118,559 on 1.1.1983). The local heavy draught horse is being improved and many specialized breeds, such as the English thoroughbred horse, Arabian horse, Light Harness horse and Mountain horse are now being raised.

Bee-keeping. The Thracians and Slavs kept bees and used great quantities of honey as food. Honey was also used in folk medicine. At the time of Khan Omurtag, honey and wax were already being exported from Bulgaria to Byzantium. Bee-keeping was one of the most profitable occupations in Medieval Bulgaria, and during the Ottoman rule Bulgaria was a supplier of big quantities of honey for the market in Constantinople. After the Liberation in 1878, bee-keeping was developed in a fast and organized way, for the most part by amateurs.

After 9 September 1944 favourable social and economic conditions were created for the rapid development of beekeeping. The number of bee-families, as well as the average yields of the products of bee-keeping, increased considerably. Bee-keeping developed in two different sectors – state-run and private farms. The state-run sector embraces the state apicultural enterprises and the apiaries of the AICs. Apicultural enterprises each keeping from 4,000–10,000 bee-families are set up in fourteen districts. There are three apiaries specializing in the production of queen-bees, which sell more than 60,000 fertilized queens. 77 per cent of the bee-families are kept on personal farms. The total number of bee-families in Bulgaria is 565,522, of which 20,118 (1.1.1983) are kept by the state-run sector. Bulgaria ranks among the first in the world in the number of bee-families per sq.km and per 1,000 people. The yield of one bee-family for one season is 10–15 kg of honey, 200–600 g of wax, 500–600 g of bee-bread and 2–3 g of bees' venom. The quantity of honey purchased in 1982 was 8,200 tons.

Veterinary Medicine. In ancient times, the people of the Bulgarian lands already knew many of the animal diseases. Instructions for their treatment are found in some written monuments – Roman and Greek in particular. Veterinary medicine began its rapid development in the fourteenth century. Infectious diseases such as rabies, glanders, rinderpest and fever were raging in the country in the years of Ottoman oppression. The sick animals were treated by self-taught doctors. It was the Russian military veterinary surgeons who initiated scientific knowledge immediately after the Liberation of 1878. The first Bulgarian veterinary bacteriological station came into being in 1901. After the First World War it became the Veterinary-Bacteriological Institute. After 1924 Bulgaria was the only capitalist country in which veterinary medicine became public and all its functions were assumed by the State.

After 9 September 1944 the existing veterinary establishments and institutes were expanded and improved. A system of veterinary units, including research, production (for serums, vaccines and diagnostic preparations), prophylactic and treatment establishments, control units, and enterprises for the collection and use of carcasses and confiscated animals, was created. Many diseases such as cattle brucellosis, sheep pox, glanders and dourine on the solidungulates, and foot-and-mouth diseases were eliminated. The spread of swine plague, erysipelas and poultry pseudo-plague was confined. The major problems of present-day veterinary medicine are the elaboration of radical measures for combating infections and parasitic diseases on farm animals; prophylactics and treatment of non-infectious diseases; curing of sterility and mastitis; production of milk, meat and other animal products of high veterinary-sanitary qualities; and the invention of new diagnostic means and medicines.

FORESTRY

Forestry is a branch of the national economy whose task is to preserve, exploit and renovate the forests. It plans and organizes the farming and exploitation of the forests with a view to meeting the economy's demands for timber and other forest products, protects and enhances their other useful functions (soil-protective, water-retaining, sanitary, hygienic, etc.), improves the quality and increases the productivity of forest plantations, and renovates and grows new forests. During the long period of Ottoman rule, many of the forests were uprooted or burnt, and others were destroyed by foreign concessionaries. This process continued after the Liberation in 1878, but on a limited scale. Forestry administration was set up for the first time after 1878. Under Capitalism and up to 1947, some 26 per cent of the forests belonged to the State, 55 per cent were owned by municipalities, 0.6 per cent by schools, 1 per cent by monasteries, 0.7 per cent by co-operatives, and 16.7 per cent were privately owned. During this period forestry was established on a scientific basis, but deforestation far exceeded afforestation, which actually began in 1893, chiefly to combat erosion and floods, but also to improve the environment of populated areas. The measurement and arrangement of forests was first introduced in 1890, and forest structuring has been the task of a state organization since 1901. Before 9 September 1944 only 28 per cent of the country's forests were structured. After 1947, the forests were nationalized and only about 3 per cent left in the care of co-operative farms, state farms, etc. The widened scope of economic activity in the domain of forestry necessitated the formation, in 1949, of a special Ministry of Forests (now the Ministry of Forests and Forestry) which managed the plantation, cultivation and preservation of the forests, wood-cutting and wood-working. With the development of the country's economy, forestry became more and more intensive. By 1957 all forests had been structured and economic activity was carried out in conformity with established forest-structure projects. The forests of economic designation (74 per cent) served chiefly as a source of timber, while the forests of special designation (26 per cent) included water-retention and forest shelter-belts, woods in the green areas, forest-reserves, the state-owned forest shelter belts and eroded forest lands (1982).

In 1980 the total area of forests in the country was 3,845,028 ha. and the afforested areas – 3,313,778 ha. The wooded areas constituted 30 per cent of the country's territory. Enclosed within the confines of the forests were 111,000 ha. of forest pasture lands which were used for the development of stock-breeding. There was an average of 0.4 ha. of forest lands per head of the population. The following kinds of forest exist in Bulgaria – coniferous forests, deciduous tall-trunk forests, forests for reconstruction, off-shoot woods to be turned into seed forests, low-trunk woods, and forests subject to thinning.

Species	Area (in %)
Coniferous	32.33
white pine	16.90
black pine	8.75
spruce	4.65
fir	0.95
white fir	0.43
black fir	0.05
Douglas firs	0.35
other	0.25
Deciduous tall-trunk	20.51
beech	10.51
oak	4.12
cerrus	0.71
hornbeam	0.83
ash	0.38
linden	0.96
aspen	0.95
walnut	0.53
other	1.52
For reconstruction	20.40
Offshoot for turning into seed forests	22.66
Low-trunk	4.05
For thinning	0.05

The total stock of wood in the forests for 1980 was 303,809,380 packed cu.m (coniferous – 103,737,717 cu.m and deciduous – 200,071,664 cu.m).

The average productivity of the forests expressed in terms of annual average growth is 2:4 cu.m per ha. The average annual growth for the forest areas as a whole was 7,857,000 cu.m (1980). Annual exploitation was 6,031,000 cu.m in 1980.

Depending on the diversity of climatic, soil and

A forest area in the Rila mountains

hydrographic conditions the plant cover is horizontally divided into three forest regions: North Bulgarian (Moesian), South Bulgarian (Thracian) and South border regions. Vertically, it is divided into three belts: (1) lower flatland-hilly and hilly fore-Balkan belt of oak forests; in the Moesian region it is up to 600 m above sea level, in the Thracian, up to 700 m and in the Southern border region, up to 800 m. (2) Medium-mountain belt of beech and coniferous woods at 600–1800 m, from 700–2,000 m and from 800–2,200 m above sea-level for the respective regions. (3) The high-mountain belt at 1,800, 2,000 and 2,200 m above sea-level respectively. Each of these belts is divided into sub-belts and regions. Bulgaria's forests are mainly in the mountain regions.

Some of the forestry programmes being carried out are for turning low-trunk forests into high-trunk ones. Forest plantations with irregular tree structures are being replaced by higher productivity species. Intensive forest cultures and plantations for the accelerated output of timber, and anti-erosion cultures are being developed and installations built to combat soil erosion. Forestry occupies an important place. During the years of the popular regime more than 1.7 million ha. of forests have been planted. Bulgaria ranks among the leading countries in the world in per capita afforested areas. Forestry is managed by the Ministry of Forests and Forestry. Fifteen forestry complexes have been formed, within the framework of which there are 165 forestry farms.

GAME FARMING

Game farming is a branch of the national economy whose task is to preserve wild life and ensure its reproduction and proper use, increase the game reserves, enrich the fauna, keep the ecological balance in the natural environment and promote the development of hunting as a sport. Wild life in Bulgaria is the property of the State: it is under the management of the Ministry of Forests and Forestry and the Bulgarian Hunters' and Anglers' Union. Hunting is organized on the basis of scientific planning on lands and waters which belong to the agricultural and forestry reserves of the country. The area of the hunting reserves is about 10.7 million ha. and falls into two sectors – state and public. The state hunting reserves amount to about 1.5 million ha. (about 14 per cent). The management and use of game on these lands is under the direct control and supervision of the Ministry of Forests and Forestry. The territory occupied by the public hunting reserves is approximately 9.2 million ha. (about 86 per cent) and is managed and used by the Hunters' and Anglers' associations on a local basis. Hunting is allowed in strictly defined seasons, taking into account the species, quantity, mating periods, etc. The hunting season is declared each year by an ordinance of the Ministry of Forests and Forestry. The Game Department organizes game-breeding stations and special game farms for the breeding, proliferation and improvement of wild life. Each year about 100 red deer, 60 does, 140 wild boar, 120 fallow deer, etc., are resettled from the game breeding bases. Best organized are the game breeding bases for red deer, does and wild boar in Sherba, in the Varna district, and Palamara, in the Shumen district; for aurochs, at the Voden Game Farm, in the Razgrad district and the Preslav Game Farm, in the Shumen district; for bears at the Kormisosh Game Farm, in the Smolyan district; for fallow deer and moufflons at

Retrieving game

the Bukovets Game Farm in Tvarditsa, the Balchik Game Farm in Balchik and the Nessebur Forest Farm. Among the bigger pheasant-breeding farms are the Dunav and Hotantsa, near Russe; Tundja, near Yambol; the forest farm at Elhovo; Aramliets, in the Sofia district, and Seslav, in the Razgrad district. In 1983 the game reserves in the country contained the following: red deer – 16,900; fallow deer – 4,250; wild boar – 32,100; does – 137,700; bears – 850; moufflons – 2,800; wild goats – 1,560; woodgrouse – 2,070; aurochs – 110; hares – 914,150; partridges – 571,950; rock partridges – 96,200; and pheasants – 389,650.

Game farming is the natural base required for hunting tourism. International hunting tourism is organized by the State Association for Tourism and Recreation through the Inter-hotels-Balkantourist complex.

FISH-FARMING ECONOMY

The water economy is the sector of the national economy whose task is to manage, use and preserve the fish and other water creatures. The organization, planning and management of this sector is based on the Law on the Water Economy (1982). The fish and other water creatures in the territorial waters are the property of the State, which lets its lands and water basins be used by state, co-operative and public organizations for fishing and fish-breeding. After 9 September 1944 fishing was placed on socialist foundations. State Fishing was founded in 1948 as the main socialist economic organization for the industrial output and processing of fish. Industrial fishing takes place in the territorial waters of Bulgaria on the Black Sea, on the Danube and in the ponds along their shores and banks, and at the big state-owned dams and other water basins. Bulgarian ocean-fishing has been developing since 1964 in all parts of the world, using trawlers, cargo- and refrigerator-ships. Angling is practised in the rivers, ponds, mountain lakes and other water basins. The regimes of industrial ocean and sports fishing are regulated by the Law on the Water Economy. The average annual catch of ocean fish is over 100,000 tons, of Black Sea fish about 20,000 tons, and of fresh-water fish about 15,000 tons (1983). Bulgarian fishing started developing in the late nineteenth and early twentieth centuries. The first carp farms were founded at the agricultural schools in Sadovo and Obraztsov Chiflik near Russe. Before 9 September 1944 there were about 40 ha. of fish-breeding basins belonging to two state farms (Samokov and Plovdiv) and 63 private farms. The foundations of industrial fishing were laid at socialist agricultural enterprises in 1950. Present day fish-breeding is carried out at 180 AICs, fifteen specialized fish-breeding enterprises of the Fresh Water Fish-Breeding Scientific Production Economic Plant, at the NAIU, at the hunting and fish-breeding enterprises of the Ministry of Forests and Forestry and the Bulgarian Hunters' and Anglers' Union, as well as at the auxiliary farms of the enterprises in a water area of about 28,300 ha. (1982). The Smolyan district has scored the highest achievements in production. The main directions in fish-breeding are in natural water basins, in basins and in cells. The species most widely bred are carp and trout. Some vegetable eating fishes, such as the white amur, the white and coloured *tolstolob,* the buffalo fish and others are being acclimatized and bred. State policies in the field of fisheries are conducted by the NAIU and the Ministry of Forests and Forestry; the Bulgarian Hunters' and Anglers' Union also plays a certain role in this respect.

A trout farm

TRANSPORT

Before the Slavs and Proto-Bulgarians settled in what is now Bulgaria, these lands were inhabited by Thracians, Illyrians, Greek colonists and Romans. They built roads which still form the basis of the road network. The road system developed particularly quickly during the period of the Roman domination, up to the end of the fifth century, and had great significance for the defence and economy of the Roman Empire. Road construction was assigned to the Emperor's deputies and the municipalities were responsible for their maintenance. The main road from Bulgaria to Istanbul passed through Bulgarian territory and was 670 Roman miles long (1 Roman mile=1,482 metres). So did many other minor roads. The old Roman roads continued to be used by the Slav-Bulgarian State, which improved them, particular attention being given to the famous Roman road from Istanbul to Edirne-Plovdiv-Sofia and from there via Nish to Belgrade, Vienna and other towns. The development of the roads during the seventh to eleventh century was carried out for military purposes, but was

Automatic track-diagram control panel at Sofia central station

An electric train on the Sofia-Plovdiv line

significant for the economy of the country as well. The main roads went through the passes of Stara Planina – the Rishki, Vesselinovski, Vurbishki and Troyanski. The road along the river Struma and along the Kresna gorge was also important. The development of economic and trade links necessitated the improvement of the existing roads and the construction of new roads and ports. Waterways were also used, for example the deep rivers Vardar, Struma, Mesta and Maritsa. The Black Sea and the River Danube were especially significant for military and commercial navigation. A Bulgarian port dating back to the ninth century was found in the fortress on a Danubian island 18 km east of Silistra. Transport was carried out mainly by caravans pulled by horses or oxen. Between the twelfth and fourteenth centuries the road from Istanbul to Sofia was used very intensively, as were the two roads which connected the fortresses on the northern slopes of the Rhodope mountains with Plovdiv and Assenovgrad as their starting points.

During Ottoman rule (1396–1878) the roads were used for military purposes and for economic exchange. Inland water transportation and sea transportation were also used. Caiques sailed on the river Maritsa up to Plovdiv and Pazardjik. The ports of Varna, Nessebar, Sozopol and Ahtopol, among others, were developed. Transport on the Black Sea and along the River Danube was provided by foreign shipping companies at a high cost. The Bulgarian shipping association Postoyanstvo, founded in Istanbul in 1892, was not succesful in its activities. At the beginning of the nineteenth century the role of the cart as a means of conveyance quickly increased. The number of inns along the roads also increased. Road construction on a large scale was begun after 1865 by Mithad Pasha, the Governor of the Danubian vilayet. The roads built during the 1860s and 1870s in the Danubian and Edirne vilayet stimulated agricultural production and facilitated its marketing. Cherna Voda to Konstanta (64 km), the first railway on the Balkan Peninsula, started operating in 1890. Later on, the railway route Shumen-Varna-Turnovo-Gabrovo-Shipka was constructed, and the building of the railway route Russe-Pleven-Lukovit-Botevgrad-Sofia began. The railway from Varna to Russe, 233 km (1866), Belovo to Lyubimets (1873) and the branch-line from Maritsa to Yambol (1874), with a total length of 309 km, were built and began to operate. The construction of the railway from Yambol to Shumen began.

After the Liberation in 1878 the Bulgarian State inherited railways with a length of 532 km and a Turkish obligation to construct the railway from Belovo via Sofia to Pirot to connect with the Serbian railways. A total of 2,570 km of old and ruined roads was all that existed. Up to the end of the nineteenth century comparatively fast progress was made in the development of transport. Considerable sums were invested in the construction of a railway system and road network. The transport policy was laid down by a special legislative system. By virtue of the Law of Railways of 1885, Bulgarian railways were constructed only with permission from the National Assembly, were state property and were used only by the State. In 1888 the National Assembly passed the first bill for the construction of the railway from Yambol to Burgas, and also for Kaspichan-Shumen-Pleven-Sofia-Kyustendil route. In 1895 the first bill for the construction of the ports at Varna and Burgas was passed. Laws were also enacted, such as the Law of Organization and Construction of the Railway System in Bulgaria (1895), the Law for Extension of the Bulgarian Railway System (1897), the Law for the Utilization of the Bulgarian State Railway and Ports (1908), the Law for Organization and Management of the Bulgarian State Railways and Ports (1929), the Law for the Extension of the Railway System and Ports (1933) and so on. In 1888 the opening of the first railway constructed by Bulgarians - Tsaribrod-Sofia-Vakarel - laid the foundation of the Bulgarian State Railways network. In 1889, through the railway from Belovo to Vakarel, Bulgaria was connected with the European railway system. With the development of foreign trade and the home market the importance of railway transport increased, and it became established as the basic means of transport for the capitalist economy.

Up to 1890 the Bulgarian railway system included the railway Russe-Varna, from the border to Sofia-Vakarel-Belovo, Yambol-Burgas, Kaspichan-Shumen, Sofia-Radomir, Roman-Shumen, Chirpan-Nova Zagora, Somovit-Yassen, Yovkovo (Razdelna)-Devnya, Russe-Turnovo - a total of 1,342 km. Some sections of the main railways were built and became operational such as the southern railway from Voluyak to Bankya (1918), Sofia-Kyustendil-Gyueshevo (1893-1910), Belassitsa-Kulata - narrow-gauge railway (1929), Stanke Dimitrov-Bobov Dol - narrow-gauge railway (1917), Septemvri-Dobrinishte (1926-45), Pazardjik-Varvara (1926), Krichim-Peshtera (1937), Plovdiv-Karlovo (1933), Plovdiv-Panagyurishte (1933), Plovdiv-Assenovgrad (1928), Plovdiv-Chirpan, Yambol-Zimnitsa-Burgas (1910) and Yambol-Elhovo (1931); of the northern railway - Mezdra-Vidin (1923), Cherven Bryag-Oryahovo (1930), Yassen-Somovit (1899), Levski-Svishtov (1909), Oresh-Belene (1921), Levski-Lovich (1927), Gorna Oryahovitsa-Lyaskovets (1930), Tsar Krum-Preslav-Shumen-Komunari (1941), Kaspichan-Todor Ikononovo - narrow-gauge railway (1917), Razdelna-Kardam (1910), Sindel-Karnobat (1943) and Yunak-Staro Oryahovo (1939); of the railway passing through the Balkan Range, which connected the capital with Burgas, Burgas-Pomorie - a narrow-gauge railway (1918) and a standard one (1939); Russe-Gorna Oryahovitsa-Dubrovo-Tulovo-Stara Zagora-Rakovski (Dimitrovgrad)-Podkova (1899-1944); and Vurbanovo-Gabrovo (1912). The extension of the railway system proceeded irregularly. The highest rates of railway construction were during the periods 1878-87 and 1901-6 (see Table 1).

In 1939 the length of the railway system was 4,426 km (main railway 3,477 km, station tracks 949 km). Railways of a 1,435 mm gauge amounted to 3,041 km; of 760 mm, 221 km; and with 600 mm, 215 km. The first

A section of the Hemus motorway

Table 1: Extension of the Network of the Bulgarian State Railways

Periods	Growth (in km)	Average annual growth (in km)
1866–78	532	44.4
1879–1913	1,576	45.0
1914–23	529	52.9
1924–33	577	57.5
1934–43	636	63.9
1878–1943	3,318	51.0

The Asparuh bridge in Varna

A viaduct on the Hemus motorway

rolling-stock of the Bulgarian State Railway was bought from England – nine steam locomotives, and from Belgium – one de luxe railcar, two mail carriages, 54 passenger carriages, and 422 freight trucks. Later on, more up-to-date locomotives, wagons and coaches were brought from Austria, Germany and Belgium, for example. The maximum speed of the trains was about 30 km per hour. Four railway workshops and 23 engine-sheds were created for the maintenance of the rolling-stock. The railway lines were technically simple – big slopes and curves with small radii. They were built through high interest loans from foreign banks. The 20,300 workers and employees who worked for the Bulgarian State Railways were heavily exploited. In 1939 12.4 million passengers and 6.1 million tons of cargo were transported. During the Second World War the volume of military transport increased. Part of the railway system and other material and technical equipment belonging to it was destroyed.

A prerequisite for the development of automobile transport was the construction of a road network. At the beginning of the twentieth century there were 3,367 km of road: 1,167 km were in comparatively good repair, 670 km were in adequate condition and 1,531 km were in bad condition. As a result of road construction, there were 22,195 km of roads in 1944: 409 km were paved, 82 km were covered with concrete and 9 km were covered with asphalt.

In 1902 the first petrol automobile was imported into Bulgaria; later on several cars were supplied to meet the needs of the royal court, and for the post office. In 1909 the first rural bus line was opened, operated by a private automobile society with one six-seat car. Later on, buses were bought, with eight, ten or twelve seats. In 1913 there were 112 passenger and five cargo lines in the country. Automobiles were more broadly used for cargo and passenger transport after the First World War and particularly in the 1930s. In 1933 there were 2,112 cars and 508 lorries. Public goods transport was carried out by private owners and co-operatives. Private automobile companies with 364 buses covered 47 permanent routes with a total length of 2,425 km. In 1939, bus lines connected 640 populated centres, and there were 2,503 cars, 780 buses, 2,415 lorries and 3,510 motor-cycles – a total of 10,013 motor vehicles, of which 9,371 were on the road. In the 1930s and 1940s automobile transport became the chief competitor of the railways. In 1935, the first Automobile Communications Law was approved. By the Automobile Communications Law of 1940, broad state participation was envisaged in the organization and management of automobile transport. The Union of Public Goods Transport was set up, making membership obligatory for owners and working automobile co-operatives owning lorries. In 1941, the first state-owned bus service between Sofia and Skopje was opened. In 1943, there were state-owned bus services from Sofia to Samokov, Sofia to Rila Monastery, from Pavlikeni via Sevlievo to Gabrovo, and, since 1944, from Sofia to Breznik to Trun. The development of city electric transport began with the operation of the first electric tram in Sofia in 1901. The effective use of Sofia's electric transport was begun by a Belgian joint stock company and continued from 1916 by the Sofia municipality. In 1939, the tram system was 21.8 km long; 131 semi-wooden tram cars were in motion, as well as 106 trailers. Trolley-bus passenger transport in Sofia was introduced in 1941, when the first trolley-bus service from Gornobanski Put to Gorna Banya was opened. This was 3,350 m. long. The beginning of the Bulgarian merchant fleet was laid by the Bulgarian Commercial Shipping Society which was established in 1892 with state participation. Several ships were delivered which served as transport between Bulgarian ports. The total dead-weight of the sea fleet reached 29,000 tons in 1939. In 1934, the State Danubian Shipping Company was set up, which at the outset of World War I had 17,600 tons, including privately-owned ships. In 1939 the sea fleet carried a total of about 53,000 tons of goods. Over 90 per cent of the imports and 85 per cent of the exports by sea were carried by foreign shipping companies. During the war, the sea and river ships were used by the German military transport. Eventually the majority were sunk.

Important Black Sea ports were Varna (1906) and Burgas (1903). The turnover of cargo was small. Important ports on the river Danube were Vidin (1908), Svishtov (1908–13), Russe (1912) and Lom (1938). The development of civil aviation began in 1927 with the opening of the international airline Paris-Belgrade-Sofia-Istanbul by the French company 'Sidna'. Later, airline companies from Romania, Poland, Hungary and Germany opened airlines which landed in Sofia. In 1939 the Soviet enterprise Aeroflot opened a service from Moscow to Sofia. During World War II, the German firm Deutsche Lufthansa was the only airline company to use the Bulgarian airlines. In the airports there was only primitive technical equipment, old navigational equipment and other technical machinery.

The popular regime established on 9 September 1944 found transport backward and almost destroyed by the war. In the first few years, great efforts were made to reconstruct, strengthen, and transform the transport system into the economy's basic connecting unit. In the period of the two-year national economic plan (1947–48), the necessary technical equipment for railway transport was reconstructed and brought in line with the requirements of industrial and agricultural production. The foundations for Bulgarian civil aviation and state automobile transport (1948) were laid, and river and sea transport were re-established. After the April Plenum of the Central Committee of the BCP (1956), considerable qualitative changes in transport took place. On the basis of public ownership of the means of production and general

Bulgarian-made Chavdar buses

state management, Bulgaria's unified transport system was established (1960) and has been operating ever since. It includes railway, automobile, water (sea and river) and air transport, plus the use of pipelines.

Railway transport was brought into line with the increased needs of the national economy. In 1948, the following railway sections began regular operations: Somovit–Cherkvitsa, Samuil–Isperich, Pernik–Voluyak, Lovech–Troyan, and Karlovo–Sopot–Klissura. The rolling-stock is being modernized by updating locomotives; equipping carriages with communication cords; expansion of repair facilities for locomotives and carriages; increasing the speed of trains; increasing the traffic, processing and transport capacity of the basic lines and important composition and loading stations; renovation of the railway by means of heavier rails, etc. During the 1949–56 period railway lines were built between Stanke Dimitrov and Bobov Dol (reconstructed from 600 mm gauge to 1,435 mm) and Lyaskovets–Zlataritsa–Makotsevo–Dolno Kamartsi–Klissura, at which point the trans-Balkan line from Sofia to Burgas was finally built. Up to 1965 new sections were built and a direct connection to Turkey was established. 215 km of narrow-gauge (600 mm) railways were dismantled along the Kaspichan–Kaolinovo–Todor Ikonomovo and Kocherinovo–Rila Monastery lines and the railway line from General Todorovo to Petrich–Kulata was re-built into a 1,435 mm one. After the construction of the bridge over the river Danube (1954), the railway network was connected with the transport systems of the Socialist Republic of Romania and the USSR. A favourable atmosphere was created for the improvement of transport and economic ties and the territorial distribution of production forces in the country.

Table 2: The Railway Network in Bulgaria

Railway lines	Years					
	1939	1948	1960	1970	1980	1982
Railway network (total in km)	4,426	4,926	5,620	6,040	6,419	6,416
Actual railways	3,477	3,786	4,111	4,196	4,267	4,273
of which						
normal (1,435 mm)	3,041	3,389	3,771	3,951	4,022	4,028
of which						
double –	—	—	25	212	535	714
electrified –	—	—	—	811	1,581	1,828
semi-normal (760 mm)	221	246	245	245	245	245
narrow-gauge (600 mm)	215	151	95	—	—	—
Station tracks	949	1,140	1,509	1,844	2,152	2,143
actual railroads on 100 sq. km territory	33.7	34.1	37.3	37.8	38.5	38.5

On the basis of the decisions of the Eighth Congress of the BCP (1961), technical reconstruction and modernization of the equipment of the Bulgarian State Railways was carried out. One basic change was the eradication of low productive and ineffective steam traction and the introduction of electric and diesel traction, automation and telemechanics etc. The use of electric traction in the Bulgarian State Railways began along the first electrified railway section from Sofia to Plovdiv (156 km) and along the section from Russe to Gorna Oryahovitsa (111 km) in 1963. The electrification of railways went ahead with a one-phase alternating current, 50 Hz, 25 KW. It was carried out with the active scientific and technical assistance of the USSR and Czechoslovakia. In 1984, the great electrified ring Sofia–Gorna Oryahovitsa–Russe–Varna–Karnobat–Burgas–Plovdiv–Sofia was completed. In 1963, diesel traction was introduced along the railway sections from Sofia to Tulovo and for the express trains along the railway lines Sofia–Russe, Sofia–Varna and Sofia–Burgas. Electric traction began to be substituted for diesel traction in 1976, and the process was completed by the end of 1979. Electric engines are supplied by Czechoslovakia, and diesel by Austria, Romania and the USSR.

The present structure of railway transport includes nine trunk-lines (as well as branch lines): (1) Kalotina–Dragoman–Sofia–Pazardjik–Plovdiv–Svilengrad–to the border (part of the international route connecting Europe with the Middle East); (2) Sofia–Mezdra–Pleven–Gorna Oryahovitsa–Kaspichan–Varna (the principal trunk-line in Northern Bulgaria); (3) the sub-Balkan railway line (the shortest route from Sofia to Burgas and Varna via Karlovo–Sliven–Karnobat); (4) Russe–Gorna Oryahovitsa–Veliko Turnovo–Dubovo–Tulovo–Stara Zagora–Dimitrovgrad–Kurdjali–Podkova (connecting the southernmost parts of the country with the centre and the Danube river); (5) Sofia–Pernik–Stanke Dimitrov–Blagoevgrad–Kulata (connecting North-Eastern and Eastern Europe with Thessaloniki); (6) Sofia–Voluyak–Pernik–Radomir–Kyustendil–Gyueshevo; (7) Mezdra–Vratsa–Vidin (the ferry line Vidin–Kalafat making it an international railway line); (8) Plovdiv–Chirpan–Stara Zagora–Yambol–Karnobat–Burgas; (9) Russe–Samuil–Kaspichan (the oldest railway line in Bulgaria).

The tram depot in Sofia

The qualitative changes in the technical development of the railway network went a long way towards meeting the transport needs of the national economy and population (see Tables 3 and 4). The country's industrialization produced significant changes in the structure of freight transport by rail. The main items are: coal, ores and ore-concentrates, liquid fuels, metals, cement, bricks and roof-tiles, timber, stone, ballast and gravel, chemical fertilizers and sugar-beet.

Table 3: Volume of Transport by Rail

	1939	1952	1960	1970	1980	1982
Loads (millions of tons)	6.1	17.9	38.4	68.1	77.8	83.9
Load turnover (millions of tons per km)	1,113	3,140	6,981	13,858	17,681	18,276
Passengers (millions)	12.4	51.8	79.0	106.0	100.0	96.9
Passenger turnover (millions of passengers per km)	711	2,489	3,617	6,223	7,055	7,092

The bus terminal in Vratsa

Table 4: Relative Share of Railway Transport according to Kind of Traction (%)

Year	Kind of traction		
	Steam	Electric	Diesel
1960	100.0	—	—
1963	92.4	6.9	0.7
1970	40.7	28.9	30.4
1980	—	60.3	39.7
1982	—	66.1	33.9

Loading and unloading operations take place at 262 railway stations, 156 of them being principal stations. Big industrial enterprises are served by 27 stations, and industrial branch-lines number over 700. The key stations concerned with the flow of goods are: Sofia, Pernik, Bobov Dol, Plovdiv, Stara Zagora, Dimitrovgrad, Burgas, Varna, Russe, Lom, Gorna Oryahovitsa and Pleven. The lines which carry the most passengers are: Sofia–Kremikovtsi–Pernik, Plovdiv–Pazardjik–Assenovgrad, Burgas–Aitos, Varna–Devnya–Provadia, Russe–Stara Zagora, Pleven–Dolni Dubnik–Dolna Mitropolya and Mezdra–Vratsa.

A container transport network has been under construction since 1972. Today, twenty container stations are in operation. The container stock amounts to 175,000 tons.

A railway-ferry line Varna (Bulgaria) to Ilichovsk (USSR) has been in operation since 1978; the distance is 433 km (247 miles). It is an integral part of the production infrastructure of the industrial transport and residential agglomerate Varna-Devnya-Provadia (see diagram). The Varna-Ilichovsk ferry-line complex is jointly operated by Bulgaria and the Soviet Union. It is served by railway ferryboats, a subsidiary fleet, port and shore facilities. Of the four ferryboats, two are Bulgarian and two are Soviet. The Bulgarian boats are called *The Heroes of Sevastopol* and *The Heroes of Odessa,* and the Soviet boats are *The Heroes of Shipka* and *The Heroes of Pleven.* All were designed in the USSR; the first two were built in Norway and the other two in Yugoslavia. They can carry 108 four-axle freight wagons, and their dimensions are: length – 185.4 m., width – 26 m., load-line – 7.4 m., height up to deck-top – 15.2 m., 13,008 tons dead weight and full displacement – 22,600 tons; maximum speed – 19.4 knots, operation speed – 18.6 knots. There are cabins on board for 40 crew members and twelve passengers. The boats are designed to function as ships of the Ro-Ro type transporting 148 long-distance TIR trucks each.

Railway transport is based on a regional structure. There are four railway boards with head offices in Sofia, Gorna Oryahovitsa, Plovdiv and Varna. They administer engine and wagon sheds, freight-operation and track-maintenance facilities, traction substations, and communications and safety equipment. The Bulgarian State Railways Economic Corporation is the central administration body.

Road transport has developed rapidly and during the years of the popular regime has become an important subsystem in the national transport complex. The state and density of the road network is of great significance. In the period of socialist construction the roads have been renovated and modernized on the basis of a Twenty-Year Plan (1961–80). Quantitative and qualitative changes

Table 5: Roads and Highways (in thousand km)

	Years					
	1939	1956	1960	1970	1980	1982
Length of roads by the end of each year with hard surface including:	19.6	25.4	27.4	30.3	32.4	32.6
Motorways	—	—	—	—	0.1	0.2
First class	—	2.0	2.0	2.4	2.4	2.9
Second class	—	4.2	4.1	4.3	4.6	5.7
Third class	—	4.0	4.2	5.9	6.1	5.9
Fourth class	—	15.2	17.1	17.7	19.6	19.9
Roads per 1,000 sq. m of territory[1]	189.7	229.1	247.1	325.9	328.6	323.6

[1] Including special-purpose unpaved roads.

The Varna-Ilichovsk ferry

have been brought about in the network. The relative share of the highways, first-class and second-class roads in the total road length is over 19 per cent, of third-class roads – 16.8 per cent, of fourth-class roads – 64.2 per cent, which provides good conditions for better transport and economic communication in the economic regions and residential areas (see Table 5).

The main highways, which form the backbone of Bulgaria's modern road network, are: (1) Kalotina–Sofia–Plovdiv–Svilengrad (part of the North–South international European highway, which starts from Gdansk [Poland] and passes through Bratislava [Czchoslovakia], Budapest [Hungary] and Sofia, before reaching Istanbul [Turkey]); (2) Kulata–Blagoevgrad–Sofia–Botevgrad–Pleven–Byala–Russe (of international importance); (3) Gyueshevo–Kyustendil–Pernik–Sofia–Botevgrad–Lovech–Veliko Turnovo–Shumen–Varna; (4) Sofia–Karlovo–Sliven–Burgas (the shortest distance between Sofia and the Black Sea coast); (5) Russe–Polski Trumbesh–Veliko Turnovo–Gabrovo–through the Shipka pass–Kazanluk–Stara Zagora–Haskovo–Kurdjali.

A 1,000 km ring-road is now under construction, including the motorways: *Trakia* (Sofia–Plovdiv–Burgas), *Hemus* (Sofia–Botevgrad–Varna), *Cherno More* (Varna–Burgas) and *Maritsa* (Popovitsa–Svilengrad). The unique Asparuh Bridge, commissioned in 1976 (2,000 m long, 21 m wide), which has a dual carriageway with two lanes each and footpaths, is considered the beginning of the construction of the Cherno More motorway.

General road transport, the main kind of short-distance transport, is developing very fast (see Table 6). In 1982 its share of the transport of goods was 31.8 per cent and in the transport of passengers (not counting transport by factory and city motor vehicles) was 88.7 per cent. The main goods transported included construction materials, grain and grain fodder, timber, coal and vegetables and fruits. Road transport between villages is developing quickly, a prerequisite furthering the good ties between town and village and the consolidation of the settlement systems. The number of towns with a bus service has also increased. There are trolleybuses in Sofia and Plovdiv, and trams in Sofia (see Table 7).

Automobile transport for general use is run by the Avtomobilen Transport Economic Corporation, which includes 28 automobile combines in all the district towns, an automotive enterprise in the town of Kazanluk and 185 automotive branches in the main residential areas. They are used for goods transport, passenger transport and combined road haulage.

International road haulage (TIR) plays an important part in the national transport complex. International

Table 6: Main Functions of Road Transport

	_____ Years _____				
	1948	1960	1970	1980	1982
Transported goods (in thousand tons)	4,467	48,647	121,871	319,354	322,114
Goods carried (in millions of tons per km)	102	889	7,902	15,893	16,727
Passengers carried (in millions)	59[1]	248[1]	1,316	1,950	2,060
Passenger transport (in million passengers per km)	620	2,319	12,246	21,622	22,860

[1] In 1952 and 1959.

Table 7: City Transport and the National Transport Network

	_____ Years _____				
	1956	1960	1970	1980	1982
Towns with bus transport (number)	44	68	94	137	146
Separate bus services between populated areas (number)	536	1,082	2,533	4,371	4,403
Total length of bus services between populated areas (in km)	23,315	50,253	154,329	169,417	165,264
Number of towns and villages linked by bus	2,140	4,434	4,861	4,658	4,822
Length of the tramway network in Sofia	92[1]	99	110	153	167
Length of trolley-bus lines in Sofia and Plovdiv (in km)	51[1]	51	89	149	153

[1] 1957 only in Sofia (1956, 1960).

transport started in 1957 with a fleet of 100 lorries. In 1982 the International Road Transport Economic Corporation, one of the largest and most respected transport organizations in Europe, did business in Europe, Asia and Africa. There are two automobile enterprises in Sofia, one in Pazardjik, one in Burgas and one in Russe operating 20- and 25-ton long-distance lorries.

Sea transport plays an important part in Bulgaria's foreign economic relations. By 1962 Bulgaria ranked among the major sea-faring nations of the world. The first ships for international navigation were bought in 1946 and 147, and the state shipping company Bulgarian Maritime Fleet was established. After 1960 the maritime merchant fleet was supplied with modern ships, to which the Bulgarian shipbuilding industry contributed several. The Georgi Dimitrov shipyards in Varna turned out the flagship of Bulgarian ship-building and navigation – the *Khan Asparuh* 100,000-ton tanker, as well as 25,000- and 38,000-ton bulk cargo ships. In addition, 50,000-ton bulk cargo ships were built in the USSR. As a result of the construction of a large number of specialized ships, such as colliers, ore-carriers, container vessels, grain-carriers and oil-tankers, the relative share of the specialized merchant fleet reached 85 per cent of its total loading capacity. Sea transport is mainly directed towards long-distance haulage.

The volume of transported goods and passengers is constantly increasing (see Table 8).

In 1982 sea transport contributed 2.6 per cent of the volume of transported freight and 60.4 per cent of the freight turnover. Passenger transport by sea depends on the number of tourist voyages during the summer months.

The Bulgarian Maritime Fleet in Varna maintains nine regular sea routes. Among them are those between the Black Sea and Cuba, the Black Sea and the Western Mediterranean, and the Black Sea and the Far East. There are regular contacts between the sea ports of Varna and Burgas and the Soviet sea ports of Odessa, Ilichovsk, Novorossiisk and Zhdanov, through which Bulgaria receives raw materials and manufactured goods from the

Table 8: Volume of Freight in Sea Transport

	\ 1948	\ 1960	Years\ 1970	\ 1980	\ 1982
Shipped cargo (in thousand tons)	173	1,064	14,518	24,693	26,436
Shipped cargo (in million tons per km)	296	2,540	38,949	62,499	59,052
Transported passengers (in thousands)	350	559	602	699	710
Transported passengers (in million passengers per km)	7	19	27	36	36

Soviet Union. Single routes are maintained to the Baltic Sea, Murmansk and some sea ports in West Africa.

Major sea ports are Varna, Varna-West and Burgas. They are continually modernized and equipped with highly efficient facilities so as to facilitate high-speed, complex handling of load operations. In 1982 mechanized work accounted for 84.4 per cent of the total volume of work done at these sea ports. A significant part of Bulgaria's imports and exports is handled at the Black Sea ports (see Table 9).

Inland Transport. Since 1960 inland navigation along the Danube has developed even more quickly. The Danube is a convenient and cheap waterway for the import and export of goods from and to the USSR, Hungary, Czechoslovakia, Poland, Austria, the GDR and Romania. Imports and exports between Bulgaria and the USSR pass mainly between the Danubian ports of Lom and Russe and the ports of Reni and Izmail in the USSR.

The passenger liner terminal in Varna

A Kometa passenger hydrofoil

The Bulgarian section of the Danube is 472 km long and there are eight ports and fourteen entrepôts along the river.

The ports of Russe and Lom are the country's major river ports. The cargo turnover of the river ports of Silistra, Svishtov, Somovit and others is continuously growing. In 1948 the cargo handled at river ports (imports, exports, transit cargo and riverside shipments) amounted to 696,000 tons; in 1960 the corresponding figure was 2,933 thousand tons, in 1970 – 9,168 thousand tons, in 1980 – 10,464 thousand tons and in 1982 – 10,367 thousand tons. In 1982 mechanized load-handling operations at river ports amounted to 96.2 per cent. Passenger cruises between Vidin and Silistra are by modern hydrofoils. Two first-class ships, *Russe* and *Sofia*, sail along the Russe–Vienna route and back, with calls at Vidin, Belgrade, Budapest and Bratislava. The fleet of river-going ships has been modernized with new vessels – tug-boats, barges and tankers. In 1982 the total loading capacity amounted to 300,000 tons. The catamarans – *Khan Asparuh, Khan Tervel, Khan Kardam, Khan Krum* carry combined land-river-land shipments along the

Table 9: Cargo Turnover of the Seaports (in thousand tons)

	1948	1960	Years 1970	1980	1982
Imports	303	1,066	13,522	28,116	24,578
Exports	177	544	2,295	3,567	4,480
Transit	16	243	18	39	13
Coastal	113	626	255	514	415
Total	**609**	**2,475**	**16,090**	**32,236**	**29,486**

Table 10: Volume of Cargo and Number of Passengers Transported by River

	1948	1960	Years 1970	1980	1982
Shipped cargo (in thousand tons)	264	1,556	3,692	4,936	4,978
Cargo turnover (in million tons per km)	58	615	1,832	2,614	2,539
Transported passengers (in thousands)	605	798	275	387	428
Transported passengers (in million passengers per km)	40				

Vidin–Passau route, into the Federal Republic of Germany. A ferryboat connects the Bulgarian river port of Vidin and the Romanian river port of Calafat. A container vessel route was started in 1976 for cargo shipments between Russe and the Soviet river port of Reni. Heavy-lift container vessels have been cruising between Russe and Bratislava since 1978. A joint Bulgarian-Soviet transport organization, Dunaitrans, with its headquarters in Russe, and a representation office in Izmail, in the USSR, has been functioning since 1977. It promotes the two countries' river transport and facilitates commodity exchange between them. The shipping agencies of Bulgaria, the Soviet Union, Czechoslovakia and Hungary are members of the Interlighter economic organization, which arranges river-land-river shipments. The headquarters of the organization is in Budapest. The Ust Dunaisk specialized port at the mouth of the Danube, in the USSR, is the basic port of the *Julius Fuchik* and the *Tibur Samueli* lighters. The organization was set up as a result of the integration of the Danubian CMEA member-countries. Its aim is to accelerate river shipments and to reduce transport costs. The lighters run along two routes: Ust Dunaisk-Bombay-Karachi, and Ust Dunaisk-Ho Chi Minh (the port of Saigon–Phnom Penh, Kampuchia), with a length of approximately 14,000 km and along the Danube – 2,900 km, i.e. the total return route is nearly 33,000–35,000 km long. Thanks to these lighters the time taken to ship cargo from the mouth of the river Danube to the Pacific and the Indian Oceans is reduced. The significance of river transport will increase after the construction of the Rhine-Main-Danube and the Oder-Danube canals. These will significantly shorten the distance between the Black Sea and the North Sea, and between the Black Sea and the Baltic Sea.

Sea and river transport is run by the Varna-based Voden Transport economic association, which is also in charge of the Bulgarski Morski Flot and the Bulgarsko Rechno Plavane shipping agencies based in Varna and Russe. Ports in Bulgaria are grouped into four complexes: the Burgas harbour complex, comprising the Black Sea ports of Burgas, Nessebur, Sozopol, Pomorie, Michurin, and Ahtopol; the Varna harbour complex which includes the Black Sea ports of Varna, Varna-West, Balchik and the Varna thermo-electric power station; the Russe harbour complex, with the Danubian ports of Russe, Svishtov, Silistra, Somovit and Tutrakan; and the Lom harbour complex comprising the Danubian ports of Lom, Vidin and Oryahovo.

Air transport, which is a relatively recent innovation, is developing at a rapid pace. Bulgarian civil air transport was only begun in the years of the popular regime. The first domestic airline, Sofia-Burgas, and the first international line, Sofia-Budapest, were inaugurated in 1947. Initially Douglas and Junkers aeroplanes were used. The IL-2 planes (speed 220 km per hour) and IL14 (speed 320 km per hour) were in operation until 1962. Over the 1962–68 period IL-18, AN-12 and AN-24 (speed 600 km per hour) turbo-prop aircraft were imported. After 1968 the jet-planes TU-134 (speed – 900

Varna airport

to 950 km per hour, maximum weight – 7.7 tons, 72 seats) and TU-154 (of the same speed, maximum weight – 20 tons, 164 seats) began to be used. Aircraft and ground control have been updated and specialized, and a modern system of airline management has been established. Air transport has become a primary method of passenger travel and cargo shipments. The all-cargo fleet facilitates the import and export of goods and above all of perishables – fresh fruit, vegetables and other foodstuffs.

Aircraft are also used in agriculture and forestry. Bulgarian agricultural aviation plays a part in fertilizing, crop-dusting and pest-control, both at home and abroad. Medical-aid aviation has been founded for emergency air services. Aircraft for use in fishing (for locating shoals), and in geological surveys, as well as charter flights, are also being developed.

There are three international airports in Bulgaria – in Sofia, Varna and Burgas. Another nine airports are functioning: in Sofia, Vidin, Gorna Oryahovitsa, Turgovishte, Russe, Silistra, Varna, Burgas and Plovdiv.

Bulgaria has a dense network of domestic air-routes. The planes of the Balkan Bulgarian Airlines fly along 45 international air routes to countries in Europe, Asia and Africa (see Table 11).

The rapid development of economic, political and cultural relations between countries and peoples and of international tourism resulted in a significant increase in civil flights and air-cargo shipments (see Table 12).

Bulgarian air transport has been developed and updated with the Soviet Union's active assistance.

Table 11: Length of Air Routes

	1952	1960	Year 1970	1980	1982
Length of air routes (in km) of which:	2,081	5,366	28,145	77,540	91,904
domestic	862	1,180	3,245	2,310	2,310
international	1,219	4,186	24,900	75,230	89,594

Table 12: Air Transport Capacity

Indicators	1952	1960	Year 1970	1980	1982
Passengers carried (in thousands):					
total	40	206	1,135	2,187	2,255
by domestic flights	38	183	625	860	980
Operation and services (in millions of passengers per km)	11.0	88.6	1,201	2,670	2,701
Aircraft shipments – total (in tons)	1,550	680	8,307	24,585	28,218
Domestic air-cargo shipments (in tons)	1,160	525	1,463	5,068	6,252

The organization and management of the country's air transport is carried out by the Balkan Bulgarian Airlines economic association which incorporates two production and economic units – Civil Aviation and Agricultural Aviation – and three airport complexes, Sofia, Varna and Burgas. The first includes the airports of Sofia, Vidin, Gorna Oryahovitsa and Plovdiv, while the other two comprise the airports of Silistra, Varna and Russe, and of Burgas and Stara Zagora respectively. Balkan Airlines has offices in 40 countries and maintains business contacts with over 60 airline companies.

Conveyance by pipeline is a progressive trend in contemporary transport. The trunk pipelines connecting the gas deposit in the village of Chiren, in the Vratsa district, with the Wielhelm Pieck cement plant in the village of Beli Izvor, in the same district (1964–65), and with the Vratsa Chemical Combines (1966) were built with the assistance of the Soviet Union. With the signing of an agreement in 1969 between Bulgaria and the USSR for an annual delivery of 3,000 million cu.m of Soviet natural gas to Bulgaria, the construction of a special pipeline was started. The northern semi-circle of the USSR–General Toshevo–Shumen–Pleven–Lukovit–Sofia–Pernik pipeline with a total length of 882.5 km (254 km through Soviet territory, 183.5 km via Romania and 445 km via Bulgaria) was completed in 1972. Its branches leading to Devnya, Varna, Pleven and Kremikovtsi provide gas, both as a raw material and energy resource, to the Devnya industrial complex, the petro-chemical works in Pleven, the Leonid Brezhnev Iron and Steel Works in Kremikovtsi and many electric power stations. The southern semi-circle General Toshevo–Burgas–Yambol–Pazardjik–Sofia and its branches to Sliven, Dimitrovgrad, Haskovo and Kurdjali links with the northern semi-circle, thus forming Bulgaria's unified gas-supply networks. In 1980 some 1,080 km of pipelines were in operation. Pipelines in Bulgaria run in various directions including Burgas–Stara Zagora–Plovdiv–Sofia and Burgas–Devnya–Varna. In 1978 pipeline transport carried 253,000 tons of liquids, in 1980 they amounted to 3,139,000 tons and in 1982 – 16,497,000. Pipeline transport is part of the national transport network and was established and is managed and used by the Ministry for Chemical Industries and the respective enterprises connected with it. With the construction of the material and technical base of pipeline transport the country's unified transport network has been finally completed. The Ministry of Transport is entrusted with its entire management, while planning, design, construction and the repair of roads is carried out by the Main Road Administration at the Ministry of Transport and its specialized auxiliary services.

COMMUNICATIONS

Communications in Bulgaria began to develop during the reign of Khan Krum (803–814). He was the first to organize a postal network which served the State. At first messages and state documents were delivered across the country by horsemen. Letters were also delivered by means of homing pigeons and falcons. The development of communications came to a halt after the invasion of the Balkan Peninsula by the Turks. The first organized post offices in the Ottoman Empire appeared as late as the sixteenth–seventeenth centuries. They engaged solely in the dispatch of government decrees and the collection of taxes. It was only at the end of the eighteenth century that they started delivering mail to the general population. The communication network was small and poor, and failed to meet the needs of the emerging trades and crafts. Russia, Austria-Hungary, France, Germany, etc. set up their own postal bureaus. In 1855 they opened the first telegraph lines in the present Bulgarian lands: Constantinople-Edirne-Shumen, Shumen-Varna and Shumen-Russe; the Russe-Varna line was opened in 1856. They were mostly used for military purposes. During the Russo-Turkish War of Liberation (1877–78) the retreating Turkish units destroyed the telegraph lines and set on fire most of the post office buildings. In the course of the war the Russians set up a field postal service, a civil postal service and military telegraph communications. Large-scale organizational and preparatory work was carried out for the construction of Bulgarian communication facilities and for the training of personnel during the Provisional Russian Administration (June 1877–July 1879). About 250 Bulgarians from the liberated areas were trained. Post offices were opened in 1878 in most of the major towns and in those located on important crossroads. The Bulgarian authorities received free two postal bureaus, 29 post offices and 26 telegraph stations. On 11 April 1879 Prince A. M. Dondukov-Korsakov, the Russian Imperial Commissioner, approved the provisional regulations for postal communications in Bulgaria which later served as the basis for the organization of communications in the country. They came into force on 1 May 1879 when the first Bulgarian postage stamp was issued. This is considered to be the birth of Bulgarian postal communications. This event also marked the beginning of the organized delivery of newspapers and magazines through subscriptions and trustees. The exclusive right of the State over the means of communications was approved in 1881. Initially, postal communications were conducted along fifteen routes. Mail was chiefly delivered by horse-driven carts. Mobile post offices were set up along the railway lines and motor vehicles were supplied. In 1939 there were 770 post offices and 354 postal routes with a length of 21,153 km. The postal network covered 53 per cent of the villages. Telegraph communications developed on the basis of the 26 stations left behind by the Russian Army of Liberation. The telegraph network gradually expanded. In 1885 there were 68 stations with 194 Morse telegraph machines. The telegraph network had 3,583 km of pole lines with 5,889 km of telegraph conductors. A Hughes letter-printing telegraph machine was introduced in 1903; teletypewriters appeared in 1931 and were used for Bulgaria's international telegraph communications. Telegraph traffic between district towns was executed by means of Hughes letter-printing machines. Telegraph services in the villages were poor, most of them using telephones for the delivery of telegrams. At the end of 1944 the telegraph network had a total length of 27,000 km; the greater number of the existing 793 machines were Morse. The first telephone calls were made near the village of Pordim, in the Pleven district (1877) and in the town of Plovdiv (1879). Regular telephone links were established after the installation of the first telephone exchange in Sofia in 1886 (a five-line upright telephone switch-board). By 1895 ten- and 50-line telephone switch-boards were introduced in three other towns. Until 1932 telephone lines connecting towns, villages, etc., were aerial double-conductor lines. High-frequency telephone devices (systems with carrier currents) began to be used during the period 1932–35. The first single-channel system of the DA-1 type was delivered by the British firm, Standard. The automation of local telephone communications began in 1935. Automatic telephone exchanges (ATE) with a total capacity of 10,120 numbers were imported and installed by 1938. Of these, 5,920 were in Sofia. As many as 18 ATEs with a capacity of 17,220 numbers were installed during the 1934–44 period. Until 1944 long-distance calls inside the country were possible only through a telephone operator and the manual operation of the connections. At the end of 1944 the country's towns and villages had 29,564 telephone subscribers and 51 coin-operated phones.

Communications

The first stage coach in Bulgaria

The development of broadcasting began with the introduction of the first spark radio telegraph transmitter in Sofia, in 1913. A law was passed in 1927, which allowed the use of radio receivers; by the end of the year as many as 427 radio sets were registered, mainly in Sofia. The year 1929 marked the beginning of radio broadcasting, with the first radio programme broadcast in Bulgarian by means of a 60-watt transmitter constructed by amateur radio technicians. The first medium-wave radio transmitter with a capacity of 1 kW was built in 1934 and the first small studio came into operation. In 1935 the monopoly of radio broadcasting was taken over by the State. Medium-wave transmitters were set up in Sofia, Varna and Stara Zagora in 1937. They were brought from the German firm Telefunken and could cover about one third of the country's territory. The Bulgarian Radio in Sofia was officially inaugurated in 1942. As many as 183,000 radio sets were registered over the 1929–44 period. In 1906 Bulgaria was represented at the constituent International Radiotelegraph Conference in Berlin. Consequently Bulgaria was regularly represented at all the international conferences concerning radio broadcasting.

The rapid economic development which started after the establishment of the popular regime (1944) promoted communications, establishing them as an important sector of the national economy. In the first years after

The central post-office in Sofia

9 September 1944 the destroyed communication facilities were restored and the construction of new ones started.

With the help of Soviet specialists, a comprehensive long-term plan was elaborated for the development of communications. Models were prepared for the design and construction of local telephone networks, trunk telephone cables and automatic telephone exchanges. The USSR delivered all types of equipment for Bulgarian telecommunications. Bulgaria imported Soviet teletypewriters, radiotelegraph transmitters, telegraph machines, etc. Considerable means were allocated to the development and modernization of the material and technical basis of information and communications. In order to provide the public production sector with adequate information, a complex communications system is being created on the basis of computers. Using the latest achievements of science and technology, scientists are developing and introducing new communications facilities. A dense network of telegraph and telephone links, radio-relay lines, radio and TV stations link the country with all parts of the world. The communications system is being refined as a result of integration with the communications systems in the USSR and the other socialist countries.

The postal services receive, process, transport and deliver letters, parcels, newspapers, magazines, books and postal orders. The post offices perform various other activities: they handle pensions payments, the money operations of the Savings Bank and the sports totalizator, collection of taxes, rates, duties, etc. They also deliver all the foreign press publications. At the end of 1944 there were 1,003 post offices, 1,718 postal agencies, and about 600 inter-town and inter-village postmen; the length of postal routes was 22,895 km, of which 4,641 km were railway routes and 18,254 km other surface routes. At present all towns and villages in the country are included in the postal network. In 1983 the number of post offices was 2,916, i.e. 290 per cent more than in 1944. 9,929 post office boxes and 922 shops and kiosks for the sale of newspapers and magazines are available to the public. The length of the one-way postal routes is 60,285 km (5,795 km – railway, 2,285 km – air, 39,500 km – automobile and 12,705 km – inter-village). The broadening of the postal network and the increase of the range of services is a prerequisite for the improvement of services and the increase of the volume of postal traffic. In 1944 a single post office used to offer services to an average of 6,380 people in a territory of 110 sq. km, while in 1983 it was offering services to 3,141 people in a territory of 39 sq. km. The population and public sector get over five million postal services daily. About 300 million letters, nearly five million parcels and about 1,100 million postal and telegraph orders, and over five million sports totalizator slips are delivered annually. The post offices perform eight million operations for the State Savings Bank. They are now introducing an automated control system for money-transfer operations, which is a new development in this kind of service. Bulgaria maintains postal contacts with almost all the countries of the world. Most of the letters and parcels are transported by the Balkan Bulgarian Airlines. There is an international postal bureau at the Sofia Post Office which is in charge of and controls all the operations connected with processing and exchange of international mail. Sealed air and surface mail and parcels, as well as newspapers and magazines, are sent from Sofia to all socialist countries and to major cities the world over.

The State has a monopoly right over the issuing of postage stamps. The first original stamp of the People's Democratic Government was issued on 3 March 1945 in connection with the Slav Gathering held in Sofia. Bulgarian postage stamps are of high artistic and polygraphic standards and are designed by the best Bulgarian artists. They reflect the heroic history of the people and the successes of socialist Bulgaria in all spheres. Stamp-collectors are organized in the Union of Bulgarian Philatelists. The existing state distribution of the press system guarantees the regular dispatch of dailies and other periodicals to readers in all corners of the country. In the towns, newspapers and magazines are delivered seven times a week and in the villages, six times. The total single circulation of publications is about twelve million and the annual, over 1,100 million. Of the total circulation, 82 per cent are delivered on the basis of the subscription system, and 18 per cent are sold over the counter. About 67 per cent of the publications are distributed in the towns and 33 per cent in the villages.

The post-office in Plovdiv

Telephone communications constitute an important sub-sector of communications. Their development is characterized by the increasing density of telephones (number of telephone subscribers, per 100 people) which has increased from 0.43 in 1944 to 18 in 1983.

The rate at which the density of telephone communications in Bulgaria is increasing is one of the highest in the world – over 10 per cent annually (See Figure 1). In terms of total telephone density Bulgaria ranks second, after Greece, among the Balkan countries, and in terms of

Development of Telephone Communications in the People's Republic of Bulgaria During the 1970–82 Period

Indicators	Types of subscriber	1970	1975	1980	1982
Local telephone communications					
– total telephone density	telephone subscribers per 100 citizens	5.6	8.9	14.1	17.0
– telephone density of main subscribers	main telephone subscribers per 100 citizens	3.9	6.3	9.9	11.0
– adequate number of private telephone subscribers (main)	main telephone subscribers per 100 households	6.6	11.9	23.2	28.8
– adequate number of telephone subscribers in the national economy	public and office telephones per 100 work places	8.1	12.1	16.3	17.6

telephone density of the main telephone subscribers and an adequate number of private telephones – first among the socialist countries. The ratio of private to office telephones is 71:29. The total telephone density in Sofia is around 37 per cent, the density of the main telephone subscribers 21 per cent, and the ratio between private and office telephones 74:26. A long-term programme for the improvement of the quality of telephone communications in terms of telephone calls efficiency, the quality of telephone traffic and the reduction of complaints on the part of subscribers has been prepared with the help of the UNDP and experts from the International Communications Union. A system is being prepared for the exploitation and maintenance of the national telephone network by means of computers.

The local ATEs function on the principle of the step system. Crosspoint exchanges are also being introduced. Varieties of the Crossbar system are used in the country regions. A long-term programme provides for the introduction (in stages) over the next few years of numerical commutation systems. Since 1962 the introduction has started of semi-automatic Bulgarian-made devices, to be used for long-distance telephone communications within the country; an inter-district automation of long-distance communications with double-conductor commutations is currently being introduced. Seven transmit Crosspoint automated telephone exchanges with four-conductor commutation were introduced in 1974. They were bought from Siemens and link about 50 major towns in the country. The level of automation is over 70 per cent. The long-distance telephone channels from the transit network are made up of radio relay lines with a capacity of 960 and 1,800 channels and a range of 2–8 GHz. Long-distance links of the lower level are mostly made up of multichannel symmetrical cables and radio-relay channels. The radio-relay lines in Bulgaria have been functioning since the beginning of 1956. The existing large network of radio-relay lines is also used for the transmission of radio and television programmes. The radio-relay and multiplex equipment have been brought from various foreign firms and are now being produced in Bulgaria. The Bulgarian-made Shipka ground-station for space communications became operational in 1977. It connects the national radio-relay network with the Intersputnik space communications system of the socialist countries.

Presently under construction is a co-axial cable trunk line for the southern part of the country with a multiplex capacity of up to 100 Mhz. Two of the tracts are being built with numerical transmission systems. About 10,000 telephone channels are now functioning. They are fitted with equipment of the NKM-30 type, mainly for connection lines between the exchanges and the regions near the cities. Systems with NKM-120 and optical tracts for 2, 8 and 34 megabytes per sec are now undergoing trials.

For the purposes of international long-distance communications a transit international centre of the T4-3 type radio relay links has been built in Sofia with a capacity of 1,800 channels, one television and four radio channels to Bucharest, Belgrade, Athens and Instanbul. This allows for telephone communications with almost all countries in the world. Traffic to America is made possible by way of hired channels from Intersputnik and Intelsat. Under construction at the moment is an international ATE of a numerical type with a capacity of 2,400 (4,800) input and output lines and 45 (60) work places. This will make it possible for all the telephone subscribers in Bulgaria to reach every country in the world. Before the new automatic exchange starts operating, international telephone communications will be performed by a manually operated international switch-board. International auto-

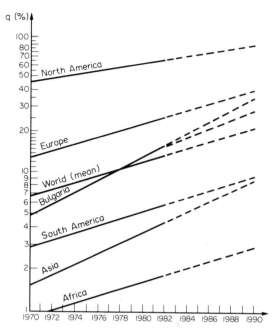

Figure 1: Telephone Density of Bulgaria and Other Regions

Automatic telephone exchange in Sofia

matic long-distance calls can be made in Sofia through the international exchange in Athens, which connects all countries in the world. The MH-60 exchange allows all telephone subscribers in Sofia to make automatic international long-distance calls to and from Moscow, Prague, Berlin and Warsaw and some other towns in the socialist countries. By using the international telephone exchange in Prague, the Crosspoint automatic international telephone exchange in Sofia allows telephone subscribers in Sofia to make international long-distance calls to thirteen European countries. Direct automatic lines with Hungary and Yugoslavia are now undergoing trials.

Telegraph communications are the second important sub-sector of long-distance communications. In 1944 the common telegraph network Genetex covered 1,003 PTTSs. Of these, eleven were serviced by teleprinters and the rest mostly by Morse and Hughes machines. A 40-line manual teletype exchange was built in Sofia in 1956, which incorporates twelve telex subscribers of import-export corporations and embassies. This introduced the Telex subscription telegraph system. The first automatic telegraph exchange was made operational in Sofia in 1959. It had a capacity of 120 subscribers – a decade-step system, supplied by the GDR. The exchange is linked to ten district exchanges, four local city PTTSs and 35 telex subscribers in the capital. Roll teleprinters were introduced in the telex network in 1970. They accelerate the processing and improve the layout of telegrams. The International Automatic Telex Exchange was built in Sofia in 1969. It has a capacity of 200 output and input lines. By 1973 all input connections were fully automated and the output connections stayed semi-automated and are operated by means of manual teletype switch-boards. The output connections were automated in 1974. Gradually, by 1977, the output connections of all district telegraph exchanges were also automated. At present all the telex subscribers included in the national telex network can reach and be reached automatically by telex subscribers in other countries. Thanks to the International Automatic Telex Exchange, the People's Republic of Bulgaria

The TV tower on Mount Snezhanka in the Rhodope mountains

Communications

Development of Telex and Genetex Density over the 1970–82 Period

Indicators	Measure	Years			
		1970	1975	1980	1982
Telex density	%	2.06	4.40	5.90	6.70
Genetex density	%	0.85	1.20	1.41	1.50
Additional telex services	number	3	3	4	12

maintains direct teletype contacts with sixteen countries and through them with all European and non-European countries. The genetex and telex density (number of teleprinters per 10,000 people) has increased considerably over the 1970–82 period.

The telegraph service meets all the needs of the national economy and the population. The first fully computerized telegraph exchange, Erikson, with 4,302 lines and an electronic telegraph multiplexor with 238 lines became operational in Sofia in 1983. It provides twelve new additional services for telex subscribers. It guarantees high quality telex lines with all countries in the world.

The latest scientific achievements and the new production technologies constitute the basis for the development of a new branch of communications – systems for data transfer between computers and computer centres (remote data processing). The development of these systems makes it possible to meet the needs of society for new types of services: videotex, teletext, telecopy, videophone, contact with moving objects, special medical services for remote medical examination observation and registering of sick patients, etc. – a total of about 30 different services (see Figure 2). Possibilities are being created for the establishment of a unified numerical data system with integrated services, using identical transfer and commutation systems and different terminal devices and, at present, different terminal commutation systems. For the purposes of data transfer Bulgaria uses hired channels and autonomous private networks. Currently under construction is a national network for data transfer intended for general use. It will be up to CCITT standards and will cover four levels. The first devices have been operating since 1976.

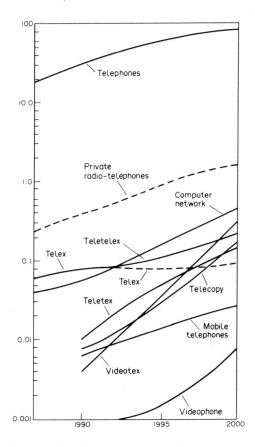

Figure 2: New Telecommunication Services

Radio and TV Broadcasting. By 1944 there were three medium-wave radio transmitters and several short-wave transmitters for radio-telegraph communications; the national radio broadcasts covered about 30 per cent of the country's territory. The national medium-wave transmitter was reconstructed and a new powerful transmitter built in North Bulgaria in 1954–55. UHF broadcasting started in 1961 with the introduction of two transmitters near Sofia. Currently the country's territory is covered by two national medium and short-wave programmes. The *Horizont* programme is on the air round the clock and is broadcast by ten transmitters with a total capacity of 1,840 kW, with a coverage of 95 per cent during the day and 75 per cent during the night. The *Hristo Botev* programme is on the air eighteen hours a day and is broadcast by ten transmitters, with a total capacity of 1,580 kW, with a coverage of 87 per cent during the day and 77 per cent during the night. Seven regions in the country broadcast local radio programmes and there is also a special programme for foreign tourists at the Black Sea coast. In the UHF range (66–73 mHz) there are two monophonic programmes (*Horizont* and *Hristo Botev*) and one stereophonic programme (*Orphei*). Each of these programmes is broadcast by twelve transmitters which

The TV tower on Mount Snezhanka in the Rhodope mountains

guarantee a coverage of about 84 per cent. The 100–108 MHz is also to be introduced and for the time being the UHF transmitter in Sofia is in operation. It broadcasts the *Horizont* stereo programme. The development of broadcasting depends chiefly on the expansion of the network of UHF transmitters. There are eleven radio sets and 22 transistor radios per every 100 citizens. There are also radio programmes transmitted by cable. There are about 11.5 cable radio sets per every 100 people. The telephone and special cable network transmit from one to six radio programmes, including stereo programmes.

Regular broadcasting of television programmes in Bulgaria began on 7 November 1959. The studio equipment and the transmitter were delivered by the British firm Pye. Experimental TV programmes have been on the air since 1954. They were made possible by an installation designed by lecturers and students from the State Polytechnic in Sofia. Colour television was introduced in 1972. Bulgaria uses the Secam system for TV broadcasting, norms D and K from two networks. The first has fourteen principal transmitters and about 300 retransmitters, which guarantee the reception of programmes on Channel One. It is on the air 90 hours a week and covers 86 per cent of the country's population. The second network has twelve principal transmitters and about 200 retransmitters, and guarantees the reception of programmes on Channel Two, covering about 70 per cent of the population. It is on the air 30 hours a week. A third TV network is due to become operational. The country's radio-relay network and the channels for satellite communications make it possible to exchange TV programmes with the Eurovision and Intervision member-countries. At the beginning of 1983 there were 27 TV sets to every 100 citizens in the country. Systems for automatic control of the broadcasting networks are being introduced, which improve the quality of broadcasts. By 1990 Channel One television programmes are expected to cover 98.5 per cent of the country's territory, and Channel Two programmes, 92.5 per cent. The cable television systems will gain greater popularity and will become multi-purpose communication systems.

Bulgaria is a member of and takes part in the work of the following international communications organizations: the World Postal Union (1879; since 1964 Bulgaria has been a member of its executive council); the International Long-distance Communications Union (1880) and its branches; the International Radio and Television Organization (1946); the Organization for Co-operation of Socialist Countries in the Field of Electric and Postal Communications (1957); the International Organization for Space Communications Intersputnik (1971); etc. The communications administration of the People's Republic of Bulgaria is taking part in the preparation of the project for the development of long-distance communications in the Middle East and the Mediterranean, and the project for Central and Eastern Europe. Bulgaria has a permanent representative on the International Committee for Frequency Distribution.

The party and state policy in the field of communications and the management of the national communications system is conducted by the Ministry of Communications. In charge of research and development is the Research Communications Institute in Sofia, and the Telecomplect Engineering Corporation is in charge of the construction and assembly of communications systems and networks.

DOMESTIC TRADE AND THE CO-OPERATIVE MOVEMENT

Domestic Trade. Commercial activities on what is today Bulgarian soil date back to antiquity. Archaeological evidence exists of barter, and copper, silver and gold coins of Thracian, Roman, Hellenic and early Byzantine origin, minted by different rulers and discovered in modern times, indicate that viable commercial activities developed during the fourth–third centuries BC and continued until the third–fifth centuries AD. The incursion of many tribes into the Balkans brought about a continuous decline of commerce, which was enlivened only after the eleventh century, when Bulgarian coins began to be minted. Weekday market places were formed in the towns, fairs were established and it became a tradition to organize them every year in the more developed settlements, or near monasteries. Bulgarian and foreign merchants were engaged in trade, and commercial middlemen appeared. Bulgaria's fall under Ottoman domination (1396–1878) resulted in a stagnation of commodity-money relations. Trade development only occurred on any significant scale in the first half of the nineteenth century, when a national market developed in Bulgaria. Distribution of goods took place through merchants in shops or in the market-place. Commercial streets were formed in the larger towns. The importance of fairs grew. The Uzundjovo, Eski Djumaya and Sliven fairs, among others, were well known throughout the Ottoman Empire and were organized on the territory of today's Bulgaria.

Only after the country's liberation from Ottoman rule (1878) did the home trade develop more fully. Trading in shops developed and gradually the importance of fairs declined. The main commercial centres in the country took shape: Russe, Plovdiv, Sofia, Varna, Burgas, Pleven, Gabrovo and Svishtov. Commercial law took shape. A Trading Floor Act (1883), Law on Weights and Measures (1889), Law on Commerce (1897), Law on the Commercial Agents in Bulgaria (1905), Law on Public Sales (1905), Law of Stock Exchanges (1907) and others were issued. In the period 1887–1911 the internal turnover of goods increased 330 per cent; in 1911 there were 230 weekly markets and 214 fairs. At the end of the nineteenth century and the beginning of the twentieth century monopolist trade organizations appeared. A system of ration distribution for staple foodstuffs was introduced during World War I and prices were controlled. These restrictions were abolished in 1924. Exploitation by commercial capital increased. The first

A toyshop in Burgas

consumer co-operatives appeared. The exchange of goods more or less doubled in the period between the two World Wars, but at the same time price instability and inflation raged. The purchasing power of the national currency – the lev – dropped by about 3,200–3,500 per cent in the 1912–29 period, compared to 1911. The concentration of trade capital increased. Commercial shareholding companies mushroomed. In 1937 there were 431, and in 1941–809 (35 of them holding practically half of the fixed capital of all companies). Private capitalist trade was established as a basic form of domestic trade. In 1939 about 78,000 private commercial firms were registered, whilst there were 100,000 by 1943. Hired workers in commercial enterprises were, on average, about 30 per cent of the work-force. The majority of commercial establishments were small shops, bars, kiosks and snack-bars, which used small premises, poorly furnished and failing to meet the most elementary requirements. Co-operative trade made a considerable development. In towns and villages it was carried out by the consumer co-operatives, which governed about 20 per cent of the country's turnover of goods by the end of the capitalist period. The significance of state-owned capitalist trade was also limited. It was performed mainly by the wholesale trade – purchasing of farm products, sales of agricultural machinery and tools, oil products and blue vitriol. In 1930 the State Commercial Enterprise, Hranoiznos Directorate, was founded and had the monopoly in the purchase and wholesale trade of cereals, flour, cotton, flax, hemp, rape, wool and other farm products. Part of the tobacco, cotton, rose flower and other agricultural products were purchased by the Bulgarian Agricultural and Co-operative Bank. In this way about half of the wholesale trade of agricultural produce was attributable to the state capitalist sector. Public catering was characterized chiefly by establishments where mainly alcoholic drinks were sold. The number of co-operative canteens for the workers, employees and students at places of work or study was small.

The central supermarket in Sofia

In the first few years after the establishment of the popular regime (1944) the efforts of the BCP and the People's Government were directed towards the restoration of the national economy, which had been ruined by the war and plundered by the nazi occupiers. The nationalization of industry, mines and banks, the establishment of a separate Ministry of Trade and Supplies and the issue of a special Decree for Prices and Supply were of great importance in setting up the socialist domestic trade. The socialist State began to take responsibility for the turnover of goods. State management and control of trade was improved and the supply of important goods to the population was reformed. The socialization of domestic trade was effected mainly between 1947 and 1952 by economic measures. Exceptionally favourable conditions were created for the development of co-operative trade,

Café at Novotel Evropa in Sofia

which in the first few years was the only public element in domestic trade. The consumer co-operatives grew in number and became more firmly established. They organized the supply of the population with consumer goods for economic needs (foodstuffs, consumer goods, building materials, seedlings, farm tools and implements, etc.), and opened shops and public catering establishments. State wholesale organizations were established in 1947–48 (tobacco and spirits monopolies, specialized

Domestic Trade and the Co-operative Movement

Table 1: Social Sectors in Domestic Trade

Social sectors	Relative Share in the Exchange of Goods (in percent)							
	1947	1948	1952	1960	1965	1975	1980	1982
Public Sector	15.9	25.31	51.51	57.72	59.89	66.04	68.26	69.00
Co-operative Sector	28.1	43.51	47.84	42.18	40.04	39.92	31.71	30.98
Private Sector	56.0	31.18	0.65	0.10	0.07	0.04	0.03	0.02

The Tihiya Kut restaurant on Mount Vitosha

purchasing organizations) and state retail trade enterprises (Naroden Magazin, Obleklo i Obuvki, Stroitelni Materiali, Hranitelni Stoki, Metalni Izdelia, Toplivo, Horemag and so on). At the beginning of 1948 there were 35 state wholesale trade enterprises. The socialization of retail trade developed more slowly.

1945 saw the beginning of organized *public catering*. The Government passed a special Decree by which the owners of enterprises, farms and heads of offices and organizations were obliged to open staff canteens and take full responsibility for their maintenance costs. In 1947 there were already 2,340 public canteens for about 230,000 members of staff. By the end of 1949 the canteens increased to 3,440 for 350,000 people; nearly half of all workers had meals once a day at the canteens at considerably reduced prices. In 1952 the canteen and the farms which supplied them became separate economic units, having their own accounting and separate management. This was the way the socialist reorganization of domestic trade developed. By the end of 1952 the private sector contributed only a small amount of the exchange of goods (see Table 1). There was a fair division of the spheres of operation between the state and the co-operative trade: state trade was active mainly in towns and industrial areas, whilst the co-operative trade operated mainly in the country and in the smaller towns. Co-

Table 2: Retail Turnover

	Years						
	1952	1957	1960	1965	1970	1975	1982
In million levs	1,155.1	1,669.8	2,383.3	3,599.1	5,627.4	8,262.1	13,315.8
1952=100	100.0	200.2	292.9	415.3	628.3	913.6	1,200

Table 3: Structure of the Retail Trade Turnover (in percentages)

Indicators	Years							
	1952	1956	1960	1965	1970	1975	1980	1982
General commodities	100.0	100.0	100.0	100.0	100.0	100.0	100.0	100.0
Foodstuffs	48.8	47.3	45.9	45.3	40.8	42.0	39.3	38.7
Non-alimentary goods	51.2	52.7	54.1	54.7	59.2	58.0	60.7	61.3

operative market trade also appeared and developed, its basic trade being carried out by the co-operative farms, which later developed into the AICs, and the personal holdings of the farmers. Under the conditions of the socialist public system development, the basic objective of domestic trade is to satisfy more fully the needs of the population. Domestic trade was established as an important factor in the promotion of the living standard of the people and the formation of a socialist way of life, in promoting and orientating material production and personal consumption, and in the consolidation of the circulation of money and the normal materialization of the reproduction process. One of the basic complex indicators for the development of trade and systematic improvement of the welfare of the people is the increase of retail trade turnover (see Table 2) and the changes in its structure in favour of increasing the production of household furnishing goods, meeting cultural needs, etc. (see Table 3). The growth of the purchasing power of the population brought about a rapid increase in the demand for and sales of high quality fashionable and luxury goods, home furnishings, and leisure and recreation items. This tendency deepened and developed on the basis of the further increase and diversification of the production of consumer goods and the growing material standards of the population.

In the period of socialist development the retail prices of personal consumer goods remained stable. Until 1952 the prices of goods sold by ration cards were very low. In 1950, together with the sales of goods by ration cards, industrial goods were freely sold on the market at relatively higher prices. In 1952 the rationing system was abolished. The retail prices of the staple foodstuffs and non-alimentary goods have been fixed by the Government.

Public catering has been developing in three main directions: supplying food for workers and other staff at their places of work and at the rest houses, at lower prices; free meals in hospitals; and in restaurants, cafeterias and so on. In 1952 the public catering turnover by current prices was 16.59 per cent of the total retail trade turnover in the country, and 17.3 per cent in 1982. In 1957 some 35.13 per cent of all foodstuffs was sold in the public catering

The Bitova Svatba restaurant at the Druzhba resort, near Varna

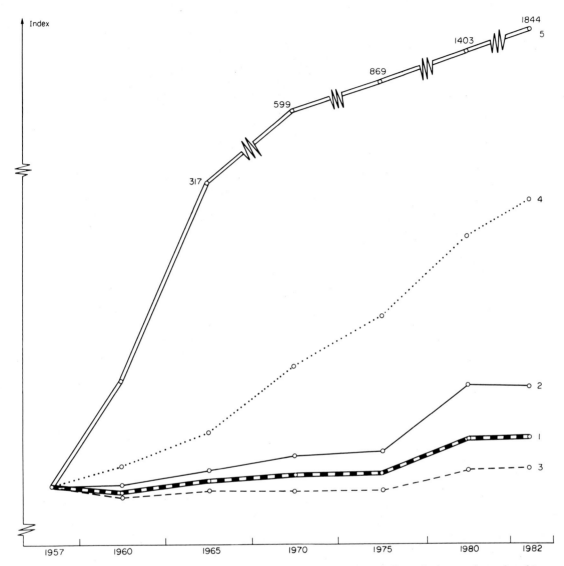

Graph: Movement of the State-Fixed Retail Price Indices (Total: 1, of foodstuffs; 2, of other goods; 3, of working wages; and 4, of old-age pensions; 5, paid in the People's Republic of Bulgaria in the 1957–82 period; 1957=100.0)

establishments and in 1982 it amounted to 39.94 per cent. The number of people engaged in domestic trade increases and turnover outlays decrease every year in order to improve the quality of customer's service and its efficiency. In 1970, one person in home trade serviced an average of 50 people, whilst in 1982 that number was 31.

Domestic trade rests on solid foundations. As a result of the continuous efforts by the socialist State the trade network is changing basically and is improving. The total number of shops was 22,396 in 1950, 33,956 in 1972, and 39,699 in 1982. The number of public catering establishments also increased considerably: from 10,194 in 1950 to 20,457 in 1972 and 25,149 in 1982. Shopping centres, department stores, trade houses, supermarkets, large specialized shops and stores, and other modern marketing outlets are being established at an increasing pace. The density of the trade network is increasing and the supply of the population with food and industrial goods is improving. Modern methods of customer service are implemented. The commercial units per 10,000 of the population increased from 38 in 1970 to 44 in 1982, and the public catering establishments – from 23 to 28.

Domestic trade is managed by the Ministry of Consumer Goods Production and Trade.

The Izvora co-operative restaurant near Shumen

Table 4: Indices for the Changes in the Development of Turnover in the Trade Network and in Public Catering

Index	Years				
	1970	1975	1978	1980	1982
Total turnover	100	145.4	165.8	174.7	190.9
Trade network turnover	100	146.0	168.0	181.0	198.4
Public catering turnover	100	142.8	157.1	150.0	161.8

The co-operative movement. The establishment of co-operatives through the free union of individuals having equal rights and duties for common activities is of great importance for social development. Under capitalist conditions it was a powerful factor for the protection of the interests of the working people and for the limitation of capitalist exploitation. The co-operative movement in Bulgaria has a long history. It stems from public and spontaneous initiative and is backed by the accumulated economic experience which is a valuable part of the cultural heritage of the country. It originated after Bulgaria's liberation from Ottoman rule (in 1878) as a result of the peasants' desire to defend their economic interests and improve their circumstances. The first co-operative, the Mirkovo Mutual Savings Agricultural Society, Oralo, was founded in 1890 in the village of Mirkovo, near Sofia. Under the specific conditions of Bulgarian rural areas, the village co-operatives developed as unique institutions of their own. Practically all of them have commercial, credit and other activities. The village co-operatives entered the national economy and were quite rightly called 'general'. They supply cheap and easy credit to farmers, collect savings, organize the supply of industrial goods, purchase and sell farm produce and open flour mills, canning factories and other production units. A distinctive feature of the Bulgarian co-operative is its democratic and popular character. The Bulgarian socialists paid great attention to the co-operative as a form of popular economic struggle. The first workers' co-operative, called *Rabotnik* (Worker), was founded in Turnovo in 1892 on the initiative of the founder of Socialism in Bulgaria, D. Blagoev. Large-scale activities were carried out by the Napred Workers' Consumer Co-operative founded in Sofia in 1903. In 1899–1902 several co-operative societies in the Russe district made their first attempts at co-operative farms. At the end of 1910 there were 981 co-operatives, comprising 721 credit co-operative societies in villages which also had commercial activities, 76 consumer co-operatives in the towns, and

Domestic Trade and the Co-operative Movement

184 specialized co-operatives. After the First World War there was great progress in the co-operative movement. The foundation of forestry co-operatives, viticulture co-operatives, artisan co-operatives, tobacco producers' co-operatives, market gardening co-operatives, animal husbandry co-operatives and popular banks, was extended. In 1919 the Osvobozhdenie Co-operative was founded under the leadership of G. Dimitrov and V. Kolarov in Sofia. Together with its economic activities, this co-operative also organized the publication of political and trade union literature and worked in close co-operation with the Trade Unions. The interest in co-operative land cultivation and management increased under the influence of the Soviet *Kolkhoz* (collective farms). In many places farm co-operatives or production departments were formed with the credit societies, following the example of the Kolkhozes. The largest number of co-operative agricultural societies were formed in 1941–44, despite the obstacles raised by the authorities. The co-operative movement became a mass movement of the poor and middle-class peasants and the urban poor. Unable to change the capitalist order, the co-operatives were able at certain times to alleviate the capitalist exploitation of the working people, carrying out much educational and cultural work among the masses, acquiring property and collecting funds, and bringing about the training of managers.

The Central Co-operative Union building in Sofia

The Georgi Kirkov co-operative market in Sofia

On 9 September 1944 the popular regime inherited a well-developed co-operative movement with a membership of 1,614,117 organized into 4,144 co-operative societies. 77 per cent of all co-operative societies were in the country. Of the 66 co-operative farms founded during the capitalist period, 29 survived and became the basis of the first co-operative farms (see the section on Agriculture, p. 408). Under the favourable conditions then prevailing, the co-operative movement became an important factor in the country's socio-economic development over a relatively short period. On Georgi Dimitrov's initiative a Central Co-operative Union (CCU) was founded at the end of 1946 as an organizational and economic centre for consumer and other co-operatives. Today, it unites all co-operative organizations in Bulgaria and has a total membership of 2.5 million. Through its 28 District Co-operative Unions it manages the consumer co-operatives (437) and opens its own enterprises for agricultural produce, and for the processing, packaging and sales of this produce. The Producer Co-operatives, of which there are 160, are organizationally subordinate to the CCU.

The consumer co-operatives, which have 17,267 shops and 9,511 public catering establishments, serve all the villages (approximately 4,500), 149 towns and some industrial, mining and building projects of national importance. The Producer Co-operatives produce a great variety of consumer goods and service the population. They have over 9,000 shops for household services and a staff of about 96,000; they render over 50 per cent of the communal services in the country. Disabled people and invalids work at home or in the production and commercial establishment of the Invalid Co-operative. The wide network of co-operative enterprises includes the production of bread and pastry products, confectionery, sugar products, canned fruit and vegetables, ice-cream, fizzy and non-alcoholic drinks, as well as industrial consumer goods.

One of the basic objectives of the CCU is the purchase of crops, vegetables, fruit and animal products, and the purchase and export of mushrooms, herbs and wild growing fruit. The Bilkoco-op enterprise, equipped with modern packaging machinery, is developing successfully. Nectarco-op purchases honey and bee-hive products. Bulgarco-op is the Foreign Trade Enterprise of the CCU and has trade relations with about 40 countries and more than 250 companies. A council for the mutual insurance of its members has been established at the CCU. It is in charge of the social insurance and medical care of the co-operative members and organizes and supervises safety of labour, the improvement of working and living conditions, and holidays for the co-operative members. The CCU provides the background for the cultural activities of its members. Active amateur artistic companies and ensembles formed at cultural centres now number 265, with a total of 65,000 participants. In its thirteen occupational training centres and two schools the CCU trains over 25,000 of its members annually.

The CCU maintains close contact and co-operation with co-operative societies in more than 60 countries. Bulgaria has been a member of the International Co-operative Alliance since 1903. An International Co-operative School on problems of the co-operative movement was set up in 1974 in Bulgaria, attended by co-operative functionaries from developing countries. The achievements of Bulgaria's co-operative movement place it among the leading nations of the world co-operative movement.

FOREIGN TRADE AND ECONOMIC CO-OPERATION

Historical Background. The development of Bulgaria's economy as a separate entity began with the country's liberation from Ottoman rule in 1878. Handicrafts, forms of land ownership, farming and commerce which had predominated in the five centuries of foreign rule were abruptly discontinued. During the first decades of national independence a large number of one-family farms and small-scale food processing and consumer goods enterprises were established, and commerce grew rapidly, fuelled by a growing monetary and banking system. These transformations gave rise to a large class of peasants, with their own piece of land, and a small but growing city proletariat, as well as to a rising number of large landowners, merchants, bankers and industrialists.

By the end of the nineteenth century there were already a number of flour mills, textile mills, distilleries, tanneries, etc. Foreign capital accounted for about one quarter of total investments in industry.

The country attained its highest pre-war level of economic development by 1939. Even then about 80 per cent of the enterprises processed farm produce, another 15 per cent belonged to the textile and wood-processing industries and only 1.5 per cent were in the metal-working industries. No heavy industry existed, with industrial production even in that peak year accounting for no more than 15 per cent of national income and 5 per cent of exports, non-agrarian industries making up merely 0.4 per cent. Four-fifths of exports were tobacco, skins and hides, grains, livestock, etc. The structure of imports was in reverse proportions – 80 per cent consisted of ready-made goods (textiles, chemicals, metals, glassware, etc.).

As to volume, Bulgaria's pre-war trade reached its highest level at the end of the 1920s. During the crisis of 1929–33 it registered a 400 per cent drop. The subsequent recovery still did not lead to a return to pre-crisis volumes, with trade in 1939 still 30 per cent lower than a decade before.

The commodity structure and volume of pre-war foreign trade reflect Bulgaria's peripheral status in relation to the major European economies, with Germany progressively gaining a dominant position both politically and in the economy. Post World War I reparations and loans, the effects of the world crisis of 1929–33, and the generally weak national industrial and merchant class provided broad possibilities for extensive foreign capital investment. This resulted in a national economic structure geared to supplying agricultural commodities and other war materials and to providing a market for mostly consumer goods, with price relationships heavily in favour of foreign companies.

During 1940–44 the economy was subjected to virtual plunder by nazi Germany. By means of an unequal clearing agreement and a policy of fixed, low prices for agricultural exports it succeeded in acquiring the predominant share of prime exports, with over 80 per cent of tobacco going to Germany by 1944, while at the same time accumulating a worthless 'debt' of over 22 billion levs to Bulgaria. Thus the revolution of 9 September 1944 found Bulgaria's economy ravaged by the war, with backward, small-scale agriculture and no real industrial base.

Post-1944 Independent Economic Development. In the early post-war years, up to 1947, the new Popular Government of the Fatherland Front faced the task of

View of the Plovdiv Trade Fair

providing the essential industrial inputs and consumer goods for restoring economic activity to more or less normal standards. In this the assistance rendered by the Soviet Union proved vital. Under the first bilateral trade agreement of 1945 it provided massive supplies of oil and oil-products, metals, cotton and other light industry raw materials, foodstuffs (wheat, etc.), as well as growing amounts of machinery and equipment.

During these years, as a result of government policy, the foreign trade system underwent successive reorganizations, thus progressively bringing it in line with real national interests and priorities. This was achieved through a new system of export and import licensing, and through enhancing the role of state trading enterprises, by granting them exclusive rights in grain trade, exports of tobacco and rose oil, etc. as well as through co-operative entities in foreign trade.

The late 1940s and the first half of the 1950s saw a succession of transformations creating the foundation for accelerated socio-economic development, international trade and co-operation on socialist economic principles and guidelines. In December 1947 the Great National Assembly, after adopting the new Constitution proclaiming Bulgaria a People's Republic, also passed a series of laws nationalizing all private industrial, mining, banking and finally trading enterprises. Thus approximately 650 export-import firms were almost wholly made public, with all but 0.5 per cent of exports and 1.2 per cent of imports passing over to publicly owned enterprises.

A successful reform of agriculture was carried out through the collectivization of arable land and the creation of co-operative farms. This allowed the introduction of mechanized farm production. By 1950 some 2,500 such farms had been created, encompassing about one half of all arable land. Collectivization was completed by about the mid-1950s.

Under the bilateral treaties concluded with all European socialist countries during 1947–48, along with political, military, etc. matters Bulgaria was also able to assure the basic conditions for mutually beneficial economic relations. Of particular importance for its all-round development was the treaty concluded with the USSR in March 1948. In 1949 Bulgaria became a fully-fledged founding member of the CMEA, thus supplementing the system of bilateral agreements with a contractual framework for multilateral trade and co-operation.

The late 1940s and early 1950s also marked the initial adoption and successive elaboration of overall planning of the national economy and the emergence of a general policy of accelerated industrialization and structural transformation of the economy. As early as April 1947 the National Assembly adopted the first National Plan – for a two-year period – aimed at completing the restoration of pre-war levels of economic activity and the creation of initial prerequisites for subsequent industrialization. As a result of the nation's effort and with the all-important Soviet inputs and assistance, pre-war levels of output were doubled. There followed two successive Five-Year National Plans, the first of which was fulfilled in less than four years (1949–53) and saw the creation of 700 new enterprises and a 230 per cent increase in industrial output at the end of 1952 over 1948 levels. The second plan, however, (1953–56) was marked by a reduction in growth rates and delays in the development of basic industrial sectors – power generation, machine-building, etc. – due to mistaken policies and concepts concerning economic priorities in general and industrialization in particular.

Tomatoes for export in the Tolbuhin district

April 1956 and Subsequent Development. Thus by 1956 the country had reached a stage where further successful growth of the economy required the formulation of a new and forward-looking overall economic and industrial strategy. This critical need, along with major concerns in the other principal political and social sectors, necessitated a thorough re-examination and reformulation of national policies, which took place at a Plenary meeting of the Central Committee of the BCP in April 1956.

The decisions taken there mark the inception of the overall economic strategy which has been followed for over two and a half decades and which has led to Bulgaria's present-day level of socio-economic achievement. The guiding features of this strategy, whose hallmark has been the optimal integration of the external economic sector into the national economy, might be summarized thus:

* full utilization of all natural resources and geographical factors, however limited, at the country's disposal – Bulgaria being relatively poor in raw material deposits; accelerated development of energy production at rates surpassing overall economic growth; development of basic extracting and processing

industries to the degree warranted by national industrial objectives and economic efficiency;
* priority development of large-scale production in chosen sectors of machine-building and other industrial goods as focal points for a progressive and forward-looking industrial and economic structure; vertical integration of intermediate echelons of production and broad end-product diversification in such sectors as means of achieving economies of scale and international competitiveness;
* external orientation of industrial policy, necessitated by the restricted national raw materials base and internal market; policy has consistently been guided by the aim of achieving optimum specialization and co-operation within the CMEA, as well as in the broader international division of labour, as a necessary prerequisite for an efficient national economy;
* consistent enhancement of agricultural production through successive stages of enlargement of individual farming units and introduction of up-to-date agro-technologies; the rural sector has also been undergoing successive stages of vertical integration into ever more comprehensive agro-industrial complexes, permitting full self-sufficiency in basic foodstuffs and increasing trade in fresh and processed agricultural commodities.

Bulgarian-made Elka-88 electronic calculator

The formulation and consistent implementation of overall national policies and economic strategy in the years since April 1956 is indelibly linked with the overall leader of the Communist Party and Head of the Bulgarian State, Todor Zhivkov. As Secretary-General of the Central Committee of the BCP and Chairman of the State Council he has consistently put forward new approaches and guidelines for adapting the economy to evolving international conditions and thus for enhancing Bulgaria's standing as an active participant in bilateral and multilateral socialist economic integration, as an active, international trading partner.

Foreign Trade and Economic Co-operation – Dynamics and Structure

Dynamics. As a function of national economic policies of the type described, foreign trade in the course of the last two and a half decades has consistently been outpacing on average by double the overall growth rates of the economy. This has led to increasing ratios of exports to national income. In 1983 this ratio reached approximately 50 per cent, with exports having grown to 11.82 billion levs (4.8 per cent over 1982) and national income registering a volume of 23.53 billion levs (3 per cent over previous year; 1 lev=$US1.002 as at 1 January 1984, official rate of the BNB). Thus today Bulgaria's is an open economy with a high degree of exposure to international market conditions and changes. This feature is of ever-growing importance as a guiding factor for shaping macro-economic policies, trade policy towards various foreign partners and, increasingly, micro-economic production and industrial organizational structures and practices.

The dynamics of development of Bulgaria's foreign trade turnover have been particularly impressive. In the course of a little over three decades it has undergone an increase of 5,400 per cent as may be seen from the following table of indices:

Table 1: Index of Growth of Foreign Trade Turnover

	1950	1960	1970	1980	1983
Trade turnover	100	490	1,700	4,400	5,400
Exports	102	540	2,000	5,900	7,300
Imports	98	450	1,400	3,100	3,700

Exports. Even more far-reaching have been the shifts in the structure of exports. The period since 1956 has seen a dramatic rise in the share of machinery, equipment and vehicles in total exports, signifying several times higher rates of increases as compared with average growth of exports. This has been paralleled by corresponding decreases in agricultural and mineral commodities and of consumer goods shares.

The figures on machine-building are a reflection of the priority given to the creation of large-scale export-orientated industries in the sectors of machine tools, forklifts and hoists, electronic data processing, communications and electro-technical equipment, farming machinery, etc. In recent years, exports under this heading have increasingly been including equipment and integrated installations in food-processing, construction materials, and continuous processing in the chemical and related industries. Recent new additions to the list have been equipment for excavation and earth-moving, metallurgy, crushing and grinding, etc.

Table 2: Commodity Structure of Exports (in percentages)

	1955	1960	1970	1980	1982
Exports – total including:	100.0	100.0	100.0	100.0	100.0
machines, equipment and transport vehicles	2.6	12.9	29.0	44.4	46.9
fuels, mineral raw materials and metals	20.7	9.2	8.1	15.0	12.7
raw materials of vegetable and animal origin, foodstuffs	60.7	56.4	43.4	24.4	22.9
industrial consumer goods	12.5	17.9	14.7	8.8	9.4
chemical products, fertilizers, rubber, building materials and other commodities	3.5	2.8	4.5	6.2	6.0

(a) Machine tools and metalworking equipment production are a direct result of the industrial policies adopted in the late 1950s. Initially, individual lathes and other metalworking equipment were produced and exported. In the course of the last decade, however, products of growing sophistication have started reaching the market: machine tools with CNC devices, specialized lathes, as well as integrated machining centres, industrial manipulators, robots and other integrated automation equipment. To these should be added hydraulics components, ball-bearings, wood-working machines, etc. A principal feature of the industry is its high rate of innovation, on average 21–22 per cent per annum, and the increasingly large degree of integration of research and development, production, marketing and servicing into integrated engineering activities for the setting-up and operation of wholly-integrated production plants for machine tools and equipment for varying applications. Such increasing sophistication has also permitted the industry to start developing sections of computer-integrated, flexible manufacturing systems which are to be the next generation of exports of the Bulgarian machine-building industry. The leading national producer is the ZMM group, with Machinoexport as the principal exporter, now supplying to over 70 markets world-wide.

(b) Fork-lifts, hoists and handling equipment in 1982 took up 8.5 per cent of overall exports. In the past 30 years Bulgaria has succeeded in developing a world-scale fork-lift industry with virtually no previous experience in this sector. Some 72,000 trucks were manufactured in 1983, with 96 per cent of production going for export. In 1978 the Balkancar national producer and marketer became the leading world manufacturer and supplier to foreign markets and has maintained its position ever since, every fifth truck on these markets being of Bulgarian origin. The fork-lift industry has a high degree of integration with local battery, electric motor and diesel engine production, as well as a full range of components – rear axles, hydraulics, etc.; both long-term specialization agreements with CMEA partners, and joint marketing and, lately, production ventures with western companies in over 65 countries; and a 25 per cent annual rate of new product introduction.

Electric hoists represent another product line of world scale dimensions of production and exports, with 123,000 pieces turned out in 1983.

Long-term contractual specialization by Bulgaria in the CMEA in fork-lifts and hoists has been of crucial importance for this rapid growth. At the beginning of the 1980s it supplied 70 per cent of the fork-lifts and 55 per cent of the hoists marketed in the CMEA countries, with the Soviet Union as by far the principal customer. Favourable opportunities for maintaining and enhancing market positions are provided by the system of bilateral and multilateral production specialization and supply arrangements with the USSR and the other socialist countries in this sector.

Bulgaria's fork-lift and mechanical handling industries are now engaged in accelerated innovation in their product lines, quality improvement and enhanced consumer support in order to utilize these opportunities and gain new world markets. New lines of trucks with greater pay-load capacity (up to 12 tons) and specialized functions are going into production. Increasingly sophisticated computerized storage and handling systems are also being developed.

(c) The electronics industry is perhaps an even more graphic example of the revolutionizing results of Bulgaria's policy of outward-oriented industrialization. The inception of this industry in the country dates back to only the late 1960s. At present a broad range of peripheral devices, central processing units and other types of equipment is produced and exported by the Izot group of enterprises. Its specialized foreign trade organization Isotimpex exports both mainframe computer systems, such as the EC 1035 (the equivalent of an IBM 370/145), and the minisystem Isot 1016 (corresponding to DEC 1130 systems); data-entry devices; memories, disc-drives, etc.

The hardware produced, along with the large pool of computer programmers and systems analysts, permits the development of data processing systems with broadly varying individualized applications. Thus today the electronics industry has an output of 1.3 billion levs, with fully 92.2 per cent going for exports.

Here again foreign trade and co-operation, above all with the USSR and other CMEA partners, have been of overriding importance. Bulgaria's electronics developed as an integral part of the CMEA-based Ryad system of electronic data processing equipment. Bulgaria was given the opportunity to build a highly sophisticated modern industrial sector from scratch as a major input for beneficial participation in intra-CMEA trade and integration. Today Bulgaria supplies a full 40 per cent of total exports to Soviet and other CMEA country markets.

The coming years will see further rapid changes in the range of products, with personal computers, new mini and micro processing systems, and new high capacity memories reaching the markets. Concurrently computerization will expand and integrate with the machine-building, communications, farming machinery and other major industries; the development of computer-aided design manufacturing and engineering systems is also rapidly progressing.

(d) Shipbuilding, communication, electrotechnical and process equipment industries are also industrial sectors with histories going back no more than two or three decades. Their development is closely linked with the intra-CMEA systems of trade and co-operation, both in setting up basic manufacturing capacities and in providing markets for the end products. In all these sectors rapid modernization, product development and marketing efforts are to be carried out.

The chemical industry comprises another major sector of the economy of prime importance for the country's exports. The beginnings of a modern industry were laid in the 1950s, with artificial fertilizers and soda ash (the Devnya soda-sash complex is today the third largest in the world). To these, after 1960, were added growing oil-processing capacities as a base for the full range of plastics and other end-products of organic chemistry. Between 1956 and 1983 nitrogen fertilizer production grew from 34 thousand to 1 million tons; plastics – from 1.3 thousand to 287 thousand tons, etc. The chemical industry today produces 11 per cent of the national income and 17 per cent of the country's industrial output.

Close economic co-operation with the Soviet Union and the other socialist countries was of crucial importance in this development. Over twenty major chemical complexes have been constructed throughout Bulgaria with direct Soviet material and technical assistance.

About 180 basic items make up the export list of the industry, with markets in 80 countries of the world. Under 38 sub-sector intra-CMEA agreements Bulgaria is supplying an increasing range of products to the socialist countries, and fully 32 per cent of total Bulgarian exports of chemicals now go to these countries. About 70 per cent of specialization-based trade in the 1981–85 period will be conducted with the Soviet Union.

Further accelerated growth rates of production and exports are envisaged for the coming years, with emphasis on speciality chemicals with low raw materials and energy intensities, such as the various agro-chemicals, process agents, pharmaceuticals, biotechnology products, cosmetics, etc.

Along with socialist markets, the chemical industry will be exporting more and more to the developing nations and industrialized western countries through the principal exporter Chimimport, while Pharmachim handles pharmaceuticals.

Agricultural commodities and processed foodstuffs are a traditional sector, both in the economy and in national exports. In the course of the last few decades Bulgaria has enhanced its self-sufficiency in staple foods and its market position as supplier of high quality fresh and processed fruits and vegetables, meat products and preserves, dairy products, tobacco, cigarettes, grapes and wine. Although decreasing in relative terms, agricultural commodities and processed products still comprise about 25 per cent of total exports, 80 per cent of which go to CMEA-country markets. Exports also go to a number of western countries – the FRG, Austria, the UK, Japan, etc – and increasingly to Middle Eastern and other developing countries. They are mainly live animals, meat and dairy products (40 per cent), canned goods (over 10 per cent), tobacco and cigarettes (over 17 per cent), etc.

The consumer goods industries are another well-established branch of the economy and trade which has been undergoing rapid transformation. During the last few years, along with growth in volume important improvements have taken place in the quality and range of consumer goods marketed, both in the country and abroad. These have affected the manufacture of clothing, footware, textiles, leatherware, sports goods, furniture, glassware, etc.

The value of exports in 1982 reached over 1 billion levs – a 300 per cent increase over 1970. A major part of these were effected under the terms of the multilateral intra-CMEA long-term agreement for the production and supply of consumer goods. The national exporter Industrialimport is also a well-known supplier to the major Western European and North American markets, as well as to a growing number of developing countries.

Imports. The change in the structure of imports reflects the need to acquire most basic raw materials from abroad and the requirements stemming from the development of new machine-building and other heavy industries. These features are easily recognizable in Table 3, on the following page.

Despite continuing growth in imports of machines and equipment their relative share shows a certain decrease due to present requirements, mainly for modernizing and reconstructing already existing plants. Besides, the relative share of fuels and mineral raw materials has been growing over the last decade due to shifts in basic price relationships. Agricultural commodities, chemicals and consumer

Table 3: Commodity Structure of Imports (in percentages)

	1955	1960	1970	1980	1982
Imports – total	100.0	100.0	100.0	100.0	100.0
including:					
machinery, equipment and transport vehicles	51.4	43.9	40.6	35.4	33.5
fuels, mineral raw materials and metals	24.3	24.3	29.1	42.9	48.2
raw materials of vegetable and animal origin, foodstuffs	13.5	16.7	15.9	9.7	8.2
industrial consumer goods	3.4	7.6	5.7	4.4	4.8
chemical products, fertilizers rubber, building materials and other commodities	7.4	7.4	8.4	7.0	6.3

goods have maintained over the last decades more or less stable relative shares of total imports.

In the energy sector prime emphasis has consistently been laid on maximum utilization of those natural resources to be found inside the country. Such utilization has naturally required large-scale imports of generating equipment and technical assistance, above all from the USSR, Czechoslovakia and the GDR. Yet through co-operation with the socialist countries, and thanks to technology developed in Bulgaria, it is now one of the very few countries in the world utilizing coal with a caloric content of less than 2,000 kcal per kg for power generation on a large scale.

Bulgaria is also implementing an amibitious nuclear power programme. The existing nuclear power-station on the Danube now generates 27 per cent of the total electric power in the country. In March 1984 a new agreement was signed with the Soviet Union for constructing a second nuclear power-station. In total, by the end of the decade the share of nuclear power should reach about 50 per cent.

About 70 per cent of total energy requirements of the

Acme of Italy pavilion at the 1983 Plovdiv International Fair

country are still met by imports. Oil and gas deposits within the country being marginal, Bulgaria has been meeting its needs through imports, above all from the Soviet Union. It has increasingly been participating in the joint construction of energy producing and transporting capacities in the USSR. Thus for its share in the construction of the Soyuz pipeline Bulgaria receives annually an additional 2.8 billion cu. m. of gas.

Annual imports of electric power from the Soviet Union amount to 4.5 billion kWh, with total imports covering about 12 per cent of the country's requirements. Since 1967 Bulgaria has been part of the combined electric power system of CMEA member-countries, which since 1979 has been operating in parallel with the Soviet power system.

The iron and steel industry has been developed since the early 1950s, entirely with equipment and technologies supplied from the Soviet Union and other socialist countries. Again maximum effort has been put into fully utilizing local iron ores. The building of the second iron and steel mill after the Lenin works in Pernik, which went into operation in 1953, was started in 1960 at the site of the Kremikovtsi deposit, with ore containing a number of valuable components. With the completion of the second steel complex the country became largely self-sufficient in basic steel products, though it still depends on imports for part of its iron ore and coking needs.

In recent years industrial co-operation has also been developing with partners from Western countries in overhauling and upgrading existing capacities. Equipment for energy saving and automated process control is being installed in the Lenin works utilizing basic Bulgarian electronics hardware with instrumentation and software from Swedish companies. Construction has also begun of a third metallurgical complex, near the Black Sea port of Burgas, which will include inputs from the socialist countries as well as from other partners.

The non-ferrous metallurgy industry also has been expanded over the years with major inputs from the Soviet Union. Bulgaria possesses industrial deposits of mainly lead, zinc and copper ore. These, however, in particular copper, have low metal content and require intensive technologies for their utilization. Over the years such technologies have been developed both with active Soviet assistance and through the efforts of Bulgarian engineers and specialists. During the 1950s and 1960s two lead and zinc plants (in Kurdjali and near Plovdiv) and a major copper mill (in Srednogorie) were constructed. Two extensive complexes for enriching non-ferrous ores were also commissioned (Medet and Elatsite).

Imports of machinery and equipment will continue to form a major proportion of imports. In 1982 their share amounted to 33.9 per cent of total import value. In future, however, the content of such imports will increasingly be shifting to uniquely high-productivity machines and automated production lines in the various

Bulgarian electric trucks for export

erally in the CMEA system of co-operation at the level of national economies and by economic sector.

The principles of trade and co-operation within the CMEA have made it possible for Bulgaria to achieve fundamental shifts in the structure of its exports to CMEA markets. Thus machinery and equipment take up over 50 per cent of exports to them – a remarkable growth considering that in 1950 virtually no machinery was included in the country's exports. In exports to the Soviet market this share is even higher: over two thirds of all machinery produced by Bulgaria under specialization agreements is marketed in the Soviet Union.

In all there are at present 30 long-term target-oriented programmes for integration by economic sector between the various Bulgarian and Soviet branch ministries. There are 146 individual agreements and contracts for specialization and co-operation covering the 1981–85 period, 118 of them stemming from multilaterally agreed programmes within the CMEA. Bulgaria has concluded 230 such agreements with the CMEA member-countries as a whole. Under them the country has specialized in the production and supply of 60 different types of machinery.

Second in importance in exports to the CMEA member-countries are foodstuffs and agricultural commodities. Bulgaria is a major supplier of fresh and processed vegetables and fruit to the Soviet Union and other socialist markets. Important exports also are wines and cigarettes.

Of growing importance are the exports of non-agricultural consumer goods. Due to the rapid modernization of Bulgaria's consumer-goods industry it is now in a position to supply increasing quantities of clothing, sports wear, cosmetics, durables, etc., which are enjoying growing consumer demand on socialist markets.

A large portion of imports from the Soviet Union and the other socialist countries consists of high-technology machinery and equipment for modernization of existing capacities and for expanding the energy, raw materials and processing base of the economy. Imports of individual machines, machine parts and components included in longer-term specialization agreements are also growing. The same is true of imports of commodities and materials received under agreements for joint investment and development of energy and raw material deposits and means for their transportation.

Another feature of prime importance in the system of trade and integration with the CMEA countries is exchange and co-operation in science and technological research and development. In the course of 1976–80 Bulgaria annually took part in joint work in about 2,500 individual research projects. One thousand of these were developed under multilateral schemes and the other 1,500 – bilaterally, 650 of them with the USSR. Over two thousand projects have been implemented in the various economic sectors. There has also been a growing exchange of trainees and experts. During this time, more than 3,600 Bulgarian specialists have spent an additional period in other CMEA countries, which have sent to Bulgaria over four thousand experts, rendering technical assistance.

Such technological co-operation with the Soviet Union alone has provided the basis for developing a full two-thirds of all modern capacities and requirements forming the basis of Bulgaria's heavy industries. Lately the emphasis in this co-operation has increasingly been shifting to more comprehensive forms of joint technological research and development. During 1976–80 direct links and co-operation were in effect between 150 Bulgarian and 200 Soviet research and development centres and institutes, with over 600 projects developed jointly. In the course of the five year period up to 1986 another 300 interdisciplinary research projects will be jointly tackled.

Multilateral intra-CMEA scientific and technological co-operation has been developing in the first half of the 1980s under a mutually agreed mid-term plan for mutual assistance on problems of common interest. These span such areas as energy, raw materials, environmental protection, automation of production and engineering etc.

This system makes possible the selective development of Bulgaria's research and development effort. It has increasingly been centring on those sectors in which the country has been enhancing its role as a specialized supplier to the CMEA countries and the other markets.

In all, the constantly evolving systems of comprehensive co-operation and integration with the Soviet Union and the other CMEA member-countries in research, manufacturing and marketing in all major branches of the economy provides a firm, long-term basis for the further accelerated expansion of mutually beneficial trade between Bulgaria and those countries.

Developed Market-Economy Countries. Bulgaria's policy on developing trade with the Western countries has consistently been founded on mutual recognition and observance of the principles of most-favoured nation treatment and non-discrimination. Trade exchanges and longer-term industrial co-operation with partners from these countries have been looked upon as areas where important potential opportunities exist for mutually beneficial business on the basis of equal treatment and respect for the rights and interests of the partners from each side.

At the same time trade and co-operation between partners from the East and the West are most important as a material contribution to the lessening of international tension and the development of constructive international relations as a whole. Such trade should play a most important role in a return to the process of *détente* and in its progressive enhancement.

In the past few decades trade with the market-economy countries has undergone steady expansion. In total imports the share of these countries in 1982 grew to 16.7 per cent. In the same year the share of exports going to them amounted to 11.4 per cent as against 7.2 per cent in 1950.

Within this group, the countries of Western Europe

have been traditional trading partners and today account for the major part of trade with the West. First among these is the FRG with over one fifth of Bulgaria's total trade with the West. Major partners are also France, Italy and the UK, as well as Switzerland, Austria and increasingly the Benelux and Scandinavian countries, and Spain.

Partners of constantly growing importance are Greece and Turkey, being direct neighbours of Bulgaria to the south. Trade with them has been one of the most dynamic sectors of the country's recent trade turnover. In the course of 1970–82 their share in Bulgaria's trade with the developed countries doubled, reaching 13 per cent. This growth is a result of geographical advantages, but above all it reflects Bulgaria's consistent policy over the years in favour of mutually beneficial co-operation in all fields of common interest and especially in trade and economic exchange. This policy has today made for stable and constructive relations in the Balkans.

Trade and other forms of business have also been growing with Japan. These have increasingly included supplies of advanced equipment and machinery for the heavy machine-building industry, for power generation, etc. Trade has also been maintained with the USA, though the withholding of most-favoured nation treatment for Bulgarian exports, and of other trade-promoting facilities, has led to limited volume and sharp reversals in its structure and trends.

During the early 1970s in line with the overall improvement in East-West relations, Bulgaria's trade with the West was supplemented by a growing range of forms of longer-term co-operation with Western partners. On the intergovernmental level a number of ten-year agreements for economic, industrial and technological co-operation were concluded with most of the West European countries. Under these agreements respective programmes for co-operation by sectors of the economy have been concluded, guiding business partners to areas of priority interest. In addition, more than 70 general framework agreements have been concluded with Western trading companies over the past decade.

Together with such forms, creating general favourable conditions for longer-term co-operation, there are at present over 200 individual industrial co-operation agreements between Bulgarian organizations and Western firms. These include all the basic forms of co-operation – supply of plants under buy-back arrangements, licensing of production with purchase of part of the end-product, product specialization with joint assembly and marketing, etc.

Since 1980 a special Decree of the State Council has been in force governing the establishment and functioning of joint equity ventures with foreign partners. Under its provisions several such ventures have been established in the engineering of automation systems, both for discrete and for continuous process industries, in consumer goods, the chemical industry, etc. with partners from Japan, USA, Switzerland and other countries.

Further growth of trade and co-operation with the developed market economy countries should be possible on the basis of mutual business interest and equality of rights and obligations. The consistent, rapid and constantly diversifying development of Bulgaria's economy undoubtedly creates broadening opportunities for trade and joint business with partners from these countries, provided basic trade policy principles are respected and general international conditions do not hamper exchanges.

Developing Countries. One of the most dynamic sectors of Bulgaria's trade has been the growth of exchanges with the developing countries. Bulgaria has been developing its trade and economic relations with them on the consistent basis of co-operation and elaboration of a mutually beneficial framework for their sustained expansion. It has adhered to the principles of providing economic assistance within the country's capacities and of developing trade on a long-term basis under bilateral arrangements with the necessary trade-promoting facilities.

Since the late 1970s this policy has been further enhanced and elaborated, integrating its various features into a comprehensive approach aimed at fully utilizing all material inputs for accelerated expansion of bilateral trade and co-operation. Much in line with the overall strategy of developing countries for rapid socio-economic progress, this approach is built on identifying priority areas in their economies, developing them jointly – thus bringing new resources to both local and foreign markets, and finally sharing in the result of such joint efforts. Thus this line corresponds to the national objectives of the partner-country and at the same time assures adequate compensation for external inputs. The result is growing complementary to economic structures, to equal advantage, with corresponding growth of mutual trade.

Over the 1976–81 period the volume of trade with the developing countries grew by over 300 per cent, surpassing 2.5 billion US dollars in 1982. Their share in Bulgaria's trade increased from 2.9 per cent in 1960 to 5.6 per cent in 1970 and 11.0 per cent in 1982. The main partners have been Middle Eastern and North African countries, with over 80 per cent of trade, followed by countries in Asia (13.3 per cent), Tropical Africa (4.3 per cent) and Latin America (2.1 per cent).

In line with development plans and with jointly agreed projects, Bulgarian exports largely consist of machinery and equipment. Commodities and foodstuffs come next, with an approximately 17 per cent share for each of these groups in exports for 1982. Imports have been increasingly diversified by including more industrial goods from the developing countries. Since 1967 Bulgaria has introduced a system of tariff preferences, in favour of imports from the developing countries, which has been successfully improved to provide full, tariff-free entry for goods exported by the least developed among them. As a result, in 1983 the share of industrial goods in total Bulgarian imports from the developing countries surpassed 45 per cent.

The various forms of trade and co-operation which have evolved over the years are in line with the principles and economic potential of Bulgaria as a partner for the developing countries. With a number of them, intergovernmental ten year agreements have been concluded setting out the basic guidelines and priority sectors for trade and co-operation.

The supply of engineering services under comprehensive long-term agreements with appropriate financing and other facilities favouring bilateral trade has rapidly been gaining prominence. Civil engineering and construction are at present most developed among these. There are now over 100 major separate projects constructed by Bulgarian organizations in the developing countries. The Technoexportstroi engineering organization has successfully built, thoughout the Middle East and Africa, hospitals, airports, sports facilities, housing projects, congress centres, municipal buildings, etc.

Bulgarian engineering organizations have increasingly undertaken the erection of installations and complete plants. Technoexport, Technoimport, Bulgargeomin, Transcomplekt, Machinoexport, Electroimpex and other Bulgarian companies have increasingly been supplying equipment, services and complete systems for such industries as food-processing, building materials, cooling and air-conditioning facilities, wood processing, machine-building, etc. In total, such projects have been executed in over 30 developing countries, including Libya, Iraq, Syria, Nigeria, Ethiopia, the Congo, Tunisia, Morocco, Mozambique, Angola, Tanzania and others.

An area of growing importance, with broad potential, is the agro-industrial sector. Bulgaria's extensive experience in the production and processing of various agricultural commodities and in organizing comprehensive agro-industrial units is finding increased and mutually beneficial application. The engineering organization Agrocomplekt has been drafting and implementing irrigation, land reclamation, fruit-growing and poultry breeding projects and complexes in a number of Middle Eastern and Latin American countries.

Technical co-operation has also been developing through the training of specialists and provision of expert assistance in various sectors. In the past years there have been over 4,000 students from developing countries being trained annually at Bulgarian universities, which is about 5 per cent of all students attending. The number of Bulgarian specialists working in developing countries amounted in the 1976–83 period to 66,000, provided mainly through the Technoimpex organization.

Table 4: Trade by Groups of Countries (in percentages)

	1950	1960	1970	1980	1982
Total	100.0	100.0	100.0	100.0	100.0
Socialist countries	88.7	83.9	77.8	74.7	74.4
Developed market-economy countries	10.3	13.2	16.6	16.5	14.0
Developing countries	1.0	2.9	5.6	8.8	11.6

Table 5: Share of Trade with the Socialist Countries in Overall Trade of Bulgaria (in percentages)

Years	Total	Socialist countries			
		All countries	Members of CMEA	USSR	Other socialist countries
1950	100	88.7	88.7	52.2	...
1960	100	83.9	81.0	53.1	2.9
1970	100	77.8	76.2	53.0	1.6
1982	100	74.4	73.0	53.8	1.4

Leather and fur articles produced by the Pirin Industrial and Commercial Organization at the 1981 Plovdiv International Fair

The PRK-3 triple attachment exhibited by the Agro-machinaimpex Foreign Trade Organization at the 1983 Plovdiv International Fair

Bulgarian-made electric hoists at the 1982 Plovdiv International Fair

The Bulgarian National Bank in Sofia

THE FISCAL AND MONETARY SYSTEM

The formation of the fiscal and monetary system of Bulgaria started at the time of the Russian-Turkish Liberation War (1877–78). A Chancellery of the Civil Government for the Bulgarian lands, headed by the Russian Prince Vladimir Alexandrovich Cherkasky (1824–78) was founded, with the task of setting up – with the help of the local national forces – a system of new state institutions to take the place of the abolished Turkish Administration. Local self-governing bodies were founded, and government and local finances were instituted. During the Provisional Russian Government in Bulgaria, under the guidance of Prince Alexander Mikhailovich Dondukov-Korsakov (1820–93), important normative acts were issued concerning the finances of the country. The Ministry of Finance and the Bulgarian National Bank (BNB) were founded. The functions of the customs were regulated. The inherited Turkish tax system with all its specific, indirect levies (*taxes corvées,* etc.) was more or less preserved, but some direct taxes were abolished. The first Budget approved by Dondukov-Korsakov covered the period from 1 March 1879 to 1 March 1880. The receipts and the expenditures were in French francs, since there was no Bulgarian national monetary unit at that time. With the adoption of the Law for Coinage, voted by the Second Ordinary National Assembly in 1880, a bimetallic monetary system was established and a Bulgarian monetary unit – the lev – instituted, subdivided into 100 stotinkas (*sto*=hundred). The parity of the lev was tied to the French franc. The first Bulgarian coins for circulation were struck in 1881, and the first banknotes with gold backing were issued in 1885, when by virtue of a law the BNB was granted the right to issue banknotes. They were printed in London.

After the Liberation in 1878 the production and financial systems typical of the Turkish feudal system were abolished. The process, already under way, of gradual consolidation of capitalist production systems, and of capital accumulation, was facilitated by the fiscal and monetary policies of the bourgeois governments. By the end of the nineteenth century a number of private banks, which were set up as joint stock companies, were already functioning. At the beginning of the twentieth century the upsurge in industry and commerce attracted a big influx of foreign capital into banking. Big joint stock banks with foreign capital were founded (the Balkan Bank, the Bulgarian General Bank, the Italian and Bulgarian Commercial Bank, the Credit Bank, etc.). In the early 1920s financial groups of related industrial, commercial and insurance companies formed around these banks. Commercial credit provided the basis for the credit systems.

The finances of the country were characterized by the rapid increase in government outlay due to the growth of expenditure on the army, the police, government administration, the building of railways, roads and ports and, first and foremost, due to government debts. This determined the respective development of government receipts. The main source of revenue for the Budget was taxation of the population. In conformity with the interests of the ruling bourgeois establishment, the tax burden was gradually and increasingly transferred to the workers, peasants and office employees. A system was introduced whereby direct taxes were gradually reduced and indirect taxes increased. Budget balances repeatedly ran into the red and deficits were financed through crippling, external loans. These loans grew into an additional burden for the Bulgarian taxpayers. The wars (1912–18) shattered the country's finances. In 1912 a new money standard was created by the BNB, and paper money was put into circulation. During the First World War (1914–18) coercive requisitions of agricultural products were introduced, which led to the total plundering of the small farmers. The reparations imposed by the victors in the war were a further great burden; they amounted to 2,250 million French francs in gold. During the short rule of the Bulgarian Agrarian Party (1920–23) some social justice in the taxation system was established; the highest levies were imposed on big capital, and an attempt was made to confiscate the profits accumulated during the war in order to alleviate the burden on earned incomes. With the re-establishment of the bourgeois Government, the crisis in government finances and monetary circulation became more acute. The wars shattered the monetary system and by 1928 the official devaluation of paper money was more than 2,800 per cent and the actual devaluation was more than 3,300 per cent. The effect of the attempt which the bourgeois Government made to consolidate monetary circulation and the Budget with the so called Stabilization Loan of 1928 was short-lived. When the world economic crisis broke in 1929 monetary circulation was characterized by the development of deflationary processes which further intensified the crises in the economy. After 1933 monetary circulation was affected by galloping inflation. The years of the Second World War (1939–45) saw the transformation of the national economy into an appendage of nazi Germany, with the latter granted credit by Bulgaria. Domestic loans, treasury bonds and the issue of paper money were used to meet the growing military expenditures. The BNB put masses of new banknotes into circulation, as well as bonds, backed by uncovered receipts from the country's clearing system, thus deepening further the process of inflation. By the end of 1944 the value of the lev dropped to one-twelfth of its 1939 parity. The treasury bonds grew from 1,837 billion levs in 1942 to 18,218 billion levs in 1944; the coins in circulation – from 1,441 billion levs in 1939 to 3,172 billion levs in 1944; the paper money – from 4,245 billion levs in 1939 to 45,838 billion levs in 1944. There was a sharp decrease in the gross domestic product and in the national income. The Bulgarian monarcho-fascist regime, linked closely with German Imperialism, completely ruined the country's finances.

Throughout the period of capitalist development, the fiscal system was used mainly in the interest of the ruling class and foreign capital. The tax system was a means of additional exploitation of the working people. Indirect taxes – excise duties, customs duties, fiscal monopolies and turnover taxes were of crucial importance. The Budget was raised mostly from taxing the population (65.4 per cent in 1944) and from internal and external government loans. The budgetary expenditures were of an unproductive, consumerist character. Budgets were not stable; deficit was chronic. The budget system featured a number of Budgets – regular, additional, extraordinary, municipal and over 420 fund budgets. The credit system encompassed three types of credit institution: *state* (the BNB, the Bulgarian Agricultural and Co-operative Bank, the Bulgarian Mortgage Bank, the Bulgarian Central Co-operative Bank and the Post Office Savings Bank); *co-operative* (popular banks, village credit co-operatives and central credit co-operatives); and *private banks* (mainly joint stock companies), which in the majority of cases were closely linked to foreign bank amalgamations. The insurance business was carried out by private joint stock companies, whose interest in the insurance was determined by the possibilities offered for profit. There were also a number of co-operative insurance corporations. Savings activities developed without a plan and the accumulated sums were used mainly to extend credits to the big capitalist enterprises and to the bourgeois Government. The share of the working people in the savings activities was limited, due to their low incomes. Bank deposits were opened mainly in the Bulgarian Agricultural and Co-operative Bank, the Post Office Savings Bank, the agricultural credit co-operatives and the popular banks.

After the popular regime was established in 1944, the fiscal and monetary system was transformed entirely to meet the needs of the socialist economy. Finances were stabilized on the basis of the planned development of the economy. The fiscal policies of the BCP were based on the understanding that the building of Socialism was inconceivable without a stable fiscal and monetary system.

The Fiscal System. One of the first tasks of the People's Democratic Government was to eliminate inflation. The first decisive step for limiting inflation was the raising of the Liberty Loan in 1945. Instead of the planned 15 billion levs, 25 billion levs (1945 lev parity) were raised. This loan was an acid test for the attitude of the working people towards the Government of the people's democracy and the policies of the BCP. One anti-inflationary measure was the intensified collection of state revenues, mainly unpaid taxes from the *haute bourgeoisie*. The end of the Bulgarian Patriotic War (1944–45) saw the restructuring of state expenditures, with the introduction of strict economic measures and priority to the productive over the unproductive. A law was voted to confiscate all property acquired by the *haute bourgeoisie* through speculation and unlawful means in the period 1935–45. The fiscal reform carried out in early 1947, together with the lump-sum progressive tax on exchanged banknotes, was decisive in combating inflation. This reform is of markedly progressive political significance, since it dealt a blow to the financial power of the politically beaten, but economically still thriving *haute bourgeoisie,* and the process of inflation in the country was eliminated once and for all. Stabilization of monetary circulation required the thorough recovery of the national economy. The Two-Year National Economic Development Plan (1947–48) was a step in this direction. The nationalization of the basic means of production (1947) followed by the promotion and strengthening of the socialist systems of production created the economic preconditions for the final stabilization of money circulation and a reform of the fiscal system. A fiscal reform was carried out in 1952. The parity of the lev was fixed to gold (0.130,687 g gold). The gold parity of the lev was raised with the reform of 1962 (0.759,548 g gold). The rates of exchange for the currencies with which Bulgaria transacted business were corrected respectively. The wages, salaries, bank accounts, receipts, etc. were brought into concurrence with the new price scale. Commodity prices were also changed. However, this reform had no effect on the income of the working people. The strengthening of the fiscal system and the setting up of the currency circulation on a planned basis transformed the lev into a powerful instrument for organization and control of the fulfilment of the Five-Year Plans for Economic Development.

Important measures for the strengthening of the country's finances were taken: overdue taxes and other sums were collected by the State from the capitalists and the big land-owners, a regime of stern economies of state expenditure was established, new taxes were introduced, etc. An important law providing for self-support by the state and state-autonomous enterprises was voted by the National Assembly in 1946. This law required the profitability of all state enterprises as a vital financial measure increasing state revenues from all economic activities. Another law which channelled receipts into the economy was the law for confiscating all properties acquired through speculation and unlawful means. The enforcement of this act brought about the return to the State of the capital accumulated by the big capitalists during the war years.

With the establishment of the popular regime the political platform of the Fatherland Front envisaged a restructuring of *taxes* with a view to a more fair distribution of the tax burden, where the true solvency of the different categories of tax-payer was taken into account. The line was to introduce progressive taxation on income, and a decrease in indirect taxes and in the number of charges paid by the working people. After the nationalization of 1947, which laid the foundation for the socialist tax system, taxes were transformed in order to bring them into compliance with the new production systems. Steps were taken to abolish some of the taxes levied on the population (railway tax, army tax, etc.) and to lower their fiscal importance. The first qualitative change in the system of taxation was introduced with the establishment of income tax in early 1947. The act provides for large tax relief for the small farmers, workers and employees. At the same time it was directed towards large unearned incomes. In this manner the fiscal system and taxation policy of the People's Democratic Government became an instrument for class struggle in the conditions of transition from Capitalism to Socialism.

With the strengthening of socialist ownership over the means of production and the introduction of planning in the management of the national economy, the revenues from state economic activities gained in importance, as also did the contribution from the profit of the self-financing enterprises and the turnover tax. The turnover tax, which was restructured in 1948 in conformity with the socialist production systems, assumed particular importance. The structure and organization of this tax was of great importance for the qualitative change in the tax structure in the People's Republic of Bulgaria. The turnover tax of 1948 established a strong, planned correlation between the State Budget and the self-financing enterprises and created the prerequisites for a simplified tax system – limited application and dropping altogether of the excise duties and the fiscal monopolies.

Income tax underwent a reform in 1950. Under the new income-tax system there were two main categories of tax-payers: urban population (workers, artisans, free-lance professionals, etc.) and rural population (farmers). Lowest among the taxes levied on the urban population was the tax on salaries and wages, where a minimum income was established as non-taxable. As to the taxation of farmers, differentiation was established between members and non-members of the co-operative farms (higher taxes were imposed on non-members in order to encourage them to join the co-operative farms, while for the co-operated farmers, a token tax was levied only on the income from their personal farms). New co-operative farms were granted big concessions, being exempted from income tax for five years. The rapid development of collective farming led to the setting up of machine-and-tractor stations as enterprises financed through the State Budget. The outlays and the receipts of the machine-and-tractor stations occupied a special place in the State Budget.

As a result of these measures, by the end of 1950 the tax system had become socialist in character. In 1956 the turnover tax and deductions from profit accounted for 65 per cent of state budget receipts. Budget revenue came mainly from turnover tax; deductions from the profits of self-supporting enterprises; receipts from income tax, including income taxes levied on co-operative farms, producers' co-operatives and consumer co-operatives; and income from machine-and-tractor stations. In general lines this tax structure is still valid today. Some changes were made in the deductions from the profit, machine-and-tractor stations were closed down, and AICs were founded. The tax system is constantly being improved to keep abreast of the development of the national economy. Price policies were also improved. In 1955 the purchasing of many agricultural products were changed; in 1956 the producers' prices of industrial goods also underwent a reform, with a view to reducing budget subsidies for planned losses in some sectors and industries and in order to create relatively similar conditions for profitable business activity in all self-supporting enterprises. During the Second Five-Year Plan (1953–57) the tax levied on the income of the co-operative farms, including on the incomes received from the personal farms of the members of the co-operatives, was greatly reduced in order to stimulate the development of collective farming and to raise the incomes of the members of the co-operatives.

In connection with the improvement of the management of the national economy which had begun in the 1960s, the tax system was also improved. Instead of deductions from profit, a number of taxes were introduced – a tax on profit, on profitability, on salaries, on differential land rent, etc. Positive changes were carried out in the taxation of the AICs and the co-operative organizations. An important element in the development of the tax system was the endeavour to curb and reduce personal taxes (taxes paid by the population), ensuing from the policy of the BCP to ensure a constant rise in the standard of living of the working people. The personal taxes decreased in number, the minimum level of salaries and wages exempted from tax was raised, and a number of taxes and charges were done away with. The share of personal income taxes in revenues dropped to less than 10 per cent. The revenues coming from state economic activities increased and their role as receipts in the State Budget grew steadily. The prevalence of the revenues from economic activity allows the socialist State to continue to increase social securities for the working people, to raise systematically the standard of living and to restrict the fiscal share of personal income taxes, with the prospects of their gradual abolition. The process is characterized by the abolition of some personal taxes and charges and by the raising of the tax-exemption minimum, which is in the 1980s already five times higher than in 1950, the tendency being for this tax-exempted minimum to catch up with the average salary. The development of the system of finance in the People's Republic of Bulgaria in the conditions of the socialist production relations has been characterized by the diminishing importance of indirect taxes (taxes on consumption).

The socialist turnover and planned price formation simplify financial relations with the State and render useless the excise and fiscal monopolies. The fiscal and economic importance of the customs duties, particularly of import duties, has been diminishing, too. The complex duty system of bourgeois Bulgaria was restructured in line with production systems. Economic systems with the socialist countries were organized in a planned manner and the complicated customs processes became a thing of the past. Import price differentiation replaced import customs duties. This made easier and more efficient the solution of the task of domestic price formation. The fiscal and economic importance of the customs duties is limited, with a tendency towards their further reduction. They still have some role to play in economic relations with capitalist and developing countries. It should be noted that the share of those two groups of countries in Bulgaria's foreign trade does not exceed 20–22 per cent.

In the period 1944–56 internal *government loans* were reduced in order to increase budget receipts. These loans contributed to a certain extent to the stability of the monetary circulation. Five internal loans were underwritten, which were expressly for production purposes, since they were closely connected with socialist construction. The productive character of the internal state loans conditioned their self-reproduction. For this reason the tax burden did not increase at all with the repayment of the annuities and the interest rates on the loans; on the contrary, it decreased. Thanks to the positive results in the building of Socialism and the consolidation of the financial system, internal state loans have not been raised since 1956. There was a delay for a few years in the repayment of the loans and interest. At the beginning of 1963 the State resumed interest and principal repayments until the full amortization of the last loan, raised in 1955. Since 1975 there have been no debts on internal state loans. Bulgaria is a country with an open economy, which is why the use of foreign credits cannot be avoided. These credits are related to the purchase of licences, machinery and equipment for new enterprises or for enterprises which are being overhauled or equipped with new advanced technologies. Bulgarian economic organizations are very cautious in their use of credits from capitalist countries because they do not want to become economically dependent on their creditors; the needs of the country for credits, and the possibilities for their reasonable and efficient utilization, as well as for their repayment, are well studied in advance.

The Budget is used as a means of solving the main political and economic tasks of the country. Steps were taken for the elimination of the inherited budget deficit from the bourgeois Government amounting to 32 billion levs (at the parity of that time). Non-productive expenditures were restricted – military expenditures, administrative outlays, etc., and budget receipts were collected in due time, this being especially true for the arrears which the capitalist class owed to the State. By the end of 1945 budget deficits were restricted and lowered considerably. During the Two-Year National Economic Development Plan (1946–47) the Budget was already fully balanced and in 1948 – the first year after the nationalization of the industrial enterprises, mines and banks – it wound up with a surplus. In the period 1945–47 great attention was paid to the budgets of the local government bodies and the State Administration – the City and Village People's Councils. The Government used to grant them financial assistance, especially during the Two-year Economic Plan, when for the first time since their establishment the local government bodies became engaged in extensive and all-round public works projects.

The beginning of the unified State Budget of the People's Republic of Bulgaria was established in 1948. The first unified Budget was in force in 1949, together with the equalization fund existing at the time. Budget unity was finally achieved fully in the 1950 Budget. The Budget of the Republic is organically tied to the local budgets and to the budget of the State Social Insurance, and subsequently to the budget of the Council for Mutual Insurance of the Members of the Producer's Co-operatives, later renamed the budget of the Council for the Mutual Insurance of Co-operative Members. A great number (over 400) of state, municipal and fund budgets were liquidated. A uniform obligatory classification of revenue and outlay at all budget levels was introduced. The receipts of the separate budgets in the aggregate budget system were differentiated and specified as to their separate origins and destinations. The budget higher up in the hierarchy was used in part to support the lower budgets in the form of subsidies and grants. From the separate entries in the Budget of the Republic it was allowed to make percentage deductions in favour of the local district and municipal budgets. By regulating the budgets in this manner conditions were created for budgetary assistance to economically underdeveloped regions of the country, which was an important factor for budgetary unity (see Table 1).

Thanks to the high and stable growth rates of the country's development and the rapid increase of the national income, the budgetary receipts and outlays rise with each passing year without an increase in the tax burden on the population. An essential feature of the State Budget of the People's Republic of Bulgaria is that for more than thirty years it has ended up with a surplus, which is an important feature of socialist budgeting. Characteristic of the development of the state budgetary receipts and outlays in the period of planned building of Socialism are the constant increase of revenue from the economic activity of the State and of

The Fiscal and Monetary System

Table 1: Revenues and Expenditures in the State Budget of the People's Republic of Bulgaria (million levs)

Periods (years) Indices	1952	2nd Five Year Plan (1953–57)	3rd Five Year Plan (1958–60)	4th Five Year Plan (1961–65)	5th Five Year Plan (1966–70)	6th Five Year Plan (1971–75)	7th Five Year Plan (1976–80)	1981	1982
1. Revenues	1,598.8	8,896.2	7,838.3	16,515.6	23,422.9	36,868.1	53,691.4	15,489.2	16,687.7
Including from the national economy									
– amount	1,205.0	6,782.7	6,121.3	12,470.7	17,494.4	27,111.0	36,710.9	10,105.2	10,691.0
– per cent	75.4	76.2	78.1	75.5	74.7	73.5	68.4	65.2	64.1
2. Expenditures	1,432.7	8,528.9	7,651.6	16,230.9	23,077.9	36,370.6	53,063.3	15,426.9	16,526.3
Including									
a) for the national economy									
– amount	822.6	4,588.7	4,531.6	9,370.0	12,195.6	18,306.5	24,407.7	8,067.7	8,519.6
– per cent	57.4	53.8	59.2	57.7	52.8	50.3	46.0	52.3	51.6
b) for social and cultural activities									
– amount	277.8	1,910.7	1,674.7	3,959.7	6,657.1	11,682.0	19,096.8	5,281.1	5,681.4
– per cent	19.4	22.4	21.9	24.4	28.8	32.1	36.0	34.2	34.4
3. Budget surplus	166.1	367.3	186.7	284.9	335.0	497.5	628.1	62.3	161.4

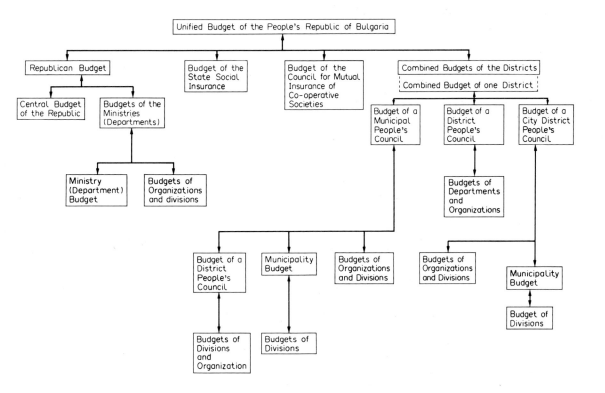

outlay for the development of the national economy and for financing social and cultural undertakings, the restriction of administrative and other non-productive expenditure and the budgetary surplus (see Table 1).

The accelerated economic development enhances the role of the local budget – districts, municipal, and city district – and of the budgets of the People's Councils. Their main purpose is to provide the necessary financial means for the development of the territorial units. The local government bodies and the people's self-government are immediately concerned with the well-being of the working people in Bulgaria. The care for man in the socialist society is most strongly felt through the budgetary outlays. The outlays in the local budgets are allocated mainly for the urbanization of inhabitated places, for housing construction and public works, for furnishing and supporting schools, hospitals, child-care establishments, etc.

The 1970s were characterized by the still wider use of the fiscal and monetary relations in economic management, which resulted in expanding the fiscal basis of the socialist economic organizations and enterprises. The fund for expansion and technical improvement and the fund for social and cultural undertakings were formed through obligatory deductions from profit and from other sources. The fund for expansion and technical improvement is allocated mainly for financing capital investment, overhauls, the increase of current assets (stocks) and for research and development, while the fund for social and cultural undertakings finances programmes for the social development of the workers' collectives. The economic organizations and enterprises also raise other funds, which are expended mainly on material incentives.

Under the operating economic mechanism capital investment is financed from the resources of the corresponding funds of the economic organizations and enterprises, and with bank funds. The investment for the development of the energy and raw materials industries for building new production facilities in industries which are decisive for the re-equipment of other industries are financed with resources coming from the State Budget. The national programmes for the implementation of scientific and technical achievements are also financed with resources of the State Budget. This change creates the impression that the expenditures from the State Budget of the People's Republic of Bulgaria are decreasing, while in fact this decrease is due to the expanded fiscal basis of the economic organizations and enterprises – an expression which is inconceivable without the indirect participation of the State Budget. The development of the unified State Budget is a proof of the stability of the financial system.

The Monetary System. After the 9 September 1944 victory the monetary system underwent qualitative changes in functional and institutional aspects. In the first years its structure was preserved, but the People's Democratic Government exercised a direct control on the functioning of the private banks. This situation continued until the end of 1947 when industries, mines and banks were nationalized. 1948 ushered in a process of socialist reconstruction of the monetary and banking system which ended in 1951.

In the period of socialist construction and of continuous economic growth, the monetary system was perfected in line with the tasks to be solved at every stage of economic development. Nowadays the monetary system consists of the Bulgarian National Bank, the Bulgarian Foreign Trade Bank, the Bank for Economic Undertakings (Mineralbank), the State Savings Bank and their district and regional branches and agencies in the country.

The Bulgarian National Bank occupies a central and leading place among the banking establishments. Its main functions are: to carry out issues and regulate the national monetary circulation; to accumulate the temporarily surplus cash in the national economy and to use it for the planned development of the economy; to grant short-and long-term credit for business activity; to organize and exercise bank control on the production and financial activity of the economic organizations and enterprises, including the wage fund; to organize and execute cheque and cash payments and bank transfers, and to implement the required bank transfer operations from the State Budget.

The Bulgarian Foreign Trade Bank is a joint-stock company and specialized institution for crediting, payments and control over foreign economic relations and payments. It settles foreign accounts and transfers to and from abroad the payments on inter-government clearing and credits; grants and uses credits to and from abroad; funds the export-import foreign trade organizations and enterprises; grants loans in foreign currency to economic organizations and enterprises; and participates in bank operations with the International Bank for Economic Co-operation and the International Investment Bank in Moscow. The Bulgarian Foreign Trade Bank is a founder of and a majority stockholder in the Lebanese Bank, Litexbank, and has representatives in London and Frankfurt am Main.

The Bank for Economic Undertakings (Mineralbank) is a joint stock company and specialized bank institute which assists in the development and contributes to the maximum utilization of local energy, mineral and other resources. To this purpose it attracts domestic and foreign currency resources and uses them for funding different initiatives in this sphere.

The State Savings Bank serves the general public and is the only savings institution to attract savings from the population and to extend it loans.

In the People's Republic of Bulgaria mobilization and investment of credit money resources is carried out according to plan. For this purpose the BNB draws up a credit plan which is an individual part of the unified plan for social and economic development. It includes the state loanable funds and their redistribution in the form of loans for the expanded socialist reproduction as

The Fiscal and Monetary System

Table 2: Ordinary Accounts

Years	Ordinary accounts total		Including					
			Savings accounts[1]		Children's savings accounts		Others	
	Thousands	Total in levs	Thousands	Total in levs	Thousands	Total in levs	Thousands	Total in levs
1932	4,353	167	3,420	136	930	31	3	0
1957	4,803	341	3,756	300	1,032	39	15	2
1960	6,091	742	4,920	676	1,155	64	16	2
1965	8,003	1,497	6,597	1,366	1,388	128	18	3
1970	7,898	3,114	6,226	2,802	1,652	336	20	6
1975	9,035	5,967	6,847	5,194	2,049	751	139	22
1980	8,755	8,442	6,183	7,065	2,301	1,318	271	59
1981	8,985	9,061	6,296	7,543	2,339	1,445	350	73
1982	9,251	9,880	6,440	8,248	2,350	1,511	461	101

[1] Up to 1975 (inclusive) the data reflects the workers', time, on demand, qualified and premium deposits; after 1975 the data are for savings accounts.

well as liabilities which lead to the formation of credit resources, creation of money and assets operations. Loans are planned in accordance with the policy of the BCP and the Government in the field of monetary circulation and of crediting the national economy. Banks allot the mobilized money resources for temporary utilization in the form of *short-term credits* for supplementing the floating assets of the economic organizations and enterprises (loans for productive assets – raw material stocks, half-finished goods, etc.) and loans for circulating assets (stocks of finished goods, forwarded goods or goods disbursement) and *long-term loans* for the building of new production facilities (basic production assets).

Under the now functioning economic mechanism and with the predominant use of intensive factors for the development of the economy, the importance of the credit system has increased significantly. The credit mechanism is one of the main economic layers for influencing the economic organizations and enterprises, and for raising the efficiency and improving the quality of their production activity. Long-term loans are allocated for the building of facilities which ensure the production of new, highly effective goods of high quality for the home and foreign markets. The credit system is an active factor in the achievement of the strategic goal – the building of a mature socialist society in the People's Republic of Bulgaria.

Savings activity. The organized savings activity is a sign of the rising living standard of the working people. The restructuring of the national economy, full employment and the ever rising national income are the objective conditions for the development of the savings activity. The protection rendered by the Socialist Government to savings plays a very important role. The savings of the people are concentrated in the *State Savings Bank*. At the

Table 3: Deposits for Housing Construction

Year	Number of accounts (in thousands)	Balance (in million levs)
1952	7.5	1.8
1957	42.2	36.3
1960	81.2	95.9
1965	149.5	259.0
1970	284.6	755.5
1975	480.9	1,333.1
1980	545.1	1,850.9
1981	539.2	1,872.0
1982	540.1	1,928.6

beginning of 1952 the permanent balance of the savings in the State Savings Bank approximated 166 million levs (at current parity) and at the end of 1982 – 9,860 million levs; the number of deposit accounts (ordinary deposits) was 9,251,000 (see Table 2). In 1982 the sum total of ordinary deposits and deposits for house construction in the State Savings Bank amounted to 11,788.8 million levs, i.e. the growth rate of organized savings activity was 7,000 per cent higher than in 1952. This is a sign of stable, sufficient and growing personal incomes.

There is a marked growth after 1970 in savings and in children's savings accounts – a result of the BCP's policy of raising the material well-being of the working people. The children's savings accounts are mostly on a long-term basis. The growth in children's savings accounts is a testimony of their assured material well-being. The average annual growth in children's savings accounts after 1970 (in capitalized interest) is approximately 100 million levs.

A twenty-lev note, 1974 issue

A ten-lev note, 1974 issue

The Fiscal and Monetary System 489

A fifty-lev note, 1923–42 issue

A ten-lev note, 1890–1907 issue

Two-lev coin, 1966 mintage
One-lev coin, 1969
50-stotinki coin, 1974
20-stotinki coin, 1974
 5-stotinki coin, 1974
10-stotinki coin, 1962

Obverse of above

Housing construction accounts form the basis for the long-term loans granted to Bulgarian citizens to buy or build homes. The maximum loan which can be obtained from the State Savings Bank is 12,000 levs, payable in 20–30 years at 2 per cent annual interest. These extremely favourable conditions for crediting private housing construction, together with the preliminary earmarked savings activity, are factors of the rapid development of construction. In 1952 the sum total of the savings of 7,500 investors in similar deposit accounts amounted to 1,800,000 levs, the corresponding figures for 1982 being 540,100 and 1,928.6 million levs (see Table 3). The average amount of a deposit account for housing construction is 3,571 levs.

Organized savings activity plays a very important role in the Bulgarian financial and credit system. With the placing of the permanent and constantly growing balances of savings accounts at the disposal of the BNB, the possibilities for crediting the national economy have been increased. The permananent balance on the savings accounts is a reliable crediting source for the BNB. Organized savings activity affects money circulation very favourably, contributing actively to a balance between consumer demand and the available stocks of goods during a given period. By drawing out of circulation the money supply corresponding to the uncovered purchasing power of the population, organized savings play the role of a stabilizing factor in money circulation and goods turnover.

Insurance. With the establishment of the People's Democratic Government the insurance companies were nationalized (1946) and the State Insurance Institution was established in place of the private and co-operative insurance companies. Its activity develops along several lines: life insurance, fire insurance, insurance of crops, livestock and other farming activities, liability insurance, etc. It operates under the direct guidance of the Ministry of Finance and is an economic, self-financing organization. Fifty per cent of its annual profit goes into the State Budget. The financial and economic importance of this institution is very great; it renders all-round insurance protection of the people and the national economy, thus contributing to the maintenance of the continuity of the production process. It constantly and significantly decreases the cost of insurance coverage and accumulates money resources which can be used for financing socialist construction.

The insurance organization Bulstrad was founded in 1961 as a joint stock company with headquarters in Sofia. Bulstrad engages in insurance activities, transportation insurance, cargo insurance, serving foreign trade, insurance of vehicles and means of transportation – ships, aircraft, liability insurance in air transport, etc. Bulstrad is governed by a Board of Managers elected from among the shareholders. A part of its net annual profit goes into the State Budget. Bulstrad has a very good reputation on the international insurance and underwriters' market. Its entire activity is based on solid financing and it is distinguished for remarkable financial stability.

The fiscal and monetary system of the People's Republic of Bulgaria plays an important role in the process of distribution and re-distribution of the national income within the framework of the set planning parameters. The existence of stable financial and credit systems in the entire national economy is an important condition for the successful construction of the socialist society. Cyclical financing and crediting of socialist construction depend to a large extent on the normal functioning of the fiscal and monetary system. The Bulgarian fiscal and monetary system successfully fulfils the tasks stemming from the economic and fiscal policies of the BCP and of the Government. The competent utilization of its distribution and control functions guarantees to a significant extent the positive results in the socialist construction and in the social policy.

Taking into consideration the main goals of the socio-economic development of the country in the years of the Eighth Five-Year Plan and up to the end of 1990, the Twelfth Congress of the BCP underlined the necessity of raising the role of the fiscal and monetary levers, so that they could contribute still more actively to the country's progress; using their control functions they should further strengthen the regime of austere measures and fianancial discipline, with the aim of transforming the State Budget into a system of earmarked funds, formed and used on the basis of long-term norms. Bank credit and interest rates should play a greater role and become an active instrument in the practical implementation of the main parameters laid down in the Plan and the Budget, in the accelerated implementation of the achievements of scientific and technical progress and in attaining a high multiplier effect.

The policies of the Party and Government in the fiscal and monetary field are entrusted to and implemented by the Ministry of Finance and the BNB, which govern the activity of all bodies and institutions in the fiscal and monetary system of the country.

TOURISM AND RESORTS

Tourism. The small territory of Bulgaria abounds in relics of the past, in remnants of many epochs and old civilizations. Nature has been bountiful to the country and has gathered together, in a picturesque conglomerate, snow-peaked mountains, meandering rivers and scenic valleys, unique natural phenomena, a beautiful Black Sea coast and curative mineral springs. Ancient museum-towns, quiet monasteries with valuable murals, icons and wood-carvings have been preserved intact for centuries. The country is the venue for picturesque festivals, renowned international competitions, and many other cultural events. History and modern life alike attract tourists to Bulgaria. The country has won recognition on the world tourist market due to the consistent and purposeful state policy in tourism. The Party and state leadership of the People's Republic of Bulgaria look on tourism as a basic factor in the development of the national economy and international relations; a factor instrumental in the proper utilization of the beneficial natural and climatic conditions, the rich cultural and historical heritage and the advantageous situation of the country at a significant geographical crossroad. Special attention is given to the development of tourism, to its constant improvement and efficiency. The development of tourism, the basic goal of which is the peaceful gathering together of people from all over the world, is a vivid,

The Shatrata restaurant at the motel near Pravets

fundamental expression of the consistent and actively peaceful policy of Bulgaria; of her aspirations to establish good-neighbourly relations and co-operation, close contacts with different nations, and friendship with different states. During the period of capitalist development, there were no conditions for the development of tourism in the economically backward country. To make the natural wealth and scenery, the ancient historical sites and monuments of culture accessible to the public, were objectives of no concern to the ruling bourgeois parties. The Central Sea Baths in the Black Sea town of Varna were not built until the beginning of the 1920s. They were opened in 1926, and were then visited by 1,100 foreign tourists. In 1929, Varna was visited by 14,700 tourists. The Bulgarian Travel Agency, which organized cruises to Istanbul and elsewhere, was founded in Sofia, and the Balkan Bureau was set up in 1937 to cater for the interests of foreign tourists – about 8,000 at that time.

The basis of international tourism in Bulgaria was laid in 1948 when the Balkantourist State Enterprise for Travel and Tourism was set up. The newly-founded agency was entrusted with the task of organizing and facilitating tourism, building homes, restaurants, etc., and attracting foreign visitors to the country. In a short time, Balkantourist built a number of new tourist sites and improved the few existing hotels and restaurants. The development of tourism during the 1960s and 1970s was marked by an acceleration in the provision of the material and technological basis of tourism, which was one of the immediate and significant results of the historic decisions of the April Plenary Session of the Central Committee of the BCP (1956); during the 1980s it is marked by intensification and a significant improvement in the quality, organization and management of tourism, which has become an important branch of the economy. During the period of socialist construction, Bulgaria has won recognition as a sought-after partner on the international tourist market; as a country which offers a quiet atmosphere, a variety of opportunities for rest and recreation, for travelling and becoming familiar with Bulgarian folklore, and with the works of Bulgarian painters, singers and musicians.

For those keen on hunting and fishing, excellent facilities have been created in the shooting grounds and breeding pools. There is a growing interest in Bulgaria as a country of congress tourism. Bulgaria is a venue for world

Tourism and Resorts

Table 1: Number of Foreign Visitors to Bulgaria

			Years		
1960	*1965*	*1970*	*1975*	*1980*	*1982*
200,600	1,084,000	2,537,000	4,049,000	5,186,000	5,647,000

congresses and symposia, of international conferences and meetings (about 400 annually) which are attended by thousands of scientists and scholars, cultural figures and artists, and are held in the big convention centres of Sofia, Varna, Russe and Burgas. The biggest convention centre in the country is the Lyudmila Zhivkova People's Palace of Culture in Sofia, a member of the International Association of Congress Palaces and of the International Association of Congresses and Conferences. The large-scale tourist building has been instrumental in gaining recognition of tourism as a highly efficient branch of the economy. Modern de luxe hotels have been erected, most of them graded as 3- to 5-star, according to international standards. Some have swimming pools, and hydropathic and sanatorial establishments provided with modern medical equipment. The hotels have comfortable restaurants and night clubs, casinos, etc. The best de luxe hotels are the Varna in the Druzhba resort near Varna, the Vitosha New Otani in Sofia, and the Leningrad and the Novotel-Plovdiv in Plovdiv. Camping sites have been built in the vicinity of resorts and natural sites. There are three categories, depending on the facilities offered. Sites with excellent facilities are the Black Sea campsites of Kavatsite and Kraimorie (south of Burgas), Europe (near Pomorie), the mountain sites Tsigov Chark (near the Batak dam), Kleptuza (Velingrad) and many others. By the end of 1980, the country had 98 campsites for a total of 49,515 campers. Motels, usually part of roadside tourist complexes, including a restaurant, are situated along the motorways and the heavily used roads. There are two categories of motel: in the first-category are the Sofia-West, Gorublyane, Sofia-East, Ihtiman, The Ninth Kilometre and Gergana motels (along the E-80 motorway); Kulata and Riltsi (E-79); and Pravets and Gerena (E-83). By the end of 1980 there were 34 motels with 1,442 beds. Whereas in 1956 there were only 14,054 beds in hotels and motels, in 1982 the total capacity of economic tourism was 103,928 beds in hotels and motels, 15,004 beds in bungalows, 152,849 beds in private houses, etc. All coastal resorts offer facilities for pastimes like yachting, scootering, pedal-floating, water-skiing and surfing, and practising underwater hunting and fishing, etc. The mountain resorts are equipped with modern ski complexes with cable-cars (gondola and double-chair), tennis courts, mini-golf links, volleyball courts, etc.

That economic tourism is an important branch of the economy is evident from the ever-growing number of foreign visitors to Bulgaria (see Table 1 above).

Most of the foreign visitors are individual tourists, but the number organized through Bulgarian tourist agencies is increasing. The travel agencies offer a number of itineraries, which include eight Bulgarian monuments of culture and interesting natural phenomena given in the Unesco list of remarkable objects of world cultural and natural interest, such as the Kazanluk Tomb, the Madara Horseman, the Rock Monastery near Ivanovo, the Boyana Church, the Rila Monastery, the town of Nessebur, the Pirin National Park and the Sreburna Reserve.

The Bulgarian Association for Tourism and Recreation was founded in 1983; it is a voluntary social and economic union, the aim of which is to organize, further and direct the creative initiative and activity of the 150 organizations and departments affiliated to it; to increase the efficiency of international tourism and to improve the conditions of rest and recreation for Bulgarian citizens. At the same time it is a state body of the Council of Ministers and in this category pursues a co-ordinated policy in the sphere of tourism and recreation. Within the system of the Association are 25 Balkantourist complexes; the Balkantourist foreign tourist corporation, the Corecom foreign trade corporation; the Touristproekt-Engineering organization for engineering and implementation; the Tourist Information and Reservation Centre, the Centre for Tourist Publicity, and the Centre for International Events economic organizations; the Touriststroi Building and

Murals in the Rila Monastery

The Rila Monastery – one of the eight unique monuments on the Unesco list

Assembly Plants; the Research and Development Institute of Tourism and Recreation; the Centre for Qualification of Managerial Staff, etc. With the constitution of the Association a unified system for management of the socio-economic processes in the sphere of tourism and rest was created, conforming to the principles and new forms of social management of the country during the stage of building a developed socialist society.

The heightened international prestige of Bulgaria is reflected in 31 inter-governmental agreements on co-operation in tourism with countries from all over the world, and in five-year agreements of the Bulgarian Association for Tourism and Recreation with the respective state bodies of tourism in the Soviet Union, the GDR, Poland, Hungary, Romania and Czechoslovakia. There is an upward trend in co-operation with the Balkan countries: Bulgaria has agreements with Greece, Romania, Turkey and Yugoslavia. In 1971 a multilateral inter-governmental agreement was concluded between the Balkan countries for co-operation in the field of tourism. The annual conferences of the governmental bodies of tourism and the tourist agencies of the Balkan countries all further the development of tourism in the Balkans. The Balkantourist foreign tourist corporation has agreements with well-known firms and agencies such as: Thomson Holidays, Global Holidays, Schools Abroad, Sunquest, Inghams and Balkanholidays from Great Britain; Touristique Union International, Neckermann and Kauhoff from the FRG; Turopa from Austria; Kuoni and Airtour from Switzerland; Transtour, Club Méditerranée, Tourisme et Travail, and Frame from France; Santourist, Lomamatka, and Matkarengas from Finland; Vingressor and Royal Tours from Sweden, etc. The specialized Bulgarian tourist bureaux are the Orbita Bureaux for Youth Tourism, co-operating with the Central Committee of the Dimitrov Young Communist League; the Shipka Motor Tourism Agency of the Central Council of the Union of Bulgarian Motorists; the Pirin Tourist Bureau of the Central Council of the Bulgarian Hikers' Association; and the Co-optourist co-operative agency attached to the CCU.

Resorts. There have been resorts in the Bulgarian lands since ancient times. The curative powers of nature, water, air and sun were known and used by primitive man, who made a cult of them and used them. Many archaeological monuments and other finds from historical times testify to the richness and widespread use of natural curative waters and their cult. As early as in primitive societies settlements appeared around mineral springs. Many of Bulgaria's resorts of today were already famous spas in the distant past.

The Thracians who inhabited the Bulgarian lands held the waters of the mineral springs in reverence. The use of spas developed from Thracian medicine, which had a natural cure basis. After colonizing the Black Sea coast (seventh–sixth centuries BC) the ancient Greeks also made the widest use of the mineral waters in the Bulgarian lands and worshipped them. This is recorded on the coins they minted. Swimming pools were built around the mineral springs and balneo-treatment featured prominently in the Hellenic concept of resorts. In the Roman era (first–seventh centuries AD) resorts, particularly spas, flourished. The Roman emperors were anxious to keep their legions in perfect physical health and hence were interested in spas. The works of Gallen (130–200 AD) and others describe the methods and values of water-cure, mud-cure and sun-cure. The Romans developed the places that had no spa waters, and in luxury and splendour they matched the resorts of the capital. Proof of this is seen in the ruins of Roman baths, reservoirs, pipes and other 2,000-year-old finds in Sofia (ancient Serdica), Kyustendil (Pautalia), Sapareva Banya (Germania), Burgas (Aqua Callide), etc. The Slavs in the Bulgarian lands also made use of local mineral waters.

Because of the ancient cult of the spas, great care was taken of the springs and the resort areas. Especially revered were the healer-god Apollo and the Hellenic gods of health Asclepius (in Roman mythology Aesculapius), Hygea and Telesphore. In honour of Asclepius, temples (asclepions) were built near the mineral springs, where care was offered to patients. In the resorts people worshipped the deities of the mineral springs – the nymphs – to whom special temples (nympheums) were dedicated. The asclepions and nympheums were the first balneo-establishments in the Bulgarian lands. They were built in places with curative springs, fresh air, and rich verdure. Ruins of Roman baths, spa installations, asclepions and nympheums are to be found in Bulgaria near the hot mineral springs of Kyustendil, Sofia, Hissar, Sapareva Banya and the Haskovo spas. Of architectural interest are the ancient baths in the region of the Burgas spas, which had ventilation for steam and light. The properties of the curative mud of Anhialo, today named Pomorie, were also known to the Thracians, the ancient Greeks and the Romans.

Bulgaria is famous for its resorts, and the number of foreigners who come to them for recreation and treatment is growing every year. Before 9 September 1944 development was unsystematic and there was a lack of qualified medical personnel. The resorts were inaccessible to the majority of people. When the Board of Health and Resorts at the Ministry of Public Health was instituted in 1947, it marked the beginning of organized balneo-therapy. The first sanatoriums, resort polyclinics, and holiday houses were opened. In the years that followed the resort facilities were considerably expanded and the number of resorts increased. The National Health Act (1973) commissioned the Ministry of Public Health to plan, organize and manage the development and protection of existing and future resorts. It laid the grounds for full use of balneological resources, established the existence of protected areas and their status, and endorsed the building of holiday houses and sporting and tourist facilities in the balneological areas. The resort network in Bulgaria is being built on the principle of socialist medical care. The country has mineral waters and mud sources of different composition, climate zones at different altitudes, and an extensive beach-line that runs along the whole Black Sea coast of Bulgaria. The medico-biological and hydrogeological aspects of the country's balneological potential, the assessment of this potentiality and its application for preventing and curing illness are being studied in depth. The balneological resources of the country include mineral waters of different composition and energy content (hydromineral resources) and several types of curative mud deposits (peloid resources).

The biological characteristics of Bulgaria are most favourable for climate prophylaxis and climate-therapy, which are practised along the Black Sea coast, in the Rhodope, Rila, Sredna Gora, and Vitosha mountains, and in the Balkan Range. The simultaneous combination of various elements of the meteorological environment (air temperature and humidity, wind velocity and thermal intensity of solar radiation, ultraviolet radiation, etc.) is also beneficial to human health.

The following groups of diseases are successfully treated at Bulgarian resorts: ailments of the respiratory system (inflammatory and allergic), ailments of the locomotor system (inflammatory, degenerative, traumatic, orthopaedic), organic and functional diseases of the central and peripheral nervous system, diseases of the gastro-intestinal system, gynaecological ailments, endocrinal and metabolic disturbances, urologic diseases and skin diseases.

The Bulgarian resorts of national and international significance are established and endorsed by decree of the Council of Ministers. They meet the needs of balneo- and resort treatment of patients from the whole of Bulgaria and from abroad. The resorts of local significance are named by the Ministry of Public Health and serve the population of certain regions of the country. According to their curative effect Bulgaria's resorts are subdivided into balneological, peloidological, mountain-climatic and seaside-climatic.

The following resorts are of national and international significance:
Balneological – Bankya; Banevo, in the Burgas district (Burgas Spas); Banya, in the Pazardjik district (Panagyurishte Spas); Banya, in the Plovdiv district (Karlovo Spas); Velingrad; Vurshets, Georgi Dimitrov, in the Sofia district (until 1950 Kostenets Spas); Gorna Banya (a suburb of Sofia); Kyustendil; Mericheri, in the Haskovo district; Momin Prohod (a quarter of Kostenets); Narechen Spas, in the Smolyan district; Ovcha Kupel (a suburb of Sofia); Pavel Banya; Sandanski; Sliven Spas; Stara Zagora Spas; Strelcha; Haskovo Spas; Hissar; Berkovitsa.
Peloidological – Balchiska Tuzla, in the Tolbuhin district; Banevo, in the Burgas district (Burgas Spas); Banya, in the Plovdiv district (Karlovo Spas); Pomorie.
Mountain-climatic – Borovets, in the Sofia district; Pamporovo (Vassil Kolarov) in the Smolyan district; Govedartsi in the Sofia district; Kotel, Yundola in the Pazardjik district.
Seaside-climatic – Ahtopol; Balchik (including the Fishfisha locality); Burgas; Varna (from Golden Sands to Galata); Michurin; Nessebur; Obzor (Burgas district); Pomorie; Primorsko (with its quarter of Kiten); Sozopol; Bal chishka Tuzla in the Tolbuhin district. Resort complexes have also been built along the Black Sea coast.

Mountain-climatic resorts. In Bulgaria most mountain resorts are in the medium-altitude mountain belt. The climatic resources of Bulgaria's mountain resorts are rich and versatile. These regions maintain a thick and stable snow cover for five to eight months a year. Spring is cooler than autumn. Ultraviolet radiation is high (particularly in winter). The biological activity of the sun's rays is some 10 per cent higher than that in the lowland regions. The ionization of the air is also high, reaching – at an altitude 1,500–2,000 m – 2,000–3,000 light positive and negative ions per 1 cu. m of air, in the close vicinity of the mountain rivers – 5,000–10,000 ions per cu. cm, and around the mountain waterfalls – an average of 20,000–50,000 ions per cu. cm (in some cases up to 100,000 ions per cu. cm). The rich afforestation and grassing of mountain resort areas also has its positive effect.

In most cases the maximum favourable effect from the mountain-climate cure in Bulgaria is to be had in resorts at an altitude of 1,200–2,000 m, because the bioclimatic factors there are in an optimum ratio. Bulgaria's mountain resorts offer a broad range of indications for climatic cure: respiratory diseases (chronic bronchitis, asthma, sinusitis, silicosis, etc.); cardiovascular diseases (hypertonic diseases of I and II degree, arteriosclerosis, etc.); nervous disorders (neuroses, cerebrasthenia, etc.); blood diseases, endocrinal and metabolic upsets (light forms of diabetic myxodema, hypothyreois, etc.); chronic skin diseases (eczemas, urticarias, etc.).

With every passing year Bulgaria's mountain resorts become increasingly established as skiing centres of world renown. The modern installations and ski-runs have prompted the International Skiing Federation to include them in its calendars. Competitions are held in Borovets in the Alpine events for the world cup; in Pamporovo laps are run for the European Cup; Vitosha was the centre of Universiad 1983. This enabled Sofia to advance its candidacy for hosting the 1992 Olympic Winter Games. The skiing season begins in December and ends in May. The snow-cover during this period is between 100 and 250 cm. Among the most important reasons for the growing prestige of Bulgaria's mountain resorts, which makes them the ideal place for skiing holidays, are their skiing schools. Teaching methods and practice there have been assessed by the world's greatest experts as leading the field. The annual polls taken by the Research Institute on Recreation and Tourism indicate that 70 per cent of Bulgaria's winter visitors come to its mountain resorts for the skiing lessons offered.

Bulgaria has some 50 mountain resorts.

Mount Vitosha rises to the south of the capital Sofia, half an hour's drive away from the city centre. It is often called 'the lungs of Sofia'. The mountain's ancient Slav name was Skopor, meaning 'steep mountain', and indeed, it looks like a huge steeply sloped mound with rounded contours. Lengthwise it stretches 23 km from north to south, its width being 18 km. The highest peak is Cherni Vruh (2,290 m). There are 27 other peaks rising above an altitude of 1,500 m. Typical of Vitosha are the impressive stone rivers of striking beauty. Winter at higher altitudes lasts about seven months and the snow-cover, which is heaviest from the middle of February to the end of March, can be as thick as 3.5 m. At Mt. Cherni Vruh there is an average of 1,884 hours of sunshine annually. The mean solar radiation is 150.1 kcal. Part of Vitosha is declared a national park (an area of 12,100 hectares.) This is the north-eastern section, which is closest to Sofia. The slopes of Vitosha are overgrown with thick forests, which are the habitat of a great variety of wild life. Because of its proximity to Sofia, the mountain is visited by thousands of hikers who can avail themselves of the services of scores of chalets, holiday houses, high-mountain shelters, hotels, restaurants and playgrounds. There is a ringroad passing through the foothills and two other radial roads connect Sofia with the localities of *Zlatni Mostové* (Golden Bridges) – with a branch to the Kopitoto (Hoof) Hotel, and Shtastlivetsa. Vitosha is criss-crossed by 300 km of foothills.

so that both beginners and expert skiers can be satisfied. Three chair-lifts and a succession of ski-tows are available, reaching as far as just below Cherni Vruh. For those who do not bring their own outfit there is a ski- and boot-lending service. Cable cars and chair-lifts operate between several high spots in the mountain and localities in the outskirts of Sofia.

Borovets is one of the oldest and biggest Bulgarian mountain resorts. It is situated on the Rila mountain at an altitude of 1,346 m, amidst vast, old, coniferous woods. The river Bistritsa, a tributary of the Iskur, flows through the resort. The mean annual temperature at Borovets is 5.4°C (the average temperature for January is 4.8°C, for July 15.2°C); the mean annual precipitation is 929 mm. The snow-cover lasts from the middle of November to the middle of April, on average (a great bio-climatic advantage because it reflects the ultraviolet rays), with a maximum thickness of 126 cm (in March). The place is aired by winds of a markedly mountain-valley character, which give a bracing atmosphere and are of great climatic and therapeutic value. The construction of Borovets began in 1896.

There are more than 200 villas, with a total of more than 2,000 beds in Borovets. Among the latest additions are several new hotels and restaurants, two villa settlements of comfortably appointed wooden bungalows, a resort polyclinic, and a campsite. The former royal

A drag-lift at the Borovets mountain resort

New hotels at Pamporovo

The Aleko tourist complex is to be found at an altitude of 1,800 m on the eastern slope, below Cherni Vruh. It includes three hotels and more than a dozen chalets. The Shtastlivetsa Hotel is within walking distance of numerous excellent ski slopes. The skiing season starts usually late in November and lasts till the middle or even the end of May. The ski runs are of various lengths and steepnesses,

palaces – Bistritsa (1,520 m above sea-level) and Sitnyakovo (1,740 m) – have been turned into holiday homes. It is connected with Sofia (71 km away) and with Kostenets (28 km away) by regular bus services. Borovets is a starting point for excursions to Mussala Peak and to other places in the Rila mountains. A modern gondola lift has been built up to the Yastrebets Peak (2,350 m high). It spans a distance of 4.8 km, and has a capacity of 1,200 persons per hour. There are also several ski lifts starting from the very doorstep of the hotels. Borovets offers fine conditions for skiing. It is also equipped with ice-hockey rinks. National and international competitions are held there.

The resort of *Govedartsi* lies on the north-western slopes of the Rila mountains at a height of 1,200 m above sea level. The locality through which the River Cherni Iskur flows is surrounded by coniferous forests. The resort is linked with Samokov (16 km away) by an asphalted road. The mean annual temperature is 6°C, the mean temperature for January 4°C and for July – 16°C. Annual precipitations are estimated at 950 mm, while the annual average cloud is about 5.5. The resort is situated above the level of the general inversion overcast, which accounts for the maximum number of sunny days during the year.

The area is shielded from aerosol pollution from industrial or other sources. In January the snow cover is about 60 cm. The ionization of the air is of the order of 1,200–1,300 ions per cu. cm (slightly positive and negative).

Treatment at the resort influences favourably patients with hypertonic diseases – first and second degree, early arterio-sclerosis – without grave afflictions of the heart muscle, light forms of diabetes, disorders in the function of the thyroid gland, functional disorders of the nervous system and conditions after mental and physical fatigue, non-specific chronic diseases of the lungs and respiratory tracts, silicosis, chronic and allergic skin diseases, bronchial asthma and recuperation after serious illnesses. The resort of Govedartsi is used as a sports and tourist centre, especially in the winter.

Pamporovo (Vassil Kolarov) is Bulgaria's biggest mountain resort and is situated at an altitude of 1,620 m in the Rhodopes, the legendary mountain where the mythical singer Orpheus is said to have lived. The town of Chepelare is 10 km away. The place is a hilly plateau, with wide meadows and ancient coniferous forests. The peaks of Murgavets and Snezhanka tower above the southern parts. The mean temperature for January is −8.0°C, for July 16.0°C, and the mean annual temperature is about 8.0°C. The annual precipitation is 900 mm. Winter is mild and relatively long lasting (80–120 days, with a thick snow blanket of 20–40 cm). The mean cloud is about 0.5. The place is protected naturally from pollution by industrial and other pollutants. The biologically active ultra-violet radiation is on average 10 per cent higher than that in the lower areas. Pamporovo is situated above the level of the general inversion overcast and is well aired. The ionization of the air is about 1,500–2,000 ions per cu. cm (slightly positive and negative). Pamporovo has many holiday-houses, restaurants and bars, and seven hotels, among which the Perelik is first class. It boasts excellent winter sports facilities, tracks for alpine and nordic skiing events, and ski jumps for 50 m, 70 m, and 90 m. It is an international ski centre, where competitions are held each year for the Pamporovo Cup, and is equally attractive in winter and summer. Conditions for skiing are particularly good: the slopes and *pistes* starting from the peak of Snezhanka (1,937 m high) will give pleasure to any skier. Two chair-lifts take tourists up to the Studenets Chalet, and the third, to Snezhanka. Nine ski-tows of various lengths are also available. A 132 m high television tower has been erected on Snezhanka Peak, with a coffee-shop at 91 metres, from which a magnificent view of the whole countryside opens up to the spectator.

Pamporovo is visited by a large number of Bulgarian and foreign tourists. It offers fine conditions for sun and air curative procedures and for curative pedestrian tourism. No less favourable is the influence of the natural factors in the treatment of hypertension, early arteriosclerosis, some endocrinological metabolic disturbances, neuroses, chronic and allergic skin diseases, states of nervous and physical fatigue, non-tubercular diseases of the lungs and respiratory tracts, etc. The resort is connected with Plovdiv (84 km), Assenovgrad (64 km), Smolyan (13 km) and other towns by bus services.

The resort of *Yundola* is situated in a mountain saddle between the Western Rhodopes and Eastern Rila at a height of 1,380 m. Its wide meadows are fringed by coniferous woods. It is situated 16 km from Velingrad and 26 km from Belovo, and is shielded from strong winds. The mean annual temperature is 5.5°C; the mean July temperature is 15.5°C and the mean January temperature is 4°C. The annual precipitation is 715 mm. The snow cover is short-lasting. The average annual number of days with snow-cover is about 80, and its medium thickness in January is about 25 cm. Annual average cloud is 0.5.

The resort is situated above the level of the general cloud inversion, which accounts for its being one of the places with the maximum annual number of sunny days. The intensity of the biologically active ultraviolet radiation is, on average, 8 per cent higher than in the plains. The ionization of the air is about 1,200–1,500 ions per cu. cm (slightly positive and negative).

The resort of Yundola is suitable for patients with hypertension – first and second degree, without grave afflictions of the heart; early arterio-sclerosis, some endocrinal-metabolic diseases (light forms of diabetes, impaired function of the thyroid gland, obesity); chronic skin and allergic diseases (orticaria, psoriasis, eczema); the state of nervous and physical fatigue; functional nervous disorders; non-tubercular chronic diseases of the lungs and respiratory tracts (bronchitis, pharyngitis, rhynitis, trachitis), silicosis; conditions of recuperation after grave illnesses; incipient emphysema of the lungs and bronchial asthma. During the winter, holidaymakers can use the fine ski tracks and ski-tow.

Balneological Resorts. Bulgaria is particularly famous for its great variety of mineral and therapeutic muds. Every type of mineral water and therapeutic mud can be found on the country's territory, in spite of its relatively small size. These valuable natural resources are now safeguarded and put into use in modern balneological resorts, some of which are already well-known. The balneo-sanatoriums and polyclinics at these resorts specialize in the treatment of many chronic diseases. Balneological resorts possess first class therapeutic facilities, in which natural factors are made use of in the light of modern requirements.

Mineral Waters. In Bulgaria over 600 natural springs (about 90 per cent of them in South-western Bulgaria) and over 370 artificial springs (boreholes) are to be found (1982). The rate of discharge varies from a few litres to many decilitres per second. About 40 per cent of the natural springs have a rate of discharge of between 1 and 10 litres per second, and many have a discharge of between 10 and 30 litres per second. The highest rates are to be found in the springs of the Rila-Rhodopes Massif (the town of Velingrad, with about 110 litres per second, and the village of Banya, near Blagoevgrad – 76 litres per second). Some artificial springs along the Black Sea coast near Varna discharge 60–80 litres per second. The total rate of discharge of all the natural and artificial springs in the country is about 5,500 litres per second, while the total volume utilized is 3,071 litres per second. Springs with a rate of discharge of 5 litres per second are most numerous, while those with a rate of discharge of over 100 litres per second are the least numerous. 80 per cent of known reserves (4,000 litres per second) are in Southern Bulgaria. Bulgarian mineral waters usually have a low content of dissolved minerals (below 1 g per litre). The highest mineral content is at the Solenata Voda spring (280 g per litre) near the Vassil Kolarov Station, near Varna and in the lower valley of the Kamchiya River. The waters with the highest mineral content are to be found in the deeper sedimentary strata: they are usually rich in iodine, bromine, boron and other minerals. The temperature of the mineral waters varies between 8°C and 101.4°C. Large quantities of hyperthermic mineral water are contained in the magmatic and metamorphic rocks in the Rila-Rhodopes Massif, as well as in the sedimentary strata in Northern Bulgaria (at a depth of over 3 km). The water with the highest temperature is that of Sapareva Banya (38.0–101.4°C), the Chepino River Basin (43–98°C), the valley of the River Struma (91°C) and Erma Reka (over 80°C). With respect to hardness, one third of the mineral waters in Bulgaria are of 1° (the upper limit being 25° on the German scale). They also differ in acidity: springs with alkaline waters (pH 7.2–8.5) are most numerous; the springs near Breznik and the village of Gorno Uino, near Kyustendil, have the highest acidity.

Mineral waters can be grouped into four hydromineral categories, according to origin and content: acratothermic (thermal waters of low mineral content, with gaseous nitrogen); thermal and cold carbonic acid nitrogen waters (with higher mineral content); chloride waters (with a mineral content of predominantly marine origin); and sodium-sulphate waters (with a mineral content of predominantly lacustrine origin). The waters of the latter two groups are mainly saline solutions from old marine and continental basins; water-bearing structures that are near to the surface have a low mineral content, yielding acratothermic or ordinary waters.

Acratotherms are the most widely-spread (they cover an area of about 450,000 sq. km; mineral content varies

Restaurant by the Kleptuza karst spring in Velingrad spa

between 0.1 and 1.5 g per litre and temperature varies between 20 and 100°C). They can be siliceous (formed in magmatic and metamorphic silicate rocks), carbonate-karstic (in limestone rocks) and territic (in terrigenous loamy and clastic rocks). The siliceous acratotherms cover an area of 19,000–20,000 sq. km and are found in the Rila-Rhodopes Massif, in the Ossogovo mountains, in the Srednogorie and the Western part of the Balkan mountains (in the Chiprovtsi and Berkovitsa area). With the exception of the Stara Zagora mineral baths, those near the village of Pancharevo (outside Sofia), and Ovcha Kupel (a district of Sofia), the balneological resorts and the mineral baths in Bulgaria are siliceous acratotherms. The total volume per second of acratothermic waters is estimated to be 4,000 litres per second at temperatures varying between 25 and 100°C (the average being 50°C); mineral content is usually between 0.1 and 1 g per litre; cation composition – sodium; anion – bicarbonate, carbonate, sulphate; the proportion of anions changes with temperature. With the exception of the mineral waters on the slopes of Lyulin, near Sofia, all Bulgarian acratothermic springs contain fluorine and metasilicic acid. The high temperature mineral waters at Kyustendil, Sapareva Banya, Blagoevgrad and Simitli contain sulphide-hydrosulphide, concentration – 15 mg per litre. Half the springs are radioactive (mainly those in the valley of Dolna Banya: at Momin Prohod, near Kostenets; at Pchelin, near Sofia; and Dolna Banya itself), and also those at Narechenski Bani in the Smolyan Area, Hissaria, Strelcha and Pavel Banya. Some of the springs with a high rate of discharge – such as those in Velingad and the village of Banya, near Blagoevgrad – have a high level of radioactivity and

Table 2: Curative Properties of Mineral Waters

Mineral springs (towns, villages and hydros)	Cardio-vascular	Of the gastro-intestinal tract, liver and bladder	Of the kidneys	Of the skin	Gynaecological	Anaemia of the lymphatic tract	Of bones and joints	Of the nervous system	Of the respiratory tract	Of the glands with internal secretion	Chemical composition and temperature characteristics
Bankya[1]		○			○		○	○			Hydrocarbonate-sulphate-sodium, radioactive (1.5 emans); hypothermal
Banya, a village in the Blagoevgrad district (Guliina Banya, Razlog hydro)	○	○					○	○			Hydrocarbonate-sulphate-sodium, fluorine (up to 9 mg per litre); hyperthermal
Banya, a village in the Pazardjik district (Panagyurishte hydro)[1]	○	○					○	○			Sulphate-hydrocarbonate-sodium, fluorine (up to 9 mg per litre); hyperthermal and hypothermal
Banya, a village in the Plovdiv district (Karlovo hydro)[1]	○	○	○		○		○	○		○	Sulphate-hydrocarbonate-sodium, fluorine (up to 9 mg per litre), strongly alkaline; hyperthermal
Banya, a village in the Sliven district (Korten hydro, Nova Zagora hydro)	○	○			○		○	○			Nitrogen, sulphate-hydrocarbonate-sodium fluorine (up to 16 mg per litre), silicon weakly alkaline; hyperthermal
Beden hydro (at the village of Beden, Smolyan district)					○		○				Carboacid, hydrocarbonate-sulphate-sodium, silicon (up to 1.87 mg per litre of metasilicic acid); hyperthermal
Blagoevgrad		○	○		○		○	○			Nitrogen, hydrosulphide, sulphate-hydrocarbonate-sodium, fluorine (up to 10 mg per litre) and silicon, alkaline; hyperthermal
Breznik			○			○	○				Sulphate, calcium-magnesium, iron (up to 430 mg per litre) and aluminium, acid; cold
Burgas hydro (near the village of Banevo, Burgas district)[1]		○	○	○	○		○			○	Nitrogen, chloride-carbonate-sodium, fluorine (8–10 mg per litre), silicon and boron (up to 7 mg per litre metaboric acid), strongly alkaline; hyperthermal
Dobrinishte, a village in the Blagoevgrad district		○	○		○		○	○			Hydrocarbonate-sulphate-sodium, fluorine, radioactive (35–61 emans); hyperthermal
Dolna Banya		○	○		○		○	○			Sulphate-sodium, fluorine, radioactive (0.5 emans on average); hyperthermal
Georgi Dimitrov (until 1950 Kostenets hydro)[1]	○	○					○	○	○	○	Hydrocarbonate-sulphate-sodium, fluorine, radioactive (35 emans); hyperthermal and hypothermal
Gorna Banya, a quarter in Sofia[1]		○	○				○	○	○	○	Slightly mineralized (hydrocarbonate-sulphate sodium); hyperthermal and hypothermal
Haskovo hydro (near the village of Mineralni Bani, Haskovo district)[1]	○	○	○	○	○		○	○			Nitrogen, sulphate-sodium-calcium fluorine (4 mg per litre), radioactive (32–54 emans); hyperthermal
Hissarya (Momina Banya springs)[1]		○	○								Hydrocarbonate-sulphate-sodium, radioactive and fluorine; hyperthermal
Knyazhevo, a quarter in Sofia		○			○		○				Hydrocarbonate-sulphate-sodium, alkaline; hypothermal
Kyustendil[1]				○			○	○			Hydrocarbonate-sulphate-sodium, sulphide, fluorine and silicon; hyperthermal
Marikostino, a village in the Blagoevgrad district				○			○	○			Hydrocarbonate-sulphate-sodium, fluorine; hyperthermal
Merichleri, a village in the Haskovo district[1]		○	○							○	
Main water-catchment											Sulphate-hydrocarbonate-chloride sodium; hyperthermal
Bore-hole 2											Sulphate-hydrocarbonate-chloride-sodium, arsenic; homeothermal
Mihalkovo, a village in the Smolyan district	○	○	○							○	Carboacid, hydrocarbonate-sodium, calcium and silicon, slightly alkaline (1.96 g per litre of carbon dioxide); cold and hypothermal

Table 2: Curative Properties of Mineral Waters (continued)

Mineral springs (towns, villages and hydros)	Cardio-vascular	Of the gastro-intestinal tract, liver and bladder	Of the kidneys	Of the skin	Gynaecological	Anaemia of the lymphatic tract	Of bones and joints	Of the nervous system	Of the respiratory tract	Of the glands with internal secretion	Chemical composition and temperature characteristics
Momin Prohod, a quarter in Kostenets[1]					O		O	O	O	O	Radioactive (562 emans), silicon and fluorine, slightly alkaline; hot
Narechenski Bani; a village in the Smolyan district[1]	O						O	O			
Bore-hole 2 and Banski Izvor											Sulphate-hydrocarbonate-sodium, radioactive, fluorine; hypothermal
Bore-hole 4 (Solenoto Izvorche)											Sulphate-hydrocarbonate-sodium, strongly radioactive (407 emans), silicon and fluorine; hypothermal
Ovcha Kupel, a quarter in Sofia		O					O	O			Sulphate-hydrocarbonate-calcium and sodium neutral; hypothermal
Pancharevo, a village near Sofia		O	O		O		O	O			Hydrocarbonate-calcium and magnesium, neutral; hyperthermal
Pavel Banya[1]							O	O	O		Hydrocarbonate-sodium; fluorine, silicon
Pchelin hydro (near the village of Pchelin, Sofia district)					O		O	O	O		Low radon, silicon, fluorine, alkaline; hyperthermal
Sandanski (Polenichka Banya)[1]							O		O		Silicon, slightly mineralized (hydrocarbonate-sulphate-sodium), slightly alkaline; hyperthermal
Sapareva Banya					O	O		O	O		
Main water-catchment											Sulphide, silicon and fluorine, slightly mineralized (hydrocarbonate-sulphate-sodium); hyperthermal
Bore-hole (Pariloto spring)											Hydrocarbonate-sulphate-sodium-calcium; hyperthermal
Sliven hydro[1] (Djinov hydro near the village of Zlati Voivoda, Sliven district)	O	O			O		O		O		Hydrocarbonate-sulphate-sodium, calcium and fluorine, carbon dioxide (316 mg per litre), hyperthermal
Sofia spa (in the city centre)	O	O	O	O			O	O			Slightly mineralized (hydrocarbonate-sulphate-sodium), alkaline; hyperthermal
Stara Zagora hydro (near the village of Sulitsa, Stara Zagora district)[1]	O	O			O		O	O			Slightly mineralized (hydrocarbonate-calcium-magnesium), neutral; hyperthermal
Strelcha[1]					O	O		O	O		
Spa-water catchment											Hydrocarbonate-sulphate-sodium, slightly radioactive; hyperthermal
Bancheto (open-air pool)											Hydrocarbonate-sulphate-sodium, slightly radioactive, silicon; hypothermal
Velingrad[1]	O				O	O		O	O		O
Chepino quarter											Hydrocarbonate- (and carbonate-) sulphate-sodium, fluorine (up to 6 mg per litre), radioactive (up to 80 emans); hyperthermal
Kamenitsa quarter											Sulphate-hydrocarbonate-sodium, fluorine and silicon (up to 137 g per litre silicic acid); hyperthermal
Vurshets[1]	O	O					O	O		O	Nitrogen, hydrocarbonate-sulphate-sodium, ultrafresh (mineralization at 0.18–20 g per litre); hyperthermal
Yagoda, a village in the Stara Zagora district					O	O		O	O		Silicon, fluorine, slightly mineralized (hydrocarbonate-sulphate-sodium); hyperthermal

[1] A hydro of national significance.

comparatively low radon content. Tungsten and germanium are usually present as trace elements. The majority of acratotherms have pronounced acid characteristics, most noticeably in hot and sulphide springs; acidity is between 7.8 and 10. The carbonate-karstic acratotherms cover an area of approximately 18,000 sq. km and are to be found in the Upper Jurassic and Lower Cretaceous layers of the artesian slopes of the Ludogorie, along the Black Sea coast; in the Triassic and Jurassic layers in the Zlatitsa-Teteven part of the Balkan mountains; the middle section of the Balkan mountains; Kraishte; the Sofia valley; the Srednogorie; and some marble strata in the Rhodopes. Total flow is estimated at 7,000–8,000 litres per second, if we include the springs of warm karstic waters, which are warmed by a deep tributary of siliceous acrotherms. Temperature ranges between 20°C and 60°C. The karstic acratotherms contain carbonates, calcium and magnesium with a mineral content of 0.3–0.8 g per litre. Some springs contain sulphates, chlorides and sulphide; mineral waters from terrigenous water-bearing layers also contain sodium. Examples include the springs near Pancharevo, outside Sofia, and the Stara Zagora Spas.

The Velingrad spa in the Western Rhodopes

Carbonic acid waters are to be found in small deposits in the Rhodopes, the Srednogorie (the village of Mihalkovo, in the neighbourhood of the village of Beden – Smolyan district, Devin, the Struma river valley, the southern part of the Sliven Plain and in the Burgas lowland. They cover a total surface of approximately 1,000 sq. km. The natural resources amount to 200–250 litres per second. *Saline chloride waters and lyes* are to be found in the hilly Danubian Plain, in many of the foothills of the Balkan Range, in the eastern part of the Balkan Range and in the eastern part of the Burgas syncline. Altogether they occupy a surface of approximately 37,00 sq. km (excluding the Ludogorie and Dobrudja). The mineral waters are thermal and slightly overheated lyes with mineralization of up to 350 g per litre. In the Post-Triassic stages thermal salt waters with mineralization of up to 35–40 g per litre predominate. They include waters rich in iodine, bromine (in the Kamchia Basin), and hydrogen sulphide. The saline chloride waters and lyes are valuable mineral waters; because of their high salt content they are highly suited to industrial application. The *sodium-sulphate mineralized waters* are connected with old deposits of lacustrine-continental waters in the neogene matrices in the Sofia Plain, the Basin of Eastern Maritsa (in the region of the village of Merichleri, Haskovo district, and the town of Harmanli). Their surface is limited, their composition is hydrocarbonate – sulphate-sodium (sodium-sulphate) and their mineralization is not greater than 8–10 g per litre. In the periphery of the depressions they are contiguous to the acratotherms. The provisions of the Bulgarian State Standard specify which mineral waters are good for drinking. They are mainly used in balneological treatment and preventive medicine. Mineral waters are also used in hot-houses in growing vegetables and flowers, in the production of soft drinks and aerated waters, and in the production of extracts for cosmetic and pharmaceutical preparations.

Pelotherapeutic Resorts. In Bulgaria therapeutic mud occurs naturally in seaside firth lakes near mineral springs in the forms of sediments consisting of organic and inorganic compounds, and gases (mainly hydrogen-sulphide). In Bulgaria, the most widespread pelotherapeutic deposits which are of greatest practical importance are found in the Black Sea firths and lagoons, such as the Shabla Lake, Ptichi Zaliv, the Varna Lake, the Beloslav Lake, the Pomorie Lake and the Atanassov Lake. Therapeutic mud is also found near the mineral springs in the village of Marikostinovo in the Blagoevgrad district, and in the village of Banya in the Plovdiv district, amongst others. The Burgas Mineral Baths, the village of Banya in the Plovdiv district (Karlovo Mineral Baths), Balchishka Tuzia, Pomorie and Shabla have been officially declared and advertised as pelotherapeutic resorts. In many resort sanatoriums and polyclinics mud treatment is applied, using the therapeutic mud from its natural deposits. Thanks to the existence of consulting rooms equipped with modern apparatus for various treatments, indoor swimming-pools and expert physicians, of cosy and comfortable hotels and recreational facilities, they rank as first-class resorts suitable for balneal treatment in all seasons.

Bankya is a town in the district of Greater Sofia. It lies 17 km west of Sofia and has a population of 8,339 (1982). It has been a national balneological and health resort since 1970. It lies in a picturesque setting at the foot of the Lyulin mountain, near the Banska river and its tributary, the Gradomanska, at an altitude of 630–750 m. The climate is temperate continental; the influence of the mountain is slightly felt – the summer is long and cool, the winter is relatively mild, with few misty days. The mineral waters spring from normal fault joints along

Tourism and Resorts

The Vurshets spa in the Western Stara Planina

which the Sofia Plain has sunk. Their temperature is 34–38°C and their combined flow is 50 litres per second. In content they are hydrocarbonate sulphate, sodium, with low mineralization (0.26–0.28 g per litre), with low radioactivity – 1.5 emans; they contain fluorine 0.6 mg per litre, and metasilicic acid – 53 mg per litre. They are recommended for cardio-vascular complaints, diseases of the locomotor system, functional disturbances of the nervous system, gynaecological disorders, hypertension, etc. There are a resort polyclinic, a balneological physiotherapy clinic, a rehabilitation hospital for children with cerebral injuries, a sanatorium for children suffering from rheumatism, a balneo sanatorium, about 30 holiday homes, hotels and a number of villas. There is a large outdoor recreation centre with swimming-pools and sunbathing area.

Velingrad is a town in the Pazardjik district, 50 km south-west of Pazardjik, with a population of 77,830 (1982 census). It has been a national balneological and health resort since 1950.

Velingrad is in the high Chepinska Valley, on the banks of the river Chepinska and its tributaries Elenka and Lukovitsa, at an altitude of 740 m. It has a transitional continental climate with Mediterranean influence, a mild winter and cool summer, and precipitation is lower than the country's average. Snow stays for about 50 days. There are foggy days. Velingrad is one of the least windy places in Bulgaria. The karst spring Kleptuza in Velingrad fills two scenic lakes and is one of the largest hot mineral water deposits in Bulgaria. The total number of springs is about 70 and they rise in the Western Rhodope Massif. The rate of discharge is about 110 litres per second and the temperature ranges from 22° to 78°C. The drilling increases the flow up to 130–140 litres per second.

All waters contain nitrogen, sodium, fluorine, and acratotherms with a variable anion composition and mineralization and a diversity of rare and trace elements. The waters are used in balneological sanatoriums, rest homes, an outdoor swimming pool and greenhouses. They cure diseases of the locomotory and peripheral nervous system, women's diseases, skin, kidney-urological, gastrointestinal and liver complaints.

Vurshets is a town in the Mihailovgrad district, 32 km south-east of Mihailovgrad, with a population of 7,146 (1982 census) and has been a national balneological and health resort since 1970. It is situated in the Vurshets valley and at the foot of Koznitsa mountain in the Western Balkan Range, along the river Botunya, at an altitude of 398 m. It has a moderate continental climate with mountainous influence, a mild winter and cool summer. Snow stays about 60 days. The hot mineral waters were of major importance for the origin and development of the town. They spring in tectonic crevices of granodiorites and consist of three springs and four bore-holes on the western bank of Botunya. The temperature ranges from 20–38°C; the rate of discharge is 20–25 litres per second. The waters are alkaline and ultra-fresh (mineralization 0.18 –0.20 g per litre). They contain nitrogen, hydrocarbonate sulphates and sodium. The mineral waters are used in the treatment of diseases of the cardio-vascular system, the locomotory system, the peripheral nervous system, functional disorders of the central nervous system, and other related illnesses. In the resort area there is a large park with a pine forest, a resort polyclinic, a balneological treatment centre, specialized balneological sanatoriums, rest houses and villas.

Narechenski Bani is a village in the Smolyan district, 65 km north of Smolyan, with a population of 1,262 (1982 census), and was declared a national balneological and health resort in 1950. It is situated in the valley of the Chepelarska river between the Chernatitsa ridge and the Radyuva mountain in the Western Rhodopes, at an altitude of 620 m. The climate is transitional continental, mild and suitable for treatments. The summers are warm but not hot; the winters are mild, moderately cold, with plenty of sunny days. Snow stays for a short time.

The Spartacus monument at the Sandanski spa

Precipitation and cloud cover are moderate; the mountain air is bracing because of its relatively high ionization.

Narechenski Bani owes its popularity to its radioactive mineral waters with a temperature of 20.5 to 31.2°C, and total flow of 6 litres per second. There are six natural springs and about 25 artificial springs. The waters contain nitrogen, radon, sulphates, hydrocarbonates, sodium, fluorine, silicon and the mineral content is between 1.46 and 1.67 g per litre. Solenoto Izvorche is one of the most radioactive springs in Bulgaria – 1,100 to 1,300 emans. Narechenski Bani is recommended for the treatment of functional disorders of the nervous system, internal diseases (hypertonicity, arterio-sclerosis), metabolic disorders and cardio-vascular complains. A balneological centre, holiday homes, pools and hotels have been built there.

Merichleri is a town in the Haskovo district, 32 km north of Haskovo, with a population of 2,830 (1982 census), and was declared a national balneological resort in 1971. It lies on the river Merichleri, a tributary to the Maritsa, in the Upper Thracian Plain, at an altitude 150 m. The

Fortress walls at the Hissarya spa (1)

climate is transitional continental. Some 3 km south-east of Merichleri there are mineral waters obtained from two bore-holes. Their temperature is 31–42°C and their rate of discharge about 1.5 litres per second. The waters contain sulphates, hydrocarbonates, chlorides, sodium, fluorine (fluorine 5 mg per litre) boron (metaboric acid 22 mg per litre), arsenic and a high content of dissolved minerals (4.8–6.4 g per litre). The Merichleri waters are similar to those of Karlovy Vary, Czechoslavakia. They won a gold medal at an international exhibition in London in 1907. Merichleri is recommended for the treatment of liver and gall complaints, gastro-intestinal and skin diseases, and metabolic malfunctions. The waters are used to obtain mineral salts for the pharmaceutical industry, for the Meri tooth paste, and for other products.

Sandanski is a town in the Blagoevgrad district, 65 km south of Blagoevgrad, with a population of 22,248 (1982

Fortress walls at the Hissarya spa (2)

census), and has been a national balneological resort since 1967. It is situated in the valley of the river Sandanska Bistritsa, at the foot of the Pirin mountains, at an altitude of 224 m. The climate is moderate continental with a strong Mediterranean influence. The winters are mild; snow stays for a short time; spring comes early; summers are dry and hot, and autumns are warm and long. Precipitation is low. The mineral waters contain silicon and have a low content of dissolved minerals (hydrocarbonates, sulphates, sodium), and are hyperthermal with a weak alkaline reaction. The climate of the town is one of the best in Europe for the treatment of bronchial asthma. The mineral waters benefit kidney complaints, gastritis, colitis, and inflammations of the skin. The construction of the new, large and attractive balneological hotel Sandanski has helped to increase the town's position as one of the most popular balneological resorts. The magnificent hotel was planned by a team of Bulgarian architects and built jointly with the Austrian firm Universale.

Hissarya is a town in the Plovdiv district, 42 km north of Plovdiv, with a population of 8,787 (1982 census), and has been a balneological resort since 1950. Situated in the southern folds of Sushtinska Sredna Gora, at an altitude of 365 m, it enjoys a moderate continental climate: the summers are hot; the winters are temperate and mild; and autumns and springs are warm. Precipitation is moderate; snow-cover is thin and melts early. The basic health benefit is the mineral water gushing from 22 springs in the town and its vicinity. The rate of discharge of the springs is 4,000 litres per minute. The water is colourless, odourless, and tastes pleasant. The mineral content is low; the components are hydrocarbonates, sulphates, sodium and fluorines, and there are radioactive elements. Hissarya is a balneological centre for the treatment of urological, gall, liver, and gastro-intestinal diseases. The first-class hotel for convalescents, Augusta, has rooms with up-to-date equipment and qualified specialists.

The Seaside Health Resorts in Bulgaria fall into three zones – the Tolbuhin (Dobrudja) Black Sea Coast includes the villages of Krapets, Durankulak and Kranevo, the towns of Kavarna and Balchik, and the resorts of Russalka, Balchishka Tuzla and Albena; the Varna Black Sea coast comprises the resorts of Zlatni Pyassatsi, Druzhba and Kamchia, the town of Varna (including the Galata suburb), and the village of Byala; the Burgas Black Sea coast takes in the villages of Obzor, Vlas, Ravda, Chernomorets, Lozenets and Sinemorets, the towns of Nessebur, Pomorie, Burgas, Sozopol, Primorsko (including the Kiten suburb), Michurin and Ahtopol, and the resorts of Slunchev Bryag and Ropotamo.

The Bulgarian Black Sea coast has been developed as a seaside resort area. It extends 378 km, out of which 130 km (34 per cent) of an approximate area of 15 million sq. m are beaches, most of them covered with fine sand. Part of the coastal area is high and rocky, with beautiful relief forms. The Black Sea water is clear and has a moderate salt content of between 16–18 per cent. The tidal variation is slight and there are no dangerous sharks or other sea creatures harmful to bathers. The sand-covered sea-bed inclines only gradually and this makes the sea safe for bathing. At the height of the summer the average temperature of the surface of the sea is 22.8°C. At times this temperature reaches 24–28°C. The bathing season begins in May and ends in October. The climate is mild. The sea air is saturated with iodine and bromine vapours and is very beneficial for health. The sea breezes temper the maximum temperature in the summer and raise the humidity; they make the air clearer and more rarefied and blow away the clouds (the average annual cloud formation is 5.0–5.8). As a result, the hours of sunshine are 70–75 per cent of the maximum possible and in Varna, for example, they have 2,236.7 hours of sunshine a year. The breezes increase the light negative ions to 1,000 per one cu. cm., which compares favourably with the interior plains of the country, where their number is 300–500 per one cu. cm. Whilst in the interior of the country during the summer there are some 40 to 50 unpleasantly hot days, on the coast this falls to an average of 10 and there is very little rain. 75 per cent of the total number of beds in Bulgaria's resorts are concentrated on the Black Sea coast. Modern resort complexes are being built. The year-round seaside resorts make it possible to provide the necessary places for day trips, or one or two week long holidays, and to offer the added benefits of therapy and preventive medicine. Each year the number of foreigners visiting the Bulgarian seaside resorts increases; it has now reached some 1.5 million per year. The seaside health resorts provide treatment for allergies; chronic non-tuberculosis pulmonary diseases; chronic diseases of the upper respiratory tract; cardio-vascular functional diseases; functional nervous diseases; gynaecological diseases; gastro-intestinal diseases; diseases of the locomotory system, of the metabolism and the endocrine glands (obesity, diabetes, etc.); tuberculosis of the bones and joints, lymphs and glands; skin tuberculosis; occupational diseases such as bronchitis, pulmonary emphysema, chronic intoxications, cardio-vascular diseases, skin diseases and vibration diseases; as well as some other afflictions.

The Estuary of the Ropotamo river on the Black Sea coast

The Russalka holiday resort

Russalka is a resort and tourist complex situated 16 km north-east of the Kaliakra Cape in the very beautiful locality of the Tauk-Liman (Bay of Birds). About 600 neat bungalows, a restaurant, a night club, an open-air theatre, some tennis-courts and yachting facilities have been constructed without spoiling the natural beauty of the rocky and picturesque sea coast. With its whiteness and its buildings, which are arranged amphitheatrically on the rocks, Russalka bears a resemblance to the Mediterranean resorts. It has a very good beach covered with fine sand. The resort is surrounded by a deposit of therapeutic mud. The charm of Russalka, which fully harmonizes with the natural, almost wild beauty of the coast, creates good conditions for a pleasant and active holiday. The resort is a place of interest to many foreign holiday-makers and bus services connect it with Kavarna.

Balchik is a town in the district of Tolbuhin, some 37 km south-east of the district centre with a population of 12,334 people (1982). The buildings of the town are situated in terraces on the white steep slopes of the Dobrudja plateau near a small Black Sea bay, at an altitude of 205 m. Balchik has been a national seaside health resort since 1967. Summer is moderately warm, winter is not

Tourism and Resorts

The holiday resort of Balchik

very cold and autumn is long, with many sunny days. The rainfall is low and mostly occurs in early spring and late autumn. Balchik has developed both as a seaside resort and as a tourist centre. Many children's camps, rest houses, hotels and camping sites have been built right next to the beach. The rich collection of cactuses, which number more than 250 varieties, in the botanical garden ranks second in Europe. Not far from Balchik is the resort of Albena. The national pelotherapeutic resort of Balchishka Tuzla lies some 5 km east of Balchik in close proximity to the small Balchishka Tuzla lagoon. The seabed is covered with therapeutic mud which contains clay particles coagulated by $Fe(Hs)_2$ (ferrous sulphite) and some organic substances, ammonia, phosphates, sulphur, etc. The pelotherapeutic sanatorium provides treatment for gynaecological diseases and diseases of the locomotory system, of the nervous system and of the spinal cord.

Albena is a resort complex in the northern part of the Black Sea coast. It is situated 9 km south-west of Balchik and 10 km north of the village of Kranevo (Varna district), very close to the Batova river. Albena's beach of fine-grain sand is 250–500 m wide and over 5 km long and its shallow sea-bed slopes only slightly. Good roads connect it with Varna (37 km), Zlatni Pyassatsi (17 km) and Balchik (9 km). Albena is visited by many Bulgarian and foreign holiday-makers. The Baltata forest plantation of 280 ha, which lies nearby, renders the resort especially beautiful. The temperature in the summer is not very high and cloud formation is low. There is plenty of sun and little rain (about 450 mm). The construction of the resort began in 1968. Today Albena impresses the visitor with its unique harmony of natural beauty, greenery and flowers, and the original design of its architecture. The 38 hotels are built on terraces leading down to the sea. Their rooms look out on the sea. Besides the beauty and facilities, Albena offers its visitors a wide range of entertainments. There is a great variety of original establishments. The snack-bars, situated on the beach itself, provide a useful service for the tourists. Young people can choose their entertainment from among the restaurants and disco-clubs, sports grounds and facilities for paragliding, windsurfing, sailing, horseriding and bowling. Furthermore, sports festivals named 'Three Days of Albena's Smile' are organized there every month. The children can go to their Sun club or to the amusement centre, the Jolly Rabbit. They can spend some hours with the Skillful Hands club, join the everyday beach games, which are fun for the children and their parents, play in the children's swimming pool and visit the kindergartens run by foreign tourist companies. In 1983 the Dobrudja

The holiday resort of Albena

big new hotel complex opened its doors to visitors. It has 500 beds, restaurants, bars, cafés and many shops and stores, as well as facilities for conferences. A balneological centre for the hotel is under construction and will be one of the biggest in Bulgaria. Besides the indoor and outdoor swimming pools, the centre will be equipped with the most up-to-date apparatus and equipment for more than 1,000 types of treatment.

It is a quite natural thing that a resort named after a young and beautiful woman (a character from a play by the writer Yordan Yovkov) should be a centre of an extremely popular beauty contest, Miss Black Sea, organized there every year. Albena is equally attractive for its sports competitions: the Albena International Boxing Tournament, organized under the sponsorship of the newspaper *Sofia News,* and the International Chess tourna-

The Zlatni Pyassatsi holiday resort complex

ment, which takes place each year. Thousands of automobile sports lovers have the chance to watch the high-speed legs of the most popular rally in Bulgaria – the International Albena-Zlatni Pyassatsi-Sliven.

Zlatni Pyassatsi (Golden Sands) is a resort complex in the northern part of the Bulgarian Black Sea coast. It is situated 17 km north-east of Varna at the foot of the Frangensko plateau on the territory of the Zlatni Pyassatsi national park. The coast is covered with forests and vineyards and leads down to the sea in steps. Nature has endowed this spot with a 3.7 km long and 7.0 m wide beach-strip of splendid golden sand. The sea-bed drops away suddenly but there are still shallow areas which are ideal for bathing. Mineral-water wells up from the boreholes on the beach and has both therapeutic and preventive uses. A scenic highway and a regular ship line connect Zlatni Pyassatsi with Varna. During the period June–September the cloud formation is low (3.6); there is little rain (about 180 mm) and an abundance of sunshine (average 2,237 hours per year).

The average annual temperature of the air in Zlatni Pyassatsi is 11.8°C (−1.4°C in January, 20.0°C in June, 22.6° in July, 22.4° in August and 18.0°C in September). The absolute maximum temperature of the air is 40.3°C (in July). The average annual rainfall is 498 mm. The average annual temperature of the sea water is 13.0°C (3.8°C in January, 23.9°C in July, 24.1°C in August). The resort is well-known and attracts Bulgarian and foreign holiday-makers from May to October.

The design and construction of Zlatni Pyassatsi started at the end of 1956. Nowadays the hot mineral spring (a Karst acratothermal formation) and the comfortable hotels, among them Glarus, Astoria, International, Ambassador, Veliko Turnovo and Shipka, guarantee all-year-round utilization of the resort. With taste and imagination the architects have situated 68 pretty hotels and numerous other establishments in a wide park. This is in fact the only sea-side resort in Bulgaria with so many high-grade entertainment places. It has two jetties, mineral-water swimming pools, playgrounds, open-air kindergartens, bowling alleys, tennis-courts, an archery ground, a horse-race course, a yacht club and other entertainment facilities. In the past two years two new attractions have appeared in the Zlatni Pyassatsi – the Aquarama built by the Rolba Company, Switzerland, and the big complex for balneological treatment at the Ambassador Hotel. The Zlatni Pyassatsi resort has long-standing traditions in balneological tourism. The them belongs the International Hotel, which has outstanding balneological facilities and has been awarded various international prizes, of which an Oscar for hotel services is the most significant. There are two camping sites, Zlatni Pyassatsi and Panorama, accommodating over 1,800 holiday-makers.

For a large number of tourists and leading cultural figures, the name Zlatni Pyassatsi is primarily connected with a number of world-known cultural events: the International Ballet Competition, the Varnensko Lyato Music Festival, the National Film Festival, the World Festival of Cartoon Films, etc. A large number of international scientific congresses and symposia are also held in Zlatni Pyassatsi.

In the immediate vicinity of the resort complex lies the ancient Aladja rock monastery.

Zlatni Pyassatsi, with its original architecture and picturesque surroundings, is justly considered to be one of the best organized seaside resorts in Bulgaria.

The seaside resort of *Druzhba* was the first to be adapted to modern recreation requirements. It lies 9 km to the north of Varna, amidst luxurious green parks. The coastline is well formed, rocky in some places, and 10–15 m above sea-level at others. The beaches are picturesque, sheltered from the winds, with fine, clean quartz sand. The sea itself is good for bathing, since the bottom slopes gradually. A hot mineral spring, recently discovered, is beneficial for the treatment and prevention of various diseases. The St. Constantine monastery (probably fourteenth or fifteenth century) which, together with the small church, forms part of the entire architectural complex, is in the central region of the resort.

The Grand Hotel Varna at Druzhba

The first hotels were built here in 1949–50. Today Druzhba has seventeen hotels, situated among cypresses, limes and fig trees. Open-air swimming pools with mineral water are hidden among the trees. In Druzhba stands the most beautiful hotel on the Bulgarian coast, the Varna De Luxe Grand Hotel (five star), which offers excellent conditions for rest and entertainment, sports and balneological treatment. The rooms are comfortable and the suites luxurious; restaurants, conference halls, a cocktail bar and night club, a folk-style restaurant and a casino-roulette, open-air and indoor swimming facilities, a squash court, tennis courts and souvenir shops are at the disposal of the hotel guests.

Recently, a new holiday village, also called the Druzhba, with 50 bungalows, was built to enlarge the facilities offered by the resort complex.

The high building of the Frederic Joliot-Curie International House of Scientists, where there is a conference hall, a modern restaurant and well-developed balneological centre, is also situated in Druzhba. A great number of scientific conferences and symposia are held in the Joliot-Curie House of Scientists.

Varna is a district town in North-eastern Bulgaria (see the article on Varna, pp. 88–89). The coastline here boasts numerous beaches with fine quartz sand and sea suitable for safe bathing, especially for children. The seashore is steep, and covered with rich verdure. Varna is extremely popular among both Bulgarians and foreign visitors, mainly for its holiday facilities. The average annual temperature there is high, about 12°C. The summer is moderately hot, the spring cool and the autumn moderately warm. Chilly winter spells are rare. The average annual hours of sunshine amount to 2,240. On drilling along the coast in the town and to the north and south, subthermal and thermal springs have been discovered. Part of the flow serves the balneological centre of the Central Beach in the town itself.

The Slunchev Bryag holiday resort

In the immediate vicinity of Varna lies Lake Varna, which is rich in healing mud deposits. The mud-baths on the town beach are used for physiotherapy, and part of the deposits are exported to other Bulgarian resorts and recreation centres.

The curative and recreational factors of Varna are beneficial for the treatment of non-specific chronic ailments, cardio-vascular conditions, metabolic and endocrine disorders, peripheral nervous diseases, locomotory disorders and chronic gynaecological diseases. There are a dispensary for tourists, a balneological treatment centre, and a group of recreation homes in the town.

There are numerous volleyball, basketball and tennis courts in the seaside park of the city, as well as outdoor and indoor swimming pools not far from the southern beaches. There are facilities for aquatic sports, and the immediate and outlying vicinities of Varna with their scenic beauty, lovely forests, parks, and historic monuments are excellent for outings.

The Black Sea coast of Varna is an attractive resort area. Hotels, recreational facilities and amenities are built along the whole stretch of coastline, from the northernmost to the southernmost point, blending into a resort ensemble which provides conditions for rest and recreation, balneological treatment and international tourism.

The *Slunchev Bryag* (Sunny Beach) complex is situated on a bow-shaped cove, 6 km north of Nessebur. To the north of it rise the easternmost parts of the Balkan Range. A good dual carriageway connects it with Burgas (36 km away) and Varna (some 100 km away), and there is a regular bus and boat service between Slunchev Bryag and almost all the resort complexes along the coast. Burgas airport receives tourists and holidaymakers from inside the country, and from Europe and the Middle East. The picturesque coastline, over 7 km long and up to 100 m wide, is covered by spacious beaches with fine golden sand and dunes. The bottom slopes gradually and the sea is shallow. The hotels and restaurants are situated on the beach itself or amid the pleasant greenery and peace of a vast park.

The mean annual temperature for the complex is 12.6°C, the mean January temperature is −2.5°C and the July temperature is 24°C. The annual sunshine duration is comparatively long, nearly 2,300 hours. Other treatment and prophylactic factors of this resort, apart from the climate, are the air, the sunbathing and the sea water. Hypothermal, strongly mineralized water has been discovered in the resort, whose 108 hotels are spread out among the Canadian poplars, sea pines, cypresses, plane trees, acacias and all sorts of decorative shrubs of the park. Slunchev Bryag is the largest Black Sea resort in Bulgaria. The hotels offer a wide range of services, including (in the majority of them) kindergartens with qualified nurses, menus for children in restaurants, and numerous children's swimming pools and playgrounds, as well as a variety of festivals, games and competitions. Slunchev Bryag is a real paradise for children.

The resort has a large, well-equipped polyclinic with specialized consulting rooms and rooms for electrotheraphy, inhalations and remedial massage.

The restaurants and places of entertainment in the resort vary a great deal in style and atmosphere. There is also a nightclub – the Bar Variété.

Slunchev Bryag has all kinds of sports and recreational facilities and amenities, including a yacht club, tennis courts, a racecourse, indoor swimming pools at the healthcare centres in the Globus and Burgas Hotels, a bowling-alley, a large stadium, volleyball and basketball grounds, mini-golf courses and cricket grounds. At nine of the hotels there are amusement arcades with billiards and slot-machines.

Slunchev Bryag is a venue for major international events such as the Golden Orpheus International Pop-

The old part of Nessebur

Song Festival and the Golden Amphora International Underwater-Fishing Tournament.

In the northern part of Slunchev Bryag lies the holiday village of Zora, providing conditions for rest and recreation closer to nature. The village consists of 52 solidly built cottages, complete with kitchen furnishings and cooking facilities. It has several restaurants, five tennis courts and a tennis-school.

Elenite, another holiday village, is under construction on the eastern slopes of the Balkan Range, adjoining the Slunchev Bryag beach. It was designed by a joint team of Bulgarian and Finnish architects and the construction is being carried out jointly by Bulgarian architects and the Finnish firm, Proeke Oy. It is a veritable oasis, covering an area of nearly 18,000 sq. m and will consist of 232 cottages, several restaurants, a café, a day bar and a night club. The camp will have a cultural centre with a multi-purpose hall, seating 220, for concerts, parties, sports competitions, congresses, etc. There will be a modern shopping centre for the needs of holidaymakers. The Elenite holiday camp is to welcome its first guests in 1985.

Nessebur is a town in the Burgas district, 38 km northeast of Burgas with a population of 6,210 (1982), which was made a national climatic and sea resort in 1954. Nessebur is among the oldest towns in Bulgaria, having sprung up in the late Bronze Age as a Thracian fortress called Messembria. The town has survived Greek, Roman and Byzantine rule. In 812 Khan Krum incorporated it into the Bulgarian State. In 1956 the ancient town was declared an architectural, town-planning and archaeological reserve. The numerous valuable historical monuments preserved here include the ancient fortress wall with gate (fifth–fourth centuries BC), the Nessebur churches (the Old Metropolitan church, St John the Baptist, St Stephen, Pantokrator, St Theodore, The Archangels Michael and Gabriel), and some 60 Revival Period houses, built at the turn of the eighteenth century, forming architectural ensembles. The narrow, cobbled streets are picturesquely overhung by bay windows supported by consoles. The town architecture is typical of the coastal areas. As a resort, Nessebur has spread outside the peninsula, on which it is situated, over to the mainland, where a new housing quarter is being developed, with holiday houses and health centres. The Slunchev Bryag resort complex lies 2.5 km north of Nessebur.

The mean annual air temperature is high – about 12.5°C. The favourable climatic conditions, the air, sunbathing and sea water are the main curative and prophylactic factors in this resort.

One and a half kilometres south of the town there is a spacious beach with high dunes spreading over a large area. The sand is fine and golden – the best to be found along the entire coast. The beach has a quay and a diving platform. Most of the holiday houses and sanatoriums have been built close to the beach. There are two sanatoriums, which are used for eight months of the year as sanatoriums and for four months in the summer, as holiday homes. They have a health centre, complete with facilities for mud-bath treatments, balneotreatment, paraffin therapy and electrolysis. The sanatoriums specialize in the treatment of coronary diseases, functional nervous disorders, diseases of the locomotor system and some gynaecological disorders. In the centre of the new residential area of Nessebur there is also a sanatorium for the treatment of obesity, open throughout the year.

Pomorie is a town in Burgas district, situated 22 km north-east of Burgas, with a population of 12,697 (1982 data), which was declared a national health seaside resort and mud-cure centre in 1953. Built on a tiny peninsula of the same name, the town is connected with the mainland by a narrow isthmus, and now spreads over in a newly-constructed quarter. Pomorie is on the site of the ancient Greek colony, Anchialo. It has a better climate than the remaining Black Sea resorts. Snow falls in Pomorie for six to seven days a year at the longest, while the average annual temperature ranges between 12.8° and 13.8°C.

Among the major factors which turn the place into a health spot are the curative mud-baths, the sea-water and climate, and the beaches. The mud comes from sedimentary alluvian deposits. It is black, greasy to the touch, moulds well, smells of hydrogen sulphide and is composed predominantly of silicates and carbonates, with a dash of gypsum. The Pomorie beach is 5 km long and about 30 m wide. It is covered with coarse-grained sand, dark-coloured from the high iron and manganese content. Because of the dark colour it absorbs a lot of sunshine and gets heated to considerable temperatures. Owing to its gradual incline the sea-shore is safe for swimming.

Indications for treatment here include bone and muscular complaints, gynaecological, vascular, skin and neurological diseases, and metabolic, endocrine and urological

Wind-surfing at a campsite near Burgas

Monument to the poet Peyo Yavorov in Pomorie

diseases. The town has an up-to-date mud-cure bath, a polyclinic, sanatoriums and the hotel complex, Pomorie, with an indoor swimming pool, a number of holiday homes, etc.

Burgas is a district city of South-east Bulgaria (see also the section on Burgas in the chapter on District Centres, pp. 93–94). Burgas started attracting holiday-makers from the very beginning of the century. It has a vast beach stretching as far as Lake Atanassovo. Its sand is fine, dark-tinted owing to the presence of titanium and magnetite and gets burning-hot in the sun, which makes it good for sand-cures. The average annual temperature stays around 12.8°C. Summer days are hot, but with refreshing sea breezes. Winter is mild, and autumn is warmer than spring. Precipitation is low. The average amount of sunshine is over 2,300 hours annually.

Apart from the climate and the sea, other medical attractions here are the unlimited reserves of curative mud in the sea-shore lakes and the existence of subterranean mineral springs. In the southern part of the beach there is a covered bath for hot sea-water therapy. Burgas has ample facilities for the development of the tourist industry. Excellently furnished hotels, de luxe private accommodation and high-class restaurants contribute to the restful holidays of both Bulgarian and foreign vacationers.

Fifteen km north-west of Burgas, hidden among the trees of a picturesque pine park, between the villages of Zhitarovo and Banevo, lies the Burgas spa resort. Winter here is mild, summer is cool and autumn is warm. It has a high average annual temperature of 12.4°C. The spa waters are hyperthermal, with a low mineral content, including chlorides, hydrocarbonates, sodium, fluorine, silicon and boron with a rate of discharge 2.280 decilitres per minute. The resort specializes in the treatment of inflammatory and degenerative diseases of the bones and

Old houses in the seaside town of Sozopol

muscles, diseases of the central and the peripheral nervous systems, of the digestive and urinary tracts, metabolic, gynaecological and skin ailments. There are balneosanatoriums, rest houses and an open-air mineral-water pool at the spa.

Not far from the Burgas spa, the Lake Atanassovo offers a wealth of curative mud, fine in structure, almost free from impurities and rich in H_2S (up to 0.133 g per 100 g of mud). It plays an important part in the balneological therapeutic programme for the treatment of a great number of diseases.

A fisherman's house in Sozopol

Sozopol is a town in Burgas district, lying 32 km southeast of Burgas with a population of 3,577 (1982 data); it was declared a national health seaside resort in 1967. Founded by Miletus Greeks in 610 BC, it bore the name of Apollonia. By 431 BC its name was changed to Sozopolis – 'salvation city'. Whole architectural ensembles from the National Revival period have survived intact to this day, and over 180 buildings in this museum town are preserved as cultural monuments. Sozopol is located on a rocky peninsula. It has a deeply indented coast-line, with three islets and a veritable necklace of spacious beaches.

The average annual temperature here is among Bulgaria's highest – about 13.3°C. Summer is hot, relieved by fresh sea breezes, while winter is mild, with few chilly days. Sunshine totals an average of 2,350 hours per annum. The sea, climate and water, sunbathing and swimming provide good medicinal and fitness-increasing opportunities. Each beach has its own micro-climate, depending on its exact exposure along the curving coast. The Harmani beach complex hires out equipment for aquatic sports. Sozopol is a starting point for sight-seeing tours to the Kavatsi resort complex, 5 km south of the town; the Alepu bay, with its breathtakingly beautiful fiords; the Ropotamo national preserve, where one can take a boat trip along the Ropotamo river; and the Arlcutino reserve, with its white water lilies in full splendour.

The Southern Black Sea Coast of Bulgaria is exremely attractive with its numerous beauty spots – romantic bays, thick sycamore, willow and oak woods and old fishermen's towns. Lovers of the romantic and the exotic can find plenty of both here. The largest camping sites in Bulgaria are built here, offering convenient bungalows, restaurants, bars, cafeterias and disco-clubs.

In one of the most eye-pleasing localities (35 km south of Burgas), an original vacationing settlement called Duni (dunes), a joint venture with the Austrian Roegner company, is now under construction. It is planned as a complex of several distinct areas. The Fisherman's village will be a one- and two-storey villa zone with a total of 1,200 beds, a sports centre, a playground, sites for children, night-clubs and restaurants and shopping-centres. Next to it, the Monastery Settlement will show the typical elements of local architecture, and will accommodate 800 guests, who will have a cafeteria, snack bars, folk-style restaurants and the like within walking distance. Perched upon the very shore will be the Sea Hamlet, complete with a yachting club and pier. It will house 300 holidaymakers and manage 150 yachts. The Duni builders have shown tremendous consideration for nature conservation concerns, taking special care not to interfere with the Via Pontica bird route over the Strandja mountain forests.

A picturesque camping site will emerge at the far side of the Fishermen's Village. The living quarters will be built in perpendicular rows to the beach, green mounds will rise over certain swimming pools and the transport services will use a sunken road, so as not to spoil the wild charm of Lake Alepu.

Apart from camping sites and bungalows, the Southern Black Sea coast offers 12,000 beds in private rooms, which can be rented in such fantastically romantic sea-side towns and villages as Sozopol – the most colourful of all fishing towns – Ahtopol, Lozen and others. Here, too, in the town of Primorsko, you can find a genuine republic of world youth at the International Youth Centre, named after the leader of the Bulgarian people – Georgi Dimitrov.

Recommended Addresses

Bulgarian Association for Recreation and Tourism
Sofia, 1 Lenin Square.
Telephone: 84131; telexes: 23553 and 23101.

Balkantourist Foreign Tourist Organization
Sofia, 1 Vitosha Blvd.
Telephone: 84131; telexes: 22583 and 22584.

Orbita Youth Travel Agency
Sofia, 45ᴬ Stamboliiski Blvd.
Telephone: 879552; telex: 221081.

Shipka Travel Agency of the Central Council of the Union of Bulgarian Motorists
Sofia, 6 Sv. Sofia Street.
Telephone: 878801; telex: 22279.

Pirin Tourist Organization of the Central Council of
Bulgarian Tourists
Sofia, 30 Stamboliiski Blvd.
Telephone: 870687; telex: 22357.

Co-optourist Travel Agency of the Central Co-operative
Union
Sofia, 99 Rakovski Street.
Telephone: 8441; telex: 22657.

Offices of the Bulgarian Association for Recreation and Tourism Abroad

Predstavitelstvo gosurdarstvennogo komiteta po turizmu
NR Bulgarii
SSSR, Moskva K-350
ploshchad Sverdlova
gostinitsa 'Metropol'
komnata 2045.
Telephone: 2218575; telex: 414560.

Predstavitelstvo gosurdarstvennogo komiteta po turizmu
NR Bulgarii
SSSR, Leningrad 191073
ul. Brodskogo 1/7 gostinitsa Evropeiskaya
komnata 336.
Telephone: 2103336; telex: 121565.

Generalnoye Konsulstvo NR Bulgarii
SSSR, Odessa
ul. Posmitnogo 9.
Telephone: 650038; telex: 457.

Predstavitelstvo gosurdarstvennogo komiteta po turismu
NR Bulgarii
SSSR, Kiev
ul. Kreshchatik 12
kv. 2, et 2.
Telephone: 290467; telex: 131339.

Bulgarisches Fremdenverkehrsamt
DDR, Berlin 108
Unter den Linden 40.
Telephone: 2292273/2292339; telex: 114074.

Zastupitelství Bulgarského Státního Výboru pro Cestovní Ruch
CSSR, Praha 1
Pařízská 3.
Telephone: 2314596; telex: 122378.

Zastupitelstvo Bulgarského Státniho Výboru Cestovní Ruch
CSSR, Bratislava 801000
Nálepkova 2.
Telephone: 336168; telex: 92366.

Balkantourist
Ungarn, Budapest 6
Nepköztarsasag utcc 16.
Telephone: 22812; telex: 224369.

Przedstawicielstwo Państwowego Komitetu do Spraw
Turystyk R. L. Bulgarii
Polska, Warszawa
ul. Marszałkowska 83.
Telephone: 298257; telex: 813550.

Ambassade de la R. P. de Bulgarie
Représentation commerciale S. F. R. Jugoslavija
Beograd
Nusiceva 3/1.
Telephone: 331-132; telex: 11665.

Office National du Tourisme Bulgare
France, Paris 75002
45, Avenue de l'Opéra.
Telephone: Opéra 2616958; telex: 211629.

Bulgarisches Fremdenverkehrsamt
BRD, Frankfurt/Main
Stephansstrasse 1–3.
Telephone: 295284; telex: 413869.

Bulgarisches Tourist Informations Bureau
Denmark
1606 Kobenhavn
V Farimagsgade 6.
Telephone: 123510; telex: 19223.

Bulgarian National Tourist Office
England, London W.1.
126 Regent Street
Telephone: 01-437 2611; telex: 28833.

Office National du Tourisme Bulgare
Belgique, Bruxelles 1000
rue Ravenstein 62.
Telephone: 5139610; telex: 25219.

Nationaal Bulgaars Verkeers Bureau
Holland, Amsterdam 1017 NV
Leidsestraat 43.
Telephone: 020/248431; telex: 16322.

Bulgarian Tourist Office
USA and Canada/New York N.Y. 10028
161 East 86th Street.
Telephone: /212/722-1110; telex: balk 429767.

Bulgarische Fremdenverkehrswerbung
Österreich/Wien 1040
Margaretenstrasse 9.
Telephone: 577762; telex: 111223.

Offizielles Bulgarisches Verkehrsbüro
Schweiz/Zürich 8023
Steinmühlerplätz 1.
Telephone: 2212777; telex: 812502.

Bulgarisches Fremdenverkehrsampt
Berlin West 15
Kurfürstendamm, 175.
Telephone: 8827418/8827419; telex: 182994.

Bulgarian Tourist Office
Finland/00120 Helsinki
Annankatu 9.
Telephone: 666244; telex: 121842.

Bulgarian Tourist Office
Sverige/Stockholm C
Kungsgatan 30.
Telephone: 115191; telex: 17523.

Bulgarian Tourist Office
España/Madrid
Princessa 12.
Telephone: 2420720/21.

Ufficio Touristico Bulgaro
00198 Roma
Viale Gorizia 14.
Telephone: 856438.

Balkan Bulgarian Airlines
Romania/Bucuresti
Str. Batistei No. 9.
Telephone: 148994; telex: 10148.

Bulgarian Tourist Office
Greece/Athens
12 Akademias Str.
Telephone: 3634675; telex: 215873.

Bulgarian National Tourist Office
Lebanon/Beirut
Mamra Street
Sabbag Centre.
Telephone office: 342265–342362, 342364–350679; telex: 41241 Le Beirut.

BULGARIAN FOREIGN TRADE ORGANIZATIONS AND OTHER ENTERPRISES CONNECTED WITH FOREIGN TRADE

1. Chambers of Commerce

Bulgarian Chamber of Commerce & Industry
11a A. Stamboliiski Blvd., Sofia 1040.
Tel.: 87-26-31
Telex: 22574
Cables: Torgpalata

The Bulgarian Chamber of Commerce Industry is a public organization. Its members are economic ministries and committees, economic organizations, corporations and combines, enterprises of foreign and home trade, of services and tourism, industrial enterprises and agro-industrial complexes.

The functions of the Chamber are:
- to promote and develop trade and economic ties between Bulgarian economic and trading organizations and foreign companies; to contribute to the activation and intensification of the country's participation in socialist economic integration with the CMEA countries, and especially with the USSR; to assist in the expansion of industrial co-operation and econonomic collaboration with developing and developed non-socialist countries;
- to receive foreign economic delegations and to send Bulgarian economic delegations abroad to study ways and means of collaboration between Bulgaria and the other countries;
- to organize Bulgaria's official participation at world expositions, international fairs and exhibitions, as well as to organize Bulgarian national exhibitions abroad;
- to organize the annual international fairs in Plovdiv and international specialized exhibitions in the country;
- to publish, in Bulgarian and in foreign languages, economic and foreign trade information materials for distribution among Bulgarian economic organizations and enterprises and interested organizations and companies abroad;
- to publish the monthly paper *Bulgarian Economic News* and also the bi-monthly magazine *Bulgarian Foreign Trade* in English, Russian, French, German and Spanish;
- to assist Bulgarian industry in the implementation of the achievements of world technical progress, with the aim of increasing the variety and improving the quality of the goods;
- to act on requests by foreign companies and persons for patenting of their inventions and registration of their trade marks and industrial models with the competent Bulgarian authorities, and on requests by Bulgarian enterprises to register their trade marks and industrial models abroad;
- to render legal assistance to Bulgarian foreign trade organizations and their foreign partners in solving disputes between them through the Arbitration Court at the Chamber;
- to keep a register of Bulgarian foreign trade organizations;
- to issue certificates of origin of goods, ATA carnets and other documents related to essential facts, necessary in the materialization of foreign trade transactions.

The BCCI system includes the following institutions and economic enterprises:
1. The Arbitration Court of the BCCI
   11a A. Stamboliiski Blvd., Sofia 1040.
2. General Average Adjuster's Office of the BCCI
   11a A. Stamboliiski Blvd., Sofia 1040.
3. Patent & Trade Mark Bureau of the BCCI
   11a A. Stamboliiski Blvd., Sofia 1040.
4. International Plovdiv Fair
   37 G. Dimitrov Blvd., Plovdiv 4018.

Arbitration Court
of the Bulgarian Chamber of Commerce & Industry
11a A. Stamboliiski Blvd.
Sofia 1040.
Tel.: 87-26-31
Telex: 22374
Cables: Torgpalata

Examines disputes arising from contractual and legal or other civil-law relations instituted between legal subjects from different countries in the materialization of foreign trade transactions or other international economic and scientific-technical ties.

Parties to such disputes can be:
(a) Bulgarian and foreign enterprises, firms and persons;
(b) foreign enterprises, firms and persons.

General Average Adjuster's Office
of the Bulgarian Chamber of
Commerce & Industry
11a A. Stamboliiski Blvd., Sofia 1040.
Tel.: 87-26-31
Telex: 22374
Cables: Torgpalata

Determines the kind of sea averages, estimates the extent of general averages and issues statements of adjustment on the request of Bulgarian and foreign interested parties (shipowners, freight owners, carriers, insurance companies, etc.).

Collects deposits and other securities related to general averages and handles their liquidation.

Collaborates with foreign adjusters in connection with general averages of Bulgarian and foreign-vessels.

Mediates for voluntary settlement of claims related to receipts due from general averages, transport, freight and insurance contracts.

Patent & Trade Mark Bureau
of the Bulgarian Chamber of
Commerce & Industry
11a A. Stamboliiski Blvd., Sofia 1040.
Tel.: 87-26-31
Telex: 22374
Cables: Torgpalata

Represents foreign nationals and juridical persons in patenting of inventions and registration of trade marks and industrial models in Bulgaria, as well as Bulgarian enterprises, public organizations and persons in registration of their trade marks and industrial models abroad.

Effects researches of Bulgarian and forign patents and trade marks.

Renders assistance related to patent rights and collaborates in administrative and law court disputes related to patents and trade marks.

2. Banks

Bulgarian National Bank
2 Sofiiska Komuna St., Sofia 1000.
Tel.: 85-51
Telex: 22031
Cables: National bank

Issuing bank and main centre for credit and payments.

Bulgarian Foreign Trade Bank
2 Sofiiska Komuna St., Sofia 1000.
Tel.: 85-51
Telex: 22031
Cables: Bulbank

Payment operations of commercial and non-commercial character with foreign countries. Foreign currency and credit operations abroad, crediting of Bulgarian foreign trade organizations and enterprises with foreign and local currencies.

Mineralbank
Bank for Economic Initiatives
17 Lege St., Sofia 1000.
Tel.: 80-17-37
Telex: 23390, 23391, 23392 Mibank
Cables: Mineralbank

Crediting and/or participation with capital in local and foreign currencies in economic activities and operations outside state planning targets of local economic organizations and enterprises, as well as in the activities of joint ventures in the country and abroad. Servicing and execution of all banking operations in national and foreign currencies in favour of these organizations.

3. Foreign Trade and Engineering Organizations

Agrocommerce
Foreign Trade Organization
86 Dondukov Blvd., Sofia 1000.
Tel.: 80-33-12
Telex: 23223
Cables: Agrocommerce

Export of produce of the economic organizations – the members of the National Agro-Industrial Union; import of industrial equipment and goods, multilateral barter and other foreign trade operations, including consumer products; organization of technical service and maintenance of imported machines and equipment.

Agrocomplect
Engineering Economic Organization
61 Vitosha Blvd., Sofia 1000.
Tel.: 87-09-62
Telex: 22021 Agroc
Cables: Agrocomplect

Survey, design, construction, delivery, erection, commissioning and technical maintenance of agro-industrial complexes, hydromeliorative, hydrotechnical and afforestation projects, semen and pedigree farms, poultry and livestock industrial farms, forage plants, greenhouses, tobacco curing and other installations, farming facilities and food industry projects abroad; research, technical, production and business collaboration with foreign economic organizations and firms; export of single machines, technological lines and complex equipment, spares and services; sale of technologies, licences and know-how; engineering services; import of separate projects, technologies, licences, know-how, machines, equipment, etc.; sending of experts abroad to render scientific and technical assistance in the field of agriculture and the food industry.

Bulgarian Foreign Trade Organizations

Agromachinaimpex
Foreign Trade Company
1 Stoyan Lepoev St., Sofia 1040.
Tel.: 23-03-91
Telex: 22563
Cables: Agromachinaimpex

Export, import and home trade in tractors, agricultural machines and equipment, complex technological lines, electronic and hydraulic equipment, pumps and spares. Engineering activity, service, re-export, barter and licence transactions, production, transport, construction and other activities in the country and abroad.

Balkancar
Research, Industrial &
Trading Corporation
48 K. Ohridski Blvd., Sofia 1040.
Tel.: 6-55-01
Telex: 23431, 23432
Cables: Balkancar

Manufacture of electric and ICE trucks, buses, lorries and cars and trailers, equipment for them and the related research, engineering and implementation activities, foreign and home trade, service in the country and abroad.

Balkancarpodem
Foreign Trade Engineering Company
48 Kliment Ohridski Blvd., Sofia 1040.
Tel.: 6-55-01
Telex: 22665, 22666
Cables: Balkancarpodem

Export, import, service and engineering activity in electric hoists, cranes and crane trolleys, lifting and material handling equipment, accessories for the production of these goods and spare parts thereof, re-export and other specific foreign trade transactions, licences, patents, know-how, etc.

Bulgarconsult
Engineering Company
130 Tsar Boris St., Sofia 1000.
Tel.: 88-48-20, 87-23-55, 80-27-29
Telex: 23201 Bulco bg

Carries out in-country and abroad studies, designing, consultations, technical assistance and other kinds of engineering services in the field of territorial and town planning, urbanization, architecture, typization and industrialization of construction, geodesy, photogrammetry, mapping, geodetic instruments, activities, power generation, water supply, sewerage, effluent treatment, engineering infrastructure, roads and road equipment, plant construction and studies of building materials and training of technical personnel.

Bulgarco-op
Foreign Trade Organization at the Central Co-operative Union
99 Rakovski St., Sofia 1000.
Tel.: 84-41
Telex: 23429
Cables: Bulgarcoop

Export: medicinal plants, essential oil seeds, wild and cultivated mushrooms, honey and bee products, game and game meat, rabbits and rabbit meat, snails, mineral waters, fresh and processed fruits and vegetables, wild fruits, consumer goods, etc.
Import: industrial consumer products and foodstuffs, concentrates for soft drinks, medicinal plants, etc.

Bulgargeomin
State Economic Corporation
3 Ogneborets St., Sofia 1619
PO Box 600.
Tel.: 57-91-71
Telex: 22859 bgm
Cables: Bulgargeomin

Comprehensive engineering services abroad; geological mapping; geophysical and geochemical surveys; complex geological prospecting and mining of ores and minerals, oil, gas, etc.; engineering–geological and hydrogeological surveys, drilling of test water wells; designing, delivery of equipment and building of open and underground mines, tunnels, ore-dressing combines, metallurgical complexes, product pipelines and equipment for oil, gas, etc.; designing, delivery and assembly of metal structures for industrial building, storehouses, etc. Renders technical assistance and trains local personnel. Delivery of drilling materials, tools and equipment, mining machinery; compensation and re-export transactions.

Bulgariafilm
Directorate
96 Rakovski St., Sofia 1000.
Tel.: 87-66-11
Telex: 22447 Filmex bg
Cables: Filmbulgaria

Export and import of films and film materials, co-productions, film services. Publicity of Bulgarian film art abroad.

Bulgarplodexport
Foreign Trade Company
22 Alabin St., Sofia 1000.
Tel.: 84-51, 88-59-51, 88-29-42
Telex: 22451, 22452, 22453, Bulgarfrukt; 23297, 23298 Bulkon
Cables: Bulgarfrukt

Export: all kinds of fresh and processed fruits and vegetables, juices, pulps, wild fruits, flowers, etc.

Import: fresh and processed citrus and other southern fruits, southern fruit concentrates, vegetables, etc.

Re-export, barter and other specific foreign trade transactions.

Engineering and implementation activities, delivery and construction of complete projects in the country and abroad in the field of market gardening, fruit growing, floriculture and canning industry.

Bulgartabac
Trading & Industrial Corporation
14 A. Stamboliiski Blvd., Sofia 1000.
Tel.: 87-52-11
Telex: 23288
Cables: Tabak

Production, export and import of tobacco, tobacco products and materials for the tobacco industry.

Chimcomplect
Engineering Economic Organization
101 A. Tolstoi Blvd., Sofia 1220.
Tel.: 38-49-07
Telex: 22785
Cables: Chimcomplect

Import and export of licences, know-how, machines and equipment, complete projects and rendering of engineering services (feasibility studies, designing, supervision of construction and assembly works, commissioning of installations) in the field of chemical industry.

Chimimport
Foreign Trade Company
2 Stefan Karadja St., Sofia 1080.
Tel.: 88-38-11/15
Telex: 22521, 22522
Cables: Chimimport

Export: *chemical fertilizers* – urea, ammonium sulphate, ammonium nitrate, sodium nitrate, etc.; soda ash; plant protection preparations – zineb; *inorganic chemicals* – caustic soda, sodium bicarbonate, sodium silicofluoride, etc.; *plastics* – polyvinylchloride, polyethylene, polystyrene, etc.; *plastic products;* synthetic rubber and rubber products, automobile tyres; *fuels* – petrol, diesel fuel, fuel oil, etc.; mineral oils, petrochemicals – acrylnitrile, monoethyleneglycol, propylene, benzene, toluene, acetone, etc.; normal paraffins, solid paraffins, propane-butane gas; photographic paper and chemical reagents; synthetic dyes.

Import: industrial chemicals – organic and inorganic salts, acids and bases, solvents; phosphorites; plant-protection preparations; raw materials for production of plastics; natural and synthetic rubber; rubber products, automobile tyres; accelerants for the rubber industry; crude oil and oil products; petrochemicals, synthetic dyes; lacquers and paints; laboratory chemicals and reagents; photographic and cinema materials.

Corecom
Foreign Trade Organization
8 Tsar Kaloyan St., Sofia 1000.
Tel.: 8-51-31
Telex: 22476
Cables: Corecom

Import and retail sale against convertible currency of imported and local goods.

Electroimpex
Foreign Trade Engineering Company
17 G. Washington St., Sofia 1000.
Tel.: 8-61-81
Telex: 22075, 22076
Cables: Electroimpex

Export and import of electric power generation units and electrical engineering machines and equipment, electric medical and pharmaceutical equipment, measuring and automation equipment, cables and conductors, household electrical appliances, electric-light fixtures, quartz and quartz products, fibre optics, components and instruments for the electrical engineering industry, etc. Import and export of licences and patents and the related single and complete equipment. Re-export, compensation and other transactions.

Engineering activity abroad – study, designing, delivery, erection, operation, know-how and maintenance connected with the export and erection of complete projects, equipment and industrial enterprises in the field of electrical engineering, automation and instrument-making, medical equipment, energy, power and relay systems, etc.

Energoimpex
Foreign Trade Organization
17a Ernst Thaelmann Blvd.
Sofia 1000
PO Box 801.
Tel.: 51-88-67
Telex: 22669
Cables: Energoimpex

Export and import of coal, electric power and other energy media.

Import of equipment and units for the needs of power generation in the country.

Study, designing, construction, assembly and maintenance of electric transmission lines up to 750 kV, of transformer substations for low, medium and high voltages. Consultant services in the field of energy generation.

Hemus
Foreign Trade Organization
6 Ruski Blvd., Sofia 1000.
Tel.: 87-03-65
Telex: 22267 Hemkik
Cables: Hemus

Import and export of books, music publications,

newspapers, periodicals, gramophone records, musical instruments, works of art, art souvenirs, museum objects, coins, photographic cameras and accessories, school aids, export of colour and black/white postcards, slides, picture albums, import and export of foreign and Bulgarian postage stamps, philatelic aids, etc.

Hranexport
Foreign Trade Company
56 Alabin St., Sofia 1000.
Tel.: 88-22-51/54
Telex: 22525, 22526
Cables: Hranexport

Export, import and other foreign trade transactions in wheat, maize, sunflower seed, beans, rice and other pulses, vegetable edible and technical oils, fodder components, sugar and confectionery, alcohol, molasses, cocoa, kernels, etc.

Hydrostroi
State Economic Corporation
30 Pozitano St., Sofia 1000.
Tel.: 88-47-11
Telex: 22644
Cables: Hydrostroi

Performs in-country and abroad study, designing, erection, reconstruction and modernization of: dam walls – concrete and earth-filled, hydropower stations, confined and non-confined tunnels, pipelines, derivations and other hydrotechnical equipment, purification stations for potable and waste waters, grouting and drilling works, gunite works, road tunnels, metropolitan reservoirs, water supply and sewerage systems, slurry ponds and slurry lines with the proper equipment, river corrections, projects related to environment protection; rendering technical assistance, supervision, resident assistance, laboratory tests and other engineering services.

Industrial Co-operation
Foreign Trade Directorate
57 Cherni Vruh Blvd., Sofia 1407.
Tel.: 66-89-01, 66-89-02
Telex: 22513
Cables: Inco

Import and export of special machines, designs and models related to implementation of licences in machine-building and electronics; means of automation; computers, instruments and apparatus for production control and management; air-conditioning equipment; peripheral devices and printed circuit boards; environment protection equipment and apparatus; hydraulic and other presses; measuring instruments and quality control equipment; machine tools, equipment and complete lines; technological equipment, etc.; engineering services.

Industrialimport
Foreign Trade Company
3 Positano St., Sofia 1040.
Tel.: 87-30-21
Telex: 22091
Cables: Indimport

Export of woollen and cotton ready-to-wear, cotton and synthetic knitwear; cotton, woollen and silk fabrics; blankets; machine-made carpets – Persian type; shoes, leather fancy goods, sports goods and appliances; synthetic, polyester fibres; polyester wool, polyamide technical silk and cord; blown and pressed glassware, sanitary porcelain, wall tiles, illuminants, etc.

Import of cotton, yarns, wool, artificial textile raw materials, textile products – cotton, woollen and silk fabrics, knitwear; artificial leather; carpets; footwear.

Information Systems & System-Engineering Services
Engineering Economic Organization
2 P. Volov St., Sofia 1000.
Tel.: 55-40-77, 55-30-11, 54-60-56
Telex: 22001 Kessi bg
Cables: Kessi

Development, import and export of information systems, program products and system-engineering services, independently or jointly with foreign companies and organizations. Setting up abroad joint ventures, system-engineering offices and computer centres for services. Rendering of all kinds of consultant and engineering services in building of information systems, training of personnel and rendering of technical assistance in the field of information, complete delivery of hardware and software systems.

Intercommerce
Foreign Trade Company
16 Lenin Sq., Sofia 1080
PO Box 676.
Tel.: 87-93-64
Telex: 22067
Cables: Intercommerce

Export and import of various goods, raw and other materials; multilateral, barter, compensation and processing operations in-country and abroad; transit commission, agent's and other foreign trade transactions, including financial operations; participation in joint ventures and in joint economic activities in-country and abroad.

Intransmash
Bulgarian-Hungarian Company
40 Bukston Blvd., Sofia 1618.
Tel.: 5-68-71; Telex: 22331
Cables: Instransmash Sofia

17 Marvani St., Budapest 1112.
Tel.: 886-360; Telex: 225762
Cables: Intransmash Budapest

Engineering organization for comprehensive solution of

problems related to material handling and warehousing in all spheres of production and consumption. Feasibility studies, designing, manufacture and delivery of 'turn-key' projects. Provision of the necessary service, spares and training of operating personnel. Independent foreign trade activity.

Isotimpex
Foreign Trade Organization
51 Chapaev St., Sofia 1113.
Tel.: 74-61-51
Telex: 22731, 22732
Cables: Isotimpex

Export and import of computing and business machines, automation and control instruments and equipment, electronic components, etc.

Kintex
State Commercial Enterprise
66 Anton Ivanov Blvd., Sofia 1407
PO Box 209.
Tel.: 66-23-11
Telex: 22471, 23243
Cables: Kintex

Import and export of special machinery and equipment, complete projects and equipment, organizational equipment, explosives, cords, detonators, apparatus and others used for industrial and mining purposes, hunting, sports and angling gear; compensation and multilateral transactions and transit operations.

Koraboimpex
Foreign Trade Organization
128 D. Blagoev Blvd., Varna 9000.
Tel.: 88-18-25, 88-56-13, 88-20-95
Telex: 77550, 77560
Cables: Koraboimpex Varna

Export: ships – sea-going, river-sea-going and river-going, container carriers, ship equipment, radar equipment, ship spares. Engineering activity – study, design and execution of complete projects. Industrial co-operation.

Import: ships, floating docks, ship equipments, drilling platforms, computer and specialized equipment for dockyards, tools, etc.

Lescomplect
Engineering Economic Organization
67 V. Poptomov St., Sofia 1000.
Tel.: 80-25-31
Telex: 23540 Lesco
Cables: Lescomplect

Feasibility study, designing, delivery of equipment and execution of projects in the fields of forest economy, timber industry, wood-processing and furniture industry, manufacture of doors and windows, dry distillation of wood, prefabricated buildings.

Consultant services rendered by highly skilled specialists; supervision on assembly and maintenance of delivered machines; training of personnel in-country and abroad.

Import of machines and spares for the wood-processing, furniture and pulp-and-paper industries.

Lessoimpex
Foreign Trade Organization
67 V. Poptomov St., Sofia 1000.
Tel.: 87-91-75, 88-02-11
Telex: 23407
Cables: Lessoimpex

Export: furniture, timber, pulp and paper, chipboard and fibreboard.

Import: pulp and paper, paper products, timber, veneers, exotic logs, furniture fittings.

Machinoexport
Foreign Trade Engineering Company
5 Aksakov St., Sofia, 1000.
Tel.: 88-53-21
Telex: 23425, 23426
Cables: Machinoexport

Export: machine tools, automatic technological lines and modules, robots, forging and pressing equipment, foundry equipment, woodworking machines and lines, bearings, hydraulic and pneumatic products, tools and measuring gauges, metal products, welding equipment, refrigeration equipment, complete projects; engineering services.

Import: machine tools, forging and pressing equipment, foundry equipment, woodworking machines, tools, bearings, hydraulic and pneumatic products and equipment, complete projects for the machine-building and wood-processing industries.

Maimex
Foreign Trade Enterprise
15 Dimiter Nestorov St., Sofia 1431.
Tel.: 5-81-01, ext. 691, 59-61-75
Telex: 22712 Maprez
Cables: Maimex

Export of endoprostheses and apparatus for osteosynthesis and instruments; specialized electronic medical equipment; contact lenses, eye prostheses and accessories; materials and preparations for hygienic control and protection; vaccines, allergens, nutrient media and other bioproducts; patents, licences and know-how in the field of medicine, medical tourism and balneological treatment of foreign nationals. Engineering activity in the country and abroad.

Import: medical apparatus, tools and instruments, ready-made pharmaceutical preparations.

Medexim
Medical Personnel Abroad Directorate
of the Ministry of Public Health
47 Kiril i Metodii St., Sofia 1202.
Tel.: 83-27-23, 83-20-47
Telex: 23698 bg
Cables: Medexim

Scientific and technical assistance in the field of medicare and medical practice. Provision of highly skilled specialists and medical teams for hospitals, out-patient clinics, dispensaries and other medical establishments abroad. Organization of emergency and specialized medical attendance centres. Provision of consultants and lecturers in higher medical institutes abroad.

Metalni Konstruktsii (Metal Structures)
Economic Combine
24 Kamen Andreev St., Sofia 1606
PO Box 46.
Tel.: 51-06-33, 51-07-12
Telex: 22771
Cables: Metalni konstruktsii

Complex research, study, designing, production and erection of buildings and equipment using metal structures in the country and abroad. Implementation of advanced building systems and sub-systems for construction of industial, agricultural, public and other facilities. Complex engineering and consultant services, foreign trade in construction materials, components and structures. Sale of patents, licences and know-how in the field of building with metal structures.

Metal Technology
Research & Production Corporation
Foreign Trade Directorate
53 Chapaev St., Sofia 1574.
Tel.: 7-14-21
Telex: 22903 Techmet
Cables: Technomet

Export: machines, dies and complete technological lines for counter-pressure casting of non-ferrous alloys, non-ferrous alloys castings obtained after the counter-pressure casting methods, thermoplast automates for counter-pressure casting of foamed thermoplastics, extruded steel sections, equipment for soldering collectors of electrical machines.
Import: industrial equipment.
Trade in technologies, licences and know-how in the field of counter-pressure casting. Industrial co-operation.

Mineralimpex
Foreign Trade Organization
44 K. Ohridski Blvd., Block 1-A,
Sofia 1156.
Tel.: 66-19-66, 66-12-61
Telex: 22973 Minex
Cables: Mineralimpex

Export: *rock-facing materials* – marble, granite and limestone blocks and slabs, granite curbstones and paving stones, mosaics, etc.; *minerals* – bentonites, perlites, kaolin, talc, talcum-magnesite, dolomite, gypsum, etc.; machines, equipment and spares for mining and processing rock-facing materials, geological equipment and mineral raw materials.

Import: *minerals* – kaolin, clays, graphite, potassium feldspar, rutile concentrate, etc.; *rock-facing materials* – granite and marble; machines for mining and processing, rock-facing materials, machines and equipment for oil extraction, geological prospecting and geophysical activities, spares; diamond tools for cutting and processing rock materials, diamond tools for metalworking, glass-processing and electronic industries, diamond powder, diamond paste; complete plant in the field of mineral resources.

Engineering and consultant activities abroad – study, design and construction related to exploration, mining and processing rock-facing materials and mineral raw materials; mounting of rock-facing materials.

Delivery of technology and know-how. Joint ventures in industry and trade with foreign partners.

Re-export and other specific commercial operations.

Mineralsouvenir
Economic Trade & Industrial
Combine
44a K. Ohridski Blvd., Sofia 1156.
Tel.: 66-19-66
Telex: 22676 Misuv
Cables: Mineralsouvenir

Manufacture, export and import of mineral souvenirs, gold, silver and mass jewelry, life-style and home decoration articles.

Montazhi (Assembly)
State Economic Corporation
3, 65th Street,
Chervena Zvezda quarter, Sofia 1156.
Tel.: 71-01-21, 72-25-64
Telex: 22142, 22143
Cables: Montazh

Performs in-country and abroad: complex assembly in modernization, reconstruction, expansion and new construction of industrial and power generation, complete projects such as – metallurgical, machine-building, metal-processing, wood-processing, petrochemical and chemical combines and plants, plants and enterprises for production and processing of building materials, agricultural produce, plants and enterprises of the food and tobacco, glass and other industries; thermo-power stations, hydro-power stations and equipment, atomic power stations, electrical sub-stations and distribution equipment; gas trunk pipelines, oil pipelines, product pipelines and related equipment, mechanical handling equipment, air-conditioning and ventilation installations, thermo and frost-resistant insulations, refractory masonry, test-loading operations related to commissioning of assembled installations. Study, designing and erection of complete projects. Import, export, re-export and other foreign trade operations in assembly components, metal structures, tanks, silos, equipment, etc. Rendering of technical assistance and collaboration, designer's supervision, resident's activity, sending of teams, personnel training, sending of workers and specialists abroad, related to the execution of the projects. Rendering of complex engineering and consultant services, sale of licences, patents and know-how, related to the assembly and commissioning of industrial and power generation enterprises.

Pharmachim
Trading & Industrial Corporation
16 Iliensko Chaussée, Sofia 1220.
Tel.: 36-55-31
Telex: 22097, 22098
Cables: Pharmachim

Manufacture, import and export of medicines, microbiological products, cosmetic and perfumery products, essential oils, veterinary preparations, dressing materials, chemical products and the related engineering, implementation and other activities.

Pirin
Industrial & Trading Corporation
19 Levski St., Sofia 1000.
Tel.: 88-14-43
Telex: 22761
Cables: Pirin

Manufacture of footwear – leather, rubber and slippers; leather and fur garments, leather fancy goods, etc. and the related research, development, designing and other activities.

Export of leather garments, leather fancy goods, shoes, sole leather, leather gloves, raw hides, auxiliary materials for the shoe industry.

Import of hides, soles, man-made leather auxiliary materials for the shoe, fancy goods and fur industries.

The corporation has its own network of shops.

Prodexim
Foreign Trade Directorate
56 Alabin St., Sofia 1000.
Tel.: 88-06-42
Telex: 22212 Prodex
Cables: Prodexim

Barter transactions with foreign organizations in consumer goods – textiles, ready-to-wear, knitwear, industrial products, foodstuffs; import-export; trade between the fraternal towns and districts; supply of goods to the Bulgarian national shops and restaurants abroad and to foreign ones in the country.

Raznoiznos
Foreign Trade Company
1 Tsar Assen St., Sofia 1040.
Tel.: 88-02-11
Telex: 23244
Cables: Raznoiznos

Export: art souvenirs, ceramic products, pokerwork, small wooden articles, bulrush and willow articles, toys, sports goods, Christmas tree decorations, hand-tied Persian-type carpets, national style Chiprovski and Kotlenski carpets, artistic life-style fabrics, illuminants, household articles – enamel ware, stainless kitchen ware, brushes, hats, suitcases, stationery, etc.

Import: organic and technical glass, laboratory glassware, optic products, insulation materials, hairdresser's equipment and accessories, metal fittings, bicycles and spares, carpets, moquettes, sanitary appliances, bathtubs, handicraft articles, souvenirs, wallpaper, illuminants, bicycle and motor-cycle tyre valves, cutlery, heat-fast and crystal kitchen utensils, metal utensils, attraction and entertainment equipment, stationery and drawing materials, gold and silver jewellery, ornaments, toys, sports, hunting and angling wear, watches and spares, musical instruments and equipment, textile fancy goods and *passementerie,* foostuffs – coffee, tea, olives, pepper, etc.

Other activities: transit operations, warehousing, consignment trade, multilateral and barter deals; participation in industrial, technical and trading co-operation; participation in consortiums and joint ventures; assignment, designing, construction, appointment and exploitation of business and warehouse projects in the country and abroad.

Ribno Stopanstvo
(Fisheries)
State Economic Corporation
Foreign Trade Directorate
42 Parchevich St., Sofia 1000.
Tel.: 80-23-03
Telex: 22796
Cables: Bulgarriba

Export and import of fish, tinned fish and fish products.

Rodopaimpex
Foreign Trade Organization
2 Gavril Genov St., Sofia 1000.
Tel.: 88-26-01, 87-26-91
Telex: 22541, 22542
Cables: Rodex

Export: meat and meat products, live animals, eggs, poultry meat, dairy products and pedigree animals.

Import: meat, pedigree animals and poultry, packing and other materials for the meat and dairy industries.

Rudmetal
Foreign Trade Company
1 Dobrudja St., Sofia 1000.
Tel.: 88-12-71
Telex: 22027, 22028
Cables: Rudmetal

Export: *ferrous metals* – steel balls, seamless and welded steel tubes, wire, coils, reinforcing bars, section steels, heavy and thin steel sheets, galvanized sheets, tinplate, plastic coated steel sheets, cold-bent sections, ferromanganese, ferrosilicon, etc.; *non-ferrous metals* – lead, zinc, cadmium, bismuth, secondary aluminium, copper, zinc, brass and aluminium rolled stock; *refractories* – chamot products; *minerals* – barytes concentrate, manganese ore, metal ores.

Import: ferrous and non-ferrous rolled stock, all kinds of metal products, steel pipes and tubes, tinplate, high grade steel, refractories, coking coal, blends, iron ore, iron concentrate, manganese ore, pyrites concentrate, steel ingots, slabs, ferrosilicon, other ferrous alloys, copper concentrate, zinc concentrate, lead concentrate, etc.

Semena & Posaduchen Material
(Seeds & Stock)
Foreign Trade Directorate
56 Alabin St., Sofia 1000.
Tel.: 87-03-89, 88-14-08, 88-21-14
Telex: 22688
Cables: Semena

Export: grass-fodder crop seeds, grain and vegetable seeds, tree and shrub seeds, biennial rose roots, stock (grafted vines on the root, vine cuttings, fruit trees, raspberry and strawberry seedlings) etc. Propagates seeds and stock supplied by the customer.

Import: grass-fodder and vegetable seeds, stock, etc.

Sportkomplekt Economic Company
67-b Bulgaria Blvd., Sofia 1142.
Tel.: 86-51, ext. 548, 511
Telex: 22737
Cables: Sportkomplekt

Designing and construction of sports projects abroad; import and export of sports articles, appliances and equipment; purchase and sale of patents, licences and know-how, barter transactions, international industrial co-operation and participation in joint companies in the field of physical culture, sport and sport totalizers.

Stroyimpex
Foreign Trade Organization
18 Chapaev St., Sofia 1040.
Tel.: 71-80-17, 70-20-57, 71-90-00
Telex: 22385 Strim
Cables: Stroyimpex

Export: cement, concrete plastificators, hydrate lime, clinker, asbestos-cement slabs and pipes, wall and floor ceramic tiles, metal structures, prefabricated houses, bungalows, arc-shaped metal hangars, industrial and warehouse hangars, vans of corrugated steel sheets, false ceilings, etc.

Import: all kinds of rubberoid and asbestos, stoneware pipes, glass-fibre, laminated and other building materials.

Technika
Foreign Trade Organization
125 Lenin Blvd., Block 2, Sofia 1113
PO Box 672.
Tel.: 70-20-41/45, 70-21-41/45
Telex: 23278
Cables: Technika

Import and export of licences, patents, equipment, technical collaboration, material and technical supply of research activities; import of chemicals and other materials (consumables) for business machines.

Technoexport
Foreign Trade Engineering Company
20 Joliot Curie St., Sofia 1113
PO Box 541.
Tel.: 73-81
Telex: 22193, 22048, 22049
Cables: Technoexport

Feasibility studies, designing, construction, export and import of complete plant, technologies, equipment, flow lines and machine systems, single machines, spares. Rendering of technical assistance in the following fields: production of building materials, mechanization of all types of building operations, printing industry, refrigeration and purification equipment, light industry, food and tobacco industries, metal structures and sheds for industrial, warehouse and general uses, cranes, greenhouse complexes.

Technoexportstroi
State Economic Corporation
11 Antim I St., Sofia 1303.
Tel.: 87-85-11
Telex: 22128, 22129
Cables: TES

Operation abroad in the field of comprehensive territorial and urban planning, study, design and construction of: housing complexes, public buildings, hotels and theatre complexes, research centres, outpatient clinics, hospitals, medical stores, sport stadiums, grain silos, factories, oil refineries, electric power stations, dams, irrigation systems, water towers, airports, motorways, roads, railway-lines, bridges, tunnels, etc.; assembly of electrical and mechanical equipment, structures, installations and machinery; designer's and investor's supervision; engineering services, technical assistance related to above activities, delivery of materials, machines and equipment for the needs of the Corporation.

Technoimpex
Directorate
10–12 Graf Ignatiev St., Sofia 1080
PO Box 932.
Tel.: 88-15-71/72, 73
Telex: 23277 Timpex bg
Cables: Technoimpex

Rendering of scientific and technical assistance to foreign countries by offering the services of specialists in the field of industry, construction and architecture, transport and communications, power generation, geological prospecting, education, etc. Rendering of study and consultant services.

Technoimport
Foreign Trade Engineering Company
20 F. Joliot Curie St., Sofia 1040.
Tel.: 73-81
Telex: 23421, 23422
Cables: Technoimport

Study, designing, construction, delivery, import and export of complete projects, technologies, licences, machines and spares for the ferrous and non-ferrous metallurgies, power generation for open and underground mining of minerals, protection equipment for atomic-power stations, pumps, compressors, ventilators and industrial apparatus.

Telecom
Foreign Trade Engineering Company
17 G. Washington St., Sofia 1040
PO Box 933.
Tel.: 8-61-81
Telex: 22077
Cables: Telecom

Study, design, delivery, assembly or assembly supervision, commissioning and service of complete communication projects and systems, know-how, technical assistance, consultations, training of specialists in communication projects and systems.

Export: radio-relay lines, multiplex systems of communications through cable lines, a range of Crosspoint automatic telephone exchanges, ultra-short wave systems and radio stations, information transmission systems and electro-acoustic equipment for hotels, TV retranslators, nuclear measuring apparatus, etc.

Import: telephone equipment, automatic telephone exchanges, communication lines, trunk-line telephone equipment, radio and TV transmitters, radio-relay lines, etc.

Telecomplect
Engineering Economic Organization
8 Totleben Blvd., Sofia 1606.
Tel.: 54-27-69, 51-70-17
Telex: 22706, 23555
Cables: Telecomplect

Feasibility studies and designing of communication, radio and TV systems and projects; design solutions for reconstruction and modernization of existing telecommunication systems; delivery of materials and equipment, construction and assembly in the country and abroad of: automatic urban telephone and telegraph exchanges, inter-urban cable trunk-lines and municipal telephone networks, USW and SW radio transmitting and receiving stations, MW radio stations, radio masts and towers, technological and telemechanical transmission systems for gas and oil pipelines.

Telerimpex
Foreign Trade Organization
4 Dragan Tsankov Blvd., Sofia 1421.
Tel.: 85-41/ext. 620, 608, 578/,
66-11-21
Telex: 22305, 22581
Cables: Telerimpex

Import and export of TV programmes and films, video-recordings; co-production with and services to foreign TV and film companies.

Tourist-Engineering
Engineering Economic
Organization
127 Rakovski St., Sofia 1000.
Tel.: 80-23-44
Telex: 23541

Research, engineering and implementation activities, rendering of technical assistance in the field of international and internal economic tourism in the country and abroad. Import and export of complete projects, machines, equipment, installations, tooling, transport vehicles, spares, goods, materials and other products for the tourist business and the related commercial and industrial co-operation with foreign and Bulgarian economic organizations, companies, etc.

Turgovsko Obzavezhdane
(Commercial Equipment)
State Economic Corporation
2 Konstantin Fotinov St., Sofia 1517.
Tel.: 45-10-46
Telex: 23293
Cables: Turgovsko obzavezhdane

Designing, implementation, production, delivery, assembly and service of commercial equipment and appointment. Designing and reconstruction of shopping centres, execution of construction and assembly works, and turn-key delivery, engineering activity in-country and abroad.

Export of technological equipment, appointment and complete projects for commercial, catering and services establishments.

Import of machines and equipment for commercial and catering establishments, including spare parts, etc.

Operation of commercial projects and centres in the country and abroad.

Bulgarian Foreign Trade Organizations

Transimpex
Foreign Trade Organization
65 Skobelev Blvd., Sofia 1606.
Tel.: 52-23-21
Telex: 22123
Cables: Transimpex

Import: locomotives, wagons, railway equipment, electronic systems for rail traffic control and spares; railway and road-construction machines and spares; floating port equipment, tugs, ship equipment and spares; service maintenance of ships, ship repair, supply of oils, fuels, paints, chemicals, foodstuffs to ships; trade in duty-free goods.

Export: wagons, railway equipment, ship repair, sale of old ships for scrap, provision of transport and technical specialists. Multilateral transactions, re-export and barter operations.

Transkomplekt
Engineering Economic Organization
35 Bulgaro-Suvetska Druzhba Blvd.,
Sofia 1156, PO Box 444.
Tel.: 44-16-18, 44-36-18, 44-28-82, 87-70-41
Telex: 23315
Cables: Transkomplekt

Study, designing and construction of transport projects. Offers equipment, systems, designs, know-how, licences, technologies for the transport infrastructure. Renders engineering services and technical assistance. Joint ventures and consultant teams.

Valentina
State Economic Corporation
5 Graf Ignatiev St., Sofia 1000.
Tel.: 80-17-26, 87-72-97, 87-10-01
Telex: 22170 Valina
Cables: Valentina

Production, delivery and retail trade in de-luxe and new varieties of goods. The corporation has its own manufacturing enterprises and 56 reperesentative shops in Sofia and other large towns in-country.
Import and export of fashionable and de-luxe articles, and of materials and equipment for their manufacture.
Barter operations with the socialist countries in consumer goods. Compensation transactions with the non-socialist countries.

Vinimpex
Foreign Trade Company
19 Lavele St., Sofia 1080.
Tel.: 88-39-21/26
Telex: 22467, 22468
Cables: Vinimpex

Export and import of wine, beer and all other kinds of alcoholic drinks, raw materials, materials, equipment and spares; industrial and technical collaboration, co-operation, engineering and other services related to its basic activity; re-export and other specific transactions.

4. International Transport, Tourism

Association of Bulgarian Enterprises
for International Automobile
Transport & Roads
56 Emil Markov Blvd.,
Block 15, Sofia 1680.
Tel.: 59-11-13, 59-21-17
Telex: 23616
Cables: Aebtri

Assistance to members in organizing and carrying out of international automobile transport of cargo and passengers; translation and issue of international normative documents related to transport activity and passengers, instructions for their application; issue of TIR carnets and AGT manifests, guarantee for transport covered by above documents; assistance in designing, construction and exploitation of roads and motorways on the territory of Bulgaria.

'Balkan' Bulgarian Airlines
Economic Corporation
Sofia Airport – 1540.
Tel.: 7-12-01
Telex: 22342, 22299
Cables: BGA Balkan

Air transport of passengers and freight over regular domestic and foreign routes. Charter transport of passengers and freight. Renting and leasing of aircraft. Maintains regular passenger lines to Abu Dhabi, Algiers, Amsterdam, Ankara, Athens, Baghdad, Beirut, Belgrade, Berlin, Bucharest, Budapest, Bratislava, Barcelona, Brussels, Cairo, Casablanca, Copenhagen, Damascus, Dresden, Frankfurt/Main, Istanbul, Khartoum, Kiev, Kuwait, La Valetta, Lagos, Luxembourg, Leningrad, London, Luanda, Madrid, Milan, Moscow, Nicosia, Paris, Prague, Rome, Salonika, Stockholm, Tripoli, Tunis, Vienna, Warsaw, Helsinki, Ho Chi Minh, Zurich, and a regular freight line to Paris.

Maintains regular passenger lines in the country to Varna, Burgas, Gorna Oryahovitsa, Russe, Turgovishte, Silistra, Vidin, Kurdjali.

Aviation services for the needs of national economy of building, environmental protection and prospecting character. Agro-chemical operations in the country and abroad.

'Balkan' has 47 representatives abroad.

Balkantourist
Foreign Trade Organization
1 Vitosha Blvd., Sofia 1000.
Tel.: 8-41-31
Telex: 22583, 22584
Cables: Balkantourist

Foreign trade activity related to international tourism.

Bulfracht
State Economic Enterprise
5 General Gurko St., Sofia 1080.
Tel.: 88-55-81
Telex: 22161
Cables: Bulfracht

Exclusive rights in organizing and chartering Bulgarian and foreign shipping tonnage for transport of Bulgarian and foreign cargo.

Bulgarinterautoservice
165 Kiril i Metodii St., Sofia 1309.
Tel.: 22-05-18
Telex: 23582
Cables: Bias

Organization and rendering of service and other kinds of assistance – technical service and repair, filling with fuel and oils, towing, storage of automobiles and cargo, loading and unloading operations and others to automobiles, buses, etc., of foreign juridical and physical persons, of their drivers and accompanying persons.

The enterprise has branches in the following towns: Sofia, Russe, Plovdiv, Vidin, Tolbuhin, Silistra, Varna, Burgas, Stara Zagora, Gorna Oryahovitsa, Haskovo, Kyustendil and Blagoevgrad.

Despred
State Forwarding Enterprise
2 Slavyanska St., Sofia 1000.
Tel.: 87-60-16
Telex: 23306
Cables: Despred

International forwarding activity related to imports and exports, re-exports and goods in transit by rail, water, road and air transport; carries out transport of exhibits for fairs and exhibitions, small consignments in group-shipment, combined transport abroad and inland transport.

Inflot
State Shipping Agency
88 V. Zaimov Blvd., Sofia 1504
PO Box 634.
Tel.: 87-25-34, 88-27-41
Telex: 22376
Cables: Bolinflot

Agent for all foreign vessels calling at Bulgarian ports. Agent for all Bulgarian vessels sailing abroad.

Branch agencies at the seaports of Varna and Burgas, and the Danubian ports of Russe, Samovit, Lom, Svishtov and Vidin.

Interhotels – Balkantourist
Tourist Complex
14a A. Stamboliiski Blvd., Sofia 1000.
Tel.: 88-29-02
Telex: 22052

Handles all types of international tourism – individual, group, therapeutic, balneological, professional in lines of interest, congresses for hunting, etc.

Offers accommodation in de luxe hotels in Sofia, Plovdiv, Burgas, Varna, Druzhba Resort, Russe and Veliko Turnovo.

Offers:
– comprehensive facilities for organization of congresses, conferences and other international events;
– round the year facilities for active rest, tourism, sport, balneological treatment;
– excursions;
– casinos at the Vitosha and New Otani hotels in Sofia, Grand Hotel Varna at the Druzhba Resort, and Bulgaria Hotel in Burgas.
– Interbalkan bureaux for booking and sale of air tickets.

International Road Transport – So Mat
Economic Corporation
Gorublyane, Sofia 1738.
Tel.: 71-21-21, 78-11-21
Telex: 22433, 22426, 22356, 22620, 22380, 22788
Cables: So Mat

International transport of Bulgarian and foreign cargo by road and in combined means of transport (truck-ship-truck, truck-river vessel-truck, railway-truck). Ferryboat transport of accompanied vehicles, semi-trailers, containers and other self-propelled wheeled and track-chained machines, of passenger cars and passengers via the Medlink ferryboat line, operated by the Corporation. Maintenance, repair and overhaul of vehicles, engines and aggregates thereof, in-country and abroad. Provision of qualified personnel to organize, execute and manage road transport operations abroad, including service, repair and maintenance of all types of automobiles. Forwarding and agent services, emergency technical assistance to foreign cargo vehicles on the territory of Bulgaria. Engineering and consultant services in the field of automobile transport.

5. Representation, Copyrights, Law Services

Interpred Association
2 A. Stamboliiski Blvd., Sofia 1040.
Tel.: 87-45-21, 87-31-91
Telex: 23284
Cables: Interpred

Association of offices for representation of foreign companies and for commercial mediation in Bulgaria. It includes the following independent offices at the above address:

Bulpharma
Tel.: 88-18-94, 87-31-91/96
Telex: 23284
Cables: Bulpharma

Kom
Tel.: 87-94-73, 87-45-21/24
Telex: 23284
Cables: Kompred

Lozen
Tel.: 87-08-27, 87-45-21/24
Telex: 23284
Cables: Lozenpred

Lyulin
Tel.: 88-16-65, 87-31-91/96
Telex: 22578
Cables: Lyulinpred

Murgash
Tel.: 87-47-65, 87-31-91/96
Telex: 23284
Cables: Murgashpred

Mussala
Tel.: 87-52-29, 87-45-21/24
Telex: 23284
Cables: Mussalapred

Pirin
Tel.: 87-14-41, 87-31-91/96
Telex: 23284
Cables: Pirinpred

Rila
Tel.: 87-42-59, 87-45-21/24
Telex: 23284
Cables: Rilapred

Ruen
Tel.: 88-48-28, 87-31-91/96
Telex: 23284
Cables: Ruenpred

Shipka
Tel.: 87-17-21, 87-45-21/24
Telex: 23284
Cables: Shipkapred

Trakiya
Tel.: 80-05-13, 87-31-91/96
Telex: 23284
Cables: Trakiyapred

Vitosha
Tel.: 88-00-52, 87-45-21/24
Telex: 23284
Cables: Vitoshapred

Jusauthor Copyright Agency
11 Slaveikov Sq., Sofia 1000.
Tel.: 87-91-11
Cables: Jusauthor

Bulgarian agency for literature, theatre and music, providing information on literary, scientific, theatrical and musical works by Bulgarian authors; exercises control of the application of copyright legislation and tariffs for fees paid to authors and other persons for creative work; mediates in conclusion of contracts for publication and use of works by Bulgarian authors abroad and in-country; ensures protection of copyrights of Bulgarian and foreign authors on Bulgarian territory, as well as copyrights of Bulgarian authors abroad in compliance with acting national legislation and international conventions.

Law Office for Foreign
Legal Matters
31 Alabin St., Sofia 1000.
Tel.: 87-77-82
Telex: 22104 lawof
Cables: Lawoffice

All legal protection of interests of foreign physical and juridical persons in Bulgaria and of Bulgarian nationals abroad in civil, administrative, criminal and other cases, legal advice and other legal assistance.

6. Insurance, Superintendence of Goods

Bulgarkontrola
Economic Enterprise
42 Parchevich St., Sofia 1080
PO Box 106.
Tel.: 8-51-51
Telex: 23318, 23653
Cables: Bulgarkontrola

Specialized independent organization for supervision of goods with head office in Sofia and representations in all ports, frontier check-points and industrial business centres in the country.

It carries out qualitative and quantitative control and expert examinations, sampling, chemical, physico-chemical and physico-mechanical analyses and tests on various types of export and import goods on customer's order.

Bulgarkontrola is Lloyd's agent, agent of P&I clubs, and average commissary on behalf of various foreign insurance companies. It issues certificates of shipping averages and files regressive claims.

Bulstrad
Bulgarian Foreign Insurance &
Reinsurance Co.
5 Dunav St., Sofia 1000.
Tel.: 8-51-91
Telex: 22564
Cables: Bulstrad

All insurance operations on import and export goods, reinsurance operations; insurance of ships, aircraft and motor vehicles; life and property insurance of foreign nationals in the country and of Bulgarian nationals abroad.

7. Advertising, International Fairs and Exhibitions, Publishing

Bulgarreklama Agency
Foreign Trade Organization
42 Parchevich St., Sofia 1040.
Tel.: 8-51-51
Telex: 23318
Cables: Bulgarreklama

Advertising Bulgarian export goods abroad through the Press, TV, radio, cinema and other mass information media. Organizes advertising for foreign companies in the country through the Press, TV, radio, direct mail, exhibitions, symposia, posters, etc. Publication in foreign languages of advertising and information printed materials for the export requirements of the foreign trade and other economic organizations. Prepares black/white and colour photos, slides, advertising prints and blow-ups for fairs and exhibitions in-country and abroad. Organizes production of foreign trade advertising films, audiovisual advertising programmes, tape recordings on orders of foreign trade organizations. Supplies local and foreign advertising gifts. Co-editor of *Messemagazin International* – an international magazine for East-West trade.

Bulgarska kniga i pechat
(Bulgarian Books and Printing)
State International and Economic Association
11 Slaveikov Sq., Sofia 1000.
Tel.: 87-91-11
Telex: 22927
Cables: DTPCC 'Bulgarska kniga i pechat'

Management of all activities related to book publishing, polygraphy, book distribution, fairs, exhibitions and book promotion in the country and abroad and co-ordination of matters connected with the structure and material and technical supply for periodicals.

International Plovdiv Fair
Economic Enterprise
37 G. Dimitrov Blvd., Plovdiv 4018.
Tel.: 5-31-91, 5-43-21
Telex: 44432 Partet bg
Cables: Panaira Plovdiv

Organization of Bulgaria's official participation in international fairs and exhibitions; preparation of structures, necessary equipment and the artistic layout of specialized and other exhibitions of the Bulgarian economic organizations abroad; organization and holding of international fairs and specialized exhibitions in Plovdiv; rendering services to foreign and local exhibitors.

Programme of international fairs and exhibitions in Plovdiv for 1984–85:

1984
7–13 May	International fair for consumer goods and foodstuffs
24 September– 1 October	International technical fair

1985
6–12 May	International fair for consumer goods and foodstuffs
30 September– 7 October	International technical fair

Sofia Press Agency
29 Slavyanska St., Sofia 1040.
Tel.: 88-58-31
Telex: 22622, 23577
Cables: Sofia press

Printing and publishing of books in foreign languages, photo and cinema promotion materials; preparation of press releases for the mass media, for enterprises and individuals abroad; publication of magazines, newspapers, bulletins, books and other printed matter in foreign languages for distribution abroad; production of films, multivision programmes; arrangement for photo exhibitions and show-windows.

BULGARIAN FOREIGN TRADE REPRESENTATIONS

AFGHANISTAN

The Embassy of the P.R. of Bulgaria
Commercial Counsellor
POB 200
Kabul

ALBANIA

La Représentation Commerciale de la R. P. de Bulgarie
7, Rue Skenderbeg
Tirane
Tel.: 2935, 6496

ALGERIA

Ambassade de la R. P. de Bulgarie
Conseiller Commercial
4, Avenue Mustapha El-Ouali
(Ex. Claude Debussy).
Alger.
BP 486
Tel.: 61-53-63, 61-53-64
Telex: 52706

ANGOLA

Embaixada da República Popular da Bulgária
Secção Comercial e Económica
Rua Da Muccáo NHO-142
1°-Ambar
C.P. 2617
Correo Central de Luanda
Luanda.
Tel.: 32-544
Telex: 3375 Bultarg AN

ARGENTINA

Embajada de la Republica Popular de Bulgaria
Departamanto Commercial
Calle Rodriguez Pena, 565, piso I
Buenos Aires 1020.
Tel.: 461284, 461285
Telex: 121314 Arlbulg

AUSTRALIA

Consulate General of the P.R. of Bulgaria
Commercial Section
4, Carlotta Road, Double Bay, NSW
Sydney 2028.
Tel.: 36-4440, 36-7591
Telex: 27755 Bulcon AA

AUSTRIA

Handelsvertretung bei der Botschaft der
Volksrepublik Bulgarien
Prinz Eugen-Straße 66
1040 Wien.
Tel.: 65-36-06
Telex: 131749

BANGLADESH, P.R. OF

Embassy of the P.R. of Bulgaria
Commercial Counsellor
Road No. 127, House No. 12
Gulshan Model Town
Dacca, 12.
Telex: 642310 bul bj

BELGIUM

Représentation Commercial de Bulgarie
36, Rue Jules Lejeune
1060 Bruxelles.
Tel.: /02/ 345-43-13, 345-52-00
Telex: 22473 Bultarg Bru

BENIN

Ambassade de la P.R. de Bulgarie
Section Commerciale
Les Cocotiers, Lot P. No. 10
B.P. 7058
Cotonou.
Tel.: 30-03-66
Telex: 21567 Bulg NG

BERLIN/WEST

Generalkonsulat der V.R. Bulgarien
Handelsabteilung
Kurfürstendamm 175
Berlin (West) 15.
Tel.: 8-82-74-40
Telex: 182994 bulg K

BRAZIL

Departamento Comercial da
Embaixada da R.P. da Bulgaria
Rua México, 21 – gr. 1601
Rio de Janeiro.
Tel.: 240-53-76
Telex: 2121567 LRPB BR

BRAZIL

Departamento Comercial da R.P. da Bulgaria
Rua Xavier de Toledo, 210/Sala 31
São Paulo.
Tel.: 37-7694
Telex: 1121918 LRPB BR

CANADA

Bulgarian Trade Commission
100 Adelaide St. West, Suite 1405
Toronto, Ontario M5H 153.
Tel.: (416) 368-1034-35
Telex: 6-23535

CHINA, P.R. of

Embassy of the P.R. of Bulgaria
Trade and Economic Section
Xiu Suit Beijie 4
Peking.
Tel.: 52-22-31

COLOMBIA

Embajada de la R.P. de Bulgaria
Departamento comercial
Calle 81 No. 7-71
Apartado Aéreo 36-36
Bogota.
Tel.: 249-8498
Telex: 41217 Bulg Co

CONGO

Ambassade de la République Populaire de Bulgarie
Section Commerciale
B.P. 28-08
Brazzaville.

CUBA

Représentacion Comercial de Bulgaria
Calle 21YO, Edificio ICP
Mezzanine Vedado
La Habana.
Tel.: 32-5586
Telex: 51354

CYPRUS

Bulgarian Trade Mission
16, A Saint Paul Street
Nicosia.
Tel.: 7-27-40
Telex: 2188 Bultarg

CZECHOSLOVAKIA

Handelswirtschaftliche Abteilung bei der Botschaft
der Volksrepublik Bulgarien
Rihtareska 1
Prag 6 Dejvice.
Tel: 37-86-16, 37-56-61
Telex: 9042

CZECHOSLOVAKIA

Generalny Konzulate BLR
Obchodne oddelenie
Obrancov Mieru 31
80100 Bratislava.
Tel.: 42312, 423539
Telex: Bultarg bva 93202

DENMARK

Botschaft der Volksrepublik Bulgarien
Handelsrat
2900 Hellerup
Kildegaardsvej 36
Kopenhagen.
Tel.: /01/ 68-22-09, /01/ 68-22-10
Telex: Bultarg dk 27020

ECUADOR

Embajada de la R.P. de Bulgaria
Departamento Económico y Comercial
Calle la Colina No. 331
POB 81-A Suc. 3
Quito.
Tel.: 23-8980
Telex: 2279 ietel ed

Bulgarian Foreign Trade Representations

EGYPT, ARAB REPUBLIC OF

Embassy of the P.R. of Bulgaria
Commercial Counsellor Office of the P.R. of Bulgaria
8, Saleh Ayoub Street – Zamalek
Cairo.
Tel.: 806654–803745
Telex: 92037 Bultarg Un

EGYPT, ARAB REPUBLIC OF

Commercial Counsellor Office
of the P.R. of Bulgaria
2, Stanbul Street, 5th Floor, Ap. No. 32
Alexandria.
Tel.: 80-12-40

ETHIOPIA

Embassy of the People's Republic of Bulgaria
Commercial Counsellor
POB 1631
Addis Ababa.
Tel.: 15-81-82

FINLAND

Handelsvertretung der Volksrepublik Bulgarien
Merikatu 3 A 4
00140 Helsinki 14.
Tel.: 170 133
Telex: 121035

FRANCE

Service Commercial près l'Ambassade de Bulgarie
1, Avenue Rapp
75007 Paris.
Tel.: 551-85-90
Telex: 201118

GERMANY, FEDERAL REPUBLIC OF

Handelsvertretung bei der Botschaft
der V.R. Bulgarien
5300 Bonn 2 Bad Godesberg
Am. Buchel 17a.
Tel.: 36-30-61/65
Telex: 885 739

GERMANY, FEDERAL REPUBLIC OF

Handelsbüro der V.R. Bulgarien
Eckenheimerland Str. 101
6000 Frankfurt/M.
Tel.: 00611/59 71 483
Telex: 4189 391

GERMANY, FEDERAL REPUBLIC OF

Handelsbüro der Volksrepublik Bulgarien
Benedictstraße 5/11
2000 Hamburg 13.
Tel.: (040) 48 6169
Telex: 2 14643

GERMANY, FEDERAL REPUBLIC OF

Handelsbüro der VR Bulgarien
Wintrichring 85
8000 München 50.
Tel.: (089) 174057
Telex: 5213128

GERMAN DEMOCRATIC REPUBLIC

Handelswirtschaftliche Abteilung bei der Botschaft
der Volksrepublik Bulgarien
Friedrichstraße 62
108 Berlin.
Tel.: 2001-321
Telex: 114034, 114035

GHANA

Embassy of the People's Republic of Bulgaria
Commercial Section
3 Kakramádu Road
East Cantonments
POB 3193
Accra.
Tel.: 21800

GREECE

Bulgarian Trade Mission
6, Loukianou-Ypsilantou Str.
Athens.
Tel.: 7214 400, 7214 830
Telex: 215873

GREECE

Bulgarian Trade Representative
12, Nikolaou Manu Str.
Thessaloniki.
Tel.: 827642
Telex: 412639

GUINEA

Ambassade de la R.P. de Bulgarie
Section Commerciale
BP 632
Conakry.
Tel.: 613-29, 415-12

HOLLAND

Handelsvertretung der Volksrepublik Bulgarien
Alexander Gogelweg 22
Den Haag.
Tel.: 46-88-72, 46-88-73
Telex: 32651 Btarg NL

HUNGARY

Handelswirtschaftliche Abteilung bei der Botschaft
der Volksrepublik Bulgarien
Marcius 15, ter, 1
Budapest V.
Tel.: 182-644
Telex: 225441

INDIA

Embassy of the People's Republic of Bulgaria
Office of the Commercial Counsellor
16/17, Chandragupta Marg, Chanakyapuri
New Delhi – 110 021.
Tel.: 697811, 697813, 698018
Telex: 31 2545

INDIA

Embassy of the People's Republic of Bulgaria
Trade Representation
Peacock Palace, Flat 11
69, Bhulabhai Desai Road
Bombay – 400 026.
Tel.: 822 86 80/822 83 58
Telex: Bultarg 011-3352

INDIA

Embassy of the People's Republic of Bulgaria
Trade Representation
6-A, Middleton Street, 5th Floor
Calcutta – 16.
Tel.: 44-2642
Telex: CA217872

INDONESIA

Bulgarian Trade Representation
Jalan Lombok, 59
Djakarta.
Tel.: 45 667, 48 650
Telex: 45106

IRAN

Embassy of the P.R. of Bulgaria
Commercial Counsellor
Ave. Dr. Fatemi
Ave. Shabnam, Ave. Eight No. 23
Tehran.
Tel.: 650012-652838
Telex: 212789

IRAQ

Bulgarian Trade Representation
POB 6103
Harithiya 14/9/23 – karkh
Baghdad.
Tel.: 5520401
Telex: 212604

ITALY

Ufficio Commerciale ed Economico Presso
l'Ambasciata della R.P. di Bulgaria
Via P.P. Rubens, 21
00197 Roma.
Tel.: 360-94-40, 360-94-43, 360-94-45, 360-94-48
Telex: 610234

ITALY

Ufficio Commerciale ed Economico della R.P. di Bulgaria
Via Vittor Pisani, 13
20124 – Milano.
Tel.: 66-43-42, 65-95-595
Telex: 311004

JAPAN

Embassy of the P.R. of Bulgaria
Commercial Counsellor's Office
36–38, Yoyogi 5-Chome
Shibuya-ku
Tokyo.
Tel.: 465-1025/8
Telex: J 22341

JORDAN

Embassy of the P.R. of Bulgaria
Trade Office
6th circle jabal
POB 950344
Amman.
Tel.: 813193, 813194
Telex: 22247 Bltrg jo

KAMPUCHEA, R.P. de

Représentation Commerciale de Bulgarie
Villa No. 32, Voie No. 232, Région 7
6-ème Quartier
Phnom-Penh.
Tel.: 2-46-81

KOREA, P.D.R. OF

Trade Representation of the P.R. of Bulgaria
Munsundon District
Pjoengjang.

SELECTED BIBLIOGRAPHY

The Development of the Economy

ANGELOV, T. *Efektivnost na kapitalnite vlozhenia i osnovnite fondove v nashata promishlenost* (Effectivity of Capital Investments and the Basic Funds in Bulgarian Industry). Sofia, 1967.

BARBOV, T. and I. LAMBEV *Organizatsiya na upravlenieto na zhelezoputniya transport v NR Bulgaria* (Organization of Railway Transport Management in the People's Republic of Bulgaria). Sofia, 1976.

BENEV, B., M. PANDEVA and N. IVANOV *Nauchno-tehnicheskata politika na NRB* (The Policy of the People's Republic on Science and Technology). Sofia, 1980.

BEROV, L., *Ikonomicheskoto razvitie na Bulgaria prez vekovete* (The Economic Development of Bulgaria through the Ages). Sofia, 1974.

BOROH, N. and S. SILVESTROV *Sutrudnichestvo na stranite-chlenki na SIV v sferata na promishlenoto proizvodstvo* (Co-operation of the CMEA Countries in the Sphere of Industrial Production). Varna, 1978.

BOZHKOV, T. *Nauchno-tehnicheska i proizvodstvena integratsiya mezhdu NRB i SSSR* (Scientific-Technological and Production Integration between the People's Republic of Bulgaria and the USSR). Sofia, 1976.

CHOLAKOV, Y. and A. SAVOV *Osnovni nasoki v razvitieto na kozharskata, kozhuharskata i obuvnata promishlenost v Bulgaria* (Basic Trends in the Tanning, Fur and Shoe Industry in Bulgaria). Sofia, 1971.

DOCHEV, I. *Ikonomicheski problemi no modernizatsiyata i rekonstruktsiyata v promishlenostta (na NRB)* (Economic Issues of Modernization and Reconstruction in Industry [in the People's Republic of Bulgaria]). Sofia, 1975.

Efektivnost na nauchno-tehnicheskiya progres (Effectivity of Scientific-Technological Progress). Sofia, 1977.

HADJIEV, S., G. DIMITROV and V. DJAMBAZOV *Sotsialisticheskata ikonomicheska integratsiya i razvitieto na stroitelstvoto v stranite-chlenki na SIV.* (Socialist Economic Integration and the Development of Construction in the CMEA countries). Varna, 1978.

Ikonomika na Bulgaria (Economy of Bulgaria). Vols 1–6. Sofia, 1969–80.

Ikonomichesko i sotsialno razvitie na NRB (Economic and Social Development of the People's Republic of Bulgaria). Sofia, 1964.

Ikonomika i obshtestveni otnoshenia v NRB na suvremenniya etap (Economy and Social Relations in the People's Republic of Bulgaria at the Present Stage). Varna, 1983.

Ikonomikata v sotsialisticheskoto obshtestvo (The Economy in the Socialist Society). Vols. 1–2. Varna, 1977.

Ikonomicheska entsiklopedia (Economic Encyclopaedia). In two volumes. Sofia, 1984.

Izgrazhdane na ikonomikata na razvitoto sotsialistichesko obshtestvo (Building the Economy of the Developed Socialist Society). Sofia, 1980.

KALIGOROV, H. *Izhgrazhdane, funktsionirane i usuvurshenstvuvane na organizatsiyata v proizvodstvoto* (Building, Functioning and Improvement of Production Organization). Sofia, 1980.

Kompleksna energetika na gradovete v NR Bulgaria (Complex Power Engineering in the towns of the People's Republic of Bulgaria). Sofia, 1966.

KUNIN, P. *Agrarno-selskiyat vupros v Bulgaria. Ot Osvobozhdenieto do kraya na Purvata svetovna voina* (The Agrarian Issue in Bulgaria. From the Liberation from Ottoman Domination to the End of World War I). Sofia, 1971.

KUNIN, P. *Vliyanieto na Oktomvriiskata revolyutsiya, na leninizma za razvitieto na agrarno-selskiya vupros v Bulgaria (1919–44)* (Impact of the October Revolution and Leninism on the Agrarian Issue in Bulgaria [1919–44]). Sofia, 1975.

MATEEV, B. *Dvizhenieto za kooperativno zemedelie v Bulgaria pri usloviyata na kapitalizma* (The Movement for Co-operative Farming in Bulgaria under Capitalism). Sofia, 1967.

MATEEV, E. *Balans na narodnoto stopanstvo* (The Balance of the Economy). 4th edn. Sofia, 1977.

MATEEV, E. *Upravlenie, efektivnost, integratsia v tursene na resheniya* (Management, Effectiveness, Integration in Seeking Decisions). Sofia, 1976.

Materialno-tehnicheskata baza no obshtestvoto v etapa na izgrazhdaneto na zreliya sotsializum (The Material and Technical Basis of Society at the Stage of Building Mature Socialism). Sofia, 1983.

Mashinostroeneto, elektronikata i elektrotehnicheskata promishlenost v Bulgaria (Machine Building, Electronics and Electrical Engineering in Bulgaria). Sofia, 1982.

Mehanizum za povishavane efektivnostta na obshtestvenoto proizvodstvo (A Mechanism for Raising the Effectiveness of Social Production). Sofia, 1984.

Modeli za perspektivno razvitie na ikonomikata (Models for the Long-Term Development of the Economy). Sofia, 1980.

Narodna Republika Bulgaria v mezhdunarodnoto ikonomichesko sutrudnichestvo i sotsialisticheskata integratsia (The People's Republic of Bulgaria in International Economic Co-operation and Socialist Integration). Sofia, 1979.

Natsionalniyat dohod v sotsialisticheskoto obshtestvo (National Income in the Socialist Society). Vols. 1–2. Varna, 1979.

PASKALEV, S. *Proizvoditelni sili i materialno-tehnicheska baza na sotsializma – NRB. Faktori v razvitieto im* (Productive Forces and the Resource and Technology Base of Socialism in the People's Republic of Bulgaria. Factors of Development). Sofia, 1976.

PENCHEV, P. and St. BANOV *Tyutyun i tsigari* (Tobacco and Cigarette Making). Sofia, 1960.

PEOVSKI, R. *Materialno-tehnicheskata baza na sotsialisticheskoto obshtestvo* (The Resource and Technology Base of the Socialist Society). Sofia, 1975.

Petdeset godini mini Pernik, 1891–1941 (Fifty years of Mining in Pernik, 1891–1941). Sofia, 1941.

Petiletkite v NR Bulgaria (Statisticheski danni) (Five-year Economic Plans in the People's Republic of Bulgaria. Statistical Data). Sofia, 1971.

PETKOV, P. *Likvidirane na kapitalisticheskata sobstvenost v promishlenostta na Bulgaria* (Abolition of Capitalism Ownership in Bulgarian Industry). Sofia, 1965.

POLYAKOV, T. *Mogusht vuzhod na nashata promishlenost v godinite na narodnata vlast* (The Powerful Upsurge of Bulgarian Industry in the Years of the Popular Regime). Sofia, 1954.

PORYAZOV, D. *Parite v sotsialisticheskoto obshetestvo* (Money in Socialist Society). Varna, 1969.

PORYAZOV, D. and T. KOTSEV *Parichno obrushtenie i credit pri sotsializma* (Money Circulation under Socialism). Varna, 1977.

Poyava i razvitie na TKZS v Bulgaria (Emergence and Development of Co-operative Farms in Bulgaria). Sofia, 1979.

RADULOV, L. *Politika na uskoreno ikonomichesko razvitie na sotsializma v Narodna Republika Bulgaria* (Policy of Accelerated Economic Development of Socialism in the People's Republic of Bulgaria). Sofia, 1976.

RUSSENOV, M. *Finansi na sotsialisticheskata durzhava* (Finances of the Socialist State). 2nd ed., Varna, 1973.

SHAPKAREV, P. *Otraslovata struktura i mezhduotraslovite vruzki na promishlenostta v NR Bulgaria* (Branch Structure and Inter-Branch Relations in Industry in the People's Republic of Bulgaria). Sofia, 1965.

SHARENKOV, S. *Ikonomicheskata pomosht na Suvetskiya Suyuz – reshavasht faktor za stroitelstvoto na sotsializma v Bulgaria* (The Economic Assistance of the USSR as a Decisive Factor of Socialist Construction in Bulgaria). Sofia, 1960.

SHARENKOV, S. *Ikonomichesko i nauchno-tehnichesko sutrudnichestvo mezhdu Narodna Republika Bulgaria i Suvetskiya Suyuz* (Economic and Scientific-Technological Co-operation between the People's Republic of Bulgaria and the USSR). Sofia, 1965.

SIMOV, B., B. BLAGOEV and O. ASLANYAN *Vuzstanovyavane i razvitie na promishlenostta v NRB. 1944–48* (Reconstruction and Development of Industry in the People's Republic of Bulgaria. 1944–48). Sofia, 1968.

Sotsialisticheska Bulgaria stroi (Socialist Bulgaria Engaged in Construction). Sofia, 1979.

Sotsialno-ikonomicheskata politika na bulgarskata durzhava. 681–1981 (The Socio-Economic Policy of the Bulgarian State. 681–1981). Varna, 1981.

Sotsialno-ikonomicheski vuzhod na NR Bulgaria (Socio-Economic Upsurge of the People's Republic of Bulgaria). Sofia, 1981.

Statisticheski godishnik na Narodna Republika Bulgaria (Statistical Year-Book of the People's Republic of Bulgaria). Sofia, 1983.

Sto godini bulgarska ikonomika (One Hundred Years of Bulgarian Economy). Sofia, 1978.

Sto godini Bulgarska industriya (One Hundred Years of Bulgarian Industry 1834–1937). Sofia, 1937.

Sto godini suobshteniya v Bulgaria (One Hundred Years of Communications in Bulgaria). Sofia, 1979.

Stopanska istoriya na Bulgaria. 681–1981. (Economic History of Bulgaria 681–1981). Sofia, 1981.

SYULEMEZOV, S. *Razvitie na kooperativnoto dvizhenie v Bulgaria* (The Co-operative Movement in Bulgaria). Sofia, 1975.

Tekstilnata promishlenost v Bulgaria. Minalo, Nastoyashte, Budeshte. (Textile Industry in Bulgaria. Past, Present, Future). Sofia, 1982.

Tehnicheskiyat vek na Bulgaria (The Technical Age of Bulgaria). Sofia, 1974.

Trideset godini ikonomika na NR Bulgaria (Thirty Years of Economy in the People's Republic of Bulgaria). Sofia, 1974.

Trideset godini sotsialisticheska Bulgaria (30th Anniversary of Socialist Bulgaria). Sofia, 1974.

Trideset i pet godini sotsialisticheska Bulgaria (The 35th Anniversary of Socialist Bulgaria). Sofia, 1979.

Vuzhod na selskoto stopanstvo v Bulgaria (The Upsurge of Agriculture in Bulgaria). Sofia, 1974.

YORDANOV, T. *Materialno-tehnicheska baza na razvitoto sotsialistichesko obshtestvo* (Resource and Technology Base of the Developed Socialist Society). Sofia, 1973.

Za uskoreno razvitie na nyakoi strategicheski napravleniya na nauchno-tehnichestiya progres v Narodna Republika Bulgaria. Materiali na Plenuma na Tsentralniya komitet na BKP, sustoyal se na 20 i 21 yuli 1978 (For an Accelerated Development of Some Strategic Branches of Scientific-Technological Progress in the People's Republic of Bulgaria. Materials of the Plenum of the Central Committee of the Bulgarian Communist Party held on 20–21 July 1978). Sofia, 1978.

ZHIVKOV, T. *Dvanadesetiyat kongres na BKP i po-natatushnoto izgrazhdane na zreliya sotsializm. Problemi, zadachi i podhodi* (The Twelfth Congress of the Bulgarian Communist Party and the Further Construction of Mature Socialism. Problems, Tasks and Approaches). 2nd enl. and rev. edn. Sofia, 1982.

ZHIVKOV, T. *Ikonomicheskata politika na Bulgarskata komunisticheska partiya* (The Economic Policy of the Bulgarian Communist Party). In 3 vols. 1982.

ZHIVKOV, T. *Problemi i podhodi na izgrazhdaneto na zreliya sotsializum v NR Bulgaria* (Problems of and Approaches to the Construction of Mature Socialism in the People's Republic of Bulgaria). Sofia, 1984.

PART VI

SOCIAL POLICIES

LIVING CONDITIONS OF THE PEOPLE

Until the socialist revolution in September 1944 the standard of living of the vast majority of the population was very low. In fact, Bulgaria held one of the last places among European countries in this respect. There was also large-scale unemployment. The earnings of the working people were insufficient to satisfy even their most basic needs. The consumption of goods and services was extremely limited. Most of the housing was primitive and unhygienic, and furnishings scarce. Community services were inadequate, and the lack of proper hygiene and safety of labour in industry made for a high morbidity and mortality rate.

The social and economic changes after 1944 created conditions for a rapid growth of the country's economic productive capacity and for a steady rise in the standard of living of the population. The overall quality of life began to improve gradually and the social, spiritual and cultural aspects of the professional and everyday life of people became richer and more diverse. A considerable role in raising the standard of living of the working people was played by the April 1956 Plenum of the Central Committee of the Bulgarian Communist Party, which placed economic and social policies on a scientific basis. The December Programme, adopted by the Party in 1972, was a concrete manifestation of that new approach. That programme was aimed above all at improving the people's living conditions. The policies adopted for social and economic development followed the motto 'Everything for man, everything for the benefit of man' thus meeting ever more fully the material, social and spiritual needs of the people.

The material basis for raising the Bulgarian people's standard of living was provided by the planned development of production and the steady increase in the national income. During the 1961–82 period the national income of Bulgaria grew at an average annual rate of 7.1 per cent. This economic growth rate generated sufficient means for further increasing economic productive capacity and for raising the standard of living. There was no unemployment. During that period about three-quarters of the national income was used for consumption and one-quarter for accumulation. The consumption fund increased at an average annual rate of 6.0 per cent. Part of the accumulation fund (resources for the construction of housing, schools, hospitals, cultural and educational establishments, sports facilities, enterprises for communal services) also contributed to raising the standard of living. Thus about 80–85 per cent of the national income was used to meet material and social needs.

The consumption fund is made available to the population mainly through wages and salaries, the size of which depends on the quantity and quality of labour invested and the results achieved. An increasingly important role in the growth of consumption is played by the centralized and decentralized social consumption funds. The two factors that determine the amount of incomes ensure social equality and justice. In Bulgaria there are no suitable economic conditions for unearned incomes. All industrial enterprises, land, commerce, financial and banking establishments, etc. are state property and the work-forces employed in the economic enterprises participate in their management. With this kind of ownership all people have an equal position. The differences in income depend solely on the differences in the quantity and quality of their labour and the different results they achieve.

The raising of the standard of living is also reflected in the substantial structural changes in the needs being met. The part of their income which the working people spend on cultural needs is growing constantly. The health service network has broadened, the number of health workers has increased and the quality of medical technology and equipment has improved considerably. All working people are socially insured in case of temporary or permanent disability (see the chapter on Social Security, pp. 608–9). Working conditions are constantly improving. Great attention is given to preventing pollution of the soil, air and water. A long-term policy has been adopted for reducing major differences between the incomes of various groups of working people and also between regions of the country. Conditions are thereby gradually being created for the versatile development of all people and for the achievement of social homogeneity.

Incomes of the Population. The changes in the incomes of the population are the result of substantial changes in the productive forces and production relations, the structure of the economy, the quality of the labour force and the social composition of the population. The sharp contrast between wages in industry and agriculture, inherited from Capitalism, has been overcome. The productivity of labour and the incomes of the population are growing at a high and steady rate (see Table 1).

Table 1: Labour Productivity, Incomes and Wages

Indices	1982 in % as compared with 1958	Average annual growth rate in %
Social productivity of labour (national income per one person employed in material production)	674	7.95
Full incomes of the population	677	8.5
working incomes	578	7.4
incomes from the social consumption funds	1,237	
Nominal income per capita	566	7.4
Real income per capita	352	5.45
Average wages	291	4.35
Real wages	207	3.0

As a result of the expansion of the material and technical basis of the improvement in productive forces, the productivity of labour in 1982 was 674 per cent higher than in 1958. i.e. it grew at an average annual rate of 7.95 per cent. This growth has contributed to the high rate of increase of the domestic product and national income, to the higher effectiveness of production, to increase in the incomes of society and of the population and to the ever fuller satisfaction of its needs.

The consistent upward trend is the most characteristic feature of the development of incomes. It must also be pointed out that the overall incomes of the three main social groups of the population – workers, farmers and employees have developed in a similar manner. Per household member the respective growth rates are: for workers – 267 per cent, for employees – 256 per cent and for co-operative farmers – 232 per cent. Significant changes have taken place in the pattern of household incomes among the social groups. Thus, for example, the relative share of the incomes from pensions, benefits and scholarships for the groups of workers and employees increased from 10–11 per cent in 1965 to 17–17.5 per cent in 1980, and for the co-operative farmers from 7.7 to 32.9 per cent. The latter change is due to the inclusion of all peasants in the social security system. The relative share of the incomes from working wages has remained relatively steady for households. The level of employment of household members is even more indicative of the sources of income. The relative percentage of working household members increased: for workers from 46.7 per cent to 48.5 per cent, for employees from 47.1 to 50.8 per cent, whereas for co-operative farmers it declined from 52.6 to 20.3 per cent. The main cause for this lies in the increased migration of young people from villages to towns under the conditions of intensified scientific and technological progress. The tendency towards reduction of the relative share of households with lower incomes manifested itself also in the process of socialist construction. There was a substantial narrowing of the gap between the incomes of workers, employees and co-operative farmers. The principles and conditions underlying the formation and regulation of incomes have become more or less similar for the various social groups. This has been facilitated by the policy of overall intensification and industrialization of agricultural production and the reduction of differences in the conditions and content of agricultural and industrial labour.

The Socialist State devotes particular attention to monitoring the changes within the middle and lower wage brackets. The average annual growth rate of minimum wages has come close to the growth rate of average wages. Over the past ten years minimum wages marked a higher growth rate (see Table 2). This resulted from the necessity to provide the means for the normal continuity of the labour force on all levels.

Social Consumption Funds. The social consumption funds have played an increasingly important role in raising the incomes and standard of living of the population. Those funds include the part of the consumption fund earmarked for education, culture and art, health services, social security, physical education and sports, tourism and other similar purposes. By means of the social consumption funds the needs of people in the areas indicated above are met, regardless of their performance at work. The largest part of those funds is used by the younger generation and by pensioners.

Owing to the low level of development of the productive forces, the possibilities for allocating funds for social consumption were rather limited after the victory of the socialist revolution. In 1952 their total volume was 248.1 million levs (87.8 million levs for education, 11.5 million levs for culture and the arts, 143.1 million levs for health services, social security, physical education, sports and tourism and 5.7 million levs for other needs). With the development of the productive forces, as the rates of economic and national income growth were stepped up, the social consumption funds increased and their structure improved. This tendency became more marked after 1956, when important corrections in the economic and social policies were made. In 1956 the social consumption funds stood at 362.5 million levs. In comparison with

Table 2: Growth Rates of Average and Minimum Wages

Indices	Periods			
	1957–75	1960–75	1965–75	1970–82
Average annual growth rate of average wages (in %)	4.15	4.00	4.25	4.20
Average annual growth rate of minimum wages (in %)	3.95	3.20	6.40	4.60

Table 3: Social Consumption Funds by Sector [1] (in million levs)

	Years					
	1952	1956	1960	1970	1980	1982
Total	248.1	362.5	700.2	2,131.2	5,288.5	5,991.9
Education	87.8	129.4	166.7	454.8	1,221.4	1,892.9
Culture and art	11.5	14.8	19.7	72.2	166.2	227.6
Public health, social security, physical education, sports and tourism	143.1	211.6	471.7	1,443.8	3,528.1	3,993.2
Others	5.7	6.7	42.1	160.4	370.8	468.2

[1] **Not including remuneration during regular paid leave and any additional leave.**

1952 the increase was 114.4 million levs (41.6 million levs for education, 3.3 million levs for culture and the arts, 68.5 million levs for public health, social security, physical education, sports and tourism, and 1 million levs for other needs).

In the period 1956–82 the social consumption funds grew very rapidly. Their total amount went up from 362.5 million levs to 5,991.9 million levs, which is more than a 1,600 per cent increase. In the different spheres of social policy the increase is as follows: education – 1,183.5 million levs; culture and the arts – 212.8 million levs; public health, social security, physical education, sports and tourism – 3,781.6 million levs; and for other needs – 461.5 million levs. The total amount of the social consumption funds doubled every two years during the period under consideration.

It is a genuine proof of the humane character of Socialism as a social system and of the fact that the development is subordinate to the process of meeting human needs ever more fully, and to the necessity of developing versatile and harmonious personalities. The structural changes in the social consumption funds are highly indicative of this trend. The funds allocated for public health, social security, physical education, sports and tourism have increased over 1,800 per cent; those set aside for culture and art – over 1,600 per cent; and those for education – about 1,000 per cent. The most significant increase (about 7,000 per cent) is that of the funds allocated for other needs – the overhead costs of canteens at enterprises and offices, the depreciation of the state housing stock not covered by rents, etc. This shows that the range of needs to be met has widened and the degree to which they are met has increased. The social consumption funds marked the highest increase in the 1960–70 period – almost 300 per cent, and during the 1970–80 period – almost 200 per cent. The increase of the funds in the different sectors is similar (see Table 3).

The average annual amount of the social consumption funds per capita rose from 34 levs in 1952 to 672 levs in 1982, which is a 2,000 per cent increase. The dynamics of the increase are as follows: 1956 – 48 levs, or 14 levs more than the 1952 level; 1960 – 89 levs or almost twice as much as in 1956; 1970 – 251 levs or almost 300 per cent over the 1960 level; 1980 – 597 levs or over 200 per cent more than in 1970; 1982 – 672 levs or by 75 levs over the 1980 level. Funds allocated for public health, social security, physical education, sports and tourism account for the greatest share of per capita social consumption funds. Funds allocated for education take second place. At the present stage we are witnessing a process of meeting the needs of the population increasingly fully at the expense of the social consumption funds; in future they will also meet part of those needs which at present are met by personal incomes.

At the current stage, under the modern conditions and possibilities for development, the social funds have

Table 4: Social Consumption Funds in Terms of Sources of Finance (in per cent)

Indices	Years					
	1952	1956	1960	1970	1980	1982
Total	100.0	100.0	100.0	100.0	100.0	100.0
From the State Budget	62.3	62.3	44.7	34.8	41.0	40.1
From insurance institutions	34.8	33.1	44.9	50.4	47.8	47.1
From enterprises, organizations and other institutions	2.9	4.6	10.4	14.8	11.2	12.8

become an important factor in raising the material well-being, in meeting ever more fully the growing needs of the population, for a further establishment of the socialist life-style. The relative share of the social consumption funds in the total nominal incomes of the population was constantly increasing – from 5.1 per cent in 1952 and 7.1 per cent in 1956 it went up to 19.4 per cent in 1975 and 21.7 per cent in 1982. The social consumption funds are developing as a unified system. The basic resources come from the insurance institutions, and from enterprises, organizations and other institutions.

For the 1975–82 period the total social consumption funds showed an average annual increase of 7.7 per cent and in 1982 their amount exceeded 700 million levs. The rise in the social consumption funds and the greater participation of enterprises, organizations and other institutions in their formation enhances their role and significance in socialist society and increases the economic and social efficiency of their use. Conditions are provided for workers whose performance is outstanding or who have more children to have priority in the use of these funds. This promotes the process of making the social consumption funds and additional material and moral incentive for raising labour productivity and increasing social production.

The amount of social consumption funds allocated for the development of public health, social security, physical education, sports and tourism has been marked by high growth rates. In the period 1961–82 they increased about 600 per cent, reaching almost 4,000 million levs in 1982. The resources used for these purposes in 1982 represent about two-thirds of the total sum of the social consumption funds (58 per cent in 1952). The free medical aid, introduced as early as 1951, is assessed as one of the most significant social gains of the Bulgarian people (see chapter on Public Health Services).

Bulgaria has to its credit one of the most humane and democratic systems of pension insurance, which is being constantly improved and developed further. The retirement age is lower than that of many developed capitalist countries. Socialist Bulgaria was the first country in the world to have introduced, by law, state old age and disability pensions for agricultural workers. The upkeep and raising of children is a prime concern in the social policy. The social consumption funds are used to pay monthly family allowances, to provide lump-sum cash grants at the birth of a child and to make up the difference between the cost of accommodating children in child-care establishments and the fees paid by the parents. Each year tens of millions of levs are spent on benefits of all kinds (see chapter on Social Security).

Considerable funds are allocated for the development of education, which is available free of charge. In 1982 1,300 million levs were spent on education i.e. over 500 per cent more compared to 1960. The sums for scholarship grants went up from 5.5 million levs in 1952 to 70.3 million levs in 1982. The system of boarding-school and semi-boarding-school education was enlarged. All elementary school pupils are given textbooks free of charge. The number of schools providing meals (breakfast and lunch) free or at token prices increased. All maintenance costs of refectories and students' canteens are covered by the State. Rents in the modern students' hostels are very low (see Education section).

The funds allocated for the development of socialist culture and art increased over 800 per cent in comparison with 1960 and in 1982 they reached almost 228 million levs. Each member of Bulgaria's socialist society has equal access to the arts and culture.

The importance of the social consumption funds is constantly growing; they have an ever more direct influence on all economic and social processes and on the further establishment of the socialist life-style. The Socialist State strives to streamline the process of their formation and use, and to achieve higher economic and social efficiency. What this actually means is that they should play an increasing role in the quantitative and qualitative continuity of manpower, and in the efforts to stabilize the labour force within the enterprises, as well as in creating more favourable conditions of work and living. Of special importance is their role in shaping such qualities within the individual as would allow his full self-realization.

Consumption. Per capita consumption is an important indicator of the living standard. It is directly related to the production of manpower. The practice in the People's Republic of Bulgaria shows that the all-round development of the individual, the development of his professional training, general knowledge, creativity and abilities, depend to a large extent on the satisfaction of the material

Table 5: Indices of the National Income, the Consumption Fund and the Per Capita Nominal Incomes

Indices	Years						
	1956	1960	1965	1970	1975	1980	1982
Per capita national income	100.0	151.3	206.8	295.2	418.7	554.0	592.1
Per capita consumption fund	100.0	132.3	174.9	238.1	326.2	391.4	421.6
Per capita nominal incomes	100.0	147.4	204.3	297.8	413.1	573.9	636.9

and spiritual needs of the people, on the volume and structure of consumption. During the 1956–82 period there was a steady increase in the per capita volume of consumption in Bulgaria, based on the increase of the domestic product (830 per cent), the national income (710 per cent), and the consumption fund (500 per cent).

Compared to the domestic product and the national income, the rise in the consumption fund was smaller, which is due to the allocation of larger sums for building the material and technical basis of society. The growth rates of the national income, the consumption fund and the per capita nominal incomes are highly indicative in this respect (see Table 5).

The growth in the country's economic potential increased the possibilities of meeting more fully and comprehensively the steadily increasing needs of the people. This was primarily reflected in the continuing expansion of the per capita volume of consumption. In 1982 the total consumption was over 400 per cent higher than in 1956. The growth was greatest between 1970 and 1975, and between 1975 and 1980.

The ever rising possibilities of satisfying the population's needs in goods and services showed qualitative, as well as quantitative changes. Characteristic of the structure of personal consumption was the shrinking share of foodstuffs in the total volume, from 49.7 per cent in 1960 to 39.4 per cent in 1982, and the increase in the share of goods other than foodstuffs, and of material services, from 50.3 to 60.6 per cent. The expectations are that the tendency will continue to develop towards a speedier expansion (in comparison with the personal consumption of foodstuffs) of personal consumption of goods other than foodstuffs, and material services, including durable goods and services. Thus, as regards the volume and structure of personal consumption, Bulgaria will be drawing nearer to the most highly developed socialist and capitalist countries.

The socialist reconstruction of agriculture, the concentration and the technical re-equipment of agricultural production resulted in further raising of the material, cultural and daily living standards of agricultural workers, and in radical changes in their social profile and psychology. This created, in its turn, the necessary prerequisites for overcoming the differences between personal consumption expenditures of the various social groups in Bulgaria. A characteristic trend was the continuous growth and improvement in structure of personal consumption expenses per household member. Also characteristic was the trend toward narrowing the gap between household expenses of farmers and those of workers and employees, for food, housing and furniture, hygiene needs etc. Greater differences existed where other needs, e.g. clothing and footwear, cultural and social life, post, transport services etc. were concerned.

In the conditions of socialist development the volume of basic foodstuff consumption quickly increased. The structure of their average annual consumption improved simultaneously (see Table 6).

This was mainly due to the dynamic economic development, including foodstuffs production, and the increasing purchasing power of the population.

Bulgaria ranks among the leading countries in the world in the per capita consumption of fruit and vegetables. The consumption of meat, milk, eggs, fresh and tinned vegetables, fresh and tinned fruit is getting steadily nearer to the annual consumption norms: 80 kg of meat, 180 kg of fresh and tinned vegetables per capita, 200 kg of fresh and tinned fruit, 260 litres of milk and dairy products and 265 eggs. It is expected that by 1990 the consumption norms will be met. This will result in an even greater improvement in the nutrition patterns. As far as the average daily consumption in terms of calories is concerned, Bulgaria is approaching the developed countries. Therefore no increase in the daily intake of calories is envisaged, but efforts will be made to improve its pattern by reducing the share of carbohydrates and increasing the share of predominantly animal protein calories.

A large share of personal consumption goes to the products of the textile, footwear, and leather and fur industry. The consumption of their products is growing quickly (see Table 7).

Table 6: Gross Consumption of Staple Foodstuffs Per Capita

Products	1960	1965	1970	1975	1980	1982	1980 as compared to 1960 (in per cent)
Bread and bakery products (kg)	261.4	285.4	238.8	217.8	216.0	210.0	99.3
Meat (kg)	29.1	39.6	41.4	58.0	61.2	68.3	284.7
Fish (kg)	2.2	3.4	5.5	6.2	6.9	7.5	340.9
Vegetable oils and animal fats (kg)	14.0	13.3	16.2	18.7	20.8	21.9	158.4
Milk and dairy products (in litres of milk)	92.3	103.6	116.6	142.8	169.4	179.2	194.1
Sugar and confectionery (kg)	17.7	22.3	32.9	32.5	34.7	35.1	198.3
Eggs (number)	84	100	122	146	204	217	258.3
Vegetables (kg)	97.2	88.8	88.9	90.1	93.8	104.9	107.9
Fresh and tinned fruit (kg)	95.3	131.4	148.2	118.6	105.8	118.9	124.8

Table 7: Gross Per Capita Consumption of Goods other than Foodstuffs

	1960	1965	1970	1975	1980	1982
Cotton fabrics (sq. m)	16.4	19.8	22.2	23.8	25.9	27.5
Woollen fabrics (sq. m)	3.0	3.2	3.8	5.5	4.6	4.7
Silk fabrics (sq. m)	0.9	1.3	2.1	3.4	4.0	4.1
Cotton and woollen knitwear (number)	4.6	6.0	8.5	10.5	12.5	13.0
Shoes (excluding rubber shoes; pairs)	0.9	1.3	1.7	2.0	2.0	2.3

Table 8: Availability of Durable Goods per 100 Households

	1965	1970	1975	1980	1982
Radio sets	59	62	76	88	91
TV sets	8	42	66	75	83
Electric washing machines	23	50	59	71	77
Household refrigerators	5	29	59	76	84
Cars	2	6	15	29	32
Telephones	—	7	12	24	30

According to the forecasts of the relevant research institutes, on the basis of present achievements, existing possibilities and the structural changes in the population's incomes, as well as other factors, it is expected that the scientifically based norms for consumption of light industry products other than foodstuffs will be attained by 1990.

The increased incomes, improved cultural and living standards of the population, and growth of the country's economic potential has increased the share of durable goods in the structure of the total personal consumption fund (see Table 8).

Conditions were created for solving the housing problem. Nowadays, a large-scale housing construction programme is being carried out in towns and villages, radically changing the look of residential areas. At the end of 1982 the available housing stock amounted to 2,970,000 dwelling units. Nearly three-quarters of them were built after the socialist revolution. Between 1958 and 1982, 1,332,000 dwellings units were built, or 45 per cent of the available housing stock. The per capita living space in 1982 was 15.4 sq. m.

Improvements are also being made in housing mainte-

Table 9: Housing and Communal Economy (1970=100)

Indices	Years				
	1965	1970	1975	1980	1981
Housing and communal economy and communal services	50.0	100.0	128.7	220.2	227.8
Housing maintenance	49.2	100.0	105.7	221.8	230.1
Sewerage system	87.5	100.0	162.5	375.0	412.5
Sanitation of inhabited places	56.5	100.0	147.1	303.5	314.1
Hotel economy	42.7	100.0	205.7	378.4	385.4
Public baths and swimming pools	70.8	100.0	113.2	90.6	94.3
Hairdressing and beauty parlours	80.6	100.0	108.3	124.8	134.5
Other facilities and communal economy, including housing stock owned by different enterprises, departments, etc.	28.3	100.0	57.6	99.0	105.4

Table 10: Communal Services for the Population

Indices	Years			
	1970	1975	1980	1982
Share of the population living in places supplied with water (in per cent of the total)	91.6	95.4	97.3	97.7
Share of the population living in places supplied with electricity (in per cent of the total)	99.6	99.8	99.9	99.97
Dwelling units supplied with central heating, per 1,000 dwelling units in towns with central heating installations (numbers)	171	242	228	249
Hotel beds per 10,000 of the population (number)	82	113	115	116
Capacity of laundries per 10,000 of the population (kg)	100	167	206	216

nance and communal services. Over 400 types of services are available nowadays in a comprehensive system of service centres, industrial enterprises, workshops, dressmaking, tailoring and other shops, and complex communal service centres (see Tables 9 and 10).

In accordance with the theses adopted by the Twelfth Congress of the BCP, the standard of living of the Bulgarian people will continue to rise. In 1984 the monthly wages and salaries of the graduates of secondary special schools are being raised by 8–13 per cent, and of the university by 10 per cent. The minimum wages are being raised by 11 per cent and the minimum old-age pensions are also being increased. Medical personnel and teachers are also receiving considerable rises. In 1985 the incomes of other categories of workers and employees will continue to increase. By the end of 1985, canteens and other catering establishments of enterprises and offices will provide meals for nearly 70 per cent of the population. Important steps are being taken towards environmental protection and reproduction; much is being done to introduce improvements in the home environment and daily living, to improve the conditions of labour and to expand and better organize leisure time. The continuous improvement of the Bulgarian people's standard of living is a cause for satisfaction in the country's level of economic development, and for optimism and faith in the future.

PUBLIC HEALTH SERVICES

Foundations of Public Health Policy. According to the programme of the BCP, adopted at its Tenth Congress (24 April 1971) 'The care of man is the main concern of the Party, the meaning and content of its entire policy'. In this connection a characteristic feature of public health policy is the satisfaction of the biological needs of man for staple foodstuffs; the provision of housing, corresponding to the fundamental hygienic norms and requirements of the socialist way of life; the building up of inhabited places in accordance with requirements for undisturbed work, living and rest and the comprehensive development of the individual; continuing efforts for the improvement of sanitary and hygienic conditions in production, everyday life and in inhabited places; protection of the natural environment against mechanical, biological, chemical and other harmful factors. For preserving the health and prolonging the life of the working people, the party programme provides for the carrying out of far-ranging socio-economic, medical and physical education undertakings, and a constant improvement in the technical and sanitary conditions of work and everyday living.

The Constitution of the People's Republic of Bulgaria (adopted on 16 May 1971) guarantees Bulgarian citizens the right to good health. The State devotes comprehensive care to the people's health by organizing therapeutic, preventive and other health establishments and services; the state and public organizations disseminate health education and culture among the people and encourage the development of physical education and hiking; every citizen has the right to free medical aid; and the State and the public organizations devote particular care to the health of children and adolescents. The Constitution guarantees the protection and preservation of nature and the natural resources of water, air and soil; the right to safe and hygienic working conditions; the right to rest; social security; a pension and assistance in case of incapacity for work due to sickness, accident, maternity, disability, old age or death, and in bringing up children. Mothers enjoy protection by being assured leave before and after childbirth, without loss of remuneration, and with free obstetric and medical care and alleviation of their type of work; and expansion of the network of child-care establishments, communal services and public catering establishments.

The provisions of the Constitution have been further elaborated in the Public Health Law, which regulates social relations in the protection of the people's health, and aims at the creation of favourable conditions for an all-round, harmonious development and a long, active life for all citizens, as well as for improving the reproduction of the population. The basic stipulations of the Law are: guaranteeing the right to free medical aid for every citizen of the country; all-round care by the State for the preservation and restoration of the citizens' health and for providing them with accessible and qualified medical aid.

The Bulgarian state system of socialist public health services is unified in structure, goals, tasks and methods of work in accordance with the principles of democratic centralism. The preservation of the people's health is based on the unity of prevention and treatment, which is free of charge, accessible to all and highly qualified. Major components of the system are medical science and medical education, supervised by the Ministry of Public Health and developed in close relationship with public health practice, with the active participation of the populace. The services are built on the basis of economic achievement and contribute, by preserving and improving the health and working capacity of the nation, to the consolidation of the Socialist State in which the people are the greatest national asset. They draw constantly on the rich experience of the public health services of the USSR, the socialist and other countries.

Development of Public Health Services in Bulgaria. Many data revealed by historical and archaeological investigations and cultural monuments from ancient Bulgaria testify to the presence of developed medicine and health care, combining the traditions and experience of the ancient Thracians, Slavs and Proto-Bulgarians. Medicine, which was then connected with religion, was distinguished for its wealth of empirical knowledge and methods and the use made of medicinal herbs, while the standards of sanitary construction, as can be judged by the water supply system of the first Bulgarian capital, Pliska, were quite high. Bulgarian medicine reached a high point in its development after the adoption of Christianity by the Bulgarians (864), when the dissemination of medical knowledge began, helped by the creation of a Slav alphabet and literature. The replies of Pope Nicholas I to the questions of the Bulgarians (106 questions were addressed by Prince Boris I to the Pope in 866) testify to the great interest of the people in matters of health, and to the medical and hygienic knowledge and practice in Bulgaria, which had reached the scientific level characteristic of the period. The disciples of Kliment of Ohrid (*c.* 840–916) studied hygiene and the treatment of diseases. Among the most significant works of this period are those of John the Exarch (ninth–tenth centuries) – *Hexameron,* in which are reflected the first Slav natural science terminology and certain medical data, and *Description of the Human Body,* the first work of its kind in the literature of the Slav peoples on the anatomy of the human body and the functions of its organs. Other valuable documents dating from those times are: *Lekarstvenik* (List of Medicines) by John of Rila (*c.* 876–880 to 947) and the book on anthropology, in the Old Slavonic language, written by Kliment of Ohrid. From these works it can be seen that methods and means were sought for treatment, but that there was also an effort to discover the causes of diseases. The medical profession was regulated; there are data on the high level attained by surgery, asepsis and antisepsis. The first hospitals in Bulgaria were set up at the monastery of St. Pantheleimon in Ohrid (ninth century) and at the Bachkovo Monastery (eleventh century). In the tenth century, at the order of Tsar Samuil, the swamps were drained near the head of the Drin river, which has its source in lake Ohrid, in order to stamp out malaria.

During the first half of the tenth century the Bogomil teaching made its appearance in Bulgaria, exerting a strong influence on the development of medicine and health care. The religious and legal norms of the Bogomil movement included a regimen of hygiene and diet, the complete exclusion of the use of alcohol and the banning of ritual practices in the treatment of diseases. Wide use was made of medicinal herbs. Priest Bogomil, the founder of the movement, enjoyed particularly wide popularity among the healers, as did Vassili Vrach (Basil the Healer; died 1111), one of the main theoreticians of Bogomilism. The Bogomil healers also compiled collections of recipes and directions for treatment. The oldest list of medicines in the territory of Bulgaria was called *Zelenik* (tenth–eleventh centuries).

The fall of Bulgaria under Ottoman rule (1398) had an adverse effect on public health. Up to the middle of the nineteenth century the poor sanitary conditions in the country led to the development of devastating epidemics of plague, cholera, typhoid fever, small pox and typhus. General and infant mortality was very high. This was a period in which nothing is mentioned of organized medical assistance. When in need, the population sought the folk healers and medical aid in the Bulgarian churches and monasteries. Popular methods of treatment were applied, herbs and natural factors being used. At the same time, the Bulgarian populations's initiative in health matters was developed. The Bulgarian communities became initiators of a number of socio-sanitary undertakings, maintaining and developing the people's traditions of cleanliness in the home, creating sanitary facilities in inhabited places and propagating rules of personal hygiene and a hygienic way of life. The first manifestation of non-professional public health activities was the so-called 'health insurance' introduced by the Bulgarian municipalities in the towns of Stara Zagora, Koprivshtitsa and elsewhere during the second half of the nineteenth century. Gradually a Bulgarian medical intelligentsia made its appearance. Young men, with their own means or supported by public benefactors, graduated in medicine in Russia, Western Europe, Athens, Bucharest and Constantinople and began to work under extremely hard conditions, some of them rising to become professors in the higher medical school in Constantinople. Side by side with providing qualified medical aid, the Bulgarian doctors embraced the progressive ideas of their times and took an active part in the struggle for the spiritual and national liberation of their people. Prominent among them were Sava Pikolo (1792–1865), Atanas Bogoridi (*c.* 1788–1826), and Ivan Seliminski (1798–1897). Peter Beron (1800–71) won European fame as a person of encyclopaedic learning. On his initiative medical knowledge began to be taught in a number of schools. In the years prior to the Liberation of Bulgaria from Ottoman rule the condition of the population's health was not studied, nor was it an object of observation on the part of the authorities. In the nineteenth century seven military and five civilian Turkish hospitals were built in the territory of Bulgaria (in the towns of Shumen, Russe, Vidin and elsewhere). The first Bulgarian hospital was built in 1867 in Turnovo with funds donated by the Bulgarian population.

In the Russo-Turkish War of Liberation (1877–78), parallel with the care for the wounded and sick in the army, the Russian Administration also organized medical aid for the population. For its part the Bulgarian population selflessly helped the Russian army medical services and with their own means set up infirmaries and field-hospitals. Even during the course of the war, medical care for the civilian population was initiated and together with the attention devoted to the development of hospital care a hygiene service was set up and prophylactic measures were undertaken. N. I. Pirogov, S. P. Botkin, N. V. Sklifassovski and other Russian doctors rendered great assistance.

After Bulgaria's Liberation from Ottoman rule (1878), the provisional Russian Administration organized medical and pharmaceutical services in the country. Services by regional and district doctors were established; in 1878 there were doctors in 28 towns. The beginning of organized public health services in Bulgaria was marked on 4 February 1874, when the Russian imperial commissioner in the country, Prince Alexander M. Dondukov-Korsakov, approved provisional rules for the organization of medical services. These consisted of provisional rules for the structure of a medical administration, a hospital statute and provisional rules for the organization of pharmacies. The supreme administrative body – the Supreme Medical Council – set up in 1879, was also guided by the provisional rules. In the first fifteen years after the Liberation of Bulgaria, nine first-class hospitals were established (in Sofia, Plovdiv, Russe, Varna, Turnovo, Pleven, Vidin, Sliven, Lovech), each one with up to 200 beds and specialized wards, and fifteen second-class hospitals, with up to 60 beds each. In certain country towns third-class hospitals were started with up to 25 beds each. According to the provisional rules, out-patient treatment and the medicines dispensed in the hospitals were free of charge, and for people producing a certificate of poverty, hospital treatment was also free of charge. Compulsory registration of those suffering from contagious diseases was introduced. A list was made of the doctors authorized to practise in the country – at the beginning of 1880 there was a total of 71 doctors in Bulgaria (53 Bulgarians and 18 foreigners). The first civil school for assistant doctors was opened in 1879 in Sofia. The pharmacies already had their own general rules. In 1880 a chemical laboratory was opened at the Supreme Medical Council, which was the first scientific medical establishment in the country. In 1888 the Provisional Rules were replaced by a Sanitary Law. The supreme body for administering the public health services became the Civil Sanitary Department under the Ministry of the Interior. The system of health services began to be increasingly penetrated by private capitalist tendencies, and the progressive traditions of Russian medicine which had been introduced after Bulgaria's Liberation by the Russian doctors who had worked in Bulgaria, were gradually abandoned. Prominent workers in the field of Bulgarian public health resisted these tendencies. In 1895 in Sofia the first school for midwives was opened and, in 1900, a school for medical nurses. In 1901 the Bulgarian Doctors' Union was founded. The Law on the Protection of Public Health (1903, amended and supplemented in 1909) permitting private practice by assistant doctors in the rural communities, gave the right to private persons to open hospitals and did away with the institution of the sanitary inspectors. In 1912 there were 640 doctors in Bulgaria. In the period of the Balkan War (1912–13) and of the First World War (1914–18), in spite of the poor sanitary conditions in the country, a movement was begun for the socialization of medicine and sanitary facilities. This was the beginning of organized industrial-sanitary supervision. Rural health services were developed. The people at large were actively involved in public health and a great number of Popular Hygiene Councils were established (in 1918 there were 2,341), which developed activities aimed at better sanitation and control of communicable diseases. Public-minded doctors studied the incidence of children's diseases and infant mortality, tuberculosis, malaria and other diseases, and working conditions (particularly of women and children) in the factories, and worked for more widespread prevention and the creation of a Ministry of Public Health. After World War I the Union of Medical and Health Workers was set up, which fought for the implementation of the progressive ideas of social preventive medicine. On 10 April 1918 the faculty of medicine at Sofia University was opened, whereby conditions were created for the systematic training of doctors in Bulgaria. In 1926 there were only seven maternity and child health centres, set up in Sofia on the initiative of the people at large. The Law on Public Health (1929) solved only some of the problems pertaining to therapy; the development of hospital care, hygiene and epidemiological services was lagging behind. In 1930 only 23 per cent of the population was supplied with water from modern water mains. Only five towns were fully supplied with sewerage systems; another eleven had such systems covering only a section of the town. In 1940 a doctor in a rural area served 13 times more people than a doctor in the town. In 1944 in Bulgaria there were 3,280 doctors, 73 state hospitals with a total of 7,265 beds, 62 private hospitals with 1,476 beds and 17 out-patient clinics providing free treatment for the poor. Medical aid at a state medical establishment was paid for; only those who had a poverty certificate were treated free of charge.

Public health was a central concern in the propaganda of socialist ideas in Bulgaria. The BCP was the initiator and bearer of progressive and revolutionary ideas concerning health services for the people and led the struggle for their implementation. As early as 1891 Dimiter Blagoev, the founder of the BCP, wrote that the health services had to become public, to be socialized. In the first party programme (1893) the demand was raised to introduce free medical and obstetric assistance. The programme of a group that called itself 'Social Doctor', established in 1931, argued the need for socializing the public health services, for unity in preventive and therapeutic care, for registering the population in out-patient clinics for permanent observation of its health and for its active participation in the public health services. In 1940 at the 25th Doctors' Rally, the Bulgarian Doctors' Union adopted a draft state health plan which expressed the doctors' progressive ideas concerning the public health services. In 1944 the Bulgarian medical community entered new socialist Bulgaria historically ripe and progressively oriented, ideologically prepared to realize the principles of socialist public health for ensuring the good health and working capacity of the people.

The building up of the system of public health services founded on the principles of socialist health started immediately after 9 September 1944. In the first programme of the Fatherland Front Government, announced on 17 September 1944, public health was treated as a paramount state concern with the main emphasis on preventive care. For the first time an independent Ministry of Public Health was set up, on 9 November 1944. At the 26th Rally of the Bulgarian Doctors' Union (1945) Georgi Dimitrov pointed out: 'The health and working capacity of the people are the nation's most precious asset... The new popular regime must not only make up quickly for what the fascist regime of traitors failed to do for the health of the people, but also must make good what they did in ruining the people's health and the working capacity of the population'. The first few years after 9 September 1944 were characterized by far-reaching activity for the health protection control of mass diseases, expansion of the social security system and training of medical personnel. Under the Law on Social Security (1945) about 1.5 million industrial and office workers and their families received the right to free medical aid. In the programme declaration of the Government of the People's Republic of Bulgaria, adopted by the Great National Assembly (13 December 1946) and in the Constitution (1947), public health is placed at the centre of the social policy. In 1949 the foundation was laid of a general reorganization of public health on the basis of the experience and principles of the Soviet health services. The hospitals and pharmacies were nationalized. The construction of health establishments was launched on a large scale. In 1950 the hospitals and polyclinics were combined and a unified network for therapeutic and prophylactic care was set up on the medical district principle, and the first specialized medical research institutes were established. A State Sanitary Inspection Board was set up which organized sanitary control, and also sanitary and epidemiological stations (after 1971 called hygiene and epidemiological inspection boards). A turning-point in the development of public health was the introduction in March 1951 of free and highly qualified medical aid for the people – one of the most important social gains of the population.

The April 1956 Plenary Session of the Central Committee of the BCP gave a fresh impetus to the development of public health. Health protection, as an inseparable part of the social policy of the State and an essential feature of the socialist way of life, is guaranteed by the Constitution of the People's Republic of Bulgaria (1971) and by the Law on Public Health (1973). The public health services are organized, guided and financed by the State. Private medical practice was abolished in 1972. The main direction in the public health service is comprehensive public and personal prophylaxis. The practical health services are developing and being improved in close connection and constant interaction with medical science. In fulfilling its many tasks, the public health system enjoys wide social support on the part of the Fatherland Front, the Bulgarian Trade Unions, the Bulgarian Red Cross and other mass organizations.

The Health Status of the Population. In the years after the Second World War since the socialist revolution of 9 September 1944, the health and working capacities of the Bulgarian people have been constantly improving. These changes, which are exceptional in character and scope, are clearly outlined against the background of the miserable state of health services which socialist Bulgaria inherited from the past. The demographic and health indicators in the period up to 9 September 1944 were extremely unfavourable, and stood at a level typical of the economically backward and dependent countries: a very high infant mortality rate (120.6 per thousand, 1944), a high death rate and a low average life expectancy (48.4 years, 1944), more than 500,000 suffering from malaria, about 200,000 from tuberculosis, a high incidence of spotted fever (22.4 per 100,000, 1944), typhoid fever, diphtheria, and other infectious diseases.

A health centre in Sofia

The socio-economic development of Bulgaria after 9 September 1944 and the success of socialist public health measures led to a rapid improvement in the health status and working capacity of the population. The indicators for the physical development of children and adolescents improved, the spread of infectious diseases was quickly controlled and some of them were completely stamped out. The incidence of tuberculosis lost its social significance. The average life expectancy of the population increased (see Table 1).

Table 1: Average Life Expectancy of the Population (in years)

Years	Men	Women	Total
1935–39	50.98	52.56	51.75
1956–57	64.17	67.65	65.89
1960–62	67.82	71.35	69.59
1965–67	68.81	72.57	70.66
1969–71	68.58	73.86	71.11
1974–78	68.68	73.91	71.31

Perinatal, neonatal and postnatal infant mortality are constantly decreasing. In 1982 the rate of stillbirths was reduced to 7.3 per thousand live births.

The main causes of death among the population are diseases of the circulatory system, neoplasms, diseases of the respiratory system, accident and poisoning. They account for more than 35 per cent of the deaths (see Table 4).

Table 2: Number of Children who Died before the Age of 1 Year (per 1,000 live-born)

Years	Towns	Villages	Total
1944	107.3	124.0	120.6
1956	35.1	79.1	72.0
1960	34.6	50.4	43.1
1965	25.6	35.0	30.8
1970	22.7	33.5	27.3
1975	19.9	29.7	23.1
1980	18.0	24.9	20.2
1982	10.4	21.7	18.2

Table 3: Perinatal mortality (per 1,000 live-born)

Years	Total
1960	23.0
1965	18.2
1970	18.8
1975	17.4
1980	15.0
1982	14.4

Table 4: Death Rate from Certain Diseases (per 100,000 of the population)

Diseases	1980	1982
Diseases of the circulatory system	612.3	648.4
Neoplasms	142.0	155.3
Diseases of the respiratory system	103.3	89.7
Accidents and poisoning	60.0	61.8

Table 5: Recorded Cases at the Health Establishments of a General Type (per 1,000)

Years	Total	0–14 years of age	15 years of age
1970	885.0	953.4	864.8
1975	943.5	1,151.3	884.8
1980	1,371.2	2,035.7	1,132.4
1982	1,467.1	2,230.4	4,244.9

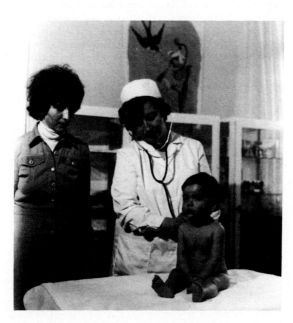

A baby and child-care consulting room

In the last ten–fifteen years, in connection with the ever fuller satisfaction of the health needs of the population, the incidence of diseases based on the records of the number of people who sought medical aid at health establishments of a general type has exhibited an upward trend, which is assessed as a major positive phenomenon, as it indicates an increasing coverage of the population by the various services of the health network.

An important achievement is the reduction in the incidence of infectious diseases and tuberculosis (see Tables 6 and 7).

The incidence of illness involving temporary or permanent disability is also decreasing. In the period 1975–80 primary disability decreased by 0.40 points.

Health Establishments and Medical Personnel. Medical care of the population is secured by a dense network of health establishments and medical personnel, whose distribution throughout the country depends upon the characteristics of the territory and the needs of medicare.

A characteristic phenomenon in the development of the network of hospitals is their concentration and the increase in their capacity, as well as the opening of new, specialized wards.

The number of hospitals and sanatorium beds is increasing. In 1944 there was an average of only nineteen beds (sixteen at hospitals and three at sanatoriums) per 10,000 of the population, whereas in 1982 the availability of beds rose to 111 (90 at hospitals and 21 at sanatoriums) per 10,000 of the population.

Parallel with the development of the network of health establishments, the number of medical personnel available to the population is also increasing.

Table 6: Recorded Cases of some Infectious Diseases (per 100,000 of the population)

Diseases	1965	1970	Years 1975	1980	1982
Diphtheria	0.3	0.1	0.0	0.0	0.0
Measles	354.1	350.9	231.2	121.5	3.2
Whooping cough	18.1	27.3	5.6	1.7	1.9
Mumps	250.9	211.2	400.8	73.6	38.3
Typhoid fever	0.4	0.2	0.2	0.0	0.0
Paratyphoid	2.3	0.0	0.0	0.0	0.0
Dysentery	127.2	228.4	184.8	96.0	80.8
Infectious hepatitis	233.5	141.4	115.7	110.8	249.2
Anthrax	1.9	0.6	0.3	0.1	0.2
Tetanus	1.4	0.6	0.3	0.2	0.3

Table 7: Recorded Cases of Active Tuberculosis (per 100,000 of the population

Form	1965	1970	Years 1975	1980	1982
Total	544.6	379.2	259.1	178.2	153.6
Including: newly-found	117.0	79.0	49.0	37.0	33.6
Tuberculosis of the respiratory organs	408.1	267.5	170.3	115.1	101.9
Including: newly-found	82.1	56.7	32.3	25.0	23.9

Table 8: Health Establishments

Establishments	1965	1970	Years 1975	1980	1982
Hospitals	1,355	200	184	184	186
Out-patient wards and polyclinics	3,074	3,588	3,597	3,758	3,818
Children's clinics	379	537	952	1,162	1,187
Sanatoriums and establishments at health resorts	150	185	189	186	195
Others	48	30	55	57	57

Therapeutic and Preventive Care

Primary medical care and sanitary aid is secured mainly at a doctor level: in the towns by district doctors – GPs for the adult population, and district pediatricians for children up to fifteen years of age; at enterprises – by workshop doctors; in villages – by doctors in the rural health services; and in schools, crèches and kindergartens – by pediatricians. In small villages and enterprises the health services are managed by assistant doctors, and in crèches and kindergartens – by medical nurses, all under the supervision of a corresponding health service. The district doctors in the towns, and the health services at enterprises, schools and children's establishments are integral components of the hospitals and polyclinics. One urban doctor's therapeutic district serves on average about 3,500 adults, and one pediatric district – 1,200 children below the age of fifteen.

The general hospital is the basic unit in the network of therapeutic and preventive establishments. It secures free in-patient and out-patient therapeutic and prophylactic treatment and is a centre for carrying out mass prophylactic measures; it also serves as a base for the training of

Table 9: Hospital and Sanatorium Beds Available to the Population

Years	Beds		Number of beds per 10,000 of the population		
	at hospitals	sanatoriums	total	at hospitals	at sanatoriums
1965	69,660	53,705	15,955	83	65	20
1970	79,736	63,426	16,310	94	75	19
1975	91,840	73,507	18,333	105	84	21
1980	98,142	79,588	18,554	111	90	21
1982	99,525	80,392	19,133	111	90	21

senior medical and paramedical personnel. There are general hospitals at three levels: at regional level, in which patients receive specialist service in internal diseases and pediatric wards, surgery, obstetrics and gynaecology wards, etc.; at district level, where services are provided by highly qualified specialists in cases for which the regional hospital has no diagnostic and therapeutic facilities; and at inter-district level, in first-class district hospitals with specialized inter-district services. Most highly-qualified specialist services are provided at the Medical Academy, which includes the higher medical schools and the medical research institutes.

The polyclinic in Pravets

A scanner

In the big urban centres, independent polyclinics are opened for every 50,000 of the population on average, with the corresponding specialized wards and doctor's offices. The in-patient, out-patient and polyclinic establishments at higher levels provide scientific and methodological guidance and assistance to establishments at lower levels. A network of strictly specialized medical establishments functions at a national and district level: including maternity hospitals, pneumophthisiatric hospitals and out-patient words, psychiatric hospitals and out-patient wards, oncological out-patient wards, skin and venereal disease wards, centres for transfusion haematology, and the Republican Centre of Sports Medicine. A great achievement in the development of specialized medical assistance is that in all districts there are artificial kidney medical centres which provide free dialysis for those who need it. This has drastically lowered mortality from chronic kidney diseases.

Medicines for home treatment of a growing number of diseases are provided free of charge. Since 1980, persons with low incomes, pensioners, patients suffering from certain chronic medical conditions, pregnant women and children under six years have been exempt from charges for prescriptions. Consistent efforts are being made to combat and limit the spread of all socially significant diseases (of the cardio-vascular and respiratory system, malignant growths, traumas, etc.).

Emergency aid, in cases of accident and sudden or acute diseases threatening the life of the patient, is provided by hospitals and polyclinics, and by the health services at

Table 10: Medical Personnel Available to the Population

Medical personnel	Number					Availability per 10,000 of the population				
	Years					Years				
	1965	1970	1975	1980	1982	1965	1970	1975	1980	1982
Doctors	13,593	13,810	18,770	21,700	23,081	16.5	18.8	21.5	24.6	25.8
Dentists	2,882	3,111	3,701	4,839	5,201	3.5	3.6	4.2	5.4	5.8
Pharmacists	1,725	2,382	3,053	3,648	3,965	2.1	2.8	3.5	4.1	4.4
Paramedical personnel	37,738	48,834	62,633	77,532	80,171	45.8	57.3	70.6	87.3	89.8
Population:										
per doctor	865	756	465	407	387	x	x	x	x	x
per dentist	2,855	2,737	2,300	1,834	1,717	x	x	x	x	x

Table 11: Maternity and Child Consulting Rooms (number)

Kinds of consultations	Years			
	1965	1970	1980	1982
Children's	187	214	389	865
Women's	203	209	256	334
Children's and women's	1,989	2,130	2,233	1,773
Total	2,379	2,553	2,878	2,972
Of these managed by a doctor	1,722	1,703	1,726	1,806

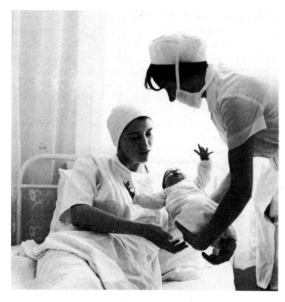

Maternity

industrial enterprises, schools and villages. In the district centres and the capital, emergency aid is provided at district hospitals, which have at their diposal specialist teams and ambulances. Specialist teams have also been formed for the treatment of physical injuries, cardiological, neurological, and gynaecological conditions of patients requiring immediate treatment. The prompt response of the teams is ensured by a wide network of radio and telephones.

In 1951 in Sofia the Republican Institute for Emergency Medical Aid named after Nikolai Ivanovich Pirogov was established as a specialized health establishment providing prompt services round the clock to patients from all over the country. When necessary, a special air service secures prompt consultation and therapeutic intervention by highly qualified specialists from the Medical Academy even in the most distant corners of the country.

Health Care for Mothers and Children. Maternity and child care is one of the most important strands in the government's health policy. The Programme for Raising the Living Standards of the People adopted in December 1972 provides that society should gradually shoulder the overall maintenance of the young generation. By a decision of the Politburo of the Central Committee of the BCP of 1973 on promoting the role of women in the construction of the developed socialist society, maternity was stressed as the foremost social function of Bulgarian women today. The decree on encouraging childbearing provided for considerable lump-sum grants, monthly family allowances and aids in cash to mothers and children. The State constantly encourages childbearing. Mothers are entitled to paid leave until children reach two years of age. They are allowed to look after a sick

Table 12: Crèches and Accommodation for Children under Three Years

	Years			
	1965	1970	1980	1982
Number of crèches	349	506	1,129	1,155
Places per 1,000 children under three years	61.6	76.1	197.0	208.0
In towns	127.7	118.1	242.6	257.3
In villages	7.7	21.2	97.3	101.5

Table 13: Pregnant Women, and Children Undergoing Regular Examinations at Maternity and Child Consulting Rooms

	Years			
	1965	1970	1980	1982
% of pregnant women visiting consulting rooms before the third calendar month	79.4	83.3	91.3	90.2
% of infants up to one month of age being visited by the district doctor and nurse	97.7	95.0	95.6	95.5

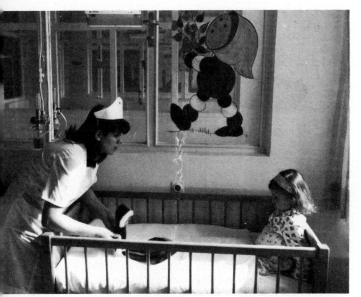

Children's unit in the Dr Racho Angelov district hospital, Sofia

child under nine without loss of labour remuneration. Drugs are supplied free of charge for children up to six years. Loans for new housing construction are partly remitted to young families with the birth of every child.

By decree of the Council of Ministers, preventive measures for protecting and promoting the physical and mental health of children have been established as the main trend in the activities of the state, economic and public bodies and organizations. Of paramount importance are the tasks of protecting the health of working women, relieving women of household chores through improving domestic standards, protecting pregnancy so as to ensure a normal childbirth, limiting abortions, and prevention of sterility. Among the priorities of the health services are: cutting down the child mortality rate and limiting the incidence of disease among children; securing, at pre-school and school age, an environment conducive to the development of the adaptability of the child's physique; promoting the child's physical and mental health; and providing correct career guidance for adolescents. Special care is devoted to the development of the network of maternity and child-consulting-rooms, securing a systematic observation of pregnant women, nursing mothers and children up to the age of seven.

A network of day-care establishments has developed for bringing up children of up to three years of age. The seasonal crèches are opened mainly in the villages during the high farming season (from May to October). For frequently sick children and for children with chronic ailments, sanatorium crèches are opened. In the crèches, conditions are provided for normal physical and mental development through systematic care, science-based methods of rearing, feeding, strengthening, disease prevention and education, and through a goal-oriented health education of parents. Crèches are opened and maintained by the People's Councils or by individual enterprises, agro-industrial complexes and institutions. Organizationally and methodologically, they are guided by the local health authorities. The Mother and Child Homes perform special functions in bringing up children of unmarried mothers, orphans and neglected children. Very popular among the population are the baby-foods kitchens which secure low-priced meals for children. They are run by the Municipal People's Councils as units under the corresponding health services, and in the villages – by the agro-industrial complexes.

The relative proportion of pregnant women visiting the maternity consulting rooms in the early stage of pregnancy is high, as is that of infants up to one month who are under the care of doctors and nurses working from the consulting room.

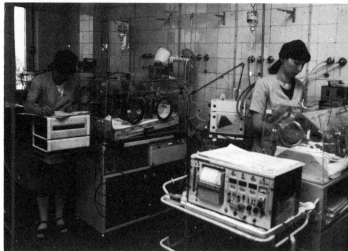

The intensive care unit at the Obstetrics and Gynaecology Research Institute, Sofia

Obstetrics and Gynaecology Research Institute of the Medical Academy, Sofia

The majority of women who are pregnant undergo regular check-ups at obstetric and gynaecological wards in hospitals, both in towns and in larger villages. As a rule, childbirth takes place at maternity hospitals. The task is, in close connection with the activity of the consulting-rooms, to raise the quality of maternity aid, to diminish childbirth traumatism and to reduce to a minimum the risks of an abnormal child delivery. Primary medical aid to women and children is implemented at doctor level: in the towns – by district obstetric and gynaecological teams, consisting of an obstetrician-gynaecologist and a midwife, and by district pediatric teams, consisting of a pediatrician, a pediatric nurse and a visiting nurse, at children's consulting rooms; in the villages – by the rural public health services; and by doctors and midwives at industrial enterprises. A higher level of pediatric and obstetric-gynaecological services, including specialist services, is provided at the obstetric and gynaecological departments and wards, by the pediatric wards at the regional and district hospitals, and – at a national level – by the specialized clinics and institutes of the Medical Academy.

Health Care for Workers. A basic principle in the organization of medical aid in Bulgaria is that industrial workers enjoy priority. This is done by bringing qualified health care closer to the place of work through the development of a dense network of workers' health establishments (workshop doctor's sector, doctor's and assistant doctor's health centre, workers' hospital and workers' polyclinic) and specialized units of occupational pathology, as well as by securing priority health care for the workers thoughout the whole health network of the country. A workshop doctor's sector, staffed by a doctor, a dentist and paramedical personnel, is provided for compact contingents of 2,000 workers and over; and in enterprises with up to 200 workers there is an assistant doctor or a trained nurse. In industrial zones, workers' polyclinics and workers' hospitals of an open type are set up for securing medical assistance for workers and their families, and for the population which lives in the industrial zone. Qualified services are also secured for the workers by the district health establishments, in which occupational pathology wards are set up. At a national level strictly specialized and highly qualified assistance is secured by the institutes and clinics of the Medical Academy, and by the Research Institute of Hygiene and Occupational Diseases of the Medical Academy in Sofia. As well as by the territorial health establishments, the health needs of the workers have been attended to by 16 workers' hospitals, 29 workers' polyclinics, 407 workshop doctor's sectors, 37 doctors' and 139 assistant doctors' health centres at enterprises (1982 data).

In the system of priority health care for workers an important place is held by the prophylactoriums. These are health establishments set up at industrial compounds, and at large industrial enterprises and groups of smaller ones, in which blue- and white-collar employees restore improve their health without being detached from the production process. The aim is the prevention of diseases among those threatened by or predisposed to them, and of recurrent and chronic ailments, with restorative treatment to cut down the time of temporary incapacity for work. In 1983 a total of 54 prophylactoriums were in operation in Bulgaria.

Of particular importance in the complex of health care is the improvement of working conditions. The theses of the Twelfth Congress of the BCP demand that particular attention be devoted to labour hygiene and the cutting down of the risk factors at places of work and in trades and professions in which incidence of occupational disease is comparatively high. More than 1,000 million levs were invested for the fulfilment of the national programme of labour safety and hygiene during the Eighth Five-Year Plan period (1981–85). Extensive rights and powers have been granted to the work-forces to solve their social and health problems by making use of the specially created fund for social welfare and cultural undertakings. Of great importance for the preservation of the health of the work-forces is the close interaction between the doctor and the sociologist and psychologist at the enterprise, both with regard to the hygienic-epidemiological, therapeutic-diagnostic and rehabilitation activity, and with regard to the measures aimed at promoting a hygienic way of life in the work-forces, diminishing the psychological and physical tension and its consequences, and the implementation on a wide front of social measures for improving the worker's performances and way of life.

Health Services for the Rural Population. The essence and the organization of the health services in the present-day Bulgarian village are determined by the health-demographic and urbanization processes, the changes in the agglomeration of the population and the dynamism of the industrial-agricultural development of the rural areas as well as by the increasing relationship between the illnesses and occupations of the rural population. This sets higher demands on the functions, structure and organization of the village health services, which should ensure immediate services to a large number of small villages and, at the same time, bring the quality of these services up to the standards of those provided to the urban population. Important trends are the obstretico-gynaecological and pediatric services, and the medical services for the active working population and the aged. Owing to the large number of small inhabited places, a dense network of transport and flexible organizations of emergency medical aid are available.

In 1982 a total of 2,216 village assistant-doctor's centres were set up, while in the larger villages 1,026 village health services and sectors, staffed by one or two doctors each, were established. A total of 83 village polyclinics secure primary and specialized medical aid in inhabited places or groups of inhabited places with a population of over 6,000. These polyclinics have therapeutic, pediatric and stomatological wards, a clinical laboratory, a physiotherapy service, maternity and child

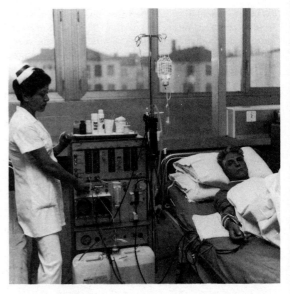

Haemodialysis

consulting-rooms. Periodically and systematically, in accordance with a timetable, the specialists from the regional or district hospitals carry out on-the-spot obstetrico-gynaecological, surgical, ophthalmological, ear, nose and throat, neurological, dermatological and venereal and X-ray diagnostic examinations. The rural population obtains specialist medical aid on a higher level at the regional and district hospitals and at the Medical Academy. Equal possibilities for hospitalization are enjoyed by the rural and the urban population. Over 98 per cent of births take place in maternity homes, while over 86.9 per cent of pregnant women are under the supervision of the village health service during their pregnancy. Mobile units visit smaller villages. There are no differences in principle between the functions of the rural and the urban health services. Since the urban health establishments of every kind also serve the rural population in the territory belonging to them, the urban health services are in the broadest sense of the word integrated with the resources and activities of the entire health network.

Medical Services for the Elderly. The growing number of elderly people in Bulgaria presents the public health services with special tasks. The medical social problems of advanced and old age are settled by means of complex health measures: the securing of medical help, as well as improving the mode of life of the elderly by creating a propitious atmosphere in the family and in society, by directing the elderly person to suitable employment if he or she is fit to work, bearing in mind that labour is a great stimulant of good health and mental balance. By a decree of the Council of Ministers of 1980 medicines are prescribed free of charge to pensioners with low incomes, and services are rendered at the homes of elderly and infirm people who live alone. Those who wish are accommodated at homes for the aged. The People's Councils organize a wide range of services for elderly persons at their homes, e.g. daily supply of warm and nourishing meals, medicines and medical attendance; cleaning and laundry services, and domestic help. Recreational workshops and cultural entertainments have been organized at cultural clubs opened for this purpose.

A pressure chamber

Testing the coronary vessels

Screening and follow-up treatment is a fundamental method of work of the socialist health services in Bulgaria. It is expressed in a dynamic observation of the health conditions of individuals and of groups of the population having a common physiological feature or practising the same profession, and of groups of patients suffering from certain diseases; of children, adolescents, pregnant women; patients suffering from chronic diseases; recurring and long-term diseases; and patients who have suffered from acute infectious diseases. Great attention is devoted to screening the population, to identify persons or groups of the population living or working under conditions involving the risk of a particular disease. They are the object of special observation and measures are taken for removing or limiting the risk. Among priority tasks are early diagnosis of such diseases, taking adequate measures for improving the working and living environment, treatment and rehabilitation of patients under special programmes, including measures for systematic observation of a hygienic and dietetic regimen, drug treatment, labour readjustment, rehabilitation at sanatoriums and resorts, etc. Increasingly large groups of the population are included in the system of screening and follow-up treatment – in 1982 alone some 1,234,886 patients. 1,360,980 women were examined for the discovery of gynaecological ailments and malignant growths. Active and early diagnosis of diseases is achieved through preventive medical examination and the everyday out-patient activities of the health establishments. For the purpose of achieving broad scope and effectiveness special teams have been set up and corresponding anamnestic, laboratory and apparatus tests are made for the early detection of the major wide-spread ailments.

Sanatoriums and Resorts. Bulgaria has a considerable number of good resorts, mostly spas; the curative properties of mineral waters have been well-known since earliest times. Before 9 September 1944 sanatoriums and resorts developed in an unplanned manner, without a scientific foundation, qualified medical assistance or specified methods and doses for sanatorium and resort treatment. Spas were used quite inadequately. After the socialist revolution reconstruction of all sanatoriums and resorts was started. In 1947, under the Ministry of Public Health, a sanatorium and resort administration was established, which assumed the management and control of resorts. In 1949 the Research Institute of Resorts and Physiotherapy was opened in the Ovcha Kupel district of Sofia (today the Research Institute of Resorts, Physiotherapy and Rehabilitation with the Medical Academy). It tackles all problems in the field of resorts and rehabilitation, common and socially significant ailments. Rehabilitation is introduced in the early stages of diseases. At the resort centres a network of sanatoriums has been developed: pneumophthysiatric and for non-tubercular patients. For outpatients and polyclinic treatment of holiday-makers outside the sanatoriums, resort polyclinics have been opened in Burgas, Varna, Hissarya, Bankya, Pomorie and many other inhabited places; and most of the mineralbaths have been transformed into balneohospitals in Sandanski, Petrich, Velingrad, Kyustendil, Hissarya and elsewhere; at many places modern mud-cure establishments have been opened. The sanatoriums and health resort establishments specialize in accordance with the curative properties of the waters and climate of the corresponding resort. Mostly the natural factors are applied – climatic treatment, mud-cure, balneotreatment, sea treatment – but also physiotherapy, massage, inhalations, remedial gynnastics, dietetic meals, etc. Sanatorium and resort treatment holds a major place in the socialist public health services. In the sanatorium and resort establishments, tremendous curative and preventive acitivy is carried out. In 1982 the number of registered visits to the doctor was 750,813, those to dentists was 72,871 and the number of patients treated at the sanatoriums was 212,868. The sanatoriums and rehabilitation centres in Bankya, Momin Prohod, Pavel Banya, Hissarya, Vurshets, Narechen, Kyustendil and elsewhere have won world fame.

At the same time effective departments of physiotherapy and rehabilitation with electro- and heliotherapy, exercise, hydro- and heliotherapy, have been set up at all district and large regional hospitals. Rehabilitation wards have been established also at workers' health centres and at out-patient wards and polyclinics.

Neuro-surgery under a microscope at the Institute for the Treatment of Foreign Citizens, Sofia

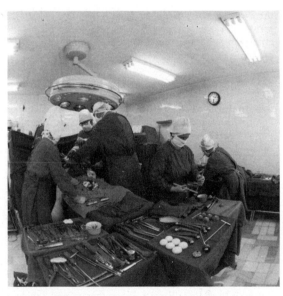

The orthopaedics and traumatology department of the Institute for the Treatment of Foreign Citizens, Sofia

Sanitary Control over the Environment, Hygienic and Epidemiological Activity. Combating Infectious Diseases.

The rapid development and intensification of the economy has led to substantial changes in the conditions and factors of the living environment because of the appearance of new industrial branches, the rapid development of non-ferrous and ferrous metallurgy, the chemical industry, machine-building and electronics, the mechanization and chemicalization of agriculture, etc. It has become necessary to step up care for the conservation of the environment by opposing and putting pressure on the state and public bodies and organizations which neglect their obligations. Primary preventive care is at the centre of the great and intricate complex of tasks with which the public health services are faced in combating diseases by removing or limiting the factors in the environment hazardous to man's health.

The far-reaching and comprehensive social activities of the State in this field have a solid normative base regulated by the Law on Public Health. In 1975 the Council of Ministers of the People's Republic of Bulgaria adopted a national programme for labour safety and hygiene and is supervising its strict fulfilment. The national industrial enterprises, the different departments, corporations and companies have preventive programmes elaborated for them. The whole production sphere is encompassed by far-reaching and goal-oriented health measures, complying with the requirements for optimizing the working environment and the working process in accordance with the physiological, ergonomical and sanitary and hygienic norms. The acititives to ensure a clean atmosphere, water and soil and to improve the hygienic conditions of inhabited places are implemented through control over pollution of the air, drinking water and reservoirs; combating noise; the construction of purifying equipment; proclamation of model towns and villages; and good sanitation in the home, at child-care establishments, schools, enterprises and public places.

The combating of infectious diseases is conducted on the basis of a five-year programme, with the participation of the People's Councils, the public organizations and the population. Management of this activity is facilitated by the automated system of epidemiological information which is secured every day by the Ministry of Public Health, containing data on the registered cases of illness throughout the country. The automated system for

Dental castings at the district dentistry, Tolbuhin

inspectorate, which exercises sanitary and anti-epidemic control in the territory of the district. The epidemiological inspection has branches and groups in other inhabited places of the district. There are 28 different hygienic and epidemiological inspectorates, of which ten perform inter-district functions with their specialized laboratories. Sanitary and control functions are also entrusted to the doctors at the enterprises and the rural health services. The hygienic and epidemiological inspectorate is supervised by the Public Health and Social Welfare department at the corresponding District People's Council, and with respect to specific activities connected with state sanitary control and the combating of infectious diseases – by the Ministry of Public Health. The Ministry is the supervising body at national level, through its Board for Sanitary and Anti-epidemiological Control, the Research Institute of Infectious and Parasitic Diseases of the Medical Academy, and the Institute for Health Education of the Ministry of Public Health.

The current problems of the hygienic and epidemiological service are: raising the effectiveness of the movement for clean and urbanized inhabited places, securing good managing preventive immunizations appoints and controls the carrying out of immunizations for every individual, and the level of immunity of the population in the districts.

For the creation and maintenance of a healthy living environment and for its development in accordance with the changes in life under the construction of developed socialist society, a paramount role is played by state sanitary control. For preliminary sanitary control, conclusions are given on the correspondence of the blueprints for the contruction of new plants for chemical and biological substances and products, and for new methods and means of production, technological processes, devices, instruments, apparatuses, etc., with hygienic norms and sanitary rules. Preliminary sanitary control orders construction to be stopped, or does not allow projects to be put into operation, when sanitary requirements have been violated. The current sanitary control makes an evaluation of projects which are in regular exploitation, and of machines to be used in production or which are newly introduced, and of the equipment, products, articles, chemical and biological means, as well as monitoring the health of workers in the catering and communal enterprises, child-care establishments, etc.

The main, specialized, public health establishment, which controls the correspondence of the factors and conditions of the living environment with the hygienic norms, is the district hygienic and epidemiological

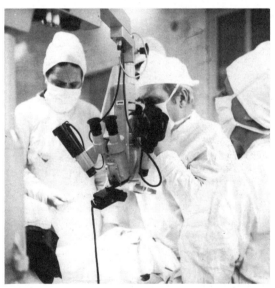

Ophthalmics Research Institute of the Medical Academy, Sofia

quality drinking water, improving hygienic conditions at resorts and tourist sites, on roadways, etc.; protecting the air, waters and soil from pollution; limiting transport and everyday noise; fulfilment of the national programme for labour safety and hygiene and of the prophylactic programmes in the districts, inhabited places and enterprises; rationalizing the diet of the population; improving hygienic conditions of child-care and educational establishments; organizing publicity and control against tobacco smoking and alcoholism.

At the present stage of the country's development, in

The mineral water pool at the Zlatni Pyassatsi holiday resort

connection with the spread of chronic and degenerative ailments and the negative consequences of plentiful but unbalanced feeding, sedentary ways, smoking, the abuse of alcohol, the increasing neuropsychic tension, there is a constantly growing role for the hygienic and epidemiological services in promoting a healthy way of life for the individual, working collectives and organized groups of the population. This task is tackled jointly with other health establishments and public organizations, enlisting also the co-operation of the community.

Health Culture and Education. The improvement of the population's health education and knowledge is the purpose of a national programme appointed by the Council of Ministers in 1976, which makes health propaganda and education the duty of all Ministries, departments, People's Councils, public organizations and the whole of society.

The chief organizing and methodological functions in this sphere are entrusted to the Ministry of Public Health, which puts them into effect through its Health Education Institute and the health education departments of the hygiene and epidemiological inspectorates. At the health-centre level these functions are performed by the heads of these centres and medical personnel appointed by them and known as health education organizers. Health education is one of the prime concerns of the staff of all health establishments at the personal, work-force and general public level. Each medical worker is duty-bound to organize and promote health-education activities.

The main emphasis falls on measures for ensuring a healthy way of life and a wholesome environment in the conditions of real Socialism.

One of the most important tasks in the field of health education and propaganda is the involvement of the population in all the campaigns and undertakings of the public health service.

The fight against smoking and drinking is one of the prime concerns of health education and propaganda, and an important state and social task. A special decree of the Central Committee of the BCP and the Council of Ministers on combating smoking and alcoholism has enforced a series of measures in this respect, including a wide range of health-education undertakings. Drug abuse is not a social problem in Bulgaria but quite an isolated phenomenon.

Management and Planning of the Health Services. The Ministry of Public Health is the central, leading body of the health services. It draws up the state plans and programmes for the development of the health services, prepares bills and other government documents relating to them and issues rules, regulations and instructions on the activity of the public health network. In the districts, the health services are guided by the public health and social welfare departments of the executive committees of the District People's Councils. In the settlement systems and the municipalities, the management of the health services is entrusted to the main health establishment in the area.

The planning of the development and activity of the health services is founded on a dynamic statutory base relating to manpower, material and financial resources, and to the volume and character of the activity performed in accordance with the population's health needs. Long-term (five-year) and operating (annual) plans are drawn up, observing certain proportions with the other social services. For the attainment of the basic goals of the health service an important role is played by the programmes for combating various diseases and for the development of public health care. They are elaborated for the whole country (such as the national programmes for labour safety and hygiene, combating cardio-vascular diseases, preserving mental health, follow-up treatment, raising the health culture of the people in order to promote a healthy way of life, combating infectious diseases, etc.), and for the public health services at district and settlement level. All state bodies, People's Councils, public bodies and organizations join in the fulfilment of the programmes.

Particular attention is devoted to ensuring the flow of information necessary for health service management. Information generated by the health establishments is directed to the public health and social welfare departments in the districts and thence to the main information and computer centre of the Ministry of Public Health and the Medical Academy, and to the Committee of the Unified System for Social Information under the Council of Ministers. An important part of this flow is the operating information about basic phenomena in the health services, which is made available periodically, and the annual information report, which sums up the data for the annual activity of the health services.

In connection with the improvement of management on the hierarchical principle, an automated system of public health information and management has been built up as an element of the unified system of social information in the country.

Automated information systems covering out-patient and polyclinic care, temporary incapacity for work, and blood donation have been functioning in all districts since 1983. Automated systems for management of preventive immunization function in 21 districts, and for screening and follow-up services – for cancer and mental illness in particular – in a few districts. Automated information systems covering the infectious diseases subject to compulsory registration, venereal diseases and invalidity, function on a national level. An automated information-retrieval system covering all personnel in the country with qualifications in medicine, dentistry and pharmacy has also been set up at national level.

Pharmacies and the Pharmaceutical Network

Pharmacies. The pharmacies and the medicine supply system implemented through them are an important part of the health service, which provides the population and health establishments with medicines, medical instruments and equipment, medical-sanitary materials, etc. The basic organizational principles governing the system were established after 9 September 1944 on the basis of socialist production relations and the people's ownership of medicines. The unity in the methods, guidance and organization was established after the final liquidation of private concessionary pharmacies and provisional pharmacies in 1949. Today medicine supply is entirely in the hands of the State. It is guided by the Ministry of Public Health and is implemented by the State Pharmacy Association in Sofia, which embraces all forms of medical supply and dispensing of drugs in Bulgaria. The association draws up long-term plans for medical supplies, forecasts and programmes for the development of pharmacies. It fulfils Bulgaria's obligations towards the United Nations in connection with control over and reporting on narcotics. As a unified body of the Ministry of Public Health, the State Pharmacy Association provides full organizational and economic co-ordination between the medicine supply units in every district.

Pharmaceutical Network. After Bulgaria's Liberation from Ottoman rule a system of concessions was introduced, under which pharmacies were personal property which could be sold, leased and inherited. Up to 9 September 1944 the distribution of medicines was handled by 636 pharmaceutical establishments – 391 private concessionary and private provisional pharmacies, 86 municipal pharmacies, 19 belonging to the social security fund, 73 in hospital establishments and 67 private chemists. Partial supply with medicines was effected by 711 district health services in the villages. In 1945 the system of concessions was established and in 1949 the nationalization of the private concessionary and provisional pharmacies was carried out and the public mode of supply with medicines established. The network of pharmacies is subordinated to the socialist principles of bringing the supply of medicines as close as possible to the population and of linking it directly with the remaining part of the care network. It is composed of district pharmaceutical enterprises – branches of the State Pharmacy Association, each one of which includes pharmacies (for the population and in hospitals); pharmaceutical warehouses (one for every branch); pharmaceutical branches and pavilions (smaller units, subordinated to the nearest pharmacy, opened mainly in inhabited places in which there is a rural health service or a rural health centre serving a population of 3,000 to 4,600); pharmaceutical stands (units accountable to, directed and supplied by the pharmacy to which they are attached; they supply the population only with medicinal preparations according to the Ministry of Public Health list, mainly in inhabited places with a population of up to 3,000, a great part being managed by the doctors and assistant doctors of the rural health services); medical supply shops (only in the towns) and medicine-packing

Table 14: Pharmaceutical Network

Establishments	1965	1970	1980	1982
Pharmacies	840	896	904	944
Pharmaceutical branches and stands	2,716	3,063	3,462	3,532
Pharmaceutical pavilions	16	28	25	31
Medical supply shops	72	103	126	107

laboratories in certain districts. Laboratories for control analyses are organized at the branches. The total number of pharmacy establishments, regardless of their departmental subordination is 4,614 (1982, see Table 14). The pharmaceutical network is staffed with pharmacists having higher, semi-higher and secondary special education.

Drug treatment in the in-patient department of all health establishments is absolutely free of charge. Free medicines for out-patient treatment and treatment at home are provided for children up to six years of age and for patients suffering from chronic and degenerative ailments, according to a list established by the Ministry of Public Health.

Social Support of Public Health. The widest participation of the population in undertakings concerning health care finds expression in the rights of the electoral bodies of the People's Government, stipulated in the Constitution of the People's Republic of Bulgaria. The permanent health-care commissions of the People's Councils constitute a basic form of organized public effort in the solution of socio-medical problems. With the assistance of a wide circle of people, they study health-care problems, control the work of the health service, submit proposals for its improvement, for urbanizing populated areas and making them more healthy, for building health establishments, etc.

The Fatherland Front, the Bulgarian Trade Unions, the Dimitrov Young Communist League, the Bulgarian Union for Physical Education and Sport, the Bulgarian Red Cross, the Bulgarian Tourist Union and other public organizations assist the voluntary efforts of the population in health care; they are concerned with improving the health education of their members, and exercise public control over the observation of sanitary-hygienic norms, rules and requirements.

The Fatherland Front is the organizer and leader of the nationwide movement for the urbanization and sanitary improvement of populated areas, for providing better sanitary-hygienic conditions in production and daily life; the drive against air, soil and water pollution, against noise and vibrations, and against the alienation of man from nature. It propagandizes, promotes, and provides the necessary conditions for a healthy way of life. The Fatherland Front organizes nationwide annual reviews on the urbanization and sanitary improvement of populated areas.

The Bulgarian Trade Unions play an important role in the defence of the working people's interests in health care. They exercise control over labour safety, actively work for the improvement of the working and living conditions of workers, organize recreation, holidays and health resort treatment, mass physical culture, sports and tourism. The trade union organizations play a major role in drawing up, organizing and controlling the fulfilment of the work-forces' social development plans; and they also exercise control over the fulfilment of the tasks stemming from the National Programme on Hygiene and Labour Safety and the concrete programmes of preventive health care at the enterprises. The trade union organizations and activists from among the workers systematically survey the incidence of diseases and injuries leading to temporary incapacity for work, take measures to reduce them and make full use of the follow-up treatment for the prevention of diseases, and to set up canteens for dietetic food and prophylactoria. They work for the attainment of high cultural standards at places of work, conducting public supervision and control over the medical attendance of workers and employees.

The Bulgarian Red Cross is a mass, voluntary, public organization with broad and varied functions in health care. It carries out wide health education, organizes and trains first-aid units and activists in the field of sanitation, participates in the movement for high standards of hygiene and sanitation countrywide, in various health care and sanitation undertakings, in the organization of public control over hygiene in industrial plants, schools, public catering and communal establishments, etc. The Red Cross conducts wide, popular propaganda and organizational work in the area of blood donation, provides relief to the victim of natural calamities, organizes mountain rescue and life saving services.

International Health Co-operation. Bulgaria conducts international scientific co-operation in the field of medicine and public health in various forms. It was one of the first member-countries of the World Health Organization (since 1948). Under an agreement with its headquarters in Geneva and the European Bureau in Copenhagen, a model health service is being built up in the district of Gabrovo, especially oriented to primary medical aid, early diagnosis and preventive care. Bulgaria takes an active part in the Conferences of Ministers of Public Health of the Socialist Countries (since 1955). Co-operation in the Permanent Commission on the problems of public health of the CMEA has been active since 1975. Bulgaria is co-operating with the USSR and the other socialist countries in the elaboration of major scientific and practical problems of public health. Scientific institutes of the Medical Academy are guiding or co-ordinating research projects carried out in the CMEA member-countries on environmental pollution control, preventive health care, etc. Scientific co-operation is implemented with a number of other countries in Europe and other continents. Bulgaria is a member of the Balkan Medical Union and maintains contacts on public health problems with its Balkan neighbours. Health co-operation with the developing countries is being constantly expanded. Bulgaria plays an active part in the work of the Unicef, FAO and other international organizations on problems connected with public health. It has on several occasions been organizer and host of international events related to public health (symposiums, conferences, seminars). Outstanding Bulgarian doctors are employed as experts of the World Health Organization, under whose auspices courses in public health are held every year for the training of doctors from the developing and other countries.

PHYSICAL EDUCATION, SPORTS AND TOURISM

Physical Education and Sports. Just like the Bulgarian State, physical education and sports in Bulgaria are both ancient and young: ancient, because they were known to the first settlers of the present day Bulgarian lands, the Thracians, who inhabited them as early as the fourth millennium BC; and young because their real upsurge and flourishing state is witnessed in the time of socialist Bulgaria.

places for field-events, wrestling, gymnastics, horse and chariot races, gladiator fights and so on. The first major sports games in the territory of present-day Bulgaria were organized in Philippopolis (now Plovdiv) and were first called Pythian and later Alexandrean and Kendrizean. There were similar games in Odessos (now Varna), in Serdica (now Sofia) and other towns. The southern Slavs, who settled on the Balkan Peninsula in the sixth and seventh centuries, were greater in number than the Thracians. Together with the Bulgars, they inherited the physical education of the Thracians and exercised some influence on each other. Physical education in the first Bulgarian State (founded 681) was thus inspired by the military training exercises, sports and games of the Thracians, the Slavs and the Proto-Bulgarians, while sports like wrestling, horse-riding, archery, hunting, rowing and so on acquired a popular character.

Bronze coin depicting wrestlers from Philippopolis (Plovdiv), minted during the reign of the Roman Emperor Elagabalus (218–222 AD)

Physical education in the time of the Thracians developed in close relation with the countries of the ancient Orient and ancient Greece. The Thracian 'hero-horseman', who was known as one of their deities, symbolized the hunting activities of the Thracians. Further evidence of the Thracians' physical education are the ancient coins of that time, depicting different sports events. Major Thracian settlements were very well laid out, and among the monumental buildings, luxurious temples, theatres, public thermal baths and numerous statues they featured special places and facilities for sports, such as gymnasiums, circuses, amphitheatres and other

Marble plaque depicting gladatorial games, from Serdica (Sofia), second–fourth century

Bulgaria's fall under Ottoman rule in the fourteenth century suspended the natural historical development of a civilization and culture, including physical education which had evolved in the Middle Ages under the protection of the powerful Bulgarian State. The Bulgarian people started an endless struggle against their oppressors.

Popular sports and training in physical education had a role to play in this struggle. They contributed to the maintenance of the people's physical and spiritual strength and to the demonstration of their courage, endurance and fighting abilities.

During the period of the Bulgarian National Revival, which was at its height in the eighteenth and the first three quarters of the nineteenth centuries, physical education was part of the comprehensive educational and training process. Educationalists and revolutionaries of the Revival Period were well aware of the necessity of a comprehensive intellectual and physical education for the contemporary generation. Prominent figures among them established a system of physical education in schools which suited the conditions in Bulgaria and defined the place of military, physical training of the population in the struggle against the Ottoman yoke. The range of this system included children's pastimes in the schools of the National Revival Period, the gymnastic classes in the schools before the Liberation, the resistance units and the Bulgarian military schools in Belgrade in 1862 and during the period 1867–68. This process was carried out in underground fraternal units, set up by the strategist, ideologist and organizer of the Bulgarian National Liberation Revolutionary Movement, Vassil Levski. Thus, even before the liberation of Bulgaria from the Ottoman rule the country already had the rudiments of a system of physical education in schools and had developed national forms of physical education and sports. There is substantial evidence of the development of some modern sports, some of which were demonstrated at fairs and public gatherings. Conditions were also created for the birth of the Bulgarian public physical education organizations.

The Bulgarian Physical Culture Movement from 1878 to 1944.

Foundation of the Bulgarian Physical Culture Movement. On 3 March 1878, as a result of the Russo-Turkish war, Bulgaria was liberated from Ottoman domination, which had lasted for five centuries. The ideas of the National Revival Period concerned with providing physical education for the people found expression in the establishment of the public gymnastics unions. The day of 28 November 1878, when the *Maritsa* newspaper published the Constitution of the Orel Public Gymnastic Union in Plovdiv, is considered as the birthday of the Bulgarian physical education movement. Such unions were also established in Burgas, Sliven, Haskovo, Yambol and other towns. The gymnastics unions resulted from the specific historical, political and social circumstances in the country. Their establishment was necessitated by the need to consolidate the foundation of the young State and to create organized units for the social education of the people. They became a socio-military factor that had its impact on the enemies.

From the very beginning the Bulgarian physical education movement was built on democratic principles with the broad participation of the people. Due credit must be given to the Russian military authorities for the assistance and support they extended in the formation and the development of the gymnastics unions. These were followed by the establishment of many gymnastics, hunting, tourist, cycling and other unions and clubs in Sofia, Russe, Turnovo and other towns. The end of the nineteenth century saw a process of amalgamation, when the separate clubs were incorporated into unions. The biggest sports' organization of the time, the Alliance of the Bulgarian Gymnastic Unions, named Yunak, was established in 1898. The Sokol Hunters' Union was established in the same year. The Bulgarian Tourists' Union in Sofia assumed the role of a central organization in 1900, while the young hikers established their separate union in 1911. In 1902 the Bulgarian cyclists had their own union too.

Organizing the Progressive Workers' Movement in Physical Education. Bulgaria's rapid capitalist development and class confrontation were reflected in the physical education movement as well. The constantly growing working class in the country became the spokesman and supporter of the country's national interests. It provided a new substance, a new purpose and a new task for the Bulgarian physical education movement, which was divided into two trends: bourgeois and proletarian.

The progressive workers' movement in physical education arose from the class struggle of the Bulgarian proletariat, guided by its Marxist Party. In its first programme in 1893, the Party declared that it was working towards the comprehensive physical education of both sexes. The beginning of the progressive workers' movement in physical education was set by the establishment in 1900 of the Borets Union in the Yuchbunar district of Sofia. Its example was followed by Russe, Gabrovo, Plovdiv and other towns. The aim and tasks of the unions were to develop physical strength, and to build up the class consciousness of the union members and strengthen their solidarity in the class struggle. The foundation of the Bulgarian Workers' Social Democratic Party (left-wing socialists) in 1903 increased the progressive influence on the workers' physical education movement. In the Resolution of the Thirteenth Party Congress (1906), submitted by Georgi Dimitrov, the party organizations were given the task of proceeding with the formation of workers' gymnastics unions, in which physical education was to complement ideological education. This Resolution resulted in the formation of a number of new unions. The workers' gymnastics union, Mladezh, in the Yuchbunar and Konjovitsa districts of Sofia, became the new centre of the movement. Following this example, other unions were formed in Varna, Vratsa, Sliven, Stara Zagora, Shumen, Gabrovo etc.

The Great October Socialist Revolution also influenced the Bulgarian progressive workers physical culture movement. The Bulgarian Communist Youth Union became the chief organizer of the workers' physical education. In 1919 some gymnastic units called Spartak were formed within the Union's system. The 1917–23 period of revolutionary upsurge in Bulgaria is characterized simultaneously by a decline in the gymnastics, sports and other organizations that were under the influence of the bourgeoisie and an all-round development of the progres-

General Vladimir Stoichev, Bulgarian participant in the Olympic Games of 1924 (Paris) and 1928 (Amsterdam) and Honorary President of the Bulgarian Olympic Committee

Third meeting of the Yunak Union of Gymnastic Societies, Sofia, 1904

sive workers' physical education movement. An expression of this development were the *spartakiads* (major competitions in several sports), organized by the Bulgarian Communist Youth Union in Plovdiv, Burgas, Dupnitsa (now Stanke Dimitrov), Kotel, in the village of Alexandrovo (now Dimovo), etc. The 'red athletes' were loyal to the interests of the working class and under the leadership of the Communist Party they also participated in the preparation of the September Uprising (1923), the world's first anti-fascist revolt. The complete failure of the uprising resulted in the banning of the workers' progressive physical education movement. However, greater efforts were made to win over the support of the young through sporting activities. The attempt of progressive sportsmen to establish a broad worker-peasant sports alliance was suppressed by the fascist police. During this period the progressive workers' physical education movement maintained relations with the international revolutionary workers' sports movement represented by the Red Sports International.

Active propaganda work was carried out by the workers' sports functionaries, the progressive media and the Workers' Youth Union for participation in the Moscow Spartakiad in 1928. The 'red athletes' were also actively preparing for the international workers' spartakiad in Berlin in 1931. During the 1930s, the monarchofascist regime in Bulgaria steered the political, economic and cultural life of the country in the direction of Hitlerite Germany. In his main report at the Seventh Congress of the Comintern in 1935, Georgi Dimitrov worked out specific tactics of creating broad anti-fascist fronts. These were adopted by the progressive sportsmen also. They became members of the legal sports clubs and organizations and began working energetically to create good conditions for sporting activities among the workers and for the denunciation and isolation of their fascist leadership.

In this way they began a struggle for the democratization of the physical education movement. During the years of armed struggle against Monarcho-Fascism (1941–44) many anti-fascist sportsmen fought in the ranks of the guerrilla and military units for the freedom of the people.

The physical education movement in Bulgaria did not acquire a broad basis until the Liberation in 1944. The efforts of the anti-fascist sportsmen to create facilities for physical education and sports for the people met with the resistance of the fascist rulers. The Monarcho-Fascist Government, pursuing its anti-national policy, left a meagre heritage of poorly developed sports scattered among small, insignificant clubs, and only a scanty material sports base.

The period 1923–44 witnessed the establishment of the Bulgarian National Sports Federation. It grew from the affiliation of several sports unions and had only about 70,000 members.

During the period 1878–1944 Bulgarian athletes participated in the Olympic Games and other international championships without much success. The successful participation of individual sportsmen like Nikola Petrov, Dan Kolov (wrestling), Vladimir Stoichev and Krum Lekarski (horse-riding), Lyuben Doichev, Vassil Venkov (track-and-field events) was mainly due to their devotion to the respective sports. The State did not care very much for the development of physical education, the medical control of sports, or improving facilities. Through some of its laws and administrative regulations the monarcho-fascist rulers curbed the development of physical education and sporting activities and attempted to use sport in the fascist ideological education of young people.

Physical Education and Sports in Socialist Bulgaria. On 9 September 1944 the Bulgarian people, under the leadership of the BCP and with the decisive help of the Soviet army, overthrew the monarcho-fascist regime and gave political power to the people. In Bulgaria's thirteen-centuries' long history, this was the first day of a new era, the era of Socialism and Communism. Under these new conditions, the physical education movement also took up a new road of development.

Building the Foundations of the Socialist Physical Education Movement in the Period 1944–56. Immediately after the victory of the 9 September Uprising in 1944, and following a decision of the Central Committee of the BCP, a temporary Central Sports Council was established which in fact was the first leading organ of the Bulgarian socialist physical education movement. At the end of 1947, on the initiative of Georgi Dimitrov the Law on Physical Education and Sports was passed, and his concern for the development of the physical education movement was reflected in the Dimitrov Constitution, adopted in December 1947. The Supreme Committee of Physical Culture and Sports of the Council of Ministers was established in 1948. The decision of the Central Committee of the BCP of August 1949 on the state of physical education and sports and the Party's future tasks ahead, was of great historical importance. In this party document concerned with physical education, Dimitrov's conception of the role and place of physical education and sports in the People's Republic of Bulgaria received further treatment.

It was during the period 1944–56 that the foundations of the socialist physical education movement were laid. Drawing on the experience of the Soviet system of physical education, the building of a socialist physical education system was started in Bulgaria and set off on the road of democratic development. Sports organizations were turned into really mass, community organizations. Sporting facilities were organized in schools, universities, enterprises, offices and in the villages, and major mass sports events, sports celebrations, trade-union youth long-distance races, relay-races and so on were also staged. Much was done to reach the standard of the sports complex of 'Ready for Labour and Defence' exercises. A comprehensive system of championships in the different sporting activities was created and the well-planned construction of a broad sports' facilities programme was promptly begun. The Bulgarian physical education movement also began active participation in international sports events.

A mass cross-country race

Accelerating the Rate of Physical Education and Sports Development in the Period after 1956. The prevailing creative atmosphere in the country after the April Plenum of the Central Committee of the BCP (1956) exercised a favourable influence on all walks of life, including the physical education movement. It boosted the further democratic development of the physical education movement and created opportunities for the real expression of its social and amateur character. The Bulgarian Union for Physical Education and Sports, which is responsible for all sports in the country, was also established.

It is a characteristic feature of the period after 1956 that the issues of physical education and sport were increas-

Spartakiad at the Vassil Levski National Stadium, Sofia

ingly ratified as a state policy in respect of the BCP's concept of the all-round intellectual and physical development of the citizens of the People's Republic of Bulgaria. The issues of physical education and sport were given special attention in a number of party and government documents. Especially important after the April Plenum (1956) were the speeches, reports and statements of Todor Zhivkov, General Secretary of the Central Committee of the BCP and State Council President.

The social basis of the physical education movement was broadened, and the role, involvement and aims of the Dimitrov Young Communist League, the Bulgarian Trade Unions, the Ministry of Education and other public organizations and state authorities were enhanced. The mutual activities of the Bulgarian Union for Physical Education and Sports and these organizations and state bodies were put on an entirely new footing. As a result of the constant concern demonstrated by the Party and the People's government and keeping in step with the accelerated socialist construction programme, the physical education movement became an inseparable part of the country's social development.

Development of Mass Physical Education and Sports. With the active participation and joint efforts of the Dimitrov Young Communist League, the Ministry of Education, the Bulgarian Trade Unions and other public organizations and state organs, the initiatives and creative potential of the masses were enlivened in order to develop the programme of physical education and sport. A number of specific events of great significance that are beneficial and attractive to children, youth and the working people are actively encouraged. Among these are the organized swimming contests across the Danube and other water locations, cycling races, jumping, running and field throwing events, group-field and ski races and winter school competitions.

The various relay-races which commemorate remarkable anniversaries in Bulgaria's history are also highly successful.

A new form of organized group sport introduced by the Dimitrov Young Communist League, is the Rodina Complex of sports exercises, with 1,500,000 holders of the special badge (1983). With the aim of popularizing swimming, youth reviews of group training and competitive-sports activities have been organized each summer since 1970. Thousands of workers and employees participate in the popular trade unions' sports events called 'Strong, Young and Able-bodied'. Every year the number taking part in the in-plant gymnastic exercises and in the remedial gymnastics groups is growing. Sports such as field events, swimming, table tennis and others have been widely developed. The system of holding mass sports competitions is being continuously perfected and enriched. Tangible evidence of the huge scale of physical education and sport are the national spartakiads, organized in the jubilee years of the armed uprising of 9 September 1944. Over 3,000,000 people participated in the Sixth Spartakiad (1979–84). The results achieved in the mass development of physical education and sport show that the sports movement is available to all and that group-organized physical education and sports in socialist Bulgaria are a reality.

Successful Development of Sport. Bulgaria has achieved remarkable successes in the field of sport. During the first decade of its socialist development only partial sporting successes were achieved. By 1955 Bulgarian athletes had won nine medals in world and European championships. It was after 1956 that sport made a big leap forward, taking on a national character, when the development of sport for children and teenagers was substantially improved and a unified system of promoting sport was set up by the Bulgarian Union for Physical Education and Sport and the Ministry of Education. In 1983 more than 158,000 young people studied sport in specialized sports schools and courses. A high standard of educational and training activities is maintained in these sports schools, which are establishing themselves as the providers of the best facilities for developing the talents of the young.

The policy of improving the standard of skills in the various domains of sport is being implemented on a long-term basis, with a definite goal in mind. Great efforts are being made to steadily improve the organization and content of education, training and the competitive activity of athletes. There is a well-organized system of planning, reviewing and control over the training of athletes. Unified training programmes for the various sports have been approved, and the classification of national sporting activities is being constantly improved, as is the system of national championships. The national and Olympic teams are trained under a special system. Specific regulations have been introduced for the development of various sports. Definite guidelines have been established for the construction of sports centres throughout the country. The number of athletes who have reached the standards for top-class competitors and the major international successes

Wrestling champion Boyan Radev receives the Olympic Flame from Greek gymnasts to carry it through Bulgaria on the way to Moscow

indicate the accelerated rate of development of Bulgarian sport. In 1983 there were 6,630 Masters of Sport, 13,089 Candidate Masters of Sport and 768 Merited Masters of Sport.

Bulgarian sports and athletes carry the unfading glory of socialist Bulgaria. At the 1952 Helsinki Olympics boxer Boris Georgiev won a bronze medal, the first ever Olympic distinction for Bulgarian sport. At the Melbourne Olympics in 1956 wrestler Nikola Stanchev took a gold medal, marking the beginning of a golden tradition in Bulgarian sport. Another wrestler, Petko Sirakov, became the first Bulgarian to win the world championship title. Bulgarian athletes have scored victories at every Olympics and at the 1980 Moscow Olympics Bulgaria ranked third in the unofficial rating by nations. During the 1952–83 period Bulgarian sportsmen and women won 27 gold, 50 silver and 40 bronze medals at Olympic Games, 120 gold, 189 silver and 202 bronze medals at world championships, and 194 gold, 233 silver and 199 bronze medals at European championships.

The BCP and the Government of the People's Republic of Bulgaria put a high value on the work done by athletes, specialists and officials in the sphere of physical education. State distinctions have been awarded to 1,729 athletes, coaches, scientific workers and sports officials. A total of 1,395 people have been made Honoured Workers of physical education and sport for their personal contributions to the development of Bulgarian sport. The high results which have been achieved at international competitions place Bulgaria among the top sporting nations in the world. The Bulgarian schools of wrestling, weight-lifting and rhythmic gymnastics have been highly rated and recognized the world over.

The Construction of Physical Education and Sport on a Scientific Basis and the Training of Cadres. The development of physical education and sport is associated with the development of sports science. There is now a unified centre for the study of science and the training of cadres for physical education and sport and an organization which handles research, scientific and applied activities. The Centre for Scientific and Applied Activity contributes greatly to improving the training standards of Bulgarian athletes for European and world championships and for the Olympic Games. The Georgi Dimitrov Higher Institute of Physical Education in Sofia trains specialists in the various sports. Bulgarian scientific workers active in the field of physical education and sport elaborate original theoretical bases and many new systems, programmes, methods, tests and types of apparatus aimed at achieving the best results in the process of training and education.

The 110,000 instructors, coaches, judges, referees and scientific workers provide a broad public basis for physical education. These are Komsomol and trade union activists, teachers and instructors of physical education, and the party and state economic executives who are actively involved in the development of mass physical education and sports.

International Activity of the Bulgarian Union for Physical Education and Sport (BSFS). The international policy of the BSFS and of the entire sports movement is aimed at promoting the struggle for peace, friendship and cooperation, and at bringing nations and youth closely together in an atmosphere of confidence and mutual knowledge. Friendship, co-operation and mutual assistance with sports movements in the USSR and the other socialist countries are expanding and being enriched. Sporting links with the countries of Europe, Asia, Africa and Latin America are likewise making good progress.

The BSFS is actively engaged in international sports organizations. It pursues a consistent policy of democratizing the international sports and Olympic movements. At present the BSFS Central Council and its units are members of the bodies of 50 international sports organizations, which include 62 Bulgarian representatives. The prestige and recognition of the BSFS in the international sports movement are above all due to the major successes scored by Bulgarian athletes at Balkan, European and world championships, as well as at Olympic Games, enhancing the country's international sporting prestige.

Bulgaria has played host to many international events: the world wrestling championships in 1963 and 1971; the world rhythmic gymnastics championships in 1956 and 1969; the world volleyball championships in 1970; the world gymnastics championships in 1974; the world weight-lifting championships in 1977; the world acrobatics championships in 1978; the world canoeing championships in 1977; the European basketball championships in 1960, 1965, 1972, 1976 and 1982; the European weight-lifting championships in 1971 and 1979; the European gymnastics championships in 1965; the European wrestling championships in 1969, 1978 and 1982; the European table-tennis championships in 1977; the ice hockey championships (group B) in 1980; the World Student Games (Universiads) in 1961 and 1977; and the Winter Universiad in 1983. It has hosted Balkan Games in various competitions 77 times.

The Olympic movement is highly popular in Bulgaria. Its noble ideals, which further the physical and spiritual growth of man, the struggle against political, racial and religious discrimination in sport, the desire for equality, peace and friendship among young people and among the athletes from all parts of the world are supported by the

The Bulgarian Olympic delegation in Moscow, 1980

entire Bulgarian public. Bulgaria was one of the thirteen countries to participate in the first modern Olympic Games in Athens in 1896. The Bulgarian Olympic Committee (BOC) was set up on 30 March 1923 and established itself as a member of the Olympic movement, for which it has organized such major events as the 53rd session of the International Olympic Committee (IOC), held in Sofia in 1957, the Tenth Olympic Congress in Varna in 1973, the 74th session of the IOC held in Varna, 1973, and the Ninth General Assembly of the Association of European National Olympic Committees (AENOC) in Sofia, 1978. One of the major events organized by the BOC was the Tenth Olympic Congress, which marked the beginning of a new stage in the international Olympic and sports movement – a stage in its development along democratic lines, updating its structure and principles, ridding the IOC of its conservatism, and consolidating its international functions. The report delivered at the Congress, the Declaration to the tripartite commission and the Appeal to athletes worldwide, adopted by the Congress, set the guidelines for the future development and perfection of the Olympic spirit. One of the most important aspects of the Tenth Olympic Congress was that it was held under the motto of 'Sports for a Peaceful World'. In his greeting to the Congress, the President of the State Council of the People's Republic of Bulgaria, Todor Zhivkov, said: 'The Olympic idea has always been linked to the desire for peace and friendship among the peoples of the world. The People's Republic of Bulgaria is making its active and constructive contribution to this lofty goal. The ethos and humane principals of the Olympic movement are in harmony with the essence of our socialist system and therefore enjoy the full support of the Bulgarian Government and of the Bulgarian people as a whole'.

Construction of Material Sports Facilities. After 9 September 1944 a large-scale programme of the construction of sports facilities began. Many state and public bodies and organizations took part in it, and the initiative of the Central Council of the Fatherland Front, the Bulgarian Trade Unions, the Dimitrov Young Communist League, and the Bulgarian Union for Physical Education and Sport, and of the Ministry of Education, for expanding facilities for light sports provided a powerful impetus. The sports movement now boasts about 20,000 sports facilities, 16,700 of them being outdoors. The most outstanding of them are the Vassil Levski stadium, the Universiad Hall, and the Septemvri Winter Sports Complex in Sofia, the Palace of Sports and Culture and the Sports Palace complex in Varna. There are similar modern sports complexes in Plovdiv, Russe, Burgas, Yambol, Kyustendil, Vidin, Vratsa, Tolbuhin, Pernik, Sozopol and Petrich. The construction of sports facilities corresponds to the rate of development of physical education and sport in the country, in terms of both quality and quantity.

Organization of Physical Education and Sports. Six specialized organizations for physical education and sports have been set up: the BSFS, the BOC, the Bulgarian

Tudor Zhivkov receives Juan Antonio Samaranch, President of the International Olympic Committee, May 1981

Tourist Union (BTU), the Bulgarian Hunting and Angling Union (BHAU), the Defence Assistance Organization (DAO), and the Union of Bulgarian Motorists.

The Bulgarian Union for Physical Education and Sports is the biggest organization of this kind. It carries out a wide range of activities. The BSFS is a public, voluntary organization which works for the popularization of physical education, promotes the development of sports and of domestic and international competitions and contributes to the training of skilful all-round athletes. In pursuance of these objectives, the BSFS co-ordinates its activities with a number of state and public bodies and organizations. It supervises the development of 35 kinds of sports: athletics, wrestling, weight-lifting, gymnastics, rhythmic gymnastics, acrobatics, swimming, diving, angling, water polo, soccer, basketball, volleyball, handball, rugby, rowing, canoeing, boxing, judo, fencing, cycling, equestrianism, modern pentathlon, lawn tennis, table tennis, badminton, bowling, shooting, archery, yachting, skiing, ice hockey, figure skating, chess and bridge. By the end of 1983 it had a membership of 1,115,048. The BSFS Congress is its supreme body and is convened at least once every five years. A Central Council is elected to administer the organization's overall activity between congresses, and a Central Auditing Committee is also elected. The Central Council elects a Bureau and a Presidency, which function as managerial bodies in the period between plenary sessions.

The BSFS Central Council carries out its varied organizing and methodological acitivities through auxiliary bodies. Sports administration is carried out through 29 sports federations and four republican committees. Particular bodies with specific functions take care of the development of amateur physical education and sports activity, the promotion of good sportsmanship, the Union's structure, the qualifications for its sections, education, propaganda, planning, science-applied sports activity, the provision of facilities and finance and the carrying out of construction and administrative activities.

The organizational structure of the BSFS covers the whole country. It has 28 district organizations, managed by elective councils and bureaux. A total of 328 physical education and sports societies had been set up by 1983. The societies are run by councils and committees elected by the ordinary members. The sections dealing with specific sports are the main structural units, tackling all issues related to membership, popularization of sports, improvement of technique and sports education. The number of sections totalled 2,850 by the end of 1983. Both the sections and sports societies promote the practice of physical exercise and sports in 2,796 sports and 4,584 councils for physical education and sports, which have been set up at secondary and higher educational establishments, businesses, offices, villages and city residential areas and jointly with other public organizations directly involved in the administration of these activities.

The Bulgarian Olympic Committee carries out its activities in close collaboration with the other physical education and sports organizations and primarily with the BSFS. It is recognized by the IOC as being the representative of the Olympic movement in the country. It contributes actively to the development of physical education and sports, the consolidation of international sports contacts and the popularization of the humane and ethical principles of the Olympic spirit. It takes part in the activities of the IOC and the other Olympic Associations. It is a member of the Association of National Olympic Committees and the AENOC. It provides assistance for the participation of representatives of Bulgarian sporting organizations in international bodies which are recognized by the IOC.

The Bulgarian Hunting and Angling Union is a public sporting and economic organization of hunters and anglers. It promotes the development of hunting and fishery, game-shooting and angling and takes care of environmental protection. The BHAU's structure is geared to the territorial-production principle. Generally speaking, the Union is divided into societies. There were 2,390 hunting and angling societies and 743 angling societies in 1983. They are incorporated into 153 Hunters' and Anglers' Associations within 27 BHAU District Councils. The Congress, convened every five years, is the supreme body. It elects a Central Council and an Auditing Committee. The Union has 169,651 regular and 35,678 auxiliary members (1983).

The Defence Assistance Organization is a mass public organization supervising the development of certain sports. It has set up a number of clubs, which carry out training and sporting activities in motor rallying, motor racing, aeronautical sports, model racing, amateur radio and shooting events. It is managed by a Central Council, a Bureau of the Central Council, District and Regional Councils and their Bureaux.

Other public and state bodies are concerned with physical education and sports. Apart from the specialized organizations, high ranking public and state bodies are actively involved in the development of Bulgarian physical education and sports. *The Bulgarian Trade Unions* introduce physical exercises at the place of work on a wide scale. They are initiators and sponsors of mass sports and tourist events among workers and employees and play an important role in constructing the necessary sports facilities. *The Dimitrov Young Communist League* actively promotes gymnastic and sports activity among young people of all ages. It is involved in the organization, management, educational process and practical activity of all physical education organizations. *The Fatherland Front* takes care of sports activity in residential areas and other population centres. *The Ministry of Education* also plays an important role, thanks to which school physical education and sports have scored enviable successes. Physical exercises and sports are an integral part of the educational process of children and adolescents, contributing to their successful development as versatile, healthy and able-bodied individuals. Other organizations which play a part in this include the *Ministry of Public Health*, the *Defence Ministry*, the *Ministry of Internal Affairs*, the *Ministry of Transport*, the *Head Department of the Engineering Army Corps* and other institutions. Set up to serve the whole nation, the organization of physical culture and sports in the People's Republic of Bulgaria produce widespread beneficial effects of great importance.

Most Popular Sports in Bulgaria. Sports organizations in the country develop more than 50 different sports, the most popular among which are:

Acrobatics. Because of its attractive nature, it developed initially as a type of circus art and was popular in Bulgaria in this form as early as the nineteenth century. Sports acrobatics made its début after 1950; the first national championship was held in 1952 within the framework of the gymnastics championship, but acrobatics championships have been held independently since 1955. Bulgaria was a co-founder of the International Acrobatic Federation set up in 1973, and Bulgarian acrobats are among the best in the world. In the period 1974–83 they won 55 medals in European championships and 96 medals in world championships. The greatest credit for the 25 championship titles won so far goes to Dimiter Minchev, Margarita Mollova, Boyanka Angelova, and Meli Mihailovg.

Basketball. Basketball made its very first steps in Bulgaria at the beginning of this century. Initially it was practised at foreign schools, but was subsequently popularized in the bigger cities. In 1925 several clubs from Sofia organized the first official championship for men. The first state championships were held in 1942. A Bulgarian men's team participated in the first European championships, held in Geneva in 1935. The greatest successes scored by Bulgarian basketball players are the silver medals won at the 1957 European Men's Championships in Sofia and the bronze medals won at the 1961 European championships in Belgrade. After their début at the European Women's Championships in 1952, the Bulgarian basketball players took their places in the European and world's élite. They were European champions in Łódź (Poland) in 1958, four times runners-up (1960, 1964, 1972, 1983) and three times bronze medallists (1954, 1962, 1976). They were runners-up to the Soviet Union at the 1959 World Championships in Moscow and won the bronze medals at the 1964 Championships. After the inclusion of women's basketball in the Olympic programmes, Bulgaria ranked third at the Munich Olympics in 1976 and second at the 1980 Moscow Games. Bulgaria has been a member of the International Basketball Federation since 1948.

Boxing. Amateur boxing has been practised since 1924. Only two state championships were held before the end of the Second World War. Fifty-four boxers competed for the titles at the first national championships in 1946. Six years later, light-heavyweight Boris Georgiev won the bronze at the 1952 Helsinki Olympic Games – the first Olympic medal in the history of Bulgarian boxing. The popularity of boxing among young people has made Bulgarian boxing a major European and world force. Bulgarian boxers have so far won a total of 56 medals at Olympic, World and European championships.

The most notable successes of Bulgarian boxers so far have been the Olympic titles of bantamweights – Georgi Kostadinov (1952) and Peter Lessov (1981) – and the World title of flyweight Ivailo Marinov (1982). Bulgaria has been a member of the International Amateur Boxing Federation since 1946.

Wrestling. Wrestling is a popular national sport with a long tradition in Bulgarian history. The organized practice of wrestling as a modern sport started in 1932. Long before that, however, Nikola Petrov of Bulgaria won the

Basketball match between Bulgaria and USSR, women's teams, 1984

Nikola Stanchev, first Bulgarian free-style wrestling champion, Melbourne Olympics, 1956

1900 world champion's title in Graeco-Roman style wrestling. Equally remarkable was the performance of Doncho Kolev (Dan Kolov), who won the European free-style wresting title in 1936. Wrestling began to develop quickly after 1944. In about a decade, the number of competitors increased from 500 to 14,000. All the necessary conditions for their systematic training were provided. They took part in scores of national and international tournaments and championships. At the Melbourne Olympics Nikola Stanchev won the first champion's title in the history of Bulgarian sport. Petko Sirakov won Bulgaria's first world title in wrestling at the 1957 Istanbul Championship.

Experienced coaches and specialists fostered the development of a Bulgarian school of wrestling. At present more than 60,000 Bulgarians take part in this sport. It has been included in the curriculum of all Bulgarian schools. Hundreds of schools and wrestling training centres of a high standard have been set up. Particular care is taken of children and teenagers. Following a Government Decree in 1969, a large number of state and public bodies and organizations became directly involved in the popularization and development of wrestling. A total of 52 national tournaments are organized in the country annually. Bulgaria has played host to four World and three European Wrestling Championships. In the 1956-83

Peter Kirov, twice Olympic champion, World and European Graeco-Roman wrestling champion

period Bulgarian wrestlers competing at Olympic, World and European Championships won a total of 134 gold, 179 silver and 158 bronze medals. The Bulgarian Graeco-Roman style wrestling team won the world title in 1971 and the European title in 1973, whereas the free-style team captured the European title in 1968, 1970, 1978 and 1983.

Bulgaria's top wrestlers are Olympic and World champions Boyan Radev, Peter Kirov, Enyo Vulchev, Prodan Gardjev, Lyndmil Damyanov – twice World champion and Alexander Tomov, five times World champion. Coaches Raiko Petrov and Filip Kriviralchev have been awarded the highest state distinction – 'Hero of Socialist Labour'. Extremely popular both in Bulgaria and abroad are Georgi Murkov and Hristo Hihailov – Olympic and World champions; Dimiter Dobrev, Levent Mizamov, Valentin Raichev and Georgi Raikov – Olympic champions; Nikola Dinev – twice World champion and scores of other medal-holders from World and European championships. Bulgaria has been a member of the International Amateur Wrestling Federation since 1954.

Weightlifting. The organized practice of this sport started after the revolution of 1944. The first official competition was held in Sofia on 21 December 1947, and the annual national championships began a year later. The first success for Bulgarian weightlifters was Ivan Vesselinov's European bronze medal and fifth placing in the world, at the 1955 Munich European and World Championships. Bulgarian weightlifting has been rapidly progressing since 1969. The modern training methods introduced by the national squad's senior coach Ivan Abadjiev brought about the creation and establishment of a Bulgarian school in this sport.

Today, Bulgarian weightlifting enjoys high prestige. Bulgarian competitors are among the world's best. They have won 75 gold, 88 silver and 57 bronze medals at Olympic, World and European championships. Bulgaria's top weightlifters are: Hero of Socialist Labour, Yanko Russev – Olympic and five times World Champion, 1981 World Cup winner and holder of 25 world records; Hero of Socialist Labour, Norair Nurikyan – twice Olympic and World champion; Blagoi Blagoev – three times World champion, twice World Cup winner (1982, 1983) and holder of twenty world records. Bulgaria's junior weightlifters have also achieved remarkable results, exceeding on many occasions those scored by the men. Bulgaria has been a member of the International Weightlifting Federation since 1951.

Volleyball. Today, more than 100,000 Bulgarians play volleyball, which became an organized sport in 1922. Only four teams competed in the first state championships held in 1942, but Bulgaria ranked third after the Soviet Union and Czechoslovakia at the first men's World championship (1949). Subsequent high placings include: silver (1970) and bronze medalists (1952) at World championships; European runners-up (1951) and bronze medalists (1955, 1981, 1983). The greatest success so far has been the second placing at the 1980 Moscow Olympics. Highest placings of Bulgaria's women's teams are: 1980 Olympic bronze medal, and 1981 European title. Bulgaria has scored successes at Club championships with four European Cup holders: Levski Spartak (Women 1964), CSKA Septemvriisko Zname (men – 1969 and women – 1979, 1984). Bulgaria has been a member of the International Volleyball Federation since 1949.

Rowing. The first official rowing competitions were organized in Burgas on 19 August 1924, and impetus was given to development of this sport after 1950. The first National Rowing Championships were held in 1953. Bulgarian rowers made their début in international competition at the 1956 European championship in Bled. They have won 27 medals at Olympic, World and European championships so far. The 1976 Montreal Olympics marked the best performance of Bulgarian rowers, who won two gold medals and one silver. Championship titles were won by the women's pairs Sika Kelbecheva and Stoyanka Grueva, and Zdravka Yordanov and Svetla Otsetova. World titles were won by the men's coxed pair (Amsterdam 1977) and the women's coxed fours and double sculls (New Zealand, 1978). Bulgaria has been a member of the International Rowing Federation since 1955.

Canoeing. Canoeing was first practised as an organized sport in 1953, and a year later the first national

Alexander Tomov, five times World champion, five times European Graeco-Roman wrestling champion, and winner of the silver medal at three consecutive Olympic games

championships were held. Bulgaria made her début at a World championship in 1958. The bulk of successes came after 1976 with a total of nineteen medals won at Olympic, World and European championships. Lyubomir Lyubenov is the men's K-I 1980 Olympic and 1978 World champion. Bulgaria has been a member of the International Canoe Federation since 1953.

Track and Field Events. The first tournament which attracted athletes from all over the country was organized in 1924. That same year four Bulgarians competed at the Paris Olympics. National championships were first held in 1926. Track and field events enjoy great popularity and are the most widely practised individual sport in Bulgaria – more than 190,000 athletes go in for them today. They form the basis of physical education at school. Relay and cross-country races and other types of competitions, involving people of all ages (from children to veterans), regularly take place.

At the 1976 Montreal Olympics, Hero of Socialist Labour, Ivanka Hristova won the first Olympic gold medal in the history of Bulgarian athletics and set up a new Olympic record. Bulgarian athletes have won a total of one gold, five silver and four bronze medals at Olympic games. In the 1944–83 period, Bulgarian athletes won a total of eleven medals at European outdoor athletics individual championships, including four gold and five silver. Testifying to the progress of Bulgarian athletics are the world records set up by: Yordanka Blagoeva – women's high jump; Svetla Zlateva – women's 800 m.; the women's 4×800 m. relay team; Ivanka Hristova – women's shotput; Maria Vergova-Petkova – women's discus; and Antoaneta Todorova – women's javelin. Bulgaria has been a member of the International Amateur Athletic Federation since 1924.

Swimming and Diving. The conditions for the development of swimming were created after 1944. Owing to special attention by the State, swimming progressed from an under-developed sport to one of the most popular in the country. Scores of swimming pools and schools were opened in a number of regions. Ample opportunities exist for systematic training for the more than 55,000 swimmers. Bulgarian swimmers made their début at the European championships in 1958. Bulgaria has been a member of the International Amateur Swimming Federation since 1953.

Skiing. The Bulgarian Skiing Union was founded in 1931. Georgi Dimitrov was the first Bulgarian alpine skier to join the world skiing élite by finishing seventh in the Alpine combination at the 1956 Cortina d'Ampezzo Winter Olympics.

The favourable conditions created for the development of skiing in Bulgaria have made it possible for more than

Norair Nurikyan, twice Olympic champion, World and European weight-lifting champion

Svetla Otsetova and Zdravka Yordanova, Olympic and world rowing champions

18,000 skiers to take up this sport. Bulgaria has been a member of the International Ski Federation since 1932.

Gymnastics made its very first steps at the schools during the Bulgarian Revival Period in the last century. Bulgarians took part in the gymnastics championships at the first Olympic Games in 1896. The Union of Gymnastics Societies, Yunak, was set up in Sofia in 1898. Mass gymnastics competitions featured predominantly in its yearly activity. In 1931 Yunak joined the International Gymnastics Federation and participated in a big international tournament in Venice. Bulgarian gymnasts became actively involved in international sports activity after 1944. Stoyan Delchev has scored the greatest successes so far. He was the first Bulgarian to top the horizontal bar record to win the first Olympic gold medal at the 1980 Games. Boryana Stoyanova won the vaulting-horse gold medal at the World Women's Championship in 1983. Over 23,000 people take part in this sport in the numerous gymnastics centres throughout the country.

Shooting. The development of this sport started after 1944. National championships have been held regularly since 1950. Bulgaria's top marksman is the World champion Anka Pelova. A total of 56 medals have been won at Olympic, World and European championships so far. A. Pelova, N. Matova and Vessela Lecheva set up four world records in 1975, 1979 and 1980. Bulgaria has been a member of the International Shooting Union since 1921.

Lawn Tennis. The first lawn tennis club in Bulgaria was founded in 1911. The Bulgarian Lawn Tennis Federation was founded in 1930 and that same year it joined the International Lawn Tennis Federation. National championships have been held since 1929. Before the Second World War Bulgarian tennis players only competed at the Balkan Games and in international matches. Nowadays they take part in scores of top-ranking international tournaments. They have been regular participants and among the best performers at the Balkan championships since their resumption in 1960. Bulgarian tennis players competed for the first time in the Davis Cup Tournament in 1964. Bulgaria has participated in the women's tournaments for the International Federation Cup since 1968.

Table Tennis. Table tennis is one of the most popular sports in the country with over 62,000 people from all age brackets actively practising it. The first state championships were held in 1942 and the first Republican section was set up in 1948. National Championships have been regularly held since 1949. Bulgarian table tennis has not yet scored any successes beyond the Balkan Games level. Bulgaria has been a member of the International Table Tennis Federation since 1950.

Fencing. Fencing gained popularity after Bulgaria's Liberation from Ottoman domination. It was primarily practised at the Military Academy and officers constituted

(Centre) Lyubomir Lyubenov, Olympic and world canoeing champion

A swimming lesson for students of the Georgi Dimitrov Higher Institute of Physical Education at the Madara swimming pool

Pioneer children preparing and launching yachts on the Iskur Dam lake

the bulk of participants in the competitions. The Bulgarian Olympic Committee sponsored the first comprehensive Balkan Games in Sofia in 1931, and fencing was among the six sports events included in the programme. The beginning of regular men's and women's national championships was laid in 1950. The best results so far have been scored by the men's sabre teams. In 1983 Vassil Etropolski was a World champion and World Cup holder. Bulgaria has been a member of the International Fencing Federation since 1950.

Association Football (Soccer). Soccer is the most popular sport in Bulgaria. The beginning of its organized development was laid down in 1909 with the formation of the first soccer teams and clubs. The first state championships were held in 1924 and that same year Bulgaria's national team participated in the Paris Olympics. Before the Second World War Bulgarian soccer's greatest successes were the winning of the 1932 and 1935 Balkan Cups. A fresh impetus was given to soccer in the post-war period. Nowadays 205,000 players go in for this sport and compete in regular championships. Bulgarian soccer teams participated in the World Cup finals in 1962, 1966, 1970 and 1974. Their greatest successes so far have been the 1956 Olympic bronze medal and the 1968 Olympic silver medal. Bulgaria has been a member of the International Football Federation since 1924.

Eurythmics. Eurythmics has as its background the group apparatus exercises included in the school curricula at the beginning of the century and the mass gymnastics shows. In its present form it began its development in 1951. National championships have been regularly held since 1952. It is a favourite sport of Bulgarian women. Today, more than 5,000 girls participate in it. For several years Bulgarian rhythmic gymnastics has been a trend-setter at European and World championships. Bulgaria has had five World champions in the individual events and four titles in the group exercises at the ten World championships

Stoyan Delchev, Olympic and European gymnastics champion

held so far. The top stars are Hero of Socialist Labour, Maria Gigova – three times absolute World champion (1969, 1971, 1973) and winner of thirteen medals at five World championships; Ilyana Raeva – 1980 absolute European champion; Anelia Ralenkova – 1981 absolute World champion; Lili Ignatova – first World Cup holder; and Dilyana Georgieva – 1983 absolute World champion. The Bulgarian school of rhythmic gymnastics was founded by Hero of Socialist Labour, Julieta Shishmanova and further developed and perfected by Hero of Socialist Labour, Neshka Robeva. This school has had its impact on the development of eurhythmics throughout the world. It has been studied and popularized in Europe, Asia and America. Bulgarian coaches work in many countries sharing their experiences and promoting the development of this graceful sport.

Chess. The setting up of the Sofia Chess Club on 1 February 1922 is generally considered to mark the birth of organized chess in Bulgaria. The first unofficial state championship was held in Turnovo in 1928 and regular championships for the country were initiated in 1933. Bulgaria made its Olympic début at the 1936 Munich Olympics. The first successes of Bulgarian chess players took place after the 1944 revolution. Bulgaria won the World students' team title in 1959. Olympic bronze medals were won by the men's team in 1968 and by the women's team in 1974. Milko Bobotsov became the first Bulgarian grandmaster. A total of eleven Bulgarian chess players have been awarded the Grandmaster title.

Training session of the Druzhba Sailing School on the Iskur Dam lake

Bulgarian rhythmic gymnasts Lili Ignatova, Anelia Ralenkova and Ilyana Raeva

Vassil Etropolski, World fencing champion and 1983 World Cup holder

Bulgaria has been a member of the International Chess Federation since 1938.

The Tourist Movement

Brief Historical Survey. The first tourist travels in Bulgaria began after her liberation from the Ottoman domination (1878). Members of the cultural and educational Slavyanska Beseda society in Sofia founded the Urvich tourist club in 1889 and began to organize trips to various beauty spots, monasteries and other places of interest round Sofia. The club was actually founded and inspired by the great Bulgarian writer and democrat, Aleko Konstantinov. It was his idea to organize, in August 1895, the first mass climbing of the Cherni Vruh Peak on Mount Vitosha by three hundred people, and this is considered as the beginning of the organized tourist movement in Bulgaria. The Club of Bulgarian tourists was then set up and the popular poet Ivan Vazov was elected president, with Aleko Konstantinov as secretary. The enthusiastic tourists made a trip to the Poganov Monastery.

The first Bulgarian tourist society, Aleko Konstantinov, on whose Board figured prominent Bulgarian patriots, was founded in Sofia in 1899. The aim was to visit and get to know the most beautiful places in Bulgaria and to make them accessible by means of path-marking, planting trees and broad publicity. From 1900 onwards affiliated tourist societies began to be founded throughout the country and the society turned into a union. The first provincial branch (society) was founded in Svishtov, followed by societies in Nova Zagora, Plovdiv, Pleven, Vratsa, Burgas, Haskovo, Turnovo and elsewhere. On 21 May 1901 the first regular assembly of the Bulgarian Tourist Society (BTS) was convened in Sofia, at which the eminent Bulgarian scholar Alexander Todorov-Balan was elected president. The first issue of the *Bulgarski Tourist* (Bulgarian Tourist) magazine (1902–43) came out in 1902. In 1908 a group of school children founded the Geolog (Geologist) high-school society in Russe with the aim of exploring their native country. The Tourist Union of Youth (TUY) was founded in 1911. After World War I the BTS became more active and directed its efforts towards path-marking, the construction of mountain huts, the publication of guide books, etc. Several smaller independent tourist societies, which were members of the Bulgarian Tourist Federation, also came into being. In 1931 the Federation united the BTS with its member-societies, and the new organization was named Bulgarian Tourist Union (BTU). This encouraged both younger

Physical Education, Sports and Tourism

Three times absolute World rhythmic gymnastics champion, Maria Gigova, now President of the Bulgarian Gymnastics Federation, awards the Julieta Shishmanova prize to Neshka Robeva, one of Bulgaria's outstanding rhythmic gymnasts, now coaching the national team

Table 1: Medals Won By Bulgarian Athletes at Olympic Games, World and European Championships over the 1944–83 Period

Sport	Olympic Games			World Championships			European Chanpionships		
	Gold	Silver	Bronze	Gold	Silver	Bronze	Gold	Silver	Bronze
Acrobatics	—	—	—	18	41	37	7	29	19
Basketball	—	1	1	—	1	1	1	5	4
Boxing	2	1	6	1	1	5	6	15	21
Wrestling	13	24	12	31	62	60	82	67	43
Wrestling–Sambo	—	—	—	2	12	28	6	14	15
Weightlifting	7	10	3	31	36	19	37	42	35
Volleyball	—	1	1	—	1	2	1	1	4
Rowing	2	2	3	3	7	7	—	1	2
Judo	—	1	1	—	—	1	1	3	1
Canoeing	1	2	3	2	3	8	—	—	—
Equestrianism	—	1	—	—	—	—	—	—	—
Athletics (individual champion)	1	5	4	—	—	3	4	5	2
Swimming	—	—	—	—	—	—	—	—	1
Angling	—	—	—	—	—	—	—	2	1
Skiing	—	—	1	—	—	1	—	—	—
Diving	—	—	—	—	—	—	1	—	—
Gymnastics	1	—	2	1	—	5	9	4	6
Shooting	—	1	2	2	6	3	10	15	17
Lawn tennis	—	—	—	—	—	—	1	2	1
Fencing	—	—	—	1	—	2	—	1	1
Football	—	1	1	—	—	—	—	—	—
Eurythmics	—	—	—	25	19	24	12	6	6
Chess	—	—	—	—	—	2	—	—	—
Total	27	50	40	117	189	208	178	212	178

Table 2: Engagement in Sports

Sports	Sports sections	Membership	Competitors
Acrobatics	27	4,080	3,527
Basketball	214	82,752	77,083
Boxing	28	4,434	3,743
Wrestling	257	65,645	60,127
Weightlifting	60	8,809	6,730
Yachting	8	817	762
Volleyball	286	108,887	101,035
Rowing	12	2,901	2,857
Judo	28	5,207	4,367
Canoeing	12	3,306	2,232
Cycling	37	7,150	6,087
Equestrianism	64	3,945	2,046
Athletics	332	200,199	190,878
Swimming	142	60,986	58,700
Skiing	44	20,252	18,328
Gymnastics	65	24,035	23,201
Shooting	42	18,503	20,129
Lawn tennis	15	5,813	5,078
Table tennis	287	71,576	62,317
Fencing	9	1,190	1,116
Football	353	268,042	204,751
Handball	134	50,984	50,684
Eurythmics	17	6,301	5,206
Chess	312	79,887	67,761

Table 3: Champions

Sport	Olympic Champion	Year
Boxing	Georgi Kostadinov	1972
	Peter Lessov	1980
Free-style wrestling	Nikola Stanchev	1956
	Enyo Vulchev	1964
	Prodan Gardjev	1964
	Hristo Mihailov	1976
	Valentin Raichev	1980
	Levent Nizamov	1980
Graeco-Roman style wrestling	Dimiter Dobrev	1960
	Boyan Radev	1964, 1968
	Peter Kirov	1968, 1972
	Georgi Murkov	1972
	Georgi Raikov	1980
Weightlifting	Norair Nurikyan	1972
	Yordan Bikov	1972
	Andon Nikolov	1972
	Norair Nurikyan	1976
	Yordan Mitkov	1976
	Assen Zlatev	1980
Rowing	Svetla Otsetova	1976
	Zdravka Yordanova	1976
	Siyka Kelbecheva	1976
	Stoyanka Gruicheva	1976
Canoeing	Lyubomir Lyubenov	1980
Athletics	Ivanka Hristova	1976
Gymnastics	Stoyan Delchev	1980

Sport	World Champion	Year
Acrobatics	Kalinka Lecheva	1974
	Stefka Spassova	1974
	Bonka Encheva	1976
	Maria Dimitrova	1976
	Milka Mateeva	1976
	Stefka Gencheva	1978
	Galya Hristova	1978
	Margarita Mollova	1978
	Dimiter Minchev	1978, 1980, 1982
	Boyanka Angelova	1980, 1982
	Galya Velinova	1980
	Valentina Kuyumdjieva	1980
	Pavlina Stankova	1980
	Meli Mihailova	1980, 1982
	Snezhana Andonova	1982
	Antoaneta Miladinova	1982
Boxing	Ivailo Marinov	1982
Free-style wrestling	Petko Sirakov	1957
	Lyndmil Damyanov	1959
	Enyo Vulchev	1962
	Prodan Gardjev	1963, 1966
	Ivan Iliev	1969
	Russi Petrov	1971
	Simeon Shterev	1981
	Hristo Makailov	1974, 1975

Sport	World Champion	Year	
Graeco-Roman style wrestling	Angel Kerezov	1966	
	Boyan Radev	1966	
	Peter Krumov	1969	
	Valentin Yordanov	1983	
	Peter Kirov	1970, 1971, 1974	
	Georgi Murkov	1971	
	Alexander Tomov	1971, 1973, 1974, 1975 1979	
	Ivan Kolev	1973	
	Kamen Goranov	1975	
	Nikola Dinev	1977, 1982	
	Stoyan Nikolov	1978	
	Yanko Shopov	1979	
	Bratan Tsenov	1983	
	Andrei Dimitrov	1983	
Wrestling – Sambo	Valentin Minev	1983	
	Georgi Petrov	1982	
Weightlifting	Norair Nurikyan	1972, 1976	
	Yordan Bikov	1972	
	Andon Nikolov	1972	
	Atanas Kirov	1973, 1974, 1975	
	Nedelcho Kolev	1973, 1974	
	Georgi Todorov	1974, 1975	
	Trendafil Stoichev	1974	
	Valentin Hristov	1975, 1977	
	Yordan Mitkov	1976	
	Yanko Russev	1978, 1979, 1980, 1981, 1982	
	Anton Kodjabashev	1979, 1981, 1982	
	Assen Zlatev	1980, 1982	
	Beloslav Manolov	1981	
	Blagoi Blagoev	1981, 1982, 1983	
	Neno Terziiski	1983	
	Alexander Vurbanov	1983	
Rowing	Todor Mrunkov	1977	
	Dimiter Yanakiev	1977	
	Stefan Stoikov	1977	
	Zdravka Yordanova	1978	
	Svetla Otsetova	1978	
	Ani Bakova	1978	
	Dolores Nakova	1978	
	Rossitsa Spasova	1978	
	Rumeliana Boncheva	1978	
	Ani Eftimova	1978	
Canoeing	Maria Mincheva	1977	
	Roza Boyanova	1977	
	Velichka Mincheva	1977	
	Natasha Yanakieva	1977	
	Lyubomir Lyubenov	1978	
Gymnastics	Boryana Stoyanova	1983	
Shooting	Anka Pelova	1974	
	Lyubcho Dyakov	1981	
	Lyubomir Vangelov	1981	team standings
	Ivan Mandov	1981	
	Jean Mihov	1981	

Table 3: Champions (continued)

Sport	World Champion	Year	
Fencing	Vassil Etropolski	1983	
Rhythmic gymnastics	Maria Gigova	1967, 1969, 1971, 1973	
	Violeta Elenska	1969	
	Vera Marinova	1969	
	Bogdana Todorova	1969	group exercises
	Sonya Peeva	1969	
	Stela Milosheva	1969	
	Despa Katalieva	1969	
	Rumyana Stefanova	1971	
	Vera Marinova	1971	
	Despa Katalieva	1971	group exercises
	Assya Kursheva	1971	
	Snezhana Decheva	1971	
	Kapka Blagoeva	1971	
	Kristina Gyurova	1979	
	Ilyana Raeva	1979	
	Anelia Ralenkova	1981, 1983	
	Lili Ignatova	1981, 1983	
	Galina Rangelova	1981	
	Kamelia Ignatova	1981	
	Diana Tabakova	1981	group exercises
	Emilia Bozhidarova	1981	
	Zdravka Chonkova	1981	
	Maria Kuzmanova	1981	
	Dilyana Georgieva	1983	
	Svetla Chobanova	1983	
	Tsvetomira Filipova	1983	
	Paulina Krusteva	1983	Group exercises
	Viktoria Dimitrova	1983	
	Ilyana Ilieva	1983	
	Mariela Spassova	1983	

and older tourists to continue with the organization of hikes, afforestation and the building of mountain huts, shelters and chalets. The first tourist hikes in winter-time were to the Cherni Vruh Peak in 1918.

In 1929 tourists from Sofia and Samokov who favoured the higher mountains, founded the Bulgarian Mountaineering Club, which was the first organization with an Alpine profile in Bulgaria. From 1931 onwards it was called the Bulgarian Alpine Club. By means of talks and showing photographs, its members and the general public became more familiar with the high mountains in Bulgaria and with Alpine practices. Despite the lack of adequate equipment, for which they compensated with enthusiasm, Bulgarian alpinists made a winter trek along the difficult karst ridge of the Pirin mountains, and they also climbed the northern face of the Malyovitsa peak in the Rila mountains. The first Bulgarian Spelaeological Society was founded in 1929, and its members launched a publicity campaign for the protection of caves in Bulgaria. The same year saw the foundation of the Union for the Protection of Nature and the BTS became a collective member. Skiing hikes and down-hill skiing became increasingly popular and the Bulgarian Skiing Union was founded in 1931. The Mountain Rescue Service, started by Bulgarian tourists and alpinists, was founded in 1933. Apart from bringing together all lovers of nature, the hikers' societies also organized large-scale building and educational activities. A number of tourist huts, shelters and chalets were built with the voluntary labour and donations of money by tourists. The first hut to be built in this way was Skakavitsa in the Rila mountains in 1922. About 70 huts, shelters and chalets were built by 1944. The members of the BTS also built the high-altitude observation posts on the Mussala peak (in the Rila mountains), on the Botev peak (in the Balkan range) and on the Cherni Vruh peak, and they took an active part in the construction of monuments on Shipka peak and other historical sites.

The Bulgarian bourgeois capitalist State did not have a positive attitude to the tourist movement and hampered its development and democratization. The attempts of reactionary forces to instil the ideology of Fascism proved futile. In its struggle against Fascism the BCP made good use of the tourist movement as a means of uniting progressive young people and as a legal platform for numerous political actions. Tourist hikes became covers for underground meetings and conferences of the Workers' Youth Union and the Bulgarian Workers' Party (Communists). Many active tourists such as Slavi Alexiev ('the Tourist'), Tsvetan Spassov, Boyan Chonos, Argir Stoilov Tomata and others perished in the armed antifascist struggle. Many tourist huts (Vladko, Makedonia, Chavdar, Uzana, Buzludja, Chumerna and others) provided shelter for the partisans. Some of those huts were burnt down and destroyed by the gendarmerie and the police.

Development in the Period after 9 September 1944. The Party and the People's Government gave special attention to the promotion of the tourist movement, to building up its new structure and to the development of new forms of tourism. The heritage of the tourist movement was popular but still developed within a rather narrow organizational framework, encompassing no more than 17,000 adults and young people. In 1945 the BTU, the TUY and the Bulgarian Skiing Union merged under the name of National Tourist Union and started the publication of the *Naroden Tourist* magazine. In order to attract more working people and young people it was decided to establish an administrative link between tourism and sports. Accordingly, towards the end of 1946 the National Tourist Union joined the National Union for Physical Culture. Proceeding from the experience of Soviet tourism and alpinism there began a thorough review of the aims, tasks and forms of tourism and mountaineering in Bulgaria and the tourist movement was placed on a much broader basis. Mass treks to partisan and other historical sites and places of interest were organized, qualified instructors were trained, a system of badges and grades was introduced, and summer holiday hiking was organized. The first mass trek, starting from a number of places and converging on

Stoletov peak, took place in 1947, and the route of the band of the Bulgarian revolutionary Hristo Botev from Kozlodui to Okolchitsa was marked and followed. Mass treks were organized to the Buzludja peak, the birth-place of the BCP and also to Oborishte, a place sacred to all Bulgarians. In 1947 the first Winter Skiing Treks of Freedom through the Rila and Pirin mountains were started. The first high-mountain relay on the occasion of the birthday of Georgi Dimitrov took place in 1948. The first one-month training course for tourist cadres was organized in the Rila mountains in 1949. With the material backing of the trade unions, holiday hiking – a new and promising form of tourism was launched in 1951. There were a number of exploratory tourist expeditions along various routes – Along the Route of Benkovski's Flying Band (1951), Along the Path of Georgi Dimitrov and Vassil Kolarov (1953), Along the Route of the Anton Ivanov Partisan Brigade (1955) and others.

The historic April Plenum of the Central Committee of the BCP (1956) marked a turning point in the development of the tourist movement too. In order to boost the tourist movement and incorporate it into the mainstream of life, the BTU was reinstated in 1957. At its Constituent Congress it incorporated 108 tourist societies with a total of 64,000 members.

The BTU developed as a voluntary public mass organization for the promotion and guidance of the tourist movement, alpinism, spelaeology and orienteering, thus continuing the progressive traditions of the BTU (1895–1945), the TUY and the Bulgarian Skiing Union. Local district councils and tourist societies were set up in towns and villages.

Tourism began to develop on a large scale. It has currently become a common activity of the BTU, the Bulgarian Trade Unions, the Komsomol Youth Organization and the Fatherland Front and is used as an active way of improving the physical fitness of the population and of installing partriotic feelings and a love of nature. The 'Weekend in the Open' movement, initiated by the workers of the Kremikovtsi Steel Plant, was welcomed by Bulgarian tourists and acquired wide popularity. On the initiative of many tourist societies and the district councils of the BTU, local celebrations and gatherings are organized in the mountains and by rivers and lakes. The 'Get to Know Your Socialist Homeland' movement has become a powerful means of patriotic education. The selected one hundred tourist sites enable nature lovers to become familiar with places of outstanding natural beauty and monuments of culture, and also with the rich and heroic past of the Bulgarian people. The 'Get to Know Your Native Place' movement is also developing successfully. Worth noting are the versatile activities of the Dimitrov tourist expeditions for young people. Thousands of expeditionary groups, which study the routes of party revolutionaries, of bands of rebels and Haiduks, as well as of individual national heroes, have been organized. These groups collect valuable materials and photos, keep diaries, organize exhibitions and visit important industrial and construction sites. Together with the tourist societies, the Trades Union organizations, the Komsomol societies, bands of Pioneer children and the Fatherland Front organizations take part in mass hikes and excursions, treks, assemblies, competitions and reviews such as 'Healthy, Strong and Fit for Work' and 'With Mother and Father in the Mountains'. As many as 248,327 excursions, hikes, gatherings, etc. with more than 26,100,000 participants took place in 1982. The BTU incorporates 911 societies with a total of 2,915,812 members.

Work teams competition in the Healthy, Strong and Fit review

The Winter Spartakiad at the Aleko Hostel on Mount Vitosha

Opening ceremony of the Student Games at the Vassil Levski National Stadium, Sofia, 1977

The various forms of tourism are becoming increasingly popular. *Skiing tourism* has become the main form of winter tourism in the mountains, and a special decree on promoting skiing was issued by the Secretariat of the Central Commitee of the BCP and the Council of Ministers. The tourist huts have been fitted out with appropriate equipment and a number of skiing centres, chair-lifts and drag-lifts have been built. Numerous instructors and skiing teachers are trained every year and all kinds of skiing hikes, down-hill races and treks widely publicized. In addition to the traditional Skiing Treks of Freedom through the Rila and Pirin mountains, treks are also organized every five years from the Kom peak (near the Bulgarian-Yugoslav border) to Cape Emine (on the Black Sea) along the ridge of the Balkan range, and also through the Rhodope mountains, etc. There are skiing treks for women, and Bulgarian tourists also take part in skiing treks and various events organized abroad.

Cycling tourism began to develop after 1952, when the first national cycling trek was organized. This kind of tourism has proved accessible and popular, especially among school children and young people. In addition to the traditional national treks to Oborishte and from Kozlodui to Okolchitsa many local cycling competitions also take place. There are festivals of cycling and mass cycling tours, e.g. the Rodina and Golden Bicycle tours. Particularly popular are the international 'Along the Path of the First Bulgarian Army' (in World War II) and 'Along the Path of the Liberators' tours and some other treks. In 1983 there were 333 cycling tourism clubs with 13,098 participants.

Boating first developed in the form of expeditions undertaken by young travellers down some of the country's rivers, later including also the Danube, from where the river Timok joins it to the town of Silistra. The Bulgarian rivers Iskur, Russenski Lom, Kamchia, Yantra, Maritsa, Tundja, Struma and others are also suitable for such purposes. A national regatta down the Danube, regattas down most of the inland rivers and water slalom competitions etc. in the Black Sea are held every year. In 1965 Bulgarian oarsmen took part in the International Danube Regatta, which starts at Ingolstadt (FRG) and finishes in Silistra (Bulgaria) and is one of the most outstanding events of its kind in the world. The participation of increasing numbers of people in boating activities is furthered by the construction of special centres, the introduction of suitable equipment and the setting up of yacht clubs. In 1975 a training centre for instructors began to operate near the town of Vidin. From 1979 onwards the first Eho maritime tourist regatta began to be held annually. By the end of 1983 there were 138 boating clubs with 7,696 members.

Alpine climbing gained popularity with the first ascents of rock faces near the village of Lakatnik, and of Mount Vitosha and the Rila mountains, and started to attract an increasing number of young people. Bulgarian alpine climbers receive training on the cliffs near Lakatnik, the Kominite peak of Mount Vitosha, the Malyovitsa peak in the Rila mountains, the Soviet Alpine camps in the Caucasus and in the Central Alpine Camp (now the 'Malyovitsa' Central Mountaineering School) in the Rila mountains, which was opened in 1952. The most difficult faces in the Rila, Pirin and the Balkan range attract increasing numbers of Bulgarian climbers.

Bulgarian alpine climbers have scored significant international successes. The 12 km-long climb across the Bezingi face in the Caucasus was performed in 1964. There followed climbs in the Alps – of Petit Dru, the western face of the Bonati tour, the Tour of the Guides, the Grands Jorasses up the Walker ridge of Mont Blanc, the Frene ridge, Matterhorn, the north face of the Eiger,

Mass marathon for the Sofia 1983 prize

the classical tour and the Japanese diretissima, and in the Caucasus the faces of the Cross of Ushba. A number of peaks with altitudes higher than 7,000 m. have been climbed – Noshak Peak (7,492) in the Hindu-Kush, Lenin (7,134), Communism (7,495) and Korzhenevska (7,105) in the Pamir. In 1981 Hristo Prodanov climbed the Lhotse Peak (8,501) in the Himalayas. In honour of the 40th anniversary of the victory of the socialist revolution in Bulgaria and the 80th anniversary of the organized revolutionary trade union movement, in 1984 H. Prodanov, I. Vulchev, M. Savov, K. Doskov and N. Petkov conquered the highest peak on earth – Mount Everest, from the difficult Western ridge. In 1983 there were 47 Alpine clubs with about 1,800 participants.

Orienteering has established itself as one of the most popular kinds of tourism. The first national competitions were held in 1955 near the town of Koprivshtitsa only a year after the introduction of that sport in Bulgaria. The first international competitions were held in 1958. The BTU is a founding-member of the International Orienteering Organization, established in 1961. The proficiency of Bulgarians engaged in orienteering has grown steadily and in Balkan and international competitions of the socialist countries Bulgarians usually rank among the first. They came second at the World Ski-Orienteering Championship in 1977 and won the bronze at the World Championship for men in Sweden (1983) and for women in Laverone (1983). In 1986 Bulgaria will host the Sixth World Ski-orienteering Championship. In 1983 there were 253 clubs with over 20,000 members.

Cave tourism and spelaeology have also developed. Special courses are organized every year. Every two years there are courses which train one hundred lifeguards, instructors, trainee-instructors and bearers of the 'Young Cave Explorer' and the 'Bulgarian Cave Explorer' badges. The Bulgarian Federation of Spelaeology and the district sections and clubs organize hundreds of expeditions for the exploration of karst terrains in Bulgaria. The data are passed over to various research institutes of the Bulgarian Academy of Science. As a result 3,970 caves and precipices have been charted and recorded. Together with various institutes of the Bulgarian Academy of Science, the BTU and its Spelaeology Federation have prepared a comprehensive programme for the preservation of caves and karst areas in Bulgaria, and have also organized scientific conferences and expeditions. Nine caves have been made accessible and are visited by about three million people annually. Two more caves will soon be open to the public. Two spelaeology centres have been built: the National Centre in the village of Karlukovo, Sofia district and the *Peshternvak* (Cave Explorer) Centre in the town of Chepelare, which houses the Rhodope Mountains Karst Museum, one of the largest in Europe. In 1983 there were 54 cave exploration clubs with 2,500 members.

Holiday hiking was intitiated in 1951 when 1,050 people went along two mountain routes. It has developed as an important joint enterprise of the Bulgarian Trade Unions, the Ministry of Education and the BTU. The facilities of the BTU and the mountain chalets of the Trade Unions

Hristo Prodanov, conqueror of the Himalayas: Llotse (1981), Everest (1984)

Fifteenth World Parachuting Championship; singles target jumping, Kazanluk, 1980

Members of a Sofia Pioneer children's group on a hike in the Rhodope mountains

are thus used to full advantage. The itineraries – on foot, by ship, train, coach, car and combined means – have gradually covered the whole country. In 1983 there were about 60,000 participants included in 53 different itineraries. There are both summer and winter routes. The successful development of holiday hiking is promoted by the constant improvement of facilities, regular supplies and the availability of qualified guides. As many as 189 tourist sites have so far been included in the holiday hiking programmes.

Nature protection holds an important place in the activities of the BTU, as a result of which the first law on the protection of the environment in Bulgaria was passed. A National Commission with the Central Council of the BTU for the protection of nature has been set up, and there are also commissions at the district councils, societies and groups of voluntary environmentalists. Much is being done for the popularization of the cause of nature protection and for the preservation of various beauty spots, rare animals and plants, the cultivation of the areas round the tourist huts and of other tourist and historical sites that are visited by large numbers of people. The BTU has emerged as a public guarantor of nature preservation in the mountains of Bulgaria.

The development of the tourist movement in Bulgaria is connected with the construction of huts, shelters and chalets. The number of huts has increased from 70 with accommodation for about 2,000 in 1944 to 375 and accommodation for 23,197 in 1983 (with about 3 million overnight registrations a year.) As many as 256 huts have restaurants. The huts have become fine places for cultural recreation and centres for patriotic education. In many of the huts there are museums and various exhibitions. At the disposal of tourists are rooms for games, libraries with more than 20,000 books, television and radio sets, film and slide projectors, etc. The areas round the huts are constantly cultivated and improved. The construction of volleyball, basketball and other playgrounds is continuing, new ski-runs are cleared through the forests, and up-to-date cable-car lines to places difficult of access. Together with the construction of huts the marking of tourist paths also continues. A special service for path-making was set up at the Bulgarian Tourist Union in 1964. At present there are permanent markings along all the major routes to the most important tourist sites. More than 4,000 km of tourist roads leading to traditional tourist sites, historical places and unique spots of nature have so far been marked. At the starting points there are route marking tables and route schemes. The routes that require special training and equipment are indicated. The marking of ski-tracks and runs is under way now. The safety measures along tourist paths, alpine ridges, steep drops, etc. are given special attention in the work on route marking.

The Bulgarian Tourist Union also boasts broad, amateur, artistic activities. The *Planinarska Pesen* (Mountaineering Songs) choir which was founded in 1959 has become one of the most important means of publicity for the amateur artistic activities of tourists. At present there are about 60 choirs, which often take part in various cultural events and tourist festivals.

The upswing of the tourist movement is also reflected in the development of the tourist Press and literature. The *Tourist* magazine has been published since 1956, the *Eho* newspaper since 1959 and the *Turisticheski Organizator* (Tourist Organizer) bulletin since 1968. They have contributed significantly to the publicizing of tourism and the beauty of Bulgaria, the involvement of thousands of new active tourists, and their education in the spirit of the love of nature and of their country. There has been a great increase in the number and size of publications in the domain of tourism. Dozens of guidebooks to the Bulgarian mountains, many towns and districts have already been published. The *Ognena Dirya* (Fiery Tracks) and *Malka Turisticheska Biblioteka* (Little Tourist Library) series have gained wide popularity. Dozens of books with a scientific and methodological slant have also been published regularly.

The Pirin Bureau at the BTU has contributed to the development of tourism in the mountains. It is in charge of economic tourism and offers numerous itineraries about the Bulgarian mountains for foreign and Bulgarian tourists.

Motorists with private cars are members of the *Bulgarian Motorists' Union,* which is a voluntary and independent public organization, founded in 1957. It offers various kinds of assistance to motorists and keeps track of the safety of roads and the infringement of regulations. The Union maintains courses and other facilities for the qualification and assistance of the owners of private cars and motor bikes. It is also engaged in developing up-to-date facilities and equipment for emergency help on the road. International motoring is sponsored by the Shipka

Buggy racing near Samokov

The Zlatni Pyassatsi car rally

Tourist Agency, which also takes care of motoring sports. It pursues its activities on a broad public basis in close co-operation with the Bulgarian Tourist Union, the Bulgarian Trade Unions, the Bulgarian Union of Physical Education and Sports, the Ministry of the Interior and other public organizations and governmental bodies. It maintains links with related organizations all over Europe. In 1982 there were 3,718 societies with 571,860 members.

REST AND RECREATION

Under the conditions of modern scientific and technological progress and rapid urbanization, the creation of favourable conditions for rest and recreation is a major economic and social question and one of the main aims of the social policy of the Bulgarian Communist Party. A harmonious balance between work and living conditions, on the one hand, and various forms of rest and recreation, on the other, are a necessary requirement for human activity.

The Albena holiday resort

The beach at the Russalka holiday resort

Depending upon the aims, recreation in Bulgaria is organized as a wide-ranging complex of activities in people's spare time, whereby (in daily, weekly and yearly cycles) the physical and mental forces of the working people and their families are restored and strengthened. Among the main factors for implementing this complex of activities in the field of rest and recreation and individual needs in this sphere are facilities available to this end and the organizational and managing structure connected with them. Building, maintenance and most effective utilization of the relevant facilities are organized by the system of rest and recreation.

Rest and recreation in Bulgaria is a social system, which satisfies the main recreational needs of the population. Unlike the international and home tourism industry, where all expenses are paid by the actual tourists, in the social system of rest and recreation the bulk of the expenses for the construction and maintenance of holiday facilities and services are covered by the public consumption funds (which come either from the State Budget or from the funds of enterprises, departments and organizations). The payment workers make for a holiday is far below the real cost of such holidays, which comes to about 100–130 levs for a fourteen-day holiday per person.

For such a holiday, a worker pays an average of 40 to 55 levs or less than 50 per cent of the real cost of the holiday. For children aged between three and fourteen the fee is 24 levs, while a third and subsequent children in a family pay nothing at all.

Prophylactoria, where workers who have suffered from some disease or who are threatened by such are sent, are

The resort of Pamporovo

Holiday houses for workers at Narechen spa

also free of charge. The expenses for such holidays are covered by the Social, Communal and Cultural Undertakings Funds of the Enterprises.

School and university students pay a minimal fee for a twenty-day holiday.

Attaching high value to the importance of holidays for the people's health and social well-being, the BCP set the holiday scheme the strategic task of creating optimum conditions for fully meeting working people's requirements for holiday facilities and services.

The holiday-scheme system in Bulgaria is conditioned by two basic groups of factors. The first comprises the factors conducive to the recreational needs of the working people and to their volume and structure. The second is those that satisfy the working people's requirements for rest and recreation, and incorporate above all the resort and tourist resources and facilities. Bulgaria's facilities in this respect are now considerable. In 1982 the total number of beds in Bulgaria in holiday camps, tourist hotels, rest houses, children's camps and sanatoriums totalled 352,000 and, taken together with those provided for the purpose in private homes and at camp-sites, the country has at its disposal more than 570,000 beds.

The variety of demands for recreation resorts and tourist resources and facilities have conditioned the development of different kinds and forms of rest and recreation: seaside holidays, mountaineering, spa holidays, study tours, sports outings, etc., fall into three basic groups depending on their duration: daily recreation, weekend holidays and annual holidays.

Daily recreation takes place in spare time between two successive workdays in various environments: the home, the place of work, special rest and recreation zones and in parks or on the outskirts of inhabited areas. Many of the new housing estates and separate buildings have additional facilities such as gymnasiums, halls for games and entertainment, libraries, reading rooms, workshops, studios, children's rooms, etc.

The daily recreation in the place of work is achieved through sport and cultural activities in nearby outdoor or indoor swimming pools, green and aquatic zones, solariums, aerariums, and special buildings intended for cultural and educational activities, study and game rooms; enterprises and plants have their own workshops and club-rooms for radio and fine mechanics enthusiasts, or the arts and crafts-rooms, halls for self-education, foreign language courses, etc. Prophylactoriums are also a particularly effective form of rest and restoring the people's working capacity. The prophylactoria are usually built and maintained by the enterprises. They are amply equipped with modern medical apparatuses and offer conditions for all-round physioprophylaxis.

The daily recreation in the open air takes place in gardens, in open-air summer theatres and cinemas, parks, etc. and is incorporated into the town development plans. Daily indoor recreation is planned and organized in accordance with the statutes on comprehensive services for the population. It is carried out through the co-ordinated efforts of the People's Councils, trade unions, Komsomol, and Fatherland Front organizations, enterprises and institutions, and in co-operation with Ministries and departments and their branches.

A holiday house for agricultural workers in Velingrad

Weekend holidays are the result of the organized hiking movement in Bulgaria. At first it covered mainly weekend outings, path-marking and construction of hostels.

Weekend recreation developed widely after the establishment of the popular regime in 1944. The organizers of mass hiking in all its forms – mountaineering, skiing, potholing, hikers' rallies, and path-finding – are the Bulgarian Hikers' Union and the Bulgarian Trade Unions (see the section on the Tourist Movement, pp. 584–93).

At weekends working people take part in a wide range of leisure activities: physically exerting activities, such as walks, outings, hiking treks, balneological procedures, gymnastics, swimming, hunting, field sports, mountaineering, pot-holing etc.

Weekend recreation is carried out in three main environments: in built-up areas – where facilities designed for daily recreation are used; everyday recreation – in suburban recreational zones, mainly wooded parks, where priority is given to the development of social forms of recreation; and in national recreational zones – national parks, forests, reservoirs, rivers, places of historical and cultural interest, etc.

The necessary facilities for weekend recreation in Bulgaria include the following main elements: accommodation at tourist hostels, hotels, motels, campsites, field sport lodges, boarding houses, private lodgings, privately and publicly owned villas, etc. – the average number of beds per 1,000 people is 170; public catering establishments – restaurants, snack bars, coffee shops, folk-style establishments, etc., with a capacity of 100 places per 1,000 holidaymakers; a specialized trade network – shops, tourist markets, kiosks, etc. for sale of foodstuff, hiking and sports equipment, which come to an average of 200 sq. m. of shopping area per 1,000 holidaymakers; sports facilities – basketball, volleyball, handball courts, ski-runs, swimming pools, etc., whose size depends on the number of beds available; cultural facilities – libraries, open-air cinemas, theatres and stages, amusement halls, etc., whose average area is stipulated to be 650 sq. m. per 1,000 beds; and service establishments – public laundries, workshops for the repair of sports and hiking equipment, etc. The development of the road network and public transport, the diverse recreation opportunities and the fact that there are two leisure days in the week has led to a constant expansion of the areas used for this purpose in more distant places. According to surveys made among working people, the proportion of weekend outings to places up to 40 km from the home has considerably decreased, while visits to places at a distance of over 60 km have greatly increased.

This trend is particularly marked in districts where the natural possibilities for organizing weekend rest and recreation are more limited. There has been a marked increase of co-operative development of zones for rest and recreation by several neighbouring inhabited areas. In Bulgaria today there are more than 4,000 specialized recreational zones near inhabited places.

In the construction of the new weekend recreation facilities, public facilities dominate: these include permanent facilities, such as boarding, hostels, etc., used all the year round; lightweight buildings and facilities – campsites, summer and winter camps etc., mainly intended for seasonal use; provisional buildings and facilities made of light materials with collapsible and transportable parts, tents, etc., mainly for temporary use; and movable structures and facilities – caravans, tents and the like, capable of being used in different environments.

There has been an increase in the use of buildings and facilities converted for the purposes of short-term recreation, such as residential areas of cultural and historic interest, places of entertainment, amusements, sports events and the like, where in certain seasons, and during fairs, folk festivals, exhibitions and celebrations, part of the sleeping accommodation and other facilities are placed at the disposal of tourists; or villages which, due to the migration to industrial centres and cities are becoming depopulated and thus leave unused a considerable proportion of their housing and other facilities, and which are in areas with a healthy climate. These facilities are placed at the disposal of industrial enterprises, institutions and public organizations; towns, villages and hamlets of historical significance and with architecturally valuable housing are being restored under the direction of the Institute for Cultural Monuments and adapted for the needs of short-term recreation and study tours; settlements of dwellings and additional facilities originally built for workers and technical staff working on construction sites of dams, weirs, roads, mines, etc., which after the completion of the project are all used for the purpose of rest and recreation.

Rest home for cultural institution workers in the Borovets tourist complex

The main advantages of the above forms are that use is made of available facilities which only have to be converted for temporary use, thus making it largely unnecessary to build up new areas, and thus preserving the existing settlement.

Weekend rest and recreation will continue to develop as a major factor in preserving and consolidating the health and the working capacity of the people. By 1990 it is expected that it will be possible for about 50 per cent of the population to make use of weekend rest and recreation facilities at one time, two thirds of these being in zones outside inhabited areas and one third being daily recreation facilities within the inhabited areas.

Annual (long-term) rest and recreation had its rather primitive beginnings before Bulgaria's Liberation from the Ottoman rule (1878), when the population built baths around thermal mineral springs. The changes in the socio-economic conditions in liberated Bulgaria brought a number of changes to recreation. Some spas, such as Hissarya, Vurshets, Kostenets and Kyustendil acquired considerable fame for their curative waters, and the number of visitors to them grew rapidly. Various forms of recreation then fashionable in the advanced capitalist countries gradually gained a foothold in Bulgaria, introduced mainly by members of the Bulgarian bourgeoisis and intelligentsia. In addition to spa resorts, attention was also focused on the Black Sea coast. In 1890 the first sea baths were opened in Varna, and places of entertainment appeared around them. Somewhat later sea baths were also opened in Burgas and the mountain resort of Borovets (Ghamkoria) started to develop. After the First World War the rapid growth of the urban population gave rise to the habit of going out into the country. The bourgeoisie built villas at the most famous spa, mountain and seaside resorts. Almost everywhere private accommodation was also used for holiday-making, although the services at the resorts were at a very low level. Certain schools set up the first school children's colonies (summer camps at suitable places in the mountains and at the seaside); many monasteries were also used by tourists and holidaymakers.

In spite of everything, the facilities for rest and recreation were few and far between, and holidaying remained the privilege of the bourgeois class.

The popular uprising of 9 September 1944 and the consolidation of people's power and of socialist production relations in Bulgaria were extremely beneficial for the development of rest and recreation. As a result of the constant care of the BCP and the Government, and the active participation of the Bulgarian Trade Unions, Ministries, etc., modern holiday houses were built in the different resort regions of the country, as were balneosanatoriums, school camps, etc., where in 1982 more than 1.7 million Bulgarian citizens spent their annual holiday subsidized by the State Budget and by the funds of enterprises. The hotel facilities, camp-sites and private lodgings were used by more than 17.6 million Bulgarians in the same year, many of them with subsidies from the social consumption funds.

The main forms of annual holidays in Bulgaria are: stationary holidays at holiday houses, treatment at balneological sanatoriums and recuperation homes, hiking holidays, holidays at specialized school and university student holiday camps and holidays abroad.

The most widespread form of annual holiday is that at holiday houses. These establishments provide all services for recreation – accommodation, meals, entertainment, sports events, etc. For a fuller combination of recreation with preventive treatment, the medical personnel at holiday houses is gradually being increased.

In 1960 some 289,000 holidaymakers visited 307 holiday houses whose beds totalled 32,815. In 1970 almost 448,000 people made use of 66,343 beds in 986 holiday homes, while in 1982 some 1,043,000 holidaymakers spent their holidays in 1,484 holiday houses with 106,804 beds.

The Central Council of the Bulgarian Trade Unions has 208 holiday houses of its own with 25,985 beds, in which in 1982 some 477,000 people spent their holidays. The

Holiday homes of the Higher Institute of Mechanical Engineers (Sofia) at Sozopol

remaining 1,276 holiday homes with 80,819 beds are managed by enterprises and economic organizations.

Nearly 66 per cent of the beds are in holiday homes situated on the Bulgarian Black Sea coast, which is the largest and most popular holiday region in the country. 25 per cent of the beds are in the mountainous regions and 9 per cent at spas. The uneven territorial distribution of beds determines to a great extent their concentration on the Black Sea coast, their seasonal nature and the fact that they are comparatively under-used. A mere 18.1 per cent of all the beds in holiday houses are used all year round, while only 2.1 per cent of those on the Black Sea coast are used throughout the year.

For this reason, during the Eighth Five-Year Plan (1981–85) period the Bulgarian Trade Unions are carrying out a large-scale development policy for the reconstruction and modernization of existing facilities and their adaptation for use all the year round. Over 165 million levs are being invested in this reconstruction and modernization programme, in which particular attention will be paid to improving the conditions for family holidays.

Treatment at resorts is the fastest and most effective way of restoring and strengthening the health of workers. Here two types of establishments are employed – clinical balneosanatoriums, where people stay for the period necessary for treatment, and non-clinical balneosanatoriums, which are used for workers who have been ill and need to restore or strengthen their health, or people threatened by ailments who require preventive treatment.

Depending on the qualities of the mineral waters (see chapter on Tourism and Resorts) and on the basis of the experience gathered so far, balneological resorts are grouped according to the ailments that can be treated there: Bankya, Vurshets, Nessebur and Druzhba for cardio-vascular diseases; Banya (Pazardjik district), Banya (Plovdiv district), the Burgas mineral springs, Pavel Banya, Sapareva Banya and the Haskovo mineral springs for ailments of the bones and joints; Bankya, Narechen and Velingrad for functional nervous diseases; Hissar, Gorna Banya and Krasnovo for urological diseases and those of the kidneys and liver; Kyustendil and Pomorie for gynaecological disorders; Sandanski, Momin Prohod and Velingrad for diseases of the respiratory tract.

Bulgaria has over 14,000 beds in non-tubercular sanatoriums, where over 205,000 people went for therapy in 1982. The bulk of the places in spa holiday houses and private accommodation are used for treatment or prevention. A recent novelty is the adaptation of seasonal holiday houses on the coast for balneological purposes during the low season. Treatment at spas in Bulgaria is organized by the Ministry of Public Health and the Bulgarian Trade Unions.

Excursion holidays have a particularly therapeutic and beneficial effect on people, at the same time helping them to increase their general knowledge. Trips are taken over a certain route in different terrains (mainly in the mountains), and usually last for four-five hours with a one-three day stay in a particular hostel. These holidays are organized for seven, ten or fourteen days, and people of all age-groups participate in them. There are a total of 35 different routes: six of them involve motor transport, one involves river and sea transport, one is combined and 27 of them are mountain hikes. In 1982 almost 81,000 people took part in them. Holiday houses, tourist hostels, hotels, ships, trains and buses are used for excursion holidays. The high degree of activity combined with the visits to beauty-spots and places of cultural and historic interest make excursion holidays one of the most popular forms of holidaymaking in Bulgaria.

The concern for improving the health and generally developing the younger generation can be seen in the constant increase in the number of beds and organized holidays for schoolchildren and students. In 1982, 433,352 students and schoolchildren spent their holidays in the 751 specialized state-subsidized children's holiday camps, whose total number of beds was 97,000. Besides these holidays, which last for twenty days, many schoolchildren and students had ten-day holidays in the winter and spring using the facilities of the trade-unions, the Bulgarian Hikers' Union and the Bulgarian tourist industry, which were converted for use by students and schoolchildren. Since 1983 a rota-system has been used for schoolchildren's winter and spring holidays, thus enabling three times as many schoolchildren to spend them in the mountains.

Over the past two decades trips abroad have become a widespread form of holiday-making. In 1965 some 207,912 people travelled abroad, while in 1970 and 1982 the figures were 305,809 and 351,398 respectively. There is a rapid increase in the number of trips organized by tourist bureaux and agencies (Balkantourist, Orbita, Pirin, Co-optourist and Shipka) and by trade unions.

Approximately 68 per cent of the total number of tourists visited the socialist countries, 75 per cent of them being for holiday purposes. The greatest number of trips are to the Soviet Union (135,759 in 1982), Romania and the GDR. The most visited capitalist countries are Greece, Turkey, the FRG and Austria. Bulgarian tourists show particular interest in places where famous Bulgarians lived and worked. The more popular holidays include combined boat trips on the Danube and Mediterranean Seas. Trade-union exchange trips between twin districts, towns and enterprises in Bulgaria and the other socialist countries are also increasing in number.

The chief organizers of workers' holidays are the Bulgarian Trade Unions. The Central Council of the Bulgarian Trade Unions takes part in deciding party and state policy on holidays and resort therapy and is responsible for its implementation. They elaborate and

Leonid Brezhnev Iron and Steel Works rest home on Mount Vitosha

propose laws, decrees and other normative acts concerning rest and recreation, supervise their observation and issue instructions for their implementation. The Central Council of the Bulgarian Trade Unions ensures co-ordination in the development and implementation of plans, organizations and regulations concerning holidaymaking. All holiday and resort therapy facilities in Bulgaria are subject to integral planning, regulation and control.

In fulfilment of the strategic goals in the sphere of holidays and resort therapy, the Central Council of the Bulgarian Trade Unions outlines the tasks of trade-union bodies and organizations, thus ensuring the participation of the workers in their organization and development. The Central Council of the Bulgarian Trade Unions also guides and aids its bodies and organizations in ensuring the inclusion of rest and recreation in their plans for the social development of work-forces and collective labour contracts and in the allocation and utilization of funds to this end.

The Central Council of the Bulgarian Trade Unions has the important and responsible task as the organizer of rest and recreation of constructing, reconstructing, modernizing and maintaining the facilities entrusted to it. Through its various bodies, it involves Ministries, authorities and enterprises in the financing of holiday construction, and organizes and directs co-operative construction.

The Central Council of the Bulgarian Trade Unions supervises and controls the development of the necessary facilities and their rational use with the aim of their most purposeful utilization. The central committees of the branch trade-unions plan the construction and development of holiday facilities for their workers and their families in conjunction with Ministries, authorities and

Relaxing in the park

plans of the branch trade unions. The central committees of the trade unions include measures for the joint efforts by the work-forces in their particular branch for the construction, reconstruction, modernization and expansion of holiday facilities when they sign contracts with the Ministries. To this end the trade-union central committees determine special limits for the purpose of holiday schemes in those they stipulate for the social development of work-forces, by proposing the inclusion of the Central Council of the Bulgarian Trade Unions in the cooperative construction of holiday facilities, which depend on the means accumulated by the particular branch and its needs. The trade-union central committees are responsible for supervision and co-ordination for the most efficient utilization of the facilities they own or use through contracts with the Ministries.

The Youth Compound of the Orbita Tourist Bureau in Sofia

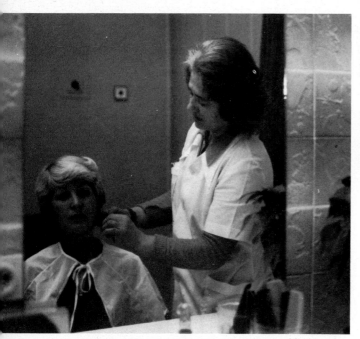

Hairdressers at the machine-tools plant in Sofia

organizations, in accordance with production development rates.

The questions of rest, recreation and resort treatment are provided for in the programmes and development

The District Councils of the Bulgarian Trade Unions are responsible for carrying out the main tasks set by the Central Council of the Bulgarian Trade Unions on a district level as far as holiday schemes are concerned. They play an active role in getting the labour forces to implement the plans for their social development in the field of recreation and resort therapy, and work in conjunction with the District People's Councils for the general implementation of party and state policy for developing holiday schemes and facilities within their districts.

The District Councils of the Bulgarian Trade Unions also work to increase the initiative of trade-union bodies and economic managements in developing holiday and

The Georgi Dimitrov International Youth Centre in Primorsko

recreation facilities by combining their means and labour resources on a district level. It is they who distribute holiday vouchers among their branches as stipulated by the Central Council of the Bulgarian Trade Unions; they also assist and collaborate with the Holiday Schemes and Resort Therapy Complexes in appointing staff and arranging supplies, transportation etc., and also in organizing cultural, sports and educational activities at resorts.

The trade-union committees of enterprises and economic organizations have the task of applying the public principle in the management of holiday schemes and of involving the workers in it. In involving the workers directly in distribution of holiday vouchers, the trade-union committees set certain requirements to make holiday vouchers serve as recognition of the workers' attitude to their labour and its level of difficulty, their health, public activity, etc. This increases the socio-economic effect of holiday schemes by increasing labour productivity.

It is among the work-force that the general organization of holiday and resort therapy schemes is realized in practice. The trade-union committees, in conjunction with economic and administrative managements, determine the labour forces' requirements and elaborate long-term programmes and plans for their satisfaction, incorporating collective contracts and work-force social development plans in this sphere, and provide them with the necessary funds. They arrange direct worker participation in the construction and maintenance of holiday facilities by using voluntary labour, economical materials, etc. The trade-union committees help to draw up annual leave plans in such a way as to ensure optimum co-ordination between production needs and the utilization of holiday vouchers all the year round.

At the same time they acquaint the workers with natural and climatic conditions in order to recommend the best ways for them to spend their holidays, depending on their state of health and interests. It should be noted that all working people in Bulgaria who are full-time members of trade unions, as well as members of their families and pensioners, can use subsidized holiday vouchers. Foreign workers, specialists and students have the same rights as trade-union members.

The right to co-ordinate the state and public organizations in the development and organization of holiday schemes is a characteristic feature of the trade unions' job as organizers in this field. To this end, an interdepartmental council was set up with the Central Council of the Bulgarian Trade Unions in 1977. It is chaired by the President of the Central Council of the Bulgarian Trade Unions. The Council is made up of representatives of twenty Ministries, departments and public organizations directly concerned with holidays and leisure.

The Inter-Departmental Council is a state and public body for the collective management of holiday schemes in Bulgaria.

The Council directs and co-ordinates the activities of Ministries, departments, District People's Councils, economic and other organizations in the field of workers' holiday schemes.

The Inter-Department Council takes decisions on the main questions concerning workers' holiday schemes and leisure, ratifying planning, normative and organizational documents which are binding for all state, economic and public bodies and organizations.

District holiday-scheme councils have been set up with the District Trade-Union Councils in order to co-ordinate activities of People's Councils, economic and public organizations in developing holiday schemes at district level.

The district holiday-scheme councils study the workers' holiday requirements in various districts and, in conjunction with the executive committees of the District People's Councils, draw up long-term and yearly plans on the construction of suburban recreation zones and facilities. They co-ordinate and help economic and public organizations to maintain, expand, reconstruct and modernize holiday facilities and to improve holiday services – catering, cultural and sports activities, transport, etc.

SOCIAL SECURITY

Social security was first introduced in Bulgaria immediately after the Liberation from Ottoman rule in 1878. At first the social security scheme covered only civil servants as there was a strong desire to consolidate the Bulgarian State. The Turnovo Constitution of 1879, the first Bulgarian Constitution, stipulated that 'civil servants appointed by the Government are entitled to pensions the size of which will be fixed in a stated manner'. With a view to legally regulating the social security scheme, laws on pensions were voted on covering disability pensions, retirement pensions and pensions paid to dependants. Those entitled to these pensions included the military, administrative and police officials, teachers, ministers, clergymen, civil servants and those holding elective posts. The bourgeois State excluded the working-class people from social security, which was the privilege of certain classes. Retirement pensions paid to the military were extremely high. The aim was to prevent civil servants and municipal workers from waging an organized struggle for better working and living conditions and from carrying out joint actions with the workers. When setting up their first trade union organization, workers from the private sector established mutual-aid funds to cover the event of unemployment, sickness and death. This voluntary trade union security failed to provide the necessary social security against labour risks.

The Bulgarian proletariat launched an organized struggle for mandatory social security for workers, a struggle which achieved the first results, whereby the State was compelled to establish mutual-aid funds and friendly societies. An important role in the proper direction of the workers' demands was played by the Bulgarian Workers' Social Democratic Party which, in the programme adopted at its first congress in 1894, demanded social security for workers as well. A law was adopted in 1905 regulating female and child labour in industrial enterprises and a social security scheme was introduced to cover sickness and industrial injuries for workers from the public building enterprises. The raising of the revolutionary spirit of the working people as a result of the victorious Great October Socialist Revolution, and of the hard conditions on the battle front during World War I led to the introduction in 1918 of compulsory social security covering industrial and office workers in the event of sickness and industrial injuries. The sickness insurance was confined only to low-paid workers, those receiving up to 2,400 levs annually. Provisions were made for treatment of up to six months with the sick person receiving, from the third day of falling sick, compensation amounting to three quarters of his or her monthly wages when treatment was given at home. In the event of hospital treatment, the corresponding proportion was 50 per cent of the respective wages. The insurance against industrial injuries came to cover, for the first time, all workers and employees at state, public and private enterprises and establishments irrespective of their age, sex, nationality and form of pay. The principle of vocational risk was recognized. It required that the insurance contributions be paid by the respective enterprise, with full compensation being given in the event of any kind of industrial injury, and there was no need to prove who was to blame for the labour accident. Those who suffered industrial injuries were allotted pensions in a statutory way rather then through the court. In 1920 the pensioning scheme covered 62,000 people but insurance was poorly organized and inadequately financed. Under the conditions of the organized struggle of the workers, grouped in powerful trade unions led by the Bulgarian Workers' Party after the suppression of the world's first anti-fascist uprising of September 1923, the bourgeois Government was compelled to initiate a workers' social security scheme. As a result of the 1924 social insurance act, an insurance scheme was introduced covering industrial injuries, occupational diseases, sickness and maternity, disability and old age. An unemployment insurance system was initiated in 1925, while from 1941 onwards insurance came to cover private craftsmen, and from 1943 onwards it included merchants as well. The contributions for the 'labour injury' insurance were paid by the employers, while the contributions for the rest of the insurances were shared equally by those covered by the insurance scheme, the employers and the State.

Bulgaria's experience in introducing insurance for private farmers in 1941 was interesting. That insurance originally covered only men, and from 1946 on women as well. The retirement age for men was 60, for women 55. The size of pensions was not very high. Under the conditions of capitalist development social security was incomplete and financially unstable, and the size of pensions was insufficient. Social security in those days was

An Old People's Home in Sofia

characterized by various financial irregularities and bureaucratic administration.

After 9 September 1944 social security became part of Bulgaria's social policy. It was aimed at promoting the working people's health and at raising their living standard. From the first years of the popular regime it was placed on solid foundations. In 1949 all social security schemes were overhauled and were administered by a unified State Social Security Institute. In 1951 free medical care was introduced throughout the whole country and a Labour Code was adopted, regulating state social provisions. Since 1957 pensioning has been regulated by a law, which is still in force today, although some amendments and additions have been made to it.

The right to social security is guaranteed by the Constitution. The main principles underlying social security may be summed up as follows: *to protect all the working people* (all people engaged in socially useful work are covered by the insurance scheme), that is, workers, employees and co-operative farmers who are employed under a labour contract, irrespective of what has necessitated such a contract, as well as part-time workers; and servicemen in the Bulgarian People's Army, re-enlisted officers and sergeants, workers and employees on the regular payroll working half a day, those vocationally rehabilitated who are on a shorter working day, Bulgarian citizens working abroad, those working in small-scale private shops and farms, members of households and private persons, foreign workers and employees based in Bulgaria, those serving in the Bulgarian Orthodox Church, masters of arts and crafts, writers, journalists, artists and sculptors; *to cover all risks against temporary or permanent incapacity for work* due to industrial injuries, occupational diseases, general sickness or injuries other than industrial ones, confinement and maternity, quarantine, infectious disease, attending a sick member of the family, disability, old age and death. *The insurance contributions are paid by the respective enterprise or establishment* (workers and employers pay no contributions, except freelances). *The size of the insurance contributions should be sufficient to enable the working people to maintain their living standard even in the event of incapacity for work. The administration of social security should be handled by the workers.* Under a 1960 law this administration was entrusted to the trade unions, the largest organization of working people in Bulgaria.

Social security in Bulgaria is based on the insurance contributions paid by enterprises, establishments, co-operatives and public organizations. These contributions are deducted from the wage fund and are paid out of the funds of enterprises and establishments. There has been a great increase in the size of these deductions. This is due to the extension of the population's rights to insurance, mainly through the increase in the size of child allowances, compensation and pensions, and through the extension of maternity leave. In 1973 these contributions amounted to 12.5 per cent, in 1974 the proportion was 20 per cent and in 1980 – 30 per cent. Social security in Bulgaria is based on the principle of the pure system of distribution. Current expenditures are met by current revenues. No reserves for meeting future payments are formed. Any surplus at the end of the year is deposited in the State Budget, thus restoring any deficit, hence the financial stability of social security in this country.

Insurance in the Event of Temporary Incapacity for Work. This covers workers who are unable to work because of sickness, industrial injuries, occupational diseases, balneological treatment, urgent medical examinations, quarantine, suspension from work on medical advice, attendance of a sick member of the family, the accompanying of a sick man to a hospital situated in another area, or treatment abroad; and for women workers and employees – in the event of pregnancy, confinement, or maternity leave. Temporary incapacity for work must be certified by the doctor in attendance when it lasts up to twenty days, and by a medical team when its duration is over twenty days. The amount of compensation for temporary incapacity for work due to sickness, accidents other than labour ones, quarantines, for attending a sick member of the family (not including children up to the age of seven) depends on the uninterrupted length of service of the person involved and is calculated as a percentage of the nominal labour remuneration during the month preceding the onset of the incapacity for work. The percentages are differentiated and depend on the number of years worked: 70 per cent of those with ten years of service, 80 per cent for people with service of between ten and fifteen years and 90 per cent for a service of over fifteen years. When temporary incapacity for work continues for more than fifteen days, the size of the compensation after the fifteenth day is 80 per cent, 90 per cent and again 90 per cent, respectively. The compensation for attending a sick child up to the age of nine is 100 per cent of the nominal wages, and for temporary incapacity for work due to industrial injuries and occupational diseases – 90 per cent. In both these cases the size of the compensation does not depend on the number of years worked. Temporary incapacity for work due to sickness and industrial injuries is considered to have begun from the first day of its onset and runs until the day the person involved is proved to be fit for work again, unless permanent disability is certified, in which case a disability pension is granted. In cases of quarantine, compensation is paid as long as the person involved is under quarantine. When someone is suspended from work for suffering from an infectious disease, for being a germ carrier, or for having been in contact with people suffering from contagious diseases, compensation is paid for 90 days. The duration of the compensation for attending a sick member of the family is ten calendar days per calendar year for each insured person. When attending a sick child under the age of sixteen, or when accompanying him or her to another area for the purpose of undergoing treatment or medical examinations, or when accompanying him or her abroad for the same purpose, the accompanying mother or father is entitled to compensation for up to 60 calendar days per calendar year. In the event of balneological treatment the compensation paid covers the whole period of treatment, plus three days travel allowance to and from the spa.

Insurance in the Event of Pregnancy, Confinement and Care of Young Children. In 1944 the number of people covered by insurance was 54,000, while in 1982 the corresponding figure was 2 million. Paid leave has been extended and the size of compensation has been raised to 100 per cent of the basic labour remuneration.

Insurance coverage may be summarized as follows:

1. *Vocational Rehabilitation of Expectant Mothers.* With a view to preserving the health of expectant mothers and to ensuring normal pregnancy, every mother-to-be working in conditions harmful or arduous to the pregnancy is transferred to another lighter job corresponding to her health condition and qualifications. A special commission made up of the respective administrative manager, the chairman of the trade union, or trade union representatives, a representative of the Inspectorate of Hygiene and Epidemiology and a doctor from the maternity health centre defines the place of work and posts expectant mothers should not occupy and those to which they can be transferred. This vocational rehabilitation begins from the day pregnancy is confirmed. When the income of expectant mothers has fallen as a result of vocational rehabilitation, the balance of the average monthly nominal wages received during the previous twelve months is restored out of the social security fund. Vocational rehabilitation is carried out on the basis of a medical certificate issued by the respective doctor. It gives the duration of that rehabilitation and the kind of job the expectant mother is capable of doing. Once back to work after the delivery, mothers are entitled to the same job they have held before their vocational rehabilitation.

2. *Maternity Leave.* All women covered by the insurance scheme are entitled to paid maternity leave. This applies to all those who have been employed, even only for one day, before going on maternity leave (which begins 45 days before confinement). They too are covered by the insurance scheme and are entitled to cash compensation. Seasonal and other women workers who are not permanently employed are also covered by the insurance scheme. They are required to have been employed for seven consecutive days before going on maternity leave. Mothers enjoy 120 days of paid maternity leave for the first child, 150 days for the second, 180 days for the third and 120 days for each subsequent child. Forty-five days of this leave are given before delivery. Throughout the maternity leave mothers receive cash grants equal to 100 per cent of the nominal wages they have received during the preceding month before going on leave (which is not less than the wage scale and not higher than 120 per cent of that same scale). Adoptive mothers, too, are entitled to the same maternity leave.

3. *Paid Leave for Looking after Young Children.* If after the paid maternity leave the child is not placed in a nursery or a child-care establishment (this is done of the parents' own free will), mothers (adoptive mothers included) enjoy an additional leave of six months for the first child, seven months for the second, eight months for the third, and six months for each subsequent child; as of 1 July 1985 this leave is extended to 24 months. During that leave mothers receive cash grants equal to their minimum wages, which are fixed throughout the country, and amount to 110 levs a month. When mothers do not wish to take such leave and prefer to return to work, as well as their wages they also receive 50 per cent of the cash grants given to mothers on leave for looking after their children. If a mother (or an adoptive mother) on maternity leave, or on leave to look after her child, is taken ill or dies, the cash grant for looking after the newborn baby is given to the father or to another relative. In addition, mothers enjoy additional unpaid leave until the child is three years old. During that leave they receive ten levs a month in the form of a cash grant. The maternity leave and the leave for looking after the child is not considered an interruption of the length of service, and their jobs are held for them until the child is three years old.

Student mothers are entitled to a 90 lev cash grant per month. This grant is given to them for ten months for the first child, 12 months for the second child, and 14 months for the third child. A decree of the Central Committee of the BCP and the Government, issued in April 1984, provides for this leave to go up to 24 months, irrespective of the number of children born, and the grants will be equal to the country's minimum wages. Only full-time students are entitled to this grant, regardless of whether they are studying in Bulgaria or abroad. Student mothers at all higher educational establishments and at special secondary schools receive a grant of up to 80 levs a month after their maternity grants have terminated. When the husband is employed, the student wife receives 50 per cent of the grant, i.e. 40 levs. Mothers who for one reason or another are not employed and do not study are likewise entitled to maternity grants, and to grants for looking after their children. The period is considered as length of service, which is twelve months for the first child, fourteen months for the second child and eighteen months for the third child.

Family Allowances are paid under the 1951 Decree for Promoting the Birth-rate and Large Families. Family income supplements were first paid in Bulgaria in 1941 to state employees only. Under legislation currently in effect, able-bodied persons are entitled to a lump-sum cash grant in the event of confinement, to monthly child allowances, and monthly grants for student families and for those of post-graduate student and conscripts. Lump-sum cash grants for live-born children at the rate of 100 levs for a first, a fourth and each additional child, 250 levs for a second and 500 levs for a third child are paid to all Bulgarian women, regardless of whether they are covered by the social security scheme. Foreign nationals giving birth to a child in Bulgaria are not entitled to cash grants. As a rule it is the wife who draws the maternity grant, though in some specific contingencies it can be drawn by

Table 1: Indemnities and Allowances Paid

Indices	1970	1975	1980	1982
Cash indemnities for temporary				
disability (in million levs)	116.1	190.3	247.8	246.4
basic indices	100.0	163.9	213.4	212.2
Sickness and injury benefits				
(in million levs)	86.4	137.7	186.3	184.4
basic indices	100.0	159.4	215.6	213.4
Pregnancy and confinement grants				
(in million levs)	29.7	52.6	61.5	62.0
basic indices	100.0	177.1	207.1	208.8
Cash benefits for raising children				
(in million levs)	—	55.3	99.2	94.9
basic indices	—	100.0	179.4	171.6
Maternity lump-sum cash grants				
(in million levs)	17.1	26.0	25.7	24.1
basic indices	100.0	152.0	150.3	140.9
Monthly child allowances				
(in million levs)	219.8	253.0	492.9	504.7
basic indices	100.0	115.1	224.2	229.6
Other benefits (in million levs)	43.3	46.8	25.9	36.8
basic indices	100.0	108.1	59.8	85.0

the husband or the guardian. During the period 1970–82 the sum total of maternity grants rose from 17.1 million levs to 24.1 million. The monthly child allowance rate is 15 levs for a first, a fourth and each additional child, 25 for a second, and 45 for a third, so that a one-child family receives monthly a 15 levs supplement, a two-child family a 40 levs supplement, and a three-child family an 85 levs supplement, amounting respectively to 15, 40 and 85 per cent of the minimum monthly pay. Child allowances are payable until the age of 16.

As of 1 July 1985 the monthly allowances for children are set as: for a second child to 30 levs; for a third child to 55 levs; after the birth of a second child, the monthly allowance for the first child increases from 15 levs to 30 levs. The monthly child-care grants for families of students, post-graduate students and conscripts are 30 levs per child. State allocations for child allowances come second after those for pensions, and have grown from 219.8 million levs in 1970 to 504.7 million in 1982.

As of 1 July 1985 young newlyweds are entitled to state loans for the construction or purchase of a flat, up to the sum of 15,000 levs, without having to pay a deposit, and are allowed up to 30 years to repay it. On the birth of a second child (four years after the birth of the first child) 3,000 levs of the loan are remitted, and of a third child – another 4,000 levs are remitted. Apart from the loan for construction or purchase of a flat, young newlywed couples are entitled to a loan of 5,000 levs for other purposes, to be paid off within ten years. For the birth of a second child within four years after the birth of the first child the entire remaining unpaid part of the loan is to be remitted.

Other Benefits. An employed person falling into a specified disability class, but not eligible for a disability pension because of incomplete length of service, and who has been re-posted to a job suitable for his health status, can qualify for financial aid starting from the date of disability, to be paid over the course of three months, if he is graded Class 3 disabled, and over the course of six months – if Class 2 or 1. In the event of the death of a member of the insured person's family, a death grant to the rate of 80 levs is payable to the person paying the funeral expenses; in the event of death through an accident at work the sum is 120 levs. Funds allocated for temporary disability indemnities, and especially those of the child allowances, are continually on the increase (see Table 1).

The Pension Scheme. Bulgaria's pension scheme is one of the most democratic in the world. Especially increased provision was made for pensions following the April 1956 Plenum of the Central Committee of the BCP. The pension scheme is being systematically improved in compliance with the extensive programme for raising the population's living standards adopted in 1972. Factory and office workers and co-operative farmers come under a unified pension scheme which covers practically all Bulgarian citizens. State expenditure on it keeps growing (see Table 2).

Retirement Pensions. Having completed the specified length of service, the socially insured person of pensionable age is eligible for a retirement pension. The length of service is defined as the time under contract in state, public or co-operative enterprises for which labour remuneration has been paid; for freelances (workers in the arts, journalists, lawyers and others) this is the term during which they have been self-employed. provided they had been paying the necessary contributions. Just as with the length of service, required pensionable age varies

Table 2: Pensions and Expenditures Thereof (by the end of the year)

Indices	1970	1975	1980	1982
Pensions:				
number (in thousands)	1,720	1,868	2,042	2,119
basic indices	100.0	108.6	118.7	123.2
Pensions paid:				
in million levs	718.9	1,044.1	1,685.2	1,974.2
basic indices	100.0	145.2	234.4	274.6
Average annual pension:				
in levs	423	566	834	938
basic indices	100.0	133.8	197.2	221.8

with the category of labour. There are three categories in Bulgaria which classify the degrees of hard and harmful labour: Class 1 – very hard and harmful labour, Class 2 – hard and harmful, and Class 3 – labour under normal conditions. In cases where an insured person has worked in more than one of the above-mentioned categories of labour, the rate of his pension is determined according to the class in which he has served the longest. A retirement pension can be claimed under the following terms: Class 1 labour – fifteen years of service, at the age of 50 for men, 45 for women; Class 2 – 20 years of service, at the age of 55 and 50 respectively; Class 3 – 25 years of service and at the age of 60 for men, and 20 years of service and at the age of 55 for women. Mothers of five or more children aged eight and over are entitled to a retirement pension after fifteen years of service and 40 years of age in Class 1 labour, and after the same length of service and 45 years of age for classes 2 and 3. Persons of pensionable age not having completed the specified length of service may qualify for a retirement pension with reduced rates, proportional to the length of service they have completed. For those under pensionable age each year of service above the term specified in the Law on Pensions brings an addition of up to 2 per cent to the basic retirement pension; the total rate of such additions cannot, however, exceed 12 per cent. Persons having reached the required pensionable age and completed the specified length of service are entitled to a 6 per cent addition to their basic pension for each extra year of service. The rate of the retirement pension is based on the average nominal monthly pay received during three consecutive years out of the last fifteen years of service as chosen by the person himself, and is deducted as a percentage of the basic remuneration – starting from 80 per cent for the lowest, down to 55 per cent for the highest labour remuneration. Retirement pensions are regularly being up-graded in line with changing economic conditions, so that the pensioner's living standards do not suffer.

Disability Security. Disability pensions can be claimed in the event of partial or complete, temporary or permanent disability. They are granted in cases where the disability was caused in the course of employment, or not later than two years after termination of contract (the former contingency not applying to the congenitally blind or to persons deprived of sight before entering employment, the term of the latter being longer for certain prescribed diseases). When, in spite of treatment, ailments do not improve and the chances of regaining capacity for work are slight or non-existent, the invalid must appear before a Medical Commission on Labour (MCL) – a specialized commission of medical experts and a representative of the trade unions, authorized to ascertain permanent disability; to pronounce on the degree of disability and the date of disablement; to establish, in cases of occupational diseases, the relation between employment and ailment; and to re-deploy persons with partially impaired capacity for work. Three disability classes have been set out in Bulgaria according to the degree of disability: Class 1 includes persons unable to perform any work (the severely disabled); Class 2 – persons unable to perform either the job they have been trained for, or any other, but who can work in specially provided conditions; Class 3 includes persons who, because of health considerations, have lowered or been induced to lower, their standard of occupation, or have changed, or been induced to change, their conditions of labour. A disability class is granted permanently when there are no indications that the disability will be cured, or temporarily – in all other cases. In the first case the disabled person can claim re-assignment to another disability class with more favourable pension terms, in cases when his condition has worsened. Disability pensions fall into the following categories: industrial injury or occupational disease pensions, pensions for general illness or injuries arising from accidents other than labour accidents, and civil and military disability pensions. Industrial injuries comprise all injuries arising out of or in the course of employment, as well as those resulting from work in emergencies or under conditions not normal for the respective occupation, if it has caused the insured person's temporary disability, invalidity or death. Industrial injuries are likewise injuries suffered in performing any service in the interest of the employer, during breaks in work-time, on the way to and from work, in saving a person's life or property, or while taking part in voluntary work-days and sports competitions. An industrial injury is ascertained by the head of the enterprise through an Industrial Injury Certificate, the disability – by the MCL. Occupational diseases fall under the class of industrial injuries. The nature of the disease and the degree of disability are established by the MCL.

The rate of the pensions granted for disability through labour accident or occupational disease is fixed according to the disability class and the basic labour remuneration. The percentage determining the rate of a disability pension grows in direct proportion to the degree of disability established, and in inverse proportion to the basic labour remuneration. Thus Class 1 disability pensions vary from 70 per cent for the highest basic remuneration to 100 per cent for the lowest, for Class 2 the percentage is respectively 55 to 85 and for Class 3 – 35 and 65. The aim is to ensure that the disabled receive approximately the same income as before the disablement.

Disability pensions are granted in the event of general illness or injuries arising from accidents other than labour accidents to disabled persons pronounced by an expert medical commission on labour to fall into one of the three disability groups and provided that they had completed the specific length of service graded in conformity to the respective age before the pronouncement of disability (up to 20 years of age – regardless of the length of service; up to 25 years of age – with three years of service, over 25 years of age – with five years of service). Disability pensions are payable to congenitally blind persons having been deprived of sight before entering employment with five years of service regardless of their age.

Disability pensions granted in the event of general illness or injuries from accidents other than labour accidents are rated in a percentage based on the basic labour remuneration. The percentage is graded according to the size of labour remuneration and the disability group as follows – for group I – rating from 73 per cent for the lowest to 55 per cent for the highest labour remuneration, for group II – from 65 per cent to 40 per cent and for group III – from 50 per cent to 25 per cent. Persons who fall into disability groups I and II due to general illness are entitled to an addition to the basic pension, the size of which depends on the length of services prior to the date of pronouncement on disability and is fixed according to the basic pension as follows: 5 per cent after ten to fifteen years of service, 10 per cent after fifteen to twenty years of service, 15 per cent after twenty years of service and over. Pensions with 25 years of service, at the age of 55 for men, and 20 years of service and at the age of 50 for women, are granted a 25 per cent addition to the basic pension.

Civil disability pensions are granted to persons injured in the performance of civil or public duties, to members of the auxiliary defence organizations, to members of sports teams injured during training sessions or competitions, to students injured during practical training or studies and to convicts injured in labour accidents. The monthly size of pensions is 65 levs for disability group I, 60 levs for disability group II and 50 levs for disability group III.

Military disability pensions are granted to persons injured while serving in the army, depending on the respective disability group, to regular privates and NCOs in the armed forces, to called-up reservists, to regular servicemen in the engineering corps or to reservists summoned to training, inspection or practical muster roll, regardless of their rank and to persons injured in rendering assistance to troops. The dead and missing also come into this category. If the injured have been socially insured before enlisting in the army, the size of their pensions is fixed on a similar scale to the rating of disability pensions that would have been granted in the event of industrial injuries and occupational diseases, if this is to their benefit.

Table 3: Monthly Size of Military Disability Pensions in Levs

	Disability group		
	I	II	III
Privates and NCOs	90	75	60
Officers	110	85	70

Dependant Pensions (Death Insurance). Dependant pensions secure an income for the heirs to the deceased if he was insured. The major requirement in Bulgaria is for the dependant to have been supported by the deceased prior to his death. The beneficiaries entitled to dependant pensions are the children, brothers, sisters and grandchildren under eighteen, and those studying at educational establishments up to 25 years of age and over if they have lost their capacity for work before reaching 18 or 25 respectively. Students are entitled to dependant pensions after 25 years of age if they have served their regular terms in the army or in the labour corps. This term is extended by as many years as served in the army or in the labour corps. The parents are also entitled to dependant pensions, the husband or wife on reaching the age of 60 for men, 50 for women, or before if disabled; one of the parents or the wife, regardless of their capacity for work or age, if they are unemployed and look after children, or the brothers and sisters of the late head of family, on reaching sixteen; the grandfather and grandmother if they have no other income or if there are no persons bound by law to support them. The size of dependant pensions is fixed on the basis of the personal pension that had been paid to the deceased: 50 per cent if there is one dependant, 75 per cent if there are two dependants and 100 per cent if there are three dependants. The size of pensions payable to the heirs of a person who has died from industrial injuries is fixed on the basis of group I disability pensions if disability was caused by an accident. Heirs of a person whose death was caused by general disease are entitled to a pension fixed on the basis of the disability pension of group II that had been granted to the deceased. If the deceased had the required length of service entitling him or her to a personal retirement pension, the dependants' pension is fixed on the basis of the former, should this be of benefit to the dependants.

Social pensions were introduced in 1973. They are insignificant in number and are granted to persons who have not been insured or do not have at least half the length of service required for retirement or disability pensions. Such pensions are granted to invalids falling into disability groups I and II (aged over sixteen), and to old people having reached 70 whose annual income is lower than the specified minimum income. These pensions are small and the same for all entitled to them. Social security has been entrusted to the Bulgarian Trade Unions. They draft bills on social security issues and, jointly with other departments concerned, submit them to the National Assembly. Jointly with the Council of Ministers the Trade Unions issue regulations, decrees and instructions.

The Trade Unions are in charge of the social security budget. The remainder of contributions deducted (after the payment of indemnities, monthly child allowances, and maternity lump-sum cash grants) is transferred by enterprises and departments to the bank account of the social security with the Trade Union's Central Council.

Sums for the payment of pensions are drawn from this bank account. The control over the correct deduction and payment of security contributions and over the lawful spending of the social security fund is exercised by the Trade Unions. At enterprises and departments, indemnities cannot be paid without the signature of the Trade Union Chairman or of the person authorized by the former. The planning of social security funds, receipts and expenditures is carried out by the Trade Unions. Pension commissions and expert medical commissions on labour must include representatives of the respective Trade Union bodies. The Trade Unions fulfil their rights and obligations through their specialized bodies: the Social Security Board with the Central Council, the social security departments with the district and city councils, the Trade Unions and the social security commissions set up with them at every enterprise, department and organization.

SELECTED BIBLIOGRAPHY

BOZHILOV, Y. and G. GOCHEV *Dohodite na naselenieto* (The Incomes of the Population). Sofia, 1983.
BOICHEV, Y. *Zhiznenoto ravnishte – nerazdelna chast ot sotsialisticheskiya nachin na zhivot* (The Standard of Living as an Inseparable Element of the Socialist Way of Life). Sofia, 1980.
Dvadeset i pet godini sotsialistichesko zdraveopazvane (Twenty-Five Years of Socialist Health Services). Sofia, 1969.
Fizicheskata kultura i sportut v Bulgaria prez vekovete (Physical Education and Sports in Bulgaria through the Ages). Sofia, 1982.
GOCHEVA, R. *Zhizneno ravnishte i nachin na zhivot* (Standard of Living and Way of Life). Sofia, 1977.
Ikonomicheski problemi na sotsialisticheskiya nachin na zhivot (Economic Problems of the Socialist Way of Life). Sofia, 1981.
Izvunproizvodstvenata sfera. Problemi na efektivnostta i ikonomicheskiyat podhod (The Extra-productive Sphere. Problems of Efficiency and the Economic Approach). Sofia, 1982.
KONSTANTINOV, N. *Sotsialistichesko preustroistvo na zdravnoto dela v Bulgaria (1944–51)* (The Socialist Reconstruction of Public Health Services in Bulgaria 1944–51). Sofia, 1983.
KOLEV, A. and T. PACHEV *Obshtestveni fondove za potreblenie* (Public Funds for Consumption). Sofia, 1978.
LYUTOV, A., B. ATANASSOV and K. STOYANOVA *Razvitie na narodnoto potreblenie* (The Development of Popular Consumption). Sofia, 1982.
MARTINSKI, T. *Vuzhodut na fizicheskata kultura i sporta v NRB* (The Upsurge of Physical Education and Sports in the People's Republic of Bulgaria). Sofia, 1976.
Meditsinskata nauka i zdraveopazvaneto v NRB (Medical Science and Public Health Services in the People's Republic of Bulgaria). Sofia, 1981.
MILANOV, D. et M. MANOLOV *La Bulgarie sportive face au Monde*. Sofia, 1983.
OPROV, K. G. *Uslugite v razvitoto sotsialistichesko obshtestvo* (Services in the Developed Socialist Society). Varna, 1982.
Otnosheniyata na razpredelenieto pri izgrazhdane na razvitoto sotsialistichesko obshtestvo u nas (Relations of Distribution in the Building of a Developed Socialist Society in Bulgaria). Sofia, 1978.
Postizhenia i problemi na zdraveopazvaneto v NRB (Achievements and Problems of Public Health Services in the People's Republic of Bulgaria). Sofia, 1974.
75 godini organiziran turizum v Bulgaria 1895–1970 (75 Years of Organized Tourism in Bulgaria 1895–1970). Sofia, 1971.
SLADKAROVA, O. *Vurhove na Bulgarskiya sport. Spravochnik.* (Peaks of Bulgarian Sports). Guide-Book. Sofia, 1981.
SLATINCHEV, Y. *Borbata za trudovo zakonodatelstvo v Bulgaria: 1878–1944.* (The Struggle for Labour Legislation in Bulgaria: 1878–1944). Sofia, 1961.
Sotsialnata politika na BKP i razvitieto na NR Bulgaria (The Social Policy of the BCP and the Development of the People's Republic of Bulgaria). Sofia, 1980.
TRENEVA, M. *Nauchna organizatsiya na potreblenieto* (The Scientific Organization of Consumption). Sofia, 1979.
TSONKOV, V. and N. PETROVA *Istoriya na fizicheskata kultura* (History of Physical Education). 2nd rev. and enl. ed. Sofia, 1976.
TSONKOV, L. and N. DESSEV *Social Security in the People's Republic of Bulgaria.* Sofia, 1983.
YANULOV, I. *Obshtestveno osiguryavane pri kapitalizma i sotsializma* (Social Security under Capitalism and Socialism). Sofia, 1964.
ZHIVKOV, T. *Ikonomicheskata politika na Bulgarskata komunisticheska partiya* (The Economic Policy of the Bulgarian Communist Party). In 3 vols. Sofia, 1982.
ZHIVKOV, T. *Za posledovatelno izpulnenie resheniyata na Dessetiya kongres na Bulgarskata komunisticheska partiya za povishavane zhiznenoto ravnishte na naroda* (For a Consistent Implementation of the Decisions of the Tenth Congress of the Bulgarian Communist Party on Raising the People's Living Standards). In *Selected Works.* Vol. 20. Sofia, 1976.
Zhivotut v sotsialisticheska Bulgaria (Life in Socialist Bulgaria). Sofia, 1980.

PART VII

EDUCATION

EDUCATION

General Development and Structure. In ancient times, official State use of the Greek script hindered the development of an independent Slav-Bulgarian culture. The work of Cyril and Methodius in the elaboration of the Slav script created favourable conditions for further political and cultural development by the Bulgarian people and State. Born out of the necessity to consolidate the State and to defend its interests, education early acquired a democratic character. The belief of Cyril the Philosopher (828/827–869) – that each people had the right to read and write in its own language – became a reality. While Latin was largely used in education in the West, in Bulgaria great efforts were made to develop a native education in the mother tongue and to create a solid body of written work in the Bulgarian language. The first measures in this respect were promulgated by the disciples Cyril and Methodius, and, above all, Kliment of Ohrid (c. 840–916), after the conversion of Bulgaria to Christianity. Countering foreign influences, he and his followers used Bulgarian to teach and preach to the ordinary people and raise their levels of culture and national consciousness. In the ninth and tenth centuries the Ohrid and Preslav schools became centres of education. The flowering of letters during the reign of Tsar Simeon (893–927 AD) and the increased material wealth of the Boyars helped to consolidate and widen the scope of new literature. During the period of the decline of Bulgarian feudalism some young people received an elementary education, chiefly at schools attached to churches and monasteries. The appearance of the Bogomil sect and the spread of their socio-religious ideas resulted in a great increase in the literary wealth of medieval Bulgaria. During the period of Byzantine dominance (1018–1186) the spread of education in the Bulgarian lands was seriously hampered, but despite the hostile attitude of the Byzantine secular and church powers, several educational centres continued their work, teaching in Bulgarian. In the largest Bulgarian monasteries, Rila and Bachkovo, Bulgarian letters and literature were created, stored and preserved, and thus the Bulgarian spirit was kept alive. The Kilifarevo and Turnovo schools, which were beginning to develop the characteristics of higher schools of theology and philosophy, were great centes of enlightenment in the period of the Second Bulgarian Kingdom (1186–1396). The 'enlighteners' Cyprian and Gregory studied there. The activity of the reformer Konstantin Kostenechki is outstanding: he introduced methods of phonetics teaching, of teaching new material only after the children had learned the previous lesson, etc. The great role played by the Church, and the domination of the Christian ethic determined the religious character of education and enlightenment in Medieval Bulgaria.

The Ottoman rule (1396–1878) slowed down the economic, cultural and political development of the Bulgarian people for a very long period of time. The only educational centres were the monastery and church schools, which kept the traditional thirst for knowledge handed down from generation to generation.

Founded and supported by the people, these schools partly answered the needs for education, helped to preserve the Bulgarian national spirit, stored the literary heritage, taught clergymen, icon-painters, psalmists, copyists, enlighteners, teachers, master-builders, etc. The mainly religious education lasted from three to four years, and the subjects taught were reading, writing, religious instruction and psalm-singing. During the sixteenth and seventeenth centuries the Sofia and Etropole schools played a great role in opposing Ottoman rule. The raising of the national consciousness during the Revival period gave a strong impetus to the development of national culture, and the necessity for secular education appeared. The original idea for the enlightenment of the ordinary people started by Paissii of Hilendar was further developed and carried into action by Sophronius of Vratsa, Neophyte of Rila, Peter Beron, Vassil Aprilov, Neophyte Bozveli of Hilendar, Emanuil Vaskidovich, etc.

In order to satisfy the requirements of the rising Bulgarian bourgeoisie for secular education and knowledge of the Greek language, Greek and Greek-Bulgarian schools were founded, at which History, Geography, Arithmetic, Greek and Bulgarian languages were taught. The Greek influence which thus came through them met with the opposition of the Bulgarian people who, in keeping with the emerging national consciousness, founded their own purely Bulgarian secular schools. During the first half of the nineteenth century, at the time of the transition from Feudalism to Capitalism, the mutual instruction schools, using the Bell-Lancaster method, were established. They were mass, people's schools, the fruit of the Bulgarian national revival. With their appearance the Bulgarian school lost its medieval character, shook off the

Dr Peter Beron

Greek influence and turned into a school of a completely new type. Grammar, the Slavonic language, Geography, Bulgarian History, Arithmetic, Natural Science and other subjects were taught in a course of education lasting from two to three years. The first Bulgarian secular school of the new type was the mutual instruction school founded in 1835 in the town of Gabrovo. Following it, many such schools (boys' and girls') were opened. The new trends in education found their place in the *Primer with Sundry Precepts* (1824) by the scholar-encyclopaedist Dr Peter Beron. It was called *The Fish Primer* by the Bulgarian people because of the illustration of two fish on the last page of the book. Encyclopaedic in form, the primer corresponded to the needs of the age and the peculiarities of children's life. It was written in colloquial Bulgarian and satisfied the need for nation-wide education and secular knowledge. It was the first children's book. In the 1830s and 1840s a rich methodological and pedagogical literature evolved. The Bulgarian school during the Revival Period underwent a strong Russian influence, the textbooks being modelled on Russian ones and the teachers being mainly Russian graduates. After the Crimean war (1853–56) education and pedagogical theory entered a new stage of development: Western influence also began to affect Bulgarian teaching methods.

A new type of school was set up where the pupils could continue their education at the mutual instruction school. Historical and philological subjects predominated. Pedagogy was taught in some of them so that the students could take it up as a future activity. The teaching-learning process was carried out in forms and classes, the material being adapted to the pupils' ages and their level of knowledge. The lesson was the basic form of instruction. Lessons followed a strict weekly timetable of the various subjects. Sunday-schools were opened for barely literate young people and adults. The objective of those schools was to widen the scope of the learners' knowledge and to prepare them for self-fulfilment in life. The first secondary schools for general education were founded. In 1859 the Sts. Cyril and Methodius secondary school was established in the town of Bolgrad, then in 1872 the Aprilov school in Gabrovo and later others in Sofia, Plovdiv, Pleven, etc. By the beginning of the 1870s there were already 1,600 primary schools and three secondary schools. The teachers' meetings organized by the town councils had a very important role in establishing unified management of the schools. They were an original and democratic forum for discussing the problems of education, such as the necessity to introduce unification in the management of the schools, in the curricula and the development plans, as well as general requirements for the training of teachers. The revolutionary ideas of the epoch permeated the school atmosphere, and a large number of teachers took part in the national liberation movement. A characteristic feature of education during the period of the Ottoman oppression was that the people themselves were concerned in the establishment, management and support of the schools and struggled against any encroachment upon them.

After the Liberation in 1878 better conditions for the development of education were created. The two political parties, conservatives and liberals, had quite different attitudes towards education. The conservatives, as representatives of the upper middle class, took measures for the founding of a great number of vocational schools to meet the needs for qualified workers and for limiting the general education of the mass of the people; the liberals, on the other hand, defended the interests of the lower classes and stood for a compulsory, free education. During the Provisional Russian Government education developed under the influence of Slavophil ideas. Schools were not only compulsory and free, but secular and with a scientific orientation. Education and schooling continued in the spirit of the democratic revival traditions. According to the provisional statute for the people's schools (1878) there were three stages; elementary (course of education – three years); middle (two years) and principal (four years). For the first time education for both sexes became compulsory. With the Education Act of 1880, the democratic principles first introduced in Marin Drinov's regulations were put into practice: the course of education in the primary schools went on for four years, free and compulsory, followed by two more stages of three- and four-year courses each. The schools were supported by the councils. The 1884 Act concerning the people's and the private schools introduced for the first time the idea of vocational training for the pupils – some of the general subjects were taught with regard to the needs of agriculture and craftsmanship. After 1885, trying to maintain its advantage, the upper middle class neglected compulsory primary education in favour of secondary education, which trained personnel for the state apparatus. A large number of foreign schools were opened, through which West European influence was increased, especially German. On the initiative of the prominent scholar Ivan Shishmanov (heading the Department of Education and Spiritual Culture at the time of the Provisional Russian Government in Bulgaria) a Higher Teachers' Training College with a department of History and Philology was founded on 1 October 1888 at the State School for Ancient Languages in Sofia. The foundations of higher education were laid. In 1891 an Education Act was passed according to which the management and control of all schools and institutions of higher education came under the Ministry of Education.

The Act also made provision for the foundation of vocational schools (teachers' training schools, schools of commerce and industry and schools of theology). The 1909 Act determined the structure of the educational system, which underwent very few changes in the course of the next few decades. Education became compulsory for children aged seven–fourteen but in fact this was only applied to the elementary school. Secondary education provided a five-year course of studies. Classical, semi-classical and ordinary schools were set up. The 1921 Education Act introduced by the Minister of Education, Stoyan Omarchevski, made considerable changes in the educational system. Especially new was an orientation towards practical training. In the years of fascist dictatorship, starting from 1923, the ruling class made an attempt to destroy the democratic traditions in the educational system. All these anti-democratic initiatives affected the structure and content of education. Entrance exams were introduced for those who wanted to receive secondary education, and in this way the admission to secondary schools of children of working-class or peasant stock was limited. Schools for gifted children were founded to educate the future members of the state apparatus. Religious instruction became a compulsory subject in secondary schools. To complete the destruction of the democratic trends in education, the management of schools came under the control of mayors, representatives of the Fascist Government.

Miserable social conditions and the way in which children were admitted to schools during this period prevented the children of poor origins from getting a primary education, let alone a secondary one, and the existing law for compulsory primary education (first–seventh classes) in fact was not put into practice. Thus thousands of children from the above-mentioned age-group remained illiterate, receiving no education whatsoever.

A completely new epoch was inaugurated by the victory of the socialist revolution of 9 September 1944. The latter provided new social and economic conditions, and the necessary equipment and organization for a rapid development of education. With Socialism under way, education exercised a decisive function in the scientific, technological and social progress of the country. It also contributed to the general social and economic uplift and spiritual morale of the nation.

Kindergarten on the Lenin housing estate in Pernik

The basic trends in the strategy of the state policy for education were incorporated as items in the Constitution of the People's Republic of Bulgaria.

Free education is secured for all citizens in all kinds of schools and institutes of higher education. All schools are run by the State. Education today keeps pace with the latest achievements of science and technology and is based on the Marxist-Leninist ideology. Education up to the age of sixteen is compulsory. The building of socialist economy and culture conditions the rapid and regular development of education. Illiteracy was quickly abolished and all children and young people have the opportunity of receiving primary and secondary education.

The Bulgarian Communist Party and the Bulgarian Government have accepted and approved documents determining the new trends in education and its democratic character.

Free education is accessible to all. Pupils are admitted to secondary schools with no restrictions. They follow their inclinations. Wider vistas are opened to those of the working people willing to improve their education by either regular or extra-mural courses of studies.

In 1948 the first Education Act in the socialist People's Republic was passed.

This Act confirmed the revolutionary changes in

Festival at the combined kindergarten and nursery in the Chervena Zvezda district of Sofia

education in the first years of the People's Government and indicated further steps in the re-organization and improvement of the educational system. A unified educational system was established. Pupils could enter secondary schools with no restrictions. A scientific approach was made to the curriculum. Universal education was finally established in Bulgaria.

The decisions of the April 1956 Plenary Session of the Central Committee of the BCP marked the turning-point of the democratization of schools, which conditioned the increase of the polytechnical basis of education and its closer ties with life. The documents adopted at the Plenary Session made schools a more decisive factor in the building of socialist society and preconditioned the realization of compulsory education and the improvement of the general and vocational training in schools, stimulating the cultural advance of the people.

The re-organization of education made closer links with economy, culture and current scientific and technical progress. Decisive measures were taken against the old-fashioned, bookish character of education, and against dogmatism and formalism, in order to help the flowering of the creative abilities of teachers and pupils. The law of 1959, which determined closer links between school and life and further improvement of education in the People's Republic of Bulgaria, is of extremely great importance in this reorganizing process. One more year has been added to the seven-year course.

The secondary school of general education has now a definitely polytechnical character. Apart from the acquisition of knowledge about nature and society, pupils acquire different skills and an appropriate trade orientation. The polytechnical education is conducted through different subjects illustrating the scientific character of production: Mathematics, Physics, Chemistry, Biology; through lessons on practical skills conducted in all forms (first–eleventh classes), pupils receive knowledge of electricity, agriculture, the machine-building industry, internal-combustion engines, etc. Extra-curricular activities flourish through various clubs organized at the schools. Displays of students' technical work are held regularly. People can attend evening and extra-mural classes whilst continuing their jobs.

The rapid pace of the socio-economic and cultural development of the country, as well as the achievements of the scientific and technical progress, requires a continuous improvement in the structure and content of education. In this context in 1969 the Central Committee of the BCP and the Bulgarian Government again discussed educational matters and took decisions to intensify the connections between school and life, to secure general secondary education for all and improve the system of vocational training of pupils. A tendency to draw the two historical educational trends together – the general and the vocational – has become predominant. A purposeful study has begun of how and under what conditions the completely new type of the unified secondary polytechnical school should be established (see Table 1).

A new epoch in education was started in 1979. A Plenary Session of the Central Committee of the BCP accepted the concept of education included in the opening speech of the General Secretary of the Central Committee of the BCP and President of the State Council, Todor Zhivkov, and in the Theses on the further development of education in the People's Republic of Bulgaria. This is in fact a scientifically worked-out strategy for the functions of the educational system in the period of building of a developed socialist society. The widely discussed Theses, on receiving general approval, turned into a programme, the realization of which is now the task of the whole nation. The general aim of the educational system is the formation of the versatile individual, and later the all-round development of the individual, able to fulfil himself completely in life. The realization of the general aim of the educational system demands the formation of socialist individuals of ample knowledge and creative abilities, a high standard of general culture, ideology and vocational training, as well as readiness to take part in socially useful labour. Education aims at making the young person a fully developed personality with a high level of education and culture, so that he can have his start in life as a conscientious citizen of Bulgaria, deeply committed to the ideas of Socialism and an active member of socialist society. He should also be a personality of versatile and accomplished abilities, aspiring towards effective realization of his acquired knowledge and skills and his further self-improvement. The general aim of education demands that every young person should participate in social labour with a skill or profession, mastered to such an extent as to suit all requirements and standards of the quality of

Table 1: Structure of the System of Education in the People's Republic of Bulgaria

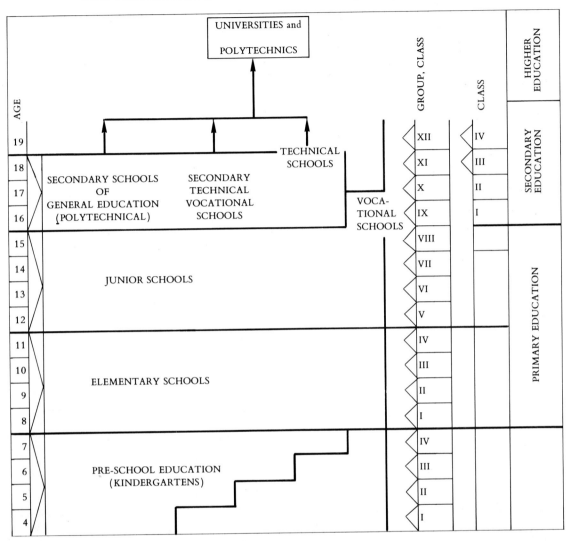

labour. He should see the chief field of his activities and source of improvement of his abilities in his work – a means of active participation in the task of the whole nation to add to the material and spiritual values of society and to the improvement of social relations.

The versatile development and highest possible fulfilment of school-leavers is based on general and technical education, vocational education and the mastering of a specific job, skill or craft.

In the course of the reorganization of the educational system all the best aspects and traditions have been preserved, together with the achievements of education in the years of Socialism in Bulgaria. A constructive attitude has been shown to the good results of the educational systems of other socialist countries and the Soviet Union.

The combined public and state principle is fundamental in the management of education and this intensifies the democratic character of education and also its efficiency. The combined public and state principle also gives better chances to state, economic and public bodies and organizations, to workers and employees, and to institutions of science and culture to take more initiatives and feel more responsible for their duties.

The reorganization of education in 1979 was of a comprehensive character. This affected the structure, contents and methods of teaching and upbringing, teacher training and the material and technical foundations of education and its management. The democratic nature of education is of paramount importance and guarantees conditions for creating equality in young people's education, and for the maximum possible development of individual talents and creativity and for the young people's complete fulfilment in life.

The two educational trends – the general and the vocational, each with a history of its own – have been drawn together and finally merged into each other to form the foundation of the new educational system, which is now starting to function through its new establishments. So all young people enjoy the right to be well educated and highly qualified in their chosen skills and trades and to raise their cultural levels.

Pre-School Upbringing. The People's Republic of Bulgaria provides a unified system of pre-school establishments – *kindergartens* – where public upbringing is successfully carried out. These take children from three to seven years of age. The first steps in character-building are taken there: an elementary course of physical training is introduced, together with a course for their intellectual development,. They are also taught how to behave socially, about their duties to and the rights of other people, and an appreciation of beauty. The kindergarten course of education also includes some skill and trade activities. Kindergartens function as state establishments accessible to all and play an important role in the development of the national economy and culture. Kindergartens assist families in bringing up their children; thus parents can take an active part in socially useful labour.

Teaching, training and all other activities in kindergartens are well planned to prepare children for school and for their future versatile development. The presence of kindergartens in the system of education is the first stage of the organized, public upbringing of children.

The idea of pre-school upbringing in Bulgaria occurred first at the end of the nineteenth century. The first kindergarten was founded in 1882 in Svishtov with the help of Nikola Zhivkov (1847–1901) – a prominent figure in the field of education – and was financed by the municipality there. Private kindergartens appeared one after another in Varna (1884), Plovdiv (1884), Sofia (1885) and other places. The first state kindergarten was founded in Sofia in 1888. The BCP since its formation has worked upon the idea of admitting all children to kindergartens. Vela Blagoeva, the wife of the first Bulgarian Marxist, Dimiter Blagoev, was one of the first pedagogues to deal with pre-school education and its organization. The bourgeois Government made no attempt to establish kindergartens for the working-class children. Up to the 1944 socialist revolution there were 243 kindergartens with 11,334 children and 292 kindergarten teachers.

The victory of the socialist revolution in 1944 turned pre-school education into a concern of the whole nation and a constituent of party and state policy. In 1948, 3,753 state kindergartens were founded with 5,580 kindergarten teachers. Kindergartens are not compulsory – children can attend them without any restrictions or privileges. How-

Construction games at a kindergarten

Swimming pool of the Dr Mara Maleeva combined kindergarten and nursery in Sofia

ever, the children of large families or those whose parents are students or post-graduates are considered first for places.

Some of the kindergartens work all the year round, some are only seasonal. The time children spend daily in the all-year-round kindergartens is from 7 a.m. till 7 p.m. They also admit children who stay there all night.

There are also part-time kindergartens, where children stay from 7 a.m. to 1 p.m. or from 1 p.m. to 7 p.m., special weekly or boarding nursery schools for mentally retarded

The Chavdartsi day kindergarten in Sofia

Education. The *seasonal kindergartens* are opened for a certain period of time, to help some groups of seasonal workers. They can be of three types: temporary, full-day kindergartens with a working time varying from four to ten months; summer, full-day kindergartens (45 days to ten months); and open-air kindergartens (30 to 50 days). Every year the Government sets aside from the State Budget the funds necessary for the maintenance of pre-school institutions. Parents pay only 20 per cent of the whole sum, while the remaining 80 per cent is undertaken by the State. The amount depends on the income of the family and the number of its members. The kindergartens are completely free for families of low income, and for children of students or men doing their national service in the army. The half-day kindergartens are free of charge. The convalescent and other kinds of special kindergartens are totally state-maintained. Each nursery or kindergarten is provided with medical care. Pediatricians visit the kindergartens at fixed hours every day, doing the necessary medical check-ups. Every kindergarten has a nurse who examines the children every morning, takes care of current disinfection, immunization and keep-fit procedures, supervises the diet and keeps a medical record of each child. To improve the health of the children, many kindergartens organize summer camps at the seaside or in the mountains. The equipment is constantly being improved and augmented, thus giving an opportunity for more children to be admitted. For example, while there were 194,280 children in the full-day kindergartens in 1975, in 1982 their number had mounted to 291,611.

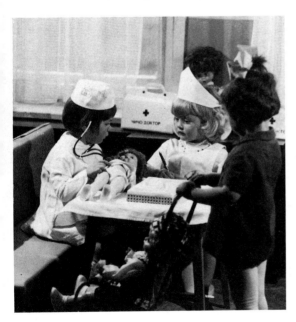

Instruction at a kindergarten

Playground at the Ho-Chi-Minh kindergarten in Sofia

children or for convalescent children recuperating after chronic diseases, and combined nurseries-kindergartens. To these, children aged between ten months and seven years are admitted. The educational process is carried out in accordance with a syllabus endorsed by the Ministry of

There is an opinion rapidly gaining ground in Bulgaria that all children should have attended a full-day or a half-day kindergarten for at least a year before entering the first class. The percentage of children attending kindergartens

In the co-operative market in Plovdiv

Classroom at the Dr Maleeva-Zhivkova kindergarten in Sofia

at the age of three, four and five is increasing every year, so that in 1985 every child can have the opportuntiy to attend a kindergarten, through which universal pre-school education will be effected. In 1982 there were 403,518 children in 5,733 kindergartens in Bulgaria. The total number of combined nurseries-kindergartens is 414, with 57,750 children. There are twelve convalescent kindergartens with 838 children, and 3,273 half-day ones with 5,495 children. 76 per cent of the children from three to six years of age attend kindergartens throughout the country.

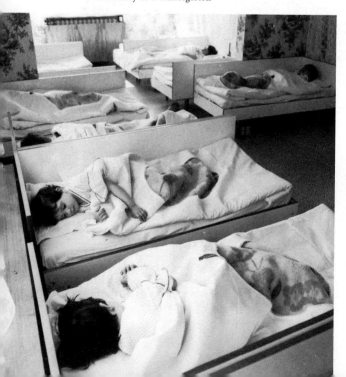

A dormitory at a kindergarten

A kindergarten in Mihailovgrad

Secondary Education. The unified secondary polytechnical school is the basic structural unit in the system of the new type of education. Its form and content reflect the new requirements of society in the socio-economic, scientific, technological and cultural spheres. The following requirements have been taken into consideration when

Table 2: Kindergartens – 1982

Kindergartens		Teachers	Children
Total	2,169	19,223	274,786
Permanent	2,073	19,064	271,042
full-day	1,408	17,414	227,243
half-day	665	1,650	43,799
Seasonal	96	159	3,744
temporary	9	24	338
summer	18	24	528
open-air	69	111	2,878

Students from the Secondary School of Art and Industrial Design in Kazanluk

A first-grade pupil from the Hristo Kurpachev Elementary School No. 122 in Sofia

forming the structure of the unified secondary polytechnical school: universal secondary education for all; solid, comprehensive training to serve as a basis of the system of professional qualification; increasing the significance of the school for developing all the natural talents and abilities of the individual, and his intellectual power and aesthetic appreciation of the world about him. This new type of school contributes to overcoming the substantial differences between physical and mental work. In its social functions, content and organization, the new type of secondary school is *unified*, since all children and young people pass through it. It unites general and vocational education in dialectical unity, and on the basis of a system of scientific knowledge it creates equal opportunities for the social fulfilment of the younger generation. The *comprehensive character* of the school is defined by the opportunity it gives to all students for a versatile intellectual, ideological, political, moral, aesthetic and physical training. The *polytechnical nature* of the new type of school originates from the close connection of the educational process with the contemporary world and the scientific-revolution. The knowledge and skills mastered in the school are linked to their efficient use in the different spheres of human activity. The students become acquainted with the scientific foundations of production and are psychologically and practically prepared for highly-organized, socially-productive labour. The vocational training of the younger generation is being established on this basis. The unified, secondary polytechnical school provides a type of training, qualification and education which fully correspond to the accepted requirements of world practice for secondary education. The full course of the unified secondary polytechnical school comprises twelve years and provides for modern vocational training of all boys and girls. The unified secondary polytechnical school includes three levels of training organically connected, having at the same time specific functions and contents. These levels are conditioned by the characteristics of the psycho-physical development of the students and by the contemporary needs of production and public services. In order to get a secondary education the students are obliged to pass through them.

The first level covers the period from the first to the tenth class and includes children from six to sixteen years of age. The fundamental purpose of the first level of the unified school is to lay the foundations of the versatile development of the student's personality. The educational process at this level provides for broad comprehensive training, forms work habits and lasting motivation for work, creates up-to-date informational needs in the students, develops their cognition, and reveals and confirms their abilities, talents and creative aspirations. The syllabus at the first level embraces the natural sciences and the humanities.

The basis of polytechnical training is the study of all subjects, through which students acquire knowledge of the

A language laboratory at Secondary School No. 114 in Sofia

A ski school on Mount Vitosha

general natural laws and natural phenomena as well as their application to social production and practice. The polytechnical training creates a broad polytechnical culture, which prepares the vocational orientation of the young people and provides for their immediate work-training in the next stages of the school. At this level a unified system for vocational training and orientation has been constructed. It reveals the social significance of every

A history lesson at the Academician Emilyan Stanev Secondary Polytechnical School in Sofia

profession and specialism, and the opportunities for professional and personal fulfilment which different jobs can offer. In the conditions of the first level of the unified school, the differentiative system of training is being introduced, which influences the individual interests and abilities. The differentiative system includes: in addition to the general compulsory syllabus, there will be education in optional subjects, from the first to the tenth class, and in the ninth and tenth classes the students may choose to train in one particular vocational polytechnical subject, which helps them to acquire some skill and prepares them to a certain extent for work in the different branches of the national economy. At the very beginning of the school year every pupil in the tenth class has to choose one or two subjects connected with the humanitarian, scientific, aesthetic and vocational polytechnical aspects of education. The overall system of differentiative education allows the pupils to choose a kind of activity which stimulates their creative abilities. Besides the general knowledge of all subjects studied, every pupil has the right to choose separate (extra) subjects, for more detailed study, which satisfy and develop their interests. The system of education including extra subjects enriches pupils' knowledge, widens the scope of their scientific outlook and develops their particular abilities.

In compliance with the specific features of the physical and psychological development of the pupils, when planning the curriculum and the organization of education at this level, the system is divided into three stages: the first stage is from the first to third class; the second, from the fourth to the seventh, and the third, from the eighth to the tenth class.

The education during the *first stage* provides for a natural transition from the kindergarten to school: the children are at school all day, five days a week; lessons are effectively

The refectory at School No. 107 in Sofia

alternated with games, recreation, an afternoon nap and a medically-supervised balanced-rational diet.

The *second stage* is crucial in the general and polytechnical education. It is during this stage that the adolescents lay the foundations of their scientific outlook.

The *third stage* extends the systematic course of general and polytechnical education. The subjects taught and the educative process are brought closer to the scientific achievements of today. This period raises the students' level of skills and qualifications.

The second level of the unified secondary polytechnical school continues for about a year and a half after the tenth class. The structure of studies at this stage covers the broad polytechnical basis of the skill/trade chosen, as well as the theoretical, technological, economic, organizational and managerial fundamentals of the respective production spheres. The system of general technical subjects is of particular importance for the students' vocational training at this stage. It includes subjects like the Foundations of Socialist Production, Mechanics and Auto-mechanics, Electronics and Electricity, Computing Machinery, Tourism and Hotel Administration, Organization and Management of the National Economy, etc.

This level continues compulsory instruction in Bulgarian Language and Literature, the integral subject called the Man and Society, and Physical Education. The system of moral, labour and aesthetic education underlies the entire educative process. The second level extends its efforts for developing the individual faculties of young people. It promotes the subjects freely opted for and extra-mural studies, and creates preconditions for raising the students' general standards and creative contribution to the various fields of public and cultural life, and scientific and technical creative work. All this provides for a thorough,

fundamental training of students and prepares for the ensuing practice.

The third level of the unified secondary polytechnical school continues for about six months. In the course of it students master one particular specialism in the circumstances of real production. This level completes the educational cycle of the unified secondary polytechnical school. School-leavers are provided with a broad general and polytechnical qualification, on the basis of which they choose a specific trade or skill. This affords opportunities for their start and complete fulfilment in life, and for improving and raising their professional standards.

The occupational training during the second and the third level is carried out in the practical-training centres. This is a completely new form of educational and practical training. The centres are part of the unified secondary polytechnical school, using the already established practical training facilities of the vocational schemes and of some technical schools. Being a labour process proper, the

The grade school in the village of Zheravna in the Sliven district

vocational practice is also carried out in the subsidiaries for vocational specialization – sections, shops, and work places of some leading enterprises and communal, health and other establishments equipped with modern machinery and technologies and employing leading workteams characterized by excellent labour organization. The number of broadly-based skills and specialisms studied at the second and the third level of the unified school reflects the dynamics and needs of material production and of other spheres of public life (see Table 3).

The new type of unified secondary polytechnical school was initiated in the school year 1981–82 and its establishment will be completed in the 1992–93 school year. This period will create all social, pedagogical, material, technical and personnel conditions for the

Table 3: Educational Institutions, Teachers and Students

Educational Institutions	Number	Teachers	Students
Total	4,238	102,285	1,498,316
Polytechnical schools for General Education	3,541	68,695	1,170,434
Schools for Children of Retarded Development	3	71	1,667
Schools for Vocational Education	129	2,371	17,237
Secondary Technical Schools	281	7,788	122,220
Technical Colleges and Schools of Arts	232	9,133	93,226
Semi-Higher Institutes of Education (Further Education Colleges)	23	973	9,899
Higher Institutes of Education (Universities and Polytechnics	29	13,254	83,633

Polytechnical Schools of General Education	Number	Teachers	Students
Elementary (I–IV)	777	24,975	430,433
Junior (V–VIII)	54	35,261	623,432
Primary (I–VIII)	2,267	—	—
Secondary (IX–XI)	94	8,459	116,569
Unified (I–XI)	349	—	—

students' education and vocational training, and will consolidate the new school as a new synthesis of the traditions of general and vocational education. Its complete form will be attained through a steady improvement of present-day educational structures and by introducing a new educative content, the tendencies typical of the unified secondary polytechnical school. These present-day

A maths lesson at the Vassil Levski Elementary School No. 8 in Sofia

The Aprilov Secondary School in Gabrovo

structures are as follows: the secondary polytechnical school of general education, having a three-year course of studies after the eighth class; the secondary technical vocational school, with a three-year course of studies starting after the eighth class; and the technical schools, with a four-year course starting after the eighth class.

At present, some 93 per cent of all students leaving the compulsory course of education (first to eighth class) continue their studies in one of these three types of secondary school. Some 50 per cent of them are controlled in the unified secondary polytechnical schools and the rest go either to secondary technical vocational schools or to technical schools.

The Lyuben Karavelov School in Koprivshtitsa

Secondary school pupils at work at the telephone plant in Sofia

vocational school, whose aim was to train technically literate workers for industry, trade, agriculture, construction, transport, communications, etc. The factory schools established after the revolution of September 1944 were analogous to the Soviet schools of that time. In 1952 they were restructured into schools for skilled labour reserves. The 1959 Act for the Close Linkage between Schooling and Life and the Further Development of Education in the People's Republic of Bulgaria established the first technical vocational schools, with a two-year term of study. The rising need for skilled rank and file personnel demanded better educational opportunities. The 1962–63 school year saw the gradual establishment of *secondary technical vocational schools,* providing for a three-year course of study after primary education. Besides the work-skills acquired, these schools give the students a secondary education and the right to continue their studies in any higher or semi-higher school. The secondary technical occupational schools are the basic type of school that train cadres for the various branches of the national economy. They specialize according to the professions acquired in them and train workers for machine-engineering, light industry, transport, construction, agriculture, services, etc. Schooling consists of general educational studies indentical to those undertaken in the unified secondary polytechnical school of general vocational training, with specialized training in the theory and practice of the trade chosen. The studies are conducted in special workshops, workrooms and laboratories built at the secondary technical vocational schools. Practical instruction also makes broad use of the material and technical facilities belonging to the enterprises the students are being trained for.

Students from the Technical School of Plastics and Rubber of the Narodna Republika state industrial enterprise during practical training

Vocational education in Bulgaria was initiated by some educational establishments opened in the late nineteenth century. The first vocational school was inaugurated in Svishtov in 1873 by Dimiter Shishmanov, a prominent public figure. The first decades of the twentieth century saw the establishment of a considerable number of agricultural, practical, economic and other types of

Table 4: Structure of the Unified Secondary Polytechnical School

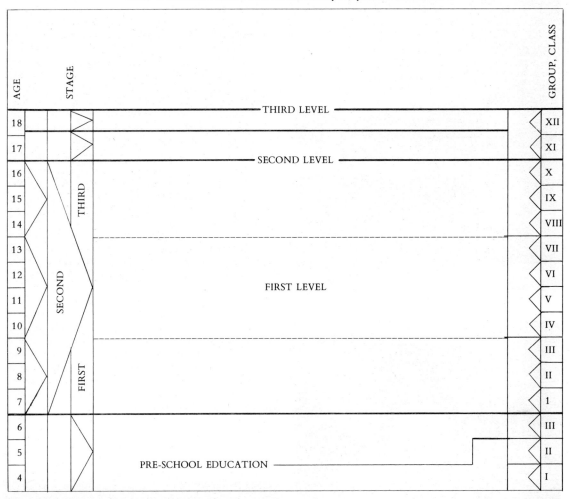

The technical schools provide secondary education for workers and executives needed by the spheres of production and services. They admit young people who have completed their primary education up to and including the eighth class. The term of studies is four years; the content provides a good theoretical background – general, vocational and specialized – for future executives in production. The theoretical and practical training of technical school students is carried out in training workshops, experimental farms, workrooms, and laboratories, well equipped with modern machinery. The professions and specialisms in technical schools relate to the basic branches of industry, agriculture, social services, construction, tourism, etc. Studies end in either sitting for a final exam or defending a diploma thesis. School-leavers have the right to continue their studies in all higher educational establishments. In the 1982–83 school year, technical schools in Bulgaria numbered 217 and enrolled 89,737 students.

The list of skilled workers' professions acquired in the unified secondary polytechnical, secondary technical vocational and technical schools is endorsed by both the Ministry of Education and the State Committee for Planning. The occupational training in these schools is to be restructured in conformity with the requirements of the broadly-based skills and specialisms taught. The list of professions and specialisms acquired in the secondary schools is revised in step with the development of the economy, science and technology. The percentage of professions relating to new technologies, automated production, electronics, biotechnologies, etc. is steadily growing.

In conformity with the principle of a gradual merger of the two types of education – general and vocational – and in the course of the establishment of a unified secondary

Education

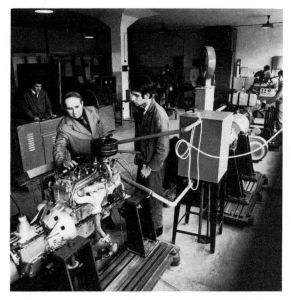

At the internal combustion engines laboratory at the Wilhelm Pieck Technical College of Energy in Sofia

A physics lesson at the S.M. Kirov Technical School of Electrical Engineering in Sofia

Practice in wood-carving at the village of Yasna Polyana in the Burgas district

polytechnical school, the relative share of secondary technical vocational schools and technical schools training pupils after the eighth grade will diminish and they will ultimately be closed down. The unified school will be the main form of providing secondary education and vocational training. A new type of technical college will be opened at a higher grade, to train leading cadres with secondary education. They will admit students who have completed either a unified secondary polytechnical school or a secondary technical vocational school. Their course of studies will continue for two years and guarantee a thorough theoretical, applied and technical knowledge of the respective occupational spheres, consisting of a general educational and cultural background, and knowledge of the socio-psychological, managerial, economic, and organizational problems of production.

The training of experts in Bulgaria is planned. It is correlated to the needs of the national economy and is an essential part of the One- and Five-Year Plans for the country's socio-economic development. The material, labour, and financial resources needed by the educational establishments are secured through planning the country's needs for trained personnel in various professions and occupations.

The process of vocational training of students is largely assisted by the enterprises and economic organizations that will employ them. To this end, the latter build and place at the disposal of schools the necessary material and technical facilities, and send engineers and technicians to help with the vocational training of students. The integration between schools and economic organizations will provide the national economy with a stream of highly-skilled workers and executives.

Bulgaria has also built a large network of specialized educational establishments for developing the individual talents and abilities of students. Thousands of young people cultivate their talents and artistic pursuits in the secondary schools of music, choreography, applied and fine arts, etc. (see Table 5).

There are also unified polytechnical schools for Russian, French, English, German, Spanish and Italian language

A concert given by the students of the Secondary Music School for Folk-Instruments and Folk-Singing at the village of Shiroka Luka in the Smolyan district

studies on the model principles of the unified school. Besides acquiring a general and vocational education, their students attain a fair proficiency in the respective language. In 1983–84 there were 74 language schools in Bulgaria. Their school-leavers either continue their studies or are employed in the occupations they have been trained for.

Sports schools. These are general secondary polytechnical schools, with the emphasis on training. The curriculum and the organization of the work is structured according to the overall plan of the general secondary polytechnical schools, with certain modifications demanded by the particular type of school. The teaching and training process takes place all the year round, with a weekly schedule varying from 16 to 24 lessons according to the types of sports and the ages of the students. Intensive sports instruction begins in the fifth or sixth class depending on the type of sport. After finishing the tenth class, the most talented sportsmen and sportswomen enrol in the eleventh and twelfth class (second and third levels) with the emphasis on training. The students in the eleventh and twelfth classes receive the same broadly-based vocational training as the students in the unified secondary schools. The school curriculum creates wonderful conditions for serious training and versatile development of the students, directed towards their fulfilment in their various spheres of life. After finishing at the sports school, students have exactly the same rights as those finishing at the general secondary polytechnical schools.

The first sports school was opened in Pleven during the school year 1968–69 for gymnastics, track and field events, basketball and volleyball. In 1971 sports schools were opened in Plovdiv and Varna, and later in Burgas, Stara Zagora and some other places. There are sixteen sports schools in Bulgaria.

Medical colleges are educational institutions for the training of secondary medical staff. The oldest are the medical colleges founded in 1878, for the training of doctors' assistants. The training of nurses is closely related to the activity of the Bulgarian Red Cross. In 1895 it opened a one year course and in 1900, a regular school in Sofia. In 1895 the first school for midwives was also opened in Sofia. Courses for the training of assistant pharmacists, dental mechanics and laboratory assistants have been organized. Since the revolution of 9 September 1944, physiotherapists, instructors, social workers and orthopaedic technicians have also been trained through the medical colleges. In 1981–82 there were fourteen medical colleges in Bulgaria.

Special schools have been established for the education of physically and mentally-handicapped children, in which special syllabuses are followed. All children in need of such care are admitted to these schools, and education and board are free.

Extra-curricular and out-of-school activities occupy a specific place in the educational system. They are organized by the Pioneers and the Communist Youth Organization under the leadership of the teacher, on a voluntary basis, and do not duplicate school activities. Widespread forms of extra-curricular activities are: mass (clubs, evenings, olympiads, quizzes, competitions, etc.), connected with hobby circles (scientific, technical, sports, amateur acting, singing, music, etc.) and individual. These forms have, in fact, proved to be one of the most effective means of producing good scientific workers, artists, sportsmen, etc. Clubs for creative scientific and technical work by young people are organized in three sections – scientific, technical and information/propaganda – in almost every school. The 'My Contemporary' clubs contribute very much to the ideological and political education of the students. The students take part in voluntary choirs, theatrical groups, sports, games, and labour initiatives following their natural interests and aspirations. 149 out-of-school institutions also operate, in the stations of young technicians, agrobiologists and others, in which the students broaden the scope of their knowledge and interests in their free time (see Tables 6 and 7).

Recreational activities at the seaside and in the mountains are annually organized for the students. They spend twenty unforgettable days in special student camps, getting fit and becoming stronger both physically and morally. For example, in 1983 some 520,000 students were included in organized recreational activities.

Teacher Training. There was hardly any specialized teacher training before the Liberation of Bulgaria from Ottoman rule in 1878. Every literate person could be a teacher in the various little schools in the country. With

Education

Table 5: Schools of the Arts – 1982

	Names of schools and the towns they are located in	Number of students
Secondary Music Schools	Sofia – L. Pipkov	579
	Plovdiv	418
	Varna – D. Hristov	385
	Russe – V. Stoyanov	337
	Burgas	350
	Pleven – P. Pipkov	368
	Stara Zagora – H. Morfova	215
	Kotel – P. Kutev – for folk-singers and musicians	189
	The village of Shiroka Luka, the Smolyan district – for folk singers and musicians	206
	Sofia – the State School of Choreography	358
Specialized Secondary Schools of the Arts	Sofia – Academy Petrov – of fine arts	344
	Sofia – of fine arts	343
	Plovdiv – Ts. Lavrenov – of fine and monumental arts	196
	Plovdiv – Scenic designers	194
	Kazanluk – of design	239
	Troyan – of applied arts	195
	Tryavna – of applied arts	338
	Sliven – of applied arts	146
	Smolyan – of applied arts	113
	Sofia – J. Fuchik – a special secondary school of polygraphy and photography	763
	Sofia – Constantine Cyril, the Philosopher – national high school for ancient languages and cultures	385
	Sofia – Gorna Banya quarter – unified secondary polytechnical experimental school	239

the establishment of mutual instruction schools and schools organized in classes, the need for well-trained teachers became apparent, though the question of specialized pedagogical training was not raised. Teachers educated abroad had the best training. Pedagogy was introduced as a school subject for the first time in 1868, into the school curriculum of the girls' school in Stara Zagora. The diocesan school, Sts. Cyril and Methodius in Plovdiv, the Aprilov high school in Gabrovo and the schools in Shumen and Turnovo acquired a definite pedagogical character. In 1869 the first boys' pedagogical school, the Pedagogical Theological School in Shtip, was opened.

In the first years after the Liberation of 1878, the training of primary teachers was not very purposeful or well organized, because no specialized schools existed at that time. The first pedagogical schools were opened in Shumen and Vratsa during the school year 1881–82. The duration of the course was three years, after the seventh class and followed a special curriculum. In 1883, the Saints Cyril and Methodius pedagogical school was opened in Kazanluk. After 1885, pedagogical schools were opened in Lom (1888) and in Silistra (1890). During the school year 1890–91, there were five boys' pedagogical schools for the training of primary teachers in Bulgaria: in Shumen, Silistra, Lom, Kazanluk and Kyustendil, with uniform syllabus, duration and content. In 1912, a girls' pedagogical school was opened in Sofia. A considerable reorganization of the pedagogical schools took place during the period of rule of the Bulgarian Agrarian Union

Students in an agriculture brigade

Table 6: Development of Extra-Curricular and Out-of-School Forms of Sports Activity

Forms	1978–79		1983–84	
	Number of sections	Number of participants	Number of sections	Number of participants
1. Sports sections	10,246	187,823	10,687	196,472
2. Groups for mass sports	9,775	194,834	11,823	436,967
3. Groups for remedial sports activities	2,037	38,808	2,188	40,226
4. Active sportsmen (in sports sections, students' sports clubs, complex and specialized clubs, sports teams)	—	355,351	—	404,377

Table 7: Artistic Amateur Activities

Forms	1978–79		1983–84	
	Groups	Participants	Groups	Participants
1. Musical groups	4,694	187,189	5,130	171,298
choirs	2,924	150,367	2,703	128,475
vocal groups	638	12,740	1,278	20,718
orchestras	1,087	24,082	1,449	26,105
2. Dancing groups	910	23,853	956	25,744
3. Ensembles	102	5,470	74	6,113
4. Theatrical troupes	1,540	28,503	2,651	38,949
5. Literature circles	246	4,975	595	10,455
6. Art	795	14,827	1,110	19,452
fine arts	491	9,437	311	6,068
applied art	304	5,390	799	13,384
7. Cine-clubs	153	2,390	197	11,193
8. Photo-clubs	1,291	20,670	1,316	19,575
9. Groups for political songs	610	8,256	1,496	17,630
10. Groups for authentic folklore	116	2,258	248	5,350
11. 'Friends of art' clubs	717	15,447	1,133	23,794

in 1920–23. Pedagogical schools were opened in many towns in the country. In 1940, five colleges for primary teachers came into existence – in Turnovo, Kyustendil, Vratsa, Stara Zagora and Silistra.

After the revolution of 9 September 1944 these teachers' training colleges were reconstructed. Secondary pedagogical schools for the training of primary teachers were opened in 1950–56. In 1956–57 they were transformed into teacher training colleges with a two-year course after secondary education. Nursery teachers were trained at first in pedagogical schools, but since 1960 colleges for the training of nursery teachers with semi-higher education have started to function. Junior-school teachers (fourth–eighth classes) are trained in teacher training colleges with four departments: Bulgarian Language and Literature and Russian; Bulgarian Language and Literature, History and a West-European language; Mathematics and Physics; Natural Sciences, Geography and Chemistry. In all departments an arts subject or Physical Education is studied as an option. The duration of the course is three years. The pedagogical and methodological management of the college is carried out by the Ministry of Education.

The high standards demanded in teachers' work required the reconstruction of the educational system (1979), including also the system for training and qualification of teachers. Gradually some colleges for the training of nursery, primary and junior-school teachers for fourth–eighth classes with a two or three-year course were transformed into higher pedagogical institutes, while others were turned into branches of universities. An in-service course for the improvement of the teachers' professional qualifications is being introduced, together with the reconstruction of the training system. In the next few years all teachers, irrespective of the type of school they work in, will have a higher education, which they

Students on field practice in the hot-houses of the Georgi Dimitrov Palace in Sofia

A classical ballet lesson at the State School of Choreography in Sofia

will receive in regular and correspondence training courses. In 1976, the first Higher Pedagogical Institute for the training of nursery and primary teachers was founded in the town of Blagoevgrad. The duration of the course is four years. Teachers with higher education are being trained in the universities of Sofia, Turnovo, Plovdiv and the Higher Pedagogical Institute in Shumen.

A central institute for the in-service training of teachers and managerial staffs of schools is in operation, as well as four regional institutes for the in-service training of teachers. Centres for the qualification and information of educational workers function in every district. This allows the scientific, methodological, and psychological training of all teachers to be updated and improved every five years in refresher courses, and their knowledge and creative pedagogical skills to be enriched.

A considerable number of pedagogical and methodological magazines are published in Bulgaria to help the teachers in their work. The professional development of Bulgarian teachers, their social importance and the solution of their social problems are concerns of the Party and the state organs and organizations.

At the first Congress of Education, which took place in 1980, a *Bulgarian Teachers' Charter* was accepted and then officially confirmed by the People's Assembly. This remarkable document reflects the enormous social functions of the teacher, and his leading role in the process of education of the young generation. The Teachers' Charter clearly defines the significance of pedagogical work and the means for solving teachers' social problems – living conditions, rest and recreation, qualifications, social position and incentives.

The system of labour incentives for teachers also includes obtaining qualification categories – first and second. Every teacher with experience and solid theoretical knowledge has the right to a qualification category, the award of which is made by the special commissions after an inspection of his work and training.

Student members of the astronomy club in the village of Avren, in the Varna district, on a visit to the observatory

The terminal training room at the Higher Institute of Mining and Geology in Sofia

Those who have acquired a first or second qualification category get additional pay.

The Government and People's Councils are responsible for settling the housing problems, rest and medical treatment of the teachers. Those working in remote regions receive additional pay, free accommodation, free heating, etc.

Teachers, and lecturers in higher institutes, together with these working at the Bulgarian Academy of Science, are united in a trade-union organization.

The Union of Bulgarian Teachers is a professional organization of Bulgarian teachers from educational institutions of all kinds and levels, including workers at the Bulgarian Academy of Science. It was founded in 1962 on the initiative of the President of the State Council and General Secretary of the BCP, Todor Zhivkov. It has inherited and continued the traditions of the Teachers' Social Democratic Organization (founded 1905), the Teachers' Professional Opposition (1930) and the Union of Workers in the Sphere of Education (1944). It is responsible for improving the Marxist-Leninist, scientific and pedagogical training of the teachers. In co-operation with the Ministry of Education, it organizes refresher courses to further the qualifications of the teachers, and investigates and popularizes valuable experience in the field of education, exploring the achievements of Soviet pedagogy and those of other countries and applying them creatively to Bulgarian education. It promotes the image of the teachers in the spirit of the Bulgarian Teachers' Charter and maintains relations with other creative and professional unions to ensure their co-operation in educational matters. Eight commissions have been formed at the Central Committee of the Union, each of them dealing with different aspects of the Union's work: organization, competition, welfare, research carried out in the higher educational establishments and the Bulgarian Academy of Science, professional skills, ideology and recreation. The Congress is the supreme body of the Union, and in the interval between congresses, the Central Committee and the Bureau elected by it act as its executives. The district committees are the local bodies and the union organization in each separate educational institution is the smallest unit. Together with the Ministry of Education, the Union publishes its newspaper *Uchitelsko Delo* (Teachers' Work) and the magazine *Narodna Prosveta* (People's Education).

The Union is a member of the Bulgarian Trade Unions and in 1946 joined the International Federation of Educational Trade Unions. It maintains relations with similar organizations from socialist and other countries.

Higher Education. Higher education provides highly qualified, highly intellectual and socially active experts and scientific workers for all spheres of social life. It originated from the literary schools of the period from the ninth to the fourteenth century in which the training of high state officials was accompanied by intense literary, cultural and educational activity, and theology, rhetoric, logic and philosophy were taught and developed.

During the time of the Bulgarian Revival, which was characterized by an upsurge in education, the formation of the national consciousness and the struggle for an independent State, the sons and daughters of the emerging bourgeoisie (tradesmen, rich landlords, craftsmen) received their higher eduation in advanced European countries like Germany, Italy, France and Russia.

After the Liberation in 1878, the resulting changes in

The Kliment Ohrid University in Sofia

the socio-economic development and the social structure in Bulgaria necessitated improvement in the levels of education and culture. A new Higher Pedagogical Course was founded in Sofia in 1888, its name being changed to Higher School in 1889 and University in 1904. At first there was a faculty of History and Philology, a faculty of Physics and Mathematics and a faculty of Law; then faculties of Medicine (1917), Agronomy (1921), Theology (1923), Veterinary Medicine (1923) and others were added. 1901–2 was the first academic year when the university admitted women among its students. The capitalist development of Bulgaria, the harsh home and foreign competition in economy and politics, and the growing need for a higher level of culture prepared the way for the foundation of the Balkan Near-East Institute of Political Sciences (1920), which, having been reformed several times, in 1952 finally became the Karl Marx Higher Institute of Economics in Sofia. Higher Schools of Economy were also established in Varna (1921) and Svishtov (1936). In 1896 a State Art School for the training of artists was founded (known as the Academy of Art since 1921, and Higher Institute of Fine Arts since 1954). Musical training could originally be received at a private school (founded in 1904), which was to become state-owned in 1912 and developed into an Academy of Music in 1921 (the Bulgarian State Conservatoire since 1954). In 1942 the Higher Technical School was founded (State Polytechnic since 1953), as well as the Higher School for Physical Instruction (the Higher Institute of Physical Education since 1953). Up to 1944 only eight higher-education institutions existed in Bulgaria.

After the victory of the socialist revolution in 1944, higher education was reformed on the basis of Marxist-Leninist principles and became widely democratic. It became free and accessible to all citizens of the People's Republic of Bulgaria without distinction of sex, nationality, religion, social or property status. In accordance with the Higher Education Acts (1947, 1958) new departments of dialectical and historical materialism and postgraduate courses were set up. Socialist higher education aims at forming workers that have profound scientific, Marxist-Leninist and social background knowledge, quick, independent thinking and the ability to make constructive theoretical and practical decisions. The differentiation of higher-education institutions and the list of subjects taught are determined by the laws and requirements of social production and its future prospects, by the level and structure of modern science and by the rapid advance of science and technology. With the Tenth Congress of the BCP, higher education started changing its content in keeping with the requirements of advancing science and technology and the needs of the well-developed socialist society in Bulgaria. The production and new technology adoption units of the Bulgarian Academy of Science and similar research institutes co-operate closely with the higher schools. This helps the efficient use of necessary equipment, teachers and research workers. To receive higher education without losing touch with the production process, extra-mural and evening courses were introduced in 1948 and 1960 respectively. Students have the right of extra, fully-paid leave to attend lectures and prepare for semester and final exams. Part-time graduates enjoy all the rights of the graduates of full-time courses.

The higher education institutions are divided into several groups: *universities,* where a wide range of subjects is taught, grouped together in faculties; *academies* – specialized or comprehensive institutions; institutes for special subjects; and *higher schools and colleges* designed to service certain Ministries and departments. Higher education is governed by the Ministry of Public Education. Besides the Academy of Social Sciences and Management in Sofia, six higher military schools and one Theological Academy, there were 29 civil higher educational institutions in Sofia during the 1982–83 academic year. Prospective workers in the spheres of education, administration and economy are trained at the Kliment Ohrid University in Sofia, the University in Veliko Turnovo named after Cyril and Methodius, the Plovdiv Paisii Hilendar University, the Higher Institutes of Pedagogy in Shumen and Blagoevgrad, and the Higher Institutes of Economics in Sofia, Varna and Svishtov. There are ten technological higher schools meeting the requirements of industry: higher institutes of electrical engineering in Sofia, Varna and Gabrovo, the Higher Technical School in Russe, the Higher Institute of Architecture and Construction in Sofia, the Higher Institute of Silviculture in Sofia, the Higher Institute of Mining and Geology in Sofia, the Higher Institutes of Chemical Technology in Sofia and Burgas, and the Higher Institute of Food Chemistry in Plovdiv. Specialists in agriculture are trained at two higher institutes: the Higher Institute of Agriculture in Plovdiv and the Higher Institute of Zootechnology and Veterinary Medicine in Stara Zagora. There are Higher Institutes at the Medical Academy in Sofia, Plovdiv, Varna, Pleven and Stara Zagora. There are also four higher institutes of the arts: the Bulgarian State Conservatoire, the Higher Institute of Theatrical Art, the Higher Institute of Fine Arts, and the Higher Pedagogical Music School in Plovdiv. The training of sports instructors, trainers and specialists in mass health gymnastics is performed at the Higher Institute of Physical Education and Sport in Sofia. Since 1951, a Theological Academy for training priests of the Bulgarian Orthodox Church has functioned independently.

Students are admitted after competitive entry examinations. The subjects of the exams differ according to the professional trend and specialization. The entry examinations include two special subjects and sociology. Besides the grades obtained in the examinations, the marks in certain subjects during the secondary course of study and the average score of the school-leaving certificate are also taken into account in deciding admission to university or higher institutes. Having been assessed according to these criteria students with the highest marks are admitted, their number having been planned beforehand. Every year approximately 13,500 students get into the higher institutions on regular study-courses, and about 4,500 for extra-mural or evening study. Among the specialized higher schools is the Institute for Foreign Students named after G.A. Nasser in Sofia, where foreigners learn the Bulgarian Language and other special subjects, in preparation for future entry into Bulgarian technical schools and higher institutions. Training lasts a year. There are similar centres attached to the universities in Plovdiv, Russe and Burgas.

Special courses, seminars and schools are organized for the foreign students after they enter the higher institute. The living conditions of foreign students are good; they share hostels with fellow-students from Bulgaria and have equal rights and obligations. They are trained according to the general syllabus, but there are special extra courses which concentrate on the characteristic climatic, socio-political and economic conditions of their own countries.

The International Friendship Club organizes mass cultural and sporting events. Here the young men and women present the history and the achievements of their countries to the other students and to the Bulgarian public. The number of foreign students is steadily increasing. During the academic-year of 1983–84, approximately 9,000 students and 750 post-graduates from 103 European, African, Asian and Latin American countries studied in Bulgaria.

The Thesis on the Development of Education in the People's Republic of Bulgaria of 1979 established a new nomenclature of the professions, skills and specializations, corresponding to the modern development of science, technology and the anticipated demands for university graduates by the various fields of public activities. Each professional line covers a group of related specialisms which require a unified, basic scientific training. The nomenclature consists of 31 professional lines with 134 specialisms. Arts students are distributed in 32 specialisms. In the *first stage* of the higher educational establishments the students acquire a uniform general-theoretical fundamental training and a general-theoretical training in a specific professional line. The purposeful mastering of practical skills and habits goes alongside mastery of the indispensable, fundamental theoretical knowledge. This first stage lasts about one and a half to two and a half years. The *second stage* lasts for two years and guarantees broad vocational training for the specialists. During this stage the students master specific theoretical, technological, constructive, economic, managerial knowledge and skills, which determine their performance in a broad vocational sphere. Their training is organized in the higher educational establishments and in the fields of science, production and services.

The course in the first and second stage is conducted in compulsory, optional and elective subjects. This contributes to the advancement of the students' talents and to their further individual improvement. The course structure in the first two stages provides mobility and opportunities for comprehensive education of the specialists in accordance with the continuously changing requirements of practical life and scientific and technical progress. Having completed the second stage, the students are assigned to specific occupations and work-places in branches of the national economy.

The *third stage* lasts for about one year. It guarantees high professional training in a specific sphere or profession, which is acquired in actual production enterprises or scientific-research establishments. The education covers compulsory and optional subjects, the latter having priority. The students receive a diploma of university education after a successful completion of the third stage and passing the state exams, and all graduates are provided with an occupation in their respective spheres. *Scientific-research* work is carried out in the higher educational establishments alongside the scientific educational activities. All higher educational establishments are equipped with task-orientated and branch scientific research laboratories, which work on fundamental and topical problems of higher education. They test and introduce new forms, methods and technical devices in the educational process, put scientific projects into practice, and supervise post-graduate qualification and the upgrading of the specialists. A significant number of the country's scientists are trained in the higher educational establishments through regular, extra-mural and optional post-graduate studies. The selection and enrolment of lecturers is decided on the basis of competitive examinations. The scientific degrees within the system of higher education are assistant lecturer, lecturer, associate professor and professor; conferring the higher degrees of Candidate of Science and Doctor of Sciences follows the defence of a thesis.

The Bulgarian higher educational establishments maintain close links with related foreign institutes. Exchange of lecturers, postgraduates and students; joint elaboration of papers; organization of international congresses, symposia

The student refectory at the Cyril and Methodius University in Veliko Turnovo

Students at the Kliment Ohrid University in Sofia

and seminars; meetings for exchange of experience, information and literature; and organization of students' brigades and excursions are among the activities organized in this connection.

The students enjoy a wide range of social benefits. About 50 per cent of the students are granted scholarships and accommodated in hostels. The value of the scholarship varies from 40 to 80 levs per month. Student-mothers are entitled to 80 levs and a supplementary monthly grant of 20 levs for each child. Half the cost of the food in the students' canteens is paid by the State. The students are entitled to a discount of 30 per cent on Bulgarian State Railways and Bulgarian River Navigation: during the summer and winter holidays they enjoy the facilities of rest homes, sports and health recreation camps at a minimum rate. They are entitled to free medical services.

Facilities and Management of Education. A special programme on the development of the material basis of education has been adopted, and a comprehensive conception of ideal classroom and architectural and environmental conditions for educational establishments is now being put into effect. This applies to all types of educational buildings; kindergartens and nurseries, schools, higher educational establishments, etc. Industrial methods are mostly applied in their construction, since they provide the opportunity for maximum unification and typification of the construction elements, and guarantee flexibility of planning. The regional and micro-regional centres to a certain extent determine the form of the school buildings. 147.9 million levs of capital investment were expended on the construction of educational buildings in 1981. In terms of urbanization, the tendency is for the school premises to be situated close to the public centre, so that the new social functions of the school may be best realized.

Amendments to the management of education were necessitated by the new educational reform now being implemented. The state-public system of management is structured on a highly democratic basis: the elected Supreme Educational Council, the Council for Higher and Secondary Special Education, the Council for Career Guidance, for Professional Training and In-Service Training, the Council for Co-ordination of Scientific Research and the Scientific Back-up of Education, function at the Ministry of Public Education. Officials from the fields of science and culture, pedagogical experts, public officials and representatives of economic organizations, state bodies and public institutes are represented on them.

District Educational Councils are elected at the District People's Councils; Municipal and Ward Educational Councils at municipalities and wards; and Boards of Trustees at schools.

These bodies fulfil co-ordinating functions and involve the participation of all social sectors, organizations and institutions in the solution of educational problems.

The Supreme Educational Council is a state-public body, which documents, discusses and solves strategic issues on the content, organization and management of education. The Council considers educational plans, curricula, scientific activities in education, construction material and technical bases of education at its periodic plenary sessions. The Supreme Educational Council is headed by a Bureau, the chairman of which is the Minister of Education. The Bureau is composed of representatives of the Ministry of Public Health, the Committee for Culture, rectors of higher educational establishments, headmasters of schools, etc., besides the members of the leaderships of

The student township in Veliko Turnovo

Table 8: Structure of Higher Education in the People's Republic of Bulgaria

Fifth courses	**THIRD STAGE** Provides high professional training in a certain specialism
Third/Fourth	**SECOND STAGE** Provides training on a broad vocational basis
First/Second	**FIRST STAGE** Provides fundamental, general-theoretical training in a specific vocational line

all other Ministries. The other councils at the Ministry of Education constitute auxiliary bodies to the Supreme Educational Council and its Bureau. They discuss and solve specific problems of secondary and higher education, teacher training and upgrading, and the scientific back-up of education and other issues.

The Educational Councils at the District and Municipal People's Councils are in charge of the implementation of state policy in the field of education. They fulfil organizational, efficiency, planning, co-ordinating and controlling functions in the course of exercising direct leadership over educational activities in the respective districts of the People's Councils (higher and semi-higher educational establishments excluded). Within the limits of their rights and authority, the Educational Councils fulfil managerial functions and adopt decisions on all problems related to the organizational-pedagogical, educational-methodological staffs and material and technical needs of education. They are also concerned with meeting the housing problems of the teachers and the communal services needed by the teachers and students from the appropriate region, for which the finances are provided by the people's council budget. Their activities and maintenance are in accordance with the requirements of the economic mechanism and its implementation in the field of education.

The Educational Councils (District or Municipal) are elected for a three-year term at meetings of representatives of the educational, state, economic, political and public bodies and organizations in the area of each People's Council. The meeting itself decides whether or not the election will be carried out by secret ballot.

There are both elected and permanently appointed members on the Educational Councils. The number of elected members is decided at a conference: 35 to 65 people for the District Educational Councils and 15 to 35 people for the municipal ones. There is a fixed number of permanently appointed members. The elected members of the Educational Councils also function as a collective management body for the organizations and their sections of the educational system in the area.

The number of teachers and workers in the field of education should not exceed 50 per cent of the elected

The Hristo Botev student township in Sofia

members of the Educational Councils. Each Educational Council elects a chairman and an executive board of five to fifteen members for a district council and of five to eleven members for a municipal one. The chairman is either a person who is working in the field of public education or a respected social, political or state figure.

The boards of trustees are public bodies attached to the kindergartens and schools. They discuss and solve any basic problems that might arise, with the active assistance of state, economic and public bodies and organizations and the public at large.

The board of trustees is elected by a show of hands at a general meeting held jointly with representatives of the local party, public, cultural, state and economic bodies and organizations, the staff, the parents' committees of each class and the Komsomol office-bearers from the school. Meetings to elect boards of trustees are called by the Municipal Educational Council with the active assistance of the staff of the schools (the school managements). The boards are elected for a three-year term and are directed by the Municipal Educational Council. Representatives of the public, educational workers, people active in the field of science and culture, managers, workers, parents and other citizens representing state bodies and public organizations are elected; the number of trustees ranges from five to fifteen.

The board of trustees elects a committee consisting of a chairman, a deputy chairman and a secretary-treasurer. It works according to an annual plan and in sessions held at least once every two months.

Ad hoc or standing committees may be constituted in order to improve educational work, facilities, financing, parent-teacher collaboration, etc. Their basic task is to discuss, take decisions and make proposals to the respective bodies and organizations on the development of the kindergartens and the schools; to improve the facilities and train teachers; to improve the conditions for educational work; to organize and carry out extra-curricular and out-of-school activities; to organize career guidance and vocational training for pupils; to manage the planning, production and sale of items produced by pupils; and to find jobs for school-leavers. It directs the activities of the parents' committees for each class, co-ordinates public activities in the area served by the school with a view to developing, maintaining and up-dating facilities for the kindergartens and schools; helps to organize canteens, transport and other services for pupils and teachers, assists the school, economic and public bodies and other organizations in making arrangements for the recreation of children and students and their participation in voluntary work brigades; and launches initiatives aimed at raising the level of pedagogical awareness of parents and trustees. Together with the headmaster or headmistress of the school, it aids the activities of the parents' committees; and together with the local public it organizes the building of sports grounds and children's playgrounds, sports halls, clubs, recreation centres and other facilities for the communist education of the pupils. The chairman of the board of trustees is an *ex officio* member of the teachers' council and has the right to put forward proposals on behalf of the board at the Municipal Educational Council so as to improve the functioning of the kindergartens and schools and stimulate their further development.

The most important school holiday – 24 May – is an expression of the Bulgarian people's respect for the founders of the Bulgarian alphabet and literature, Cyril and Methodius.

24 May – Day of Bulgarian Education and Culture, Day of the Slav Alphabet and its Creators Cyril and Methodius and Day of the Bulgarian Press. This day is an official holiday, celebrated by all Bulgarian people. The work of Cyril and Methodius was honoured as early as the ninth century when they were canonized. Though initially it was honoured by the Church, 11 May (24 May on the Gregorian calendar) has never been a purely religious holiday; above all the two brothers have always been famous as the founders of Slav letters. Having started in Bulgaria, this festival of the Slav enlighteners spread to the other Slav peoples as well, especially the Russians and the Serbs. During the periods of Byzantine and Ottoman domination the tradition was preserved and helped to keep alive the national consciousness of the Bulgarians. During the Revival the celebration of the first Slav teachers Cyril and Methodius was an expression of thanksgiving for their work. The first celebration of 11 May as a school holiday took place in 1851 in the Sts. Cyril and Methodius eparchial school in Plovdiv. For four to five years 11 May was a nationwide and school holiday. On that day the portraits of Cyril and Methodius and the school buildings were decorated with flowers, and festivities at which the teachers made stirring speeches were held. After the Liberation in 1878, the commemoration of the work of Cyril and Methodius became associated with the struggles of the Bulgarian people for the democratization of

Schoolgirls gather for the 24 May Parade, Sofia, 1968

education and culture. In 1892 Stoyan N. Mihailovski wrote the words of the anthem, 'Onward, nation reborn', and in 1901 Panayot Pipkov composed the music. During the years of Fascism, the progressive forces in Bulgaria were inspired by the words of Georgi Dimitrov, who in 1933 pointed out the importance of what Cyril and Methodius had done by saying: 'At a period of history when the German Emperor Karl V vowed that he would talk German only to his horses, at a time when the nobility and intellectual circles of Germany wrote only Latin and were ashamed of their mother tongue, in "barbarous" Bulgaria the apostles Cyril and Methodius invented and spread the use of the old Bulgarian script'.

Today the 24 May is a very important day, since in 1957 it was declared an official holiday, and many exhibitions, competitions and recitals are organized to commemorate it. Every year, on 24 May, people who have distinguished themselves in the arts, science, culture and education are honoured officially, by being given the order of Cyril and Methodius, or receiving the titles *naroden* (people's) or *zasluzhil* (honoured).

SELECTED BIBLIOGRAPHY

ANGELOV, T. *The development of Secondary Education in the People's Republic of Bulgaria.* Sofia, 1980.

Istoria na obrazovanieto i pedagogicheskata misul v Bulgaria (The History of Education and Pedagogical Thought in Bulgaria), Vols. 1–2. Sofia, 1975–82.

KONEV, I. *Bulgarskoto Vuzrazhdane i Prosveshenieto* (The Bulgarian Revival Period and the Enlightenment). Sofia, 1983.

ZHIVKOVA, L. *Visokata misiya na obrazovanieto – otgovornost pred budeshteto* (The High Mission of Education as a Responsibility to the Future). Report to the First Congress of People's Education, 12 May 1980. Sofia, 1981.

ZHIVKOV, T. *Obrazovatelnoto delo – delo vsenarodno* (Education as an All-National Task). In two vols. Sofia, 1980.

PART VIII

SCIENCE

THE DEVELOPMENT OF SCIENCE PRIOR TO 1944

The first elements of scientific knowledge in Bulgaria are contained in the oral tradition, heroic epics and the myths about the earliest history of the Thracians, Slavs and Proto-Bulgarians inhabiting what is present-day Bulgaria. After the Bulgarian State was founded (681), public life also found expression in written monuments. Scientific thought and activities developed after Cyril and Methodius created the Slav script and after the appearance of Bulgarian literature (mid-ninth century). Their disciples Kliment of Ohrid, Naum of Ohrid and others founded the Ohrid and Preslav Literary Schools, which marked the beginning of teaching and the dissemination of knowledge. A number of scientific achievements were popularized and penetrated also into other Slav countries (Russia, Serbia) through the Old Bulgarian literature, the oldest Slav literature. Different domains of knowledge – theological, philosophical, historical, literary and natural-scientific – began to develop. Constantine Cyril the Philosopher gave the first definition of philosophy among the Slav peoples. Together with his brother Methodius he created the first philosophical terminology in the Old Bulgarian language, which comprises many generic concepts and categories. The works of Konstantin of Preslav, Chernorizets Hrabr, John the Exarch, Presbyter Kozma and others reflect the consciousness and self-confidence of the already formed Bulgarian ethos. The first Bulgarian chronicle *The Name-List of the Bulgarian Khans* – contains historical information. Evidence about the history of the Bulgarian State and on legislation is synthesized in one of the oldest Bulgarian acts of legislation: the Law on Justice for the People. Individual works of the Old Bulgarian literature testify to the interest of the educated circles of Bulgarian society in secular knowledge. The *Hexameron* (tenth century) by John the Exarch is one of the most remarkable achievements in Bulgarian medieval culture. This is the first encyclopaedic philosophical-theoretical book in the Old Bulgarian language, comprising knowledge about the world and about the essence and properties of the cosmic powers and atmospheric phenomena. The author created and established a large number of the categories, concepts and terms of the modern philosophical-theoretical and scientific language. Bulgarian medieval science developed remarkably during the fourteenth century, when secular work began to be disseminated parallel with the religious literature. Books containing knowledge about the nature surrounding man were translated from Greek. One of the most remarkable translated works is Manasius' *Chronicle* (fourteenth century), in which the nineteen comments added by the Bulgarian translator to elucidate events in Bulgarian history up to the reign of the Assens are of extreme interest and importance. In the second half of the fourteenth century Patriarch Euthymius of Turnovo founded the Turnovo Literary School in which he and his disciples discussed theological problems on a Hesychast basis, introduced a language reform, developed extensive literary and educational activities and raised the prestige of Bulgarian medieval culture to a high level. The character of philosophical thought during this age was shaped by the social processes and by Eastern Orthodox Christian theology. The works of the great thinkers of that time formulate, interpret and solve cosmological, ontological, epistemological, ethical and natural-scientific problems. Economic views on feudal property, taxes, trade and money, took shape. The first attempts at musical-theoretical generalizations were made by Yoan Kukuzel, one of the most remarkable singers, musicians, composers and theoreticians of his time. The fall of Bulgaria under Ottoman domination (1396) dealt a heavy blow to science and culture. Monasteries were the only places in which the created literature was hidden and preserved, and which continued the tradition of writing and translating books. The centres of spiritual life were the Rila and Bachkovo monasteries and the Sofia Literary School.

The Bulgarian National Revival period (eighteenth–nineteenth centuries) was the time of comprehensive socio-economic and political processes which brought about the great upsurge of national power and the systematic development of scientific knowledge. The beginning of Bulgarian historiography was marked by *Istoriya slavenobolgarskaya* (Slav-Bulgarian History; 1762) by Paissii of Hilendar, which played a major role in the formation of the national consciousness and in kindling the national church and political struggle of the Bulgarians. Secular education emerged and developed. European science began to be actively mastered and put into practice. The most intelligent Bulgarian youths went to study in the universities of Europe. Some of the eminent figures of the Bulgarian National Revival period popularized the achievements of the nation by translating articles and books and by enriching the Bulgarian language with new scientific concepts and terms. A young intelligentsia was born and represented the emerging Bulgarian progressive scientific thought to other countries and nations. Since conditions for scientific research were non-existent in the country, Bulgarian scientists at that time worked predominantly outside the confines of the Ottoman Empire. Most of them received doctor's degrees, joined various scientific circles, collaborated in various research institutes and published scientific works in other countries. Stefan Bogoridi (1775–1859) became corresponding member of the Turkish Academy when it was founded in 1851, Nikola S. Piccolo (1772–1865) was elected a foreign member of the French Academy of Sciences in Nancy (1853). Contributions to the history of Bulgarian science include the first dissertations of Atanas Bogoridi (1816), Peter Beron (1831), Dimiter Mutev (1842), and others. The most outstanding representative of Bulgarian scientific thought during the Bulgarian National Revival period was the scientist-encyclopaedist Peter Beron (1800–71), who was a member of many European scientific societies (Paris, Athens, etc.). As a result of the generalization and classification of the achievements of the natural and humanitarian sciences, he created his own natural-philosophic system, presented in *Panepistemia* (in French, 7 volumes, 1861–67). P. Beron took an active part in international scientific life: he delivered scientific lectures in Athens, Paris, before the Royal Society of Arts in London in 1850, and elsewhere. In 1824 he published the *Riben Bukvar* (Fish Primer) – the first secular textbook, which had a reforming influence in public education. The most important work on natural-scientific subjects, published in the first half of the nineteenth century, was the book *Conclusions from Physics* (1849) by Naiden Gerov. Other independent publications of scientific character also appeared. The first Bulgarian bibliographic edition appeared in 1852 – *List of Bulgarian Books in the Newly-Revised Bulgarian Literature of the Nineteenth Century* by Ivan Shopov. In 1853 Nikola Ikonomov published the first Bulgarian book on agronomy – *Agriculture*. The work of Nikola Toshkovich, *Practical Notes on Ships* (in French, 1860) was the first major manifestation of Bulgarian technological thinking. The works of Spiridon Palauzov marked the beginning of modern Bulgarian historical science. Marin Drinov contributed publications on the origin of the Bulgarians, on the Slav settlement on the Balkan Peninsula and on other historical matters. The first Bulgarian military book – *Manual for Successful Fighting against the Turks,* by General Ivan P. Kishelski – appeared in 1876 in Bucharest. Studies started on the scientific creativity and the ethnic characteristics of the Bulgarians. Several large folkloric volumes were published, the most important among them being *Bulgarian Folk Songs* (1861) by the brothers Dimiter and Konstantin Miladinov. During the National Revival period there appeared publications on the history of music, as well as information and reviews about Bulgarian and foreign events and phenomena. Nikolai Pavlovich wrote articles on Bulgarian fine arts. Specialized translated and compiled editions also appeared: *Experimental Physics* (1869, translated from French by Yoakim Gruev), *Animal Medicine* (1869, translated from Turkish), *Agricultural Inorganic and Organic Chemistry,* (1871, translated from Russian by D. Enchev), *Elementary Literacy in Two Courses* (1873, translated from Russian by Todor Shishkov), and *Literacy Manual* (1874, compiled after French models by Dobri Voinikov).

Marin Drinov

The striving for encyclopaedic knowledge, typical of the National Revival period, found expression in the periodic Press as well. Many outstanding figures of that time emphasized in their writings the great social significance of the Press. Original philosophic ideas, freed of religious elements, were created during the National Revival period. Materialistic philosophy became the basis for the ideology of the progressive social forces. The eminent figures of the Bulgarian national-liberation movement, Georgi Rakovski, Lyuben Karavelov, Vassil Levski and Hristo Botev created a system of views on the economic and political liberation and development of Bulgaria. The philosophical and sociological ideas of L. Karavelov are related to the social problems of the national-liberation struggle and the cultural-educational upsurge of the people.

Foreign scientists, travellers and researchers had a strong impact on Bulgarian scientific thought in the eighteenth and nineteenth centuries. They published numerous articles in the European periodic Press and wrote scientific works and travel notes on Bulgarian themes. The French natural scientists A. Boue, O. Viquenelle, the Austrian geologists F. Hochstedter, F. Toula and others undertook the first geological prospecting. The Hungarian scholar I. Friewaldski, the German botanists A. Griesebach and H. Dingler and the Swiss botanist P. Boissier were the first to study Bulgarian flora. The Hungarian scholar Felix Kanic gave information about the physical geography of Bulgaria, about its antiquities and the social relations and the way of life of the Bulgarians. The Czech researchers Konstantin Jireček, the Skorpil brothers, as well as Václav Dobruský laid the foundation of archaeological study in Bulgaria and of the preservation of the antiquities. The Slav scholars Vuk Karadžić, Yuri Venelin, Stefan Verković and I. Grigorovich evoked scientific interest in Bulgarian folklore and written heritage. The Austrian scholar W. Tomaschek, the Slav scientists J. Kopitar, F. Miklošič, A. M. Selishchev, Nikolai S. Trubetskoi and others were the first to tackle a number of problems connected with the Bulgarian language. The studies of the foreign scholars contributed in bringing Bulgarian culture closer to European culture.

The development of Bulgarian culture gave rise to the need to create a centralized organization for education and science, such as the Bulgarian Literary Society (founded 1869, since 1911 – the Bulgarian Academy of Science). The main aim of the Society was to study the history of the Bulgarians and the Bulgarian territory, the Bulgarian language and literature, flora and fauna, and Bulgarian geography, geology, mathematics and physics. From the very beginning of its existence, the Society contributed to the development of scientific thought and became a stronghold of Bulgarian nationalism. Professor Marin Drinov was the first President of the Bulgarian Literary Society.

After the Liberation from Ottoman rule (1878), scientific thought began to develop rapidly. The Bulgarian Academy of Science became a national scientific centre with three branches: historical-philosophical, philosophical-social and natural-mathematical. Its main objectives were to create the scientific and ideological basis for the unification of the Bulgarian nation, to study scientifically its history, way of life and culture, to develop the modern Bulgarian language and literature, and to carry out research in the field of geology and the flora and fauna of Bulgaria. Scientific societies were formed on the principle of public service and have contributed to the development of various branches of science. They organized discussion of scientific achievements. Research activities also flourished in the Higher School founded in 1888 – the present-day Kliment Ohrid University of Sofia. Other centres of scientific activities were also formed: the Balkan Near-East Institute for Political Sciences (1920) in Sofia, higher business schools in Varna (1921) and Svishtov (1936), the State Music Academy and State Arts Academy in Sofia (1921) and the State Higher Technical School in Sofia (1942). The lecturers in these educational institutions were involved in research in all spheres of science.

Nikola Obreshkov

Capitalist Bulgaria offered very limited possibilities for the development of science. In spite of the difficulties, the Bulgarian scholars succeeded through their efforts, enthusiasm and love of science, in creating Bulgarian schools in the fields of mathematics, physics, chemistry and the arts, laying the foundations of many branches of science.

Research in the field of mathematics concentrated on the composition of quadratic forms, mathematical problems of external ballistics and differential equations. Nikola Obreshkov defined new methods for the summation of divergent series and expanded the Cartesian theorem on complex roots. Problems of the roots of polynomial geometry were also investigated. In the 1920s and 1930s the greatest achievements can be credited to Georgi Nadjakov (1896–1981), who studied photoelectric phenomena in broad-band conductors, electrometry and discovered the photo-electret state of matter (the first Bulgarian discovery). The foundations were also laid for research in the field of radio-activity and nuclear physics, and of some new high-frequency electric oscillations in electronic lamps and magnetrons and of colour centres in ionic crystals. The first publications in Bulgaria on the theory of relativity and quantum mechanics appeared. Ivan N. Stranski (1906–72), Lyubomir Krustanov (1908–77) and Rostislav Kaishev (1908) laid the foundations of the molecular-kinetic theory of crystal growth. Pencho Raikov (1864–1940) was the founder in Bulgaria of the oldest branch of chemical science – organic chemistry. Assen Zlatarov (1885–1936) worked in the sphere of bromatology and biochemistry. Dimiter I. Popov (1891–1975) and Alexander V. Spassov (1905) discovered and developed a new type of mixed organomagnesium reagents. The composition of Bulgarian essential oils and waxes was studied. Research was carried out on the analysis and geochemistry of rare gases. Dimiter Balarev (1885–1961) created the theory of the thermodynamic foundations of the disperse structure of real-crystal systems. Dimiter Ivanov (1894–1975) devised an original method for obtaining a wide spectrum of organic compounds which are difficult to synthesize, using polyfunctional organomagnesium and organolithic reagents, known in scientific literature as 'Ivanov's reagents'. In 1935 Dragomir Mateev (1902–71), Hristo Petrov and V. Schwarz discovered gravitational shock in man. Atanas Ishirkov (1868–1937) developed the study of the historical, political and settlement geography of Bulgaria. Anastas Beshkov (1896–1961) was the first to criticize the reactionary nature of anthropogeography, absolute chorologism and geopolitics. Seismology, meteorology, climatology and synoptic meteorology also underwent development. Research was carried out on atmospheric physics and atmospheric turbulence. Georgi Zlatarski (1854–1909) was the author of the first Bulgarian geological work on ores in Bulgaria (1882) and of the first geological map of Bulgaria. Stefan Bonchev (1870–1947) was the founder of the Bulgarian geological school. The works of Georgi Bonchev (1866–1955) on the magmatic, metamorphic and sedimentary rocks in Bulgaria are of great scientific value. Problems of mineralogy, stratigraphy, petrology, geotectonics and palaeontology were investigated. The first significant achievement in the sphere of general biological sciences was the research on experimental anabiosis in insects, carried out in 1895 by Porfiri I. Bahmetiev (1860–1913). Metodi A. Popov (1881–1954) was the author of an important textbook of general biology (1919) and also created stimulation theory. Differentiated studies on groups of plants began. Works of a generalized nature appeared in the field of ecology and geobotany. The first Bulgarian work in zoology – *Materials on the Study of Bulgarian Fauna* by Georgi Hristovich (1863–1926) – was published in 1892.

Research in the agricultural sciences was concentrated mainly on studying the local plant species and varieties, the acclimatization of imported species and varieties and their protection against diseases and pests, investigation of the soil, and the working out of systems for its cultivation and fertilization. Work was being done on problems in the diagnostics and prevention of infectious diseases, as well as on the production of sera and vaccines. Partial studies were carried out in forest meteorology and phenology, forest taxation and forest cultures. Differentiation in health care also contributed to the development of clinical medicine, but the research was not related to the co-ordinated solving of vital problems.

Rostislav Kaishev

The capitalist production relations after the Liberation from Ottoman rule defined the character and trends in the development of the social sciences. Bourgeois philosophical, sociological, economic, psychological and aesthetic thought was shaped under the impact of the idealistic theories and schools of Western Europe. At the end of the nineteenth century Marxism began to spread and Dimiter Blagoev (1856–1924) laid the foundations of Bulgarian Marxist philosophy, economic sciences, history, sociology and theory of literature. His scientific and theoretical activities were closely connected with the creative application of Marxist ideas to explain Bulgarian social development. Eminent Marxists of this period are: Georgi Kirkov (1867–1919), Georgi Dimitrov (1882–1949), Vassil Kolarov (1877–1950) and others. The philosophical sciences were characterized by the formation toward the end of the nineteenth century of various idealistic trends, such as Neo-Kantianism, Berkeley's philosophy and psychophysical parallelism and, after the First World War, Bergsonism and Freudism. The philosophy of Johannes Rehmke, represented by Dimiter Mihalchev (1880–1957), exercised the strongest impact. Todor Pavlov (1890–1977), Georgi Bakalov (1873–1939), Sava Ganovski (1897) and others propagated Marxist-Leninist principles in explaining the ideological life of society. The main problem treated was the epistemological one, on the basis of the theory of reflection (T. Pavlov).

Metodi Popov

Sava Ganovski

Problems of aesthetics and social development were also dealt with.

Sociology also made the first attempt at concrete empirical studies (Iliya Yanulov), parallel with general idealistic theoretical concepts. Marxist sociologists took a creative approach and advocated the scientific theory of society, guiding the Bulgarian revolutionary movement along the path of the struggle for Socialism. In the 1930s Todor Pavlov was to define the specific domain of Marxist sociology as the most general, non-philosophical science of society, developed on the basis of historical materialism and closely related to it (*Theory of Reflection. Basic Problems of the Theory of Knowledge,* 1938). The development of Marxist philosophy and the struggle for the creation and propagation of scientific psychology date back to the first half of the twentieth century (Ivan Hadjiski; 1907–44). Historic documents from the National Revival period were sought out and studied. The study of historical memoirs started to be developed. Historiography contributed to the study of facts and events in the political and cultural development of the Bulgarian people. The first important historical works appeared (Vassil Zlatarski (1886–1933). The consolidation of Marxism as a theoretical basis in the study of historical matters resulted in the correct elucidation of the socio-economic development and history of the national-liberation and revolutionary workers' movement in Bulgaria, and repudiated unscientific theories.

Studies in the field of ethnography and folklore appeared. Extensive studies on folk culture from all parts of the country were published in *Miscellany of Folklore and Ethnography* (1889; until 1913 entitled *Miscellany of Folklore, Science and Literature*). Theoretical economics formed the academic wing of bourgeois economic thought. The propagation of Marxism gave a new direction to economic scientific thought. The first translation of works by Karl Marx (1886) and Friedrich Engels (1890) were printed. The first Bulgarian Marxist publication entitled *What is Socialism and Can It Thrive On Bulgarian Soil?* by Dimiter Blagoev appeared in 1891.

Research was in progress in the fields of history of the Bulgarian State and legislation, constitutional and administrative law, civil law and procedural law. In spite of the wide scope and individual achievements of bourgeois law, the publications were predominantly of an informative, comparative and commentary nature. The formal dogmatic, normativist, positivist or eclectic methods of knowledge predominated in most of the studies. The representatives of Marxist thought in Bulgaria provided a scientific characterization of the Bulgarian Bourgeois State and developed socialist ideas, views and principles on the State and law.

The history and present-day state of the literary language and dialects, problems of Thracian studies, Slavonic linguistics, Indo-European studies, the modern analysis of the Balkan linguistic union and the central role of the Bulgarian language in it, were investigated by

Vassil Zlatarski

Alexander Todorov-Balan

Benyu Tsonev (1863–1926), Lyubomir Miletich (1863–1937), Stefan Mladenov (1880–1963), Alexander Todorov-Balan (1859–1959) and others. In the 1880s and 1890s literary criticism was predominantly cultural-publicistic in character. The first literary histories were published: *History of Bulgarian Literature* (1884) by A. N. Pipin and V. D. Spassovich, *History of Bulgarian Literature* (1887) by Dimiter Marinov (1866–1940) and *Bulgarian Literature* (1896) by A. Todorov-Balan. The demonstration of the original metrorhythmic structure of Bulgarian folk music, for which Dobri Hristov (1875–1911) created a consistent system and which he guided composers toward recreating was a great success. The notion of Bulgarian rhythm gained popularity abroad, and the formation of a Bulgarian national musical style was of primary significance. Until the 1940s publications on the science of dramatic art were mainly in the field of theatre criticism, which gradually became orientated towards theoretical-scientific analyses and evaluations of contemporary practices in the theatre. Critical and theoretical publications appeared, assessing possibilities for the development of the Bulgarian cinema, its character, and the interrelations of the cinema and the theatre, and pointing out the ideological superiority of Soviet cinema. Bibliographical references appeared during this period on the literature published in the fields of history, ethnography, law, economics and other sciences. Studies in bibliography were also published.

Technological thought developed, prompted by the demands of everyday practice. The Bulgarian Society of Engineers and Architects contributed greatly to the popularization of technological knowledge.

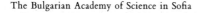

The Bulgarian Academy of Science in Sofia

SCIENTIFIC POLICY. THE ORGANIZATIONAL STRUCTURE AND MANAGEMENT OF THE SCIENTIFIC UNITS

The socialist revolution of 1944 opened up new vistas for Bulgarian science, and presented new tasks of unprecedented scope and dimensions, stemming directly from the objectives of the new stage in the country's development. The scientific front in the People's Republic of Bulgaria was not built overnight. It followed a complex and long process which started with the new Charter for the Bulgarian Academy of Science of 1947 and passed through a number of government decisions and acts which laid the foundations and formulated the basic principles of state policy in the sphere of scientific and technological progress.

The fundamental principles for the development of the system of management of science in Bulgaria and ways of its integration with production, outline the following main guidelines in scientific development: adherence to a selective strategy and the utilization of progressive foreign achievements; the concentration of the forces of the scientific front for the solution of important scientific problems and for the introduction of new equipment, advanced technology and the optimum organization of all branches of the national economy; the rapid introduction and utilization of advanced scientific and technological achievements from the rest of the world, the accelerated implementation of research findings and a shortening of the 'science-production' cycle; the intensive development of the scientific front and the improvement of the qualifications of the individuals engaged in scientific activities; the improvement of the organization and management of research activities and of the material and technological base of research; the dynamic planning of the correlation between fundamental and applied research and development projects, with priority to applied research and development projects and at the same time attaching great importance to fundamental research; closer contacts with the Soviet Union in all spheres; co-operation and economic integration with the CMEA member-countries in the fields of scientific and technological progress through intensified joint research and development activities in spheres of vital importance for economic and social development; and vigorous scientific co-operation with other countries and peoples.

The establishment of state bodies responsible for the unified management of the scientific and technological front is a prime concern in Bulgaria. The State Committee for Science and Technological Progress occupies an important position among these bodies. The Committee was founded in 1962 and developed as an organization putting into practice the national scientific and technological policy. The President of the Committee is a member of the Government. The Committee is responsible for a vast range of tasks: it defines the main trends in the development of science and technology, creates conditions for higher efficiency in research, guarantees rapid integration of scientific and technological achievements into the national economy, and organizes scientific and technological information. The Committee co-ordinates the solving of scientific and technological problems of national importance, working out complex national programmes for these problems. The decisions of the State Committee for Science and Technological Progress within its jurisdiction are mandatory for all Ministries and other organizations.

The Committee carries out its activities through scientific and technical councils. It comprises more than twenty co-ordination councils. The make up of the Committee includes a number of organizations which are instrumental in the implementation of the scientific and technological policy: the Central Institute for Scientific and Technical Information, the Central Institute for Inventions and Technological Innovations, the Progress Centre for the Intensive Implementation of Technology, the Technika Foreign Trade Organization and the Scientific Centre for the Organization, Management and Economy of Scientific and Technological Progress.

The Progress Centre for the Intensive Implementation of Technology was set up in 1977 with the aim of effective and rapid implementation of Bulgarian and foreign scientific and technological achievements. The achievements to be introduced are innovations which have not been implemented by the respective economic organizations either because no resources have been provided or due to the great economic risk. The Centre can also be defined as a specialized risk organization for the implementation of scientific and technological achievements. The activities of the Centre are planned on the basis of proposals submitted for implementation. The decision for the inclusion of these projects in the plan is made by the administration of the Centre in accordance with the criteria stipulated by the State Committee for Science and Technological Progress. The Progress Centre functions on a self-financing principle, the funds being supplied on the basis of contracts with the organization in which the development project is to be implemented. During the 1977–82 period the Centre signed more than 140 contracts for innovations which have been of great significance in technological progress.

The Central Institute for Inventions and Technological Innovations is responsible for all activities connected with inventions and technological innovations. A special concern of the Institute is the consolidation of inventions as a specific resource of scientific and technological progress for the effective updating of production. About 3,300 Bulgarian inventions were implemented during the five years 1976–80 and 750 in 1982. This demonstrates the ability of qualified Bulgarian researchers, engineers and technicians to generate their own scientific and technological ideas. As regards the number of inventions, Bulgaria ranks twelfth in the world and fourth among the socialist countries. Inventors in the fields of electronics, electrical engineering and the chemical industry are particularly active. For example, a new trend in Bulgarian electrical engineering developed on the basis of ten inventions, namely the production of high-torque electric drives, is a major export item. A system of five innovations made it possible to design small-sized oil circuit breakers for high voltages, corresponding to the highest technological standards. A number of Bulgarian inventions guarantee the technological superiority in the production of a wire-feeding device of the Isoplan type, which Bulgaria successfully exports to many countries.

The Technika Foreign Trade Organization is responsible for the export and import of licence rights, equipment and technology for research engineering activities.

Research and production enterprises with promising technological developments, such as the Polimerstroi Research and Industrial Complex, as well as the Centromet and Vacuumterm research and industrial enterprises, also form part of the system of the State Committee for Science and Technological Progress.

The State Committee for Science and Technological Progress provides guidance in the activities connected with standardization and metrology, exercises state supervision over the implementation and adherence to standards, and co-ordinates activities related to the quality of production. The Committee works in close co-operation with a number of state and public institutions on the problems of planning, financing and the stimulation of scientific and technological progress.

Sectoral Ministries play an important part in the state system of decision-making in the sphere of science and technology, and they are gradually being transformed into the headquarters of scientific and technological progress. They are granted the authority and right to finance research on the respective problems and possess their own information bases. The Ministries are responsible for the technological standards of production. They are in charge of the elaboration and implementation of the scientific and technological programmes, and they control the state of the technological-material and experimental facilities for their engineering and innovative activities.

Instrumental in the organizational development of research, engineering and innovatory activities is the system of scientific institutions, consisting of academic, higher educational and specialized organizations, directly involved in the solution of scientific and technological problems.

The academic form of the organization of research activities in the country is represented by the Bulgarian Academy of Science and by two specialized academies: the Medical Academy and the Agricultural Academy.

The Bulgarian Academy of Science is the largest scientific organization in Bulgaria. It unites the most outstanding scientists and carries out research. Nearly 11,000 people are employed by the Academy, more than 3,000 of them being researchers. It has assumed its present-day form during the years of socialist construction.

In 1982 the Bulgarian Academy of Science had 47 full members (academicians), 67 corresponding members,

The Frederic Joliot-Curie holiday house for scientists in Varna

Table 1: Staff of the Bulgarian Academy of Science

Staff	1944	1956	1965	1974	1982
Total	16	2,008	4,421	6,432	10,112
Number of research workers	6	495	954	1,720	3,055

171 Doctors of Science and 1,540 Candidates of Science (approximately equivalent to PhD).

The Academy develops the most effective and the most promising trends, carrying out fundamental research in the natural, technical and social sciences related to scientific and technological progress in the national economy, the public and cultural life of the country and the continuity of science. The Academy participates actively and directly in scientific and technological decision-making in the various branches of the national economy and in public life. In close co-operation with the Kliment Ohrid University of Sofia and with other higher educational institutions, the Academy takes part in the training of highly specialized experts. It guides and co-ordinates fundamental research on basic trends and problems in the natural and social sciences, develops, guides and participates in the implementation of programmes for the scientific development of the most important, complex and fundamental problems, and in the implementation of the results in social practice. Scientific societies, associations and national committees on various fields of science and on complex scientific domains exist within the system of the Academy, which also organizes international and national congresses, conferences and symposia.

The supreme administrative organ of the Bulgarian Academy of Science is its General Assembly, which consists of the academicians and corresponding members of the Academy, as well as a certain number of outstanding scholars and representatives of various institutions, elected for a five-year period. The general activities of the Academy are guided by the Presidium, headed by the President of the Bulgarian Academy of Science, elected by the General Assembly.

The basic form for the organization of research in the Bulgarian Academy of Science is in the hands of the scientific units, which consist of permanent institutions: research institutes, central laboratories, centres, independent departments and sectors, problem groups, and research and development bases. When the scientific unit includes faculties of the University of Sofia, it takes the form of a unified centre which works in accordance with a common research plan and participates in the training of graduate and postgraduate students through the relevant university faculty included in the Centre.

The Bulgarian Academy of Science consists of basic structural units which comprise the majority of the research institutes and laboratories: the Centre for Mathematics and Mechanics, the Centre for Physics, the Centre for Chemistry, the Centre for Geosciences, the Centre for Biology, the Centre for Philosophy and Sociology, the Centre for the Sciences of the State and Law, the Centre for Language and Literature, the Centre for History, the Scientific Unit for the Science of Art, and the Scientific Unit on the Basic Problems of the Technical Sciences. These scientific units and centres comprise a total of 80 academic research institutes and laboratories. In addition to them, directly subordinated to the Presidium of the Bulgarian Academy of Science, are the Institute of Economics, the Institute of International Relations and Socialist Economic Integration, the Institute of Contemporary Social Theories, the Centre for the Science of Science, the National Natural History Museum, the Centre for Bulgarian Studies, the Centre for Scientific Information, the Central Library and Scientific Archives, the Main Office of Hydrology and Meteorology (Institute of Hydrology and Meteorology) and the Bulgarian Encyclopaedia.

For outstanding achievements in science, the Bulgarian Academy of Science grants every year the following awards: 'Nikola Obreshkov' – for mathematics and physics; 'Assen Zlatarov' – for chemistry and biology; 'Dimiter Blagoev' – for the social sciences; 'Paissii of Hilendar – for Bulgarian studies; the 'Cyril and Methodius' prize –

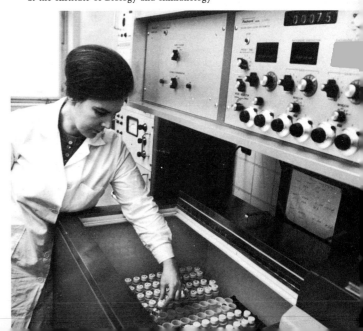

The Packard scintillation counter for measuring radioactivity of the reproduction and development of organisms at the Institute of Biology and Immunology

for achievements in the study of the Old Bulgarian script, literature and culture; as well as other awards for merits in the geosciences and in the technical sciences. The Bulgarian Academy of Science co-ordinates, co-operates and integrates its research activities with the Academy of Science of the USSR and Academies of Science of the other socialist countries, engages in co-operation with academies and scientific organizations in the USA, Great Britain, France, the FRG, Italy and Austria, and participates in international scientific organizations and institutes.

The Publishing House of the Bulgarian Academy of Science releases more than 200 books annually, covering studies in the field of natural, social and technical sciences, monographs, collections of articles, documents connected with the history and culture of the Bulgarian nation and general and specialized encyclopaedias. Its objective is to make known scientific and technological achievements. The Bulgarian Academy of Science publishes 31 scientific journals: the *Journal of the Bulgarian Academy of Sciences;* the *Journal of the Bulgarian Geological Society;* the *Historical Review; Philosophical Thought; Comptes rendus de l'Academie bulgare des Sciences; Bulgarski Ezik* (Bulgarian Language); *Nature; Hydrology and Meteorology; Economic Thought; Literaturna Misul* (Literary Thought); *Pravna Misul* (Journal of Law); the *Physico-Mathematical Journal; Archaeology; Linguistique Balkanique; Forestry Science; Études Balkaniques; Technical Thought, Genetics and Plant Breeding;* the *Journal of Chemistry; Problems of Art; Sociological Problems; Theoretical and Applied Mechanics;* the *Bulgarian Historical Review; Acta physiologica et pharmacologica bulgarica;* the *Bulgarian Journal of Physics; Plant Physiology;* the *Bulgarian Geophysical Journal; Geologica Balkanica; Problems of Geography; Serdica; Bulgaricae mathematicae publicationes;* and *Palaeo-bulgarica.*

The Agricultural Academy is a unified scientific and educational centre for the guidance and development of agricultural science, for the training of specialists and for the implementation of scientific and technological progress in the agro-industrial production. It was founded in 1982 as a collective member of the National Agro-industrial Union.

The Agricultural Academy guides, co-ordinates and is responsible for the general research, educational, engineering and development activities of all scientific and educational units within the system of the National Agro-Industrial Union. Its governing bodies are the General Assembly and the Presidium. The organizational structure of the Agricultural Academy directly comprises thirteen research institutes: the Institute of Economics and the Organization of Agriculture and Farming; the Nikola Pushkarov Soil Sciences and Yield Programming Institute; the Wheat and Sunflower Institute; the Institute of Introproduction and Plant Resources; the Institute for Plant Protection; the Soyabean Institute; the Cotton Institute; the Institute of Horticulture in Kyustendil; the Institute of Oleaginous Roses, Essential Oil and Medicinal Plants; the Institute of Animal Breeding; the Fodder Institute; the Institute of Cattle and Sheep Breeding; the Institute of Mountain Animal Breeding and Agriculture; as well as two higher educational institutions: the Vassil

Mass-analysis laboratory at the Nikola Pushkarov Institute of Soil Sciences and Yield Programming in Sofia

Kolarov Higher Institute of Agriculture and Farming, and the Higher Institute of Zootechnics and Veterinary Medicine. There is also an Institute for the Postgraduate Specialization of Agricultural Experts in Sofia. Moreover, the Agricultural Academy offers scientific and methodological guidance, as well as planning and co-ordinating the research, engineering and development activities of 26 research institutes and of about ten research and industrial organizations and engineering units. Research in the Agricultural Academy is channelled by the respective scientific-co-ordination centres. They organize, guide and co-ordinate the activities of the research organizations in the development of guide-lines, forecasts, plans, complex-target and co-ordination programmes for the scientific servicing of the respective sub-branches of agriculture and farming. Other organizations functioning within the Agricultural Academy are: the Editorial and Publishing Board, a Centre for Scientific, Technical and Economic Information, the Central Agricultural Library and the National Museum of Agriculture.

The Agricultural Academy actively co-operates with similar institutions in the socialist countries and also in a number of capitalist countries. It takes an active part in the work of the Standing Committee on Agriculture at the CMEA. The Agricultural Academy participates with direct membership or through its subdivisions and scientific personnel in a number of international agricultural organizations, above all in the Food and Agriculture Organization (FAO).

The Centre for Scientific, Technical and Economic Information of the Agricultural Academy publishes sixteen journals: nine scientific – *Veterinary-Medical Sciences,*

Horticulture and *Viticulture, Animal Breeding, Economy of Agriculture, International Agricultural Journal, Soil Science and Agro-chemistry, Plant-Growing Sciences, Agricultural Science* and *Agricultural Technology*, and seven popular-scientific journals – *The Veterinary Journal, Horticulture, Animal Breeding, Mechanization of Agriculture, Horticulture, Apiculture,* and *Plant Production.* The Agricultural Academy also releases monthly bulletins *Innovations in Agricultural Practice* with twelve issues on various branches.

The Medical Academy is a complex scientific organization for the development of the medical sciences and higher medical education, and for therapeutic and diagnostic work. It offers consultative and organizational methodological assistance to the health-care system of the country. It is established as a scientific complex which guides the formation of a unified scientific medical front on a national scale. The Medical Academy is laying the scientific foundations for more rapid solutions of the outstanding problems of the health services. For the solutions of the basic structural scientific problems it elaborates comprehensive programmes which concentrate the research activities. The long-term development of medicine is based on fundamental medico-biological research which studies the intimate structure and the function of the various components of the human organism at molecular, cellular and organismal levels. In the research activities of the Medical Academy there is a prevalence of applied research, which is of immediate value and help to the objectives of health care. An efficient organization for the implementation of experimental results has been created.

Rational organizational and functional integration is achieved in the most important and promising scientific areas of research on the fundamental medico-biological problems between the Medical Academy and the Bulgarian Academy of Science.

According to its Statute of 1979, the supreme body of the Medical Academy is its General Assembly, consisting of scientists with academic rank in the Medical Academy, while by decisions of its Presidium it is also possible to include outstanding scientists from other institutions and scientific organizations. All the activities of the Medical Academy are guided by the Presidium, headed by its President.

The basic structural units of the Medical Academy are the institutes which consist of departments, clinics and laboratories. In 1981 there were five Higher Medical Institutes at the Medical Academy: the Higher Medical Institute in Sofia, with Departments of Medicine, Stomatology and Pharmacology; the Ivan Petrovich Pavlov Medical Institute in Plovdiv, with Departments of Medicine and Stomatology, and a branch Medical Department in Pazardjik; the Higher Medical Institute in Varna, with a branch Medical Department in Tolbuhin; the Higher Medical Institute in Pleven; and the Higher Medical Institute in Stara Zagora, as well as 24 research institutes: Research Institute of Obstetrics and Gynaecology; Research Institute on Internal Diseases; Research Institute of Pharmacology; Research Institute on Gastroenterology and Nutrition; Research Institute on Dermatology and Venereology; Research Institute on Endocrinology, Gerontology and Geriatrics; Research Institute on Infectious and Parasitic Diseases; Research Institute on Ideological Problems; Research Institute on Balneology, Physiotheraphy and Rehabilitation; Medico-Biological Institute; Research Institute on Neurology, Psychiatry and Neurosurgery; Research Institute on Nephrology, Urology, Haemodialysis and Kidney Transplantation; Research Institute on Oncology; Research Institute of Orthopaedics and Traumatology; Research Institute of Ophthalmology; Research Institute of Paediatrics; Research Institute of Pneumology and Phthisiatry; Research Institute of Roentgenology and Radiobiology, Institute of Social Medicine; Research Institute of Cardio-Vascular Diseases; Research Institute on Oto-Rhino-Laryngology; Research Institute of Haematology and Blood Transfusion; Research Institute of Hygiene and Occupational Diseases; and Research Institute of Surgery.

The Medical Academy organizes and co-ordinates international co-operation in various forms, the most important of which is the joint research on scientific projects and themes. Co-operation takes place on the broadest scale through the Standing Committee on Health Care Problems under the CMEA, as well as on a bilateral basis with the socialist countries: the USSR (the Academy of Medical Sciences of the USSR), the GDR, Czechoslovakia, Hungary, Poland and Romania. Wide co-operation exists with the World Health Organization, the International Atomic Energy Agency, and the Intercosmos Space Research Programme of the socialist countries. Some Bulgarian research institutes play a leading role or act as co-ordinators on problems raised by the international organizations. Scientific co-operation also takes place with some capitalist countries: Great Britain, the USA, France, Austria, the FRG, Spain, Italy, and other countries.

The central publication of the Medical Academy is the periodic journal *Acta Medica Bulgarica* (published in English). The Institutes of the Medical Academy publish the following scientific journals: *Ideological Problems of Medicine; Medico-Biological Problems; Problems of Obstetrics and Gynaecology; Problems of Internal Medicine; Problems of Infectious and Parasitic Diseases; Problems of Neurology, Psychiatry and Neurosurgery; Problems of Oncology; Problems of Paediatrics; Problems of Pneumology and Phthisiatry; Problems of Roentgenology and Radiobiology; Problems of Stomatology; Problems of the Cardio-Vascular Diseases; Problems of Pharmacy; Problems of Hygiene;* and *Problems of Surgery.* The editions of the Higher Medical Institutes are: *Scripta scientifica medica* at the Higher Medical Insitute of the Bulgarian Academy of Science, *Folia medica* of the Higher Medical Institute in Plovdiv, and *Scientific Works* of the Higher Medical Institute in Pleven. The Centre for Scientific Information on Medicine and Health Care with the Central Medical Library at the Medical Academy publishes 24 information bulletins, seven of which are general and seventeen specialized.

The Academy of Social Sciences and Management, of the Central Committee of the Bulgarian Communist Party

Testing circuit boards for control devices at the Institute of Cybernetics and Robotics Engineering

(BCP), is a centre for training specialists in the field of the social sciences and social management, for research and for applied scientific activities. Created in 1969 on the basis of the then existing Higher Party School at the Central Committee of the BCP, the Institute for the Organization of Management of the Council of Ministers and the Centre for Postgraduate Specialization of Managerial Personnel at the Ministry of Labour and Social Welfare, the Academy maintains close contacts with similar institutes and centres in the USSR and the other socialist countries, organizing joint research projects, symposia, seminars and other forms of exchange of information and experience. The scientific publications of the Academy started to appear after 1970 in four branches: Philosophy, Economics, History and Social Management.

The Georgi Stoikov Rakovski Military Academy is a higher military educational institution for training highly qualified command, political, engineering and military research staff. The Academy was founded in 1912. Its publications are: *The Works of the G. S. Rakovski Academy,* and *Social Disciplines* (1958–   ).

The St. Kliment of Ohrid Theological Academy is a higher Orthodox educational institution. It is engaged in research and publishes many works, monographs and studies, printed at the Publishing House of the Holy Synod. An *Annual of the Theological Academy* (1924–   ) is also published.

Science in the Higher Educational Institutions. The higher educational institutions in Bulgaria developed as complex educational and research centres. They are characterized by the combination of research work with teaching and with the needs of social practice (for the structure and tasks of the higher educational institutions and further details see Part VII: Education). The higher educational institutions represent a considerable portion of the personnel and scientific potential in the country. This scientific and personnel potential has been formed almost exclusively from highly qualified experts in all fields of science, technology and higher education. The proportion of members with doctorates is 33 per cent, and of those with scientific degrees 41 per cent. At present, two thirds of the postgraduate training takes place at the higher educational institutions and one out of five postgraduate students takes part in work on research projects. The higher educational institutions are reliable and valued partners of the economic organizations and the branch Ministries in the solving of their numerous and varied scientific and technological problems.

In accordance with the main trends in the economic development and with the scientific strategy and policy of Bulgaria, the forces of the scientific front in the higher educational institutions are concentrated predominantly in the following basic branches: machine-building, electronics, chemical industry and biotechnologies.

The higher educational institutions work for the expansion of the raw materials base, for the complex and rational utilization of raw materials, and for raising the effectiveness of the energy resources. In the sphere of fundamental research the efforts of the representatives of the natural and mathematical sciences are aimed at maintaining the high standards of scientific knowledge, whereas the representatives of the social sciences are concerned with solving the problems stemming from the formation of the developed socialist society. The problems of the scientific servicing of higher education attract increasing attention. Many of the scientific and technological achievements of the higher educational insitutions are models of creativity and talent and have won justified recognition at home and abroad.

The sectoral form of organization of the research, engineering and development activities involves a wide variety of organizations and activities in which large numbers of specialists are concentrated. In accordance with their specialization, the engineering and industrial transfer organizations which are structural subdivisions of economic organizations and Ministries, established in the order stipulated by the State Committee for Science and Technological Progress, are responsible for raising the technical-economic, technological and organizational levels of production. The engineering organizations master new technologies; study and introduce the techno-scientific experiences of other countries and also participate in the reconstruction and modernization of existing capacities; offer technical assistance to the production subdivisions; take part in the purchasing and selling of licences; patent rights and technical documentation; and study the market and the demand for the available types of production and technology.

In 1982 there were about 380 engineering organizations. Nearly 50 per cent of them are bases for research and development. These bases are the main bodies of the

Machine-tool plant and Research Institute in Sofia

economic organizations, making it possible to constantly raise the technical-economic, technological and organizational levels of production. The bases independently work out technologies, projects, and documentation for the future technological updating and re-equipping of production.

Of particular significance in the engineering activities of the People's Republic of Bulgaria are the research and industrial organizations. They unite production and economic units, and engineering organizations. Different organizational forms of research and industrial organizations are used in Bulgaria: research and production associations, enterprises, laboratories and factories. The research and production organizations are a new form in the structure of the engineering activities in the country.

In the established research and production associations, work on the generation and implementation of the innovations in science and technology proceeds much faster. Engineering units are being formed primarily in the economic organizations and their subdivisions, which contribute to the consolidation of the organizational foundations of scientific and technological progress.

In recent years there has been a tendency to create units which comprise academic, university and branch organizations for research and engineering activities. The higher technical institutes and some engineering organizations have recently created reseach-educational-industrial complexes which are engaged in research, industrial engineering and educational work in a given branch or in its sub-branches.

Systems have been created for the territorial management of scientific and technological progress at district, municipal and regional levels. At the centre of these territorial systems are the councils for scientific and technological progress, which function on the public-state principle. These councils work out district programmes for the management of scientific and technological progress.

A considerable role in the general scientific and technological progress of Bulgaria is played by the organizations formed on a public principle, such as the Union of Scientific Workers and the Scientific and Technical Unions.

The Union of Scientific Workers in Bulgaria is a public creative organization of scientific workers, founded in 1944. Its main objectives are: to contribute to the construction of the developed socialist society by making use of the achievements and advanced experience of Bulgarian and foreign science for elaboration of research problems and for giving assistance to state and public organizations in the implementation of these research projects, following signed contracts for engineering and development work; to organize congresses, symposia, and colloquia, and to establish and maintain links with progressive international organizations. In 1983 the Union of Scientific Workers had 7,100 individual members, mainly people with academic degrees, whereas through its scientific societies it unites about 56,000 researchers and specialists. The Union has eighteen district branches and comprises eighteen departments: Physical-Mathematical; Technical Sciences; Geological-Geographic; Chemistry and Pharmacy; Biology; Medicine and Stomatology; Veterinary Medicine and Zootechnics; Agrobiology; Microbiology; Biochemistry; Biophysics; Philosphical-Pedagogical; Economics; Law; Historical-Linguistic; Military History; Scientific Problems and Physical Culture; and The Assen Zlatarov Department of Nutrition Forestry Sciences. The Union publishes the journals *Nauchen zhivot* (Scientific life) and *Scienca mondo* (in Esperanto). The Union of Scientific Workers is a member of the World Federation of Scientific Workers (since 1948). The F. Joliot-Curie International House of Scientists in Varna offers facilities for scientific congresses and conferences, as well as for recreation.

The Scientific and Technical Unions are a united voluntary creative mass public organization of engineers, technicians, economists, agricultural specialists, scientific workers, students in the higher educational institutions and in the secondary vocational technical schools, and worker-innovators and inventors in Bulgaria. The basic task of the Scientific and Technical Unions is to contribute to the accelerated implementation of scientific and technological progress through the mobilization of the creativity of their members.

In 1983 the Scientific and Technical Unions had a membership of 350,000. The governing body is the

Table 2: Research Workers in General and Specific Branches of Science

Branches of Science	1965	1970	1975	1980	1982
Technical sciences	2,262	4,072	5,764	7,641	8,200
Medical sciences	1,701	2,164	3,430	4,785	4,550
Natural sciences	1,734	2,127	3,443	3,769	3,950
Agricultural sciences	1,367	1,739	2,003	1,965	2,320
Social sciences	1,834	2,632	3,724	4,441	4,580
Other sciences	49	72	—	—	—
Total	**8,973**	**12,765**	**18,436**	**22,601**	**23,600**

Central Union; its bodies in the districts are the District Councils which unite and guide about 4,374 scientific and technical societies. The Central Union, the District Councils of the Scientific and Technical Unions, and the Scientific and Technical Unions themselves are juridical bodies with their own structure and organs. Subdivisions of the Central Union and District Councils of the Scientific and Technical Unions are: the Unions of Geodesy and Land Management; Water Economy, Power Production, Electrical Engineering and Communications, Economics, Forestry, Machine-building; Mining, Geology and Metallurgy; Agriculture and Farming; Constructions; Textiles and Clothing; Transport; Chemistry and the Chemical Industry; and the Food Industry. A total of 160 specialized sections are working on specific problems and in specific branches within the Unions.

Nuclear reactor at the Institute of Nuclear Research and Nuclear Power Engineering

The Scientific and Technical Unions organize and conduct scientific and technical activities related to the improvement of scientific and technological propaganda: the providing of prompt information to specialists about the latest achievements in science and technology; the expansion and improvement of the system of postgraduate training and the improvement of the qualifications of staff with secondary education; the development and intensification of co-operation and interaction with other public, state and economic bodies and organizations, with the aim of accelerating scientific and technological progress and consolidating the public principle in its guidance and management. The Scientific and Technical Unions publish the *Tehnichesko Delo* – a weekly newspaper of scientific and technical information – and a series of booklets on topical scientific and technical problems; and they also participate in the publication of other scientific and technical journals. The Scientific and Technical Unions are members of almost all international engineering organizations, and maintain relations with similar organizations in other countries.

For nearly thirty years after the victory of the socialist revolution in Bulgaria in 1944, one of the main characteristics of the system of research and development was the *rapid quantitative growth of scientific personnel.* As a result of this the necessary personnel was provided for fundamental and applied research in the main trends of scientific and technological progress. In 1982 the number of specialists engaged in research and engineering activities was 72,000, of which 23,600 were research workers, i.e. there were 162 research workers and specialists per 10,000 people employed in the national economy.

Attention is focused mainly on the qualitative growth of the scientific personnel. The most important factor in this respect is the constant raising of their qualifications to a higher level.

A unified system for improving the qualifications of research and engineering personnel has been in operation in Bulgaria since 1980. Various forms of improving qualifications within the research and engineering organizations are also widely used. Periodic personnel certification and evaluation of the work of each research worker and specialist involved with research and engineering activities acquires an ever increasing importance.

Considerable changes have recently taken place in the

Table 3: Research Workers According to Scientific Degrees

Scientific Degrees	1965	1970	1975	1980	1982
Academicians	31	34	38	36	38
Corresponding members	56	44	42	33	41
Professors	452	718	883	986	1,025
Associate professors	661	885	1,482	2,008	2,180
Senior research associates	960	1,542	2,178	3,468	3,240
Doctors of science	66	103	244	645	706
Candidates of science (approximately PhD)	1,189	2,469	4,814	7,945	8,380
Total number of research workers	8,973	12,765	18,436	22,601	23,600

qualification structure of the personnel. The number of highly qualified specialists has grown considerably.

Parallel with this there was an increase in the number of builders, technologists and designers. In 1982 they amounted to nearly 30 per cent of all people engaged in research and engineering activities. The Ministry of Electronics and Machine-building employs the largest number of researchers and engineers: nearly 27,000 in 1982, which is equal to 40 per cent of the research and engineering personnel. This figure was 5,000 for the Ministry of Metallurgy and Mineral Resources and the Ministry of the Chemical Industry, i.e. in the organizations working on the problems of raw materials.

An important trend in the development of scientific and technological potential is the *rapid increase of basic funds for research and engineering activities*. The material and technological base has a vital importance for the effectiveness of the scientific and technological policy and for the intensification of research and development activities. The fixed capital increased twofold in the 1975–82 period, whereas the number of personnel grew 120 per cent.

This process is parallelled by constant modernization of the machinery, equipment and facilities of research, experimenting and designing activities. Instrument-building began to be developed on a scientific basis. The scope of experimental production was expanded and now accounts for nearly 30 per cent of all funds. Alongside the progressive tendencies in the development of the material and technological base should be noted the lines of its most effective and complete utilization. A gradual transition has taken place in the establishment of specialized forms for the use of expensive scientific and technological equipment, such as collective use through temporary loans of equipment from other organizations or through co-operation among the various units. An improvement can also be noted in the forms and methods of material and technological supply to the research and engineering organizations. A long-term programme until the year 2000 has been elaborated with a view to solving the complex problems of the material and technological base for research and experimentation in the country.

Table 4: Fixed Capital Put into Operation (million levs)

1965	1970	1975	1980	1982
10.1	25.8	39.8	53.5	62.4

The funds allocated for research and engineering activities have grown rapidly. A 140 per cent increase has been planned for the 1980–85 period. In 1985 these funds amount to nearly 2.8 per cent of the national income. In the year 2000 this figure is expected to reach 4.6 per cent.

Extensive information and patent files have been created over the past years. The Central Institute for Scientific and Technical Information (Cinti) is now

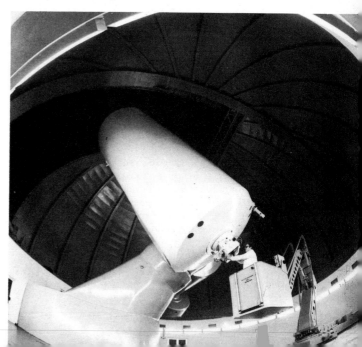

Two-metre telescope at the National Astronomical Observatory in the Rhodope mountains

Table 5: Capital Investments (million levs)

1965	1970	1975	1980	1982
14.9	30.1	53.6	63.0	70.4

functioning and possesses a modern computer centre, highly qualified staff and a good technical library.

Essential changes took place in 1980: the number of documents exceeded 2.7 million, grouped in fifteen automated systems for scientific and technical information. They are used by more than 4,500 clients from 200 organizations. The Computer Centre stores and processes more than 800,000 records of different documents annually. Links have been established between Cinti and the European regional network for scientific and technical information, Euronet.

There is wide-scale co-operation with the system of information services in the USSR.

Co-operation with other socialist countries within the International System for Scientific and Technical Information of the CMEA member-countries is also developing successfully. In the past years co-operation has been established with 21 other countries, among them the USA, Japan, Great Britain, and France.

A national *system for management of scientific and technological progress* has been set up.

A unified plan for the socio-economic development of Bulgaria over five-year periods defines the basic tasks concerning the development of science and technology. The section on scientific and technological progress stipulates concrete demands for innovations. The plan concerning scientific and technological progress has been developed in three main directions and in accordance with the scope offered by the material and technological base. The first is strategic planning, concerning planned long-term scientific and technological prospects. The second comprises the Five-Year Plans through which the tactical planning of scientific and technological progress takes place. The third comprises operational plans.

The strategic objectives of the national scientific and technological policy are formed on the basis of long-term planned prospects. These contain scientific and technological forecasts, complex programmes for scientific and technological development and analysis and evaluation of the problem areas of scientific and technological progress.

A system of scientific and technological forecasting has been created in Bulgaria and is functioning successfully. More than twenty scientific and technological forecasts have been made for determining scientific policy for the period until the year 2000 and into the more distant future. The forecasts are grouped together in a unified document. The State Committee for Science and Technological Progress is the methodological leader and organizer of this project.

The National Complex Programme for the Development of Science and for Technological Progress has a place of particular importance in the structure of long-term scientific and technological prospects. It is a complex of sub-programmes for the development of different strategic trends which determine the scientific and technological development of Bulgaria over a longer period of time. Concrete solutions are being worked out for the implementation of this national programme.

The sub-programmes of the National Programme are grouped in four main blocks.

The first block comprises programmes for the development of the strategic trends of scientific and technological progress. The aim of these programmes is to outline future qualitative structural changes in the material and technological base of the national economy and mainly to develop new equipment and technology, which are of vital importance for the intensification of the national economy.

The second block of programmes comprises the long-term development of the national economic branches faced with important technological tasks.

The third block of programmes comprises the development of fundamental research as a basis for future applied research and engineering activities. They outline the strategy in the fundamental scientific domains: mathematics, physics, chemistry, biology and geosciences.

The fourth block includes the programmes guaranteeing the development of the basic elements of the national scientific and technological potential: personnel, material and technological base, financing, international scientific and technological co-operation and scientific and technical information.

The Interprograma Bulgarian-Soviet Research and Design Institute

The programmes are used as basic preliminary documents in drafting the Five-Year Plans for the socio-economic development of the People's Republic of Bulgaria. The tasks ensuing from the national complex programmes are endorsed by the Council of Ministers. For the sake of more efficient organization and management in the implementation of the tasks formulated in the national programmes, Programme Councils have been set up within the State Committee for Science and Technological Progress, headed by Deputy-Presidents of the Committee. The councils comprise managerial personnel, outstanding scientists and specialists from Ministries and various establishments, from academies and institutions of higher education, from economic and engineering organizations.

Strategic planning takes into account potential and actual problems which are to be dealt with over a long period. The scientific and technological objectives, being subject to the Five-Year Plan, are classified in three groups:

(a) State scientific and technological tasks included in the national complex programme. They are related to the most important problems of the national economy and for this reason are mandatory. About 900 of these are included every year in the Five-Year Plan.
(b) Scientific and technological tasks through which the branch strategy is put into practice. They reflect the scientific and technological policy of the Ministries.
(c) Scientific and technological tasks of the economic organizations, comprising the immediate technological needs of the units and subdivisions.

The technological level, terms of fulfilment of the task, economic effect and organization responsible for the implementation of the task, are each determined for every innovation forming part of the plan.

Particular importance in state strategy concerning the implementation of scientific and technological policy is attached to the *economic methods and means of management* through which the economic approach in the management of scientific and technological progress is applied. The administrative methods and means are insufficiently effective for the conditions of the intensively developing the Bulgarian economy. The economic mechanism is a unified system of legally and mutually bound economic methods, means and indices, capable of effectively influencing scientific and technological progress.

A major characteristic feature of the national system for the financing of scientific and technological progress is its flexibility. Financing can be authorized at different levels: national, branch and enterprise level; the object of financing is not the individual research engineering organization itself, but rather the scientific and technological task; the basic source for the financing of these funds is the Budget and the profit of the economic organizations.

An important aspect in the economic policy aimed at stimulating the economic organizations in the sphere of technological progress is to guarantee compensation for possible losses incurred in the course of the implementation of the new technology. For this purpose the economic organizations have a special fund with compensating functions, available since 1979. This fund has been successfully introduced into the general structure of the economic mechanism. The constant updating of the range of production causes each economic organization to constantly introduce the latest technological innovations, as they are the basic and decisive source of progress.

The stimulation of the economic interest of the engineering organizations in the practical applications of the results of their activities is the next main line of intensification. A special economic regime has been established in these organizations for this purpose, its essence being that it makes the income of the organization dependent on its research, design and engineering activities. The basic source of income for engineering organizations is provided by revenues from the implementation of scientific and technological innovations. After deducting expenditures, the organizations form an income fund which is distributed among the internal funds and added to the labour remuneration fund.

The activities of the engineering organizations are assessed every year, using a special system of parameters which define its goal and efficiency. After receiving the evaluation, special state certification committees evaluate the activities of the organization. The reports of these committees are endorsed by the managerial body of the respective ministry, as well as by the State Committee for Science and Technological Progress. The individual scientific and technological projects are assessed by estimating the economic effect which includes the revenues from the production of the new article and the economy of resources ensuing from the innovation. The evaluation of the performance of the individual worker in the respective working group characterizes the professional, working and social aspects of his activities. This individual evaluation and certification has a direct effect on labour remuneration.

The implementation of scientific and technological achievements in practice is a task of prime inportance in the national scientific and technological strategy.

This process has acquired decisive importance in the dynamic and planned development of science and technology in Bulgaria. Great attention is attached to the further improvement of the planning and organization of the innovation processes with a view to minimizing the duration of time from research to implementation. About 12,000 scientific and technological achievements were put into practice in the 1981–82 period alone; in most cases these were designed for the further improvement of the technological processes. The tendency at present is to implement inventions and innovations which would lead to qualitative changes in the structure of the national economy and would guarantee success on international markets. In this respect implementation work is planned and strictly takes into account the particulars of each branch. In branches with a high fixed capital share (e.g. in power production, heavy machine-building and the chemical industry) the implementation of scientific and technological achievements is aimed at a highly effective utilization of these funds. In branches with a high proportion of the total percentage of raw materials,

M. A. Markov addresses an international seminar on the theory of elementary particles in Varna, 1968

implementation is oriented towards economy of raw materials and metals. In branches with a high relative share of labour costs (in the building industry, agriculture, coal-mining etc.), the emphasis is placed on the mechanization, automation and scientifically-based organization of production.

Engineering projects consist basically of scientific and technological tasks which the working groups propose for implementation. These projects reflect above all the initiative and the creativity of workers and specialists, of inventors and researchers, and of all members of the working team. Engineering projects are a specific programme for technological renovation of the list of articles produced and for the updating of the production capacities.

Various measures are aimed at systematically and permanently raising the interest of industrial enterprises in guaranteeing technological progress. These comprise both economic and organizational methods. The economic measures are designed to evoke in the producer interest in new production. More specifically, use is being made of the stimulating effect of prices for the implementation of such production which also requires new technologies. The enterprises create special target funds which guarantee economic stability and interest in the implementation of the innovations. With a view to increasing the collective and individual interest, special funds are allocated for bonuses for successfully completed tasks.

The economic organizations work out annual and five-year plans for technological development, including reconstruction and modernization, production of new articles and materials, implementation of new technologies, and the purchase and implementation of foreign scientific and technological achievements. Through their own engineering organizations they study and analyse the trends in the technological development of their branch.

Consumers participate in certain technical gatherings to elucidate the structural characteristics of the innovations, and they determine the volume or amount of the production released both to the home and the international market. Special organizations are set up for the operation of the most complex innovations. These organizations provide assistance to the users in collecting data on the possible uses, and in the assembly and servicing of the implemented innovations.

The international scientific and technological co-operation of the People's Republic of Bulgaria represents a system of contacts and interrelations with other states and international organizations, guaranteeing wide and effective implementation of the achievements of science and technological progress, and optimum utilization of the national scientific potential. Under the conditions of the scientific and technological revolution, of the intensifying trend towards science, scientific and technological co-operation is a means of accelerating scientific development, and of raising productive forces and the efficiency of social production. As a whole, scientific and technological co-operation is an important and promising trend in international economic co-operation by Bulgaria. It comprises the exchange of scientific and technical documentation (either free of charge or against payment) and production experience, co-operation, co-ordination and joint organization of research and international scientific and technical events and the training of personnel in the sphere of science and technology.

The scientific and technological co-operation of the People's Republic of Bulgaria is developing in accordance with the planned strategic trends in the country's economic and scientific-technological policy. Attention is focused mainly on the further improvement of the target orientation of this type of co-operation in all its forms. The target orientation is expressed in all activities related to the exchange of experience, specialization and the receiving of technical documentation.

The main international scientific links are with the socialist countries and above all with the USSR. During the first years after World War II bilateral contracts for economic and scientific-technological co-operation were signed with the USSR (1947), and Poland and Romania (1948). With the founding of the CMEA in 1949, the way was paved for extensive and planned multilateral and bilateral international scientific co-operation among the socialist countries. Since 1949, co-operation between Bulgaria and the other socialist states has undergone considerable development. From the original exchange of scientific and technical documentation, training of research specialists and aid with practical experience, Bulgaria embarked upon new modern foundations for co-operation with the socialist countries. At present the guiding principles and objectives of this co-operation are formulated in the Complex Programme for intensification of the socialist economic and scientific-technological co-operation with the CMEA member-countries, as well as the General Plan for specialization and co-operation in the sphere of material production between the People's Republic of Bulgaria and the USSR. The scientific and technological co-operation among the socialist states is being built at present on the basis of long-term target programmes.

On the basis of multilateral scientific and technological co-operation, Bulgaria plays an active part in the work of the international organizations of the CMEA in the field of science and technology. Bulgarian scientists participate actively in joint institutes, international laboratories and centres, as well as in the International Programme, Intercosmos, as a result of which the spaceship *Soyuz-33* carrying on board the first Bulgarian cosmonaut was launched in 1979.

In multilateral scientific and technological co-operation Bulgaria participates in the implementation of more than two thirds of the signed contracts and agreements. As a result of the joint efforts of scientists and specialists from the CMEA member-countries, highly effective power plants have been built, new industrial catalysts have been synthesized and new varieties and hybrids of agricultural crops have been created.

The scientific and technological co-operation of Bulgaria with the capitalist countries is based on intergovernmental agreements and programmes for joint research predominantly in the field of electronics, atomic power, machine-building, the chemical industry, pharmaceutical and biological industries. The main forms of this co-operation are: the exchange of documentation and information, joint research, the organization of specializations and training of specialists, and the sale of licence rights.

Licences are bought above all from the FRG, Great Britain, Italy and France, and are sold mainly in Japan, the USA and Great Britain. The licences sold are in the field of metallurgy, machine-building and the food industry. Good results have also been attained in some joint research projects with research organizations in France, the FRG and the USA, in co-operation with which Bulgaria is working on more than 150 projects. Co-operation with the capitalist countries grew after the special decree of the State Council (1980) concerning economic co-operation between Bulgarian juridical persons and foreign physical and juridical persons. On the basis of this act of administration, a number of joint ventures were set up, one of their tasks being to pursue scientific and technological co-operation.

Scientific co-operation with the developing countries is oriented toward offering assistance in surmounting their economic and technological backwardness, as well as towards the discovery and utilization of their internal resources. The basic forms of scientific and technological co-operation with these countries is in the training of graduate and postgraduate students, specialists, the offering of licences and the dispatching of Bulgarian specialists to work in these countries. Students from the developing countries are trained mainly in engineering departments (about 40 per cent of them). For the majority of them education is free of charge. Approximately 6,000 Bulgarian physicians, engineers, builders, teachers and university lecturers and agronomists are working in Libya, Mozambique, Angola, Morocco, Syria and Iraq.

The Bulgarian Academy of Science occupies a special position in the international scientific co-operation of the People's Republic of Bulgaria. Jointly with the Academy of Science of the USSR and the Academies of Science of the other socialist countries, as well as with related academic organizations in the non-socialist countries, the Bulgarian Academy of Science is responsible for the major part in bilateral and multilateral co-operation in the spheres of fundamental and applied research. In 1983 the Bulgarian Academy of Science was engaged in co-operation with thirteen socialist and twelve non-socialist countries. One of the main forms of this bilateral co-operation is the joint work on problems and themes of mutual interest. Nearly 87 per cent of the 486 joint projects in various scientific branches are with the socialist countries, the remaining 13 per cent represent co-operation with Austria, Great Britain, the FRG, Greece, Italy, Mexico, the USA, Finland, France, Switzerland and Sweden.

Co-operation with the Academy of Science of the USSR has made a considerable contribution to the development of Bulgarian science. For the past three years co-operation with the Soviet institutes has been channelled into priority problem areas. Co-operation between the Bulgarian Academy of Science and other academies of science in the socialist countries is proceeding also in accordance with plans co-ordinated in advance and is yielding very good mutually beneficial results for both academies.

In 1981 the People's Republic of Bulgaria was either a collective or an individual member of about 140 international non-governmental scientific organizations. The Bulgarian Academy of Science is represented in 52 such organizations, a large number of which form the networks of the great international unions: the International Council of Scientific Unions (ICSU) and the International Social Sciences Council (ISSC), both organizations having very close working contacts with Unesco and also with governmental organizations. The membership of the Bulgarian Academy of Science in non-governmental organizations has proved particularly useful for the establishment of scientific contacts and for participation in international scientific forums, thus gaining valuable information about the state and development of modern science and at the same time making known the achievements of Bulgarian scientific thought.

The co-operation of Bulgarian science develops along the lines of international governmental organizations as well. The Bulgarian Academy of Science is one of the most active Bulgarian scientific institutions, maintaining close links with Unesco. National committees have been set up for the intergovernmental programmes 'Man and the Biosphere' and 'The International Hydrological Programme'. Bulgarian scientists occupy leading positions in this world organization. Another considerable part of the scientific contacts of Bulgarian science with the world scientific governmental organizations is seen in the International Atomic Energy Agency and the World Meteorological Organization. Contacts with the latter organization resulted in the construction of a regional telecommunications centre in Sofia, for which modern equipment was imported and opportunities were provided for Bulgarian specialists to aquire knowledge and experience abroad.

Along the lines of the UNDP projects in Bulgaria the Centre of Phytochemistry and Chemistry of Natural Substances was formed in Bulgaria and the Institute of Cybernetics and Robotics Engineering of the Bulgarian Academy of Science was modernized.

International scientific co-operation is a basic component in the scientific and technological policy of Bulgaria. There are long-term plans for the development of this co-operation and this has a particularly favourable influence on the co-operation of Bulgaria with the socialist countries within the CMEA system, as well as with the non-socialist countries along the lines of the agreements signed with them.

BASIC TRENDS AND ACHIEVEMENTS OF THE NATURAL AND TECHNICAL SCIENCES

Mathematics and Mechanics. Mathematical science has sound and lasting traditions in Bulgaria. The Bulgarian mathematical school has earned international prestige and followers. Many high international distinctions and titles have been granted to Bulgarian scientists for their considerable contribution in the field of mathematics.

In 1962 Vitosha, the first Bulgarian computer, was designed and produced at the Institute of Mathematics (founded 1951) of the Bulgarian Academy of Science, whereas the first Bulgarian electronic calculator Elca was designed and put into production in 1964. The main areas of mathematics developed rapidly. Fundamental and applied research is united in three complex problems: mathematical structures, mathematical informatics, and mathematical modelling.

Creative work is in progress in the field of universal mathematical structures and considerable results have been obtained in the following fields:

(a) in algebra (Nikola Obreshkov) – classical algebraic structures, algebraic manifolds, commutative algebra and algebraic geometry, polynomials, combinatorial algebra and theory of coding;

(b) in topology – general topology, algebraic topology, topological algebra;

(c) in mathematical logic – generalized theory of recursive functions, abstract computational models, non-classical logics and their application in theoretical informatics, foundations of the constructive and intuitionistic mathematics, theory of proof;

(d) in real and functional analysis (Yaroslav Tagamlitski) – diagonal principle, minimal Abelian groups, kernel and analytical spaces, theory of operators, classical and modern analysis, theory of graphs;

(e) in complex analysis (Lyubomir Chakalov, Lyubomir Iliev) – theory of analytical functions, geometric theory of analytical functions, analytical manifolds and spaces, operational methods in the analysis;

(f) in differential equations – boundary problems for mixed-type equations, boundary problems for linear and non-linear equations with non-negative quadratic form, microlocal investigation of the specificities of differential equations and their application for the study of boundary problems, spectral properties in the theory of dispersion, topological methods in differential equations;

(g) in geometry (Boyan Petkanchin) – differential manifolds with geometric structures, axiomatic, combinatorial, algebraic and synthetic studies of geometric structures and graphs.

Lyubomir Iliev

In the field of mathematical informatics successful research has been carried out in a number of trends: automation of programming, algorithmic languages and translation methods, operational systems, systems for the automation of information services, computer graphics, artificial intelligence, computer networks, control systems for data bases, research and designing of discrete processes and systems, machine translation, systems for linguistic communication with computers, theory of the formal languages and grammar, software for microcomputer systems, and application of informatics in the solving of scientific-technological and economic problems.

Georgi Nadjakov

In the field of modelling successful work is being done in:
(a) mathematical modelling (Blagovest Sendov, Vassil Popov) with the following trends: theory of approximations, numerical methods, mathematical methods in biology and ecology, methodology of mathematical modelling of natural and social processes;
(b) operations research with the following trends: modelling and optimization of real processes, theoretical foundations of optimization, working out of algorithms and programme products for optimization;
(c) probabilities and statistics with the following trends: theory of probabilities, mathematical statistics, application in other sciences, industry, economy and agriculture, computer stochastics.

Some of the results obtained by Bulgarian mathematicians are published in the form of monographs. Many of them are widely cited and used in the world mathematical literature, others have been translated into foreign languages.

Physical Sciences. Georgi Nadjakov made the first Bulgarian discovery of scientific significance: he formulated the photo-electret state of matter. Research in the field of photo-electrets and the phenomena associated with them is still in progress, and particularly fruitful.

Certain principal inherent regularities in the induction and preservation of the photo-electret state have been established. In collaboration with Soviet scientists a theory of effect has been elaborated on the basis of the present-day understanding of the electronic processes in the crystals. The theory has found wide application in electrophotography. Research is still in progress today, mainly with the aim of clarifying the mode of induction of this state under the effect of X-rays and its practical applications in radiography and X-ray dosimetry.

Research for many years on photo-electret phenomena in dialectrics has gradually comprised some high-resistance semi-conductors as well, mainly compounds between the elements of groups two and three in the periodic system of elements. Some of the results obtained are of pioneering significance for physics and have gained justified recognition in world science. These activities are continuing to be developed, some of the materials being used for the production of various photo-receivers and for nuclear radiation detectors.

Photo-electric and photo-electret phenomena in amorphous semi-conductors is another important sphere of research. On the basis of the valuable results obtained, which have long enjoyed justified recognition abroad, research is now under way on the use of amorphous semi-conductors in electro-photography. Parallel with this, some inorganic photo-resistors are being used in photo-lithography and in electronic lithography.

After 1956 a powerful impetus was given through Soviet aid to the development of nuclear physics in Bulgaria. A research nuclear reactor was built and put into operation. It marked the beginning in the development of neutron physics and nuclear reactor physics. These studies brought about the following important scientific and applied results: new methods have been devised for obtaining ultracold neutrons through the principle of their mechanical deceleration by reflection in a moving reflector; new experimental data have been obtained in the sections of uranium, thorium and plutonium fusion, as well as a number of parameters of great significance for the operation of nuclear reactors, and with great accuracy; new programmes with improved precision have been devised for rapid calculation of the neutron-physical characteristics of WWER-type reactors; methods and equipment have been introduced in the first Bulgarian nuclear-power plant, which have improved the regime of operation of the reactors and permit the monitoring of the process of the 'burning' of the nuclear fuel and the accumulation of secondary fuel.

Research and development have started in the spheres of nuclear methods, nuclear instrument-building and nuclear electronics.

The setting up of the Joint Institute for Nuclear Research in Dubna (USSR) in 1956 initiated research in Bulgaria in the quantum field theory, high-energy physics and elementary particles. In a relatively short time Bulgarian physicists achieved scientific results which put them on an equal footing with the most outstanding scientists from the socialist countries, with whom they worked jointly on common themes and subjects of research.

The work of Bulgarian physicists related to the development of the conformational-invariant field theory and the asymptotic approach in this theory enjoys high recognition. Of particular note are the elaborated variants of the higher symmetries theory, of paraalgebras as an apparatus for more general quantification, of superalgebras and their application in supergravitation, of the calculation of the effects of weak interactions in a higher degree of approximation and of the quantum theory including a new universal constant in elementary length.

Research on some important trends in high-energy physics was carried out jointly with the Institute of Physics of the Academy of the USSR. Wide atmospheric torrents in cosmic radiation were studied and light was shed on the nature of the particles of primary cosmic radiation with energy exceeding 10^{14} eV, as well as on the character of their interaction. The Bulgarian space research station at Mussala peak in the Rila mountains is an important prerequisite for research.

With the aid of the Joint Institute for Nuclear Research in Dubna research has been developed in Bulgaria on the theory of the atomic nucleus, nuclear reactions and nuclear spectroscopy. These studies further consolidated the prestige of Bulgarian physics, raising it to the scientific level of the great schools. The programme complex created by Bulgarian physicists for experimentation and estimation of basic monoparticle characteristics for the heavy nuclei is being used by all specialists interested in the microscopic theory of the nucleus.

The period after 9 September 1944 saw the beginnings of research in other important areas of physics as well. In the sphere of atmospheric physics the studies of Lyubomir Krastanov on the phase transitions of water in the atmosphere and its turbulence continued. The scientific school of Lyubomir Krastanov is well known abroad. The works of his scientific collaborators are often cited and used in atmospheric science and in investigations of atmospheric processes. Some of the earlier studies are considered to be classics; the more recent findings are of great practical significance.

In the sphere of technologies applied in microelectronics, high-resolution photographic plates have been developed in Bulgaria. They can be used to produce photographic patterns for the needs of microelectronics. Some theoretical and experimental studies have been carried out, as well as evaluation of the factors limiting resolution in electronic lithography. Investigations on electronic resistances and the possibilities for their practical application have also been made. Ionic implantation is used to create a technology for the production of silicon dosimetric detectors for nuclear radiation and a technology for synthesis of SiO_2 layers on silicon. The defects occurring during this process in the structure of silicon and germanium have been studied, as well as the changes taking place in the properties of the metal alloys layer under the effect of nitrogen implantation.

The new phenomenon discovered – hypercannellation – offers new possibilities for studying crystal surfaces. Systems have been worked out to obtain a mono-crystal layer of alphaquartz, several tens of microns thick, on the silicon surface. Technologies have been developed for obtaining effective protective layers of silicon nitrate at low temperatures and a method for dry plasmic etching of aluminium.

Multilayer interference films have been investigated with a view to producing in Bulgaria thin-layer interference filters with a high coefficient (above 80 per cent) and a narrow pass-band, as well as laser mirrors (narrow- and broad-band) with a very high reflection coefficient for the different spectral regions.

A new and promising trend – cryogenic radiophysics and electronics – has been developed in recent years. A superconducting monocontact quantum interferometer has been designed for the first time in Bulgaria. The sensitivity attained of the magnetic flux is among the major achievements in the world. The interferometer can be used for high-sensitivity electromagnetic measurements, by recording the changes in the magnetic field with the sensitivity which is necessary in important practical and biomagnetic measurements, for detecting magnetic anomalies and for establishing the magnetic properties of the materials.

Flexoelectricity in liquid crystals is a phenomenon similar to piezo-electricity in solid bodies. The molecular-static theory on the direct and inverse flexoeffect in nematic crystals has been developed in Bulgaria. Expressions have been obtained for the two flexo-electric coefficients in the case of real orientation order of the medium. The influence of steric molecular asymmetry and flexo-electricity on the dielectric properties of liquid crystals has been demonstrated. A general theory of dielectric-flexo-electric deformations induced in a homogeneous electric field under static and dynamic conditions has been assessed, introducing the notion of surface molecular field.

The approach used in the study of the general theory of liquid crystals includes constant monitoring of the concrete molecular conformation in explaining the macroscopic properties of liquid crystals. This approach has led to the explanation of some electro-optical phenomena in thermotropic liquid crystals, as well as to a fruitful transfer of the notion about steric and electrical molecular asymmetry and flexo-electricity in the physics of lyotrophic liquid crystals, in biological membranes included. These studies confirmed the notion about the unity of the liquid-crystal state of matter, about the joint action of the

The Bulgaria-1300 satellite made under the Intercosmos programme and launched in 1981

The Bulgarian cosmonaut Georgi Ivanov

three types of molecular asymmetry – electrical, steric and amphiphilic – and on this basis about the effectiveness of the complex investigation of some problems, ostensibly very different, of liquid crystals physics.

Space Science was born in Bulgaria as a result of the country's participation in the space research programme of the socialist countries – Intercosmos. Bulgarian devices were launched into outer space on board the satellites Intercosmos -8, -12, -14 and -19, of the heavy rockets Vertical-3, -4, -6 and -7, and six meteorological rockets M-100. An expression of the high regard in which Bulgarian space research activities are held was the invitation for Bulgarian apparatus to be launched in India in the containers of two Indian rockets, Centaur-II, in 1978 and 1979. Twenty different systems and apparatuses, designed and produced in Bulgaria, have been launched in these and other types of spacecraft. Bulgaria specializes in two directions: devices for measuring the basic plasmic characteristics of the upper atmosphere (electronic and ionic concentration, and their respective temperatures) and electro-photometric systems for measuring the natural optical irradiation in outer space.

Control and measuring devices and equipment, ground-based devices for calibration and testing of space equipment, ground-based spectrometers for measuring the spectral reflection characteristics of various mineralogical, plant, soil and other formations, have been designed and produced.

Research has resulted in original approximate expressions for the distribution of the electronic and ionic concentrations and of the electronic temperature at a height of 1,300 km above the Earth. New data have been found on the so-called mid-latitude drop of the electronic concentration (a drop in the concentration of the electrons observed at moderate geographic latitude). For the first time in world space practice, full vertical profiles have been obtained for a number of daytime basic light emissions from the high atmosphere and their sources have been clarified. A large number of theoretical and applied studies resulting from the operations of the space apparatus designed by Bulgarian specialists, new ideas and suggestions for future research have been reported at international meetings on space research.

Bulgarian scientists made a considerable contribution to science and technology in the working out of a varied and detailed programme for the first Bulgarian cosmonaut – Georgi Ivanov. Three original apparatuses were worked out: the Duga electro-photometric system, the Spektur-15 multi-channel chamber and the Sredets apparatus for psychophysical tests. Twenty-seven different experiments in physics, orbital technologies, remote sensing of the Earth, biology and medicine in space were included in the cosmonaut's programme. Four 'Pirin' ampoules with new semiconductor materials and alloys were designed and produced for the orbital technologies. Bulgarian foods were used during the spaceflights in fresh and lyophilized form. Bulgaria is a valued partner for space co-operation not only for CMEA member countries, but also for the USA, Italy, India and Greece.

Chemical Sciences. Physico-chemistry represents one of the most outstanding trends in the chemical sciences, with firmly established traditions. It comprises the theory of crystallization phenomena and the fundamental research associated with them. The foundations of the molecular-kinetic theory of crystal growth were laid in the early 1930s; these made it possible to gain an insight for the first time into the molecular nature of these phenomena, to elucidate the basic problems involved in the formation of crystal nuclei, problems of the kinetics of the growth of crystal walls and of the equilibrium forms of crystals. This trend developed under the scientific guidance of Ivan N. Stranski and Rostislav Kaishev, and it is one of the central domains in Bulgarian physico-chemistry. Considerable results have been obtained in this field: the effect of external factors on crystal growth and crystal nucleus formation has been made clear, e.g. the influence of foreign substrata and nuclei, of adsorption and inclusion of admixtures and the electric fields. Important contributions have been made to the modern atomist theory about phase formation and crystal growth, vitrification, thin-layer formation and the processes of electro-crystallization. Original methods have been worked out for the experimental investigation of crystallization processes, more specifically of electro-crystallization problems, which have made it possible to study these processes for the first time at molecular level, to confirm the basic theoretical formulations and accumulate considerable experience on the formation phase composition, texture and other basic characteristics of the thin layers, obtained by means of electro-crystallization or through condensation from the gas phase.

Important results have been obtained in the last three decades in another subdivision of physico-chemistry as well – electro-chemistry – more specifically in the field of electro-chemical power sources. Research on the influence of the crystallographic structure on electro-adsorption and electro-catalysis, on the theory of gas-diffusion electrodes, on the electro-chemistry of lead batteries, on some electro-chemical systems which look promising for the development of new electro-chemical power sources, as well as on some methods for studying the electrode processes, is considered to be fundamental. Purely practical results have been obtained on the basis of this research, e.g. metal-air elements for light signalling in the sea have been designed, and new progressive technologies and new designs of lead batteries have been developed. The results of these studies are extensively cited in world scientific literature; they are applied in solving concrete research tasks as a starting-point for new research in this field.

Bulgarian physico-chemical science has scored great successes in the study of surface phenomena and disperse systems – one of the important subdivisions of colloid chemistry. Methods have been created for studying the thermodynamic and kinetic behaviour of these thin liquid films. The basic colloid systems – emulsions and foams – have been modelled in this way. Important aspects of the primary process during flotation have been elucidated. This achievement, together with the results of the application of original methods to study light diffusion in colloid systems, to a considerable extent solves the problem concerning the stability of disperse systems.

Bulgarian physico-chemistry is a leading science in the study of photographic processes in solid-state bodies. Studies have been carried out for the first time on the properties and characteristics of the so-called holes (defect electrons) formed in solid bodies during the primary photoprocess. Original methods have been used to clarify their role in the formation of the photographic image in the photomaterials. The theory of photographic processes, previously based only on the electron stage, acquired its present-day form. A basically new method for photography has been created in Bulgaria on the basis of the accumulated experience and research findings. Based on the hole stage of the primary process, it possesses a number of advantages over conventional photography and it can use compounds not containing silver.

The quantum mechanical theory of the elementary transitions through energy barriers is another trend in physico-chemistry and has been successfully investigated under the guidance of Stefan Hristov. It comprises: electron and proton transitions during the electrode processes in electro-chemistry – Bulgaria is a pioneer in this field; general theory of energy barriers; theory about the velocity of chemical reactions, applied recently also in connection with the photosynthesis process; the theory about electron transitions in solid bodies which explains a wide range of phenomena: electron emission of metals and semi-conductors in a vacuum, and electron currents through thin layers of metals and semi-conductors.

Adsorbents have been worked out on the basis of natural raw materials, as well as new types of highly active and highly selective oxide catalysts for decontamination of gases released from industry and the exhaust gases of motor vehicles. A highly effective method for the production of low-temperature catalysts for conversion of carbon oxide with water vapours has been created and put into practice on the basis of theoretical knowledge. The parameters of the newly-implemented catalysts are superior in many cases to other catalysts used in world practice.

Traditional studies on the effect of the gloss-forming admixtures on the mechanism of deposition and on the physico-mechanical properties of metal coatings have made it possible to create modern technologies for the production of shiny zinc layers from slightly acid or slightly alkaline electrolytes.

Bulgarian analytical chemistry has long-standing traditions and enjoys undisputed international prestige. Great successes have been scored in the fields of analytical chemistry of rhenium, in complexometry, in raising the sensitivity of spectroanalytical determinations and in atom-absorption analysis.

Organic chemistry helped consolidate the prestige of Bulgarian science many years ago. The studies initiated half a century ago by Dimiter Ivanov on the application of magnesium and on organic synthesis are already considered classics. He and his school should be credited with the detailed studies of Bulgarian essential oils. 'Ivanov's reaction' and 'Ivanov's reagents' are the most important achievements of Bulgarian organic chemistry.

On the basis of its own experience the Bulgarian pharmaceutical industry worked out a technology for synthesizing the medicines Furantril, Cinnarizin and Venoruton, which are ranked among the highest achievements in the world in this field, on account of their excellent properties. Particularly efficient are the new Bulgarian medicines Biocarbosin, which treats some early cancer localizations, as well as Simethylin (Tagamet) which is recognized as the best medicine for the treatment of gastric ulcers. Under the guidance of Lyubomir Zhelyazkov, the Reasearch Chemical-Pharmaceutical Institute developed a technology for the production of Furantril – a synthetic salidiuretic with weak toxicity, prescribed for treatment of oedematous states and slight hypertension. Owing to its therapeutic properties, it has proved to be an indispensable medicine among salidiuretics and cardio-vascular agents.

New organomagnesium reagents have been obtained as a result of research on the reactivity of organic compounds under different conditions (e.g. naphthalene in liquid ammonia). The phenomenon of metallotrony was discovered, which is an important contribution to modern theoretical organic chemistry. The beginnings of organic photochemistry in Bulgaria were laid with the studying of the effect of UV-light on the prototrophic and metallotrophic equilibrium.

The Centre for Hygiene in Sofia

Biological Sciences. Biologization is one of the main trends along which the economy develops under the conditions of the accelerated construction of the developed socialist society, i.e. orientation of the biological sciences towards the solution of such important social problems as the preservation of human health and the prolonging of the period of human activity, raising the activity of the natural biological resources and their rational utilization and protection of the environment.

The first important contributions in the sphere of genetics are credited to Doncho Kostov. His studies on the ultrastructure of chromosomes and on the cytogenetics of interspecies tobacco hybrids are internationally known. Among the important achievements of Bulgarian genetics and plant breeding are the hybrid tomatoes No. 10 Bison, Triumph and Ogosta and the combinations Pioneer-1 and Pioneer-2 which are known for their high quality and suitability for industrial technologies. The utilization of male sterility in the heterosis breeding of tomatoes, tobacco and maize has yielded valuable practical results. Special importance in the studies is attached to investigating the specific action, the physical mutagenic factors and the differential sensitivity of the genetic material.

Many of the results obtained, especially over the past twenty years, have been put into practice, e.g.: the discovery of the way in which the genes are localized in the chromosomes, the existence of genetic systems controlling the meiotic conjugation of the chromosomes in the species, determination of new sources of tobacco resistance to the causes of the mildew, black mould and wildfire and against aphids, of wheat against powdery and black rust and of tomatoes against bacterial fungal diseases and the white greenhouse fly.

The immunogenetic studies have resulted in tobacco, wheat and tomato hybrid varieties which are resistant to diseases. Some of them are used in breeding practice. The hybridization of heteroploid Triticale forms is one of the most effective methods for creating highly fertile and highly productive Triticale hexaploid varieties, possessing a number of valuable properties.

The possibilities for building a highly productive and profitable stock-breeding industry have been opened up through application of cryobiology, cryoconservation of sex-cells, transplantation of ovocytes and zygotes, intensification of the reproductive process and hybridization.

Original methods have been developed for immunodiagnostics, immunotherapy and immunoprophylaxis of immunologically conditioned infertility in cows, as well as methods for the early diagnosis of pregnancy in cows, methods for diagnostics and prophylaxis of immunologically conditioned (sterile) mastites in farm animals, for evaluation of the seminal fluid in inseminating stations and for the effect of its cryoconservation. An important

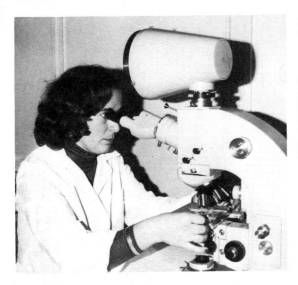

The State Institute for Control of Medicaments in Sofia

The Institute of Genetics

practical problem has been solved, namely the prolonged conservation of the seminal fluid of some farm animals, more specifically bulls, at ultra-low temperatures of 196°C below zero. Cryobiology of sex cells is a new branch of the biological sciences.

Complex research for many years on reproductive biology has resulted in vital contributions to the theoretical work on the regulation and intensification of the reproductive process in sheep and cows. Biotechnical methods have been implemented on the basis of these studies, permitting the fuller utilization of the biological parameters of fertility, which would result in increased production of meat and milk.

Bulgarian biochemists are working on an original theory about the cellular and tissue organization of animal organisms. Sex and blood tissues are differentiated as separate tissue systems. The studies on the cellular and molecular mechanisms of auto-immune processes and diseases have generated great interest. Lasting resistance against auto-immune experimental allergic enciphalomyelitis has been successfully induced.

Considerable interest is being shown in the data obtained concerning the binary structure of the nucleosomes in the chromatin, the elucidation of the mode of histone distribution during DNA replication and the methods devised for obtaining antitoxic sera on the basis of a plasmid containing the gene of the thermostable enterotoxins of *Escherichia coli*.

Studies on the molecular mechanisms of inflammatory proliferative processes in the tissues and cellular differentiation have been carried out for many years under the guidance of Rumen Tsanev. It has been proved that cell damage leads to rapid DNA degradation and the resulting degradation products are the basic factor which triggers important defence processes after injury: expansion of the capillaries, attracting of leucocytes and stimulation of phagocytosis. These data reveal one of the biochemical mechanisms of the reactive processes after injury and result in the practical utilization of the RNA hydrolysis for the treatment of slow-healing wounds.

The studies on the mechanisms activating cell proliferation after injury suggest the idea that this process is not induced by substances stimulating cell division, but by substances which normally suppress it. On the basis of this idea and of current notions on the regulations of the gene activity the first mathematical model for the regulation of cell proliferation has been worked out, based on intracellular molecular mechanisms. Through this model it has become possible to trace the process of malignant degradation of the tissues, which has led to the idea that this process may have underlying mechanisms which do not affect DNA, but only disturb some regulatory links with the genome.

The use of the mathematical modelling method led to the basically new idea that in addition to the genetic information of the protein structure, contained in the DNA, the cell should also contain another type of information which defines the programme according to which the genetic information will be used in the course of the embryonal development. This programme is not contained in the DNA, but in the protein complexes which DNA forms in the nucleus of higher organisms, and it is epigenetic in nature. Recently a number of foreign authors have accepted this idea.

Of great practical significance in experimental biology is the two-wave spectrophotometric method worked out for determining the amount of RNA and DNA in animal tissues. Methods have also been elaborated for electrophoretic fractionation of ribonucleic acids and ribosomes in gel; the methods have a higher resolution capacity compared with ultracentrifugation.

Medical Sciences. The primary task facing medical sciences is the solution of topical questions of preventive treatment in the conditions of rapid industrialization, intensive chemicalization of agriculture, the revolution in science and technology and urbanization. It studies environmental factors which have a harmful effect on the population's health and activity and the hygienic aspects of the conservation and reproduction of the environment.

Most of the scientific medical workers are concerned with the solving of important questions of preventive treatment, diagnostics and the treatment of the most widespread diseases. The comprehensive programmes for the study of cardiovascular diseases, malignant tumours, and the like, occupy a particular place in the plans of medical researchers.

New methods are being worked out at the Medical Academy for the prevention, diagnosis and treatment of diseases. They are being implemented in the health care institutions of the country through the direct participation of scientific personnel and specialists. The Bulgarian State Register of Discoveries has as its second entry the phenomenon 'sixth heart sound'. The author of this scientific discovery is Ivan Mitev from the Medical Academy's Institute of Paediatrics. The discovery offers better opportunities for functional diagnosis of cardiac activity and more specifically for closer studying of the low-frequency sound phenomena occurring in the heart. Of special interest are the investigations of the sixth heart sound in children, where its appearance is almost always a pathological finding. A great number of these sounds cannot be heard, they can only be recorded phonocardiographically. The studies of the sixth heart sound and its origin open up new vistas before cardiac diagnostics and create possibilities for new studies in the field of cardiac physiology and functional diagnostics.

Investigations on the clinical immunology of hepatic diseases have resulted in new breakthroughs in the field of internal diseases. A working team led by Atanas Maleev applied, for the first time in Bulgaria, parallel study of the hepatic B-virus: HBsAg, HBcAg and HBeAg. New original methods have been devised for determining the system e-antigen (e-antibodies), as well as an immunoenzyme method for anti-actin antibodies. These studies have served as a basis for the concept of the forms of chronic hepatic diseases having the HB-virus as etiological factor. This concept is important for differentiated treatment of the patients.

A method has been created for treatment of diabetic ketoacidosis with long-term intravenous infusion of small insulin doses. With this therapy, fast-action crystal insulin is infused in physiological serum in doses of 2–8 units per hour. The following parameters are controlled: the patient's state, his blood sugar, glycosuria, acetonuria, the level of erytocytes in the serum and the alkaline reserve. In this way ketoacidosis is surmounted for a period of three to

Atanas Maleev

twelve hours, using an average insulin dose of 5–6 units per hour.

A working team under Iliya Tomov has elaborated and is applying in clinical practice a comprehensive unidimensional (M-type) and bi-dimensional echocardiographic system for functional assessment of the left and right cardiac ventricles. Echocardiographic indicators have been established for early diagnosis of cardiac insufficiency of the left ventricle.

In Bulgaria the problem of producing artificial human organs is placed on a sound scientific basis. An Experimental and Production Base for Artificial Human Organs has been organized. Gamma-oxygenators, gamma-reinfusion devices, artificial kidneys, artificial livers, and artificial cardiac valves, have been developed and put into practice. Bulgarian inventions in this field serve as the basis for the production of artificial organs in the country. The tests have proved that the devices created by Bulgarian researchers and designers – the oxygenators of the Oxymil type and the reinfusion device of the Reinmil type – comply with the highest standards in the world. The reinfusion device Reinmil is a cardiogenic microfilter reservoir through which the blood released into the body cavities as a result of injuries or operation is processed, filtered and then reinfused to the same patient, thus achieving great economy of blood for transfusion needs.

Without this device the body rejects the blood. Sometimes the amount of this blood is 4–5 litres, or even more. It is usually replaced with blood from donors, which is a very expensive and slow process, and it often leads to complications resulting from incompatibility, pulmonary complications and pulmonary shock.

Counterpressure casting by the Angel Balevski method at the Research Institute of Metal Casting in Pleven

Angel Balevski

The original Bulgarian endoprosthesis for the hip-joint, Etropal-DMP, designed by Alexander Gerchev, is widely used both at home and abroad, and has earned high recognition. The endoprosthesis has been recognized as an invention and it is covered by patent rights in more than twenty countries. The prosthesis is made of titanium alloy and has a plastic bed in the acetabular part. Its main advantage consists in the existence of a set of acetabular capsules and femural stems, which are fixed either mechanically or are glued with bone cement. This completeness ensures the possibility to choose the most suitable method for fixing the prosthesis during the implantation in accordance with the concrete indications. Another original achievement in orthopaedic surgery is the method of Elena Planeva-Holevich of bilateral tendinous plastic surgery for injuries of the tendons in the so-called 'no man's land' of the hand. This method enjoys high recognition abroad.

The production units of the Medical Academy have achieved major successes in the development of new medicines, anti-tumour preparations included.

Ivan Matev was the first to introduce the distraction method for extension in the case of loss of the whole thumb or of part of it in hand injuries, with a view to restoring or at least improving its functions. The method has been recognized as an invention and is widely applied both in Bulgaria and abroad.

Technical Sciences. Peak achievement of the Institute of Metal Sciences and Technology of Metals of the Bulgarian Academy of Science is the method of casting under counterpressure and the processing of materials with gas counterpressure. This is known as the Balevski-Dimov method and is recognized as an invention.

The method finds application in several fields: casting of machine parts from metal alloys, casting of machine parts from non-metal materials, gas alloying and the production of new alloys. The main feature of this method is the production of gas counterpressure in the casting mould, acting in the opposite direction to the movement of the melt filling the mould. Filling with the molten metal takes place under the action of another pressure which is higher than the counterpressure. In this way the processes of casting and crystallization are controlled and regulated, and additional opportunities are created for improving the quality of the casting. One of the main applications of the counterpressure casting method is the casting of parts for which there are high requirements with respect to strength and quality. Bulgaria has supplied machines and technologies for counterpressure casting, and know-how, related to the Balevski-Dimov method, to the USSR, Poland, Hungary, Austria, Romania, India, Canada and Italy, which is of considerable economic benefit to Bulgaria. The principle of casting with gas counterpressure is very effective for the processing of foaming thermoplastic melts. This serves as a basis for new ideas about developing new methods and technologies for casting foam rubber and for making machinery and equipment for their industrial production.

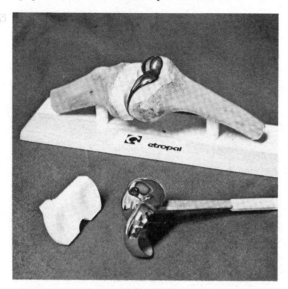

Etropal – an artificial knee joint

Great opportunities are offered by the counterpressure casting method. It can be applied not only to the casting of machine parts, but also to the gas alloying of metal alloys. The system for metal casting in closed premises under conditions of high pressure, designed by a working team led by Ivan Dimov, develops the counterpressure casting method in a new aspect. The system is designed for the production of high-quality stainless and nitrided tool steels. It can be used to produce casts and blocks intended for further rolling or hammering, as well as for different types of profile casting. A working team led by Ivan Dimov has also created a new method for the electroslag melting of metals and alloys with counterpressure or under pressure. The method is applied for the production of austenitic nitrided steels either nickel-free or with reduced nickel content. Such steels are welcome from a technological and economic point of view, because they successfully take the place of the more expensive chromium-nickel steels type 18–8.

More and more foreign firms are displaying interest in the method of casting with counterpressure.

One of the strategic trends of scientific and technological progress is the complex automation and electronization of the national economy. On the basis of microprocessor technology, the Institute of Cybernetics and Robotic Engineering of the Bulgarian Academy of Science designed the control devices and the mechanical aspect of the first Bulgarian industrial robots, produced in large numbers, for transport, welding and assembly operations, as well as the digital programming control of some metal-cutting tools. The robot-operated welding complex RB-251 provides effective arc welding within preset parameters and with high-quality results. A class of original planetary devices – Isplan – for feeding welding wire has been designed and put into practice, which makes it possible to conduct high-quality continuous welding operations in various protective media.

In the sphere of engineering cybernetics, microprocessor hardware and software have been designed for hierarchic decentralized control of continuous processes – the MIC-2000 system. Methods and software have been worked out for control of completely centralized and hierarchic systems with pure delay, as well as 16-bit modular microprocessor systems for CAD/CAM systems and for operative systems in industrial transport – the systems for operative management, control and recording from Trasi class in agriculture, the Eco systems for automation of greenhouse production, and Delta-Chrome for tobacco sorting; in the cement industry – the system Betoncontrol. General methods and software have been elaborated for the multi-criterial analysis of control systems.

The first Bulgarian personal computers have been created at the Institute and are already in production.

The Central Laboratory of Physico-Chemical Mechanics of the Bulgarian Academy of Science is engaged in research on the changes in the structure formation of cement stone. Inclusion of new and more active components in the composition of the complex admixtures has been found to have a favourable effect on hydration and hence on the structure-forming of the cement stone. Accelerated setting is guaranteed and also parallel with it control over the expansion of the cement stone. The new methods and development projects, for which four authors' certificates and eight patents have been granted in the USA, Sweden, the GDR and Spain, cover all technologies in construction. The results obtained form part of the comprehensive objective for further improvement of the methods for designing and optimization of the composition of concretes and concrete solutions, for prediction of their properties and behaviour under operational conditions and for producing concrete with the desired structure and properties. In recent years specialists from the laboratory have been working on the possibility of introducing polyamide phenolformaldehyde foam rubber materials in construction. These materials have several important advantages: increased bending strength, high thermo-insulating capacity and they have no corrosive effect on the surface to which they are glued. The working group at the laboratory has applied phenolformaldehyde foam rubbers in construction in heat and sound-insulating materials in different forms (panels, insulations for pipelines and containers), for the production of thermo-insulating and carrying elements with steel casing, in railway car-building, shipbuilding and refrigeration technology.

A research team led by Nikola Todoriev worked out a new technology for burning low-calorie fuels with high ash content and high humidity, without preliminary drying in special drying installations. With the active help of institutes in the USSR, and after surmounting many difficulties, the problem was successfully solved. At

The Geophysical Institute

present the four sectors of the Maritsa-Iztok 3 Thermoelectric Power Plant produce power using this highly effective technology. Future extensions of the power plant will also burn the low-calorie fuels of the region without preliminary drying.

An important achievement of prospecting geophysics is the decoding of the in-depth geophysical structure of Bulgaria and the eastern part of the Balkan Peninsula. The schemes worked out reflect the behaviour of the surface after Mohorovich and Conrad, as well as the spatial distribution of the large fault structures. The analysis of geophysical fields offers an opportunity for working out a preliminary model for the block tectonics of the territory of Bulgaria. It is closely related to the problem of the pathways of the deep sources of substances and energy, and contributes to the more comprehensive elucidation of the inherent regularities in the accumulation of ore and mineral deposits.

A major achievement in the processing of mineral resources in Bulgaria is the mining of the Medet copper-porphyry deposit. In its scale and technology it is a large mine, even by world standards. The technology implemented for the first time in the Medet factory guarantees maximum complex extraction of copper and other valuable components (pyrites and molybdenum) in high-quality concentrates, with minimum capital investment and operational costs. It represents a major scientific and technological achievement.

A new trend has been developed in recent years, namely the extraction of halogen and rare elements from strongly mineralized natural waters, oil-accompanying waters and effluent lye after salt production from sea water.

Quartz sands obtained by enrichment of the kaolin raw material are the main source for supplying quartz raw materials for the needs of industry. The technology designed for deferrization of quartz sands approaches the physico-chemical properties of the processed sands to the ideal raw material for the production of potash and lead crystal glass. In this respect Bulgaria ranks among the leading countries: the USSR, the USA and the FRG.

The growth and expansion of the material base for the non-ferrous industry has been paralleled by the development and implementation of a number of applied scientific projects which are the work of Bulgarian specialists.

Many new processes and apparatuses have been designed for zinc metallurgy, which makes it possible to further improve the entire technological process. The method and device worked out for the separation of the solid phase from zinc sulphate solution during purification from admixtures are considerable technological achievements. The development of the new technological variant and the application of chamber disc filters provides a solution to one of the most labour-consuming processes, extremely difficult to control, in zinc hydrometallurgy. Parallel with this, a much higher quality of electrolyte and solid semiproducts is attained.

The application of continuous fuming of lead slags with black mineral oil is a remarkable scientific and technological achievement in the field of lead metallurgy, guaranteeing complex utilization of the components of the raw material. The technology has been patented in many countries. It has a number of advantages over the periodic system, namely: higher productivity, better labour conditions, lower specific fuel expenditure and fuller extraction of the lead and zinc.

Device for laying boride coatings on cutting instruments

Owing to their high quality, Bulgarian non-ferrous metals are accepted without certificate at the London Metal Stock Exchange.

The iron ore from Kremikovtsi is unique in its composition. With a view to its fuller utilization, pyrometallurgical technologies have been worked out for lead separation, as well as an original technology for the production of metallized products through the extraction of lead, zinc, silver and manganese. A new technology for the production of cast iron with high and moderate manganese content has been introduced at the L. I. Brezhnev Combine in Kremikovtsi. For the first time in world practice, quantitative data have been obtained on the distribution of lead and on its behaviour in the blast furnaces during processing of lead-containing raw iron materials. Complex studies have also been carried out on the distribution of manganese and sulphur between the cast-iron and the blast-furnace slag in cases of increased manganese and barium oxides in the raw materials.

The graphite electrodes with protective coatings, known all over the world, are the creation of a working team led by Alexander Vulchev. More than 50 per cent of the electrode consumption results from lateral oxidation at temperatures of 1,600–1,700°C during the steel-making process. Numerous attempts at creating a suitable protective coating to reduce lateral oxidation, carried out in many countries, were unsuccessful. Only the protective coating made in Bulgaria met with the approval of industry. This coating is distinguished by two main properties: a low melting temperature (about 600°C) with high aluminium content and excellent cohesion to the graphite surface, which is achieved after the processing of the electrode with an electric arc. Specialized machines producing the protective coating with the minimum of labour, energy and investments have been designed in Bulgaria. The method is covered by patent rights. Its innovations and improvements, the machines for the production of the protective coating and the reconstruction of the electric furnaces, have been patented in more than twenty countries.

A working group led by D. Petrov has elaborated a method for electrolytic refining of copper using increased current density of up to 400 A/m^2, unlike the classical method adopted throughout the world which uses about 230 A/m^2. This is achieved by applying alternating current. The technology can be introduced in any plant for copper refining. The method has been patented in France, Belgium, Great Britain, Chile and Canada. It has been or is being successfully implemented by firms in Italy, Japan, the USA, Spain, Turkey, Finland, Iran and Mexico.

Alternating current is also used successfully in the new method and electrical circuit for the electro-extraction of zinc, developed by a research team led by Ivan Enchev. They ensure the optimum utilization of alternating rectifiers. Labour productivity increases by about 30–40 per cent, electric power input decreases by about 100 kW/h per ton of metal, the quality of the deposited cathode metal is improved and capital investment is reduced by 20–30 per cent. The technology permits the increase of the current density, whereas the duration of the deposition may last 96 hours, depending on the requirements. High awards have been given to the project: gold medals at the International Trade Fair in Plovdiv and at the International Exhibition of Inventions in Brussels, a silver medal at the International Exhibition of Inventions in Brussels, a silver medal at the International Exhibition of Inventions and Discoveries in Nuremberg (FRG). The invention has been patented in Canada, the USA, the FRG, Belgium, Spain, Great Britain, Czechoslovakia and France.

One of the remarkable achievements of research and development activities in light industry has been the creation of the Prenomit technologies for the production of smooth and textured yarns – the work of a research team led by Georgi Mitov.

Three basic modifications have been introduced for machines for combined yarns, for spinning smooth and textured yarns and for twisting textured yarns with and without loops. The machines designed on the basis of the

Prenomit technologies have been patented in the USA, Japan, France, Great Britain, the GDR and other countries. These machines are produced in Britain under patent rights from Bulgaria.

Agricultural Sciences. Agricultural sciences in Bulgaria have a long history and sound traditions. Considerable results have been obtained in the breeding and introduction of cultivated plants. A total of 728 varieties and hybrids from 67 species of field, vegetable, viticultural and other crops have been created since the time of the first agricultural experimental stations established at the turn of the century. More than 613 of them have been bred during the years following the socialist revolution in Bulgaria. A considerable number of Bulgarian varieties have attained recognition as the best examples in plant breeding practice in the world.

A research group led by Todor Rachinski has been working on the wheat varieties Dobrudja-1, Slavyanka, Levent, Jubilee and Ludogorka. Dobrudja-1 has been obtained by hybridization between the varieties Liebululla Bezostaya-1. It has large, very heavy and perfectly growing grain. This is a high-quality and strong variety, with excellent baking properties. The productive capacity of the variety under field conditions is 7,100–7,300 kg per hectare. The Slavyanka variety has been obtained through inter-varietal hybridization from the cross NS-313 x Bezostaya-1, and is the highest quality and strongest variety grown in Bulgaria.

The variety Levent has been obtained through inter-varietal hybridization from the cross Bezostaya-1 x Fiorello. It is suitable for cultivation in moderately fertile soils in the plains and semi-mountainous regions. The productive capacities of the variety under field conditions are 7,100–7,300 kg per hectare. Among the varieties cultivated in Bulgaria, Jubilee is the most resistant variety to stem bending. It is classified as a low-stem intensive variety. It is obtained through inter-varietal hybridization from the cross Fleischmann x Bezostaya-1, being suitable for highly intensive cultivation and under irrigation. In 1977 and 1978 this variety ranked first in the world contest for wheat varieties, organized by the FAO in 48 countries. The productive capacities of the variety under field conditions are 7,900–8,000 kg per hectare. Ludogorka is the first Bulgarian high-quality strong wheat variety. It is obtained by the method of inter-varietal hybridization from the cross Bezostaya-1 x Étoile de Choisir. The wheat grain is medium-sized, of high quality, with increased protein content – up to 14.5 per cent, and wet glutene content exceeding 32 per cent. The productive capacities of the variety under field conditions are 6,500–6,800 kg per hectare.

The varieties Sadovo-1 and Sadovo Super are a great achievement of wheat breeding and have made a considerable contribution to the increased wheat grain. The varieties were created by a research team led by Pavel Popov. The common winter wheat variety Sadovo-1 has been obtained as a result of hybridization between the Bulgarian variety Yubileina-3 and the Soviet variety

Laboratory of plant cytoembrology at the Institute of Genetics

Bezostaya-1. In 1977 and 1978 Sadovo-1 ranked second at the world testing of 30 varieties in 60 regions of the Northern and Southern hemispheres, which proves its high productivity class and its qualities of a genotype having wide ecological adaptability. In both years the variety ranked first in the world with respect to grain size. The Sadovo Super variety is the result of complex hybridization between the French variety Moisson, the variety Yubileina-3 and the Soviet variety Pshenichno Pireen hybrid No. 186. Since 1978 the use of this variety has been widespread throughout the country. It is characterized by high resistance to droughts and satisfactory resistance to cold; its vegetational period is approximately equal to that of the Sadovo-1 variety and it is highly resistant to fungal diseases. This is the only Bulgarian variety which is resistant to bacteriosis.

The hybrid tomato varieties No. 10 x Biton, Triumph, Ogosta and Kristi are the work of a research team led by Hristo Daskalov. In Bulgaria the entire early and late field production of tomatoes and about 40 per cent of the moderately early production is provided by the heterosis varieties. Heterosis seed production reaches 10,000 to 12,000 kg seed annually, at an average yield per hectare of 200–300 kg. Bulgaria ranks first in the world in the production of hybrid tomato seeds. Comprehensive research has been conducted for more than forty years at the Maritsa Institute for Vegetable Crops in Plovdiv and at the Institute of Genetics of the Bulgarian Academy of Science – by a working team led by Hristo Daskalov, with

A series of maize hybrids distinguished by their high resistance to droughts and good productivity have been created following planned breeding and genetic research at the Maize Institute in Knezha. Dry biomass yield under irrigation reaches 89,450 kg per hectare for some varieties.

Bulgarian stock-breeding has made important contributions to the improvement and creation of breeds of milk cows, and especially fine-fleeced and semi-fine-fleeced sheep.

A research team led by Ivan Tonev created the North-Bulgarian fine-fleeced sheep breed with its two intra-bred types: Dobrudja and Shumen. The sheep of the new breed are characterized by the following basic breeding parameters: live weight 67–68 kg for the ewes, compared with 70–76 kg for the local breed; wool yield 7.2–7.9 kg versus 2.1–2.3 kg for the ewes, and 14–15 kg versus 4.8–5.2 kg for the rams. The quality of the wool is radically different. It is fine, its evenness coresponding to quality 64. Staple length for the ewes is 8.0–8.5 cm. The milk yield is 95–98 litres for the whole lactation period, compared with 72–78 litres for the local sheep breed. Fertility is also high: 124–125 lambs per 100 ewes, compared with 105–107 for the local breed.

The Thracian fine-fleeced sheep breed has been created by a group of researchers led by Rada Balevska, applying

Hristo Daskalov

The Maritsa Institute of Vegetable Culture in Plovdiv

a view to elucidating the nature of the heterosis phenomenon and its practical utilization. A number of inherent regularities have been discovered, contributing to the extensive and successful utilization of heterosis in tomato breeding. For more than twenty years the variety No. 10 x Bison has been the only variety of early field growing tomatoes. Owing to this, Bulgaria is one of the leading exporters of fresh tomatoes. The variety is also widespread in the USSR, Romania, Yugoslavia, Albania and elsewhere. Today the Triumph variety accounts for about 60 per cent of the early production of tomatoes for export. This hybrid variety is one of the most highly productive and high-quality tomato varieties. With good agrotechnical care it yields 60,000–80,000 kg per hectare. The Ogosta variety is widespread in Bulgaria and is used in the production of high-quality canned peeled tomatoes for export; between 15,000 and 20,000 tons are produced and exported annually. The promising hybrid variety Kristi was recently created: it has complex resistance to tobacco mosaic, Ventriculum, Fusarium and nematosis, being of high fertility and excellent quality. With good agrotechnical measures it produces 80,000–100,000 kg high-quality tomatoes per hectare, 90 per cent of which are exported. The wide introduction of this variety in early field production contributes to raising its effectiveness and competitiveness on markets abroad.

the method of complex reproductive crossing of the local coarse-fleeced and semi-coarse-fleeced sheep (from the Maritsa and Stara Zagora regions) bred in the Thracian lowlands (Southern Bulgaria), with rams from the following breeds: merino, fleisch and Caucasian fine-fleeced breed. The sheep from the Thracian breed have fine fleeces, good milk, butter and meat yield. The wool is white, with fineness of quality 60–64, with good length and staple structure of the wool fibre. Average wool yield is 13–14 kg for the rams and 6.5–7 kg for the ewes. The average live weight of the rams is 110–112 kg, and 65–70 kg for the ewes. Young lambs grow very well, reaching 28–30 kg towards the 90th day, with a daily growth rate of 280–300 g. Ewes from the Thracian breed have the highest milk yield among all fine-fleeced and semi-fine-fleeced breeds in Bulgaria. Their milk yield is 120–125 litres for a 200-day lactation period. Fertility is also high: 123–125 lambs per 100 ewes. Twinning ewes reach 40–45 per cent in the different breeding flocks.

Strategic Trends. The efforts of the country's scientific potential are aimed at solving a number of important problems related to the further increase in the social labour productivity, the efficient utilization of the raw materials and energy resources and improved quality.

Automation of the discrete production will be attained by designing series of highly productive machines, robots, transmanipulators and other equipment based on the modular principle for building flexible automated production systems. Efforts will continue for the automation of the most widespread assembly operations.

The main trends in the automation of continuous production for the next five to ten years are: complete transition to decentralized and hierarchic digital control and management of the technological processes, operative dispatcher control of production and integrated management systems for the different enterprises. In the near future, regular production of the basic technological means and systems will be implemented, with a view to automating the design, experimental and research activities, comprising the entire cycle from the idea to the preparation for production.

Automation of management will be oriented towards the technical preparation, technological-economic planning and operative control of production. Integrated systems will be created for the management of enterprises, combines and complexes in material production.

Fundamental and applied research of automation, computer technology and microelectronics is concentrated above all on the elaboration of methods and algorithms for decentralized and hierarchic management systems, as well as optimization, recognition of patterns and situations, elaboration and implementation of new software and hardware for high-speed processing of large data arrays, as well as dialogue between man and computer; implementation of electronic-radiation and X-ray lithography, ionic implantation, laser methods for the processing of materials and other promising technologies in microelectronics; carrying out supporting research in the sphere of the

Data processing at the Georgi Dimitrov Research and Production Complex in Plovdiv

Card-punching room at the Institute of Techno-Scientific Information in Sofia

Todor Pavlov

cyrogenic, molecular, bio- and magnetoelectronics, with a view to their industrial utilization after 1990.

The creation and application of basically new technologies is considered to be the basis for radical changes in the formation of new sub-branches in the national economy and a fundamental factor for considerable economies in all types of resources. The development of new technologies will be an important task facing the research and design institute working on the production and implementation of new articles. Preference will be given to the development and implementation of plasmic, electron-ray, laser, vibrational and other technologies, geobiotechnologies and new technologies for power production from non-traditional sources.

Biotechnologies occupy an independent position in the general development of technologies. The main tasks facing science and technological progress in this sphere consist of the elaboration and implementation of a wide range of biological products for meeting the country's needs and for export, mass introduction of the biological processes into the various sectors of the national economy, resulting in radical improvement of the quality of production and an increase in the scientific and technological level and effectiveness of these processes. Fundamental and applied research is focused on the development of gene and cell engineering, and its application in biotechnologies, genetics and plant breeding.

The creation and utilization of new materials is one of the most important trends of scientific and technological progress, aimed at better quality, expansion of supplies of raw materials and economical utilization of resources. The efforts of scientists are orientated towards devising methods and technologies for the production of new types of materials out of Bulgarian raw materials, production and ample utilization of non-traditional materials, mainly through combining raw materials of different composition and properties, programming on this basis their qualities depending on the needs. In the field of construction materials there will be a substitution of ordinary steels and rolled stock with new materials processing properties such as high strength and lower metal and energy consumption; production of new non-traditional materials; and complex utilization of local raw materials for the production of new materials with high operational characteristics. Wide use will be made of Bulgarian methods for processing with gas counterpressure in powder metallurgy, hydroplastic deformation and composite materials.

BASIC TRENDS AND ACHIEVEMENTS IN SOCIAL SCIENCES AND HUMANITIES

After the socialist revolution the main trends in social sciences and humanities were defined as the study of the development of society, its laws, structure, social relations, intellectual life, culture and formation of the personality; study and evaluation of the historical past of the nation and State; language, literature, culture, and the anaylsis of the spiritual and material values created in the course of the thirteen centuries-long existence of the Bulgarian State.

In **philosophy** are elaborated: general methodology of science and problems of ontology, epistemology and the history of philosophy. From its early days, philosophy took an orientation toward a reconstruction of the ideological front on the principles of dialectical and historical materialism (T. Pavlov, S. Ganovski, A. Kisselinchev, A. Polikarov, N. Iribadjakov, P. Gindev, and others). Parallel with this, important results have been obtained in the study of the problems of the developed socialist society, the socialist way of life, the philosophical and methodological problems of scientific knowledge and the history of philosophy.

An essential contribution to the enrichment and further development of the Leninist views on the nature, genesis and varied forms of reflection as a universal property of matter is made in a number of special studies and in the two-volume work *The Leninist Theory of Reflection in the Light of the Development of Science and Practice* (Editor-in-Chief: Todor Pavlov) on the basis of a detailed and comprehensive analysis and generalization of the latest achievements of science and scientific and technological thought.

Serious research has also been carried out in the spheres of the history of world philosophy and the history of philosophical thought in Bulgaria.

The analysis of dialectics as theory and method introduces further clarification of the structural details of materialist dialectics, and of the theoretical aspects of dialectical categories and dialectical method. Certain more concrete problems of logic and the methodology of scientific knowledge have been analysed.

A conception of the socialist way of life has been formulated, which reveals the changes in relations between people in society and the impact on man's development as a social subject. The intrinisic unity of the

Fifteenth World Congress of Philosophy in Varna, 1973

socialist way of life and real humanism, moral relations as an aspect of the way of life, and the formation of a standard of living in the social realization of the individual are clarified.

Many varied studies have been performed in historical materialism (Zh. Oshavkov, T. Stoichev), logic (A. Bunkov, D. Spassov), aesthetics (T. Pavlov, A. Stoikov, I. Passi), atheism (N. Mizov), philosophical problems of the natural sciences (N. Iribadjakov, A. Polikarov, I. Kalaikov) and history of philosophy (R. Karakolov, M. Buchvarov).

Research in **sociology** is orientated mainly toward the macrosocial processes and the fundamental changes in society (Zh. Oshavkov, N. Yahiel, S. Mihailov). Studies have been successfully carried out on the periods, structure and functioning of the socialist society as a whole system, particularly the regularities in the construction of the developed socialist society. Attention is focused on the changes in the social-class and demographic structure, the social problems of working groups, science and technological progress, the social characteristics of art and mass communications, the problems of social management, different social groups, such as youth and women, the family and reproductive behaviour. Empirical sociological investigations, on a nationwide or more limited scale, have been carried out, with extensive use of mathematical and statistical methods and computer technology. Information from these investigations is also widely used for the purposes of social management.

Jacques Natan

Research on the labour capacities and the working realization of the Bulgarian population has been carried out, prompted by studies on the problem of Population and Social-Class Structure of the Socialist Society.

A characterization of the population of the People's Republic of Bulgaria is the first study in which modern sociological and statistical methods are used to determine the quantitative parameters of the capacity, ability and readiness for labour characteristic of the different social groups and strata of the country's population. The study represents a contribution to the methodology of empirical research. A sociological investigation of religious belief in Bulgaria was carried out in 1952. The results and conclusions of this investigation are reflected in the collective publication *The Process of Surmounting Religion in Bulgaria* (1968). Special attention is paid to the role of the sociology of science in the development of present-day society and in the formation of the scientific organizations (N. Yahiel).

In research on the contemporary social theories significant results worthwhile mentioning are the two series *Methodology and Criticism* (in five volumes), and *The Arms Race and Ideological Struggle,* and the collective work *Problems of Peace and War.*

In **economics** the main concern is the problem of the country's socio-economic development. Research is carried out on problems of contemporary Capitalism, and on the political economy of Capitalism and the developing countries (T. Vladigerov, E. Kamenov, T. Trendafilov, and others). Detailed and comprehensive studies exist in the field of economic history and criticism of modern bourgeois political economy and revisionism (J. Natan, K. Grigorov, and others). Investigations are also being conducted in the political economy of Socialism, on fundamental problems of the transition period, the inherent features of socialist development and the stage of formation of the developed socialist society (K. Dobrev, V. Nikiforov, and others), the management of the national economy and the effectiveness of social production, related mainly to the methodology of planning. E. Mateev, I. Iliev, I. Stefanov, A. Totev, and others have contributed greatly to the development of statistics and to the emergence of the economic-mathematical trend in economic sciences. A number of works have been published on the territorial distribution of the productive forces and on territorial planning. Important scientific results are reported in the collective publication of Bulgarian scientists and economists, the multi-volume edition *The Economy of Bulgaria,* the monograph *The Formation of the Economy of the Developed Socialist Society,* and the work *The Socio-Economic Effectiveness of Scientific and Technological Progress.* These publications discuss the problems of all-round intensification of the economy, providing a qualitative characterization of the economy of developed Socialism and proposing methods of evaluating the socio-economic effect of the generation and implementation of science and technology.

A co-ordinated system of economic and mathematical methods for the long-term and medium-term prediction of the economic development of the People's Republic of Bulgaria and for characterizing the basic properties and factors of economic growth and further improvement of the economy has been set up with a view to modelling the long-term development of the economy.

Considerable research has been carried out in connection with the country's socio-economic development. Guidelines for the socio-economic development of the different regions for the 1981–2000 period have been worked out. A system incorporating qualitatively new methods of long and medium-term prediction has been designed.

Two publications commissioned by organizations belonging to the UN (Unido) – *Industrial Co-operation between Bulgaria and the Developing Countries* and *The Development of*

Dimiter Kossev

Hristo Hristov

Industry for the Basic Means of Production – investigate the fundamental principles and requirements of Bulgaria's industrial co-operation with the developing countries, financial and legal aspects of industrial co-operation, and also evaluate their mutual benefit and effectiveness.

Plans and forecasts have been worked out for the development of the Bulgarian economy up to the year 2000, offering variants of this development relative to possible trends in the world economy.

Considerable successes have been achieved by historians in the study of Bulgaria's past, in the search for and publication of documents about **Bulgarian history and culture.**

Bulgarian historians have published a three-volume *History of Bulgaria* in two editions, as well as short histories of the Bulgarian people and State (D. Angelov, I. Duichev, H. M. Danov, A. Burmov, D. Kossev, H. Gandev). On the occasion of the thirteen hundredth anniversary of the founding of the Bulgarian State, publication of *A History of Bulgaria* (in 14 volumes), edited by D. Kossev, H. Hristov, V. Vassilev, et al. has begun. This is the first comprehensive fundamental study of Bulgarian history. The volumes cover the historical development of the Bulgarian lands from the Palaeolithic Age to the present day. Four volumes had appeared up to 1983.

The cultural and historical heritage of the Thracians is being intensively studied (H. M. Danov, A. Fol, D. Dechev). In the monographs *The Historical Geography of the Thracian Tribes before the Third Century BC* and *Thracian Monuments* (in three volumes) all existing material is systematized, much of it newly-discovered.

The elucidation of the cultural and historical heritage has shed light on various problems related to the formation of the Old-Bulgarian culture, on the basis of many years of archaeological excavation and historical research into the culture and way of life of Bulgaria during the Early Middle Ages (I. Duichev, S. Vaklinov, V. Beshevliev). Proto-Bulgarian inscriptions have been studied, making a valuable contribution to historiography (V. Beshevliev).

Over the past few decades Bulgarian historians have studied in great detail the process of the consolidation of the Bulgarian State in the seventh–fourteenth centuries, the achievements of its material and spiritual culture, as well as its relations with the European states (I. Duichev, H. Gandev, A. Burmov, D. Angelov, P. H. Petrov, V. Gyuzelev, B. Tsvetkova, N. Todorov).

The problems of the Bulgarian National Revival period: the formation of the nation, its struggle for national liberation and state sovereignty, the role of the great European powers in the process, and the problems of Bulgaria's economic, social and cultural development during the period, have received a scientific interpretation in a number of monographs and collections (K. Sharova, V. Paskaleva, K. Kossev, S. Damyanov, H. Hristov, H. Gandev).

Special attention has been paid to the study of the ethnic unity of the Bulgarian people. Of great significance in this respect is the work *Macedonia. Documents and Materials* – a large collection of primary sources elucidating the history of the Bulgarians in Macedonia from the Early Middle Ages to the beginning of World War II. Special emphasis is placed on the period after the Berlin Congress (1878). The study of *The Kresna-Razlog Uprising of 1878–79* is of interest, as is the monograph *The National-Liberation Movement in Macedonia and the Adrianople Region (1878–1903)*.

Vladimir Georgiev

Parallel to the development of historiography, careful research has been carried out in the field of archaeology, ethnography, archival science and numismatics. Fundamental works have been published on the ethnography of Bulgaria, on the country's historical role as a centre and bearer of cultural values and traditions that have contributed, sometimes decisively, to the formation of European civilization.

The Institute of Bulgarian Communist Party History conducts research and co-ordinates studies on historical and present-day problems of Balkan and international communist and working class movements. Its scientific organ is *Bulletin of the Institute of History of the Bulgarian Communist Party* (since 1957).

The study of **linguistics** has notched up considerable successes in the following language fields: Bulgarian (K. Mirchev, L. Andreichin), Proto-Bulgarian (V. Georgiev, V. Beshevliev), Slavonic (V. Georgiev, I. Lekov, S. Ivanchev), Balkan (V. Georgiev, I. Duridanov) and Thracian (D. Dechev, V. Beshevliev, I. Duridanov). The book *The Thracians and their Language* (1977) by V. Georgiev has initiated a new phase in research on the Thracian language. With his work on Indo-European linguistics, V. Georgiev raised Bulgarian linguistics to world standard. In their works on general linguistics, Bulgarian scholars confirm their dialectical-materialist view of linguistic reality, of the nature and functions of language and its systems and structure (V. Georgiev, I. Duridanov, and others). In *Historical Grammar of the Bulgarian Language* (1958), K. Mirchev compared a number of forms and phenomena in the Bulgarian language with those from other Balkan languages for the first time. *Basic Bulgarian Grammar* (1944) by L. Andreichin was the first Bulgarian theoretical grammar free from the traditions of the Latin-Greek model. Research is in progress on problems of lexicology, stylistics and dialectology.

A number of recent works have made a valuable contribution to diachronic and synochronic linguistics.

The work *The Unity of the Bulgarian Language in the Past and Today* provides a rigorous and scientifically grounded solution of the problem of the unity of the Bulgarian language, both diachronically and synchronically. By its nature this study is a theoretical and methodological programme for Bulgarian linguistics.

Some lexicographic works are also very valuable, such as *Dictionary of the Bulgarian Language* which covers Bulgarian lexical material from the National Revival Period to the present day, and the *Bulgarian Etymological Dictionary*, volumes I and II. This is a fundamental study of the origin of the words of the Bulgarian language, covering every stage in the development of the language in all areas where Bulgarian is spoken. The work is of great linguistic and social importance, containing valuable information not only for linguists, but also for historians, ethnographers, folklorists, writers and students. The *Dictionary of Synonyms in Modern Standard Bulgarian* was compiled on the basis of a large corpus of lexical material and demonstrates the wealth of lexical synonymity in standard Bulgarian.

The book *Foreign Scholars on the South-Bulgarian Dialects* contains articles by world-famous linguists who have confirmed Bulgarian views on the geographical distribution of the Bulgarian language. The results of systematic and comprehensive research into all Bulgarian dialects have been generalized in the *Bulgarian Dialect Atlas* (4 volumes, 1964–81).

Theory of Literature is a survey of the development and

Evlogi Georgiev

specific features of **Bulgarian literature** during the Middle Ages, the National Revival period, the Monarchy and the period of Socialism. The work *General Theory of Art* (1938) by Todor Pavlov is of fundamental importance for the theory of literature, and for the theory of art in general. P. Zarev, P. Dashev and others have also made valuable contributions to literary theory. Work is being done on the periodization of Bulgarian literature, styles, genres and the problems of Socialist Realism (A. M. Arnaudov, P. Dinekov, E. Georgiev, P. Zarev, G. Tsanev, K. Kuev, B. Angelov, E. Karanfilov, P. Danchev, S. Karolev, L. Tenev, S. Karakostov). Studies have also been made of problems in Russian and Soviet literature, as well as Bulgarian-Russian and Bulgarian-Soviet literary contacts (S. Russakiev, V. Velchev, V. Kolevski, and others), of Slav literatures (E. Georgiev, K. Kuev, B. Nichev), of Western European literature (A. Peshev, K. Gulubov and others). The development and achievements of Bulgarian literature have been traced in *History of Bulgarian Literature* (4 volumes, 1962–76), *Essays on the History of Bulgarian Literature after 9 September 1944* (2 books, 1979–80), *Panorama of Bulgarian Literature* (5 volumes, 1966–76) by P. Zarev, et al.

The three-volume *Lexicon of Bulgarian Literature*, a valuable contribution to the theory of literature, is the first reference book to be published in Bulgaria on the development of Bulgarian literature over the past eleven centuries. Old Bulgarian literature is represented by Kliment's works, which have been published under the title *Kliment Ohridski. Collected Works* (in three volumes). *The Cyril and Methodius Encyclopaedia* is a very important publication, describing the significance of the work of Cyril and Methodius and the scientific interest that it has aroused.

The two-volume *Essays on the History of Bulgarian Literature after 9 September 1944* are devoted to modern literature, examining the development of Bulgarian literature during the first thirty years of Socialism.

Many collections of literary essays have also been published, for example: *Elin Pelin. 100th Anniversary of His Birth. New Studies*; *Nikola Vaptsarov. New Studies and Materials*; and *Yavorov – Divided and United. New Studies*. These collections of essays have cast new light on the work of these classics of Bulgarian literature. Pantelei Zarev's *The Bulgarian National Mentality and Literature* (1983) is a landmark in research into the theory of literature.

Folklore studies concentrate on theoretical, sociological and aesthetic research into folklore, on the basis of primary sources. Various problems in the theory and history of creativity in folk poetry have been discussed, e.g. *Bulgarian Folklore* (volume 1, 1972) by P. Dinekov and *Problems of Bulgarian Folklore* (1976) by Ts. Romanska. Folklore is being studied by G. Keremidchiev, S. Stoikova, T. I. Zhivkov and others, by musicologists and folklorists and by some literary critics. The series *Bulgarian Folklore* (13 volumes, 1961–65) is the result of organized collection and systematization of the folkloric material.

The Scientific Study of Art deals with the history and topical problems of the fine arts, revealing the character and roots of progressive realistic traditions, and examines specific periods in art and the work of different artists. Theoeretical problems are investigated and criticism finds expression in the work of T. Pavlov, M. Tsonchev, A. Obretenov, N. Mavrodinov, A. Vassiliev, A. Stoikov, A. Bozhkov and others. The multi-volume edition *History of Bulgarian Fine Arts* is the result of the work of several specialists.

The research of Bulgarian musicologists (V. Krusstev, B. Sturshenov, P. Stoyanov, A. Karastoyanov, D. Hristov, Z. Manolov, G. Gaitandjiev, and others) covers historical, theoretical and comparative musicology and Palaeobalkan Studies, musical sociology, musical psychology, bibliography and informatics. Considerable generalizations are being made in musical folklore studies on the basis of regional investigations and the recording of more than 270,000 Bulgarian folk songs, instrumental melodies and dances. The musical folklorists (R. Katsarova, S. Djudjev, N. Kaufman, E. Stoin, A. Motsev, and others) are studying the musical folklore genres, instrumental music, children's musical folklore and folk musical instruments. Bulgarian musicologists have also made a contribution to the development of Marxist-Leninist aesthetics.

S. Russakiev addresses a meeting of the International Association of Teachers of Russian Language and Literature in Sofia, 1968

Outstanding figures in the theory and history of **dramatic art** include G. Gochev, P. Penev, S. Karakostov, L. Tenev, D. Kanushev, D. B. Mitov, A. Natev, and Ch. Dobrev. Research deals with such areas as ideological content, artistic methods, the form and means of expression of the theatre, the influence of social aspirations, the historical development, the genre and stylistic trends in stage direction, and actors' creativity.

Film Studies have been stimulated by the upsurge of the Bulgarian film industry. Film critics (Y. Molhov, G. S. Bigor, E. Petrov, A. Alexandrov, A. Grozev, L. Tenev and others) consider problems relating to the specificity of the art of cinema, its genres, its relationship to other arts, and socialist realism in the cinema, and study the history of the Bulgarian and foreign cinema.

In the **science of art** recent achievements in research into the problems of the historical development and theory of modern Bulgarian art, architecture and folklore, include such works as: *Encyclopaedia of Pictorial Arts in Bulgaria, Bulgarian Artistic Heritage, Between Folklore and Literature, The Art of the USA, Social Function of Architecture,* and *Contemporary Aesthetic Problems of Housing Architecture.*

In the domain of **law** detailed studies have appeared on the nature, objectives, functions and structure of the people's democratic State and law; the development of the feudal and bourgeois State and law, of the State and legal system under Socialism; and the building of developed socialist society, of socialist democracy and legality, of the constitutional foundations of the socialist State (M. Genovski, B. Spassov, Y. Radev, M. Andreev), the foundations of the theory of administrative and financial law (P. S. Stainov, A. Angelov, and others). L. Vassilev, D. Silyanovski, L. Raduilski, A. Kozhuharov, Zh. Stalev, V. Tadjer and others have contributed to the development of the regulation of the economic mechanisms by law, civil law, the civil procedural law for guaranteeing the free exercise of the rights and legitimate interests of citizens and socialist organizations, while N. Mevorah and others have worked on problems of the legal protection of children and the family. Research has also been carried out on problems in socialist criminal law.

Books have been dedicated to the improvement of socialist democracy. These publications reveal the theoretical and practical significance of the problem of democracy.

Research is in progress on problems related to the further improvement of the legal forms for the further intensification of economic co-operation and the development of economic relations between CMEA member-countries. The study *The Courts of Arbitration of the Chambers of Commerce of the CMEA Member-States* is a considerable contribution to the unification of the legal framework of international economic co-operation between the CMEA member-states. From the standpoint of comparative law, this work examines the competence, procedure, and other matters related to the work of these courts.

The role of legislation in further improving socialist social relations in developed socialist society is dealt with in a number of publications: *Legal Regulation of Relations in the National Economy, Development of the Penal Procedure Code of the People's Republic of Bulgaria at the Present-Day Stage, The Legal Status of the Foreigner in the People's Republic of Bulgaria, Legal Problems of State, People's and Public Control, Guaranteeing the Right of Defence to Citizens in the Penal Procedural Code and in the Penal Procedural Practice of the People's Republic of Bulgaria, Fundamental Problems of the Effectiveness of Civil Procedural Law in the People's Republic of Bulgaria.*

The science of science developed as a new scientific branch in Bulgaria after 1966. The first body to deal with the subject – The Centre for the Science of Science of the Bulgarian Academy of Science – was founded in 1968. The research carried out at the Centre includes problems of scientific policy, methodological management of research activities (N. Stefanov), sociology (N. Yahiel), economy of science (I. Yordanov), prediction and planning of scientific activities, organization of interdisciplinary research, the problems of science personnel, and the interaction of social, natural and technical sciences.

The results are contained in several works, the more important of which are: *Science and Management* (1970), *Social Sciences and Social Technology* (1973), *The Small World of 'Great' Science* (1981), *Multiplying Approach and Effectiveness* (1983), *Along the Way of Great Science* (1976), *Scientists in Great Science* (1975), *Sociology and Science* (1975), *An Introduction to the Management of Scientific Research* (1982), *Development Projects and Innovations* (1978), *Interdisciplinary Movement and the Organization of Research* (1973), *Unanimity and Difference of Opinion in Science* (1982), *Forecasting in Science* (1982), and *Adaptive Planning of Scientific Research* (1982).

SELECTED BIBLIOGRAPHY

BALEVSKI, N. *Problemi na naukata i obrazovanieto* (Problems of Science and Education). Sofia, 1974.

BENEV, B., M. PANDEVA and N. IVANOV *Nauchno-tehnicheskata politika v Narodna Republika Bulgaria* (Scientific-Technological Policies in the People's Republic of Bulgaria). Sofia, 1980.

BRADINOV, B. *Adaptivno planirane na nauchnata deinost* (Adaptive Planning of Scientific Research). Sofia, 1982.

BUCHVAROVA, N. *Prirodonauchnite znania i knizhninata prez bulgarskoto vuzrazhdane* (Natural Science and Scholarship in the National Revival Period). Sofia, 1982.

KACHAUNOV, S. *Prognoznata deinost v naukata* (Forecasting in Science). Sofia, 1982.

KOSTOV, K. *Uvod v upravlenieto na nauchnoto izsledvane* (Introduction to the Management of Scientific Research). Sofia, 1982.

MINKOV, Y. *Naukata i chovekut* (Science and Man). Sofia, 1973.

MONCHEV, N. *Razrabotki i novovuvedeniya* (Development and Innovations). Sofia, 1976.

Po putya na golymata nauka (Along the Way of Great Science). Sofia, 1976.

POLIKAROV, A. *Naukata i suvremenniya sviyat* (Science and the Modern World). Sofia, 1981.

POLIKAROV, A. *Ochertsi po metodologiya na naukata* (Essays on Scientific Methodology). Sofia, 1981.

SIMEONOVA, K. *Mezhdudistsiplinno dvizhenie i organizatsiya na nauchnite izsledvaniya* (Interdisciplinary Movement and the Organization of Scientific Research). Sofia, 1973.

STEFANOV, N. *Malkiyat Sviyat na 'golyamata' nauka* (The Small World of 'Great' Science). Sofia, 1981.

STEFANOV, N. *Nauka i upravlenie* (Science and Management). 2nd rev. and enl. edn. Sofia, 1970.

STEFANOV, N. *Obshtestveni nauki i sotsialna tehnologia* (Social Science and Social Technology). Sofia, 1973.

TODOROV, I. Ml., E. NIKOLOV and K. ILIEV *Postizheniya na naukata i tehnicheskiya progres v Narodna Republika Bulgaria* (Scientific Achievements and Technological Progress in the People's Republic of Bulgaria). Sofia, 1983.

YAHIEL, N. *Sotsiologiya i sotsialna praktika* (Sociology and Social Practice). Sofia, 1982.

Za uskoreno razvitie na nyakoi strategicheski napravleniya na nauchno-tehnicheskiya progres v Narodna Republika Bulgaria. Materiali ot Plenum na Tsentralniya komitet na BKP, sustoyal se na 20 i 21 yuli (For the Accelerated Development of Some Strategic Branches of Scientific-Technological Progress in the People's Republic of Bulgaria. Materials of the Plenum of the Central Committee of the Bulgarian Communist Party held on 20–21 July). Sofia, 1978.

PART IX

CULTURE

THE DEVELOPMENT OF CULTURE AND CULTURAL POLICY IN SOCIALIST BULGARIA

Bulgaria has a millennial culture. Its very rich and original cultural traditions and values reveal the creative genius of the Bulgarian people, their aspirations to freedom, social justice and cultural advancement, their humanism, love of peace and active interest in the cultural values of other peoples. A country with a rich cultural past and a dynamic many-sided and fruitful modern cultural life, the People's Republic of Bulgaria ranks among the most developed states with regard to culture and makes its own contribution to the treasure-store of world culture.

The turbulent and contradictory political history of Bulgaria led to the relatively early formation and spread of deeply rooted national, democratic and revolutionary traditions. Under the conditions of existence as an independent state and almost seven centuries of Byzantine and Ottoman domination there ripened the seeds of a truly national and democratic culture, which became one of the main strongholds of the national spirit and contributed significantly to the formation of a national consciousness and to the powerful spiritual manifestations of the Bulgarian people. This markedly democratic and progressive line in the development of Bulgarian culture found its continuation in the revolutionary and democratic culture of the National Revival Period, which, for its part, provided a natural basis for the relatively early and wide spread of socialist ideas as early as the end of the nineteenth century, as well as for the development of proletarian revolutionary culture and art in bourgeois Bulgaria.

In the dawn of the socialist movement in Bulgaria, Dimiter Blagoev and his associates, the left-wing socialists, adopted as a supreme value the deeply democratic, progressive educational and cultural achievements and tradition of the people. They combined in an inseparable unity the centuries-long struggle of the Bulgarian people for cultural advancement with the Marxist ideas of the revolutionary transformation of society. This is a major historic merit of the Bulgarian Communist Party (BCP) and one of the basic reasons for the enormous prestige and unreserved support it already enjoyed in the years of Capitalism, among the widest strata of the progressive intelligentsia. The Party had sown its socialist ideas among the Bulgarian intelligentsia comparatively early, preserving and developing the spirit it had inherited from the National Revival, and within a short historical period formed around itself a solid detachment of men active in literature and art, in education and culture and the Press, who became outstanding figures in Bulgarian culture in the first half of the century. It was a really wide social basis for influence on cultural processes, and after the victory of the socialist revolution in 1944 it grew extensively.

From its foundation down to the present day the BCP has been following, developing and enriching the valuable tradition of its cultural policy: a marked concern for the successful development of progressive literature and art; comradely respect and recognition of all their talented representatives, attention and care for their fulfilment; active personal participation of the most outstanding party functionaries in the theoretical and practical elucidation and guidance of the development of processes in the field of culture.

The foundations of this tradition were laid by Dimiter Blagoev, who was not only an outstanding revolutionary and Marxist, but also a leading theoretician of Bulgarian culture. He made an enormous contribution to the formation of correct Marxist views by the Party on the social nature and class functions of culture, and on its role in the life of society and the education of the people. Georgi Dimitrov performed an invaluable service in the further development of these views, bringing the patriotic, democratic and revolutionary intelligentsia closer to the party line and policy and to the Leninist ideas of culture. These ideas struck deep roots in the consciousness of broad strata of the intelligentsia and the people and were embedded in the foundations of the Party's cultural policy. After the victory of the revolution of 9 September 1944 the BCP embarked upon a radical socialist renewal of cultural life.

In the years of socialist construction the Party developed and enriched its cultural policy in conformity with the new historical conditions and needs. It gave extremely great attention and care to the development of socialist literature and art as an important constituent of social life. The guaranteeing of the accelerated and all-round cultural progress of Bulgaria has become a basic concern of the Socialist State and of all its bodies and institutions. Socialism turned culture from the privilege of an élite into a possession and field of creative activity of the broadest popular strata, thus multiplying the possibilities both for its own development and high achievements and for its role in the education of the individual, as well as its effect on the development of all spheres of social life.

The time since the April 1956 Plenum of the Central Committee of the BCP has been a particularly characteristic, rich and dynamic period in the cultural flowering of the People's Republic of Bulgaria. The historic importance of the April Plenum consists above all in the fact that it put a decisive and irreversible end to the deviations from the creative principles in the guidance of culture and art that had appeared after the death of Georgi Dimitrov. It rejected and overcame the manifestations of subjectivism in the guidance and evaluation of cultural processes and phenomena and created an exceptionally favourable social climate for the flourishing of culture and art on the unified basis of socialist realism, releasing the enormous creative energy of the intelligentsia and the whole people. It also introduced a number of new elements in the democratization of culture by involving broad strata of the people and above all of the intelligentsia itself in the creation of cultural values and in the direct control of cultural processes.

A very important role in the formation and implementation of the April line in the cultural policy was played by the personal contribution of Todor Zhivkov, General Secretary of the Central Committee of the BCP and President of the State Council. Continuing the traditions of the Party, he became the mastermind of the most important ideas and processes in the realm of cultural policy. His personal style of leadership is characterized by the great attention he devotes in his daily work to the problems of art and culture, the intimate knowledge of cultural facts and processes, his close contacts with a broad circle of personalities of all generations in the area of culture, his exceptional skill in maintaining contacts with them, helping them in their work and daily needs, supporting the real talents and uncompromisingly criticizing all shortcomings. By his example he instils a similar approach and style of work among all officials. This is from where his great popularity, prestige and influence among the intelligentsia stems.

The implementation of the cultural policy over the past decade and its creative development are inseparable also from the life and work of Lyudmila Zhivkova, who played an outstanding role in the political, cultural and scientific life of Bulgaria. As a member of the Politburo of the Central Committee of the BCP, as Chairman of the Committee for Culture and as a scholar she showed the exceptional qualities of a versatile personality, initiating and organizing a number of important cultural events, which received due recognition within the country and elsewhere. Her untimely death was a great loss, but the heritage of her ideas and projects lives on.

During the post-April period and particularly in the past decade Bulgarian socialist culture has developed fruitfully, scored major key successes and reached maturity. Today, both in Bulgaria and abroad, people speak about a new Golden Age of Bulgarian culture, about Bulgaria being 'a state of the spirit', about the confidence and concern of the Communist Party and the Socialist State with regard to men of culture, and about the moral and material conditions being provided for the efflorescence of art and creativity.

The main feature of the cultural scene in Bulgaria today is the unity and cohesion of the intelligentsia round the April policy of the Party, its great responsibilities for the development and realization of talented people, the consistent, unswerving adherence to socialist principles and criteria in all creative artistic and cultural activities, the affirmation and creative development of the method of socialist realism, and the active participation in the life of the country of all generations of writers, artists, scholars, etc.

This stimulating, creative climate has made it possible for literature and art to achieve a higher ideological and artistic standard. Significant works of art have been created in all spheres of culture. There are notable successes not only in the orientation of writers and artists to contemporary themes, which provide the main impetus to their creative energies, but also in the truthful and highly artistic rendering of socialist reality. The capacity of writers and artists to reflect life in all its diversity, and to probe the complexity of the processes which characterize the building of developed Socialism has grown. Significant works of art on themes and plots from older and more recent history, which constituted a valuable contribution on the part of the cultural front to the celebration in 1981 of the 1,300th anniversary of the foundation of the Bulgarian State, constituted also an important contribution to the patriotic education of the people. The artistic skill of writers, artists, musicians, etc., has also increased. The variety of genres, styles and creative idioms, and the presence of new quests and trends are permanent and constantly developing features of modern Bulgarian art.

One of the most characteristic features of the socialist way of life, the spiritual contacts and mutual enrichment of men of culture and working people, is growing in prominence. The use of the achievements of the techno-scientific revolution in the mass media opened up wide opportunities for expanding the scope and impact of culture by creating new, hitherto unknown possibilities of reaching an audience of millions. More and more culture enters the plants, farms and homes of working people, and appreciation of the values of literature and art is becoming more and more a vital need for the widest masses of people, a source of optimism and vitality, inspiration for work, creativity and life.

Over the past years the overall many-sided domestic and international activities of the national cultural complex have been based on a long-term programme and concentrated round the following national programmes of the Committee for Culture:

- The National Programme of nationwide aesthetic education and its transformation into a nationwide movement;

- The long-term complex programme 'Harmoniously Developed Personalities', together with the 'Illustrious Bulgarians' programme.
- The International Children's Assembly and the 'Banner of Peace' movement.
- The programme of the Committee for Culture and of other state and public bodies for the education and stimulation of young intellectuals.
- The national programme for the celebration of the 1,300th anniversary of the foundation of the Bulgarian State.

The experience that has been accumulated in designing and implementing the national and regional programmes for nationwide aesthetic education, introduced on the initiative of and with the active participation of Lyudmila Zhivkova, is particularly valuable. The achievements in the realization of this idea during the 1970s and early 1980s have fully confirmed its correctness and viability. The problems of aesthetic education have become the focus of attention of state policies and of the whole of society. All regional programmes for aesthetic education in the various districts were approved at plenums of the district party committees, together with the District People's Councils and the governing boards of public organizations. The control of the implementation of the programmes and their continuous updating are also carried out by the respective state and public bodies, the Committee for Culture and the local bodies of cultural management, exercising mainly co-ordinating and some other more specific functions.

The aims of the nationwide aesthetic education are closely connected with the Harmoniously Developed Personalities programme, the Nikolai Konstantinovich Roerich (1978), Leonardo da Vinci (1979), V. I. Lenin (1980), Constantine Cyril the Philosopher (1981), Georgi Dimitrov (1982), K. Marx and A. Einstein (1983), stages of which have already been carried out. The F. Engels (1948–85) and Mikhail V. Lomonosov (1985–86) stages are forthcoming. The programme provides also for the special commemoration of the life and work of some of the great figures of Bulgarian national history, such as Georgi Rakovski, Vassil Levski and Hristo Botev. Through the example of the life and work of such eminent representatives of the human spirit, who have set themselves the task of striving towards the perfection of man and furthering the struggle for the improvement of society, these programmes aim to reveal the crucial role of art and culture for the harmonious development and education of the socialist personality.

Lyudmila Zhivkova's wonderful idea of the Banner of Peace movement and the International Children's Assembly under the slogan 'Unity, Creativity, Beauty', held in 1979 and 1982 produced basically new and significant results both in Bulgaria and abroad. The Banner of Peace movement in Bulgaria has unfolded on a large scale mainly as a system for the development of the creative talents of children and adolescents. National and district Banner of Peace centres have been set up, which are very active. Children from 105 countries from all over the world and a large number of prominent figures of art, culture, science, politics, etc., have taken part in the two International Children's Assemblies held so far. More than 30,000 children's works of art, literature, music, photography and technology have so far been deposited in the funds of the Assembly.

A lot of work is being done in Bulgaria for the further improvement of the system for the dissemination of cultural values. A relatively good level of provision of cultural benefits for the population as a whole has already been achieved and the social possibilities of every citizen for enjoying the values of culture and art and developing and expressing his talents have increased considerably. The country has a well developed network of cultural and educational institutions – cinemas, theatres, musical institutes, libraries, museums, houses of culture, educational establishments of the arts, as well as more than 4,000 traditional community centres, a typically Bulgarian cultural institution. There exists a relatively good material basis for cultural activities, which is being constantly improved. Every Five-Year Plan allocates substantial funds from the national income for the development of culture. In fact, these funds have doubled over the past ten years. Measures are taken for the most effective use of the cultural potential that has been created so as to achieve a better social effect and to raise the quality of cultural activities in accordance with the tasks set by the National Party Conference (1984) and the Fourth Congress of Bulgarian Culture (1983).

The theoretical motivation and practical application of the public-state principle in the management of cultural processes on the national and local level is one of the most characteristic and original features of cultural policy in socialist Bulgaria. This innovative idea is the creative continuation under new conditions of a number of national traditions and forms, which originate from the time of the Ottoman domination in the nineteenth century, of involving broader social circles in the management of spiritual life. It is the result of the positive processes in the development of Bulgarian culture during the years of Socialism and is the expression of great popular confidence in the officials and intellectuals of Bulgarian socialist culture. The public-state principle is intimately linked with the high dynamism and complexity of the economic, social, political and cultural processes, with the further improvement of social relations, the heightening of the social consciousness and activity of the working people, the development of socialist democracy and the further improvement of the political system of society.

The First Congress of Bulgarian Culture, at which delegates from all over the country and the authorized representatives of the artistic and creative intelligentsia elected the public-state body for the management of culture – the Committee for Culture – was held in 1967. In accordance with the principle of the administrative and territorial structure of the country, district and local councils of culture, including representatives of all social and age groups, of the creative intelligentsia, state departments, public organizations, artists' and writers' unions, as well as of cultural institutions, were elected on a broad democratic basis. In 1974, by a decree of the Central Committee of the BCP and the Council of Ministers, the Arts, Culture and Mass Media National Complex was set up, which brought under its structural and functional control the entire cultural activity of all cultural institutions, unions for the arts, and governmental and public organizations. In 1977, by a decision of the Government, the territorial Culture complexes were set up on the basis of the public-state principle. Another official document delegated the functions and rights of bodies for the management of culture to the community centres in the smaller towns and villages.

Four congresses of Bulgarian culture have been held so far – in 1967, 1972, 1977 and 1983. According to the Statutes of the Committee for Culture, the congresses are held at five-year intervals, whereas the conferences of the councils for culture are held every two years. If necessary, a national conference for culture can be called between congresses. The elected bodies are the Committee for Culture, its Bureau and its Board of Chairmen. The Chairman of the Committee for Culture is elected by the Committee itself and is then endorsed by the supreme organ of state power, the National Assembly, as a Minister, thus becoming a member of the Government. The chairmen of the district and local councils for culture are also elected and become members of the local authorities. There are subsidiary bodies of the Committee for Culture, such as its co-ordinating councils, for the various types of activities – the Council of the Chairmen of the Cultural Unions, the Council of the Representatives of the public organizations, the Council of the Representatives of government organizations involved in cultural activities, and others. Thus in a democratic manner 217 people were elected to the Committee for Culture at the Fourth Congress of Bulgarian Culture, 2,250 people were elected to the district councils for culture and 2,262 were elected to the smaller local councils for culture at the district and muncipal conferences.

By involving representatives of all strata of society and of the cultural community directly in the guidance of cultural processes, the application of the public-state principle broadened significantly the social basis of the guidance of culture. The advantages of that social experiment justified all expectations and the system of public and state bodies for the guidance of culture is already bearing fruit. The possibilities of the State and the capabilities and interests of the whole people are combined on a higher level and the real democratic basis is broadened, thus affording a real opportunity to the broadest strata of the working people not only to be consumers of cultural values, but also to have a share in the management of cultural processes. An increasing number of international organizations, representatives of governments and public institutions in various countries show interest in the public-state bodies for the guidance of Bulgarian culture.

As an inseparable part of the overall process of the building of a mature socialist society in Bulgaria, and as a result of the consistent implementation of the April 1956 cultural policy, of the high achievements in literature and art, and of the consolidation and enrichment of the public-state principle in the management of cultural processes, during the 1970s Bulgarian culture gradually entered a new, still more dynamic and intensive period of its development, characterized by many valuable features and promising tendencies. The Twelfth Congress of the BCP highly appreciated the achievements during that period, the new considerable step forward in the construction of a highly developed system of cultural life, and the heightened social role and creative possibilities of culture. The dynamic efflorescence of socialist culture increasingly becomes an objective need and an active factor in social development. Arts and culture in the People's Republic of Bulgaria increasingly become an irreplaceable instrument for the education of man and, at the same time, one of the basic spheres in which the new man will have a full-blooded realization of his creative nature and of his pursuit of the beautiful and the sublime.

The cultural policy of Bulgaria is based on the concept of the international nature of the processes developing in socialist culture. Taking into consideration the circumstance that the building of society's cultural sphere cannot be successful if the experience and achievements of other peoples are not used, this policy is orientated toward the promotion of wider cultural co-operation with the socialist countries and above all with the Soviet Union, and to studying and utilizing the world cultural achievements. In this way Bulgarian literature and art entered into still more direct contact with world literature and art. This helped find correct criteria for development and self-assessment, to enrich cultural values and judge the true worth of what has been achieved in Bulgaria. At the same time the contacts enhanced the international prestige of Bulgarian culture.

The development of the contemporary culture of the People's Republic of Bulgaria is effected in the spirit of Socialism and serves social and human progress, peace and understanding among nations.

FOLK ART (FOLKLORE)

In early times, Bulgarian folklore arose, existed and developed within the framework of the family and the home. The formation of the Bulgarian nation contributed to the ethnic individualization of the cultural processes. The people created an artistic culture of their own, which for a long time preserved its regional character. Bulgarian folklore was formed as a synthesis of the artistic traditions of the Slavs, Proto-Bulgarians and Thracians. Cultural threads were preserved, interweaving pagan and Christian ideas and images. This expressed the people's thoughts and was a means of its continuity. At times of foreign domination it ensured the preservation of the Bulgarian nationality. Under the impact of the historical conditions its type and purpose changed. In the everyday life of the old peasant society folklore was interwoven with the economic cycle of the year. The word, the song, the dance, the clothing, the ritual objects and the participating persons in the ritual are of equal value to the action and express its social meaning. The development of the expressiveness and the use of images in folklore arose out of its practical application and the ideological meaning lent to it in life. The new urban culture, which made its appearance during the National Revival Period (eighteenth–nineteenth centuries), developed on the basis of the existing regional forms of folklore and within the framework of the nation. This gave rise to new artistic works, which, without losing their link with the old creative traditions, used an expressiveness and imagery of their own. This folklore was no longer regional and today it is an inseparable part of the national culture.

After the Liberation in 1878, **oral folklore** lost its place as the main artistic reflection of the people's life and the diminished role of folklore led to a loss of some folk traditions. In spite of this, almost until the middle of the twentieth century it continued to be one of the most important components of artistic culture in the countryside. Its vitality was reaffirmed by the anti-fascist folk songs and by their most important part – the partisan songs, which reflected the armed struggle of the people against Fascism. During the first half of the twentieth century the influence of personal poetry and of group song was felt more deeply.

The political and historical changes in life gradually diminished the importance of folklore in the artistic life of the modern Bulgarian. Today in Bulgaria serious efforts are made for the preservation of tradition, for lending it

Martenitsas

new life. Conditions were created for the preservation and stimulation of the people's artistic initiatives. New forms arose for the existence and functioning of folklore. Its performance and popularization was also taken up by amateur artistic groups, and by the Press, radio and television. Local festivals of folk art became a tradition in cultural life. District rallies were organized, as well as festivals and folklore celebrations, a National Folklore Rally in Koprivshtitsa, a National Festival of the *Kukeri* (Mummers) and *Survakari* (New Year Well-Wishers) in Pernik, a National Festival of Humour and Satires in Gabrovo, exhibitions of folk art, a National Fair of the Arts and Crafts and Applied Art in the village of Oreshak in the Lovech district, etc. Craftsmen's workshops have been restored, as well as National Revival streets and houses. Folk plastic art has been lent a new meaning in the works of the members of the Labour Society of the Masters of Folk Crafts.

The folk-songs were the richest and best developed part of Bulgarian folklore. In them the people reveal their life and soul in vivid and live colours and with an astonishing poetical mastery. The folk-songs are notable for a great

Celebration of Midwive's Day in the Sofia region

variety of subject matter and motives. Most ancient are the ritual songs connected with the calendar and the family. They are linked with primitive agriculture and stock-breeding and with patriarchal relations within the family. In the course of time they were subjected to great changes, but have preserved traces of old, religious, pagan and Christian ideas reflecting the past, naive world outlook of the people, their customs and rituals performed in connection with the calendar and on different family events – childbirth, weddings, and deaths. Particularly rich in motifs are the songs for Christmas and for St. Lazarus's Day. The songs enjoy wide popularity, especially those connected with the customs for bringing rain: *Peperuda* (Butterfly) and Gherman. Similar to the ritual songs are those for mythical beings – wood nymphs, fairies, bad fairies, dragons, lamias, and celestial bodies. The religious and legendary songs are few; they were created under the influence of the canonical church literature and more often of the apocrypha – legendary and religious works, which the Church, as official guide of cultural life in the Middle Ages, denied and persecuted because of the 'heretical' revolutionary and social elements in them. Among the oldest works in folklore are the *yunak* (heroic) songs. They also have a historical character, extolling heroes and heroic events from the second half of the fourteenth century. Their system of images connected with mythological ideas, certain old subject matter and motifs give ground to suppose that they arose in prehistoric times and reflect an earlier stage in the development of the people's consciousness. The heroic poems were subjected to a profound transformation in connection with the Ottoman invasion and the fateful events for the Balkan people resulting from it. The heroic songs reflect the ideas of a people with a strong and heroic personality, their aspirations for heroic deeds, their love for their country and hatred of the enslavers, their conceptions of duty, justice and honour. The main heroes in the heroic epic poems are Momchil (the beginning of the fourteenth century–1345) – a military chieftain and independent ruler in the Rhodope mountains, one of the first fighters against Ottoman invasion, and Krali Marko, whose prototype was the feudal ruler Marko (1371–95), for whom a great cycle (over 120 subjects) of epic songs was created, containing Bulgarian and international motifs and many legends. The haiduk songs are connected with the haiduk movement; there is a kind of continuity between them and the heroic songs, expressed in their common ideological orientation, in the adaptation of some subject matter from one theme into another, and in part also in their artistic structure. The haiduk songs most vividly reflect the resistance of the people against the political oppression and outrages of the Ottoman Turks, the increased self-consciousness of the people, their faith in the future and the democratic ideas of the National Revival Period. Their hero is the haiduk (rebel captain), revenger and defender of the oppressed people. The historical folk-songs were created over a long period (fourteenth to twentieth centuries), which is why they do not have a unified artistic style. They are united into one genre on the basis of their content, which relives actual historical events and persons – the people's heavy plight under the Ottoman yoke, the national liberation struggle, Russian and Russo-Turkish wars, the wars and political developments after the Liberation in 1878, etc. The largest and most vivid group of songs are those about everyday life and the social folk songs – love, family, labour, social and humorous songs. They reflect the personal, family and social life of the people, their mores and mutual relations and class conflicts. The everyday and social folk-songs constitute the most diverse part in subject matter and themes. Many of the everyday songs have the character of ballads. In addition to everyday songs there are also mythical and historical folk ballads about unusual and strongly dramatic events in family and social life. In the course of time the fabulous gave way to subject matter closer to real life.

Bulgarian folk-tales reflect the age-long experience of the people, their world outlook, and depict life in the Bulgarian village. Most ancient are the tales about animals, but these are comparatively few. They include also the folk-fables, which have a didactic function. The fairy-tales treat many specific national subjects as well as those common to other nations. The legendary tales are fewer; some of their subjects bring them closer to the fairy-tales and others to the folk stories and narratives. They treat ethical problems, with strong satirical elements and social commitment. The folk stories reflect the personal and social life of the people and are much like the fairytales, but the supernatural element in them has lost its magical character and has been transformed into hyberbole of the possibilities of man. Most interesting are the tales of everyday life which depict typical Bulgarian characters and situations, where human shortcomings and

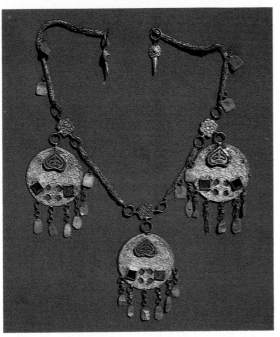

Necklace from the Yambol region

A bracelet from the Burgas region

Shepherd's crook from the Smolyan region

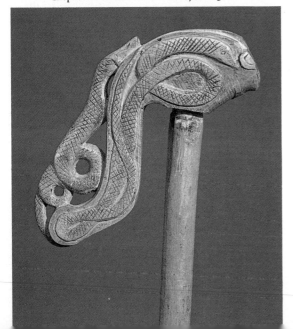

Pottery jug, bowl and brandy flask from Samokov

Belt buckles from the Plovdiv region

Embroidered pillow from the Russe region

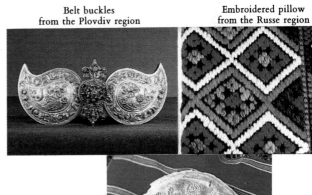

Ritual loaf from the Burgas region

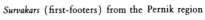

Survakars (first-footers) from the Pernik region

Baking dish with stand from the Smolyan region

Interior of a peasant home

Appliqué ornament on a dress from the Stanke Dimitrov region

Floral pattern embroidery from the Sliven region

Shirt-sleeve embroidery from the Stanke Dimitrov region

Coppersmith's workshop

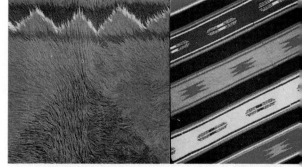

Tufted rug from Kotel   Carpet from the Blagoevgrad

Copper bells

Shepherds from the Sofia region

Bridegroom's costume from the Pleven region

Bride's costume from the Pleven region

Shepherd from the Haskovo region

Bagpipe player

Harvest festival

Homespun national costumes from the Sofia region

St. Lazar's day celebration of health, happiness and fertility

Town costumes from Kotel

Young girls' festive costumes from the Tolbuhin region

Man's harvest-time costume from the Dobrudja region

Costumes from the Vidin region

St. Lazar's day costumes from the Pomorie region

Women's costumes from the Smolyan r

Homespun costumes from the Burgas region

National costume from the Sofia region

Saiyana-type costume from the Gotse Delchev region

vices are ridiculed and social injustice is unmasked. Anecdotes make up one of the most vital forms of prose narrative. They are distinguished for their realism, satirical sharpness and class orientation. A favourite hero connected with Nasredin Hodja, is sly Peter, a popular character in the folk-tales of other peoples. The legends are about historical events, historical personalities, geographical and contemporary developments, but most of all about the anti-fascist struggle. Legends are also about mythical beings (demonological); Christian (from the Old and New Testament), created mainly under the impact of the canonical church and apocryphal literature about the creation of the world and the origin of man; about different peoples, animals, birds, and plants; and peculiarities of natural phenomena. They are connected with old pagan traditions and in part also later transferences of dualistic conceptions. The proverbs and sayings vividly express the wisdom and philosophy of the people and manifest their centuries-long socio-historical experience *in highly artistic, synthesized form*. The incantations, in verse or in prose, expressing faith in the magic power of the words, are among the oldest genres of oral folklore. Through them people have tried to influence human happenings or natural phenomena. Children's folklore contains works of oral, musical and dance folklore, created by adults and performed by them or by children.

Feast on Vine-Growers' Day (Trifon Zarezan)

Musical folklore includes vocal and instrumental folk music. In the ideological and the artistic composition of the songs, folk music and folk poetry play an equal part. The songs connected with different aspects of life have common musical features which unite them into song cycles characteristic of folklore. The melodies of a given cycle accompany only a certain kind of theme, and they have very strong bonds between meaning and function. The performance of songs of a certain cycle demonstrates the emotional atmosphere and mood of the people. Within the framework of the music cycle, the dialectical traits are revealed, typical of every region, and one can trace the different phases of historical development. A widely representative sample of old music is encountered among the melodies of the ritual cycles. There are melodies which have no independent musical meaning. They constrain the free melodies within the framework of the ritual words. The re-interpretation of the words and their liberation from their ritual orientation is accompanied by the formation and development of new musical means of expression. Most developed musically are the songs for festive dinners and for occasions combining work with pleasure. In accordance with the metrical groups of the stanza, the melody is divided into motifs, parts and sentences. Some of the melodies reach a certain independence through purely musical means – tone durations, pitch, ornament, etc.

Side by side with the regular 2/4 time (very popular), 3/4, 3/8 and 4/4 times are used, as well as irregular metres, formed by irregular times or bars. Each bar consists of 2 or 3 primary values, which as a whole are accepted as ordinary and lengthened secondary times in a ratio of 1:1, 1:2 or 2:3. The most wide-spread irregular times are 5/8 (5/16), 7/8 (7/16), 8/8 (8/16), 9/8 (9/16), 10/8 (10/16), 11/8 (11/16), 12/8 (12/16), and 13/8 (13/16). The hemiolically lengthened values of 12- and 13-time bars are often changed, becoming diaplastic, and then the measure 7/8 is determined. There are also larger metro-rhythmic formations in Bulgarian folk music, formed from combinations of several irregular bars, which are periodically repeated. They are combined metric groups composed of 7/16+5/16, 9/16+5/16, 7/16+11/16, 12/16+9/16, 11/16+5/16+5/16, etc. A considerable number of folk-songs cannot be recorded in such measures. They are called measureless songs or songs without bars. They have a smooth movement, a broad melodic line, prolonged tones and rich ornamentation. Most of the Bulgarian folk-songs are in one part. In Western and South-western Bulgaria two-part folk singing is extensively used in several variants. Two-part melodies with persistent tonic drone in the second part predominate. Chiefly women sing two-part songs. One singer sings the first part of the song, and two, three or four singers the second part, as a drone accompaniment. Another variant of two-part songs has a broken line for the second part, which in certain conditions departs from the tonic to the sub-tonic tone. The second part can also move more freely, setting on the II, III, IV or V degree and temporarily forming a second tonal centre. Many folk songs are traditionally sung antiphonally. They are sung and responded to by solo singing, of two by two, or groups of singers.

In instrument folk-music the melody obtains a fully independent musical development. Its entry into the instrumental repertory proves the capability of music to express independently with its own means and methods a certain ideational and emotional content. The appearance of instrumental music is a sign of the disintegration of syncretism in folklore. Instrumental folk-music in its late kind is not related to rituals; rather is it 'attached' to well-known rituals, at weddings, carnivals and the like. It develops in two directions – as accompaniment and as an independent kind. The folk music instruments are divided into four main groups, according to the kind of the vibrator and according to the way of obtaining the tone: idiophones, membranophones, aerophones, and chordophones. They are most often made by the players themselves. Most widespread in almost all Bulgaria are the shepherd's pipe, the bagpipe, the rebec and drum. Instrumental folk music replaced the song as accompanying the *horo* (chain dance). The text loses its ritual sense and encompasses the chain dances as a whole as winter, spring, summer and autumn dances. The development of the artistic and technical means of expression of the folk-dances depends upon their liberation from the necessities of ritual performance. Characteristic of the Bulgarian folk-dances is the great variety of forms, metro-rhythm, ways of holding one's partner and combination of steps and movements, which are danced by the feet. Upon this rests the vivid, ornamental character and general expression of the folk-dances. They lack what is typical of the ancient Greek, Indian and other eastern dance systems, namely gestures and movements – an expression of merriment, sorrow, love, hate and other feelings. The folk dances are mainly group and chain dances, danced in an open or closed chain. The chain dances which are more complex and more diverse in steps are danced by a small number of people (up to sixteen), in a curved or straight line. Smaller is the percentage of the solo and pair dances (*Peshachkata, Kasumskité, Treperushkité* and the like). Typical of all Bulgaria are the straight *horo*, the Paidushko, Daichovo, Elenino horo, Ruchenitsa, etc. Besides these, local folk-dances are widespread in the different regions. Today Bulgarian folk-music and dances underlie the modern festivals revealing the rich traditions in the development of folklore.

To the **applied arts** belongs the making of embroideries, fabrics (bedspreads, covers and clothing), folk costumes, paintings (on household utensils, buildings, Easter eggs), shaping of ceramic articles, pottery, stone carving, ritual loaves of bread, ritual objects, metalwork (coppersmith's articles, wrought iron, goldsmith's articles), wood-carving, architecture (types of houses), etc. The development of folk plastic art has led to the appearance of the arts and crafts whose mass production in the period of the National Revival had long-standing traditions.

Architecture creates material objects, but has also an indisputably artistic character, both an aesthetic and a practical function. It is the work of ordinary master-builders who have acquired their professional skills from personal experience and from practical life. They created and designed the people's houses with their picturesque courtyards, fountains, clock towers, bridges and churches. The talent and skill of the master-builder found its most

Folk-song and dance festival at Rozhen

Wooden-pipe players from the Sofia region

vivid expression in interior design. Connected wth it are the popular arts and crafts – pottery, coppersmith's articles, woodcarvings, woven articles, embroidery, etc. They developed on the basis of the inherited old cultures on the territory of Bulgaria and in the course of time became established as original, highly artistic manifestations. The cosiness of the Bulgarian home is partly a result of the production of home-made rugs, carpets, etc., the artistic effect of which is achieved by the regular pattern of stripes in one direction in vivid colours (red, green, etc.). During the National Revival Period rug and carpet making centres were established in the towns of Chiprovtsi, Kotel, Panagyurishte, etc. The Chiprovtsi carpets are produced by a fibre technique with predominating floral ornamentation; those of Kotel by a square technique, with geometrical motifs; and the Panagyurishte carpets use the ornaments and colours of the eastern carpets. Typical of Bulgarian weaving art are the tufted rugs and the woollen and goat-hair rugs from the Rhodope region. Their production continues to this day. Ceramics and pottery were already flourishing in the Middle Ages, when decorative slates and tiles were produced for ornamental purposes and household utensils for every day living. During the National Revival Period well-known ceramic centres were established in the village of Bussentsi in the Pernik district, and in the town of Troyan. Very typical is the decoration of the Troyan household utensils – coloured, painted and glazed. Woodcarving has a long tradition. Primitive shallow and flat woodcarving satisfied the household needs; craftsmen elaborated more sophisticated techniques for the more exquisitely made objects such as shepherd's staffs, distaffs, ladles, spoons, cups, etc. Ceilings, doors and cupboards in the houses of the National Revival Period were decorated with woodcarvings. From among the craftsmen, professional master-woodcarvers appeared, who created masterpieces of woodcarving (iconostases, altar gates, bishop's thrones and the like). Poker-work assumed a national character and its products today have mainly a souvenir character. The painting of objects also developed as a trade. All kinds of things were painted: eggs, trunks, carts and houses. Church mural paintings reached artistic perfection and developed into schools of icon painting during the National Revival Period. During the same period the production of wrought iron and copper articles also developed. The coppersmith's trade developed in close connection with the main handicrafts and the increased needs for household utensils during the National Revival Period. In many towns, such as Sofia, Shumen and Plovdiv, whole market places and streets with workshops and shops were created. In Sliven, Gabrovo, Samokov and other towns cutlery and the gunsmith's trade developed. The goldsmith's trade has developed ever since remotest antiquity in the territory of present-day Bulgaria. Under the Ottoman rule and particularly during the National Revival Period goldsmith's centres were set up in Chiprovtsi, Vratsa, Sofia, Turnovo and elsewhere. In the production of jewellery, filigree work was mainly employed which, enriched and up-dated, is applied in the goldsmith's industry to this very day. The making of folk-costumes was distinguished for its variety and high mastery. The male costume is of two basic kinds, according to the colour of the upper clothes, white or black, with a different line and cut. The female folk-costume is richer than that of the men. It may have a single apron or two aprons, made as a tunic or as a low-cut sleeveless dress. The main elements in both the male and female costumes are the shirt, richly decorated with embroidery, and the sash or girdle – in the female costumes with belt buckles – usually richly ornamented. The aprons are woven with a keen sense of decoration and colour combinations, in which the Bulgarian woman uses all her skill and experience accumulated through the ages. Costumes for major festivals and celebrations are completed with special pieces of clothing and accessories.

After the Liberation in 1878, considerable changes were introduced in folk dress under the impact of town wear. Particularly after the First World War (1914–18), with the stepping up of the production and import of industrial goods, there was a decline in the arts and crafts trades and they suffered a period of crisis. After 9 September 1944 some of the home crafts disappeared. The arts and crafts practised at home were preserved and developed mainly as a cultural and historical heritage. A new national style was created which developed on the basis of the traditions that were established through the ages in Bulgarian plastic folk art, which has been preserved and recreated in many architectural and ethnographic preserves, such as Koprivshtitsa, Old Plovdiv, Varosha in Lovech, the village of Bozhentsi in the Gabrovo district, the village of Zheravna in the Sliven district and elsewhere; in the Etur architectural and ethnographic museum in the town of Gabrovo, and in the ethnographic museums.

Mihail Arnaudov

The beginning of the Bulgarian **folklore studies** was laid during the National Revival Period. In general, collecting work was done and for the first time folklore collections were published. The aim of folklore studies was patriotic, closely connected with the educative and revolutionary activity of the Bulgarian intelligentsia. The earliest records of folklore works belong to the first quarter of the eighteenth centry. A lasting scientific and social interest in folklore was aroused in the first decades of the nineteenth century. Credit for this goes to the outstanding Serbian scholar, Vuk Karadžić, who first published samples of Bulgarian folk songs, and the Ukrainian, Yuri Ivanovich Venelin, who pointed out to the Bulgarians the invaluable significance of folklore for the awakening and consolidation of their national awareness. Folklore works were recorded by Vassil Aprilov, Naiden Gerov, Ivan Bogorov, Petko Slaveikov, Georgi Rakovski, Lyuben Karavelov, the brothers Dimiter and Konstantin Miladinov, and many others. Several large folklore collections were published, most important among which was *Bulgarian Folk Songs* (1861) by the brothers Dimiter and Konstantin Miladinov. It was in this

Dimiter Miladinov

Facsimile of *Bulgarian Folk Songs,* collected by the Miladinov brothers

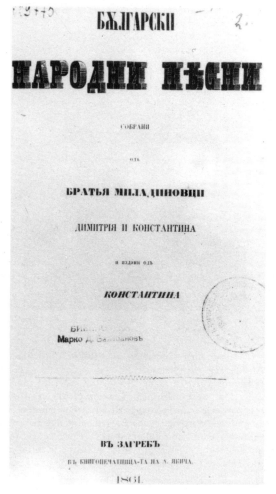

Konstantin Miladinov

that for the first time Bulgarian folklore was comprehensively fully presented. A testimony of the great interest on the part of the Bulgarian people at large was the great number of copies made in longhand, despite its great volume – 542 pages. Also a substantial contribution to the science of Bulgarian folklore is the collection *Narodne Pesme Makedonski Bugara* (1860) by Stefan Verković.

The first scientifically trained Bulgarian folklorists appeared in the 1860s and 1870s. After Bulgaria's liberation from Ottoman rule in 1878 the problems of folklore were tackled by reaffirmed scholars whose investigations were at the level of European science: Ivan Shishmanov (1862–1928), Mihail Dragomanov (1841–95) and others. The research work was concentrated mainly in Sofia University. The collection and publication of folklore materials was organized and guided by the Ministry of Education, the Bulgarian Literary Society (after 1911 the Bulgarian Academy of Science), and the National Ethnography Museum (established in 1906). Personal initiative continued to play a major role.

Folk Art (Folklore)

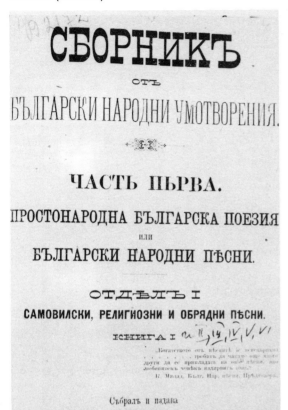

Facsimile of a collection of folklore, by Kuzman Shapkarev

Kuzman Shapkarev

The collection of the more unusual items was developed by Kuzman Shapkarev – *Collection of Bulgarian Folklore Works* (6 volumes, 1891–92) and many others. Of great significance for stepping up collecting and publishing was the series *Collection of Folklore Works, Science and Literature* (of 1889). In the 1920s and 1930s a tremendous collection was accomplished by the musicologist and folklorist Vassil Stoin, who together with his associates recorded more than 15,000 folk songs from the whole country – text and melody. Songs written down in notes were published also by Angel Bukureshliev, Raina Katsarova and others. Investigations of folk music were written by Karel Mahan, Dobri Hristov, Vassil Stoin, Stoyan Djudjev, A. Bukureshliev and others. A major figure in Bulgarian folklore studies between the two world wars was Mihail Arnaudov (1878–1978).

After 9 September 1944 Bulgarian folklore studies were placed on new Marxist methodological foundations. The publishing and collecting work was organized by the Ethnography Institute and Museum of the Bulgarian Academy of Science, its Institute of Musicology and, in part, by its Institute of Literature, and by Sofia University. After 1973 it was concentrated mainly in the Institute of Folklore of the Bulgarian Academy of Science. For the first time interest was manifested in the contemporary phenomena in folklore, in the anti-fascist, city and workers' folklore. The organized collecting, systematizing and publishing of folklore materials continues. The *Bulgarski Folklor* magazine, of the Folklore Institute of the Bulgarian Academy of Science, has been published since 1974.

LITERATURE

Bulgarian literature is the oldest Slavonic literature, appearing in the second half of the ninth century. Its beginning is related to the educationalist cause of the brothers Cyril and Methodius, who invented the Slavonic alphabet and laid the foundations of a Bulgarian literary language, Old Bulgarian, and Slavic literature. That cause was furthered by their disciples (Kliment, Naum and Angelarius), who in the middle of the ninth century promoted the south-western Old Bulgarian Thessaloniki dialect to a literary language, and translated the first Old Bulgarian books from Greek. This moment of crucial significance for the whole of Slavdom, preconditioned by the needs of official Christian religion, meant the admission of the Slav States to the bosom of European civilization. As a result, Bulgaria was to play the leading role in the complex processes of committing the southern and eastern Slavs to Byzantine cultural traditions; she was to experience unprecedented spiritual growth and was to enjoy immense literary fame in the course of centuries.

By the middle of the ninth century when Cyril and Methodius were doing their missionary work among the western Slavs, conditions were already ripe in Bulgaria for a wide reaching literary and enlightenment movement. This is testified by the achievements in building, in the arts and crafts, and by the Old Bulgarian inscriptions in Greek (eighth–ninth centuries) on stone, of which ninety have been found and which record events from state and public life. Though short (and some of them partially destroyed), they shed light on Bulgarian reality at that time, give us an idea of social relations and serve as a clue to what was considered at the time worth being perpetuated on stone. Individual inscriptions were the fruit of deep reflection about the value of human life and human ethical standards.

The Old Bulgarian inscriptions present the chronicle genre in its oldest form and could justifiably be regarded as the earliest stage, the 'pre-history' of Bulgarian literature. After the overthrow of Slavic religious service in Great Moravia and Pannonia, the work of Cyril and Methodius was treasured and continued by their disciples, who managed to escape the persecution of the German-Latin clergy and arrived in Bulgaria in 885. Prince Boris I welcomed Kliment, Naum and Angelarius with honours and ensured for them excellent conditions to engage in

Cyril and Methodius: a mural from the Church of the Holy Virgin in Smolyan

literary activities. Old Bulgarian literature became classical medieval literature in the early decades of its impetuous development.

Mosaic featuring St. Cyril from the Church of San Clemente, Rome

The flourishing education and culture in Bulgaria during the ninth and tenth centuries, referred to as the Golden Age of Bulgarian Culture, is associated with two remarkable literary schools of the day – the Ohrid literary school (in South-western Bulgaria) and the Preslav school (Eastern Bulgaria). The founder and most eminent representative of the Ohrid school was Kliment of Ohrid (c. 840–916) who in the course of thirty years, helped by Naum of Ohrid (c. 830–916), engaged in brisk educationalist and literary activities in the Ohrid region – activities which were crucial for the establishment of Christianity in Bulgaria, the consolidation of the Bulgarian State and the affirmation of a Bulgarian national identity. Kliment and Naum trained for literary work a considerable number of well-versed scholars, among whom were the Bishop of Devol – Marco, and the unknown author of the *Lives of Kliment and Naum*. Kliment was a translator, the author of original works and founder of new literary genres; he also developed further the Old Bulgarian literary language created by Cyril and Methodius. His poetic skill is demonstrated by his *Sermons of Cyril and Methodius* and his translation of the Floral Triodion. Kliment wrote a few dozen didactic sermons, and econoniums of all the important Christian holidays,

Apostles, prophets and martyrs. In writing them he was prompted by the need to find a way to instil in the hearts and minds of his newly baptized compatriots a new dogma and ethical code. His *Eulogy of Cyril the Philosopher* marked a peak in the development of Old Bulgarian literature and his overall work was a high point in the history of Slavdom.

Literary creative work and the training of scholars at the Preslav school was concentrated mainly in the capital of Preslav and the Monastery of St. Panteleimon at the mouth of the River Ticha. Literary life was headed by Naum (until he moved to Ohrid), Prince Boris, his brother Doks and by Tsar Simeon who issued an order for the writing of encyclopaedic collections, which were to contain the wisdom of his time. Well known are the names of the enlighteners Todor Doksov, who copied *Four Homilies against the Arians* (translated by Konstantin of Preslav), Presbyter Gregori and Presbyter John – translators of the biblical books and of lives of the saints – a medieval literary genre which related the life of a person who had been canonized by the Christian Church. The most brilliant representatives of the Preslav school were Konstantin of Preslav – a poet and priest, clerical translator and chronicler, John the Exarch of Bulgaria – scholar, philosopher, orator and poet, and Monk Hrabr – author of a fiery apology of the Slavonic alphabet, *On Letters*. It is to them and Kliment of Ohrid that Old Bulgarian literature owes its international renown. Their works were a vivid testimony of the high level of uniformity and consistency attained by the Bulgarian literary language in its early stage of development and of the genre variety of young Bulgarian literature. It aspired after lofty heights in the solemn panegyric art, in the emotional verse form and in the philosophical, polemic and publicistic works. The two schools took shape and functioned on just slight differences in dialects of the Old Bulgarian language of the ninth–tenth centuries, and flourished while keeping in touch and being all the time dependent on each other. They were organically linked by the philological experience in translation and original literary creation remarkable for its time by the profound and steady belief of the Old Bulgarian writers in the social service performed by the written word, and the passionate support by Cyril and Methodius' humanitarian principles, the ardent defence of the cause of Slav enlightenment – the latter being the most significant original theme of Old Bulgarian literature.

Initially the Glagolitic alphabet was used at both the Ohrid and Preslav literary schools, but during the tenth century the Cyrillic alphabet began to be used together with the Glagolitic. Even in its early period Old Bulgarian literature was not confined to narrow national aspirations and was characterized by a desire for monumentality of ideas and problems, thus becoming accessible to other nations. Old Bulgarian literature played a historic role in the development of the Russian, Serbian and Romanian literatures. Not only did Russia receive from Bulgaria the script and divine service books it needed after the adoption of Christianity (988), but the Old Bulgarian language was furthermore the official literary language of Old

Russian literature for quite a long time. The Old Bulgarian translations of works by prominent Byzantine authors – John Chrysostom, Gregory the Theologian, Basil the Great, Athanassius of Alexandria, Cyril of Alexandria and Ephraim the Syrian, among others, reached Russian soil.

The Old Bulgarian enlighteners opened up before Eastern Orthodox Slavdom a new world of ideas, themes, genres and literary skill. Old Bulgarian literature was one of great genre variety: hagiography, hymnography and declamatory poetry, rhetoric prose (instructive, panegyric and polemic literature), annals and chronicles – typical historiographic works of that time, *belles-letters* – the secular genre of the epoch, natural science books, philosophical and grammar works, epistles, marginal notes, question-and-answer literature – all of which flourished. The genre variety of Old Bulgarian literature was directly dependent on its contacts with the literatures of the other peoples and on foreign artistic experience. Through the medium of Byzantium it was infiltrated at a rather early date and in a felicitous atmosphere of open-mindedness by works of Oriental literature and of Greek classicism; some works, as for instance *The Parable of Troy*, came to be known through the mediation of the literature of the West.

A specific place in Old Bulgarian literature was occupied by the Apocrypha – legendary-religious works denounced and persecuted by the Church in its capacity as leader of cultural life in the Middle Ages for containing 'heretical' elements. The Apocrypha combined elements of fiction with philosophical reflections, fantastic description, legends, and poetry. They gave expression to popular cosmogenic, eschatological, astrological and social concepts and elements of criticism of official dogmatism.

Bogomil literature, which served as the ideological weapon of the Bogomils against the dogmas of the official Church and the feudal order, was among the earliest anti-canonical works in medieval Europe. It propagated intolerance of the injustices of social order. Its form was religious, as typical of the times, and its sources were the Bible, apocryphal legends and the oral tradition. Bogomil literature was anonymous and was labelled heretic because it interpreted rather freely the Christian myths and notions sanctified by the Church, expounding in them theological, cosmogonic and social ideas which were contrary to the official religion. A number of written sources give as the author of 'heretic books' Priest Bogomil (tenth century) – founder and chief exponent of Bogomilism. *The Secret Book* is the most significant Bogomil work. Irreconcilability with the traditional notions, though expressed in a conventional religious form, characterizes Bogomilism as rebellious, and its literature was persecuted and destroyed by the Church. One of the most interesting phenomena in tenth century Bulgarian literature was the original polemic work *A Homily Against the Bogomils* by Presbyter Kozma. From the position of official ideology, the author sought out the causes of the moral dissipation he witnessed around him. Juxtaposed with the emotionally charged Bogomil literature, Presbyter Kozma's work rounded off the dramatic picture of social clashes and moral collisions.

During the period of Byzantine domination (1018–1186) literary life suffered a decline. The most famous of the classical Old Bulgarian literary masterpieces of the tenth and eleventh centuries were the *Zograph Gospel*, the *Marianus Tetraevangelia*, the *Assemanius Gospel*, the *Sinai Psalter*, the *Sinai Prayer-Book*, the *Clozianus manuscript*, *Codex Supraciliensis* – the oldest treasured texts in the history of the Bulgarian language. The thirteenth and especially the fourteenth century, witnessed a new upsurge of Old Bulgarian literature. Among the most valuable works of the thirteenth century are Boril's *Synodikon*, which included the 'Story about the Restoration of the Turnovo Patriarchate', the 'Story of the Zograph Martyrs', 'Dragan's Minaios', the 'Berlin collection', the 'Bologna Psalter' and others. Several manuscripts date back to the reign of Tsar Ivan Alexander (1331–71) – the *Sofia Psalter* (1337), the *Gospel of Ivan-Alexander* (1356) and the *Tomich Psalter* (*c.* 1360). The growing interest in history books, the newly translated short stories, novellas and natural science books, and the appearance of voluminous collections of a most varied content testify to the tastes of the Bulgarian cultural community at that time. The Turnovo literary school founded in the middle of the fourteenth century by Theodosius of Turnovo and Patriarch Euthymius of

Facsimile of the Tomich Psalter (fourteenth century)

Turnovo was the most outstanding cultural manifestation of the late medieval period, which brought resplendence and popularity to Bulgarian mediaeval literature.

The most lively centres of literary life were the town of Turnovo, the Kilifarevo Monastery (south of the town of Kilifarevo, on the right bank of the river Bregalnitsa), Vidin, the Rila Monastery, Lesnovo Monastery and the Mount Athos sanctuaries – above all the Zograph Monastery and the St. Athanassius privileged monastery. The most active period of the Turnovo literary school was from 1371 to 1393, when Patriarch Euthymius of Turnovo and his pupils worked there. In a broader sense the phenomenon is known to encompass the preparations for the establishment of the school thoughout the thirteenth and in the first half of the fourteenth centuries, the life's work of Theodossius of Turnovo and the Kilifarevo literary school (at Kilifarevo Monastery) headed by him, the expansion of cultural relations with the Mount Athos monasteries and the influence of the Turnovo literary traditions outside the boundaries of Bulgaria. The Turnovo literary school was closely linked with the overall material and spiritual upsurge of Bulgarian society on the eve of the Ottoman domination. Ideologically it rested on the theological and philosophical position of Hesychasm and represented an important element of the final period of efflorescence of the Eastern Orthodox spiritual community. There are serious grounds to suppose that Euthymius' school was not just a monk's community, but a broadly conceived organization for literary and cultural activities. Euythmius had applied himself to leading it with rare skill and purpose.

Making use of the experience of the Bulgarian men of letters on Mount Athos, he passed a linguistic and orthographic reform which also concerned the practice of the divine service. An important aspect of the work of Euthymius and his associates was the revision and complementing of the translation stock of Old Bulgarian literature. The literary standards became much higher and some genres typical of the Turnovo literary school flourished – lives of saints, panegyrics and polemic literature (part of it are the epistles).

Hymns occupy an essential place in the overall literary upsurge. The translations of polemic treatises – the works of Gregori Palama, Nil Kovassila and David Dissipat among others – popularize some of the most outstanding exponents of the Eastern Orthodox religion and Hesychasm.

Collections of hagiographies, panegyrics and homilies were thoroughly edited, and new translations of lives and eulogies appeared. These were subordinated to uniform artistic principles and high aesthetic requirements. From Hesychasm, in which the contemporary science traced portents of the Eastern pre-Renaissance, Euthymius of Turnovo borrowed the conviction about the divine predestination of the written word. For him the literary work was an original icon, a prayer, which afforded unsuspected opportunities for moral impact. By force of dramatic socio-political circumstances he associated that philosphical idea with the call of the historical moment and created remarkable works – lives and sermons – in which he defended his concept about the noble purpose of literature, while at the same time voicing his disquieting forebodings, references to the country's glorious past, and preoccupation with the consolidation of the moral unity of his compatriots. Euthymius' talent as a writer and his literary idiom, which Slavic scholars refer to as *pletenie sloves* (weaving words), are among the highlights in the history of mediaeval Slavic literature.

Euthymius enjoyed prestige beyond the boundaries of Bulgaria. He had a large number of students and disciples who popularized his orthographic rules and literary methods in Serbia, Russia and the Romanian principalities. The most talented writers among them – Cyprian, Yoasaf of Bdin, Konstantin of Kostenets and especially Grigori Tsamblak – continued the traditions of the Turnovo literary school even after Bulgaria lost its political independence. The Bulgarian manuscripts found their way far into the North, beyond the Danube, to the Romanian monasteries. They laid the foundations upon which the Slav-Romanian literature flourished almost up to the eighteenth century. The principles established at the Turnovo Literary School exerted an influence on the Old Russian literature known as 'the second South Slav influence in Russia'. For the second time after the Bulgarian Golden Age, Old Bulgarian literature reached a zenith and enjoyed enormous prestige.

Facsimile of the Rila Book of Homilies (seventeenth century)

The literary activity did not die down during the Ottoman domination (1396–1878). The literary tradition was kept alive in churches and monasteries – mainly by copying literary works created during the previous centuries. New works also appeared, like the *Bulgarian Chronicle* of the fifteenth century, which in its ideology was directed against the Ottoman invaders, and new translated works such as the fifteenth century *Story of the Fall of Constantinople*. The marginal notes in the different manuscripts contain valuable information about the life of

Facsimile of the *Panegyric of Rila* (1479) by Vladislav the Grammarian

the Bulgarian people under foreign rule. Very proliferous during the fifteenth century was Vladislav the Grammarian, who compiled four encyclopaedic collections of a total of about 4,300 pages and wrote the *Rila Story* about the transfer of the relics of John of Rila, the most worshipped saint in Bulgaria, from the city of Turnovo to the Rila Monastery. The works of Dimiter Kantakuzin are indicative of the sentiments of the intelligentsia during the first decades of Ottoman domination.

Literature blossomed in Sofia and its district during the sixteenth century when the Sofia literary school was established and became the most important educational and literary centre in the period after Bulgaria's fall under Ottoman domination. Its most outstanding representatives were the writers Priest Peyo and Matei the Grammarian, who in their works extolled the feats of Georgi Novi of Sofia and Nikola of Sofia. Both had died as martyrs for the Christian faith. By praising the self-sacrifice of the heroes, the authors pointed out the advantages of Christianity over Islam, and gave encouragement to people's resistance against the enslavers and to the Bulgarians' religious staunchness.

A new spirit was felt and new ideals emerged in the literature of the seventeenth and the eighteenth centuries.

The writers were inspired by political, social and genre themes, approaching them individually from their personal point of view rather than following a stereotype. Some 450 hand-written collections are known from this period, most of them treating various subjects and representing different genres – from sermons and lives to imaginative literature and chronicles. They were written in a popular language and were destined for the broad masses. Many monasteries kept the religious traditions alive by preserving and disseminating the liturgical and other books; some of these monasteries became literary centres. The Damascenes were an interesting literary phenomenon; these were collections with religious and didactic content, which first appeared in Bulgaria in the late sixteenth century and continued to be created up until the mid-nineteenth century. They consisted mainly of translated works from the *Treasure* collection of the Greek writer Damascene Studit. The Damascenes were written in the vernacular and raised questions of a social and ethical character. The Damascene writers were not ordinary compilers and translators; they proceeded from definite civic and moral principles, which often diverted them from the original, and led them to work their own reflections into the translation, thus becoming co-authors. Their contributions were indicative of the changes which were taking place in the life and the consciousness of the Bulgarian people. Yossif the Bearded (1714–c. 1759) was an outstanding Damascene writer. He championed the idea of a language which would commit Bulgarians to the cultural values of the time. In his writings he dealt with social and everyday questions. The vernacular made its way gradually and steadily into the literature. Chronicle writing became a literary trend. Priest Methodius Draginov (seventeenth century) wrote a short story about the Bulgarians' conversion to Islam. A remarkable historical document in itself, it is a literary work of exceptional power of expression. Priest Peyo, of the village of Mirkovo in the Sofia district, has left chronicle notes about the Turkish-Austrian War of 1690, in which one witnesses the rudiments of political consciousness. The chronicle-writing tradition continued throughout the nineteenth century – represented by the notes of Todor of Pirdop, a teacher, priest Yovsho of Tryavna and many others. Catholic literature – in translation and in the original – occupied a place of prominence in the overall development of literature (seventeenth–nineteenth centuries). Its representatives were highly educated men who had graduated in Western Europe – Peter Bogdan (1601–74), Philip Stanislavov (1608–74), Krustyo Peikich (late seventeenth–early eighteenth centuries), Franz Xavier Peyachevich and others. It attracts interest with its secular problems, genre variety and most of all with its national spirit. Works include theological and philosophical treatises, historical notes, poems, and elements of the autobiographical genre. Thematically they draw their subject matter from the political bondage and the historical past, reflecting the unflinching Bulgarian spirit of the Catholic writers.

During the seventeenth and the eighteenth centuries

some writers worked abroad. They committed the Bulgarian people to the spirit of the European Enlightenment. The most outstanding exponents of the ideas of the Renaissance were Hristofor Zhefarovich (end of the seventeenth century–1753) – a writer, artist and clergyman, and Partenii Pavlovich (end of the sixteenth century–1760). In 1741 Zhefarovich published his book *Stemmatigraphion*, the first printed Bulgarian book treating secular subjects and containing twenty portraits of Bulgarian and Serbian rulers and saints and 56 coats of arms of Slav and other countries, with explanatory quatrains below each of them. With its arguments and lessons drawn from the past, the book influenced the formation of national consciousness amongst Bulgarians, Serbs and other Slav peoples. It is a source of Bulgarian heraldry. With his unfinished autobiography (1757) Partenii Pavlovich laid the beginning of the autobiographical genre in the literature of the Southern Slavs. He stands out as a politician who wished to attract the interest of the ruling circle abroad to the political destinies of the Bulgarian people.

Facsimile of a page from the *Stemmatigraphion* (1741) by Hristofor Zhefarovich

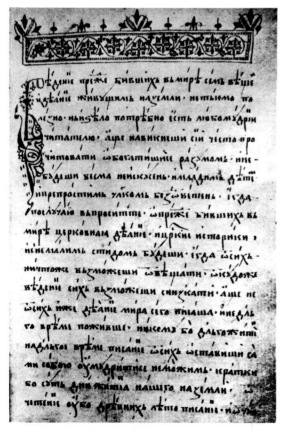

Facsimile of the *Slav-Bulgarian History* made by Sophronius of Vratsa in 1765

The new tendencies in the Bulgarian literature which appeared in the late seventeenth and the first half of the eighteenth centuries are most brilliantly and fully illustrated in the work of Paissii of Hilendar – a thinker of a new type, a writer with a strong national awareness. His *Slav-Bulgarian History* (1762) is the first work of the literature of the National Revival Period and reflects the topical problems of that epoch; in fact it is its literary and political manifesto. The *History* is a continuation of some of the preceding literary traditions, mainly of the damascenes and chronicles; yet it is an original phenomenon in Bulgarian literature, the fruit of the historical conditions obtaining at that time. It is a viable literary work, the creation of an illustrious personality. Paisssii had a profound knowledge of the tendencies in life, and developed further the ideas of the Renaissance and the suppositions of his predecessors and contemporaries about the tasks and prospects of the political and cultural life of the Bulgarian people. He outlined the programme of the Bulgarian National Revival and blazed a new trail for literature, linking it to life, to the sorrows and the struggles of the people, as a literature of civic commitment. His followers pointed out the patriotic motives which served them as loadstars in their literary activities. The literary work of Paissii was continued by the author of the Zograph history and by Hieromonk Spiridon (first half of the eighteenth century–1815), who wrote a short history of the Bulgarian people. Though they did not measure up to Paissii's clear and bright thought or vivid, temperamental style, his disciples shared with him the inner necessity of influencing their contemporaries through the lessons drawn from the past. Paissii's own pupil and most brilliant disciple was Sophronius of Vratsa (1739–1813) – the first exponent of the ideas of the Enlightenment and rationalism in Bulgaria. With his literary works he defended the right of the vernacular to become the official language of the new Bulgarian literature. He was the author of the first printed book of the new Bulgarian literature *Kiriakodromion* or *Sunday Prayer-book* – 1806, and of an original autobiography *The Life and Suffering of Sinful Sophronius* (1806–11) published in 1861, which was translated into a number of European languages. It gave expression to the literary concepts and aspirations of the new Renaissance personality that was taking shape and reflected a whole epoch in the life of the Bulgarian people. The influence of Sophronius' literary life was most palpably felt during the early decades of the nineteenth century. His most zealous disciples were called 'Sophroniotes'.

Neophyte Bozveli of Hilendar

Neophyte of Rila

Facing the burning issues of the epoch, the literature of the National Revival Period reflected the aspirations towards education and enlightenment, the struggle against the Phanariot clergy, and the sharp social contradictions. Works of publicist, didactic, moral and philosophical character predominated in the initial stage of its development. The literature of the second half of the eighteenth and the early decades of the nineteenth centuries was syncretistic, moralizing and educative in character.

The new social forces which began a struggle for the formation of a Bulgarian nation also sponsored a number of literary and cultural initiatives. The educational system was restructured. In 1824 Peter Beron (1800–71) published his *Fish Primer* which, owing to its diverse encyclopaedic content, became an educational programme of the National Revival Period. Secular education in Bulgaria was initiated with the opening of the mutual education school in Gabrovo in 1835. Its founder was Vassil Aprilov (1789–1847), an enlightener and ideologist of the Bulgarian educational movement. The flowering of education created a boom in the writing of textbooks and pedagogical manuals which popularized the principles of secular education. Their authors were the most prominent teachers and writers of the period – Hristaki Pavlovich (1804–48), Neophyte of Rila (*c.* 1793–1881), Neophyte Bozveli of Hilendar (*c.* 1785–1848), Emanuil Vaskidovich (1795–1875), Raino Popovich (*c.* 1773–1858), Konstantin Ognyanovich (1789–1858) and others.

The 1840s marked a new stage in the development of Bulgarian literature and culture. The periodical Press played a significant role in the promotion of literature. The *Lyuboslovie* (1844–46) magazine, *Bulgarski Orel* (Bulgarian Eagle; 1846–47), *Tsarigradski Vestnik* (Constantinople Gazette; 1848–62) were among the first periodicals which raised problems of literature, culture and education. The newspapers and magazines, around which groups of writers were formed, united by their common ideas, became the arena of severe socio-political and literary debates. The greater part of the Bulgarian periodicals were printed outside Bulgaria and the Ottoman Empire. From the early 1850s up to the Liberation in 1878 a total of 30 Bulgarian newspapers and magazines were published within the Empire and more than 60 in Romania, Russia, Serbia and Austria. The most progressive Bulgarian newspapers printed within the Empire were *Gaida* (Bagpipe; 1863–67) and *Makedonia* (1866–72). They were edited by Petko Slaveikov and carried literary articles too.

Of the *émigré* revolutionary Press most outstanding

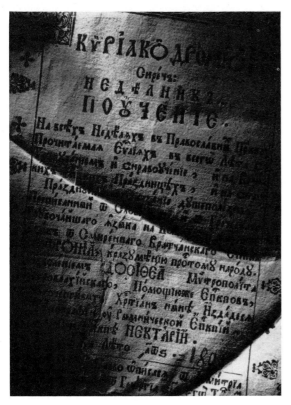

Facsimile of a page from *Kiriakodromion* (1806) by Sophronius of Vratsa

were the newspapers *Dunavski Lebed* (Danubian Swan; 1860–61) of Georgi Rakovski, *Svoboda* (Liberty; 1863–74) which assumed the title of *Nezavissimost*

Hristo Botev

(Independence) as of February 1873, of Lyuben Karavelov, *Duma na Bulgarskite Emigranti* (Word of the Bulgarian Emigrés; 1871), *Budilnik* (Alarm; 1873), and *Zname* (Banner; 1874–75) of Hristo Botev. They were concerned with the problems of political slavery and the national-liberation movement and fought for an original democratic literature and culture. Publicist works with educational purposes appeared, and caustic pamphlets against the local landlords were written. Men of letters turned to the literature of other nations, too. The first translations of pedagogical, morally instructive and literary works were made. Some translators adapted foreign literary works to the Bulgarian language, playing the role of co-authors; they modified the plot, the characters and sometimes even the composition, thus making the work suitable for the specific national conditions.

The first timid attempts in poetry were made. The beginnings were laid of book publishing – chiefly with funds provided by subscribers who paid a part of the sum or the whole amount in advance. The creation of literary works brought to the fore the question of the national literary language. A struggle began for an official literary language based on the spoken language.

Dobri Chintulov

The activities of Yuri Venelin (1802–39), a Ukrainian historian and Slavonic scholar who performed great services for the Bulgarian National Revival, stirred up interest in folklore as a consistent manifestation of the emerging national awareness. The Bulgarian educationists and cultural figures turned to folklore as a model of artistic excellence, a treasure-house of national traditions and an exponent of the people's strivings and hopes. Vassil Aprilov, Nikola Katranov, Naiden Gerov, Raino Popovich, Zahari Knyazheski, Ivan Bogorov and many other men of letters began to record and publish folklore works. In 1842 Ivan Bogorov (c. 1820–92) published the first collection of *Bulgarian Folk Songs and Proverbs*. Folklore materials were also published by Georgi Rakovski, the brothers Dimiter and Konstantin Miladinov (*Bulgarian Folk Songs*, 1861), Petko Slaveikov, and Lyuben Karavelov. Folklore became a mentor and adviser of Bulgarian writers. It played an important role in enriching Bulgarian literature.

The Greek influence, which had been dominant in education, literature and culture until the 1840s, gradually declined during the 1850s. The impact of Russian social and literary thought was ever more tangibly felt. Several Bulgarian poets of talent were active in Odessa in the 1840s. There Naiden Gerov (1823–1900), a writer, linguist, folklorist and active public figure, wrote his first lyrical works, which stood out sharply from the mass of earlier poetic endeavours. He also published the first epic poem in modern Bulgarian literature – *Stoyan and Rada* (1845) written in the spirit of the folk-song tradition. Gerov distinguished himself as one of the pioneers of modern Bulgarian poetry, writing verse with a new spirit, prosody and imagery. On his way from Russia to Bulgaria in 1846 he wrote the first letters in the form of travel notes, thus pioneering this genre in Bulgarian literature. Between 1840 and 1849, in Odessa, Dobri Chintulov (1823–86) wrote revolutionary songs. Gerov and Chintulov used the vocabulary and artistic richness of folklore in an original way, establishing the use of mixed syllabic and tonic versification in Bulgarian poetry. The first Bulgarian poetesses started writing in the same period: Elena Muteva (1825–1954) in Russia and Stanka Nikolitsa (the penname of Spasso Elenina; 1835–1920) in Bulgaria.

In the 1840s literature left the confines of the school and was linked with the social struggles; the writer became an active creative public figure. Neophyte Bozveli of Hilendar, who stood at the centre of the ecclesiastic and national struggle, because the pioneer of a new genre in Bulgarian literature, the dialogue.

Gradually the didactic stance in literature gave way to a more intimate, confessional tone. Along with the problems of national identity and duty, love and the attitude to women became an important subject. The first attempts at theoretical studies of some literary categories were undertaken at that time. Prose and drama works on secular subjects, mainly moralizing and sentimental in character, were translated in the 1840s and 1850s.

The Crimean War of 1853–56 helped to differentiate the class forces in Bulgarian society and created the prerequisites for the emergence of a revolutionary ideology and the appearance of works of markedly social and militant motives. The two main trends in socio-political life, revolutionary and educationist, were also manifested in literature. Sometimes their representatives defended the same ideological and aesthetic views and concepts, while in other cases there was a clear distinction. Literature evolved as profoundly democratic, with strong civic overtones, and aimed at serving the people's interests. The revolutionary line in poetry, initiated by Dobri Chintulov, was continued by Georgi Rakovski (1821–67), the founder of the Bulgarian national revolutionary movement and of the Bulgarian revolutionary Press, a publicist, historian and ethnographer, the first Bulgarian poet who deliberately, in the spirit of romanticism, used historical subjects as an ideological weapon. His verse and the narrative poem *Forest Traveller* (1854, published 1857), an ardent apotheosis of the haiduks' heroism, are an expression of his view of history as an important factor in rousing the national consciousness and keeping it alive. Rakovski reinterpreted the folk-song tradition and raised the poetry of the National Revival Period to a new ideological level. His poems are of lasting importance, mainly for the ideas that inspire them and the traditions they established in the treatment of revolutionary subjects. The most versatile poet of the time, who can also be credited with the highest artistic achievements, is Petko Slaveikov (1827–95) the first versatile talent in Bulgarian literature, a teacher, an active journalist and publicist, folklorist, translator and public figure. He, together with Dobri Chintulov, accomplished the transition to genuine, aesthetically significant art. Having started with imitative, intimate verse in the 1840s, Slaveikov wrote original, civic, lyrical poetry, reflecting the essence of that age, the complex and contradictory processes that took place in the Bulgarian consciousness during the National Revival. He made his literary work an integral part of his public activities. This was particularly evident in the 1860s and 1870s, when he wrote his finest lyrical verse. His poetic confessions show a keen sense of civic duty, and are permeated by historical optimism, a moving patriotism and sharp criticism of social ills.

Raiko Zhinzifov

Directly linked with the tendencies in lyrical verse of the 1860s and 1870s is the poetry written by Bulgarians living abroad – Konstantin Milanov (1830–62), born in Struga, and Raiko Zhinzifov (1839–77), born in Veles. Their keynote is one of nostalgia, anguish and sympathy

Grigor Purlichev

with the work of the world's most outstanding poets of the 1860s and 1870s.

The narrative poem became established as a literary genre in the period following the Crimean War. Rakovski's *Forest Traveller* and *Haiduk Sider and the Blackamoor* (1868) by Nikola Kozlev (1824–1902) directly reflect the social problems of that period. The genre recorded its highest achievements in terms of ideas and artistic standard in Botev's *Haiduks* (1871) and Petko Slaveikov's *The Spring of the White-footed Maiden* (1873). Other notable works in the genre are the poems of Grigor Purlichev (1830–93), born in Ohrid – *The Sirda* (in Greek, it received the first prize at a poetry competition in Athens in 1860 and was translated into Bulgarian in 1930), and *Iskandur Bey* (1861 in Greek, translated into Bulgarian in 1967), which are important landmarks in the literary process during the National Revival.

with the suffering of the enslaved Bulgarian people. Revolutionary motifs in poetry found a most forceful expression in the writings of Lyuben Karavelov and Hristo Botev. Karavelov wrote poems rich in revolutionary ideas, having a strong patriotic impact; some of them had much in common with folk-songs and were also sung. A place of prominence in the development of lyrical poetry in the National Revival Period belongs to the poems of Stefan Stambolov (1854–95). They appealed to the reader's feelings with the force of slogans and exerted a strong impact on the public mind.

The highest summit in National Revival poetry was the lyrical verse of Hristo Botev (1847–76), born in Kalofer, a poet of remarkable talent, a revolutionary, a national hero and a brilliant publicist. He developed further the national literary and poetic tradition, raising it to the highest level it ever attained. Botev's poetry is rich in ideas; it is active, mobilizing, imbued with the romanticism, pathos and the heroic spirit of the revolutionary struggle. Botev reflected reality in vivid, strong and synthesized poetic images, depicted brilliantly the national-liberation struggle of the Bulgarian people and immortalized its heroes. His poems mirror the ideas, sensibility and talent of an extraordinary creative personality, and a new attitude to beauty, to the heroic and the sublime. Botev created brilliant examples of political and revolutionary verse, eulogies of the revolution and of the exploits of the fighter for freedom and justice. His poetry is national and original and at the same time international and universally human. His poems are intimate confessions, a philosophical and ethical credo of the poet, and inspired hymns of the revolution. While his verse is close in rhythm to the folk-song, his overall prosodic system is highly original and entirely new in Bulgarian poetry. Botev's poetry, the world of ideas, images and emotions that it creates, its novelty of content and form, can vie

Vassil Drumev

The first works of fiction appeared in the 1860s and early 1870s. The pioneer in the field was Vassil Drumev (1841–1901), born in Shumen, with his short novel *Unfortunate Family* (1860). The very first prose works that appeared at the time reflected the problems of the day, and the political and social relations in Bulgaria during the National Revival. Their authors at first tended to copy Western European literary models, and were influenced by the methods of sentimentalism and the romantic literature – thrillers and adventure stories, notably in *Unfortunate Family* and *Stanka Perdue* (1865) and *Ill-starred Krustinka* (1870) by Iliya Bluskov (1839–1913).

Familiarization with Russian literature channelled the creative development of some writers into a new direction. Most illustrative of the ideological, thematic and artistic originality of the National Revival prose are the short stories and novellas by Lyuben Karavelov (1834/35–79) – organizer and ideologue of the Bulgarian national

Lyuben Karavelov

revolution, one of the founders of the new Bulgarian literature, a talented journalist, publicist and folklorist. The impact of his works on his contemporaries was determined by their topical ideas and artistic significance. In *Voivode* (1860), *Bulgarians of Yore* (1867), *Hadji Nicho* (1870), *The Rich Poor Man* (1872), and *Mother's Darling* (1875) among others, the talented critical realist depicted the panorama of Bulgaria's political and social life, of the national-liberation struggle to which he dedicated his entire literary work. Karavelov manifested his artistic skill to perceive life as it is, to narrate skillfully, to create vivid and original images and characters, and to search and reflect the original features and nuances. With his novellas *Is Fate to Blame?* (1868), *Soka* (1869) and others, Karavelov laid the beginnings of critical realism. Novelists of the National Revival Period posed and tried to solve important ideological and artistic problems. Iliya Bluskov displayed great talent in depicting landscapes, battle scenes and popular customs. Vassil Drumev sought to gain an insight into his characters' mentality, to find out the socio-political motives of their behaviour and to achieve a diverse and plastic, realistic portrayal. Lyuben Karavelov recreated the character of the epoch's hero and posed a number of problems: about the link between *belles-lettres* and folklore, the role of the vernacular in the literary process, the diversity of styles and the great possibilities provided by the short story and the novella as belletristic forms.

The 1860s and 1870s saw a corresponding development of drama. French, Russian, Greek and German comedies and tragedies were translated and often Bulgarianized according to the taste and requirements of both translators and readers. Initially dramatic works were regarded as a reading matter. The interest in and taste for theatrical performances developed gradually – first through the dialogue and later on through drama. Petko Slaveikov and others wrote dialogues whose aim was to arouse the Bulgarians' interest in works for the stage. The first theatrical performances date back to 1856, and the appearance of original Bulgarian dramatic works made it possible to attempt to establish an original Bulgarian repertoire. Dobri Voinikov (1833–78) is the founder of the original Bulgarian drama and theatre, as author of the dramas *Princess Raina* (1866), *Conversion to Christianity of the Preslav Court* (1868), *Velislava, Bulgarian Princess* (1870), the comedy *Civilization Misunderstood* (1871) and others. In spite of some compositional and artistic shortcomings, Voinikov's dramas appealed to audiences, developed their taste and educated them in the spirit of patriotism. Drumev furthered Voinikov's traditions in the historical drama, employing some of the same artistic devices, yet his drama *Ivanko, the Killer of Assen I* (1872) elevated Bulgarian drama, counting on psychological effect and involving his characters in contradictory situations. Todor Peev (1842–1904) tried to carry on Drumev's creative achievement with his tragedy *Fudulescu, Hadji Stefaniya's Profligate Son-in-Law* (1875, published in 1976). Dramatic works were written in the 1870s also by Atanas Uzunov (1851–1907), Nikola Zhivkov (1847–1901) and others. Lyuben Karavelov published his revolutionary drama *Hadji Dimiter Yassenov* (1872).

Dobri Voinikov

Being witnesses to great historical events as well as participants in ecclesiastical struggles and in the national liberation movement, the writers of the National Revival Period produced memoirs, too. In their works of that genre Georgi Rakovski, Petko Slaveikov and Lyuben Karavelov described Bulgarian customs, spiritual and material life during the National Revival, and the formation and consolidation of the Bulgarian's national consciousness.

The social commitment of National Revival literature

Naiden Gerov

Petko Slaveikov

provided a favourable ground for the writing of travel notes as well. Having begun as descriptions of towns, villages and localities in geography textbooks, travel notes became established as a separate genre with traditions of its own through Naiden Gerov's letters and travel notes, Ivan Bogorov's *Several Days' Journey in the Bulgarian Lands* (1886), Lyuben Karavelov's *Notes on Bulgaria and the Bulgarians* (1867) and others.

The participation of writers and publicists in the social life and their high sense of civic duty prompted the appearance of a new artistic and publicistic genre, the *feuilleton*, whose founder was Petko Slaveikov. Through the newspapers *Tsarigradski Vestnik* (1859–63), *Shutosh* (1873–74) and mainly *Gaida*, he revealed his great talent as a humorist and satirist, and his skill in influencing the reader through an intriguing plot, lively dialogues and wealth of dialect. The *feuilleton* reached its summit on the pages of the revolutionary, *émigré*, periodical Press. Karavelov and Botev turned this genre into a powerful weapon of the liberation struggle and enriched it with new ideas.

Being the product of the public and creative work of National Revival writers, non-fiction became a powerful ideological and educational factor in the ecclesiastical struggles and the national-liberation movement of the 1860s and 1870s. Through periodical publications journalists promoted the development of literary genres and of literary taste. The talented articles and reviews written for the *Gaida* and *Makedonia* newspapers, the *Bulgarski Knizhitsi* (1858–62) and *Chitalishte* (1870–75) journals, and for other periodicals published in the Ottoman Empire, as well as for the newspapers *Dunavska Zora* (1868–70), *Svoboda (Nezavissimost)*, *Duma na Bulgarskite Emigranti*, *Zname*, the *Periodical Review of the Bulgarian Literary Society* (1870–76, 1882–1910) and others published by the Bulgarian *émigrés* in Romania and other countries, raised and elucidated topical political, social, and general cultural problems and significantly contributed to the literary life of the National Revival Period. The periodicals featured short stories, poems, drama, travel notes, translations, critical reviews and essays on literary problems, among others.

The appearance of the first more significant literary works predetermined the emergence and development of literary criticism. The periodical Press carried articles on the essence and purpose of literature, on literary trends and schools, and on foreign writers. Many Bulgarian men of letters were concerned with the problem of creating a literature modern in content and with specific national characteristics. Poets and novelists, playwrights and journalists such as Petko Slaveikov and Todor Ikonomov (1838–92) expressed original views about the essence and functions of *belles-lettres*. Individual articles and manuals dealt with the problem of the genre variety of literature. Talented critics and writers such as Nesho Bonchev (1839–78), Vassil Drumev, Lyuben Karavelov and Hristo Botev affirmed the criteria and principles of Russian revolutionary democratic aesthetics and used criticism as a powerful instrument of aesthetic education and social impact. In the *Periodical Review* N. Bonchev and V. Drumev dwelt on the problem of literature as an activity, its purpose and tasks. N. Bonchev examined the problems of translated works, spoke against literature as an end in itself, and advised writers to work out contemporary subjects. Agreeing with the assessments of Bonchev and Drumev made in their book reviews, both Karavelov and Botev introduced a novel feature into literary criticism of the National Revival Period – that of a literature subordinated to the imperatives of revolutionary struggle.

After the Crimean War, readers showed a growing interest in and demanded higher requirements of translated literary works. The latter included fiction, poetry, drama and popular science works. Greater attention was paid to the subject matter and aesthetic qualities of the translations. Both translators and readers showed a preference for Russian and the other Slavonic literatures because they were closest to the Bulgarian national character. Most recommended for translation were such great writers as Gogol, Pushkin, Shakespeare, Cervantes and Balzac. Translations into Bulgarian included *Taras Bulba* by Gogol, *The Captain's Daughter* by Pushkin, *The Robbers* by Schiller, and works by Herzen, Belinsky and others.

National Revival literature travelled a long and complex path. As a result of its accelerated development many a talented poet, novelist, playwright, critic and publicist left his mark on it within a little over a century, and new genres and forms appeared. What it took other literatures centuries to create, was formed and developed in Bulgarian literature of the National Revival Period within decades only.

After Bulgaria's liberation from Ottoman domination in 1878 literature started developing under new socio-political conditions. It mirrored the phenomena characterizing the formation and evolution of capitalist society in Bulgaria and the working people's struggle against Capitalism. During the 1880s literature continued the progressive ideological and creative traditions of the National Revival, acquiring at the same time new ideological and aesthetic features. Its development was influenced by a reality with rapidly changing social characteristics, marked by clashes between liberals and conservatives, between democratic forces striving to consolidate and expand people's freedoms, and reactionary forces which sought to restrict and exterminate them. The nationwide cultural upsurge also made its impact on the development of literature. Bulgarian intellectuals, particularly those who participated actively in socio-political life, were mostly influenced by Russian democratic ideas. The most prominent Bulgarian writers were bearers and exponents of these ideas, siding with the people and responding to their needs and aspirations.

Cultural and literary life prior to the Unification of Eastern Rumelia and the Principality of Bulgaria in 1885 was concentrated in two centres – Plovdiv, where the most prominent men of letters were Ivan Vazov (1850–1921), Konstantin Velichkov (1855–1907), Zahari Stoyanov (1850–89), Naiden Gerov and Petko Slaveikov (who worked there for a while in 1881 after the *coup d'état*); and Sofia, with Petko Slaveikov, Stoyan Mihailovski (1856–1927) and Marin Drinov (1856–1906) as the most eminent figures. In the first few years following the Liberation Plovdiv had priority over Sofia, but in both cities the Bulgarian intelligentsia carried out brisk and diverse cultural work. The periodical Press was a major factor in the development of literature: 1881 saw the publication of the *Nauka* (Science) scientific literary review (1881–84), the publication of the *Periodical Review* was renewed in 1882, and the *Collection of Folk Wisdom, Science and Literature* started coming out in 1889. The first

Stoyan Mihailovski

purely literary Bulgarian journals were *Zora* (Dawn; 1885) and *Dennitsa* (Morning Star; 1890–91), published by Ivan Vazov. These were democratic in their essence and dealt with topical issues of social and literary life. The 1890s saw the appearance of the journals *Missul* (Thought; 1892–1907), *Bulgarian Review* (1893–1900), and *Bulgarian Collection* (1894–1915). They did away with the profusion of ethnographic and folklore material typical of publications in the 1880s; instead they featured a great number of original works of art, theoretical articles on philosophy, aesthetics and history of literature. These changes came as a result of the advance of national culture and literature, and of the development of creative thought of writers and critics.

The salient feature of literature during the 1880s was patriotism, inherited from writers and public figures of the National Revival Period. The literary works expressed feelings of gratitude to and love for the Russian liberators, mirrored moments of the national liberation movement and the Russo-Turkish War of Liberation of 1877–78, enthusiastically extolled the heroism of revolutionary fighters and brought back to life bright moments of the recent revolutionary past. The most characteristic feature which determined the ideological and emotional content of literature was love for the homeland, which permeated the most outstanding works of art; the main positive character was the national freedom fighter. Parallel with this, other works were created, mainly poetry, whose leitmotif was a critical attitude towards socio-political reality. They exposed the moral and political corruption of the ruling class, the lack of ideals, greed and mercenariness. Literature embarked on the path of critical realism.

The 1890s ushered in stormy changes in literature, influenced by Populism and Socialism – new ideological trends which had made their appearance already in the 1880s. New literary genres emerged; the novelette and the

Aleko Konstantinov

short story gained ground, thus enabling writers to cover a greater range of socio-political problems and to introduce more characters into their works. Authors showed greater interest in the emotional experiences of their characters, and their language became richer and more poetic. The critical attitude towards the bourgeois reality was enhanced. Critical realism became the main creative method and was to remain so until the middle of the first decade of the twentieth century. Critical realists exposed the predatory mentality of the nascent bourgeoisie, its moral and political degradation. Writers focused their attention on everyday life – on socio-political problems and the universal human traits of the Bulgarian. Their works reflected the discord in social life, and castigated socio-political and moral corruption, the social cankers of capitalist reality. They depicted these negative phenomena with bitter irony, opposed violence and sympathized with the underprivileged. The critical attitude was particularly strong in some writers, lying at the core of their works; in others it was just a component part. There were also those who coupled it with populist or populist-social feelings.

The poet Ivan Vazov, the patriarch of modern Bulgarian literature, is the first and most prominent representative of post-Liberation literature. He created significant works in all literary forms and genres. Having matured as a poet under the direct impact of the patriotic upsurge of the April 1876 Uprising and under the favourable influence of National Revival and Russian literatures, Vazov upheld and affirmed throughout his life his view of the writer's great civic responsibility and the social mission of art. The homeland is the central image in his works, which are imbued with patriotic feeling. In his most significant works (the books of poetry *Banner and Rebec*, 1876; *Bulgaria's Sorrows*, 1877; the *Epopee of the Forgotten* cycle, 1881–84; the novella *Outcasts*, 1883–84 and others) Vazov extolled the national liberation struggles and their heroes. Filled with indignation at Christian Europe's indifference toward the Bulgarian question prior to the Liberation he sharply denounced the falseness and hypocrisy of the western capitalist world, while praising the Russian people's humanism and liberation mission with enthusiasm and profound gratitude. Vazov's most remarkable and popular work is the novel *Under the Yoke* (1889–90) in which he recreated with vivid realism and artistic power the preparations for the outbreak and the suppression of the April Uprising, as seen through the eyes of the common people. The novel is in fact a national epic, presenting a broad panorama of the people's struggles, life, customs and spiritual aspirations during Ottoman domination and prior to the April Uprising. In many of his poems and short stories extolling the social idealism of National Revival figures Vazov rebuked the selfishness of the new bourgeois society.

Zahari Stoyanov

The heroic spirit of the national liberation struggles drew the attention of other writers, too, who left a lasting imprint on the history of Bulgarian literature. The 1880s and the 1890s saw the flowering of memoirs which were mainly written by participants in historical events. Their most prominent representative was Zahari Stoyanov, the author of *Notes on the Bulgarian Uprising* in three volumes (1884–92). This is a historical and artistic document of the Stara Zagora Uprising of 1875 and the April 1876 Uprising. Of interest and value also are the memoirs *The Past* (in four books, 1884–88) by Stoyan Zaimov (1853–1932), *Reminiscences from Constantinople Gaols* (1881) by Svetoslav Milarov (1850–92) and *In Gaol* (1899) by Konstantin Velichkov (1855–1907).

The exposing, realistic line in literature initiated by Vazov through his satirical verses, short stories and essays was taken up by Aleko Konstantinov (1863–97) – the most brilliant and most consistent critical realist of the

Facsimile of the cover of Ivan Vazov's novel *Under the Yoke* (1894)

Anton Strashimirov

Petko Todorov

1890s. His overall literary work is a reflection of sociopolitical life. His works deal with topical problems and reflect the attitudes of the most progressive and democratic part of the Bulgarian intelligentsia. In his *feuilletons* Aleko Konstantinov mercilessly denounced the rulers' violence and high-handedness, political time-serving, careerism and corruption. With courage and bitter sarcasm, rare for those times, he held King Ferdinand I up to ridicule and exposed his anti-popular activities. In his classical work *Bai Ganyo. Tall Tales about a Modern Bulgarian* (1895) the writer built up the most popular satirical image in Bulgarian literature, a generalization of the upstart bourgeoisie from the period of the initial accumulation of capital in Bulgaria during the first decades after the Liberation. In his travel notes *To Chicago and Back* (1893-94) Aleko Konstantinov also revealed in a realistic manner the nature of the capitalist world, both in Europe and in the US. In satires, fables, and poems in dialogue, the satirical poet Stoyan Mihailovski expressed bitter truths about people's freedoms that were trampled underfoot, stigmatized the lackeys of the court, and unmasked political machinations and social corruption. His lyrical works are imbued with reflections about evil, about the political situation in Bulgaria and man's never-ending discontent with his own imperfection. The allegoric poem *A Book about the Bulgarian People* (1897), which is a biting criticism of social ills and the monarchy, marked the summit of his satirical works.

During the 1880s and 1890s a great proportion of writers recreated the personal and social life of the Bulgarian peasants, and studied their mentality and customs. Some of them dreamed of bringing back to life the patriarchal past; others pinned their hopes on education and the peculiar role the intelligentsia had to play. Populist writers adhered to traditions in the people's life and strove to oppose the new social factors which destroyed patriarchal, idyllic life. In some of his novelettes Todor Vlaikov (1865-1943) idealized the petty bourgeois patriarchal life, while in others he described, by the method of critical realism, its decay and the heavy lot of the ruined craftsmen and farm-hands. In his short stories Hristo Maximov (1867-1902) revealed the gloomy side of peasant life, the cruel exploitation of the people. The critical realistic short stories of Mihalaki Georgiev (1854-1916) dwelt on the sorry plight of toiling peasants and the social crisis in the Bulgarian village. In some of them he idealized the far-off patriarchal world, opposed past morality to the amoral essence of his own times, depicted with deep compassion the lot of ordinary man and his dramatic clash with the bureaucratic state apparatus.

Parallel with the emergence of Socialism the beginnings were laid of socialist literature. Socialism attracted prominent intellectuals who enthusiastically embraced its ideas. Some young writers became its convinced champions, while others were more or less influenced by socialist ideology and focused their creative attention on social problems. Socialism enhanced the critical attitude in art and culture, levelled at the depressive, grim life, full of injustices. Critical realism was coupled with populist-socialist feelings, which was a typical Bulgarian national phenomenon. Anton Strashimirov's (1872–1937) works reveal the painful transition toward capitalist relations in the Bulgarian village; they are a protest against social injustices. Petko Todorov (1879–1916) wrote sentimental revolutionary short stories full of indignation at the cruel capitalist world. Elin Pelin alias Dimiter Ivanov Stoyanov (1877–1949), born in the village of Bailovo, Sofia district, created characters which came close to the socialist ideal. In the beginning of his literary career Tsanko Tserkovski (1869–1926), bard of the miseries, toil and struggles of the peasantry, also shared and popularized socialist ideas. Ivan Andreichin (1872–1934), who wrote socialist mass songs, poems, short stories and essays on social subjects, was also prominent as a socialist writer at the end of the nineteenth century. Peyo Yavorov (1878–1914), born in Chirpan, matured and grew as a poet under the influence of the socialist movement. In the works of his first period, when he was an outstanding representative of critical realism, the poet voiced his civic protest against social injustice and national oppression, as well as his love and sympathy for the deprived people (*Poems*, 1901). The first poems by Kiril Hristov (1875–1944) were also characterized by civic and social sentiments. The fiction of Vela Blagoeva (1858–1921), Ana Karima (1871–1949), and of some other writers was also linked with the socialist movement. Strashimir Krinchev (1884–1913), Dimiter Boyadjiev (1880–1911), Dimcho Debelyanov (1887–1916), and Dimiter Podvurzachov (1881–1937), were influenced by Socialism to a different degree.

Proletarian revolutionary literature came into being as an exponent of the ideas, feelings and struggles of the Bulgarian working class. Dimiter Polyanov (1876–1953) was a pioneer of the proletarian poetry. He was born in Karnobat, and his works extolled the emergence and development of revolutionary consciousness in the worker – the new positive hero in Bulgarian literature. The poet decisively rejected the false idols of the old world – religion, militarism, big business, and voiced his faith in the creative powers of the working class. He was the first to foresee the future that was to be built by the working class. His poetry is propagandist, distinguished more for its ideological and didactic message than for its concrete vital content. However, it did contribute a great deal to the creation and consolidation of the proletarian literary front. The poems of Vassil Karagyozov (1889–1925) and of others were in the spirit of Dimiter Polyanov's poetry.

The founding father of Bulgarian proletarian fiction was Georgi Kirkov (1867–1919), born in Pleven, a talented publicist and satirist, a master of the *feuilleton* and the pamphlet. His *fueilletons* stand out by their witty criticism of the bourgeois state and politics, and their ardent propaganda of Socialism and the uncompromising struggle waged against the 'common-cause' revision of Marxism. Georgi Kirkov was the first writer in Bulgarian literature to create a truthful image of the socialist worker of the late 1890s. His literary work is closely linked with the emergence of Bulgarian Marxist journalism.

Pencho Slaveikov

After Bulgaria's Liberation from Ottoman domination, favourable conditions were created for the development of literature for children and teenagers – Vassil Popovich (1883–97), Stoyan Popov, known under the pen-name of Uncle Stoyan (1865–1939), Tsonyu Kalchev (1870–1942), Vassil Stoyanov (1880–1962), and Stoyan Drinov (1883–1922). The names of some of the most talented Bulgarian writers of books for adults, such as Konstantin Velichkov, Ivan Vazov, and later Tsanko Tserkovski and Elin Pelin, are also linked with that literature.

In the 1880s and particularly in the 1890s, poetry came to occupy an important place in literature. Civic, lyrical poetry, represented by Ivan Vazov, Stoyan Mihailovski, Tsanko Tserkovski and Peyo Yavorov, was flourishing, as was epic poetry. Ivan Vazov, Stoyan Mihailovski, and later Pencho Slaveikov (1856–1912; born in Tryavna) produced outstanding epic poems which entered the golden fund of Bulgarian literature. Prose developed in two directions: one was heroic and emotional in tone, associated with the depiction of national liberation struggles, and the other was of a satirical and social

Elin Pelin

Ivan Shishmanov

character, reflecting the problems and conflicts in contemporary life. During this period, drama was poorly represented; many genre comedies were published in which the problems of love and the family were treated in an unimaginative way. Realistic comedies and dramas were written in the late nineteenth and early twentieth centuries. The first eminent playwright in the post-Liberation period was Ivan Vazov, the author of historical dramas and of social comedies of everyday life.

In the early years after the Liberation, literary criticism was mainly of a cultural and publicistic character. Writers like Ivan Vazov, Konstantin Velichkov, Zahari Stoyanov and Aleko Konstantinov frequently wrote critical articles and reviews in response to the new literary works, either in the original or in translation, evaluating folklore collections and making critical reviews of periodicals. Their articles showed a deep concern for the development of Bulgarian literature. The most active literary critics in those days were Ivan Shishmanov (1862–1928; born in Svishtov) and Krustyo Krustev (1866–1919; born in Pirot). Shishmanov was a well-known specialist in West European literature and in folklore, while Krustev made a study of German idealistic philosophy and aesthetics. Both were actively involved in the country's literary life, writing critical articles, engaging in literary historical research and editing important periodicals. Ivan Shishmanov showed a preference for realistic art. He is the founder of the cultural-historical school in Bulgarian theory of literature and folklore studies, and left behind valuable studies of Bulgarian culture of the National Revival Period. Krustyo Krustev began his activity as a literary critic with the belief that the writer and literature have a major role to play in the life of society, and that realism leads to the creation of truthful works of high artistic value. However, he did not remain consistently true to these views.

A new period in the history of Bulgarian literature began at the turn of the century, when new elements made their appearance in the development of critical realism. The life of the toiling peasants and their spontaneous protest against social injustice was more extensively depicted. Elin Pelin wrote some of his most critical short stories, imbued with affection and sympathy for the peasants, and portraying their rich inner life. Georgi Stamatov (1869–1942) published in 1905 his first collection of short stories, which were distinguished for their original critical style. Anton Strashimirov published his realistic novel *Autumn Days* (1902), the short novel

Peyo Yavorov

Crossroads (1904), and *Vampire* (1902), a drama of everyday life. Peyo Yavorov was still writing about the hardships of village life, at times giving way to individualistic sentiments. Mihalaki Georgiev continued to castigate wealthy village landowners, usurers, and mayors.

Noticeable during this period was a decrease in the civic concern and critical attitudes of some writers, as well as a toning down in their condemnation of contemporay life. This was felt even in the works of Ivan Vazov. He looked back to history in order to turn obliquely to the present. Stoyan Mihailovski indulged in abstract philosophical contemplations, pervaded by mysticism.

A deeper interest was shown in man's personal life, and his inner world, moral drama, spiritual quests, thoughts and feelings were elucidated. A wider range of ethical and aesthetic problems were explored in art. The way was paved for new artistic quests.

Nikolai Liliev

Dimcho Debelyanov

The national liberation struggle of the Bulgarians in Macedonia and Adrianople Thrace strengthened the social and patriotic feeling of some of the Bulgarian intelligentsia, impelling them to self-sacrifice and heroic suffering. It had a favourable effect on Bulgarian literature and consolidated its militant and humanistic trends. Anton Strashimirov and Peyo Yavorov devoted their works to the Macedonian revolutionary movement.

During the latter half of the 1890s and the early years of the twentieth century, some writers, influenced by West European culture and literature, began looking for new aesthetic values. They familiarized themselves with individualism as a philosophical school and with the symbolism of poetics, endeavouring to embody them in their literary works. Pencho Slaveikov became the initiator of the 'new quests' in Bulgarian literature. Having rid himself of the sentimental experience that underlay the poems in his first collection he propagated the aesthetics of individualism, but inwardly he did not remain true to it all the way, nor did he fall entirely under the influence of Nietzsche. His ideal was intricate and contradictory. He advocated a new art which ennobled man, perfecting him ethnically through beauty, and created a 'new' realism, mostly a psychological realism of daily life in which the social element, the direct disclosure of class conflicts and struggles, remains in the background. The ideological and emotional essence of his most important works in the collections *A Dream of Happiness* (1906), *Epic Songs* (1907), and *On the Island of the Blessed* (1910), is in most general terms characterized by humanism. Pencho Slaveikov strove to depict only what was of lasting value for man, for society; he strove to combine what was national with universal humanistic ideas. He accentuated the personal life of his characters, trying to probe their moral essence and their spiritual longings. He recreated the heroism of the April 1876 Uprising in his unfinished poem *Song of Blood* (1893, published 1911–13). In it the author made an attempt to create a national epic, synthesizing the philosophical meaning of Bulgaria's national development.

Pencho Slaveikov's efforts to introduce the major philosphical and moral problems of the time into Bulgarian literature, to renew and to 'Europeanize' it, were supported by his followers from the *Missul* literary circle (who published a review of the same name) – Krustyo Krustev, Petko Todorov and Peyo Yavorov. They popularized modernism. The ideological and aesthetic programme of the circle envisaged an ideological and thematic enrichment of Bulgarian literature, raising the problems it treated to the level of the most topical questions of the time, its artistic renewal, and elevating its standards to match the highest achievements of world literature. Failing to appreciate, for a variety of reasons, the revolutionary meaning of the socialist movement, the representatives of the *Missul* circle at the same time opposed the official regime and held anti-monarchist and democratic views. In one degree or another they lost touch with social realities and sought salvation in the temple of 'pure' art, away from the bustle of the 'market-place'. They placed the problems of the individual at the centre of attention in their works, attaching prime importance to the 'moral revival' of the individual. Along with this they strove to express the national, Bulgarian element, to seek out and reveal the beauty and the profound meaning of Bulgarian folklore. The contradiction between the individual and the social element is the basis of the atmosphere of conflict in most of Petko Todorov's idylls and dramas. His characters, who in the majority of cases are village people, true to their natural striving for what is unusual and poetic, for all that stands above the monotony of daily life, enter into conflict with their surroundings. Their impulses are most frequently a source of suffering and unhappiness.

Todor Trayanov

Geo Milev

After the defeat of the 1903 Ilinden-Preobrazhenie Uprising, Peyo Yavorov went through a spiritual crisis – his social views and spiritual dreams collapsed. At that time Yavorov also experienced the influence of Nietzsche's philosophy and the aesthetics of the French modern poets. He gave poetic expression to the crisis and to his new 'creed' in his *Song of My Songs* (1906), which can be regarded as a manifesto of the symbolism that was making its way into Bulgarian literature. Detached from the people, Yavorov became an introvert and adopted the philosophy of extreme subjectivism. The poet's own personality became the subject of his poetry, a personality which was alien to society and its struggles and whose basic tone was pessimism. The lyrical hero in Yavorov's works experienced the contradiction between the moral ideal and the reality of life very dramatically and tragically. His torments reflect the disharmony between modern man and the period of decadence. With his philosophical conception of the world expressed in his collection of poems *Sleepless Nights* (1907), Yavorov was the first to introduce symbolism in Bulgarian literature. He legitimized the theme of the alienation of the individual and was later to hand it down to the younger symbolist poets.

Bulgarian modernism is a medley of the poetics of various trends – symbolism, impressionism and expressionism. Typical features of some of its representatives are the distancing from the national democratic traditions, the break with realism, the avoidance of social themes, and isolation in the world of subjective experiences. Some of them adopt the escape from reality as a peculiar protest against society, as an opportunity to preserve their independence in the world of 'pure art'. The most persistent adherent to the tenets of symbolism was Todor Trayanov (1882–1945). In his lyrical poetry, Nikolai Liliev (1885–1960) shows himself as a lonely introvert, a sad poet of deep feelings. His individualism intertwines with a warm sympathy for all sufferers. His poems emanate noble humanism, revealing the image of a highly ethical personality. The poetry of Dimcho Debelyanov, one of the tenderest and warmest Bulgarian lyric poets, expresses complicated and contradictory human emotions. He adopted some of the aesthetic views of the symbolists but did not detach his unaffected and sincere poetry from life. His lyric poems are a personal confession of unfulfilled dreams and romantic yearnings, reflecting the poet's emotional conflict, generated by the tragic contradiction between reality and the ideal. There is a strong outburst of social feeling in the secret sighs of Debelyanov's poems. Poets Hristo Yassenov (1889–1925), Lyudmil Stoyanov (1888–1973), and Emanuil Popdimitrov (1885–1943), among others, were strongly influenced by symbolism early in their creative careers. Geo Milev (1895–1925) was a follower and advocate of symbolism and expressionism. Another poet who to a certain extent came close to symbolism was Dimiter Boyadjiev (1880–1911). His poetry, however, is always down to earth and his sadness results from definite experiences of life rather than from the principle. In prose, Bulgarian symbolism is chiefly represented by the literary works of Nikolai Rainov (1889–1954).

Although fleeing from reality into the world of dreams, the symbolists were unable to escape from it entirely. Along with his symbolist works Yavorov wrote a cycle of realistic poems devoted to the national-liberation struggles of the Bulgarians in Macedonia, while his dramas expose the flaws in bourgeois morality. In 1913 Todor Trayanov broke the symbolist dogmas and under the influence of the 1912–13 Balkan War produced his finest works, permeated with a sincere sense of patriotism. Dimcho Debelyanov produced tender and intimate lyric verse. The poems he wrote during World War I (1914–18) are realistic and concrete in their imagery, and humanistic in ideological and emotional content. He also wrote critical realistic, humorous and satirical works.

Hristo Smirnenski

Kiril Hristov

Late in the 1890s and early in the twentieth century Kiril Hristov established himself as a poet of the impulse of life. He rid love of philistine sentimentalism, laying emphasis on its sensual power. His vitalism is combined with Epicurism, his overcoming of restrictions resulted in unrestrained eroticism and a longing for a wanderer's life. Kiril Hristov's individualism is not associated with the struggle for the spiritual emancipation of man or the yearnings of the fighting individual, but with the ego, his supreme credo in life.

During the second decade of the twentieth century Bulgaria experienced three wars. Very few of the Bulgarian writers assessed properly the character of the 1913 Second Balkan War or of World War I. At times some of them succumbed to a mood of national exaltation – Kiril Hristov, Elin Pelin, Ivan Vazov, Anton Strashimirov, to mention but a few. But in Bulgarian literature of that period there also sounded deeply humane notes – dejection and concern at the relentless extermination of people, at the senseless shedding of blood. This could be felt in the poetry of Dimcho Debelyanov and Nikolai Liliev, and in the essays *Stains of Blood* (1921) by Vladimir Mussakov (1887–1916). The many faces of the war were likewise recreated by Yordan Yovkov (1880–1937), one of the most remarkable Bulgarian short story writers. During the above-mentioned wars there was a sharp drop in the intensity of spiritual and cultural life in the country. Compelled to experience the horrors of the war and of defeat, some of the most important representatives of symbolism abandoned their ivory towers and came close to the problems of life. In poems written on the battlefront, Dimcho Debelyanov came increasingly closer to realistic poetry. Hristo Yassenov wrote *Petrograd* in 1917, a moving civic poem, through which he greeted the 1917 October Revolution in Russia. The poets Geo Milev and Lyudmil Stoyanov experienced confusion and ideological hesitations. The first poems of Hristo Smirnenski (born in Kukush; 1898–1923) appeared in humorous magazines.

The new artistic trends late in the 1890s and early in the twentieth century influenced the variety of genres in literature. The content of lyric poetry changed. Poetry of abstract dreams, of striving for the unattainable was created. Beauty and perfection became a cult. From Yavorov's *Sleepless Nights* to Liliev's verse, poetry endeavoured to come close to music. Poets painstakingly sought the metric patterns and rhyme schemes capable of conveying their states of experience. Public sentiments had a note of ethical dissatisfaction, a longing for a pure, perfect world, an embodiment of what is beautiful and good. Peyo Yavorov, Dimcho Debelyanov and Nikolai Liliev enriched and perfected the artistic form of poetry, revealing the entire wealth and music of the Bulgarian language.

Lyudmil Stoyanov

Ivan Vazov

A new and original phenomenon in Bulgarian poetry was the appearance of humour in lyrical verse. Many burlesques were written, deriding and denying the literary tradition, and many humorous weeklies were published.

The early years of the twentieth century saw the appearance of rich and varied love lyrics by Pencho Slaveikov, Peyo Yavorov, Dimcho Debelyanov, Kiril Hristov, Dimiter Boyadjiev and others.

In fiction, the short story and the novel figured prominently, while the short novel was less well represented. Late in the nineteenth century and early in the twentieth century the short story depicted the many aspects of life and became increasingly varied in the themes and characters. Short stories were written by Ivan Vazov, Elin Pelin, Anton Strashimirov, Georgi Stamatov, Georgi Kirkov, and others. The novels were social and psychological, treating problems of everyday life. The most important work produced during this period was Ivan Vazov's novel *The Kazalar Queen* (1903). Vazov was the author who laid the beginnings of the truly artistic historical novel with *Svetoslav Terter,* written in 1907. The high-water mark in the development of the short novel treating problems of everyday life was Elin Pelin's *The Geraks* (1911), which shows the advent of Capitalism and the disintegration of patriarchal relations in the Bulgarian village. Georgi Kirkov continued the traditions of Hristo Botev and Aleko Konstantinov, adding a new content and new character to the *feuilleton* and the satirical short story, imbuing them with revolutionary optimism. The idylls of Petko Todorov are a highly original phenomenon in Bulgarian fiction.

In the early years of the twentieth century Bulgarian drama, too, was on the upswing. The founding of the National Theatre in Sofia in 1907 gave an impetus to dramatic literary output. Many political, psychological, historical, satirical and genre plays were written. Ivan Vazov was the most outstanding Bulgarian playwright in the post-Liberation period and in the first decade of the twentieth century. His satirical, social and political comedies dealing with problems of daily life posed topical social problems, exposing the political corruption, morality and customs in bourgeois Bulgaria: *Duel* (1892), *A Reporter, Did You Say?* (1900), *Candidates for Fame* (1901), and *Office Seekers* (1903).

Ivan Vazov's lasting popularity as a dramatist is due to his play *Exiles* (1894), a dramatization of his short novel *The Outcasts,* and to his historical dramas. Petko Todorov's plays occupy a particular place in Bulgaria's national dramaturgy as bold attempts at interpreting the Bulgarian way of life in the spirit of interesting philosophical conceptions. For Petko Todorov social morality and social conflicts are of prime importance, as in *Bricklayers* (1899, published in 1902) and *The First Ones* (1907). His lyrical and realistic plays are based on traditional beliefs and folkloric motifs. The realistic dramas of everyday life and the legendary and symbolic plays of Anton Strashimirov occupied an important place in Bulgarian dramaturgy during this period. He tackled social questions, sought for national self-determination for Bulgarian dramaturgy and at the same time was attracted by various short-lived fashionable trends and emulated them. His most important dramatic works are the tragedy *The Vampire* (1902) and the comedy *Mother-in-Law* (1906). The biggest success in the composition of civic and psychological dramas was achieved by Peyo Yavorov with his plays *At the Foot of Mount Vitosha* (1911) and *When Thunder Strikes, How the Echo Dies Down* (1912). The main themes in Yavorov's dramatury are tragic love and the powerlessness of social idealism. Unlike his predecessors he skilfully highlights the principal moments of the action, introduces an element of psychological drama in social and personal conflict, makes dialogue more detailed and gives it a poetic ring. The dramatic work of Kiril Hristov is uneven as regards ideological and artistic standards, and in most general terms continues the romantic and historical traditions. His most successful play is *Boyan the Magician*.

In the first two decades of the twentieth century Bulgarian comedy made a decisive leap forward as a characteristic manifestation of the critical realism prevailing in literature at the time. Along with those of Vazov and Strashimirov, a variety of other comedies and farces were written, enriching that dramatic genre in terms of themes, ideas and style.

Literary criticism in the late nineteenth century and the early years of the twentieth century expressed sharply and clearly the conflict between the materialist and idealist aesthetics. This was most fully reflected in the fierce polemics between the critics Krustyo Krustev, an idealist, and Dimiter Blagoev (1856–1924), the first Marxist critic in Bulgaria. In many of his articles K. Krustev, founder and editor of the *Missul* magazine, took a stand in favour of realism, but in the late 1890s and the beginning of the twentieth century the magazine became a centre of the aesthetics of individualism and formalism in literature. Krustev's alienation from the social struggles of the people led him to a renunciation of social humanism in art; he advocated and preached decadent theories and propounded the principle of 'pure' art. The aesthetics of the *Missul* circle influenced Boyan Penev (1882–1927), a highly erudite scholar and exacting aesthete, an inspired literary historian and critic. He strove to promote the artistic standards of Bulgarian literature, advising writers to pose and solve problems of a universal human nature. Having evolved his critical standards on the basis of foreign models, Boyan Penev did not always give sufficient consideration to the realities of Bulgarian life. His studies of National Revival literature was particularly valuable, marking a summit in Bulgarian literary scholarship in the past, eg. his *History of Bulgarian Literature* (4 volumes, 1930–36). At the outset of his literary career Vladimir Vassilev (1883–1963) also kept in touch with writers contributing to the *Missul* magazine. A follower of bourgeois-aesthetic literary theory, he often argued with Marxist critics.

Through the *Novo Vreme* magazine, Dimiter Blagoev tried to bring home the ideas of scientific Socialism to the workers, impoverished craftsmen and working intellectuals. A born polemicist, with a keen insight into the class nature of art, Blagoev used persuasive arguments in his polemics with the populists, utopian socialists and anarchists, and with the individualists and 'supermen' around the *Missul* magazine, trying to explain various literary questions from socialist positions. Blagoev regarded art as

Georgi Bakalov

Boyan Penev

a reflection of reality, as a mirror of public life. He expected creative artists to gain an insight into the characteristic, typical aspects and manifestations of life, and judged writers according to the vitality and truthfulness of their work, its social significance and impact. Marxist aesthetics, literary theory and criticism in Bulgaria as a whole continued and developed Blagoev's work. One of his followers was Georgi Bakalov (1873–1939), whose interest in Bulgaria's cultural and intellectual life and in world progressive culture was clearly reflected in the magazine he published, *Suvremennik* (Contemporary; 1908–10). Bakalov made his mark as a talented literary critic with a trenchant critical and publicistic style. He tried to elucidate artistic phenomena from Marxist positions. However, following the Communist Party's left sectarian course, he failed to give a scientifically based and fair assessment of the outstanding realist writers (Vazov, Mihailovski, Konstantinov, Strashimirov, Elin Pelin and others) or of those who showed a preference for the poetic style of modernism (Pencho Slaveikov, Yavorov and Liliev). He was convinced that political outlook played a decisive role in the creation of socially active, democratic works. Another follower of Blagoev in literary criticism was Dimiter Dimitrov (1876–1902). From the positions of Marxist aesthetics he argued against the modernist view, advocating a realist literature, rich in progressive ideas, and making a positive effort to establish the social-revolutionary line in literature and to emphasize the mounting influence of Socialism. Strashimir Krinchev likewise was a talented literary critic working from the position of Socialism, and sharply attacked the advocates of modernism and decadence.

Modernism in literature was defended by the magazine *Is Nov Put* (On a New Course; 1907–10) and to a certain extent by the magazines *Nash Zhivot* (Our Life; 1901–12), *Hudozhnik* (Artist; 1905–9) and *Slunchogled* (Sunflower; 1909). Its first theorist was Ivan Andreichin. Dimo Kyorchev (1884–1928) and Ivan Radoslavov (1880–1969) also proclaimed themselves followers and theorists of modernism.

Literary critics were clearly aware of the impending disturbances in literary development due to the profound crisis in the public consciousness. The social orientation caused the formation of literary groups following definite socio-political tendencies, and a loss of direction for writers who remained without socio-political support.

In the period from the First World War to 9 September 1944 literature reflected the constantly intensifying social contradictions, the onslaught of Fascism, the revolutionary upsurge of the proletariat, the bolshevization of the Bulgarian Communist Party (BCP) and its establishment as the main political force in the country, organizer and leader of the struggle of the working class, peasants and intelligentsia. Authors responded to all major events in world and Bulgarian history. Though the literary process was complex, contradictory and uneven, its most salient characteristic gained increasing prominence: in a ruthless struggle against reactionary bourgois ideas and aesthetic theories the proletarian literary front was growing strong, literature was turning left and socialist realism was increasingly established as a new creative method. Bulgarian writers, with very few exceptions, did not accept Fascism or become its advocates.

The political maturing of the masses during the First World War, the revolutionary upsurge in its wake, the mass demonstrations, strikes and revolts and particularly the September 1923 Anti-fascist Uprising profoundly shook the Bulgarian intelligentsia, changing its outlook and focusing its attention on the main problems in the life of the people. The example of the Russian writers who had embraced the ideas of the October Revolution exerted a strong impact on writers in Bulgaria. The experimental efforts of some cultural figures, such as Geo Milev in his *Vezni* magazine, and the belated dreams of a creative artist independent of society, and of aesthetes like the Vladimir Vassilev circle of the *Zlatorog* (Golden Horn) magazine and Ivan Radoslavov and the adherents of his *Hyperion* magazine, proved unable to give modernism a new lease of life. The attempts to win popularity for the pseudo-patriotic literary output which had been churned out profusely during the war, to defend and revive the bourgeois 'national ideal' of unification in the form of a 'unified mentality', 'self-knowledge', 'creativity' and 'beauty', were equally futile. The period following 1918 saw the publication of remarkable books by the Bulgarian symbolists: *Evening Mirages* (1920) and *Free Verse* (1921) by Emanuil Popdimitrov; the posthumous edition of Dimcho Debelyanov's works, *Poems* (1920); *Bulgarian Ballads* (1921) and *Song of Songs* (1923) by Todor Trayanov; *The Knight's Castle* by Hristo Yassenov; *Moon Spots* (1922) by Nikolai Liliev; and *Land* (1923) by Lyudmil Stoyanov. They marked summits in the development of Bulgarian poetic language and demonstrated its musical and rhythmic potential, but nevertheless it can be felt that the traditional spirit of symbolism in them is shaken by the surge of more objective, earthy feelings and moods, by realistic pictures and sober thoughts. Many of the symbolist poets underwent an evolution of ideas and creative style and wrote realistic and democratic works. A case in point is the change in Hristo Yassenov, who adopted communist ideas; the strivings of the crowds were still alive in the mind of Nikolai Liliev; in their prose, Lyudmil Stoyanov and Nikolai Rainov opted for a realistic portrayal of life; Geo Milev experienced a complex ideological and aesthetic evolution – while his collection of expressionist verse *The Cruel Ring* (1920) was still in print, the harrowing realism of the cycle *Ugly Prose* was already taking shape in his mind. Nikolai Hrelkov (1894–1950) also outgrew the symbolist tradition.

The revolutionary writers found a creative platform in the party Press – the *Cherven Smyah* (Red Laughter) magazine (1919–23), the *Mladezh* (Youth) newspaper (1921–23) and other publications. The poetry of Hristo Smirnenski provided the fullest artistic expression of the revolutionization of the masses following the October Revolution. From 1920 onwards, he published poems directly associated with the struggles of the working class. Collected in the book *Let Daylight Come* (1922), they marked the second stage in Bulgarian revolutionary-proletarian literature and the formulation of socialist realism as its artistic method. Hristo Smirnenski heralded the end of the old world, glorified the new hero of the age – the revolutionary worker – and praised the triumphant march of the red squadrons and the grandeur of the young Soviet capital. Many of his poems, though devoted to specific events in the heroic struggle of the world proletariat, reflect the essence of the historical process, the irresistible progress of the proletarian revolution and the leading role of the Communist Party in the revolutionary upsurge of the labouring masses. With remarkable political foresight the poet reveals the pioneering mission of the

Russian working class in the building of the future socialist society; his poems passionately defend the Soviet regime, mercilessly condemn the counter-revolution and the intervention against the Soviet State and express the Bulgarian people's love of the Soviet Union and their belief in its ultimate victory. His work combined communist devotion to the Party with bright artistic imagery and a rich orchestration of poetic rhythm.

The democratic and critical realistic trend in literature gained momentum after the First World War. Many authors, humanists and democrats, wrote works full of humanity, love of the people and faith in its strength and its future, which made a profound artistic impact. Deeper psychological insight was shown in the portrayal of literary characters and in rendering their thoughts and feelings. The people's poet, Ivan Vazov, published his finest intimate verse – the collection *I Smell the Scent of Lilac* (1919) and his last collection of poems – *She Will Not Die!* (1920). Yordan Yovkov printed his last war stories, *Last Joy* (1920) and others, and the long story *The Harvester* (1920). A number of interesting works appeared by Georgi Raichev (1882–1947), Anton Strashimirov, Nikolai Rainov, Svetoslav Minkov (1902–66) and other writers. Important new works representative of critical realism included short stories by Georgi Stamatov and the long story *Zemya* (1922) by Elin Pelin.

The military fascist *coup d'état* of 9 June 1923, the September 1923 Uprising and the events of April 1925 were a trial of the writers' civic and creative conscience. The progressive authors, linked with the people, took their side without hesitation. The so-called September literature, sharply condemning Fascism and singing inspired praise of the heroism of the insurgents, came into being. Anton Strashimirov branded the fascist hangmen in his novel *Horo* (1926). In his poem *September* (1924) Geo Milev combined powerful condemnation with a graphic portrayal of the revolutionary storm and the horror of defeat. The people's tragedy and their power of recovery were movingly depicted in the September verse and short stories of Assen Raztsvetnikov (a pen-name of Assen Kolarov, 1897–1951) – *Sacrificial Pyres* (1924); Nikolai Furnadjiev – *Spring Wind* (1925); Angel Karaliichev (1902–72) – *Rye* (1925); Georgi Karaslavov (1904–80) – *The Reed-Pipe Weeps* (1927); Nikolai Hrelkov – *A Ballad of Three Sisters* (1935); Krum Kyulyavkov (1893–1955) – the ballad *Nikodim* (1944); and Krum Velkov (1902–60) – the novel *The Village of Borovo* (1932).

The socio-political events of the late 1920s and the 1930s deepened not only the social but also the literary differentiation and reflected favourably on the work of the democratically-minded realist writers who were not directly linked to any political party. The general political and economic crisis finally destroyed all illusions of apolitical and non-committed art or purely emotional and aesthetic joys. They were replaced by the idea of man's duty to contemporary life. Critical realism evolved new modern forms and became richer in terms of genre and style. Humanism and democratism in realistic literature gained ground and there was a persistent effort to portray inner human drama. Authors underwent a complex and contradictory development. Progressive tendencies found expression in different ways: in an individual revolt against the conventional values of bourgeois ethics, in a direct criticism of capitalist social relations, or in retreat into the past, where genuine moral purity and humanity were sought. The development of democratic literature was assisted by the newspapers *Razvigor* (Spring Wind; 1921–27, 1937), *Literaturen Glas* (Literary Voice;

Svetoslav Minkov

Facsimile of the cover of Hristo Smirnenski's *Let Daylight Come* (1922)

Atanas Dalchev

Elissaveta Bagryana

Nikola Furnadjiev

Yordan Yovkov

1928–46), the magazines *Listopad* (Autumn; 1913–35), *Bulgarska Missul* (Bulgarian Thought; 1925–44), *Izkustvo i Kritika* (Art and Criticism; 1938–43) and others. Atanas Dalchev (1904–78) wrote lyrical poetry with rich philosophical undertones: *Window* (1926), *Paris* (1930), and *The Angel of Chartres* (1943). The poet conveyed his impressions of the social contradictions in the capitalist city and of the dissatisfaction of the intelligentsia in concrete and tangible images, and depicted with sympathy the life of the working people. A sharp and implacable revolt against the philistine restrictions imposed on women, a thirst for unknown distances and new frontiers, deep patriotic feelings and earthly vitality are characteristics of the fresh and colourful verse of Elissaveta Bagryana (a pen-name of Elissaveta Belcheva, b. 1893) – *The Eternal and the Sacred* (1927), *The Sailor's Star* (1932), and *Human Heart* (1936). The poetry of Dora Gabe (1888–1983) is rich in humanistic motifs – *Earthly Path* (1928) and *The Sleep-Walker* (1933). Nikola Furnadjiev wrote some of his finest verse, marked by a keen sense of civic duty, longing for social change and hope of overcoming isolation in life. The poems of Slavcho

Krassinski (a pen-name of Ventseslav Krustev, 1908–84) are a hymn to the land and to work – *Spring Visitor* (1932). Realism and humane ideas are characteristic of

Stefan Kostov

the poetry of Dimiter Panteleev (b. 1901), Nikola Rakitin (1885–1934), Alexander Vutimski (a pen-name of Alexander Vutov, 1919–43), Blenika (a pen-name of Penka Tsaneva, 1899–1978). Magda Petkanova (1900–70) and others. In a series of collections of short stories the great prose writer Yordan Yovkov revealed the rich personality, patriotism and moral heroism of Bulgarian men and women, defending the people's moral and aesthetic standards, its striving for humanity and beauty, social justice and moral nobility – *Last Joy* (1926), *Balkan Legends* (1927), *Evenings at the Antimovo Inn* (1928), *A Woman's Heart* (1935), and *If They Could Speak* (1936). In some of his works Yovkov is more directly critical of his time – the dramas *The Millionaire* (1930), *Boryana* (1932), the novels *The Farm by the Border* (1934), *The Adventures of Gorolomov* (1938) and others. Realistic stories were written by Angel Karaliichev – *Deceitful World* (1932), *Whirlwind* (1938), Konstantin Konstantinov (1890–1970) – *To My Fellow-man* (1920), *Third Class* (1936) and others. Historical prose recorded some notable successes. In his novel *The Last Day, the Day of God* (3 parts, 1931–34), the first significant historical novel in Bulgarian literature, written with patriotic inspiration, a good knowledge of history and in fine language, Stoyan Zagorchinov (1889–1969) treated the subject of the people's masses as motive forces of historical progress, and of the strength of their patriotism and heroism. Anton Strashimirov wrote a novel about the national liberation movement in Macedonia, *Slaves* (1929). Many historical novels and long stories were published by Konstantin Petkanov (1891–1952), Fani Popova-Mutafova (1902–73) – *Kaloyan's Daughter* (1936), Dimiter Talev (1898–1966) and other writers. The realistic portrayal of the recent historical past (the wars of 1912–18) conveyed a protest against the new world war being prepared by Nazism – *Midnight Congress* (1932) by Nikolai Hrelkov, *A Purple Star, a Bloody Star* (1934) by Konstantin Petkanov, *Cholera* (1935) by Lyudmil Stoyanov, some short stories by Svetoslav Minkov and others. The satirical short stories by Georgi Stamatov are characterized by their caustic criticism – *Short Stories* in two volumes (1929–30), and *Motes* (1934); the same applies to those written by Svetoslav Minkov – *Automations* (1932), *The Lady with the X-ray Eyes* (1934), *Tales in a Hedgehog Hide* (1936); the psychological fiction of Georgi Raichev – *A Legend about Money* (1931), *The Golden Key* (1942) and to the stories about peasant life by Iliya Volen (a pen-name of Marin Stamenov, 1905–82) written in the spirit of critical realism – *Black Fallow Lands* (1928), *Shocks* (1931), and *Blessed People* (1937). The authors penetrate the essence of the bourgeois world, unmasking its moral degradation; they present the drama of alienation, the philosophy of egotism and misanthropy, and depict man and society in their troubled existence. The civil servant deprived of his individuality, the promiscuous woman, the corrupt policy-maker and vile speculator are typical literary characters. They are portrayed (by Svetoslav Minkov in particular) with new, modern means of expression against the background of a cold, urban landscape and in severe, sometimes even drastic language, which is in full contradiction to the melodious nature of Bulgarian, as praised by Ivan Vazov, and with the harmony reigning in the world of Yovkov's characters. The critical style of Chudomir (a pen-name of Dimiter Chorbadjiiski, 1890–1967) is quite different, however. In his humorous short stories, the embodiment of the people's healthy sense of humour the author portrays ordinary man gaily touching upon primitive aspirations and ambitions, upon the comic side of everyday life – *I Am Not One of Them* (1935), *Countrymen* (1936), and *A la minute* (1938). Stefan Kostov (1879–1939), Bulgaria's most prominent comedy writer, was also skilled in using humour and satirical devices. He depicts incisively the Bulgarian bourgeoisie, revealing the unfounded nature of its prestige, and exposes its detrimental effect on the national spirit, thus creating well-rounded, comic images and turning them into literary generalizations. The character of Golemanov, a prototype of the megalomaniac middle-class person and political demagogue, is finely drawn. The comic effects, conflicts and catastrophes in Kostov's works are rooted mainly in the conduct of the characters. *The Golden Mine* (1925), *Golemanov* (1928), *Grasshoppers* (1931) and *The Healer* (1933) are fine examples of this. In general, drama lagged behind in its development as compared with lyrics and fiction in the period between the two world wars. Racho Stoyanov's

(1883–1951) play, *Master-builders* (1927) stands out in the genre of psychological drama. The work is marked by sharp conflicts and compositional integrity, with problems and characters of national significance, and deep psychological insight.

The 1920s and 1930s saw major successes in literature for children and adolescents. The magazines *Svetulka* (Firefly; 1904–47), *Detska radost* (Children's Joy; 1910–13 and 1919–47), *Venets* (Wreath; 1911–43) and others of this type largely contributed to its development. The first to promote the proletarian trend were the party magazines *Drugarche* (Little Comrade; 1919–20) and *Svetlina* (Light; 1923), followed by *Detski Glas* (Children's Voice; 1932–34) newspaper and the *Rossitsa* (Dew; 1934–45) magazine. Among those who wrote works for children and adolescents were Georgi Karaslavov – *The Sophisticated Mice* (1935), *Winners* (1935); Hristo Radevski (b. 1903) – *Toothy Subko* (1938), *The Tricks of Ivan Boyan* (1938); and Nikola Vaptsarov (1909–42) – *Train*, published posthumously in 1947. Many prominent democratic writers also created literary

Dimiter Polyanov

Dimiter Talev

Pavel Vezhinov

works for children and adolescents. For example, Elin Pelin's lyrics and fiction for children combine poetic and romantic elements with comic ones; they abound in worldly wisdom typical of the Bulgarian people: *Dense Forests* (1919), *The Smooth-tongued Granny* (1919), *Good and Wrong* (1920), *Brooklets Clear* (1931), *Jan Bibijan* (1933), and *Jan Bibijan on the Moon* (1934). The treasure-house of Bulgarian classic literature for children and adolescents contains many works by Ran Bossilek (a pen-name of Gencho Negentsov, 1886–1958), Dora Gabe, Elissaveta Bagryana, Angel Karaliichev, Assen Raztsvetnikov, Kalina Malina (a pen-name of Raina Radeva Mitova, 1898–1979), Emil Koralov (a pen-name of Emil Stanchev, b. 1906), and Emiliyan Stanev (a pen-name of Nikola Stanev, 1907–79).

The 1920s and 1930s saw the creative work of prominent literary scholars who, although they did profess revolutionary ideas, strove to achieve objectivity in their appraisals and conclusions (Marxist literary studies had neutralized bourgeois aesthetics to a great extent). The philosophical and literary apology of reaction, Fascism and war is but a short episode in literary theory and criticism. Vladimir Vassilev showed a negative attitude to revolutionary and social literature and approved individualistic aestheticism, yet he helped discover talented writers and raise the standard of literary creativity. Ivan Shishmanov published new studies; some of Boyan Penev's works and his most significant studies appeared posthumously during the 1930s. Mihail Arnaudov (1878–1978) achieved substantial success, mainly in the history of literature and the psychology of literary creativity. Dimiter Mitov (1898–1962), Georgi Konstantinov (1902–70), Malcho Nikolov (1883–1965), and Alexander Balabanov (1879–1955) made efforts to attain objectivity in the sphere of theory of literature.

The leftist periodicals, such as the magazines *Zvezda* (Star; 1932–34) published by Georgi Bakalov, *Nakovalnya* (Anvil; 1925–33) by Dimiter Polyanov, the newspapers *Vedrina* (1926–27) by Anton Strashimirov, *RLF* (Workers' Literary Front; 1929–34), *Pogled* (Outlook; 1930–34), *Shtit* (Shield, 1933–34), *Literaturen pregled* (Literary Review; 1934–35), and *Kormilo* (Rudder; 1935–36), and the comic and satirical publications *Zhupel* (Fiend; 1931–34), *Gorchiv Smyah* (Bitter Laughter; 1936–37) and *Horovod* (Round Dance; 1937–39) were unifying centres of proletarian literature during the second half of the 1920s and in the 1930s. Marxist literary thought increasingly gained ground. In their articles and books Georgi Bakalov, Sava Ganovski, Todor Pavlov and other men of letters set themselves the task of purging proletarian and revolutionary aesthetics of 'all kinds of opportunist and revisionist theories' i.e. of placing it on Leninist foundations, elaborating the theoretical principles of socialist realism and treating various problems of art. Despite some deviations and wrong opinions, their theory of literature channelled proletarian art in the right direction and influenced a large number of creative workers. In this respect credit goes to the progressive literary scholars Georgi Tsanev (1895), Ivan Meshekov (1891–1970) and many others. Todor Pavlov, Georgi Bakalov, Georgi Karaslavov, Hristo Radevski, Lyudmil Stoyanov, Mladen Issaev (1907), Nikolai Lankov (1902–65), Kamen Zidarov (a pen-name of Todor Manev, 1902), Angel Todorov (1906), Krum Kyulyavkov, Orlin Vassilev (a pen-name of Hristo Vassilev, 1904–77) and Dimiter Polyanov were amongst those who organized the proletarian front in literature. In December 1931 the Union of Proletarian Writers (from the beginning of 1932 the Union of Labour Militant Writers in Bulgaria) with its organ, the newspaper *Front of the Labour Militant Writers in Bulgaria*, was constituted in Sofia. The Union existed till the end of 1936 and united about 50 writers – communists, anarchists and anti-fascists. It was the first broader organization of creative intelligentsia, with an anti-fascist political and aesthetic platform. In 1934 No. 38 of *Shtit* (Shield) newspaper carried a *Literary Manifesto, Goals and Tasks of Literature*, Like the Programme of the Union of Labour Militant Writers it rejected the principle of art for art's sake as reactionary and non-historical, appealed for struggle against Fascism and war, and for recognition and defence of Soviet culture as custodian of the most valuable possessions of human thought. Numerous articles in the party Press demanded that art should consciously and devotedly serve the Party as an organized vanguard of the workers' movement. The range of social themes in literature expanded and became enriched in genre and style, reflecting revolutionary ideals, the pursuit of freedom and the dreams of social justice. The literary character was the man in the street: the worker, striker, participant in street demonstrations, or the revolutionized peasant in his daily life, in the process of overcoming philistine prejudices. He is portrayed as a staunch revolutionary, alien to the illusions and temptations of the intellectual. Georgi Karaslavov published his novel *Sporzhilov* (1931) – a narrative about the exploitation of Czech workers; and the collection of short stories *Sentry* (1932) about the heroism of revolutionary workers. In his novella *Village Correspondent* (1933) the writer revealed (as Georgi Bakalov put it) 'the nature of class struggle in the contemporary Bulgarian village in flesh-and-blood characters'. In his novels *Thornapple* (1938) and *Daughter-in-Law* (1942) he described in a new way the drama of private ownership. New problems and ideas are to be found in the prose works of Orlin Vassilev, Lyudmil Stoyanov, Krustyo Belev (1908–78), Gyoncho Belev (1889–1963), and others.

Lyrical verse, narrative and satirical poems were written. The lyrical verse of Hristo Radevski was imbued with militant and party irreconcilability: *To the Party* (1932), *Ours Is a Constitutional State* (1933), *Pulse* (1936), etc. The poems of Mladen Issaev were permeated with romanticism and charged with emotion: the collections *Fires* (1932), *Victims* (1934), *Serenity* (1936), *Anxious Planet* (1938), etc. The ballad element is strong in the work of Nikolai Hrelkov. In *Poems Without a Lie* (1936) and *Useful Poems* (1938), Angel Todorov searched the world and the epochs for plots to express revolutionary ideas. Emotional poems were written by Nikola Lankov, Pantelei Mateev (1897–1957), Krum Penev (b. 1901), Ivan Burin (pen-name of Ivan Alexiev, b. 1903), Lyudmil Stoyanov, Maria Grubeshlieva (1900–70) and others. In the second half of the 1930s talented young prose writers and poets made a name for themselves and linked their work with the proletarian literary front: Andrei Gulyashki (b. 1914), Dimiter Chavdarov-Chelkash (pen-name of Dimiter Kotsev, 1912–72), Armand Baruh (b. 1908), Ivan Martinov (pen-name of Ivan Georgiev, b. 1912), Pavel Vezhinov (pen-name of Nikola Gugov, 1914–83), Nikolai Marangozov (pen-name of Nikolai Tsanev, 1900–67), Bogomil Rainov (b. 1919), Valeri Petrov (Valeri Mevorah, b. 1920), Alexander Gerov (b, 1919), Venko Markovski (pen-name of Venyamin Toshev, b. 1915), and others. The poetry of Nikola Vaptsarov marked the peak in the development of revolutionary literature. It was deeply related to the most dramatic period of the workers' and anti-fascist struggle in Bulgaria and in the world, and drew inspiration from the literary traditions, folklore, the homeland and working people. In his poems (the collection *Motor Songs*, 1940) the poet depicts the revolutionary movement as a deed of the ordinary people who create history with their blood, suffering and heroism, and glorifies their revolutionary devotion and invulnerable faith in the victory of the socialist revolution. Nikola Vaptsarov reveals the arduous and dramatic path of the working class, the most revolutionary and most powerful class, united in the struggle for the attainment of the most noble, human ideals. The solidarity of workers from all countries and continents, the maturity and the strength of the working class are the source of Vaptsarov's faith in the triumph of socialist humanism, the underlying pathos in his lyrical verse. The poet makes a vivid class characterization of the proletarian revolution and at the same time reveals its scope as a great struggle in the name of man. Vaptsarov's poetry is an expression of a new romanticism – the poet sees the future harmony between man and machines, and draws concrete pictures of human material and spiritual progress. Vaptsarov's poetry is innova-

Nikola Vaptsarov

tory in form. His poetical style is characterized by simplicity, lyricism, colloquialism and blank verse. He has achieved the deepest synthesis between the turbulent social movements of the time and the dramatic experiences of the lyrical hero. The great conflicts of the epoch stand out in subtle psychological detail in the 'simple human drama'. The individual and his fate are invariably brought to the foreground. Vaptsarov's works established the method of socialist realism and rank among the finest achievements of the world revolutionary poetry. Like Vaptsarov, a number of young writers and poets laid down their lives in the armed anti-fascist struggle of 1941–44, loyal to Botev's tradition of the oneness of ideas and cause: Tsvetan Spassov (1919–44),

Hristo Kurpachev (1911–43), Marko Angelov (1921–44), Ivan Nivyanin (pen-name of Ivan Vurbanov, 1919–44), Atanas Manchev (1921–44), Vassil Vodenicharski (1918–44), and others.

After 9 September 1944 literature developed under highly favourable socio-political conditions. The majority of writers had been closely connected, since the time of the anti-fascist struggle, with the BCP, with the socialist ideas and the working people's aspirations for a new, more just society. Other writers gravitated to the Party and the Fatherland Front movement, and a third group tended to readjust their ideas. The victory of the socialist revolution

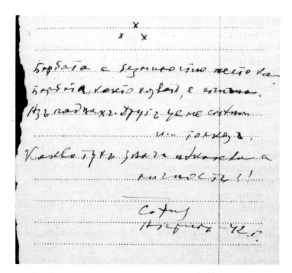

Facsimile of Nikola Vaptsarov's poem On Parting, written on 23 July 1943, shortly before the poet's execution

catalysed the evolution of many creative minds – realist writers and democrats, and inspired the communist and anti-fascist writers. Two main periods can be outlined in the literary process: from the People's Uprising on 9 September 1944 to the April Plenum of the Central Committee of the BCP in 1956, and from the April Plenum to the present day.

In the first period, literature followed and developed the revolutionary traditions created by proletarian literature between World Wars I and II. The personal, creative quests of the writers went in parallel with the transformation of the socio-political reality. No. 1 of Literaturen Front newspaper, which rallied writers of different creative inclinations but with a common ideological and aesthetic platform, came out on 30 September 1944. It printed an Appeal of the Writers from the Fatherland Front in which they expressed their readiness to continue the cause of the people's liberation and invest their energies in building a new Bulgarian culture. In his letter to the Union of Bulgarian Writers of

14 May 1945, Georgi Dimitrov (1882–1949), stressed that the people needed 'genuine popular artistic literature like bread and air'. The communists – writers and party functionaries – organized and guided the literary process. The literary development was complicated and controversial. In 1945–46 heated debates were conducted among the intelligentsia. The most argued question was about the freedom of the writer, directly connected with the artist's mission, the response of his artistic conscience, and the inculcation of a new and correct sense of history. Gyoncho Belev, Georgi Karaslavov, Boris Delchev (b. 1910), Alexander Nichev (b. 1922), Dimiter Mitov, Pantelei Zarev (pen-name of Pantelei Pantev, b. 1911), Nikolai Rainov and others resolutely opposed the illusions of non-committed art upheld by some intellectuals, who made use of the newspapers *Izgrev* (Dawn) and *Svoboden Narod* (A Free People) to disseminate them. Despite the complexity of the problems and the situation, those writers who related directly with the people found the correct path. The main themes in literature were the anti-fascist struggle, the Patriotic War of Bulgaria 1944–45, socialist construction, and the radical changes that were taking place in all spheres of life. Immediately after the people's victory newspapers and magazines gave much space to poems. A product of enthusiasm and spiritual renewal they expressed the people's joy, the moving welcome given to the partisan fighters and the Red Army, the impressive departure of the first echelons for the front, and the *élan* of the youth construction brigades. The earliest works were not particularly good poetry and their value is to be sought in the writer's orientation to life, to the working people and against reaction. The topical genres – march, propaganda play, song and report were widespread. Gradually, more significant works appeared and shaped the image of literature. In 1945 Hristo Radevski published his collection of poems, *Air Was Scanty*. Written in the style of emotional reflection characteristic of this poet, it told about the time of ferocious fascist terror and was a testimony to the author's maturity as a poet. In 1946 Mladen Issaev's best collections of poems came out – *Fire* – containing works dedicated to the Red Army, to concentration camp inmates and partisan fighters, and to the feats of the Bulgarian soldiers. The poems sang praise to the unbending courage and heroism of the Bulgarians, and showed the ruthlessness of the struggle; nevertheless they sound like intimate lyrical confessions. *Partisan Songs* (1947) by Vesselin Andreev (pen-name of Georgi Andreev, b. 1918) is a collection of poems in which life and militant experience were crystallized, partisan life was seen 'from within', with all its severity and romanticism; so these poems are an apotheosis of the heroes of the partisan struggle. In his partisan poems David Ovadia (b. 1923) achieved a great simplicity of poetic expression, as in *Partisan Diary* (1948). *Thirst* (1951), by Dobri Zhotev (pen-name of Dobri Dimitrov, b. 1921), was written with more emotion. His comrades from prison and from the partisan battalion are the prototypes of the heroes of his poems. In her poem *Vela* (1946), Dora Gabe glorified the heroic partisan fighters and sought the grandeur and beauty of the near past. Georgi Djagarov (b. 1925) published poems about the resistance; his collection *My Songs* (1954) was entirely about the anti-fascist struggle of the communist youth and extolled the moral feat of the heroes who perished at the hands of the keepers of public order but saved their human dignity and communist honour. Prose and drama also dealt with the theme of anti-fascist struggle. Emil Manov (1919–82) wrote documentary sketches and stories; Angel Karaliichev published *Defender of the People* (1949), a collection of documentary short studies about Georgi Dimitrov. Writers and poets who had been partisans wrote memoir pieces: Vesselin Andreev's *In the Lopyan Forest* (1947), Kosta Lambrev's *Partisans in the Sredna Gora* (1948), Ivan Zurlov's *Notes of the Political Commissar* (1949), the work of Mitko Yavorski (pen-name of Gatyo Gatev, b. 1925) *Loyal Sons to the People* (1954), Zhelyazko Kolev's *Along Haiduks' Paths* (1954), the writings of Slavcho Trunski (pen-name of Slavcho Savov, b. 1914) – *Partisan Time Memoirs* (1955) and many others. In the most significant works of guerrilla memoir literature, the documentary and the literary elements are interwoven. Documentary writers like Vesselin Andreev, David Ovadia and Emil Manov had a knack for the aesthetic organization of the material according to the author's vision. Characteristic in this respect is Emil Manov's documentary short novel *Birds in Captivity. A Prison Chronicle* (1947), in which the individual destinies of the political prisoners were synthesized to portray the hero of the resistance movement with his most typical features. Vesselin Andreev's memoir sketches in *In the Lopyan Forest* are documentary stories about exceptional moments in the life of the Chavdar partisan detachment. In the choice of documentary facts the narrator integrates his own humane ideas, makes moral evaluations, and expresses his philosophical and ethical concepts.

Georgi Karaslavov, Orlin Vassilev, Gyoncho Belev, Ivan Martinov, Stoyan Daskalov, Zdravko Srebrov (b. 1902), Marko Marchevski (pen-name of Marko Markov, 1898–1962), Armand Baruh, Dimiter Hadjiliev (1895–1960) and many others wrote short stories about the popular resistance. The best works outline the tendencies of anti-fascist prose towards epic breadth and psychological insight, revealing the writers' aesthetic attitude to the near past. In his treatment of the resistance in the short novel *On a Quiet Evening* (1948) Emiliyan Stanev remained true to the intimate psychology of the narrative and took an analytical approach to the hero's emotional experience. The structure and the composition of that short novel – departure from the traditional narrative, employment of cinema techniques, colourful visualization, etc. – are innovations in the genre. Pavel Vezhinov's short novel *In the Field* (1950) is an excellent piece of writing. The writer conveys his message – the historically inevitable end of Fascism and the glory of the partisan exploits – by two parallel plots: an exposing, satirical plot (the fascists) and a romantic lyrical plot (the partisans). Georgi Karaslovov's short novel *Tango* (1946) was the beginning of the epic portrayal of class warfare in

Bulgaria in the years of Fascism. It is a transition not just to his epic work *Ordinary People* (6 parts, 1952–75) but also to the broad depiction of life in contemporary fiction and a direct predecessor of the flowering of the Bulgarian novel in the 1950s. The writer outlined the vast scope of the class duel revealed in his heroes – young communist peasants – revolutionary changes in folk psychology.

The time of Fascism, the struggle, the victory – these were usually the subjects with which novels about the resistance were preoccupied. The novelists turned to events explaining the new Bulgarian history: World War I, the socialist movement and the Bolshevization of the Party, the September Anti-fascist Uprising of 1923. The third edition of Krustyo Belev's novel *The Breakthrough* (1937), much talked about among the progressive

Dimiter Dimov

intelligentsia even before 9 September 1944, came out in a complete form in 1952. In his novel *September Fighters* (1945) Emil Koralov treated the familiar September theme. The first, fine, literary achievement of the period in question was *Doomed Souls* (1945) by Dimiter Dimov (1909–66), one of the most successful novels about the Civil War in Spain (1937–39). It introduced a new poetry into the Bulgarian novel. Although the subject-matter is international, the work is related to the most recent Bulgarian history. The characters in the novel are typical of Dimov: products of a philosopher writer, slaves of one passion, exponents of the spiritual and moral features of their class. A characteristic feature of the novels of the first period is their closeness to the didactic novel which studies character-formation and the obstacles the hero overcomes in his life. This model appears as early as Kamen Kalchev's book *At the End of Summer* (1945).

Stoyan Daskalov's novel *Road* (3 vols., 1945–54), which demonstrated the opportunities for the traditional narrative technique under new conditions to deal with new problems, is a typical example. *Road* manifests one of the major peculiarities in the development of the Bulgarian novel in this period – the tendency towards an epic novel as a proper form to recreate the turbulent epoch that brought about radical social changes. With *New People* (1945), *Red Dawn* (1952) and *Act One* (1958), Gyoncho Belev continued his cycle of novels, *Happenings from the Life of Minko Minin,* begun in 1940 with *The Needle Broke.* The writer made an interesting attempt to present the vast social picture through the eyes of one character, a peculiar contradiction between the large scale of the author's intention and his ability to 'subjectivize' the epic. Lyudmil Stoyanov's *At Daybreak* (1945) was meant to be the first book of a trilogy. The novel by Peter Slavinski (pen-name of Peter Cholakov, b. 1909) entitled *The Last Attack* (1951) is part of a series which later came out under the title *The End of the Masters.* The epic scope is also characteristic of the novel *Life and Death Struggle* (1953) by Dimiter Angelov (1904–77), dedicated to the armed anti-fascist struggle. During 1946 the *Literaturen Front* newspaper began publishing the first chapters of Dimiter Dimov's novel *Tobacco* (1951, second edition 1954). What was new in the novel was above all Dimov's highly intellectual, philosophical and moralistic approach in treating the problems of the most recent Bulgarian history. Also new was the aesthetic system of the novel: character-drawing rested on the romantic principle, accentuating one salient trait and concentrating on the characteristics of the type rather than the individual character. It was the first time in the history of the Bulgarian novel that the romantic principle of contrast, based on actual historical material, had been applied on such a scale; each noble character has his negative counterpart, each heroic scene – its parody, and each particular plot – its counterplot. 1952 saw the publication of the first novel, *The Iron Candlestick,* of Dimiter Talev's tetralogy, followed by *Ilinden* (1953), *The Bells of Prespa* (1954), and *I Hear Your Voices* (1966). Although it is more closely linked with the traditions of National Revival realism, it is also akin to the epic novels on subjects of the anti-fascist struggle, judging from its genre peculiarities, the problems it treats concerning the destiny of the people, and the struggle for liberation of Macedonia in particular. The first novel, entitled *Ordinary People,* in the series of six volumes by Georgi Karaslavov, also appeared in 1952, giving an artistic picture of the most important events over the three decades from 1914 to the socialist revolution of 9 September 1944. The writer was preoccupied with the social and political processes developing in the Bulgarian village.

The anti-fascist theme provided the initial great impetus to the development of Bulgarian drama. The period in history it covered abounded in intense conflicts and drama which had to be rendered in artistic form – the surging revolutionary wave, the growing self-awareness of the people's masses as the true makers of history. Many works which reflected the fundamental revolutionary political conflict of the times appeared in 1944–49. Some scholars refer to the period as the era of the anti-fascist play, and others as that of storms and pressure. Chronologically, and by importance as pioneer, first came Krum Kyulyavkov's play, *The Struggle Goes On*, written on the eve of 9 September 1944 and staged by the National Theatre in Sofia on 5 April 1945. It was followed in 1948 by *Alarm*, by Orlin Vassilev and *Royal Mercy*, by Kamen Zidarov, and further, in 1949, by *Reconnaissance* by Lozan Strelkov (pen-name of Lozan Bogdanov, 1912–81). They portrayed events in the recent history of the Bulgarian people, when communist ideas gained in scope, giving a new interpretation of the historic subject-matter and showing the heroism of the fighters for Socialism. Since the past was surveyed from a short distance, there was a tendency to depict events from their external side by explicit illustration and adherence to ready-made theses and not by sufficient psychological insight, while on the other hand these plays were remarkable for the wide scope of artistic, historical vision, sincere emotion and inspiration. The founders of Bulgarian socialist drama laid the foundations for its further development. The theme of anti-fascist struggle, to which the Bulgarian writers are committed to this day, was central in literature during the 1944–56 period. Though many of the works written at that time are today of interest only to the specialist in the history of literature, in their day they had a particular socio-political, cognitive and educative significance in explaining the role of the class struggles for the future destinies of the Bulgarian people as well as in its recent history, promoting new moral criteria and reaffirming the character of the anti-fascist militant. This theme is treated with great literary skill in the novels *Tango, Tobacco, Ordinary People* and a number of others.

The anti-fascist theme had its natural continuation in works dedicated to the Patriotic War of 1944–45, perceived by Bulgarian authors as a further stage in the national anti-fascist movement. Many of them took part in that war as soldiers, editors of front publications, correspondents and feature-writers; Matvei Vulev (pen-name of Dimiter Vulev; 1902–44) and Ivan Hadjiiski (1907–44) were killed in action. Moving poems dedicated to the Bulgarian soldier were written by Mladen Issaev, Lamar (pen-name of Lalyo Marinov Ponchev, 1898–1974), Angel Todorov, Radoi Ralin (pen-name of Dimiter Stoyanov, b. 1932), Ivan Peichev (1916–76), Krum Penev and others. The collection *Cartridge-box Poems* (1954) by Vesselin Hanchev (1919–66) is one of the finest achievements of modern Bulgarian lyrical poetry. As well as seeing the heroic and epic character of the struggle, the poet described touching details from the life and experiences of the soldiers at the front, their intimate human emotions expressed with manly restraint. Because of the fact that those events had taken place but a short time before, no epic works were created. Ivan Martinov's novel *The Drava Flows through Slav Lands* (1946) is the only major attempt at encompassing in full the events on the war front in Hungary, depicting the process of re-education of the Bulgarian soldier and his new image. Most significant in the shorter genres of literature are Pavel Vezhinov's short novel – *Second Company* (1949); the short stories by Ivailo Petrov (pen-name of Prodan Kichukov, b. 1923) – *Baptism* (1953); by Rangel Ignatov (b. 1927) – *Mityo of Ravnogore* (1954); and by Dimiter Chavdarov-Chelkash – *Ivan and His Comrades* (1955). Later on Bogdan Gloginski (pen-name of Bogdan Dimov, b. 1924) published the novel *Six Hundred Unshaven Men* (1967) outlining the dimensions of the new socialist morality with psychological insight and acumen in constructing a plot.

A number of simplified concepts and methods of work, alien to the approach of the Communist Party towards the arts, began to appear in the guidance of literature in the 1950–56 period. These weaknesses did not change the essence of party policy.

Penyo Penev

In the first half of the 1950s many talented poets emerged – Valeri Petrov, Bogomil Rainov, Alexander Gerov, Pavel Matev (b. 1924) and others. The romantic vision of constructive work and sincerity of feeling are characteristic of the 'poetry of the youth-brigade movement' in the works of Ivan Radoev, Kliment Tsachev (b. 1925), Blaga Dimitrova (b. 1922), Stanka Pencheva (b. 1929), Vladimir Golev (b. 1922) and others. A great many works of poetry were dedicated to socialist construction, of the greatest value among them being the poems of Penyo Penev (1930–59). This 'poet in a quilted jacket' evokes in the minds of the generations the image of the new industrial and construction sites of Dimitrovgrad. Though somewhat crude and raucous, they exude the charm of unadulterated truth about life and convey the feeling of the grand scale of the times and high spirits of the class of builders, of the optimism of the working masses and of their nobility and simplicity. In the works of fiction, dogmatic thinking gained the upper hand mostly in works on the subject of 'production'. Owing to the long-standing fine traditions developed in the shorter genres, 'rustic' literature of that period is more convincing. Though the works about the new village were also marked by excessive descriptiveness the set of political and psychological problems were treated in greater depth and the conflicts in life were more clearly expressed, often in original lyrical intonations. Angel Karaliichev wrote one of the first short stories about the labour upsurge in the village following the socialist revolution – *Sokolov's Cornfield* (1946). Stoyan Daskalov tackled the problems arising in the new village in his collection of short stories *Deleted Boundaries* (1948) and in his story *One's Own Land* (1947). They give a truthful picture of social life in the countryside but do not treat the great moral problems concerning the Bulgarian peasants. Other works of a similar character include the collection of short stories *Chairman of the Village Council* (1950) and the novella *Iglichevo* (1958) by Krum Grigorov (b. 1909). The dramas on contemporary themes from the period up to 1956 are characterized by sketchiness and lack of conflict. Among those distinguished by a measure of achievement are *Years of Great Effort* (1954) by Kamen Zidarov and the *The Gaberovs* (1955) by Georgi Karaslavov, depicting the initial difficulties in the construction of Socialism in the villages. Among the fine works in dramaturgy are those on subjects from the past – *Daughter-in-Law* (1951) by Georgi Karaslavov after his novel of the same name, *Time of Evil* (1955) by Iliya Volen, and *Land Riot* (1956) by Magda Petkanova. The central character in the drama *The First Blow* (1952), by Krum Kyulyavkov is Georgi Dimitrov. Based on authentic material and documents the play throws abundant light on his activity as a leader of the international proletariat and the revolutionizing influence exercised by him over ordinary people. A breakthrough in the prevailing sketchiness was made by the considerable artistry and philosphic insight into the conflicts in reality and the inner world of characters in several plays on subjects of the day and of the anti-fascist struggle, such as *Love* (1952) and *Happiness* (1954) by Orlin Vassilev, and *Faith* (1954) by Todor Genov (b. 1903), among others.

In spite of the difficulties experienced from the end of the 1940s to 1956, the objective laws in the development of society and culture emerged clearly in the authors' continued searches. The main efforts were channelled toward the artistic portrayal of the positive character as the principal hero of our times. Both the positive experience of writers and their failures have been of significance in this respect.

The April 1956 Plenum of the Central Committee of the BCP ushered in a new period in the development of literature. An editorial entitled *Upsurge of Creative Powers* (*Literaturen Front* No. 15, 1956) read: 'The Plenum's resolution... is an appeal by the Party for closer and immediate links of the creators of our culture with the life of the people; for giving full play to initiative, independent thinking, individual creative quests and original style in literary works'. The successes achieved in literature since the Plenum are due to the constructive adoption and application of the party principles. The appeal made to the writers in 1958 by the party leadership in the person of Todor Zhivkov 'Closer to the People, Closer to Life' was an appeal to apply themselves to the real problems of life, and above all, to contemporary themes, and to uphold realism in literature.

A number of established poets achieved new dimensions in their work, applying their talents with fresh vigour. There was general striving to sum up what had been done, shedding delusions, penetrating to the core of the subject-matter. Poetry gained in philosophic substance, associations, dramatic force and earth-bound vitality. Nikola Furnadjiev's collections of poems, *Along the Paths You Charted* (1958) and *Sunshine over the Mountains* (1961), are distinguished by maturity, wisdom and rich context. Original poems were created by Nikola Lankov, Hristo

Yordan Radichkov

Radevski, Dimiter Panteleev, Mladen Issaev, Lamar and Atanas Dalchev. Condemnation of the cult of personality is conspicuously present in the works of Vesselin Hanchev. Profound works, an expression of a new outburst of their poetic gift, were written by Elissaveta Bagryana – *From Coast to Coast* (1963), *Counterparts* (1972) and *Chiaroscuro* (1977), and also by her contemporary Dora Gabe – *Wait, Sun* (1967), *Thickened Silence* (1973) and *Depths. Conversations with the Sea* (1976). Pavel Matev's works were enriched with new civic motifs and feelings, interpreting the inner world of the communist of today in all its aspects. Dobri Zhotev, David Ovadia, Vladimir

Golev and others remained true to the revolutionary traditions, to which new emotions and thoughts were added. The revolution is also the leitmotif in the April poems of Georgi Djagarov, which are imbued with optimism and communist purity – *Lyrics* (1956), *In Minutes of Silence* (1958) and in the monologue poems of Dimiter Metodiev (b. 1922) – *Song to Communism* (1959), *Song to the General Line* (1969) and *Song to the Great Migration* (1970). The years after 1956 marked a new stage in the poetic development of Blaga Dimitrova, Stanka Pencheva, Alexander Gerov – *A Child in the Window* (1959), *Free Verse* (1967), *Specks of Dust* (1973), Radoi Ralin, Bozhidar Bozhilov (b. 1923), Ivan Davidkov (b. 1926) and Venko Markovski. Among the successes of poetry was Valeri Petrov's long poem *In the Mild Autumn* (1961), imbued with a profound feeling of patriotism and written in light and playful verse with resonant rhymes. The stimulating atmosphere following the April 1956 Plenum of the Party proved propitious for the growth of poets like Purvan Stefanov (b. 1931), Ivan Peichev, Liliana Stefanova (b. 1929), Anastas Stoyanov (b. 1931), Naiden Vulchev (b. 1925) and others who created a sincere emotional poetry, combining lyricism with civic sensitivity.

Dora Gabe

The new trends in the short story and the short novel after the April Plenum were expressed in a greater psychological depth of character drawing, in more pronounced moral and philosophical undertones, in a more lyrical and subjective narrative and in an elimination of outward pomposity and preconceived ideas. The assertion of new artistic methods was a difficult and controversial process. Authors like Georgi Karaslavov, Emiliyan Stanev, Pavel Vezhinov, Iliya Volen, Kamen Kalchev, Andrei Gulyashki, Stoyan Daskalov, Ivan Martinov, Krum Grigorov, Emil Manov, Orlin Vassilev and others kept writing, and a new generation of story-tellers appeared on the literary scene – Bogomil Rainov (till then an established poet, art expert and critic), Yordan Radichkov (b. 1929), Nikolai Haitov (b. 1919), Vassil Popov (1930–80), Atanas Nakovski (b. 1925), Diko Fuchedjiev (b. 1928), Dragomir Assenov (a pen-name of Jacques Melamed (1926–81), Lyuben Dilov (b. 1927), Lada Galina (b. 1934), Kliment Tsachev, Gencho Stoev (b. 1925), Georgi Mishev (b. 1935), Dimiter Vulev (b. 1929), Dimiter Gulev (b. 1927), Doncho Tsonchev (b. 1933), Nikolai Kirilov (b. 1922), Kolyo Georgiev (b. 1926), Lyuben Stanev (b. 1924) and others. The new content of reality and its artistic interpretation called to life new forms and styles – the grotesque of Yordan Radichkov, the associativeness and essayistic narrative of Bogomil Rainov, the monologue-like short stories of Diko Fuchedjiev and Nikolai Haitov, etc.; the lyrical-impressionist narrative and cyclic short stories were revived. The traditions of classical Bulgarian prose were continued in Iliya Volen's short novels, such as *Between Two Worlds* (1958), in the short novels *Nonka's Love* (1956) by Ivailo Petrov, *The People of Razdol* (1958) by Krum Grigorov, etc. Many achieved remarkable psychological depth in portraying salient features of the contemporary literary hero: Serafim Severnyak (b. 1930), Tsilia Lacheva (b. 1922), and Peter Bonev (b. 1928) are examples. Very good prose works were created by Yordan Radichkov, Milcho Radev (b. 1925), Nevena Stefanova, Kliment Tsachev, Lyuben Dilov, and Lada Galina, among others. The great possibilities of the intellectual trend in prose were demonstrated by Bogomil Rainov's collections of short stories *The Man at the Corner* and *Blue Flowers* (1967). With the moral problems they treat, with the simplicity of expression, flawlessness of structure and natural ease of the narrative they are a considerable achievement in the artistic recreation of reality.

Mature novellas, short stories and novelettes appeared in the late 1960s, a period during which one of the main subjects of fiction was the destiny of the Bulgarian countryside. The artistic and intellectual maturity of literature was expressed in the overcoming of the genre and ethnographic treatment of subjects and of description for its own sake, and in the profound exploration of the ideological problems stemming from such revolutionary cataclysms in rural life as the disintegration of its age-old ethic and value system, the moral formation of the individual in the conditions of rapid technological advancement, urbanization and the shortage of moral ideas against the background of the dynamics of modern life. Yordan Radichkov is a prolific and original writer whose books include *Fierce Mood* (1965), *High Noon* (1965), *The Goat's Beard* (1967), *Aquarius* (1967), *We, Sparrows* (1968), *Memories of Horses* (1975), *An Inch of Land* (1980), *The Gentle Spiral* (1983), etc. By artistic means Radichkov recreates the process of social and human readjustment over the past couple of decades, the collision of the peasants' sense of family relationships, their community spirit and customs with modern times and with the new life in socialist Bulgaria. The new relationships usually prevail and impose their own logic, which is in accord with the justice of life. The charm of Radichkov's prose lies in the fusion between the funny and the tragic, the lofty and the comic, the serious and the frivolous. The problems of rural life are also reflected in the works of Ivailo Petrov, such as the short novel *Before I Was Born* (1968), and in Georgi Mishev's short stories and novels, which are characterized by their ethical and social thrust, unaffected humour, critical edge and clarity of moral assessments. Their characters are peasants of natural and simple reaction, with an inborn sense of justice, mistrustful and uncommunicative, yet naive and helpless. The novella *Matriarchate* (1967) is the epitome of Georgi Mishev's artistic position. The author used the anecdotal and the *feuilleton* style to lend meaning to an individual human fate as exemplifying the fate of the village. In the short stories from the collection *The Roots* (1967), Vassil Popov likewise paints a colourful and true picture of Bulgarian rural life during the first two decades following the collectivization of farming. The cyclical structure of the narrative is subordinated to the author's

Mladen Issaev

idea of revealing the whole life of a small and almost depopulated mountain village, which mirrors an important aspect of the country's reconstruction. Vassil Popov's short stories in the collection *Eternal Times* (1973) have a similar subject-matter and setting. The short stories, novellas and novelettes of Dimiter Vulev are distinguished for their austere narrative style and concentration of social problems.

A new, contemporary literary hero was introduced in the short stories and novelettes during the second half of the 1960s. There was more emphasis on the analytical element, irony and observation from a distance, introspection and concentration. The authors distanced themselves from the traditional narrative techniques, broke up the classical compositional pattern of the short genre and began using other artistic methods, including associations, retrospection and sharp twists of plot called for by the inner movement of thoughts and feelings, etc. The quests in the field of style and language vary from a conciseness and clarity of phrase to the use of rough-and-ready turns of speech to enhance the authenticity of observations. Even though Nikolai Haitov's short stories – *Hornbeam Leaves* (1965) and *Wild Tales* (1967) – appear to be unconcerned with the problems of the day in terms of plot and subject-matter, his characters fit perfectly into our time. His characters are primitive, yet they are the offspring of an age making an abrupt jump from a patriarchal and semi-feudal environment straight into modern times. Despite all the humour, anecdotal and highly entertaining descriptions, the author has a clear preference for romantic traits, strong characters, noble human qualities, myths and legends. More often than not, the moral values of his characters are out of place in this changing world, and this is their tragedy; their state of mind is conveyed in its most conspicuous characteristics, without any subtle analyses and minute details.

The characters in the short stories and novellas set in the period of the anti-fascist resistance also exhibit new human and moral features. Notable among such works are Yordan Radichkov's *Gunpowder Primer* (1969), Kolyo Georgiev's *Likely and Unlikely Confessions* (1970), Dobri Zhotev's *Stories of Experience* (1973), Vassil Akyov's (b. 1924) *Quiet Sunsets* (1973), some works by Kostadin Kyulyumov (b. 1925) and David Ovadia, among others. The heroic struggle and personalities are depicted in a profoundly humane and moving way. The characters, ordinary people with no ambitions for great exploits and glory, never hesitate in deciding their moral behaviour when confronted with the appalling trials of Fascism.

In the development of the short story and the novella one can also speak of an urban theme with its specific heroes, plots and conflicts in the works of authors like Pavel Vezhinov, Bogomil Rainov, Alexander Gerov, Kamen Kalchev, Andrei Gulyashki and others. They deal with problems which were engendered by the country's socio-political development during the 1960s – the moulding of the young individual in the conditions of dynamic revolutionary changes, modern urban living and the related problems of morality and human relationships, the overcoming of the style of thinking associated with the personality cult in political and social life, the evolution of new ideals and interrelations, the danger of the appearance of a consumer mentality in modern philistines, etc. Bogomil Rainov's *Tobacco Man* (1969) provides plenty of food for thought on the subject of fathers and sons (a theme continued in *That Strange Craft,* 1976), and on the dramatic life path of the Bulgarian intelligentsia. In *Roads to Nowhere* (1966), within the framework of a drama which is almost a genre today, Bogomil Rainov also poses the question of man's eternal struggle against injustice. Despite his personal catastrophe, the main character remains undaunted and preserves his dignity. The moral problems of everyday life under Socialism are also at the centre of Kamen Kalchev's attention. In *Sofia Short Stories* (1967) he recreates and castigates a world which, shaken by the revolution, has not yet overcome completely the conservation of philistine life, and gives vent to his affection for those ordinary people who carry out the revolutionary changes. The philosophic and moralistic novelettes and short stories of Pavel Vezhinov show his evolution towards ever more detailed depiction of life: the short stories in the collections *The Boy with the Violin* (1963) and *A Smell of Almonds* (1966) which are concerned with the upbringing of the youth, the generation gap and the ethical values of the man of today, the short novels *The Barrier* (1976), *The White Lizard* (1977), etc. The issues arising from the man-machine conflict dominate the novelettes and short stories of Andrei Gulyashki. The author does not go to extremes in his apology for man and human reason; he reveals modern urban life soberly and objectively. Present-day urban problems are also treated by Atanas Nakovski, Lada Galina, Lilyana Mihailova (b. 1939), Lyuben Dilov, Stefan Poptonev (b. 1928), Emil Koralov, Ivan Davidkov, Viktor Baruh (b. 1921), Ivan Ostrikov (1929–81), Radoslav Mihailov (b. 1928), Ivan Arzhentinski (b. 1910), Rangel Ignatov, Dragomir Assenov, Atanas Mandadjiev (b. 1926) and Kliment Tsachev, among others. Working-class life is the artistic concern of a still limited number of eminent prose-writers. One of them who is closely linked with the life of the working class is Gencho Stoev. In his collection of novellas and short-stories *Like Swallows* (1970) he portrays characters from major construction sites in various parts of Bulgaria, personalities of strong appeal and manly behaviour who believe in the possibilities of the class to which they belong. They are part of their work-force, masters of their trades and artistic personalities with a rich inner life; work gives them a sense of fulfilment and joy. Another work based on working-class life is Doncho Tsonchev's short novel *Men Without Neckties* (1966), in which the vitality of the characters and the sharp moral crises and conflicts accompanying their social and private lives are recreated directly and convincingly. A new view of the character and inner world of contemporary workers is clearly discernible in the novelettes and short stories of Kosta Strandjev (b. 1929).

Considerable success has also been achieved in the writing of short stories and novelettes on historical subjects. Not only did the most talented writers convey the colourful historical atmosphere in their works, they recreated the past refracted through the emotional and philosophical prism of the contemporary world, through the way of thinking of the modern intellectual – *Job* (1964) by Iliya Volen, *The Price of Gold* (1965) and *A Journey Home* (1976) by Gencho Stoev, *Legend of Sibin – the Prince of Preslav* (1968) by Emiliyan Stanev, *The Squadron* (1968) by Stefan Dichev, the historical sketches and essays by Vera Mutafchieva (b. 1929), etc.

Georgi Karaslavov

Contemporary science fiction also dwells on topical philosophical and psychological problems. Here, too, the plot and general line of narration merely serve as the background against which contemporary fates and ideas unravel. Although the works of Bulgarian science fiction are not numerous, they are characterized by a wide range of ideas and genres. Satirical science fiction continues the traditions of the popular fairy tale, of diabolical literature of the period following World War I, and of Svetoslav Minkov's satirical prose. This new type of literature is marked rather by the life-like portrayal of the characters and conflicts depicted and not by the credibility of scientific hypotheses and crime and adventure action – *Fantastic Novelettes* (1966) by Alexander Gerov, *Vanya and the Figurine* (1967) and *A Trip in Wibrobia* (1975) by Emil Manov, *The Blue Butterflies* (1968) and *The Death of Ajax* (1973) by Pavel Vezhinov, *My Strange Friend – the Astronomer* (1971) and *The Path of Icarus* (1974) by Lyuben Dilov, *Without Shades* (1970) by Atanas Nakovski, etc.

The novel continued to make headway. In the 1960s Bulgaria entered the stage of building the advanced socialist society, which paved the way for success in this major epic form. *Ivan Kondarev* (a four-part novel, 1958–64) by Emiliyan Stanev is related in structure and form to the novels-epopees of the early 1950s, and in problems and ideas to the period following the April Plenum (1956). Monumental in structure, *Ivan Kondarev* has an emotional charge stemming from its treatment of the problems of reinstating human rights in history, of communist transition from sectarian abstractions to truths about life and of surmounting left-wing dogmatism. A basic theme in the novel is the life-path of the communist Ivan Kondarev, which passes through the temptations of hegemonistic ambition, Nietzscheism and disregard for ordinary man, and then takes a turn toward achieving harmony between the goals and means of political struggle, toward greater human warmth and compassion, toward discovering life-tested truths about life and 'indifferent nature'.

Compared with the epic trend of the early 1950s, the novel on contemporary subjects developed at a slower rate, gradually, and sometimes painfully. There could be no doubt, however, that writers took up the problems of the new reality, achieving considerable success in portraying daily life and the new type of hero. *The Family of Weavers* (a two-part novel, 1956–60) by Kamen Kalchev is closely related to the 'production novel' of the preceding period; yet the realistic approach and psychological insight with which the individual characters have been drawn, also affiliate the novel to the anti-schematic trend of the following period. The complex changes brought about by the April Plenum are treated by the author in *Two in the New Town* (1964), a novel which clearly shows the tendency during the 1960s towards subjective and lyrical representation in prose. The lyrical tone is also characteristic of the novel *At Life's Source* (1964). In the documentary novels *Fiery Summer* (1973) and *The Uprising* (1975) the writer draws more directly on the sources of national history and on contemporary facts. *The Golden Fleece* (1958) by Andrei Gulyashki, which is preoccupied with the spiritual world of the characters, is a novel characteristic of the post-April Plenum period. The highly polemical novel-allegory by Gulyashki, *The Seven Days of Our Life* (1964), reflects on contemporary life and its heroes. On the one hand, the integrated, unflinching personalities, who pursue their ends with 'a marshal's baton in their knapsack', without taking heed of ordinary people, are defended and shown in a romantic light; on the other hand, the author reveals their life drama and spiritual conflict, their unfitness to live in a socialist society. The radical transformations brought about in the countryside, the tremendous efforts of the Communist Party and the socialist State for modernizing agriculture are closely followed and reported in Stoyan Daskalov's novels *Village near the Plant* (1963), *Republic of Rains* (1968), and *Earthly Light* (1968). In his novels *Groundsea* (1961) and *Wolf Hunt* (Part One, 1983), Ivailo Petrov gives a strictly realistic account of the pooling of the land into co-operative farms and optimistically discusses the question of the goals and means used in the social struggle. The range of problems of the advanced socialist society is tackled in Emil Manov's novels too. He dwells mainly on the problems of conscience and morality. His trilogy *Chasing the Blue-bird*, for instance, traces the life of Bulgarian communists, and the complex process of their readjustment, as they turn from underground functionaries into managers and workers in the new socialist reality. The contemporary ring of *Diverging Ways* (1959) is the greatest merit of Dragomir Assenov's novel; his series of novels, from *Brown Horizons* (1961) to *The Greatest Sin* (1980), fuse the chronicle, memoir and free story-telling into one. The difficulties in searching for and discovering the new types of hero are particularly evident in the novel by Todor Monov (b. 1928), *Death Does Not Exist* (1980), an attempt to draw the character of the modern builder in bold lines and harsh colours. In the series of novels by Atanas Nakovski – from *Maria versus Piralkov* (1962) to *The Accident* (1982) – the theme of construction and the technological intelligentsia is linked with questions of the conduct, moral virtues and spiritual wealth of people of today. Pavel Vezhinov's novels are outstanding creative achievements: *The Stars above Us* (1966) gives us a general idea of his notion about the Bulgarian people, or of popular culture and morality; *With the White Horses at Night* (1975) and *Balance* (1982) are an expression of the writer's urge for stock-taking and confession, for seeking harmony between feeling and reason in contemporary man, or unravelling the unfathomable secrets of life and death. Works focused on ecological problems are in step with the latest trend in modern world literature. Included here are *The River* (1974) and *Cold Estrangement* (1981) by Diko Fuchedjiev, and *The House with the Mahogany Staircase* (1975) by Andrei Gulyashki. Literature turned to major present-day problems connected with the great social and natural transformations, with the scientific and technical revolution, with the positive and negative aspects of the transformation of nature and our spiritual bonds with the heritage of our ancestors. Novels on working class and genealogical subjects also gained ground with *The Lowland* (1977) by Vassil Popov, *Life* (1978) by Vladimir Zarev (b. 1947) and other novels. At the core of contemporary novels also lies the interesting problem of the creative personality, of the 'eccentricty' of artistic *élan*, a theme mainly elaborated by Ivan Davidkov in *Distant Fords* (1967), and *Ballad of the Lonely Seafarers* (1979). Also devoted to this theme are *This Strange Trade* (1976) and *The Magic Lantern* (1983) by Bogomil Rainov, and *The Green Grass of the Desert* (1978) by Diko Fuchedjiev, etc.

The development of the detective novel, with plots about political intelligence, the struggle against the counter-revolution and the enemies of Socialism, the ideological duel between the two worlds and the attempts at ideological diversion testify to the widening range of novel genres. Andrei Gulyashki blazed a trail along this line with his novel *Counter-intelligence* (1959), and a part of the series *The Life and Adventures of Avakum Zahov* (in two volumes, 1977). Bogomil Rainov has also become established as a master of this type of novel. Under the form of spy novel, and drawing on the traditions created by prominent world writers, Rainov launches an ideological dispute between the bourgeois and socialist views of the world and people.

Often parodying the form he has chosen, he imbues it

with a new, unusual content and ironic wit and has raised the detective novel to a high professional level.

Emiliyan Stanev

The illustrative approach to history predominated in the earlier development of the historical novel. By the mid 1960s the focus had moved from events to people; the illustrative approach gave way to the treatment of psychological problems, and authors strove to build up their own conception of historical development and to impart their own philosophical view of Bulgarian and world history. Typical representatives of the historical novel are: *Time of Parting* (1964) and *Legend of Khan Asparuh, Prince Slav and the Priest Teres* (Book One, 1982) by Anton Donchev (b. 1930), *Chronicle of Troubled Times* (in two parts, 1965–66), *The Djem Case* (1967) and *A Book about Sophronius* (1978) by Vera Mutafchieva, *The Blue Amethyst* (1968) by Peter Konstantinov (b. 1928), *Liturgy for St. Elijah's Day* (1969) by Svoboda Buchvarova (b. 1925), *Haiduk Blood* (1969) by Dimiter Mantov (b. 1930), and *Antichrist* (1970) by Emiliyan Stanev. At the same time the historical novel shows a strong tendency toward documentary narration. The authors have written their historical works bearing in mind the burning issues of the contemporary world.

The development of memoir and documentary literature about the anti-fascist struggle has continued after 1956. On average, some 30–35 books of this genre have been published per year. The main trend of memoir literature has continued to be the depiction of the partisan struggle. Memoirs of combat groups were published, describing all major actions. Outstanding among the numerous publications are Mitka Grubcheva's memoirs *In the Name of the People* (1962), and those of Elena and Dobri Djurov – *Murgash* (1976). There are a number of books about the underground and militant activities in towns, life and struggle in prison, concentration camps, the 'black companies', the activities of progressive officers and soldiers in the tsarist army, the life of the Bulgarian political *émigrés* who, having survived the trials of the Spanish Civil War, or of underground struggle in other countries, or of the war on the 'quiet front', came back to Bulgaria as submariners and paratroopers to take part in the 'final battle'. The struggle against Fascism is depicted in great detail by Tsola Dragoicheva (b. 1898) in her memoir trilogy *The Call of Duty* (1972), *The Storm Attack* (1975) and *The Victory* (1979). The memoir literature overcame excessive descriptiveness and piling up of facts, and orientated itself towards meaningful interpretation of events and summary of experiences, featuring the drama of the struggle and the fighter's inner problems. Memoir writers gained in mastery; a main characteristic of their style, irrespective of the diversity, is the lively human narration, the unity of recollections and reflections and the profound depiction of the motives of human actions. Significant successes have been achieved by documentary literature. There is a tendency not only to expand this genre, which is characteristic of the overall world literary process, but also to improve the documentary writers' artistic skill and to enrich the range of problems. A lasting place in literature is held by works such as *On the Track of Those Missing* (1972) and *The Shooting Range* (1983) by Nikolai Hristozov; *They Died the Death of the Immortal* (2 vols., 1973–82) by Vesselin Andreev; *Impressions from Bulgaria* (1974) by Ivan Dinkov; *Levanevski* (1978) by David Ovadia and others.

Young writers tackle the current socio-psychological problems in an original way, taking a fresh attitude and offering topical ideological and philosophical interpretations. Among these are Stanislav Stratiev (pen-name of Stanko Stratiev, b. 1941), Lyuben Petkov (b. 1939), Vladimir Zarev, Georgi Markovski (b. 1941), Rossen Bossev (b. 1946), Dimiter Yarumov (b. 1942), Simeon Yanev (b. 1942), Dimiter Korudjiev (b. 1941), Rashko Sugarev (b. 1941), Yanko Stanoev (b. 1944) and others. The young prose writers feature the image of the youth of today – attractive and at the same time inspiring respect with its drive for many-sided and harmonious self-expression, and for creating new criteria for humaneness and spiritual beauty. There is an obvious endeavour to surmount narrowness of depiction, to avoid being preoccupied with a limited range of intimate youthful experiences, scarcity of subject-matter and confinement within the recollections of childhood and adolescence, and to become involved in tackling significant social problems, to search for new themes and synthetic structural forms and means of expression.

The April 1956 spirit renovated Bulgarian drama also. Poets, prose writers, documentary writers and journalists displayed a creative interest in drama and introduced new styles, doing away with dogmatic narrow-mindedness and creating plays in which ethical problems come to the foreground. The 'lyric drama' emerged in the late 1950s and early 1960s. The poets (mainly of the middle generation) created dramatic works distinguished by strong lyricism and attempting a philosophical summary of moral conflicts. Greater attention was paid to the heroes' intimate world and to psychological collisions, and major problems of communist morals were tackled. Greater emotionality, laconicism, expressiveness and inner drama were attained in the dialogue. In Ivan Peichev's play *Every Autumn Evening* (1958) a unity between the poetic and dramatic principles has been achieved, whereby the psychological sketches and poetic confessions do not dampen the dramatic tension. Outstanding is the drama of the individual, though each character has a general symbolic meaning as well. Events take place in heroes' souls – this is the innovatory essence of the play, a representative of the poetic wave in drama. The play *The World Is Small* (1959) by Ivan Radoev is actually a dramatic poem about love and patriotism, the class essence of the concept of patriotism, the love-duty conflict. Ivan Radoev's plays, *The Great Return* (1963), *Red and Brown* (1972), his dramatic miniatures, the satirical comedy *The She-Cannibal* (1978) testify to a growing mastery of the dramatic genres. In Valeri Petrov's lyrical comedy *When Roses Dance* (1959) the problems of love, youth and old age, of egoism and the greatness of the human spirit are presented wittily, with bold improvizations, through carnival interludes, details of lifestyle, variety technique and at the same time by tender, soft lyricism. The names of Dobri Zhotev and Georgi Djagarov are also associated with the 'lyrical drama'. Djagorov's plays are the work of a publicist moved by civic sentiments; the dispute of socially important issues, the clash between opposing outlooks and conduct come to the foreground. *The Public Prosecutor* (1965) is an innovatory play owing to its anti-cult content, its architectonic style – a concise plot developing on two planes, implied lyrical meaning, metaphorical dialogue, and the main hero's inner flights of thoughts and feelings. The play entitled *This Small Land* (1974) is devoted to topical problems of our time. The problem of conservation of nature enables the author to give universal scope to the theme of the drive for safeguarding harmony in man, his moral purity. The experience of poet-playwrights has a positive effect on the development of drama. They have a contribution of their own to the whole range of problems and styles of dramatic works, thereby making the illustrative approach and mere recording of events quite unacceptable in drama.

Many prose writers have devoted their attention to drama. Dimiter Dimov's plays stand out by their brilliant dialogue, intellectual atmosphere, satirical and ironical implications and original characters – as in *Women with a Past* (1959), *The Guilty* (1961) and *Holiday in Arco Iris* (1983). Georgi Karaslavov made his debut in the sphere of comedy with *A Stone in the Bog* (1959), in which he denounces philistinism. In his play *Vox Populi* (1963) Karaslavov developed further the ideas and conflicts embedded in *Thornapple* and *Daughter-in-Law*, and the play *Lenin In Our Home* (1970) is built on ample subject-matter based on life itself. The dramas *Tango* (1967) and *Mother for All* (1973) present the opposing forces in class society in a life and death struggle. Dragomir Assenov has added his share to the development of drama. By recreating private cases, he reveals socially significant dramas and proves a master of the dramatic dispute in his plays *Roses for Dr Shomov* (1967), *A Trade for Angels* (1971), *The Gold Holding* (1973), and *The End of the Day* (1976). Original modern plays have been written by Nikolai Haitov, such as the dramatic minatures *Dogs, A Boat in the Woods, Paths* (1966) and the play *The Celebration* (1984); and by Yordan Radichkov – *Hullabaloo* (1967), *January* (1974), *Lazaritsa* (1977), *Attempt to Fly* (1979) and *Baskets* (1982). Dramas were also written by Kolyo Georgiev, Bogomil Rainov, Rangel Ignatov, Georgi Mishev, Lada Galina and Georgi Danailov.

The April 1956 atmosphere renewed and raised to a still higher level the works by a number of distinguished playwrights. In his historical plays, Kamen Zidarov developed the realistic tradition of the genre and infused new, modern elements into it. He has a detailed mastery of dramatic structure, the intricate portrayal of character attaining success in staging the conflict and lending universal meaning to the dramatic action. This is evident in *Immortal Song* (1959, revised in 1962 under the title *Ivan Shishman*), *Kaloyan* (1969) and *Boyan – the Magician* (1972). Nadezhda Dragova (b. 1931), Purvan Stefanov and Stefan Tsanev (b. 1936), who has also written plays devoted to our times, have contributed to the development of the historical genre. Their plays illustrate an endeavour to interpret the remote past from a contemporary standpoint, and to interpret history in a philosophic light. Lozan Strelkov's development as a playwright was ideosyncratic. He had a preference for modern drama, written with publicistic zeal. His plays focus on the problem of communist morality. The characters of the positive heroes are true to life and imbued with inner dramatic quality – *The Meeting* (1961), *Not Subject to Appeal* (1972), and *In Search of Truth* (1972). Nikola Russev (b. 1932) wrote an interesting cycle of plays depicting complex moral conflicts from the past and the present – *What an Abundance of Poppies...* (1968), *The 18-Carat Gold Cigarette Case* (1967), *The Old Man and the Arrow* (1968), and *From Earth to Heaven* (1978). Moral conflicts likewise prevail in plays by Mihail Velichkov (b. 1917), Velichko Neshkov (b. 1914), Boyan Balabanov (b. 1912), Nedyalko Yordanov and Stanislav Stratiev. Modern Bulgarian drama has entered its period of maturity, which finds expression in the wide range of artistic quests and the polemic, innovatory spirit of play writing rather than by the great quantity of plays written and staged.

The rich, realistic and democratic traditions of literature for children and adolescents have been continued efficiently and creatively after the socialist revolution and the April 1956 Plenum in particular. The main set of problems treated in this literary genre is linked with the Bulgarian people's past, the Bulgarian-Russian and Bulgarian-Soviet friendship, and life in the socialist society. Works have been created in all genres, the language and stylistic means as well as those of expression and composition, have been enriched. Works for children and adolescents have been written by many outstanding writers, such as Georgi Karaslavov, Assen Raztsvetnikov, Marko Marchevski, Emil Koralov, Emiliyan Stanev, Pavel Vezhinov, Elissaveta Bagryana, Dora Gabe, Hristo Radevski, Mladen Issaev and Kamen Kalchev. Works by Angel Karaliichev, Kalina Malina, Luchezar Stanchev (b. 1908), Assen Bossev (b. 1913), Nikolai Zindarov (b. 1921), Tsvetan Angelov (1922–82), Leda Mileva, (b. 1920), Mihail Lukatnik (pen-name of Mihail Arnaudov, 1920–74), Georgi Strumski (b. 1932), and others are dedicated mainly to children and adolescents.

The stupendous successes scored by literature after the socialist revolution are also due to the work of men of letters – theoreticians, historians and critics. Relying on the rich traditions set before 9 September 1944 by Dimiter Blagoev, Georgi Bakalov and Todor Pavlov, under the leadership of the Communist Party, the theory of literature was restructured with regard to the methods used. Under the new socio-political conditions the way was paved for free application of Marxist-Leninist principles in the sphere of literary theory, history and criticism. Close ties with Soviet literary scholars have contributed to this end. The Institute of Literature of the Bulgarian Academy of Science was set up in 1948, new philological subjects were introduced in the University of Sofia, new literary periodicals began to be published – the *Pulse* and *ABC* newspapers in 1963 and 1979 respectively, the magazines *Septemvri* (September) in 1948, *Plamuk* (Flame) in 1957, *Literaturna Missul* (Literary Thought) in 1957, *Suvremennik* (Contemporary) in 1973, *Literaturna Istoria* (History of Literature) in 1977. This fact helped to increase the number of men of letters and was conducive to their advancement. The Marxist-Leninist assessment of classical literary heritage, the correct attitude to democratic literature, was a great success for literary science. The progress made by the theory of literature stands out in bold relief in the *History of Bulgarian Literature* (4 vols, 1962–78), a publication of the Bulgarian Academy of Science; *Pages from the History of Bulgarian Literature* (4 vols, 1967–75) by Georgi Tsanev; *A Companion to Bulgarian Literature* (3 vols, 1976–82); studies of various periods of the development of literature, of individual writers or various literary problems by Peter Dinekov, Emil Georgiev (1910–82), Kuyu Kuev (b. 1909), Bonyo Angelov (b. 1915), Ivan Dujchev (b. 1907), Georgi Tsanev, Pantelei Zarev, Georgi Dimov (b. 1918), Stoyan Karolev (b. 1921), Pencho Danchev (b. 1915), Efrem Karanfilov (b. 1915), Boris Delchev, Simeon Sultanov (b. 1927), Peter Pondev (b. 1914), Minko Nikolov (1929–66), Zhelyo Avdjiev (b. 1920), Rozalia Likova (b. 1922), Georgi Konstantinov, Toncho Zhechev (b. 1929), Krustyo Kuyumdjiev (b. 1933), Zdravko Petrov (b. 1928), Boyan Nichev (b. 1930), Milena Tsaneva (b. 1930), Docho Lekov (b. 1928), Ivan Sarandev (b. 1934) and Svetozar Igov (b. 1945). Combining the literary historical and literary critical approach with theoretical interpretation has been a characteristic phenomenon over the past years. Pantelei Zarev's study *Panorama of Bulgarian Literature* (5 vols, 1966–76) is an outstanding work in this respect, a successful attempt to elucidate the development of literature from the National Revival up to the present,

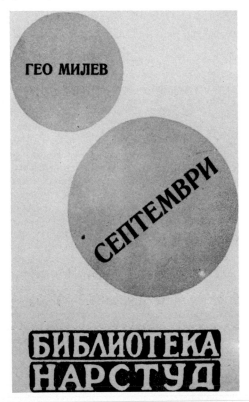

Facsimile of the title page of Geo Milev's *Septemvri* (September); 1927 Paris edition

Kamen Zidarov

Lyubomir Levchev

taking into account the overall cultural development, the lasting features of national mentality and the vicissitudes in the history of the people. Todor Pavlov, Pantelei Zarev, Georgi Dimitrov-Goshkin (b. 1912), Vassil Kolevski (b. 1925), Atanas Natev (b. 1929), Georgi Markov (b. 1927) and Ivan Popivanov (b. 1927) lent their share to the development of theory of literature. Literary criticism was taken up by Pantelei Zarev, Boris Delchev, Pencho Danchev, Efrem Karanfilov, Emil Petrov (b. 1924), Stoyan Karolev, Vesselin Yossifov, Ivan Ruzh (pen-name for Hrisim Koedjikov (b. 1909), Zdravko Petrov, Lyuben Georgiev (b. 1933), Mihail Vassilev (b. 1935) and Atanas Svilenov (b. 1937). Problems of Russian and Soviet literature were studied by Simeon Russakiev (b. 1910), Velcho Velchev (b. 1933), Georgi Germanov (b. 1926), Vassil Kolevski, Ivan Tsvetkov (b. 1924), Georgi Penchev; of Slav literatures – by Emil Georgiev, Boyan Nichev, Iliya Konev (b. 1928); of West European literatures – by Konstantin Gulabov (b. 1892), Lyubomir Ognyanov-Rizor (b. 1910), Marko Minkov (b. 1909), Stefan Stanchev (b. 1906), Atanas Natev, Nikolai Donchev (b. 1898), Simeon Hadjikossev (b. 1941). Since the 1960s comparative studies of literary phenomena, of literary relations (Bulgarian-Russian, Bulgarian-Soviet, Bulgarian-Czech, Bulgarian-Polish, etc.) have been under way, in addition to attempts in the sphere of comparative theory of literature. Problems of literature for children and adolescents have been tacked by Nikolai Yankov (1925–82), Georgi Vesselinov (1909–78), Peter Dimitrov-Rudar (a pen name of Peter Kelov, b. 1906) and others.

The Party's activities after the April 1956 Plenum provided conditions for the comprehensive development of literature and the theory of literature. Issues of literature and arts were raised and discussed at party congresses, conferences, plenums, meetings and other forums. The General Secretary of the BCP, Todor Zhivkov, has repeatedly drawn the attention of men of letters to topical issues of the literary process and to Marxist-Leninist formulations on the organic link between art and society, literature and people, science and socio-economic and cultural development. The Programme of the BCP (1971) reads in part: 'Portraying artistically the profound transformations in society, the intricate spiritual life of the man of today and the germs of the new communist consciousness and mentality by the method of socialist realism, Bulgarian literature and arts have attained the real historic opportunity of advancing to the forefront of mankind's artistic creativity and enhancing their ideological-educational and transforming role'. Enormous work has been carried out by the Union of Bulgarian Writers, set up in 1913, which is a professional, creative organization having a membership of 368 and five sections – Poetry, Prose, Literature for Children and Adolescents, Drama and Criticism – in addition to seventeen district societies of writers having a membership of 609 and the Young Writers' Club with a membership of 102 (1984). The Union's Governing Board (comprising 57 members) organizes its congresses and guides work between congresses. The Congress, which is the supreme governing body of the Union, is held every four years, reviews the Union's overall activities over the past period and adopts decisions on its future work, makes amendments and supplements to the statute of the organization and elects a governing board. The Union of Bulgarian Writers organizes various meetings to discuss the major theoretical and practical issues related to the state and development of literature, international literary co-operation and the training and education of literary cadres. It is responsible for improving the living and working conditions of writers, and for creative maturing, as well as for the early discovery and encouragement of young talent. National literary conferences are held, in addition to celebrations of writers' anniversaries, literary discussion, scientific sessions, talks and discussion on literary issues in the party Press. A particularly useful forum is the annual April literary discussion, held regularly since 1980, which in the words of the Chairman of the Union of Bulgarian Writers, Lyubomir Levchev, 'has become a measure of the spirit of criticism and self-criticism, evidence of the development of socialist democracy in literature and in the life of society, as well as a symbol of supreme ideological and creative unity based on the free emulation among various creative personalities rallied by the common communist ideal and by the leading April policy line of our Party "Quality – Our Foremost Duty"' (in *Literaturen Front*, No. 14, 1984). The Union of Bulgarian Writers likewise organizes the regular international writers meeting 'Peace – the Hope of the Planet'. Four meetings have been held (1977, 1979, 1980 and 1981) at which prominent writers from various countries and continents, with various political and aesthetic views, discussed issues related to the role of men of letters in the drive for safe-guarding and strengthening world peace and for the further development of the world democratic and humanistic literature. The rich traditions and achievements of Bulgarian literature, the leading role of the BCP and the staunch ideological unity of Bulgarian writers, as well as of the care lavished on them by the socialist State and the public, guarantee the further successful development of the literary process in the People's Republic of Bulgaria.

FINE AND APPLIED ARTS THROUGH THE AGES

The territory of the First Bulgarian State (681–1018) was inhabited by various peoples and tribes that have handed down a rich and varied artistic heritage. The beginnings of fine art can be seen in the pre-historic societies. A fine example of pre-historic workmanship on Bulgarian soil was unearthed in a burial mound near the village of Karanovo, near Sliven. Valuable finds have also been made in Southern Bulgaria and identified as belonging to the *Maritsa Culture,* which in the period between 6000 BC and 3000 BC produced work of a high aesthetic standard, revealing a subtle sense of proportion and decorative effect in carving and painting, and embossed ornamentation. Along with geometrical patterns, anthropomorphic representations began to appear on earthenware, e.g. carved hands and faces. Clay figurines of idols belonging to the neolithic culture (Karanovo levels II and III) have been discovered, which are the earliest specimens of their kind to have been found in Bulgarian soil. The figurines are female fertility symbols and testify to the existence of a cult of maternity and womanhood. During the late Neolithic Age (Karanovo level IV) these figurines gradually became more naturalistic. Specimens made of marble and bone have also been found. Zoomorphic cult vessels, models of dwellings, jewellery and ornaments are also among the early works of art that have been found in Bulgaria. The development of metallurgy and metalwork (gold, copper, bronze) stimulated the development of the fine arts, and craftsmanship became more elaborate and refined. Other fine specimens from the period include a hieratic gold chain from the treasure unearthed in the burial mound near the village of Hotnitsa, near Turnovo; several gold plaques with symbolic ornamentation from the Copper age necropolis near Varna and an exquisite bronze fawn from Sevlievo. These artefacts were the work of various more or less differentiated Thracian ethnic groups. During the Bronze Age the aesthetic standard of pottery tended to decline and idol figurines disappeared altogether. The drawings in the Magura cave near the village of Rabisha, near Vidin (second millennium BC) are a notable example of the art of the end of the Bronze Age. The human and animal drawings are primitively executed in ochre and in all probability were of magical significance. The gold treasure from the village of Vulchitrun, near Pleven (thirteenth–twelfth centuries BC) is unique. It weighs a total of 12.425 kg and consists of thirteen ritual vessels with silver encrustation in an archaistic geometrical style.

From the end of the Bronze Age and well into the Iron Age, another kind of art developed, that of the Thracians, with a distinctive zoomorphic style (according to Greek sources the Thracians inhabited the eastern part of the Balkan Peninsula). The art of the Greek colonies on the Black Sea coast also belongs to this period (seventh–sixth centuries BC). A number of unique masterpieces bear witness to the high standard of Thracian pottery, gold and silver work, mural painting, sculpture and coin minting. The remains of the ancient Thracian towns of Seuthopolis and Kabile; the remarkable discoveries in the tombs near the village of Duvanlii, near Plovdiv and Mezek, near Haskovo; the hoards discovered in the town of Vratsa and the village of Letnitsa, near Lovech; the murals in the Kazanluk tomb; the coins minted by successive Thracian rulers, together with many other discoveries reveal the nature and aesthetic achievements of Thracian art. One of the most remarkable discoveries in recent years has been the gold treasure from the town of Panagyurishte. It weighs a total of 6.164 kg, and consists of nine vessels (four zoomorphic rhytons, three anthropomorphic jugs, a large phial and amphora with handles shaped like centaurs). The vessels are decorated with designs from Greek mythology. Thracian sculpture in its pure form has survived in a few works, e.g. the archaistic bronze wild boar (fourth century BC) found in the village of Mezek, portrait sculptures on the graves of gladiators and about 2,000 votive tablets dedicated to the Thracian deity, Heros. During the period of Greek colonization of the Black Sea coast and the later period of Roman rule in Bulgaria a number of works were produced which influenced Thracian sculpture. The latter is represented by bronze and silver portrait-mask helmets (first century BC). The Hellenistic period of Thracian culture (fourth–first century BC) is also represented by the fine examples of Thracian mural-painting in the Kazanluk tomb. The murals are completely original, both in their composition and in their handling of individual figures and groups. They are clearly the work of an excellent painter of the Early Hellenistic Period and reveal the best traditions of fourth century BC Greek painting, enriched by local artistic concepts. The murals provide important data for the study of both Thracian painting and the life-style of the Thracian nobility of the fourth–third centuries BC. They are the best surviving examples of Thracian art of the period. The Kazanluk tomb figures on the Unesco list of the world's natural and cultural heritage (1979).

The Hellenistic Thracian tomb, one of the major achievements of Bulgarian archaeology, discovered near the village of Sveshtari, in the Razgrad district in 1982 dates back to the fourth–third centuries BC. It is distinguished by its rich pictorial and relief ornamentation and perfect synthesis between architecture and sculpture.

During the Roman and early Christian period of Thracian history in Bulgaria, the art of mosaic flourished: examples include the mosaics from the Roman villa near Ivailovgrad; the mosaic representation of a scene from Menander's *The Acheans* in Oescus, now the village of Gigen, near Pleven; and the mosaics in the early Christian necropolis in the Church of St. Sophia in Sofia. The fourth century has bequeathed some fine murals, richly ornamented with Christian symbols, in the vaults of the St. Sophia necropolis, and also some well-preserved frescoes, decorated with ornaments and figures, from the tomb in the town of Silistra. The vaulted ceiling is especially rich in ornamentation.

The Slavs and Proto-Bulgarians made their own distinctive contribution to the growth of fine art. Some remnants of the Middle Proto-Bulgarian culture (seventh–thirteenth century) in the Don region and Volga-Kama Bulgaria testify to the considerable achievements of Khan Asparuh's Bulgarians in monumental architecture (e.g. the archeological excavations of the palaces in Pliska, the first Bulgarian capital), stone reliefs (of horsemen and hunting scenes) in the Old Bulgarian capitals Pliska and Preslav, and also pottery. The territorial, political and economic growth of the First Bulgarian State (681–1018) went hand in hand with the development of monumental art (especially in the ninth–tenth centuries). In Pliska, art still preserved its heathen, barbarian and primitive form. Under Khan Omurtag (814–831) and Khan Pressiyan (836–852) building and architecture (Pliska, Madara, Ticha, Chatalar, Preslavets) reached a new peak. Toreutics was highly developed, as evidenced by the Madara buckles – two gold belt buckles discovered in 1926 and 1934 in the necropolis on the cliffs not far from the village of Madara, near Shumen and belonging to the eighth or ninth century; the Nagy Szent Miklos treasure of the eighth–ninth centuries – a Proto-Bulgarian gold hoard discovered in the village of Nagy Szent Miklos (in present-day Hungary), housed in the Kunst-Historisches Museum in Vienna); the torques from the Sheremet treasure dug up in the village of Sheremet, near Veliko Turnovo, which date back to the tenth–eleventh centuries. One of the most striking works of sculpture is the Madara Horseman (eighth century), which is the only rock relief of its kind in Europe. It is cut into the northwestern face of the Madara plateau at a height of 75 feet, and represents a life-sized man on a horse, cantering with a dog running behind. At the horse's feet a lion lies empaled by a spear. The scene symbolizes the grandeur of the Bulgarian ruler and the Bulgarian State, as is evident from the inscriptions around the relief. The monument figures on the Unesco list of the world's natural and cultural heritage (1979).

In Preslav, the second Bulgarian capital, philosophical humanism, arts and letters reached a peak. The use of stone and ornamentation in Preslav architecture gave rise to the decoration of cornices, facings and floors with colourful ceramic embellishment. The pottery workshops also evolved an original tradition of icon-painting on ceramic tiles, e.g. the ceramic icon of St. Theodore Stratilatus from the Patleina district near Preslav (ninth–tenth centuries). A large number of bas-reliefs bearing stylized animal representations appeared on stone capitals (in Preslav) and on stone slabs used for facing and partitioning in churches and palaces (in Stara Zagora, Nova Zagora, Preslav and Ohrid). The level of sculpture is attested to by the animal statues on the buildings in the Avradaka area (Preslav).

The art of painting developed under the influence of Byzantium after the adoption of Christianity by the Bulgarians in 864 and was largely subordinated to the needs of the Christian religion. Gradually it acquired a local character. The tenth century saw the rise of large-scale mural-painting in the archaistic style, with a touch of eastern influence (representations of prophets and flying angels in St. George's Rotunda in Sofia, and in the churches in the town of Kostur and the village of Vodocha, near Struma). According to John the Exarch the palace of Tsar Simeon was decorated with secular (portrait) murals.

The illumination of manuscripts reached a very high level, some of the decorative material having not only aesthetic but also great historical value. The earliest extant Bulgarian portraits are copies of miniatures of Knyaz (Prince) Boris I (852–889) in the Homiletic Gospel of Konstantin of Preslav (c. 893), miniatures of Tsar Simeon (893–927) in the Hippolitus Anthology (1076), copies of miniatures of the Evangelists from the Ostromir Gospel (1056–1057) and the Simeon (Svetoslav) Anthology (1073) found in Russian manuscripts dating from the twelfth–thirteenth centuries. One fairly widespread design is an interlacing, with teratological elements. This was an original variation on the geometrical Byzantine illumination and was subsequently to provide the basis for Slavonic illumination.

Byzantine domination (1018–1186) led to periods of Byzantine neo-classicism (tenth–eleventh centuries) and intensely religious painting (eleventh–twelfth centuries). This development can be seen in the murals in the Boyana Church near Sofia (first layer, tenth–eleventh centuries), the murals in the ossuary at the Bachkovo monastery near Plovdiv (these eleventh–twelfth century paintings are considered to be among the most striking examples of the painting of the period in Eastern Europe), the murals in the Church of St. Panteleimon in the village of Nerezi, near Skopje (1164) and the Church of St. George in the village of Kurbinovo near Lake Prespa (1191). In these works the influence of the canons of Constantinople and Salonika are transported into a more realistic and emotional key. During the Second Bulgarian State (1186–1396), stimulated by rapid economic and cultural development, the fine arts flourished in a variety of directions. The new capital of Turnovo became the centre of a fast-growing culture, with its own school of painting, the Turnovo school, which was founded during the thirteenth century. It went through two peak periods, the first in the thirteenth and the second in the fourteenth century. The Turnovo school of painting maintained its independence and originality due to the fall of Byzantium under Latin rule in 1204. Using earlier classical models and local traditions, slavonicizing the Byzantine canon and borrowing freely from real life, the Bulgarian painters created a classical tradition of religious mural painting exemplified by the Church of St. Nicholas near the town of Melnik (late twelfth and early thirteenth centuries), the murals in the churches on Mount Trapezitsa, the Church of Forty Martyrs (1230), the first layer of murals in Sts. Peter and Paul Church in Turnovo, the Boyana church (the second layer, dating from 1259, is a high point in Bulgarian medieval painting), the rock church in the Gospodev Dol area and the buried church of the Archangel Michael near the village of Ivanovo, near Russe. A new development in the painting of the Second Bulgarian Kingdom was the departure from the symbolic and liturgical style of Byzantine painting and mosaics, and the creation (according to the French art critic André Grabard) of a completely new style and conception of painting, by developing 'pictorial' and narrative-realistic compositions derived from miniature painting. The Bulgarian mural masters employed a combination of fresco and tempera techniques which achieved increasingly colourful and realistic effects. The portraits of church donors and ecclesiastical figures in the Boyana church possess a psychological depth achieved by purely technical means: softer lines and a subtle gradation of tone. The portraits of Tsar Konstantin-Assen (1257–77) and Tsaritsa Irina and of Sebastocrator Kaloyan and his wife Dessislava painted in the narthex are of special interest in terms of both artistic and historical value. Although the iconography adheres to the Byzantine tradition, the vitality of its humanistic realism strikingly prefigures the Renaissance. The church is listed by Unesco. During the thirteenth and the fourteenth centuries the Black Sea-Nessebur school of icon-painting emerged in the towns of Nessebur, Sozopol and elsewhere. In Plovdiv, the Bachkovo monastery, the Rila monastery, Sofia, Turnovo and Ohrid the first classical Bulgarian icons were painted. A local tradition was established in icon painting with a degree of slavonicization of the canon and a more pictorial execution. The earliest classical icons of the period are *The Virgin Hodigitria with Child* from Sozopol (thirteenth century) and the diptych *Virgin Eleusa* and *Christ Pantocrator* from Nessebur (thirteenth century). As to the literary works of the thirteenth century, the Dragan's Minaios and the Radomir Psalter (housed in the Zograph Monastery, Mount Athos) are the most richly illuminated. In the Dobreisha Gospel (first half of the thirteenth century) the interlaced decoration and teratological motifs in the portrait miniatures of the Evangelists display considerable originality.

In the fourteenth century, Bulgarian painting followed the style of Byzantine neo-classical and the Paleologue Renaiassance. This receptivity was a result of the realist achievements in Bulgarian painting in the thirteenth century, which were surpassed in the fourteenth century by several unquestionably masterly murals – in the Transfiguration Chapel in the Hrelyo Tower in the Rila Monastery (1335); in the Church near the village of Ivanovo, near Russe, the Church of the Holy Virgin in the village of Dolna Kamenitsa, near Nish, now in Yugoslavia (1330), and the church in the village of Kalotina, near Sofia; at St. Todor's Church near the town of Bobosnevo, St. John the Baptist's Church in Nessebur, and the mortuary chapel of St. Peter in the village of Berende, near Sofia; the portraits of the Bakurinai brothers, donors to Bachkovo Monastery and of Tsar Ivan Alexander in the narthex of the ossuary; and the sublime and monumental representations of 22 prophets of the dome of the Church of St. George in Sofia. The images in these churches express anxiety which can clearly be attributed to the Ottoman invasion. No less remarkable in execution and impact are two icons with silver repoussé covers: one features the *Virgin Eleusa* (in the Bachkovo Monastery, dated 1311) and the other is from Nessebur dated 1342; the *Virgin Hodigitria*, the *Virgin Pantobasilissa* and the *Virgin Life-Giving Fountain* from Sozopol; the markedly pictorial and psychological individualized icons of St. Ivan of Rila and St. Arsenius in the old Hrelyo church in the Rila Monastery, and the exquisite diptych icon featuring the Virgin with John the Divine and the Vision of Ezekiel (the Latom Miracle) of *c.* 1395 housed in Poganovo Monastery. Alongside official painting, a popular artistic trend developed in the fourteenth century in West Bulgaria. Its immediacy, primitivism and conventional colour schemes were influenced by local archaism and the tradition of monastic painting from Asia Minor (the monastery of St. John the Divine in the town of Zemen; the rock churches near the village of Karlukovo, near Lovech; the church near the village of Lyutibrod, near Vratsa; the upper part of the narthex of the church in the village of Kalotina, near Sofia).

The Turnovo school of miniature painting produced outstanding achievements in manuscript illumination. The manuscripts illuminated for the book-lovers of the royal court were richly ornamented with miniatures whose themes and style were sometimes original and sometimes revealed Byzantine influence (the Manassius Chronicle, the Gospel of Ivan Alexander, the Tomich Psalter, and the Dragan Minaios). Some of the most original examples of manuscipt illumination are to be found in depictions of Bulgarian historical events which decorate the Manassius Chronicle (a Bulgarian translation of *Chronicle of the World* by the twelfth century Byzantine writer Constantine Manasses, commissioned by Tsar Ivan Alexander in 1344–47) and the portraits of the Tsar's family in the Ivan Alexander (London) Gospel (1331–71).

Although there was some fine stone carving on sarcophagi, the plastic arts did not make great progress during the period. In the thirteenth and fourteenth centuries, and later under Ottoman domination, the Turnovo school produced a large amount of table pottery elaborately decorated with carved or painted designs. Fine examples of carving include: the wooden icon of Sts. George and Dimiter in the Church of the Holy Virgin in Sozopol (twelfth–thirteenth centuries, or sixteenth century according to other authors); the stone reliefs of the Apostles in the village of Dorkovo, near Pazardjik; the carved capitals in Turnovo; the statue of Christ in Nessebur; the wooden figure of Kliment of Ohrid in the Church of St. Kliment in Ohrid; several entrance gates – that of the Hrelyo Church (fourteenth century) and of the Church of St. Nicholas Bolnichki in Ohrid; Hrelyo's throne in the Rila Monastery; and a part of a choir loft in the Nativity Church in the village of Arbanassi, Veliko Turnovo. There are also fine examples of the goldsmith's craft, such as chainmail, repoussé covers of icons and ecclesiastical books, crosses, ornaments and jewellery.

During the period of Ottoman domination (1396–1878) the conquerors destroyed a sizeable number of cultural treasures and temporarily retarded the growth of fine arts in medieval Bulgaria. During the second half of the fifteenth century the solid traditions of painting established in the thirteenth–fourteenth centuries were still alive, as can be seen from a number of extant works. It is justifiable to speak of a late Bulgarian classical period in Turnovo (the second layer mural paintings in the Church of Sts. Peter and Paul) and more particularly in Western Bulgaria in the monasteries near the village of Dragalevtsi and Kremikovtsi (today residential districts of Sofia), the town of Zemen, the Convent of Orlitsa at the Rila Monastery, and the Church St. Petka Samardjiiska in Sofia. In the sixteenth–seventeenth centuries icon and mural painting passed into the hands of monks and local masters, which accounts for some new aesthetic and stylistic characteristics evident in the execution. In some instances the latter becomes more schematic and conventional, more ascetic and primitivistic, and in others, more graphic and decorative. On the whole, however, the artistic process was not characterized by any great uniformity and yet at times rose to new heights, as in the churches around Sofia painted by Pimen of Sofia, the first named icon painter in the late fifteenth and early sixteenth centuries; the murals of the monastery in the village of Rozhen, in the Blagoevgrad district (1535, 1611, 1662, 1728); the churches in the village of Vukovo in the Kyustendil district (1598), the Church of St. Stephen in Nessebur (1599), of St. George in Turnovo (1616), in the village of Dobursko in the Blagoevgrad district, (1614), in the Bachkovo Monastery (the refectory, 1623 and the church, 1643), in churches and monasteries in the Sofia district (c. 1626), in Vidin (the churches of St. Panteleimon, 1646 and St. Petka 1682); and in the village of Arbanassi (seventeenth–eighteenth centuries). During the seventeenth and eighteenth centuries a number of masters appeared, whose names we know. They had learned from the icon-painting traditions of the monasteries on Mount Athos, and from the icon-painting murals used in eastern-orthodox church painting. Splendid examples of icon-painting from the period of Ottoman domination are on display in the crypt of the Alexander Nevski Memorial Church in Sofia (a branch of the National Art Gallery), the Museum of National Revival Art in Varna and in other icon collections in Bulgaria. As far as miniature art is concerned, during this particular period it failed to reach the level it had attained prior to the fall of Bulgaria under Ottoman domination; however, interesting examples of monastic and secular book illumination and illustration include the Kremikovtsi Tetraevanagelia and the Slepchen Gospel of the fifteenth century; the Gospel of Priest John of Kratovo, the Krupnik Gospel, the Dragalevtsi Psalter of the sixteenth century; the Troyan Homily, the Prayer-book of Master Philip of the seventeenth century; and Priest Puncho's Anthology (with self-portraits) of the eighteenth century. Stone reliefs on churches and fountains and stone crosses for graves were also made. Wood-carving was the main form of plastic art, impressive decorative work being exemplified by church iconostases (from the sixteenth and seventeenth centuries and the National Revival Period) in the monasteries in the Rila Mountains and the town of Troyan; in churches and private houses (where the decoration is applied to ceilings, doors, chests, columns, etc.). The art of the goldsmith flourished again in the seventeenth century. In the centres of the goldsmith's trade such as Chiprovtsi, Vratsa, Sofia, Plovdiv and Pazardjik, a variety of church plate was beaten and decorated with figures and scenes: repoussé covers of gospels, gift repositories, crosses, etc.

It was not until the National Revival Period (eighteenth century–1878) that the national spirit could fully express itself in Bulgaria and closer contact could be established with European art. In the late eighteenth and the early nineteenth centuries a few schools emerged with their own distinctive characteristics: the Tryavna school (the Vitanov and Zahariev families of icon-painters), the Samokov school (the Dospeis, the Valyovs, the Obrazopisovs), the Bansko school (the Vishan-Molers), and the Debur school (the Dichovs, Filipovs, Fruchkovs, etc.). They played an important role in the growth of mural painting, icon painting and wood-carving. The promotion of these schools created a new art voicing the hopes and optimism of the Bulgarian people. Art sloughed off religious mysticism and became more earthly and life-like. Religious themes and subjects were still used, but the colours, the composition, the introduction of Bulgarian national symbols (such as the two-headed eagle and the lion), the images of Bulgarian saints and rulers, coats of arms, and scenes featuring subjects from Bulgarian history suggest a passion for freedom. They affirmed the Bulgarian name and national consciousness and raised the self-confidence of the enslaved people. Between 1817 and 1847 builders, wood-carvers and painters from Samokov built the Rila Monastery – a complex ensemble of buildings that was the greatest Bulgarian artistic and architectural achievement of the National Revival Period. However it had been founded as early as the tenth century and had functioned as a major centre of national culture, education and learning. It had been repeatedly destroyed by fire and rebuilt, thus becoming a symbol of the invincibility of the Bulgarian spirit. In 1961 it was declared a national musem and in 1983 came under the protection of Unesco.

Icon painting and other forms of religious painting flourished, the graphic arts began to develop in the monasteries and in the first printing house in Bulgaria, which had been set up by Nikola Karastoyanov and his sons in the town of Samokov. Wood-cuts and etchings were widely used for prints and book illuminations. Lithographs of a patriotic nature were also created. In the latter half of the nineteenth century and especially after the Crimean War (1853–56) religious murals and icons began to depart from the Byzantine tradition and gradually became more academic. Real life rushed into art and a complex ideological-aesthetic process began of shaking off the medieval artistic system and adopting a new system evolved during the National Revival Period. Zahari Zograph (1810–53) is a unique exponent of these new tendencies whose works effected the transition from religious to secular and realistic painting. He painted icons and frescos of a genre character, mural portraits of donors and also self-portraits in the Bachkovo Monastery, the Monastery of the Transfiguration, the Troyan, Dolno-Beshovish and Rila Monasteries; in the Church of Sts. Konstantin and Elena in Plovdiv; and in the narthex of the Velika Lavra Monastery of St. Athanasius on Mount Athos. Zahari Zograph also tried his hand at easel painting, mainly portraits, landscapes and flowers painted from life.

Father Puncho. A self-portrait in a collection compiled by him in 1796. National History Museum, Sofia

Zahari Zograph – Self Portrait. Mural in the Troyan Monastery

Various genres of painting developed vigorously during the National Revival Period, although portraiture took the lead. Secular tendencies also appeared in the portraits by the first Bulgarian painters to receive academic training abroad, in Moscow, St. Petersburg, Vienna and Munich: Stanislav Dospevski (1823–78), Nikolai Pavlovich (1835–96) and Hristo Tsokev (1847–83). They returned to Bulgaria between the 1850s and early 1870s. Along with Georgi Danchov (1846–1908) they evolved a realistic method of representation, which they used in secular and religious works alike, introducing academicism into icon-painting. Genre drawing and painting was likewise developed in the works of such painters as Nikolai Pavlovich, Georgi Danchov, Hristo Tsokev, and Nikola Obrazopisov (1828–1915).

Secular painting and the graphic arts reached new heights in the historical compositions of Nikolai Pavlovich, Henrik Dembitski (1830–1906) and the political allegories of Nikolai Pavlovich and Georgi Danchov. Graphic art ranging from the religious print to secular figure compositions gained wide popularity among political emigrants in Wallachia. Henrik Dembitski promoted the caricature. These genres were entrusted with an explicit political mission and expressed the aspirations of the enslaved people for freedom. They were created in a period of turbulent national-liberation movements and are infused with a romantic revolutionary spirit.

The study of art became an automonous subject in the 1860s and was closely associated with the national-liberation movement and Russian revolutionary-democratic aesthetics. Its first representatives included Neophyte Rilski (1793–1881), Vassil Popovich (1832–97), Nikolai Pavlovich, Peter Odjakov (1834–1906), Lyuben Karavelov (1834–1879), and Marin Drinov (1838–1906).

Following the Liberation from Ottoman domination in

Ivan Murkvichka

Ivan Angelov

Nikolai Pavlovich

Tsanko Lavrenov

1878 Bulgarian painting became based entirely on democratic ideology and the techniques of academic realism, partly as a result of the influence of foreign painters working in Bulgaria. In 1896 the Art School opened in Sofia – today the Nikolai Pavlovich Higher Institute of Fine Arts – which afforded opportunities for the development of all the art forms: painting, sculpture, graphic art and applied art. The first teachers were Ivan Murkvichka (principal in 1896–1909), Anton Mitov (1862–1930), and sculptor Boris Schatz (1866–1932). By 1932 seven art societies had been set up, the earliest being the Society for the Promotion of Art, established in 1893, to unite artists of disparate ideological activities and aesthetic programmes. The aesthetic education of the public was catered for by art exhibitions from 1885, art criticism, art magazines (the first was *Izkustvo* [Art], 1895–99) and by scholarly monographs on various aspects of art. The study of art was furthered by the contributions of the Russian scholars Fyodor Uspensky and Nikolai Kondakov; the Czech scholars Konstantin Ireček and the brothers Karel and Hermingild Skorpil, the French scholars André Grabard and Gabriel Millet and the Bulgarians Andrei Protich (1875–1959) and Dimiter Daskalov (1874–1914). During the early 1880s genre painting became dominant. Initially it was ethnographic and documentary in character, but it subsequently became

Yaroslav Veshin

Vladimir Dimitrov – the Master

more critical of social reality. The major representatives were Ivan Murkvichka (1856–1938) and Yaroslav Veshin (1860–1915), both of Czech origin; Ivan Dimitrov (1850–1944); Anton Mitov (1862–1930); Ivan Angelov (1864–1924) and other painters who had received their education abroad (Prague, Munich, Paris, Florence). They upheld the democratic traditions of the National Revival Period in their realistic works. During the late nineteenth century and the first two decades of the twentieth century formal portraiture thrived in the works of Georgi Mitov (1875–1900), Stefan Ivanov (1875–1951), Nikolai Mihailov (1876–1960), Tseno Todorov (1877–1953), Dragan Danailov (1873–1948), Boris Mitov (1891–1963) and the sculptures of Marin Vassilev (1867–1931), Zheko Spiridonov (1867–1945), Andrei Nikolov (1878–1959), and Marko Markov (1889–1966). Monumental sculpture also emerged. Impressionist landscape painting was taken up by Atanas Mihov (1879–1975), Hristo Stanchev (1870–1950), Alexander Mutafov (1879–1957), Nikola Petrov (1881–1916), Marin Georgiev-Ustagenov (1872–1937), Yordan Kyuvliev (1877–1910), Boris Denev (1883–1969), Nikola Tanev (1890–1962), and Danail Dechev (1891–1962). The artistic idiom became more subjective, lyrical and metaphorical in the impressionist landscapes and emotionally coloured genre and figure paintings of Elena Kara-Mihailova (1875–1961), Nikola Marinov (1879–1948), and Boris Georgiev (1888–1962). The early works of Stefan Ivanov, Hristo Stanchev, Sirak Skitnik (1883–1943) deal with sezession subjects and sentiments. Goshka Datsov (1894–1917), Nikola Kozhuharov (1892–1971) and others betray the symbolist influence that was creeping in at the time and which was most tangibly felt in Bulgarian literature.

The feelings provoked by successive wars between 1912 and 1918 (the First Balkan, Second Balkan and the First World Wars) found expression in painting battle scenes and realistic historical pictures e.g. by Yaraslov Veshim, Anton Mitov, Otto Horeishi (1857–1937), Dimiter Gyudjenov (1891–1979), Boris Denev, and Vladimir Dimitrov – the Master (1882–1960). After the war a broad and fairly original movement for a Bulgarian national art was launched (the Rodno Izkustvo society was founded in 1919) expressing the national spirit, life and destiny, as well as the romantic atmosphere of the National Revival Period in a decorative style based on folk art. It was joined by Vladimir Dimitrov – the Master, Ivan Milev (1897–1927), Nikolai Rainov (1889–1954), Sirak Skitnik, Tsanko Lavrenov (1896–1978), Vassil

Dechko Uzunov

Zahariev (1895–1978), Pencho Georgiev (1900–40), Dechko Uzunov (b. 1899), Zlatyo Boyadjiev (1903–76), Stoyan Venev (b. 1904), Vassil Stoilov (b. 1904), sculptor Ivan Lazarov (1889–1952). Due to increasing social movements and socio-class contradictions and tensions, the art of the 1930s became more socially concerned and artists gradually evolved what was known as new artistic realism (socialist realism) and its innovative forms, influenced by Cézanne and post-impressionistic painting (the movement was inspired by the Society of New Artists which was founded in 1931). Among its members were painters Stoyan Sotirov (1903–84), Boris Ivanov (b. 1904), Alexander Zhendov (1901–53), Kiril Petrov (1897–1979), Bencho Obreshkov (1899–1970), Kiril Tsonev (1896–1961), Ivan Nenov (b. 1902), Vassil Barakov (b. 1902), Iliya Petrov (1903–75), David Perets (1906–82), Vera Nedkova (b. 1905), Mara Tsoncheva (b. 1910); the graphic artists Boris Angelushev (1902–66), Iliya Beshkov (1901–58), Marko Behar (1914–73); sculptors Ivan Funev (1900–83), Mara Georgieva (b. 1905), Vaska Emanuilova (b. 1905), and Nikolai Shmirgela (b. 1911). Some innovators like Nenko Balkanski (1907–77) and Georgi Popov (1906–60) joined the Rodno Izkustvo society, although in their subjects and stylistic inventory they tended towards the New Artists. In the 1920s and 1930s the theoretical basis of socialist realism was reinforced by the research of Todor Pavlov (1890–1977). Art criticism was invigorated by the writings of Nikolai Shmirgela, Sirak Skitnik, Nikola Mavrodinov (1904–58) and Alexander Obretenov (b. 1903).

Boris Angelushev

After the socialist revolution of 9 September 1944 very favourable conditions were created for the development of the fine arts. Artists were both morally and financially supported by the State. They adopted the method of socialist realism whose beginnings can be traced back to the art of the New Artists of the 1930s. Art was brought closer to the life of the people and mirrored the revolutionary struggles, socialist construction and the life and work of the Bulgarian nation, thereby promoting ideological-aesthetic education. The great majority of Bulgarian artists had already been progressively-minded before 1944 and now they pooled their efforts to create a socially committed and truly progressive art. They were united in a union (the Union of Artists in Bulgaria, which was renamed the Bulgarian Artists' Union in 1959). Art thrived and a large number of young artists were successfully trained. The patriotic upsurge now called for

Nenko Balkanski

romantic commitment and revolutionary verve in painting, sculpture, and the graphic arts, and for the expansion of previous artistic activities. Art was no longer confined to private studios and instead addressed the broader masses in the idiom of monumental panel paintings, allegorical sculptures and political posters. Between 1944 and 1946 the pictorial arts played an active role in the struggle for building a people's republic. Such art-forms as the socialist political poster, the graphic arts, the caricature and the satirical drawing (Alexander Zhendov, Iliya Beshkov, Boris Ivanov, Stoyan Sotirov, Nikolai Shmirgela) all became prominent. Charged with patriotic appeal was the Eighteenth General Art Exhibition 'Front and Rear' which was organized in Sofia late in 1944 to commemorate those who had fallen during the Second World War and Fascism. Taking part in it were a number of progressive artists, such as Boris Angelushev, Iliya Petrov, Alexander Zhendov, Vesselin Staikov (1906–70), Stoyan Venev, Stoyan Sotirov, Alexander Poplilov (b. 1916), Alexander Stamenov (1905–71) and Marko Behar. After 1946 the constructive creative work, the moulding of a new type of mentality, became a major artistic concern for the next two decades. The first decade following the establishment of Socialism in Bulgaria was marked by creative searches for a new form and content congruent with the new life-style and radical changes in reality. The

Alexander Zhendov

Nikolai Rainov

Iliya Petrov

Stoyan Sotirov

group compositions or individual portraits tell of the new reality, the new type of socialist man and the new relations between people. Special attention is given to the new positive hero. Settings tend to become more concrete and characters minutely individualized. The individual character and subtle psychological states are depicted in the canvases of Iliya Petrov, Vladimir Goev (b. 1925), Nenko Balkanski and Peter Mihailov (b. 1920).

Memorial sculpture and monumental ensembles commemorating recent historic events and anti-fascist heroes began to be erected throughout the country. Notable works in this genre were created by Ivan Lazarov, Ivan Funev, Vaska Emanuilova, Mara Georgieva, Lyubomir Dalchev (b. 1902), Vladimir Ginovski (b. 1927). Freestanding sculpture was often combined with large man-figured high-relief friezes. Job satisfaction and the new socialist gains were primarily expressed in painting in the work of Panayot Panayotov (b. 1909), Stoyan Venev, Dechko Uzunov, Marko Markov, Elena Paneva (b. 1917), Peter Doev (b. 1914), Elza Goeva and graphic artist Petrana Klissurova (b. 1908).

The general upsurge of socialist society in Bulgaria that resulted from the 'April line' – the policy adopted by the BCP in April 1956 – has greatly stimulated the growth of the fine arts. The prescriptive and dogmatic interpretation of socialist realism was soon surmounted and the artist was encouraged to seek with variety and individuality. The expressive range of the fine arts was greatly broadened: metaphor, symbolism, expressiveness, associativeness were adopted as legitimate artistic devices, and a rich variety of styles, genres and individual artistic idioms were thus created. The plastic arts grew in strength and vitality, bridging the gap between the Bulgarian nation's past and present, between historical traditions and socialist transformations, by using the finest achievements of the country's artistic heritage.

Stoyan Venev

Actively engaged in research into the theory, history and criticism of fine arts, are Alexander Obretenov (b. 1903), Mara Tsoncheva (b. 1910), Atanas Stoikov (b. 1919), Bogomil Rainov (b. 1919), Vera Dinova-Russeva (b. 1924), Kiril Krustev (b. 1904), Dimiter Avramov (b. 1929), Atanas Bozhkov (b. 1929) and Maximilian Kirov (b. 1930). The theory and practice of the fine arts are promoted by such bodies as the Institute of Art Studies of the Bulgarian Academy of Science (established in 1949), the Bulgarian Artists' Union, the Nikolai Pavlovich Higher Institute of Fine Arts and the National Art Gallery in Sofia.

Between the 1960s and the 1980s Bulgarian painting has greatly broadened its range of content and expressive form. The period has been typified by large-scale, multi-figured compositions depicting the revolutionary struggles and daily work of the Bulgarian nation. Heroic subjects have been treated in a national context. Heroic passion and dramatic dynamics have been blended with subtle psychological insights. Portraiture and landscape painting made great advances also. However, the preference for thematic compositions that is promoted by party policy and the leadership of the Bulgarian Artist's Union does not inhibit the development of smaller art forms. Along with the multi-figured ideological compositions and monumental portraits of heroes of the anti-capitalist and anti-fascist resistance or of earlier national historical themes, a number of outstanding portraits of distinguished people in the arts, and leading workers in industry and agriculture have been created by Stoyan Venev, Alexander

Zlatyu Boyadjiev

Stamenov, Mara Tsoncheva, Boyan Petrov (1902–71) and Naiden Petkov (b. 1918). The transition from an agrarian to an industrial-agrarian economy characteristically affected the landscape genre. Industrial subjects reflecting intensive socialist construction successfully vied with pure nature and won a legitimate place in the works of Vassil Barakov (1902), Zlatyu Boyadjiev, Peter Dochev (b. 1934), Maria Stolarova (b. 1925) and Hristo Donkov (b. 1930).

The continuity with earlier artistic traditions is maintained by artists who began their career in the 1920s and 1930s, such as Dechko Uzunov, Kiril Tsonev, Stoyan Venev, Iliya Beshkov, Boris Angelushev, Stoyan Sotirov, Ivan Funev, Mara Tsoncheva, Vaska Emanuilova, Mara Georgieva, Vera Nedkova, Nenko Balkanski, Vassil Barakov, Alexander Stamenov, Bencho Obreshkov and Vassil Stoilov. They were soon joined by a younger generation: Alexander Popilov, Alexander Petrov (1916–83), Naiden Petkov (b. 1918), Nikola Mirchev (b. 1921–73), Vladimir Goev, Kalina Tasseva (b. 1927), Svetlin Russev (b. 1933), Ivan Kirkov (b. 1932), Liliana Russeva (b. 1932), Yoan Leviev (b. 1934), Dimiter Kirov (b. 1935), Bissera Prahova (b. 1937), Hristo Stefanov (b. 1931), Dora Boneva (b. 1936) and Toma Vurbanov (b. 1943). They offer artistic syntheses of impressive origi-

Nikola Mirchev

Ivan Funev

Svetlin Russev

Ivan Lazavov

nality and great evocative potential. There is another group of artists who make a wider use of free metaphors, associations and other stylistic ingenuities; well-known names include Alexander Petrov, Atanas Patsev (b. 1926), Grigor Spiridonov (b. 1930), Georgi Bozhilov (b. 1935), Georgi Baev (b. 1924), Emil Stoichev (b. 1935), Rumen Gasharov (b. 1936), Atanas Yaranov (b. 1940), Teofan Sokerov (b. 1943), Vanko Urumov (b. 1941), Hristo Simeonov (b. 1935), Nikolai Maistorov (b. 1943), and Aneta Dragushinu (b. 1937). In the 1970s another original group was formed: the very meticulous 'New Realists' with such names as Dimiter Buyukliiski (b. 1943), Ivan Dimov (b. 1943), Maya Gorova (b. 1946), Nadezhda Kuteva (b. 1946), and Tekla Aleksieva (b. 1944). The primitivists Dimiter Kazakov (b. 1933), Radi Nedelchev (b. 1928), Vasselin Parushev (b. 1920) have infused colour and vigour into present-day Bulgarian art and at the same time have revived the folk decorative traditions of the Rodno Izkustvo movement which flourished in the 1920s. The strong traditions of Bulgarian landscape painting from the beginning of the century have been continued by a number of painters who strive to suffuse the traditional impressionist or expressionist methods of depicting nature with a deeper subjective and lyrical tone and dramatic intensity. The most notable landscape painters include

Zdravko Alexandrov (b. 1918), Vladimir Kavaldjiev (b. 1908), Petko Abadjiev (b. 1913), Boicho Grigorov (b. 1916), Genko Genkov (b. 1923), Naiden Petkov, Georgi Baev, Svetlin Russev, Kalina Tasseva, Vanya Decheva (b. 1926), Dilo Dilov (1914–84).

The graphic arts and the art of illustration have also made progress. Graphic techniques have proliferated in step with growing artistic skills. The outstanding graphic artists Iliya Beshkov, Boris Angelushev, Marko Behar, Vesselin Staikov, Sidonia Atanasova (b. 1909), Dimiter Draganov (b. 1908), Anna Kramer (1899–1976), Evtim

Alexander Bozhinov

Tomov (b. 1919), and also Todor Panayotov (b. 1927) and Hristo Neikov (b. 1929) of the younger generation, uphold the traditions of graphic realism. A fairly large group of folk-style and decorative artists have been recruited from the younger generation: Anastasia Panayotova (b. 1931), Zlatka Dubova (b. 1927), Mana Parpulova (b. 1925), Zhana Kosturkova (b. 1927), Zafir Yonchev (b. 1924). Another group has veered toward modern and more expressive subjects and techniques: its members include Yuli Minchev (b. 1923), Atanas Neikov (b. 1924), Todor Atanassov (b. 1928), Maria Redkova (b. 1925), and Hristo Gradechliev (b. 1935). Some of the group have a taste for complex stylistic, compositional and pictorial problems, e.g. Peter Chuhovski (b. 1922), Rumen Skorchev (b. 1932), and Peter Chuklev (b. 1936). Illustrations for juvenile and adult fiction have reached a high level of perfection. Among the indisputable masters of the medium are Boris Angelushev, Alexander Poplilov, Vassil Yonchev (b. 1916), Ivan Kyosev (b. 1933), Neva Tussuzova (b. 1908), and Lyuben Zidarov (1923). The caricature tradition established by Boris Angelushev, Iliya Beshkov, Alexander Bozhinov (1878–1968) and Stoyan Venev has been continued in the work of Assen Grozev (b. 1916), Tenyu Pindarev (b. 1921), Boris Dimovski (b. 1925), and Tsvetan Tsekov-Karandash (b. 1924), etc.

Prior to the Liberation from Ottoman domination sculpture as an autonomous art-form was non-existent. In the works of the classics Zheko Spiridonov, Vassil Marinov, Alexander Andreev (1878–1971), Andrei Nikolov, Marko Markov and Ivan Lazarov and the socially

Iliya Beshkov

committed works of Ivan Funev, Mara Georgieva, Vaska Emanuilova and Nikolai Shmirgela, as well as those of the representatives of modern trends, the classical and traditional view of sculpture was largely retained. Attempts at a more autonomous sculpture have been made by Stoyo Todorov (b. 1919), Sekul Krumov (b. 1922), Hristo Simeonov (b. 1927), Dimiter Ostoich (b. 1928), Dimiter Boikov (b. 1927), Velichko Minekov (b. 1928), Georgi Apostolov (b. 1923) and Valentin Starchev (b. 1935).

The applied arts have undergone vigorous development too: pottery – Georgy Bakurdjiev (1899–1972), Venko Kolev (b. 1915), Zdravko Manolov (b. 1920), Georgi Kolarov (b. 1908); textiles – Mara Yossifova (b. 1905), Marin Vurbanov (b. 1932); wood-carving – Assen Vassilev (b. 1909); and goldsmithing – Violeta Dunin (b. 1933). Among the masters of the poster should be mentioned Alexander Poplilov, Assen Stareishinski (b. 1936), Ognyan Funev (b. 1945), Dimiter Serezliev (b. 1931), Lyudmil Chehlarov (b. 1938); theatrical and operatic stage design – Alexander Milenkov (1882–1971), Assen Popov (1895–1976), Ivan Penkov (1897–1957), Georgi Karakashev (1899–1970) and Mariana Popova (1914–1982); film design and animation – Todor Dinov (b. 1919, Donyu Donev (b. 1929), and Maria Ivanova (b. 1929); monumental decorative painting – Georgi Bogdanov (1910–74), Mito Ganovski (b. 1925), Hristo Stefanov and Yoan Leviev; industrial design – Dimiter Mehandjiiski (b. 1916), Georgi Petrov (b. 1924); book design and illustration – Vladislav Paskalev (b. 1933) and Borislav Stoev (b. 1927).

Since 1970 the Bulgarian Artists' Union has held congresses every four years and in every district artists' groups have been set up. The following sections function within the Bulgarian Artists' Union: Painting, Sculpture, Graphic Art, Applied Graphic Art, Applied and Decorative Arts, Stage Design, Design, Art Criticism, Caricature, Space Design, and Decorative and Monumental Art. A board for aesthetic education and ten specialized boards for different branches of the fine arts have been set up by the State Board for Fine and Applied Arts and Architecture of the Committee for Culture, the Bulgarian Artists' Union and the Union of Architects in Bulgaria; these organizations work in close co-operation. The Bulgarian Artists' Union is a permanent member of the Association Internationale des Arts Plastiques. There is likewise a Bulgarian section of the Association Internationale des Critiques d'Art.

Over the last decade the life of the Bulgarian Artists' Union and Bulgarian art in general has been enlivened by several authoritative international forums – the International Triennial of Realistic Painting held regularly every three years and the International Competition – Exhibition of Young Painters in Sofia; the International Biennial of Caricature and Satirical Sculpture and the International Biennial of Humour and Satire in Painting held in the town of Gabrovo; the Biennial of Graphic Art held in Varna; and the International Symposium on Sculpture held annually in Burgas. Open air events are also organized, in which artists from abroad take part.

The general art exhibitions arranged locally in districts are an attractive challenge for artists and an interesting event for art-lovers. Most of these exhibitions are held every two, three or five years and are devoted to a single theme; *Glimpses into History* in Veliko Turnovo, *Man and Labour* in Gabrovo, *The Rhodope Mountains* in Smolyan, *Men and Their Land* in Tolbuhin. So far a total of 21 general art exhibitions have been arranged. Genre exhibitions, one-man shows, commemorative exhibitions and other minor events are organized in the country's art galleries. The local thematic and genre exhibitions (*The Land of Botev*, Vratsa, 1976; *Landscape* Vidin, 1982; *Spring in Kyustendil*, etc.) have given a fresh boost to modern Bulgarian art.

Bulgarian painting is motivated by an explicit desire to create works of great ideological-aesthetic and professional value in all genres – thematic group portraits, individual portraits, landscapes, and still-lifes. The nude is another genre that has been encouraged and which has evoked a desire to explore new approaches. Bulgarian painting is searching for new spiritual achievements. The richness and variety of present-day Bulgarian plastic arts is an unquestionable fact, reflecting the country's busy artistic life. In 1983 alone some 300 exhibitions were arranged, with the increasing participation of young artists born in the 1950s. They are organized by the Young Artists' Studio of the Bulgarian Artists' Union. These exhibitions reveal high artistic standards, new trends and fresh artistic potential. They are charged with patriotic, socialist energy and emotion, which has been a source of richness and variety.

The exhibitions organized in Bulgaria, like the galleries, aim to bring Bulgarian fine arts not just to the cultural community but also to people of all walks of life, to enrich their spiritual world, introduce them to cultural values and promote aesthetic and all-round education.

The following magazines on the fine and the applied arts are published in Bulgaria: *Izkustvo* (Art), *Problemi na Izkustvoto* (Problems of Art), and *Dekorativno Izkustvo i Promishlen Dizain* (Decorative Art and Industrial Design). Bulgarian artists are encouraged and supported by state purchases, contracts and working trips abroad, along with the provision of studios and other facilities. The comprehensive concern of the BCP and the socialist State, and of the Bulgarian Artists' Union promotes modern art and provides it with a wide range of opportunities for further growth.

Over the last few decades Bulgarian Art in all its forms has become known world-wide as a result of a number of exhibitions arranged abroad, for example, *2,500 Years of Art in the Bulgarian Lands* in Vienna and Paris (1960–61), the *Bulgarian Medieval and Modern Art* exhibition at the Charpentier Gallery in Paris (1963), large exhibitions of art from the Bulgarian lands in London, New York, Boston, Tokyo, Mexico, Munich, Stuttgart, Moscow, Leningrad, Warsaw and other cities; the exhibition of *Modern Bulgarian Art* held in Moscow and Budapest (1969), the exhibition of *Selected Paintings by Vladimir Dimitrov – the Master* in India, Austria, USSR, Czechoslovakia, Portugal (1982–84) etc. The works of Bulgarian artists receive high awards at an increasing number of prestigious international exhibitions, for example, the Biennial in Venice, Youth Biennial in Paris, Triennial of Stage Design in Novi Sad, the Quadriennial in Stage Design in Prague, etc. One-man shows are likewise arranged both in socialist and in non-socialist countries. Bulgarian art critics and scholars regularly take an active part in national and international congresses, symposia and conferences.

ARCHITECTURE

The architectural and archaeological monuments in Bulgaria are under state protection. The National Institute for Cultural Monuments set up in 1957 is the authority responsible for the identification, conservation and popularization of these monuments.

The Boyana Church. Colour woodcut by Vassil Zahariev, 1950. National Art Gallery, Sofia

Murals depecting a chariot race and funeral feast, from the dome of the Kazanluk tomb, fourth–third centures BC

By 1982 in Bulgaria over 500 settlement mounds, over 10,500 necropolis mounds, about 4,000 monuments from Antiquity and the Middle Ages, over 7,000 residential buildings and architectural monuments from the National Revival Period, about 750 architectural monuments from the period of Capitalism, 300 frescos and mosaics, and about 3,800 historical sites and places were declared monuments of culture. The Unesco list of the world natural and cultural heritage includes the Kazanluk Tomb, the Boyana Church, the rock churches near the village of Ivanovo in the Russe district, the Madara Horseman, the Rila Monastery and the architectural reserve in the old part of Nessebur.

Among the earliest historical monuments found on Bulgarian soil are the prehistoric dwellings in the mounds near the village of Karanovo in the Sliven district, fourth and third millennium BC, and dolmens (prehistoric tombs made of megaliths) from the first millennium BC in the Rhodope, Strandja and Saka mountains.

Information about Thracian architecture is provided by the excavations carried out in the Thracian towns of Kabile (near Yambol), Seuthopolis (West of Kazanluk) and Pulpudeva (Plovdiv).

The Thracian house in Seuthopolis had foundations of broken stones, rickety walls (wooden frames, filled with adobe) and wooden roofing covered with ancient Greek tiles (*tegulas*).

Better preserved Thracian tombs have been found near the village of Mezek in the Haskovo district and in Kazanluk. They are beehive tombs with ante-rooms (*dromos*) of different lengths, covered with earth and forming a Thracian mound. The cupolas are false, being made of stone. An exception is the Kazanluk tomb, where one of the earliest uses of brickwork in Europe has been established.

An abundance of architectural monuments were left by the Romans, including towns, suburban villas, sanctuaries and necropolises.

The Roman towns in the Bulgarian lands fall into three groups: Greek colonies along the Black Sea coast, further

Caryatids from a Thracian tomb near Sveshtari, in the Razgrad district. Fourth century

Bulgaria's present towns came into being as early as Roman times. A large complex of public buildings from the second half of the fourth century has been found in the centre of Sofia, as well as a stadium and a theatre in Plovdiv, therma in Varna, Roman villas near Madara in the Shumen district and near Ivailovgrad, buildings and architectural fragments from Oeskus, Nikopolis ad Istrum, Augusta Trajana, etc. Ruins of pagan sanctuaries are to be found near the village of Patalenitsa in the Pazardjik district, near the village of Kopilovtsi in the Kyustendil district, and elsewhere. Unique patterns of tomb architecture are the beehive tomb near Pomorie, with a ring-shaped vault, and the Silistra tomb of Late Antiquity, covered with a semi-cylindrical vault and richly ornamented with frescos (fourth century).

developed by the Romans (Apollonia, Messambria, Odessos); preserved Thracian citadels in the hinterland, such as Kabile and Philipopol (Pulpudeva), as well as extended Thracian settlements in Moesia and Thrace such as Martianopol (Devnya), Nikopolis ad Istrum (near the village of Nikyup in the Veliko Turnovo district), Ulpia Serdika (Sofia), Augusta Trajana (Stara Zagora), Pautalia (Kyustendil) and Nikopolis ad Nestum (near the town of Gotse Delchev); and new towns, developed in the civilian suburbs near the frontier Roman military camps – Ratsiaria (the village of Archar in the Vidin district), Oeskus (the village of Gigen in the Pleven district), Nove (near Svishtov), Durostorum (near Silistra), etc. Most of

The Madara Horseman. A relief carved in rock near Madara in the Shumen district. Eighth century

Mural composition in the rock church near Ivanovo in the Veliko Turnovo district, 1360s

Remnants from the heyday of Early Byzantine architecture in the country are the cruciform domed St. Sophia church (fifth–sixth centuries) which has given its name to the capital. The Old Bishopric (second half of the fifth century) in Nessebur, the vaulted basilicas in the village of Golyamo Belovo in the Pazardjik district (fifth century), the four-apsed Basilica near Pirdop (fifth–sixth centuries), the Red Church near Perushtitsa, etc. Most of the basilicas are big, lavishly adorned with murals inside and without any external decoration.

St. Theodore Stratilates. A ceramic icon from the Patleina locality near Preslav. National History Museum, Sofia

The centralized and compact organization of the first Bulgarian State found expression in the layout of the newly built metropolises of Pliska and Veliki Preslav, as well as in the impressive architecture – powerful fortifications, monumental throne halls and royal palaces, and majestic temples. The rich ornamentation of the buildings stimulated the development of the monumental-decorative art – stone sculptures, frescos, mosaics, incrustations and uniquely Bulgarian monumental ceramics.

The plans of both capitals conform with the peculiarities of the natural surroundings. In the outer city, which had the shape of an irregular polygon and was surrounded by a protective belt, were the modest dwellings of the common people, the workshops of the potters and other craftsmen as well as many churches. The second, or inner fortified wall was designed to protect the 'inner city' where the partly preserved throne hall, royal palace, court church, boyars' residences, as well as the premises of the guards and some administrative buildings stand to this day. During the period of Byzantine rule and in the time of the Second Bulgarian State (eleventh–fourteenth centuries) church construction was one of the main building activities, with single-nave cruciform vaulted structures and their variations prevailing. The so-called 'picturesque style', one of the most outstanding achievements of Bulgarian church architecture in the Middle Ages, developed in the twelfth–fourteenth centuries. Its distinguishing features are the colourful composition of the façades, contrasting alternating belts of stones and bricks on whitewash, big and small blind niches, jagged cornices, particoloured ceramic rosettes and geometrical brick ornaments. Inside the churches are covered with frescos. Specimens of high artistic value can be found mainly in Nessebur – the churches of St. John the Unconsecrated Pantokrator, the New Metropolis, St. John Alyturgetos, etc.

The Roman amphiteatre at Plovdiv

During the period of Ottoman rule all the representative buildings (palaces, castles, fortresses) and cultural centres (monastery schools and libraries, churches, etc). were plundered or completely destroyed; only a small number of churches were left and these were reconstructed and turned into mosques. The building of big stone churches and belfries was forbidden; Buyuk Mosque in Sofia (fifteenth century), Cuma Mosque in Plovdiv

Murals from an early Christian tomb near Silistra. Fourth century

The Bachkovo Monastery

(sixteenth century), Ibrahim Pasha Mosque in Razgrad (seventeenth century), Tombul Mosque in Shumen (eighteenth century), and other mosques were built, together with *medreses, tekes* and *kales* (restored Bulgarian citadels in Belogradchik and Baba Vida in Vidin), caravanserais, bazaars (Yambol) and baths (*hamams*). Bulgarian architecture was isolated from West European culture and did not follow the styles it went through – Gothic, Renaissance, Baroque, Rococo, Classicism. But Bulgarian architecture did not die altogether, though construction was confined mainly to churches and houses. Most of the newly built churches were concentrated in the western parts of the country, where Turkish colonization was weaker. They were modest, low buildings with one nave and without bell-towers, very much like the houses of the poor. An interesting seventeeenth century phenomenon are the so-called 'Arbanassi' churches, while the most significant work of sacral architecture of that time is St. Mary's Church in the Bachkovo Monastery (1604). The 'Arbanassi' churches have one nave, vaulted, with gable roofs and overhanging eaves. The masonry is of stone (the walls are up to a metre thick), the windows are small, the doors are covered with iron plates or studded with wedges, and the churches themselves are surrounded with high stone walls; while there are almost no external ornaments, the interior is richly decorated. The walls and the vaults are decorated with murals and the ritual objects, the iconostases and the stalls are all covered with exquisitely gilded carvings.

Bulgarian Revival architecture (from the end of the eighteenth century until 1877) reflects the entire economic political and cultural upswing in the life of the people. The merchant and crafts centres of Gabrovo, Kotel, Kalofer, Karlovo, Koprivshtitsa and Panagyurishte which developed at that time had an informal and picturesque layout, with the merchant and artisan shops concentrated in the main street. Splendid architectural ensembles of the Bulgarian Revival determining the specific image of each settlement appeared at that time. Many cultural monuments of the period are the work of Bulgarian master-builders, famous among whom are Nikola Fichev (1800–81), Gencho Kunev (1829–90) and Alexi Rilev (late eighteenth–early nineteenth centuries). Each had his own creative style.

Residential buildings constituted a considerable part of Revival architecture. Various types of houses appeared in the different regions of the country. The basic form of compositional plan was the single-room dwelling with a fireplace. As the people's needs and affluence grew, new premises were added to the houses: *chardaks* (broad roofed

Architecture

Sarafka's House in Samokov

rich wooden fretwork and multicoloured painted walls, they were the product of a new vision of the world and of the increased economic and cultural resources of the emerging Bulgarian bourgeoisie. Marvellously well-preserved examples are to be found in Plovdiv, Koprivshtitsa, Smolyan and elsewhere in Bulgaria.

During the revival period architecture was also enriched with various Bulgarian public buildings, until then a monopoly of the conquerors. *Maazas* (commercial buildings and storehouses), schools (for example, the Aprilov secondary school), clock-towers and bridges (the Roofed

The Central Mineral Baths in Sofia

balconies), larders, *sobas* (bedroom and parlour), and *prusts* (vestibules). The plan depended on local tradition. The further development of National Revival architecture was linked with the transition from a rural to an urban way of life. More premises were added, the *chardak* was moved to the centre and glazed, and there appeared a marked tendency towards more elaborate ornamentation and representativeness. The final product and growing achievement of the century-long development of Revival architecture are the Plovdiv symmetrical houses. With their original plans, size, design and façades, with their

Old Houses in Tryavna

The Central Army Club in Sofia

Bridge in Lovech and the bridge near the town of Byala) were built at that time.

During the Revival the churches grew bigger in size, moving from simple to more elaborate and monumental design and decoration. Over the simple roof a dome would be erected, often combined with a bell-tower in the west. Remarkable achievements of the Revival architecture are the monastery compounds: Rila Monastery, Rozhen Monastery, Troyan Monastery, and Transfiguration Monastery. The monastery wings encircle open spaces around the main church forming an irregular polygon, which is imposed by the peculiarities of the site. The stern external appearance contrasts with the gay colours of the inner yards and the warmth of the wooden balconies.

After the Liberation in 1878 the Revival building tradition was preserved in the small towns and in the villages. In the capital and in the bigger cities (Plovdiv, Russe, Varna, etc.) the local master-builders were replaced by educated, foreign architects and techniciams who began applying European techniques in their designs and building practice. The Austrian architects Gruenanger

The National Art Gallery, Sofia

Sofia: Byzantine Church of St George in the city centre

and Rumpelmeyer, the Swiss architect Hermann Meyer, the Czech architect Antonin Kolar, and others worked in Sofia. In terms of style their work can be referred to the West European architecture from the end of the nineteenth century and was in sharp contrast with the architectural environment inherited from the Revival. Until 1900 Neo-classicism prevailed but was later taken over by the then modern Sezession styles. In the 1890s the first Bulgarian architects to complete their studies abroad returned home – Petko Momchilov (1868–1923), Nikola Lazarov (1872–1942), Pencho Koichev (1877–1958) *et al* – and a new current known as National Romanticism appeared in an attempt to apply architectural forms and details from the national heritage. Among their works are the Church of Saint Cyril, Saint Methodius and Their Five Disciples (1901), the Synodical Palace (1906–8) and the Sofia Mineral Baths (1910–11), the Mausoleum in Pleven (1903–5) and the Covered Market in Sofia (1910).

Even some foreigners joined this current. The Russian architect Alexander Pomerantsev used old Bulgarian and Byzantine forms in the construction of the Alexander Nevsky Cathedral in Sofia, which was erected to commemorate the liberation of Bulgaria from Ottoman domination. The cathedral is a monumental, five-nave basilica with domes arranged crossways, a narthex and a bell-tower. The façades are decorated with stone carvings, the domes are gold-plated, and the interior decoration is distinguished for its harmony of forms and colours. The wall-paintings and icons are the work of many eminent Bulgarian and Russian artists.

After 1910 National Romanticism gradually died out. Its most remarkable achievements up to the end of the First World War were the reconstruction of the *konak* (town-hall) into a royal palace (architects Gruenanger and Rumpelmeyer), the State Printing House (architect F. Schwamberg), the National Assembly building (architect Konstantin Yovanovich), and the National Theatre (architects F. Felner and H. Helmer), all in Sofia.

There are also some in Russe – the Rental House (architect Paul Brank), the Secondary School for Boys (architect Petko Momchilov) and the palace of Prince Alexander (architect F. Gruenanger), as well as in

The National Theatre, Sofia

The Alexander Nevsky Memorial Church, Sofia

Varna – the Popular Theatre (architect Nikola Lazarov), the Secondary School for Girls (architect Petko Momchilov), etc. By the end of the 1920s architectural activity was entirely taken over by Bulgarians who had received their training abroad and were organized in private, municipal and public societies. The most remarkable architectural achievements from that time are those of architects Ivan Vassilyov (b. 1893), Ivan Danchov (1893–1972), Yordan Yordanov (1888–1969), Georgi Menov (1862–1935), and Georgi Ovcharov (1889–1958).

Some of the masterpieces of Bulgarian architecture are to be found in Sofia – the Sofia University (designed by French architect Nicolas Breasson, working with two Bulgarian architects, Nikola Lazarov and Yordan Milanov), the Law Courts (architect Pencho Koichev), the Teachers's Fund (architects Yordan Yordanov and Sava Ovcharov), the Bulgarian National Bank (architects Ivan Vassilyov and Dimiter Tsolov), Hotel Bulgaria (architects Stoyan Belkovski and Ivan Danchov), the University Library (architects Ivan Vassilyov and Dimiter Tsolov), the National Library (Dimiter Tsolov and Ivan Vassilyov), etc.

The Black Mosque, Sofia

Monument to Tsar Alexander II

For all the diversity of styles and specific atmosphere, the architectural ensembles of public, commercial and residential buildings in Sofia, Russe, Plovdiv, Varna and elsewhere have something in common in the impression they make.

After the victory of the socialist revolution, Bulgarian architecture broadened its scope. Its development has passed through two stages: up to the April Plenum of the Central Committee of the Bulgarian Communist Party in 1956, and from 1956 to the present day. During the first stage (from 1944 up to the April Plenum), restoration of residential and public buildings destroyed in air-raids was

Contemporary monumental sculpture, Sofia
(*Photographs by Vaughan James*)

carried out and some projects, started before the war, were completed. Urban plans were worked out for the larger residential and industrial centres, and the building of new towns like Dimitrovgrad, Madan and Rudozem began. Prerequisites for the development of socialist architecture in the People's Republic of Bulgaria were created: in 1948 the architectural department of the Civil Engineering Faculty of the State Polytechnic developed into a separate faculty, with all the necessary professorial chairs, and the first designers' bureaux, which later became architectural schools, were founded; and in 1949 the Institute of Urban Planning and Architecture of the Bulgarian Academy of Science was set up.

Towards the end of the 1940s the striving for representativeness increased the tendency towards a recreation of historical styles, mainly in public buildings. In this spirit the Georgi Dimitrov Mausoleum (1949, architects Georgi Ovcharov and Racho Ribarov and team), the Cyril and Methodius National Library (1949, architects Ivan Vassilyov and Dimiter Tsolov), and the extensions of the Kliment Ohrid University of Sofia (1952, architect Lyuben Konstantinov and team) were designed and built in Sofia. After 1952 the tendency towards embellishment and formalism spread even further. Typical examples are some of the representative buildings in Sofia; the Party House (1953, architect Peter Zlatev and team), the Alexander Stamboliiski Memorial House with a theatre hall (1954, architect Lazar Parashkevanov), the Ministry of Electrification, today's State Council (1956, architect Ivan Danchov and team), the Hotel Balkan (1957, architect Dimiter Tsolov and team), and the Central Department Store and the Ministry of Heavy Industry, now the Council of Ministers (1957, architect Kosta Nikolov and team). Eclecticism made its way into the architecture of residential buildings as well, with details from the National Revival architecture often preferred.

After 1956 a new period began in Bulgarian architecture, when an aspiration towards modern and functional forms was shown. Representative architectural ensembles

Restored houses of the National Revival Period, Plovdiv
(*Photographs by Vaughan James*)

The Ethnographic Museum, Plovdiv

were built in district city centres, changing their image completely. They included theatres, cinemas, museums, administrative buildings, sports halls, etc. New public buildings reshaped the centres of small towns and villages – community cultural centres, schools, commercial and administrative buildings. Their designers sought to achieve a modern architectural image, abandoning altogether the striving after embellishment. The administrative buildings stood out as vertical dominants in the townscape or were subordinated to the scale of the surrounding architectural environment, without losing their monumental appearance. The high-rise became a preferred form in hotel construction, too: the Maritsa Hotel in Plovdiv (1964, architects Vladimir Rangelov and K. Antonova), the Tundja Hotel in Yambol (1971, architects Dimiter Yarumov and Dimiter Krustev), the Pliska Hotel (1964, architects Hristo Tsvetkov and Lozan

Lozanov), the Rila Hotel (1962, architect G. Stoilov) and the Moscow Hotel (1974, architect A. Belokapov) – in Sofia. The hotels from the Interhotels chain: the Novotel Plovdiv (1977, French architect J. S. Loubet) in Plovdiv, the Grand Hotel Varna at Druzhba resort (1977, architect Georgi Ganev), and the Novotel Europa (1977, French architect Pierre Cochin) and Vitosha-New Otani (1979, Japanese architect Kisho Kurokawa) in Sofia are on a high architectural level. The exterior design of these buildings and the furnishing of their premises (rooms, restaurants, playing halls, etc.) correspond to their representative character).

The flowering of the Bulgarian socialist culture has been an impetus to the construction of buildings for cultural purposes. The large cubic bodies of new theatrical buildings add to the diversity of the city ensembles in Pazardjik (1967), Veliko Turnovo (1971), Stara Zagora (1971), Vratsa (1976), Kyustendil (1978), Shumen (1980), Smolyan (1983) and elsewhere. The National Theatre in Sofia has been completely reconstructed (1977, architect Ivan Tomov and team), and the ancient theatre in Plovdiv has been under reconstruction (1981, architects V. Kolarova and G. Todorov). Many new libraries, museums, art galleries and community cultural centres have been under construction. The Ivan Vazov National Library in Plovdiv (1974, architect Maria Mileva) and the library-museum in Shumen (1980, architect B. Bozduganova) have been designed in an interesting way.

Architectural studios under eminent architects have been organized for the building of unique structures, in which the synthesis between architecture and monumental arts is extensively used; the Boyana Government Residence in Sofia (1974, architect Alexander Barov and team), the Pantheon of the National Revival Figures in Russe (1978, architect Nikola Nikolov and team), the Party Memorial on Mt. Buzludja (1981, architect Georgi Stoilov), and the Lyudmila Zhivkova People's Palace of Culture in Sofia (1981, architect Alexander Barov and team). The People's Palace is a centre of major socio-political and cultural events of local, national and international significance. It is a member of the Association of International Congresses and Conferences. Located in a picturesque part against the background of the Vitosha mountain, its architectural and artistic image is modern and compact. It is a multi-functional complex.

Memorial ensembles have been built in commemoration of important historical events and heroic deeds. They are a combination of architectural and sculptural elements, united ideologically, spatially and functionally and blending organically with the surrounding environment. They owe their impact to their proportions and size and to the

(Photographs by Vaughan James)

Above, right above, right below: Restored houses of the National Revival Period, Plovdiv

The Bulgarian Communist Party Memorial on Mount Buzludja

The Ministry of Foreign Trade in Sofia

expressive combinations between symbolically united elements and a specific artistic image – the Common Grave in Plovdiv (1974, architects Vladimir Rangelov and Lyubomir Shinkov, sculptor Lyubomir Dalchev), the Defenders of Stara Zagora Memorial in Stara Zagora (1977, sculptor Krum Damyanov, architect Boris David-kov), the April Uprising Memorial in Panagyurishte (1976, sculptors Sekul Krumov, Velichko Minekov, D. Daskalov, architects Ivan Nikolov and Bogdan Tomalevski), the 1300 Years of Bulgaria Memorial near Shumen (1981, sculptor Krum Damyanov and team), etc. The 1877 Pleven Epopee Panorama was built to mark the 100th anniversary of the liberation of Pleven from Ottoman rule (1977, architects Ivo Petrov and Plamena Tsacheva with a team of artists under M. V. Ovechkin). Episodes from the battles for liberation of the town have been recreated in the panorama hall.

The development of physical culture and sports have necessitated the construction of sports halls, stadiums and whole sports complexes. Those intended for big sports events and shows have a universal character. To overcome the problem of supporting large expanses, steel constructions are used. Among the outstanding achievements in this field are the Palace of Sports and Culture in Varna (1968, architect S. Kunchev), the Sofia Tennis Hall in Sofia (1968, architect Stefka Georgieva), the Druzhba Sports Complex on Rila mountain (1968, architect Encho Balukchiev), the sports complexes Bonsist (1977, architects Kristian Milenov and Alexander Konakliev) and October (1977, architects T. Gospodinev and D. Savov) in Sofia, the Hristo Botev Hall (1977, architect Alexander Konakliev and team), the Palace of Winter Sports (1983, architects Alexander Konakliev, Rumen Mladjov) in Sofia, etc.

The revolutionary leap in the country's industrialization has determined the quantitative parameters and the industrial methods of housing construction. Bulgarian residential architecture after 9 September 1944 has developed from separate buildings to large residential complexes, from small houses to high-rise blocks of flats, to better planning and technical equipment of houses, from ordinary masons' building methods to industrial ones. This development is reflected in modern housing construction in the capital, Sofia, and elsewhere in the country. The mass use of large prefabs, sliding shuttering and other industrial building methods and the necessity to overcome the housing shortage in the country quickly, created a danger of uniformity and conventionality. The residential complexes designed and built after 1970 show a tendency towards the use of smaller structural units (quarters), the centralized location of public utilities has been replaced by a linear one (along pedestrian zones), and efforts are made to isolate the automobile traffic. The use of compositional elements from Revival residential architecture has led to the creation of picturesque ensembles in Veliko Turnovo – the Triangle residential complex, in Plovdiv – the Trakia complex, in Smolyan – residential groups in the centre, and elsewhere. Together with the new housing construction, the existing residential quarters in the central parts of many towns have been renovated. This renovation is connected with the problem of the future of urban town-planning and the architectural heritage and explains the desire to preserve the existing streets, groups of buildings, individual structures and entire complexes. Successful examples in this respect are the National Revival ensembles, the old part of Plovdiv (which was awarded a European gold medal in 1979 for the preservation of cultural monuments), the Varosha quarter in Lovech, quarters built at the beginning of this century in Sofia, Shumen, Russe, Vidin and elsewhere. The old houses have been reconstructed and adapted to the contemporary way of life, with some of them turned into catering establishments or museums. But many are still occupied and function as living organisms.

The rapid growth of international tourism called for intensive construction of resort facilities. Bulgaria was the first in Europe to build whole resort complexes, designed and built as independent economic settlement units for short and long vacations. The varied landscape and good climate, the beautiful and easily accessible mountains, the warm and calm sea and long sunny summers offer ideal opportunities for building modern resort complexes, both at the seaside and in the mountains. The first resort complex built in Bulgaria was Druzhba, the construction of which began in 1949–50 (architect Georgi Ganev and team). The Frederic Joliot-Curie International Home for Scientists is located there. A number of resorts sprang up along the Black Sea coast within a short time: Zlatni Pyassatsi (architect Georgi Ganev and team), Slunchev Bryag (architect Nikola Nikolov and team), Albena (architect Nikola Nenov and team) and Russalka (architect Marin Marinov and team); also developed were Pamporovo resort in the Rhodope mountains and Borovets in the Rila mountains. These resort complexes include hotels, holiday houses, bungalows, camp-sites, services, shops, polyclinics, restaurants, night clubs, casinos, bars, etc. Particular attention is devoted to sports facilities. Having begun on a small scale, resort construction later passed through the stage of large and tall buildings; then purism was abandoned, in favour of broken-up forms.

Up to 9 September 1944 there were about 4,001 industrial enterprises in Bulgaria, most of them small and poorly equipped. The real development of industrial construction began after 1944. The first specialized designers' bureaux for industrial projects were set up in 1948 and the first large industrial enterprises were built with Soviet assistance during the 1950s. A concentration of capital investments began after 1956 and priority was given to the construction of heavy-industry facilities, while preserving the high rates in the other branches of industry. Complete industrial complexes were formed and the principles of the general planning of industrial territories were mastered. Industrial architecture was subordinated to technological requirements and developed according to the ensemble principle and under the influence of large volumes and forms. The construction of compound industrial facilities began, the relative share of metal constructions increased and new, effective materials were introduced. Cases in point are the Leonid Ilyich Brezhnev Metallurgical Works, the petrochemical plant in Burgas, the Maritsa-East state mining enterprise, the Kozlodui atomic power station, the plants for heavy investment machine-building in Russe and Radomir, and electronic engineering, light and food industry enterprises.

Many facilities have been designed and built by Bulgarian civil engineering organizations in more than 40 countries of Europe, Africa, Asia and Latin America. Bulgarian designers have won major international competitions and tenders, including urban development projects in Tunis (1960, a team under architect Lyuben Tonev), a project for the centre of Berlin (1962, a team under architect Vladimir Romanski), a project for the centre of Karlsruhe in the FRG (1963, architect Evgeni Zidarov and team), and a project for a sports complex in Tunis

The monument *1300 Years of Bulgaria,* by Krum Damyanov, in Shumen

(1964, architect Chudomir Pavlov and team). The design for the National Theatre in Lagos, Nigeria (1973, architect Stefan Kolchev and team) brought wide recognition for the talent of the Bulgarian architects.

The theoretical problems in the field of architecture are tackled at the Institute for the Theory and History of Town Planning and Architecture of the Bulgarian Academy of Science, the Faculty of Architecture of the Higher Education of Architecture and Civil Engineering, and many departmental research institutes, centres and sections. Eminent architects with rich professional experience are involved in these studies.

The Bulgarian architects are united in a voluntary principle in the Union of Bulgarian Architects. It was founded in 1965 as an independent public organization. Under its leadership is the Studio of Young Architects, set up in 1977 for creative work and the qualification of youth. The main objective of the Union of Bulgarian Architects is to render competent, timely and active assistance to the Party and the Government, and to the departments and institutes in solving basic town-planning architectural and building problems; to enhance the prestige of architects; to develop the theory of modern architecture and architectural criticism; and to work for the improvement of the professional mastery of its members. The Union of Bulgarian Architects is a member of the International Union of Architects and maintains contacts with similar organizations all over the world. The Union has organized International biennials of architecture (1981 and 1983).

The Tombul Mosque in Shumen

MUSIC

The written tradition in Bulgarian music started in the ninth century with translations of Greek liturgical chants into Old Bulgarian by the two Slavonic 'enlighteners', the brothers Cyril and Methodius (826/7–869; c. 815–885) and existed throughout the centuries alongside the folk tradition. Until the appearance of the secular (urban) musical culture of the Bulgarian Revival in the middle of the nineteenth century, Bulgarian music was mainly vocal and monodic. In genre and style it belonged to the musical tradition of the Byzantine-Slavonic cultural community. The two brothers are assumed to have made a creative selection by referring to Byzantine, Gregorian and Slavonic chants when compiling the Slavonic Liturgy (the Byzantine and Gregorian chants were based on the early Christian model). The result was the formation of a new, Slavonic branch of the Eastern Orthodox musical tradition. A large number of ninth–tenth century hymnic compositions have been preserved, and the Bulgarian and early Russian musical monuments are a proof of the notational and musical professional practice, originating from Old Bulgarian sources.

The socio-political and socio-cultural changes in Bulgaria after it had fallen under Byzantine rule in 1018 brought about changes in its church singing. Bulgaria broke away from Byzantine musical practice. Musical manuscripts from the twelfth–fourteenth centuries testify to the development of church singing during this period. There are seventeen manuscripts written in what is called theta-notation, related to a strong oral tradition. It is named after its basic sign – *theta*, which represents a coded melodic stereotype. The musical significance of this sign has not yet been decoded. The Minaios of Dragan, a Trephologion of Zograph from the latter half of the thirteenth century, is of particular interest. It contains seven notated chants, and sign combinations of this kind have not been found in Byzantine manuscripts.

At the end of the thirteenth century and the beginning of the fourteenth centuries after the transition from the Tyticon of the Studios Monastery to the Tyticon of Jerusalem, some changes took place in church music to bring it into accord with the new liturgical practice. Byzantine sources ascribe major participation in those changes to the Bulgarian Yoan Kukuzel, one of the most prominent singers, musicians and theoreticians of his time. His creative work was substantial; manuscript copies of his compositions exist from right up to the nineteenth century. His famous Hieronymic chant (the earliest copy dating from 1336) contains melodic formulae which were the basis for religious chants, and he is the most likely author of the chants marked 'Bulgarian' in Byzantine musical manuscripts from the fourteenth–fifteenth centuries. It is the name of his pupil Dimitrios Dokeanos which is associated with the chant (*kratema*) bearing the unique inscription 'imitation of a Bulgarian lamentation', which has an intonation structure similar to that of Bulgarian epic folk songs. The large reformatory work of the renowned writer and patriarch Euthymius of Turnovo (c. 1327–c. 1401/2) is supposed to have included official liturgical music as well. Proof of this is found in a musical monument of that time – the Palauzov copy of the Synodikon of Boril – which is considered to have been edited by Euthymius. Its four musical texts accompanying Greek doxologies are written in the new, late-Byzantine notation, introduced after the reforms made by Kukuzel and his contemporaries at the beginning of the fourteenth century. The reform carried out by the Turnovo School of Learning gave a strong impetus to the development of the art of church singing in the Eastern Orthodox countries Serbia, Moldavia and Russia.

The first Bulgarian instrumental ensemble, founded by Mihai Safran in Shumen in 1851

Between the end of the sixteenth and the eighteenth centuries, the South-western Russian hierologies contained chants inscribed *Bolgarskii Rospev*, *Bolgarskoe Penie*, *Tropari napelu Bolgarskogo* (Bulgarian chants), etc. They

Dobri Hristov

Petko Stainov

Georgi Atanassov – the Master

Lyubomir Pipkov

are written in the Kievan staff notation, and some of them, although rarely, in sematic notation. They are considered to have originated in Bulgaria and existed in an oral form. The practice of the 'Bulgarian' chants discloses a simplified compositional and structural system, which supposedly took shape during their oral transmission. There are three hierologies inscribed chiefly *Bolgarskii Rospev*, which give reasons to believe that there was a local school, and investigations show that such a school did exist at the Monastery of Skit Mare near the village of Manyava, today the Ivan Franco district of the Ukrainian SSR. There are no Bulgarian manuscripts dating from the fifteenth–eighteenth centuries in late-Byzantine notation. The manuscripts in Bulgarian libraries, which were meant to be sung and contain singing instructions for voice, genre and stylistic features, etc. are not notated. This also gives grounds for believing that the art of church music in the Bulgarian language existed mainly in the oral form.

During the period of the Bulgarian Revival (eighteenth century–1878) the Chant School of the Rila Monastery acquired new musical models. In the 1820s the Modern Greek notation system of *Churmuzios-Chrysanthos*, established in 1814, was introduced, and Neophyte of Rila (1793–1881) – a man of great learning, Averkii Popstoyanov (1803–81) and others were engaged on its incorporation into Bulgarian liturgical practice. They succeeded in creating a written singing practice in Bulgarian for the entire year-round liturgical cycle.

The second half of the nineteenth century saw the creation of a new, secular (urban) musical culture. New forms of music-making were sought, to reflect the new outlook on life. The first Bulgarian school songs were composed towards the middle of the century, and different strata of town populations created what is now called town folklore. 1851 saw the formation of the first instrumental ensemble in Shumen, led by the Hungarian musician Mihai Safran. Their first concerts included only foreign compositions, but gradually they began to perform Bulgarian town folklore tunes as well. In 1861 the writer Dobri Voinikov (1833–78) became their conductor, and they also performed his own works, written in the style of European music. The foundations of polyphonic choral singing in Bulgaria were laid by Dobri Voinikov's school choir in Shumen (1859) and by the church choir of Yanko Mustakov (1842–81) in Svishtov (1868), which also participated in secular events. In 1870 Todor Hadjistanchev, a schoolmaster, founded a Bulgarian Church Singing Society in Russe, which became active in the events of the Bulgarian reading clubs and theatre, as well as in the struggle for church independence and political liberation. The greater number of the participants were also members of the local revolutionary committee. The Society's choir sang in church in the Bulgarian language; it also gave theatrical performances.

Bulgaria's liberation from Ottoman rule in 1878 gave impetus to the development of professional secular musical culture, whose foundations were laid by compositions in the choral genre. The reasons were the choral singing tradition and the constant formation of amateur choirs and music societies in need of repertory, as well as the composers' preference for songs, since they had not completely mastered the vehicles of musical expression. The choral works were pervaded with intonations from folk songs and town folklore; the harmonic code was that of classical harmony, and the subject matter was multifarious – domestic and lyrical, humorous, patriotic, proletarian. Emanuil Manolov (1860–1902) is considered the first Bulgarian professional composer, whose greatest achievements are in choral and children's songs. In most of his choral works the folk tunes are only harmonized, but in *What a Girl I Saw*, which belongs to Bulgarian choral classics, Emanuil Manolov composed the first original, Bulgarian-sounding tune. Dobri Hristov (1875–1941), composer, student of musical folklore, publicist, conductor, teacher and public figure, made a multi-faceted contribution to the formation of Bulgarian musical culture. By recreating thematic material from different musical folkloric fields, he created the Bulgarian national choral song. His creative style has its disciples up to the present day. Other composers who have contributed to the choral genre are Angel Bukoreshtliev (1870–1950), Alexander Krustev (1879–1945), Alexander Morfov (1880–1934) and Panayot Pipkov (1871–1942). The first Bulgarian opera was *Poor Woman*, by Emanuil Manolov, and was first performed in 1900. The foundations of composing for the stage were laid by Maestro Georgi Atanassov (1882–1931). In his first operas he employs simple musical means of expression, based on tunes from national musical folklore, but in his later works he uses a more elaborate language, employing the leitmotif compositional technique. The first steps in composing operettas were connected with amateur musical and theatrical activities in schools at the beginning of the twentieth century. Maestro Georgi Atanassov, Panayot Pipkov and others wrote children's operettas specifically for such activities. Some pieces for wind instruments by Emanuil Manolov, Panayot Pipkov and others are considered to be the first Bulgarian orchestral works, and in 1912 Nikola Atanassov (1886–1969) wrote the first Bulgarian symphony. The first works for solo instruments were harmonizations of folk songs and dances; dance music was also written in the form of waltzes, polkas, mazurkas, etc.

The October Revolution of 1917 in Russia, the Antifascist September 1923 Uprising in Bulgaria, the revolutionary worker's movement and the armed anti-fascist struggle in 1941–44 were important influences on the ideological and thematic range of musical compositions in the years between the mid-1930s and 9 September 1944. The main creative task of Bulgarian composers was the creation of a national musical style and national school of composition, by achieving Bulgarian-sounding intonational and stylistic unity, by mastering the principles of musical dramaturgy in the symphonic and chamber genres – as well as music for the stage, and by employing contemporary vehicles of expression for the problems of the day. This enormous task was achieved by the composers Pancho Vladigerov (1899–1978), Petko Stainov (1890–1972), Vesselin Stoyanov (1902–69), Lyubomir Pipkov (1904–74), Marin Goleminov (b. 1908), Filip Kutev (1903–83), Parashkev Hadjiev (b. 1912), Svetoslav Obretenov (1909–55), Dimiter Nenov (1901–53), Boyan Ikonomov (1900–73) and others. As members of the Modern Music Society, founded in 1933 and presided over by Petko Stainov, they laid the foundations of the Bulgarian national school of composition. Most of them made considerable contributions to choral music. The harmonic vein of the most prominent works drew on the modal and melodic structures of the respective regional folk songs. The ballads of Petko Stainov and Svetoslav Obretenov contributed in a new and important way with their subject matter, musical form and roots in the national musical tradition. The composers of symphony music were mastering the different genre forms in order to express Bulgarian ideas and imagery. For the most part, symphony music was of a programme nature; it praised the beauty of the Bulgarian land and re-created features of the Bulgarian spiritual character. The fragmentary nature of the orchestral pieces from the previous period was gradually overcome to achieve a neat musical structure through a much richer compositional technique. Symphonies, instrumental concertos, overtures, suites and

Pancho Vladigerov

Ivan Vulpe

The Svetoslav Obretenov Bulgarian a cappella choir

tone poems were written. Highlights in the national music include Pancho Vladigerov's rhapsody *Vardar* and the Third Piano Concerto, Petko Stainov's *Thracian Dances* suite and Vesselin Stoyanov's *Bai Ganyo* suite. Lyubomir Pipkov's tone poem *The Wedding* and Svetoslav Obretenov's cantata *At Parting* led the way to the intensive development of the oratorio-cantata genre in later years. The beginnings of instrumental chamber composition were laid by Lyubomir Pipkov and Pancho Vladigerov, the latter making a major contribution to piano, violin and cello music for solo instruments. The operas of Pancho Vladigerov, Lyubomir Pipkov and Vesselin Stoyanov quickly became highlights in the genre and were proofs of its maturity. The 1930s saw the first attempt in ballet composition – *Yana and the Dragon* by Hristo Manolov (1900–53). Operetta composing is scantily developed, though some operettas, with themes of domestic idylls, have started a promising trend in this genre. Bulgarian 'pop music' is almost entirely imitative.

The foundations of musical stage performances were laid at the end of the nineteenth century. In general the origin and early development of Bulgarian operatic theatre followed that of most other theatres with a short history (for instance, the Balkan operatic theatres). Its forerunners were the visits of touring theatres, mainly Italian, as well as some events typical of musical culture in towns. During the first period of initial growth and strengthening of Bulgarian opera, from the late nineteenth century until 1921, events were concentrated entirely in the capital city of Sofia. The development of cultural life in the country at the end of the nineteenth century permitted the creation of new contributions by the Operatic Division of the Metropolitan Drama and Opera Company, founded by Dragomir Kazakov (1866–1948). The first performance was given on 2 January 1891 and consisted of excerpts from famous operas. In September 1892 the Operatic Division ceased to exist, but on 18 October 1908 the Bulgarian Operatic Society was inaugurated in Sofia. Among its founders were the singers Kiril Mihailov-Stoyan (1853–1914), Ivan Vulpe (1876–1929), Bogdana Gyuzeleva-Vulpe (1878–1932), Dragomir Kazakov; the conductors Heinrich Wisner (1864–1951), Todor Hadjiev (1881–1956); and the choirmasters Konstantin Ramadanov (1875–1961) and Dobri Hristov. The members of the Bulgarian Operatic Society and many an eminent public figure, supporters of opera, fought to overcome the severe material and moral obstacles. There

Assen Russkov

Hristina Morfova

was a lack of well-trained local cadres, especially stage directors. For nearly fifteen years of almost self-financed existence the Bulgarian Opera Society performed mainly newly-written Bulgarian operas and works of undoubted artistic value by Slavonic and Western European composers. Some of the characteristic features of the national operatic theatre appeared, a type of organization most pertinent for Bulgarian conditions was established, facilitating compliance with the requirements of realism in musical comedy and there was a constant striving for balance between singing and acting. The nationalization of the Bulgarian Operatic Society in 1921 (named the People's Opera since 1922) brought an end to the most difficult period in the development of the Bulgarian operatic theatre, in that it ended the struggle for its very existence, aggravated by the lack of organization and the inherent contradictions within it. The character of Bulgarian operatic theatre was determined by the works performed by the People's Opera in Sofia – Giacomo Puccini, Giuseppe Verdi, Richard Wagner, Jacques Offenbach, Alexander Borodin, Pyotr Tchaikovsky, Nikolai Rimsky-Korsakov, Mikhail Glinka, Bedrich Smetana, Pancho Vladigerov, Lyubomir Pipkov, etc. The mid-1930s witnessed the emergence of top-rank soloists Peter Raichev (1887–1960), Kristina Morfova (1889–1936), Maria Zolotovich (1898–1983), Ana Todorova (1892–1973), Subcho Subev (1899–1950), Mihail Popov (1899–1978), Hristo Brumbarov (1905–74), Tsvetana Tabakova (1905–36), Katya Spiridonova (1902), Lyuben Minchev (1904–59) and others. Among the eminent conductors were Moisei Zlatin (1882–1953), Assen Naidenov (b. 1899) and Assen Dimitrov (1894–1960), and among stage directors, Hristo Popov (1891–1970), Dragan Kurdjiev (1896–1968), Hristan Tsankov (1890–1971), Yuri Yakovlev (1888–1936) and Boyan Danovski (1899–1976). Musicians directed their efforts towards the establishment of operatic theatres in the provinces; thus Stara Zagora saw the setting up of an operatic company in 1925, and Varna saw performances by an operatic company (1928–29) and by Communal Opera (1929–31). 1918 saw the formation of the first professional operetta company, led by Angel Sladkarov, whose première of Dellinger's *Marquis Bonelli* laid the foundations of professional operetta theatre in Bulgaria. The first permanent operetta theatre was the Free Theatre

A scene from Parashkev Hadjiev's opera *Madcap*, Sofia National Opera, 1975

Mimi Balkanska

The Dimov Quartet

in Sofia (1918–24), with Peter Stoichev (1879–1945) as Director. The theatre was purpose-built, with up-to-date stage conditions and an auditorium seating 1,200 people. The opening night was on 5 December 1918, with Emmerich Kálmán's *The Gipsy Princess*. The main centre of activity in 1920–22 was the Rennaissance operetta theatre in Sofia. Despite the artistic achievements, the working conditions in the theatre were constantly worsening, with salary reduction and personnel dismissals. In defence of their labour interests, in 1921 the theatre went on the first strike of actors and stage workers in the country, led by the General Federation of Trade Unions. On 17 December 1922 the Co-operative Theatre opened in Sofia; this was the first artistic co-operative society on an equal, democratic footing in the world, a fact that was of decisive importance for the historical development of the national operetta theatre. Among the founding co-operators were the eminent operetta singers Mimi Balkanska (b. 1902), Assen Russkov (1896–1958), Angel Sladkarov and Vera Sulplieva (1903–78), with Stoil Stoilov (1893–1944) as Director, and Iliya Stoyanov Chancheto (1892–1975) as Conductor. The season 1933–34 saw the first performances in the private, Odeon theatre, owned by the actor Stefan Penchev (b. 1898), which became the third most important operetta in Bulgaria before 9 September 1944, and employed the most prominent operetta singers of the capital. The first period in the history of operetta theatre in Bulgaria was the setting up, development and consolidation of the operetta genre as a professional art-form offering great opportunities for comprehensive display of

The Sofia Philharmonic Orchestra

Scene from Alexander Raichev's ballet *Haiduk Song*. Sofia National Opera, 1976

Elena Nikolai

the actor's talent and having a broadly educative influence on the audience. 1928 saw the beginning of Bulgarian ballet with the performance of *Coppelia* by Leo Delibes, staged by Anastas Petrov (1899–1978), ballet-master, stage director and dancer.

Choral singing also developed. Distinguished ensembles were assembled, including the Choral Society of Plovdiv, the *Rodna Pesen* (Native Song), the *Gusla* (Rebeck), the *Kaval* (Shepherd's Pipe) and the Sofia Women Teachers' Choir. In 1926, the Union of Bulgarian Popular Choirs was set up. Among the eminent concert performers were the pianists Dimiter Nenov (1901–53), Tamara Yankova (1907–80), Donka Kurteva (b. 1906), Lyuba Encheva (b. 1914) and Otto Libich (1912–60); the violinists Nedyalka Simeonova (1901–59), Vladimir Avramov (b. 1909) and Vassil Chernaev (1912–75); the cellist Konstantin Popov (b. 1904); and the contrabass player Assen Vapordjiev (b. 1901). The development of chamber and symphony music stimulated the setting up of instrumental ensembles; the 1920s and 1930s saw the formation of the Bulgarian Popular Philharmonic, the Academic Symphony Orchestra, the Royal Military Symphony Orchestra, the Lechev Quartet and the Bulgarian String Quartet (later the Avramov Radio Quartet). Musical education in Bulgaria started in 1904 with the foundation of a private school of music in Sofia. In 1912 the school was nationalized, and in 1921 it was turned into a State Academy of Music, today's Bulgarian State Conservatoire.

9 September 1944 promoted the development of all spheres of the nation's musical culture. The relationship of music to life, its democratic nature and its popularity among the broadest circles of listeners show its new social function. The interrelationship with Soviet music widens the range of themes, ideas, intonations and genres in Bulgarian music. The broader employment of modern vehicles of musical expression was stimulated by the decision of the 1956 April Plenum Session of the Central Committee of the BCP. The musical works are dedicated to the motherland and to the BCP, they echo moments of the heroic past of the Bulgarian people, Bulgarian-Soviet friendship, socialist construction, the inner life of modern man, etc. The significance of the folk-song as a source of tunes has not diminished; it is mostly the typical elements and motifs that are taken over. The melodies of town folklore, workers' and resistance fighters' songs and of Soviet popular marching songs are also represented; certain cross-national melodic sources have been domesticated. Genres novel to Bulgarian music have also achieved popularity. Composition techniques have grown in diversity. Not merely were the traditions of the national music reincarnated, but the process of radical renovation of musical language became particularly fruitful after 1956. In the 1940s and 1950s, the composers of the older generation were in their creative maturity and composed

Todor Mazarov

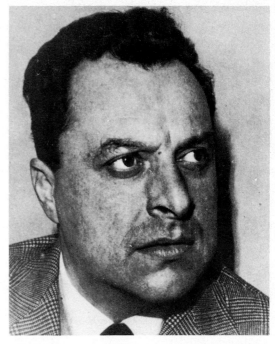
Boris Hristov

some of their most impressive works. Most of them expanded their ideological and thematic range, while maintaining, in varying degrees, the distinctive traits of their personal style. Alexander Raichev (b. 1922), Todor Popov (b. 1921), Dimiter Petkov (b. 1919), Konstantin Iliev (b. 1924), Lazar Nikolov (b. 1922) and other talented composers began composing in the 1940s. Composers of several generations, with varying preferences as to theme, genre, musical techniques, etc., were working simultaneously, which resulted in the richness of contemporary music and maintained its close links with musical traditions. Shortly after 9 September 1944 a new genre of choral singing emerged in the form of the popular marching songs, into which the new themes were woven through the influence of the national vocal melodies of Soviet revolutionary songs and popular marches. This grew to be a substantial genre in music and affected all other genres. Popular marches were written by almost all outstanding composers.

By instituting the State Ensemble for Folk Songs and Dances Filip Kutev pioneered music-making for folk choirs. The traditions of the concert works for academic choirs continue and are evolving further. Songs for young pioneers' choirs form a representative share of the national choral music heritage. Notable masters of stature in all varieties of choral singing are Todor Popov (b. 1921) and Dimiter Petkov (b. 1919). Nowadays, choral music is among the most specific and original phenomena in Bulgarian music. The high standard of Bulgarian choral singing is another feature that has spurred on its intensive development. The cantata and oratorio genres have entered a period of maturity. Regardless of the composition techniques used, the works in these genres feature contemporary socialist subject-matter, a distinctly Bulgarian sound and a marked affinity with the democratic vein. The first outstanding cantata was *The Ninth of September,* by Filip Kutev (1946) and the first oratorio, *Partisans,* by Svetoslav Obretenov (1949).

In the first years after 1944 the fruitful trends of the preceding period were continued in symphony music. Epic symphonies predominated; the works synthesized the victorious revolutionary upsurge of the times. After 1956, there was a desire to embody the modern world view, expressed in the quest of the appropriate musical means of expression. Symphony No. 2 (*The New Prometheus,* 1958) by Alexander Raichev shows, besides certain marks of the epic symphony, altogether modern architectonics. In the 1970s, the symphony compositions of some masters displayed a propensity to lyrical, psychological meditations. The concerto for orchestra is a new genre in Bulgarian music, which combines the fruitful traditions of national music with a modern musical language. Chamber

Nikolai Gyuzelev as Mephistopheles in Gounod's *Faust*

Ghena Dimitrova in the title role of Puccini's *Tosca*. Sofia National Opera, 1981

music, which has followed the trends of the symphony, has also had substantial successes. Among the new arrivals are chamber works for folk instruments.

In the 1960s, and especially the 1970s, 'pop music' has set in a distinctive mould. Certain works have tended to interpret the modern theme or to absorb elements of Bulgarian national music, for example the music of Toncho Russev (b. 1931), Boris Karadimchev (b. 1933), Angel Zaberski (b. 1931) and Villi Kazassyan (b. 1934). The political song was also ushered in as another sociocultural phenomenon. Operatic compositions treat historical, domestic and modern themes. The first Bulgarian opera on a contemporary subject was *Antigone '43* by Lyubomir Pipkov. Ballet music has been mostly geared to historical subjects; the musical and expressive means are very often close to those employed by the opera on a historical or genre subject. The first Bulgarian ballet written after 9 September 1944 – *The Haiduk Song*, by Alexander Raichev – was produced in 1953. The principles underlying light opera are democraticism and commitment to present-day topicality. Composers tend to favour the larger operatic form with an advanced musical dramaturgy. Since the 1970s, the musical has become very popular. The popular regime has ensured all necessary preconditions for the development of musical and stage performance. In 1946 Stara Zagora founded the first State Opera Theatre outside the capital. Varna followed suit in 1947, Russe in 1949, Plovdiv in 1953, Burgas in 1972 and Pleven in 1975. The first Chamber Opera House in Bulgaria was formed in the 1970s. The vocal ensembles of all theatres are recruited for the most part among professionally trained singers, graduates of the State Musical Schools, the Bulgarian State Conservatoire or the operatic schools, where they are taught to aim at the full evolution of the idea and theme in the score, with singing and acting to match.

High professionalism and rare artistic skill are exhibited on the stages of the most famous opera theatres in the world by Nikolai Ghiaurov (b. 1929), Dimiter Uzunov (b. 1922), Raina Kabaivanska (b. 1934), Ghena Dimitrova (b. 1941), Anna Tomova-Sintova, Nikola Gyuzelev (b. 1936), Nikola Nikolov (b. 1925), Alexandrina Milcheva (b. 1934) and others. A good share of the credit for the international fame of Bulgarian opera must go to the older generation of singers such as Boris Hristov (b. 1914), Elena Nikolai (b. 1905), Todor Mazarov (b. 1905), Lyuba Velichkova-Velich (b. 1916), Lyubomir Vishegonov (b. 1913) and others, who live and work outside Bulgaria and belong to the casts of world opera stages of the highest repute. Also to be cited for their part in the development of Bulgaria's operatic art are directors

Nikolai Ghiaurov as Philip II in Verdi's *Don Carlos*. Sofia National Opera, 1980

Anna Tomova-Sintova being presented with the Hristna Morfova Opera Prize in Stara Zagora

Misho Hadjimishev (b. 1914) and Svetozar Donev (b. 1933); conductors Russlan Raichev (b. 1922), Konstantin Iliev (b. 1924), Dobrin Petkov (b. 1923), Emil Glavanakov (1918–78) and others; stage-designers Assen Popov (1895–1976), Evgeni Vashchenko (1887–1979), Mariana Popova (1914–82). Soviet guest-directors Pavel Rumyantsev, Evgenii Sokovnin, Boris Pokrovsky and others have made exceptional contributions to the solution of problems of staging in Bulgaria. Opera orchestras are in most cases quite independent of symphony orchestras and form separate theatre ensembles. The choruses in the opera theatres have inherited the excellent tradition of Bulgarian choral singing and successfully perform difficult modern music, together with the acting often demanded by modern operatic repertory. Artists are also preoccupied with problems of modern life. The revival of old operas and the staging of new ones have been of permanent concern. Some 50 new Bulgarian works appeared between 1948 and 1984: Lyubomir Pipkov's operas – *Momchil* and *Antigone '43*, Parashkev Hadjiev's *Once Upon a Time*, *The Madcap*, *Albena*, *A Night in July*, *Masters*, *The Millionaire*, *The Golden Apple*, *The Knight*, *Anno Domini 893*, *Maria Desislava*, *Johannes Rex*, *Paradoxes*; by Marin Goleminov (b. 1908) – *The Golden Fowl*, *Ivailo*, *Zahari Zograph* and *Thracian Idols*; by Vesselin Stoyanov – *Sly Peter*; by Boyan Ikonomov – *Indje Voivoda*; by Konstantin Iliev – *The Boyana Master* and *The Realm of the Deer*; by Simeon Pironkov (b. 1927) – *The Good Woman of Sechuan* and *The Bright-Plumed Fowl*; by Alexander Raichev – *The Bridge*, *Alarm*, and *Khan Asparuh*; by Krassimir Kyurkchiiski (b. 1936) – *Yulla*; by Ivan Dimov – *They've Kidnapped the Consul* and *The Émigrée*; by Alexander Yossifov (b. 1940) – *Khan Krum of the Yuvigs*, etc. Of the classical operas staged those of Verdi, Puccini, Tchaikovsky, Bizet, Gounod, Mussorgsky, Borodin, Rossini and Donizetti are the most frequently performed. In the socialist period the Stefan Makedonski Musical Theare in Sofia has become the leading theatre for operetta. It was founded in 1948 and was so named, in 1952, after the singer Stefan Makedonski (1885–1952). Succeeding to and investing with a new meaning the best traditions of the past, the theatre of operetta is being gradually transformed. Its artists seek to transcend the conventional bounds of stock repertory and obsolescent aesthetic concepts of the art of light opera. It has been the tradition of the musical theatre to stage one Bulgarian opus a season – the light operas *Aika*, *Dilyana*, and *Cyrano de Bergerac* by Parashkev Hadjiev; *The Maestro's Green Years*, *Madam Saint-Jean*, *The Lucky One*, *House of Cards* and *Encounters* by Victor Raichev; *Song of Rebellion* by Georgi Zlatev-Cherkin (1905–77); *Bulgarians of Olden Times* by

Raina Kabaivanska as Queen Elizabeth in Verdi's *Don Carlos*. Sofia National Opera, 1980

Scene from Alexander Raichev's opera *Khan Asparuh*. Sofia National Opera, 1983

Assen Karastoyanov; *The Golden Fowl* by Marin Goleminov; *Review with Smiles* and *Duel* by Dimiter Vulchev; *Restless Hearts* by Dimiter Petkov; *The Straw Hat* by Peter Stupel; the musical drama *The Stone Pile* by Alexander Tanev; the musicals *The Girl I Loved* and *It's a Small World* by Jules Levi (1930); *A Time to Love* and *Twelfth Night* by Dimiter Vulchev; *Careerists* by Parashkev Hadjiev; *The Golden Turnip* by Peter Stupel; *The Orchid's Nightingale* by Alexander Raichev, etc. Soviet light operas, musicals by Western composers and classical operas have also been successfully staged. With the eminent soloists of the pre-1944 period Mimi Balkanska, Assen Russkov, Tinka Kraeva, etc. there also shine the stars of the younger generation – Vidin Daskalov (b. 1929), Liliana Kissyova (b. 1928), Minko Bossev (b. 1927), Liliana Koshlukova (b. 1925), etc. Among the promoters and developers of operettas are directors Hristan Tsankov, Georgi Markov, Svetozar Donev; and conductors Rossitsa Batalova, Boris Leviev, Iliya Stoyanov, Viktor Raichev, etc.

The socialist society also creates favourable conditions for the development of ballet. Ballet-dancing ensembles have been formed at the light opera theatres in Stara Zagora, Varna, Russe, Plovdiv, Burgas and Pleven. Notable representatives of contemporary Bulgarian ballet are Vera Kirova (b. 1940), Krassimira Koldamova (b. 1938), Kalina Bogoeva (b. 1937) and Tsveta Djumalieva (b. 1938). The *Arabesque* group, formed in 1967, stages mostly modern ballets. The ballet repertory includes the Bulgarian ballets: *Haiduk Song* and *The Fountain of the White-Limbed Maid* by Alexander Raichev, *Kurdjali* by Ivan Dimov, *Orpheus and Rhodopa* by Tsvetan Tsvetanov, *Legend of the Lake* by Pancho Vladigerov, *The Daughter of Kaloyan* by Marin Goleminov and others.

Eminent musical performers are the instrumentalists Boyan Lechev (b. 1926), Stoika Milanova, Mincho Minchev, Georgi Badev (b. 1939), Martha Deyanova, Peter Radev, Simeon Shterev; the conductors Assen Naidenov (b. 1899), Dobrin Petkov (b. 1923), Konstantin Iliev (b. 1924), Vassil Stefanov (b. 1913), Emil Chakurov and the choir conductors Vassil Arnaudov (b. 1933), Georgi Robev (b. 1934) and Mihail Milkov (b. 1923).

In Bulgaria there are ten symphony orchestras, eight opera theatres, one operetta theatre and a number of state choir and dancing ensembles. The Bulgarian a cappella choir *Svetoslav Obretenov* is the country's first professional mixed choir; it was founded on 4 November 1944 at Radio Sofia. The members of the choir are professional vocalists or choral singers of solid singing experience in the amateur choirs of *Gusla, Kaval*, etc. Besides its chief function – singing on Radio Sofia, the choir also gives regular concerts. Its repertory is mostly Bulgarian choral songs. Almost all the major works by contemporary Bulgarian composers have been sung for the first time by this a cappella choir; the first Bulgarian cantata *The Ninth of September* by Filip Kutev, the first Bulgarian oratorio *Partisans* by Svetoslav Obretenov and other works by Vesselin Stoyanov, Georgi Zlatev-Cherkin, Lyubomir Pipkov, Alexander Raichev, Dimiter Tupkov, etc.

Scene from Dimiter Vulchev's operetta *Duel*. Sofia State Music Theatre, 1977

Violinist Stoika Milanova at a musical poetry performance of verse by Dimiter Metodiev

The State Folk-Song and Dance Ensemble, founded in 1951, which now bears the name of its founder and chief art director, Filip Kutev was the first of its kind formed in Bulgaria. After his death in 1984, he was succeeded as conductor by Mikhail Bukureshliev. The chief choreographer and director of the dancing ensemble is Margarita Dikova (b. 1916). The ensemble's repertory in the main consists of Bulgarian folklore material, appropriately arranged, presenting the folklore heritage of various musical regions. The musical arrangements are those of Filip Kutev and eminent composers such as Krassimir

The Pirin Folk Song and Dance Ensemble

Music

Violinist Mincho Minchev

Filip Kutev

Kyurkchiiski (b. 1936), Dimiter Hristov and others. The State Ensemble sets a high artistic standard for folk choirs and ensembles, the number of which has markedly increased since 1944, when the State Ensembles for Folk songs and Dances – Rhodopa, Pirin, Trakia and others – were founded.

The network of musical education includes the Bulgarian State Conservatoire in Sofia, the Higher Pedagogical Institute for Music in Plovdiv, five musical schools and many amateur societies. A great number of Bulgaria's composers, performers, music critics and teachers are

Filip Kutev State Folk Song and Dance Ensemble

Scene from *Moods*, performed by the Arabesque Ballet Company, 1982

Scene from *Ballet Pieces*. Russe National Opera, 1983

graduates of the Bulgarian State Conservatoire. The Conservatoire students give concerts, instruct amateur groups, and organize lecture cycles with concerts and talks on music. In 1964 the Learners' Opera Theatre came into being. Among the outstanding teachers are professors Vladimir Avramov, Alexander Neinski (b. 1920), Sava Dimitrov (b. 1919), Lyubomir Pipkov, Marin Goleminov, Benzion Elizer, Dimiter Sagaev (b. 1915), Pencho Stoyanov (b. 1931), Georgi Zlatev-Cherkin, Hristo Brumbarov, Dimiter Russkov (b. 1925), Assen Diamandiev (b. 1915) and many others. In addition to music teachers, folk singers and instrumentalists the Higher Pedagogical Institute of Music in Plovdiv trains leaders of choirs and orchestras in the local style, directors and leaders of dance groups and the like. Folk singers and instrumentalists are trained at the secondary schools of folk music in Kotel and Shiroka Luka (in the Smolyan district).

A key role in the musical life of Bulgaria is played by the Union of Bulgarian Composers and the Union of

Performance at the Eighth International Ballet Competition, Varna

Bulgarian Musicians. The basis of the Union of Bulgarian Composers was laid in 1933 by the Contemporary Music Society. The Union has affiliated bodies, the sections 'Composers' and 'Musicologists', the Commission for work in the field of genres, the Council of the Board's Delegates for work with the districts, and standing commissions for the consideration and adoption of new works. The Studio of Young Composers seeks to stimulate the creative work of young composers and musicologists. Each year the Union organizes a review of new Bulgarian music, which makes a summing up of past achievement and charts the new trends and directions in the development of musical composition. The organ of the Union of Bulgarian Composers and the Committee of Culture is the magazine *Bulgarska Musica,* founded in 1948. The Union

Performance at the Fifth International Competition for Young Opera Singers, Sofia

of Bulgarian Musicians carries on the work of the first professional union of musicians in Bulgaria, founded in 1901. The Union has affiliated sections of ballet, 'pop music', chamber music, musical and choreographic education, opera and light opera, symphony music, choir and ensemble art, and wind instruments. The Studio of the Young Musicians, of the Union of Bulgarian Musicians, engages in a wide range of activities for the musical education of Bulgaria's youth and working people.

Under the direct guidance of, and with the participation of the two creative unions, a number of international and national competitions are organized in Bulgaria; these prove most helpful for the promotion and popularization of Bulgarian musical performance. The international ballet competition is the first of its kind in the world: it has been an annual event since 1964 and is organized in Varna during the Varna Summer musical festival. It is most important in stimulating the progress and creative growth of young performers, many of whom embark on their international career after first gaining recognition in Varna. The competition demonstrates the artistic achievement of various pedagogical schools in modern choreography; it includes conferences on problems of choreography, theory, criticism, etc. Since 1961, Sofia has held annually the world's only International Competition of Young Opera Singers, within the larger framework of the international music festival – Sofia Music Weeks. The main criterion in the final classification is the competitors' maturity in singing and stage-acting. This defines the special character of the competition, which has come to command high international prestige. Prize-winning competitors achieve international fame and sing on the most famous opera stages. Other musical festivals and competitions of international renown are: the International May Choir Competition in Varna; the International Festival-March Music Days in Russe; the International Chamber Music Festival in Plovdiv; the Golden Orpheus International Pop Song Festival, the Alen Mak International Festival of Political Songs, and others.

The Institute of Musicology at the Bulgarian Academy of Science does research in the field of music. It was set up in 1948; its first Director was composer Petko Stainov, and after his death, Professor Venelin Krustev, the eminent musicologist and prolific author of books, monographs and studies on the history of Bulgarian music and on the creative work of individual composers and musicians. The institute studies basic problems of Bulgarian musical culture, elucidates topical theoretical and methodological matters, studies the historical development of the separate kinds of performers' art, musical genres, the creative record of Bulgarian composers, the problem of musical education and other questions. The institute issues the *Bulgarian Musicology* series (1977–80; since 1981 – a magazine). Since 1975 the specialized publishing house, Music, has published not only music, but also musical-theoretical and musical-historical literature.

In the history of Bulgarian music, the period after 9 September 1944 has been the most significant and most fruitful, providing a wealth of ideas and artistic achievement. The triumph of the socialist revolution removed all the barriers which were erected in the past between artists and people, between performers and audiences, and the Government has pursued with zest and vigour the task of democratizing culture and art, in which music has been given due attention. The Government has created, and continues to create, favourable conditions for the expression and development of musical talent, which Bulgaria has in abundance.

THEATRICAL ARTS

The theatre in Bulgaria came into being in the nineteenth century, when in Europe it was already an established cultural institution with a centuries-old history. The origins of Bulgarian theatre date from the 1840s and are connected with school celebrations, at which the pupils' skill in recital and in writing and performing dialogues was demonstrated. Within a very short time the theatre emerged as an independent cultural activity; in 1856 performances of complete works of drama were already being staged, with elements of stage direction, acting skill, and stage and costume design. Theatre became part of the activities of the cultural clubs, in which the entire cultural life of towns and villages was concentrated during the National Revival Period. Bulgarian theatrical activities also developed outside Bulgarian territory. The Bulgarian pupils of the Robert College and the Sultan's Medical School in Istanbul staged classical works from elsewhere in the world and from the Bulgarian dramatic canon then developing. Various theatrical activities were also launched by the Bulgarian immigrants in Braila, Romania, where Dobri Voinikov (1833–78) worked. He was the most dedicated figure in the Bulgarian National Revival theatre, and was actually the founder of national dramaturgy, theatrical criticism and theory. The Bulgarian Theatrical Society and the Bulgarian National Theatre in Braila were founded under his direction, and a company was set up which included actresses for the first time. The great Bulgarian revolutionary and poet Hristo Botev (1847–76) was also a member of the company. Voinikov also formed theatre companies in Bucharest and Gyurgevo.

The Bulgarian theatre did not develop without influence from world theatre. The leading works of drama were made accessible to the Bulgarian public both by means of translation and through Bulgarian adaptations that brought them closer to the national way of life and mentality. But the Bulgarian theatre performed its mission of propagating the ideas of the Revival Period, especially through works of the new Bulgarian drama. The first performances on the stage appeared simultaneously with the first dramas which, within a very short period – from the end of the 1850s to the 1870s – emerged as one of the most developed genres of the literature of the Revival Period and achieved thematic and stylistic variety. The first Bulgarian dramas – the plays of Dobri Voinikov and Vassil Drumev (1841–1901) – were based on plots from Bulgarian history. This interest in the past was of particular importance at that time because it was to provide the psychological basis for the process of national reawakening, the accumulation of sentiments, evaluations and knowledge, which were to lead the Bulgarians to a new consciousness. Through the overall ideological and emotional orientation of their works those authors strove to create a feeling of national pride, thus making their contribution to the process of spiritual liberation and the rejection of the slave mentality. Lyuben Karavelov (1834–79), Atanas Uzunov (1851–1907) and Bacho Kiro (1835–76), who were among the leaders, ideologists and active figures of the revolutionary movement, also contributed to the development of Bulgarian theatre. Their plays combined creative work with the artistic reinterpretation of their time, with the striving to reveal the destinies of the Bulgarian people during that period when it was deprived of national independence and was weighed down by social injustice. Their works were full of hatred for the rich, who kept the people under social bondage, and of open protest against the tyranny and violence of the Ottoman Empire. They helped the conscious resistance and defiance of the people. The first comedies *Misconceived Civilization* by Dobri Voinikov, *Malakova* by Petko Slaveikov (1827–95) and *The Bishop of Lovech* by Todor Ikonomov (1838–92), in which irony and criticism were used to promote the establishment of lasting spiritual values, were also focused on the pressing problems of the times.

The most educated people, who were in the ideological forefront of their time, took an active part in the organization of the theatre as it developed during the Revival Period and in the staging of performances. To a great extent the theatre of the Bulgarian Revival Period was the spiritual offspring of the revolutionary intelligentsia. This fact determined the basic lines of its artistic development, and from its very beginnings involved the theatre with the crucial problems of the nation's destiny. Public commitment thus became a basic principle of the theatre. Almost all the leading figures of the Bulgarian Revival theatre were prompted by the striving to express above all their attitude to the problems of their time. Sava Dobroplodni (1820–94) attributed the spiritual crisis of his time to the decline in morality and stressed the moral purity of the people. According to Dobri Voinikov the crisis was due to lack of national confidence, and he tried

Scene from Nikolai Pogodin's *Aristocrats*. Stara Zagora Drama Theatre, 1980

to restore this by resurrecting heroic scenes from Bulgaria's past on the stage. Bacho Kiro saw revolution as the only way out for the oppressed people and he considered that the theatre was a means for further stimulating the revolutionary mood.

During the National Revival Period the Bulgarians developed a taste for a new kind of theatrical performance, turning away from the stock of folklore with its stylized imagery, metaphors and synthetic structure. The public looked instead for more realistic reflections of life. The purpose of theatrical performances changed, too. They abandoned their dependence on folklore and the suggestion of the magical impact of unknown forces. The aim now was to show real life and to reveal the true greatness of the past of the Bulgarian people, and to contribute to the formation of definite public ideals.

After the Liberation from Ottoman domination in 1878 the Bulgarian theatre developed as an institution with a clearer cultural purpose and began to build up its professional and artistic structure. Between 1881 and 1883 the first state-subsidized company was formed in Plovdiv, Eastern Rumelia. The writers Konstantin Velichkov (1855–1907) and Ivan Vazov (1850–1921) were active in its formation. In a number of articles, they stressed the importance of the theatre for the spiritual development of the people and under the pressure of public opinion which they had brought about, the Regional Assembly voted for the allocation of a fund amounting to 75,000 groshes for the setting up of a theatre company. The first Bulgarian theatre director was Stefan Popov (1854–1920), a journalist. The company was recruited from participants in the amateur theatricals of the Plovdiv Printers Society, schoolboys and schoolgirls, teachers and employees. Vassil Kirkov (1870–1931) and Ivan Popov (1865–1966), who later were recognized as being the best Bulgarian actors, were accepted as apprentice actors. In June 1888, the company visited Sofia with the play *La Tour de Nesle* by Alexander Dumas *père* and stayed there. The actors erected the building of the theatre themselves: this was the wooden Osnova theatre on the site of today's National Theatre. On 4 December 1888 the newly-built theatre was opened with the play *Sidonia the Beautiful* by Zocke. That date marked the beginning of a permanent theatre company in Sofia. Its repertory included Nikolai Gogol's *The Wedding* and *The Government Inspector*, Victor Hugo's *Hernani*, and Schiller's *Kaballe und Liebe*, which gave wider scope to the performing abilities of the actors. In 1892 the company was named Sulza i Smyah (Tears and Laughter), and continued to function under that name until 1904. The activities of the company were closely linked with the creative development of the Bulgarian theatre. One of its best actors was Vassil Kirkov;

Vassil Kirkov

Krustyu Sarafov

Adriana Budevska

Radul Kaneli (1868–1913), who had graduated from a theatrical course in St. Petersburg, was director and producer. Krustyo Sarafov (1876–1952), Adriana Budevska (1878–1955), Geno Kirov (1866–1944), Hristo Ganchev (1876–1912), Vera Ignatieva (1877–1972), Atanas Kirchev (1879–1912) and Nedelcho Shturbanov (1875–1913) all made their appearance on the stage in 1899. They received the first state scholarships for studying dramatic art. Some of them studied and specialized in Moscow and St. Petersburg, others in Vienna, Paris, Berlin and Prague, under the supervision of the greatest figures of the theatre in the world at that time. In addition to their personal, high professional skill they brought to the Bulgarian theatre also a new approach to the artistic image, formed under Russian influence. The existence of the Sulza i Smyah company was pervaded throughout by its efforts to create a theatre whose importance and public mission would be equal to that of the leading cultural institutions in the country. As a result the company was renamed the Bulgarian National Theatre in 1904, and on 3 January the National Theatre building was solemnly opened; it has preserved its basic architecture even to this day. The period from 1904 to 1912 included years which were particularly fruitful for the development of Bulgarian theatrical art. The theatre produced plays from the world classical repertory – works by Shakespeare, Molière, and Schiller and plays by contemporary playwrights, such as Alexei Tolstoy, Henrik Ibsen, and Gerhardt Hauptmann, were mounted. The repertoire extended the range of Bulgarian actors and stimulated them to master new and unexplored spheres of dramatic characters and get deeper insights into man's fate. National drama also flourished during the period. Social commitment, characteristic of the National Revival drama, acquired a new meaning; under the impact of modern European drama its philosophical range was expanded and its imagery was diversified and given new colour. Some of the greatest Bulgarian poets and fiction writers became involved in the theatre; they introduced their strict artistic criteria, their vivid keenness of observation and experience of life, and the power and profundity of their lyrical expression. Faithful to the traditions of the National Revival, Ivan Vazov turned to history once again. In the dramas *On the Road to Ruin* and *Ivailo* Vazov made a straightforward comparison between the past and the present as a means of propaganda, seeking the direct suggestiveness of the tragic and memorable lessons of history. His characters are marked by idealistic romanticism and selflessness and fervently expose the moral degradation and petty ideals of the Post-Liberation bourgeois society. One of his works dealing with the past is *Outcasts*, a dramatic version of his novella of the same name. In this play Vazov depicted the life of the Bulgarian revolutionary exiles in Romania during the National Revival; he gave a vivid description of their deep love for their motherland by portraying realistic, flexible, tragic and at the same time comic characters, who gave a new and more humane interpretation of the heroic element. In plays such as *Careerists*, Vazov turned his attention to the monstrous nature of social relations; he unmasked the ugliness of careerism and the humiliating scramble for high posts and jobs. Anton Strashimirov (1872–1937) in *Vampire* and *Mother-in-law* revolted against the over-severe moral standards and against the legitimized violence and brutality of the period, which restrained the individual's aspirations, depriving him of the inalienable human right to choose whom he will marry, thus making his life completely senseless. Petko Todorov (1879–1916) showed deep insight into the national character; he probed into the human soul, seeking its instinctive reverence for freedom and beauty. This reverence for beauty is shown against the background of the highly dramatic spiritual crisis of his contemporaries; it is related to the philosophical revolt of the individualist, embodied in poetically idealistic characters, who are the incarnation of frustrated, unhappy, and at the same time searching characters, tragically incompatible with the pettiness of life, and with the gloomy daily struggle for money and bread. The summit of the national drama during this period were the dramatic works of Peyo Yavorov (1878–1914). His play *In the Foothills of Vitosha* gripped the audiences with its lyrical portrayal of human destiny, confessions of innermost feelings, analysis of social relations, social inequality and violent political struggles. Yavorov showed deep insight into the tragedy of human existence in a world of spiritual stagnation and revealed the most personal and intimate human feeling, love, as a protest against violence and brutality; love is depicted as an urge, a yearning to attain a just and more humane world, morally and socially different from that in which he lived.

During that period, too, the theatre was the hub of Bulgaria's cultural life. There were heated discussions on the ideological and aesthetic trends in drama among the most idealistic members of the intelligentsia. For Vazov, the landmark in the development of the Bulgarian theatre was the National Revival theatre with its democratic ideals and commitment to the problems of everyday life.

Scene from Molière's *Le Malade Imaginaire*. Sofia National Theatre, 1967

Scene from Peyo Yavorov's *When Thunder Strikes*. Sofia National Theatre, 1977

Dimiter Blagoev (1836–1924) and other Marxists raised the question of the influence of the theatre in the pursuit of a more politically explicit and definite social analysis. According to Pencho Slaveikov (1866–1912), the theatre was among the strongest forces of spiritual impact – a supreme cultural institution, the artistic verbal expression of the nation, a university ranking above all other

Sava Ognyanov

universities, able even to supplement an academic education since it was the school of life. The heightened interest in the theatre stimulated the development of dramatic theory and criticism. Among the authors of reviews, articles and theoretical studies on the theatre were eminent writers such as Pencho Slaveikov, Peyo Yavorov, Petko Todorov, Ivan Vazov, Anton Strashimirov; critics such as Dr Krustyo Krustev (1866–1919), scholars like Boyan Penev (1882–1927), connoisseurs of theatrical art like Ivan Andreichin (1872–1934) and Andrei Protich (1875–1959). A number of them actively participated in the life of the Bulgarian theatre. For a certain period of time Anton Strashimirov was a repertory director at the National Theatre and founded a touring company. In 1908 the director of the National Theatre was Pencho Slaveikov and the repertory director, Peyo Yavorov. Their ambition to completely reform the structure of the national theatre left a deep mark on its overall development.

From its foundation, the Bulgarian Theatre was faced with the significant problem of stage directing. Stage directors were actors. There were also foreign actors who worked in the Bulgarian theatre as stage directors: in 1900, Adam Mandrovich (1839–1912) from the Zagreb Theatre; and Joseph Shmaha (1848–1915) from the Prague Theatre. Their stage directing created a definite type of theatre where the actor was the sole carrier and interpreter of the ideas and purport of the play. This sovereignty of the actor over stage and audience and, to a certain extent, over the play itself, stimulated the process of giving free rein to the actor's own personal means of expression. It was on this basis that the first generation of unforgettable and vivid actors and actresses flourished. The Bulgarian theatre discovered and established the priority of the individual character; he became not only the focus in the structure of the performance, imposed by the all-powerful actor, but the centre of the play's ideas and intentions. The ideas of the play were suggested by the emotionality and insight of the individual, and by the incompatibility of the personal, social and moral ideals with the moral standards of the existing social system.

During this period only the National Theatre was subsidized by the State. The most talented actors were concentrated in this theatre; their artistry turned it into an original creative laboratory of Bulgarian theatre. Touring companies familiarized the rest of the country with the drama. The most important among them, the Suvremenen Teater (Modern Theatre) and the Rosa Popova Theatre, performed the humanistic and social dramas of Maxim Gorky and Henrik Ibsen. At the end of the last century, amateur theatricals were common in the clubs of the Bulgarian Workers' Social Democratic Party in almost every town and village in the country. There, they were used to promote socialist ideas, which intensified the people's awareness of social injustice and the class character of society. Not infrequently, they filled the gap created by the lack of professional theatres, and many of the great Bulgarian actors started their careers as amateur actors.

During the 1920s and 1930s the Bulgarian theatre experienced the historic large-scale transition from one kind of theatrical system to another, radically different one, where the actor yields precedence to the director. This turning-point was the result of both the development of theatrical art, which had already built up its basic artistic unit, the actor, and of the social and ideological conditions of the post-war period which followed World War I, a period marked by national tragedy and social clashes, by the elation of the artists with the ideals of the October Revolution in Russia, by the growing confidence in the people's forces and by the deeper impact of art which was trying to find its bearings amidst the spiritual and social cataclysms of the times. Though its history had been rather brief, at this decisive moment the Bulgarian theatre swung towards the innovations and discoveries of world theatre. Geo Milev (1895–1925) upheld the view of the supremacy of the director, of his right and possibility to create original works of art which obey their own specific laws. Under the influence of Reinhardt and of the theatre of expressionism, Geo Milev formulated a concept of the performance which was new for the Bulgarian theatre. Its underlying principle was the requirement for 'the spectator not to be blinded by the illusion of reality, but to be raised, in a dazzling way, to the higher and loftier reality of art'. Thus, Geo Milev gave a new meaning to the functions of the components in the theatre. He directed them to the poetic and metaphorical imagery of what was happening on the stage. In 1932 Geo Milev staged *Man and the Masses* by Ernst Toller with politically committed actors from the party theatrical companies and music ensembles. Here his innovative approach is manifested to its fullest. The innovative concept of the play's imagery has been rendered profound and meaningful by his progressive views by a new, more vigorous and more explicitly political attitude to the great

Scene from Orlin Vassilev's *Happiness*. Sofia National Theatre, 1951

issues of the day. This performance placed Geo Milev among the directors who, consciously, and guided by deep conviction, turn the stage into a rostrum from which political ideas were voiced. From 1924 to 1944, Hristan Tsankov (1890–1971) worked as a stage director at the National Theatre. In his productions he sought to achieve an organic synthesis of all arts, weaving dance, music and mime into the texture of certain plays. The greatest credit for the development and establishment of Bulgarian stage direction in the 1920s and 1930s goes to Nikolai Ossipovich Massalitinov (1880–1961), an actor from the Moscow Art Theatre. In the long years he spent as a director in the Bulgarian theatre, he assumed the difficult task of effecting the historically-determined change – of connecting the epoch of actors' theatre with that of director's theatre. Massalitinov brought home to Bulgarian actors and directors some of the principles of the theatrical reform of Konstantin Stanislavsky and Vladimir Nemirovich-Danchenko and elevated stage production, turning it into a versatile image of life. Unlike the actor's and the director's theatre, his theatre is one in which the author's philosophical and psychological interpretation of life and human fate predominate. Massalitinov deepened the realistic content in the art of the Bulgarian directors, guiding them to an analysis of the interaction between man and environment, to the portrayal of the complex relationship between the individual human fate and the spiritual and social forces of society. Deep insight into the complexity and uniqueness of human nature and a realistic stage interpretation true to nature – these are the criteria which Massalitinov promoted among Bulgarian actors. In

Konstantin Kissimov

his numerous productions of Bulgarian, Russian and West-European plays Massalitinov added new, greater depth to the realism of the Bulgarian theatre, enhancing its humanism to higher levels of potency and vigour.

Ivan Dimov

In their attempt to deepen the ideological and moral insights and to expand the range of images of theatrical art the directors were greatly assisted by the inspired work of the second generation of Bulgarian actors. Vivid, original individuals stand out among these actors, unique in terms of both intellect and performance. The most famous among them include: Ivan Dimov (1897–1965), an actor of great psychological insight and intellectual depth; Zorka Yordanova (1904–70), known for her graceful

Nikolai Massalitinov

and sophisticated acting, and the vivid portrayal of moral beauty and idealism; the amazing stage presence of Konstantin Kissimov (1897–1965) – equally earnest and convincing in the representation both of the tragic and of the comic; Vladimir Trendafilov with his compelling emotionality and explosive passions; Boris Mihailov (1893–1963) with his intellectual lucidity and analytical mind; and Georgi Stamatov (1893–1963) known for the broad strokes of his acting and his compelling stage presence.

The widening scope of philosphical and aesthetic ideas of the Bulgarian theatre and its stronger humanism can further be attributed to some new stages in the development of the national drama, characterized by a more profound analysis of life and a greater insight into its moral and social contradictions. Among the most outstanding works of this period are the comedies of Stefan Kostov (1879–1939), a direct continuator of Vazov's vigorous use of ridicule, of the angry protest against all social evils. Ignorance and the deforming effect of miserliness, the feverish craving for money and the obsequious creeping up the social ladder, the corruption and baseness – these are the butts of ridicule in Kostov's comedies *The Gold Mine*, *The Grasshoppers*, *The New Haven*, and *Vrazhalets*. In *Golemanov* the author portrays a generalized character of the political demagogue, working his way to a ministerial post at the price of humiliation and hypocrisy by machinations and bribes; in the exaggerated craving for power, the author synthesizes the moral and social deformation of bourgeois society. During this period, the national theatre conquered new territories of the Bulgarian mentality. In his plays *Albena* and *Boryana,* Yordan Yovkov (1880–1937) asserts the ethical values of the people – their moral purity, their taste for beauty and harmony, and their poetic perception of the world around them. At the same time the author reveals the stoicism with which they preserve intact their human dignity in the face of the ugliness and deformity of social injustices. In *Master* by Racho Stoyanov (1883–1951) the Bulgarian mentality has been refracted philosophically and poetically through the dramatic world of the creative personality, in the moral trials and tribulations accompanying the artist's craving for new, unknown horizons.

The revolutionary art of the political theatre constituted an inherent element of the cultural atmosphere of the 1920s and 1930s, of the movement of philosophical and aesthetic ideas, of the hectic pattern of artistic endeavour and the discovery of new artistic horizons. In 1930 a theatre for political propaganda was set up in Sofia calling itself 'Blue Shirts'. Although, owing to constant police threats and bans the troupe had to change names more than once – from 'Young Actors Theatre' through 'Tribune' and 'People's Stage', to 'Realistic Theatre', it maintained its class and political commitment. The most outstanding personality in this theatre was Boyan Danovski (1899–1976), who found the true answers to his questions as an artist and a citizen in the political theatre of Piscator, in Bertolt Brecht's aesthetics and the innova-

Scene from Brecht's *Mother Courage*. Sofia National Theatre, 1962

tions of the Soviet revolutionary theatre. Boyan Danovski's entire artistic and organizing activity in 'Tribune' and 'People's Stage' was characterized by the struggle to link acting closely with the revolutionary philosophy and aesthetics of the politically vigorous theatre. The penetration of ideas from revolutionary art into the professional theatre is an element inherent in the artistic atmosphere of the 1920s and 1930s. This process was most explicit in the actitivy of theatres headed by Georgi Kostov (1895–1961) with repertoires including mainly Soviet plays.

The Union of Actors, Musicologists and Theatrical Stage established in 1919 on the initiative of Georgi Dimitrov (1882–1949), the then Secretary of the General Workers' Trade Union, was another major factor in the ideological and artistic development of dramatic art. The Union was a kind of a pillar of democracy and social awareness of the Bulgarian actors and stage directors, a token of the increasing political character of their art.

The experimental theatres – Issak Daniel Theatre Workshop, where actors like Zorka Yordanova and Konstantin Kissimov began their careers, and the experimental troupe of Stefan Surchadjiev (1912 65), which gave the Bulgarian theatre some of its most outstanding people after the revolution, also contributed to the rich intellectual atmosphere of this period. Workshops developed at this time along with the traditional methods of teaching in drama schools. Thus Nikolai Massalitinov headed several workshops at the National Theatre, teaching Bulgarian actors high professionalism. Boyan Danovski set up a couple of workshops, one of them based entirely on Stanislavsky's principles, developed in his book *An Actor Prepares*.

The Bulgarian theatre's interest in the latest theatrical achievements on the world stage was maintained by the critics, especially by Geo Milev, Lyudmil Stoyanov (1888–1973), Sirak Skitnik (1883–1943) and Dimiter Mitov (1898–1962). At the same time, through the articles and studies of Todor Pavlov (1890–1977), Bulgarian Marxist aesthetics revealed the scope of philosophical generalization and mapped out some methodological principles on which the socialist theatre could base its realistic and poetically metaphoric, communist and humanistic art.

After 9 September 1944 the Bulgarian theatre underwent radical ideological and aesthetic, as well as organizational transformations. The socialist revolution changed the very nature of the theatre as a social institution. It developed in a cultural context where the theatre was no longer an art for the élite and a class monopoly, in a cultural system created to build a new relationship between man and culture.

The first expression of this new context was the widely-deployed process of bringing the mass of the people closer

Scene from Stefan Kostov's *Golemanov*. Sofia National Theatre, 1977

Boyan Danovski

where there are also two genre companies, the 'National Youth Theatre' with a special section for children's and teenagers' theatre, formed in 1945, and the 'State Satirical Theatre', formed in 1957. Every theatre has a separate small stage and some companies have stages given over entirely to young actors. The first state-financed puppet theatre was set up in 1946. Today, there are in the country eighteen state puppet theatres which have performed in many countries abroad. The radio and TV theatres also contribute to the development of drama. The Higher Institute of Theatrical Art, known as VITIZ, was opened in 1948. It trains actors, puppeteers, stage directors, drama and film critics and specialists. From its very foundation, the teaching of practical and theoretical disciplines was taken up by prominent actors, directors, drama specialists and critics. Immediately after the socialist revolution a group of Bulgarian producers graduated from the Moscow and Leningrad Higher Institutes of Drama. Bulgarian students studied in Berlin, Prague and Warsaw. On the other hand, students from many other countries came to Bulgaria to study at VITIZ. The professional standards of the Bulgarian actors are kept up to the mark, through their participation in international exchanges involving guest performances. Of particularly great significance for the development of Bulgarian theatre are the numerous drama festivals and reviews – National Review of the Bulgarian Drama and Theatre, National Review of Plays and Theatres for Children and Teenagers, National Review of Historic Drama and Theatre 'Panorama of the Centuries', National Review of Bulgarian Puppet Theatres, National Review of Small Stage Productions, National Review of Poetry Reading and Small Theatrical Forms, Review of Plays on Subjects Involving the Younger Generation, and the Golden Dolphin International Festival of Bulgarian Puppet Theatre. They help strengthen the unity between the dramatic and production process, provide incentives for Bulgarian playwrights, stimulate the development of the different theatrical forms, and sustain an atmosphere of creative competition.

Scene from *Puss in Boots*, after C. Perrault. Plovdiv Puppet Theatre, 1954

to culture and creative activity. Before the establishment of popular power there were in the country only four state theatres. Immediately after the 9 September 1944 Socialist Revolution 24 professional companies appeared. Today there are 36 drama theatres, nine of them are in Sofia

During this period drama criticism also had better opportunities for development. The magazine *Theatre* started publication in 1948 and the newspaper, *Narodna Kultura* (National Culture) specializing in the field of culture, appeared in 1957. Apart from the Stagecraft Department at VITIZ, a new section on the theory and history of the Bulgarian theatre was set up in the Institute of the Theory of Art within the Bulgarian Academy of Art, which specializes in drama research. In the 1940s and 1950s drama criticism found its own place in cultural life and this period produced a group of drama critics – Gocho Gochev, Lyubomir Tenev, Pencho Penev, and Stefan Karakostov – who laid the foundations of modern Bulgarian stagecraft. They had all worked for a long time as drama teachers and had written a great many works on problems pertaining to the history and recent development of Bulgarian stagecraft and drama.

In the period following the 9 September 1944 socialist revolution the functions of the Union of Bulgarian Actors increased. Together with the state institutions, it took part in the organization of theatre life, and did a great deal to promote and refine the art of Bulgarian actors, stage directors, etc.

The first contributions of the Bulgarian socialist theatre in the 1940s and the 1950s were closely associated with the challenges of revolutionary reconstruction, with the struggle for a new kind of art, ideologically and methodologically different from that of the past. In those years the Bulgarian theatre broadened its creative scope, displaying an interest in the epic development of history – the fate of the class, the people and mankind. It removed the restrictions on the dramatic impact of the individual, pulled it out of the closed world of the intimate and tried to establish its relationship with society. Bulgarian drama became orientated towards the life of society with its ever-growing political awareness and social clashes, and with

Krustyu Mirski

the circulation of ideas, so as to transform it into an ever-present subject for creative observation and analysis, for judging the individual's level of spiritual development. These years were marked by an intensive endeavour to create a methodology of aesthetics of art capable of revealing and promoting the social, political and moral ideals of the new society. During the period of ideological and artistic reorientation, when the new creative problems had to be solved, the Bulgarian theatre fell back on the theoretical and practical experience of Konstantin Sergeyevich Stanislavsky. In the development of the young socialist theatre, Stanislavsky's theory was a kind of defence of its actual substance – a realistic, profound and ideologically purposeful interpretation of life. The process of aesthetic reorientation was of a national character and found an expression in subjects varied in scope and content. In the first years after the socialist revolution the National Theatre became the place where Stanislavsky's theory was actually put to the test. It was here that the creation of a new methodology became integrated with the attempt to preserve and develop the valuable achievements of the past. In those years, the National Theatre was noted for its outstanding professional actors. Working at the Theatre were also the most prominent directors: Nikolai Massalitinov, Boyan Danovski, Stefan Surchadjiev, Mois Baniesh, Filip Filipov and Krustyo Mirski. Associating the high dramatic skill of the actors at the National Theatre with the active class position of those who promoted the party line, as did Boyan Danovski, for instance, is in a way a recognition of the contribution of actors to national theatrical culture, and a directly expressed desire to accelerate the process of their ideological and aesthetic renewal.

The first productions of the socialist theatre recreated the greatest historical event – the revolution. The theatre found expression in the latest Bulgarian drama, in Soviet

Scene from Kuzman Krustev's *The Merry Little Boy*. Mihailovgrad Puppet Theatre, 1983

Façade of the Varna Drama Theatre

plays and in the Western classical plays. Along with plays in which sincere and powerful ardour predominated, the Bulgarian playwrights were writing plays in which the political element was enriched by the thorough analysis of life. In Nikolai Massalitinov's production of Maxim Gorky's *Enemies* and Kamen Zidarov's *Royal Pardon* the individual is interpreted from a social point of view but the explicitness of the social truth is combined with an individual characterization which is psychologically orientated to the experience itself. The intellectualization of art is present in Boyan Danovski's productions of *Yegor Bulychov and the Others* by Maxim Gorky and in *Armoured Train* by Vsevolod Ivanov. In this approach the generalization of the phenomena of life reaches a point at which profound social and moral processes governed by objective laws are emphatically expressed.

In those early years the theory of drama became fully integrated with the theatrical process. It influenced and was itself influenced by its dynamic development. It was linked with the dialectics of the developing values within it, with its endeavour to interpret problems from a social point of view, to make more thorough moral generalizations. This was the time when the first major plays appeared, those of Krum Kyulyavkov (1893–1955), Lozan Strelkov, Orlin Vassilev (1904–77) and Kamen Zidarov (b. 1902). They were influenced by the breathless pace of the crucial historical events which were affecting the life of the entire nation. The playwrights were in search of the new meaning of life, trying to interpret it from the viewpoint of its revolutionary development; and from the dynamism of the complex social and psychological processes they derived the dynamism of the historical events. Outstanding among these works are *Royal Pardon* by Kamen Zidarov and *Alarm* by Orlin Vassilev, in which the scope of the historic elements finds an expression in the magnitude of the characters, the profoundness of the thought, and the power of the experience. In the first socialist plays the aesthetic and moral categories were interpreted through revolutionary truths. They were expressed through the conduct of the man who had devoted himself to the triumph of the communist ideal. The heroism lost its exclusiveness, and was expressed through the belief of ordinary man in the historical justice of his cause.

The Bulgarian stage directors played a key role in the reformist processes of the 1940s and 1950s. Being among the most active motive forces in the development of the theatre, they concentrated their creative efforts on a philosphical approach to the phenomena of life. Persistently and anxiously, they sought to portray the multiformity of theatrical imagery and to depict interrelations which would eventually become established theatrical laws.

In the years following 1956 an even greater importance was attached to spiritual culture. The highly humane principles of the 1956 April Plenum of the Central Committee of the BCP gave a powerful impetus to the multiform development of theatre art, to its striving to reflect as fully as possible the problems of society and human life. The urge to reform and renew became paramount.

The example of creative and civic attitudes towards art set by the directors of the 1940s and 1950s was closely followed by the younger generation. Sasho Stoyanov and Grisha Ostrovski, Metodi Andonov (1932–74) and Villi Tsankov, Leon Daniel, Nikolai Lyutskanov (b. 1927), Julia Ognyanova and Stancho Stanchev, the generation of Krikor Azaryan (b. 1934), Encho Halachev and Assen Shopov, combined what they had learned from Konstantin Stanislavsky's theory of acting, with the ideas of Bertolt Brecht and the enchanting discoveries of Eugene Vakhtangov and Vesvolod Meyerhold. They were anxious to seek and learn, to know more about the achievement of world progressive theatrical culture. Aware of the claims and responsibilities of their time, they sought to adapt to them the theatrical experience which had accumulated throughout the centuries. Thus, in the 1960s, the Bulgarian theatre became a centre of the combined efforts and creative energy of several generations of directors; it proved to be a useful experience and opened the way to new aesthetic values. The Bulgarian directors of the 1960s were not afraid of experimenting in both ideas and concrete stage work; they desperately wanted to demonstrate their ability to reach the heights of stagecraft and their willingness to make the cultural imperatives, formulated after 1956, cherished imperatives of their own work. They earnestly desired to establish their own image as capable and knowing creators of deeply original concepts of the human soul.

Due to the efforts of stage directors like Boyan Danovski (*The Bedbug* by Vladimir Mayakovsky), Metodi Andonov (*The Government Inspector* and *Tarelkin's Death* by

Nikolai Gogol), Filip Filipov (*Woe from Wit* by Alexander Griboyedov), the theatre ceased to be an instrument for mere entertainment and became instead a means of analysing life in its essence, and of unmasking unseemly aspects. In the 1960s the Bulgarian directors were already working exclusively to the standards of modern theatre; they set great store by the witty, inventive, seemingly spontaneous play, whose climax is always directly proportional to its depth of vision, its clear and unquestionable conclusion.

The fruitful efforts of the directors of the 1960s were favourably influenced by the interplay of ideas. This also contributed to the new ideas regarding the strict concepts of theatrical illusion. This particular type of theatre did not radically change its style. Nevertheless it nowadays combines truthful portrayal of life with artistic generalization, psychological depth with convincing theatrical forms. Emotional revelations were also given new justification. The conventional observations on the individual's innermost world gave way to a passionate expression of the man-world relationship. The modern director thus led the individual out of the universe of his own soul into a new world of political and social values, while preserving his faith in the sincerity and truthfulness of emotion. The delicate probing into the human soul, the lyrical appreciation of its impulses, inner tension and emotional burdens, became essential to the director's everyday work on the plays. The actors responded by creating stage characters which were long remembered. The actresses Ruya Delcheva, Olga Kircheva and Ivanka Dimitrova; the actors Apostol Karamitev (1923–74), Stefan Getsov (b. 1932), Lyubomir Kabakchiev (b. 1925), Assen Milanov (b. 1922), Spass Yonev (1927–66); the actresses Mila Pavlova (1929–63), Tanya Massalitinova (b. 1922), Slavka Slavova – to mention but a few, became well-known to the theatrical public for their fine acting. The same held good also for the younger actors who strove to show the poetic, spiritual world of their contemporaries in their purely lyrical or passionately bitter interpretations: actors Kosta Tsonev, Georgi Cherkelov, Georgi Kaloyanchev, Naum Shopov; actresses like Emilia Radeva, Tsvetana Maneva (b. 1944), Nevena Kokanova (b. 1938), Vancha Doicheva (b. 1942), Maria Stefanova, and Nevena Mandajieva (b. 1941). The poetic metaphor and the convincing expressiveness which were characteristic of the work of directors like Anastas Mihailov, Nadia Seikova, Dimitrina Gyurova, Hristo Krachmarov, gradually became characteristic also of all theatrical forms, since they reinforced their moral and aesthetic impact. The intellectual message of the theatre became direct and outspoken; it preserved its appeal to the audience while getting rid of extreme rationalism and imposed ambiguity. It is also preserved, despite other conventional theatrical elements, in the uncompromising vitality of the acting. Direction thus added new dimensions to the traditional actor's wish to regard with respect the hero's emotions and struggle for fulfilment, and tried to find an adequate expression of this respect by experimenting with new forms. The actors were asked to combine their search for

Apostol Karamitev

aesthetic values with vivid grotesque, and to give sincere psychological motivation to their action, thus transforming conventional grotesque into moral categories of unusual intellectual and emotional strength.

In the 1960s Bulgarian stage design showed considerable progress. It had its predecessors in the art of Alexander Milenkov (1882–1971), Pencho Georgiev (1900–40), Assen Popov (1895–1976), Boris Angelushev (1902–66), Ivan Penkov (1897–1957) and Georgi Karakashev (1899–1970), who laid the foundations of theatrical design in Bulgaria in the 1930s, 1940s and 1950s. However, stage design became indispensable to theatrical art in the years of rapid cultural development after the 1956 April Plenum. In this period artists like Mihail Mihailov (1924–82), Assen Mitev, Svetoslav Genev (b. 1932), Assen Stoichev, Angel Achryanov and Mladen Mladenov were outstanding.

The Bulgarian lyrical drama was also directly related to the desire to introduce the spirit of renewal into Bulgarian theatre. The poets who wrote for the theatre based their criteria of human values on their sense of personal responsibility to society and its ideal, humanism. In *Every Autumn Night* Ivan Peichev tried to show the moral grounds of the communist ideal in the situation of conflicting revolutionary events, to show that the struggle for Communism is a deeply personal mechanism for connecting essentially individual problems with a responsible attitude to human problems in general; the play related the superior goals of the revolution to the establishment of sublime moral values. In *The Prosecutor* Georgi Djagarov portrayed his contemporary through the conflict between the individual and his ideal.

Viewed from this angle, the dramatic fate of the hero turns into a reflection of one's own responsibility to the communist ideal and its humanism. A brilliant expression of this impulse and striving for new philosophical and

Scene from Kamen Zidarov's *Boyan the Magician*. Gabrovo Drama Theatre, 1976

poetic horizons, so typical of this type of drama, is Valeri Petrov's play *When the Roses Dance*. It introduces a rich gamut of lyrical and ironical notes into modern Bulgarian

Scene from Yordan Radichkov's *January*. Sofia Theatre of Satire, 1975

comedy, and the capacity for simultaneously arriving at philosophical generalizations and gently touching the tender world of love. This malleable alloy of thought, judgement and innermost ideals is characteristic also of the works of Ivan Radoev and Nedyalko Yordanov. Concentration of the dramatic action within the personality and increased interest in its spiritual world are also found in another group of dramatic works, of a more analytical nature. In his *Women with a Past* and *A Holiday in Arko Irris* Dimiter Dimov reveals the complex interaction between human nature and the laws of society and synthesizes his multi-layered analysis into a philosophical inference about the attitude of the personality to the world, its belief in, or estrangement from, revolutionary and humanistic ideals. The plays of Dragomir Assenov are devoted entirely to an analysis of everyday life. He focuses his attention on the social dimensions of everyday life; he investigates the social significance of resignation and inertia, and reveals the social meaning of compromise and its incompatibility with a firm stand on matters of revolutionary principle and personal exigency. The various metamorphoses of the disparity between the ideal and the actual behaviour of the personality find their expression in the works of Lozan Strelkov; they are also the basis on which the dramaturgical texture of the works of Nikolai Haitov, Kolyo Georgiev, Ivan Radoev, Konstantin Iliev and Georgi Danailov, is built. These playwrights analyse the spiritual motives which set in motion the destructive, or constructive, powers of the hero and study the content of the personality's social thinking. The untiring search for the spiritual needs of our contemporaries is explicitly revealed in the development of Bulgarian historical plays. Drawing a parallel with the present, the playwright takes the dimensions of the national character, studies the moral values the Bulgarian has treasured and preserved intact through the ages, and analyses the remote or recent past from a contemporary point of view, trying to find his ideological landmarks in a Marxist explanation of the historical process. This comprehensive approach of the modern writer to the past finds its first categorical artistic defences in the historical plays of Kamen Zidarov, *Ivan Shishman, Kaloyan, Cain and the Magician,* while Nikola Russev's *The Man and the Arrow* outlines the trends leading to sweeping intellectual generalizations, to a revelation and philosophical rationalization of the long-lasting conformity in social existence. Bulgarian historical dramaturgy has developed greatly in terms of subject-matter and style. It has given rise to a number of plays treating some of the most significant moments in the political struggle of Georgi Dimitrov – *The First Strike* by Krum Kyulyavkov, *The Unfinished Monologue* by Todor Genov, and *Red and Brown* by Ivan Radoev.

The rich and varied creative life of the 1960s brought to the fore a number of new drama critics – Chavdar Dobrev, Vladimir Karakashev, Vassil Stefanov, Julian Vuchkov, Sevelina Gyorova, Nevyana Indjeva, Dimiter Kanushev, and Snezhina Panova, many of whom proved their worth as researchers in the field of the history and theory of dramaturgy and theatrical art.

The development of the Bulgarian theatre – of stage art and of dramaturgy in general use in the 1960s and 1970s is characterized by a harmony of ideas; its underlying principles are the principles of socialist art, which have thus found their most outstanding artistic confirmation and have been further enriched with new fruitful ideas. The study by the dramatists of the specific features of the people's mentality, of their social memory, aesthetic notions and moral principles is becoming more prominent in the Bulgarian theatre (for example, in the productions of Nikolai Lyutskanov and Encho Halachev); author and producer are more and more persistent in the unceasing search for the right answer to the question, 'What features from the national character or from the social experience of the past are in harmony with the present?' Many theatre-workers, such as Kriktor Azaryan, Mladen Kisselov and others, are looking for ways to restore the traditions of the folk-shows and find them a place in the modern theatre, and have thus produced stage performances rich in carnival elements, opening up unlimited scope for the imagination, full of unrestrained enjoyment and entertainment, witticism, ingenuity and fantastic invention. A more penetrating study of the people's psychology can be found in the staging of plays from the Bulgarian dramaturgical heritage, characterized by a search for the dialectical unity between the nationally unique and the universal. This point of intersection is the basis of many theatrical performances, for instance Villy Tsankov's staging of *The First* and *Builders* by Petko Todorov, etc., which apart from being typically Bulgarian in inspiration and execution also reveal some universal human features and outline the dramatic rise of mankind towards new ideological and moral principles. This process can be most easily traced in the plays of Yordan Radichkov, which strive towards the depths of the dialectical movement of the primal impulse towards the social and moral structures of human relationships of today. His plays *Turmoil, January, Lazars' Story, An Attempt at Flying* and *Baskets* build a profound picture of the strange cataclysms in the spiritual layers of the human mind, triggered off by modern life's encounter with social and moral experience accumulated through the centuries. Radichkov's work gives flesh and blood to the conflicts and clashes between a fantastic view of the world and the rationality and the discoveries of the modern world; it outlines the paradoxical ways of their interpretation, studies the dramatic tension within man, stretched between the boundaries of this span of time; tries to find the vital truths of the past in the manifestations of the present and suggests ideas about the dramatic movement of mankind through the ages. The tendency towards a deeper study of the essence of a phenomenon can be observed also in the development of the comedy. Accumulating a great deal of experience in the study of the paradoxical and the absurd, in the discrepancy between the personality and the challenges of life and the testing of new methods for grotesque generalizations, it has now found expression in the satirical generalization of Stanislav Stratiev. He is engrossed in the study of social evil, trying to find out the scale of the deformity of its consequences. In *Roman Bath, The Suede Jacket* and *The Maximalist* he studies the matamorphoses of the 'automated' human personality, which turns its dogma into commands of the day and is capable of hampering the timely realization of a useful social idea. Stratiev traces the manifestations of this behaviour to the point where it ceases to be a feature of a character and becomes more like a view of life. To this character, Stratiev opposes the monolithic nature of a

Scene from Lada Mileva's *The Magic Spectacles*. Sofia Central Puppet Theatre, 1975

Scene from Rada Moskova's *Adventure in the Attic*. People's Youth Theatre, Sofia, 1975

morality, which is an expression of the constructive significance of life. Its positive features are its civic demands and its incompatibility with compromises and resignation.

The fruitful unity of dogmatic literature and theatrical art is the soil on which a variety of forms of stagecraft have grown. The interest of the dramatist in the roots of acting, and his desire to penetrate yet deeper lend a modern ring to the world of fantasy and find expression in a number of dramatic works. Among the most distinguished authors of plays for children and adolescents are Valeri Petrov, Yordan Radichkov, Pancho Panchev, Nedyalko Yordanov and Boris Aprilov. Initially designed to offer solutions to some specific dramatic problems, the small theatrical forms have transcended the narrow confines of the experimental exercise by the transforming power of stagecraft. Today the genre of closet drama is typified by works like *Easter Wine* and *Nirvana* by Konstantin Iliev or *Flight* by Mihail Velichkov, which rise to philosophic syntheses of national and universal issues, full of insight.

What distinguishes the latest productions of the Bulgarian theatre is an emphatic search for creative variety in terms of style and subject-matter. The rich gamut of expressive means and devices, the panoramic presentation of conflicts and events, the subtle analysis of emotions, and the depths of poetic experience bring the Bulgarian theatre of today close to contemporary man and his social commitments and stimulate it to keep even more closely in touch with social dynamics and the changing interaction of the individual and society. Artistic vision in the modern Bulgarian theatre draws on the ideas and principles of socialist art, on the great theatrical inventions of the century and on the national experience in this field. Producers and actors now have the talent to find their embodiment in the fantastic world of the various dramatic works, to look for and discover what is particular in them, to transform theatrical performance into a brilliant, unique expression of their artistic civic credo, and to exert an influence through theatre art on the mentality, values and social position of the individual.

CINEMA

The first film shows in Bulgaria were held in 1897, most of them organized by foreigners. The first cinema theatre, Modern Theatre, was built in Sofia in 1908, and in 1910 several newsreels and documentaries were produced. During the First and Second Balkan Wars (1912–13) and the First World War a number of militaristic films were likewise released. With the encouragement of the Government of the Bulgarian Agrarian Party the Duna joint-stock company made several films in 1921–23, recording political and economic activities under the administration of the outstanding figure of the revolutionary peasant movement in Bulgaria, Alexander Stamboliiski (1879–1923). Following the fascist *coup d'état* of 9 June 1923, newsreels showed military celebrations and marches, life at court, diplomatic events, religious ceremonies, sports events, etc. In 1941 the regular weekly film review *Bulgarsko Delo* (Bulgarian Affairs) was launched which, however, only rarely covered historic events of national importance, let alone the plight of the working people in Bulgaria.

The first Bulgarian popular science film entitled *Rustic Life* was shown on 21 August 1910 in Sofia. Several dozen other films were also produced, treating medical, geographic, historical, agricultural and other subjects. At the time almost all producers and cameramen set about making 'scientific-educational' or 'cultural' films, as the popular science films were then termed. A large part of these are marked by shallow and sporadic amateurism.

The first attempts at making animated films date from 1934, when some private commercial firms produced a few specimens combining cartoons and feature episodes, all for advertising purposes.

The first Bulgarian feature film *Bulgar Men Are Gallant* was released on 13 January 1915, starring Vassil Gendov (1891–1970). Notable among the few feature films then made are *Under the Old Skies* (1922) directed by Nikolai Larin, *Crossless Graves* (1931) by Boris Grezhov (1899–1963), *The Stone Pile* (1936) by Alexander Vazov (1900–72), *Strahil Voivode* (1938) by Yossip Novak. They present a true picture of the life and struggles of the Bulgarian people and show remarkably high ideological and artistic professional standards.

In bourgeois Bulgaria the art of film-making was poorly developed. Film attendance in 1944 amounted to some 13 million and towards 9 September 1944 the cinemas across the country were no more than 213. The facilities were

Still from *Stars,* directed by K. Wolf, 1959

inadequate and there were few experts. The total output included 42 feature films and several hundred shorts and newsreels. Most of these promoted reactionary ideas and their professional-artistic level was rather low.

The victory of the revolution of 9 September 1944 ushered in a new stage in the development of Bulgarian film art. The foundations were laid by the first issue of the *Fatherland Newsreel,* with documentary stills featuring popular demonstrations, cordial welcome of resistance fighters and other film records of the first days of freedom. In 1948 cinematography was nationalized. The new factors which determined the further development of the Bulgarian cinema were: the transformation of the cinema into a major component of culture and the great care devoted by the State to its development; rapid expansion of the network of cinemas and audiences; expansion of the repertory of Bulgarian films and introducing into it Soviet films and those made in the other socialist countries, along with progressive and realistic works created by cinematographers in non-socialist countries; training professionals in all cinematic specialities at the higher cinematographic establishments in the USSR, Czechoslovakia, Poland and elsewhere; and building up a solid technical basis and boosting the output of the Bulgarian cinema. The themes and ideological content of films radically changed and professional artistic standards rose, too. The principal characters in feature films were the popular fighters for freedom and builders of Socialism. The first feature film of the socialist cinema was *Kalin the Eagle* (1950) directed by Boris Borozanov (1897–1951). In the early period in the development of the contemporary feature film (1950–56) some fourteen films were released, the best among them being: *Alarm* (1951) and *The September Insurgents* (1954) directed by Zahari Zhandov (1911), *Under the Yoke* (1952) and *The Stormy Path* (1955) directed by Dako Dakovski (1919–62), *Song of Man* (1954) directed by Borislav Sharaliev (1922), the first Soviet-Bulgarian co-production *The Heroes of Shipka* (1955) directed by Sergei Vassiliev (1900–59), and *It Happened in the Street* (1956) directed by Yanko Yankov (1924). *Dawn over the Homeland* (1951) directed by Anton Marinovich (1907–76) and Stefan Surchadjiev (1912–65) introduced the contemporary theme into the feature film. 1956 saw the first Bulgarian colour feature *Item One on the Agenda* directed by Boyan Danovski (1899–1976).

Still from *Alarm*, directed by Zahari Zhandov, 1951

In 1945 the first documentaries and popular science films of the Bulgarian socialist cinema were born, and in 1949 – the first animated films. Among the most successful documentaries released up to 1948, were *A Village Wedding* (1946) by Stefan Hristov, *Men in the Clouds* (1946) directed by Zahari Zhandov (b. 1911) and *Bright Days* (1947) by Boncho Karastoyanov (1899–1962). *The Power of the Weak* (1945) directed by Emil Karastoyanov, with cameraman Boncho Karastoyanov, was the first popular science film produced after 9 September 1944, which was a contribution to the popular persuasion work on the advantages of the co-operative cultivation of land. Young film-makers came to the fore in the ensuing years which brought the first international successes and recognition for the Bulgarian cinema: in 1947 *Village Wedding* and *Men in the Clouds* were awarded prizes at the Ninth International Film Festival in Venice. The thematic range was broadened to include agriculture, natural science, medicine, art, science and technology, etc., and technical facilities gradually became more sophisticated. The year 1953 saw the release of the first colour, popular science film *The Fruit of Our Land* directed by Stefan Hristov, with cameraman Dimiter Kitanov (1921). Under the supervision of Dimiter Todorov-Zharava the first cartoons were also made after 9 September 1944: *Serves Him Right* (1949), *Lamb and Wolf* (1950), and *The Forest Republic* (1952) the first Bulgarian colour cartoon. The first puppet film, *The Frightful Bomb,* directed by Dimo Lingurski, was made in 1951.

The far-reaching and thorough changes that took place in Bulgaria as a result of the April 1956 Plenum of the Central Committee of the BCP had a beneficial effect on the development of the Bulgarian cinema as well. During the 1956–61 period the feature film was gradually freed from the influence of dogmatism and, while preserving its ideological pathos, acquired greater philosophical depth and artistic subtlety. Notwithstanding the growing number of films treating contemporary themes, the best works of this period were inspired by historical-revolutionary subjects: *On the Little Island* (1958) and *First Lesson* (1960) a co-production with DEFA directed by Conrad Wolf, *Paupers' Street* (1960) directed by Hristo Piskov (b. 1927), and *When We Were Young* (1961) directed by Binka Zhelyazkova (b. 1923). Film-makers tried their hand at all genres, producing children's musical films, comedies etc. The thematic range of newsreels and documentaries was expanded and diversified as their numbers increased. Dry information gradually yielded to a more original, individual interpretation of facts and events. Film reportages become more incisive; along with critical commentaries on reproachful facts (assigned a special place in the newsreels) they included amusing reports and exciting profiles of outstanding personalities. The range of themes of the popular science films expanded and their producers addressed themselves to more serious issues and won international recognition. The cartoon series *Cine-Hornet* produced by artist Todor Dinov (b. 1918) in 1956–58 steered the genre toward the problems of the day. Being a disciple of Soviet cartoonist Ivan Ivanov-Vano, Dinov helped establish the aesthetic platform of the Bulgarian animated film on the basis of the traditions and experience of the Soviet one. In 1959 the Bulgarian animated films received their first international distinction at the Thirteenth International Film Festival in Edinburgh for the puppet film *The Invisible Mirko,* directed by Stefan Topaldjikov (b. 1909).

After 1961, a five-year period of relative stability set in, in which achievements were consolidated and promoted further, and the general level of feature films was upgraded. The Bulgarian film-makers mastered the wide variety of contemporary cinematographic means of expression. Successful works appeared on all major themes, which were developed in the basic genres. Film comedies, children's films, historical and detective films became

Still from *Ballad of the Horse*, a popular science film, directed by Konstantin Grigoriev, 1979

increasingly popular. An adequate qualitative and quantitative balance was attained between historical films and films about present-day Bulgaria. The landmarks of this period are: *Sun and Shadow* (1962) directed by Rangel Vulchanov, *Tobacco* (1962) by Nikola Korabov (b. 1928), *Birds in Captivity* (1962) by Ducho Mundrov (b. 1920), *The Captain* (1963) by Dimiter Petrov (b. 1924), *The Chain* (1964), by Lyubomir Sharlandjiev (1931–79), *Between the Rails* (1964) by Villi Tsankov, *The Peach Thief* (1964) by Vulo Radev (b. 1923), etc. In the second half of the 1960s the conditions were created for a fresh upsurge in the feature film industry. It was the ambition of film-makers to reify modern ideas by modern means of artistic expression. The view that the cinema has primarily an illustrative function gave way to that of its analytical and heuristic function. The success of the feature film now depended on the artistic solutions given to some specific problem, viz. ideological and philosophical reinterpretation of the concept of heroism, new ways of expressing the connection and continuity between the revolutionary past and the socialist present, enhancing the critical spirit, and the establishment of new creative relationships with the literature, theatre and the other arts. Noteworthy are the following feature films: *The Side Track* (1967) directed by Grisha Ostrovski and Todor Stoyanov (b. 1930); *The White Room* (1968) by Metodi Andonov; *The Last Summer* (1974) by Hristo Hristov; *Villa Zone* (1975) by Eduard Zahariev; *Advantage* (1977) by Georgi Dyulgerov; *The Barrier* (1979) by Hristo Hristov; *Equilibrium* (1983) by Lyudmil Kirkov; the remarkable historical films *The Goat's Horn* (1972) directed by Metodi Andonov (1932–74) and *Khan Asparuh* (1982) directed by Lyudmil Staikov; the historical-revolutionary films *Birds in Captivity* (1962) by Ducho Mundrov; and *Tsar and General* (1965) by Vulo Radev; *The Longest Night* (1967), *Black Angels* (1970) by Vulo Radev; *The Last Word* (1973) by Binka Zhelyazkova, *And There Came the Day* (1973) by Georgi Dyulgerov; *Ivan Kondarev* (1974) by Nikola Korabov; *Amendment of the State Security Law* (1976) by Lyudmil Staikov; and the children's films *Knight Without Armour* (1966) directed by Borislav Sharaliev, *Porcupines Are Born without Spikes* (1971) by Dimiter Petrov, and *Ill-timed Examinations* (1974) by Ivanka Grubcheva. The headway made by the Bulgarian feature film has been the result of the concerted efforts of producers and creative workers in other arts. Especially fruitful is the work of script writers Valeri Petrov (b. 1920), Angel Wagenstein (b. 1922), Pavel Vezhinov (1914–83), Hristo Ganev (b. 1924), Georgi Mishev (b. 1935), etc.; cameramen Vassil Holyochev (b. 1908), Atanas Tassev (b. 1931), Borislav Punchev (b. 1928), Dimo Kolarov (b. 1924), Boris Yanakiev (b. 1929), etc.; film artists Maria Ivanova (b. 1929), Konstantin Russakov (b. 1929); and composers Simeon Pironkov (b. 1927), Boris Karadimchev (b.

Still from *When We Were Young,* directed by Binka Zhelyazkova, 1961

1933), Kiril Tsibulka, etc. The conspicuous achievements in film acting contributed a great deal to the fruitful development of the Bulgarian feature film. Notable among the actors and actresses acclaimed for their remarkable achievements are: Apostol Karamitev (1923–74), Nevena Kokanova (b. 1938), Tsvetana Maneva, Georgi Georgiev-Gets (b. 1926), Grigor Vachkov (1932–80), Georgi Cherkelov (b. 1930), Kosta Tsonev (b. 1929), Naum Shopov, Stefan Danailov (b. 1942), etc.

Employing a variety of innovative methods and means of expression, science film makers have created outstanding works in the basic genres and thematic areas. The

Still from *The Wise Village,* a cartoon film directed by Donyo Donev, 1972

Still from *The Last Word,* directed by Binka Zhelyazkova, 1973

Still from *Khan Asparuh*, directed by Lyudmil Staikov, 1982

Still from *The Goat's Horn*, directed by Metodi Andonov, 1972

Still from *The Warning*, directed by J. A. Bardem, 1982

Bulgarian popular science films disseminate scientific knowledge, inculcate a materialistic world-view and cultivate the aesthetic taste of the public. The rapid development of modern science, the proliferation of scientific disciplines, the establishment of borderline sciences (biophysics, biochemistry, physico-chemistry etc.) and synthesizing sciences (cybernetics, ecology, etc.), the mathematizing and philosophizing of science, all these factors have a material impact on popular science films, broadening their range and scope. The efforts to penetrate into the innermost secrets of Nature, the social relations and the theory and practice of socialist construction are being pursued in close connection with the interest in man. Along with the new orientation toward serious new scientific issues, popular science films tend to acquire more humanitarian aspects. These developments have faced producers with the requirement for higher scientific and professional standards, and have stimulated artistic individuality and originality. Among the most distinguished science-film makers are directors Konstantin Kostov (1916–82), Konstantin Obreshkov (1934–81), Avram Ignatov (b. 1928), Konstantin Grigoriev (b. 1927), and Stiliyan Parushev.

Newsreels and documentaries now give more attention to the interpretation rather than the mere presentation of facts and events, to thematic and genre variety, and the stronger analytical, individual presence. Ciné-realism shows man in the context of a new world. Special credit for the achievements made in this film genre goes to producers Hristo Kovachev (b. 1929), Nevena Tosheva (b. 1922), Yuri Arnaudov (1917–76), Yuli Stoyanov (b. 1930), etc.

Animated-film makers have won world-wide recognition and established a 'Bulgarian school' in film animation marked by a wealth of themes and genres. These films go beyond the boundaries of subject-matter for children and encompass a wide range of topical problems by seeking new associations and meaning in their treatment. The major representatives of present-day Bulgarian animated film makers are Todor Dinov, Donyo Donev (b. 1929), Zdenka Doicheva (b. 1926), Radka Buchvarova (b. 1918), Georgi Chavdarov, Pencho Bogdanov (b. 1923), Hristo Topuzanov (b. 1930), Stoyan Dukov (b. 1931), Ivan Vesselinov, etc.

The intensive growth of the film industry is paralleled by the development of film distribution. After 9 September 1944 a network of cinemas began to be built systematically throughout the country. Present-day socialist Bulgaria had (on 1 January 1983) a total of 3,302 cinemas seating 748,000 viewers, 520 to them in the towns and 2,782 in the villages. The cinemas in the larger towns are supplied with most up-to-date equipment and facilities and constant concern is shown for improving the interiors of cinema theatres. New cinemas are put into operation using the latest scientific-technological achievements in the field – Dolby stereo sound, three-dimensional image, magnetic-carrier video cinema, etc. The electronization of the cinema network, i.e. the employment of video technology opens up good prospects for video-clubs in the larger towns as well as for video cinemas in the smaller places. The world-wide tendency towards building multi-hall cinemas, with each seating 250 to 400 viewers, is also followed in Bulgaria. Through the operation of the wide cinema network the number of cinema-goers over recent years has ranged between 90,000 and 100,000 annually (94,499 in 1983). According to statistics, each citizen goes to the cinema 10.6 times annually. By this indicator Bulgaria ranks among the foremost countries in the world.

No less attention is being given to the diversification and enrichment of the film repertory. About 170 new Bulgarian and foreign films are released annually. The ambition is constantly to increase their number and thematic variety. Repertory policies aim at broader popularization of Bulgarian films alongside the best achievements of the socialist, western and developing countries. Viewers show the greatest interest in Bulgarian and Soviet films – 21.9 per cent and 27.7 per cent respectively of the total of cinema-goers for the year (1983). The films produced by the other socialist countries were seen by 16.6 per cent, and those by western and developing countries, by 33.8 per cent of the total of Bulgarian cinema-goers.

Various forms of film distribution are used in order to meet the specific interests of film audiences – studio distribution for film devotees and experts, the Kultura cinemas for viewers showing preference for documentaries, popular science and animated films, and distribution of scientific-technological films by the scientific-technological unions to help raise the level of training of research workers, engineers and technicians.

The feature and short-length films made in socialist Bulgaria have also attracted viewers abroad. In 1983, some 96 feature films and 122 shorts were sold in the socialist countries and 26 and 195 respectively in the non-socialist countries.

Bulgarian film-makers (directors, cameramen and film critics) are trained (since 1973) at the Higher Institute of Theatrical Art in Sofia. Along with TV film-makers, they are united in the Bulgarian Film-Makers' Union in the following sections: Feature Films, Popular Science Films, Documentaries, Animated Films, and Film Criticism. The Young Film-Makers' Studio has been set up by this Union to organize and channel the creative activities of young film artists. The Cinema and Television Department of the Institute of Art Studies under the Bulgarian Academy of Science carries out research into the history and theory of the cinema and television. The specialized publications on film art are *Kinoizkustvo* (Fine Art) since 1946, *Filmovi Novini* (Film News) since 1955, *Bulgarski Film* (Bulgarian Film) since 1959, also published in foreign languages, *Kinorabotnik* (Cinema-worker) since 1973, *Kino i vreme* Bulletin (The Cinema and Our Times) since 1972, etc. In the late 1940s the foundations were laid for the Bulgarian National Film Library. Bulgarian Cinema regularly and actively participates in international reviews and festivals of feature, popular science, documentary and animated films and has won many awards and distinctions. Among the regular film events organized in Bulgaria is the Festival of Bulgarian Feature Films in Varna (since 1961) and the Festival of Bulgarian Short Films in Plovdiv (since 1975). International film festivals are likewise held, e.g. the International Festival of Red Cross and Health Films in Varna (since 1965), the International Film Festival on the Organization and Automation of Production and Management in Sofia (since 1969), the World Festival of Animated Films (since 1979), and the International Festival of the Film Comedy and Satire in Gabrovo (since 1981).

THE CIRCUS

The setting up of the Bulgarian circus was preceded by the appearance of circus-type acts performed by professionals *(karagyozchii* and *ipdjambazi)*, and by the tours of foreign circuses. The *karagiozchii* were actors who gave puppet performances, showed tricks, and performed comic scenes from everyday life. After Bulgaria's Liberation from the Ottoman yoke in 1878, the *ipdjambazi* appeared; they did tightrope-walking, sometimes with elaborate props. Visits by the circuses of the Japanese owner Yamamoto and the Italian Angello Pisi generated widespread interest in the art and contributed to the training of the first Bulgarian circus artistes. In 1897 the gymnast Peter Panayotov (1878–1926), who was a pupil of Angello Pisi, established the first Bulgarian circus known as 'The First Bulgarian Gymnastic, Acrobatic and Comic Troupe – Bulgarian Banner'. Panayotov realized that the time was ripe for the development of gymnastics and acrobatics in Bulgaria and most of the items in his circus programmes were of that type. Thus Panayotov had a great influence on the future of the Bulgarian circus, for ever since then its greatest achievements have been in gymnastics and acrobatics. Gradually small circuses of seasonal duration appeared, such as the Tango circus run by Angel Dimitrov (1890–1965) and Nikola Dimitrov-Poshtata (1886–1969), the Bulgaria circus, run by Nencho Tsankov, and the Arena run by Alexander Dobrich (1879–1958) and the athlete Zebich. The pupils of Peter Panayotov also became popular, and were the first generation of professional circus artistes in Bulgaria – Yanko Panayotov, Elena Purvanova (1892–1971), Assen Purvanov (1885–1946), Pavlina Pencheva, Georgi Penchev (1885–1947), Todor Pironkov (1891–1962), Maria Pavlova, Angel Dimitrov (1890–1965), the clowns Gancho Pehlivanov (Gancho; 1878–1950) and Dimiter Maitapov (Maitapa; 1884–1940). One of the most famous Bulgarian circus artistes was Lazar Dobrich (1881–1970), who in 1905 presented 'The Trapeze of Death' for the first time in the Schumann circus in Berlin. He himself designed the device that he used, and it came to be widely used in the circus. Together with his brother Alexander, Lazar Dobrich built a permanent circus in Sofia known as the Coliseum (1919–26). The Coliseum also ran a school for circus artistes which was attended by Hristo Pavlov, Kiril Mihailov, Nikola Panov (1907–83), Georgi Sillagi (1902–63), and Mihail Sillagi, to mention but a few.

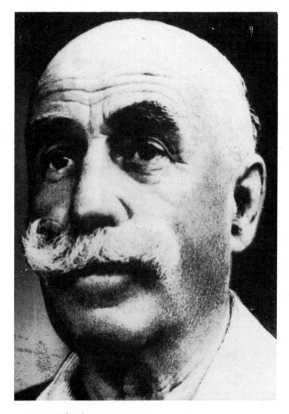

Lazar Dobrich

Nikola Panov was a gymnast of world repute. At a competition in Berlin, organized in 1931 by the International Artistic Lodge, he ranked second after the world-famous Mexican circus gymnast, Codonas. After 1930 there was a tendency for circuses to merge and form co-operatives in order to solve financial difficulties and unemployment. In 1931 the Co-operative Circus was founded, to be followed by the First Bulgarian Co-operative Circus and Co-operative United Circuses in 1933, and the Circus of United Artistes in 1934. The programmes of the co-operative circuses devoted more time to vaudeville and variety, and the clowns often directed their humour at topical issues of the day. The uniting of circuses was stimulated by the Bulgarian Circus Artistic Organization (BCAO), a professional organization of circus artistes in Bulgaria which was established in 1931 with Stoyan Milenkov (1889–1953) as President. Its aim was to defend the professional interests of artistes, circus directors and owners of entertainment establishments, and to improve their everyday life and working conditions. Membership was not limited to Bulgarians, many foreign circus artistes were also members, because of the protection and rights that the BCAO gave them. In 1945, the BCAO discontinued its activity, since the Union of Performing Artistes had set up a Circus Art section. Between 1933 and 1941 the organization published a magazine called *Circus Voice,* which was edited by Lazar Dobrich. The development of the Bulgarian circus before the revolution of 9 September 1944 was greatly influenced by the work of Lazar Dobrich and Nikola Panov, as well as that of such artistes as Vassil Kozarov (1886–1961), Angel Dimitrov, the brothers Georgi and Michail Sillagi, and Nikola Dimitrov-Poshtata (1886–1969), all of whose artistry had acquired international popularity. In spite of the successes of the circus artistes up to 9 September 1944, the Bulgarian circus as a whole had been in a precarious state. Successive bourgeois Governments had neglected the circus as an art form, and had underestimated its role and significance. Denied any

The clown, Dimiter Sedoi

Tightrope performance by Angel Bozhilov

financial support by the State, circus artistes worked and lived in extremely bad conditions. Apart from some individual achievements, circus acts were not notable for their artistic value; organization was chaotic, and as a result the genre balance of performances was upset. For financial reasons some acts, such as those involving wild animals, were hardly represented at all.

The Bulgarian people's victory on 9 September 1944 opened up new possibilities for the development of the circus. The main step forward was the social recognition that this profoundly democratic art received from the Popular Government. The old view that circus was an inferior form of art and of minor importance in cultural life was discarded. Eminent writers, painters, scholars and public figures expressed their views on circus problems, noting its progressive traditions, and emphasizing its significance for the setting up of the new socialist culture. Public discussions on the tasks and problems of the circus were organized nationwide, and articles on circus matters were published in the Press. The nationalization of circuses in 1948 marked the beginning of the rapid development of the socialist circus in Bulgaria. Three circuses were set up, Globus, Rodina and Republika, to be followed by a fourth, the Bulgaria circus, which travelled round the smaller towns and villages. The prospects for future development were outlined by a Decree of the

The Circus

Tightrope act by the Kondov Troupe

technique for the artistic expression of a comprehensive plot, theme or idea.

Thematic performances went on to become a tradition in the Sofia State Circus repertoire – *Laughter Parade, The Floor is Given to Women, Arena of the Brave, Northern Lights, A la Minute,* and others. The first serious test of what had been achieved was in 1955 when a Bulgarian circus programme took part in the First International Circus Festival in Warsaw. The artistes Angel Bozhilov (b. 1926) and Andon Kovachev were awarded gold medals, and Lazar Dobrich was the first circus artiste in the world to receive the title of 'Seigneur de l'Art du Cirque'. By 1957 the time was ripe for uniting the individual circuses under a single managing body that would co-ordinate efforts and direct creative activity. In that year the Chief Direction of Bulgarian circuses was set up. The creative artistic and managing personnel grew. Circus acts and performances were placed under the direct control of teachers, trainers, producers, choreographers, artistes and musicians. Props and costumes were made in workshops at the circus in Sofia. The introduction of new acts into performances was carefully worked out so as to maintain a balance between the different genres. Care was taken to ensure that circus artistes had professional training and qualifications. Prominent masters of the circus taught in a special circus school, and courses and seminars on the

The Boichanov Acrobatic Troupe making synchronized somersaults on two quadruple poles

Council of Ministers on the state, work and tasks of the Bulgarian circus (15 November 1950). This decree assigned various important tasks to Bulgarian circus artistes: to develop the best traditions of classical circus; to add more ideological and artistic content to circus entertainment; to create new acts that would reveal the best virtues of man, encourage optimism and stimulate a sense of beauty. The decree also emphasized the importance of clowning and pointed out the possibilities that it affords for people's ideological education in the spirit of Socialism. Todor Kozarov (b. 1926) and Dimiter Sedoi (b. 1926) were the first to entirely do away with the classical clown and replace him with a modern one. Todor Kozarov developed Hitar (Sly) Peter, a famous character from Bulgarian folklore, and Dimiter Sedoi presented him as the builder of the new society – the working man. Gradually, humour-for-humour's sake and non-ideological humour disappeared from the repertoire; political satire became a part of the circus. The foundations of political satire in clowning were laid by Dimiter Sedoi. A new generation of young circus artistes appeared, as the old masters went on to become directors and teachers. In 1954 thematic circus performances were introduced (*Our Happy Decade,* written by Dimiter Sedoi and directed by Todor Kozarov). This provided a new and better structure for circus performances. Individual circus acts became a

circus and matters of general cultural interest were regularly held. The Soviet circus also contributed a great deal: Bulgarian artistes were trained at the Moscow Circus School; directors and producers were exchanged; and guest performances of entire programmes or individual acts acquainted Bulgarians with the achievements and experience of Soviet circus artistes. As members of the Union of Artistes, Bulgarian circus performers have taken considerable steps forward. Decrees have been promulgated which give them pension rights at ages lower than those for other professions. As a result of the careful encouragement by the Popular Government, the Bulgarian circus has risen rapidly to the level of the best circuses in the world. The achievements on the horizontal bars, and in acrobatics and juggling, are known world-wide. In 1956, at the Jugglers' Festival in Italy, the Bulgarian juggler Angel Bozhilov was awarded the title of 'Best Juggler in the World'. Bulgarian circuses and individual performances hold numerous gold and silver awards from the World Festival of Youth and Students in Moscow, from the First Festival of European Circus Humour in Sofia, from the Clowns' Festival in Italy, from the Festival de l'Art du Cirque in Monte Carlo, and from the Circus Championship in London. Among the prominent circus artistes that have made the Bulgarian socialist circus what it is, the names of Lazar Dobrich, Nikola Panov, Dimiter Pironkov (b. 1925), Todor Kozarov, Kiril Panov (1915–83), Angel Bozhilov, Nikola Boichanov (b. 1925), Dimiter Sedoi, Peter Purvanov (b. 1914), Alexander Balkanski, and Temenuzhka Boichanova are outstanding.

AMATEUR ART ACTIVITIES

Amateur art activities have an impressive history and a long-standing tradition, establishing them as an important component of present-day Bulgarian culture. It was in the National Revival Period that, drawing on folklore, they first started to emerge as independent forms of organized, amateur artistic expression to mark the shift towards a new, non-folklore type of national culture. The process was triggered off by a performance staged by a teacher named Stefan Izvorski on 11 August 1846 at the Slav-Bulgarian school in the town of Shumen. The development of amateur art in the period was a direct result of the tendencies prevailing in the socio-political and cultural climate accompanying the transition to Capitalism, the consolidation of the Bulgarian nation and the start of the national liberation and social struggle. Stifled in the iron grip of national and political oppression under the Ottoman rule, with no state or cultural institution of their own, the Bulgarians still felt a craving for a new culture so strong that it found an outlet in a manner dictated by the requirements of the day. Enthusiastic educationists engaged in setting up theatrical groups and choirs at schools and library clubs, which duly proceeded to give public performances. Mihai Safran (c. 1824–1905) founded the first orchestra of amateur musicians in 1851 in Shumen, where, as well as in Lom, the first drama shows were held in 1856. In 1868 the first polyphonic choir was born in Svishtov, while in 1869 Bacho Kiro (1835–76), a highly learned man and revolutionary, initiated the first rural theatrical company at the village of Byala Cherkva. Up to Bulgaria's Liberation from Ottoman bondage in 1878 the major feature of amateur art activities was their close link with the revolutionary struggles of the people, with the social and cultural processes of the day. Most plays, both original and translated, selected for theatricals, dealt with the struggle against oppression and ended with revolutionary songs by the Bulgarian poet Dobri Chintulov (1823–86). Among those acting on the amateur stage were the most eminent leaders of the revolutionary movement Hristo Botev (1848–76), Bacho Kiro, Vassil Petleshkov (1845–76), Panayot Volov (1850–76), Pavel Bobekov (1852–77) and others.

The overall socio-political, economic and cultural upsurge brought on by Bulgaria's Liberation from Ottoman domination in 1878 had an extremely stimulating effect on the development of amateur art activities. Choirs mushroomed. In 1891 the Lira music society in Russe organized its own polyphonic choir, the Rodni Zvutsi (Nature Sounds) choir was founded in Shumen in 1898, as were the Plovdiv singing society choir (1893) and a number of army choirs throughout the country. In 1905 small singing groups became particularly abundant at the Bulgarian schools. The expanding workers' and socialist movement, with the formation of the Bulgarian Workers' Social Democratic Party in 1891 also gave a tremendous boost to amateur art. Fully appreciating the substantial part it played in the cultural improvement of the people and in spreading the socialist ideas, the Party was quick to offer a helping hand and guidance to a whole range of amateur groups. Choirs sprang up at the workers' clubs in Plovdiv, Gabrovo, Kazanluk, Sofia and elsewhere. The social orientation of amateur art became plainly manifest after the triumph of the October Socialist Revolution of 1917 in Russia and for this reason quite a few of the workers' amateur choirs and drama groups were suppressed by the bourgeois authorities. Yet the network of amateur choirs had spread so wide that in 1926 the Union of People's Choirs was set up. During the 1930s the number of workers' and community centre theatrical groups grew particularly fast. Their repertoire included progressive works by Russian, Soviet and Western writers. Bulgarian folk-dance ensembles began to make their appearance at sports clubs and schools in 1930. Up to 1934 they continued emerging at the tailors' guilds, and the metal workers and railwaymen's associations in Sofia; in 1936 they all joined together to form the dance ensemble of the Bulgarian Workers' Union. Greatest popularity in that period was earned by the Sinebluzi (Blue Collar) theatrical company, the Georgi Kirkov choir and the Mensa Academica students' choir. Persecuted rather than supported by the bourgeois authorities under the regime of police terror, severe censorship and cruel demagogy, amateur art managed to keep intact its democratic and progressive nature.

The victory of the socialist revolution in Bulgaria on 9 September 1944 opened up unlimited prospects to amateur artistic expression. In the first years of popular rule it served as a powerful means of fighting against the lingering prejudices of the past, of mobilizing the forces for the victory of the Patriotic War Bulgaria waged in 1944–45, and then for overcoming the ravages of war and rebuilding the economy in the First Two-Year Plan period. The close link with the vital issues of the day and

The Kaba Gaida Children's and Youth Music Ensemble, Smolyan

the distinctly political character of amateur art materialized in the activities of the most dynamic and successful among its performers – the propaganda groups. These made use of all the genres, singing, dancing, reciting, showing plays and comic acts, self-prepared on most occasions. The socialist State and the public organizations contributed lavishly to the development of amateur art activities. As early as 1947 the Central Committee of the People's Youth League and the Supreme Library Clubs' Council launched the first national training course for artistic directors. The Ministry of Information and the Chamber of People's Culture saw to it that there was a sufficient supply of repertory material and methodological handbooks for amateur artists. In 1947 the first national review of amateur art activities was carried out, its achievements earning the warm approval of Georgi Dimitrov (1882–1949) – the prominent leader of the Bulgarian and the international revolutionary workers' movement. With the foundation of the Committee for Science, Art and Culture in 1948 and its specialized department for amateur art activities the state apparatus was properly equipped to render methodological assistance to amateur art. In 1948 amateur groups already numbered around 4,000, with over 200,000 members. The propaganda groups gave way to art collectives with higher achievement potentials. In this favourable atmosphere a wider scope and variety in genre was bound to be attained. In 1954 a Central Club for Folk Art was set up at the Ministry of Education and Culture as a specialized organ providing creative and methodological guidance to amateur art activities, and in 1955 methodological departments were established at the District People's Councils. The decisions of the April 1956 Plenum of the Central Committee of the BCP marked a new stage in the further expansion of mass amateur art activities, leading also to higher ideological and artistic standards. The First and Second Republican Festivals of Amateur Art Activities, held in 1959 and 1964 respectively and entered by all amateur art groups without exception, turned into memorable events in the socio-political and cultural life of Bulgaria. Of particular significance for the further development of amateur art activities have been the series of

The Trovante Ensemble, from Portugal, at the Alen Mak (Scarlet Poppy) International Political Song Festival

The Naiden Kirov Dance Ensemble, Russe

documents of ruling bodies of the BCP and the Council of Ministers of the People's Republic of Bulgaria, adopted in the post-April 1956 period, among which were the special Decree on Amateur Art Activities of 1968 and the Ordinance for transforming the Institute of Amateur Art Activities into a Centre for Amateur Art Activities in 1972.

At the current stage of building the developed socialist society, amateur art is an important area of the national culture of Bulgaria for creative, rewarding activity and self-realization by all members of society, an indispensable factor in the aesthetic education of the people and of the younger generation in particular. Taking part in these activities are hundreds of thousands from all social and age groups. Statistics show that in 1982 amateur art groups numbered 29,484, with a total of 623,816 members. Their annual artistic output amounts to 168,415 concerts, theatre and art shows and the like before an audience of 46,576,000. All interests can find fulfilment in the wide gamut of amateur art genres. The musical talent of the Bulgarian people and its enduring folk-song traditions give predominance to musical activities, engaging nearly half of the country's amateur groups (13,755 in 1982). In 1982 choirs and vocal groups were top of the list (8,285), followed by accordion, mandolin, symphony and light music orchestras, etc. (2,550) and pop groups (1,509). There are no less than 36 amateur opera and operetta companies in Bulgaria. At the initiative of the Central Committee of the Dimitrov Young Communist League a mass movement was launched in 1975 for the creation, artistic interpretation and popularization of political songs. Since that same year the Alen Mak Festival of Political Songs has been held annually in Blagoevgrad. The artistic achievements of a number of amateur choirs, such as Kaval, Gusla, Sofia's Lyubomir Pipkov Chamber Choir, Madrigal, Bodra Smyana, the Children's Choir of Bulgarian Radio, Hristina Morfova, Rodna Pessen, Morski Zvutsi, Rodina, and various choirs from Plovdiv, Tolbuhin and other towns, are on a high professional level. These are conducted by such distinguished Bulgarian musicians as Vassil Stefanov (b. 1913), Angel Manolov (b. 1912), Vassil Arnaudov (b. 1933), Mihail Angelov (b. 1932), Zahari Mednikarov (b. 1924), Marin Chonev, Stoyan Kralev and others. Their repertoires comprise world and Bulgarian classical songs, as well as contemporary pieces specially written by Bulgarian composers. 1982 statistics give second place, according to the number of participating groups, to theatrical activities, of which 1,581 are drama groups, 391 are puppet theatres, 480 engage in variety and comic shows and 4,234 prepare poetry and literature recitals. These latter have obviously been preferred of late to the classic modes of histrionic expression, as have been the political theatre, the documentary theatre, satirical shows, etc.

Third on the list are dance ensembles doing classical and rhythmic, popular and folk dances (2,966 ensembles in 1982). Prominent among them are the Sredets and Mayakovsky groups, the ensembles at the Naiden Kirov works in Russe, the District People's Council in Varna, the Devnya chemical works and the Central Students' Club in Sofia. Fine arts circles have been rapidly multiplying of late. In 1982 they amounted to 1,565, while there were 1,957 photography clubs and circles, 383 film-makers' clubs and circles, and 917 literary circles. The works of

Mosaics made by members of the Children's Mosaics Society at the Lyuben Karavelov Community Centre in Burgas

The Seventeenth International Folklore Festival, Burgas

amateur painters and masters of the applied arts show an enviably high level of artistic achievement.

Along with the major events in amateur art activities – the national festivals and the folklore competitions at Koprivshtitsa – regular national festivals, competitions, reviews and celebrations are also held in the various genres. Most important among them are: the May Choir Festival in Varna; the Male-Voice Choir Festival in Gabrovo; the Golden Diana Chamber Choir Festival in Yambol; the Thracian Lyre Festival of Amateur Popular Music in Stara Zagora; the Competition of Choirs from the Danubian Towns in Silistra; the Pioneer Children's Choir Festival in Shumen; the festival of amateur theatricals in Blagoevgrad; the days of satire and humour in Gabrovo; the puppet-show days in Haskovo; the Competition for Chamber Dances in Gorna Oryahovitsa; the Festival of Soviet National Dances in Pleven; the photo competitions in Sliven and Troyan; and the festivals of amateur historic films in Shumen and of tourist films in Veliko Turnovo, etc. Every year district reviews, festivals and days of amateur art activities take place in all districts of the country. Notable among them are the periodic folklore festivals held in some districts: Pirin in Blagoevgrad, Beautiful Thrace Sings and Dances in Haskovo, events in Burgas, Kyustendil, Razgrad and Smolyan, and the festival devoted to various popular customs in Pernik. Bulgaria is an active participant in international festivals of amateur art. Scores of Bulgarian amateur groups from all genres, and above all choirs, orchestras, folk and dance ensembles, take part every year in such renowned international choir competitions as those in Arezzo and Gorizia (Italy), Cork (Ireland), Llangollen (North Wales), Neerpelt (Belgium), Tallin (USSR) and others; and in the folklore competitions and festivals in Zagreb and Ohrid (Yugoslavia), Zakopane (Poland), Nice (France), Greece, Turkey, Tunisia, Italy, Great Britain, the USSR, Czechoslovakia and others. Bulgaria plays host to traditional international competitions and festivals of amateur art: the International May Choir Competition in Varna, the International Folklore Festival in Burgas, the International Festival of the Laureates of Folklore Competitions and Festivals in Blagoevgrad and the International Student Film Days in Sofia. Bulgaria is a member of several international associations of amateur art, such as the International Association of Amateur Theatres, and the International Association of the Organizers of Folk Festivals, etc.

Amateur art activities are directed by the Committee for Culture and its specialized organ for unified ideological artistic, methodological and organizational guidance – the Centre for Amateur Art Activities, set up in 1972. A state-public body, it integrates and co-ordinates the activities of the state and public organs aimed at developing amateur art. Guidance at district level is the responsibility of the district centres for amateur art activities as special departments of the district councils for culture. The councils for culture run amateur art activities in the towns and villages, and the trade union and Komsomol committees – those at the enterprises, offices and schools.

Amateur artistes in national costume

INTERNATIONAL CULTURAL CO-OPERATION

The International cultural activities of the People's Republic of Bulgaria are a component of the country's general socio-economic development and its consistent, peaceful foreign policy. They constitute a worthy contribution of the Bulgarian Communist Party and the Bulgarian Government's aspirations for mutual acquaintance, understanding and cultural communion among peoples, for strengthening world peace.

The People's Republic of Bulgaria maintains cultural relations with over 130 foreign countries. International cultural co-operation is regulated by normative acts such as bilateral treaties of friendship, co-operation and mutual assistance with the socialist countries, as well as by bilateral inter-governmental cultural exchange agreements with the other countries, bilateral plans for cultural exchange, direct agreements among cultural unions, public and state institutions, etc.

The peaceful mission of Bulgarian socialist culture, its human ideals and creative spirit have advanced the country to the front ranks of international cultural co-operation.

The co-operation between Bulgarian and world culture has increased considerably in the years since the Third Congress of Bulgarian culture of 1977. Under the direction of Lyudmila Zhivkova, President of the Bulgarian Committee for Culture from 1974 to 1981, Bulgaria's international cultural activities entered a higher phase.

Fourth Congress of Bulgarian Culture, Sofia, 1983

The Mayakovsky Dance Ensemble at the Bulgarian-Soviet Friendship House, Sofia

International Cultural Co-operation

Cultural questions constituted a major part of the theoretical and organizational work of Lyudmila Zhivkova, who considered culture a powerful instrument for bringing about progressive changes in people's minds and for strengthening the positive international processes. A respected leader and scholar, an inspired champion of cultural intercourse in the name of peace and progress, Lyudmila Zhivkova did a great deal for the promotion of Bulgarian culture.

At the fourth Congress of Bulgarian Culture in 1983 Bulgarian international cultural co-operation was highly appraised by the visiting official delegations from twelve socialist countries, 28 eminent cultural figures from 24 different countries, and by the Deputy Director General of Unesco, M. Henry Lopez.

Bulgaria maintains the most vigorous cultural relations with the USSR, forming an inherent part of the country's international cultural co-operation. The kinship between the Bulgarian and Soviet culture is based on the ethnic, linguistic and cultural similarities between the two peoples, whose supreme goal is the education of man as a universally-developed being. The foundations of the active cultural exchange between Bulgaria and the USSR were laid by the Bulgarian-Soviet Treaty for Friendship, Co-operation and Mutual Assistance of 1948. The present-day statutes governing this exchange include various types of basic international treaties: the stipulations of the 1967 Bulgarian-Soviet Treaty for Friendship, Co-operation and Mutual Assistance, and the 1976 Agreement for Scientific and Cultural Co-operation; Five-Year Plans for cultural co-operation with concrete measures envisaged in annual protocols; and direct treaties between cultural institutions and public organizations of the two countries. Of considerable importance in the enrichment of the process of cultural intercourse is the Programme for the Further Development and Strengthening of the Co-operation and for a Closer Relationship between Bulgaria and the USSR in the Cultural Sphere adopted by the Committees for Culture in 1975. 1979 saw the establishment of an intergovernmental Bulgarian-Soviet commission for cultural co-operation at ministerial level. Cultural co-operation between Bulgaria and the USSR covers all aspects of artistic creativity, efforts to preserve and disseminate cultural wealth, and cultural research.

Telling manifestations of the Bulgarian-Soviet festivals of cultural intercourse between the two peoples are the regularly-held days of culture, theatre and drama, literature and music, amateur art, etc. Prominent among these are the large-scale events to commemorate the centenary of Bulgaria's Liberation from Ottoman rule, the 1,300th anniversary of the foundation of the Bulgarian State, the 35th anniversary of the socialist revolution in Bulgaria, the centenary of the birth of Georgi Dimitrov, the 60th and the 65th anniversaries of the Great October Socialist Revolution, the 60th anniversary of the foundation of the USSR and the 110th anniversary of Lenin's birth. In recent years guest tours of Bulgaria have been made by some of the most outstanding Soviet musical, theatrical and ballet troupes such as the Bolshoi Theatre, the Moscow and Leningrad Philharmonic Orchestras, the Latvian and the Ukrainian symphony orchestras, the Academic Choir of the All-Union Radio, the Beryozka Dance Ensemble, and the Perm Ballet; and visits by eminent Soviet composers, singers and instrumentalists etc. Works by Russian and Soviet playwrights have been presented through guest performances by the Moscow Academic Arts Theatre, the Mossoviet Theatre, the Leningrad Maxim Gorky Academic Theatre, the Mayakovsky Academic Theatre in Moscow, the Vakhtangov Theatre in Moscow, the Lenin Komsomol Theatre in Moscow, the Malaya Bronnaya Theatre in Moscow, the Shota Rustavelli Theatre of Georgia, the Tashkent and Ukrainian State Puppet Theatres and others. Bulgaria has been visited by Soviet Exhibitions of Ancient Scythian Art, Decorative Glassware and Tapestry in the Soviet Union, Applied Arts of the Soviet Peoples, Lenin in the Works of the Soviet Graphic Artists, and Monuments of Old Bulgarian Culture in Soviet Libraries and Museums.

The best Bulgarian artistic groups have visited the Soviet Union, among them the Sofia National Opera, the Svetoslav Obretenov Bulgarian a cappella choir, the Stefan Makedonski Operetta Theatre, the Ivan Vazov National Theatre, the Sofia Theatre, the People's Army Theatre, the Sulza i Smyah (Tears and Laughter) Drama Theatre, and others. Representative Exhibitions of

Prague Wind Quintet at a Music Week, Sofia, 1982

The West Berlin Philharmonic Orchestra conducted by Herbert von Karajan at the Lyudmila Zhivkova People's Palace of Culture, Sofia, 1983

Bulgarian Manuscript Books, Bulgarian Medieval Civilization, the Ethnographic Treasures of Bulgaria and the Troy-Thrace Exhibition organized jointly with the GDR have visited the USSR.

Bulgaria carries out a vigorous cultural exchange programme with the other countries of the socialist community. The socialist states' common foreign policy aims and cultural goals greatly contribute to the development of cultural co-operation while preserving their national characteristics and individual cultural goals.

Prominent in this cultural exchange are the large-scale events on a bilateral and multilateral basis, with the increasing involvement of state and public artistic institutions and organizations. A significant place in Bulgaria's cultural links with the other socialist states is taken by the days of culture and various other events in the field of music, drama, the cinema, literature, architecture, arts and amateur arts.

Days of culture have been organized on a mutual basis with Czechoslovakia, Hungary, Poland, the GDR, Romania, Cuba, Mongolia and Vietnam; days of music with Czechoslovakia, Hungary and Poland; and days of literature with Czechoslovakia.

Bulgarian art has been represented by the Exhibition of Thracian Art in the GDR, Hungary, Czechoslovakia, Poland and Romania; the Exhibition of 1,000 Years of Bulgarian Iconographic Art and the Exhibition of Painting, Graphic Art and Sculpture in Poland. The Ethno-

The Bodra Smyana choir

graphic Treasures of Bulgaria and the Lights of Medieval Sofia have visited Hungary, the Troy-Thrace and Bulgarian Arts and Crafts Exhibitions have visited the German Democratic Republic, an exhibition of the works of Vladimir Dimitrov – the Master, has visited Czecholovakia and Poland, and an exhibition of Bulgarian icons has been to Cuba.

Eighteenth Congress of the International Theatre Institute, Sofia, 1979

The children's choir of Bulgarian Radio

The Sofia National Opera has given guest performances in Poland and Czechoslovakia; the Sofia State Philharmonic Orchestra in Czechoslovakia, Poland and Hungary; the Ivan Vazov National Theatre in the GDR and Poland; the Sofia Drama Theatre has visited Hungary; the Rhodopa State Folk Song and Dance Ensemble has performed in the People's Democratic Republic of Korea; and the Kaval Choir has toured Yugoslavia. Many other guest tours have also been organized.

The Bulgarian cultural public shows great interest in visiting exhibitions: Huns, Germans and Avars from Hungary; Czech Ceramics, Majolica and Porcelain; Modern Romanian Decorative Art; Glassware and Porcelain, and The Theatre in Art from the GDR; Drawings and Crafts from Cuba; Polish Artistic Textile Design from the eighteenth century to the present day; Mongolian Art; Yugoslav Art (1900–50); etc. The Budapest and Dresden Philharmonic Orchestras; the Ballet of the Prague Opera; the Berlin Symphony Orchestra; the Belgrade Madrigal Choir; the Volksbühne and the Metropol Theatre of the GDR; the Fialka Theatre of Czechoslovakia; the Drama Theatre and the Mazowsze Folklore Ensemble of Poland; the Opera of the Korean People's Democratic Republic; and the National Folklore Ensemble from Kampuchea are just some of the groups that have visited Bulgaria.

Bulgaria attaches significance to cultural co-operation with the countries of the Balkan region. Her activities in this field comply with the aims of the BCP and the Bulgarian Government's foreign policy, which are to maintain peace, good neighbourly relations and mutually advantageous co-operation. The Appeal by Todor Zhivkov, Secretary-General of the Central Committee of the BCP and State Council President, to turn the Balkans into a nuclear-free zone has been of primary importance for the promotion of the dialogue in the field of art and culture.

Bulgaria strives to expand its co-operation with all Balkan states, to strengthen and broaden its cultural contacts, and to enrich its experience and traditions.

Bulgaria has initiated and hosted a number of regional cultural events. In 1983 the Balkan Film Festival took place in Sofia. Among the more important Bulgarian cultural events in Greece, Turkey and on Cyprus have been the guest performances of the Ivan Vazov National Theatre, the Lyudmila Zhivkova National Youth Theatre, the Sofia State Philharmonic Orchestra, the Svetoslav Obretenov, State a cappella Choir, the Bodra Smyana Children's Choir, and Sofia Soloists' Chamber Music Ensemble.

The more significant cultural events organized by Balkan states in Bulgaria include the greatest performances of the Elsa Vergi Drama Theatre of Athens, the State Opera and Ballet Theatre from Ankara, and the Ankara State Folklore Ensemble.

Bulgaria attaches great importance to its cultural relations with the newly-liberated and developing countries of Asia, Africa and Latin America. The speedy development of Bulgaria's cultural relations with those states are due to its active peaceful foreign policy, which supports them in their struggle against neo-colonialism, racism and all other forms of oppression.

Todor Zhivkov with participants at the International Writers' Meeting, Sofia, 1979

Among the cultural events organized by Bulgaria in the developing countries are, for example, the Days of Bulgarian Culture in India, Angola, Tunisia, Syria, Sri Lanka, Malta and Libya; the Exhibitions of Thracian Art, the Works of Vladimir Dimitrov – the Master, and Bulgarian Icons in India and Venezuela; the concerts of the Sofia State Folk-Song and Dance Ensemble in the People's Democratic Republic of Yemen, Syria and Jordan, of the Pirin State Folk-Song and Dance Ensemble in Indonesia and the Philippines, and of the Dimov String Quartet in India. Bulgarian puppet theatres, folklore ensembles, symphony orchestras and individual artists regularly visit various developing countries.

Major cultural events organized in Bulgaria by the developing countries include the Exhibitions of Modern Indian Painting, Traditional and Modern Nigerian Art, Buddhist Paintings from Sri Lanka, Miniatures from Algeria, Makonde Ritual Masks from Tanzania, the works of O. Guayasmin from Ecuador, and 2,000 Years of Nigerian Art.

Bulgaria's cultural relations with the countries of Western Europe, the USA, Canada, Japan, Australia and others are developing in the spirit of the Helsinki Final Act. It is Bulgaria's aim to make these relations serve the causes of mutual trust and friendship among nations, the struggle for disarmament, *détente* and peaceful coexistence among nations with different social systems.

Second International Writers' Meeting, Sofia, 1979

Bulgarian cultural initiatives in the Western European countries, the USA, Canada, Japan and Australia help to increase the awareness of the public in those countries of the Bulgarian contribution to mankind's cultural development. The Exhibitions of Thracian Art, Bulgarian Medieval Civilization, 7,000 Years of Sofia, the Neolithic and Aeneolithic Ages in Bulgaria have visited Great Britain, France, the USA, the FRG, Austria, Mexico, Japan, Belgium, Sweden, Portugal, Spain and other countries.

They became the focal points of general cultural initiatives incorporating scientific forums, musical events, film weeks and literary recitals, which on many occasions became days of Bulgarian culture. Great successes in a number of capitalist states such as France, Great Britain, Holland, Spain, the USA and Japan were the concerts given by the Sofia National Opera, the Sofia State Philharmonic Orchestra, the Pioneer Philharmonic Orchestra, the Pirin, Thrace and Filip Kutev State Folk-Song and Dance Ensembles, and others. The Bulgarian cultural public for its part displayed great interest in the

International Cultural Co-operation

The Bulgarian actor Georgi Georgiev-Gets receiving the silver medal awarded to the Bulgarian feature film *Equilibrium* at the Thirteenth International Moscow Film Festival, 1983

guest performances of La Scala of Milan, the Viennese Philharmonic Orchestra, the London Philharmonic Orchestra, the French Symphony Orchestra, the Renaud-Barrault Theatre Group, the Hikossen and Noh Theatres, the Salon d'Automne de Paris, the Gold of El Dorado, Aztec Art, American Impressionists, the Exhibition of Five Centuries of Masterpieces from the collection of Dr Armand Hammer, as well as the exhibitions of works by Turner, Hundertwasser, Orozco, and Henry Moore.

The Bulgarian centres for cultural information abroad are authoritative institutions which actively assist Bulgaria's international cultural activities by popularizing its achievements in socio-economic, political and cultural life. They organize numerous events such as Days of Bulgarian Culture in the capitals and the larger cities of the relevant countries, as well as exhibitions, scientific forums, competitions and quizzes, film shows, literary events, meetings with Bulgarian writers and poets, and press conferences. They also maintain close contacts with cultural institutions, public and state organizations, educational institutions, mass media and publishing houses. Some of the centres have clubs for Bulgarian scholars, translators and friends of Bulgarian literature.

Bulgaria considers her cultural activities in Unesco and non-governmental cultural organizations to be particularly important. Thanks to the initiative and energy on the

First International Children's Assembly, Banner of Peace, Sofia, 1979

Children's Carnival during the Second International Assembly, Banner of Peace, Sofia, 1982

Second International Children's Assembly, Banner of Peace, Sofia, 1982

part of Bulgaria, and personally to Lyudmila Zhivkova's activities at the 21st Session of the Unesco General Conference in Belgrade, eight resolutions were adopted in the field of culture, as well as a Unesco special resolution for the member-states to commemorate the 1,300th anniversary of the Bulgarian State and its inclusion in the Unesco calender for outstanding personalities and events. The Days of Bulgaria in Unesco in 1981, the inclusion of six Bulgarian historical monuments on the World Cultural Heritage list, Bulgarian fiction published in conjunction with Unesco and the organization of a number of joint, scholarly forums have raised the prestige of Bulgaria's cultural heritage and of her modern socialist culture. The Unesco *Courier* is now published in Bulgarian, and there is a special bookshop for Unesco publications in Sofia. Through its international cultural activities the People's Republic of Bulgaria supports the progressive trends in Unesco and contributes toward its more active participation, through the specific means at its disposal, in the struggle for peace.

The Bulgarian contribution to international cultural cooperation has been highly praised at the 1982 World Conference on Cultural Policies in Mexico, at the Fourth Extraordinary General Conference, and at the recent 22nd Session of the General Conference in Paris in 1983. The 23rd Session of the Unesco General Conference (which will commemorate the organization's 40th anniversary) will take place in Bulgaria in 1985.

Bulgaria has initiated and hosted a number of significant international cultural events. In 1979, the International Year of the Child, on the initiative of Bulgaria's Banner of Peace Movement, noted for its deeply humane and internationalist nature, was launched. It was another great achievement of Lyudmila Zhivkova's public, scientific and creative activities. An expression of the BCP's great support for international recognition of this initiative is the fact that the First (1979) and Second (1982) Banner of Peace Assemblies were held under the honorary patronage of the General Secretary of the Central Committee of the BCP and Bulgarian State Council

Leonard Bernstein rehearses the Pioneer Philharmonic Orchestra for a concert at the United Nations

President, Todor Zhivkov, and the Unesco Director-General, Amadou Mahter M'Bow. Within a very short period the Banner of Peace Movement won thousands of adherents all over the world. High points in the movement's development were the First International Children's Assembly in 1979, the Sofia '80 and Sofia '81 meetings, and the Second International Children's Assembly in 1982, to which 116 countries sent over 4,500 children and young people with outstanding achievements in literature, art, music, science, technology and sports. A significant proportion of the participants were the holders of awards and distinctions from international and national competitions and festivals, which gave the forums in Sofia the character of world festivals of young people's creativity.

Bulgaria's cultural activities have helped to expand the Banner of Peace Centre in Sofia into a centre for research and theoretical and practical work in children's art education, in the spirit of peace, mutual understanding and fruitful co-operation among the younger generation. Bulgaria has played host to numerous meetings and discussions of progressive cultural figures from all over the world. Four international writers' meetings under the slogan 'Peace – the Hope of the Planet' have been held in Sofia in 1977, 1979, 1980 and 1982, with the participation of eminent writers from all over the world; the Fourth Meeting, in 1982, was attended by 154 writers from 57 countries, who expressed their optimism and

Lyudmila Zhivkova opens the International Children's Art exhibition at the First International Children's Assembly, Banner of Peace, Sofia, 1979

First International Congress on Bulgarian Studies, Sofia, 1981

confidence that the aim of art is to serve man's magnanimous aspirations for moral improvement and peaceful and creative labour. In 1982 there was a Theatre of Nations' Season involving the participation of renowned drama groups from many countries. The great interest shown in the First International Exhibition of Realist Paintings in 1973 led to the decision to hold it regularly every three years. At the latest fourth triennial (1982), 110 artists from 23 countries took part. The People's Republic of Bulgaria has been host to the World Biennial of Architecture, the International Exhibition entitled *Satire and the Struggle for Peace,* the World Philatelic Exhibition under the slogan of 'Philately in Service of Peace' (1979), the First International Bulgarian Studies Congress (1981), the International Theatre Poster Exhibition (1982), the Seventh Regional Congress of INSEA (1983), and a number of other events.

On the personal initiative of Lyudmila Zhivkova, Bulgaria played host to the International Meeting and Dialogue on the Importance of Culture for the Development of Man and Society, in 1980, and the International Meeting of the Heads of National Committees for Children's Affairs, in 1981. Other important international meetings held in Bulgaria include the symposium on 'Humanism and the Development of Culture. The Historical Fate of Humanism in the Eastern Orthodox Church Culture and in the Western European Renaissance', the international creative meeting and talk on Lenin and the Contemporary World in 1980, the international meeting of film-makers on the Cinema as Art and as a Means for Communication in the Contemporary World, etc.

World-famous musicians and performers, and well-

A parade on 24 May

Vladimir Zhivkov, General Director of the Banner of Peace Centre, closing the Second International Children's Assembly, Sofia, 1982

known works are presented at the prestigious international festivals and competitions in the territory of Bulgaria, such as the Sofia Weeks of Music symphonic music festival, the Varna Summer and the March Musical Days festivals, the Chamber Music Festival in Plovdiv, the International Folklore Festival in Burgas, the International Festival of Red Cross and Medical Films in Varna, the International Festival of Animated Cartoons in Varna, and the International Book-Fair in Sofia.

Bulgaria awards prizes and distinctions to foreign cultural figures for their progressive works and their contribution to the cultural wealth of mankind. Among those awarded the Dimitrov Prize for Arts, Literature and Science are Gabriel Garcia Marquez of Columbia, Charles Percy Snow of Great Britain, Henry Winston of the USA and Karl Bleha of Austria.

The Nikola Vaptsarov International Award has been given to Amrita Prittam of India, Eduardis Mejelaitis of the USSR, Ignasio Buttita of Italy, William Meredith of the USA, and Jarosław Iwaszkiewicz of Poland.

The Hristo Botev International Award has been presented to Ahmad Sulayman al-Ahmad of Syria, Miroslav Karleza of Yugoslavia, Rasul Gamzatov of the USSR and Rafael Alberti of Spain. A great number of other foreign cultural figures have been distinguished with the International Brothers Cyril and Methodius and Patriarch Euthymius Awards, the Sofia Grand Prix for Literature and the Hiter Peter International Award for Humour and Satire.

In 1981 Bulgaria celebrated the 1,300th anniversary of the foundation of the Bulgarian State. The anniversary was celebrated in more than 100 countries by approximately 4,100 cultural events held between 1978 and 1981. The most prestigious international governmental and non-governmental organizations, heads of state, parliaments and political parties, as well as eminent public, scientific and cultural figures expressed their respect for the Bulgarian people.

Bulgaria's international cultural activity in the jubilee year was in harmony with the persistently peaceful foreign policy aims of the BCP and the Bulgarian State.

The jubilee was an opportunity to expand and deepen Bulgaria's cultural relations and co-operation with many countries and to increase and strengthen the direct contacts between Bulgarian and foreign cultural institutions and organizations, thus winning new friends for Bulgaria. The jubilee provided the occasion for a closer study of the role of Bulgarian culture in the development of human civilization and for popularizing modern socialist cultural achievements.

The 62 jubilee committees set up in 54 countries with the participation of eminent foreign public figures,

International Cultural Co-operation

The Lyudmila Zhivkova People's Palace of Culture, Sofia

Lyudmila Zhivkova and Svetoslav Roerich, Sofia, 1978

scholars and cultural figures, also contributed to the celebration of the 1,300th anniversary of the foundation of the Bulgarian State.

Along with the new experience accumulated and the new ideals and forms to which it gave birth, the jubilee brought back some valuable Bulgarian traditions such as voluntary donations for public causes. The Thirteenth-Century Bulgaria Fund, a combined state and public organization, was set up with the aim of stimulating and expanding the donation movement.

The sincere desire to promote and further popularize the highly humane cause whose champion was Lyudmila Zhivkova, brought into existence a new type of organization, the Lyudmila Zhivkova International Foundation. The very essence of multifarious activities of the patron of this foundation determines its international nature. Its aims have been defined in its Constituent Statutes as follows:

(a) to study, develop and popularize the magnanimous ideas and initiatives of Lyudmila Zhivkova for the harmonious development of children and young people, the development of man's creative gifts and the strengthening of peace and understanding among peoples;

(b) to encourage the creative spirit and to promote the dissemination of achievements in the fields of culture, art, science and education;

(c) to discover gifted children and young people and to assist their development as future creative personalities for the sake of human peace and progress.

Apart from the Bulgarian organizations represented at the inauguration of the Foundation, eminent foreign representatives of academic, cultural and business circles abroad took part, among them Sergei Mikhalkov, Dr Armand Hammer, Rafael Alberti, Robert Maxwell, Hervé Bazin and Lorenzo Gallo.

Eighteenth Golden Orpheus international festival of Bulgarian pop songs, Slunchev Bryag, 1982

International Ballet Competition, Varna, 1983

The foundation achieves its aims by organizing cultural events and academic forums, publishing books for young people and children, and granting scholarships for outstanding achievements in the field of culture, art, science and education.

From the very first, the Lyudmila Zhivkova Foundation has enjoyed considerable renown, as is shown by the numerous donations and legacies it has received from Bulgarian and foreign organizations and individuals.

The international cultural programme of the People's Republic of Bulgaria is supported by closest co-operation on the part of creative and production organizations and the creative unions in Bulgaria. This includes the activities of such public organizations as the Central Committee of the Dimitrov Young Communist League, the National Council of the Fatherland Front, the Central Council of the Bulgarian Trade Unions, and the District Councils of Culture.

Such is the broad public basis that guarantees the widest participation of the representatives and creators of Bulgarian culture in cultural exchange. Bulgaria's comprehensive, international, cultural activities and the wide response abroad to its initiatives all confirm the historical and modern achievements of Bulgarian culture and establish socialist Bulgaria's place in world culture.

SELECTED BIBLIOGRAPHY

Literature

ANGELOV, B. *Suvremennitsi na Paissii* (Contemporaries of Paissii). Books 1–2. Sofia, 1963–64.
ANGELOV, B. *V zorata na bulgarskata vuzrozhdenska literatura* (The Dawn of the Bulgarian Literature of the Revival Period). Sofia, 1969.
BAKALOV, T. *Literaturni statii i izsledvaniya* (Articles and Studies on Literature). Sofia, 1973.
DELCHEV, B. *Rodeni mezhdu dve voini* (Born Between Two Wars). Sofia, 1963.
DIMITROV, Georgi *Za literaturata i izkustvoto* (On Literature and Art). Sofia, 1982.
DIMOV, G. *Bulgarskata literaturna kritika prez Vuzrazhdaneto* (Bulgarian Literary Criticism During the Revival Period). Sofia, 1965.
DIMOV, G. *Bulgarskata marksicheska kritika i razvitieto na natsionalnata ni literatura* (Bulgarian Marxist Criticism and the Development of Bulgarian National Literature). Sofia, 1980.
DINEKOV, P. *Iz istoriyata na bulgarskata literatura* (On the History of Bulgarian Literature). Sofia, 1969.
DINEKOV, P. *Mezhdu folklora i literaturata* (Between Folklore and Literature). Sofia, 1978.
DINEKOV, P. *Vuzrozhdenski pisateli* (Writers of the Revival Period). 2nd ed. Sofia, 1964.
GACHEV, G. *Uskorenoto razvitie na kulturata* (The Accelerated Development of Culture). Sofia, 1979.
Istoriya na bulgarskata literatura (A History of Bulgarian Literature). In 4 vols. Sofia, 1962–76.
KOLAROV, S. *Po aprilskata magistrala na vremeto* (The New Vistas of Literature Opened up by the April 1956 Party Line). Sofia, 1983.
KOLEVSKI, V. *Literatura na svobodata* (The Literature of Freedom). Sofia, 1983.
KONEV, I. *Literaturni vzaimootnosheniya i literaturen protses* (Literary Relations and Literary Process). Sofia, 1974.
KONSTANTINOV, G. *Bulgarskata literatura sled voinata* (Bulgarian Post-War Literature). Vol. 1. Sofia, 1933.
KONSTANTINOV, G. *Pisateli-realisti ot Drumev do nashi dni* (Realistic Writers Since V. Drumev). Sofia, 1965.
LEKOV, D. *Literatura, obshtestvo, kultura* (Literature, Society, and Culture). Sofia, 1982.
LEKOV, D. *Problemi no bulgarskata belatristika prez Vuzrazhdaneto* (Problems of Bulgarian Fiction During the Revival Period). Sofia, 1970.
NICHEV, B. *Suvremenniyat Bulgarski roman* (The Modern Bulgarian Novel). Sofia, 1981.
Ochertsi po istoriya na bulgarskata literatura sled Deveti septemvri 1944 godina (An Outline of the History of Bulgarian Literature after 9 September 1944). Books 1–2. Sofia, 1979–80.
PAVLOV, T. *Izbrani proizvedeniya* (Selected Works). Vol. 7. Sofia, 1964.
PENCHEV, G. *Suvremennost i minalo* (The Present and the Past). Sofia, 1975.
PENEV, B. *Istoriya na novata bulgarska literatura* (A History of Modern Bulgarian Literature). Vols. 1–4. Sofia, 1976–78.
STANCHEV, K. *Poetika na starobulgarskata literatura* (The Poetics of Old Bulgarian Literature). Sofia, 1982.
TSANEV, G. *Stranitsi ot istoriyata na bulgarskata literatura* (Essays on the History of Bulgarian Literature). In 4 vols. Sofia, 1967–75.
TSANEVA, M. *Pisateli i tvorbi* (Writers and Works). Sofia, 1980.
VUCHKOV, Y. *Bulgarska dramaturgiya 1878–1944* (Bulgarian Dramaturgy in 1878–1944). Sofia, 1983.
VUCHKOV, Y. *Bulgarska dramaturgiya 1944–79* (Bulgarian Dramaturgy in 1944–79). Sofia, 1981.
Za literaturnite zhanrove prez Bulgarskoto vuzrahdane (On the Literary Genres During the Bulgarian Revival Period). Sofia, 1979.
ZAREV, P. *Panorama na bulgarskata literatura* (A Panorama of Bulgarian Literature). In 5 vols. Sofia, 1969–76.
ZAREV, P. *Preobrazena literatura* (A New Literature). Sofia, 1969.
ZAREV, P. *Suchineniya v 3 toma. T.1 Literaturna istoriya* (Works in three volumes. Vol. 1. Literary History). Sofia, 1981.
ZHIVKOV, T. *Za literaturata* (On Literature). Sofia, 1981.

Culture

ARBALIEV, G. *Bulgarskata vuzrozhdenska kushta i neinata ukrasa* (The Bulgarian House of the Revival Period and its Ornamentation). Sofia, 1974.
Arhitekturata v Bulgaria, 1878–1944 (Bulgarian Architecture, 1878–1944). Sofia, 1975.
Arhitekturata na bulgarskoto vuzrazhdane (The Architecture of the Bulgarian Revival Period). Sofia, 1975.
Arhitekturata na Narodna Republika Bulgaria (The Architecture of the People's Republic of Bulgaria). Sofia, 1975.
ARNAUDOV, M. *Ochertsi po bulgarskiya folklor* (Essays on Bulgarian Folklore). Vols. 1–2. Sofia, 1968–69.
BOSILKOV, S. *Bulgarskiyat plakat. Predvestnitsi, ranni proyavi, suvremenno razvitie* (The Bulgarian Poster. First Steps and Modern Development). Sofia, 1973.
BOYADJIEV, S., N. CHANEVA-DECHEVSKA and L. ZAHARIEVA *Izsledvaniya vurhu arhitekturata na Bulgarskoto srednovekovie* (Studies on the Architecture of Medieval Bulgaria). Sofia, 1982.
BOZHKOV, A. *Die Bulgarische Malerei.* Recklinghausen, 1969.
BOZHKOV, A. *Die Bulgarische Volkskunst.* Recklinghausen, 1972.
BOZHKOV, A. *Bulgarskata istoricheska zhivopis* (Bulgarian Historical Painting). Parts 1–2. Sofia, 1972–78.
BOZHKOV, A. *Trevnenskata zhivopisna shkola* (The Tryavna School of Painting). Sofia, 1967.
Bulgarska narodna kultura (The Bulgarian People's Culture). Sofia, 1981.
Bulgarskata arhitektura prez vekovete (Bulgarian Architecture Through the Ages). Sofia, 1982.
Bulgarskata literatura i narodnoto tvorchestvo (Bulgarian Literature and Folklore). Ed. D. Lekov. Sofia, 1977.
CHUHOVSKI, P. *Suvremenna bulgarska grafika* (Contemporary Graphic Arts). An Album. Sofia, 1971.
DINEKOV, P. *Bulgarski folklor* (Bulgarian Folklore). Part 1. 3rd edn. Sofia, 1980.
DINOVA-RUSSEVA, V. *Bulgarska Stsenografiya* (Bulgarian Stage Design). Sofia, 1975.

Entsiklopediya na bulgarskata muzikalna kultura (An Encyclopaedia of Bulgarian Music). Sofia, 1967.
Etnografiya na Bulgaria (Bulgarian Ethnography). In 3 vols. Sofia, 1980.
FILOV, B. *Starobulgarsko izkustvo* (Old Bulgarian Art). Sofia, 1924.
FOL, A. and I. MARAZOV *Thrace and the Thracians*. London, 1977.
Folklor i obshtestvo (Folklore and Society). Sofia, 1977.
Folklorut i narodnite traditsii v suvremennata natsionalna kultura (Folk and Popular Traditions in Contemporary Bulgarian Culture). Sofia, 1976.
GRABAR, A. *La peinture religieuse en Bulgarie*. Pref. de G. Millet. Vols. 1–2. Paris, 1928.
Istoriya na bulgarskoto izobrazitelno izkustvo (A History of Bulgarian Fine Arts). Vol. 1. Sofia, 1976.
IVANOVA, V. *Suvremenna bulgarska skulptura* (Contemporary Bulgarian Sculpture). An Album. Sofia, 1971.
KARAKOSTOV, S. *Bulgariskiyat teatur. Srednovekovie, renesans, prosveshtenie* (Bulgarian Theatre, Middle Ages, Rennaissance, Enlightenment). Sofia, 1972.
KARAKOSTOV, S. *Bulgarskiyat vuzrozhdenski teatur na osvoboditelnata borba* (The Bulgarian Theatre During the Period of the National Revival Struggle). Sofia, 1973.
KOLAROVSKI, V. *Bulgarska peizazhna zhivopis* (Bulgarian Landscape Painting). Sofia, 1979.
KOZHUHAROV, G. *Bulgarskata kushta prez pet stoletiya* (The Bulgarian House Through Five Ages). Sofia, 1967.
Kratka istoriya na bulgarskata arhitektura (A Short History of Bulgarian Architecture). Sofia, 1965.
KRUSTEV, K. *Nachenki na Renesans v Srednovekovna Bulgaria* (The Beginnings of Rennaissance in Medieval Bulgaria). Sofia, 1971.
KRUSTEV, K. and V. ZAHARIEV *Stara bulgarska zhivopis* (Old Bulgarian Painting). An Album. 2nd edn. Sofia, 1961.
KRUSTEV, V. *Ochertsi po istoriya na bulgarskata muzika* (Essays on the History of Bulgarian Music). 2nd enl. edn. Sofia, 1977.
MAVRODINOV, N. *Izkustvoto na bulagarskoto vuzrazhdane* (The Art of the Bulgarian Revival Period). Sofia, 1957.
MAVRODINOV, N. *Novata bulgarska zhivopis* (The New Bulgarian Painting). Sofia, 1947.
MAVRODINOV, N. *Starobulgarskata zhivopis* (Old Bulgarian Painting). Sofia, 1946.
MAVRODINOV, N. *Starobulgarskoto izkustvo. Izkustvoto na Purvoto bulgarsko tsarstvo* (Old Bulgarian Art. The Art of the First Bulgarian Kingdom). Sofia, 1959.
MAVRODINOV, N. *Starobulgarskoto izkustvo. XI–XIII v* (Old Bulgarian Art in the Eleventh–Thirteenth Centuries). Sofia, 1966.
MIHALCHEVA, I. *Portretut v bulgarskata zhivopis* (The Portrait in Bulgarian Painting). Parts 1–2. Sofia, 1968–71.
MIYATEV, K. *Die Mittelalterliche Baukunst in Bulgarien*. Sofia, 1974.
MIYATEV, K. 'Ikonite v Bulgaria'. V: *Ikoni ot Balkanite, Sinai, Gurtsia, Bulgaria, Yugoslavia* ('Icons in Bulgaria'. In: Icons from the Balkans, Sinai, Greece, Bulgaria and Yugoslavia). Sofia and Belgrade, 1966.
Obredi i obreden folklor (Rites and Ritual Folklore). Sofia, 1981.
PENEV, P. *Istoriya na bulgarskiya dramaticheski teatur* (A History of Bulgarian Drama). Sofia, 1975.
PROTICH, A. *Petdeset godini bulgarsko izkustvo* (Fifty Years of Bulgarian Art). Books 1–2. Sofia, 1933–34.
SHMIRGELA, N. *Skulpturata po nashite zemi* (Sculpture in the Bulgarian Lands). Sofia, 1961.
STOIKOV, A. *Bulgarskata karikatura. Kratuk ocherk za neiniya put i za tvorchestvoto na A. Bozhinov, I. Beshkov, A. Zhendov, B. Angelushev i S. Venev* (The Caricature in Bulgarian Art. A Short Outline of its Development and of the Work of A. Bozhinov, I. Beshkov, A. Zhendov, B. Angelushev, and S. Venev). Sofia, 1970.
Suvremenno bulgarsko izkustvo (Contemporary Bulgarian Art). An Album. Sofia, 1982.
TODOROV, T. *Suvremenni problemi v izuchavaneto na bulgarskoto narodno muzikalno tvorchestvo* (Present-Day Problems in the Study of Bulgarian Musical Folklore). Sofia, 1974.
TODOROV, T. *Suvremennost i narodna pesen* (Modern Times and the Folk Song). Sofia, 1978.
TOMOV, E. *Bulgarska grafika. Gravyura.* (Bulgarian Graphic Art. Engravings). An Album. Sofia, 1955.
TSONCHEVA, M. *Bulgarsko vuzrazhdane. Zhivopis i grafika* (The Bulgarian Revival Period. Oil Painting and Graphic Arts). Sofia, 1962.
TSONCHEVA, M. *Hudozhestvenoto nasledstvo na trakiiskite zemi* (The Artistic Heritage of the Thracian Lands). Sofia, 1971.
VAKARELSKI, H. *Etnografiya na Bulgaria* (Bulgarian Ethnography). 2nd edn. Sofia, 1977.
VAKLINOV, S. *Formirane na starobulgarskata kultura VI–XI vek.* (The Formation of Old Bulgarian Culture in the Sixth–Eleventh Centuries). Sofia, 1977.
VASILEV, A. *Bulgarski vuzrozhdenski maistori. Zhivopistsi, rezbari, stroiteli* (Bulgarian Masters of the Revival Period. Painters, Wood-carvers, Builders). Sofia, 1965.
VENEDIKOV, I. and T. GERASIMOV *Trakiiskoto izkustvo* (Thracian Art). Sofia, 1973.
ZHIVKOV, T. I. *Folklor i suvremennost* (Folklore and the Modern Times). Sofia, 1981.
ZHIVKOV, T. I. *Narod i pesen* (People and Songs). Sofia, 1977.

PART X

CULTURAL AND EDUCATIONAL ESTABLISHMENTS

THE LIBRARY CLUBS

The Library Clubs are originally Bulgarian mass cultural and educational institutions which emerged in the middle of the nineteenth century. In their structure, objectives and forms of activity and development they were an unprecedented phenomenon in the field of education and culture and have made a historical contribution to the formation and consolidation of the Bulgarian nation, the affirmation and development of the Bulgarian literary language and spiritual culture.

Having appeared during the conditions of Ottoman rule they set themselves the major task of raising the educational standards of the population and working for its moral perfection, for the spread of Bulgarian literature and for strengthening the awakened national consciousness. The first three library clubs were founded in 1856 in Svishtov, Lom and Shumen on the initiative of enlightened Bulgarians. The idea was adopted also by other towns and the number of library clubs rapidly increased. In 1861 Bulgarian patriots who had emigrated to Bucharest founded a library club which they named Bratska Lyubov (Brotherly Love) and in 1866 the Bulgarian colony in Constantinople set up a library club, which published the *Chitalishte* magazine (1870–75).

From their very emergence the library clubs became powerful cultural and educational centres of the community, places of self-education of the population, public expression and many other such initiatives and forms of activity. Libraries and reading rooms were set up, lectures were held, and amateur dramatic and musical activities underwent rapid development. Choirs and orchestras were formed, and some of the community centres also set up museum collections and published newspapers. They also became centres for propagating progressive views and revolutionary ideas; a great many library clubs played a significant role in the national liberation movement, especially in the preparation of the April 1876 Uprising.

Characteristic of the library clubs was their joint activity with the schools. The library club board of trustees was often the most important factor in choosing the school board, in the opening of new schools and in finding capable teachers.

As many as 186 library clubs were founded before the Liberation from Ottoman rule in 1878. Dimiter Blagoev wrote that 'prior to the Liberation the library club was the favourite place for every Bulgarian intellectual and patriot. For the adults it was a school and a forum

The Videlina Community Centre in Pancharevo, Sofia

for the settlement of public questions and for celebrating patriotic festivals, a promoter of revolutionary spirit, of freedom and justice . . . it reflected the alert social spirit'.

In the centuries-long history of Bulgaria the schools and community centres were among the institutions with the most creditable record for conspicuous contribution to the spiritual growth of the Bulgarian people. In them flourished the national art and culture and through them the most positive traditions of the Bulgarian heritage were created, preserved and transferred to the coming generations.

After the Liberation of Bulgaria from Ottoman rule the number of community centres increased still more rapidly, the forms of mass cultural and educational activity were diversified, the importance of the libraries and reading rooms increased, and new tasks were posed. Alongside the propagation of progressive views and revolutionary ideas they also awakened the consciousness of the working class.

The Third Congress of the Bulgarian Workers' Social-Democratic Party in 1893 recorded its decision that 'Municipal councils must organize free library clubs and libararies for the workers'.

Lectures and discussions on a wide range of subjects were also organized by the 'library club popular universities'. The collection of exhibits was the nucleus in the setting up of museum collections.

In 1911 a General Union of Popular Library Clubs (renamed Supreme Library Club-Union in 1923) was set up in order to consolidate the popular library clubs. Library club laws were passed to settle questions connected with their material and technical facilities, the organization of courses for librarians, the holding of library club conferences, the perfection of the structure and other problems of decisive significance for the development of clubs.

In the years of Monarcho-Fascism the community centres became legal centres for the anti-fascist activity of the Party and for rallying progressive forces. Under the leadership of the Bulgarian Communist Party, the underground central body of library club activitists played a major role, directing the activity of the library clubs throughout the country. Along with books by prominent Russian and West-European authors, Soviet books were also procured for the club libraries.

A significant role was also played by the library cinemas, in which Soviet films were welcomed with particularly great interest.

As many as 480 community centres were closed down for anti-fascist activities in the 1941–44 period, and many library club workers were subjected to persecution.

After the victory of the socialist revolution in 1944, along with the enhancement of the functions and goals of the popular library clubs, their tasks changed fundamentally in keeping with the new requirements. Their democratic traditions were enriched and their activity in the cause of the education of the people and the development of the individual and society intensified. The managements of the clubs (ruling councils) assumed new and higher functions, becoming unified public-state bodies for the administration and leadership of cultural processes and cultural life in centres of population. The organizational, ideological and creative activity of the popular library club was enriched particularly at the centres of settlement systems, where they developed their educational, aesthetic and mass cultural work into a broad social function.

The Dobri Voinikov Community Centre, Shumen

An amateur artistic group at the Bozhurishte Community Centre in the Sofia District

The principle of the organization of cultural activities, the detection of and concern with the most topical problems of the day, the fostering of close relations between the creative elements and the common people, and unreserved service to the country are traditions born in the popular library clubs. The socialist revolution in Bulgaria preserved, enriched and developed these traditions in the new circumstances.

The library clubs in the district centres and some of the larger towns began looking for and applying new forms of cultural and educational activity conducive to the establishment and strengthening of social contacts, making every community centre a place of interesting and

The Library Clubs

attractive cultural recreaction. Particularly valuable are the original forms and initiatives for mass work with children, designed to stimulate the creative resources of the young generation. They have created as a natural development the basis on which the Programme for Nationwide Aesthetic Education and the Banner of Peace movement was developed. All the necessary prerequisites are being created in the popular library clubs for the awakening, manifestation and development of the creative resources and gifts of every individual, of every member of society.

An amateur artistic group at a community centre in Stara Zagora

On the basis of the Programme for Nation-wide Aesthetic Education, the community centres are carrying out rich and varied creative, cultural activities. While in 1939 the libraries at the clubs possessed about 2,000,000 books, their stock of books now is nearly 31,500,000 volumes. They serve about 1,750,000 readers annually and lend 24,000,000 volumes. Nearly 200,000 amateur artistes, who give 40,000 to 45,000 shows and concerts, are united in more than 11,000 amateur groups creating original work in the spirit of the healthy folk traditions. Some 1,200 popular universities with general and specialized themes, in which more the 17,000 lectures are given every year, are also functioning on the basis of the community centres. Lively activities are carried out by more than 1,500 special interest clubs and societies with nearly 26,500 participants, and by more than 400 schools and courses for foreign languages with about 26,000 participants; the museum collections have now exceeded 500. Over 29,000 children attend the community centre schools, 2,500 the ballet schools, and over 20,000 the special interest clubs and societies; some 90,000 Chavdarche and Pioneer children take part in the different forms of amateur art activities. In the preparations for, and during the Banner of Peace Second International Assembly, exhibitions of drawings by 60,000 children were organized; more than 9,000 children manifested their abilities in the sphere of music, performing art and composition, and 9,200 presented their works to the literary reviews.

With every passing year the material foundations of the popular library clubs are being enriched and expanded. More than half of the total of 4,252 community centres have modern buildings, which afford opportunities for rich and varied cultural and creative activity.

The Twelfth Congress of the BCP sets as a fundamental task for the entire cultural front to continue to foster the growth of culture even more as a basic factor in the development of man and society, in the formation of a many-sided, balanced, versatile and harmoniously developed personality. This high requirement applies with particular urgency to the popular library clubs, which are most directly connected with the working people where they actually live. The clubs are required to enhance their role and constantly to enrich their activities. With all their diverse activity they contribute to the active involvement of working people in the reaffirmation of the socialist way of life, and the increasing satisfaction of the constantly growing cultural needs and interests of the population in the People's Republic of Bulgaria.

LIBRARIES

The first libraries in Bulgaria appeared in the monasteries, the centres of cultural life in the Middle Ages. There are records of the existence of libraries in the towns of Preslav and Ohrid, the two chief centres of Bulgarian culture, as early as the ninth century. The stocks of books at the monastery libraries consisted mainly of religious literature and allegories and short stories of a moralizing nature.

During the Ottoman domination (1396–1878) more libraries came into being, the major characteristic feature of which was their democratism. The libraries founded at the schools and community centres were the major lever in the spiritual revival of the Bulgarian people. The first school library was founded at the grade school in the town of Gabrovo in 1837, and later, in 1856, libraries were opened at the community centres in the towns of Lom, Svishtov and Shumen. The year 1869 saw the foundation of the first scientific library at the Bulgarian Learned Society, which later grew into what is known today as the Central Library of the Bulgarian Academy of Science.

Immediately after the Liberation in 1878 a public library was founded in Sofia (the Cyril and Methodius National Library of today), and in 1888 the University Library, from which the present Central Medical Library branched off in 1918, a number of departmental and municipal libraries, and many school and community centre libraries were instituted. The Publications Depositing Act of 1897 ensured the supply and enrichment of the acquisition of the National Library, as well as of a number of other libraries, with newly published books. In 1939 Bulgaria had four national libraries (in Sofia, Plovdiv, Turnovo and Shumen), four municipal libraries and 2,608 school, community and departmental libraries.

The speedier development of the libraries in Bulgaria along more modern lines is the direct result of the victory of the 9 September 1944 socialist revolution. The libraries now had important tasks to fulfil in the consolidation and further development of socialist culture. The beginnings were laid for the planned construction of a network of territorial and departmental libraries. The year 1953 saw the foundation of the district libraries, and the number of trade union, community centre and school libraries was also increased. The Central Agricultural Library was set up in 1961 and the Central Scientific and Technological Library in 1962. The development of higher education and the opening of new higher educational establishments brought new libraries at each of them. Special priority was

The library of the Dimiter Ganev non-ferrous metal works, Sofia

given to the development of the Cyril and Methodius National Library.

The growth of the network of libraries made it necessary to create a unified library system, legal provision for which was made by Decree No. 2 of the Council of Ministers of 20 January 1970. The unification and standardization of book processing gained in scope, and the development of research and its application in practice advanced further. Work was started on the centralization of some of the major processing operations in the community centre libraries. In 1975 the district libraries in Plovdiv, Sofia, Russe and Varna rose to the rank of regional libraries. The year 1978 marked the beginning of the building up of district Library Departments on the basis of the already existing district libraries.

Libraries play an important role in all the spheres of public and academic pursuits. By their specific means and methods they contribute to the formation of the socialist personality and the establishment of the socialist way of life, to the introduction of scientific and technological progress and a higher efficiency of science in production, and to the enhancement of the cultural and spiritual life of the country.

process and research work, and to provide library services for the academic and administrative staff and for the students.

Specialized libraries have been set up at scientific, technological and medical institutes, central departments, party and public organizations, theatres, museums and archives. In most cases they constitute a part of the above-mentioned institutes and organizations and operate as independent units or as parts of their information centres.

A study room with audio-visual equipment in the State Institute of Librarianship, Sofia

The library of the Pravets Cultural Centre

The unified library system of the People's Republic of Bulgaria has the following types of libraries, according to the types of service they provide: a national library, central specialized libraries, district libraries and libraries at higher educational establishments, specialized libraries, community centre libraries, school libraries and libraries at different enterprises and offices.

The central specialized libraries are: the Central Library of the Bulgarian Academy of Science, the Central Scientific and Technological Library, the Central Medical Library and the Central Agricultural Library. These are the major research libraries in the respective fields, centres of specialized bibliographies, which provide methodological guidance for the network of specialized libraries.

There are 27 district libraries. They are the major, general popular libraries on the territory of a district, depositories of the archives of regional ethnographic studies and of the local Press, and centres which provide all types of library services. Ten of them enjoy the privilege of receiving deposit copies free of charge.

There are 26 libraries at higher education establishments. Their major task is to facilitate the educational

The community centre libraries play a major role in providing territorial library services, accessible to all strata of the population. Among the biggest community club libraries are the Georgi Dimitrov Library in Pleven, Rodina in Stara Zagora, Iskura in Kazanluk, Zora in Sliven, and Yordan Yovkov in Tolbuhin.

The libraries of enterprises and offices are united in an independent network. Their basic task is to contribute to the further advancement of the commmunist education of the working people and their professional training and skills, to promote the development of technological thought and to enhance the general culture and knowledge of workers and employees. The libraries at the trade union houses of culture and holiday houses also belong here.

Libraries have also been set up at all types of schools in the People's Republic of Bulgaria, to assist teachers and pupils in the educational process.

The State guides the work of libraries through the Committee for Culture, which has supra-departmental functions. This guidance is exercised directly in respect of the district and community centre libraries, while for the central departments and organizations, which have library networks of their own, it is done through the Inter-departmental Co-ordination Council. The separate central departments and organizations guide and control the work of the libraries within their jurisdiction; they are responsible for their activities and ensure their planned development and the interaction between all the libraries within the respective network. The work of the district and community centre libraries is guided by the respective district councils for culture, under which district library departments are set up. Methodological guidance is ensured on a national territorial and departmental level.

A reading room in the Cyril and Methodius National Library, Sofia

The Cyril and Methodius National Library, Sofia

The Cyril and Methodius National Library is the body which provides methodological guidance for, and coordinates the work of, the whole library system in the country through its specialized centre for scientific and methodological guidance and through the various functional sub-systems constituting its structure.

Two councils have been set up to ensure a better organization of the Unified Library System: the Board of Directors of the Central Scientific Libraries and the Board of Directors of the District Libraries. Their major task is to consider and find solutions to the problems of developing libraries on a national and territorial level.

There were very few qualified librarians in Bulgaria before 1944 and those worked mainly in the big libraries. After the socialist revolution the Popular Government did its best for the training of library personnel. At first a training course was organized at the National Library, which in 1950 was transformed into the State Library Institute. The Institute provided a regular two-year course of training for librarians and information officers, graduates having the qualifications of specialists with semi-higher education.

In 1953 a department of library science, bibliography and scientific information was set up at Sofia University. The course is three and half years and can be taken in parallel with the major specialism. The department trains librarians and information officers with higher education.

In spite of the important role these educational establishments play, the need to train more librarians has also made it necessary to set up special courses at the library science department of Sofia University and at the Cyril and Methodius National Library.

The activities of the Cyril and Methodius National Library, the central specialized scientific libraries and the regional and district libraries play an important role in the training of qualified librarians.

The Cyril and Methodius Library is the national library of the People's Republic of Bulgaria. It was founded in Sofia on 10 December 1878 and was one of the first cultural institutions after the Liberation of Bulgaria from Ottoman bondage.

The primary stock of books of the National Library was accumulated through donations by various institutions, educational establishments and individuals from Russia and Bulgaria. Under the Publications Depositing Act of 1897, the National Library received a copy of every book published in Bulgaria. With this it took over the functions of a national depository of books and archives of Bulgarian letters, assuming also the obligation to prepare and publish Bulgarian national bibliographies. Many eminent Bulgar-

State and Activities of the Major Types of Libraries (1982)

Type	Number	Book stock (in thousands)	Readers (in thousands)	Number of borrowed books (in thousands)
National library	1	2,299	23	885
Libraries at higher educational establishments	27	5,040	107	2,615
Specialized libraries	677	24,783	207	6,183
District libraries	27	9,157	325	6,277
Community centre libraries	3,864	32,632	1,399	22,626
Libraries at enterprises and offices	1,867	8,882	462	5,008
School libraries	3,587	15,684	867	8,850
Total	10,050	98,477	3,390	52,444

ian men of letters contributed to the development of the National Library. Some of them were also its directors, for example Petko R. Slaveikov, Konstantin Jireček, Pencho Slaveikov, Stiliyan Chilingirov, Veliko Yordanov, and many others.

Before the socialist revolution in 1944 the National Library was not in a position to fulfil its tasks as national library adequately because of limited financial resources, and lack of a suitable building and qualified library personnel.

During the air raids over Sofia in the time of World War II the library building was completely demolished and all the catalogues and part of the book stocks were destroyed.

The victory of the socialist revolution of 9 September 1944 marked a turning-point in the development of the National Library. With the assistance of the socialist State it was freed from the financial difficulties that had cramped its development and for the last forty years it has been making giant strides forward. When a new modern building was constructed and inaugurated in 1953, the funds allocated for the library and its staff were increased. The tasks of the library are now fulfilled in harmony with the socialist development of the country. The National Library is successfully implementing its duty as an indispensable assistant of all men of science, art and culture, as an important factor for cultural progress and for the all-round advancement of ideological life in Bulgaria.

The National Library is also the centre of research in library science and bibliography, and is engaged in extensive publishing activities. Today it is the archive of Bulgarian letters and a national bibliographic centre, the depository of manuscripts dating back to the eleventh and up to the nineteenth century, of historical documents from the epoch of Ottoman bondage, feudalism and the Bulgarian National Revival; it is the central scientific library of the country, with a rich stock of foreign journals, books and reference works; an information centre which renders great assistance to the bodies of social management; and finally it is the institute which provides methodological guidance, thus helping the Committee for Culture in its work for the development of the unified national library system. It has a special laboratory which takes care of the conservation and restoration of the book stock, and manuscripts and documents in particular, read mainly by researchers, specialists, students and the general reading public.

In 1982 the Library had 2,299,000 library entries; 5,100 manuscripts dating back to between the eleventh and the nineteenth centuries; special collections of rare and valuable books, among which are old Bulgarian printed books from the nineteenth century; more than 1,500,000 documents from the Bulgarian National Revival; oriental manuscripts and 900,000 Turkish documents from the time of Ottoman rule (from the fourteenth up to the nineteenth century); graphic and cartographic publications; and scores of Bulgarian and foreign musical works, official publications and microfilms.

In its capacity as bibliographic and information centre the National Library publishes: National Current Bibliography (in eight series); the directories of Bulgarian books and Press *Bulgarian National Revival Literature, 1806–78* (in 2 volumes, 1957–59); *Bulgarian Periodicals, 1844–1944* (in 3 volumes, 1962–69); *Bulgarian Periodicals, 1944–69* (in 3 volumes, 1975); and *Bulgarian Books, 1878–1944* (in 6 volumes). Other more important publications are: *Proceedings of the Cyril and Methodius National Library, Bibliography of Bulgarian Bibliographies* since 1852, many special and recommended bibliographies, research papers in the field of palaeography, archive science, librarianship and bibliography, compendiums of catalogues of foreign periodicals, current information bulletins, etc.

The National Library is a member of the International Federation of Library Associations and of the International Association of Music Libraries.

The Central Library of the Bulgarian Academy of Science was founded in 1869, as a library of the Bulgarian Learned Society. It received its name in 1911. At the beginning the major sources of accumulation of books were the donations of the members of the Academy, their own publications or books from their private libraries. The first contacts and exchanges of books were established with the learned societies of the Slav countries, and later on with France, Germany, Hungary and Great Britain.

After 9 September 1944 the Library became the leading methodological centre in a network which consists of the central library and 46 branches, and assumed a number of functions and obligations in the accumulation of the unified stock of books. It supplies Bulgarian men of science with most valuable books, papers and documents, and the latest scientific literature in a variety of research fields. In 1967 it became the centre of special bibliographies for mathematics and the natural and social sciences. In 1983 its stock consisted of 1,486,028 volumes, 788,612 of which were books, 639,425 volumes of periodicals and 57,991 copies of some specialized publications. It also has a collection of 5,230 microfilms of valuable manuscripts. Through the International Book Exchange it maintains contacts with 3,495 scientific institutions in 92 countries. It publishes two bibliographic series (bibliographies of Bulgarian scientists and specialized bibliographies), as well as monographs and reference books.

The University Library, Sofia

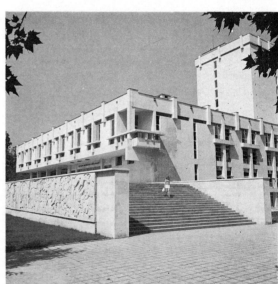

The Ivan Vazov Library, Plovdiv

The Central Library of the Bulgarian Academy of Science, Sofia

In 1949 it became a scientific institute and is now developing research and publishing activities.

The University Library in Sofia was set up in 1888, on the foundation of Sofia University. The primary stock of books consisted of extra copies sent from the national libraries in Sofia and Plovdiv, to which were added 500 volumes by the Brockhaus company and other donations. Its stock was then gradually enriched through a regular acquisition of new publications. In 1904 the beginnings were laid of exchanges of books with foreign university libraries, whose number today has reached 1,200.

The University Library is now integrated with the Central Library of the Bulgarian Academy of Science. In

1983 its stock consisted of more than 1,300,000 library entries, 830,000 of which were books, and 470,000 periodicals. It has in its possession twenty manuscripts and a comprehensive collection of old Bulgarian printed books and periodicals.

Today it represents a network comprising the central library and eighteen specialized branches at the respective faculties and departments. It is a centre of research in the field of library science and ensures methodological guidance for the libraries at the higher educational establishments under the authority of the Ministry of Education.

The Ivan Vazov National Library in Plovdiv is a general scientific library, the fourth largest in Bulgaria. It was founded in the second half of 1879 as the national book depository of Eastern Rumelia and was opened for the reading public on 15 September 1882. The library was started on the basis of Venelinov's library of about 1,850 volumes, which was the property of the Bulgarian Pupil's Society and was transported from Odessa to Plovdiv in 1876. After the Unification of Eastern Rumelia with the Bulgarian Principality in 1885 it became a national library and developed as the second in size in the country. In 1983 it had 1,029,841 volumes of books, graphic, musical and other specialized periodicals and audio-visual materials. It also has a valuable collection of manuscripts of over 400 volumes on parchment and paper, archive materials of eminent Bulgarians of the National Revival Period and of outstanding party functionaries (more than 11,000 items), rare and valuable publications and a stock of feature films.

It has published the first printed catalogue in Bulgaria (1885), *Slavonic Manuscripts and Old Printed Books of the National Library in Plovdiv* (1920), *The Year-Book of the National Library in Plovdiv* (since 1905), and many others. Apart from current bibliographies it also publishes individual reference books, for instance *Dimiter Blagoev in Plovdiv, 1893–1902* (1975), *The April 1876 Uprising in the Plovdiv Revolutionary District as Reflected in the Literature of the National Revival* (1978), etc.

In 1975 it acquired the rank of a regional library, which is expected to play a co-ordinating role in the acquisition of book stocks, and in bibliographic and information activities. The Ivan Vazov National Library has also been assigned a number of functions and obligations in the field of ethnographic and regional studies, the preparation of recommended bibliographies and the dissemination of aesthetic values.

ARCHIVES

The first archive collections in Bulgaria appeared immediately after the Liberation from Ottoman occupation (1878) and the foundation of the bourgeois State and its institutions. Their main function was to trace and pool the Bulgarian people's documentary heritage scattered in monasteries, churches, local libraries and private collections. To this end the Oriental Department of the Sofia National Library and the Bulgarian Revival Archives Collection of the National Museum Ethnographical Department were established in 1879 and 1904 respectively. By the Law on Public Education of 1921 the Archives Department of the National Library (renamed later Bulgarian Historical Archives) was created to combine the collections of the Ethnographical Museum and the Oriental Department. The objective of the Bulgarian Historical Archives to collect governmental records and records of state institutions in general was not realized, as it did not have the functions of a state Public Records Office. In 1914, a Military History Commission of the Staff of the army-in-the-field was set up to trace and carefully preserve the documents of the Ministry of War, the army staffs and military units. A considerable number of documents have been stored in the Bulgarian Academy of Science. The collection started as far back as 1881 with the revival of the Bulgarian Literary Society in Sofia. In 1911 this Society, already called the Bulgarian Academy of Science, made its aim to pool and store the stocks of the Academy and its members. The City Archives Office of the Sofia Municipal Library was set up in 1942. However, until 1944 neither a national depository of archive documents nor a network of local archive offices had been created.

The victory of the socialist revolution on 9 September 1944 and the establishment of popular democratic power in the country provided the conditions for a breakthrough with regard to the collection and storage of archive documents.

On 10 October 1951 the Presidium of the National Assembly, by virtue of Decree No. 515 on the State Archives Stock, laid the basis of a centralized archive system in Bulgaria. The Decree determined the composition of the State Archive Stock and introduced strict centralization of activity relevant to the registering, storing, processing and use of documents of historical, scientific and practical value.

The management bodies of the State Archives Stock (State Public Records Office) were built up after the issuance of Ordinance No. 344 of 18 April 1952 of the Council of Ministers, laying down the system of state archives, their classification and management, and regulating the setting up of archives offices in the state institutions and training of personnel for the archives system. Gradually, as a result of the new administrative and territorial divisions of the country and by virtue of supplementary decisions, the original archives network expanded and its functions broadened considerably. Its historical development was boosted after the April 1956 Plenum of the CC of the BCP.

The composition and management of the State Archives Stock of the People's Republic of Bulgaria have been set by the Law on the State Archives Stock adopted by the National Assembly in 1974. According to this Law the State Archives Stock is an accumulation of documents which have come into being as a result of the activities of the Bulgarian state institutions, political, economic, scientific, cultural, religious and other public organizations, and of prominent Bulgarian political, economic, scientific, cultural and other functionaries and figures, regardless of time, place and technique of creation of the documents.

The State Archives Stock also includes documents and copies of documents of interest to Bulgarian science and social management, formerly owned by foreign institutions and citizens, donated to, or purchased for, the Bulgarian archives.

The management of archives in Bulgaria as well as the guidance of and control over activity ensuring the supply, storage and use of the State Archive Stock is effected by the Central Board of Archives – a state body with functional competence and the rank of a Committee of the Council of Ministers. The decisions and directives of the Central Board of Archives are binding for Ministries, institutions, local state bodies, public organizations, their sub-divisions and individual citizens. The Central Board of Archives effects the direct structural, administrative, economic and personnel management of the central state archives as well as the functional and methodological management of the other governing bodies of the State Archives Stock.

The system of governing bodies of the State Archives Stock (the network of state archives) is based on the principles of distribution of documents by time, national and local importance, pertinence to individual spheres of social and state activity, and techniques of creation of the documents, as well as on the principle of indivisibility of the archives stocks and their complexes. This system embraces three central, 27 district and seven specialized (institutional) archives.

Each of these bodies – central, district or specialized – registers, compiles, processes, stores, files and offers for use the documents within its competence, defined as follows:

The Central State Archives of the People's Republic of Bulgaria were established in 1952. They store the documents of institutions, organizations and prominent figures of national importance from the period after the socialist revolution of 9 September 1944.

The Central State Historical Archives were set up in 1952. They store the documents of institutions, organizations and distinguished personalities on a national level in the period from the Liberation of Bulgaria from Ottoman domination until the victory of the socialist revolution (1878–9 September 1944).

The Council State Technical Archives came into existence in 1974. They store the valuable scientific and technical documentation of institutions, organizations and eminent personalities of national significance in the period after 9 September 1944.

The Manuscript Department of the Cyril and Methodius Library was set up in 1921. It contains the documents of institutions, organizations and important figures from the time of feudalism, the whole period of Ottoman domination and national liberation struggle until the Liberation of Bulgaria in 1878.

The Central Military Archives were established in 1947. They store the documents of the Ministry of Defence, the army units and outstanding military functionaries regardless of the date of the documents.

The Bulgarian Academy of Science Archives were established in 1911. They store documents pertaining to the activities of the Bulgarian Academy of Science (BAS) and its subdivisions, the Academy members, professors and senior research workers employed in the BAS system, regardless of the time of creation of the documents.

The Archives and Documentation Department of the Ministry of Foreign Affairs dates back to 1963 and stores documents of the Ministry of Foreign Affairs and its missions abroad.

The Central Party Archives and the District Party Archives were set up in 1946. They contain the documents of the Central Committee of the BCP and of the district committees regardless of the date and form of the documents.

The District State Archives and the Sofia City and District Archives were established in 1952. They keep the documents of the insititutions, organizations and functionaries whose activities bear upon the respective districts and the city of Sofia regardless of the date and kind of the documents.

The historically valuable documentation of the institutions and organizations of national importance located in the territory of the districts is also compiled by the district state archives. For their part, the state archives supply methodological guidance of and control over the work of the book-holding offices and archives of the institutions. Thus, a common approach has been achieved in the documentation work, and continuity has been ensured between the processes in book-holding and archives. The archives of the institutions, although the basic element, are not the sole object of the compilation activities of the state archives. By way of donation or purchase, the archives also receive private documents and collections from outstanding socio-political and cultural figures, scientists, Heroes of Socialist Labour, etc. Long-term programmes have been worked out for the collection of memoirs and search for documents on Bulgarian history in foreign archives. By the beginning of 1984, the State Archive Stock of the People's Republic of Bulgaria was known to contain 50,954 documents of institutional and private origin and over 2,500,000 frames of documents on Bulgarian history, stored in foreign archive depositories.

In addition to the functions of tracing, storing and ensuring the use of archive documents, the system of state archives carries out information, research and methodology actitivities, and international co-operation in the relevant fields and guidance in the training of personnel.

The information work of the Bulgarian state system runs in two main directions: first, the elaboration of a system of reference books (guidebooks, inventories, catalogues, etc.) to be used by each archives office and researcher in the archives reading room; and second, the publication of collected documents, anthologies and individual documents.

Documentary publications are prepared jointly by the state archives and Bulgarian institutes of history and universities, as well as by foreign archive institutes.

The Bulgarian archives have two periodicals: *The Archives Review* (since 1955) and *The State Archives Bulletin* (since 1957) for current information, research and methods of work in the field of archival science, sources of knowledge and other specialized historical subjects.

Since 1959, the Bulgarian archives have been a member of the International Archives Council and its committees.

The rich documentation stored in the Bulgarian archives has been used to advantage in social management, economy, science, literature, arts and other spheres of socio-political life.

MUSEUMS

The development of museums in Bulgaria began during the National Revival Period (eighteenth century–1878) and may be divided into three periods. The first coincides with the development of Bulgarian culture, the struggle for national independence and the formation of Bulgarian science in the second half of the eighteenth and the early nineteenth century. Great care was taken in the preservation of the cultural heritage. During the feudal period works of art of that time gradually accumulated in the monasteries of Rila, Bachkovo, Troyan, Dragalevtsi and Zograph, as well as in other monasteries. These medieval collections later assumed significance as the first Bulgarian museums. Outstanding public figures actively involved

After the Liberation (1878) the collections were donated to public organizations, cultural community centres and town museums. The first museum collection was set up in 1856 in the Svishtov community centre. From its very foundation (1869), the Bulgarian Literary Society set itself the task of searching out and collecting antiques for the purpose of gradually creating a collection of books, manuscripts, coins, and other objects. The society also undertook to acquire private museum collections from all over the country. Objects of great historical value began to arrive as donations to the future museum. Of great significance for the development of the musuems were also the temporary exhibitions organized in Moscow (1867) and Constantinople (1873), where for the first time works of Bulgarian applied art and genre paintings began to be put on show.

The second period in the development of Bulgarian museums (1878–1944) was characterized by wide-ranging voluntary public activity for organizing museum

The Etura architectural and ethnographic reserve in Gabrovo

The Ethnographic Museum in Smolyan

themselves in the endeavour to discover Bulgarian historical monuments. Private museum collections were set up, the oldest being that of Todor Tsenovich from Vratsa.

Ivan Mrkvicka, *Ruchenitsa*, 1894

Vladimir Dimitrov the Master,
Maiden from the Village of Kalotino, Radomir County, c. 1929

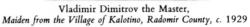

Jaroslav Vesin, *Back from the Market*, 1898

Hristo Stanchev, *In the Field*, 1937

Nikola Marinov, *Back from the Well*, 1915–20

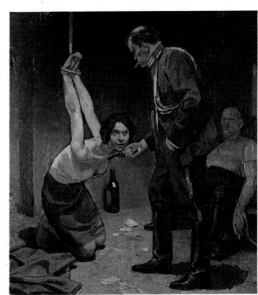

Nikola Mirchev, *Komsomol Member*, 1964
(Museum of the Revolutionary Movement)

Anton Mitov, *Peasants in a Marketplace in Sofia*, 1903

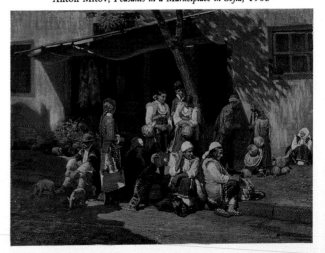

Boris Denev, *Landscape*, c. 1937

Alexander Zhendov, *Card Players*, 1946

Danail Dechev, *Spring Landscape*, 1936

Tsanko Levrenov, *Old Plovdiv*, 1938

Zlatyo Boyadjiev, *Bachkovo*, 1962

Nenko Balkanski, *A Working Man and his Wife*, 1937

Stoyan Sotirov, *Rest*, 1936

Toma Vurbanov, *Portrait*, 1983

Tseno Todorov, *Portrait of My Mother*, 1910

Kiril Tsonev, *Portrait of Professor Hutner*, 1929

Bissera Prahova, *Anton*, 1977

Dechko Uzunov, *Portrait of Krustyu Sarafov as Falstaff*, 1932

Stefan Ivanov, *Portrait of the Poetess Dora Gabe*, 1930

Mara Yossifova, *Composition* (tapestry), 1980–81
(National Museum of Decorative and Applied Arts, Sofia)

Bencho Obreshkov, *Still-Life (Fish and Grapes)*, c. 1935

Ivan Nenov, *Plasticity*, 1952 (National Museum of Decorative and Applied Arts, Sofia)

Ivan Funev, *Third Class* (relief), 1935

Dimiter Kirov, *Chamber Music*, 1980

Kalina Tasseva, *Matei Preobrazhenski, Mitkaloto*, 1969

Unless otherwise stated, all the above may be seen in the National Art Gallery in Sofia

Iconostasis of the Church of the Holy Trinity, Svishtov

Detail of wood-carving from the iconostasis in the Church of the Holy Virgin, Pazardjik, c.1840

Gabrovo in the Past (emulsion sgrafitto) by G. Bogdanov, 1962. Gabrovo House of Culture.

A Peasant Woman (lithograph), by Todor Panayotov, 1962

Head of a Girl (gypsum), by Sekul Krumov, 1965

Spirit and Matter (marble), by Andrei Nikolov, c.1922

Nymph (bronze), by Zheko Spiridonov, 1897

Mother (limestone), by Ivan Lazarov, 1934: detail from the monument at the grave of the poet Dimcho Debelyanov at his home town of Koprivishtitsa.

Arrest, by Georgi Apostolov, 1959

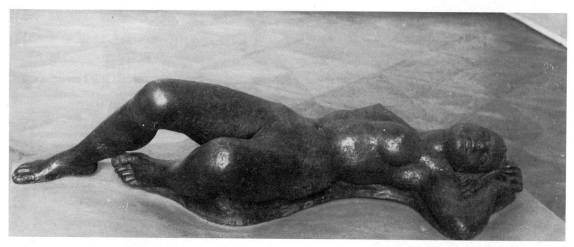

Recumbent Nude (terracotta), by Vaska Emanuilova, 1938

Blockade (ink and tempera), by Iliya Beshkov, 1942

Melnik (lithograph), by Vesselin Staikov, 1967

Holiday (woodcut) by Zlatka Dubova, 1970

Village Women (colour lithograph),
by Mana Parpulova, 1969

Unless otherwise stated, all the works of art illustrated may be seen at
the National Art Gallery, Sofia

A potter's workshop in the Etura reserve, Gabrovo

ceramic icon of St. Theodore from the Patleina locality near Preslav, various works of art of Thracian, Greek and Roman origin – sculptures, jewellery, coins, etc.). The museums carry out systematic research work and constitute local centres for scientific research – organizing scientific expeditions and excavations, publishing annuals, monographs and producing scientific publications. The main aim of the museums is to carry out scientific and educational and propaganda activity through the organization of permanent and temporary reviews and thematic exhibitions. The museums also prepare popular science publications – guidebooks, illustrated volumes, folders, almanacs, etc. They are methodologically guided by the Committee for Culture, scientifically by the Bulgarian Academy of Science and administratively by the authority to which they are attached and which directs them (District and Municipal People's Councils, public organizations, etc.). Bulgaria has been a member of the Unesco International Council of Museums since 1958; 18 May is celebrated as International

collections and studying the historical heritage and by the first manifestations of state-organized museums. On the initiative of the first Bulgarian civil governor of Sofia, Pavel V. Alabin, the foundations of a museum which later developed into a national museum were laid adjoining the Sofia Public Library (today the Cyril and Methodius National Library). A Natural Science and History Museum was set up in 1889 (today the National Natural Science Museum), and an Ethnographical Museum (1892; today's National Ethnographical Museum). On the initiative of a special national committee headed by Stoyan Zaimov (1853–1932), the Military History Parks and Museums in Pleven and its district and in Byala were set up. A spontaneous movement began for setting up museum collections at community centres and archaeological, geographical, historical, natural sciences and other societies. By 1944 there were thirteen state museums in existence.

The third period in the development of museums began after 9 September 1944. Museums hold an important place in the country's ideological, scientific and public life. The State secures considerable sums for their organization, maintenance and activities. In 1952 work began on setting up a uniform state museum network, with the museum collections of community centres, schools and churches serving as the basis for the state musuems; a great number of new museums and museum-houses were also set up. In 1983 there were 227 state museums in Bulgaria and some 500 museum collections at community centres, schools, enterprises, military units, monasteries, etc. They contain more than 3,500,000 valuable museum exhibits in the form of documents and objects. Some of the articles they contain are unique and of world cultural significance (specimens of prehistoric ceramics and idol figurines, the Vulchi Trun, Panagyurishte and Varna gold treasures, the

A coppersmith's workshop in the Etura reserve, Gabrovo

Museum Day. The museums can be divided into the following categories: national and local (in significance and territorial scope), historical (district and town), memorial (museum-houses and museums concerned with specific historical events), specialized (according to the aims and nature of exhibits – archaeological museums, ethnographical museums, military history museums, the museum of socialist construction, the museum for the history of physical culture and sport, the museum of the September 1923 uprising, and art galleries).

The following national museums exist in the country: **The Georgi Dimitrov Mausoleum.** This is situated in the Sofia Municipal Gardens and contains a sarcophagus

The Georgi Dimitrov Mausoleum, Sofia

The Freedom Monument on Mt. Shipka

The Georgi Dimitrov National Museum, Sofia

The memorial church in Shipka

with the embalmed body of Georgi Dimitrov. The design is the work of a team of architects headed by Georgi Ovcharov and Racho Ribarov. The building was erected in six days and the solemn burial of Georgi Dimitrov took place on 10 July 1949. The Mausoleum consists of a single building with an open rostrum on its northern façade. The mourning hall is of almost cubic form surrounded by an isolation corridor. The sarcophagus is cast in bronze made in the USSR. Behind the eastern part of the rostrum is built the grave of Vassil Kolarov (1950; architect V. Tiholov, sculptor K. Todorov).

The National Museum of the Revolutionary Movement. It was opened in 1950 and contains more than 175,000 exhibits divided into two departments. In the first exhibition department (1876–9 September 1944) documents and photos trace the first steps of the workers' and socialist movement, the foundation of the Marxist Party in 1891 and its struggle against the opportunism, militarism and chauvinism of the bourgeois parties. The second exhibition department (from 9 September 1944) follows up the years of national economic reconstruction and the construction of the socialist society. Tables and graphs show the success of socialist Bulgaria. Branches of the museum are the Workers' Club in Sofia, the underground headquarters of the Politburo, and the collection entitled the Struggles of the Bulgarian Students against Capitalism and Fascism Led by the Bulgarian National Students' Union. The museum works in cooperation with kindred museums from the socialist countries.

The Bulgarian Communist Party Memorial on Mt. Buzludja

The Georgi Dimitrov National Museum. This was set up in 1972. A decision of the Politburo of the Central Committee of the BCP on 10 July 1949 declared the house in which the leader and teacher of the Bulgarian people, Georgi Dimitov, lived with his family from 1888 to 1923 a museum-house, which was opened to the public in 1951. In 1972 a new building was erected nearby in which a documentary exhibition of the heroic life of Georgi Dimitrov was organized. The Georgi Dimitrov National Museum and the Dimiter Blagoev, Georgi Kirkov and Vassil Kolarov museum-houses form a united research, ideological, cultural and educational institution, whose task is to investigate and popularize the life and revolutionary heritage of the first leaders of the Party and the people.

The Shipka-Buzludja National Memorial Park. Set up in 1964, it incorporates the Shipka and Buzludja National Parks which contain architectural and sculptural monuments, among them the Liberty Monument on Mt. Stoletov, the Memorial Church in Shipka, the monuments near the village of Sheinovo and those on Mt. Buzludja.

One of the most impressive monuments in Bulgaria raised in honour of the Russian soldiers who died during the War of Liberation of 1877–78 is the Memorial Church above the town of Shipka, built to a design by the architect A. Y. Tomishko in 1885–1902 in the style of seventeenth-century Russian church architecture. Under the central part of the church is the crypt (the ossuary), which contains seventeen stone sarcophagi with bones of the heroes from the battle of Shipka. Another of the remarkable features of the Memorial Church is its seventeen bells.

Another of the most impressive monuments is the Liberty Monument on Mt. Stoletov, designed by the architect Atanas Donkov and the sculptor Alexander Andreev. It was erected with funds donated by the people and represents a regular four-sided truncated stone pyramid. Above the central entrance stands an eight-metre bronze lion and an inscription: 'To the Fighters for Freedom'.

On Mt. Buzludja stands the Monument to the Bulgarian national hero Hadji Dimiter and his comrades who fell in battle with the Ottoman enslavers. Here also is the monument commemorating the Buzludja Congress held in 1891, when the Bulgarian Marxist Workers' Party was founded. The third monument on Buzludja was erected in honour of the partisans of the Gabrovo-Sevlievo detachment who lost their lives here. There is also a Monument to Victory near the village of Sheinovo. The Communist Party Memorial House on Mt. Buzludja was erected between 1974 and 1981 after a design by the architect Georgi Stoilov. The low building symbolizes a sacrificial altar and the 70-metre tall column an unfurled communist banner. The ceremonial hall, which occupies the central part of the building, has interior walls covered with a monumental mosaic depicting the struggles of the Party from its foundation to the construction of the socialist society. In the gallery surrounding the hall there are fourteen artistic compositions reflecting the years of peaceful construction. Many well-known artists and sculptors took part in the shaping of the Monument.

The National History Museum. Founded in 1973, the museum contains more than 106,000 exhibits, distributed in the following departments: Prehistory, Antiquity, Middle Ages, National Revival, Capitalism, Revolutionary and Workers' Movement, Socialist Construction, and Historical Illustrations. Through temporary exhibitions and lectures – such as Pre-History Art in the Bulgarian Lands, Bulgarian Womens's Jewellery through the Ages, A Thousand Years of Bulgarian Icons, Art and Culture in the Medieval Bulgarian State, the Centennial of the April 1876 Uprising, the Centennial of the Russo-Turkish War of Liberation – the museum exhibits and popularizes the most valuable monuments of the country's millennial history at home and abroad. A permanent exhibition of the museum is to be opened in stages in the former Palace of Justice after its reconstruction. The first stage was opened in 1984 and covers the period from antiquity to the 1877 Russo-Turkish War of Liberation. It publishes a Bulletin (*Izvestia*) of the National History Museum (since 1975).

The Bulgarian Academy of Science National Archaeological Museum. Founded in 1878 as a museum collection the Public Library (today's Cyril and Methodius National Library), in 1892 it was established as an independent national museum. Its first exhibition

The National History Museum, Sofia

was opened on 18 May 1905. In 1909 it was re-named the National Archaeological Museum. Its departments are early mediaeval, late mediaeval, prehistoric and ancient. At the end of 1948 the museum was merged with the archaeological institute and became the Archaeological Institute and Museum of the Bulgarian Academy of Science, and after 1977 the Archaeological Institute and Museum. The museum has organized excavations and studies at Pliska, Preslav, Madara and other historical sites. Its fund contains some of the earliest tools of the Early

The National History Museum, Sofia

The National Archaeology Museum, Sofia

The Troy-Thrace exhibition at the National Archaeology Museum, Sofia

The National Art Gallery, Sofia

The Palaeontology Museum in the Kliment Ohrid University, Sofia

The National Military History Museum, Sofia

Stone Age; tools, arms, utensils, figurines, jewellery and other objects from the Neolithic Age, the Stone, Copper, Bronze and Early Iron Ages; collections of Greek painted ceramics; gold jewellery, articles of the Thracian arts and crafts, from the pre-Roman and Roman periods; items from the early Christian and Medieval periods; Old Bulgarian inscriptions, icons, collections of coins, seals, etc. The museum has contributed to the exhibitions of Thracian Art, a Thousand Years of Bulgarian Icons, and the Art and Culture of the Medieval Bulgarian State (Medieval Bulgarian Civilization), which have been shown in Bulgaria and also in Austria, the GDR, the FRG, Italy, Mexico, Poland, the USA, the USSR, Hungary, France, Switzerland, Sweden, etc. It has organized visiting exhibitions from museums in the USSR (Scythian Art – 1976); Mexico (Three Thousand Years of Mexican Art, 1976 and the Art of the Aztecs, 1978); France (Celtic Art – 1977); Poland (the Baltic Peoples, Neighbours of the Slavs, 1978); Yugoslavia (Graeco-Illyrian Treasures, 1979); Colombia (the Gold of El Dorado, 1980); Sweden (the Art of Vikings and Their Predecessors, 1980); the FRG (Roman Treasures from the Rhineland, 1980); and Great Britain (Ancient Egyptian Art from the British Museum, 1981). The museum publishes *Excavations and Studies* (1948–50 and since 1975) and the *Archaeology* magazine.

The National Ethnographical Museum. Founded in 1906 as an independent National Ethnographical Museum, in 1949 it was merged with the Institute of Ethnic Science (founded in 1947 at the Bulgarian Academy of Science) and was set up as an Ethnographical Institute and Museum. The museum owns about 100,000 exhibits, which cover the period from the sixteenth century to the present day and are organized in the following departments: Bulgarian National Dress and Embroidery, Bulgarian Folk Fabrics, Home-Made Articles and Household Utensils, Bulgarian Folk Customs and Rites, Bulgarian

The Museum of the Liberation of Pleven, 1877. The room in which Osman Pasha had his sword returned to him by the Russian Emperor, Alexander II

The 1877 Pleven Epopee historical panorama

Folk Handicrafts (Ceramics, Gold and Copper Working, Wood Carving, Wrought Iron, etc.), Agriculture and Stockbreeding, National Architecture, Musical Instruments and Foreign Department. The popularizing activity is carried out by organizing exhibitions at home and abroad (Europe, America, Asia). It has been publishing the magazine *Bulgarian Ethnography* since 1975.

National Natural Science Museum. This was founded in 1889. To enrich its exhibition it purchased collections from abroad, including: a collection of birds and choleoptera of the Balkan Peninsula and Asia Minor from the Frenchman, A. Aleone; a collection of South African birds and mammals from the Czech traveller, E. Holub; and a collection of plumage of Indian birds from the English ornithologist, S. Baker. Specimens have been collected from the fauna of all continents. In 1907 the museum was opened to the public. A Botanical Department was set up in 1919 and a department of Mineralogy and Geology in 1920. In a short time the three departments became scientific centres for the work of the prominent Bulgarian national scientists, who completed its exhibits with their own collections. Co-operation was arranged with foreign scientific centres and scientists, some of whom donated their collections to the museum. The exhibition contains some 8,300 exhibits, 720 of which are entomological cases, each one containing on average several dozen insects. The scientific fund is several times larger, amounting to hundreds of thousands of items. It has three departments – Zoology, Geology and Botany. The exhibits of the Zoology Department contain invertebrates, fish, amphibians and reptiles, birds and mammals; the Geology Department – cosmology, mineralogy, petrography and rock formation, and historical geology. The museum maintains contacts with kindred museums and scientific institutes in the socialist countries and elsewhere.

National Military History Museum. Set up in 1916, its activity began with the collection and preservation of objects and documents connected with Bulgarian military history. More than 105,000 exhibits and war relics are kept in the museum, which covers several historical periods: the Middle Ages (seventh–fourteenth centuries), the Struggle of the Bulgarian People Against Ottoman Rule (fourteenth-century–1878)), the War Period (1885–1913), the 1915–23 Period, the 1924–39 Period, the Armed Anti-Fascist Struggle (1941–44), the Patriotic War (1944–45), and the Post-War Period of the Bulgarian People's Army (1945–present day). The exhibitions include chain-mail, helmets and military standards from the earliest period, personal effects of prominent soldiers, orders, uniforms and so forth. Branches of the National Military History Museum are the Naval Museum (opened in 1956) and the museum park of Comradeship in Arms of 1944, with the Mausoleum of Wladislas Jagiello (opened in 1963 in Varna). The museum maintains contacts with the Museum of the Armed Forces of the USSR and with kindred museums in the Warsaw Treaty member-states. Since 1973 it has been publishing a *Bulletin of the National Military History Museum*.

National Museum of Bulgarian-Soviet Friendship. This was set up in 1953, and started its activities in 1954.

Museums

The Rila Monastery

More than 45,000 archive documents, photos, microfilms and other materials are kept in the museum, reflecting political, economic and cultural relations between Bulgaria and Russia from the earliest times to present-day Bulgarian-Soviet relations. The characteristic events and facts of different periods are illustrated in several departments: Bulgarian-Russian relations and contacts from the fifth-fourteenth centuries, Russian Aid in the National Liberation of the Bulgarian People – 1393–1878, Bulgarian-Russian Relations from the Liberation to the Great October Socialist Revolution, the October Revolution and the Triumph of the Socialist Revolution in Bulgaria – 1917–44, Bulgarian-Soviet Friendship and our Socialist Development. There is also a Department of Friendship and a Hall of Distinction with a collection of orders and medals of the USSR and Bulgaria. In conjunction with the National Committee for Bulgarian-Soviet Friendship the museum has been publishing *Annals of Friendship* since 1969.

The National Museum of Bulgarian Literature. It was founded in 1976 and has four departments: Old Bulgarian Literature (ninth–eighteenth centuries), Literature of the National Revival Period (eighteenth–twentieth centuries), Modern Bulgarian Literature (1900–9 September 1944), and Contemporary Bulgarian Literature (post-1944). Branches of the museum are: the Literary museum-houses in Sofia and the museum-houses of Emilyan Stanev in Veliko Turnovo and of Georgi

The Rila Monastery National Museum

The Mausoleum in Pleven

Karaslavov in Purvomai. The museum provides methodological guidance also for all literary museums in Bulgaria. It establishes contacts and concludes long-term contracts for co-operation with kindred museums abroad – the National Museum for Letters and Literature in Prague, the National Literature Museums in Moscow, Warsaw, Budapest, and others. Since 1982 it has been publishing the Bulletin of the National Museum of Bulgarian Literature – *Literary Sources*.

The National Technical Science Museum. Founded in 1957, it contains more than 10,000 exhibits. It shows the development of technology through collections of measuring devices, electrical engineering, etc. The museum maintains contacts with technical science museums in Moscow, Prague, Brno and Dresden, and with the Palace of Discovery in Paris. Its publication, *National Science Museum Year-Book,* was first issued in 1971. It also publishes journals on special subjects.

The National Agricultural Museum. It was originally founded as part of the Georgi Dimitrov Higher Agricultural Institute (established in 1956), but in 1978 became a national museum. It contains more than 17,000 articles and has two departments, Agriculture and Stockbreeding, which cover all farming activities, their economy, organization, mechanization, servicing and processing branches. It publishes the *Bulletin of the National Agricultural Museum* (first published 1971).

The National Museum of Church History and Archaeology. Set up in 1921 under the Holy Synod of the Bulgarian Orthodox Church, it now has more than 10,000 exhibits. It has four departments: Manuscripts and *Incunabula,* Iconographic Department, Arts and Crafts Department, and History and Archives Department. The exhibits include manuscripts on parchment and paper from the tenth–sixteenth centuries, and icons and documents on the ecclesiastical and national struggles.

Bulgaria, possessing 32,000 volumes of which 9,000 are *incunabula.* It contains Slav and Bulgarian manuscripts, Bulgarian *incunabula* and foreign editions.

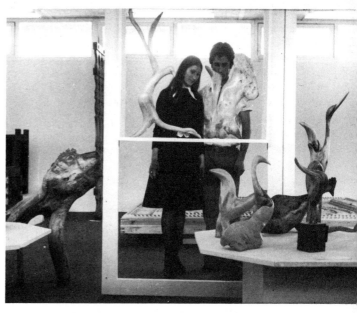

The museum in Troyan

The Museum of Folk Arts and Crafts in Troyan

The Rila Monastery National Museum. Set up in 1961, it includes the Monastery Architectural Complex consisting of the Hrelyo Tower and the Church of the Nativity of the Virgin; the Pchelino Nunnery and the nearby Church of the Dormition of the Virgin; the Orlitsa Nunnery with the Church of Sts. Peter and Paul; the St. Luke Hermitage with the Churches of St. Luke and of the Virgin's Shroud and the school of Neophyte of Rila; the church near the grave of St. John of Rila; the Church of the Presentation of the Blessed Virgin near the Monastery Ossuary, and the nineteenth-century Watch-tower which stands about 10 km south-west of the Monastery. In 1966 the Zemen Monastery was made a branch of the Rila Monastery museum. The museum's fund includes a rich collection of church vessels. The collection of icons and prints contains specimens dating from the fourteenth–nineteenth centuries. Ethnographical exhibits include fabrics, jewellery, metal-work, etc. from different parts of Bulgaria, mainly of the nineteenth century. The Monastery Library is one of the oldest and most valuable in

The National Museum of Education in Gabrovo. Set up in 1973, it has three departments: Fund, Scientific Exhibits and Cultural and Educational. It keeps books, periodicals, rare publications in Bulgarian, Russian and West European languages, and reports from the first secondary educational establishments after the Liberation of 1878.

Other museums of interest are:

Etur. An architectural and ethnographical open-air museum park, the first and only one of its kind in Bulgaria, it is situated in the Etur district of Gabrovo. Open to the public since 1964, it illustrates the economy, architecture and old handicrafts of the Gabrovo region during the National Revival Period (nineteenth century). Its exhibits include houses, operating artisans workshops, sheds for technical implements, etc.

The Museum of Woodcarving and Icon Painting in Tryavna. Founded in 1963, the exhibition shows masterpieces of wood-carving and icon painting art of the Tryavna area.

The Museum of Folk Arts and Crafts in Troyan. Opened in 1968, it depicts the development of ceramics in Bulgaria and of the Troyan ceramic school. On show are the achievements of the Troyan master-craftsmen in all branches.

The Museum-Houses are memorial museums of prominent Bulgarian public, political and cultural figures. Their

Sava Filaretov's house in Zheravna, in the Sliven district

exhibits are usually divided into two departments: personal – re-creating wholly or in part the atmosphere in which the patron of the museum-house lived, and documentary – exhibiting archive materials depicting his life and work. In 1983 there were 74 museum-houses in Bulgaria.

Museum-Houses of Heroes of the National Liberation Struggle. In Karlovo is the Vassil Levski Museum-House; in Kalofer – the house of Hristo Botev; in Koprivshtitsa – the houses of Todor Kableshkov, Lyuben

The Museum of Woodcarving and Icon Painting in Tryavna

Karavelov and Georgi Benkovski; in Sliven that of Hadji Dimiter; in Pangyurishte that of Raina Knyagina (the Princess); in Russe that of Granny Tonka; in Shumen – that of Lajos Kossuth; and in Pleven – that of Stoyan and Vladimir Zaimov.

Museum-Houses of Outstanding Figures of the Bulgarian Communist Party, and Heroes of the Anti-Fascist Struggle. In Sofia are the museum-houses of Georgi Dimitrov, Dimiter Blagoev, Vassil Kolarov, Alexander Stamboliiski and Georgi Kirkov; in Velingrad that of Vela Peeva; in Stanke Dimitrov – that of Stanke Dimitrov; in Mihailovgrad that of Hristo Mihailov; in Razlog that of Nikola Parapunov; in the village of Krun, Stara Zagora district, that of Tsvyatko Radoinov; in the town of Sliven that of Subi Dimitrov; in the village of Georgi Damyanovo, Mihailovgrad district, that of Georgi Damyanov; in Plovdiv that of Yordanka Nikolova; and in Gabrovo that of Mitko Palauzov.

Museum-Houses of Outstanding Bulgarian Writers and Artists. In Sofia there are the museum-houses of Ivan Vazov, Nikola Vaptsarov, Hristo Smirnenski, Peyo Yavorov, Dimiter Dimov; in Svishtov is that of Aleko Konstantinov; in Stara Zagora that of Geo Milev; in

The Ivan Vazov Museum in Sopot

Karnobat that of Dimiter Polyanov; in the village of Byala Cherkva, in the Veliko Turnovo district, that of Tsanko Tserkovski; in the village of Zheravna, in the Sliven district, that of Yordan Yovkov; in Koprivshtitsa that of Dimicho Debelyanov; in Pazardjik that of Stanislav Dospevski; in Provadia that of Svetoslav Obretenov; in the village of Draganovo, in the Veliko Turnovo district, that of Assen Raztsvetnikov; and in Kazanluk, that of Chudomir.

SELECTED BIBLIOGRAPHY

CHILINGIROV, S. *Bulgarski chitalishta predi osvobozdenieto* (The Bulgarian Cultural Clubs Before the Liberation of 1878). Sofia, 1930.

KISYOV, I., M. RAICHEV, and Z. MILANOV *Suvremenni problemi na bulgarskoto Muzeeznanie. V: Purvi kongres na Bulgarskoto istorichesko druzhestvo* ('Problems of the Organization and Management of Museums in Bulgaria Today'. In: *The First Congress of the Bulgarian Historical Society*). Vol. 2. Sofia, 1972.

KONDAREV, N., S. SIRAKOV and P. CHOLOV *Narodnite chitalishta v Bulgaria* (The Bulgarian Cultural Clubs). Vols. 1–2. Sofia, 1972–79.

KUZMANOVA, M. P. *Istoriya na arhivite i organizatsiya na arhivnoto delo v Bulgaria* (Archive History and Organization of Archival Work in Bulgaria). Sofia, 1966.

SILYANOVSKA-NOVIKOVA, T. *Osnovi na muzeeznanieto* (The Basis of the Organization and Management of Museums). Sofia, 1972.

PART XI
MASS MEDIA

NEWSPAPERS AND MAGAZINES

The People's Republic of Bulgaria possesses a wide variety of mass media – newspapers, periodicals, television, radio, books and other forms of information and publicity.

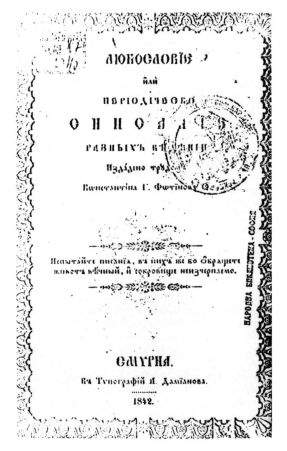

Facsimile of the title page of the magazine *Lyuboslovie* (1844–46)

Newspapers first appeared in Bulgaria during the National Revival Period, and were closely associated with three main factors: the struggle to overcome the last vestiges of feudalism, the establishment of capitalist relations within the Ottoman Empire, and the development of the Bulgarian national liberation movement. The first Bulgarian magazine *Lyuboslovie* (Love of Letters), a trial issue of which had been published in 1842, appeared in Smyrna between 1844 and 1846. It was founded and edited by Konstantin Fotinov. Ivan Bogorov published the first Bulgarian newspaper *Bulgarski Orel* (Bulgarian Eagle; 1846–47) in Leipzig, and then went to Istanbul where he founded the *Tsarigradski Vestnik* (Istanbul Newspaper; 1848–62). Prior to Bulgaria's Liberation in 1878, more than 60 newspapers and magazines were published outside the Ottoman Empire, while only 30 – in Istanbul, Russe and some other towns – were published within the Empire.

Above all, the Press appealed to people's thirst for knowledge, for closer contact with the world, and provided information about Bulgaria's past. In this way it stimulated the aspiration for cultural and national independence. Soon two basic trends appeared in the National Revival Press – evolutionary and revolutionary. The newspapers that advocated an evolutionary line and educative ideas were *Bulgaria* (1859–63), *Turtsiya* (Turkey; 1864–78), *Pravo* (Law; 1869–78), *Vek* (Century; 1874–76), *Gaida* (Bagpipe; 1863–67), and *Makedonia* (1866–72). The newspapers edited by the great writer and journalist Petko Slaveikov took a more radical position, often close to the line pursued by the revolutionary party.

The revolutionary Press propagated the ideas of national liberation and social emancipation and drew attention to the Bulgarian Church question, the problems of the educational system and economic and cultural development, from a revolutionary point of view. Georgi Rakovski is considered the founder of the revolutionary Press with the newspapers *Bulgarska Dnevnitsa* (Bulgarian Report; 1857) and *Dunavski Lebed* (Danube Swan; 1860–61). His work was taken up and continued by Lyuben Karavelov and Hristo Botev, whose newspapers are a high point in the history of Bulgarian journalism and publicistic writing: *Svoboda* (Freedom; 1869–72) and *Nezavissimost* (Independence; 1873–74) were published and edited by Karavelov, while *Duma na Bulgarskite Emigranti* (Cause of the Bulgarian Émigrées; 1871), *Budilnik* (Alarm Clock; 1873), *Zname* (Banner; 1874–75) and *Nova Bulgaria* (New Bulgaria; 1876–77) – were published and edited by Botev.

In spite of the difficult conditions of Ottoman bondage more than 95 newspapers and magazines were published.

Facsimile of the newspaper *Bulgarski Orel* (1846–47)

Following Bulgaria's Liberation in 1878 there was a press boom which reflected two main political tendencies: conservative – the newspapers *Maritsa* (1878–85) and *Vitosha* (1879–80), and liberal – the newspapers *Tselokupna Bulgaria* (Integrated Bulgaria, 1879–80), *Nezavissimost* (Independence; 1880–81), and *Borba* (Struggle; 1885).

Facsimile of the newspaper *Svoboda* (1869–74)

The country's rapid economic development played an important role in the process of stratification of Bulgarian society and also contributed to the differentiation of the Press and the popularization of different approaches to the main domestic international issues. 150 dailies came out in the period 1900–32. The end of the nineteenth century saw the publication of some important newspapers which became the mouthpieces of various parties: they included *Mir* (Peace; 1894–1944), the newspaper of the Popular Party; *Pryaporets* (Banner; 1898–1932), the newspaper of the Democratic Party, and *Bulgaria* (1898–192), the newspaper of the Progressive Liberal Party. Reputable magazines of a sensationalist type also appeared – *Vecherna Poshta* (Evening Post; 1900–24), *Dnevnik* (Diary; 1902–44), and *Utro* (Morning; 1911–44).

The turn of the century saw the founding of various newspapers and periodicals belonging to the Agrarian Party: *Selski Vestnik* (Peasant Newspaper; 1893–97, 1901–8), the magazine *Seyach* (Sower; 1896–1900), the newspapers *Zemedelska Borba* (Agrarian Struggle; 1898–1901) and *Zemedelska Zashtita* (Agrarian Defence; 1899–1902) and most important of all *Zemedelsko Zname* (Agrarian Banner), which first appeared in 1902 and has been appearing ever since, although there have been a few gaps. As the official newspaper of the Bulgarian Agrarian Party it has contributed a lot to the Party's development as an independent agrarian political organization.

Following the Congress of Berlin in 1878, the question of the liberation of the Bulgarian population of Macedonia, Aegean Thrace and Edirne (Adrianople) Thrace from Ottoman domination became a permanent topic of discussion in the Bulgarian newspapers, and in some of them even occupied a central place. The question of the strategy and tactics of the liberation movement was discussed in many newspapers including *Makedonets* (The Macedonian; 1880–85, 1894–95), *Makedonskii Glas* (Macedonian Voice; 1885–87), *Makedonia* (1888–93), *Pravo* (Law; 1894–96, 1901–3), *Revolution* (1895), *Makedonia* (1896–97), *Politicheska Svoboda* (Political Freedom; 1898–99), *Autonomy* (1898–99, 1901–2), *Vardar* (1898), *Borba za Svobodata na Makedonia i Odrinsko* (Struggle for the Liberation of Macedonia and the Edirne Region; 1899), *Reformi* (Reforms; 1899–1905), *Delo* (Cause; 1901–2), and *Svoboda ili Smurt* (Freedom or Death; 1903).

Facsimile of the newspaper *Zname* (1874–75)

Facsimile of the newspaper *Dunavskii Lebed* (1860–61)

During that period more than 8,000 newspapers and periodicals appeared in Bulgaria.

Suvremenni Pokazatel (Modern Indicator), the first socialist magazine, came out in 1885 and was edited by Dimiter Blagoev, who founded the party of the Bulgarian socialists and Marxists. The first socialist newspaper *Rossitsa* came out in 1886. *Rabotnik* (Worker; 1892–94) became the official newspaper of the Bulgarian Social Democratic Party which had been founded in 1891. *Rabotnik* conducted polemics with the newspaper *Drugar* (Comrade; 1893–94) and the magazine *Sotsialdemokrat* (1892–93), which belonged to the Bulgarian Social Democratic Union. The revolutionary Marxists created the first purely workers' newspaper – *Rabotnicheski Vestnik* (1897–1937, 1939), and the theoretical journal *Novo Vreme* (1897–1923), which has been the official journal of the Communist Party Central Committee since 1947. Its founder and editor was Dimiter Blagoev. Socialist and communist periodicals exerted a great influence on the people both before and after the Bulgarian Uprising of September 1923, an important event since it was the first anti-fascist uprising in the world. During the 1920s a large number of trade union, literary, co-operative, temperance, youth and other publications openly upheld the anti-fascist cause: they included the magazine *Zvaner* (Bell-Ringer), the newspapers *Narodna Zashtita* (People's Defence; 1923–24), *Rodina* (Motherland; 1923–24), *Narodovlastie* (People's Rule; 1924), *Oranzhevo Zhame* (Orange Flag; 1924), *Vuzrazhdane* (Revival; 1926–32); the magazine *Vik na Svobodni Hora* (Appeal of the Free People; 1923–24) edited by Yossif Herbst; *ABC* (1924–25); *Eko* (Echo; 1924–25); *Dnes* (Today; 1925); and *Plamuk* (Flame; 1924–25) edited by the writer and poet Geo Milev.

After the *coup d'état* of 19 May 1934, in the preparation of which the magazine *Zveno* (Circle; 1928–34) played a key role, all political newspapers and magazines were banned.

During the Second World War the Press was placed entirely under the control of the Directorate of National Propaganda, a newly established body that was organized on nazi lines. Only *Mir* (Peace), *Zarya* (Dawn; 1914–20, 1929–44) and *Daga* (Rainbow; 1934–44) carried any information about the progress of the war, along with occasional daring commentaries. Not surprisingly, the underground anti-fascist and youth Press developed considerably.

The socialist revolution of 9 September 1944 saw the beginning of free socialist journalism in Bulgaria. Fascist newspapers and periodicals were immediately banned. More than 100 newspapers and magazines began to be published. *Otechestven Front* (Fatherland Front), the newspaper of the new Government, and the newspapers *Rabotnichesko Delo*, *Zemedelsko Zname* (Agrarian Banner), *Izgrev* (Dawn) and *Narod* (People) – publications of individual parties, the *Literaturen Front* newspaper, the trade union newspaper *Zname na Truda* (Banner of Labour), the youth newspapers *Mladezhka Iskra* (Youth Spark) and *Narodna Mladezh* (People's Youth) appeared legally for the first time, along with new local newspapers.

Facsimile of the newspaper *Makedoniya* (1866–72)

Facsimile of the newspaper *Suvremennii Pokazatel* (1885)

The rapid increase in newspaper circulation was indicative of the great interest aroused by the Press. By the end of 1944, newspaper circulation had reached record figures: *Rabotnichesko Delo* – 157,000 issues, *Otechestven Front* – 146,000 issues, *Svoboda* (Freedom) – 75,000 issues, and *Zemedelsko Zname* – 28,000 issues. The total daily circulation of newspapers came to 778,500.

This was a period of radical restructuring for Bulgarian journalism. The main themes in the Press during the early years of the revolution included the consolidation of the Fatherland Front Government, the Patriotic War against Nazism, the rebuilding and restructuring of the national economy, increasing Bulgaria's prestige internationally, creating a people's republic and the adoption of a new constitution, and the nationalization of industry.

Bulgarian took part in the war against nazi Germany as from 9 September 1944 and fought until V-day, 9 May 1945. During that period 120 different war newspapers were published for the soldiers at the front.

The Union of Bulgarian Journalists was founded on 24 December 1944 and Bulgarian journalism was particularly prolific between 1944 and 1948.

The party and state leader Georgi Dimitrov, a talented journalist with vast experience who had founded and directed many anti-fascist newspapers and magazines, devoted considerable attention to raising the standards of the Bulgarian Press, and to the training of first-class journalists.

The period following the April 1956 Plenum of the Party's Central Committee was marked by a new creative upsurge, and journalism took rapid steps forward. The decisions taken at the Plenum were of historic significance for Bulgarian journalism and opened up a new period in the role of the mass media in the country's development.

Problems related to the country's economic, social and cultural development, to people – workers, artists and citizens – and to international relations, dominated the work of Bulgarian journalists. The professional standards of journalism became higher; it became more thorough, more analytical and more expressive. A talented younger generation of journalists grew up. Successive party congresses discussed and adopted resolutions aimed at perfecting the Press, television, radio and other mass media, and the further development of the professional abilities of Bulgarian journalists.

The mass media developed rapidly. New weeklies began to appear: *Pogled* (Review), *Anteni* (Antennas), *Orbita*, *Puls* (Pulse), the youth newspapers *Komsomolska Iskra* (Komsomol Spark) published in Plovdiv, and *Polet* (Flight) published in Varna; trade union periodicals also appeared and the network of local and special interest periodicals also expanded.

The number of newspapers and periodicals trebled over the 1956–65 period. By 1970 the total annual circulation of newspapers had reached 816,720,000 and that of magazines was 45,195,000, i.e. a 600 per cent and 1,000 per cent increase respectively as compared with the 1944 figures.

Facsimile of the newspaper *Novo Vreme* (1897–1923, 1947–)

The Structure of the Press. Responding to the high level of political thinking of the Bulgarian man in the street, and his desire for knowledge, education and self-improvement, the Press continues to develop rapidly, together with radio and television. At present over 1,400 newspapers and periodicals come out in Bulgaria with a total circulation of more than 11,000,000. There are 463 newspapers, 255 magazines in addition to 51 foreign language editions, and over 600 bulletins, many of which are virtually magazines.

The dynamic development of the Press can be seen in the following table, which compares newspaper circulation for 1944 (the year of the socialist revolution), 1956 (the year of the April Plenum) and 1983:

Circulation of Newspapers

Year	1944	1956	1983
Number of titles	322	80	463
Annual circulation (in thousands)	122,000	498,782	1,046,492

The rapid increase in the number of magazines is even more eloquent. During 1956, 105 titles were published – a total of 9,445 issues. By 1983, if we include foreign language editions, 306 magazines were published, with a total annual circulation of 60,150.

In spite of Bulgaria's size in comparison with a number of other developed countries, she can now boast of a well developed, rich and varied range of periodicals. In 1982 as many as 103 copies of newspapers were printed per head of the population. In 1983 there were 1,245 copies of periodicals per thousand of the population out of a single circulation of 11,126,140.

Collage of central Bulgarian newspapers

The past four decades have seen great improvements in the content and level of the Press, and in its level of professionalism. As the Bulgarian head of state, Todor Zhivkov, put it, 'by its very nature, journalism is an art that deals with topical problems and, as in all forms of art, the main focus of attention should be man, the education of the new individual'.

Wide-ranging nation-wide discussions and surveys are conducted by means of the Press. These enable a vast number of people to express their opinions on important problems of the country's social, economic and cultural development. The leading newspaper in this respect is *Rabotnichesko Delo,* which has the largest circulation in Bulgaria.

Bulgarian periodicals are published both in the capital,

Sofia, and in the provinces. As a result of historical tradition, and also because of the compactness of the country's territory, many periodicals are published in Sofia. There are 62 central newspapers with a total circulation of 5,689,500 copies per issue out of the total national circulation per issue of 7,082, 500. The total annual circulation of the central daily and weekly newspapers comes to 886,817,000 copies. There are eight dailies:

Rabotnichesko Delo, launched in 1927, incorporates the *Rabotnicheski Vestnik,* which was founded in 1897. It is the newspaper of the BCP Central Committee. Its circulation is 820,000 copies.

Kooperativno Selo, launched in 1951, is the newspaper of the Central Council of the National Agro-Industrial Union and the Central Committee of the Trade Union of Workers Engaged in Agriculture and the Food Industry. Its circulation is 195,000 copies.

Narodna Armiya, launched in 1944, is the newspaper of the Ministry of National Defence; its circulation is 70,000 copies.

Vecherni Novini, launched in 1951, is the newspaper of the Sofia City Committee of the BCP. It is an evening newspaper with a circulation of 122,000 copies.

There are 33 central weeklies, which cover a wide range of themes and interests and are aimed at different sections of the reading public. Some of these weeklies have a very large circulation: *Sturshel* (Hornet), a humorous and satirical weekly has a circulation of 375,000 copies; *Pogled,* which is run by the Union of Bulgarian Journalists has a circulation of nearly 300,000; *Anteni* (Antennas), a political and cultural weekly, has a circulation of 180,000; *Napravi si Sam* (Do-It-Yourself), a monthly, is published by the Komsomol for do-it-yourself enthusiasts and has a monthly circulation of 180,000.

Almost all the magazines in Bulgaria are published in Sofia, although most of them are aimed at special interest groups. General information magazines are very popular

Collage of Bulgarian magazines

Zemedelsko Zname, launched in 1902, is the newspaper of the Standing Committee of the Bulgarian Agrarian Party (BZNS). Its circulation is 170,000 copies.

Otechestven Front, launched in 1942, is the newspaper of the National Council of the Fatherland Front. It comes out in the afternoon in two editions and has a circulation of 280,000 copies.

Trud, launched in 1956, is the newspaper of the Central Council of the Bulgarian Trade Unions. Its circulation is 295,000 copies.

Narodna Mladezh, launched in 1944, is the Komsomol newspaper. It has a circulation of 235,000 copies.

Facsimile of the newspaper *Rabotnichesko Delo* (1927)

and a total of 45 titles are published, with a total print of 2,300,000. Most of them are monthly. They vary in size, though the number of pages usually runs to about 60. The magazine with the largest circulation, around 450,000 copies, is *Woman Today*. Other leading monthlies include *Otechestvo* (Homeland) with 95,000 copies, published by the National Council of the Fatherland Front; *Bulgaro-Sovetska Druzhba* (Bulgarian-Soviet Friendship) published by the National Bulgarian-Soviet Friendship Committee with a circulation of 65,000; *Paraleli* (Parallels), 220,000 copies, published by the Bulgarian News Agency (BTA); and *Obshtestvo i Pravo* (Society and Law), 60,000 copies, published by the Union of Bulgarian Jurists. A total of 29 literary and arts magazines are published, dealing with various aspects of literature, art, culture and aesthetic education. Average circulation usually totals about 30,000, and magazines of this kind usually contain about 130 pages. The 68 scientific and technical magazines and reviews are mostly specialized publications appearing four to six times annually, with an average circulation of about 2,000 copies. Most of them are published by the Bulgarian Academy of Science. They include the reviews *Arheologiya*, (Archeology), *Ekonomicheska Missul* (Economic Thought), *Istoricheski Pregled* (Historical Review), *Filosofska Missul* (Philosophical Thought), and *Sotsiologischeski Problemi* (Sociological Problems), as well as the publications of the Bulgarian Academy and Agricultural Academy, among others. There are twelve popular science magazines, which are among the most sought-after periodicals. These include *Zdrave* (Health), circulation 240,000; *Kosmos* – 160,000; *Nauka i Tehnika za Mladezha* (Science and Technology for Youth) – over 70,000; *Matematika* – 82,000; *Radio, Televiziya i Elektronika* (Radio, Television and Electronics) – 60,000; *Mlad Konstruktor* (Young Designer) – 40,000 copies. There are more than 80 publications on methodological, specialist and production problems. These are highly specialized and therefore have a smaller ciculation.

A total of 403 newspapers and a fairly large number of magazines are published in the provinces. The 27 newspapers published in the various regional centres form the basis of the regional Press. These newspapers supply general information on problems of the social, economic and cultural life in the country. They also cover events and developments in the region and are mainly distributed within the region itself. District youth weeklies are published in Plovdiv and Varna, and a weekly for school students comes out in Sofia. The circulation of the various regional newspapers varies between 20,000 and 70,000 copies; the total circulation comes to over 700,000, and is rising steadily. Five of the regional newspapers are dailies. Twenty appear three times a week while two appear twice a week. Sixteen have literary supplements and four have supplements on the problems of young people.

Around 180 newspapers are published for specific territorial and settlement systems, and about 190 in such places as factories, enterprises, educational establishments, etc.

Facsimile of the newspaper *Rabotnicheski Vestnik* (1899–1939)

In addition, three newspapers and nineteen magazines are published in foreign languages, for readers abroad. Taken as a whole they have a circulation of 800,000 copies.

Foreign readers also show an interest in magazines published in Bulgarian. The total circulation per issue of all the Bulgarian-language magazines distributed abroad comes to 110,000 copies.

Some 14,000 foreign newspapers and magazines in Russian, English, French, German and other languages are available in Bulgaria. Their total per issue circulation amounts to 2,930,950 copies, i.e. 325 copies per thousand of the population. The total annual circulation of foreign newspapers and magazines on sale in Bulgaria comes to 149,722,452 copies.

There are no privately-owned newspapers or magazines in Bulgaria. Periodicals are published by political parties, public organizations, trade unions, cultural unions, scientific institutes, companies, corporations, state bodies and organizations. One of the leading publishers is the Bulgarian Academy of Science, which is responsible for 34 magazines, one newspaper and 44 bulletins. The National Agro-Industrial Union and the Agricultural Academy bring out 23 magazines, 30 bulletins and one daily. The Committee for Culture publishes eighteen magazines, 24 bulletins and two weeklies. The Central Council for Physical Education and Sport publishes two newspapers, two magazines and ten bulletins. The periodicals produced by the Central Committee of the Dimitrov Young Communist League are notable for their wide range of themes and styles. Apart from *Narodna Mladezh,* a daily newspaper which provides general information, the Young Communists publish weeklies: *Puls* (Pulse), a

Facsimile of the newspaper *Otechestven Front* (1942)

literature and art newspaper; *Orbita* (Orbit) dealing with science and technology; and the magazines *Matematika* and *Posoki* (Directions), which give career guidance; *Rodna Rech* (Native Language); *Kosmos* (Outer Space); *Nauka i Tehnika za Mladezhta* (Science and Technology for Young People); and *Septemvriiche,* a newspaper for children.

The Law Relating to the Press. The free development and functioning of the Press is guaranteed by law. There is no censorship in Bulgaria. Article 54 of the Constitution of the People's Republic of Bulgaria states that 'citizens enjoy freedom of speech, Press, meetings, association and demonstration. These freedoms are guaranteed by placing the necessary material prerequisites for the purpose at the disposal of citizens'. Many institutions and bodies make use of these basic rights: apart from the two political parties, newspapers and periodicals are published by all the public organizations, professional and cultural unions, some state corporations and associations, the principal scientific institutions, all the regional centres, as well as a number of other towns, villages and enterprises. At the same time the Constitution allows for the banning of any abuse of the freedom of speech and the Press. Under Article 9, for example, these rights and liberties cannot be exercised to the detriment of the public interest. Under Article 63, incitement to war and war-propaganda are prohibited and punishable by law as grave offences against peace and mankind.

In 1977, the State Council of the People's Republic of Bulgaria issued decree No. 1086, which has been and will continue to be of great importance for the development of Bulgarian society. The decree concerns the increasing role of published criticism. After pointing out its importance for the implementation of state policy, the elimination of shortcomings and violations, and the extension of working people's participation in administration, Article 9, paragraph 1 of the Decree states that 'Ministries and managers of other departments, establishments and organizations, and the presidents of the Executive Committees of the People's Councils are to communicate their point of view, and the measures they have taken to eliminate weaknesses and violations, to the respective editorial board not later than one month after receiving the published criticism or information about it'. Article 19, paragraph 1 of the same Decree specifies that 'a manager who for no valid reason fails to reply to the critical publication within the set period will be fined between 20 and 100 levs. If the violation is repeated the fine will be doubled and other sanctions will be imposed'.

Special provisions ensure access to all areas of the

administrative, political, industrial, economic and cultural life of society. An official who hinders the work of representatives of the mass media, or refuses to help them, or gives inaccurate information is liable to a fine of between 20 and 200 levs, if he has not committed an offence which makes him liable to more severe punishment. The same punishment can be imposed on officials who persecute someone in connection with published criticism, or someone who has supplied information for or provided assistance to someone else in connection with published criticism.

At the same time, the Decree upholds the interests of bodies, institutions and persons that have been criticized. Under Article 12, they are entitled to refute published criticism that is based on inaccurate information. In these cases the members of the respective editorial boards have to ensure the timely publication of the refutation, while those who have been responsible for the publication of the inaccurate information are punishable by law. An editor-in-chief who fails to ensure the publication of a refutation in good time is liable to a fine of between 20 and 100 levs, and if the violation is repeated the fine is doubled and other sanctions may also be imposed.

Thanks to Article 16 of the Decree there is adequate mutual consultation between the Government and the mass media as regards published criticism. At least once every six months the Committee and the Union of Bulgarian Journalists submit information to the Council of Ministers on published criticism in the mass media and the response of the bodies and organizations concerned. The Council of Ministers examines this information carefully and makes decisions on the basis of it. This practice greatly increases the effectiveness of published criticism and enhances the role and prestige of the Press.

Facsimile of the *Periodical Journal of the Bulgarian Literary Society* (1870–76, 1882–1910)

BROADCASTING

The Committee for Television and Radio is a state organization whose task is to produce, broadcast, organize international exchange and keep archives of TV and radio programmes. It is part of the National Complex of Art, Culture and Mass Media, as a separate corporate body with its headquarters in Sofia and branch divisions in other parts of the country.

The Committee for Television and Radio is headed by a President and collective management bodies – a College and Board of Directors.

The public-state principle is widely applied in the committee's activities. Representatives of various social and professional strata and different age groups take an active part in the production of radio and TV programmes.

Bulgarian Television and Radio receives over 150,000 letters annually from viewers and listeners containing recommendations for new programmes and improvements in existing ones, as well as complaints and proposals regarding various shortcomings and weaknesses in the life of various work-forces and organizations.

The programming activities of radio and television are carried out by specialized departments, each with its own editor-in-chief.

The Bulgarian radio building, Sofia

Bulgarian Radio. The first Bulgarian radio broadcast was made in Sofia in November 1929. The Union of Radio Amateurs in Bulgaria was set up in 1930 and was responsible for the first broadcasting on a regular basis.

At first there were only two or three broadcasts weekly with a duration of about four hours, the programmes consisting of cultural and educational talks, news bulletins from the Bulgarian Telegraph Agency (BTA), advertisements and stock market reports, as well as concerts of Bulgarian folk, dance and light music. The first programme for children was broadcast on 8 February 1931 and the Countryside Hour was introduced in late 1933. Daily radio broadcasts in Bulgaria began in 1932, when live broadcasting was also started. The State took over radio broadcasting on 26 January 1935. A new 100-kW transmitter commissioned in 1937 made it possible to increase broadcasts up to nine hours daily.

The regional radio stations in Varna and Stara Zagora came into operation in 1936. They broadcast two hours of their own programmes and re-broadcast the Radio Sofia Programme during the rest of the day.

During the Second World War (1939–45), on a decision by the Bureau-in-Exile of the Bulgarian Communist Party and under the direction of Georgi Dimitrov, the underground radio stations *Hristo Botev* and *Naroden Glas* (People's Voice) started broadcasting from the territory of the USSR on the wave-lengths of Radio Sofia (on 23 July 1944 and 7 October 1941 respectively). The two radio stations challenged the official fascist radio propaganda and helped organize the Bulgarian people's armed resistance against the monarcho-fascist dictatorship. The first broadcasts of the *Hristo Botev* and *Naroden Glas* radio stations constituted the beginnings of the present-day socialist Bulgarian radio.

After the victory of the socialist revolution in Bulgaria on 9 September 1944 radio broadcasting made rapid progress. New technical facilities were built and the radio-relay network was extended. Radio Sofia's Second Programme on medium wave was launched in 1946, and a short-wave transmitter centre for broadcasts to foreign countries started operating in 1951. The duration of broadcasting increased: an all-day programme was introduced in 1958, a Third Programme came on the air in 1962 on VHF, and the regional radio network was also extended. New radio stations were opened in the cities of Plovdiv (1962), Shumen (1972) and Blagoevgrad (1973).

The country's transmission facilities expanded on a large scale, and the number of radio sets was also growing rapidly. In 1944 there were 183,000 radio receivers registered in Bulgaria, and by 1970 their number had reached 1,554,873. In 1983 the number of mains radio sets fell to 952,976 on account of the growing use of portable transistor radios (whose number now runs into millions) and the existence of over a million radio rediffusion sets.

Since 1966 Radio Varna has been broadcasting a special programme in Russian, Polish, Czech, German, French and English for foreign holidaymakers on the Bulgarian Black Sea coast.

In 1974 the First Programme of the Bulgarian Radio *Horizont* (Horizon) became a round-the-clock programme. The fourth national radio programme, *Znanie* (Knowledge), was introduced in 1977.

Today Bulgarian Radio's standards of programming are up-to-date and high. The yearly total of its programmes put on the air comes to about 45,000 hours, comprising a wide variety of programmes in different networks to suit different interests. It transmits four different national daily programmes, a system of broadcasts to foreign countries in twelve languages and a wide network of regional radio services, with six regional radio stations.

The national programmes of the Bulgarian radio are:

The *Horizont* programme – a 24-hour information and musical programme whose ratio of music to speech is about 3:1. It is the chief source of information about events at home and abroad, with more than 30 newscasts in 24 hours. The programme's guiding principles are promptness and topicality, with live broadcasts, wide use of on-the-spot reporting, and running commentaries. Phone-ins are also used.

The *Hristo Botev* programme – a cultural and publicistic programme with a duration of 18 hours a day (from 06.00 to 24.00 hours) addressed to different social and age groups. It employs all the modern programming forms, with the main accent on commentary and discussion of problems and topics by specialists in different areas of socio-political, economic and cultural life. The programme follows the main trends in social, moral and artistic life on a national and global scale. This programme also makes use of block phone-ins. Radio plays are a prominent feature.

The *Orphei* (Orpheus) programme – a specialized arts programme – broadcasts on VHF from 16.00 to 24.00 hours, and on Fridays and Saturdays until 01.00. The programme presents listeners with samples of Bulgarian and foreign classical culture and contemporary world art. The commentary-type broadcasts present the development of aesthetic thought in the various schools of art or in individual works. Stereo braoadcasts are a common feature of this programme.

The *Znanie* programme – a specialized scientific-educational programme transmitted on medium wave and VHF

Table 1: Radio Broadcasting in 1983

Radio Programmes	Horizont		H. Botev		Orphei		Znanie		Total	
	Hours	%	Hours	%	Hours	%	Hours	%	Hours	%
Information	1,366	9.4	473	4.7	126	2.5	—	—	1,965	6.4
Problems-Commentary	1,295	8.9	1,619	15.9	25	0.5	—	—	2,939	9.5
Educational	—	—	—	—	—	—	1,098	100	1,098	3.6
Cultural	64	0.4	619	6.1	643	13.7	—	—	1,326	4.2
Radio Plays	—	—	142	1.4	194	3.8	—	—	336	1.1
Musical and Entertainment – Total	5,799	39.8	3,589	35.3	2,001	39.3	—	—	11,389	36.9
Concerts (classical, pop and folk music)	5,116	35.2	2,130	21.0	1,624	32.0	—	—	8,870	28.7
Speech and Music	674	4.6	1,426	14.0	349	6.9	—	—	2,449	7.9
Humour and Satire	9	0.1	33	0.3	28	0.6	—	—	70	0.2
Sports	186	1.3	—	—	—	—	—	—	186	0.6
Other	50	0.3	134	1.3	88	1.7	—	—	272	0.8
Total	14,559	100	10,165	100	5,078	100	1,098	100	30,900	100

from 10.00 to 12.00 hours and repeated on the same day from 14.00 to 16.00 hours. It is a kind of national radio university, incorporating about 50 individual programmes in the sphere of the social and the pure sciences. The programme also broadcasts foreign language courses (Russian, English, German, French and Spanish).

Bulgarian Radio has correspondents in all districts of the country.

The regional radio in Bulgaria comprises a system of six radio-stations. Radio Varna and Radio Plovdiv broadcast daily five and a half hours of their own programmes, Radio Stara Zagora – five hours and Radio Shumen and Radio Blagoevgrad – two and a half hours each. There is a daily one hour programme for the capital.

The audience of Bulgarian radio comprises 70 per cent of the country's population above the age of seven.

The transmissions to foreign countries are broadcast in twelve languages to a total of 40 hours a day. They include broadcasts to countries in the Balkan Peninsula, Western Europe, North America, Latin America, Africa and the Near and Middle East. There is a special programme for Bulgarians living or working abroad.

The volume and structure of the four national programmes of Bulgarian radio are illustrated in Table 1.

The Technical Facilities of Bulgarian radio have been built on national and regional levels. At first all the programmes were broadcast live 'before the microphone'. Tape recordings were first introduced in the early 1940s, and the professional use of studio tape recorders and players began in 1948. A specialized technical unit for making radio programmes was formed in 1954. The use of wireless radio communications by means of VHF devices in outside studio broadcasts was introduced in 1962, stereo-recording in 1968, and multi-channel technology in sound recording – in 1975.

The television tower, Sofia

Bulgarian Television. The first TV signal in Bulgaria was emitted by wire from the building of the Higher Institute of Mechanical and Electrical Engineering in Sofia in 1952. Wireless television broadcasts began in 1954. Regular broadcasts were made twice weekly. In those days the TV signal could only cover the capital city and its surroundings. These broadcasts virtually laid the foundations of the first TV institution in the Balkans.

Bulgarian National Television was officially inaugurated on 26 December 1959. In 1960 there were 2,573 TV receivers in Bulgaria and in 1983 their number was 1,699,000, not counting portable TV sets.

Colour TV broadcasts date from 1972. A regional TV centre was established in Plovdiv in 1970, Varna followed suit in 1971, Russe in 1972, and Blagoevgrad in 1975. Bulgarian Television has studios in the towns of Vratsa, Veliko Turnovo and Burgas and correspondents' bureaux in all district centres.

Since 1972 Bulgarian Television has been broadcasting specialized programmes for foreign holidaymakers and tourists in Russian, German, French and English. Since 1982 there has been a specialized programme for the hard of hearing. Bulgarian Television broadcasts on two national channels with an annual total of 5,191 hours or an average weekly total of 100 hours, 96 per cent of them in colour.

Channel One, with a wide range of feature programmes, is devised to satisfy the interests of the widest audience. It is on the air for a total of 3,657 hours annually (70 hours a week. See Table 2). On weekdays Channel One has the following broadcasting schedule:
morning programme – from 08.30 to 12.50 hours
early evening programme – from 17.30 to 20.00 hours
main evening programme – from 20.00 to about 23.00 hours.

The morning programme, adapted to the specific requirements of its audience (school children, students, shift workers, housewives and pensioners), is composed of educational and children's broadcasts, feature films and news. All television genres are represented on the evening programme. Up to 20.00 hours television broadcasts are mainly programmes for the children, popular science films, economics broadcasts, documentary films and folklore. The main newscast 'Around the World and at Home' is on at 20.00 hours. The artistic broadcasts – films, dramas, shows – are mainly timed for the period after the news, i.e. after 20.30 hours. On Saturdays, Sundays and holidays, Channel One lays the accent on arts, entertainment and sports broadcasts. Predominant in Channel One are arts broadcasts (37 per cent) and news, current affairs and documentary programmes (28 per cent). Special attention is paid to the broadcasts for the youngest audience. Young people's programmes comprise 20 per cent of the total programme time. On weekdays the Second Channel broadcasts from 19.00 to 23.30 hours, and during holidays from 18.30 to 22.00 hours. Its total annual air-time is 1,534 hours. (See Table 2).

Channel Two is addressed primarily to a more specialized audience of schoolchildren and students, viewers

Table 2: Annual Television Broadcasts by Kinds of Programme

Kinds of Programmes	Channel One		Channel Two		Channels One and Two – Total	
	Hours	%	Hours	%	Hours	%
News, Current Affairs and Documentary Programmes	1,010	27.6	225	14.7	1,235	23.8
Arts Programmes	1,343	36.7	693	45.1	2,036	39.2
Educational Programmes	368	10.1	312	20.4	680	13.1
Children's and Youth Programmes	823	22.5	156	10.2	979	18.9
Sports Broadcasts	78	2.1	135	8.8	213	4.1
Other Programmes	35	1.0	13	0.8	48	0.9
Total	3,657	100	1,534	100	5,191	100

with wider intellectual interests and sports fans. 60 per cent of the broadcasts on the channel are addressed to a specialized audience – 20 per cent educational, 20 per cent films, theatre, music and folklore and 9 per cent sports. At the same time the Second Channel complements the First by providing a wider choice for audiences in the most popular programmes – feature films (21.5 per cent) and news, current affairs and documentary broadcasts (15 per cent).

Four television centres and three television studios produce programmes exclusively for the national television network only.

The audience of Bulgarian Television consists of 80 per cent of the population over seven.

The technical facilities of Bulgarian Television, like those of Bulgarian Radio have been built at national and regional levels.

Bulgarian Television uses the SECAM-III V option, Colour TV System, standard 625 lines (50 half-frames).

Film production. To meet its own programme demands, Bulgarian Television has organized its own documentary film production (90 films a year) as well as the production of feature films and serials (a total of twenty a year). Some of them are made as co-productions or contracted to Bulgarian film-making organizations.

Musical productions. The Committee for Television and Radio is the organizer and producer of a great portion of the nation's musical production. Besides recording music it takes an active part in the promotion of musical composition in all genres. The committee has its own symphony orchestra, big band, mixed choir, folk-song ensemble, children's choir and the Sofia String Quartet. These ensembles give many concerts in the country and abroad. The musical ensembles record annually over 8,000 minutes of music in all genres.

International relations. The Committee for Television and Radio has contracts and agreements with 57 national broadcasting corporations and maintains close co-operation with forty others. Its own TV and radio programmes participate in established international reviews, festivals and markets.

The Committee is a Member of the International Organization for Radio and Television, through which it maintains close co-operation with EBU and all regional broadcasting unions in the world. It is a member of the International Federation of Television Archives and of the International Association of Women in Broadcasting.

The Committee for Television and Radio has correpondents in Moscow, Budapest, Warsaw, Prague, Berlin, Bonn, Vienna, Helsinki, Beirut and Havana.

Each year the Committee for Television and Radio holds an International Festival of Television Plays in the city of Plovdiv. It is co-sponsor of other international events in Bulgaria – the Festival of Red Cross Films and the World Festival of Cartoon Films held in Varna, the International Golden Orpheus Pop Song Festival at the Black Sea resort of Sunny Beach, the International Festival of Humour and Satire in Gabrovo and other cultural and sports events.

The Committee for Television and Radio participates in the co-production of films with a number of foreign TV organizations.

The Radio and TV in Bulgaria are included in the Intervision and Eurovision networks via the Sofia-Moscow, Sofia-Belgrade, Sofia-Istanbul and Sofia-Athens radio-relay lines, and through the ground station for space communications of the Intersputnik system. Conditions have been created for the interchange of programmes with all neighbouring countries, as well as for transit links between the countries of Europe and the Middle and the Near East.

With their multifarious activities Bulgarian Television and Bulgarian Radio contribute to increasing the people's political, cultural and educational awareness, and to the promotion of better understanding among nations and the cause of world peace.

THE BULGARIAN TELEGRAPH AGENCY

The Bulgarian Telegraph Agency (BTA) is the central news agency of the People's Republic of Bulgaria; it has the status of a committee attached to the Council of Ministers (1 August 1973). It was founded at the beginning of 1896; the Agency's first news bulletin was handwritten and appeared on 16 February of that year. Until the 1912–13 Balkan War, the BTA department functioned as part of the Ministry of Foreign Affairs.

The first director of the BTA, Oscar Iskendur, had for a long time been editor-in-chief of *La Bulgarie,* the official French language Bulgarian newspaper. He established relations with some of the world news agencies of the day – Reuters and Havas – as well as with the German agency Wolfbüro, thus providing Bulgarian bulletins with more and more international news.

After the Balkan War, the BTA was reorganized as a press agency with two departments: the BTA and the Press Department, the key task of the latter being to study the foreign Press and translate excerpts of important articles, especially those dealing with Bulgaria and the Balkan countries. A talented Bulgarian journalist, Yossif Herbst (1875–1925), was appointed head of the Press Agency. Later he fought courageously against the fascist regime that had been established as a result of the 9 June coup in 1923 and was killed by the fascists in 1925.

In 1936 a third section of the Press Agency was set up to deal with research and documentation. Between 1925 and 1939, the BTA bulletins increased considerably – from 687 pages in 1925 (two issues daily) to 5,700 in 1939 (three issues daily). In 1928 the BTA acquired a radio-receiver by means of which it received daily 147 regular radio broadcasts from abroad.

After 9 September 1944, when power in the country was taken over by the Fatherland Front, the Press Agency underwent a thorough reform and a new stage was thus initiated in its development. The tasks and activity of the Press Agency rapidly increased. Sources of information increased also: the news unit became a separate office with a news bulletin of its own; a special office was set up for the transmissions of news abroad; the radio-telegraph service was extended and equipped with modern facilities. A news bureau was established to write articles and publish brochures, magazines and other materials in foreign languages for distribution abroad.

In 1951 the Press Agency, with all its editorial offices and services, was renamed the Bulgarian Telegraph Agency and thus became an independent information institute of the Council of Ministers. Its basic tasks are: to collect and provide news on events in the country and abroad, on the development of the People's Republic of Bulgaria and other socialist countries, on their peace-promoting foreign policy, on the struggles of peoples for national liberation and social progress; to provide the Press, radio and television, state and public bodies, institutions and establishments, foreign news agencies, Press and publications with news; to distribute official documents and announcements for publication within Bulgaria and abroad; and to publish news bulletins and other journalistic material.

Since 1965, the BTA has been publishing four weeklies: *Paraleli* (Parallels) – an illustrated magazine with a weekly circulation of 220,000; *Po Sveta* (Around the World) – a magazine with articles on events and problems in foreign countries with a circulation of 18,000; *LIK* (Literature, Art and Culture) with a circulation of 19,000; and *Nauka i Tehnika* (Science and Technology) – with accounts of the latest foreign scientific and technological achievements, and with a circulation of 18,000.

The BTA maintains telecommunication contact with five world agencies – Tass, Reuters, Associated Press, United Press International and Agence France Presse, and about sixty national news agencies. In 24 hours BTA receives 4,000 pages (30 lines a page) of foreign news reports and transmits political, economic, cultural, sports and other news on ten channels, using the most sophisticated telecommunications equipment. Photos are exchanged via wirephoto links. The Agency is included in the permanent wirephoto network of the Photo International and in the wirephoto network of Associated Press and United Press International.

All the central newspapers, the radio and the television and the five provincial dailies receive news via a direct teletype-communications system with the Agency.

The BTA is headed by a Director-General and three deputies. The Agency has twelve editorial offices, two departments and a production base. The editorial offices are: Foreign News, Domestic News, External Services, Political Analyses and Commentaries, Press Photo, Local Press, Sports Office, Reference Office, *Nauka i Tehnika* (Science and Technology), *LIK* (Literature, Arts and Culture), *Paraleli* (Parallels), and *Po Sveta* (Around the World).

The Bulgarian Telegraph Agency

The BTA has corespondents inside the country and abroad. It maintains 23 bureaux staffed with 26 correspondents in Moscow, New York, Paris, Berlin, London, Bonn, Rome, Athens, Belgrade, Tokyo, Peking, Ankara, Madrid, Hanoi, Cairo, Budapest, Prague, Bucharest, Havana, Warsaw, Beirut, Delhi and Mexico. Within Bulgaria itself, the BTA has 25 correspondents' bureaux.

The Agency publishes 27 bulletins: ten daily bulletins for Press, radio and television, two daily bulletins in Russian and English, and five weekly bulletins in five foreign languages. The BTA also publishes a *Spravochnik NRB* (Bulgarian Year-Book) with a monthly addendum and a *Spravochnik na stranite chlenki na SIV* (Reference Book on the CMEA countries) with a quarterly addendum. Twice a week the BTA publishes sets of photos entitled *Around the World and at Home* and *Physical Culture and Sport*. It provides daily supplies of pictures for the Press.

The BTA takes an active part in the Alliance of European News Agencies, and has signed international co-operation agreements with 48 agencies in Europe, Asia, Africa and America.

The Bulgarian News Agency building, Sofia

BOOK PUBLISHING

'Void is a nation without books', is written in one of the earliest Bulgarian poems, dating from the ninth century.

The Bulgarian people wrote their first books more than a thousand years ago. Rare copies of these books can be seen to this day in the world's largest libraries such as the British Museum, the Vatican Library and the Lenin Library in Moscow.

Until the end of the eighteenth century, owing to the hard conditions that prevailed during the five centuries of Ottoman rule, books in Bulgaria were mainly handwritten, although as early as the sixteenth century patriotic Bulgarians brought out the first books printed in the Cyrillic alphabet outside their subjugated country: Yakov Kraikov's *Book of Hours* (1566), *Psalter* (1569), and *Prayer Book* (1570), Filip Stanislavov's *Abagar* (1651) and Hristofor Zhefarovich's *Stemmatographion* (1741), amongst others. The publication in 1806 of *Nedelnik* (Sunday Prayer Book) of Sophronius of Vratsa, marked the beginning of new Bulgarian literature.

By the mid-nineteenth century the number of printed books had reached 170, quite an achievement for enlighteners of the National Revival period.

The first attempts at establishing independent book publishing facilities inside the country resulted in the setting up of the Bulgarian Publisher, Ephoria, in Bucharest and of the Bulgarian Publishing Society in Brasov. The most active publisher before the liberation from Ottoman rule was Hristo G. Danov, who founded a publishing house in Plovdiv which brought out 145 books, calendars, and textbooks between 1835 and 1878.

During the National Revival Period Bulgarian publishing played a major role in increasing the people's national consciousness. Books helped progressive national forces in their struggle against the feudal order and the assimilatory aspirations of the Greek clergy. They furthered Bulgarian education and the preparations for a national revolution.

In 1878, the year Bulgaria was liberated from Ottoman rule, the country had a modest but valuable literary heritage. Whereas during the first decade following the Liberation a mere 200 books were printed per year, by the turn of the century their number reached between 800 and 1,000. In addition, major works of world and Bulgarian authors and scholars were published.

In the 1890s the first books on Socialism were published. The first Congress of the Bulgarian Workers' Social Democratic Party, held on Mt. Buzludja in 1891,

Title page of the 1569 *Psalter*

decided that the publication of the *Bulgarian Socialist Democracy Library* should be started in Turnovo. The first book in that series was Dimiter Blagoev's *What Is Socialism and Can It Thrive on Bulgarian Soil?* The Socialist Bookshop and Printing House set up in Sofia in 1903 published and distributed socialist books.

In the early years of the twentieth century the publication of books centred mainly in Sofia, although it suffered setbacks during the 1912–18 wars. Some publishing houses specialized in the printing of fiction and popular science works. The *Globus* and the *Nov Svyat* (New World) publishing houses published books of a progressive nature. From 1918 to 1944 around 2,000 titles were published each year, which was a considerable increase. Book publishing, printing and distribution became established professions and a science in themselves.

Following the victory of the socialist revolution on 9 September 1944 book publishing became the right and obligation of the State and of public organizations. The number of books published and their total print increased greatly, their thematic range was broadened and their layout was improved. New publishing houses were founded, either state-owned or belonging to public organizations.

St. Demetrios the Martyr (wood-cut), by Atanas Karastoyanov, 1852

Pizho and Pendo Shaking Hands (water-colour and ink), by Alexander Bozhinov, 1944

Land and People (lino-cut), by Yuli Minchev, 1960

Since 1982 a State association – Bulgarska Kniga i Pechat (Bulgarian Books and Press) has been functioning. It is in charge of book publication, printing, book distribution, book fairs, exhibitions and advertising at home and abroad, as well as of questions related to the structure and material and technical facilities of the periodical Press. The state-public principle is applied in directing the work of that extremely important sphere by enlisting the broad support of experts in various fields – production, science, literature and the arts – as members of consultative and editorial councils. Various forms of co-ordinating the efforts of cultural unions, public organizations departtions, departments and institutions directly involved in book publishing were employed.

Book publishing is now an inseparable part of the national Art, Culture and Mass Media Complex. Book publishing is carried out in an organized and planned way by public and departmental publishing houses, of which there were 27 in 1983. Thirteen of them are state-controlled and are completely managed by state bodies as regards to theme, staff and finance: *Narodna Prosveta, Nauka i Iskustvo, Tehnika, Narodna Kultura, Zemizdat, Meditsina i Fizkultoura, Muzika, Septemvri, Otechestvo, Dr Peter Beron* and *Svyat* publishing houses in Sofia, *Hristo G. Danov* in Plovdiv and *Georgi Bakalov* in Varna. There are fourteen publishing houses run by public organizations, cultural unions, Ministries or departments: *Partizdat* (the publishing house of the Bulgarian Communist Party), *Narodna Mladezh, Voenno Izdatelstvo, Bulgarski Pisatel, Profizdat,* the Fatherland Front Publishing House, the publishing house of the Bulgarian Agrarian Party, the publishing house of the Bulgarian Academy of Science, *Bulgarski Hudozhnik, Sofia Press, Standartizatsia, Bulgarreklama* and the publishing house of the Holy Synod, all based in Sofia. Publishing is mainly concentrated in Sofia, which has 25 publishing houses.

While 737 titles were published in 4,140,000 copies in 1944, 5,070 titles came out in 1982 with a total print of 59,662,000, i.e. an increase of 690 per cent as regards titles, and an increase of 1,440 per cent in the total print.

Apart from the wealth of books available to the Bulgarian reader, each year between 15,000 and 16,000 titles in Russian are imported from the Soviet Union.

Illustration for the book *Panchatantra* (ink), by Ivan Kyossev, 1975

Illustration for Ilse and Vilmos Korn's *Meister Hans Röckle und Mister Flammenfuss*, by Rumen Skorchev, 1971

These are widely read in the original because of the similarity of the two languages.

A characteristic feature of publishing in Bulgaria is the tendency to give priority to growth rates in the total print of books as against the number of titles. The aim is to satisfy readers' demands more fully. In 1944 there were 106.6 titles per 1,000,000 people, while in 1955 the corresponding figure was 401.7, in 1975 420.2, and in 1982 567.8 titles. There has also been an increase in the average titles published per person per year: 0.6 titles in 1944, 2.7 titles in 1955, 5.6 titles in 1975 and 6.7 titles in 1982. Sociological surveys conducted in 1972 and 1981 indicated that the average Bulgarian reads five books a year, while 27 per cent of the population reads over ten books annually. In this respect Bulgaria is among those countries with a well-developed publishing trade. According to Unesco, Bulgaria is ahead of the United States, Japan, Italy and others in the per capita number of titles published each year, and is on a par with France in this respect.

Bulgaria is one of the countries which publishes a large number of translations from foreign languages – over 17 per cent. Books by classical and modern Russian and Soviet authors account for the greatest number of translations, followed by books by American, British, French, German and Spanish authors.

As a result of the intensive development of international cultural co-operation in the spirit of the Helsinki Final Act, Bulgarian publishers are now acquainting their readers with the works of outstanding writers from all parts of the world.

Foreign language publications are of particular importance. An average of about 350 titles are translated from Bulgarian each year. These comprise fiction as well as books on social, political, scientific and technical subjects by well-known Bulgarian authors. They are translated into the most widely-spoken foreign languages and are aimed at familiarizing readers abroad with the achievements and culture of the Bulgarian people. A new state publishing house, *Svyat*, was set up in 1982 with the aim of publishing Bulgarian books in foreign languages.

In Bulgaria sociological studies are regularly conducted to research reader demand, interest and opinion and to invite proposals. These studies are carried out by a specialized unit of the respective state body, the Cyril and Methodius National Library, and other cultural institutions.

Books and Brochures Published in 1939–82

Year	Population	Number of titles published	Percentage increase on 1939	Number of titles per 1,000,000 people	Total print (in thousands)	Percentage increase as against 1939	Copies of books per one million people
1939	6,139,000	2,169	100.0	353.3	6,484	100.0	1,056.2
1944	6,913,300	737	36.3	120.1	4,140	63.8	598.8
1955	7,537,800	3,028	139.6	401.7	20,702	319.3	2,746.4
1982	8,929,300	5,070	233.7	567.8	59,662	920.1	6,681.6

PRINTING AND PUBLISHING

The first printing house on Bulgarian soil was set up by Nikola Karastoyanov in the town of Samokov. He imported a printing-press in 1828 and sets of Cyrillic typefaces in 1835 to start the publication of books in Bulgarian.

Prior to the Liberation in 1878, books in Bulgarian were published by printing-presses equipped with Cyrillic typesets in Turgovishte (Romania), Rome, Vienna, Brasov, Thessaloniki, Istanbul, Braila, Leipzig and elsewhere.

In the post-Liberation period printing houses were opened in Svishtov, Russe, Shumen and Turnovo, and their number was constantly increasing. The State Printing House, set up in Sofia in 1881, emerged as the country's major institution of this kind.

About 400 small-scale printing-presses were in operation in Bulgaria in 1943, 200 of which were in Sofia, where 80 per cent of the total number of publications in the country were printed. The printing facilities in the capital city were almost completely destroyed in the heavy bombing raids in 1943 and 1944.

The deplorable state of the country's printing facilities after 1945 necessitated the construction of new printing houses, as well as the restoration of the destroyed ones. In the 1947–49 period the industry was nationalized and all printing houses were placed under state administration. New printing houses came into operation in the 1952–82 period, as a result of which fixed capital formations increased by more than 230 per cent. The share of

Third National Folk Art Festival. Poster by Alexander Poplivov, 1976

Illustration of the book *Treasure Island* (tempera), by Lyuben Zidarov, 1976

different publications in the overall printing output is as follows:

Books	24.2 per cent
Magazines	7.9 per cent
Newspapers	40.4 per cent
Blanks	13.3 per cent
Advertising materials	7.9 per cent
Miscellaneous	6.3 per cent

The development of printing and publishing in Bulgaria is connected with the wider use of electronic photocopying systems, the priority use of offset printing and the integration of printing and binding processes. Plants for aluminium offset plates and printing ink as well as a new complex of the *Rabotnichesko Delo* newspaper with phototyping and an editorial-publishing system, roller-offset printing and automated dispatch, are under construction.

Today the capacity of Bulgaria's printing and publishing industry has greatly increased: the total output in 1982 exceeded the 1939 figure by 2,200 per cent, and the total print of books in 1982 marked a nearly 1,000 per cent increase over 1939.

The Bulgarska Poligrafia State Association, which publishes all books, newspapers and magazines in the country includes 33 major printing houses (seven in Sofia and 26 in the district centres) with 112 branches.

The Bulgarian Academy of Science Publishing House, Sofia

Uprising (lithograph), by Hristo Neikov, 1976

BOOK DISTRIBUTION

The traditional method of book distribution in medieval Bulgaria and in the early years of Ottoman rule was by hand copying.

The first Bulgarian bookshop was opened in Veliko Turnovo in 1809 by Velcho Atanassov (the Glazier). The founder of modern book distribution is Hristo G. Danov, who opened his own bookshops in Plovdiv, Veles, Russe, Svishtov, Bitola and Thessaloniki.

After Bulgaria's liberation from Ottoman rule in 1878 the number of booksellers rose rapidly. In most cases these were well-known publishers.

A Union of Bookpublishers and Booksellers was founded, with its own press organ, *Knizharska Duma*, which appeared from 1934 to 1943.

After the socialist revolution of 9 September 1944 book distribution in Bulgaria made rapid progress and became an important cultural activity.

In 1952 the abolition of private publishing and bookshops began and the foundations of the socialist book trade were laid. The years that followed saw the establishment of a new, socialist book trade.

In 1972 the state book corporation, Knigorazprostranenie, was set up. A special programme was adopted in 1977 with the aim of promoting and improving Bulgarian book distribution, its concentration, specialization and modernization, and developing various up-to-date forms of book trading.

Knigorazprostranenie runs 28 district enterprises and 1,800 bookshops selling Bulgarian books, textbooks, stationery and foreign literature. Being the main link between readers and publishers, they endeavour to meet reader's demands and play an active role in helping to plan the titles to be issued by publishing houses.

Bookshops account for 70 per cent of total book sales. Books are also sold by schools, higher education establishments, scientific institutions, enterprises, and offices, where there are about 500 bookstores with paid regular staff as well as part-time salesmen.

Book sales and exhibitions are regularly organized, as are meetings with poets and writers.

A significant proportion of books are also sold through the recently established Book Friends' Movement. This is done by people who work on a voluntary basis, in advertising and distributing them at offices, institutions, educational establishments, plants and enterprises. Their numbers are constantly growing. The movement now has 60,000 members in 1,500 different clubs. An International Book Fair is held in Sofia every two years, bringing together representatives of publishing houses and book trading firms from different countries. Its motto is 'Books in the Service of Peace and Progress'. Bulgaria is a regular participant at almost all established international book fairs.

THE SOFIA PRESS AGENCY

The Sofia Press Agency is a publishing house which provides publications for foreign countries and for foreigners visiting the People's Republic of Bulgaria. It was founded on 16 December 1967 in Sofia by the Union of Bulgarian Writers, the Union of Bulgarian Journalists and the Union of Bulgarian Artists. Subsequently other public organizations affiliated themselves to Sofia Press: the Committee for Culture, the Bulgarian Book and Press Publishing Company, the Committee for Television and Radio, the Union of Bulgarian Film Makers, the Union of Research Workers in Bulgaria, the Union of Translators in Bulgaria, the Union of Jurists in Bulgaria, the Union of Architects in Bulgaria, the Union of Bulgarian Composers, the Committee of Bulgarians Living Abroad, the Bulgarian Chamber of Commerce and Industry, the Bulgarian Association for Tourism and Recreation, the Bulgarian Olympic Committee and the Bulgarian Union for Physical Culture and Sport. Affiliation to Sofia Press is open to cultural unions, public organizations, departments and institutes.

laws of the country. Under the Statute adopted at the Fifth Conference in 1983 the Agency is an independent legal entity with offices in Sofia and as a public institution it is not responsible for carrying out the functions of the State, nor is the State responsible for carrying out the functions of the Agency.

Since its creation Sofia Press has been developing as a specialized publisher, whose main purpose is to familiarize foreign readers in Bulgaria and abroad, with the political, public, economic, social and cultural life of the Bulgarian people, as well as the home and foreign policy of the country. One of the Agency's most important tasks is to encourage mutual acquaintance, understanding, confidence and friendship between peoples, and to make the Bulgarian public familiar with the life of other peoples and countries.

The Sofia Press building, Sofia

The actitivity of Sofia Press is in accordance with the Constitution of the People's Republic of Bulgaria and the

The Sofia Press logo, by Stefan Kunchev

An important aspect of the activity of the Sofia Press Agency is the preparation and distribution of articles, commentaries, interviews, discussions, surveys and other kinds of information reflecting the attitude of the Bulgarian public to significant domestic and international events. Sofia Press's publications are intended for the daily, periodical and specialist Press, and for news agencies, publishing houses, radio and television companies, research institutes, companies and private persons abroad.

The publication of newspapers and magazines in foreign languages is one of the Agency's basic functions. *Sofia News* is a weekly newspaper mainly intended for foreigners visiting the country. It reports all the major events in the home and foreign policy of the People's Republic of Bulgaria, her achievements in the development of the economy, culture, technology, the arts, education, international tourism and sport. It is published in four languages (Russian, English, French and German) and more than a million copies are printed annually. The Agency's magazines are illustrated publications intended for distribution abroad. They carry accounts of significant events in the life of the country, report on important historical developments and dates, and popularize international attractions. The Sofia Press monthlies are: *Bulgaria* (in Russian), *Bulgaria Today* (in English, French, German, Spanish, Italian and Portuguese), *New Bulgaria* (in Arabic), *Mai Bulgaria* (in Hungarian for Hungary), and *Noticias de Bulgarie* (in Spanish for Cuba); the bi-monthlies are: *Bulgaria in Construction* (in Serbo-Croatian) and *Co-operative Review* (in English, French and Spanish). The total annual circulation of these magazines exceeds 2,500,000.

The publishing activity of Sofia Press Agency is aimed at popularizing translated Bulgarian material. The Agency's books are mainly socio-political, literary and illustrated publications. In the last three years the Agency has published more than 1,200 books, albums, folders and other publications covering current affairs, literature and science-fiction in more than fifteen languages. The number of copies printed comes to more than 5,500,000. The Agency prints its books either at its own printers (distributing them independently), or makes arrangements with foreign agencies, authors, publishers and publishing houses to have them printed in other countries.

Sofia Press also makes films for cinema and TV specially for foreign audiences with dubbing in foreign languages. Films are either made independently by the Agency, or are co-produced with Bulgarian and foreign organizations, television companies, film companies and firms. Annually Sofia Press turns out more than sixty short, medium, and full-length films.

Photographs are another area in which Sofia Press is active. Photos are used to illustrate the Agency's own publications – books, brochures, albums, newspapers, magazines, and articles. The Agency also arranges various kinds of photo exhibition, as well as photo-displays and multiple-slide shows.

Sofia Press maintains contacts with many foreign writers, artists, journalists, etc. visiting Bulgaria. The Agency arranges programmes, meetings with statesmen and politicians, interviews, etc., for foreign newsmen and film crews. Annually Sofia Press co-operates with more than two hundred journalists, some of whom are invited by the Agency. The Agency has bureaux in the provinces and abroad.

Part of the Agency's work is to establish contact and conclude contracts with foreign agencies, newspapers, magazines, publishers, film studios, radio and television companies, firms and individuals, and to supply them with material. The Agency takes part in international conferences and meetings on co-operation in the sphere of information and organizes Bulgarian representations, either on its own account or on behalf of the country at international events, such as fairs and exhibitions.

BULGARIAN PHOTOGRAPHY

The popularization of photography in Europe coincided with the cultural revival of the Bulgarian nation and its national liberation struggles. The founder of Bulgarian photography was Anastas Karastoyanov (1822–80). The nation's most active and public-spirited educationists – men of letters, teachers and revolutionaries, such as Toma Hitrov, Georgi Danchov and Ivan Zografov – were among the pioneers of photography in Bulgaria. The important social changes in the country gave rise to reportage photography. The discovery of the auto-type process furthered the development of photography by making possible the unlimited printing duplication of photos. A Society of Photo-Journalists was set up in Bulgaria in 1924. The gradual development and improvement of photography made it an independent art based on authenticity and bearing an artistic character.

The socialist revolution on 9 September 1944 gave a powerful impetus to the development of photography both as regards its scale and organization and in its ideological, aesthetic, socio-political and thematic direction. Georgi Dimitrov rated highly the significance of photography, and ordered the creation of 'State photo-archives' to 'register and preserve for the coming generations all major events in our political, public, economic and cultural life'. The nationalization of photo activities in 1951 resulted in the establishment of the state enterprise, Bulgarian Photography. After 1956, photography made great progress. In 1963 the Bulgarian Photography state enterprise became an independent department, and in 1975 it was established as a cultural, propaganda and ideological institute and developed into an artistic-production corporation with the Committee for Culture – the Bulgarian Photography Corporation. Within the system of the national complex, Artistic Creativity, Cultural Activity and Mass Media, the corporation has been assigned the important task of photographing the country's advance, of propagating the successes scored in socialist construction, of developing photography, of promoting the fulfilment of the programme for aesthetic education and for the versatile development of the personality and of ensuring photo-services. The Bulgarian Photography artistic-production corporation implements these tasks through its branch offices – the National Photographic Agency, district photo-propaganda, photo-information and photo-services centres, the Bulgarian Photographers' Club and the *Bulgarian Photo* magazine. The National Agency and 27 district centres organize photo-exhibitions, periodical and thematically arranged show-cases of photos, multivision programmes, photo slogans and posters, etc. They publicize the successes of Bulgarian culture, economy and industry, the achievements of science, technological progress and protection of health, show the care of children, and furnish information about social and economic problems, about the development of sports and tourism inside the country and abroad. Over 100 photo periodicals and thematic photo publications dealing with more than 1,600 subjects are issued every year. The volume of this activity is constantly expanding.

The Dimiter Blagoev printing house, Sofia

The growth of the necessary equipment over the past few years has resulted in an increase of colour photography in the overall productions.

The Photographers' Club was started in 1971. It has been entrusted with the task of raising the level of art photography and its role in the country's cultural life, of popularizing the successes of Bulgarian photographic art, and of enhancing the professional artistry of photographers and their theoretical and aesthetic training. The club arranges national and international photographic exhibitions, encourages the participation of Bulgarian photographers in international photographic exhibitions, organizes visits and meetings with foreign photographers, and theoretical conferences, seminars and consultations on artistic and creative, techno-scientific and juridical problems of photography. More than 400 experts in photography have joined the club, its district centres and the Studio of the Young Photographer on the principle of social-state leadership of culture. Many of them have had the honorary title of Artistic-Photographer conferred on them. Bulgarian photographers are members of the International Federation of Photographic Art (FIAP). The honorary titles conferred on them by the Federation – AFIAP, Honoris AFIAP, EFIAP, Honoris EFIAP are a well-deserved recognition for their professional work.

In the years following the socialist revolution, amateur photography became a popular form of amateur art activity. Jointly with the Centre of Amateur Activities, the Bulgarian Photography Artistic-Production Corporation directs and assists over 40,000 photographers organized in photographic circles and clubs.

The problems emerging in the country's socio-political and cultural life, the tasks and successes of Bulgarian photography, the best achievements of the world theoretical, practical and technological experience, and the history of world and Bulgarian photography are given pride of place in the only specialized publication *Bulgarian Photo,* issued since 1966. There are 320 photographic studios which develop films and print photos snapped by amateur photographers, give them professional consultations and render them other kinds of photographic services.

THE COPYRIGHT AGENCY

The Copyright Agency protects the rights and interests of authors on Bulgarian territory and represents Bulgarian authors abroad and foreign authors in Bulgaria. It was founded in 1962, but following an amendment to Bulgarian copyright legislation in 1972, the Agency was reorganized. It is now a subsidized body – a legal entity with offices in Sofia. The Agency uses the name Jusautor in its international work.

The structure of the Agency is similar to that of foreign copyright bodies. The arts unions (well-known writers and composers) and all the departments concerned are represented on the Board of Managers, which is the governing body of the Agency. The Director-General is the administrative chief and representative of the Agency.

The activity of the Agency is channelled in several directions. It acts as a literary, musical and theatrical agency which popularizes the work of Bulgarians abroad through publishing, performing, staging, etc. The agency acts as an intermediary when foreign authors' works are used on the territory of the People's Republic of Bulgaria. The Agency concludes contracts and negotiates the terms under which Bulgarian works are published or otherwise used abroad, and performs the same functions when foreign works are published or otherwise used in Bulgaria. All payments to authors must be made through the Agency.

The Agency carries out various formalities. It collects and pays remunerations for the use of works by Bulgarian authors abroad and Bulgarian and foreign authors in Bulgaria. In return for payment by the authors, the Agency acts on their behalf.

The Agency ensures that copyright regulations are legally observed. It represents the State where copyright is concerned and plays a leading role in the drafting of legislation. It drafts mandatory regulations for the implementation of the Bulgarian copyright legislation.

In 1976 the Agency became a member of the International Conference of Societies of Authors and Composers (CISAC); as a result of the Agency's membership of CISAC and Bulgaria's joining the Berne Convention for the protection of literary and artistic works in 1921 and the Universal Copyright Convention in 1975, many bilateral contracts allowing for reciprocal representation have been concluded with copyright societies from all parts of the world. The Agency maintains active contacts with hundreds of foreign publishing houses, theatrical and musical agencies, as well as with individual translators

The Georgi Dimitrov photographic type-setting shop

of the Bulgarian language. The extent of the Agency's activity can be illustrated if one compares the number of books published before and after the socialist revolution. The period between Bulgaria's liberation from Ottoman bondage (1878) and the socialist revolution of 9 September 1944, saw the publication of 347 Bulgarian books abroad, an average of five books a year, whereas the period between 9 September 1944 and 1983 saw the publication abroad of more than 4,500 books, or more than 100 books a year. Between 1944 and 1956, 710 Bulgarian books were published abroad, an average of 59 titles a year. For the period 1956–83 the total was 3,790, or an average of 150 books a year. During the last six or seven years the number of Bulgarian books appearing abroad has gone up to two hundred a year.

HEMUS FOREIGN TRADE ORGANIZATION

The foreign-trade organization, Hemus, exports and imports books, periodicals, gramophone records, musical instruments, sheet-music publications, stamps, banknotes and coins, picture postcards, works of fine and applied art, reproductions, antiques, etc.

Set up in 1967 with its main office in Sofia, the organization has exclusive rights to the import and export of these items. It maintains trade relations with many companies all over the world; their number is constantly growing.

Hemus offers its foreign partners – bookselling companies, publishing houses, libraries and institutes – books by Bulgarian authors translated into foreign languages of all literary genres: socio-political, scientific, fiction, music publications, Bulgarian dictionaries and textbooks for foreign learners, children's literature, tourist publications, etc. Through the export of Bulgarian periodicals the organization provides its numerous foreign readers with varied information about socio-political and cultural life in this country.

In the field of music, it exports gramophone records and cassettes with recorded performances of Bulgarian opera singers, chamber, symphony, and folklore ensembles, pop groups, instrumentalists, etc.

The works of the Union of Bulgarian Artists, of the Creative Fund and the Union of Craftsmen and Handicrafts occupy a prominent place among exports.

Hemus exports paintings, black-and-white drawings, plastic artefacts, replicas of Bulgarian icons, household pottery, wood-carving, jewellery, artefacts, wrought-iron articles, handmade folk-style fabrics, embroideries, and dolls, tapestries, souvenirs, and other articles.

It offers its foreign clients a rich assortment of Bulgarian postage stamps, first-day and special covers, Bulgarian and foreign collections, etc.

It exports all kinds of Bulgarian and foreign banknotes and coins.

The organization participates regularly in international sample fairs, in book fairs, in exhibitions of art, musical instruments and philately, etc. and arranges sales-exhibitions.

With its foreign-trade activity in works of art and culture Hemus is instrumental in the expansion of international cultural co-operation.

UNION OF BULGARIAN JOURNALISTS

The Union of Bulgarian Journalists was founded in 1894 when journalists and writers gathered at their first congress. Three organizations of journalists were established next – the Society of Journalists from Sofia, the Union of Provincial Professional Journalists and the Union of Publicists, which united journalists writing for the magazines. On 24 December 1944 the existing Society of Journalists from Sofia and the Union of Provincial Professional Journalists merged into the Union of Journalists in Bulgaria. It existed until 12 June 1955, when the present-day Union of Bulgarian Journalists was established.

The Union of Bulgarian Journalists is a broad, creative professional organization. Over 4,400 journalists from the mass media belong to it, and its Statute is the guiding document for its work. With its all-embracing cultural and creative, socio-political, organizational, professional and social activity, the Union of Bulgarian Journalists contributes to improving the standard of Bulgarian journalism, in the Press, radio, television, documentaries and photo propaganda. The Union represents Bulgarian journalism in the state, public and cultural bodies and organizations. It identifies, documents, itself solves or

The House of Books, Pleven

The International House of Journalists near Varna

submits for solution issues relevant to the journalistic profession.

The Union has entered into contract and maintains co-operation with organizations of journalists from 40 countries. The Union of Bulgarian Journalists is an active member of the International Organization of Journalists whose main office has been in Prague since 1947. The headquarters of the International Organization of Journalists' Social Commission is in Sofia. The Commission is presided over by the Chairman of the Union of Bulgarian Journalists, of the International Committee for the Defence of Journalists and of the International Club of Agrarian Journalists. In 1978, jointly with the International Organization of Journalists, the Union of Bulgarian

Journalists set up the Georgi Dimitrov International Institute of Journalism, where agrarian journalists from the developing countries are trained.

The supreme governing body of the Union is the congress which is held every five years. In between sessions of congress the Council of Management assumes the leadership of the Union of Bulgarian Journalists.

The Union of Bulgarian Journalists is self-supporting, the sources of its income being membership dues, publications, percentages from royalties of journalists, cultural initiatives and others.

The Union's members are united into journalistic societies (91 in number), creative sections (23 in number) and clubs which are centres for the exchange of creative and professional experience. A scientific-information centre of journalism has been set up at the Union of Bulgarian Journalists.

The Union of Bulgarian Journalists has a creative assistance fund whose resources come from deductions from journalists' royalties. It allocates means for the creative and social activity of the union. The International House of Journalists in Varna, opened in 1959 by a decision of the International Organization of Journalists, is an acknowledged centre for recreation, creative meetings and contacts among people working in the sphere of mass media from all countries.

The Union of Bulgarian Journalists has five rest-houses where Bulgarian journalists and their families can spend their holidays.

It issues several publications: the newspaper *Pogled* (Review) – an illustrated socio-political weekly; the magazine *Bulgarski Zhurnalist* (Bulgarian Journalist); the magazine *Survremenna Zhurnalistika* (Modern Journalism); the newspaper *Duma na Bulgarskiya Zhurnalist* (Bulgarian Jounalistic Opinion); the Cyril and Methodius newspaper *O Pismeneh* (On Letters) and the newspaper *Plovdivski Panair* (Plovdiv Fair).

SELECTED BIBLIOGRAPHY

ATANASSOV, P. *Nachalo na bulgarskoto knigopechatane* (The Beginning of Bulgarian Printing). Sofia, 1959.

BORSHUKOV, G. *Istoriya na bulgarskata zhurnalistika. 1844–47, 1878–85* (A History of Bulgarian Journalism in 1844–77 and 1878–85). 2nd enl. edn. Sofia, 1976.

IVANCHEV, D. *Bulgarski periodichen pechat. 1844–1944* (The Bulgarian Periodic Press: 1844–1944). Vols. 1–3. Sofia, 1962–69.

IVANOV, Y. *Bulgarski periodicheski pechat ot vuzrazhdaneto mu do dnes* (The Bulgarian Periodic Press Since Its Revival). Books 1–3. Sofia, 1891–1982.

KUTINCHEV, S. *Pechatarstvoto v Bulgaria do Osvobozhdenieto* (Printing in Bulgaria Before the Liberation of 1878). Sofia, 1920.

SPASSOVA, M. *Bulgarski periodichen pechat. 1944–69* (The Bulgarian Periodic Press: 1944–69). Vols. 1–3. Sofia, 1975.

TOPENCHAROV, V. *Bulgarskata zhurnalistika. 1885–1903* (Bulgarian Journalism: 1885–1903). 2nd edn. Sofia, 1983.

TOPENCHAROV, V. *Bulgarskata zhurnalistika. 1903–17* (Bulgarian Journalism: 1903–17). Sofia, 1981.

PART XII

FOREIGN POLICY OF THE PEOPLE'S REPUBLIC OF BULGARIA

FOREIGN POLICY OF THE PEOPLE'S REPUBLIC OF BULGARIA

The foreign policy of the People's Republic of Bulgaria (PRB) is the realization of the external functions of the Bulgarian Socialist State; it is the activity through which the PRB expresses its attitude to the phenomena and events in international life, defends its national and international interests, and guides and directs relations with other states and nations. Bulgaria's foreign policy corresponds to its internal policy and is its continuation in the sphere of international relations. Its profoundly democratic and peaceful essence is the result of its socialist social and state system. Its basic directions and tasks are determined by the Bulgarian Communist Party – the leading force in society. These directions and tasks were formulated in the Programme of the BCP (1971) and were enshrined in the Constitution of the PRB (1971). The foreign policy of the PRB defends the revolutionary gains of the people, national independence, state sovereignty and the territorial integrity of the country, contributes to the construction of the developed socialist society in Bulgaria, and assists the strengthening of the unity of the world socialist system and the development and promotion of friendship, co-operation and mutual assistance with the USSR and the other socialist states. The PRB supports the just struggle of nations for independence and social progress, against aggression, violence, Colonialism and Apartheid, and against the violation of fundamental human rights and freedoms. It pursues a policy of peace and mutual understanding in the spirit of the principles of peaceful co-existence between states with different social systems. The main objective of the foreign policy of the PRB is 'to ensure the most favourable international conditions for the construction of the socialist society in this country, to contribute to strengthening the position of Socialism, progress and peace throughout the world' (Todor Zhivkov, *Selected Works*, Vol. 18. Sofia, 1976, p. 378).

Bulgaria's foreign policy is based on the Marxist-Leninist theory of the development of society. It is a creative combination of the generally accepted features and principles of socialist foreign policy with the specific conditions in Bulgaria, with the progressive traditions of its people and the revolutionary experience they have acquired. It is a class-orientated, socialist, internationalist policy. This determines its characteristic features: it is democratic, peace-loving, equitable, open, realistic, flexible, principled, active and scientifically-grounded. Any covert and self-serving aims are alien to it. Bulgaria's foreign policy is unconditionally and consistently faithful to the international obligations undertaken by the PRB.

The socialist content of the foreign policy of the PRB determines its fundamental guiding principles: proletarian and socialist internationalism; peaceful coexistence between states with different social systems; support for the struggle of the developing countries for complete independence, equality and social progress; and active struggle for the safeguarding of peace and security in the world.

Bulgaria bases its policy on the fraternal relations with the socialist countries and develops them, proceeding from Socialist Internationalism and participating in the historical process of creating a new type of international and inter-state relations – those of all-round and fruitful co-operation.

In its relations with the capitalist countries the PRB is guided by the Leninist principle of peaceful coexistence, and on the basis of this principle it develops with these countries relations of mutually advantageous and equitable co-operation. The PRB holds that it is the inalienable right of every nation to decide its own internal affairs and to choose the social system under which it will live. The historic advantages of social systems are judged from the point of view of the opportunities they create for the optimal satisfaction of human needs, for freedom in conditions of security and prosperity, and this can only be evaluated under the conditions of lasting peace.

In conformity with the principles of its foreign policy, the PRB develops equitable relations with the newly-liberated countries, supports their struggle for a new, just and democratic international order, and renders them assistance and aid for their economic and social development and in overcoming the age-old backwardness.

The PRB observes and upholds the principles of international law which have been reflected and enshrined in the UN Charter (1945) and have found concrete expression and further development in the 1975 Final Act of the European Conference on Security and Co-operation in Helsinki – sovereign equality, non-use of force or the threat of force, inviolability of frontiers, territorial

The ceremonial meeting of the 28th Jubilee Session of the CMEA, Sofia, 1964

integrity of states, peaceful settlement of disputes, non-interference in internal affairs, respect for basic human rights and freedoms (including the freedom of thought, conscience, religion and conviction), equal rights and the right of nations to self-determination, co-operation among states, and fulfilment of the obligations of international law in good faith.

To accomplish its foreign-policy aims, the PRB uses the means accepted in international practice – intergovernmental negotiations, multilateral and bilateral treaties, accords and conventions. Bulgarian diplomacy attaches particular importance to meetings at the highest level of state and political leaders. The foreign policy of socialist Bulgaria resolutely opposes subversive acts and any form of international terrorism. It intensively develops economic co-operation, and maintains active relations in the spheres of foreign trade, culture, science, technology and education, international tourism, etc. Bulgaria now has diplomatic relations with 118 countries (before 1944 with only 32 countries) and participates in intensive co-operation in various spheres with about 120 countries; is it a member of more than 350 international organizations.

The bodies through which foreign policy is effected function at home and abroad. The main bodies are determined by the country's Constitution. The supreme state guidance is effected by the National Assembly, which decides issues of peace and war and the frontiers of the country, and ratifies or abrogates the most important international treaties. The President of the National Assembly administers international relations. The State Council, as the supreme body of the National Assembly acting on a permanent basis, ensures the general guidance of foreign policy and represents Bulgaria in international relations. The President of the State Council discharges representative functions and prerogatives in foreign policy, which have been delegated to him by the State Council in accordance with the Constitution of the country. As the supreme executive and managing body, the Council of Ministers organizes the effectuation of foreign policy, guarantees the security of the country, concludes international treaties, endorses or abrogates international treaties which do not require ratification, etc. The Ministry of Foreign Affairs, on behalf of the Government, directs all bodies concerned with international relations in which other Ministries, administrations and services participate. The bodies abroad carrying out foreign policy are the diplomatic and consular missions, the permanent missions to the international organizations and their specialized agencies (e.g. the UN, the CMEA, the Warsaw Treaty Organization, Unesco, etc.), the provisional diplomatic, economic, technological, cultural and other delegations to international congresses, conferences, meetings, etc., and the party and governmental delegations.

BULGARIA AND THE OTHER SOCIALIST COUNTRIES

Bulgaria's relations of friendship and co-operation with the socialist countries hold a foremost place in its foreign policy. These relations are based on the principles of Socialist Internationalism, on international socialist solidarity, and on the unity of the common interests and goals on the path to Socialism and Communism. The basic aim of these relations is the further consolidation of their unity of purpose and action – a process which reflects the objective need for international co-operation and corresponds to the vital necessities of the development of the individual socialist countries and of the world socialist system as a whole.

From the very first days after the victory of the People's Uprising on 9 September 1944 Bulgaria established bilateral and multilateral co-operation with the other socialist countries. The Treaties of Friendship, Co-operation and Mutual Assistance between the PRB and the socialist countries formulate the principles, aims and basic trends of present-day socialist international relations. The contracting parties undertake to act for strengthening world peace and security, to consolidate friendship between the socialist countries, and to make efforts to settle the most important international problems on the basis of peaceful co-existence and of the United Nations Charter. Under these treaties the parties are to render each other assistance and to consult each other on all problems which may be of interest to them. The treaties successfully serve as a political basis and as a foundation under international law for Bulgaria's all-round co-operation with these countries. It is on the basis of these treaties that numerous agreements are signed in the spheres of the economy, science, technology, culture, law, etc.

The summit meetings between the state leaders of Bulgaria and the other countries of the socialist community are of decisive importance for the development of relations between these countries. A major role is also played by the numerous bilateral and multilateral meetings and talks at different levels and the traditional visits of party and government delegations, at which a wide range of problems are discussed and the joint political line is co-ordinated. Bulgarian delegations participate actively in the meetings of the Ministers of Foreign Affairs of the socialist countries, which work out the most important diplomatic actions or specify their joint line in the United Nations and in the other inter-governmental organizations. Both the bilateral treaties of friendship, co-operation and mutual assistance and the Warsaw Treaty provide for co-ordination of the common line and for more concerted actions by the socialist countries in international life.

The successes of Bulgarian foreign policy depend largely on unity of action with the other socialist countries. The active co-operation among the socialist countries is a guarantee of the realization of the goals of their foreign policy and of securing the best possible conditions for peaceful socialist construction.

The beginnings of the better organized multilateral political co-operation among the socialist countries were laid with the signing of the Warsaw Treaty (14 May 1955). Thus parties to the treaty are guided by the United Nations Charter and by the principles of mutual respect for the independence and sovereignty of the states and of non-interference in each other's internal affairs. In conformity with the UN Charter, Articles 1 and 2 of the Warsaw Treaty envisage that the member-states will refrain in their international relations from the use of force or from the threat to use force; that they will settle their international disputes by peaceful means; that they will participate in all international actions, aimed at the consolidation of world peace and security; that they will contribute to the general reduction of armaments, and to the banning of the atomic, hydrogen and other weapons of mass destruction. The member-states are obliged to consult on all important international problems which affect their common interests. In the case of an armed attack against any of the member-states, the others assume the obligation to render to it immediate support with all means they find necessary, including armed forces – on the basis of the right to individual and collective self-defence, provided for under Article 51 of the UN Charter, and to co-ordinate their measures against the aggressor (Article 3).

The Warsaw Treaty not only provides for co-operation in the sphere of defence, but also for the further development and strengthening of political, economic and cultural relations between the member-states. Bulgaria also takes an active part in the main political body of the Warsaw Treaty Organization – the Political Consultative Committee (PCC), composed of the member-states on the basis of equal representation and on a completely equal footing. The PCC discusses the most important question of foreign policy and problems related to the consolidation of the defensive capacity of the contracting parties, and formulates decisions on the problems under discussion. The member-states of the Warsaw Treaty make collective efforts to avoid armed confrontation in Europe and the world and to ensure that the peoples live in an atmosphere of peace and *détente*. The member-states set themselves the noble task of enlisting the support of peace-loving forces the world over to make the process of *détente* irreversible, and of securing the complete normalization of relations among states. The member-states have reiterated on many occasions their readiness to dissolve the Warsaw Treaty Organization together with the disbanding of the North Atlantic Treaty Organization (Nato), and to assume the joint obligation not to extend the two military and political blocs. With its active and constructive participation in the Warsaw Treaty Organization Bulgaria contributes to the strengthening of the relations of friendship, co-operation and mutual assistance among the countries of the socialist community and of understanding and peaceful co-existence among countries with different social systems.

The President of the State Council of the PRB, Todor Zhivkov, has pointed out: 'The People's Republic of Bulgaria has always been and will aways be a loyal member of the socialist community. Our freedom and independence, our all-round development are most intricately linked with the destinies of the countries fraternal to us – our loyal and selfless friends. We are doing and shall be doing our level best to strengthen our political and defence alliance – the Warsaw Treaty Organization' (*Report of the Central Committee of the Bulgarian Communist Party to the Twelfth Congress and the Forthcoming Tasks of the Party*. Sofia, 1981. p. 142).

The socialist states hold the main place in Bulgaria's economic, scientific and technical co-operation with the other countries. During the first years after the socialist revolution the country signed bilateral agreements on economic and technical co-operation with the USSR, the Polish People's Republic, Romania (1948), etc. The multilateral, international, economic co-operation with the socialist countries started with the establishment of the Council for Mutual Economic Assistance (CMEA) in 1949. The initial forms and principles of scientific and technical co-operation were outlined at the second session of CMEA (1949), held in Sofia. They provide for the mutual free transfer of scientific and technical documentation, for the exchange of scientific experts and of practical experience. It is largely on this basis that Bulgaria was able to stabilize its economy and successfully develop it further. Permanent commissions organizing multilateral co-operation in various branches of the economy have been functioning in CMEA on a regular basis since 1956. Bulgaria takes a direct part in them. The socialist countries account for 80 per cent of Bulgarian trade and the CMEA member countries alone – for 78 per cent.

Bulgaria participates in international, economic, socialist co-operation both on a bilateral basis and through its activities in CMEA. In the 1960s the economic relations among the CMEA member countries entered a new stage – the stage of economic integration, which acquired new forms and a new scope in the 1970s. Bulgaria actively participates in the drafting and realization of the Comprehensive Programme for the further strengthening and improvement of co-operation and for the development of socialist economic integration of the CMEA member countries (1971), and in the long-range special-purpose programmes for co-operation in the major sections of material production, worked out for a period of fifteen to twenty years.

Extended socialist production in Bulgaria is achieved also with the aid of broad co-operation and integration with the CMEA member countries, through the co-ordination of their national economic plans and through joint planning activities, through the international specialization and co-operation of production (in the sphere of power production and power engineering, machine-building, chemical production, mining, agriculture, etc.), through foreign trade, mutual crediting and participation in international economic organizations, and through joint enterprises between member countries.

The President of the State Council of the People's Republic of Bulgaria, Todor Zhivkov, signing the final documents at the Conference on Security and Co-operation in Europe, Helsinki, 1975

Bulgaria participates actively in the work of CMEA's international organizations in the spheres of agriculture, science and technology: the Joint Institute for Nuclear Research; the Institute of Standardization; the Institutional Laboratory for Strong Magnetic Fields and Low Temperatures; the International Centre for Scientific and Technical Information; the Intersputnik organization for space communications; the scientific and production associations Interatominstrument, Interatomenergo, Interelectrotest, Interetalonpribor, Interhimvlakno, Intervodoochistka; the economic organizations Intermetal, Interhim and Interelectro; and the joint rolling-stock of freight cars, etc. The PRB participates in both the international banks of the CMEA member countries – the International Bank for Economic Co-operation (founded in 1963) and the International Investments' Bank (founded in 1970). Bulgarian scholars participate with their own equipment and their own scientific works in the Intercosmos programme of the CMEA member countries, which was launched in 1967; they also participate in the processing of the results of the experiments in outer space. On 10 April 1979 the *Soyuz-33* spaceship was put into orbit with the first Bulgarian cosmonaut, Georgi Ivanov, on board.

Bulgaria also engages in all-round cultural co-operation with the socialist countries on a long-range and planned basis, in conformity with the bilateral treaties of friendship, co-operation and mutual assistance; the arrangements on co-operation in the fields of culture, science and education; and accords and agreements between different government offices and public organizations. Agreements on co-operation in the spheres of culture, science and education have been signed with the Socialist Republic of Vietnam, the GDR, the People's Democratic Republic of Korea, the Laotian People's Democratic Republic, the Mongolian People's Republic, the Polish People's Republic, the Republic of Cuba, the Socialist Republic of Romania, the Socialist Federal Republic of Yugoslavia, the Hungarian People's Republic and the Czechoslovak Socialist Republic. Since 1976 cultural exchanges with the socialist countries have also been planned through Five-Year Plans.

International cultural co-operation with the socialist states in the sphere of fine arts finds its expression in the exchange of general and individual exhibitions. Dramatic, symphonic, dance, folklore and other ensembles are also exchanged.

Days, weeks and fortnights of the culture of a given socialist country in Bulgaria, and of Bulgaria in the other socialist countries, are also organized. Co-operation between the broadcasting and television networks, the Press, etc. is also widespread. The specialized cultural centres play a substantial role in cultural exchanges with the socialist countries. There are Bulgarian cultural centres in the GDR, the Mongolian People's Republic, the Polish People's Republic, the Socialist Federal Republic of Yugoslavia, and the Czechoslovak Socialist Republic (in Prague and Bratislava). All these countries (with the exception of Mongolia) also have cultural centres in Sofia. In Sofia there is a House of Soviet Science and Culture.

Bulgaria's policy towards the **Soviet Union** is fundamental in its mutual relations with the countries of the socialist community. The political and legal foundation of Bulgarian-Soviet relations is the 1967 Treaty of Friendship, Co-operation and Mutual Assistance between the PRB and the USSR which, under the new conditions, further develops and specifies the principles and trends of co-operation of the first treaty of alliance, signed in 1948. This treaty is in full harmony with the principles laid down in the UN Charter. In conformity with the treaty the two countries undertake to pursue a policy of peaceful co-existence among states with different social systems and consistently to work for peace and security in the Balkans, in Europe and the world.

The treaty serves the development of mutually beneficial economic, scientific and technical co-operation, helping to link the most important sections of the national economies of the two countries through the specialization and co-operation of production and contributing to the extension of co-operation in the spheres of science, education, culture, health, the arts, etc.

The summit meetings between the leaders of the two countries are of decisive importance for the development of Bulgarian-Soviet relations. In 1977, at the talks between the party and government delegations of Bulgaria and the Soviet Union in Moscow, new possibilities were tapped and practical measures mapped out for the extension of co-operation between the two countries in all economic spheres. It was decided to elaborate a General Plan for specialization and co-operation in Bulgarian and Soviet material production until 1990. Measures were also outlined for the further deepening of political and ideological co-operation, and for the all-round development of cultural ties and the exchange of spiritual values. The Twelfth Congress of the BCP (1981) found that Bulgarian-Soviet relations 'have acquired an even richer content and have been raised to a qualitatively new level' (*Report of the Central Committee of the Bulgarian Communist Party to the Twelfth Congress and the Forthcoming Tasks of the Party.* Sofia, 1981. pp. 143–4). These relations correspond not only to the basic interests of the PRB and the USSR but also to the interests of the whole socialist community. With its policy towards the USSR, Bulgaria endeavours to make relations between the two countries a model of relations among socialist countries.

The contacts and co-operation between the Central Committees of the BCP and the Communist Party of the Soviet Union (CPSU) play an important role in Bulgarian-Soviet relations. Business contacts are maintained between the supreme bodies of state power, the Governments, the different Ministries and government offices, the districts and regions, cities and towns, work-forces, public organizations, etc. The National Committee for Bulgarian-Soviet Friendship, in close co-operation with the Society for Soviet-Bulgarian Friendship in the USSR, works very actively for the implementation of the policy of all-round co-operation between the two countries, for the further development and consolidation of their political alliance, for the patriotic and internationalist education of the Bulgarian working people and youth, etc. For this purpose the National Committee and the local

Committees for Bulgarian-Soviet Friendship co-operate with the other public organizations in Bulgaria.

The intensive and comprehensive economic, scientific and technical co-operation between Bulgaria and the Soviet Union encompasses the whole production process in both countries. It comprises joint, long-range planning and co-ordination of the national economic plans, the specialization, concentration and co-operation of production, the building of joint enterprises and associations with common production programmes, the steady increase of trade, etc. By 1982, as a result of the scientific, technical and economic aid and co-operation of the USSR, enterprises and other economic projects have been built or modernized in Bulgaria which now produce approximately 65 per cent of its total industrial output. Trade between the two countries has reached 11,750 million levs (1982). The joint scientific, technical and development efforts of Bulgaria and the Soviet Union are directed at the promotion of the strategic sectors of science and technology – the chemicization of the national economy, the development of robotics, automation, pneumatic and hydraulic equipment, etc. The provisional bilateral teams make a major contribution to this end. A number of joint institutes, designers' and other organizations have been set up. In the past few years co-operation with the USSR in the spheres of science and technology has been focused mainly on engineering, computer, electronic, laser, robot and other equipment. The Soviet Union also helps greatly in the training of highly qualified cadres for Bulgaria's national economy.

Scientific and cultural co-operation between the two countries gets steadily richer. Mutual relations in the fields of science and culture are maintained on the basis of five-year inter-governmental plans. Joint programmes and studies are prepared on the problems of forecasting cultural development, on the organization of the guidance of culture and the arts, on problems of education – including aesthetic education, etc. There are also joint studies in the fundamental sciences. A system of close business contacts has been established and is maintained between the mass media and the institutions of science, culture, university and other education.

Bulgaria's relations with the **German Democratic Republic** are also developing successfully on the basis of the treaties of friendship, co-operation and mutual assistance (1967, 1977). The mutual co-operation, regulated by various treaties, accords and agreements, embraces all spheres and develops at all levels. A major role in this respect is played by the joint document *Guidelines for Economic, Scientific and Technical Co-operation and for the Further Development of Socialist Economic Integration between the People's Republic of Bulgaria and the German Democratic Republic for the Period after 1980*, signed in 1977 and brought up-to-date in June 1983 to cover the period up to 1990 and even further. The bilateral inter-governmental commission for economic, scientific and technical co-operation is doing useful work. Mutual trade is growing, and the output of the specialized and co-operated enterprises already accounts for more than 30 per cent.

The basic trends in scientific and technological co-operation are related to the elaboration of problems and studies pertaining to the modernization and reconstruction of the enterprises and to the increase of the efficiency of production. The enterprises and industrial plants of the two countries maintain direct economic, scientific and technical co-operation. Cultural co-operation is also active and planned.

Relations between Bulgaria and the **Polish People's Republic** are developing successfully in all spheres of social, political and economic life in conformity with the Bulgarian-Polish Treaty of Alliance. There are lively and mutually beneficial contacts and exchanges of experience between the political parties, the Parliaments, public and mass organizations and working people of the two countries. Parallel with the growth of mutual trade, the specialization and co-operation of production on a planned and long-range basis in machine-building, ship-building, electronics, chemical production, etc. accounts for an ever greater share of the bilateral economic relations. In April 1984 a Long-Range Programme was signed in Warsaw for the further development and extension of economic, scientific and technical co-operation between Bulgaria and Poland. Bulgaria gives 'full support to the policy of the Polish United Workers' Party, to the efforts of the Government of the Polish People's Republic and to all patriotic forces in Poland which are aimed at the building of Socialism, the consolidation of society and the political and economic stabilization of the country' (Joint Declaration, *Rabotnichesko Delo*, 6 April 1984). Scientific and technical contacts with Poland are maintained on a great number of major problems and subjects, covering the basic spheres of economic co-operation, so as to introduce highly effective technologies, to increase the quality and the degree of mechanization and automation of production. Cultural exchange between the two countries is also steadily growing.

Relations between Bulgaria and the **Socialist Republic of Romania** are developing on the basis of the treaties of friendship, co-operation and mutual assistance of 1948 and 1970. The traditional summit meetings are of prime importance in this respect. They map out the guidelines for the further extension of the co-operation between the two neighbouring and friendly countries. The participation of Bulgaria and Romania in CMEA and in the Warsaw Treaty Organization is a major factor in their mutual relations. Active contacts are maintained between the Parliaments and the other state bodies, and between the public and mass organizations. Mutual economic relations are put on an organized and long-range plan basis. The inter-governmental Commission for Economic, Scientific and Technical Co-operation is also active. The main forms of Bulgarian-Romanian economic co-operation are regulated by numerous bilateral agreements, covering the different economic sectors of the two countries – engineering, power generation and production, electronics, electrical equipment, metallurgy, chemical and petrochemical production, the light industries, etc. The building of a joint enterprise for heavy engineering in

A meeting between the General Secretary of the Central Committee of the Communist Party of the Soviet Union and Chairman of the Presidium of the Supreme Soviet of the USSR, Konstantin Chernenko, and the General Secretary of the Central Committee of the Bulgarian Communist Party and President of the State Council of the People's Republic of Bulgaria, Todor Zhivkov, Moscow, 31 May 1984

Russe and Giurgiu is of particular importance. Trade between the two countries is steadily growing. Relations in the sphere of science, culture and the arts are developing on the basis of long-range plans.

Bulgaria's relations with the **Hungarian People's Republic** mark a steady advance in all spheres in conformity with the treaties of friendship, co-operation and mutual assistance of 1948 and 1969. The document *Guidelines for the Economic, Scientific and Technical Co-operation and for the Socialist Economic Integration between the People's Republic of Bulgaria and the Hungarian People's Republic up to 1990*, signed in June 1979 in Sofia and brought up to date in June 1983 in Budapest, is of major importance in the economic field. The inter-governmental Commission for Economic and Scientific and Technical Co-operation between the People's Republic of Bulgaria and the Hungarian People's Republic, established in 1960, is working successfully, and trade grows annually. The processes of specialization and co-operation of production are extending. Through the Intransmash Mixed Society the two countries are co-operating in the design and improvement of the machinery and the systems for the mechanization of factory transport and warehouse equipment. Cultural relations between the two countries are also steadily developing.

Relations between Bulgaria and the **Czechoslovak Socialist Republic** are developing at a stable and dynamic pace in all fields of life on the basis of the treaties of friendship, co-operation and mutual assistance of 1948 and 1968. In the past few years relations have been developing at a particularly rapid pace in the sphere of the economy and, especially, in the automobile industry, heavy engineering, computer equipment, chemical production and other sectors. In September 1982 the two countries signed *Guidelines for the Extension of the Economic, Scientific and Technical Co-operation after 1985*. This document gives importance to the principle of the co-ordination of plans and of co-operation in the economic sphere. In 1983 the share of specialized and co-operated production amounted to nearly 40 per cent and is one of the highest among the CMEA member countries. Broad scientific and technical co-operation is developing and joint studies are made, mainly in the fields of automobile production, engineering, electronics, power and chemical production, agriculture, etc. Co-operation in the fields of culture, science, and education, and among the mass media, is also active.

The contacts between Bulgaria and the **Republic of Cuba** are steadily extending at all levels, both governmental and public, and favourable conditions are created for the regular exchange of information, experience and consultations of mutual interest. The *Programme for the Further Development of Economic, Scientific and Technical Co-operation and for Socialist Integration between the People's Republic of Bulgaria and the Republic of Cuba after 1980*, signed in April 1979 by Todor Zhivkov and Fidel Castro, provides for a steadier, long-range development of economic ties and of relations in the spheres of science and technology, by making greater use of the possibilities, offered by the co-ordination of the national economic plans on the basis of the specialization and co-operation of production, the expansion of mutual trade and the improvement of its structure, and by co-operation in the scientific and technical fields and with third countries. The mutual relations in the spheres of culture, science, education and the mass media are active and contribute to the better acquaintance of the two peoples.

Bulgaria's all-round relations with the **Mongolian People's Republic** is developing on the basis of the Treaty of Friendship and Co-operation of 1967. The participation of the two countries in international socialist economic integration within the framework of the CMEA is an important factor for furthering Bulgarian-Mongolian economic ties. The exchange of experience in socialist construction between districts and towns which directly co-operate with each other is also of substantial importance. The PRB helps Mongolia in the development of its national economy by the construction of major economic projects, the granting of credits, the supply of equipment, etc.

Bulgaria maintains relations of sincere friendship with **the Socialist Republic of Vietnam, the Laotian People's Democratic Republic** and **the People's Republic of Kampuchea.** In 1975 the Bulgarian people warmly hailed the historic victory of the Vietnamese people and, in 1976, the unification of the country and the formation of the Socialist Republic of Vietnam. On the international arena Bulgaria fully supports the struggle of the three states of Indo-China in defence of their independence, territorial integrity and revolutionary achievements. In October 1979 Bulgaria signed treaties of friendship and co-operation with the Socialist Republic of Vietnam and with the Laotian People's Democratic Republic, and in November 1980 – with the People's Republic of Kampuchea. Together with the other countries of the socialist community, Bulgaria makes its contribution to their political and economic stabilization.

Bulgaria's relations with the **People's Democratic Republic of Korea** are steadily developing in the spirit of the agreement reached between the leaders of the two countries, Todor Zhivkov and Kim Il Sung at their meetings for talks in Pyongyang in 1973 and in Sofia in 1975. Bulgaria fully supports the struggle of the Korean people for peaceful and democratic unification of their country, without foreign interference.

BULGARIA AND THE OTHER BALKAN COUNTRIES

Situated in the centre of the Balkan Peninsula, Bulgaria pays particular attention to the political climate in this region of Europe. It makes great efforts to contribute actively to its transformation into an area of peace, security, good neighbourly relations and co-operation. Bulgaria pursues a principled, constructive and peace-loving policy, and considers the development of its political relations with the Balkan countries in direct connection with the problems of peace and security in the region. 'Our country does not threaten anybody with war, it does not intend to impose its way of thinking and of living on any Balkan state. All we propose ... is peace, friendship and mutual understanding, the establishment of mutual confidence among our countries, so as to relieve the atmosphere in the Balkans and to enable our peoples to work and live in peace. Let us improve our relations, let us trade, let us develop cultural, scientific, technical and sports relations among the Balkan peoples and countries!' (T. Zhivkov, *Selected Works*, Vol. 4. Sofia, 1976. p. 47). The PRB supported the Soviet initiatives in 1959 and 1963 for turning the Balkans, the Adriatic and the Mediterranean into a nuclear-free zone, and in 1981 it proposed the convening in Sofia of a conference of state leaders of the Balkan countries to discuss the problem of creating a nuclear-free zone on the peninsula. This initiative was adopted by other Balkan countries, and in 1984 meetings of experts were held in Athens. In accordance with the democratic principles of the Final Act of the Conference on Security and Co-operation in Europe (1975), in 1981 Bulgaria voiced its readiness to sign bilateral agreements with its neighbours, which would include a code of good neighbourly relations, renunciation of territorial claims, and prohibition of the use of the territories of the parties to the agreement for purposes and actions hostile to one another. The Bulgarian Government regards the realization of this idea as a new, important, confidence-building measure and as a major step towards the development of co-operation among the Balkan nations. The proposal also provided for continued co-operation on a multilateral basis on matters of common interest.

Economic co-operation between Bulgaria and the other Balkan countries is developing intensively. Particularly indicative in this respect in the past decade are the high rates of growth of Bulgaria's trade with Greece and Turkey, which is in full conformity with the tendency towards more dynamic foreign trade relations among Balkan countries with different social systems. Apart from the traditional foreign trade, relations in the other economic fields – tourism, communications, transport, the use of water resources, etc. – are also increasing. Since 1965, on the initiative of the Bulgarian side, the official tourist organizations of the Balkan countries hold annual conferences. The Bulgarian side contributes concrete and constructive proposals at different Balkan meetings on problems of plant protection, the common tasks of co-operation in industry, transport, agriculture and forestry, communications and telecommunications, etc. Bulgarian trade and other economic relations with the Balkan countries develop on the basis of long-range bilateral agreements on economic, technical and scientific co-operation, on trade and payments, specified and supplemented by annually signed protocols. Bulgaria also extends and develops its economic co-operation with the states in the region through the application of more advanced forms of economic relations – industrial co-operation, the exchange of materials, technologies and goods, the creation of mixed enterprises and the joint development and utilization of natural resources. The broader use of such forms may help extend the international division of labour among the Balkan countries and put to better use the advantages of geographical proximity for the development of equal and mutually beneficial economic co-operation, corresponding to the interests of peace and understanding.

The forms of scientific and cultural co-operation, regulated by long-term bilateral agreements, plans and programmes, are also being steadily extended. As an active participant in and bearer of new initiatives in cultural co-operation with the Balkan countries, Bulgaria is guided by the view that the main results of the development of contacts in this sphere of inter-state relations are the better acquaintance and *rapprochement* of the Balkan nations, the development of the exchange of cultural values and achievements in the scientific field, and the creation of an atmosphere propitious to co-operation among the Balkan countries in different spheres. Bulgarian representatives regularly participate in the traditional meetings of architects, engineers, journalists and mathematicians of the Balkan countries, and in the conferences of the Red Cross and the Red Crescent; and Bulgaria organizes Balkan art and philatelic exhibitions, etc. By a resolution of the

Eighteenth General Conference of Unesco an International Documentation Centre for Balkan Studies (Cibal) was created. In 1975 the International Association of South-East European Studies chose Sofia as the seat of Cibal. Co-operation in the field of sports has long-established traditions, one manifestation of which are the Balkan Games in Athletics and the Balkan tournaments and championships in tennis, wrestling, boxing, basketball, fencing, weight-lifting, swimming, etc.

The National Council of the Fatherland Front, the Standing Committee of the Bulgarian Agrarian Party, the Central Council of the Bulgarian Trade Unions, the Central Committee of the Dimitrov Young Communist League, the Committee of the Movement of Bulgarian Women, the Bulgarian Union of Physical Culture and Sport, the Holy Synod of the Bulgarian Orthodox Church and others contribute to the implementation of Bulgaria's peaceful and consistent policy in the Balkans. The Bulgarian Committee for Balkan Understanding and Co-operation, founded on 26 December 1959 in Sofia on the initiative of prominent public figures, scholars, artists and writers, also makes a contribution to the strengthening of friendly relations and co-operation among the countries in the region. Its activity is linked with the Movement for Balkan Understanding and Co-operation. The Committee supports the broad programme of the Bulgarian Government for the assertion of peaceful co-existence in the relations among the Balkan states with different social systems, for the settlement of disputes by peaceful means, for the development of economic, cultural and scientific contacts between the Balkan countries and peoples, and for turning the Balkans into a nuclear-free zone – a region of peace and co-operation. As Todor Zhivkov points out, the Bulgarian Government is deeply confident that 'along with the great states, all other countries can do a great deal for the strengthening of peace. This is what our own experience, accumulated here, in the Balkans, has taught us. It is a well-known fact that today there are still quite a few controversial and unresolved issues at this sensitive crossroads, where since millennia peoples, civilizations and cultures have clashed and influenced each other, where Bulgaria was founded thirteen centuries ago. Regardless of all this, today we can boldly say that the Balkans at the present moment are not the "powder keg" of Europe, and despite the existing complexity and difficulties, in practice they are manifesting themselves ever more energetically as a constructive factor for peace, security and co-operation in Europe' (T. Zhivkov, *The Borderline between the two Decades, between the 1970s and the 1980s, Should Not Be a Borderline between Détente and Confrontation*. Speech at the Opening of the International Meeting and Dialogue 'For *Détente*, Peace and Social Progress', Sofia, 1981. p. 34).

Bulgaria attaches particular importance to the development of its relations with the neighbouring socialist states. In its policy towards the **Socialist Federal Republic of Yugoslavia** it is guided by its desire for an all-round development of political, economic and cultural relations. The contacts between the state leaders and the leaders of the public organizations are of major importance for co-operation between the two counries. There is a regular exchange of visits by delegations and working groups between the Parliaments of the two countries, between the National Council of the Fatherland Front and the Socialist Union of the Working People of Yugoslavia, between trade unions, youth, women's and other public and mass organizations. In the sphere of inter-governmental relations a number of bilateral bodies (general, and by sectors of the economy, mixed commissions, working groups, border commissions, etc.) have been established and are functioning, discussing and settling concrete matters of co-operation in specific fields.

There also exist differences between the two countries, and there are complex problems, inherited from history. The Bulgarian Govenment defends consistently and on a principled basis the view that certain differences in the positions of the two countries, including those on the so-called Macedonian Question, should not be brought to the fore and their settlement should not be considered as a preliminary condition for co-operation. They can be successfully solved only in an atmosphere of all-round co-operation, and in the spirit of mutual respect, provided that independence, equality, sovereignty and territorial integrity are strictly observed, that there is no interference in each other's internal affairs and that the scientifically established historical and present-day realities are unconditionally observed.

Economic, scientific and technical co-operation, which are on a long-range basis, are making particular progress. Trade between the two States actually doubles every five years. In 1983 it amounted to 252 million dollars. At the beginning of the 1970s production co-operation emerged as a superior form of Bulgarian-Yugoslav economic relations. During the 1972–83 period the two neighbouring countries signed long-term agreements on industrial co-operation for the production of electro-technical equipment and household electrical appliances, diesel motors, pumps and pump aggregates, electric trucks and motor trucks, electric hoisting machinery, railway carriages, etc. In 1982 a Mixed Industrial and Trade Chamber was established, with headquarters in Sofia and Belgrade.

Cultural ties are also developing, and the historical language kinship between the two peoples is a contributory factor in this sphere. Since 1979 the co-operation between the academies of science, universities, national libraries, individual research institutes and the cultural unions has also been developing. The two countries exchange art exhibitions, musical and dramatic ensembles and individual performers.

Bulgaria pursues a consistent policy with regard to the **Peoples' Socialist Republic of Albania** for the development of mutually beneficial co-operation in the interest of both nations. In spite of the interruption of party and government contacts between the two countries and the freezing of cultural, scientific and technical ties during the 1960s there is a certain advance in the development of economic co-operation. Foreign trade is carried on in accordance with five-year trade agreements

The Prime Minister of the Hellenic Republic, Andreas Papandreou, being awarded the Order of Stara Planina, with ribbon, at the Boyana residence of the State Council of the People's Republic of Bulgaria, 1982

and one-year agreements on commodity exchange. The further development of relations between the two socialist countries is of mutual benefit and corresponds to the vital interests of the two peoples.

The development of Bulgaria's relations with Greece and Turkey is characterized by the efforts of the Bulgarian Government to seek, extend and develop the sphere of the common, coinciding and parallel interests in their mutual relations. A concrete expression of this policy are the series of Bulgarian initiatives for the assertion of peaceful co-existence between the Balkan states with different social systems – the signing of bilateral political declarations, which formulate the guidelines in the development of relations, statements and declaration of the meetings of state leaders, parliamentary groups, parties, public organizations and others, the unconditional recognition of the territorial status quo on the peninsula after the Second World War, etc.

Post-War relations between Bulgaria and the **Republic of Greece** demonstrate the possibilities, opened up by the policy of peaceful co-existence, pursued between two relatively small States with different social systems. Bulgaria resumed its diplomatic relations with Greece in 1954. At the same time trade relations between the two countries were also resumed. On 9 July 1964 twelve Bulgarian-Greek agreements were simultaneously signed in Athens. They settled the financial and property problems between the two countries and agreement was reached on co-operation in the fields of transport and tourism, navigation and the utilization of the rivers flowing through the territories of the two countries. Important bilateral documents were signed in 1973 – a Declaratation on the Principles of Good Neighbourly Relations, Understanding and Co-operation, a Consular Convention, an agreement on the opening of a Bulgarian Consulate General in Salonika and of a Greek Consulate in Plovdiv, and a cultural agreement. These were good prerequisites for new opportunities for the development of Bulgarian-Greek relations. The practice was established of periodic talks at the highest state and government level and between the Ministers of the two countries. Bilateral and international problems were discussed, on a wide range of topics on which the countries arrived at the same or similar positions, such as: the need to supplement political *détente* with *détente* in the military field (1976), co-operation for the consolidation of peace and security in Europe and the world (1979), the necessity of signing a world treaty on the non-use of force in international relations (1980), the recognition of the right of the Palestinian people to a homeland of their own, etc.

The discussion of bilateral problems brought to the fore the common striving of the two States to extend and deepen their economic ties, to raise their co-operation in the cultural sphere and scientific and technological exchange to a new level, etc. After 1981 Bulgarian-Greek relations and the relations between the Bulgarian Government and the Government of the All-Greek Socialist Movement, led by Andreas Papandreou, made a further step forward. A new element is the *rapprochement* of the positions of Bulgaria and Greece on the overall assessment of the international situation and the mutual support of each other's efforts for turning the Balkan Peninsula into a nuclear-free zone in the spirit of the proposals of Todor Zhivkov, President of the State Council of the PRB. Relations between the two Parliaments, between the Ministries and government offices, the economic and cultural institutions and organizations, hold an increasingly important place in bilateral relations. In addition to inter-state relations the contacts between the political parties, trade unions, youth, women's, co-operative, scientific, church and other organizations and circles give a broader and more mass character to the political relations, thereby creating a favourable climate for the development of bilateral co-operation. Economic relations are on the upswing. Mutual trade, which amounted to 37 million dollars in 1970, rose to 84.4 million dollars in 1975 and to 382.7 million dollars in 1980; in 1982 it amounted to 283.4 million dollars. Current forms of economic co-operation which are producing concrete results include the joint production of machinery and equipment and the joint participation of Bulgarian and Greek firms in the markets of third countries. The development of rail and road transport, and of air and sea transport for passengers and freight continues. Cultural co-operation is being extended in several directions. Joint exhibitions are organized, and visits of symphony and chamber orchestras and of drama companies are exchanged. The exchange of scholars and lecturers from the universities and research institutes of the two countries is growing. The work of the Bulgarian-Greek and of the Greek-Bulgarian committees of friendship, led by prominent public figures and cultural workers in Bulgaria and Greece, contributes to the lively cultural and scientific exchange and co-operation.

Relations between the PRB and the **Republic of Turkey** grew more active after 1964. In 1964–65 the Ministers of Foreign Trade of the two countries exchanged visits and the Ministers of Foreign Affairs and the Prime Ministers did likewise in the 1966–70 period. These contacts laid the beginnings of planned and systematic meetings of representatives of Bulgaria and Turkey at different levels – of public and political organizations, of cultural and business meetings between parliamentarians, etc. The meetings between the Heads of State of the two countries were of decisive importance for the extension of relations. They signed important documents such as the Declaration on the Principles of Good Neighbourly Relations between the PRB and the Republic of Turkey (1975), joint communiqués, agreements, conventions,

The welcoming ceremony for the President of the People's Republic of Bulgaria, Todor Zhivkov, at Esenboa airport, near Ankara, during his visit to the Republic of Turkey, 1983

etc. The scope of the bilateral and multilateral problems jointly discussed was enlarged and the positions of the two countries grew closer on a number of problems, including the need to extend co-operation and to transform the Balkans into a region of peace and good neighbourly relations.

By 1984 Bulgaria and Turkey had signed some 30 agreements, conventions, protocols and programmes for economic, technical and scientific co-operation, trade, international transport of passengers and goods by road, air and railway communications, linking the electric-power grids of the two countries, prevention and settlement of border incidents, etc., for co-operation in the field of agriculture, the food industry, on veterinarian matters and plant protection, for legal assistance in civil cases and criminal actions, a consular convention, etc.

The relations between the parliamentarians of the two countries, between the National Council of the Fatherland Front and the Popular Republican Party, between the Bulgarian Agrarian Party and the Party of Justice, etc. were useful for mutual acquaintance and for manifesting good neighbourly relations in the 1970s.

Bulgaria's consistent and stable policy finds understanding in Turkey, and this is also confirmed by the results of the visits of Gen. Kenan Evren, Head of State of the Republic of Turkey, to Bulgaria in 1982, and of Todor Zhivkov to Turkey in 1983. The Bulgarian-Turkish Committee for Economic, Industrial, Scientific and Technical Co-operation, founded in 1975, plays a major role in the extension of economic relations.

In 1983 mutual trade amounted to 152 million dollars as against 117.8 million dollars in 1982, 106 million dollars in 1980, 77 million dollars in 1977 and 10.1 million dollars in 1970. Economic co-operation includes agreements on industrial co-operation, the supply of complete installations, etc.

Cultural and scientific co-operation is developed on the basis of two-year programmes. The exchange of scientists, writers, journalists and public figures, orchestras, dance companies, etc. is being extended.

The People's Republic of Bulgaria maintains friendly relations with the **Republic of Cyprus** since its establishment (1960); the position of the Bulgarian Government on the Cyprus question contributes to this end. Bulgaria consistently stands for the respect for the independence, sovereignty and territorial integrity of Cyprus, and for the peaceful solution of the Cyprus question in accordance with the decisions of the United Nations, by observing the just interests of the Cypriot Greeks and the Cypriot Turks. This position was reiterated by the Bulgarian Government at the time of the visit of the President of Cyprus, Spyros Kyprianou, to Bulgaria in 1981. At that meeting unanimous views were expressed on the most topical international problems related to peace, *détente* and disarmament, and the Cypriot side supported the idea of the establishment of a nuclear-free zone in the Balkans. Economic relations are expanding as a result of the long-term agreements on trade and payments, and on economic, scientific and technical co-operation, signed in 1976. Cultural relations are also successfully developing on the basis of two-year plans. The two countries exchange exhibitions of fine arts, visits of orchestras, drama companies, etc.

BULGARIA AND THE DEVELOPING COUNTRIES

Since it came into existence, the PRB has rendered consistent support to the nations which have taken the path of national independence. The relations of the PRB with the developing countries are based on sovereign equality and mutual advantage, on the support for their just struggle against Imperialism, Colonialism, and Neo-colonialism. At the same time, the PRB grants material and moral aid to the developing countries and maintains all-round relations with them. A graphic testimony to the active foreign policy of the PRB is the fact that it has established diplomatic relations with 80 countries (1983) in Asia, Africa and Latin America. Of these, fourteen are Asian countries, thirteen Arab, fifteen are Latin American and 38 are African states situated south of the Sahara Desert.

The constructive policy of Bulgaria finds eloquent expression in the meetings and talks held in recent years by the President of the State Council of the PRB, Todor Zhivkov, with the leaders of the People's Republic of Angola, the Democratic Republic of Afghanistan, the Algerian Democratic and People's Republic, the People's Revolutionary Republic of Guinea, the Republic of Guinea-Bissau, Grenada, Socialist Ethiopia, the Republic of Zambia, Zimbabwe, the People's Democratic Republic of Yemen, the Republic of Iraq, the Republic of India, Kuwait, the People's Republic of Mozambique, the Mexican United States, the Islamic Republic of Mauritania, the Federal Republic of Nigeria, Nicaragua, the Syrian Arab Republic, the United Republic of Tanzania and the Republic of Tunisia.

The principled character of the foreign policy pursued by the PRB toward the developing countries is also apparent in the bilateral treaties for friendship and co-operation concluded with Angola, Mozambique, Ethiopia, the People's Democratic Republic of Yemen, Afghanistan and the Socialist People's Libyan Arab Jamahiriyya. These documents laid the foundations of a qualitatively new stage in the foreign-political and foreign-economic co-operation of the PRB with the developing countries.

In the overall relations of Bulgaria with the newly-liberated states, an important role belongs to the various public and political organizations – the Bulgarian Agrarian Union, the Fatherland Front, etc. – which maintain useful contacts with parties and organizations in these countries.

The PRB stands against the imperialist policy of 'spheres of interest' and 'spheres of influence' which include sovereign states, and supports all efforts for establishing a durable and secure peace in Asia, Africa and Latin America. Notable within the framework of these efforts is the support which Bulgaria gives to the practical steps and concrete proposals of the developing countries aimed at reducing tensions in the world, at finding peaceful and just solutions to the urgent international problems, and at creating zones of peace, security and co-operation in different areas of the world. Bulgaria supports the idea of declaring the Indian Ocean a zone of peace and of creating a zone of peace and co-operation in the Mediterranean. It welcomes the idea launched by the Organization of African Unity (OAU) to declare Africa a nuclear-free zone, and approves the eventual turning of Latin America into such a zone. The PRB supports the idea of the non-extension of the sphere of action of the Warsaw Treaty Organization and Nato over Asia, Africa and Latin America. It is firmly against the setting-up of new military bases and for the dismantling of the existing bases in the developing countries, and against all economic and military pressures on them.

The PRB holds the opinion that the Non-Aligned Movement is a serious factor in international relations. It supports the struggle of the non-aligned nations against Colonialism and Neo-colonialism, Apartheid and racial discrimination, for improving the international situation.

The PRB holds a positive attitude to the efforts of the non-aligned and developing countries to establish a new, just, economic order under which they would freely and equitably dispose of their natural resources and have the opportunity of transforming their international economic relations onto a fair and equitable basis. The PRB supports the demands of the developing countries in the struggle for eliminating discrimination and overcoming the artificial obstacles created in world trade and economic relations, and the elimination of all forms of inequality, exploitation and domination in international economic relations. The Government of the PRB has expressed its support for the declaration, the programme of action and the charter on economic rights and obligations of nations adopted by the UN.

An important direction in the relations of the PRB with the developing countries is their economic, scientific and technological co-operation. The relations of the PRB with the developing countries are the most dynamic element in the economic ties of the country with the non-socialist

The late Indira Gandhi, Prime Minister of the Republic of India, being accorded the title of Doctor Honoris Causa of the Kliment Ohrid University, Sofia, 1981

states. Bulgaria maintains economic and trade relations with more than 58 developing countries and makes efforts to develop these relations on a long-term and comprehensive basis. New, more effective forms of economic relations are also used, such as production and scientific co-operation, specialization and co-operation in production, the construction and functioning of up-to-date enterprises and companies. The supply of complete plants has become an important and promising item. In the last twenty years, the PRB has supplied, built and put into operation more than 580 plants, technological lines, installations and other projects in more than 40 developing countries.

A considerable share in this multi-faceted work belongs to construction. With the assistance of the PRB, more than 80 major projects have been built in developing countries – in the food industries, the production of construction materials, refrigeration equipment, air-conditioning, mining and ore-dressing plants, chemical and pharmaceutical, wood-processing, engineering and electrical-appliances plants.

The PRB has extended the practice of using various forms of direct co-operation and specialization between Bulgarian enterprises and firms in the developing countries in the spheres of the production of means of transportation, agricultural machinery, elevating gear and machine-tools, and in the chemical and pharmaceutical industries. The development of direct co-operation in other fields has led to the re-assertion in recent years of the joint venture as a promising form of economic collaboration. These new forms have successfully complemented the traditional forms of international trade.

The structure of Bulgarian trade with the developing states is co-ordinated with their interests; it also assists and stimulates the diversification of their exports. Taking into account the needs of the developing countries and the recommendations of UNCTAD, Bulgaria has introduced a preferential customs tariff for industrial and agricultural commodities originating and imported from the developing countries. The preferential treatment varies from a 50 per cent reduction of the value in the case of the 'most favoured nation clause' to the full exemption from customs duties. Resulting from these conditions, the share of industrial goods from the developing countries in the imports of the PRB has exceeded 45 per cent in 1983.

The PRB carries out scientific and technological co-operation with 50 developing countries; official contracts have been signed with more than half of them. Among its basic forms, significant importance belongs to the sending of Bulgarian specialists who help to train cadres, and the exchange of know-how, technologies, production experience, research workers, etc. At the end of 1983, the number of Bulgarian specialists sent to the developing countries reached 6,600 people.

The intensive development of the various forms of trade and economic co-operation is accompanied by the improvement of the institutional, organizational and legal foundations of bilateral relations. Resulting from the active inter-governmental contacts, 86 documents have been signed, including agreements on co-operation, bilateral comprehensive programmes and guidelines for the development of economic relations, programmes for the planned co-ordination of economic relations, provisions for establishing joint ventures, for granting credits, etc. Important bodies regulating bilateral trade and economic relations are the inter-governmental commissions for economic, scientific and technological co-operation established with more the 30 developing countries.

The PRB renders concrete assistance in training national cadres for the developing countries. About 10,000 students are being educated and trained annually in the higher and specialized secondary educational establishments of the PRB. More than 11,000 young specialists from developing countries have completed their higher education in Bulgaria. A number of initiatives are being carried out under FAO auspices for training experts from developing countries in the fields of agriculture, veterinary medicine, etc.

Bulgaria maintains active relations with the developing countries. At the state level, long-term agreements have been signed between the artistic unions, envisaging exchanges of painters, actors, musicians, artistic ensembles, etc. The weeks of mutual visits and meetings of outstanding scholars, e.g. between the PRB and India, have become a tradition. An exchange is made of feature-films and films made for TV; fruitful relations exist between the cultural unions in Bulgaria and those in India, Mexico, Tunisia, Algeria and other countries. The importance is growing of the participation of children from the developing countries in events related to the Banner of Peace Assemblies, which are held regularly in Bulgaria.

Bulgaria maintains particularly active and traditional political, economic and cultural relations with the developing countries of the geographic area closest to it – the Arab countries.

In 1956 the PRB expressed its resolute support for the nationalization of the Suez Canal and the struggle of the Egyptian people for full sovereignty over it. Together with the other socialist countries, Bulgaria helped to avert the threat to Syria's independence in 1957 and supported the progressive reforms in Iraq in 1958. Bulgaria rendered effective assistance to the Algerian people during the hard years of their struggle for national independence. Bulgaria hailed the independence of North Yemen and the successes of the South-Yemeni people in their struggle against colonial occupation, which was crowned with success in 1967. The PRB resolutely supports the struggle of the Lebanese people against the Israeli occupation, to restore the unity, sovereignty and territorial integrity of Lebanon. It expresses its solidarity with the struggle of the Arab people of Palestine to create a State of their own, and renders them all-round support, recognizing the Palestine Liberation Organization as their only legitimate representative.

The PRB makes a constructive contribution to the search to find the correct way to a comprehensive solution of problems related to the Middle-East conflict. The position of the PRB on the settlement of the Middle-East issue is unequivocal and is aimed at the establishment of a just and durable peace, which requires the withdrawal of Israeli forces from all territories occupied in 1967 and after, including the eastern part of Jerusalem; the realization of the legitimate rights of the Arab people of Palestine, including their right to create an independent State, the guaranteeing of the right to all states in that area to a secure and independent existence and development under the necessary international guarantees.

The PRB develops active economic relations with the Arab countries. In 1983 their share in Bulgarian trade with the developing countries reached 78 per cent. Particularly active are the economic exchanges with Iraq, Libya, Algeria, Lebanon, Syria and Jordan. Trade is also increasing with Morocco, Tunisia, Kuwait and other Arab countries.

Bulgaria decisively condemns the attempts of racist South Africa to perpetuate the illegal colonial occupation of Namibia and supports the just cause of the Namibian people who are fighting for national independence under the leadership of their only legitimate representative – the South West Africa People's Organization (SWAPO). The Bulgarian Government insists on the earliest settlement of the Namibian question on the basis of Resolution 435 of the UN Security Council and taking account of the vital interests of that country, without infringing its territorial integrity. The PRB condemns Apartheid in South Africa and supports the African National Congress of South Africa, the genuine spokesman for the interests of the South-African people in their struggle against the apartheid system, for national liberation and social justice. At various international forums the Bulgarian representatives have decisively spoken against the racist regime in South Africa, against the policy of Apartheid which is the main threat to peace and security in that area of the world.

Bulgaria maintains active political, economic and cultural relations with the Latin American countries, and at present has diplomatic relations with fifteen of them. In recent years, there has been a considerable extension of the relations on a legal basis – about 70 treaties, accords and protocols have been signed on co-operation with Nicaragua, the Republic of Venezuela, the Federal Republic of Brazil, Colombia, etc. With Nicaragua, Mexico, the Republic of Colombia, the Republic of Peru, the Republic of Costa Rica and the Republic of Ecuador, agreements have also been signed on cultural co-operation. In addition, the PRB has established broad economic relations with a number of Latin American countries, and has signed trade agreements with thirteen of them. With Nicaragua, Colombia, Brazil, Mexico and Peru mixed commissions for economic, scientific and technological co-operation have been set up, and with the Republic of Argentina – a trading company. At international forums,

Bulgaria and the Developing Countries

A meeting of the President of the State Council of the People's Republic of Bulgaria, Todor Zhivkov, with the General Secretary of the Arab League, Chedli Klibi, Sofia, 1983

the Latin American countries have enjoyed and are enjoying the support of the PRB in their struggle for consolidating their political and economic independence. The PRB supports the initiatives of the Contadora Group aimed at the peaceful settlement of conflicts in Central America, at establishing good-neighbourly relations and ensuring the development of these states along the road they have chosen.

BULGARIA AND THE CAPITALIST COUNTRIES

The foreign policy of the PRB toward the capitalist countries is based on the principles of peaceful coexistence in relations between states with different social systems which include such principles of general democracy as: respect for sovereignty and territorial integrity; equality and non-interference in the internal affairs of other countries; recognition of the right of each state to freely choose its social, economic and political systems; and the solution of existing problems and contradictions through negotiations on a bilateral and multilateral basis.

Guided by these principles, the chief aim of Bulgaria's foreign policy is the consolidation of peace and the development of mutually advantageous relations. Todor Zhivkov, President of the State Council of the PRB, has said that there is not and cannot be any task more important than safeguarding peace. Peace is a necessary condition for the construction of the developed socialist society in the PRB, and socialist Bulgaria makes its active contribution, within the limits of its possibilities, to the enhancement of mutual trust and development of co-operation between the socialist and the capitalist countries, to the reduction of tension, and to the promotion of *détente* and mutual understanding in Europe and in the world.

The PRB holds the opinion that the principles of peaceful coexistence in relations between states with different social systems are the only reasonable and reliable basis for normalizing relations, overcoming the dangerous tendencies in the world during the period of the 'cold war', and creating lasting conditions for the consolidation of security and development of co-operation in all spheres, which is in the interest of both the socialist and the capitalist countries.

Steadily pursuing the course of peaceful coexistence with the capitalist countries, displaying realism, a constructive approach, energy, readiness for negotiation and reasonable compromise, the PRB has gradually succeeded in normalizing and begun to promote its relations with those capitalist countries which have manifested their desire to do so. The 1960s and 1970s saw the creation of the appropriate negotiated basis for mutually advantageous co-operation in all fields.

Through its consistent policy of improving relations between East and West, as well as generally within Europe and the world, the PRB has contributed to the process of *détente* which started in the early 1970s, to the successful holding and conclusion of the European Conference on Security and Co-operation (Helsinki, 1975) and to the progress achieved in reducing international tension. An important role in this respect belongs to the meetings of the President of the State Council of the PRB, Todor Zhivkov, with the leaders of France, Italy, Japan, the FRG, Austria, Portugal, the Scandinavian countries, etc. Contacts with state and political figures of most of these countries have become regularized, and Bulgaria has no major unsolved issues with any capitalist country, which makes possible the dynamic development of relations in all spheres, as well as discussion and the search for opportunities to solve the global problems of our time. As Todor Zhivkov has said, only those leaders of Western states who do not wish to sit at the round table for a frank dialogue may have some objection to Bulgaria's foreign policy.

Useful contacts are also maintained between the PRB and Western countries at the Minister of Foreign Affairs level. They became particularly intensive and regular in the 1970s and early 1980s with countries like Austria, Spain, Finland, France, the FRG, Belgium, the Netherlands and Japan. An effective form of contact with the capitalist countries are the regular political consultations between the Ministries of Foreign Affairs. They help considerably to clarify a number of problems and stimulate the development of bilateral relations. For the time being, these consultations are the best established and most fruitful form of systematic official dialogue with many countries.

A significant place in pursuing the foreign policy of the PRB belongs to the relations with capitalist countries at the level of political parties, and of public trade-union, youth, sports and other organizations. The results in these fields are encouraging and provide concrete evidence of the political prestige of Bulgaria, which is becoming an increasingly desired, sought-after and respected partner. The role of the Bulgarian Agrarian Party should be particularly emphasized. It maintains traditionally good relations with agrarian parties and kindred organizations in the capitalist countries.

The relations of the National Assembly of the PRB with the parliaments of a number of capitalist countries are becoming increasingly active. A notable expression of the recognition of Bulgaria's activities in inter-parliamentary work was the Congress of the Inter-parliamentary Union held in Bulgaria in 1977.

A very favourable situation for the all-round development of relations was created by the successful conclusion of the European Conference on Security and Co-operation in 1975. However, the exacerbation of the international situation towards the end of the 1970s and the beginning of the 1980s has had a negative and restricting impact on the relations of the PRB with many of the capitalist countries.

The consistent foreign policy pursued by the PRB toward the capitalist countries, which is based on its profoundly peaceful nature, becomes particularly apparent in the present-day international conditions when the situation in the world has reached a dangerously tense stage and the very existence of mankind is threatened. In the strained international environment of today, the PRB is striving to foster political good will and realism and to take all possible steps to avert the threat of nuclear war in the world. It is to this aim that all foreign policy initiatives of the Bulgarian Government are subordinated, both on a bilateral and on a multilateral basis. In the name of peace the PRB, together with the other countries of the socialist community, puts forward initiatives and supports all proposals aimed against the danger of war and the policy of discrimination in international relations.

The PRB resolutely condemns the policy of nuclear arms build-up, which has sharply exacerbated the military and political situation in Europe and in the world and is a factor with direct negative repercussions in the relations between the two world systems.

In its policy toward the capitalist countries, the PRB is guided by the unquestionable fact that a considerable number of leading statesmen in the West have preserved their ability to soberly and realistically assess events, that peace and *détente* have supporters among all classes, strata and circles in the capitalist countries. Evidence of this are the relations of the PRB with the majority of the capitalist countries. Although they have been negatively influenced by the deterioration of the international situation during the recent years, marked progress has been achieved in some spheres.

The expansion of relations with the capitalist states in the fields of economy, science and technology is an important direction of the foreign policy of the PRB, and it is in these spheres that the most significant results have been achieved. The tremendous successes achieved in socialist development, the open character of the Bulgarian economy, together with the legislative facilities, have made Bulgaria a desired and respected partner of the capitalist countries. During no other period of history has it maintained so diverse, mutually advantageous relations with the developed capitalist states in the spheres of economy, science and technology. These relations cover both classic trade exchanges and industrial co-operation, as well as the realization of joint technological and economic ventures and co-operation in science and technology. During the period 1955–83, the volume of trade of the PRB within the capitalist countries has grown more than 3,000 per cent. The PRB has signed long-term governmental agreements for trade, industrial, scientific and technological co-operation with nearly all the developed capitalist countries. Dozens of contracts and agreements have been signed between Bulgarian trade and economic organizations and foreign firms; most of them were concluded after the European Conference on Security and Co-operation.

Co-operation was facilitated after the legislative measures introduced in the PRB and especially after the adoption of Decree No. 535 of the State Council in 1980 on economic co-operation between Bulgarian juridical persons and foreign firms and juridical persons. This Decree provides for the establishment of favourable legal, financial, credit, customs and other conditions for organizing a mutually advantageous economic relationship. Economic co-operation is being carried out by means of contracts on industrial direct co-operation and the setting-up of companies. In the case of industrial co-operation, the parties agree on long-term, mutually advantageous collaboration in the spheres of science, technology, production and trade. In setting up such companies, joint economic activity is carried out on the basis of payments by the contracting parties, joint management and distribution of profits and losses.

The PRB wishes to establish stable and mutually advantageous economic relations with the developed capitalist countries and to extend them, while strictly observing the undertaken mutual obligations. The intensive development of the modern material and technological base has become a pre-requisite for the constant qualitative improvement and intensification of the foreign-trade relations of the PRB with the Western countries. At present Bulgaria's volume of trade with the countries of Western Europe, the USA and Canada exceeds 3 billion currency levs (1982). The volume of Bulgarian exports to these countries is growing constantly, and industrial goods, including engineering equipment, have an increasingly significant place among them.

The PRB consistently stands for the elimination of the discriminatory limitations and for the improvement of trade and political conditions in the economic relations between the socialist and the capitalist countries; unfortunately, this does not meet with a positive response from all Western states. Showing a constructive spirit of initiative the PRB wishes to materialize conditions for full and mutually advantageous co-operation, not only on a bilateral basis, but also between the CMEA and the EEC.

In its stand against the discriminatory political restrictions in the foreign-economic policy of some capitalist countries, the PRB is confident that the efforts for mutually advantageous and equitable trade are not only of economic significance – they are efforts for *détente*, for creating a favourable climate facilitating the all-round development of relations between states with different social systems.

A substantial place in the foreign policy of the PRB toward the capitalist countries belongs to international cultural co-operation. Bulgarian socialist culture makes its contribution to the struggle for peace and *détente*, for the social and spiritual prosperity of mankind. Cultural policy helps to popularize the success of Bulgaria, the rich history and the cultural traditions of the Bulgarian people; it helps the mutual knowledge of nations, the strengthening of trust among them and the safe-guarding of peace.

Bulgarian socialist culture and arts continue to gain international recognition. There is now a regular practice of organizing weeks of the Bulgarian cinema, music, and theatre, which add to the image of Bulgarian culture in the individual countries. Representatives of the PRB take an active part in international cultural events organized in Western countries and win prestigious prizes. Intensive contacts have been established between writers, critics, actors, etc. Bulgaria has become well known around the world as the initiator and sponsor of international competitions, festivals, meetings, exhibitions, and of artistic ensembles, etc., opening its doors wide to the genuine, authentic values created by human genius, to cultural co-operation in the name of peace and progress.

The celebration of the 1,300th anniversary of the Bulgarian State in 1981 opened up new, still greater and lasting opportunities for various large-scale cultural events, for cultural exchanges between the PRB and the capitalist countries. The realization of new ideas, programmes and events which will be the Bulgarian contributions to international cultural co-operation, has now been envisaged.

One of the fundamental tasks of Bulgarian foreign policy which is acquiring increased importance and topicality is the creation of a broad, legal basis for bilateral relations with the capitalist countries. The agreements reached in various spheres outline the directions and prospects for the long-term, stable development of relations with these countries along the principles of peaceful coexistence. Their aim is to secure the most favourable foreign political and economic conditions and pre-requisites for the construction of the developed socialist society in Bulgaria. In the mid-1980s the PRB has concluded more than 150 agreements which are in force with these countries. The treaties, agreements and other official documents signed between the PRB and individual Western states are the result of the initiative and strenuous efforts of Bulgaria and the understanding shown by the other contracting party.

Particularly favourable is the development of bilateral relations between the PRB and the **Republic of Austria**. The President and the Federal Chancellor of the Republic of Austria have visited the PRB several times, and their visits have been returned by the President of the State Council of the PRB, Todor Zhivkov. In frank dialogue at the highest level they reached the conclusion that bilateral relations are extended and enhanced with every passing year and that economic co-operation between the two countries is gaining strength. The exchange of opinions is not limited to the issues of bilateral relations alone; it bears upon major international matters, on a number of which the PRB and the Republic of Austria have similar stances.

An expression of friendly feelings for the Bulgarian

A meeting between the President of the State Council of the People's Republic of Bulgaria, Todor Zhivkov, and the President of the German Social Democratic Party, Willy Brandt, Sofia, 1978

people was the large-scale celebration of the 13th centennial of the Bulgarian State in Austria, with numerous regional and national events. According to the President of the State Council of the PRB, Todor Zhivkov, the relations between Bulgaria and Austria are to a great extent a model of relations between states with different social and political systems.

Since the establishment of diplomatic relations between the PRB and the **Federal Republic of Germany** in December 1973, contacts and relations between the two countries have advanced considerably. A substantive role in this respect was played by the visit of the President of the State Council of the PRB, Todor Zhivkov, to the FRG in 1975 and the visit of the Federal Chancellor of the FRG, Helmut Schmidt, to the PRB in 1979. The documents signed during these visits and especially the declaration on relations between the PRB and the FRG of 1975, are a stable basis for all-round co-operation between the two countries.

The development of relations betweeen the PRB and the FRG continued after the change of Government in Bonn and the advent of the coalition of the Christian Democratic Union/Christian Socialist Union and the Free Democratic Party. The Federal Minister of Foreign Affairs of the FRG, Hans Dietrich Genscher, visited Sofia in 1983, and his Bulgarian counterpart, Peter Mladenov, visited the FRG in 1984. There are good pre-requisites for their extension, particularly in the field of economy, where a number of joint economic and technological projects are being realized. The FRG is the major economic partner of the PRB among the capitalist states.

Good relations exist between the PRB and the **Republic of France.** They were given a powerful impetus after Todor Zhivkov's official visit to France in 1966 at the head of a governmental delegation. The healthy state of Bulgarian-French relations was also pointed out during the several visits of the Bulgarian Minister of Foreign Affairs to France, the latest of which was in April 1984. Just as with the other capitalist countries, the main direction in Bulgaria's relations with France is economic co-operation, which is developing on an increasingly large scale. This is being effectuated on the basis of the inter-governmental agreement and long-term programme for economic, industrial and scientific and technological co-operation, signed in the mid-1970s. France is among the main Western trade partners of the PRB. At the summit meetings the need was pointed out to search for new forms and ways to make bilateral relations more active.

Also successful and fruitful is the development of bilateral relations between the PRB and the Empire of **Japan.** The President of the State Council of the PRB, Todor Zhivkov, visited that country in 1970 and 1978. Also positive were a number of other initiatives in Bulgarian-Japanese co-operation, among them the visit to Japan in 1976 of Bulgaria's Foreign Minister and the visit to Bulgaria in 1978 of the Crown Prince Akihito – the first visit of this nature to a socialist country. Economic, scientific, technological and cultural co-operation between the two countries is developing intensively. During the visit of Shintaro Abe, Minister for Foreign Affairs of Japan, to Bulgaria in 1983, the two Ministers noted the good state of relations and the considerable prospects; they also stressed their common concern over the deterioration of the international situation and their conviction that world peace and security must be safeguarded.

The relations of the PRB with the Republic of **Italy** are developing actively. A number of new projects have been envisaged and their implementation begun, thus creating a good foundation for the further extension of economic, industrial, scientific and technological co-operation.

The PRB has good relations with the **Benelux** countries the Kingdom of **Belgium,** the Kingdom of **The Netherlands** and the Grand Duchy of **Luxemburg.** The main direction in the development of relations is the trade and economic co-operation realized on the basis of long-term agreements for economic, industrial and technological co-operation; certain results have also been achieved in the field of foreign policy. Particularly important is the role of the visits of the Minister of Foreign Affairs of the PRB to the Netherlands in 1976 and to Belgium in 1978. The practice of mutual political consultation both on problems of bilateral relations and on matters of European security and co-operation is gaining ground. There has been an exchange of parliamentary delegations, and cultural ex-

The President of the World Peace Council, Romesh Chandra, delivering the closing speech, Sofia, 1974

changes have also grown considerably. As Todor Zhivkov put it, 'the good relations of the PRB with the Benelux countries are developing in many spheres, and there are no serious unresolved problems about them; it is our desire, by strengthening our co-operation with these countries, to help to improve the climate in Europe and in the world'.

The PRB maintains traditionally friendly relations with the **Scandinavian countries.** A fruitful exchange of opinions has been effectuated at the top level: the President of the State Council of the PRB, Todor Zhivkov, visited Norway and Denmark in 1970, and the Chairman of the Council of Ministers of the PRB visited Finland in 1973 and Sweden in 1974. Business and trade relations and cultural exchanges are expanding.

The visit of the President of the Republic of **Portugal,** Gen. Eanes, to the PRB in 1979 and the talks held at that time gave a very strong impetus to the further development of Bulgarian-Portugese relations.

Diplomatic relations between the PRB and the Kingdom of **Spain** were established in the late 1970s. Bilateral co-operation has a solid legal foundation.

The PRB makes efforts to extend its co-operation with **Great Britain,** the **USA** and **Canada.** A fruitful form to this end are the consultations between the Ministries for Foreign Affairs on matters of mutual interest. Possibilities exist for the advancement of economic relations with these countries which, unfortunately, have not been fully taken up, through no fault of the PRB.

Guided by its peaceful foreign policy, the PRB is developing and will continue actively to develop its relations with the capitalist states on the basis of the principles of peaceful coexistence. The economic, scientific and technological co-operation as well as the cultural exchanges with these states are not only of mutual interest, they also contribute to the improvement of the general climate in international relations.

BULGARIA AND THE STRUGGLE FOR PEACE, SECURITY AND CO-OPERATION

In Bulgaria, efforts for peace are a state policy raised to the level of a constitutional principle. The Bulgarian Constitution of 1971 stipulates: 'Every citizen is duty-bound to help preserve and consolidate peace. Incitement to war and war-propaganda are prohibited and punishable by law as grave crimes against peace and mankind'. As far back as December 1950, the National Assembly voted for the Defence of Peace Act, which determines the content of the crimes of that nature and foresees severe penalities for them. The appropriate texts have been included in the Penal Code of the PRB, and there are many state documents on foreign policy which illustrate its active struggle for peace and security, for co-operation between states with different social and economic systems, its firm protest against all aggression and injustice, against all acts of violence, and interference in the internal affairs of others and against all threats to peace.

The PRB takes part in the co-ordination of foreign policy and the formation of positions of unity on the issues of security and co-operation in the Warsaw Treaty Organization. Bulgaria uses the rostrums of the international organizations and above all the UN, to express its active support for the cause of peace, and to re-affirm the idea of convening the European Conference on Security and Co-operation (ECSC). The PRB took an active part at all stages of the preparations for and the work of the ECSC: during the first stage of the Conference, Bulgaria tabled, together with Poland, a draft document on the Basic Directions in the Development of Cultural Co-operation, Contacts and Exchanges of Information. It was one of main working documents of the Commission on Co-operation in the Humanitarian Spheres. Bulgarian diplomacy made its contribution to the adoption of the proposal to hold the final stage of the ECSC at the summit level.

In July 1974, the National Assembly of the PRB sent an appeal to the parliaments of the participants in the ECSC calling on them to strive with all their energies to overcome the difficulties encountered at the second stage of the Conference and to support its earlier and fruitful conclusion. The importance of this document was borne out by the numerous letters sent to the National Assembly of the PRB by a number of parliaments of European countries, expressing their unreserved support for the appeal of the supreme body of state power of the PRB.

The realistic and constructive policy of the PRB in the field of bilateral relations with the Western European countries, in combination with its active participation in the joint actions and initiatives of the Warsaw Treaty member-states, characterizes Bulgaria as one of the active architects of security and co-operation in Europe. The consistent policy of good-neighbourly relations and mutual understanding in the Balkans and in the world is widely assessed as a considerable contribution to efforts for reducing tensions in international relations.

Détente is a living tendency in present-day international relations. Its success, however, necessitates still greater efforts, a still closer rallying of the forces of peace and an even more dynamic, constructive and intensive foreign policy for the implementation of peace initiatives and the extension of their scope. In the autumn of 1976, at its regular session, the Political Consultative Committee of the Warsaw Treaty member-states adopted a declaration appealing to the ECSC participant-states for active moves to strengthen security and co-operation in Europe. At that session they formulated the proposal to all participants in the ECSC to conclude a treaty for the renunciation of the first use against each other of nuclear arms on land, at sea, in the atmosphere or in outer space, and to refrain from undertaking actions which could lead to the enlargement of existing military and political groupings and alliances or the establishment of new ones. After the signing of the Final Act in Helsinki (1975), the realization of the agreed text and the accepted principles is one of the major tasks in the foreign policy of the PRB, which is fully in harmony with the course of foreign policy it has been carrying out since the first days of its existence as a socialist state. The PRB assists efforts to fill the agreements at the Conference with concrete meaning and to re-affirm them as a long-term programme of action. There is no field of international life in which the PRB has not been making its worthy contribution in the interests of peace, co-operation and mutual understanding.

In pursuance of the obligations undertaken in Helsinki, Bulgaria immediately started to improve its legislation and to widely popularize the results of the ECSC. In accordance with the Final Act, the Ministries, administrations and public organizations in the PRB have worked out plans, programmes and initiatives for their activities. For the complete realization of the recommendations from Helsinki, a Co-ordination Commission on the Problems of Security and Co-operation in Europe was established in Bulgaria. A number of measures were adopted for the development of even more effective economic co-operation. By a Decree of the Council of Ministers of the PRB (1975), the issues of the establishment and functioning of representations of foreign firms in Bulgaria were regulated. Problems related to the text of the Helsinki Final Act in the humanitarian sphere are being solved in the same spirit. Bulgaria has introduced visa-free regulations for all persons who wish to visit the PRB as tourists. When a visa is necessary, the formalities for issuing it are minimal. Since 1972, the Sojourn of Foreigners in the PRB Act has been in force, which establishes a very simplified procedure. At the initiative of the PRB, visa-free regulations have been established with a number of West European countries – Austria, Sweden, Finland, Denmark, Norway, Iceland, etc. In their practice, the state bodies of the PRB strictly abide by the meaning of the moral and political norms of the Final Act in settling cases of contacts on the basis of family ties. The status of foreign journalists accredited to Bulgaria is similar. An instruction has been endorsed on their status and they have been facilitated in obtaining entry visas. Since May 1977, a Foreign Journalists Club has functioned in Sofia. The national authorities of the PRB create favourable conditions for their work. Press conferences, travel by crews around the country, meetings and interviews are organized. The work of the respective authorities in the PRB aimed at materializing the spirit of Helsinki assists to a great extent in the realization of the basic aims of the country's foreign policy. This is an expression of its willingness to discharge the undertaken obligations strictly and, whenever necessary, to make internal legislation conform with them. However, this willingness does not in the least mean that any form of interference into the internal affairs of the PRB would be tolerated.

In the efforts to apply the provisions of the Final Act strictly in matters of economic co-operation, the PRB undeviatingly follows the course of democratization of international economic relations, creating real opportunities for the full utilization of the advantages offered by the international division of labour. At the same time, strenuous efforts are made for the constant economic development of the country, for improving the quality of production, in order to meet the growing requirements set by the intensified international economic relations. The PRB strictly observes the obligations it has undertaken in Helsinki for ensuring good conditions for business contacts and economic information.

To conform with the agreements reached in Helsinki, the broad public in Bulgaria is being acquainted with the highest achievements of world culture by means of exhibitions, lectures, symposia and with the participation of the mass media.

Films and programmes of western countries are shown on Bulgarian television. Western periodicals are distributed, and Bulgarian publications reprint articles and other materials from Western countries. The number of foreign tourists visiting Bulgaria is growing, and each day confirms the words of the President of the State Council of the PRB, Todor Zhivkov, at the signature ceremony of the Final Act in Helsinki: 'The PRB also attaches special importance to international co-operation in the spheres of culture and education, of information and contacts among people. Open doors are a symbol of trust and hospitality. Our doors will be open to all people with open hearts, with good and honest intentions, who respect the laws, traditions and customs of the house whose guests they are'.

The PRB fully supports efforts for the steady advancement of the ideas contained in the Final Act at Helsinki. Bulgarian diplomacy is making its contribution to the continuation of the constructive dialogue in Belgrade (1977–78), at the meetings of experts in Bonn, Valletta, Montreux and Athens, at the difficult but fruitful work in Madrid (1980–83) and the Stockholm Conference on Confidence-Building Measures, Security and Disarmament in Europe (1984) which was opened in pursuance of the agreements reached at the Madrid Meeting.

The Madrid Meeting ended with a substantive and balanced final document in spite of the acute exacerbation of the international situation. The strenuous efforts of the socialist community countries contributed to the greatest extent to the successful conclusion of the Madrid Meeting. The PRB also had its worthy share in this work – throughout the Meeting it consistently upheld the constructive line. Bulgaria also made a number of proposals which found a place in the Final Document of the Meeting (e.g. the texts on the inclusion of the principles of the Document in international treaties, on the ratification of the human-rights pacts, on studying the problems of the younger generation, on the participation of small and medium-size firms in trade and industrial co-operation, etc.). At the final stage of the Madrid Meeting the Minister of Foreign Affairs of the PRB, Peter Mladenov, pointed out that 'the process of *détente* is vital and dynamic' and 'deep-rooted', and went on to say: 'The PRB will continue working to reach a solution of the main issue – preventing the further deterioration of the international situation and the slide toward nuclear destruction; ensuring to every man his basic right – the right to live'.

In his speech to the Stockholm Conference, the Minister of Foreign Affairs of the PRB emphasized: 'We came to Stockholm with the awareness that we would be working on crucial problems of European relations'. He pointed out Bulgaria's stand on the specific political importance of this forum in the present-day, dangerous international situation, which can and must give an impetus to *détente* and result in a constructive political dialogue in order to create an atmosphere of trust in Europe and to reduce the risk of a military confrontation in the continent. The PRB and the Warsaw Treaty member-states proposed a broad constructive programme:

the countries of the socialist community hold the opinion that a particularly important confidence-building measure will be the realization of their proposal put forward at the Prague Session of the Political Consultative Committee (1983) for the conclusion of a treaty between the member-states of the Warsaw Treaty Organization and the Nato countries for mutual non-use of military force and for maintaining peaceful relations. Bulgaria insists that the nuclear states which have not done so must follow the USSR in undertaking the obligation not to use nuclear weapons first.

The results from Madrid, the beginning of the Stockholm Conference, the decisions to carry out a number of future steps – in Athens (1984), Venice (1984), Ottawa (1985), Budapest (1985), Vienna (1986) and Berne (1986) and the commemoration of the tenth anniversary of the signing of the Helsinki Final Act (1985), bear out the fact that the objective basis of *détente* has not been destroyed and can serve for the development of a new stage of positive shifts in the international scene. In this connection Todor Zhivkov said: 'We are optimists. We are doing and will continue to do everything possible, to turn the world back to *détente* and co-operation. We are convinced champions of mutually advantageous co-operation with all countries and in all spheres – science, technology, production, and trade. This co-operation is necessary not only for us, and we believe that it will make increasingly broad breakthroughs, ignoring all artificial barriers'. This optimism is based on the profound knowledge of the laws of social development.

While realizing its programme aims and tasks, the PRB, guided by the progressive principles of its foreign policy, works toward the formation of the broadest possible peace front and combines its active diplomatic work flexibly with other effective forms. This is borne out by holding in the PRB such important forums as the International Meeting-Dialogue on *Détente*, organized on the occasion of the 80th anniversary of the founding of the Bulgarian Agrarian Party (8 July 1980), the World Parliament of Peoples for Peace (September 1980), the International Symposium on East-West Industrial Co-operation (May 1982), the international conference on the Life's Work of Georgi Dimitrov and Our Time (June 1982), and the international trade-union meeting-dialogue on Peace and Trade Unions (October 1983). The National Committee for European Security and Co-operation, established in 1969, is active in Bulgaria, where it numbers among its ranks figures of various public and political organizations and cultural unions and expresses the will of the Bulgarian public. The Committee organizes actions, meetings, international conferences, etc., and since 1971, it has been a member of the International European Security and Co-operation Committee, seated in Brussels. The governing body of the peace movement in Bulgaria is the National Peace Committee created in 1948. The Committee rallies all political forces, social layers and religious circles. Peace committees have been set up in the district towns.

The International Meeting-Dialogue on *Détente*, Sofia, 1980

In its foreign policy the PRB renders full support to efforts for curbing the arms race and achieving general and complete disarmament and is active in all initiatives leading step by step to this aim.

The PRB highly values the efforts of the USSR for reaching real results in the Soviet-US dialogue in the field of disarmament.

By its consistent course in foreign policy, the PRB actively supports the realization of the proposals made by the USSR to discontinue the production of nuclear weaponry in all its forms; to discontinue production and ban all weapons of mass destruction; to discontinue the development of new types of conventional weapons with great destructive power; to renounce the enlargement of armed forces and the increase of conventional weapons by states which are permanent members of the UN Security Council and by member-states of military alliances. The PRB insists on the gradual destruction of stockpiles of nuclear weapons. It fully supports the initiative to accompany the ban on nuclear weapons with a world treaty for the non-use of force in international relations and hails the declaration of the USSR that it will never use nuclear weapons against states which refuse to produce, to possess or deploy them on their territories.

Being a Balkan country situated close to the Mediterranean, the PRB joins its voice to the proposals made by the USSR for the withdrawal of all Soviet and US ships with nuclear weapons on board from that area of the world.

The PRB has signed and ratified a number of treaties, e.g.: the treaty banning nuclear-weapon tests in the atmosphere, in outer space and under water (1963); the treaty on principles governing the activities of states in the exploration and use of outer space, including the moon and other celestial bodies (1967); the treaty on the non-proliferation of nuclear weapons (1968); the treaty on the prohibition of the emplacement of nuclear weapons and other weapons of mass destruction on the sea-bed and the ocean floor and in the subsoil thereof (1971); the convention on the prohibition of the development, production and stockpiling of bacteriological (biological) and toxic weapons and on their destruction (1972); and the convention on the prohibition of military or any other hostile use of environmental modification techniques (1977). The PRB took an active part in the preparatory work for the two special sessions on disarmament at the UN General Assembly, as well as in the work of the multilateral bodies on disarmament – the Disarmament Conference in Geneva, the Special Committee for the World Conference on Disarmament, etc. It is a co-author of the Prague declaration of the Political Consultative Committee of the Warsaw Treaty (January 1983), which contains a realistic programme for averting a nuclear holocaust, for curbing and putting an end to the arms race, and for disarmament.

The PRB also attaches great importance to regional initiatives aimed at reducing international tension, particularly proposals to create nuclear-free zones in different areas of Europe, as well as the freeing of the old continent from nuclear weapons. The PRB and the socialist community countries have always been guided by the realistic opinion that nuclear weapons are an intricate matter whose solution makes it necessary to rally the efforts of all nuclear and non-nuclear states, at multilateral and bilateral levels, on a global and regional scale. They are well aware of all difficulties but at the same time they resolutely refute the theory that this problem is insoluble.

BULGARIA'S PARTICIPATION IN THE UNITED NATIONS AND IN OTHER INTERNATIONAL ORGANIZATIONS

Bulgaria's participation in the work of international governmental and non-governmental organizations and above all of the UN and its specialized agencies is one of the main spheres of foreign political activities of the People's Republic of Bulgaria. The Bulgarian Government took official steps for Bulgaria's admission to the UN immediately after the signing of the Paris Peace Treaty (10 February 1947). On 26 July 1947 the Bulgarian Government sent the Secretary-General of the UN an application for the admission of Bulgaria to the UN. It stressed that Bulgaria accepted the fundamental aims and principles laid down in the UN Charter, and was prepared to assume the obligations stemming from its admission to the organization. The application was renewed in 1948, 1952, 1954 and 1955. The United States and Great Britain vetoed Bulgaria's request in the Security Council, and its admission to the UN was delayed. By Resolution 995, passed on 14 December 1955, thanks to the support of the USSR and many other states, the Tenth Session of the UN General Assembly admitted Bulgaria and fifteen other countries to the world organization. On that occasion the Bulgarian Government issued a declaration in which it stated that Bulgaria would make every effort for the maintenance of peace and security, for the development of friendly relations among nations and for the achievement of co-operation in the settlement of international problems of a political, economic, social, cultural and humanitarian nature. A Permanent Mission of the PRB to the UN was opened in New York in 1956.

Bulgaria's admission to the UN was a major achievement of its principled policy of peace and international co-operation. It opened up new possibilities and gave new scope to its foreign political activities. Bulgaria's admission to the UN coincided with the April 1956 Plenary Session of the Central Committee of the BCP, after which Bulgaria's foreign policy became much more active.

From the very beginning Bulgaria's participation in UN activities was marked by its active involvement in the discussion and solution of international problems. In its work in the UN and other organizations within the UN system Bulgaria is unswervingly guided by its concern for the maintenance of world peace and security and for the promotion of friendship and co-operation among nations. Bulgaria's active contribution to the enhancement of the prestige and possibilities of the UN has found broad international recognition. As UN Secretary-General, Javier Perez de Cuellar, stressed Bulgaria 'invariably firmly supports the efforts of the World Organization for peace and international security' (*Rabotnichesko Delo,* 25 February 1984).

The First International Congress of Unesco associated schools, Sofia, 1983

The PRB takes an active part in the work of the UN for the attainment of the aims laid down in its Charter. By their participation in the General Assembly and in other main bodies of the UN, the Bulgarian delegates contribute to the strengthening of peace and security in the world, to the ending of the arms race, to disarmament; they champion the cause of the peoples fighting for freedom and independence and stand for the abolition of racism and discrimination, for the development of equal and mutually beneficial international co-operation. Bulgaria also stands for the promotion of the role of the organization and for the strict observance of its Charter. After

Bulgaria was elected to the Security Council (1966–67) it participated actively in its work. The Bulgarian delegation contributed effectively to the elaboration of the Declaration for the Strengthening of World Security, adopted by the 25th Jubilee Session of the UN General Assembly (1970) and was a co-sponsor of the resolution with which the Declaration was adopted by the General Assembly. Bulgaria holds the view that in accordance with the stipulations of the UN Charter the Organization should take all necessary steps, including the use of armed force (UN operations) for the maintenance of peace. Bulgaria participates in the work of almost all the multilateral bodies in the field of disarmament. It is a member of the Disarmament Conference (which began its work in 1960 in Geneva as the Ten Nations' Disarmament Committee, in 1962 became the Eighteen Nations' Disarmament Committee, and since 1978 consists of 40 states), of the Special UN Committee for the Indian Ocean, of the Special UN Committee for a World Disarmament Conference, etc. Bulgarian representatives have been elected to the leadership of bodies related to the problems of international security and disarmament.

Bulgaria also participates with a special status in the Vienna talks on the mutual reduction of armed forces and armaments in Central Europe. It is a party to all the multilateral treaties in the sphere of disarmament (the Treaty Banning Nuclear Tests in the Atmosphere, in Outer Space and under Water of 1963, the Treaty on the Non-Proliferation of Nuclear Weapons of 1968, etc.). Within the framework of the co-ordinated foreign policy of the Warsaw Treaty member countries, together with the USSR and the other socialist countries, Bulgaria has been the co-sponsor of a number of initiatives, the most important of which were the proposals for the founding principles of general and complete disarmament (1960, in the Disarmament Committee), the banning of chemical and bacteriological weapons (1969, in the Disarmament Committee, draft convention), the banning of neutron nuclear weapons (1978, in the Disarmament Committee, draft convention), the beginning of talks on the cessation of the production of all types of nuclear weapons and the gradual reduction of their stocks until their complete destruction (1979, in the Disarmament Committee). On Bulgaria's initiative the UN General Assembly has adopted a number of resolutions on disarmament problems. At the 36th Session of the UN General Assembly (1981) the Bulgarian delegation tabled three draft resolutions: on strengthening the guarantees for the security of non-nuclear states; on the carrying out of a world campaign for the collection of signatures in support of measures for the prevention of a nuclear war, for the restriction of the armaments drive and for disarmament; and on the state of the multilateral agreements on disarmament. At the 37th Session of the UN General Assembly (1982) the First (Political) Committee of the General Assembly adopted the resolution, tabled on Bulgaria's initiative and co-sponsored by Mongolia and Romania, on the organization of a World Disarmament Campaign. In this resolution it is more specifically proposed to launch a campaign for the collection of signatures in support of measures for the prevention of a nuclear war, for the restriction of the armaments drive and for disarmament.

Bulgaria actively participates in various conferences on disarmament held under the auspices of the World Organization. It took part in the UN Conference on the banning or restriction of the use of certain concrete types of conventional weapons which may be considered extremely brutal or have an indiscriminate action (the Conference formulated a corresponding convention), etc. Together with the USSR and the other socialist countries Bulgaria was among the first states to sign this convention (in 1981) and to ratify it (in 1982). On Bulgaria's initiative the 38th Session of the UN General Assembly adopted a resolution on restricting naval activities and armaments, on guaranteeing the security of non-nuclear states and on the World Disarmament Campaign. The PRB also takes an active part in a number of other major spheres of UN activities.

Liquidation of Colonialism. Bulgaria is active in all UN bodies in defence of the national-liberation struggles of the peoples of Asia, Africa and Latin America. It wages a resolute struggle against the attempts of world Imperialism to check the popular drive for liberation and to maintain its positions in the former dependent and colonial countries. Bulgaria participated in drawing up the Declaration on the Granting of Independence to Colonial Countries and Peoples, adopted on the initiative of the USSR at the Fifteenth Session of the UN General Assembly (1960) and actively works for its application. It was elected to the Special Committee on Decolonization (the Committee of the 24 [States]; 1961), formed for the implementation of the Declaration. At Bulgaria's proposal in 1967 the UN General Assembly included on its agenda the question of the implementation of the Declaration on the Granting of Independence to Colonial Countries and Peoples by the specialized agencies and by the international institutions, associated with the UN. In Resolution 2311/XXII (1967), adopted by the UN General Assembly, the specialized agencies of the UN – Unesco, the World Health Organization (WHO), the International Labour Organization (ILO) and others – were called upon to co-operate with the Organization of African Unity (OAU) and to render direct material support to the national-liberation movement. This question is still on the agenda of the sessions of the UN General Assembly. At its 25th Jubilee Session (1970), with Bulgaria's active participation, the General Assembly adopted a Programme of Action for the Full Implementation of the Declaration on the Granting of Independence to Colonial Countries and Peoples. In 1973 the Special Committee on Decolonization sent a mission to the headquarters of the main specialized agencies to discuss the basic forms of application of Resolution 2311/XXII. The mission included representatives of Bulgaria, Iraq, Sweden and Tunisia. The 28th Session of the General Assembly (1973) made a positive assessment of the mission's work. Bulgaria is also a member of the Council for Namibia (since 1978) and works for the realization of the right of the Namibian people to self-determination and independence.

Defence of the Rights of Man. Bulgaria helps towards the promotion of international co-operation in the sphere of the assertion of the rights of man and of fundamental liberties for all, irrespective of race, sex, language or religion. It takes part in the work of various UN bodies dealing with these matters, and in the drafting of pacts and other international acts on human rights, and it assumes obligations under the international agreements in this field. Bulgaria has ratified twelve of the twenty international agreements on human rights signed under the auspices of the UN. It was the first country in the world to ratify the Convention for the Abolition of All Forms of Racial Discrimination (1965, in force since 1969). Under Article 8 of the Convention, a Committee of eighteen experts was established with the task of observing its implementation on the basis of reports submitted by the Governments. A Bulgarian expert also sits on this Committee. By 1982 Bulgaria had submitted six reports which were favourably assessed by the Committee.

In 1975 the PRB ratified the International Pact for Civil and Political Rights (in force since 1976). The fundamental principles of these pacts are laid down in the Constitution of the People's Republic of Bulgaria (1971, Chapter III). In conformity with the Pact for Civil and Political Rights, a Committee on the Rights of Man was established (1976). Bulgaria is a member of the Session working group of fifteen countries at the Economic and Social Council (Ecosoc) which observes the implementation of the Pact for Economic, Social and Cultural Rights. It regularly presents reports on the implementation of the international human rights pacts which are discussed by the Human Rights Committee (1979) and by the working group of the Ecosoc (1982). In 1973 the UN General Assembly adopted a Convention for the Prosecution and Punishment of the Crime of Apartheid (in force since 1976). Bulgaria participated in the drafting of this Convention and was one of the first countries to sign and ratify it (1974). After the Convention came into force, Bulgaria was several times elected to the group of three countries to review the implementation of the Convention, and it submitted two reports for discussion. It supported the proclamation of the Decade for Struggle against Racism and Racial Discrimination (1973–83) and participated in the implementation of the Decade's programme and in the World Conference for Struggle Against Racism and Racial Discrimination (Geneva, 1978). It also contributed to the fulfilment of the aims of the UN Decade for Women; it participated in the world conferences in Mexico (1975) and Copenhagen (1980) and in the drafting of the Convention for the Abolition of All Forms of Discrimination Against Women; and it was co-sponsor of the resolution with which the UN General Assembly approved and adopted the Convention (1979, in force since 1981), ratified by Bulgaria in 1981. The country's active participation in the humanitarian activities of the UN finds recognition in its repeated election since 1973 to the Human Rights Commission. At the Commission's session in 1973 and 1979 in Geneva the representative of Bulgaria was elected Vice-Chairman and in 1982 – Chairman. In 1980 the Bulgarian representative was elected to chair the Third Committee of the 35th regular session of the UN General Assembly.

The Use of Outer Space for Peaceful Purposes. Bulgaria is a member of the Committee on the Use of Outer Space for Peaceful Purposes (1959). It is a party to all treaties on outer space adopted by the UN, and to the Convention between Bulgaria and the Soviet Union on the transmission and utilization of data from the remote sounding of the earth from outer space (1968). According to the UN list Bulgaria became the eighteenth space country in the world. On 10 April 1979 a crew consisting of the Soviet cosmonaut Nikolai Rukavishnikov and the Bulgarian cosmonaut Georgi Ivanov (born in 1940) orbited the earth aboard the *Soyuz 33* spaceship, and Bulgaria thus became the sixth country in the world to send a cosmonaut into outer space.

Utilization of the Sea-Bed and the Ocean Floor. A Special Committee, consisting of 35 states, including Bulgaria, was established at the 22nd Session of the UN General Assembly (1967). At the 23rd Session (1968) it was renamed Committee for the Peaceful Uses of the Sea-Bed and the Ocean Floor beyond the Boundaries of National Jurisdiction (also known as the Sea-Bed Committee). Its membership grew to 42 states and later on – to 90 states. Together with the other socialist countries, Bulgaria actively participates in the work of the Committee and of its Sub-Committees. It also submitted for discussion a special document on determining the width of the territorial sea. At the proposal of the USSR and with the active support of the socialist countries the UN General Assembly adopted a Treaty Banning the Deployment of Nuclear and Other Weapons for Mass Destruction on the Sea-Bed and the Ocean Floor and in their Recesses (1970, in force since 1972). Bulgaria signed and ratified the Treaty in 1971. At the Third Conference on the Law of the Sea (1974) the Bulgarian representative was elected to chair one of the Conference's three committees. At the Ninth and Tenth Conferences on the Law of the Sea (1980, 1981) the PRB insisted on the adoption of a comprehensive convention on the law of the sea, including the establishment of an equitable international regime for the exploration and utilization of the resources of the sea-bed and the ocean floor beyond the boundaries of national jurisdiction.

Participation in International Bodies, Organizations and Conferences of the United Nations. Bulgaria takes an active part in the work of the UN and of the organizations within its system for the progressive development of international economic relations, for their reorganization on a just and democratic basis, and for the broadest possible utilization of the advantages of the

The Ninth World Youth Festival held under the motto 'For Solidarity, Peace and Friendship', Sofia, 1968

international division of labour. During the 1981–82 period Bulgaria was elected to the Ecosoc, the main economic and social body of the UN, and at the 37th Session of the General Assembly (1982) it was re-elected for another three years. Bulgaria is also active in the United Nations Industrial Development Organization (Unido, established in 1967 with its seat in Vienna), the basic aim of which is to contribute to the industrial development of its member countries and, more particularly, to help speed up the industrialization of the developing countries. Bulgaria was a founding member of Unido and it has established a National Commission for Work with Unido. During the 1977–80 and 1983–85 periods it was and still is a member of Unido's Industrial Development Board. In the past few years seminars and meetings on industrial development and co-operation with representatives of developing countries have been held in Bulgaria with the co-operation of Unido. The practical results of Bulgaria's participation in the organization are a number of projects, built in the country and financed by the United Nations Development Programme (UNDP). Bulgaria has been a member of the UNDP since its establishment (1965) and actively participates in its work by sending specialists, experts and consultants to carry out surveys and studies in developing countries and by receiving representatives of such countries for training and specialization. Many Bulgarian specialists improve their professional qualifications in foreign institutes within the framework of the UNDP. Bulgaria was elected to the UNDP's Governing Council for the 1981–83 period. A number of institutes and centres are being erected in Bulgaria with UNDP funds.

Bulgaria takes an active part in the work of the UN Economic Commission for Europe (ECE), its representatives participate in the work of all the auxiliary bodies of the Commission and it attaches major importance to the ECE plans: the North-South Highway, the linking of the electric power systems of the Balkan countries, joint research on the utilization of low-calorie coal, etc. A representative of Bulgaria was elected to chair the ECE's 39th Session (April 1984). Since 1967 Bulgaria has been participating as an observer in the General Agreement on Tariffs and Trade (GATT). Bulgaria has also been taking an active part in the work of the International Atomic Energy Agency (IAEA) since its Statute entered into force (1957). A National Committee for the Peaceful Uses of Atomic Energy has been established, and during the 1967–69 period Bulgaria was a member of the Board of Governors – the agency's executive body. Bulgarian scholars have given papers at IAEA conferences and the agency contributes to the improvement of the qualifications of Bulgarian scientists working in this field as well as making use of Bulgarian expertise. Bulgaria has also been participating in the UN Children's Fund (Unicef) since its establishment (1946) and has a National Committee for Co-operation with Unicef. Since 1972 it has been on the Executive Board of the Fund. In 1982 the 28th Annual Session of the European National Committees of the Unicef was held in Sofia. The participants in the session unveiled the memorial plaque, built into the walls, of the Banner of Peace Monument in Bulgaria and bearing the inscription: 'For the Happiness and Well-Being of Children All Over the World'. Bulgarian scholars – jurists and diplomats – have contributed to the drafting of a number of international conventions adopted by international conferences: the 1961 and 1963 Vienna Conventions on diplomatic and consular relations, on contractual law, conventions in the spheres of maritime law, shipping, the safety of civil aviation, etc., as well as on the protection of intellectual property within the framework of the World Intellectual Property Organization.

A meeting of the second session of the World Children's Parliament, Sofia, 1982

Participation in the Work of the International Specialized Agencies. The PRB values the role of the specialized states. Bulgaria is a member of ten of the fifteen specialized agencies within the UN system (in 1982): the International Tele-Communications Union (ITU), the International Labour Organization (ILO), the International Civil Aviation Organization (ICAO), the International Maritime Organization (IMO; formerly the Inter-governmental Maritime Consultative Organization [IMCO]), the United Nations Educational, Scientific and Cultural Organization (Unesco), the World Meteorological Organization (WMO), the World Intellectual Property Organization (WIPO), the World Health Organization (WHO), the Food and Agriculture Organization of the UN (FAO) and the Universal Postal Union (UPU). Bulgaria is also a member of the United Nations Industrial Development Organization (Unido), the United Nations Conference on Trade and Development (UNCTAD) and the World Organization on Tourism (WOT).

Bulgaria is not a member of the following specialized agencies: the International Bank for Reconstruction and Development (IBRD), the International Monetary Fund (IMF), the International Finance Corporation (IFC) or the International Development Association (IDA).

Bulgaria participates in the International Telecommunications Union (with headquarters in Geneva). As a member of the ILO based in Geneva, it was elected to the Governing Body in 1972 for a period of three years and re-elected for the 1981–84 period. It has ratified 77 conventions, and by the number of the recommendations it has introduced into ILO internal legislation Bulgaria ranks among the first in the world. With the co-operation of the ILO and of the UNDP, a Centre for the improvement of leading trade union and other cadres, mainly from the developing countries, has been established in Sofia. Bulgaria is a member of the ICAO with headquarters in Montreal (since 1967) and contributes actively to the work of the organization, and more particularly to the regulation of the conditions for the conveyance of passengers and commodities. Bulgaria participates in the FAO, with headquartes in Rome, and Bulgarian specialists and experts have participated in a number of FAP projects in developing countries. Since 1972 Bulgaria has been a member of the International Union for the Conservation of Nature and Natural Resources (UCN), which is closely affiliated with the FAO. Bulgarian representatives take an active part in the work of scientific conferences and symposia organized by FAO. With the co-operation of that organization and the financial assistance of UNDP, the Nikola Pushkarov Institute of Soil Sciences and Yield Programming, the Research and Experimental Station for the Biological Testing of Fodder Mixtures in the town of Kostinbrod, and the Experimental Farm in the town of Breznik have been established in Bulgaria. The FAO organizes seminars and symposia in Bulgaria on plant-growing and stock-raising, attended with interest by representatives of the developing countries. The Thirteenth Regional Conference of the FAO for Europe was held in October 1982 in Sofia. Since 1961 Bulgaria has been a member of the IMO (formerly the IMCO) with headquarters in London. In 1984 it was elected to its Council. It co-operates in the drafting of the international conventions adopted by the IMO. The Bulgarian representative chaired the Assembly of the organization in 1973–74. In 1976 the conference of the IMCO member countries adopted a convention for the establishment of an international Organization for Maritime Telecommunications via Satellites (INMARSAT), with headquarters in London. In 1979 Bulgaria was elected to the Council of the organization and it was re-elected in 1981.

The International Meeting-Dialogue on Peace and the Trade Unions, Sofia, 1983

President of the State Assembly of the People's Republic of Bulgaria, Todor Zhivkov, addressing the Fourteenth Congress of the International Union of Students, Sofia, 1984

Bulgaria has been a member of Unesco since 1956 and has a permanent mission to the organization. A National Commission of the PRB for Unesco was founded, also in 1956. Bulgarian representatives actively participate in the work of the General Conferences and Executive Board, in drafting Unesco programmes, and in seminars and symposia.

Bulgaria has been a member of the WMO since 1952 and participates through the Main Board of Hydrology and Meteorology at the Bulgarian Academy of Science. It contributes to all the activities of the organization. In September 1982 a conference on countering damage caused by hailstorms was held in Sofia; it was organized by an international initiative committee with the co-operation of the WMO. Bulgaria has been a member of the WIPO, with headquarters in Geneva, since 1970 and contributes to the development of international co-operation in the sphere of authors' and inventors' rights. In 1975 Bulgaria also joined the Universal Copyright Convention, signed under the auspices of Unesco. Bulgaria has been a member of the WHO, with headquarters in Geneva, since 1949. During the 1969–72 period it was a member of the WHO Executive Board; the Minister of Public Health of the PRB was a vice-president of the 25th World Health Assembly. The organization has held some of its official meetings in Bulgaria (seminars on cardiovascular diseases in 1971 and 1973; session of the working group, studying the effectiveness of the programmes worked out by the European Regional Office of WHO [1972], etc.). The organization sends many Bulgarian physicians and other medical personnel to help the developing countries. With the aid of WHO, an International Centre of Hygiene for the professional improvement of medical personnel from all parts of the world has been established in Sofia. Since 1879 Bulgaria has been a member of the UPU with headquarters in Berne. It has been elected to the Union's Executive Council.

Bulgaria and International Non-Governmental Organizations (NGOs). Bulgaria realizes that the international non-governmental organizations as associations of the public or of individuals of three or more countries play an increasing role in international life. A considerable number of them are becoming important factors for the reorganization of international relations on a democratic basis. The rapid growth of the number of NGOs (in 1981 there were more than 4,400) shows vast support for the idea of founding organizations for co-operation among national social, political, youth, trade union, sports or other public organizations to forward specific common purposes and aims. Bulgaria takes a very positive view and renders all-round co-operation with the NGOs which serve the ideals and the cause of peace, disarmament, friendship, good neighbourliness and co-operation among nations, as well as the economic and social progress of mankind.

The statute of each NGO determines the specific character of the forms of co-operation and especially of the support afforded by Bulgaria to many of them. Bulgaria realizes that the role and importance of the NGOs will continue to grow with the advance of the progress in science and technology and with the rapid increase in the number and scope of problems, the solution of which will need the co-operation of representatives of the public in most varied spheres of spiritual life, material production, etc.; it also realizes that the process of the formation of new, world and regional NGOs in the sphere of political, scientific, technical, social, sports and other international relations will continue to develop at an accelerated pace. Bulgaria and its Government encourage membership of democratic NGOs; it is significant that by 1981 Bulgarian public organizations and individuals were collective or individual members of nearly 370. Bulgarian representatives are members of the governing and executive bodies of many NGOs or work in their secretariats. Bulgarian delegations play an active role in their annual conferences with their initiatives and views on the problems under discussion.

Conclusion. Bulgaria's foreign political activities on a bilateral and multilateral basis are aimed at helping the struggle for social progress, for the dynamic development of the newly liberated states, for equal and mutually beneficial co-operation among the countries with different social systems and for peaceful co-existence in Europe and the world, for disarmament, for the transformation of the Balkans into a zone of peace and good neighbourly relations.

Bulgaria wants to live in peace with all peoples of the world. Precisely for this reason the foreign policy of the People's Republic of Bulgaria is and will continue to be principled and consistent; it will continue to contribute within its powers to a real advance in the all-round development of this planet, to the assertion of the principles of peaceful co-existence, to disarmament, to the maintenance of *détente* and to its transformation into an irreversible process, to guarantee world peace.

SELECTED BIBLIOGRAPHY

GANCHEV, T. *Balkanite – zona i faktor na evropeiskiya mir* (The Balkans – a Zone and Factor of European Peace). Sofia, 1979.

Georgi Dimitrov i nyakoi problemi na mezhdunarodnoto pravo i mezhdunarodnite otnosheniya (Georgi Dimitrov and Some Problems of International Law and International Relations). A Collection of Papers. Sofia, 1977.

Mezhdunarodni otnosheniya i vunshna politika na Bulgaria sled Vtorata svetovna voina (International Relations and Foreign Policy of Bulgaria after the Second World War). Sofia, 1982.

Mezhdunarodni otnosheniya i vunshna politika na Narodna Republika Bulgari. Za sistemata na partiinata prosveta (International Relations and Foreign Policy of the People's Republic of Bulgaria. For the System of Party Education). Sofia, 1982.

MLADENOV, P. 'Aprilskata vunshna politika na Narodna Republika Bulgaria' (The April 1956 Line of the Party in the Foreign Policy of Bulgaria). In *Mezhdunarodni otnosheniya (International Relations)*, No. 2, 1981.

MLADENOV, P. 'Narodna Republika Bulgaria – aktiven uchastnik v borbata za mir i razoruzhavane' (The People's Republic of Bulgaria – an Active Participant in the Struggle for Peace and Disarmament). In *Novo Vreme (New Times Journal)*, No. 3, 1981.

MLADENOV, P. 'Osnovopolozhnik na sotsialisticheskata vunshna politika na Bulgaria' (Founder of the Socialist Foreign Policy of Bulgaria). In *Novo Vreme (New Times Journal)*, No. 6, 1982.

Narodna Republika Bulgaria i preustroistvoto na mezhdunarodnite otnosheniya vurhu demokratichna osnova (The People's Republic of Bulgaria and the Reshaping of International Relations on a Democratic Basis). Sofia, 1984.

NRB – SSSR. Sutrudnichestvo i vzaimodeistvie v imeto na mira i progresa (The People's Republic of Bulgaria and the USSR. Cooperation and Interrelations in the Name of Peace and Progress). Sofia, 1983.

Sotsialisticheskata vunshna politika na Narodna Republika Bulgaria, 1944–74 (The Socialist Foreign Policy of the People's Republic of Bulgaria, 1944–74). Sofia, 1974.

STEFANOV, G. *Mezhdunarodni otnosheniya i vunshna politika na Bulgaria 1789-1970* (International Relations and Foreign Policy of Bulgaria in 1789–1970). 2nd rev. edn. Sofia, 1974.

Vunshna politika na Narodna republika Bulgaria (Foreign Policy of the People's Republic of Bulgaria. Collected documents and materials). Vols 1–3. Sofia, 1970-83.

Za vsestranno razvitie na bulgaro-yugoslavskite otnosheniya (On an All-Round Development of Bulgarian-Yugoslav Relations). A Declaration by the Ministry of Foreign Affairs of the People's Republic of Bulgaria, Sofia, 24 July 1978. Sofia, 1978.

ZHIVKOV, T. *The borderline between the two decades, between the 1970s and the 1980s, should not be a borderline between détente and confrontation*. Speech delivered at the International Meeting and Dialogue on *Détente, Peace and Social Progress*, organized by the Bulgarian Agrarian Party, 21 May 1981.

ZHIVKOV, T. *Bulgaria – Ancient and Socialist*. Toronto, 1975.

ZHIVKOV, T. *Bulgaria is for Peace, for Détente, for Co-operation*. Speech delivered at the Session of the National Council of the Fatherland Front. Sofia, 11 February 1980.

ZHIVKOV, T. *Edinodeistvieto v imeto na mira – povela na zhivota* (Unity of Action in the Name of Peace – The Challenge of Life). A Speech delivered at the International Trade Union Meeting and Dialogue – *Peace and the Trade Unions*, 27 October 1983. Sofia, 1983.

ZHIVKOV, T. *Internal and Foreign Policy of the People's Republic of Bulgaria*. Speech delivered at the Tenth Session of the Seventh National Assembly, Sofia, 27 April 1979. Sofia. 1979.

ZHIVKOV, T. *Otchet na Tsentralniya komitet na Bulgarskata komunisticheska partiya pred Dvanadesitiya kongres i predstoyashtite zadachi na partiyata* (A Report of the Central Committee of the Bulgarian Communist Party before the Twelfth Party Congress and the Forthcoming Tasks of the Party), 31 March 1981. Sofia, 1981.

ZHIVKOV, T. *Otchet na Tsentralniya komitet na Bulgarskata komunisticheska partiya za perioda mezhdu Desetiya i Edinadesetiya kongres i predstoyashtite zadachi. Doklad pred Edinadesetiya kongres na BKP, 29 mart 1976. Rech pri zakrivaneto na Edinadesetiya kongres na Bulgarskata komunisticheska partiya. 2.4.1976* (Report of the Central Committee of the Bulgarian Communist Party for the Period between the Tenth and the Eleventh Party Congresses and on the Forthcoming Tasks, Delivered at the Eleventh Party Congress, 29 March 1976. Speech delivered at the Closing Session of the Eleventh Party Congress. 2 April 1976). Sofia, 1976.

ZHIVKOV, T. *Otcheten doklad na Tsentralniya komitet na Bulgarskata komunisticheska partiya pred Desetiya kongres na partiyata* (Report of the Central Committee of the Bulgarian Communist Party Delivered at the Tenth Party Congress), 20 April 1971.

ZHIVKOV, T. *Selected Works – Todor Zhivkov* New Delhi, 1982.

ZHIVKOV, T. *Socialism and Peace are Inseparable: Todor Zhivkov Answers Questions Put to Him by Robert Maxwell – President of the Anglo-American Publishing House, Pergamon Press*. Sofia, 1982.

ZHIVKOV, T. *Speech about Bulgaria*. Delivered at the Ceremonial Meeting Dedicated to the 1,300th Anniversary of the Founding of the Bulgarian State, Sofia, 20 October 1981. Sofia, 1981.

ZHIVKOV, T. *Statesman and Builder of New Bulgaria* Translated selections from various sources. Oxford, 1982.

ZHIVKOV, T. *Stroitelstvoto na sotsializma i komunizma i svetovnoto razvitie* (The Building of Socialism and Communism and World Development). Speech delivered at the Opening of the International Theoretical Conference 'The Building of Socialism, Communism and World Development', 12 December 1978. Sofia, 1978.

ZHIVKOV, T. *Through the Efforts and Will of all Peoples a Steadfast Struggle for Just and Lasting Peace*. Speech delivered at the Opening of the World Parliament of the Peoples for Peace in Sofia, 23 September 1980. Sofia, 1980.

ZHIVKOV, T. *Today Humanity Has No Task More Important and More Urgent than to Safeguard and Strengthen Peace*. Speech delivered at the International Meeting and Dialogue on *Détente*, organized on the occasion of the 80th anniversary of the Bulgarian Agrarian Party, 8 July 1980. Sofia Press, 1980.

ZHIVKOV, T. *A Word about the Party of Lenin*. Sofia, 1981.

APPENDIX:
BULGARIA IN FIGURES

Geography of the People's Republic of Bulgaria

Territory	Square kilometres
Total – land and river frontier waters	110,911.5
Area included between the land frontiers, the sea-shore and river frontier banks	110,548.6
Area of frontier river and sea islands	95.2
Area of territorial waters of frontier rivers	267.7

Boundaries (kilometres)	Total	Land	River	Coastal
Total	**2,245**	**1,181**	**686**	**378**
Northern – with the Socialist Republic of Romania	609	139	470	—
Eastern – the Black Sea	378	—	—	378
Southern – with the Republic of Turkey	259	133	126	—
– with the Republic of Greece	493	429	64	—
Western – with the Socialist Federal Republic of Yugoslavia	506	480	26	—

Geographic co-ordinates of Bulgaria's endmost points

Direction	District	Northern Latitude	Eastern Longitude[1]
North	Vidin – river Timok mouth	44°13′	22°40′
South	Kurdjali – Veikata peak	41°14′	25°17′
East	Tolbuhin – cape Shabla	43°32′	28°36′
West	Kyustendil – north-west of Kitka peak	42°19′	22°21′

[1] Greenwich longitude.

Land Use in 1980 (thousand hectares)

Countries	Total area	Dry-land area	Cultivable land		Forests	Other lands
			Arable land plantations of perrenials	Natural meadows and pastures		
World total	13,392,148	13,075,248	1,452,215	3,116,685	4,093,547	4,412,801
1. Austria	8,385	8,273	1,635	2,040	3,282	1,316
2. Bulgaria	11,091	11,055	4,181	2,004	3,845	1,025
3. Czechoslovakia	12,787	12,549	5,169	1,682	4,578	1,120
4. France	54,703	54,563	18,643	12,883	14,582	8,455
5. FRG	24,858	24,434	7,494	4,754	7,320	4,866
6. GDR	10,833	10,610	5,034	1,235	2,955	1,386
7. Great Britain	24,482	24,160	6,969	11,473	2,102	3,589
8. Greece	13,194	13,080	3,926	5,255	2,619	1,280
9. Hungary	9,303	9,234	5,333	1,294	1,610	997
10. Italy	30,123	29,402	12,465	5,136	6,346	5,455
11. Poland	31,268	30,454	14,901	4,046	8,684	2,823
12. Romania	23,750	23,034	10,497	4,467	6,337	1,733
13. Turkey	78,058	77,076	28,479	9,700	20,199	18,698
14. USSR	2,240,220	2,227,200	231,966	373,700	920,000	701,534
15. Yugoslavia	25,580	25,540	7,884	6,401	9,290	1,965

Distribution of the Inhabited Places in Terms of Their Population

Inhabited places with population	Inhabited places								Population							
	Number				Relative share				Number				Relative share			
	1975	1980	1982	1975	1980	1982			1975	1980	1982	1975	1980	1982		

Total

	1975	1980	1982	1975	1980	1982	1975	1980	1982
Total[1]	5,373	5,368	5,379	100.0	100.0	100.0	8,727,771	8,876,652	8,929,332	100.0	100.0	100.0
Below 200	1,483	1,626	1,684	27.6	30.2	31.2	114,872	124,837	127,821	1.3	1.4	1.4
200–499	1,239	1,302	1,318	23.0	24.3	24.6	416,252	436,740	440,001	4.8	4.9	4.9
500–999	1,184	1,120	1,121	22.0	20.9	21.2	850,677	809,095	819,945	9.7	9.2	9.2
1,000–1,999	882	803	754	16.4	15.0	14.0	1,214,960	1,110,604	1,046,567	13.9	12.5	11.7
2,000–4,999	412	347	321	7.7	6.5	6.0	1,178,851	999,820	925,575	13.5	11.3	10.3
5,000–9,999	85	76	77	1.6	1.4	1.4	581,170	506,062	514,552	6.7	5.7	5.8
10,000–24,999	52	56	56	1.0	1.0	1.0	779,900	844,770	864,005	8.9	9.5	9.7
25,000–99,999	29	31	30	0.6	0.6	0.5	1,537,766	1,741,030	1,710,901	17.6	19.6	19.2
100,000–499,999	6	6	7	0.1	0.1	0.1	1,086,109	1,246,749	1,397,650	12.5	14.0	15.7
500,000 and over	1	1	1	0.0	0.0	0.0	967,214	1,056,945	1,082,315	11.1	11.9	12.1

Towns

	1975	1980	1982	1975	1980	1982	1975	1980	1982	1975	1980	1982
Total	214	221	227	100.0	100.0	100.0	5,061,087	5,546,022	5,735,834	100.0	100.0	100.0
Below 1,000	2	3	2	1.0	0.9	0.9	1,362	1,241	1,025	0.0	0.0	0.0
1,000–1,999	3	6	7	1.4	2.7	3.1	5,128	10,284	11,092	0.1	0.1	0.2
2,000–4,999	48	53	56	22.4	23.9	24.7	176,535	200,199	206,395	3.5	3.6	3.6
5,000–9,999	73	66	68	34.1	30.0	30.0	507,073	444,804	462,451	10.0	8.0	8.1
10,000–24,999	52	56	56	24.3	25.3	24.7	779,900	844,770	864,005	15.4	15.2	15.1
25,000–99,999	29	31	30	13.5	14.0	13.1	1,537,766	1,741,030	1,710,901	30.4	31.5	29.7
100,000–499,999	6	6	7	2.8	2.7	3.1	1,086,109	1,246,749	1,397,650	21.5	22.4	24.4
500,000 and more	1	1	1	0.5	0.5	0.4	967,214	1,056,945	1,082,315	19.1	19.2	18.9

Villages

	1975	1980	1982	1975	1980	1982	1975	1980	1982	1975	1980	1982
Total[1]	5,159	5,147	5,152	100.0	100.0	100.0	3,666,684	3,330,630	3,193,498	100.0	100.0	100.0
Below 200	1,483	1,626	1,684	28.8	31.6	32.7	114,872	124,837	127,821	3.1	3.7	4.0
200–499	1,238	1,301	1,317	24.0	25.3	25.6	415,835	436,285	439,562	11.3	13.1	13.8
500–999	1,183	1,119	1,130	22.9	21.8	21.9	849,732	808,309	819,359	23.3	24.3	25.7
1,000–1,999	879	797	747	17.0	15.4	14.5	1,209,832	1,100,320	1,035,475	33.0	33.0	32.4
2,000–4,999	364	294	265	7.1	5.7	5.1	1,002,316	799,621	719,180	27.3	24.1	22.5
5,000–9,999	12	10	9	0.2	0.2	0.2	74,097	61,258	52,101	2.0	1.8	1.6
10,000–24,999	—	—	—	—	—	—	—	—	—	—	—	—

[1] Excluded are 64 inhabited places which around 2 December 1975 had no population. A State Council Decree has not deleted them from the list of inhabited places.

Towns with over 50,000 Inhabitants

Towns	1982	Towns	1982
1. Sofia	1,082,315	14. Plovdiv	367,195
2. Blagoevgrad	64,442	15. Razgrad	51,761
3. Burgas	178,239	16. Russe	178,920
4. Dimitrovgrad	50,688	17. Shumen	99,642
5. Gabrovo	80,901	18. Silistra	57,670
6. Haskovo	87,639	19. Sliven	100,637
7. Kazanluk	59,568	20. Stara Zagora	141,722
8. Kurdjali	55,762	21. Tolbuhin	98,857
9. Kyustendil	54,657	22. Varna	295,038
10. Mihailovgrad	54,240	23. Veliko Turnovo	64,985
11. Pazardjik	77,830	24. Vidin	60,877
12. Pernik	94,859	25. Vratsa	73,014
13. Pleven	135,899	26. Yambol	86,216

Main Rivers in the People's Republic of Bulgaria

River	Length (km)	Water-catching area (sq.km)	River	Length (km)	Water-catching area (sq.km)
Danube	471.0	46,930.0	Vurbitsa	98.1	1,203.0
Iskur	368.0	8,646.0	Dospat	96.2	633.5
Tundja	349.5	7,884.0	Lom	92.5	1,140.0
Maritsa	321.6	21,084.0	Harmanliiska	91.9	956.3
Ossum	314.0	2,824.0	Lefedja	91.8	2,424.0
Struma	290.0	10,797.0	Hursovska	90.8	997.3
Yantra	285.5	7,862.0	Topchiiska	88.6	659.8
Kamchia	244.5	5,358.0	Tsibritsa	87.5	933.6
Arda	241.3	5,201.0	Fakiiska	87.3	641.0
Luda Kamchia	200.9	1,612.0	Muchuritsa	85.9	1,278.0
Golyama Kamchia	198.5		Chepelarska	85.9	1,010.0
Russenski Lom	196.9	2,947.0	Malki Iskur	85.5	1,284.0
Vit	188.6	3,225.0	Djyulyunitsa	85.3	891.6
Rossitsa	164.3	2,265.0	Chepinska	82.7	899.6
Topolnitsa	154.8	1,789.0	Golyma Reka	74.7	663.3
Veleka	147.0	994.8	Luda Yana	74.0	685.3
Sazliika	145.4	3,293.0	Neikovsko Dere	72.0	430.8
Ogosta	144.1	3,157.0	Kalnitsa	71.9	577.3
Beli Lom	140.7	1,279.0	Popovska	71.6	532.9
Skut	134.0	1,074.0	Pyasuchnik	71.5	662.9
Cherni Lom	130.3	1,549.0	Ovcharitsa	71.5	636.1
Mesta	125.9	2,767.0	Tolbuhinska	70.4	516.3
Spuha Reka	126.0	2,404.0	Gospodarevska	70.2	422.0
Provadiiska	119.0	2,132.0	Dragovishtitsa	70.0	867.0
Strumeshnitsa	114.0	1,900.0	Byala	69.6	593.8
Rezovska	112.0	183.4	Vesselina	69.6	381.9
Vucha	111.5	1,645.0	Sredetska	69.0	985.3
Stryama	110.1	1,394.0	Botunya	68.9	731.7
Kanagyol	109.6	1,745.0	Vrana	67.6	937.6
Tsaratsar	108.0	1,062.0	Vidima	67.6	560.4
Senkovets	101.6	553.0	Topolovets	67.6	582.8

Mountains in Bulgaria, Main Peaks and Altitude

Mountains and Peaks	Altitude (m)	Mountains and Peaks	Altitude (m)
Rila		Shipka	1,326
Mussala	2,925	Chetiri Polyani	1,244
Dimitrov	2,902	Bulgarka	1,181
Malyovitsa	2,729	Razboina	1,128
Cherna Polyana	2,716	Vedernik	1,124
Golyam Skakavets	2,706	Babin Nos	1,108
Ibur	2,666	Lissets	1,073
Kovach	2,634	Gavanite	1,034
Kolarov	2,627	Ushite	1,011
Golyam Mechi Vruh	2,618		
Golyam Polich	2,615	**Vitosha**	
Kabul	2,531	Cherni Vruh	2,290
Chaushka Chuka	2,515	**Ossogoco**	
Maluk Polich	2,342	Ruen	2,251
Slavov Vruh	2,306	Beltok	1,524
Balabanitsa	2,157	Choveka	1,334
Zekiritsa	1,734		
		Rhodopes	
Pirin		Golyam Perelik	2,194
Vihren	2,915	Golyam Snezhnik	2,188
Polezhan	2,850	Golyama Syutka	2,186
Kamenitsa	2,816	Golyam Persenk	2,091
Uleven	2,645	Batashki Snezhnik	2,082
Sinanitsa	2,516	Malka Syutka	2,079
Gotsev Vruh	2,212	Persenk	2,074
Orelek	2,099	Prespa	2,000
Sveshtnik	1,973	Modur	1,992
		Sredniya Vruh	1,950
Balkan Range		Beslet	1,938
Botev	2,376	Srebren	1,901
Triglav	2,276	Devriklin	1,897
Vezhen	2,198	Tuzlata	1,891
Golyam Kupen	2,169	Kurbanieri	1,864
Midjur	2,168	Chernovets	1,834
Levski	2,166	Tsigansko Gradishte	1,827
Baba	2,071	Cherkovniya Vruh	1,816
Paskal	2,029	Ardin Vruh	1,730
Ravnets	2,021	Elvarnika	1,715
Kom	2,016	Velnitsa	1,712
Svishtiplaz	1,888	Djenevra	1,674
Todorini Kukli	1,785	Videnitsa	1,652
Haidushki Kamuk	1,721	Gyozchukarakol	1,637
Murgash	1,687	Milevi Skali	1,593
Koznitsa	1,637	Cherni Rid	1,545
Chukava	1,588	Sini Vruh	1,537
Gorno Yazovo	1,573	Snezhnik	1,482
Chumerna	1,536	Muglenik	1,266
Ispolin	1,524		
Atovo Padalo	1,495	**Belassitsa**	
Izdrimets	1,493	Radomir	2,029
Vassilyov Vruh	1,490	Tumba	1,880
Bedek	1,488		
Beglichka Mogila	1,482	**Vlahina**	
Surbenitsa	1,479	Kadiitsa	1,924
Hadji Dimiter	1,441	Ruen	1,134
Manyakov Kamik	1,439		
Golyam Klimash	1,355	**Ograzhden**	
		Markovi Kladentsi	1,523

Mountains in Bulgaria, Main Peaks and Altitude *(continued)*

Mountains and Peaks	Altitude (m)	Mountains and Peaks	Altitude (m)
Sredna Gora		Kaleto	1,263
Golyam Bogdan	1,604	Bratan	1,236
Bunaya	1,572	Golyamata Ikuna	1,221
Bratya	1,519	Pleshivets	1,204
Lissets	1,386	Popov Dol	1,190
Ostra	1,289	Benkovski	1,186
Truna	1,275	Brunkova Kitka	1,037

Indices of Total Industrial Output (1975=100)

Countries	1976	1980	1982
Austria	106	125	122
Bulgaria	107	134	146
Czechoslovakia	106	126	130
France	109	117	112
FRG	109	118	115
GDR	106	127	137
Great Britain	103	107	104
Hungary	104	118	124
Italy	112	130	125
Poland	109	126	110
Romania	111	158	163
USSR	105	124	132
Yugoslavia	104	139	—

Per Capita Production of Some Industrial Commodities in 1982

Countries	Electric power (kWh)	Cast iron and ferrous alloys (kg)	Steel (kg)	Nitrogen fertilizers (100 per cent N)[1] (kg)	Sulphuric acid (kg)	Cement (kg)
Austria	5,867	412	620	31.8		663
Bulgaria	4,531	181	289	84.7	102.6	629
Czechoslovakia	4,860	620	975	44.0	77.0	672
France	4,837	279	340	29.4	76.3	483
FRG	5,980[1]	450	590	18.0	71.8	487
GDR	6,147	128	428	57.8	55.0	700
Great Britain	4,853	152	162	22.7	46.2	214
Greece	2,150	36[2]	113[2]	31.8	102.3	1,350
Hungary	2,290	205	346	63.9	55.6	408
Italy	3,269	206	425	20.9	42.0[1]	736[1]
Poland	3,246	200	400	35.2	74.0	443
Romania	3,065	396	581	81.5	81.2[1]	620
Turkey	550[1]	45[2]	50	15.3	15.2	339
USSR	5,059	395	545	39.3	88.3	458
Yugoslavia	2,734	127	104[1]	18.6	52.1	428

[1] 1981.
[2] 1980.

Statistics of the Country's Social and Economic Development

Industrial Production	1939	1948	1952	1957	1960	1965	1970	1975	1980	1982
Electricity (million kWh)	266	550	1,352	2,656	4,657	10,244	19,513	25,235	34,831	40,455
Coal: total output (thousand tons)	2,214	4,266	7,410	11,889	17,147	26,254	31,411	28,849	31,571	33,550
Iron ore: content of iron (thousand tons)	11	10	54	134	188	585	792	775	590	474
Manganese ore (thousand tons)	2	—	13	81	25	42	33	35	49	45
Pig-iron and ferrous alloys (thousand tons)	—	1	9	54	192	695	1,251	1,590	1,583	1,617
Steel (thousand tons)	6	5	6	159	253	588	1,800	2,265	2,565	2,584
Rolled ferrous metals (thousand tons)	4	3	11	117	193	431	1,420	2,495	3,213	3,253
Internal combustion engines (number)	—	54	1,059	2,410	10,657	23,073	40,958	19,969	14,200	24,104
in terms of hp units	—	0.4	10.9	61.4	154.5	172.7	228.9	365.5	814.3	1,600.5
Electric engines (thousand sets)	—	3.0	13.5	111.0	136.0	497.1	750.5	947.3	1,251.4	1,337.5
Thousand kWh	—	5	125	200	919	2,864	4,194	5,117	7,111	6,446
Power transformers: number	17	187	1,087	1,798	3,924	4,224	4,807	6,179	7,645	8,621
thousand kWh	—	33	303	635	1,172	2,672	3,256	3,924	4,136	4,459
Storage battries — (thousand items)	—	6	14	59	364	2,271	3,274	4,088	4,003	4,431
Synthetic ammonium: 100% nitrogen (thousand tons)	—	—	—	53.2	110.8	337.7	788.2	934.6	1,008.8	1,032.6
Nitrogen fertilizers: without urea; 100% nitrogen (thousand tons)	—	—	24.0	40.1	83.6	245.8	286.8	379.9	436.0	459.0
Phosphates: 100% P$_2$O$_5$ (thousand tons)	—	—	18.2	3.2	40.9	93.7	147.6	245.9	216.6	238.9
Soda ash: 98% (thousand tons)	—	—	—	92	131	227	306	1,009	1,479	1,459
Sulphuric acid: monohydrate (thousand tons)	—	—	12.1	40.4	122.6	317.6	502.5	853.6	851.9	915.9
Caustic soda: 96% (tons)	8	64	148	12,934	17,783	33,749	47,977	88,018	168,369	161,292
Automobile tyres (thousand items)	2.9	13.8	50.3	103.9	172.1	327.0	546.2	1,227.3	1,531.7	1,576.9
Cement (thousand tons)	225	378	672	880	1,586	2,681	3,668	4,358	5,359	5,614

Statistics of the Country's Social and Economic Development

Indices of Production of some industrial articles (1939=100)	1939	1948	1952	1957	1960	1965	1970	1975	1980	1982
Pulp and paper industry	100	208	325	629	932	1,600	3,300	5,800	7,100	7,800
Glassware, porcelain and faience industry	100	325	817	2,000	4,200	10,100	18,700	28,000	38,400	42,100
Textile and tailoring industry	100	188	405	775	1,100	1,500	2,400	3,300	4,100	4,500
Fur, leather and shoe industry	100	275	765	1,000	1,200	1,700	3,000	4,100	4,600	5,100
Printing and publishing industry	100	165	229	348	471	595	1,000	1,300	2,100	2,200
Food and beverages industry	100	164	329	466	653	1,100	1,400	1,900	2,200	2,500
Indices of total per capita production	100	178	370	640	976	1,600	2,600	4,000	5,200	5,700

Production of Coal (Brown and Lignite; thousands of tons)

Countries	1970	1975	1982
Albania	606	850	n.a.
Austria	3,670	3,397	3,300
Bulgaria	28,854	27,515	31,974
Czechoslovakia	81,298	86,272	97,100
France	2,785	3,186	n.a.
FRG	107,766	123,377	127,308
GDR	261,482	246,706	276,000
Greece	8,081	18,408	27,288
Hungary	23,679	21,867	23,000
Italy	1,393	1,213	1,908
Poland	32,766	39,865	37,600
USSR	144,745	160,216	159,000
Turkey	4,437	6,939	n.a.
Yugoslavia	27,779	34,939	n.a.

Steel Production (thousands of tons)

Countries	1970	1975	1982
Austria	4,079	4,066	4,692
Bulgaria	1,800	2,265	2,584
Czechoslovakia	11,480	14,323	14,991
France	23,773	21,530	18,420
FRG	45,040	40,414	36,348
GDR	5,053	6,472	7,169
Great Britain	28,316	20,098	9,084
Hungary	3,108	3,673	3,703
Poland	11,795	15,004	14,477
Romania	6,517	9,549	13,055
Turkey	1,312	1,457	2,316[1]
USSR	115,889	141,344	147,145
Yugoslavia	2,228	2,916	n.a.

[1] Only ingots.

Power Output (millions of kWh)

Countries	1970	1975	1982
Albania	944	1,800	n.a.
Austria	30,036	35,205	44,412
Bulgaria	19,513	25,235	40,455
Czechoslovakia	45,163	59,277	74,700
France[1, 3]	146,966	185,312	262,248
FRG	242,605	301,802	n.a.
GDR	67,650	84,505	102,900
Great Britain	249,016	271,987	272,328
Greece	9,820	16,147	21,048
Hungary	14,542	20,465	24,500
Italy[2]	117,423	147,333	184,000
Poland	64,532	97,169	117,600
Romania	35,088	53,721	68,900
Turkey	8,624	15,569	26,496
USSR	740,926	1,038,607	1,366,000
Yugoslavia	26,024	40,040	62,100

[1] Net output (producer consumption and transmission losses excluded).
[2] Including San Marino.
[3] Including Monaco.

Production of Iron Ore (Metal Content; thousand tons)

Countries	1970	1975	1981
Austria	1,304	1,201	945
Bulgaria	792	775	537
Czechoslovakia	447	468	522
France	17,759	15,309	6,473
FRG	1,773	1,053	515
Great Britain	3,365	1,212	190
Greece	380	817	547
Hungary	160	153	88
Italy	318	259	53
Poland	707	376	32
Romania	881	794	624
Turkey	1,663	1,300	1,642
USSR	127,870	155,039	166,335
Yugoslavia	1,301	1,928	1,676

Bulgaria in Figures

Production of Cast Iron and Ferrous-Alloys (thousand tons)

Countries	1970	1975	1982
Austria	2,970	3,098	3,120
Bulgaria	1,251	1,560	1,617
Czechoslovakia	7,548	9,281	9,525
France	19,575	18,381	15,132
FRG	33,897	30,330	27,744
GDR	1,994	2,456	2,149
Great Britain[1]	19,023	12,262	8,508
Hungary	1,828	2,219	2,191
Italy	8,525	11,941	11,604
Poland	6,984	7,752	7,244
Romania	4,210	6,602	n.a.
Turkey	1,176	1,367	n.a.
USSR	85,933	102,968	106,724
Yugoslavia	1,377	2,196	2,880

[1] Ferrous-alloys produced in electric furnaces are excluded.

Production of Manganese Ore (Metal Content; thousand tons)

Countries	1970	1975	1980
Bulgaria	10.3	10.0	14.2
Greece[1]	3.3	6.1	2.5
Hungary	34.8	27.8	23.3
Italy	12.5	n.a.	2.7
Romania	26.5	27.8	53.0
Turkey	5.5	11.6	11.9
USSR	2,446.1	2,951.4	3,039.6
Yugoslavia	4.6	5.9	10.6

[1] Including concentrates.

Production of Sulphuric Acid (100% H_2SO_4; thousand tons)

Countries	1970	1975	1982
Bulgaria	502	854	916
Czechoslavakia	1,110	1,245	1,183
France	3,682	3,758	4,136
FRG	4,435	4,157	4,428
GDR	1,099	1,002	920
Great Britain	3,352	3,166	2,592
Greece	623	920	1,002
Hungary	471	647	595
Italy	3,327	3,006	n.a.
Poland	1,901	3,413	2,682
Romania	994	1,448	n.a.
Turkey	28	299	n.a.
USSR	12,059	18,645	23,841
Yugoslavia	696	871	1,183

Production of Nitrogen Fertilizers (Nitrogen Content; thousand tons)

Countries	1970	1975	1981
Austria	219	256	239
Bulgaria	602	673	753
Czechoslovakia	352	496	674
France	1,351	1,361	1,588
FRG	1,505	1,259	1,108
GDR	395	538	967
Great Britain	748	1,055	1,270
Hungary	350	436	684
Italy	956	1,000	1,195
Poland	1,030	1,533	1,274
Romania	647	1,292	1,822
Turkey	82	171	695
USSR	5,423	8,535	10,705
Yugoslavia	266	358	419

Production of Caustic Soda (100% NaON; thousand tons)

Countries	1970	1975	1982
Bulgaria	46	85	155
Czechoslovakia	189	257	324
France	1,094	1,120	1,402
FRG	1,682	2,489	3,208
GDR	413	442	695
Hungary	63	81	167
Italy	999	1,002	n.a.
Poland	313	392	363
Romania	330	566	n.a.
Turkey	4[1]	25	41
USSR	1,783	2,395	2,783
Yugoslavia	90	94	169

[1] Production of state-owned plants only.

Production of Soda Ash (100% Na_2CO_3; thousand tons)

Countries	1970	1975	1980
Bulgaria	300	989	1,449
Czechoslovakia	104	121	123
France	1,419	1,279	1,560
FRG	1,334	1,249	1,411
GDR	676	818	866
Poland	644	716	762
Romania	582	693	937
USSR	3,485	4,962	4,780
Yugoslavia	113	147	129

Consumption of Chemical Fertilizers (in Pure Substance; thousand tons)

Countries	1970–71	1975–76	1981–82 Total	Kg per ha of arable landscape
Austria	408	313	394	241
Bulgaria	639	679	1,044	250
Czechoslovakia	1,282	1,684	1,720	333
France	4,651	4,686	5,789	311
FRG	3,228	3,107	3,131	418
GDR	1,554	1,826	1,726	343
Great Britain	1,878	1,820	2,301	329
Greece	337	463	525	134
Hungary	837	1,518	1,485	279
Italy	1,338	1,490	2,029	163
Poland	2,572	3,671	3,686	247
Romania	594	1,197	1,618	154
Turkey	430	891	1,294	45
USSR	10,368	17,243	19,167	83
Yugoslavia	632	720	1,010	128

Production of Plastics and Artificial Resins (thousand tons)

Countries	1970	1975	1982
Bulgaria	89	156	337
France	1,548	2,016	n.a.
FRG	4,321	5,046	6,800
GDR	370	605	985
Great Britain	1,469	2,057	2,090
Hungary	56	124	326
Italy	1,567	2,053	2,100
Poland	224	431	466
Romania	206	347	591
USSR	1,670	2,838	3,317
Yugoslavia	79	160	409

Production of Synthetic Fibres (thousand tons)

Countries	1970	1975	1981
Austria	79.3	84.2	n.a.
Bulgaria	n.a.	21.5	42.8
Czechoslovakia	69.2	71.0	55.1
France[1]	130.7	83.7	66.0
FRG[1]	226.3	115.7	148.8
GDR	167.5	170.1	157.0
Great Britain	262.5	201.3	147.6
Greece	4.1	5.9	7.4
Hungary	4.0	8.8	7.5
Italy	182.8	89.8	60.1
Poland	84.3	97.9	65.7
Romania	34.3	48.8	44.5
Turkey	1.4	4.0	n.a.
USSR	456.4	590.4	604.0
Yugoslavia	32.1	63.9	64.1

[1] Including cigarette filters.

Production of Cotton Yarn (Pure and Mixed; thousand tons)

Countries	1970	1975	1981
Austria	20.8	16.4	18.0
Bulgaria	73.7	78.9	87.3
Czechoslovakia	113.9	129.1	137.1
France[1]	269.8	183.9	217.2
FRG	238.6	192.3	146.4
GDR	132.6	123.0	134.5
Great Britain	184.0	125.4	84.0
Greece	41.9	79.4[1]	110.4
Hungary	65.1	67.6	65.3
Italy	215.1	200.4	160.8
Poland	207.8	212.3	196.3
Romania	115.7	153.4	198.9
Turkey[2]	178.0	127.0	127.3
USSR	1,434.8	1,572.6	1,645.3
Yugoslavia	102.3	107.2	117.6

[1] Pure cotton yarn.
[2] Since 1975 – only factory-made.

Production of Woollen Yarns (Pure and Mixed; thousand tons)

Countries	1970	1975	1981
Austria	7.8	5.2	9.6
Bulgaria	24.3	31.5	36.9
Czechoslovakia	45.5	52.6	57.3
France[1]	75.3	58.8	n.a.
FRG	78.8	51.4	51.8
GDR	68.2	62.4	74.8
Great Britain	227.0	187.5	130.8
Greece[1]	16.6	15.4	n.a.
Hungary	20.6	22.0	27.0
Poland	84.4	102.7	88.2
Romania	35.7	50.8	75.0
Turkey	30.0	10.5	4.0[2]
USSR	350.2	416.9	452.7
Yugoslavia	37.9	41.8	51.2

[1] Since 1975 pure woollen yarns.
[2] Production in the state sector only.

Production of Cotton Fabrics (millions of square metres)

Countries	1970	1975	1980
Austria	102	71	84
Bulgaria[1]	279	341	340
Czechoslovakia[1]	518	551	548
France	1,418	1,088	986
FRG[1]	930	781	791
GDR[1]	445	473	486
Great Britain	733	462	370
Greece	137	217	n.a.
Hungary[1]	304	352	332
Italy	831	1,153	1,469
Poland[1]	847	953	960
Romania[1]	437	572	732
USSR[1]	6,152	6,634	7,068
Yugoslavia[2]	390	376	386

[1] Finished textiles.
[2] Including fabrics from artificial fibres.

Production of Woollen Textiles (million cubic metres)

Countries	1970	1975	1980
Austria	21.1	20.2	19.9
Bulgaria[2]	37.2	54.8	59.6
Czechoslovakia[2]	80.9	84.7	90.0
France	94.4	82.6	62.4
FRG[2]	126.2	99.4	109.2
GDR[2]	123.7	106.7	107
Great Britain	215.2	151.4	118.2
Greece	10.9	8.9	n.a.
Hungary[2]	36.7	36.2	39.3
Poland[2]	138.3	181.0	181.9
Romania[2]	58.3	78.1	95.9
Turkey[1]	7.7	27.9	n.a.
USSR[2]	642.9	740.0	761.9
Yugoslavia[3]	56.9	66.1	91.9

[1] Production in the state sector only.
[2] Finished textiles.
[3] Including fabrics of synthetic fibres.

Statistics of the Country's Social and Economic Development

	1939	1948	1952	1957	1960	1965	1970	1975	1980	1982
Dressed and cleft timber (thousand cu.m)	1,002	1,772	2,081	2,808	3,858	4,343	3,950	3,720	3,280	3,126
Plywood (thousand cu.m)	7.4	17.0	28.4	42.3	64.3	81.4	70.7	65.0	56.8	56.5
Cellulose (thousand tons)	—	—	0.1	15.9	21.3	65.9	77.3	179.0	215.6	212.3
Paper (thousand tons)	14.5	22.0	30.0	35.7	53.9	85.2	199.7	283.0	316.1	353.6
Pane glass 2 mm (million sq.m.)	1.0	1.4	3.1	3.6	7.6	17.4	20.0	22.5	24.5	24.9
Cotton textiles (million m.)	34.1	58.3	105.9	153.0	218.4	291.3	318.8	371.8	346.1	366.6
Woollen textiles (million m.)	5.3	5.4	9.2	13.3	18.7	20.1	26.8	37.9	39.5	37.3
Silk textiles (million m.)	1.1	1.5	3.9	6.4	10.7	15.1	21.3	31.4	33.6	35.9
Hides and skins (thousands tons)	3.1	2.1	3.3	3.8	4.5	4.1	3.4	2.5	1.6	1.4
Upper leathers (million sq.dm.)	12	42	96	137	184	250	358	472	505	539
Leathers (million sq.dm.)	2	27	60	85	113	169	168	216	205	213
Meat (thousand tons)	49.7	25.5	66.6	118.0	148.0	207.2	270.5	399.5	462.3	491.0
Butter (thousand tons)	0.7	0.6	3.1	5.4	11.0	11.9	14.1	16.0	19.8	23.0
White cheese (thousand tons)	13.3	8.0	18.2	27.5	42.8	63.2	70.4	76.6	92.5	98.3
Yellow cheese (thousand tons)	3.5	3.0	3.6	3.4	9.6	8.9	13.9	16.8	16.8	25.3
Vegetable oils (thousand tons)	39.8	29.0	64.1	60.6	88.3	115.4	160.6	149.1	175.1	167.8
Incl. edible oils	34.6	22.7	52.3	52.8	78.1	98.3	146.9	136.7	156.4	161.8
Tobacco products (thousand tons)	3.9	7.4	8.1	12.5	13.6	32.1	55.1	71.4	85.2	88.1
Indices of per capita labour productivity in the state and co-operative industrial enterprises (of blue and white collar employees; 1948=100)	100	100	148	191	224	312	434	601	775	829
Total production in construction – in million levs according to prices of the respective year	364.0	206.5	420.1	527.5	913.0	1,310.6	2,368.2	3,124.9	5,311.7	5,792.8
Indices of total production of the building industry (1952=100)	31	49	100	141	249	356	628	828	1,100	1,200
Indices of total agricultural production (1939=100)										
Crop raising	100	99	95	129	154	173	205	222	214	254
Stock-breeding	100	109	106	120	149	189	222	284	328	344

Statistics of the Country's Social and Economic Development

	1939	1948	1952	1957	1960	1965	1970	1975	1980	1982
Livestock (thousands)										
Cattle. including cows	1,495	1,783	1,638	1,442	1,452	1,450	1,279	1,656	1,796	1,783
Pigs	253	703	592	547	547	581	589	670	702	703
Sheep	743	1,078	1,337	1,993	2,553	2,408	2,369	3,889	3,808	3,810
Poultry	9,028	9,266	7,759	7,742	9,333	10,312	9,678	10,014	10,433	10,761
	—	11,380	12,610	14,302	23,366	20,845	33,706	38,061	41,636	42,853
Animal products:										
Meat (animals sold for slaughtering and slaughtered on the farms in carcass weight including edible offal; thousands of tons)	202	208	199	270	307	464	476	657	781	807
Milk (millions of litres)	669	733	619	885	1,081	1,346	1,583	1,749	2,151	2,350
Unwashed wool (thousands of tons)	13	14	13	15	21	26	29	34	35	35
Eggs (million of items)	744	633	750	871	1,221	1,449	1,617	1,851	2,434	2,489
Productivity of livestock:										
Average milk yield –										
per milk cow	450	403	438	922	1,402	1,689	2,147	2,198	2,589	2,859
per milk ewe	42	42	43	41	41	39	46	45	45	47
Average wool yield per sheep (kg)	1,511	1,607	1,784	1,980	2,417	2,486	3,345	3,820	3,881	3,906
Average number of eggs per hen (items)	73	74	75	77	91	98	112	126	138	147

Statistics of the Country's Social and Economic Development

Per capita consumption of some alimentary and non-alimentary products per annum	1952	1957	1960	1965	1970	1975	1980	1982
Meat (kg)	21.3	27.5	29.1	39.6	41.4	58.0	61.2	68.3
Vegetable and animal fats (kg)	9.3	11.4	14.0	15.3	16.2	18.7	20.8	21.9
Milk (litres)	80.2	84.9	92.3	103.6	116.6	142.8	169.4	79.2
Eggs (number)	68	76	84	100	122	146	204	217

Indices of Total Agricultural Produce (1969–71=100)

Countries	1970	1975	1982
Austria	99	108	123
Bulgaria	99	109	122
Czechoslovakia	97	114	122
France	100	108	131
FRG	100	102	117
GDR	100	118	135
Great Britain	102	107	124
Greece	103	127	143
Hungary	91	127	146
Italy	100	109	122
Poland	102	115	101
Romania	90	127	164
Turkey	99	121	148
USSR	102	108	113
Yugoslavia	93	116	136

Meat[1] (thousand tons)

Countries	1969–71	1975	1982
Austria	475	544	653
Bulgaria	413	553	673
Czechoslovakia	1,043	1,304	1,367
France	4,079	4,832	5,586
FRG	3,823	4,064	4,459
GDR	1,300	1,721	1,846
Great Britain	2,654	2,933	2,890
Greece	313	466	531
Hungary	950	1,297	1,537
Italy	2,470	2,843	3,656
Poland	2,026	2,786	2,059
Romania	883	1,337	1,788
Turkey	649	799	967
USSR	12,438	15,060	15,337
Yugoslavia	961	1,268	1,465

[1] All kinds in slaughtered weight, excluding viscera and fat.

Milk (thousand tons)

Countries	1969–71	1975	1982
Austria	3,248	3,190	3,547
Bulgaria	1,634	1,803	2,411
Czechoslovakia	4,978	5,562	5,986
France[2]	28,514	30,910	35,956
FRG[2]	21,784	21,628	25,372
GDR	6,830	7,458	6,821
Great Britain[1]	13,007	13,937	16,400
Greece	1,353	1,691	1,733
Hungary	1,726	1,835	2,733
Italy	10,080	10,032	11,414
Poland[2]	14,988	16,395	15,203
Romania	3,912	4,581	4,361
Turkey	4,308	4,817	5,884
USSR	83,016	90,804	90,056
Yugoslavia	2,759	3,803	4,601

[1] Cow's milk.
[2] Including milk suckled by young animals.

Sheep[1] (thousands)

Countries	1969–71	1975	1982
Albania	1,249	1,163	n.a.
Bulgaria	9,518	10,014	10,761
Czechoslovakia	955	805	990
France	10,023	10,568	13,121
FRG	838	1,040	1,108
GDR	1,696	1,883	2,198
Great Britain	26,332	28,364	33,049
Greece	7,646	8,274	7,900
Hungary	2,986	2,039	3,203
Italy	8,097	7,995	9,632
Poland	3,236	3,178	3,676
Romania	13,984	13,865	16,828
Turkey	36,470	40,539	49,598
USSR	136,434	141,436	142,059
Yugoslavia	9,136	8,175	7,398

[1] The statistics refer to twelve-month periods ending on 30 September of the year indicated, and for the socialist countries – at the end of the year.

Bulgaria in Figures

Pigs[1] (thousands)

Countries	1969-71	1975	1982
Austria	3,245	3,517	4,010
Bulgaria	2,159	3,889	3,810
Czechoslovakia	5,234	6,683	7,126
France	10,516	12,031	11,859
FRG	19,675	20,234	22,310
GDR	9,481	11,501	12,107
Great Britain	8,205	7,540	8,082
Greece	407	761	n.a.
Hungary	6,271	6,953	9,035
Italy	8,501	8,814	9,132
Poland	14,348	21,647	17,558
Romania	6,061	8,813	n.a.
USSR	57,528	57,899	76,523
Yugoslavia	5,733	7,683	8,431

[1] Per twelve-month periods ending on 30 September of the year indicated (by the year's end for the socialist countries).

Cattle[1] (in thousands)

Countries	1969-71	1975	1982
Albania	416	427	n.a.
Austria	2,440	2,581	2,530
Bulgaria	1,357	1,725	1,822
Czechoslovakia	4,253	4,555	5,131
France	21,669	24,119	23,605
FRG	14,124	14,430	14,992
GDR	5,157	5,532	5,690
Great Britain	12,632	14,764	13,275
Greece	1,014	1,247	n.a.
Hungary	1,952	1,904	1,922
Italy	9,486	8,242	9,012
Poland	10,990	12,764	11,037
Romania	5,129	6,126	6,082
Turkey	14,419	14,410	16,983
USSR	97,172	111,034	117,066
Yugoslavia	5,194	5,938	5,222

[1] The statistics refer to twelve-month periods ending on 30 September of the year indicated, and for the socialist countries – at the end of the year.

Eggs (thousand tons)

Countries	1969-71	1975	1982
Austria	86	88	98
Bulgaria	89	102	133
Czechoslovakia	186	225	245
France	644	761	906
FRG	884	893	800
GDR	258	278	342
Great Britain	878	799	750
Greece	100	103	122
Hungary	176	222	260
Italy	549	671	685
Poland	387	449	448
Romania	165	249	310
Turkey	98	130	245
USSR	2,250	3,176	4,000
Yugoslavia	137	180	232

Poultry[1] (thousands)

Countries	1969-71	1975	1981
1. Albania			
Total of poultry	2,023	2,255	2,484
2. Austria			
Total of poultry	11,861	12,445	14,430
including hens	11,658	12,250	14,160
3. Bulgaria			
Total of poultry	28,695	38,061	40,563
including hens	27,020	35,891	38,765
4. Czechoslovakia			
Total of poultry	34,797	40,130	47,388
including hens	34,007	38,700	45,300
5. France			
Total of poultry	175,047	202,514	211,039
including hens	164,264	n.a.	185,965
6. FRG			
Total of poultry	96,914	91,239	88,165
including hens	94,673	89,398	85,461
7. GDR			
Total of poultry	41,467	47,122	54,392
8. Great Britain			
Total of poultry	136,401	136,676	137,280
including hens	130,487	130,471	122,639
9. Greece			
Total of poultry	24,412	30,366	30,265
including hens	23,981	30,053	30,000
10. Hungary			
Total of poultry	60,165	56,313	66,348
including hens	56,883	53,389	63,629
11. Italy			
Total of poultry	109,000	113,776	110,000
12. Poland			
Total of poultry	87,304	97,130	69,780
including hens	74,931	88,767	65,482
13. Romania			
Total of poultry[2]	51,948	78,626	109,200
14. Turkey			
Total of poultry	34,200	41,006	61,449
including hens	32,213	38,660	53,584
15. USSR			
Total of poultry[2]	575,899	734,400	1,071,200
16. Yugoslavia			
Total of poultry	38,952	53,396	68,262
including hens	36,576	50,591	65,187

[1] The statistics refer to a twelve-month period, ending on 30 September of the the year indicated and for the socialist countries – towards the end of the year. They cover the number of hens, ducks and turkeys.
[2] Including geese.

Statistics of the Country's Social and Economic Development

Average yields of some crops – kg per ha	1939	1948	1952	1957	1960	1965	1970	1975	1980	1982
Wheat	1,310	1,155	1,427	1,659	1,900	2,548	2,990	3,287	3,973	4,634
Rye	1,086	984	1,142	972	1,042	1,103	1,255	1,087	1,410	1,454
Barley	1,492	1,273	1,602	1,874	2,087	2,347	2,891	2,955	3,229	4,077
Maize – grain	1,359	1,118	688	1,943	2,359	2,207	3,721	3,416	3,845	5,484
Dry beans	825	641	331	681	1,156	503	967	1,024	691	960
Sunflower seed	960	900	800	1,118	1,452	1,337	1,458	1,790	1,534	2,010
Raw cotton	689	453	253	699	812	827	859	1,194	947	1,578
Oriental tobacco	956	587	442	829	698	994	1,029	1,320	1,115	1,430
Sugar beet	17,668	16,016	8,968	23,574	24,525	20,640	30,951	23,050	26,692	27,263
Tomatoes	20,493	20,018	23,122	25,701	29,049	33,907	27,616	20,392	28,606	29,413
Potatoes	7,795	6,381	6,373	9,371	9,979	6,959	11,782	10,457	8,541	11,512
Apples	7,974	10,440	8,310	2,544	8,653	6,796	7,237	8,948	11,375	12,612
Dessert grapes	6,597	5,386	4,512	5,883	4,754	10,217	5,690	5,541	5,435	8,697
Wine grapes	5,116	3,170	3,001	4,220	3,994	6,319	5,038	4,270	5,611	7,569

Rice (Unhusked)

Countries	1969–71	1975	1981
Harvested area (thousands of hectares)			
Bulgaria[1]	17.2	16.8	16.4
France	22.0	10.0	5.0
Greece	17.0	20.0	16.0
Hungary[1]	24.0	27.0	13.1
Italy	172.0	174.0	169.0
Romania[1]	28.0	21.9	20.0
Turkey	63.0	55.0	70.0
USSR[1]	356.0	500.0	634.3
Yugoslavia	7.0	8.0	8.0
Output (thousands of tons)			
Bulgaria	66	68	74
France	88	42	21
Greece	84	102	74
Hungary	54	69	39
Italy	858	1,010	837
Romania	67	69	49
Turkey	257	240	290
USSR	1,272	2,010	2,400
Yugoslavia	32	37	40
Average yield (kg per hectare)			
Bulgaria	3,850	4,030	4,470
France	4,060	4,160	4,000
Greece	4,830	5,100	4,630
Hungary	2,250	2,540	2,990
Italy	4,980	5,810	4,950
Romania	2,370	3,130	2,450
Turkey	4,100	4,380	4,140
USSR	3,570	4,020	3,790
Yugoslavia	4,260	4,630	5,000

[1] Sown area.

Sunflower Seed

Countries	1969–71	1975	1982
Harvested area (thousands of hectares)			
Bulgaria[1]	277	238	253
Czechoslovakia[1]	2	3.8	25
France	31	72	269
Greece	2	2	3
Hungary[1]	98	129	290
Italy	4	25	45
Romania[1]	562	511.1	496
Turkey	347	418	425
Yugoslavia	199	194	140
Output (thousands of tons)			
Bulgaria	471	426	511
Czechoslovakia	4	5	40
France	55	110	608
Greece	2	3	5
Hungary	122	155	530
Romania	769	728	700
Turkey	383	488	620
USSR	6,055	4,993	5,200
Yugoslavia	334	272	227
Average yield (kg per hectare)			
Bulgaria	1,700	1,790	2,010
Czechoslovakia	1,620	1,430	1,600
France	1,770	1,530	2,260
Greece	1,040	1,500	1,580
Hungary	1,240	1,190	1,830
Italy	2,010	1,800	1,860
Romania	1,370	1,430	1,410
Turkey	1,100	1,170	1,460
USSR	1,290	1,230	1,180
Yugoslavia	1,680	1,400	1,620

[1] Sown area.

Wheat

Countries	1969–71	1971	1982
	Harvested area (thousands of hectares)		
Austria	279	270	289
Bulgaria[1]	1,013.9	911.5	1,059.5
Czechoslovakia[1]	1,076	1,182.8	1,073
France	3,892	3,876	4,844
FRG	1,511	1,569	1,578
GDR[1]	597	688.5	591
Great Britain	980	1,035	1,664
Greece	1,010	920	1,030
Hungary[1]	1,289	1,251	1,310
Italy	4,089	3,545	3,327
Poland[1]	2,004	1,842.1	1,456
Romania[1,2]	2,527	2,385.7	2,175
Turkey	8,732	9,309	9,250
USSR[1]	65,230	61,985	57,278
Yugoslavia	1,928	1,616	1,559
	Output (thousands of tons)		
Austria	912	945	1,236
Bulgaria	3,032	2,996	4,913
Czechoslovakia	3,436	4,202	4,606
France	14,112	15,013	25,342
FRG	6,268	7,014	8,632
GDR	2,203	2,736	2,734
Greece	1,867	2,140	2,992
Hungary	3,410	4,005	5,747
Italy	9,756	9,610	8,998
Poland	4,925	5,207	4,476
Romania[2]	4,433	4,912	6,505
Turkey	11,423	14,830	17,650
USSR	92,804	66,224	84,000
Yugoslavia	4,760	4,408	5,239
	Average yield (kg per hectare)		
Austria	3,270	3,510	4,480
Bulgaria	2,990	3,290	4,630
Czechoslovakia	3,190	3,570	4,290
France	3,630	3,870	5,230
FRG	4,150	4,470	5,480
GDR	3,690	3,970	4,630
Great Britain	4,220	4,530	6,170
Greece	1,850	2,330	2,910
Hungary	2,640	3,200	4,390
Italy	2,390	2,710	2,710
Poland	2,460	2,830	3,070
Romania	1,750	2,060	2,970
Turkey	1,310	1,590	1,910
USSR	1,420	1,070	1,500
Yugoslavia	2,470	2,730	3,360

[1] Sown area.
[2] Including rye.

Potatoes

Countries	1969–71	1975	1982
	Harvested area (thousands of hectares)		
Austria	109	69	46
Bulgaria[1]	30.8	30	40.3
Czechoslovakia[1]	331	251.4	199
France	376	311	215
FRG	580	415	265
GDR[1]	643	574.2	504
Great Britain	270	214	191
Greece	55	57	49
Hungary[1]	169	100	84
Italy	277	179	148
Poland	2,707	2,580.9	2,178
Romania[1]	310	288.9	286
Turkey	160	178	180
USSR[1]	8,019	7,912	6,856
Yugoslavia	328	314	291
	Output (thousands of tons)		
Austria	2,787	1,579	1,121
Bulgaria	374	318	469
Czechoslovakia	4,864	3,565	3,608
France	8,569	6,642	6,750
FRG	15,804	10,853	7,821
GDR	10,432	7,673	8,820
Great Britain	7,359	4,551	6,550
Greece	700	878	888
Hungary	1,874	1,630	1,464
Italy	3,632	2,943	2,680
Poland	45,012	46,429	31,951
Romania	2,671	2,716	4,533[2]
Turkey	1,984	2,490	2,992
USSR	93,739	88,703	78,047
Yugoslavia	3,020	2,394	2,774
	Average yield (kg per hectare)		
Austria	25,510	22,850	24,550
Bulgaria	11,780	10,460	11,510
Czechoslovakia	14,680	14,240	18,100
France	22,780	21,380	31,400
FRG	27,250	26,130	29,540
GDR	16,220	13,360	17,500
Great Britain	27,220	21,270	34,290
Greece	12,730	15,400	17,990
Hungary	11,100	12,640	17,500
Italy	13,130	16,470	18,070
Poland	16,630	18,000	14,700
Romania	8,630	8,910	16,700
Turkey	12,380	13,990	16,670
USSR	11,690	11,200	11,400
Yugoslavia	9,200	7,620	9,530

[1] Sown area.
[2] Autumn potatoes.

Maize

Countries	1969–71	1975	1982
	Harvested area (thousands of hectares)		
Austria	122	144	198
Bulgaria[1]	635.3	652.1	620.9
Czechoslovakia	127	158.5	180
France	1,436	1,960	1,630
FRG	99	96	160
Greece	162	127	164
Hungary[1]	1,272	1,423	1,155
Italy	986	897	1,009
Romania[1]	3,170	3,305	n.a.
Turkey	646	597	590
USSR[1]	3,617	2,652	4,162
Yugoslavia	2,391	2,363	2,236
	Output (thousands of tons)		
Austria	677	981	1,446
Bulgaria	2,375	2,822	3,418
Czechoslovakia	511	843	941
France	7,394	8,209	9,833
FRG	500	531	1,054
Greece	498	488	1,310
Hungary	4,542	7,172	7,910
Italy	4,601	5,326	6,864
Romania	7,354	9,241	12,640[2]
Turkey	1,058	1,200	1,400
USSR	9,993	7,328	11,000
Yugoslavia	7,399	9,389	11,200
	Average yield (kg per hectare)		
Austria	555	682	732
Bulgaria	373	432	548
Czechoslovakia	402	549	533
France	515	419	603
FRG	505	552	658
Hungary	357	502	583
Italy	466	594	680
Romania	232	278	411[2]
Turkey	164	201	237
USSR	276	274	290
Yugoslavia	310	397	501

[1] Sown area.
[2] Including sorghum.

Tomatoes

Countries	1969–71	1975	1982
	Harvested area (thousands of hectares)		
1. Bulgaria[1]	24.0	27.2	27.0
2. Czechoslovakia[1]	4.0	3.4	4.0
3. France	18.0	23.0	18.0
4. GDR	1.0	1.3	n.a.
5. Great Britain[2]	1.0	3.0	1.0
6. Greece	33.0	42.0	44.0
7. Hungary[1]	19.0	15.0	15.0
8. Italy	127.0	113.0	116.0
9. Poland[1]	27.0	27.0	31.0
10. Romania[1]	57.0	56.1	75.0
11. Turkey	74.0	82.0	114.0
12. Yugoslavia	33.0	37.0	38.0
	Production (thousands of tons)		
1. Bulgaria	716	599	853
2. Czechoslovakia	105	97	122
3. France	514	639	876
4. GDR	18	34	n.a.
5. Great Britain[2]	108	181	145
6. Greece	976	1,627	1,918
7. Hungary	359	313	365
8. Italy	3,571	3,512	4,075
9. Poland	339	418	442
10. Romania	780	858	1,400
11. Turkey	1,756	2,300	3,824
12. Yugoslavia	332	335	465
	Average yields (kg per ha)		
1. Bulgaria	27,620	20,390	29,410
2. France	29,140	27,770	48,670
3. GDR	19,470	21,670	n.a.
4. Great Britain[2]	101,560	57,600	163,820
5. Greece	29,850	38,740	44,070
6. Hungary	19,100	18,000	25,000
7. Italy	28,110	31,030	35,180
8. Poland	12,610	13,800	14,660
9. Romania	13,760	11,740	18,670
10. Turkey	23,680	28,050	33,650
11. Yugoslavia	10,150	9,050	12,130

[1] Planted area.
[2] Mainly hothouse production.

Apple Production (in thousand tons)

Countries	1969–71	1975	1982
Austria	293	304	429
Bulgaria	363	329	426
Czechoslovakia	205	182	504
France	3,895	3,285	3,016
FRG	2,110	2,035	2,775
GDR	289	454	577
Great Britain	524	386	373
Greece	230	264	257
Hungary	700	809	1,000
Italy	1,923	2,127	2,200
Poland	636	841	1,893
Romania	265	315	410
Turkey	716	900	1,257
Yugoslavia	362	370	667

Grapes

Countries	1969–71	1975	1982
	Area planted – thousands of hectares		
Bulgaria[1]	169.2	164.9	148
Czechoslovakia[1]	24	31.5	38.4
France	1,242	1,308	n.a.
Greece	222	204	n.a.
Hungary[1]	229	199.2	129.7
Italy	1,432	1,450	n.a.
Romania[1]	346	295.7	n.a.
Turkey	843	790	n.a.
USSR[1]	1,095	817.8	973.4
Yugoslavia	254	247	n.a.
	Output – thousands of tons		
Bulgaria	1,040	885	1,246
Czechoslovakia	128	209	275
France	9,706	11,627	11,230
Greece	1,569	1,480	1,617
Hungary	809	813	1,000
Italy	10,638	10,917	11,150
Romania	1,020	1,182	2,192
Turkey	3,779	3,274	3,741
USSR	4,220	5,400	7,300
Yugoslavia	1,232	3,961	1,450

[1] Only fruit-yielding area.

Tobacco

Countries	1969–71	1975	1982
	Harvested area (thousands of hectares)		
Albania	18	20	103.3
Bulgaria[1]	118	127	103.3
including Oriental tobacco	108.1	106.4	88.2
France	20	20	16
Greece	99	94	94
Hungary[1]	17	15.6	15
Italy	44	57	61
Poland[1]	45	50.4	49
Romania[1]	34	57.3	45
Turkey	326	242	154
USSR[1]	168	173.1	180
Yugoslavia	52	63	56
	Output (thousands of tons)		
Albania	12	14	n.a.
Bulgaria	120	157	145
France	45	51	40
Greece	88	117	115
Hungary	20	17	25
Italy	79	113	133
Poland	82	102	80
Romania	26	40	39
Turkey	157	200	180
USSR	256	289	288
Yugoslavia	46	70	63

[1] Sown area.

Railways (length of railways used by year's end in km)

Countries	1970				1981			
	Total	Including electrified	Per 100 sq.km	Per 10,000 of the population	Total	Including electrified	Per 100 sq.km	Per 10,000 of the population
1. Austria	6,546	2,628	7.8	8.8	6,409	3,266	7.6	8.5
2. Bulgaria	4,196	811	3.8	4.9	4,267	1,730	3.9	4.8
3. Czechoslovakia	13,308	2,510	10.4	9.3	13,130	3,081	10.3	8.6
4. France	36,532	9,359	6.6	7.1	34,595	10,477	6.3	6.4
5. FRG	33,123	8,883	13.4	5.6	31,482	11,501	12.6	5.1
6. Great Britain	19,327	3,162	7.9	3.4	17,769	3,729	7.3	3.2
7. GDR	14,658	1,357	13.5	8.6	14,222	1,788	13.1	8.5
8. Greece	2,571	—	1.9	2.9	2,479	—	1.9	2.6
9. Hungary	9,168	935	9.9	8.9	7,867	1,619	8.5	7.3
10. Italy	20,089	9,330	6.7	3.7	16,162[1]	8,758[1]	5.4[1]	2.8
11. Poland	26,678	3,872	8.5	8.2	27,172	7,091	8.7	7.6
12. Romania	11,012	494	4.6	5.4	11,093	2,706	4.7	5.0
13. Turkey	7,985	72	1.0	2.3	8,193	392	1.1	1.8
14. USSR	135,190	33,861	0.6	5.6	142,806	44,832	0.6	5.3
15. Yugoslavia	10,289	1,510	4.0	5.0	9,393	3,320	3.7	4.2

[1] Only state-owned railways.

Transport and Communication Services (in numbers)

	1970	1975	1980	1982
Per capita passenger trips[1]	167.7	224.8	293.6	308.9
Including				
By train	12.5	11.9	11.3	10.9
By road	155.0	212.5	220.1	231.0
Per capita passenger trips by train (for Sofia)	427.9	406.7	396.7	387.3
Per capita passenger trips by trolleybus (Sofia and Plovdiv)	109.4	84.4	85.5	94.7
Telephones per 1,000 of the population	55.6	89.0	141.5	169.5

[1] Including passengers using city transport.

Merchant Fleet (Sea-Going; thousands GRT)

Countries	1970	1975	1981
Bulgaria	651	793	1,074
FRG	7,881	8,517	7,068
France	6,458	10,746	10,807
Italy	7,448	10,137	9,804
GDR	989	1,389	1,385
Great Britain	25,825	33,157	22,879
Greece	10,952	22,527	39,679
Poland	1,580	2,817	3,201
Romania	341	777	1,811
Turkey	697	994	1,720
Yugoslavia	1,516	1,873	2,494

Bulgaria in Figures

Export of Some Commodities by Countries (thousands of tons)

Countries	1970	1975	1981
Cast iron, steel and their products			
Austria	806[1]	1,983	2,728
Bulgaria	757	992	—
France	8,060	9,076	12,616
FRG	12,681	16,988	20,434
USSR	11,800	—	—
Wheat			
Bulgaria	211	116	319
France	3,446	6,329	12,785
FRG	1,450	679	496
Hungary	580	952	1,298
Romania	14	705	200
USSR	4,852	2,681	1,600
Tobacco			
Bulgaria	58	71	69
Greece	63	51	58
Italy	11	59	76
Turkey	74	66	131
Wines			
Bulgaria	196	250	269
France	515	635	920
FRG	35	82	206
Greece	98	110	27
Hungary	98	161	255
Italy	554	1,413	2,041

[1] Cast and rolled iron.

Imports and Exports ($US millions)

Countries	1970		1982	
	Imports	Exports	Imports	Exports
	cif	fob	cif	fob
Austria	3,549	2,856	19,557	15,685
Bulgaria[1]	1,831	2,004	11,554	11,453
Czechoslovakia[1]	3,695	3,792	18,724	18,443
France	19,114	17,935	115,743	92,756
FRG	29,947	34,228	155,856	176,428
GDR[1]	4,847	4,581	21,743	20,196
Great Britain	21,695	19,382	99,654	96,994
Greece	1,958	642	10,023	4,297
Hungary[1]	2,506	2,317	8,825	8,795
Italy	14,970	13,206	86,213	73,490
Poland[1]	3,608	3,548	10,248	11,208
Romania[1]	1,960	1,851	n.a.	n.a.
Turkey	894	588	8,753	5,701
USSR[1]	11,732	12,800	77,793	86,949
Yugoslavia	2,874	1,679	n.a.	n.a.

[1] Imports fob.

Number of Foreign Visitors (thousands)

Countries	Code	1970	1975	1981
Austria	R	8,867	11,540	14,241
Bulgaria	F	2,537	4,049	6,046
Czechoslovakia	F	3,546	13,863	17,731
FRG	R	8,468	7,403	11,600
Great Britain	F	6,730	8,844	11,486
Greece	T	1,408	2,840	5,094
Hungary	T	4,040	4,995	10,450
Italy	T	—	—	20,036
Poland	F	1,889	9,320	2,172
Romania	F	2,290	3,206	7,002
Turkey	F	446	1,541	1,405
USSR	F	2,059	3,691	5,870
Yugoslavia	R	4,748	5,835	6,616

Code:
T – Tourists having visited the country.
F – Foreigners having visited the country.
R – Foreigners registered in hotels and other similar establishments.

Indices of Retail Trade (at prices in the respective year; 1970=100)

Countries	1975	1982
Austria	168.9	259.0
Bulgaria	146.8	236.6
Czechoslovakia	130.0	n.a.
FRG	149.2	208.0
GDR	127.9	160.0
Great Britain[1]	n.a.	155.0[2]
Greece	192.9	626.0
Hungary	157.0	288.0
Italy	185.2	n.a.
Poland	182.0	556.0
Romania	151.0	n.a.
USSR	136.0	n.a.
Yugoslavia	327.0	1,800.0

[1] Excluding Northern Ireland.
[2] 1978=100.

Indices of Retail Prices (1970=100)

Countries	1975	1982
Austria	142.2	207.1
Bulgaria	101.0	123.9
Czechoslovakia	100.9	—
FRG	134.7	183.3
France	152.8	318.7
GDR	98.4	—
Great Britain	184.4	438.3
Greece	178.6	571.5
Hungary	114.6	175.1
Italy	171.1	513.7
Poland	113.3	—
USSR	99.7	—
Yugoslavia	242.7	1,000

Population Censuses

Year	Total				In Towns	
	Total	Men	Women	Women per 1,000 men	Total	Men
1880[1]	2,007,919	1,027,803	980,116	954	336,102	179,490
1884[2]	942,680	476,462	466,218	978	224,105	112,280
1887	3,154,375	1,605,389	1,548,986	965	593,547	307,310
1892	3,310,713	1,690,626	1,620,087	958	652,328	341,109
1900	3,744,283	1,909,567	1,834,716	961	742,435	384,041
1905	4,035,575	2,057,092	1,978,483	962	789,689	409,992
1910	4,337,513	2,206,685	2,130,828	966	829,522	428,668
1920	4,846,971	2,420,784	2,426,187	1,002	966,375	496,635
1926	5,478,741	2,743,025	2,735,716	997	1,130,131	574,961
1934	6,077,939	3,053,893	3,024,046	990	1,302,551	660,914
1946	7,029,349	3,516,774	3,512,575	999	1,735,188	888,538
1956	7,613,709	3,799,356	3,814,353	1,004	2,556,071	1,276,670
1965	8,227,866	4,114,167	4,113,699	1,000	3,822,824	1,911,860
1975	8,727,771	4,357,820	4,369,951	1,003	5,061,087	2,517,708

[1] North Bulgaria.
[2] Eastern Rumelia.

Statistics of the Country's Social and Economic Development

	1948	1952	1957	1960	1965	1970	1975	1980	1982
Population of active age (thousands)	4,243	4,405	4,519	4,608	4,788	4,938	5,057	5,088	5,059
Men (16–59 years)	2,194	2,291	2,363	2,418	2,527	2,603	2,633	2,704	2,691
Women (16–54 years)	2,049	2,114	2,156	2,190	2,261	2,335	2,424	2,388	2,368
Persons employed structure – %	100.0	100.0	100.0	100.0	100.0	100.0	100.0	100.0	100.0
In material production	95.7	93.6	92.5	90.8	89.2	86.9	84.3	83.0	82.8
In the non-productive sphere	4.3	6.4	7.5	9.2	10.8	13.1	15.7	17.0	17.2
Workers and employees average annual number (thousands)	629	1,016	1,339	1,774	2,197	2,749	3,677	4,025	4,100
including workers[1]	362	580	808	1,128	1,406	1,737	2,292	2,620	2,625

[1] The category of workers was designated in the Communications sector in 1964.

Average Longevity by Periods (in years)

Sex	1916–25[1]	1927–34[1]	1935–39[1]	1956–57	1960–62	1965–67	1969–71	1974–76	1978–80
Total	44.64	48.40	51.75	65.89	69.59	70.66	71.11	71.31	71.14
Men	44.35	47.81	50.98	64.17	67.82	68.81	68.58	68.68	68.35
Women	44.98	49.09	52.56	67.65	71.35	72.67	73.86	73.91	73.55

[1] According to statistics of the Bulgarian Academy of Science.

		In Villages				Relative share of the urban population in relation to the rural
Women	Women per 1,000 men	Total	Men	Women	Women per 1,000 men	
156,612	873	1,671,817	848,313	823,504	971	16.7
111,825	996	718,575	364,182	354,393	973	23.8
286,237	931	2,560,828	1,298,079	1,262,749	973	18.8
311,219	912	2,658,385	1,349,517	1,308,868	970	19.7
358,394	933	3,001,848	1,525,526	1,476,322	968	19.8
379,697	926	3,245,886	1,647,100	1,598,786	971	19.6
400,854	935	3,507,991	1,778,017	1,729,974	973	19.1
469,740	946	3,880,596	1,924,149	1,956,447	1,017	19.9
555,170	966	4,348,610	2,168,064	2,180,546	1,006	20.6
641,637	971	4,775,388	2,392,979	2,382,409	996	21.4
846,650	953	5,294,161	2,628,236	2,665,925	1,014	24.7
1,279,401	1,002	5,057,638	2,522,686	2,534,952	1,005	33.6
1,910,964	1,000	4,405,042	2,202,307	2,202,735	1,000	46.5
2,543,379	1,010	3,666,684	1,840,112	1,826,572	993	58.0

Population, Territory, Density and Capitals

Countries	Population counted		1981		Capitals or administrative centres		
	Date of Census	Thousands	Territory 1,000 sq.km	Density[3] per sq.km	Name	Population Date	Thousands
Albania	2.10.1960	1,626	29	97	Tirana	1970	200
Austria[1]	12.5.1981	7,555	84	90	Vienna	1.7.1980	1,568
Bulgaria	2.12.1975	8,728	111	80	Sofia	31.12.1982	1,140
Czechoslovakia	1.11.1980	15,283	128	120	Prague	31.12.1981	1,182
France[1] [4]	4.3.1982	54,257	547	99	Paris	20.2.1975	8,613
FRG[1]	27.5.1970	60,651	249	248	Bonn	30.6.1981	286
GDR	3.12.1981	16,733	108	155	Berlin	31.12.1981	1,166
Greece	5.4.1981	9,740	132	74	Athens	14.3.1971	2,101
Hungary	1.1.1980	10,709	93	115	Budapest	31.12.1981	2,062
Italy	25.10.1981	56,244	301	190	Rome	31.7.1980	2,914
Poland	7.12.1978	35,061	313	115	Warsaw	31.12.1981	1,612
Romania	5.1.1977	21,560	238	94	Bucharest	1.7.1981	1,929
USSR	17.1.1979	262,436	22,402	12	Moscow	31.12.1981	8,302
UK[2]	5.4.1981	55,671	244	229	London	30.6.1980	6,849
Yugoslavia	31.3.1971	20,523	256	88	Belgrade	31.3.1981	1,455

[1] *De jure* population.
[2] Excluding Isle of Man.
[3] Population count at mid-year.
[4] Excluding overseas territories.

Population by Sex and Age

Countries	Sex	Population thousands	Age (years)					
			0–4	5–14	15–24	25–49	50–59	60 and over
			Percentage of the population					
Austria	Total	7,505	5.8	14.7	16.3	32.1	11.9	19.2
1.7.1980 (E)	Male	3,551	6.3	15.9	17.6	34.1	11.1	15.0
	Female	3,954	5.3	13.7	15.2	30.3	12.6	22.9
Albania	Total	1,391	15.9	22.9	18.5	27.5	6.3	8.9
2.10.1965 (C)	Male	713	16.1	23.5	19.4	27.4	5.9	7.7
	Female	678	15.8	22.2	17.6	27.5	6.7	10.2
Bulgaria	Total	8,929	7.1	15.0	13.7	34.3	13.4	16.5
31.12.1982 (E)	Male	4,443	7.3	15.5	14.1	34.5	13.3	15.3
	Female	4,486	6.9	14.5	13.3	34.0	13.5	17.8
Czechoslovakia	Total	15,289	8.8	15.6	14.5	33.7	11.7	15.7
31.12.1980 (E)	Male	7,444	9.2	16.3	15.3	34.7	11.3	13.2
	Female	7,845	8.4	14.8	13.8	32.8	12.1	18.1
France	Total	54,085	7.0	14.9	15.7	33.2	11.6	17.6
1.1.1982 (E)	Male	26,490	7.3	15.6	16.3	34.7	11.5	14.6
	Female	27,595	6.7	14.3	15.1	31.8	11.6	20.5
FRG	Total	61,566	4.8	13.4	16.0	34.6	11.9	19.3
30.6.1980 (E)	Male	29,417	5.1	14.4	17.3	37.1	11.2	14.9
	Female	32,149	4.4	12.5	14.9	32.2	12.5	23.5
Great Britain	Total	55,852	6.3	16.1	14.8	31.2	11.8	19.8
30.6.1977 (E)	Male	27,184	6.7	17.1	15.5	32.2	11.8	16.7
	Female	28,668	6.0	15.3	14.0	30.2	11.8	22.7
GDR	Total	16,736	6.8	12.7	16.5	34.2	10.7	19.1
30.6.1981 (E)	Male	7,864	7.4	13.8	18.0	36.8	9.8	14.2
	Female	8,872	6.2	11.7	15.1	30.5	11.6	24.9
Greece	Total	9,449	7.5	15.7	14.8	33.1	11.4	17.5
1.7.1979 (E)	Male	4,633	7.9	16.5	15.5	32.9	11.2	16.0
	Female	4,816	7.1	14.9	14.1	33.4	11.7	18.8
Hungary	Total	10,711	7.3	14.8	12.6	35.2	12.6	17.5
1.1.1982 (E)	Male	5,185	7.7	15.7	13.4	36.2	12.2	14.8
	Female	5,526	6.9	13.9	11.9	34.4	13.0	19.9
Italy	Total	57,070	6.4	15.6	15.2	33.3	12.3	17.2
1.7.1980 (E)	Male	27,871	6.7	16.4	15.9	34.0	12.0	15.0
	Female	29,199	6.1	14.8	14.6	32.6	12.5	19.4
Poland[1]	Total	36,062	9.1	15.5	25.5	25.4	11.2	13.3
31.12.1981 (E)	Male	17,572	9.6	16.2	26.8	25.9	10.7	10.8
	Female	18,490	8.7	14.7	24.4	24.9	11.7	15.6
Romania	Total	22,201	9.0	17.6	14.4	34.0	11.7	13.3
1.7.1980 (E)	Male	10,953	9.4	18.3	14.8	34.6	11.3	11.6
	Female	11,248	8.7	17.0	13.9	33.5	12.0	14.9
Turkey	Total	40,198	13.4	26.5	19.9	27.6	5.2	7.4
26.10.1975 (C)	Male	20,417	13.5	27.0	20.3	27.2	5.1	6.9
	Female	19,781	13.4	26.0	19.6	27.9	5.2	7.9
USSR[2]	Total	250,869	—	16.7	19.8	42.6	8.0	12.9
1.1.1974 (E)	Male	116,225	—	18.4	21.7	44.6	6.4	8.9
	Female	134,644	—	15.3	18.1	40.8	9.4	16.4
Yugoslavia	Total	21,974	8.4	16.5	17.4	35.5	10.2	12.0
30.6.1978 (E)	Male	10,821	8.8	17.1	18.1	36.3	9.3	10.4
	Female	11,153	8.0	15.8	16.8	34.8	11.1	13.5

[1] The groups are: 0–4, 5–14, 15–29, 30–49, 50–59, 60 years and over.
[2] The groups are: 0–9, 10–19, 20–49 years.
E=estimates; C=Census.

Bulgaria in Figures

Average Life Expectancy

Countries	Periods Years	Men	Women
Albania	1965–66	64.9	67.0
	1975–80	68.00	70.70
Austria	1959–61	65.60	72.03
	1970	66.34	73.52
	1980	68.97	76.15
Bulgaria	1965–67	68.81	72.67
	1969–71	68.58	73.86
	1978–80	68.35	73.55
Czechoslovakia	1964	67.76	73.56
	1970	67.2	73.6
	1981	67.6	72.6
France	1965	67.8	75.0
	1970	68.6	76.1
	1978–80	70.05	78.20
FRG	1964–65	67.59	73.45
	1968–70	67.24	73.44
	1978–80	69.60	76.36
GDR	1967	68.35	73.43
	1972	68.5	73.9
	1980	68.7	74.6
Great Britain[2]	1963–65	68.30	74.40
	1968–70	67.81	73.81
	1977–79	70.20	76.40
Greece	1960–62	67.46	70.70
	1970	70.13	73.64
Hungary	1964	67.00	71.83
	1970	66.28	72.1
	1981	66.0	73.4
Italy	1960–62	67.24	72.27
	1970–72	68.97	74.88
	1974–77	69.69	75.91
Poland	1965–66	66.8	72.8
	1970–72	66.8	73.8
	1980–81	66.9	75.4
Romania	1964–67	66.45	70.51
	1970–72	66.27	70.85
	1976–78	67.42	72.18
Turkey	1966	53.7[1]	—
	1975–80	60.30	61.60
USSR	1965–66	66	74
	1971–72	64	74
Yugoslavia	1961–62	62.41	65.58
	1970–72	65.42	70.22
	1977	67.6	72.6

[1] Statistics are for both sexes.
[2] Statistics for the periods 1963–65 and 1977–79 only for England and Wales.

Birth Rate

Countries	Live-born per 1,000 of the population		
	1970	1975	1981
Albania	32.5	31.9[1]	n.a.
Austria	15.1	12.5	12.4
Bulgaria	16.3	16.6	14.0
Czechoslovakia	15.9	19.6	15.5
France	16.7	14.1	14.9
FRG	13.4	9.7	10.1
GDR	13.9	10.8	14.2
Great Britain	16.3	12.5	13.1
Greece	16.5	15.7	n.a.
Hungary	14.7	18.4	13.3
Italy	16.8	14.8	10.9
Poland	16.6	18.9	18.9
Romania	21.1	19.7	17.0
USSR	17.4	18.1	18.7
Yugoslavia	17.8	18.2	16.7

[1] For 1970–75 period – UN Population Department estimates.

Mortality Rate

Countries	Per 1,000 of the population		
	1970	1975	1981
Albania	9.3	6.9[1]	n.a.
Austria	13.4	12.8	12.3
Bulgaria	9.1	10.3	10.7
Czechoslovakia	11.6	11.5	11.7
France	10.7	10.6	10.3
FRG	12.1	12.1	11.7
GDR	14.1	14.3	13.9
Great Britain	11.8	11.9	11.8
Greece	8.4	8.9	n.a.
Hungary	11.6	12.4	13.5
Italy	9.7	9.9	9.5
Poland	8.1	8.7	9.2
Romania	9.5	9.3	10.0
USSR	8.2	9.3	10.3
Yugoslavia	8.9	8.7	9.0

[1] For the 1970–75 period, according to estimates of the UN Population Department.

Children's Mortality Rate

Countries	For infants up to one year of age per 1,000 live-born		
	1970	1975	1981
Austria	25.9	20.5	12.6
Bulgaria	27.3	23.1	18.9
Czechoslavakia	22.1	20.8	16.7
France	18.2	13.8	9.6
FRG	23.6	19.8	11.6
GDR	18.5	15.9	12.3
Great Britain	18.4	16.0	11.2
Greece	29.6	24.0	n.a.
Hungary	35.9	32.8	20.6
Italy	29.5	21.2	14.1
Poland	36.7	29.1	24.7
Romania	49.4	34.7	28.6
USSR	24.7	n.a.	n.a.
Yugoslavia	55.5	39.7	30.7

Natural Population Growth (per 1,000 of the population)

Countries	1970	1975	1980
Albania	23.2[1]	25.0	—
Austria	1.7	−0.3	0.1
Bulgaria	7.2	6.3	3.3
Czechoslovakia	4.3	8.1	3.8
FRG	1.3	−2.4	−1.6
GDR	−0.2	−3.5	0.3
Great Britain	4.5	0.6	1.3
Greece	8.1	6.8	—
Hungary	3.1	6.0	−0.2
Italy	7.1	4.9	1.4
Poland	8.5	10.2	9.7
Romania	11.6	10.4	7.0
USSR	9.2	8.8	8.4
Yugoslavia	8.9	9.5	7.7

[1] For the 1970–75 period – UN Population Department estimates.

Marriages and Divorces (per 1,000 of the population)

Countries	Marriages			Divorces		
	1970	1975	1981	1970	1975	1981
Austria	7.1	6.2	6.4	1.40	1.43	1.79
Bulgaria	8.6	8.6	7.5	1.17	1.26	1.49
Czechoslovakia	8.8	9.5	7.6	1.74	2.18	2.3
France	7.8	7.4	5.8	0.79	1.16	n.a.
FRG	7.3	6.3	5.8	1.26	1.73	n.a.
GDR	7.7	8.4	7.7	1.6	2.5	2.9
Great Britain	8.5	7.7	n.a.	1.18	2.43	n.a.
Greece	7.7	8.5	n.a.	0.40	0.41	n.a.
Hungary	9.3	9.9	7.2	2.2	2.5	2.5
Italy	7.4	6.7	5.5	n.a.	0.19	0.19
Poland	8.5	9.7	9.0	1.06	1.20	1.1
Romania	7.2	8.9	8.2	0.39	1.62	1.5
USSR	9.7	10.7	10.4	2.6	3.1	3.5
Yugoslavia	9.0	8.4	7.7	1.01	1.18	n.a.

Employment by Sectors (per cent)

Sectors	1970	1975	1980	1982
Total	**100.0**	**100.0**	**100.0**	**100.0**
In material production	*86.9*	*84.3*	*83.0*	*82.8*
Industry	30.3	33.5	35.2	36.3
Construction	8.4	8.0	8.2	8.2
Agriculture	35.2	27.7	23.8	22.3
Forestry	0.5	0.5	0.4	0.4
Transport	5.2	5.5	5.9	5.8
Communications	0.8	0.8	0.9	0.9
Trade, material and technical supply and purchasing	6.1	7.8	8.0	8.2
Other sectors of material production	0.4	0.5	0.6	0.7
In the non-productive sphere	*13.1*	*15.7*	*17.0*	*17.2*
Housing and communal services	2.1	2.1	2.1	2.1
Science	1.1	1.4	1.5	1.6
Education	4.3	5.2	5.7	5.8
Culture and the arts	0.7	1.0	1.0	1.0
Health services, social insurance, sports	2.7	3.6	4.3	4.4
Finance, loans and insurance	0.4	0.5	0.5	0.5
Management	1.5	1.5	1.5	1.4
Other sectors of the non-productive sphere	0.3	0.4	0.4	0.4

Administrative Managerial Staff in the National Economy per 100 (percentage)

	1970	1975	1980	1982
Relative share of administrative managerial staff in the total number of workers and employees	12.5	13.1	10.9	11.0
Indices at base (1970=100)	100.0	130.1	141.4	143.0
Administrative managerial staff	100.0	136.0	122.6	125.1

Workers and Office Employees by Sectors

Sectors	1970	1975	1980	1982
	Average annual number			
	Workers and office employees			
Total	**2,748,720**	**3,676,632**	**4,024,823**	**4,100,259**
Industry	1,155,266	1,296,632	1,368,920	1,401,997
Construction[1]	303,761	316,818	341,213	349,029
Agriculture	272,477	788,097	955,830	937,609
Forestry	21,858	22,568	17,771	16,879
Transport	192,725	231,529	258,684	260,809
Communications	32,514	37,920	39,076	40,463
Trade, material and technical supply and purchasing	232,050	308,464	331,648	352,731
Other sectors of material production	14,324	23,140	27,879	30,386
Housing and communal services	77,077	74,295	50,962	51,498
Science	46,681	60,538	62,401	73,602
Education	173,795	223,127	243,076	254,199
Culture and the arts	29,628	42,618	45,034	44,989
Health services, social insurance, sports	110,600	153,723	185,503	195,644
Finance, loans and insurance	16,182	19,888	20,421	21,528
Management	58,597	62,344	61,057	56,356
Other sectors of the non-productive sphere	11,185	14,931	15,348	12,540
	Workers			
Total	**1,737,473**	**2,291,514**	**2,619,722**	**2,624,829**
Industry	917,148	1,019,274	1,111,403	1,133,551
Construction[1]	237,790	227,794	256,956	263,037
Agriculture	225,971	593,385	742,762	712,843
Forestry	19,446	19,325	14,484	13,555
Transport	161,975	195,874	224,625	225,943
Communications	28,693	33,460	34,425	35,367
Trade, material and technical supply and purchasing	140,205	193,330	219,816	223,479
Other sectors of material production	6,245	9,072	15,251	17,054

[1] Since 1977, including those employed by the building enterprises with the Water Economy State Corporation, and the Town Planning and Public Utilities enterprises with the District People's Councils.

Bulgaria in Figures

Basic Assets in the National Economy (million levs to complete initial value at the end of the year)

Sectors	1970	1975	1977[1]	1978	1979	1980	1981	1982
Total[2]	32,916.6	48,125.9	62,663.5	67,175.7	71,805.6	76,286.4	81,812.3	88,250.9
Basic production assets	21,212.1	32,537.0	42,190.7	45,410.4	48,729.9	51,868.3	55,596.9	60,216.4
Industry[3]	11,092.2	17,390.8	21,558.9	23,402.6	25,317.9	27,906.0	30,158.0	33,102.7
Construction	613.3	1,203.1	1,571.6	1,761.7	1,928.3	2,054.0	2,158.2	2,332.9
Agriculture[2]	4,352.5	5,983.4	7,425.7	7,774.9	8,122.5	7,933.1	8,356.1	8,799.3
Transport	4,156.7	6,115.9	9,045.3	9,715.8	10,329.1	10,782.1	11,542.2	12,342.0
Communications	245.5	455.5	733.3	777.3	862.3	954.8	1,040.3	1,128.3
Trade, material and technical supply and purchasing	730.5	1,274.7	1,703.0	1,790.1	1,940.6	1,999.0	2,152.8	2,298.2
Other sectors of material production	21.4	113.6	152.9	188.0	229.2	239.3	189.3	213.0
Non-productive basic assets	11,704.5	15,588.9	20,472.8	21,765.3	23,075.7	24,418.1	26,215.4	28,034.5
Housing and communal services	9,522.5	12,483.7	16,244.8	16,948.1	17,863.6	18,899.9	20,148.1	21,693.2
Including housing stock	7,766.0	9,581.2	11,912.1	12,400.4	13,026.0	13,815.8	14,830.5	15,766.5
Housing stock inhabited by people	6,926.6	8,238.2	10,073.1	10,412.3	10,765.7	11,211.2	11,754.6	12,562.8
Science and scientific service	201.6	423.9	483.3	500.8	533.5	574.0	614.0	661.9
Education	894.9	1,251.9	1,837.8	2,169.5	2,292.7	2,101.9	2,252.9	2,350.7
Culture and art						293.5	523.8	496.4
Medical care, social security, physical education, sport and tourism	399.1	768.0	1,027.6	1,124.4	1,219.8	1,320.0	1,337.4	1,437.2
Other sectors of the non-productive sphere	686.4	661.4	899.3	1,022.5	1,166.1	1,228.8	1,339.2	1,395.1

[1] Replacement value. The balance sheets of enterprises, departments and organizations reflect the results of the general inventory and re-evaluation of the fixed assets around 31 December 1974.
[2] Until 1979 sheep, pigs and goats were included in the fixed assets, and since 1980 have been included in the material current assets.
[3] In 1965 the fixed assets of the Forestry sector were included in Industry sector.

Fixed Assets in the National Economy (annual average availability – million levs)

	1970	1975	1977	1978	1979	1980	1981	1982
Total	31,724.6	48,186.4	60,156.1	64,179.5	68,104.1	72,942.2	78,729.2	85,194.3
Fixed assets	20,180.4	31,170.9	40,401.0	43,643.0	47,074.2	49,845.8	53,738.0	57,875.2
Material current assets	11,544.2	17,015.5	19,755.1	20,536.5	21,029.9	23,096.4	24,991.2	27,319.1

Average Annual Growth Rate of Some Basic Indicators by Periods and Five-Year Plans

	1949–82 period	1949–52 First	1953–57 Second	1958–60 Third	1961–65 Fourth	1966–70 Fifth	1971–75 Sixth	1976–80 Seventh	1981–82 Eighth
Capital investments	9.9	14.1	5.1	29.2	7.9	12.5	8.6	4.0	7.9
Fixed capital put into operation	11.6	16.7	8.2	15.3	10.3	13.7	7.4	6.8	9.9
Social product	8.65	10.7	8.8	13.5	8.6	9.5	7.75	5.9	4.6
National income	7.75	8.4	7.8	11.6	6.7	8.75	7.8	6.1	4.6
Per capita national income	7.0	7.8	6.7	10.5	5.8	8.0	7.25	5.75	4.25
National income per person employed in material production	7.95	9.8	6.5	15.2	7.1	8.25	7.7	6.1	3.7
Total industrial output	11.4	20.7	12.7	16.2	11.7	10.9	9.1	6.0	4.6
Labour productivity per blue and white-collar worker in industrial production, in the state and co-operative industrial enterprises	6.4	10.3	5.2	5.6	6.8	6.8	6.75	5.25	3.4
Total production of the building industry	8.7	19.6	7.1	20.8	7.4	12.0	5.7	5.9	4.5
Labour productivity per person employed in building industry	5.4	n.a.	9.3	4.4	5.4	5.7	6.6	4.4	4.4
Total output of agriculture	3.1	0.9	4.9	6.6	3.2	3.5	2.9	0.9	5.55
Labour productivity in farming	6.8	2.7	5.0	17.4	7.4	7.5	7.3	3.7	7.9
Consumption fund	6.05	n.a.	7.2	8.5	5.75	6.4	6.5	3.7	4.2
Nominal work pay	4.8	n.a.	4.8	4.8	3.4	6.0	3.4	4.5	4.0
Real work pay	3.25	n.a.	11.3	4.5	2.0	5.3	3.0	0.5	3.4
Nominal incomes	6.2	n.a.	1.4	9.5	6.7	7.8	6.75	6.8	5.4
Real incomes	5.45	n.a.	7.1	8.3	4.6	6.0	5.75	2.6	4.9

Fixed Capital Commissioned According to Five-Year Plans, and Forms of Ownership (millions of levs)

Five-Year Plans Years	Total	State-owned enterprises	Co-operative enterprises	Population	Others
First 5-Year Plan (1949–52)	1,577.0	1,166.6	118.4	291.3	0.7
Average per year	394.3	291.7	29.6	72.8	0.2
Second 5-Year Plan (1955–57)	3,015.9	2,194.1	317.2	503.2	1.4
Average per year	603.2	438.8	63.4	100.7	0.3
Third 5-Year Plan (1958–60)	2,728.2	1,739.3	493.0	495.2	0.7
Average per year	909.4	579.8	164.3	165.1	0.2
Fourth 5-Year Plan (1961–65)	7,562.1	5,410.7	1,294.8	856.5	0.1
Average per year	1,512.4	1,082.1	259.0	171.3	0.0
Fifth 5-Year Plan (1966–70)	13,360.0	10,559.8	1,765.1	1,035.9	0.1
Average per year	2,672.2	2,112.0	353.0	207.2	0.0
Sixth 5-Year Plan (1971–75)	18,474.3	15,757.0	1,481.4	1,235.7	0.2
Average per year	3,694.9	3,151.4	296.3	247.1	0.1
Seventh 5-Year Plan (1976–80)	27,878.6	25,710.9	485.8	1,681.2	0.7
Average per year	5,575.7	5,142.2	97.2	336.2	0.1
Eighth 5-Year Plan (1981–82)	14,520.7	13,228.7	189.2	1,102.7	0.1
Average per year	7,260.3	6,614.3	94.6	551.4	0.0

Size of Saving Accounts[1] (at year's end)

Size of accounts	1970	1975	1980	1982
	\multicolumn{4}{c}{Account (in thousands)}			
Total	**7,878**	**9,013**	**8,733**	**9,229**
Up to 50 levs	2,933	2,360	2,527	2,586
51–100 levs	1,593	1,584		
101–500 levs	1,707	2,059	2,169	2,249
501–1,000 levs	883	1,378	1,641	1,689
1,001–2,000 levs	492	923	1,231	1,339
2,001–4,000 levs	214	512	784	890
Over 4,000 levs	56	197	381	476
	\multicolumn{4}{c}{Sums (millions of levs)}			
Total	**3,138.7**	**5,950.4**	**8,405.5**	**9,822.5**
Up to 50 levs	76.2	72.4	176.8	188.3
51–100 levs	131.9	128.8		
101–500 levs	579.5	682.0	723.4	759.0
501–1,000 levs	677.2	1,023.7	1,169.5	1,269.7
1,001–2,000 levs	742.5	1,350.7	1,760.7	1,932.9
2,001–4,000 levs	611.6	1,477.2	2,138.5	2,465.2
Over 4,000 levs	319.8	1,215.6	2,436.6	3,207.5

[1] Without the accounts of foreign nationals and, since 1975, without blocked accounts.

Housing Construction Deposits (at year's end)

	1970	1975	1980	1982
Deposits (in thousands)	284.6	480.6	545.1	540.1
Sums (millions of levs)	755.5	1,633.1	1,850.9	1,928.6

Savings (at end of year)

	1970	1975	1980	1982
Deposits in thousands	7,898	9,035	8,755	9,251
In towns	4,950	5,965	6,075	6,553
In villages	2,928	3,048	2,658	2,676
Undeclared	20	22	22	22
Sums in million levs	3,144.0	5,967.1	8,441.7	9,860.2
In towns	1,763.1	3,578.4	5,407.8	6,418.6
In villages	1,375.6	2,372.0	2,997.7	3,403.9
Undeclared	5.3	7.4	11.3	14.2
Blocked	—	9.3	24.9	23.5

Expenditure and Savings of the Population in Terms of Sources of Means

	1970	1975	1980	1982
	\multicolumn{4}{c}{Million levs}			
Total	9,501.1	13,581.8	19,506.2	21,888.5
From earned incomes	7,369.9	10,316.5	14,219.7	15,896.6
From public consumption funds[1]	2,131.2	3,265.3	5,286.5	5,991.9
	Relative share			
Total	100.0	100.0	100.0	100.0
From earned incomes	77.6	76.0	72.9	72.6
From public consumption funds[1]	22.4	24.0	27.1	27.4

[1] Excluding remuneration during regular and additional annual leave.

Pensions

Kinds of pensions	1970	1975	1980	1982
Number of pensions as of 31 December (in thousands)	1,720	1,868	2,042	2,119
Record of service, disability due to general illness or employment accident	540	720	1,165	1,360
Old-age-war veterans'	2	2	1	1
War invalids'	41	32	25	23
For special services	52	57	61	59
Civil invalids'	2	1	1	1
Farmers'	1,072	1,008	737	622
Craftsmen's, merchants', freelances'	11	12	14	14
Social	—	36	38	39
Pensions paid (millions of levs)	718.9	1,044.1	1,685.2	1,974.2
Record of service, disability due to general illness or employment accident	356.4	593.7	1,196.1	1,478.4
Old-age-war veterans'	1.2	1.5	0.9	0.8
War invalids'	17.2	14.5	16.7	14.3
For special services	30.0	34.1	44.4	41.4
Civil invalids'	0.7	0.7	0.8	0.7
Farmers'	307.4	382.9	400.3	411.0
Craftsmen's, merchants', freelances'	6.0	8.1	11.6	12.8
Social	—	8.6	14.4	14.8
Average annual size of pension (in levs)	423	566	834	938
Record of service, disability due to general illness or employment accident	680	858	1,077	1,121
Old-age-war veterans'	576	749	783	932
War invalids'	406	447	655	612
For special services	594	599	721	688
Civil invalids'	439	467	588	550
Farmers'	288	378	521	636
Craftsmen's, merchants', freelances'	549	691	871	912
Social	—	251	380	382

Pensions According to Annual Size of Pension (thousands at year's end)

Size of pensions	1970	1975	1980	1982
Total	1,720	1,868	2,042	2,119
Up to 300 levs	710	353	7	5
301–360 levs	333	195	79	67
361–480 levs	254	500	305	110
481–720 levs	229	378	667	590
721–960 levs	122	194	337	506
961–1,200 levs	40	134	348	418
1,201–1,440 levs	13	50	140	189
Above 1,440 levs	12	61	157	233
Undeclared	7	3	2	1

State Savings Bank Network

Divisions	1970	1975	1980
Branches	211	254	196
Sub-branches	98	107	195
Representatives	3,038	2,991	2,939
Including those with post offices	2,062	2,109	2,194
Consumer co-operatives	57	17	7
Other enterprises and establishments (co-operative farms included)	898	752	633

Insurance Instalments and Insurance Benefits Paid (millions of levs)

Kinds of insurance	1970	1975	1980	1982
Insurance instalments				
Total	**205.9**	**369.6**	**538.1**	**629.1**
Compulsory insurance	78.9	144.9	348.5	390.8
Property	72.7	135.6	334.8	376.1
Passengers	6.2	9.3	13.7	14.7
Voluntary insurance	127.0	224.7	189.6	238.3
Agricultural crops	39.4	55.8	0.1	0.1
Livestock	16.7	28.6	3.3	4.0
Transport, fire, etc.	18.9	47.7	39.8	46.1
Life	40.4	75.8	129.6	167.4
Accident	2.7	5.4	9.7	13.1
Reinsurance	8.9	11.4	7.1	7.6
Insurance benefits paid				
Total	**137.8**	**256.4**	**481.2**	**366.9**
Compulsory insurance	28.4	83.3	373.9	243.4
Property	27.7	82.6	373.0	242.4
Passengers	0.7	0.7	0.9	1.0
Voluntary insurance	109.4	173.1	107.3	123.5
Agricultural crops	61.4	80.7	0.1	0.1
Livestock	9.7	24.6	3.8	4.6
Transport, fire, etc.	5.1	13.3	19.7	12.0
Life	27.2	43.5	75.3	96.6
Accident	1.1	2.0	3.0	4.1
Reinsurance	4.9	9.0	5.4	6.1

Bulgaria in Figures

Statistics of the Country's Social and Economic Development

	1939	1948	1952	1957	1960	1965	1970	1975	1980	1982
Foreign trade indices (1939—100)										
Exports	100	78	123	275	495	977	1,700	2,900	4,400	5,200
Imports	100	66	140	325	544	1,100	2,000	3,200	5,900	7,200
	100	88	109	233	453	859	1,400	2,700	3,100	3,500
Trade structure by countries in %	100.0	100.0	100.0	100.0	100.0	100.0	100.0	100.0	100.0	100.0
including the socialist countries	0.0	82.8	88.7	84.5	83.9	76.8	77.8	75.8	74.7	74.4
with the USSR alone	0.0	55.3	57.1	53.5	53.1	51.1	53.0	52.5	53.5	53.8
Per capita consumption funds (levs)			279	324	412	589	874	1,214	1,772	1,966
Per capita consumption funds (levs)			256	308	394	569	839	1,142	1,685	1,871
Indices of per capita consumption (1952=100)			100	145	186	250	339	451	544	591.4
Nominal wages (levs)		485	646	815	939	1,109	1,486	1,757	2,185	2,363
Indices of nominal work pay (1952=100)		75	100	126	145	172	230	272	339	367
Indices of real work pay (1952=100)			100	171	195	215	278	321	331	354
Indices of real per capita income (1952=100)			100	141	179	224	300	396	449	494
Indices of social consumption funds (1960=100)					100	145	274	415	535	601
Indices of social consumption funds per capita (1960=100)					100	138	254	374	474	529
Retail trade turnover per capita (levs)			159	218	304	439	663	947	1,364	493
Indices of per capita agricultural production (1932–38=100)	110	99	94	113	134	150	171	193	199	220
Crop raising	120	105	99	126	148	159	182	192	182	215
Stock-breeding	94	90	86	93	112	137	150	193	220	229
Production of some agricultural produce (thousand tons)										
Wheat	2,003	1,688	2,041	2,395	2,379	2,921	3,032	2,996	3,847	4,913
Barley	414	270	429	478	622	876	1,167	1,699	1,375	1,436
Maize-grain	1,077	802	487	1,492	1,505	1,238	2,375	2,822	2,256	3,418
Sunflower seed	171	166	184	208	344	357	407	426	380	511
Sugar beet	234	560	381	1,434	1,650	1,392	1,714	1,758	1,414	1,583
Tomatoes	44	95	215	366	634	775	716	599	838	853
Potatoes	136	159	268	313	478	285	374	318	301	469

Statistics of the Country's Social and Economic Development

	1939	1948	1952	1957	1960	1965	1970	1975	1980	1982
Benefits paid (million levs)			22.6	85.0	134.0	157.8	396.3	571.4	891.5	906.9
Scholarships paid (million levs)			5.5	6.1	8.1	14.5	19.6	30.0	75.9	70.3
Sales staff per 1,000 of the population					139	164	200	261	288	294
Relative share of population in water-supplied towns and villages (%)					69.8	82.7	91.6	95.4	97.3	97.7
Relative share of population in towns and villages supplied with electricity					91.7	97.5	99.6	99.8	99.9	99.97
Theatre attendance per thousand of the population	242	260	426	633	764	604	637	691	723	696
Cinema attendance per capita	2	4	5	10	14	15	13	13	11	10
Radio subscribers per thousand of the population	10	30	51	117	182	250	269	261	242	234
TV subscribers per 1,000 of the population					0	23	121	173	186	188
Hospital beds per 10,000 of the population	17	26	43	51	56	65	75	84	90	90
Sanatorium beds per 10,000 of the population	2	3	8	11	16	20	19	21	21	21
Population per doctor	2,021	1,460	1,149	774	715	605	538	465	407	387
Population per dentist	5,240	5,535	4,093	3,542	3,304	2,855	2,737	2,360	1,834	1,717
Dwelling units (thousand)						2,085	2,283	2,510	2,839	2,970
Floor space (thousand sq.m)						86,354	96,021	116,394	131,675	137,894
Per capita floor space (sq.m)						10.5	11.3	13.3	14.8	15.4
Newly-built dwelling units				43,462	49,786	45,211	45,656	57,151	74,308	80,246
Newly-built dwelling units (thousand sq.m)				1,687	1,759	2,112	2,243	2,741	3,328	3,255
Sugar (kg per capita)			6.5	14.2	17.7	22.3	32.9	32.5	34.7	31.1
Vegetables (kg per capita)			79.6	90.7	97.2	88.8	88.9	90.1	93.8	104.9

Statistics of the Country's Social and Economic Development (continued)

	1939	1948	1952	1957	1960	1965	1970	1975	1980	1982
Fruit (kg per capita)			91.8	102.0	95.3	131.4	148.2	118.6	105.8	118.9
Cotton material (m per capita)			8.6	13.1	16.4	19.8	22.2	23.8	25.9	27.5
Woollen material (m per capita)			1.7	2.3	3.0	3.2	3.8	5.5	4.6	4.7
Silk material (m per capita)			0.3	0.4	0.9	1.3	2.1	3.4	3.9	4.1
Shoes (excluding rubber soles – pairs)			0.3	0.6	0.9	1.3	1.7	2.0	2.0	2.3
Accommodation in crèches (number)		3,425	8,961	19,908	24,164	42,994	43,695	66,424	77,369	76,735
Accommodation in crèches per 10,000 children, aged under 3 (number)		77	204	495	613	1,137	1,064	1,604	1,986	2,080
Children in kindergartens – (in thousands)	13	166	253	270	284	324	332	393	421	404
Accommodation in kindergartens per 1,000 children aged 3–6		332	448	494	542	628	680	747	751	777
Pensions paid (millions of levs)		45.3	67.3	120.1	207.5	380.3	718.9	1,044.1	1,685.2	1,974.2
National income (million levs at prices in respective year)	5,009.0	1,576.9	2,580.6	3,208.9	4,488.8	6,635.6	10,527.4	14,288.6	20,508.6	22,849.5
National income indices (1939=100)	100	101	140	203	282	390	593	864	1,200	1,300
Per capita national income indices (1939=100)	100	89	121	167	226	299	440	624	825	897
National income structure by sectors (%)	100	100	100	100	100	100	100	100	100	100
Industry	15	23	29	41	48	49	55	54	51	54
Construction	3	4	7	7	7	8	9	9	9	10
Agriculture	65	58	39	34	27	27	17	18	19	19
Forestry	0	1	1	0	1	1	1	1	0	0
Transport	2	2	2	3	4	5	6	7	7	7
Communications			0	0	0	0	1	1	1	1
Trade, material–technical supply and purchase of produce	12	8	19	12	11	8	9	8	10	6
Other sectors of material production	3	4	3	3	2	2	2	2	3	3

Statistics of the Country's Social and Economic Development (continued)

	1939	1948	1952	1957	1960	1965	1970	1975	1980	1982
Indices of utilized national income (1952=100)			100	142	205	288	429	648	745	818
Accumulation fund			100	120	237	353	590	1,100	1,100	1,200
Consumption fund			100	149	195	269	380	533	650	709
Structure of utilized national income (%)			100.0	100.0	100.0	100.0	100.0	100.0	100.0	100.0
Accumulation fund			23.8	20.1	27.5	28.3	30.8	32.8	25.0	25.3
Consumption fund			76.2	79.9	72.5	71.7	69.2	67.2	75.0	74.7
Indices of national income per person employed in material production (1956=100)			83	114	174	245	363	528	71	765
Indices of total industrial output (1939=100)	100	203	430	782	1,200	2,100	3,600	5,500	7,400	8,100
Capital goods (group A)[1] production	100	329	821	1,500	2,700	5,200	9,200	14,900	21,000	22,700
Consumer goods (group B)[2] production	100	166	315	557	793	1,200	2,000	2,900	3,500	4,000
Electric- and thermo-power production	100	200	490	1,000	1,800	3,600	6,700	8,600	12,700	14,800
Fuel industry	100	184	299	572	766	1,800	3,700	5,600	7,300	7,200
Ferrous metallurgy (including ore-mining)	100	56	333	3,100	5,300	17,600	41,100	69,000	90,700	95,000
Machine-building and metal-working industry	100	622	1,800	4,000	8,600	20,000	41,200	81,800	129,700	150,500
Chemical and rubber industry	100	220	688	1,800	3,700	8,000	21,200	36,700	57,700	64,000
Building material industry	100	199	535	1,300	2,400	5,200	8,500	13,200	19,000	20,200
Timber and woodworking industry	100	227	317	488	653	871	1,100	1,500	1,700	1,800
Indices of labour productivity in agriculture (1939=100)	100	95	106	135	218	312	448	637	762	886
Total area of forests (thousand ha.)			4,032	3,704	3,635	3,612	3,709	3,797	3,845	3,859
Afforested area (thousand ha.)			3,276	3,194	3,190	3,144	3,162	3,228	3,293	3,305
Freight carried (million tons)	8.9	14.7	39.2	82.8	169.3	345.0	579.2	826.7	926.7	1,012.1
(million tons per km)	1,162	2,173	4,253	7,885	12,175	24,736	62,555	74,271	99,476	97,721

Statistics of the Country's Social and Economic Development *(continued)*

	1939	1948	1952	1957	1960	1965	1970	1975	1980	1982
Passengers carried (million)	17.8	60.2	69.1	92.6	153.1	256.6	442.6	683.5	833.6	890.8
(million passengers per km)		2,885	2,963	3,625	5,403	8,581	14,756	23,069	25,205	26,174
Annual average carriage per person employed in transport										
Railway transport (thousand tons per km)		145.9	184.5	217.9	246.6	341.4	421.0	488.7	445.1	490.1
Motor transport (levs)						3,575	4,764	6,671	11,863	12,879
Water (sea) transport (thousand tons per km)		707.4	869.7	865.2	2,243.2	3,720.2	7,550.7	7,815.0	11,095.5	10,737.5
Water (river) transport (thousand tons per km)		219.4	498.3	577.7	640.6	889.7	1,246.6	1,466.0	1,531.5	1,560.6
Air transport	—	—	1,673	2,780	4,389	8,720	15,862	25,331	34,324	39,613
Capacity of the telephone exchanges in towns and villages at year's end (thousand sets)	35.0	67.5	111.4	127.2	150.9	239.4	387.8	617.5	1,181.7	1,388.6
Telephone sets at year's end (thousand sets)		57.9	94.2	127.6	171.7	279.2	473.0	777.1	1,255.8	1,513.4
Retail trade (million levs at prices in respective year			1,155.1	669.8	2,388.3	3,599.1	5,627.4	8,262.1	12,083.9	13,315.8
Retail trade index (1952=100)			100	200	293	415	628	914	1,100	1,200
Foreign trade (million currency levs at prices in respective year)	116.7	301.9	384.9	821.6	408.7	2,753.6	4,486.8	9,777.0	17,184.4	21,855.9
Exports	63.4	149.4	199.8	433.0	668.6	1,375.7	2,344.5	4,541.4	8,901.5	10,880.0
Imports	53.3	152.5	185.1	388.6	740.1	1,377.9	2,142.3	5,235.6	8,282.9	10,975.9
Capital investment indices (1949=100)			149	191	412	602	1,100	1,600	2,000	2,300
Mixed capital put into operation (million levs)			451.5	670.5	1,029.1	1,681.9	3,190.5	4,565.8	6,353.1	7,670.9

Statistics of the Country's Social and Economic Development (continued)

	1939	1948	1952	1957	1960	1965	1970	1975	1980	1982
Domestic product (million levs at prices in respective year)	7,316.2	2,431.0	4,945.2	6,823.4	10,310.1	15,976.5	26,243.0	37,051.1	57,902.0	66,509.3
Domestic product indices (1939=100)	100	123	185	282	412	621	980	1,400	1,900	2,100
Domestic product structure by sectors (%)	100	100	100	100	100	100	100	100	100	100
Industry	19	32	45	55	58	63	65	66	66	67
Construction	5	6	9	8	9	9	9	9	9	9
Agriculture	59	49	31	26	22	19	14	14	13	13
Forestry	0	0	0	0	0	0	0	0	0	0

[1] Group A=heavy industry.
[2] Group B=light industry.

Power Consumption

	1970	1975	1980	1982
Power consumption per capita (kWh)	290.7	526.0	772.3	908.9
Indices (1970=100)	100.0	180.9	265.7	312.6

Food Consumption per Capita (calories per day)

Countries	Total			Vegetable products			Animal products			
	1969–71	1975–77	1978–80	1969–71	1975–77	1978–80	1969–71	1975–77	1978–80
Austria	3,431	3,435	3,495	2,153	2,074	2,084	1,279	1,361	1,411
Bulgaria	3,512	3,590	3,638	2,982	2,860	2,870	530	729	768
France	3,371	3,355	3,390	2,197	2,130	2,162	1,174	1,225	1,228
FRG	3,351	3,376	3,537	2,059	2,076	2,161	1,292	1,300	1,376
Great Britain	3,352	3,247	3,316	2,048	2,016	2,088	1,305	1,231	1,228
GDR	3,448	3,635	3,746	2,229	2,302	2,355	1,219	1,333	1,391
Greece[1]	3,087	3,424	3,365	2,521	2,725	2,608	566	699	757
Italy	3,496	3,463	3,650	2,819	2,669	2,754	678	794	896
Poland	3,382	3,550	3,520	2,312	2,318	2,286	1,070	1,232	1,233
Romania	3,027	3,358	3,395	2,499	2,611	2,562	528	747	833
Turkey	2,808	2,934	2,965	2,517	2,634	2,663	292	300	303
Yugoslavia	3,336	3,526	3,511	2,716	2,750	2,684	602	776	826

[1] In periods 1969–71; 1974–76; 1977–79 (fish consumption excluded).

Communal Services Per Capita (prices in levs as of 1 January 1982)

	1981	1982
Metalworking services	2.53	2.35
Maintenance and repair of household appliances	2.09	2.36
Maintenance and repair of motor vehicles	6.07	7.31
Maintenance and repair of radio and TV sets	2.89	3.03
Optician's services	0.11	0.12
Cabinet-making and joiner's services	8.52	8.89
Maintenance and repair of lifts	0.61	0.51
Textile and knitwear services	5.67	5.77
Tailoring services	7.02	7.04
Repair of footwear and leather haberdashery	2.23	2.27
Dry cleaning	0.96	1.04
Washing	1.03	1.16
Barber's and hairdresser's services	3.10	3.32
Miscellaneous services	1.77	1.94
Public bath services	1.13	1.04
Mason's services	2.80	2.90
Photographic services	1.62	1.67

Servicing by Retail-Sale and Public-Catering Establishments

	1970	1975	1980	1982
Number of retail-sale per 10,000 of the population	38	41	42	44
Population per retail-sale establishment[1]	260	247	238	225
Commercial area per capita (sq.m)	0.15	0.17	0.19	0.20
Number of public-eating establishments per 10,000 of the population	23	25	26	28
Population per public-eating establishment	431	403	381	355
Seats in public-eating establishments per 10,000 of the population	1,110	1,305	1,528	1,668

[1] Shops, pavilions, stands and kiosks, not commercial organizations.

Medical Services

Countries	Years	Doctors	Population per doctor	Years	Hospital beds	Population per bed
Albania	1977	2,641	991	1977	16,313	156
Austria	1981	19,157	392	1981	83,687	90
Bulgaria	1981	22,088	402	1981	80,791	110
Czechoslovakia	1981	43,246	354	1981	156,167	98
France	1976	86,306	613	1976	540,800	97
FRG	1980	139,431	442	1980	707,710	87
GDR	1981	34,626	483	1981	171,157	98
Great Britain	1977	85,552	653	1977	462,108	121
Greece	1980	23,469	410	1980	60,067	160
Hungary	1981	28,658	374	1981	96,882	111
Italy	1980	167,213	341	1980	542,260	105
Poland	1981	64,896	525	1981	203,453	177
Romania	1980	32,762	678	1980	208,213	107
Turkey	1980	27,241	1,631	1980	99,117	448
USSR	1981	1,033,900	259	1981	3,384,000	79
Yugoslavia	1977	28,528	768	1976	129,983	166

Water Economy (at end of year)

	Measure	1970	1975	1977	1978	1979	1980	1981	1982
Fixed assets of complete initial value[1]	thousand levs	373,370	613,247	778,827	844,945	912,646	958,758	1,027,300	1,089,410
Workers and employees	average annual	5,577	5,740	5,332	5,042	5,314	5,726	5,478	5,656
including workers[2]	number	2,809	2,957	3,999	3,698	3,904	4,714	4,031	4,198
Annual average salary of workers and employees	levs	1,443	1,678	1,769	1,651	2,120	2,549	2,292	2,222
including workers[2]	levs	1,230	1,459	1,635	1,531	1,846	2,503	2,084	2,018
State dams used for irrigation	number	99	125	136	139	144	151	154	161
Holding capacity	million cu.m	1,358	2,266	2,481	2,434	2,356	2,364	2,421	2,561
State pumping stations	number	438	547	567	609	594	605	664	695
Installations in them	thousand kWh	219	354	384	414	427	552	512	527
Dams owned by agricultural enterprises	number	1,893	2,000	2,001	2,009	1,957	1,957	2,030	1,903
Holding capacity	million cu.m	589	638	643	645	641	641	665	633
Spraying machines and installations	number	8,999	12,766	12,962	13,787	14,881	14,280	8,193	9,820

[1] For 1977 – according to replacement value.
[2] Only those employed in basic activity.

Environmental Protection and Restoration Investment by Sectors

Sectors	1975	1977	1978	1979	1980	1981	1982
Total	**47,296**	**83,067**	**95,991**	**116,449**	**142,877**	**141,823**	**146,233**
Water	21,135	45,842	49,971	59,824	76,395	76,857	70,504
Soil	3,497	10,246	17,858	22,472	30,173	28,226	34,326
Air	22,664	26,979	28,162	33,584	35,317	26,243	17,060
Waste-free technologies and use of recycled water	—	—	—	569	992	4,298	19,427
Noise protection	—	—	—	—	—	6,199	4,916
Total	**27,237**	**51,210**	**60,651**	**74,785**	**100,453**	**97,713**	**94,658**
Water	18,783	31,509	38,096	44,883	56,723	55,749	52,961
Soil	993	5,241	10,092	11,356	16,512	15,008	19,917
Air	7,461	14,460	12,463	18,138	26,275	19,485	10,453
Waste-free technologies and use of recycled water	—	—	—	408	943	1,700	6,847
Noise protection	—	—	—	—	—	5,771	4,480

Spending on Protection and Restoration of the Environment[1]

	1975	1977	1978	1979	1980	1981	1982
				Millions of levs			
Total	88.2	141.1	157.6	172.3	214.9	217.4	223.8
Including capital investments	47.3	83.1	96.0	116.4	142.9	141.8	146.2
				Indices (1975=100)			
Total	100.0	160.0	178.7	195.4	243.7	246.5	253.7
Including capital investments	100.0	175.7	203.0	246.1	302.1	299.8	309.1

[1] Unplanned spending excluded.

Statistics of the Country's Social and Economic Development

	1939	1948	1952	1957	1960	1965	1970	1975	1980	1982
Pupils and students per 10,000 of the population	1,697	1,604	1,550	1,682	1,792	1,981	1,825	1,752	1,642	1,678
University students per 10,000 of the population	16	55	41	53	70	103	105	121	96	94
Book stock of libraries per 1,000 of the population			1,119	2,002	2,736	4,395	5,527	6,818	8,125	8,504
Annual total print of books per capita – items	1.0	2.8	2.7	3.2	3.8	4.8	4.8	5.6	6.1	6.7
Annual total print of magazines per capita – items	1.8	1.5	0.8	1.9	2.7	3.1	5.3	5.7	7.3	6.9
Annual total print of newspapers per capita – items	20.7	48.5	44.0	31.3	76.6	73.4	96.2	96.7	101.8	102.9

Schoolchildren and Students

Countries	Years	Pre-school establishments	Elementary schools	Secondary general schools	Secondary vocational schools	Secondary teacher-training schools	Higher and semi-higher educational establishments — Students	Percentage of women	Number of students per 10,000 of the population
Austria	1980	160,948	400,397	583,382	150,981	5,339	127,746[1]	42[1]	171[1]
Bulgaria	1980	420,804	994,018	91,863	222,890	—	99,354[2]	56[2]	112[2]
Czechoslovakia	1980	694,720	1,904,476	145,395	228,709	14,457	190,929[1,2]	41[1,2]	126[1,2]
France	1980	2,383,465	4,621,670	3,911,054	1,104,393	—	1,060,412[1]	46[1]	199[1]
FRG	1980	1,535,959	5,044,424	3,690,340	610,400	—	151,978[1]	41[1]	189[1]
GDR	1980	663,491	2,203,991	46,927	459,485	—	399,204[1,2]	57[1,2]	237[1,2]
Great Britain	1979	346,687	5,133,710	5,116,354	243,743	—	799,462	36	143
Greece	1978	137,146	922,698[3]	585,130[3]	125,039	—	117,407[1]	39[1]	127[1]
Hungary	1980	478,100	1,162,203	89,400	107,491	5,897	103,469[1,2]	50[1,2]	97[1,2]
Italy	1980	1,840,555	4,435,217	3,493,003	1,579,374	236,218	1,097,954[1]	42[1]	194[1]
Poland	1980	1,349,528	4,167,313	345,214	1,309,952	18,703	609,997[1,2]	56[1,2]	172[1,2]
Romania	1980	935,711	3,236,808	80,879	784,061	6,317	192,769[2]	43[4]	87[2]
Turkey	1980	4,691[6]	5,656,494	1,710,610[1]	497,114[1]	17,809[1]	369,864[1]	25[1]	61[1]
USSR	1980	10,212,000	21,714,000	17,356,000	2,919,000	433,300[3,5]	5,236,000[2]	52[2]	197[2]
Yugoslavia	1980	290,870	1,431,582	1,835,636	587,895	2,633	448,755[1]	40[1]	203[1]

[1] 1979.
[2] Including extra-mural and night classes.
[3] Including pupils at night schools.
[4] 1978.
[5] Including private pupils.
[6] Kindergartens only.

Bulgaria in Figures

Persons and Work-forces Awarded Titles, Orders and Medals

	1961–81			1982		
	Men	Women	Work-forces	Men	Women	Work-forces
Title of 'Hero of the People's Republic of Bulgaria'	31	1	—	5	1	—
Title of 'Hero of Socialist Labour'	909	107	—	75	15	—
Order of Georgi Dimitrov	1,716	191	263	185	17	10
Order of People's Republic of Bulgaria, 1st–3rd class	6,929	527	338	966	118	34
Order of 9 September, 1944, 1st–3rd class	33,478	6,917	47	3,142	938	3
Order of People's Freedom 1941–44, 1st–2nd class	40,700	13,244	2	126	37	—
Order of Red Banner of Labour	14,533	1,360	1,335	1,148	118	72
Order of Red Banner	1,246	5	7	—	—	—
People's Order of Labour	115,902	20,750	1,216	3,963	938	58
Order of Cyril and Methodius, 1st–3rd class	19,071	10,566	2,723	858	468	157
Medal of Labour Distinction	18,505	6,603	1	—	—	—
Honoured with other titles and orders (People's, Merited, Honorary, etc.)	29,143	4,837	28	2,323	442	—
Honoured Mothers of Many Children	—	323,531	—	—	10,649	—
Including those awarded the title of 'Heroine Mother'	—	495	—	—	19	—
'1300 Years of Bulgaria' Jubilee Medal	100,017			10,937		
'Georgi Dimitrov Birth Centenary' Jubilee Medal	—			107,933		

INDEX

This index does not cover the Appendix on pp. 917–63 nor participants in certain activities who are named in lists or dense textual sequences. For these latter see 'personalities' under the following entries: architecture, art, cinema, circus, music, sport, theatre.
Portraits are noted ('port.') only when they are on a different page from the text.

Academy of Social Sciences and Management, 288, 633, 651–2
acrobatics, 576
administrative and economic districts, 3, 81–134, 237, 239, 241
advertising, 529–30
Agrarian Association of Farmers, 223
Agrarian Defence, 97
Agricultural Academy, 648, 650–1, 857
agricultural land, 32, 47, 54, 57, 62; arable, 23, 38, 47, 52–4, 246, 344, 346
agriculture, 23, 32, 75, 179, 194, 200, 221, 246–7, 290, 297, 342–4, 408–34; chemicals in, 37, 61; research, 673–5
agro-industrial complexes (AIC), 247, 277, 349, 411–13, 483. See also National Agro-Industrial Union *and under* District Centres, 81–134
air force, 329
air transport, 83, 88, 91, 443, 449–51
Akyov, V., 732
Alabin, P., 837
Aladja Monastery, 65
Albania, 251, 253, 892
Albena, 22, 505, 507, 763
Albert, R., 820
Alexander I, 194–9
Alexander, *Exarch,* 186
Alexander Nevski Church, 84, 743
Alexander of Macedon, 142
Alexandrov, Ch., 287
Alexandrov, T., 220
Alexius I, 160–1
Alexius III, 163
Allusian, 159–61
alphabet, 77, 80, 153, 155, 550, 613, 641, 698–9
Andonov, M., 790, 797
Andreev, V., 727, 735
Andreichin, I., 713, 720, 784
Andrew II, 164
Angelarius, 698
Angelov, D. (historian), 679
Angelov, D. (novelist), 728

Angelov, I, 746; port. 745
Angelov, Marko, 726
Angelov, Mihail, 807
Angelov, Ts., 737
Antim I, 119, 185
antimony, 27
Aprilov, B., 794
Aprilov, V, 80, 180, 613, 696, 704–5
Arabs, 148, 154; Palestinian, 898
Arbitration Court, 517
architecture, 753–64; personalities, 758–63. See also under District Centres, 81–134
archives, 834–5
Arda, river valley, 20–1, 31, 40–1, 43
area, 3
aristocracy, 152–3, 155, 158, 163–4
armed forces, 193, 203, 233–4, 250–1, 253, 283, 323–36
Armenians, 86; newspaper, 856
Arnaudov, M., 697, 725, 737; port. 695
Arnaudov, V., 807
Arnulf, *King,* 153
art, 740–51; amateur, 805–9; exhibitions, 752, 814–15, 817; personalities, 745–52
Arzhentinski, I., 733
asbestos, 25, 28
Asparuh, *Khan,* 146, 148
Assen I, 81, 88, 115, 117, 129, 161–3
Assenov, D, 731, 733–4, 736, 792
Atanassov, G, 767; port. 766
Atanassov, G.I., 288
Atanassov, N., 767
Atanassov, V., 872
Athens (Ancient), 142
athletics, 579
Athos, Mount, 700
Austria, 186, 195, 198–9, 206–7, 209–10, 251–2, 902; wars with Turkey, 176–7
automation, 675
automotive industry, 369
Avars, 95, 97, 124, 145–6, 148–9, 152
Average Adjuster, 518

Avramov, Ts., 301
awards: cultural, 818; state, 4, 287
Azaryan, K., 790, 793

Bagryana, E., 722, 724, 730, 737
Bagryanov, I., 229
Bahmetiev, P. 644
Bakalov, G., 88, 644, 719, 725, 737
Balabanov, B., 736
Balarev, D., 644
Balchik, 507
Baldwin I, 163
Balev, M., 287
Balevska, R., 674
Balevski, A., 669–70
Balkan Airlines, 83, 527
Balkan Alliance/Wars, 207–8, 290, 299
Balkan federation idea, 206, 290
Balkan mountains, 3, 5, 7, 9–10, 14–16, 18, 22, 31–4, 36–43, 45–7, 51–4, 57, 59, 66
Balkan relations, 252–4, 891–5, 905
Balkan studies, 892
Balkantourist, 492–3, 495, 514, 527–8
ballet, 772–3
Baniesh, M., 789
banks and credit, 177, 201, 206, 343–5, 481, 486–7, 518
Barbusse, H., 220
barium, 27
Baruh, A., 725, 727
Baruh, V., 733
barytes, 25–8
Basil II, 124, 157–9
basketball, 576
Bayer, L., 96
Bazim, H., 820
beaches, 21–2, 89, 94, 496, 505–14
bee-keeping, 434
Belassitsa mountains, 3, 8, 19, 43, 47, 51, 53, 59
Belev, G., 725, 727–8
Belev, K., 725, 728
Belgium, 903
Benderev, A., 198–9
Benevolent Society, 188–9

965

Berlin Treaty (1878), 86, 118, 122, 133, 192, 194–7, 199, 205, 343
Bernsteinism, 289–90
Beron, P., 80, 180, 182, 551, 613–14, 642, 704
Beshevliev, V., 679–80
Beshkov, A., 644
bibliography, 831–3
biology, 666–8
bismuth, 25, 27
Bitola, 158; uprising (1903), 206
bitumens, 24
Black Sea, 3, 6, 30, 36, 38, 60, 440; coast, 21–2, 30–3, 36–8, 48, 53–5, 57–8, 61, 89, 93, 505–14
Blagoev, D: and Communist International, 213; as editor, 853; on health services, 552; as Marxist theoretician, 644–5, 687, 719, 737; on library clubs, 825; local activities, 95, 101, 115, 119; party activity, 201, 288–90; in local areas, 95, 101, 115, 119; port. 200; schooling, 95; and teachers' associations, 315; on theatre, 783; *What is Socialism?* 645, 866; on World War I, 210
Blagoeva, V., 101, 618, 713
Blagoevgrad, 3, 118
Bluskov, I., 707–8
boating, 590
Bobekov, P., 805
Boboshevski, Ts., 233
Bogomils, 155–6, 161, 164, 171, 551, 613, 700
Bogoridi, Alexander, 194, 196
Bogoridi, Atanas, 551, 642
Bogoridi, S., 642
Bogorov, I., 80, 696, 705, 709, 851
Bonchev, G., 644
Bonchev, N., 709
Bonchev, S., 644
Bonev, P., 731
bookselling, 867, 872, 878
borders, 3; guards, 335
Boril *Tsar*, 163–4
Boris I, 152–4, 551, 698–9
Boris II, 156–7
Boris III, 212, 223, 226, 229
Borovets, 497, 763
Bosnia, 172, 189, 207
Bossev, A., 737
Bossev, R., 735
Botev, H., 108, 182, 189–90, 642, 689, 705, 707, 709, 780, 851
Boue, A., 72, 643
Bozhilov, B., 731
Bozhilov, D., 229
Bozhinov, T., 287
bourgeoisie, 177–9, 180, 187, 190, 193, 195, 198, 201, 343–4
boxing, 573, 577
boyars *see* aristocracy
Bozveli, Neophyte, 183, 613, 704, 706
Bucharest Peace Treaty: (1886), 198; (1913), 103, 122, 208, 210

Buchvarov, M., 677
Buchvarova, S., 735
Budevska, A., 782
budget, 281, 481–6
building materials, 379, 670
Bukoreshtliev, A., 767
Bulgaria Today, 874
Bulgarian Academy of Science, 83, 633; Archaeological Institute, 101; archives, 834; and Bulgarian language, 80; Centre for the Science of Science, 683; co-operative research, 659–60; and environmental protection, 63, 66, 71; and folklore, 697; foundation, 182, 643; functions, 647–51; Institute of Art Studies, 749; Institute of Literature, 737; Institute of Metal Sciences, 669; Institute of Musicology, 779; Institute for Town Planning, 763; Laboratory of Physico-Chemical Mechanics, 670; library, 828–9, 832; museum collections, 836–7; publications, 857, 869
Bulgarian Agrarian Party/Union (BAP/BAU), 275, 299–303; foundation, 97, 201; and World War I, 209–10; and Soldiers' Mutiny, 212; Government, 212–17, 344, 481; and cinema, 795; and Communist Party, 220, 291, 294, 346; in elections, 220, 222, 224, 242, 290; in Fatherland Front, 229, 233, 235–7, 295, 305–6; publications, 852, 856, 869; in foreign relations, 894, 900
Bulgarian Artists' Union, 747, 749, 752
Bulgarian Association for Recreation and Tourism, 514
Bulgarian Central Charity Society (BCCS), 190, 192
Bulgarian Communist Party (BCP): Central Committee, 81, 84, 287–8; Congress, 287; General Secretary, 237–8, 242, 296–7; Politburo, 287; Secretariat, 63, 287; archives, 835; and culture, 687–8; and the environment, 62; foreign policy, 883; and health services, 552; publications, 853, 856, 869; relations with Soviet Party, 887; 1971 Programme, 241, 297; April 1956 Plenum, 239, 245–7, 249–50, 253, 296, 302, 307, 317, 347, 361, 443, 468, 492, 543, 553, 517–2, 616, 688, 690, 726, 730, 731, 736, 771, 790, 796, 806, 827, 834, 854; history, 287–99; foundation, 213, 290; and Agrarian Party, 214–17, 301; and 1923 Uprising, 219; in elections, 212; in 1920s and 1930s, 219–24, 302; transformation into Bulgarian Workers' Party, 224; since World War II, 238–42, 244,

246, 248, 252; local organizations: Kurdjali Regional Committee, 122; Silistra organization, 121; Smolyan groups, 133; Vratsa District Committee, 113
Bulgarian Communist Youth Union, 290, 311
Bulgarian Foreign Trade Bank, 486, 518
Bulgarian language, 4, 75, 77–80, 180, 182, 613, 680, 699
Bulgarian Literary Society *see* Bulgarian Academy of Science
Bulgarian National Bank (BNB), 481–2, 486, 491, 518
'Bulgarian Question', 185–92, 197–8
Bulgarian Revolutionary Central Committee (BRCC), 90, 189–90; *for local committees see under* District Centres, 81–134
Bulgarian River Shipping Co., 91
Bulgarian Social Democratic Party (BSDP), 201, 212, 215, 217, 224, 288–9, 294, 307, 315; district organizations, 82, 86, 99, 115, 128
Bulgarian Social Democratic Union (BSDU), 201, 288–9; newspapers, 853
Bulgarian Telegraph Agency (BTA), 83, 860, 864–5
Bulgarian Travel Agency, 492; *see also* Balkantourist
Bulgarian Typographic Society, 201, 315
Bulgarian Workers' Party, 118, 224, 226–7, 229, 231, 233, 294, 304–7
Bulgarian Workers' Social Democratic Party (BWSDP), 201–2, 233, 235–7, 289–90, 306, 569, 784, 805; Central Committee, 82, 88; district organizations, 95, 97, 104, 106, 111, 113, 118, 119, 121, 126, 129; newspaper, 853
Bulgarian Worker's Social Democratic Party (left-wing socialists), 202, 209–12, 215, 315, 569
Bulgarian Workers' Union (BWU), 223, 316
Bulgarska Dnevnitsa, 851
Bulgarski Orel, 704, 851
Bunkov, A., 677
Burgas, 3, 5, 93–4, 505, 512
Burmov, A., 677
Burmov, T., 193, 195
Byala Cherkva, 20
Byzantium: conflicts and conquests, 81, 85, 88, 95, 97, 101, 107, 115, 119, 124, 129, 133, 141, 144, 145, 146, 148–63, 166–7, 169–70, 172; cultural influence, 613, 700, 741–2, 754

cadmium, 25, 27
camping sites, 492–3
Canada, 904

Index

Canetti, E., 92
canoeing, 578–9
capital cities, 3, 81–2, 115, 154, 157–8, 165, 193
Capitalism, 177–9, 187, 193–232, 342–4, 359; control of, 244
Carpathian mountains, 5–6, 8, 30, 37
catering, public, 461–3, 549
cattle see stock-breeding
caves, 588, 591
Celts, 104, 143
cement, 28
Central Co-operative Union, 241, 277, 282, 345, 465–6
Central Council of Trade Unions, 241
Central Institute for Scientific and Technical Information (Cinti), 655–6
Chakalov, L., 661
Chamber of Commerce, 517
Chavdarov-Chelkash, D., 725, 729
Chelebi, Aha, 133, 172
chemical industry, 38, 375–8
chemicals in agriculture, 37, 61
chemistry, 665–6
Cherevin, P. A., 133
Cherkasky, *Prince* Vladimir A., 193, 481
Chernyshevsky, A.G., 188
Chervenkov, Vulko, 238, 296
chess, 583–4
Chilingirov, S., 831
Chintulov, D., 706, 805; port. 705
Chonev, M., 807
Christianity, adoption of, 151–5, 613, 699, 741
chromium, 25
Chudomir, 723
Church, Eastern Orthodox: adherents, 75; Greek dominance,159, 179–81, 183–4; independent Bulgarian, 153–5, 157, 163, 165, 175, 177, 185–6; and early literature, 700; in Macedonia, 200. *See also under* District Centres, 81–134
Church Roman Catholic *see* Roman Catholic Church
church music, 765–7
Church Slavonic, 79–80
cinema, 682, 795–800, 874; personalities, 795–800
circus/personalities, 801–4
Civil Defence, 334–5
civil obligations, 279
civil rights, 277–8, 279, 911
clay, 28
climate, 21, 22–3, 30–3, 38, 42, 506–14. *See also under* District Centres, 81–134
climbing, 590
Closier, M. de, 209–10
coal, 24, 104, 364
coast *see* Black Sea coast
cobalt, 25, 27

collective farms, 76, 234, 239, 246–7, 302, 345, 347–8, 408–10, 465, 483; personal plots, 413
colonial independence, 910–11
Commission for State Control, 237
Committee for Culture, 276, 809, 811, 837, 857, 873
Committee for Environmental Protection (CEP); 62–4
Committee for Science, Art and Culture, 237, 248, 806
Committee for Scientific and Technical Progress, 276
Committee for State and Public Control, 276
Communist International, 213, 217, 224, 290–1, 294
Communist Manifesto, 90
Communist Party *see* Bulgarian Communist Party; Bulgarian Social Democratic Party; Bulgarian Workers' Party; Bulgarian Workers' Social Democratic Party; Workers' Party
commuting, 76
computing equipment, 371, 661, 670
conference venues, 89, 492
construction industry, 399–407
Conservative Party, 193, 195–6, 201, 614
Constantine IV, 146
Constantine V, 148
Constantine VI, 149
Constitution: 'Turnovo', 115, 193–6, 207; 1947 'Dimitrov', 233, 236–7, 253, 296; of 1971, 238, 241, 297; provisions, 275–80; and education, 615; and environment, 62; and foreign policy, 883–4; and freedom of speech, 858; and health care, 550, 553; and peace, 905; and public organizations, 287; and religion, 75
Constitutional bloc, 215
consumption, 546–8
co-operatives, 277, 343, 408, 459–66
copper, 25–7
copyright, 528–9, 877
Council for Mutual Economic Assistance (CMEA), 4, 64, 247, 253–5, 276, 350, 360, 362, 468–71, 474, 476, 567, 647, 650–1, 656, 659, 682–3, 884, 886–8
Council of Ministers, 62, 71, 81, 84, 237, 246, 280–6, 345, 859, 884; Chairman, 241, 242
coup d'état (1934), 223
courts, 282, 284–5, 307
Craiova, Treaty of, (1940), 103, 121, 251
credit *see* banks and credit
crest, state, 4
Crimean War, influence of, 186
Croats, 152, 154

crops, 32, 46, 47, 52, 53, 54, 341, 416–26, 673–4
Crusades, 85, 159, 162–3, 165, 172, 176
Cuba, 887, 890
cultural activities, 154, 166, 171–2, 176, 200, 248–50, 279, 687–90; in the armed forces, 332–4; international, 810–20, 902; *see also under* District Centres, 81–134
cultural sites *see* monuments
cycling, 590
Cyprus, 895
Cyril and Methodius, 77, 181, 613, 637–8, 641, 681, 698–99, 765
Czechoslovakia, 215, 251, 253–4, 887, 889

Dalchev, A., 722, 730
dams, 41, 346
Damyan, *Patriarch*, 157
Damyanov, S., 680
Danailov, G., 736, 792
Danchov, G., 744, 875
Danev, S., 201, 203, 207–8, 214
Daniel, L., 790
Danov, H.G., 866, 872
Danov, H.M., 679
Danovski, B., 786–7, 789–90, 796; port.788
Danube, river, 3, 6, 13–15, 36, 38–40, 43, 45–8, 53, 91, 121, 440–1, 448
Daskalov, H., 673–4
Daskalov, R., 212, 214, 300
Daskalov, S., 727–8, 730–1, 734
Davidkov, I., 731, 733–4
Debelyanov, D., 716–18, 720; port. 715
Dechev, D., 679, 680
Delchev, B. 727, 738
Delchev, G., 205, 205; port. 203
Dembitski, H., 744
Democratic Alliance, 217, 219–20, 222, 224, 301
democratic centralism, 280, 287, 289
Democratic Party, 201, 203, 206, 209, 221, 224, 229, 235, 852
Den, 288
Denmark, 904
developing countries, 477–8, 659, 679, 763, 814, 883, 896–9
Deyan, *Despot*, 124, 170
Dichev, S., 733
Dilov, L., 731, 733
Dimiter, H., 108, 189; port. 187
Dimitriev, R., 198–9
Dimitrov, D., 720
Dimitrov, G.: and the 1923 Uprising, 217–20, 291; Reichstag Fire trial, 223–3, 293; in Comintern, 224, 294; and Fatherland Front, 227, 235, 295, 304–5, 307; as Prime Minister, 236–7, 252, 296; and

amateur art, 806; and army, 326; and culture, 250; on the economy, 346; on health care, 552; as journalist, 854; on literature, 726–7; as Marxist theoretician, 644; on photography, 875; on physical education, 569–71; and trade unions, 315–16; and Union of Actors, 787; and war-time broadcasting, 860; and youth organizations, 311, 313; local activities, 104, 126; ports. 251, 267, 282, 308
Dimitrov, G.: Mausoleum, 84, 837–8; museum, 839; order of, 4
Dimitrov, I., 746
Dimitrov, S., 220
Dimitrova, B., 729, 731
Dimitrovgrad, 76, 759
Dimov, D., 728, 736, 792
Dimov, I. (actor), 786
Dimov, I. (metallurgist), 669–70
Dimov, S., 133
Dimov, V., 133
Dinekov, P., 681, 737
Dinkov, I., 735
disarmament, 254–5, 813, 891–2, 896, 901, 907–8, 910, 915
Djagarov, G., 272, 731, 736, 791
Djurov, Dobri and Elena, 735
Djurov, Dobri M., 287
Dobrev, K., 678
Dobrolyubov, N. A., 188
Dobroplodni, S., 101, 780
Dobrudja, 3, 13–14, 36, 38–41, 43, 46, 53, 57, 103, 121, 172, 192, 208, 211–12, 226, 251; Revolutionary Organization (DRO), 222
doctors, 551–2, 554–7, 567
Doinov, O., 287
Dokeanos, D., 765
Doksov, T., 699
dolomites, 28
Donchev, A., 735
Dondukov-Korsakov, A.M., 128, 193, 481, 551; port. 194
Dorotheus, *Bishop*, 97
Dospevski, S., 114, 133, 182, 744
Dragiev, D., 299, 300
Draginov, M., 702
Dragoicheva, Ts., 735
Dragomanov, M., 697
Dragova, N., 736
Dramaliev, K., 305
Drinov, M., 182, 193, 614, 642, 643, 710, 744
Drugar, 289, 853
Drumev, V., 182, 195, 198, 707–9, 780
Druzhba, 22, 89, 505, 509–10, 763
Duichev, I., 679
Dunavski Lebed, 182, 705, 851
Dyulgerov, P., 288, 315
Dyustabanov, Ts., 108

Eastern Rumelia, 122, 192–4; Unification with Bulgaria (1885), 86, 196–8
economics, 678–9
economy, 341–57; in Depression (1929–34), 221, 293, 344; in World War II, 227, 250, 344; post-war reconstruction, 244; introduction of socialist reform (1944), 234, 276, 344–5, 348, 350, 361. *See also* plans
Edinstvo (Unity) Committee, 118
education, 180–3, 193, 200, 279, 613–38; administration, 635–7; *see also under* District Centres, 81–134
elderly, health care of, 560–1
electoral law, 193, 196, 278, 280
electric power *see* power
electrical and electronics industries, 367–75
Elenite, 511
emigration: demographic, 73–4, 173; literary, 706, 709; political, 179, 181–2, 184–5, 187–90, 192, 220
employment, 76, 344, 544
Enchev, 672
Enravotha, 151
environment: destruction, 23; pollution, 31, 37, 61; protection, 55, 62–4, 549, 592
Ernrot, K., 196
ethnic minorities, 86
European Conference on Security and Co-operation (ECSC), 255, 883, 891, 900–1, 905–7
Euthymius of Turnovo, 79, 115, 641, 700–1, 765

fairs, 99, 111, 132, 177, 530
family allowances, 606–7
Fascism, 215, 217, 219, 222–4, 291, 301; eradication, 248
Fatherland Front, 94, 227, 229, 231–2, 233–7, 243, 248, 250–2, 282, 287, 295, 302, 304–9, 345, 566; National Committee, 82, 229, 233, 237, 244, 295, 305–6; National Council, 63, 241, 307, 309; publications, 856, 869; newspaper, 856; publishing house, 869
fauna, 56–61
fencing, 581
Ferdinand I, 199–200, 202–3, 207–8, 210, 212, 299, 712
festivals: cultural, 691, 752, 778, 788, 800, 806–8, 817–18; in district centres, 84, 87, 89, 94, 96, 98, 99, 103, 105, 108, 112, 117–18, 121, 123, 127, 128–9, 132
feudalism, 145, 150, 152, 155, 159, 167, 173, 175, 177, 341–2
Ficheto, K., 117
Fichev, N., 131, 183, 756
Filipov, F., 789, 791

Filipov, G., 242, 287
Filov, B., 226–7, 229
Finland, 904
fishing, 61, 62, 439
flag, 4
fluorite, 25, 27
Fol, A., 679
folk arts/lore, 176, 681, 691–7, 706–7, 806–8
food industry, 394–8
football, 582
footwear, 394
foreign capital investment, 201, 206, 343–4, 481
foreign firms: joint ventures, 477, 659, 901; representation in Bulgaria, 528, 906
Foreign Journalists' Club, 906
foreign loans, 200, 203, 209, 221, 442, 481, 484
foreign relations, 4, 64, 198, 207, 215, 219, 223–4, 239, 250–5, 883–915; *see also* 'Bulgarian Question'
foreign students, 634
forests, 22, 32, 34, 46–7, 52–5, 61–2, 69–71, 435–7
Fotinov, K., 80, 851
France, 185–6, 199, 209, 215, 223–4, 250–1, 476, 903
Franks, 148–2
fruit, 423–5
Fuchedjiev, D., 731, 734
fur, 393
Furnadjiev, N., 721–2, 730

Gabe, D., 722, 724, 727, 730, 737; port. 731
Gabrovo, 3, 31–2, 108–10; Humour and Satire Festival, 108
Gabrovski, N., 115
Gaida, 704, 709, 851
Galina, L., 731, 733, 736
Gallo, L., 820
game farming 61–2, 438
Ganchev, H., 782
Gandev, H., 679–80
Ganev, V., 233
Ganovski, S., 644, 677, 725
Garibaldi, G., 188
gas, 24–5
Gavril-Radomir, *Tsar*, 157
General Workers' Trade Union (GWTU), 202, 213, 215, 220–1, 235, 289–91
General Youth Front, 224
Genoa, 169
Genov, G., 217
Genov, T., 730, 792
geographical position, 3, 30, 32–3, 492
geology, 5–12, 661
Georgi Terter I, 167, 169
Georgi Terter II, 169
Georgi Voiteh, 160
Georgiev, B., 573, 577

Index

Georgiev, E., 681
Georgiev, G., 289
Georgiev, Kimon, 222-3, 231, 233, 235, 305
Georgiev, Kolyo, 731-2, 736, 792
Georgiev, M., 712, 715
Georgiev, S., 287
Georgiev, V., 680
Gerchev, A., 669
German Democratic Republic, 254, 887-8
Germanic tribes, 145, 152-3
Germany, 195, 199, 209-10; economic dependence on, 224, 344; in World War II, 229; declaration of war on, 231, 250
Germany (Federal Republic of), 253, 476, 902
Gerov, A., 725, 729, 731-3
Gerov, N., 80, 193, 642, 696, 705, 706, 709, 710
Geshov, I., 196, 201, 207, 214
Gichev, D., 222, 301-2
Gindev, P., 677
Giurgiu revolutionary committee, 190
glass, 383-4
Gloginski, B., 729
gold, 25-7, 743
Golden Sands *see* Zlatni Pyassatsi
Goleminov, M., 767
Golev, V., 729-31
Gorky, M., 220
Goths, 143
Govedartsi, 498
governments, list of, 266
granite, 26
Great Britain: and the Bulgarian Question, 185-6, 190, 192; declaration of war on, 227; relations with after 1923, 219; in the 1930s, 224; in World War II, 229, 231; Peace Treaty with, 250; trade with, 199-200, 476, 904; currently 904
Greece: alliance with (1912), 207; border, 3, 251; liberation war, 179; occupation of Macedonia, 208-9; relations with, 252-4, 893; territorial claims by 198, 212, 221; trade with, 47-6, 891; war with (1913), 208; in World War I, 210; in World War II, 226
Greek clergy, 183-5
Greek colonies (classical period), 141, 740
Greek language, 80, 613
Greeks: Ancient, 144; in Bulgaria, 86, 180; in Eastern Rumelia, 196; in emigration, 185
Grekov, A., 126
Grigorov, K. (economist), 678
Grigorov, K. (Novelist), 730-1
Grubcheva, M., 735
Gruev, D., 205-6; port. 203
Gulev, D., 731
Gulyashki, A., 725, 731-4

Gurko, I.V., 95, 111, 115
gymnastics, 573, 581
gypsum, 28
Gyuzelev, V., 679

Hadjiev, P., 767
Hadjiiski, I., 645, 729
Hadjiliev, D., 727
Hadjioglu, A., 90
Hadjistanchev, T., 767
haiduks, 175-6, 323; songs, 692
Haitov, N, 731-2, 736, 792
Halachev, E., 790, 793
Hammer, A., 820
Hanchev, V., 729-30
Haskovo, 3, 106
Hazaras, 146
health resorts, 133, 496-8, 505-14, 561-2, 598; *see also* spas
health services, 279, 550-67
Hemus, 878
Herbst, Y., 864
hermitism, 155, 161
Herodotus, 141
Hertzen, A. I., 188
higher education, 632-35, 652
hiking, 588, 591
historical sites *see* monuments
history, study of, 679-80
Hitov, P., 189; port. 186
Hitrov, T., 875
holiday resorts, 20, 22, 81, 88, 93-4, 492-516, 594-5, 597-8, 763
Homer, 140
horses, 434
hospitals, 551-6
hotels, 20, 65, 84, 89, 492-3, 761
housing, 548-9, 762; *see also under* District Centres, 81-134
Hrabr, *Monk*, 78, 641, 699
Hrelkov, N., 720, 721, 723, 725
Hristov, D., 646, 697, 767-8; port. 766
Hristov, H., 679-80
Hristov, K., 713, 717-19
Hristov, S., 665
Hristovich, G., 644
Hristozov, N., 735
Hungarians: early, 150, 152, 154-7, 159-60, 162-7; in political emigration, 101; cultural relations, 887
Hungary: invasion by (1365), 172; relations with, 251-4, 889; trade, 889
Huns, 95, 144-5
hunting, 61-2, 438, 492, 576
Hunyadi, Jan, 176

icon-painting, 742-4
Idrisi, 99
Ignatieva, V., 782
Ignatov, R., 729, 733, 736

Ikonomicheski Zhivot, 287
Ikonomov, B., 767
Ikonomov, N., 642
Ikonomov, T., 709, 780
Iliad, 140
Iliev, I., 678
Iliev, K., 772, 792, 794
Iliev, L., 661
Iliev, N., 131
Ilinden-Preobrazhenie Uprising (1903), 118, 205-6
Ilichovsk-Varna ferry, 88, 349
immigration, 72, 74, 82, 123
incomes, 277, 543-4, 548
Independent Workers' Trade Unions, 221-2, 292-4, 316
industrial: construction, 763; research, 653
industry, 62, 75-6, 177-8, 195, 199, 201, 203, 206, 221, 244-5, 247, 343-4, 358-98; *see also under* District Centres, 81-134
inland navigation, 91, 111, 121, 440, 443, 448-9
Innocent III, *Pope*, 163
Institute for Foreign Students, 634
Institute of Bulgarian Communist Party History, 288, 680
Institute of Public Administration, 288
insurance, 481, 482, 491, 529
Internal Macedonian-Adrianople Revolutionary Organization (IMARO), 205-6, 209
Internal Macedonian Revolutionary Organization (IMRO), 215, 217, 220, 222-3
intelligentsia, 248, 250, 295, 641-2
Internal Revolutionary Organization, 129
international co-operation: cultural, 810-20, 887; economic, 245, 247, 253, 255, 276, 351, 360, 362, 467-8, 682-3, 886, 891: scientific and technical, 647, 656, 658-60, 663, 664-5, 670, 886-7
international non-governmental organizations, 914-15
inventions, 648, 669
Iribadjakov, N., 677
Irina, *Empress*, 95
iron, 25-7
irrigation, 37, 47, 408
Isaac II, 162
Isbul, *Kav-Khan*, 151
Ishirkov, A., 644
Iskandur, O., 864
Iskra, 88, 289
Iskur, river, 7, 15, 38-9, 47, 50, 67
Islam, 75, 122, 133, 173, 175
Israel, 253
Issaev, M., 725, 727, 729-30, 737; port. 732
Italy, 188, 215, 219, 226, 250-1, 476, 903
Ivailo, *Tsar*, 167

Ivan Alexander, 170–2
Ivan Assen II, 85, 117, 163–6
Ivan Assen III, 167
Ivan Rilski *see* John of Rila
Ivan Shishman, 172
Ivan Stratsimir, 171–3
Ivan Vazov National Library, 833
Ivan Vladislav, 158
Ivanko, *Despot*, 85, 163, 172
Ivanov, D., 644, 665
Ivanov, G., 131, 664–5, 887
Ivats, *Voivode*, 158
Izvorski, S., 805

janizaries, 173, 177
Japan, 477, 903
Jews, 86, 227; newspaper, 856
Jireček, K., 72, 643, 831
John I Tsimisces, 157
John III Ducas Vatatses, 133, 165
John X, *Pope*, 154
John Cantacuzene, 170–1
John the Exarch, 78, 551, 641, 699, 741
John of Rila, 55, 166, 551
Joseph, *Archbishop*, 153
journalism *see* Press
Justinian I, 85
Justinian II, 148

Kaishev, R., 643, 665; port. 646
Kalaikov, I., 677
Kalchev, K., 728, 731–4, 737
Kalfa, H., 93
Kaliman I, 166
Kaliman II, 166
Kaloyan, *Tsar*, 85, 88, 124, 133, 162–3
Kamenov, E., 678
Kampuchea, 890
Kaneli, R., 782
kaolin, 28
Karadimchev, B., 773
Karadja, S., 108, 189; port. 188
Karadjov, S., 288
Karadžić, V., 643, 696
Karakolov, R., 677
Karaliichev, A., 724, 727, 730, 737
Karaslavov, G., 724–5, 727–8, 730–1, 736–7; port. 733
Karastoyanov, A., 875
Karastoyanov, N., 744, 870
Karavelov, L., 182, 188–9, 642, 696, 705–9, 744, 780, 851
Karevelov, P., 195–6, 198, 202
Kardam, *Khan*, 148–9
Karima, A., 713
Kaulbars, A., 196, 199
Kazakov, D., 768
Kazanluk Tomb, 142, 493, 740–1, 753
Kazassov, D., 222, 305
Kazassyan,V., 773
Kersebleptis, 142

Khazars, 150, 152
Kiev (Ancient), 156–7
kindergartens, 618–20, 621
Kirchev, A., 782
Kirilov, N., 731
Kirkov, G., 289, 315, 644, 713, 718
Kirkov, V., 781; port. 782
Kiro, B., 780–1, 805
Kirov, G., 782
Kishelski, I., 642
Kisselinchev, A., 677
Kisselov, M., 793
Kissimov, K., 786–7; port. 785
Kissyov, H. and V., 134
Kliment of Ohrid, 78, 551, 613, 641, 681, 698–9
Kliment, *Bishop*, *see* Drimev, V.
Knyaginya, R., 95
Kolarov, V., 101, 126, 217, 220, 235, 238, 253, 291, 293, 296, 644; port. 267
Kolev, Z., 727
Kolov, Dan, 571, 577
Komuna circle, 90
Konstantin Assen, 166
Konstantin Bodin, 160
Konstantin of Preslav, 78, 641, 699
Konstantinov, A., 80, 584, 711–12, 714
Konstantinov, P., 735
Koralov, E., 724, 728, 733, 737
Korea, North, 253, 887, 890
Kormisosh, *Khan*, 148
Korndjiev, D., 735
Kossev, D., 679
Kossev, K., 680
Kossuth, L., 101
Kostenechki, K., 613
Kostov, D., 666
Kostov, G., 787
Kostov, S., 723, 786
Kostov, T., 235
Kosturkov, S., 306
Kotys I, 142
Kozlev, N., 707
Kozma the Presbyter, 78, 641, 700
Kraikov, Y., 866
Kralev, S., 807
Kremikovtsi, 672
Kresna-Razlog Uprising (1878–9), 118, 192, 205
Krum, *Khan*, 81, 149–50
Krustev, A., 767
Krustev, K., 714–5, 719, 784
Krustev, V., 779
Krustevich, G., 196–7
Kubadinski, P., 287, 304
Krustanov, L., 643
Kuber, *Khan*, 146
Kubrat, *Khan*, 146–7
Kukuzel, Yoan, 641, 765
Kumanians, 162, 164
Kumluk desert, 21–2
Kunchev, A., 128
Kumev, G., 110, 756

Kurdjali, 4, 73, 122–3
Kurpachev, H., 726
Kutev, F., 767, 772, 776; port. 777
Kyoseivanov, G., 223, 226
Kyulyavkov, K., 721, 725, 729–30, 790, 792
Kyulyomov, K., 732
Kyustendil, 4, 31, 124–5

Lacheva, Ts., 731
lakes, 34–5, 37–8, 48, 52, 61, 93
Lamur, 729–30,
Lambrev, K., 727
land ownership, 75, 175, 177, 179, 193–4, 215, 237, 246, 277, 342, 345, 408–9, 413
language schools, 627–8
Lankov, N., 725, 730
Laos, 887, 890
Latin Empire, 163–5
Lavrenov, Ts.: port. 745
law, 281–3, 285; studies, 682
Lazar, *King*, 172
lead, 26–7
League of Nations, 215
leather, 393
leisure activities, 32, 278, 594–602
Lenin, V. I., 289. *See also* Marxism-Leninism
Leo VI, 154
Levchev, L., 738
Levski, V., 189, 197, 569, 642, 689; establishment of revolutionary committees by, 82, 86, 95, 97, 99, 106, 108, 111, 115, 129; Monument, 84; port. 185
Liberal Party, 193, 195–6, 200–1, 614
libraries, 828, 833
library clubs, 182, 825–7
licences, 659
Liliev, N., 716–18, 720; port. 715
limestone, 28
linguistics, 680
literature, 182–3, 681, 704–7; study of, 737–9; early, 78–9, 153–5, 171–2, 176, 550, 613, 641, 698–703
Livy, 124
Lovech, 4, 129–31; Treaty of (1187), 129
Ludwig, *King*, 152
Lukanov, A., 287
Lukatnik, M. *see* Arnaudov, M.
Lukyanov, S., 193
Luxembourg, 903
Lyapchev A., 220–1
Lyuboslovie, 704, 851
Lyutskanov, N., 790, 793

Macedonia: Greek influence in, 181; study of history, 680; territorial claims, 192, 196–7, 207–8, 210, 226, 852, 892
Macedonia (Ancient), 142–3

Index

Madan, 76, 759
magnesites, 25, 28
magnetite, 25–6
Makariopolski, I., 183–5
Makedonia, 704, 709, 851–2
Malamir, *Khan*, 85, 151
Maleshevska mountain, 3, 8, 19
Maleev, A., 668
Malina, K., 724, 737
Malinov, A., 201, 206, 211–2, 214, 221–2, 300
management: economic, 239, 348, 362, 657; social, 244
Manasius, 641
Manchev, A., 726
Mandadjiev, A., 733
manganese, 25–7
Manolov, A., 807
Manolov, E., 767
Manolov, H., 768
Manov, E., 727, 731, 733–4
Mantov, D., 735
manuscripts, 741–2, 766, 866
Marangozov, N., 725
marble, 26, 28
Marchevski, M., 727, 737
Marinov, D., 646
Maritsa, river, 3, 5, 15, 18, 21, 37–8, 40–1, 47
Markovski, G., 735
Markovski, V., 725, 731
marls, 28
Martinov, I., 727, 729, 731
Marxist circles, 104
Marxism, 201–2, 288–90, 644–5; -Leninism, 287, 290–1, 293–4, 296–8, 883; and culture, 248–9; in education, 615, 633; in literature, 725, 737–8; in philosophy, 677; in the theatre, 787
Massalitinov, N. O., 785, 787, 789–90; port. 786
Mateev, D., 644
Mateev, E., 678
maternity and child-care, 557–9, 606
Matev, I., 669
Matev, P., 729–30
mathematics, 661–2
Maximov, H., 712
Maxwell, R., 820
Mazzini, G., 188
mechanical engineering, 367–75
mechanics, 661–2
medicine, 550–3, 668–9
medicines, 666
Medical Academy, 648, 651, 668, 669
Mednikarov, Z., 807
megmetite, 25
mercury, 27
metallurgy, 366–7, 669, 671–2
Metodiev, D., 731
Michael IV, 160
Michael VII, 160
Michael, VIII, 166, 167
Mihail II Assen, 133, 166

Mihail III Shishman, 119, 124, 169–70
Mihailov, B., 786
Mihailov, H., 126, 226
Mihailov, I., 126
Mihailov, R., 733
Mihailov, S., 677
Mihailova, L., 733
Mihailovgrad, 4, 126–7
Mihailovski, S., 638, 710, 712–13, 715
Mihalchev, D., 644
Mikhalkov, S., 820
Milanov, K., 706
Miladinov, D. and K., 642, 696, 706
Milarov, S., 711
Miletich, L., 80, 646
Milev, G., 716–17, 720–1, 784–5, 787
Mileva, L., 737
Military Academy, 652
Military League, 215, 222–3, 301
Milutin, *King*, 124
minerals, 24–9, 37, 42
ministries, 280–1, 284
Ministry of Foreign Affairs, 884–5
Ministry of National Defence, 328
Ministry of Public Education, 248
Minkov, S., 721, 723, 733
Mirski, K., 789
Mishev, G., 731–2, 736
Missul, 710, 715–6, 719
Mitev, I., 668
Mithad Pasha, 440
Mitov, A., 745–6
Mitov, D., 725, 727, 787
Mizov, N., 677
Mladenov, P., 287, 903, 906
Mladenov, S., 80, 646
molybdenum, 25–7
Momchev, G., 288
Momchil, *Rubr*, 133, 170–1
monarcho-fascist dictatorship, 223–6
monarchy: establishment, 193–4; abolition, 235, 275, 296
monasteries, 84, 99, 153, 162, 165, 176, 641, 701; libraries, 828; schools, 180, 613; *see also under* District Centres, 81–134
money, 4, 115, 165, 341–5, 481–91
Mongolia, 887, 890
Monov, T., 734
Montenegro, 207–8
monuments, cultural/historical, 64, 753, 816; *see also under* District Centres, 81–134
Moravia, 152–3
Morfov, A., 767
mortality, 553–4
Moscow Peace Treaty (1944), 250
mosques, 84, 96, 101, 120, 128, 755–6
motels, 493
motoring, 592–3
mountains, 3, 5–23, 30–2, 41, 49, 51, 56, 496–8, 588, 593

municipalities, 3
Murad I, 172
Murad II, 176
Muraviev, K., 229, 301–2
Murvichka, I., 746
museums, 332, 836–46; in district centres, 83, 87, 89, 91, 94, 96, 98, 99, 101, 103, 105, 106, 108, 110, 111, 117, 121, 123, 127, 128, 129, 132, 133
Mushanov, N., 222
music, 83–4, 87, 89, 91, 94–6, 98, 101, 103, 107, 112, 123, 765–79, 805, 807; personalities, 768–76; sales agency, 878
musicology, 681
Muslims *see* Islam
Mutakov, Y., 767
Mutafchieva, V., 733, 735
Mutev, D., 642
Muteva, E., 706
Mutkurov, S., 198–9

Nadjakov, G., 643, 662
Nakovski, A., 731, 733–4
Narechen, 20
Natan, J., 678
National Agro-Industrial Union, 244, 349, 412–13, 650, 857
national anthem, 3–4
National Art Gallery, 749; photo, 841
National Assembly, 62, 81, 84, 115, 193–6, 199, 202–3, 207, 209, 213, 215, 222–4, 226, 229, 233, 235–7, 241–3, 275–6, 280–1, 283–6, 290, 293–4, 296, 299, 301, 307, 328, 441, 884, 900, 905
national holidays, 4, 181, 637–8
national income, 247, 344, 346–50, 547
National Liberal Party, 215, 222, 224
National Liberation Insurgent Army (NOVA), 229; General Staff, 82, 229, 231
National Library, 83, 828, 830–1, 834, 837, 869
National Opera, 83
national parks, 61, 64–71, 84, 101
national revolutionary movement, 185–90
National Revival, 85–6, 95, 176, 177–92, 569, 613, 641–2, 687
National Social Movement, 222, 301
National Theatre, 83, 101, 718, 761, 782, 785, 787, 789; demonstration at, 203
National Youth Theatre, 788
nationalization, 75, 234, 237, 244, 277, 296, 345, 360
natural sites, 67–71
nature conservation *see* environment: protection
nature reserves, 61, 64, 66
Naum of Ohrid, 78, 641, 698–9

navy, 329
Nedelnik, 179
Nenov, D., 767
Neophyte of Rila, 180, 613, 704, 744, 766
Neshkov, V., 736
Nessebur, 493, 505, 511-12, 742, 753-4
Netherlands, 903
Neuilly Peace Treaty (1919), 212-13, 220-1, 343
new towns, 759
news agencies *see* Bulgarian Telegraph Agency; Sofia Press Agency
newspapers *see* Press
Nicean Empire, 165-6
Nicephorus I, 149
Nicholas I, *Pope*, 153, 550
nickel, 25, 27
Nikiforov, V., 678
Nikola, 'Dyado', 108, 115
Nikolai I, 186
Nikolitsa, S., 706
Nikolov, L., 771
Nikulitsa, *Voivode*, 158
Nikulitsa Delphinus, 160
Nivyanin, I, 726
non-alignment, 896
Normans, 159-60, 162
North Atlantic Treaty Organization (Nato), 253, 328, 886
Norway, 904
Novo Vreme, 287, 289, 719, 853

Obbov, A., 235, 302
Obrazopisov, N., 744
Obreshkov, N., 643, 661
Obretenov, S., 767-8, 772
Obretenova, 'Baba' Tonka, 90
Obshto Delo, 289
Odjakov, P., 744
office equipment, 371
Ognyanov, S., port. 784
Ognyanova, Y., 790
Ograzhden mountain, 3, 8, 19, 38
Ohrid, 157-8; school, 153, 613, 641, 699
oil, 24-5
Olympic Games, 571-83
Omarchevski, S., 615
Omurtag, *Khan*, 101, 150-1
opera, 83-4, 87, 89, 94, 95, 96, 98, 767-9
Orbita Youth Travel Agency, 514
orienteering, 591
Oshavkov, Zh., 677
Osman Pasha, 97
Ossogovo mountain, 3, 9, 19, 38, 46-7, 51, 59
Ostrikov, I., 733
Ostrogoths, 144-5
Ostrovski, G., 790, 797
Otechestven Front, 305, 853-4, 856
Ovadia, D., 727, 730, 732, 735

painting, 741-7
Paissii of Hilendar, 179-80, 613, 641, 703-4
Palauzov, Spiridon, 182, 642
Palestine Arabs, 898
Pamporovo, 20, 498, 763
Panayotov, S., 288
Panchev, P., 794
Pannonian Principality, 152
Panteleev, D., 730
Papandreou, A., 894
paper, 382-3
Paris Agreement (1954), 253
Paris Peace Treaty (1947), 236, 251-2
Parliament *see* National Assembly
Partien Zhivot, 287
Paskaleva, V., 680
Passi, I., 677
Patent and Trade Mark Bureau, 518
Paulicians, 156, 160-1
Pavlov, T., 233, 644-5, 677, 681, 725, 737-8, 747, 787; port. 676
Pavlovich, H., 80, 704
Pavlovich, N., 182-3, 642, 744; port. 745
Pazardjik, 3, 111-12
Pazvantoglu, 119
pearlite, 28
peaceful co-existence, 883, 892, 896, 900, 904-15
peasant uprisings, 167, 175-6, 289, 299, 343
peasants, 179, 194-5, 201, 214, 237, 246-7, 299, 342
Pechenegs, 152, 154-7, 159-60, 162
Peev, T., 708
pegmatites, 28
Peichev, I., 729, 731, 736, 791
Pelin, Elin, 681, 713-14, 717-18, 721, 724
Pencheva, S., 729, 731
Penev, B., 719, 725, 784
Penev, K., 729
Penev, P., 729
pensions, 279, 546, 603, 607-10
People's Councils, 63, 237, 241, 275, 280, 281, 283, 284, 286-7, 307; and amateur art, 806
People's Liberal Party, 200-1, 212
People's Youth Union, 311
Permanent Committee for Environmental Protection, 62
Pernik, 3, 104-5
personality cult, 238, 249, 296
Peter I, 155-6
Peter II, 115, 117, 129, 163
Peter Delyan, 159-60, 161
Peter of Mirkovo, 702
Peter the Woodcutter, 131
Petkanchin, B., 661
Petkanov, K., 723
Petkanova, M., 730
Petkov, D. (composer), 772
Petkov, D. (politician), 201, 203

Petkov, L., 735
Petkov, N., 302, 305
Petkov, P. D., 301
Petleshkov, V., 805
Petrich, 31-2
Petrov, D., 672
Petrov, G., 222
Petrov, H., 644
Petrov, I., 729, 731-2, 734
Petrov, N., 571, 577
Petrov, P., 679
Petrov, R., 206
Petrov, V., 725, 729, 731, 736, 792, 794
pharmacies, 565-6
Philes, M., 93
Philip II of Macedon, 85, 142
philosophy, 677
Photius, *Patriarch*, 153
photographic agencies, 874-5
photography, 875-6
physical education, 568-84
physics, 662-4
Piccolo, N., 642
Pikolo, S., 551
pipelines, 443, 451
Pipin, A., 646
Pioneer Children's Organization, 313
Pipkov, L., 768, 772; port. 766
Pipkov, P., 638, 767
Pirin mountains, 5, 9, 19-22, 34, 36-8, 41, 43-4, 46-7, 49, 51-3, 56-7, 59-60, 66
Planeva-Holevich, E., 669
plans, 236, 242, 244-5, 247, 281, 296-7, 345-9, 360, 364, 656-7; international co-ordination, 886
platinum, 25
Plekhanov, G. P., 210, 290
Pleven, 3, 31, 97-8
Pliny, 124
Pliska, 741, 755
Plovdiv, 3, 31, 85-7, 161, 710, 742, 762; University, 80, 633
Poland, 215, 250-1, 253-4, 886-8
Polikarov, A., 677
Polish political *émigrés*, 101, 186
Politicheska Agitatsia, 287
Politicheska Prosveta, 287
Polyanov, D., 713, 725; port, 724
Pomorie, 512
pop music, 768, 772, 779, 807
Popdimitrov, E., 716, 720
Popov, D., 644
Popov, I., 781
Popov, M., 644
Popov, S., 781
Popov, T., 771-2
Popov, V. (mathematician), 662
Popov, V. (novelist), 731-2, 734
Popov, Z., 126
Popovich, R., 704-5
Popovich, V., 744
Popstoyanov, A., 766
Poptonev, A., 733

Popular Alliance, 301
Popular Bloc, 222–3, 301–2
Popular Front Movement, 224, 294, 302
Popular Party, 200–1, 203, 207, 209, 229, 300; newspaper, 852
population, 3, 72–6, 553–4, 678; *see also under* District Centres, 81–134
Popzlatev, P., 235
porcelain, 384
ports, 88, 90, 93–4, 111, 121, 199, 440, 443
Portugal, 904
postal service, 343, 452, 454
pottery, 743
poultry, 433
power production/consumption, 24, 41, 365–6, 670–1; imports 472
pre-history, 139–40, 740; *see also under* District Centres, 81–134
Presidency of the Republic, 235, 296
President of the State Council, 4, 238, 242, 884
Preslav, 154, 157; architecture, 741, 755; school, 153, 613, 641, 699
Press, 182, 203, 392–3, 530, 642–3, 704, 709–10, 725, 851–7, 879–80; law, 858–9
Pressian, *Khan*, 133, 151
Pressian, *Tsar-Presumptive*, 158
printing, 392–3, 744, 866, 870–1
Progressive Liberal Party, 201, 203, 207, 209, 212, 300, 852
property, 276–7, 350
Protestantism, 75, 186
Protich, A., 784
proto-Bulgarians, 144–50, 741
Prussia, 186
public organizations, 244, 276–7, 282, 287, 304–17
Public Prosecutor, 81, 237, 276, 280–2, 285–6
publishing, 86–7, 392–3, 530, 650, 866–9, 870–1; in English, 874; legal deposit, 828, 830
puppet theatre, 788, 807
Purlichev, G., 707

quartz, 28

Rabotnicheski Vestnik, 289–90, 853, 856
Rabotnichesko Delo, 221, 287, 292, 853–6, 871
Rabotnik, 115, 288–9, 853
Rachinski, T., 673
Radev, M., 731
Radevski, H., 724, 725, 727, 730, 737
Radical Democratic Party, 203, 209
Radical Party, 222, 224, 235, 306, 307
Radichkov, Y., 731–2, 736, 793–4
radio and television, 453, 457–8, 860–3, 874, 879; in English, 861

Radoev, I., 729, 736, 792
Radoslavov, V., 200–1, 209, 211, 214
Raichev, A., 771–3
Raichev, G., 721, 723
Raikov, P., 648
railways, 199–200, 206, 343, 440–2, 443–5; *see also under* District Centres, 81–134
Rainov, B., 725, 729, 731–4, 736
Rainov, N., 717, 720, 721, 727
Rakovski, G., 88, 182, 188, 323, 642, 689, 696, 705–8, 851; port. 186
Ralin, R., 729, 731
Razgrad, 4, 128
Raztsvetnikov, A., 721, 737
Red Book, 71
Reichstag Fire, 222, 293
religion, 75, 279, 678
Relits, A., 183
rhenium, 25, 27
Rhodope mountains, 3, 5–6, 9–10, 16, 19–21, 23, 32, 34, 36–8, 41–3, 46–7, 52–4, 57, 59–60, 139
Rifat Pasha, 129
Rila: Monastery, 20, 493–4, 613, 641, 700, 742–4, 753, 766, 836, 845; mountain, 9, 19, 23, 34, 36–40, 43–4, 46–7, 49, 51–2, 56–7, 59–60, 497
Rilev, A., 756
rivers, 3, 5–23, 32, 34, 36–7, 38–44, 48, 54–5, 61, 62, 440; shipping *see* inland navigation
road transport, 20, 199, 206, 343, 440–3, 445–7; *see also under* District Centres, 81–134
Rolland, R., 220
Roman, *Tsar*, 156–7
Roman Catholic Church: adherents, 75; conversion attempts, 186; early relations with, 165; and literature, 176, 702; temporary jurisdiction in Bulgaria, 153, 163
Roman period, 81, 85, 90, 95–7, 99, 101, 103, 108, 119, 124, 126, 128–9, 133, 143–5, 495, 740–1, 753
Romania, 188–90, 208, 210–12, 226, 253–4, 886–9; border, 3; occupation by, 103, 121; in Russo-Turkish Wars, 97
Romanus Lecapenus, 154–5
roses, 18, 418
Rossitsa, 108, 853
Rousset, A., 90
rowing, 578
rubber industry, 378
Rudozem, 76, 759
rulers, list of, 265
Russalka, 506, 763
Russe, 3, 90–2
Russev, N., 736, 792
Russev, T., 773
Russian language, 80
Russinov, G., 126

Russo-Turkish wars: early, 93, 129, 176–7; (1877–78), 190–2, 193, 323, 343, 551; local actions, 82, 88, 93, 95, 97, 99, 101, 103, 108, 113, 115, 119, 121, 124, 126, 128, 129, 133

Safran, M., 101, 767, 805
St. Sophia Church, 81, 84
Sakuzov, Y., 201, 288–90
salt, 28, 38
Samuil, *Tsar*, 157, 551
San Stefano Peace Treaty (1878), 191–2, 193
sanatoriums, 561–2
Sarafov, K., 782
savings, 486–7, 491
Savoyard incursion, 172
Schutz, B., 745
Schwartz, V., 644
science, 250, 279, 297, 641–6, 661–76; organization and administration, 647–60; personnel, 653–5; science of, 683
Scientific and Technical Unions, 653–4
sea-bed, 911
sea-going transport, 88, 90, 93–4, 440, 443, 447–8
Second International, 210, 290
secondary education, 620–4
Secret Central Bulgarian Committee (SCBC), 188
selenium, 27
Selim I, 133
Seliminski, I., 182, 551
Sendov, B., 662
Serafimov, V., 133
Serbia: alliance with against Turks, 172, 179, 190, 207; early conflicts with, 151, 154–5, 165–7, 169–70; émigrés in, 188–9; territorial claims, 186, 206; to Macedonia, 181, 207, 212; war with (1885), 198; (1913), 208–9; (1915), 210
Serbs, 179
Serdica Ecumenical Council, 81
services, 549
settlements (local government), 3, 241
Seuthes, I, III, 142
Severians, 146
Severnyak, S., 731
Shapkarev, K., 697
Sharova, K., 680
sheep *see* stock-breeding
Shishmanov, D., 625
Shishmanov, I., 615, 697, 714, 725
shooting, 581
Shopov, A., 790
Shopov, I., 642
Shopova, S., 311
Shturbanov, N., 782
Shumen, 3, 101–2; Higher Pedagogical Institute, 80
Sigismund, *King*, 173, 176

Silistra, 3, 4, 121
silver, 26–7
Simeon I, 153–5, 613, 699
Simeon II, 229
Simeonov, T., 108–10
Sitalkes, 141–2
sites: cultural/historical *see* monuments; natural *see* natural sites
skiing, 20, 496–8, 579, 588–90
Skitnik, S., 787
Slaveikov, Pencho, 713, 715, 718, 783–4, 831
Slaveikov, Petko, 80, 182, 195–6, 696, 704, 706–10, 780, 831, 851
Slavinski, P., 728
Slavs' invasion/settlement, 85, 90, 95, 97, 101, 124, 126, 133, 144–50
Sliven, 3, 99–100
Slunchev Bryag, 22, 63, 505, 510–11, 763
Smilets, *Tsar*, 169
Smirnenski, H., 717, 720–1
Smolyan, 4, 133–4, 762
Sobolev, L., 196
social consumption funds, 544–6
social sciences, 677
social security, 278, 543–4, 546, 603–10
Socialism, 4, 62, 76, 80, 90, 93, 115, 124, 129, 200, 232–55, 276, 280, 295
sociology, 677–8
Sofia, 3, 31, 76, 81–4, 193, 496–7, 613, 641, 702, 710; archives, 835; underground railway, 83; University, 80, 83, 200, 203, 206, 552, 633, 643, 649, 697, 737; library, 828, 832
Sofia News, 874
Sofia Press Agency, 530, 873–4
soils, 45–8, 420; erosion, 23, 62
Soldiers' Mutiny (1918), 211–12, 290, 300
Sophronius of Vratsa, 113, 119, 179, 613, 704, 866
Sotsialdemokrat, 853
South Africa, 898
Sozopol, 31, 505, 513–14, 742
space research, 663–5, 887, 911
Spain, 904
Spartacus, 140
spas, 20, 32, 81, 93, 95, 118, 124–5, 132, 495–6, 498–505, 561
Spassov, A., 644
Spassov, D., 677
Spassov, Ts., 726
Spassovich, V., 646
sport, 32, 568–84, 586–8, 892; personalities, 578–83, 587–8; schools, 628; *see also under* District Centres, 81–134
Srebov, Z., 727
Sredna Gora mountains, 5–10, 16–18, 34, 43, 45–6
Stainov, P., 767–8, 779; port. 766

Stamatov, G., 714, 718, 721, 723, 786
Stamboliiski, A., 201, 210, 212, 213–15, 299–301, 344, 795; port. 211
Stambolov, S., 198–201, 203, 206, 707
Stanchev, L., 737
Stanchev, N., 573, 577
Stanchev, S., 790
standard of living, 543
Stanev, E., 727, 731, 733–5, 737
Stanev, Lazar, 222, 301
Stanev, Lyuben, 731
Stanislavov, F., 866
Stanoev, Y., 735
Stara Zagora, 3, 31, 95–6
State: crest, 4; foundation, 4, 146–52; anniversary, 250, 818, 902; offical independence from Turkey, 206; restoration (1878), 193–8; foundation of socialist, 233–7, 242, 244, 275, 296
State Committee for Science and Technological Progress, 64, 647, 652, 656
State Council, 62, 75, 81, 84, 241, 276, 281–7, 328, 884; President, 4, 238, 242, 884
State Folk-Song and Dance Ensemble, 776
State Planning Commission/Committee, 237, 626
state public organizations, 244, 276
Stefan Dechanski, 124, 169
Stefan Dushan, 170
Stefanov, I., 678
Stefanov, P., 731, 736
Stefanov, V., 807
Stefanova, L., 731
Stefanova, N., 731
stock-breeding, 97, 426–33, 674–5
Stoev, G., 731, 733
Stoichev, T., 677
Stoichev, V., 251, 571; port. 570
Stoichkov, G., 288
Stoikov, A., 677
Stoilov, K., 200–1
Stoin, V., 697
Stoyanov, A., 731
Stoyanov, D., 288
Stoyanov, L., 716–17, 720, 723, 725, 728, 787; port. 718
Stoyanov, R., 723, 786
Stoyanov, S., 790
Stoyanov, V., 767–8
Stoyanov, Z., 196, 710–11, 714
Strandja mountain, 3, 5, 8–10, 18–19, 32, 38, 40, 43, 46–7, 53, 57
Strandjev, K., 733
Stranski, I., 643, 665
Strashimirov, A., 713–15, 717–18, 721, 723, 725, 783–4; port. 712
Stratiev, S., 735–6, 793
Strelkov, L., 729, 736, 790, 792
strikes, 99, 104, 201–3, 213, 221–2, 224, 290, 293–4, 300, 316

Struma river, 5–6, 10, 19, 31, 40, 45–7, 53–5, 59
Strumski, G., 737
Sugarev, R., 735
Suleiman Pasha, 95
sulphur, 25, 27
Supreme Court, 81, 237, 276, 282, 284–5
Supreme Economic Council, 345
Supreme Macedonian-Adrianople Committee (SMAC), 205–6
Surchadjiev, S., 787, 789, 796
Sursuvul, G., 155
Suvremenni Pokazatel, 853
Svetlina circle, 90
Svoboda (Nezavissimost), 182, 705, 709, 851, 854
Svyatoslav Igorevich, 156–7
swamps, 38, 48, 61, 551
Sweden, 904
swimming, 579
Switzerland, 251
Synesius, *Bishop*, 90

table tennis, 581
Tagamlitskii, Y., 661
tailoring, 391
Talev, D., 723, 728; port. 724
Tanchev, P., 299
Tartars, 88, 119, 165–9
taxes, 177–8, 187, 200–1, 203, 215, 244, 289, 299, 342–4, 481–4
teachers: associations, 119, 315, 632; training 628–32
technical schools, 626
telegraph, 452, 456
telephones, 452, 454–5
television *see* radio and television
Telerig, *Khan*, 158
telex, 456–7
tennis, 581
Teres, 141
territory, 3, 13, 30, 32–3, 42–3, 212, 224, 251, 343, 883
Tervel, *Khan*, 99, 148
textiles, 108, 385–91, 672
theatre, 682, 780–94; amateur, 807; personalities, 791–2
Theodosius III, 148
Theodosius of Turnovo, 115, 700–1
Theological Academy, 633
Theopompus, 124
Thomas the Slav, 150
Thracians, 81, 85, 90, 93, 95, 97, 101, 104, 106–8, 111, 113, 118–19, 121–2, 124, 126, 128–9, 133, 140–4, 495, 568, 740–1, 753; language, 680; study of, 679
Thucydides, 142
titanium, 25–7
tobacco, 122–3, 395–6, 420
Todor Comnenus, 164–5
Todor Svetoslav, 169
Todoriev, N., 670

Index

Todorov, A., 725, 729
Todorov, N., 679
Todorov, P., 713, 715–16, 718, 783–4, 793; port. 712
Todorov, S., 241
Todorov, T., 212, 300
Todorov-Balan, A., 80, 584, 646
Tolbuhin (town), 3, 103
Tolbukhin, *General*, 325
Tomov, K., 300, 301
topography, 13–23
Toshkovich, N., 642
Totev, A., 678
Totyu, F., 189; port. 187
tourism, 20, 32, 64–5, 438, 492–516, 584, 588–93, 763, 891; see also under District Centres, 81–134
towns, 3, 61, 73–6, 81–134, 200
trade, 86, 96, 296, 440, 459–66; foreign, 4, 86, 90, 97, 108, 128, 129, 177, 187, 195, 199–200, 221, 224, 226, 253, 296, 341–2, 344, 467–8, 886, 901; organizations, 517–30, 878; representation abroad, 531–7. *See also* international co-operation, economic
trade unions, 99, 104–5, 201–2, 215, 223, 235, 244, 276, 282, 287, 291–2, 315–17, 345, 360, 566, 598–601, 610; newspaper, 856
Traikov, G., 235, 302; port. 303
translations, 869, 874, 877–8
transport, 440–51; organizations, 527–8
travel agencies' addresses, 514–16, 527–8
Trayanov, T., 716–17, 720
treaties, *see under* Berlin (1878); Bucharest (1886, 1913); Craiova (1940); Lovech (1187); Moscow (1944); Neuilly (1919); Paris (1947); San Stefano (1878)
Trendafilov, T., 678
Trendafilov, V., 786
Trud, 315, 856
Trunski, S, 727
Tsachev, K., 729, 731, 733
Tsanev, G., 737
Tsanev, R., 667
Tsanev, S., 736
Tsankov, A., 215, 220, 222, 301
Tsankov, D., 195, 196, 198, 200, 201
Tsankov, H., 785
Tsankov, V., 790, 793, 797
Tsarigradski Vestnik, 704, 851
Tsenovich, T., 836
Tserkovski, Ts., 299, 713
Tsokev, H., 744
Tsonchev, D., 731, 733
Tsonev, B., 80, 646
Tsvetkova, B., 679
Tundja, river, 19, 38, 40–1, 47, 53–4
tungsten, 25–7
Turgovishte, 4, 132

Turkey, 3, 199, 207, 251, 252–4, 476, 891, 894–5; war with (1912), 207–8
Turkish conquest, 72–4, 172–6, 342–3, 440, 551, 613, 641; *see also under* District Centres, 81–134
Turkish language, 80
Turks, 74, 86, 146, 152, 160, 170–1, 180, 194, 196, 856
Turlakov, M., 300–1
Turnovo *see* Veliko Turnovo

Ubaydallah al-Mahdi, 154
Unesco, 66, 660, 811, 815–16, 837, 884, 914
Uniate Church, 186
Union of Actors, 787
Union of Bulgarian Writers, 738, 873, 879–80
Union of Scientific Workers, 653
Union of Workers' Social Democratic Youth, 290
United Nations, 4, 64, 252–3, 883–5, 896, 905, 908–15
United Workers' Syndicate, 315–16
universities, 633
Uprising (1923), 97, 113, 118–19, 126, 133, 215–19, 291, 293, 301, 323
Uprising (1944), 231–2, 295, 296, 325
USA, 227, 229, 231, 250–2, 477, 904
USSR/Russia: relations with, 195–6, 197–200, 206, 207, 208, 215, 223, 237, 239, 250–3, 255, 294, 298, 301, 323, 883, 886–8; Treaty of Friendship, 253, 254, 811, 887; cultural relations, 80, 248, 288, 574, 614, 690, 699, 701, 706, 811, 886–8; economic relations, 245, 247, 345–6, 351, 360, 408, 468, 656, 659–60, 886–8; declaration of war and invasion of Bulgaria, 229–30, 295; liberation from Turks by, 98, 178–9, 185–6, 188, 190, 193, *see also* Russo-Turkish War of Liberation; Nazi invasion (1941), 226, 294; relations during World War II, 226, 229; Revolution (1917), 290, 569
Uzi, 159
Uzunov, A., 780

Vaklinov, S., 679
Vaptsarov, N., 227, 681, 724–6
Varna, 3, 5, 31, 88–9, 492, 505
Vaskidovich, E., 613, 704
Vassilev, O., 725, 727, 729–31, 790
Vassilev, V., 719–20, 724–5
Vazov, I., 80, 182, 584, 710–11, 713–15, 717–18, 721, 723, 781, 783–4
Vazov National Library, 833

vegetables, 421–3
vegetation, 31, 49–55
Velichkov, K., 80, 710–11, 713–14, 781
Velichkov, M., 736, 794
Veliki Preslav *see* Preslav
Veliko Turnovo, 3, 115–16, 162, 167, 172, 176; architecture, 165–6, 762; parliament, 193–4; school, 79, 171–2, 613, 641, 700–1, 742–3; university, 80, 633
Venelin, Yuri, 72, 643, 696, 705
Venice, 169
Verković, S., 696
vermiculite, 25
Veshin, Y., 746
veterinary medicine, 434
Vezhinov, P., 725, 727, 729, 731–4, 737; port. 724
Vidin, 3, 31, 73, 119–20, 700
Vietnam, 887, 890
villages, 3, 73, 74–6; free communities, 145; health services, 560
Vitosha, Mount, 8, 10, 17, 30, 40, 44, 46–7, 51–2, 57, 60, 65, 81, 84, 496–7, 584
Vladigerov, P., 767–8
Vladigerov, T., 678
Vladimir, *Tsar*, 153
Vlahina mountain, 3, 19, 38, 43
Vlaikov, T., 712
vocational training, 623–5
Vodenicharski, V., 726
Voinikov, D., 101, 182, 642, 708, 767, 780
Volen, I., 723, 730.1, 733
volleyball, 578
Volov, P., 850
Vratsa, 3, 31–2, 113–114
Vulchev, A., 672, 731
Vulev, D., 729, 731–2

wages *see* incomes
Wallachians, 160, 172
Warsaw Treaty Organization, 4, 253–4, 327–8, 884–7, 905–8
water resources, 34–44, 62, 439, 552
weight-lifting, 573, 578
Willard, M., 220
wine, 97–8, 395
Wladyslaw III, 176
women, employment, 76
wood-processing, 62, 381
wool *see* textiles
work-forces, 276–7
workers, 76, 201–3, 221, 237, 278, 295, 343; health care, 559–60, 566
Workers' Party, 108, 124, 126, 221–2, 224, 292–4
Workers' Social-Democratic Youth Union, 311
Workers' Youth Union, 82, 118, 221–2, 224, 227, 234–5, 292–3, 311, 570

World Health Organization, 566, 651
World War I, 209–12, 290, 299–300; peace treaty, 212–13, 220–1, 343
World War II, 226–32, 234, 250–2, 325–6; anti-fascist insurgency, 226–32, 294–5, 323–4; peace treaties, 231, 250–2; *see also under* District Centres, 81–134
Wrangel, *Baron*, group, 215, 290
wrestling, 573, 577

Yahiel, N., 677, 678, 683
Yambol, 3, 107
Yanev, S., 735
Yanulov, I., 645
Yarumov, D., 735
Yassenov, H., 716–17, 720
Yavorov, P., 713, 715–19, 783–4; port. 714
Yavorski, M., 727
Yordanov, N., 736, 792, 794
Yordanov, V., 831
Yordanova, Z., 786–7
Yorov, Y., 287

Young Communist League, 237, 241, 282, 287, 309–13, 572, 807, 857–8. *See also* Bulgarian Communist Youth Union; Workers' Youth Union
Young Turks, 206, 208
Yovkov, Y., 717, 721, 723, 786; port. 722
Yugoslavia, 3, 215, 219, 226, 253–4, 326, 887, 892
Yugov, A., 238
Yundola, 498

Zaberski, A., 773
Zabunov, Y., 299
Zagorchinov, S., 723
Zahariev, Ts., 108
Zaimov, S., 711, 837
Zarev, P., 681, 727, 737–8
Zarev, V., 734–5
Zemedelsko Zname, 299, 852–4, 856
Zhefarovich, H., 703, 866
Zhelyazkov, D., 99, 358
Zhelyazkov, L., 666
Zhinzhifov, R., 706
Zhivkov, G., 199
Zhivkov, N., 618

Zhivkov, T., 4, 231, 238–9, 241–2, 246–7, 249, 254, 275, 287, 297–8, 312, 317, 328, 347, 469, 572, 575, 616, 688, 730, 738, 813, 816–17, 855, 883, 886, 890–2, 894–6, 900, 902–4, 906–7; ports. 268, 308, 310, 814, 889, 899
Zhivkov, T.H., 287
Zhivkova, L., 688–9, 810–11, 816–17; port. 819
Zhotev, D., 727, 730, 732, 736
Zidarov, K., 729–30, 736, 790, 792
Zimmerwald Movement, 290
zinc, 26–7
Zindarov, N., 737
Zlatarov, A., 646
Zlatarski, G., 644
Zlatarski, V., 645; port. 646
Zlatev, P., 223
Zlatni Pyassatsi (Golden Sands), 22, 65, 89, 508–9, 763
Zname, 182, 705, 709, 851
Zografov, I., 875
Zograph, D., 133
Zograph, Z., 182, 744
Zurlov, I., 727
Zveno, 222–3, 229, 295, 305–7; Popular Union (PU), 233, 235–7
Zveno, 853

Foreign Trade Engineering Company
Exporter of Electrical Engineering Products
and Engineering Services

Electroimpex

Our Export List includes:
- Complete electrical engineering plant
- Transformers
- Electric Motors
- Low and high voltage apparatus
- Cables and conductors
- Insulators
- Electric wires and insulating materials
- Manual electric tools
- Electro-medical instruments
- Lighting fixtures
- Lamps and auto-lamps
- Quartz products
- Household electrical appliances

Electroimpex have extensive experience in the study, design, supply, assembly, commissioning and servicing of complete hydroelectric plant and technological lines for electrical engineering products. Our experts are also on hand offering their technical assistance in consultation or training.

17 George Washington Street,
Sofia, Bulgaria.

Telephone: 8-61-81
Telex: 022075 / 022076

balkancar

48 Kliment Ohridski Boulevard,
1000 Sofia, Bulgaria.
Telephone: 655-01
Telex: 023431/2

**POWERFUL • EFFICIENT
TECHNICALLY ADVANCED**

Bulgar plod export

22 Alabin Street, Sofia, Bulgaria.
Telephone: 88-59-51
Telex: 22451/3 / 23297/8
Cables: Bulgarfrukt.

Exporting over 1 million tons of food products to over 70 nations, each and every year.

Our export lists bulge with over 300 different foods, carefully selected for taste and nutritiousness. We export fresh or preserved fruit and vegetables, pulps, wild fruit and flowers, and can handle re-export, specific foreign trade operations and engineering activities.

In addition to this we import fresh or preserved Citrus and other Southern fruit and vegetables.

FOREIGN TRADE COMPANY

Agromachinaimpex

For the export and import of:
- Tractors
- Farm Machinery and Equipment
- Complex Technological Lines
- Electronic and Hydraulic Equipment
- Pumps and Spare Parts

- Compensation and Licence Deals
- Production
- Transport
- Construction
- And many other related activities

Services available abroad:
- Engineering
- Servicing
- Re-exporting

1 Stoyan Lepoev Boulevard,
Sofia, Bulgaria.
Telephone: 23-03-91
Telex: 022563

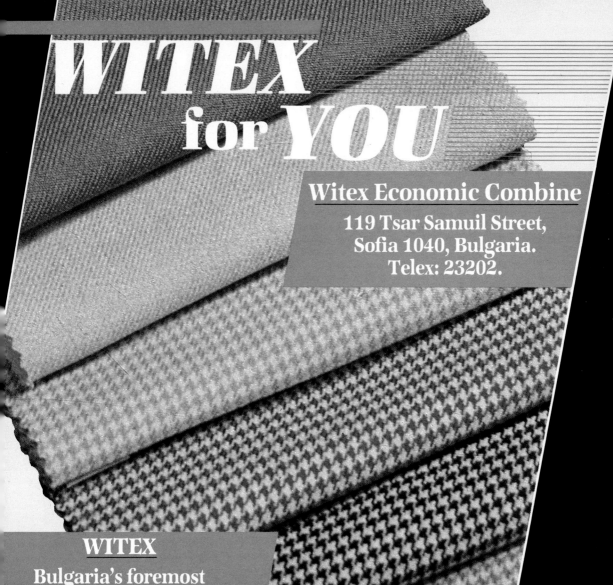

WITEX for YOU

Witex Economic Combine
119 Tsar Samuil Street,
Sofia 1040, Bulgaria.
Telex: 23202.

WITEX
Bulgaria's foremost manufacturers of woollen and silk textiles.

WITEX
A gorgeous collection of classic styles and the latest fashion designs.

WITEX
Beauty and quality to satisfy the most discriminating tastes.

WITEX
TRADITION REFINEMENT STYLE

Exporter:
**Industrialimport,
3 Positano Street,
Sofia, Bulgaria.
Telex: 22090.**

TECHNOIMPORT
Engineering Economic Association

Comprehensive range of services:
- ▼ Feasibility studies
- ▼ Design
- ▼ Selection and supply of equipment
- ▼ Erection
- ▼ Supervision and technical assistance
- ▼ Training of personnel in service and operation
- ▼ Import and export of complete projects
- ▼ Technology
- ▼ Licences
- ▼ Machines and spare parts

Comprehensive range of applications:
- ▼ Power generating industries
- ▼ Ferrous and non-ferrous metallurgy
- ▼ Mining and ore dressing
- ▼ Oil storage tanks and depots
- ▼ Waste product salvage
- ▼ Any other related field

Comprehensive range of contracts:
- ▼ Product in hand
- ▼ Turnkey
- ▼ Cost plus

Technoimport
20 Joliot Curie Street,
Sofia 1040, Bulgaria.
Telephone: 7381 Telex: 23421

bulgartabac

As one of the world's foremost tobacco associations, Bulgartabac has established itself as a reliable and popular business partner. Our services cover all commercial and production operations: from the selection of suitable soils, regionalization and standardization, through agrotechnics, manipulation and fermentation, to cigarette manufacture, import and export, international collaboration and research. Bulgartabac exports not only the classic Oriental tobacco but also Virginia and Burley, and we can supply large tobacco lots. But whatever the type and whatever the quantity, the emphasis remains firmly on quality.

BULGARTABAC
Commercial-Industrial Association
14 Stamboliiski Boulevard,
Sofia, Bulgaria.
Telephone: 87-52-11/16 Telex: 23288.

bulgartabac

ENGINEERING ECONOMIC ORGANISATION

A comprehensive service covering the study, design, delivery, assembly and maintenance of complete plant or single machines, in the fields of:

Food and Tobacco Industry:
- Abbatoirs and meat processing plants
- Dairy installations
- Fruit and vegetables processing installations
- Bottling equipment
- Tobacco processing lines
- And other associated installations

Refrigeration, Air-conditioning and Purification Equipment:
- Refrigeration storehouses for meat, fish, fruit and vegetables
- Ice plants
- Air-conditioning installations
- Purification equipment
- And other associated installations

Light Industry:
- Leather and fur manufacturing enterprises
- Textile machines
- Essential oils installations

Construction Industry:
- Ceramic Enterprises
- Installations for the manufacture of concrete elements
- Construction equipment
- Road construction machines
- Machines and equipment for the glass industry
- Glasshouses, plant halls, metal structures
- And other associated installations

PO Box 541, 20 Joliot Curie Street, Sofia, Bulgaria. Tel.: 73-81
Telex: 22193 / 22048 / 22049